Handbook of NANOTECHNOLOGY

Handbook of NANOTECHNOLOGY

Edited by

Dr. Parag Diwan

PENTAGON PRESS

Handbook of Nanotechnology / Dr. Parag Diwan (Ed)
Includes preface, Bibliography and index

ISBN 978-81-8274-415-8

First Indian Edition 2009

Published by

PENTAGON PRESS
206, Peacock Lane, Shahpur Jat, New Delhi-110049
Phones: 011-64706243, 26491568
Telefax: 011-26490600
email: rajan@pentagon-press.com
website: www.pentagon-press.com

Printed at

Swan Press
Naraina Ind. Area, Ph. II, New Delhi, India.

Preface

One of the biggest scientific trends of the 21st century has been focused on something incredibly small: i.e. of the order of 10^{-9} metre. You have read about it in the papers, heard glowing reports on television, and perhaps you have enjoyed reading about the possibilities of nanotechnology in science fiction. But what is nanotechnology? That is the most difficult question to answer, even though it's all over the news these days. To understand what nanotechnology is, one needs to accurately encapsulate the information explosion in this area that is going on as we read.

The word nanotechnology means different things to different people. For some, it conjures up images of fabulous new materials, lighter and stronger than steel. Others envision microscopic robots that clean plaque from our arteries and tartar from our teeth. In the purest sense, nanotechnology is the science on small—very small things measured in units called nanometers, which are one-billionth of a meter. In other words, a nanometer is so small that if one meter were stretched from Kashmir to Kanyakumari, each nanometer would still be the size of an aspirin.

Experts believe nanotechnology will have a more far-reaching impact than the silicon integrated chip with many more applications than electronics. Nanotechnology is expected to revolutionize certain areas, such as semiconductors, pharmaceuticals and materials. The total societal impact of nanotechnology is expected to be much greater than that of the silicon circuit because it s applicable in many more fields than electronics. Nanotechnology experts believe that because of nanotechnology, we would see more changes in the next 30 years than we saw in the entire last century.

The foundations of nanotechnology have emerged over many decades of research in many different fields. Computer circuits have been getting smaller, chemicals have been getting more complex, biochemists have learned more about how to study and control the molecular basis of organisms and mechanical engineering has been getting more precise.

In 1959, great physicists Richard Feynman suggested that it should be possible to build machines small enough to manufacture objects with atomic precision. His talk "there's plenty of room at the bottom.' Is widely considered to be the foreshadowing of nanotechnology. Among other things, he predicted that information could be stored with amazing density.

In the late 1970's, Eric Drexler began to invent what would become molecular manufacturing. He quickly realized that molecular machines could control the chemical manufacture of complex products, including additional manufacturing systems which would be very powerful technology. In 1986, he introduced the term nanotechnology in his book Engines of Creation to describe this approach to manufacturing and some of its consequences. Subsequent research showed that

Taniguchi had previously used the word in Japan to describe precision micromachining.

The term 'nanotechnology' rapidly became popular, and almost immediately its meaning began to shift. By 1992, Drexler was using 'molecular nanotechnology' or 'molecular manufacturing' to distinguish his manufacturing ideas from the simpler product-focused research that was borrowing the word.

By 2010, technology will be 2000 times more advanced than in 1900. Therefore, we will experience an equivalent level of technological change in the first decade of the 21st century to whole what we experienced in the 20th century! Extrapolating to the end of the 21st century, one finds technology will have progressed to become a million times more advanced than the technology that existed in 1900. In other words, we will experience a thousand times the technological progress achieved in the 20th century, during the 21st century.

The Handbook of nanotechnology seeks to explain this complex field, and provide an understanding of what is happening now in the field, as well as what to expect in the short and long terms. The Handbook consists of following distinct sections.

Section I	Nanotechnology Revolution
Section II	Nanostructured Materials
Section III	Nanocomposites
Section IV	Nanoelectronics
Section V	Nanorobotics
Section VI	Nanomedicine

Section I: **Nanotechnology Revolution** delineates a number of alternative scenarios to develop our understanding of nanotechnology. These scenarios cannot represent what will happen, because no one knows. They can, however, show how post-breakthrough capabilities could mesh with human life and Earth's environment. The result will appear to be quaintly conservative from a future perspective, however most of it seem like science fiction today. It also describes some other important components of nanotechnology namely, Molecular Nanotechnology, Fullernes, Nanotubes and Superlattices.

Section II: **Nanostructured Materials** details recent advances in the synthesis and investigation of functional materials, focusing on the novel size-dependent physics and chemistry that results when atoms and electrons are confined within nanoscale semiconductors and metal clusters. Nanostructure science and technology is a broad and interdisciplinary area of research and development activity that has been growing explosively worldwide in the past few years. It has the potential for revolutionizing the ways in which materials and products are created and the range and nature of functionalities that can be accessed. It is already having a significant commercial impact, which surely will increase in the future and this has been discussed at length in this section.

Section III: **Nanocomposites** provides a perfect insight into the latest technologies, emerging markets and commercial opportunities for nanocomposites. It examines the key drivers behind this interest and commercialization opportunities for nanocomposite technology in the various macro and micropolymer marketplaces.

Section IV: **Nanoelectronics and Nanooptics** is a topic that is attracting a lot of interest in scientific and nonscientific corners alike. Researchers are driven to explore the field because further miniaturising today's small electronic circuits will lead to faster, more sophisticated, and portable devices. As you will see in this volume nanoelectronics also encompasses what is known as molecular electronics, which uses individual molecules in electronics. This section enables readers to better understand drivers and demand patterns for nanoelectronics and nano-optical products, such as the trend in computing, communications, optical satellite telecommunications or sensor technology and IT.

Section V: **Nanorobotics** focuses on the state of the art in the field of nanorobotics by describing various molecular level systems and associated design and control issues. Following that some of the man-made and naturally occurring molecular machines are discussed. This section also describes SPM principles, surveys SPM uses in Nanomanipulation and looks at the SPM as a robot.

Section VI: **Nanomedicine** touches upon some of the leading areas to which nanotechnology tools, materials, devices and intelligent materials and machines are currently applied in medical research.

Realising the Potential of Ten Raised to the Power minus Nine Scientists studying molecular nanotechnology are interested in ways in which we might manipulate individual atoms and molecules to build things with almost unimaginable precision. Building things with atomic precision is an amazing contrast to the manufacturing techniques of today, which are very crude when examined at the molecular level. This handbook would be of use to such researchers as a compilation of things attempted and future that beholds us.

Today, making a smooth edge requires manufactures to grind away and discard trillions of atoms. Much in the way Michelangelo created statues from blocks of marble, manufactures today frequently create objects by first creating larger objects and then removing excess material by grinding, chiseling or sanding. Components that seem to fit together precisely are really billions of atoms out of alignment. Poorly fitting components wear faster, require costly lubrication and eventually lead to mechanical breakdown.

From a molecular perceptive, even the most precise of today's mechanical components fit poorly. These factors plague modern manufacturing techniques causing enormous waste, pollution and imprecision.

Enter molecular nanotechnology. The ability to precisely pick and place individual atoms and molecules will allow us to create objects with almost efficiency. Right now, scientists are picking up and moving individual atoms of silicon and other materials using scanning tunneling microscopy. STM uses an amazingly sharp tip, usually made of tungsten, to push atoms around on a surface. This handbook provides details of this technology and further prospects as the technology improves, we will be able to move hundreds, thousands and then millions of atoms at will, to precisely construct objects of virtually any size, shape and material.

Objects created using molecular nanotechnology will be precise to within the size of a molecule. Machined components will no longer abrade one another because they will fit together with molecular precision. Moreover, they will have been made one atom at a time with readily available components like carbon, oxygen and nitrogen as their building blocks.

The purpose of this handbook is to introduce new entrants into the world of nanotechnology by focusing on certain key application areas. A substantial amount of literature exists regarding the applications of nanotechnology. The material presented here is a compilation of literature available in the public domain. It has been compiled keeping the average reader in mind. Specific company and university research activities should be interpreted as examples and are only a subset of the vast scope of Nanoscience and nanotechnology research currently underway. As such a basic knowledge of science is assumed. However, no prior exposure to nanotechnology concepts is required.

The quote "The Next Big Thing is Really Small" by Jack Uldrich and Deb Newberry sums it up nicely. Nanotechnology is no longer a far-off, fuzzy, futuristic technology. It has already established a beachhead in today's world economy. We hope that the materials, concepts and ideas r presented in this handbook would be useful for all those who would interface with this emerging technology.

New Delhi
10th May 2009

Dr. Parag Diwan

Contents

SECTION II
NANOSTRUCTURED MATERIALS

SECTION III
NANOCOMPOSITES

SECTION IV
NANOELECTRONICS

SECTION V
NANOROBOTICS

SECTION VI
NANOMEDICINE

APPENDICES

SECTION A

NANOTECHNOLOGY REVOLUTION

Unit I–A

Nanotechnology—Looking Ahead

Molecular nanotechnology ("MNT") is an anticipated manufacturing technology that would allow precise control and positional assembly of molecule-sized building blocks through the use of nanoscale manipulator arms. Molecular nanotechnology is usually considered distinct from the more inclusive term "nanotechnology", which is now used to refer to a wide range of scientific or technological projects that focus on phenomena or properties of the nanometer scale (around 0.1-100nm). Nanotechnology is already a blossoming field, but molecular nanotechnology—the goal of productive, molecular-scale machine systems—is still in the preliminary research stage.

Nanotechnology was first introduced in 1959, in a talk by the Nobel Prize-winning physicist Richard Feynman, entitled "There's Plenty of Room at the Bottom". Feynman proposed using a set of conventional-sized robot arms to construct a replica of themselves, but one-tenth the original size, then using that new set of arms to manufacture an even smaller set, and so on, until the molecular scale is reached. If we had many millions or billions of such molecular-scale arms, we could program them to work together to create macroscale products built from individual molecules—a "bottom-up manufacturing" technique, as opposed to the usual technique of cutting away material until you have a completed component or product "top-down manufacturing".

Feynman's idea remained largely undiscussed until the mid-80s, when the MIT-educated engineer K. Eric Drexler published *Engines of Creation*, a book to popularize the potential of molecular nanotechnology. Because MNT would allow manufacturers to fabricate products from the bottom up with precise molecular control, a very wide range of chemically possible structures could be created. Since MNT systems could put every molecule in its specific place, molecular manufacturing processes could be very clean and efficient. Also, because every little bit of matter in an molecular nanotechnology system would be part of a nano-scale manipulator, nanotechnological systems could be far more productive and maintain much higher throughputs than modern manufacturing techniques, which use macro-scale manipulators to fabricate products.

To initiate an MNT revolution would require an "assembler"—a reprogrammable nanoscale manipulator capable of creating a wide range of molecular structures, including a complete copy of itself. The first assemblers will only function effectively in lab-controlled environments, such as a vacuum. The advent of self-replicating molecular nanomachines could quickly lead to "desktop nanofactories", tabletop appliances that consume modest amounts of power and contain the software required to manufacture an interesting range of useful products. The arrival of MNT would revolutionize wide sectors of human activity, including manufacturing, medicine, scientific research, communication, computing, and warfare. When full-blown molecular nanotechnology will arrive is currently unknown, but many experts foresee its arrival between 2010 and 2020.

A Japanese professor and his American visitor paused in the rain to look at a rising concrete structure on a university campus in the Tokyo suburbs near Higashikoganei Station. "This is for our Nanotechnology Center," Professor Kobayashi said. The professor's guest complimented the work as he wondered to himself, when would an American professor be able to say the same?

This Nanotechnology Center was being built in the spring of 1990, as Eric Drexler was midway through a hectic eight-day trip, giving talks on nanotechnology to researchers and seeing dozens of university and consortium research laboratories. A Japanese research society had sponsored the trip, and the Ministry of International Trade and Industry MITI) had organized a symposium around the visit—a symposium on molecular machines and nanotechnology. Japanese research was forging ahead, aiming to develop "new modes of science and technology in harmony with nature and human society," a new technology for the twenty-first century.

There is a view of the future that doesn't fit with the view in the newspapers. Think of it as an alternative, a turn in the road of future history that leads to a different world. In that world, cancer follows polio, petroleum follows whale oil, and industrial technology follows chipped flint—all healed or replaced. Old problems vanish, new problems appear: down the road are many alternative worlds, some fit to live in, some not. We aim to survey this road and the alternatives, because to arrive at a world fit to live in, we will all need a better view of the open paths.

How does one begin to describe a process that can replace the industrial system of the world? Physical possibilities, research trends, future technologies, human consequences, political challenges: this is the logical sequence, but none of these makes a satisfactory starting point. The story might begin with research at places like IBM, Du Pont, and the ERATO projects at Tsukuba and RIKEN, but this would begin with molecules, seemingly remote from human concerns. At the core of the story is a kind of technology—"molecular nanotechnology" or "molecular manufacturing"—that appears destined to replace most of technology as we know it today, but it seems best not to begin in the middle. Instead, it seems best to begin with a little of each topic, briefly sketching consequences, technologies, trends, and principles before diving into whole units on one aspect or another. This unit provides those sketches and sets the stage for what follows.

All this can be read as posing a grand "What if?" Question: *What if molecular manufacturing and its products replace modern technology?* If they don't, then the question merely invites an entertaining and mind-stretching exercise. But if they do, then working out good answers in advance may tip the balance in making decisions that determine the fate of the world. Later units will show why we see molecular manufacturing as being almost inevitable, yet for now it will suffice if enough people give enough thought to the question "What if?"

A Sketch of Technologies

Molecular nanotechnology: Thorough, inexpensive control of the structure of matter based on molecule-by-molecule control of products and byproducts; the products and processes of molecular manufacturing.

Technology as we know it is a product of industry, of manufacturing and chemical engineering. Industry as we know it takes things from nature—ore from mountains, trees from forests—and coerces them into forms that someone considers useful. Trees become lumber, then houses. Mountains become rubble, then molten iron, then steel, then cars. Sand becomes a purified gas, then silicon, then chips. And so it goes. Each process is crude, based on cutting, stirring, baking, spraying, etching, grinding, and the like.

Trees, though, are not crude: to make wood and leaves, they neither cut, grind, stir, bake, spray, etch, nor grind. Instead, they gather solar energy using molecular electronic devices, the photosynthetic reaction centers of chloroplasts. They use that energy to drive molecular machines—active devices with moving parts of precise, molecular structure—which process carbon dioxide and water into oxygen and molecular building blocks. They use other molecular machines to join these molecular building blocks to form roots, trunks, branches, twigs, solar collectors, and more molecular machinery. Every tree makes leaves, and each leaf is more sophisticated than a spacecraft, more finely patterned than the latest chip from Silicon Valley. They do all this without noise, heat, toxic fumes, or human labour, and they consume pollutants as they go. Viewed this way, trees are high technology. Chips and rockets aren't.

Trees give a hint of what molecular nanotechnology will be like, but nanotechnology won't be biotechnology because it won't rely on altering life. Biotechnology is a further stage in the domestication of living things. Like selective breeding, it reshapes the genetic heritage of a species to produce varieties more useful to people. Unlike selective breeding, it inserts new genes. Like biotechnology—or ordinary trees—molecular nanotechnology will use molecular machinery, but unlike biotechnology, it will not rely on genetic meddling. It will be not an extension of biotechnology, but an alternative or a replacement.

Molecular nanotechnology could have been conceived and analyzed—though not built—based on scientific knowledge available forty years ago. Even today, as development accelerates, understanding grows slowly because molecular nanotechnology merges fields that have been strangers: the molecular sciences, working at the threshold of the quantum realm, and mechanical engineering, still mired in the grease and crudity of conventional technology. Nanotechnology will be a technology of new molecular machines, of gears and shafts and bearings that move and work with parts shaped in accord with the wave equations at the foundations of natural law. Mechanical engineers don't design molecules. Molecular scientists seldom design machines. Yet a new field will grow—is growing today—in the gap between. That field will replace both chemistry as we know it and mechanical engineering as we know it. And what is manufacturing today, or modern technology itself, but a patchwork of crude chemistry and crude machines?

Next unit will paint a concrete picture of molecular machines and molecular manufacturing, but for now analogy will serve. Picture an automated factory, full of conveyor belts, computers, rollers, stampers, and swinging robot arms. Now imagine something like that factory, but a million times smaller and working a million times faster, with parts and workpieces of molecular size. In this factory, a "pollutant" would be a loose molecule, like a ricocheting bolt or washer, and loose molecules aren't tolerated. In many ways, the factory is utterly unlike a living cell: not fluid, flexible, adaptable, and fertile, but rigid, preprogrammed and specialized. And yet for all of that, this microscopic molecular factory emulates life in its clean, precise molecular construction.

Advanced molecular manufacturing will be able to make almost anything. Unlike crude mechanical and chemical technologies, molecular manufacturing will work from the bottom up, assembling intricate products from the molecular building blocks that underlie everything in the physical world.

Nanotechnology will bring new capabilities, giving us new ways to make things, heal our bodies, and care for the environment. It will also bring unwelcome advances in weaponry and give us yet more ways to foul up the world on an enormous scale. It won't automatically solve our problems: even powerful technologies merely give us more power. As usual, we have a lot of work ahead of us and a lot of hard decisions to make if we hope to harness new developments to good ends. The main reason to pay attention to nanotechnology now, before it exists, is to get a head start on understanding it and what to do about it.

A Sketch of Consequences

The United States may have become famous for its obsession with the elections and the next quarter's profits, and the future be damned. Nonetheless, we are writing for normal human beings who feel that the future matters–ten, twenty, perhaps even thirty years from now—for people who care enough to try to shift the odds for the better. Making wise choices with an eye to the future requires a realistic picture of what the future can hold. What if most pictures of the future today are based on the wrong assumptions?

Here are a few of today's common assumptions, some so familiar that they are seldom stated:

- Industrial development is the only alternative to poverty.
- Many people must work in factories.
- Greater wealth means greater resource consumption.
- Logging, mining, and fossil fuel burning must continue.
- Manufacturing means polluting.
- Third World development would doom the environment.

These all depend on a more basic assumption:

- *Industry as we know it cannot be replaced.*

Some further common assumptions:

- Today's economic trends will define tomorrow's problems.
- Spaceflight will never be affordable for most people.
- Forests will never grow beyond Earth.
- More advanced medicine will always be more expensive.
- Even highly advanced medicine won't be able to keep people healthy.
- Solar energy will never become really inexpensive.
- Toxic wastes will never be gathered and eliminated.
- Developed land will never be returned to wilderness.
- There will never be weapons worse than nuclear missiles.
- Pollution and resource depletion will eventually bring war or collapse.

These, too, depend on a more basic assumption:

- *Technology as we know it will never be replaced.*

These commonplace assumptions paint a future full of terrible dilemmas, and the notion that a technological change will let us escape from them smacks of the idea that some technological fix can save the industrial system. The prospect, though, is quite different: The industrial system won't be fixed, it will be junked and recycled. The prospect isn't more industrial wealth ripped from the flesh of the Earth, but green wealth unfolding from processes as clean as a growing tree. Today, our industrial technologies force us to choose better quality or lower cost or greater safety or a cleaner environment. Molecular manufacturing, however, can be used to improve quality *and* lower costs *and* increase safety *and* clean the environment. The coming revolutions in technology will transcend many of the old, familiar dilemmas. And yes, they will bring fresh, equally terrible dilemmas.

Molecular nanotechnology will bring thorough and inexpensive control of the structure of matter. We need to understand molecular nanotechnology in order to understand the future capabilities of the human race. This will help us see the challenges ahead, and help us plan how best to conserve values, traditions, and ecosystems through effective policies and institutions. Likewise, it can help us see what today's events mean, including business opportunities and possibilities for action. We need a vision of where technology is leading because technology is a part of what human beings are, and will affect what we and our societies can become.

The consequences of the coming revolutions will depend on human actions. As always, new abilities will create new possibilities both for good and for ill. We will discuss both, focusing on how political and economic pressures can best be harnessed to achieve good ends. Our answers will not be satisfactory, but they are at least a beginning.

A Sketch of Trends

Technology has been moving toward greater control of the structure of matter for millennia. For decades, microtechnology has been building ever-smaller devices, working toward the molecularsize-scale from the top down. For a century or more, chemistry has been building ever-larger molecules, working up toward molecules large enough to serve as machines. The research is global, and the competition is heating up.

Since the concept of molecular nanotechnology was first laid out, scientists have developed more powerful capabilities in chemistry and molecular manipulation. There is now a better picture of how those capabilities can come together in the next steps, and of how advanced molecular manufacturing can work. Nanotechnology has arrived as an idea and as a research direction, though not yet as a reality.

Naturally occurring molecular machines exist already. Researchers are learning to design new ones. The trend is clear, and it will accelerate because better molecular machines can help build even better molecular machines. By the standards of daily life, the development of molecular nanotechnology will be gradual, spanning years or decades, yet by the ponderous standards of human history it will happen in an eyeblink. In retrospect, the wholesale replacement of twentieth-century technologies will surely be seen as a technological revolution, as a process encompassing a great breakthrough.

Today, we live in the end of the pre-breakthrough era, with pre-breakthrough technologies, hopes, fears, and preoccupations that often seem permanent, as did the Cold War. Yet it seems that the breakthrough era is not a matter for some future generation, but for our own. These developments are taking shape right now, and it would be rash to assume that their consequences will be many years delayed.

In later units, we'll say more about what researchers are doing today, about where their work is leading, and about the problems and choices ahead. To get a sense of the consequences, though, requires a picture of what nanotechnology can do. This can be hard to grasp because past advanced technologies–microwave tubes, lasers, superconductors, satellites, robots, and the like–have come trickling out of factories, at first with high price tags and narrow applications. Molecular manufacturing, though, will be more like computers: a flexible technology with a huge range of applications. And molecular manufacturing won't come trickling out of conventional factories as computers did: it will replace factories and replace or upgrade their products. This is something new and basic, not just another twentieth-century gadget. It will arise out of twentieth-century trends in science, but it will break the trend-lines in technology, economics, and environmental affairs.

Calculators were once thousand-dollar desktop clunkers, but microelectronics made them fast and efficient, sized to a child's pocket and priced to a child's budget. Now imagine a revolution of similar magnitude, but applied to everything else.

More Consequences: Scenes from a Post-breakthrough World

What nanotechnology will mean for human life is beyond our predicting, but a good way to understand what it could mean is to paint scenarios. A good scenario brings together different aspects of the world (technologies, environments, human concerns) into a coherent whole. Major corporations use scenarios to help envision the paths that the future may take–not as forecasts, but as tools for thinking. In playing the "What if?" game, scenarios present trial answers and pose new questions.

The following scenarios can't represent what will happen, because no one knows. They can, however, show how post-breakthrough capabilities could mesh with human life and Earth's environment. The results will likely seem quaintly conservative from a future perspective, however much they seem like science fiction today. The issues behind these scenarios will be discussed later.

Scenario: Solar Energy

In Fairbanks, Alaska, Linda Hoover yawns and flips a switch on a dark winter morning. The light comes on, powered by stored solar electricity. The Alaska oil pipeline shut down years ago, and tanker traffic is gone for good.

Nanotechnology can make solar cells efficient, as cheap as newspaper, and as tough as asphalt–tough enough to use for resurfacing roads, collecting energy without displacing any more grass and trees. Together with efficient, inexpensive storage cells, this will yield low-cost power (but no, not "too cheap to meter").

Scenario: Medicine that Cures

A lady has been a bit hoarse for weeks, and just came down with a horrid head cold. For the past six months, she's been seeing ads for some cure for the Common Cold, so she spends her five dollars and takes the nosespray and throatspray doses. Within three hours, 99 percent of the viruses in her nose and throat are gone, and the rest are on the run. Within six hours, the medical mechanisms have become inactive, like a pinch of inhaled but biodegradable dust, soon cleared from the body. She feels much better and won't infect her friends at dinner.

The human immune system is an intricate molecular mechanism, patrolling the body for viruses and other invaders, recognizing them by their foreign molecular coats. The immune system, though, is slow to recognize something new. For her five dollars, Sue bought 10 billion molecular mechanisms primed to recognize not just the viruses she had already encountered, but each of the five hundred most common viruses that cause colds, influenza, and the like.

Weeks have passed, but the hoarseness Sue had before her cold still hasn't gone away; it gets worse. She ignores it through a long vacation, but once she's back and caught up, Sue finally goes to see her doctor. He looks down her throat and says, "Hmmm." He asks her to inhale an aerosol, cough, spit in a cup, and go read a magazine. The diagnosis pops up on a screen five minutes after he pours the sample into his cell analyzer. Despite his knowledge, his training and tools, he feels chilled to read the diagnosis: a malignant cancer of the throat, the same disease that has cropped up all too often in his own mother's family.

He touches the "Proceed" button. In twenty minutes, he looks at the screen to check progress. Yes, Sue's cancerous cells are all of one basic kind, displaying one of the 16,314 known molecular markers for malignancy. They can be recognized, and since they can be recognized, they can be destroyed by standard molecular machines primed to react to those markers. The doctor instructs the cell analyzer to prime some "immune machines" to go after her cancer cells. He tests them on cells from the sample, watches, and sees that they work as expected, so he has the analyzer prime up some more.

Sue puts the magazine down and looks up. "Well, doc, what's the word?" she asks. "I found some suspicious cells, but this should clear it up," he says. He gives her a throat spray and an injection. "I'd like you to come back in three weeks, just to be sure." "Do I have to?" she asks.

"You know," he lectures her, "we need to make sure it's gone. You really shouldn't let things like this go so far before coming in."

"Yes, fine, I'll make the appointment," she says. Leaving the office, Sue thinks fondly of how old-fashioned and conservative Dr. Fujima is.

The molecular mechanisms of the immune system already destroy most potential cancers before they grow large enough to detect. With nanotechnology, we will build molecular mechanisms to destroy those that the immune system misses.

Scenario: Cleansing the Soil

California Scout Troop 9731 has hiked for six days, deep in the second-wilderness forests of the Pacific Northwest.

"I bet we're the first people ever to walk here," says one of the youngest scouts.

"Well, maybe you're right about walking," says Scoutmaster Jackson, "but look up ahead–what do you see, scouts?"

Twenty paces ahead runs a strip of younger trees, stretching left and right until it vanishes among the trunks of the surrounding forest.

"Hey, guys! Another old logging road!" shouts an older scout. Several scouts pull probes from their pockets and fit them to the ends of their walking sticks. Jackson smiles: It's been ten years since a California troop found anything this way, but the kids keep trying.

The scouts fan out, angling their path along the scar of the old road, poking at the ground and watching the readouts on the stick handles. Suddenly, unexpectedly, comes a call: "I've got a signal! Wow–I've got PCBs!"

In a moment, grinning scouts are mapping and tracing the spill. Decades ago, a truck with a leaking load of chemical waste snuck down the old logging road, leaving a thin toxic trail. That trail leads them to a deep ravine, some rusted drums, and a nice wide patch of invisible filth. The excitement is electrifying.

Setting aside their maps and orienteering practice, they unseal a satellite locator to log the exact latitude and longitude of the site, then send a message that registers their cleanup claim on the ravine. The survey done, they head off again, eagerly planning a return trip to earn the now-rare Toxic Waste Cleanup Merit Badge.

Today, tree farms are replacing wilderness. Tomorrow, the slow return to wilderness may begin, when nature need no longer be seen as a storehouse of natural resources to be plundered. Later we will discuss just how little need be taken from nature to provide humans with wealth, and how post-breakthrough technologies can remove from nature the toxic residues of twentieth-century mistakes.

Scenario: Pocket Supercomputers

At the University of Michigan, Joel Gregory grabs a molecular rod with both hands and twists. It feels a bit weak, and a ripple of red reveals too much stress in a strained molecular bond halfway down its length. He adds two atoms and twists the rod again: all greens and blues, much better. Joel plugs the rod into the mechanical arm he's designing, turns up the temperature, and sets the whole thing in motion. A million atoms dance in thermal vibration, gears spin, and the arm swings to and fro in programed motion. It looks good. A few parts are still mock-ups, but doing a thesis takes time, and he'll work out the rest of the molecular details later. Joel strips off the computer display goggles and gloves and blinks at the real world. It's time for a sandwich and a cup of coffee. He grabs the computer itself, stuffs it into his pocket, and heads for the student center.

Researchers already use computers to build models of molecules, and "virtual reality systems" have begun to appear, enabling a user to walk around the image of a molecule and "touch" it, using computer-controlled gloves and goggles. We can't build a supercomputer able to model a million-atom machine yet–much less build a pocket supercomputer–but computers keep shrinking in size and cost. With nanotechnology to make molecular parts, a computer like Joel's will become easy to build. Today's supercomputers will seem like hand-cranked adding machines by comparison. The following units take a closer look at a simulated molecular world.

Scenario: Global Wealth

Behind a village school in the forest a stone's throw from the Congo River, a desktop computer with a thousand times the power of an early 1990s supercomputer lies half-buried in a recycling bin. Indoors, Joseph Adoula and his friends have finished their day's studies; now they are playing together in a vivid game universe using personal computers each a million times more powerful than the clunker in the trash. They stay late in air-conditioned comfort.

Trees use air, soil, and sunlight to make wood, and wood is cheap enough to burn. Nanotechnology can do likewise, making products as cheap as wood–even products like supercomputers, air conditioners, and solar cells to power them. The resulting economics may even keep tropical forests from being burned. In a later unit we will discuss how costs can fall low enough to make material wealth for the Third World easy to achieve.

Scenario: Cleansing the Air

In Earth's atmosphere, the twentieth-century rise in carbondioxide levels has halted and reversed. Fossil fuels are obsolete, so pollution rates have lessened. Efficient agriculture has freed fertile land for reforestation, so growing trees are cleansing the atmosphere. Surplus solar power from the world's repaved roads is being used to break down excess carbon dioxide at a rate of 5 billion tons per year. Climates are returning to normal, the seas are receding to their historical shores, and ecosystems are beginning the slow process of recovery. In another twenty years, the atmosphere will be back to the pre-industrial composition it had in the year 1800.

Scenario: Transportation Outward

Jim Salin's afternoon flight from Dulles International is on the ground, late for departure. Impatiently, Jim checks the time: any later, and he'll miss his connecting flight.

At last, the glassy-surfaced craft rolls down the runway. With gliderlike wings, it lifts its fat body and climbs steeply toward the east. A few pages into his novel, Jim is interrupted by a second recitation of safety instructions and the captain's announcement that they'll try to make up for lost time. Jim settles back in his seat as the main engines kick in, the wings retract, the acceleration builds, and the sky darkens to black. Like the highest-performance rockets of the 1980s, Jim's liner produces an exhaust of pure water vapor. Spaceflight has become clean, safe, and routine. And every year, more people go up than come down.

The cost of spaceflight is mostly the cost of high-performance, reliable hardware. Molecular manufacturing will make aerospace structures from nearly flawless, superstrong materials at low cost. Add inexpensive fuel, and space will become more accessible than the other side of the ocean is today.

Scenario: Restoring Species

Restoration Day Ceremonies are always moving events. For some reason, the old people always cry, even though they say they're happy.

Crying, Tracy Stiegler thinks, doesn't make any sense. She looks again through the camouflage screen over the sandy Triangle Keys beach, gazing across the Caribbean toward the Yucatán Peninsula. Soon this will be theirs again, and that's all to the good.

Tracy and the other scientists from BioArchive have positions of honor in today's Restoration Day Ceremony. Since the mid-twentieth century there had been no living Caribbean monk seals, only grisly relics of the years of their slaughter: seal furs and dry museum specimens. Tracy's team struggled for years, gathering these relics and studying them with molecular instruments. It had been known for decades—since the 1980s—that genes are tough enough to survive in dried skin, bone, horn, and eggshell. Tracy's team had collected genes and rebuilt cells.

They worked for years, and gave thanks to the strict protection—late, but good enough—that saved one related species. At last, a Hawaiian monk seal had given birth to a genetically-pure Caribbean monk seal, twin to a seal long dead. And now there were five hundred, some young, some middle-aged, with decent genetic diversity and five years' experience living in the confines of a coastal ecological station.

Today, with raucous voices, they are moving out into the world to reclaim their ecological niche. As Tracy watches, she thinks of the voices that will never be heard again: of the species, known and unknown, that left not a even a bloody scrap to be cherished and restored. Thousands (millions?) of species had simply been brushed into extinction as habitats were destroyed by farming and logging. People knew–for years they had known–that freezing or drying would save genes. And they knew of the ecological destruction, and they knew they weren't stopping it. And the ignorant bastards didn't even keep samples.

Tracy discovers that she, too, cries at Restoration Day Ceremonies.

People will surely push biomedical applications of nanotechnology far and fast for human healthcare. With a bit more pushing, this technology base will be good enough to restore some species now thought lost forever, to repair some of the damage human beings have done to the web of life. It would be better to preserve ecosystems and species intact, but restoration, even of a few species, will be far better than nothing. Some samples from endangered species are being kept today, but not enough, and mostly for the wrong reasons.

Scenario: An Unstable Arms Race

Disputes over technology development and trade had soured relationships between Singapore and the Japan-United States alliance. Diplomatic inquiries regarding peculiar seismic and sonar readings in the South China Sea had just begun when they suddenly became irrelevant: an estimated one billion tons of unfamiliar, highly-automated military hardware appeared in coastal waters around the world. Accusations began to fly between Congress and PeaceWatch personnel, "If you'd done your jobs—" "If you'd let us do our jobs—".

And so, in late February, Singapore emerged as a military superpower.

Low cost, high quality, high-speed production can be applied to many purposes, not all attractive. Nanotechnology has enormous potential for abuse.

Technologies Revisited

Molecules matter because matter is made of molecules, and everything from air to flesh to spacecraft is made of matter. When we learn how to arrange molecules in new ways, we can make new things, and make old things in new ways. Perhaps this is why Japan's MITI has identified "control technologies for the precision arrangement of molecules" as a basic industrial technology for the twenty-first century. Molecular nanotechnology will give thorough control of matter on a large scale at low cost, shattering a whole set of technological and economic barriers more or less at one stroke.

A molecule is an object consisting of a collection of atoms held together by strong bonds (one-atom molecules are a special case). "Molecule" usually refers to an object with a number of atoms small enough to be counted (a few to a few thousand), but strictly speaking a truck tire (for instance) is mostly one big molecule, containing something like 1,000,000,000,000,000,000,000,000,000 atoms. Counting this many atoms aloud would take about 10,000,000,000 billion years.

Scientists and engineers still have no direct, convenient way to control molecules, basically because human hands are about 10 million times too large. Today, chemists and materials scientists make molecular structures indirectly, by mixing, heating, and the like. The idea of molecular nanotechnology begins with the idea of a molecular assembler, a device resembling an industrial robot arm but built on a microscopic scale. A general-purpose molecular assembler will be a jointed mechanism built from rigid molecular parts, driven by motors, controlled by computers, and able to grasp and apply molecular-scale tools. Molecular assemblers can be used to build other molecular machines–they can even build more molecular assemblers. Assemblers and other machines in molecular manufacturing systems will be able to make almost anything, if given the right raw materials. In effect, molecular assemblers will provide the microscopic "hands" that we lack today. (Chemists are asked to forgive this literary license; the specific details of molecular binding and bonding don't change the conclusion.)

Nanotechnology will give better control of molecular building blocks, of how they move and go together to form more complex objects. Molecular manufacturing will make things by building from the bottom up, starting with the smallest possible building blocks. The *nano* in nanotechnology comes from *nanos*, the Greek word for dwarf. In science, the prefix *nano-* means one-billionth of something, as in nanometer and nanosecond, which are typical units of size and time in the world of molecular manufacturing. When you see it tacked onto the name of an object, it means that the object is made by patterning matter with molecular control: nanomachine, nanomotor, nanocomputer. These are the smallest, most precise devices that make sense based on today's science.

(Be cautious of other usages, though—some researchers have begun to use the *nano-* prefix to refer to other small-scale technologies in the laboratory today. In this book *nanotechnology* means the *precise, molecular* nanotechnology of the future. British usage also applies the term to the small-scale and high precision technologies of today—even to precision grinding and measurement. The latter are useful, but hardly revolutionary.)

Digital electronics brought an information-processing revolution by handling information quickly and controllably in perfect, discrete pieces: bits and bytes. Likewise, nanotechnology will bring a matter-processing revolution by handling matter quickly and controllably in perfect, discrete pieces: atoms and molecules. The digital revolution has centered on a device able to make any desired pattern of bits: the programable computer. Likewise, the nanotechnological revolution will center on a device able to make (almost) any desired pattern of atoms: the programmable assembler. The technologies that plague us today suffer from the messiness and wear of an old phonograph record. Nanotechnology, in contrast, will bring the crisp, digital perfection of a compact disc.

A Road Map

The next two sections say a bit more about why molecular nanotechnology is already worth your attention and about whether it's possible to understand anything about the future. Later units answer questions like the following:

- Who is working on nanotechnology? What are they doing, and why?
- How can this work come together to provide breakthrough capabilities? When might this happen? What developments should we watch for?
- How will nanotechnology work? Who will be able to use it?
- What will it mean for the economy? For medicine? For the environment?
- What are its risks? What basic regulations will we need? What will it mean for the global arms race?
- What might go wrong as this technology emerges, and what can we do about it?

In a democratic society, only a few people need an in-depth understanding of how a technology works, but many people need to understand what it can do. In the next unit, we'll lead off by describing the molecular world and how it works–after all, everything around us and inside us is made of molecules—but the main story is about what this technology will mean for human beings and the biosphere.

Why Talk About It?

It is these concerns–the implications of nanotechnology for our lives, the environment, and the future–that guided the writing of this book. Nanotechnology can bring great achievements and solve great problems, but it will likewise present opportunities for enormous abuse. Research progress is necessary, but so is an informed and cautious public.

Our motivation in presenting these ideas is as much a fear of potential harm, and a wish to avoid it, as a longing for the potential good and a wish to seek it. Even so, we will dwell on the good that nanotechnology can bring and give only an outline of the obvious potential harm. The coming revolution can best be managed by people who share not only a picture of what they wish to avoid, but of what they can achieve. If we as a society have a clear view of a route to follow, we won't need a precise catalog of every cliff and mine field to the side of the road.

Some will hear this emphasis and call us optimistic. But would it really be wise to dwell on exactly how a technology can be abused? Or to draw up blueprints, perhaps?

Still, sitting here, preparing to tell this story, is an uncomfortable place for a researcher to be. In his book *How Superstition Won and Science Lost*, historian John C. Burnham tells of the century-long retreat of scientists from what they once saw as their responsibility: presenting the content and methods of science to a broad audience, for the public good. Today, the culture of science takes a dim view of "popularization." If you can write in plain English, this is taken as evidence that you can't do math, and vice versa. Robert Pool, a member of the news staff of the most prestigious American scientific journal, *Science*, acknowledges this negative attitude in writing that "some researchers, either by choice or just by being in the wrong place at the wrong time, make it into the public eye." So how can a researcher keep out of trouble? If you stumble on something important, wrap it in jargon. If people realize that it's important, run and hide. Robert Pool gently urges scientists to become more involved, but the social pressures in the research community are heavily in the other direction.

In response to this negative attitude toward "popularization," we can only ask that scientists and engineers try to act in a thoroughly professional fashion when judging a given proposal–which is to say, that they pay scrupulous attention to the scientific and technical facts. This means judging the validity of technical ideas based on their factual merits, and not on their (occasionally readable) style of presentation, or on the emotional response they may stir up. Nanotechnology matters to people, and they deserve to know about its flesh-and-blood human consequences, its impact on society and nature. We urge scientifically inclined readers to consult the Technical Bibliography at the end of the book, and then to point out any major errors they can find in the technical papers on this topic. We urge nonscientists who encounter scientifically knowledgeable critics to ask for specific, technical criticisms. Years of discussion with scientists and engineers—in public, in private, at conferences, and through the press—indicate that the case for nanotechnology is solid. Japanese and European industry, government, and academic researchers are forging ahead on the road to nanotechnology, and more and more U.S. research is applicable. Some researchers have even begun to call it an obvious goal.

Words that Block Thinking

Americans, so often in the forefront of science and technology, have a curious difficulty in thinking about the future. Language seems to have something to do with it.

If something sounds futurelike, we call it "futuristic." If that doesn't stop the conversation, we say that it "sounds like science fiction." These descriptions remind listeners of laughable 1950s fantasies like rockets to the Moon, video telephones, ray guns, robots, and the like. Of course, all these became real in the 1960s, because the science wasn't fiction. Today, we can see not only how to build additional science-fictional devices, but–more important, for better or worse–how to make them cheap and abundant. We need to think about the future, and name-calling won't help.

Curiously, the Japanese language seems to lack a disparaging word for "futurelike." Ideas for future technologies may be termed *mirai no* ("of the future," a hope or a goal), *shōrai-teki* (an expected development, which might be twenty years away), or *kûsõ no* ("imaginary" only, because contrary to physical law or economics). To think about the future, we need to distinguish *mirai no* and *shōrai-teki*, like nanotechnology, from mere *kûsõ no*, like antigravity boots.

A serious problem (Calvin and Hobbes).

A final objection is the claim that there's no point in trying to think about the future, because it is all too complex and unpredictable. This is too sweeping, but has more than a little truth. It deserves a considered response.

The Difficulty of Looking Forward

If our future will include nanotechnology, then it would be useful to understand what it can do, so that we can make more sensible plans for our families, careers, companies, and society. But many intelligent people will respond that understanding is impossible, that the future is just too unpredictable. This depends, of course, on what you're trying to predict:

The weather a month from now? Forget it; weather is too chaotic.

The position of the Moon a century from now? Easy; the Moon's orbit is like clockwork.

Which personal-computer company will lead twenty years from now? Good luck; major companies today didn't even exist twenty years ago.

That personal computers will become enormously more powerful? A virtual certainty. And so on. If you aim to say something sensible about the future of technology, the trick is to ask the right questions and to avoid the standard pitfalls. In his book *Megamistakes: Forecasting and the Myth of Rapid Technological Change*, Steven Schnaars surveys these pitfalls and their effects on past predictions. Borrowing and adapting some of his generalizations, here are our suggestions for how to blunder into a Megamistake in forecasting:

- Ignore the scientific facts, or guess.
- Forget to ask whether anyone wants the projected product or situation.
- Ignore the costs.
- Try to predict which company or technology will win.

In looking at what to expect from nanotechnology—or any technology—all of these must be avoided, since they can lead to some grand absurdities. In a classic demonstration of the first error, someone once concocted the notion that pills would someday replace food. But people need energy to live, and energy means calories, which means fuel, which takes up room. To subsist on pills, you'd need to gobble them by the fistful. This would be like eating a tasteless kibbled dog food, which was hardly the idea. In short, the pills-for-food prediction ignored the scientific facts. In a similar vein, we once heard promises of a cure for cancer—but this was based on a guess about scientific facts, a guess that "cancer" was in some sense a single disease, which might have a single point of vulnerability and a single cure. This guess was wrong, and progress against cancer has been slow.

Earlier, we presented a scenario that includes the routine cure of a cancer using nanotechnology. This scenario takes account of the currently known facts: cancers differ, but each kind can be recognized by its molecular markers. Molecular machines can recognize molecular markers, and so can be primed to recognize and destroy specific kinds of cancer cells as they turn up.

Even nanotechnology can't cram a meal into a pill, but this is just as well. The pills-for-food proposal didn't just ignore the facts, it also ignored what people want—things like dinner conversation and novel ethnic cuisines. Magazines once promised cities beneath the sea, but who wants to live in the ultimate damp, chilly climate? California and the Sunbelt have somehow proved more popular. And again, we were promised talking cars, but after giving them a try, people prefer luxury cars from companies that promise silence.

Many human wants are easy to predict, because they are old and stable: people want better medical care, housing, consumer goods, transportation, education, and so forth, preferably at lower costs, with greater safety, in a cleaner environment. When our limited abilities force us to choose better quality *or* lower cost *or* greater safety *or* a cleaner environment, decisions become sticky. Molecular manufacturing will allow a big step in the direction of better quality and lower costs and increased safety and a cleaner environment. (Choices of how much of each will remain.) There is no existing market demand for "nanotechnology," as such, but a great demand for what it can do.

Neglecting costs has also been popular among prognosticators: Building cities under the sea would be expensive, with few benefits. Building in space has more benefits, but would be far more expensive, using past or present technologies. Many bold projections gather dust on shelves because development or manufacturing costs are too high. Some examples include personal robots, flying cars, and Moon colonies–they still sound more like 1950s science fiction than practical possibilities, and cost is one major reason.

Molecular manufacturing is, in part, about cost reduction. As mentioned above, molecular machines in nature make things cheaply, like wood, potatoes, and hay. Trees are more complex than spacecraft, so why should spacecraft stay more expensive? Gordon Tullock, professor of economics and political science at the University of Arizona, says of molecular nanotechnology, "Its economic effect is that we will all be much richer." The prospect of building sophisticated products for the price of potatoes gives reason to pull a lot of old projections down from the shelf.

Even staying within the bounds of known science, focusing on things people want, and paying attention to costs, it's still hard to pick a specific winner. Technology development is like a horse race: everyone knows that some horse will win, but knowing which horse is harder (and worth big bucks). Both corporate managers betting money and researchers betting their careers have to play this game, and they often lose. A technology may work, provide something useful, and be less expensive than last year's alternative, yet still be clobbered in the market by something unexpected but better. To know which technologies will win, you'd have to know all the alternatives, whether they've been invented yet or not. Good luck!

We won't try to play that game here. "Nanotechnology" (like "modern industry") describes a huge range of technologies. Nonetheless, nanotechnology in one form or another is a monumentally obvious idea: it will be the culmination of an age-old trend toward more thorough control of the structure of matter. Predicting that some form of nanotechnology will win most technology races is like predicting that some horse will win a horse race (as opposed to, say, a dachshund). A technology based on thorough control of the structure of matter will almost always beat one based on crude control of the structure of matter. Other technologies have already won races in the literal sense of being first. Few, however, will win in the sense of being best.

Exploratory Engineering

Studies of molecular nanotechnology are today in the exploratory engineering phase, and just beginning to move into engineering development. The basic idea of exploratory engineering is simple: combine engineering principles with known scientific facts to form a picture of future technological possibilities. Exploratory engineering looks at future possibilities to help guide our attention in the present. Science–especially molecular science–has moved fast in recent decades. There is no need to wait for more scientific breakthroughs in order to make engineering breakthroughs in nanotechnology.

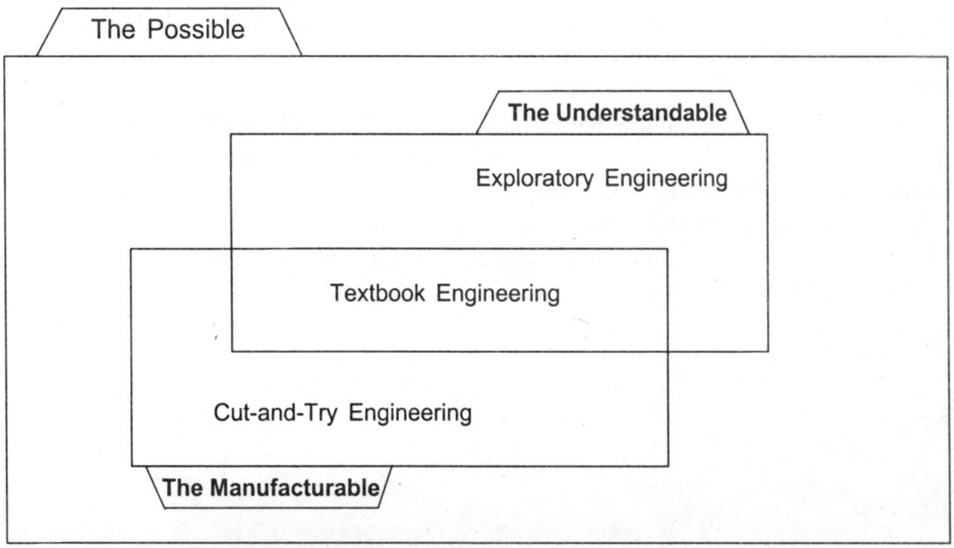

Exploratory Engineering Venn Diagram

The outer tagged rectangle represents the set of all technologies permitted by the laws of nature, whether they exist or not, whether they have been imagined or not. Within this set are those technologies that are manufacturable with today's technology, and those that are understandable with today's science. Textbooks teach what is understandable (hence teachable) and manufacturable (hence immediately practical). Practical engineers achieve many successes by cut-and-try methods and put them into production. Exploratory engineers study what will become practical as manufacturing abilities expand to embrace more of the possible.

The above illustration shows how exploratory engineering relates to more familiar kinds of engineering. Each works within the limits of the possible, which are set by the known and unknown laws of nature. The most familiar kind is the engineering taught in schools: this "textbook engineering" covers technologies that can be both understood (so they can be taught) and manufactured (so they can be used). Bridge-building and gearbox design fall in this category. Other technologies, however, can be manufactured but aren't understood—any engineer can give examples of things that work when similar things don't, and for no obvious reason. But as long as they do work, and work consistently, they can be used with confidence. This is the world of "cut-and-try engineering," so important to modern industry. Bearing lubrication, adhesives, and many manufacturing technologies advance by cut-and-try methods.

Exploratory engineering covers technologies that can be understood but not manufactured–yet. Technologies in this category are also familiar to engineers, although normally they design such things only for fun. So much is known about mechanics, thermodynamics, electronics, and so forth that engineers can often calculate what something will do, just from a description of it. Yet there is no reason why everything that can be correctly described must be manufacturable—the constraints are different. Exploratory engineering is as simple as textbook engineering, but neither military planners nor corporate executives see much profit in it, so it hasn't received much attention.

The concepts of molecular manufacturing and molecular nanotechnology are straightforward results of exploratory-engineering research applied to molecular systems. As we observed above, the basic ideas could have been worked out forty years ago, if anyone had bothered. Naturally enough, both scientists and engineers were preoccupied with more immediate concerns. But now, with the threshold of nanotechnology approaching, attention is beginning to focus on where the next steps lead.

Nanotechnology seems to be where the world is headed if technology keeps advancing, and competition practically guarantees that advances will continue. It will open both a huge range of opportunities for benefit and a huge range of opportunities for misuse. We will paint scenarios to give a sense of the prospects and possibilities, but we don't offer predictions of what will happen. Actual human choices and blunders will depend on a range of factors and alternatives beyond what we can hope to anticipate.

Unit I–B

The Molecular World

Nanotechnology will be a bottom-up technology, building upward from the molecular scale. It will bring a revolution in human abilities like that brought by agriculture or power machinery. It can even be used to reverse many of the changes brought by agriculture or power machinery. But we humans are huge creations with no direct experience of the molecular world, and this can make nanotechnology hard to visualize, hence hard to understand.

Scientists working with moleculas face this problem today. They can often calculate how molecules will behave, but to understand this behavior, they need more than heaps of numbers: they need pictures, movies, and interactive simulations, and so they are producing them at an ever-increasing pace.

Molecules are objects that exert forces on one another. If your hands were small enough, you could grab them, squeeze them, and bash them together. Understanding the molecular world is much like understanding any other physical world: it is a matter of understanding size, shape, strength, force, motion, and the like—a matter of understanding the differences between sand, water, and rock, or between steel and soap bubbles. Today's visualization tools give a taste of what will become possible with tomorrow's faster computers and better "virtual realities," simulated environments that let you tour a world that "exists" only as a model inside the computer. Before discussing nanotechnology and how it relates to the technologies of today, let's try to get a more concrete understanding of the molecular world by describing a simulation embedded in a scenario. In this scenario, events and technologies described as dating from 1990 or before are historically accurate; those with later dates are either projections or mere scenario elements. The descriptive details in the simulation are written to fit designs and calculations based on standard scientific data, so the science isn't fiction.

Exploring the Molecular World

In a scenario in the Unit I—A, we saw Joel Gregory manipulating molecules in the virtual reality of a simulated world using video goggles, tactile gloves, and a supercomputer. The early twenty-first century should be able to do even better. Imagine, then, that today you were to take a really long nap, oversleep, and wake up decades later in a nanotechnological world.

In the twenty-first century, even more than in the twentieth, it's easy to make things work without understanding them, but to a newcomer much of the technology seems like magic, which is dissatisfying. After a few days, you want to understand what nanotechnology is, on a gut level. Back in the late twentieth century, most teaching used dry words and simple pictures, but now—for a topic like this—it's easier to explore a simulated world. And so you decide to explore a simulation of the molecular world.

Looking through the brochure, you read many tedious facts about the simulation: how accurate it is in describing sizes, forces, motions, and the like; how similar it is to working tools used by both engineering students and professionals; how you can buy one for your very own home, and so forth. It explains how you can tour the human body, see state-of-the-art nanotechnology in action, climb a bacterium, etc. For starters, you decide to take an introductory tour: simulations of real twentieth-century objects alongside quaint twentieth-century concepts of nanotechnology.

After paying a small fee and memorizing a few key phrases (any variation of "Get me out of here!" will do the most important job), you pull on a powersuit, pocket a Talking Tourguide, step into the simulation chamber, and strap the video goggles over your eyes. Looking through the goggles, you seem to be in a room with a table you know isn't really there and walls that seem too far away to fit in the simulation chamber. But trickery with a treadmill floor makes the walk to the walls seem far enough, and when you walk back and thump the table, it feels solid because the powersuit stops your hand sharply at just the right place. You can even feel the texture of the carvings on the table leg, because the suit's gloves press against your fingertips in the right patterns as you move. The simulation isn't perfect, but it's easy to ignore the defects. On the table is (or seems to be) an old 1990s silicon computer chip. When you pick it up, as the beginners' instructions suggest, it looks like Figure I-B.1A. Then you say, "Shrink me!", and the world seems to expand.

Vision and Motion

You feel as though you're falling toward the chip's surface, shrinking rapidly. In a moment, it looks roughly like Figure II-B.1B, with your thumb still there holding it. The world grows blurrier, then everything seems to go wrong as you approach the molecular level. First, your vision blurs to uselessness—there is light, but it becomes a featureless fog. Your skin is tickled by small impacts, then battered by what feel like hard-thrown marbles. Your arms and legs feel as though they are caught in turbulence, pulling to and fro, harder and harder. The ground hits your feet, you stumble and stick to the ground like a fly on flypaper, battered so hard that it almost hurts. You asked for realism, and only the built-in safety limits in the suit keep the simulated thermal motions of air molecules and of your own arms from beating you senseless.

"Stop!" gives you a rest from the suit's yanking and thumping, and "Standard settings!" makes the world around you become more reasonable. The simulation changes, introducing the standard cheats. Your simulated eyes are now smaller than a light wave, making focus impossible, but the goggles snap your vision into sharpness and show the atoms around you as small spheres. (Real nanomachines are as blind as you were a moment ago, and can't cheat.) You are on the surface of the computer chip, between a cell and two blocky nanocomputers like the ones in Figure I-B.1D. Your simulated body is 50 nanometers tall, about 1/40,000,000 your real size, and the smaller nanocomputer is twice your height. At that size, you can "see" atoms and molecules, as in Figure I-B.1E.

The simulation keeps bombarding you with air molecules, but the standard settings leave out the sensation of being pelted with marbles. A moment ago you were stuck tight to the ground by molecular stickiness, but the standard settings give your muscles the effective strength of steel—at

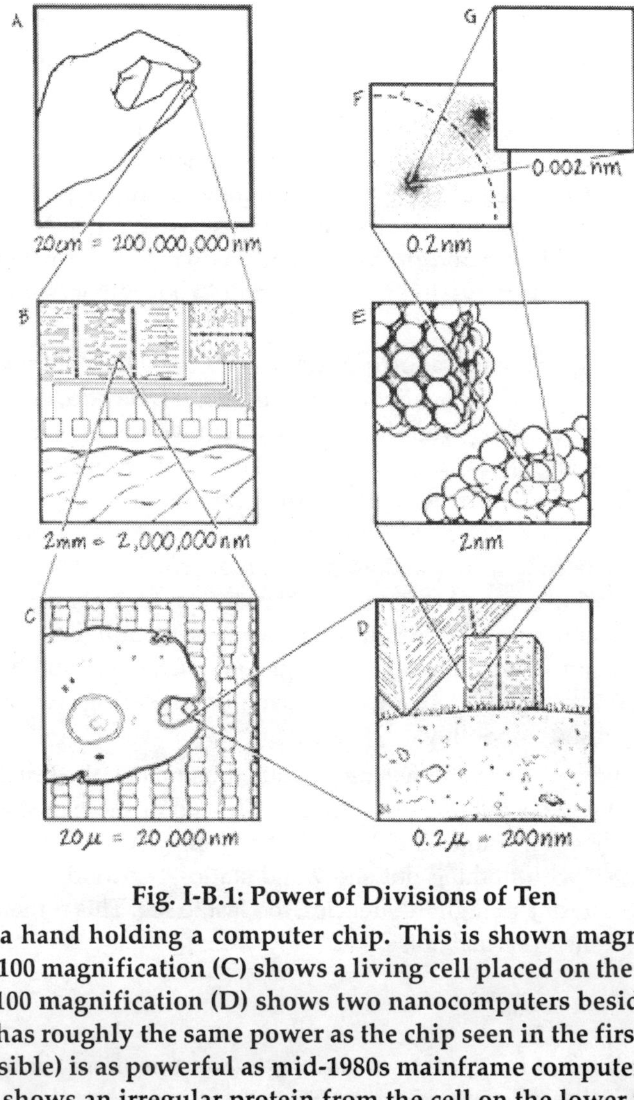

Fig. I-B.1: Power of Divisions of Ten

Frame (A) shows a hand holding a computer chip. This is shown magnified 100 times in (B). Another factor of 100 magnification (C) shows a living cell placed on the chip to show scale. Yet another factor of 100 magnification (D) shows two nanocomputers beside the cell. The smaller (shown as block) has roughly the same power as the chip seen in the first view; the larger (with only the corner visible) is as powerful as mid-1980s mainframe computer. Another factor of 100 magnification (E) shows an irregular protein from the cell on the lower right, and a cylindrical gear made by molecular manufacturing at top left. Taking a smaller factor of 10 jump, (F) shows two atoms in the protein, with electron clouds represented by stippling. A final factor of 100 magnification (G) reveals the nucleus of the atom as a tiny speck.

least in simulation—by making everything around you much softer and weaker. The tourguide says that the only unreal features of the simulation have to do with you—not just your ability to see and to ignore thermal shaking and bombardment, but also your sheer existence at a size too small for anything so complex as a human being. It also explains why you can see things move, something about slowing down everything around you by a factor of 10 for every factor of 10 enlargement, and by another factor to allow for your being made stronger and hence faster. And so, with your greater strength and some adjustments to make your arms, legs, and torso less sticky, you can stand, see, feel, and take stock of the situation.

Molecular Texture

The ground underfoot, like everything around you, is pebbly with atom-sized bumps the size of your fingertips. Objects look like bunches of transparent grapes or fused marbles in a variety of pretty but imaginary colors. The simulation displays a view of atoms and molecules much like those used by chemists in the 1980s, but with a sharper 3-D image and a better way to move them and to feel the forces they exert. Actually, the whole simulation setup is nothing but an improved version of systems built in the late 1980s—the computer is faster, but it is calculating the same things. The video goggles are better and the whole-body powersuit is a major change, but even in the 1980s there were 3-D displays for molecules and crude devices that gave a sense of touching them.

The gloves on this suit give the sensation of touching whatever the computer simulates. When you run a fingertip over the side of the smaller nanocomputer, it feels odd, hard to describe. It is as if the surface were magnetic–it pulls on your fingertip if you move close enough. But the result isn't a sharp click of contact, because the surface isn't hard like a magnet, but strangely soft. Touching the surface is like touching a film of fog that grades smoothly into foam rubber, then hard rubber, then steel, all within the thickness of a sheet of corrugated cardboard. Moving sideways, your fingertip feels no texture, no friction, just smooth bumps more slippery than oil, and a tendency to get pulled into hollows. Pulling free of the surface takes a firm tug. The simulation makes your atom-sized fingertips feel the same forces that an atom would. It is strange how slippery the surface is—and it can't have been lubricated, since even a single oil molecule would be a lump the size of your thumb. This slipperiness makes it obvious how nano-scale bearings can work, how the parts of molecular machines can slide smoothly.

But on top of this, there is a tingling feeling in your fingers, like the sensation of touching a working loudspeaker. When you put your ear against the wall of the nanocomputer, you flinch back: for a moment, you heard a sound like the hiss of a twentieth century television tuned to a channel with no broadcast, with nothing but snow and static—but loud, painfully loud. All the atoms in the surface are vibrating at high frequencies, too fast to see. This is thermal vibration, and it's obvious why it's also called thermal noise.

Gas and Liquid

Individual molecules still move too quickly to see. So, to add one more cheat to the simulation, you issue the command "Whoa!", and everything around seems to slow down by a factor of ten.

On the surface, you now can see thermal vibrations that had been too quick to follow. All around, air molecules become easier to watch. They whiz about as thick as raindrops in a storm, but they are the size of marbles and bounce in all directions. They're also sticky in a magnetlike way, and some are skidding around on the wall of the nanocomputer. When you grab one, it slips away. Most are like two fused spheres, but you spot one that is perfectly round—it is an argon atom, and these are fairly rare. With a firm grip on all sides to keep it from shooting away like a watermelon seed, you pinch it between your steel-strong fingers. It compresses by about 10 percent before the resistance is more than you can overcome. It springs back perfectly and instantly when you relax, then bounces free of your grip. Atoms have an unfamiliar perfection about them, resilient and unchanging, and they surround you in thick swarms.

At the base of the wall is a churning blob that can only be a droplet of water. Scooping up a handful for a closer look yields a swarm of molecules, hundreds, all tumbling and bumbling over one another, but clinging in a coherent mass. As you watch, though, one breaks free of the liquid and flies off into the freer chaos of the surrounding air: the water is evaporating. Some slide up your arm and lodge in the armpit, but eventually skitter away. Getting rid of all the water molecules takes too much scraping, so you command "Clean me!" to dry off.

Too Small and Too Large

Beside you, the smaller nanocomputer is a block twice your height, but it's easy to climb up onto it as the tourguide suggests. Gravity is less important on a small-scale: even a fly can defy gravity to walk on a ceiling, and an ant can lift what would be a truck to us. At a simulated size of fifty nanometers, gravity counts for nothing. Materials keep their strength, and are just as hard to bend or break, but the weight of an object becomes negligible. Even without the strength-enhancement that lets you overcome molecular stickiness, you could lift an object with 40 million times your mass–like a person of normal size lifting a box containing a half-dozen fully loaded oil tankers. To simulate this weak gravity, the powersuit cradles your body's weight, making you feel as if you were floating. This is almost like a vacation in an orbital theme park, walking with stickyboots on walls, ceilings, and whatnot, but with no need for antinausea medication.

On top of the nanocomputer is a stray protein molecule, like the one in Figure I-B.1E. This looks like a cluster of grapes and is about the same size. It even feels a bit like a bunch of grapes, soft and loose. The parts don't fly free like a gas or tumble and wander like a liquid, but they do quiver like gelatin and sometimes flop or twist. It is solid enough, but the folded structure is not as strong as your steel fingers. In the 1990s, people began to build molecular machinery out of proteins, copying biology. It worked, but it's easy to see why they moved on to better materials.

From a simulated pocket, you pull out a simulated magnifying glass and look at the simulated protein. This shows a pair of bonded atoms on the surface at 10 times magnification, looking like Figure I-B.1F. The atoms are almost transparent, but even a close look doesn't reveal a nucleus inside, because it's too small to see. It would take 1,000 times magnification to be able to see it, even with the head start of being able to see atoms with your naked eye. How could people ever confuse big, plump atoms with tiny specks like nuclei? Remembering how your steel-strong fingers couldn't press more than a fraction of the way toward the nucleus of an argon atom from the air, it's clear why nuclear fusion is so difficult. In fact, the tourguide said that it would take a real-world projectile over a hundred times faster than a high-powered rifle bullet to penetrate into the atomic core and let two nuclei fuse. Try as you might, there just isn't anything you could find in the molecular world that could reach into the middle of an atom to meddle with its nucleus. You can't touch it and you can't see it, so you stop squinting though the magnifying glass. Nuclei just aren't of much interest in nanotechnology.

Puzzle Chains

Taking the advice of the tourguide, you grab two molecular knobs on the protein and pull. It resists for a moment, but then a loop comes free, letting other loops flop around more, and the whole structure seems to melt into a writhing coil. After a bit of pulling and wrestling, the protein's structure becomes obvious: It is a long chain–longer than you are tall, if you could get it straight—and each segment of the chain has one of several kinds of knobs sticking off to the side. With the multicolored, glassy-bead portrayal of atoms, the protein chain resembles a flamboyant necklace. This may be decorative, but how does it all go back together? The chain flops and twists and thrashes, and you pull and push and twist, but the original tight, solid packing is lost. There are more ways to go wrong in folding up the chain than there are in solving Rubik's Cube, and now that the folded structure is gone, it isn't even clear what the result should look like. How did those twentieth-century researchers ever solve the notorious "protein folding problem"? It's a matter of record that they started building protein objects in the late 1980s.

This protein molecule won't go back together, so you try to break it. A firm grip and a powerful yank straightens a section a bit, but the chain holds together and snaps back. Though unfolding it was easy, even muscles with the strength of steel—the strength of Superman—can't break the chain itself. Chemical bonds are amazingly strong, so it's time to cheat again. When you say, "Flimsy world–one second!" while pulling, your hands easily move apart, splitting the chain in two before its strength returns to normal. You've forced a chemical change, but there must be easier ways since chemists do their work without tiny superhands. While you compare the broken ends, they thrash around and bump together. The third time this happens, the chain rejoins, as strong as before. This is like having snap-together parts, but the snaps are far stronger than welded steel. Modern assembler chemistry usually uses other approaches, but seeing this happen makes the idea of molecular assembly more understandable: Put the right pieces together in the right positions, and they snap together to make a bigger structure.

Remembering the "Whoa!" command, you decide to go back to the properly scaled speed for your size and strength. Saying "Standard settings!," you see the thrashing of the protein chain speed up to hard-to-follow blur.

Nanomachines

At your feet is a ribbed, ringed cylindrical object about the size of a soup can—not a messy, loosely folded strand like the protein (before it fell apart), but a solid piece of modern nanotechnology. It's a gear like the one in Figure I-B.1E. Picking it up, you can immediately feel how different it is from a protein. In the gear, everything is held in place by bonds as strong as those that strung together the beads of the protein chain. It can't unfold, and you'd have to cheat again to break its perfect symmetry. Like those in the wall of the nanocomputer, its solidly attached atoms vibrate only slightly. There's another gear nearby, so you fit them together and make the atomic teeth mesh, with bumps on one fitting into hollows on the other. They stick together, and the soft, slick atomic surfaces let them roll smoothly.

Underfoot is the nanocomputer itself, a huge mechanism built in the same rigid style. Climbing down from it, you can see through the transparent layers of the wall to watch the inner works. An electric motor an arm-span wide spins inside, turning a crank that drives a set of oscillating rods, which in turn drive smaller rods. This doesn't look like a computer; it looks more like an engineer's fantasy from the nineteenth century. But then, it is an antique design–the tourguide said that the original proposal was a piece of exploratory engineering dating from the mid-1980s, a mechanical design that was superseded by improved electronic designs before anyone had the tools to build even a prototype. This simulation is based on a version built by a hobbyist many years later.

The mechanical nanocomputer may be crude, but it does work, and it's a lot smaller and more efficient than the electronic computers of the early 1990s. It's even somewhat faster. The rods slide back and forth in a blur of motion, blocking and unblocking each other in changing patterns, weaving patterns of logic. This nanocomputer is a stripped-down model with almost no memory, useless by itself. Looking beyond it, you see the other block–the one on the left in Figure I-B.1D–which contains a machine powerful enough to compete with most computers built in 1990. This computer is a millionth of a meter on a side, but from where you stand, it looks like a blocky building looming over ten stories tall. The tourguide says that it contains over 100 billion atoms and stores as much data as a room full of books. You can see some of the storage system inside: row upon row of racks containing spools of molecular tape somewhat like the protein chain, but with simple bumps representing the 1s and 0s of computer data.

These nanocomputers seem big and crude, but the ground you're now standing on is also a computer—a single chip roughly as powerful as the smaller, stripped-down nanocomputer at your side. As you gaze out over the chip, you get a better sense for just how crude things were a few decades ago. At your feet, on the smallest scale, the chip is an irregular mess. Although the wall of the nanocomputer is pebbly with atomic-scale bumps, the bumps are as regular as tile. The chip's surface, though, is a jumble of lumps and mounds. This pattern spreads for dozens of paces in all directions, ending in an irregular cliff marking the edge of a single transistor. Beyond, you can see other ridges and plateaus stretching off to the horizon. These form grand, regular patterns, the circuits of the computer. The horizon—the edge of the chip—is so distant that walking there from the center would (as the tourguide warns) take days. And these vast pieces of landscaping were considered twentieth-century miracles of miniaturization?

Cells and Bodies

Even back then, research in molecular biology had revealed the existence of smaller, more perfect machines such as the protein molecules in cells. A simulated human cell–put here because earlier visitors wanted to see the size comparisons–its on the chip next to the smaller nanocomputer. The tourguide points out that the simulation cheats a bit at this point, making the cell act as though it were in a watery environment instead of air. The cell dwarfs the nanocomputer, sprawling across the chip surface and rearing into the sky like a small mountain. Walking the nature trail around its edge would lead across many transistor-plateaus and take about an hour. A glance is enough to show how different it is from a nanocomputer or a gear: it looks organic, it bulges and curves like a blob of liver, but its surface is shaggy with waving molecular chains.

Walking up to its edge, you can see that the membrane wrapping the cell is fluid (cell walls are for stiff things like plants), and the membrane molecules are in constant motion. On an impulse, you thrust your arm through the membrane and poke around inside. You can feel many proteins bumping and tumbling around in the cell's interior fluid, and a crisscrossing network of protein cables and beams. Somewhere inside are the molecular machines that made all these proteins, but such bits of machinery are embedded in a roiling, organic mass. When you pull your arm out, the membrane flows closed behind. The fluid, dynamic structure of the cell is largely self healing. That's what let scientists perform experimental surgery on cells with the old, crude tools of the 20th century: they didn't need to stitch up the holes they made when they poked around inside.

Even a single human cell is huge and complex. No real thinking being could be as small as you are in the simulation: A simple computer without any memory is twice your height, and the larger nanocomputer, the size of an apartment complex, is no smarter than one of the submoronic computers of 1990. Not even a bendable finger could be as small as your simulated fingers: in the simulation, your fingers are only one atom wide, leaving no room for the slimmest possible tendon, to say nothing of nerves.

For a last look at the organic world, you gaze out past the horizon and see the image of your own, full-sized thumb holding the chip on which you stand. The bulge of your thumb rises ten times higher than Mount Everest. Above, filling the sky, is a face looming like the Earth seen from orbit, gazing down. It is your own face, with cheeks the size of continents. The eyes are motionless. Thinking of the tourguide's data, you remember:

The simulation uses the standard mechanical scaling rules, so being 40 million times smaller has made you 40 million times faster. To let you pull free of surfaces, it increased your strength by more than a factor of 100, which increased your speed by more than a factor of 10. So one second in the ordinary world corresponds to over 400 million here in the simulation. It would take years to see that huge face in the sky complete a single eyeblink.

Enough. At the command "Get me out!", the molecular world vanishes, and your feeling of weight returns as the suit goes slack. You strip off the video goggles—and hugely, slowly, blink.

Unit I–C

Bottom-up Technology

The Unit I-B showed the sizes, forces, and general nature of objects in the molecular world. Building on this, we can get a better picture of where developments seem to be leading, a better picture of molecular manufacturing itself. To show the sizes, forces, and general nature of things in molecular manufacturing, we first invite the reader (and the reader's inquisitive alter ego) to take a second and final tour before returning to the world of present-day research. As before, the pre-1990 history is accurate, and the science isn't fiction.

The Silicon Valley Faire

The tour of the molecular world showed some products of molecular manufacturing, but didn't show how they were made. The technologies you remember from the old days have mostly been replaced—but how did this happen? The Silicon Valley Faire is advertised as "An authentic theme park capturing life, work, and play in the early Breakthrough years." Since "work" must include manufacturing, it seems worth a visit.

A broad dome caps the park—"To fully capture the authentic sights, sounds, and smells of the era," the tourguide politely says. Inside, the clothes and hairstyles, the newspaper headlines, the bumper-to-bumper traffic, all look much as they did before your long nap. A light haze obscures the buildings on the far side of the dome, your eyes burn slightly, and the air smells truly authentic.

Pocket Libraries

The Nanofabricators, Inc., plant offers the main display of early nanotechnology. As you near the building, the tourguide mentions that this is indeed the original manufacturing plant, given landmark status over twenty years ago, and then made the centerpiece of the Silicon Valley Faire ten years later, when . . . With a few taps, you reset the pocket tourguide to speak up less often.

As people file into the Nanofabricator plant, there's a moment of hushed quiet, a sense of walking into history. Nanofabricators: home of the Superchip, the first mass-market product of nanotechnology. It was the huge memory capacity of Superchips that made possible the first Pocket Library.

This section of the plant now houses a series of displays, including working replicas of early products. Picking up a pocket library, you find that it's not only the size of a wallet, but about the same weight. Yet it has enough memory to record every volume in the Library of Congress–something like a million times the capacity of a personal computer from 1990. It opens with a flip, the two-panel screen lights up, and a world of written knowledge is at your fingertips. Impressive.

"Wow, can you believe these things?" says another tourist as he fingers a pocket library. "Hardly any video, no 3-D–just words, sound, and flat pictures. And the cost! I wouldn't've bought `em for my kids at that price!"

Your tourguide quietly states the price: about what you remember for a top-of-the-line TV set. This isn't the cheap manufacturing promised by mature nanotechnology, but it seems like a pretty good price for a library. Hmm . . . how did they work out the copyrights and royalties? There's a lot more to this product than just the technology . . .

Nanofabrication

The next room displays more technology. Here in the workroom where Superchips were first made, early nanotech manufacturing is spread out on display. The whole setup is surprisingly quiet and ordinary. Back in the 1980s and 1990s, chip plants had carefully controlled clean rooms with gowns and masks on workers and visitors, special workstations, and carefully crafted air flows to keep dust away from products. This room has none of that. It's even a little grubby.

In the middle of a big square table are a half-dozen steel tanks, about the size and shape of old-fashioned milk cans. Each can has a different label identifying its contents: MEMORY BLOCKS, DATA-TRANSMISSION BLOCKS, INTERFACE BLOCKS. These are the parts needed for building up the chip. Clear plastic tubes, carrying clear and tea-colored liquids, emerge from the mouths of the milk cans and drape across the table. The tubes end in fist-sized boxes mounted above shallow dishes sitting in a ring around the cans. As the different liquids drip into each dish, a beater like a kitchen mixer swirls the liquid. In each dish, nanomachines are building Superchips.

A nanofab "engineer," dressed in period clothing complete with name badge, is setting up a dish to begin building a new chip. "This," he says, holding up a blank with a pair of tweezers, "is a silicon chip like the ones made with pre-breakthrough technology. Companies here in this valley made chips like these by melting silicon, freezing it into lumps, sawing the lumps into slices, polishing the slices, and then going through a long series of chemical and photographic steps. When they were done, they had a pattern of lines and blobs of different materials on the surface. Even the smallest of these blobs contained billions of atoms, and it took several blobs working together to store a single bit of information. A chip this size, the size of your fingernail, could store only a fraction of a billion bits. Here at Nanofab, we used bare silicon chips as a base for building up nanomemory. The picture on the wall here shows the surface of a blank chip: no transistors, no memory circuits, just fine wires to connect up with the nanomemory we built on top. The nanomemory, even in the early days, stored thousands of billions of bits. And we made them like this, but a thousand at a time–" He places the chip in the dish, presses a button, and the dish begins to fill with liquid.

"A few years latter," he adds, "we got rid of the silicon chips entirely"–he props up a sign saying, THIS CHIP BUILD BEGAN AT 2:15 P.M., ESTIMATED COMPLETION TIME: 1:00 A.M." and we sped up the construction process by a factor of a thousand."

The chips in the dishes all look pretty much the same except for color. The new chip looks like dull metal. The only difference you can see in the older chips, further along in the process, is a smooth rectangular patch covered by a film of darker material. An animated flowchart on the wall shows how layer upon layer of nanomemory building blocks are grabbed from solution and laid down on the surface to make that film. The tourguide explains that the energy for this process, like the energy for molecular machines within cells, comes from dissolved chemicals–from oxygen and fuel molecules. The total amount of energy needed here is trivial, because the amount of product is trivial: at the end of the process, the total thickness of nanomemory structure–the memory store for a Pocket Library–amounts to one-tenth the thickness of a sheet of paper, spread over an area smaller than a postage stamp.

Molecular Assembly

The animated flowchart showed nanomemory building blocks as big things containing about a hundred thousand atoms apiece (it takes a moment to remember that this is still submicroscopic). The build process in the dishes stacked these blocks to make the memory film on the SuperChip, but how were the blocks themselves built? The hard part in this molecular-manufacturing business has got to be at the bottom of the whole process, at the stage where molecules are put together to make large, complex parts.

The Silicon Valley Faire offers simulations of this molecular assembly process, and at no extra charge. From the tourguide, you learn that modern assembly processes are complex; that earlier processes–like those used by Nanofabricators, Inc. –used clever-but-obscure engineering tricks; and that the simplest, earliest concepts were never built. Why not begin at the beginning? A short walk takes you to the Museum of Antique Concepts, the first wing of the Museum of Molecular Manufacturing.

A peek inside the first hall shows several people strolling around wearing loosely fitting jumpsuits with attached goggles and gloves, staring at nothing and playing mime with invisible objects. Oh well, why not join the fools' parade? Stepping through the doorway while wearing the suit is entirely different. The goggles show a normal world outside the door and a molecular world inside. Now you, too, can see and feel the exhibit that fills the hall. It's much like the earlier simulated molecular world: it shares the standard settings for size, strength, and speed. Again, atoms seem 40 million times larger, about the size of your fingertips. This simulation is a bit less thorough than the last was–you can feel simulated objects, but only with your gloved hands. Again, everything seems to be made of quivering masses of fused marbles, each an atom.

"Welcome," says the tourguide, "to a concept for a molecular-manufacturing plant. These exploratory engineering designs were never intended for actual use, yet they demonstrate the basics of molecular manufacturing: making parts, testing them, and assembling them."

Machinery fills the hall. Overall, the sight is reminiscent of an automated factory of the 1980s or 1990s. It seems clear enough what must be going on: Big machines stand beside a conveyor belt loaded with half-finished-looking blocks of some material (this setup looks much like Figure I-C.1); the machines must do some sort of work on the blocks. Judging by the conveyor belt, the blocks eventually move from one arm to the next until they turn a corner and enter the next hall.

Since nothing is real, the exhibit can't be damaged, so you walk up to a machine and give it a poke. It seems as solid as the wall of the nanocomputer in the previous tour. Suddenly, you notice something odd: no bombarding air molecules and no droplets of water–in fact, no loose molecules anywhere. Every atom seems to be part of a mechanical system, quivering thermal vibration, but otherwise perfectly controlled. Everything here is like the nanocomputer or like the tough little gear; none of it resembles the loosely coiled protein or the roiling mass of the living cell.

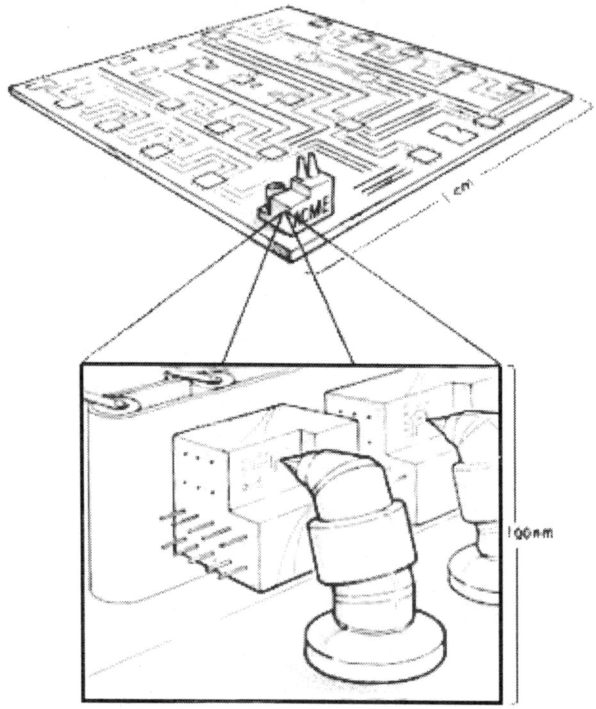

Fig. I-C.1: Assembler with Factory on Chip
A factory–large enough to make over 10 million nanocomputers per day would fit on the edge one of today's integrated circuits. Inset shows an assembler arm together with workpiece on a conveyor belt.

The conveyor belt seems motionless. At regular intervals along the belt are blocks of material under construction: workpieces. The nearest block is about a hundred marble-bumps wide, so it must contain something like 100 × 100 × 100 atoms, a full million. This block looks strangely familiar, with its rods, crank, and the rest. It's a nanocomputer–or rather, a blocklike part of a nanocomputer still under construction.

Standing alongside the pieces of nanocomputer on the conveyor belt, dominating the hall, is a row of huge mechanisms. Their trunks rise from the floor, as thick as old oaks. Even though they bend over, they rear overhead. "Each machine," your tourguide says, "is the arm of a general-purpose molecular assembler."

One assembler arm is bent over with its tip pressed to a block on the conveyor belt. Walking closer, you see molecular assembly in action. The arm ends in a fist-sized knob with a few protruding marbles, like knuckles. Right now, two quivering marbles—atoms—are pressed into a small hollow in the block. As you watch, the two spheres shift, snapping into place in the block with a quick twitch of motion: a chemical reaction. The assembler arm just stands there, nearly motionless. The fist has lost two knuckles, and the block of nanocomputer is two atoms larger.

The tourguide holds forth: "This general-purpose assembler concept resembles, in essence, the factory robots of the 1980s. It is a computer-controlled mechanical arm that moves molecular tools according to a series of instructions. Each tool is like a single-shot stapler or rivet gun. It has a handle for the assembler to grab and comes loaded with a little bit of matter–a few atoms–which it attaches to the workpiece by a chemical reaction." This is like the rejoining of the protein chain in the earlier tour.

Molecular Precision

The atoms seemed to jump into place easily enough; can they jump out of place just as easily? By now the assembler arm has crept back from the surface, leaving a small gap, so you can reach in and poke at the newly added atoms. Poking and prying do no good: When you push as hard as you can (with your simulated fingers as strong as steel), the atoms don't budge by a visible amount. Strong molecular bonds hold them in place.

Your pocket tourguide–which has been applying the power of a thousand supercomputers to the task of deciding when to speak up–remarks, "Molecular bonds hold things together. In strong, stable materials atoms are either bonded, or they aren't, with no possibilities in between. Assemblers work by making and breaking bonds, so each step either succeeds perfectly or fails completely. In pre-breakthrough manufacturing, parts were always made and put together with small inaccuracies. These could add up to wreck product quality. At the molecular scale, these problems vanish. Since each step is perfectly precise, little errors can't add up. The process either works, or it doesn't."

But what about those definite, complete failures? Fired by scientific curiosity, you walk to the next assembler, grab the tip, and shake it. Almost nothing happens. When you shove as hard as you can, the tip moves by about one-tenth of an atomic diameter, then springs back. "Thermal vibrations can cause mistakes by causing parts to come together and form bonds in the wrong place," the tourguide remarks. "Thermal vibrations make floppy objects bend further than stiff ones, and so these assembler arms were designed to be thick and stubby to make them very stiff. Error rates can be kept to one in a trillion, and so small products can be perfectly regular and perfectly identical. Large products can be almost perfect, having just a few atoms out of place." This should mean high reliability. Oddly, most of the things you've been seeing outside have looked pretty ordinary–not slick, shiny, and perfect, but rough and homey. They must have been manufactured that way, or made by hand. Slick, shiny things must not impress anyone anymore.

Molecular Robotics

By now, the assembler arm has moved by several atom-widths. Through the translucent sides of the arm you can see that the arm is full of mechanisms: twirling shafts, gears, and large, slowly turning rings that drive the rotation and extension of joints along the trunk. The whole system is a huge, articulated robot arm. The arm is big because the smallest parts are the size of marbles, and the machinery inside that makes it move and bend has many, many parts. Inside, another mechanism is at work: The arm now ends in a hole, and you can see the old, spent molecular tool being retracted through a tube down the middle.

Patience, patience. Within a few minutes, a new tool is on its way back up the tube. Eventually, it reaches the end. Shafts twirl, gears turn, and clamps lock the tool in position. Other shafts twirl, and the arm slowly leans up against the workpiece again at a new site. Finally, with a twitch of motion, more atoms jump across, and the block is again just a little bit bigger. The cycle begins again. This huge arm seems amazingly slow, but the standard simulation settings have shifted speeds by a factor of over 400 million. A few minutes of simulation time correspond to less than a millionth of a second of real time, so this stiff, sluggish arm is completing about a million operations per second.

Peering down at the very base of the assembler arm, you can get a glimpse of yet more assembler-arm machinery underneath the floor: Electric motors spin, and a nanocomputer chugs away, rods pumping furiously. All these rods and gears move quickly, sliding and turning many times for every cycle of the ponderous arm. This seems inefficient; the mechanical vibrations must generate a lot of heat, so the electric motors must draw a lot of power. Having a computer control each arm is a lot more awkward now than it was in pre-breakthrough years. Back then, a robot arm was big and expensive and a computer was a cheap chip; now the computer is bigger than the arm. There must be a better way–but then, this is the Museum of Antique Concepts.

Building-Blocks into Buildings

Where do the blocks go, once the assemblers have finished with them? Following the conveyor belt past a dozen arms, you stroll to the end of the hall, turn the corner, and find yourself on a balcony overlooking a vaster hall beyond. Here, just off the conveyor belt, a block sits in a complex fixture. Its parts are moving, and an enormous arm looms over it like a construction crane. After a moment, the tourguide speaks up and confirms your suspicion: "After manufacturing, each block is tested. Large arms pick up properly made blocks. In this hall, the larger arms assemble almost a thousand blocks of various kinds to make a complete nanocomputer.

The grand hall has its own conveyor belt, bearing a series of partially completed nanocomputers. Arrayed along this grand belt is a row of grand arms, able to swing to and fro, to reach down to lesser conveyor belts, pluck million-atom blocks from testing stations, and plug them into the grand workpieces, the nanocomputers under construction. The belt runs the length of the hall, and at the end, finished nanocomputers turn a corner–to a yet-grander hall beyond?

After gazing at the final-assembly hall for several minutes, you notice that nothing seems to have moved. Mere patience won't do: at the rate the smaller arms moved in the hall behind you, each block must take months to complete, and the grand block-handling arms are taking full advantage of the leisure this provides. Building a computer, start to finish, might take a terribly long time. Perhaps as long as the blink of an eye.

Molecular assemblers build blocks that go to block assemblers. The block assemblers build computers, which go to system assemblers, which build systems, which—at least one path from molecules to large products seems clear enough. If a car were assembled by normal-sized robots from a thousand pieces, each piece having been assembled by smaller robots from a thousand smaller pieces, and so on, down and down, then only ten levels of assembly process would separate cars from molecules. Perhaps, around a few more corners and down a few more ever-larger halls, you would see a post-breakthrough car in the making, with unrecognizable engine parts and comfortable seating being snapped together in a century-long process in a hall so vast that the Pacific Ocean would be a puddle in the corner .

Just ten steps in size; eight, starting with blocks as big as the ones made in the hall behind you. The molecular world seems closer, viewed this way.

Molecular Processing

Stepping back into that hall, you wonder how the process begins. In every cycle of their sluggish motion, each molecular assembler gets a fresh tool through a tube from somewhere beneath the floor, and that somewhere is where the story of molecular precision begins. And so you ask, "Where do the tools come from?", and the tourguide replies, "You might want to take the elevator to your left."

Stepping out of the elevator and into the basement, you see a wide hall full of small conveyor belts and pulleys; a large pipe runs down the middle. A plaque on the wall says, "Mechanochemical processing concept, circa 1990." As usual, all the motions seem rather slow, but in this hall everything that seems designed to move is visibly in motion. The general flow seems to be away from the pipe, through several steps, and then up through the ceiling toward the hall of assemblers above.

After walking over to the pipe, you can see that it is nearly transparent. Inside is a seething chaos of small molecules: the wall of the pipe is the boundary between loose molecules and controlled ones, but the loose molecules are well confined. In this simulation, your fingertips are like small molecules. No matter how hard you push, there's no way to drive your finger through the wall of the pipe. Every few paces along the pipe a fitting juts out, a housing with a mechanically driven rotating thing, exposed to the liquid inside the pipe, but also exposed to a belt running over one of the pulleys, embedded in the housing. It's hard to see exactly what is happening.

The tourguide speaks up, saying, "Pockets on the rotor capture single molecules from the liquid in the pipe. Each rotor pocket has a size and shape that fits just one of the several different kinds of molecule in the liquid, so the process is rather selective. Captured molecules are then pushed into the pockets on the belt that's wrapped over the pulley there, then–"

"Enough," you say. Fine, it singles out molecules and sticks them into this maze of machinery. Presumably, the machines can sort the molecules to make sure the right kinds go to the right places.

The belts loop back and forth carrying big, knobby masses of molecules. Many of the pulleys–rollers?–press two belts together inside a housing with auxiliary rollers. While you are looking at one of these, the tourguide says, "Each knob on a belt is a mechanochemical-processing device. When two knobs on different belts are pressed together in the right way, they are designed to transfer molecular fragments from one to another by means of a mechanically forced chemical reaction. In this way, small molecules are broken down, recombined, and finally joined to molecular tools of the sort used in the assemblers in the hall above. In this device here, the rollers create a pressure equal to the pressure found halfway to the center of the Earth, speeding a reaction that–"

"Fine, fine," you say. Chemists in the old days managed to make amazingly complex molecules just by mixing different chemicals together in solution in the right order under the right conditions. Here, molecules can certainly be brought together in the right order, and the conditions are much better controlled. It stands to reason that this carefully designed maze of pulleys and belts can do a better job of molecule processing than a test tube full of disorganized liquid ever could. From a liquid, through a sorter, into a mill, and out as tools: this seems to be the story of molecule processing. All the belts are loops, so the machinery just goes around and around, carrying and transforming molecular parts.

Beyond Antiques

This system of belts seems terribly simple and efficient, compared to the ponderous arms driven by frantic computers in the hall above. Why stop with making simple tools? You must have muttered this, because the tourguide speaks up again and says, "The Special-Assembler Exhibit shows another early molecular-manufacturing concept that uses the principles of this molecule-processing system to build large, complex objects. If a system is building only a single product, there is no need to have computers and flexible arms move parts around. It is far more efficient to build a machine in which everything just moves on belts at a constant speed, adding small parts to larger ones and then bringing the larger ones together as you saw at the end of the hall above."

This does seem like a more sensible way to churn out a lot of identical products, but it sounds like just more of the same. Gears like fused marbles, belts like coarse beadwork, drive shafts, pulleys, machines and more machines. In a few places, marbles snap into new patterns to prepare a tool or make a product. Roll, roll, chug, chug, pop, snap, then roll and chug some more.

As you leave the simulation hall, you ask, "Is there anything important I've missed in this molecular manufacturing tour?"

The tourguide launches into a list: "Yes–the inner workings of assembler arms, with drive shafts, worm gears, and harmonic drives; the use of Diels-Alder reactions, interfacial free-radial chain reactions, and dative-bond formation to join blocks together in the larger-scale stages of assembly; different kinds of mechanochemical processing for preparing reactive molecular tools; the use of staged-cascade methods in providing feed-molecules of the right kinds with near-perfect reliability; the differences between efficient and inefficient steps in molecular processing; the use of redundancy to ensure reliability in large systems despite sporadic damage; modern methods of building large objects from smaller blocks; modern electronic nanocomputers; modern methods for–"

"Enough!" you say, and the tourguide falls silent as you pitch it into a recycling bin. A course in molecular manufacturing isn't what you're looking for right now; the general idea seems clear enough. It's time to take another look at the world on a more normal scale. Houses, roads, buildings, even the landscape looked different out there beyond the Faire domeless crowded, paved, and plowed than you remember. But why? The history books (well, they're more than just books) say that molecular manufacturing made a big difference; perhaps now the changes will make more sense. Yes, it's time to leave.

As you toss your goggled, gloved jumpsuit into another bin, a striking dark-haired woman is taking a fresh one from a rack. She wears a jacket emblazoned with the name "Desert Rose NanoManufacturing."

"How'd you like it?" she asks with a smile.

"Pretty amazing," you say.

"Yes," she agrees. "I saw this sim back when I was taking my first molecular-manufacturing class. I swore I'd never design anything so clunky! This whole setup really brings back the memories–I can't wait to see if it's as crude as I remember." She steps into the simulation hall and closes the door.

Crude Technology

As the Silicon Valley Faire Scenario shows, molecular manufacturing will work much like ordinary manufacturing, but with devices built so small that a single loose molecule of pollutant would be like a brick heaved into a machine tool. John Walker of Autodesk, a leading company in computer-aided design, observes that nanotechnology and today's crude methods are very different: "Technology has never had this kind of precise control; all of our technologies today are bulk technologies. We take a big chunk of stuff and hack away at it until we're left with the object we want, or we assemble parts from components without regard to structure at the molecular level."

Molecular manufacturing will orchestrate atoms into products of symphonic complexity, but modern manufacturing mostly makes loud noises. These figurative noises are sometimes all too literal: A crack in a metal forging grows under stress, a wing fails, and a passenger jet crashes from

the sky. A chemical reaction goes out of control, heat and pressure build, and a poisonous blast shakes the countryside. A lifesaving product cannot be made, a heart fails, and a hospital's heart-monitoring machine signals the end with a high-pitched wail.

Today, we make many things from metal, by machining. From the perspective of our standard, simulated molecular world, a typical metal part is a piece of terrain many days' journey across. The metal itself is weak compared to the bonds of the protein chain or other tough nanomechanisms: solid steel is no stronger than your simulated fingers, and the atoms on its surface can be pushed around with your bare hands. Standing on a piece of metal being machined in a lathe, you would see a cutting blade crawl past a few times per year, like a majestic plough the size of a mountain range. Each pass would rip up a strip of the metal landscape, leaving a rugged valley broad enough to hold a town. This is machining from a nanotechnological perspective: a process that hacks crude shapes from intrinsically weak materials.

Today, electronics are made from silicon chips. We have already seen the landscape of a finished chip. During manufacturing, metal features would be built up by a centuries-long drizzle of metal-atom rain, and hollows would be formed by a centuries-long submergence in an acid sea. From the perspective of our simulation, the whole process would resemble geology as much as manufacturing, with the slow layering of sedimentary deposits alternating with ages of erosion. The term *nanotechnology* is sometimes used as a name for small-scale microtechnology, but the difference between molecular manufacturing and this sort of microlandscaping is like the difference between watchmaking and bulldozing.

Today, chemists make molecules by solution chemistry. We have seen what a liquid looks like in our first simulation, with molecules bumping and tumbling and wandering around. Just as assemblers can make chemical reactions occur by bringing molecules together mechanically, so reactions can occur when molecules bump at random through thermal vibration and motion in a liquid. Indeed, much of what we know today about chemical reactions comes from observing this process. Chemists make large molecules by mixing small molecules in a liquid. By choosing the right molecules and conditions, they can get a surprising measure of control over the results: only some pairs of molecules will react, and then only in certain ways.

Doing chemistry this way, though, is like trying to assemble a model car by putting the pieces in a box and shaking. This will only work with cleverly shaped pieces, and it is hard to make anything very complex. Chemists today consider it challenging to make a precise, three-dimensional structure having a hundred atoms, and making one with a thousand atoms is a great accomplishment. Molecular manufacturing, in contrast, will routinely assemble millions or billions. The basic chemical principles will be the same, but control and reliability will be vastly greater. It is the difference between throwing things together blindly and putting them together with a watchmaker's care.

Technology today doesn't permit thorough control of the structure of matter. Molecular manufacturing will. Today's technologies have given us computers, spacecraft, indoor plumbing, and the other wonders of the modern age. Tomorrow's will do much more, bringing change and choices.

Simple Matter, Smart Matter

Today's technology mostly works with matter in a few basic forms: gases, liquids, and solids. Though each form has many varieties, all are comparatively simple.

Gases, as we've seen, consist of molecules ricocheting through space. A volume of gas will push against its walls and, if not walled in, expand without limit. Gases can supply certain raw materials for nanomachines, and nanomachines can be used to remove pollutants from air and turn them into something else. Gases lack structure, so they will remain simple.

Liquids are somewhat like gases, but their molecules cling together to form a coherent blob that won't expand beyond a certain limit. Liquids will be good sources of raw materials for nanomachines because they are denser and can carry a wide range of fuels and raw materials in solution (the pipe in the molecular-processing hall contained liquid). Nanomachines can clean up polluted water as easily as air, removing and transforming noxious molecules. Liquids have more structure than gases, but nanotechnology will have its greatest application to solids.

Solids are diverse. Solid butter consists of molecules stronger than steel, but the molecules cling to one another by the weaker forces of molecular stickiness. A little heat increases thermal vibrations and makes the solid structure disintegrate into a blob of liquid. Butterlike materials would make poor nanomachines. Metals consist of atoms held together by stronger forces, and so they can be structurally stronger and able to withstand higher temperatures. The forces are not very directional, though, and so planes of metal atoms can slip past one another under pressure; this is why spoons bend, rather than break. This ability to slip makes metals less brittle and easier to shape (with crude technology), but it also weakens them. Only the strongest, hardest, highest-melting-point metals are worth considering as parts of nanomachines.

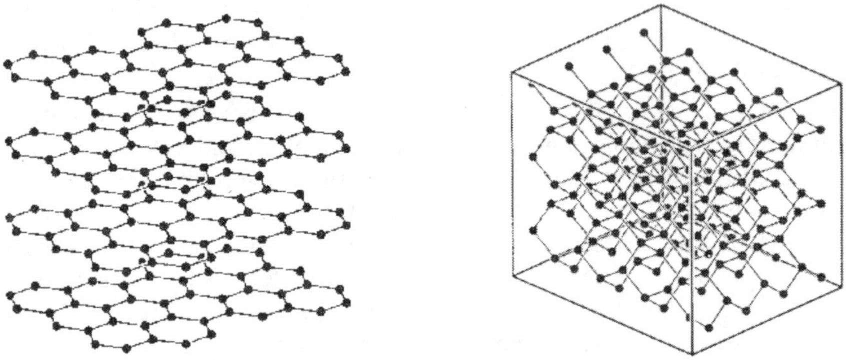

Fig. I-C.2: Carbon-Soft and Hard.
On the left is graphite, the material called "lead" in pencils—made of carbon atoms. On right is diamond, the same atoms arranged in a different pattern.

Diamond consists of carbon atoms held together by strong, directional bonds, like the bonds down the axis of a protein chain. (See Figure I-C.2.) These directional bonds make it hard for planes of atoms to slip past one another, making diamond (and similar materials) very strong indeed–ten to a hundred times stronger than steel. But the planes can't easily slip, so when the material fails, it doesn't bend, it breaks. Tiny cracks can easily grow, making a large object seem weak. Glass is a similar material: glass windows seem weak–and a scratch makes glass far weaker–yet thin, perfect glass fibers are widely used to make composite materials stronger and lighter than steel. Nanotechnology will be able to build with diamond and similar strong materials, making small, flawless fibers and components.

In engineering today, diamond is just beginning to be used. Japan has pioneered a technology for making diamond at low pressure, and a Japanese company sells a speaker with excellent high-frequency response–the speaker cone is reinforced with a light, stiff film of diamond. Diamond is extraordinary stuff, made from cheap materials like natural gas. U.S. companies are scrambling to catch up.

All these materials are simple. More complex structures lead to more complex properties, and begin to give some hint of what molecular manufacturing will mean for materials.

What if you strung carbon atoms in long chains with side-groups, a bit like a protein chain, and linked them into a big three-dimensional mesh? If the chains were kinked so that they couldn't pack tightly, they would coil up and flop around almost like molecules in a liquid, yet the strong bonds would keep the overall mesh intact. Pulling the whole network would tend to straighten the chains, but their writhing motions would tend to coil them back up. This sort of network has been made: it is called rubber.

Rubber is weak mostly because the network is irregular. When stretched, first one chain breaks, then another, because they don't all become taut at the same time to share and divide the load. A more regular mesh would be as soft as rubber at first, but when stretched to the limit would become stronger than steel. Molecular manufacturing could make such stuff.

The natural world contains a host of good materials–cellulose and lignin in wood, stronger-than-steel proteins in spider's silk, hard ceramics in grains of sand, and more. Many products of molecular manufacturing will be designed for great durability, like sand. Others will be designed to fall apart easily for easy recycling, like wood. Some may be designed for uses where they may be thrown away. In this last category, nanotailored biodegradables will shine. With care, almost any sort of product from a shoe to computer-driven nanomachines can be made to last for a good long time, and then unzip fairly rapidly and very thoroughly into molecules and other bits of stuff all of kinds normally found in the soil.

This gives only a hint of what molecular manufacturing will make possible by giving better control of the structure of solid matter. The most impressive applications will not be superstrong structural materials, improved rubber, and simple biodegradable materials: these are uniform, repetitive structures not greatly different from ordinary materials. These materials are "stupid." When pushed, they resist, or they stretch and bounce back. If you shine light on them, they transmit it, reflect it, or absorb it. But molecular manufacturing can do much more. Rather than heaping up simple molecules, it can build materials from trillions of motors, ratchets, light-emitters, and computers.

Muscle is smarter than rubber because it contains molecular machines: it can be told to contract. The products of molecular manufacturing can include materials able to change shape, color, and other properties on command. When a dust mote can contain a supercomputer, materials can be made smart, medicine can be made sophisticated, and the world will be a different place. Smart materials will be discussed later.

Ideas and Criticisms

We've just seen a picture of molecular manufacturing (of one sort) and of what it can do (in sketchy outline). Now let's look at the idea of nanotechnology itself: Where did it come from, and what do the experts think of it? The next unit will have more to say on the latter point, presenting the thoughts of researchers who are advancing the field through their own work.

Origins

The idea of molecular nanotechnology, like most ideas, has roots stretching far back in time. In ancient Greece, Democritus suggested that the world was built of durable, invisible particles–atoms, the building blocks of solid objects, liquids, and gases. In the last hundred years, scientists have learned more and more about these building blocks, and chemists have learned more and more ways to combine them to make new things. Decades ago, biologists found molecules that do complex things; they termed them "molecular machines."

Physicist Richard Feynman was a visionary of miniaturization who pointed toward something like molecular nanotechnology: on December 29, 1959, in an after-dinner talk at the annual meeting of the American Physical Society, he proposed that large machines could be used to make smaller machines, which could make still smaller ones, working in a top-down fashion from the macroscale to the microscale. At the end of his talk, he painted a vision of moving individual atoms, pointing out, "The principles of physics, as far as I can see, do not speak against the possibility of maneuvering things atom by atom." He pictured making molecules, pointing clearly in the direction taken by the modern concept of nanotechnology: "But it is interesting that it would be, in principle, possible (I think) for a physicist to synthesize any chemical substance the chemist writes down. Give the orders, and the physicist synthesizes it. How? Put the atoms down where the chemist says, and so you make the substance."

Despite this clear signpost pointing to a potentially revolutionary area, no one filled the conceptual gap between miniature machines and chemical substances. There was no clear concept of making molecular machines able to build more such machines, no notion of controllable molecular manufacturing. With hindsight, one wonders why the gap took so long to fill. Feynman himself didn't follow it up, saying that the ability to maneuver atoms one by one "will really be useless" since chemists would come up with traditional, bulk-process ways to make new chemical substances. For a researcher whose main interest was physics, he had contributed much just by placing the signpost: it was up to others to move forward. Instead, the idea of molecular machines for molecular manufacturing didn't appear for decades.

From today's viewpoint, molecular nanotechnology looks more like an extension of chemistry than like an extension of miniaturization. A mechanical engineer, looking at nanotechnology, might ask, "How can machines be made so small?" A chemist, though, would ask, "How can molecules be made so large?" The chemist has the better question. Nanotechnology isn't primarily about miniaturizing machines, but about extending precise control of molecular structure to larger and larger scales. Nanotechnology is about making (precise) things *big*.

Nature gives the most obvious clues to how this can be done, and it was the growing scientific literature on natural molecular machines that led one of the present authors (Drexler) to propose molecular nanotechnology of the sort described here. A strategy to reach the goal was part of the concept: Build increasingly complex molecular machinery from simpler pieces, including molecular machines able to build more molecular machines. And the motivation for studying this, and publishing? Largely the fear of living in a world that might rush into the new technology blindly, with ugly consequences.

The first book on the topic, *Engines of Creation*, was published in 1986. In 1988, Stanford University became the first to offer a course in molecular nanotechnology, sponsored by the Department of Computer Science. In 1989, this department hosted the first major conference on the subject, cosponsored by the Foresight Institute and Global Business Network. With the upcoming publication of a technical book describing nanotechnology–from molecular mechanical and quantum-mechanical principles up to assembly systems and products–the subject will be easier to teach, and more college courses will become available.

Macroscopic and Molecular Components

Technology	*Function*	*Molecular Examples*
struts, beams, casins	transmit force, hold positions	cell walls, microtubules
cables	transmit tension	collagen, silk
fasteners, glue	Connect parts	intermolecular forces
solenoids, actuators	move things	muscle actin, myosin
motors	turn shafts	flagellar motor
drive shafts	Transmit torque	bacterial flagella
bearings	support moving parts	single bonds
clamps	hold workpieces	enzymatic binding sites
tools	modify workpieces	enzymes, reactive molecules
production lines	control devices	enzyme systems, ribosomes
numerical control systems	store and read programs	genetic system

In parallel with the development and spread of ideas about nanotechnology and molecular manufacturing–ideas that remain pure theory, however well grounded–scientists and engineers, working in laboratories to build real tools and capabilities, have been pioneering roads to nanotechnology. Research has come a long way since the mid-1980s, as we'll see in the next unit. But, as one might expect with a complex new idea that, if true, disrupts a lot of existing plans and expectations, some objections have been heard.

"It Won't Work"

Life might be much simpler if these ideas about nanotechnology had some fatal flaw. If only molecules couldn't be used to form machines, or the machines couldn't be used to build things, then we might be able to keep right on going with our crude technologies: our medicine that doesn't heal, our spacecraft that don't open a new frontier, our oil crises, our pollution, and all the limits that keep us from trading familiar problems for strange ones. Most new ideas are wrong, especially if they purport to bring radical changes. It is not unreasonable to hope that these are wrong. From years of discussions with chemists, physicists, and engineers, it is possible to compile what seems to be a complete list of basic, critical questions about whether nanotechnology will work. The questioners generally seem satisfied with the answers.

"Will Thermal Vibrations Mess Things Up?"

The earlier scenarios describe the nature of thermal vibration and the problems it can cause. Designing nanomachines strong enough and stiff enough to operate reliably despite thermal vibration is a genuine engineering challenge. But calculating the design requirements usually requires only simple textbook principles, and these requirements can be met for everything described in this book.

"Will Quantum Uncertainty Mess Things Up?"

Quantum mechanics says that particles must be described as small smears of probability, not as points with perfectly defined locations. This is, in fact, why the atoms and molecules in the simulations felt so soft and smooth: their electrons are smeared out over the whole volume of the molecule, and these electron clouds taper off smoothly and softly toward the edges. Atoms themselves are a bit uncertain in position, but this is a small effect compared to thermal vibrations. Again, simple textbook principles apply, and well-designed molecular machines will work.

"Will Loose Molecules Mess Things Up?"

Chemists work with loose molecules in liquids, and they naturally tend to picture molecules as flying around loose. It is possible to build nanomachines and molecular-manufacturing systems that work in this sort of environment (biological mechanisms are an existence proof), but in the long run, there will be no need to do so. The Silicon Valley Faire simulation gives the right idea: systems can be built with no loose molecules, making nanomechanical design much easier. If no molecules are loose inside a machine, then loose molecules can't cause problems there.

"Will Chemical Instability Mess Things Up?"

Chemists perform chemical reactions, which means that they tend to work with reactive, unstable molecules. Many molecules, though, can sit around in peace with their neighbors for millions of years, as is known both from chemical theory and from the study of molecules trapped in ancient rock. Nanomachines can be built from the more stable sorts of structure. The only necessary exception is in molecular assembly, where molecules must react, but even here the reactive molecules need not be turned loose. They can be applied just when and where they are needed in the construction process.

"Is It Too Complex, Like Biology?"

An easy way to explain molecular manufacturing is to say that it is somewhat like molecular biology: small, complex molecular devices working together to build things and do various jobs. The next point, however, is that molecular manufacturing is different in every detail and different in overall structure: compare the nanocomputers, assembler arms, and conveyor belts described above to the shaggy, seething living cell described in the Unit I—B. Biology is complex in a strange and wonderful way. Engineers need not even understand life, much less duplicate it, merely to build a molecular-scale factory.

"I don't see anything wrong with it. But it's so interdisciplinary–couldn't there be a problem I can't see?"

Nanotechnology is basically a shotgun marriage of chemistry and mechanical engineering, with physics (as always) presiding. This makes a complete evaluation difficult for most of today's specialists, because each of these fields is taught separately and usually practiced separately. Many specialists, having highly focused backgrounds, find themselves unequipped to evaluate proposals that overlap other disciplines. When asked to do so, they will state feelings of discomfort, because although they can't identify any particular problems, they can't verify the entire concept as sound. Scientists and engineers with multidisciplinary backgrounds, or with access to specialists from other fields, can evaluate the idea from all sides.

It Will Work

When physicists, chemists, biologists, engineers, and computer scientists evaluate those parts of nanotechnology that fall within their disciplines, they agree: At no point would it require new principles or violate a physical law. There may for many years be some experts offering negative off-the-cuff opinions, but the consensus among those who have taken the time to examine the facts is clear. Molecular nanotechnology falls entirely within the realm of the possible.

"It Would Work, but Isn't It a Bad Idea to Implement It?"

If this means, "These new technologies could easily do far more harm than good," then there is no argument, because no one seems to disagree.

If this means, "These new technologies will certainly do more harm than good," then we disagree: much good is possible, much harm is avoidable, and it would be too bold to declare any such outcome "certain."

If this means, "These new technologies should be avoided," then we reply, "How, with what risks, and with what consequences?"

If this means, "Don't think about it or describe it," then we reply, "How else are we to understand it or make decisions?"

Increased human abilities have routinely been used to damage the environment and to make war. Even the crude technologies of the twentieth century have taken us to the brink. It is natural to feel exhilarated (or terrified) by a prospect that promises (or threatens) to extend human abilities beyond most past dreams (or nightmares). It is better to feel both, to meld and moderate these feelings, and to set out on a course of action that makes bad outcomes less likely. We're convinced that the best course is to focus on the potential good while warning of the potential evils.

"But Isn't It Unlikely to Arrive Within Our Lifetimes?"

Those in failing health may be justified in saying this; others are expressing an opinion that may well be wrong. It would be optimistic to assume that benefits are around the corner, and prudent to assume that they will be long delayed. Conversely, it would be optimistic to assume that dangers will be long delayed, and prudent to assume that they will arrive promptly. Whatever good or ill may come of post-breakthrough capabilities, the turbulence of the coming transition will present a real danger. While we invite readers to take a "What if?" stance toward these technologies, it would be imprudent to listen to the lulling sound of the promise "not in our lifetimes."

Even today, public acceptance of man's exploration of space is slow. It is considered an event our children may experience, but certainly not one that we shall see.

Perspective

We are still many years away from nanotechnology based on molecular manufacturing. It might even seem that such vast, slow giants as ourselves could never make such small, quick machines. The following sections will describe how advances in science and technology are leading toward these abilities. We'll try to get some feel for the road ahead, for its length, and for how fast we're moving. We are already surprisingly close to developing a crude molecular manufacturing technology, and getting visibly closer every week. The first, crude technology will enable the construction of molecular machines that can be used to build better molecular machines, climbing a ladder of capabilities that leads to general-purpose molecular assemblers as good or better than those described here.

The opportunities then will be enormous. If we haven't prepared, the dangers, too, will be enormous. Whether we're ready or not, the resulting changes will be disruptive, sweeping industries aside, upending military strategies, and transforming our ways of life.

Unit I–D

Fundamental Issues of Nanotechnology

A basic question about nanotechnology is, "When will it be achieved?" The answer is simple: No one knows. How molecular machines will behave is a matter for calculation, but how long it will take us to develop them is a separate issue. Technology timetables can't be calculated from the laws of nature, they can only be guessed at. In this unit, we examine different paths to nanotechnology, hear what some of the pioneers have to say, and describe the progress already made. This will not answer our basic question, but it will educate our guesses.

Molecular nanotechnology could be developed in any of several basically different ways. Each of these basic alternatives itself includes further alternatives. Researchers will be asking, "How can we make the fastest progress?" To understand the answers they may come to, we need to ask the same question here, adopting (for the moment) a gung-ho, let's go, how do we get the job done? attitude. We give some of the researchers' answers in their own words.

Will It Ever Be Achieved?

Like "when will it be achieved?", this is a basic question with an answer beyond calculation. Here, though, the answer seems fairly clear. Throughout history, people have worked to achieve better control of matter, to convince atoms to do what we want them to do. This has gone on since before people learned that atoms exist, and has accelerated ever since. Although different industries use different materials and different tools and methods, the basic aim is always the same. They seek to make better things, and make them more consistently, and that means better control of the structure of matter. From this perspective, nanotechnology is just the next, natural step in a progression that has been under way for millennia.

Consider the compact discs now replacing older stereo records: both the old and the new technologies stamp patterns into plastic, but for CDs, the bumps on the stamping surface are only about 130 by 600 nanometers in size, versus 100,000 nanometers or so for the width of the groove on an old-style record. Or look at a personal computer. John Foster, a physicist at IBM's Almaden Research Center, points to a hard disk and says that "inside that box are a bunch of whirring disks, and every one of those disks has got a metal layer where the information is stored. The last thing on top of the metal layer is a monolayer that's the lubricant between the disk and the head that flies over it. The monolayer is not fifteen angstroms [15 angstroms = 1.5 nanometers] and it's not three, because fifteen won't work and neither will three. So it has to be ten plus or minus a few angstroms. This is definitely working in the nanometer regime. We're at that level: We ship it every day and make money on it every day."

The transistors on computer chips are heading down in size on an exponential curve. Foster's colleague at IBM, Patrick Arnett, expects the trend to continue: "If you stay on that curve, then you end up at the atomic scale at 2020 or so. That's the nature of technology now. You expect to follow that curve as far as you can go." The trend is clear, and at least some of the results can be foreseen, but the precise path and timetable for the development of nanotechnology is unpredictable. This unpredictability goes to the heart of important questions: "How will this technology be developed? Who will do it? Where? When? In ten years? Fifty? A hundred? Will this happen in my lifetime?" The answers will depend on what people do with their time and resources, which in turn will depend on what goals they think are most promising. Human attitudes, understanding, and goals will make all the difference.

What Decisions Most Affect the Rate of Advance?

Decisions about research directions are central. Researchers are already pouring effort into chemical synthesis, molecular engineering, and related fields. The same amount of effort could produce more impressive results in molecular nanotechnology if a fraction of it were differently directed. The research funders—corporate executives, and decision makers in science funding agencies like the National Science Foundation in the United States and Japan's Ministry of International Trade and Industry—all have a large influence on research directions, but so do the researchers working in the labs. They submit proposals to potential funders (and often spend time on personally chosen projects, regardless of funding), so their opinions also shape what happens. Where public money is involved, politicians' impressions of public opinion can have a huge influence, and public opinion depends on what all of us think and say.

Still, researchers play a central role. They tend to work on what they think is interesting, which depends on what they think is possible, which depends on the tools they have or—among the most creative researchers—on the tools they can see how to make. Our tools shape how we think: as the saying goes, when all you have is a hammer, everything looks like a nail. New tools encourage new thoughts and enable new achievements, and decisions about tool development will pace advances in nanotechnology. To understand the challenges ahead, we need to take a look at ideas about the tools that will be needed.

Why Are Tools So Important?

Throughout history, limited tools have limited achievement. Leonardo da Vinci's sixteenth century chain drives and ball bearings were theoretically workable, yet never worked in their inventor's lifetime. Charles Babbage's nineteenth century mechanical computer suffered the same fate. The problem? Both inventors needed precisely machined parts that (though readily available today) were beyond the manufacturing technology of their times. Physicist David Miller recounts how a sophisticated integrated circuit design project at TRW hit similar limits in the early 1980s: "It all came down to whether a German company could cool their glass lenses slowly enough to give us the accuracy we needed. They couldn't."

In the molecular world, tool development again paces progress, and new tools can bring breathtaking advances. Mark Pearson, director of molecular biology for Du Pont, has observed this in action: "When I was a graduate student back in the 1950s, it was a multiyear problem to determine the molecular structure of a single protein. We used to say, 'one protein, one career.' Yet now the time has shrunk from a career to a decade to a year—and in optimal cases to a few months." Protein structures can be mapped atom by atom by studying X-ray reflections from layers in protein crystals. Pearson observes that "Characterizing a protein was a career-long endeavor in part because it was so difficult to get crystals, and just getting the material was a big constraint. With new technologies, we can get our hands on the material now—that may sound mundane, but it's a great advance. To the people in the field, it makes all the difference in the world." Improved tools for making and studying proteins are of special importance because proteins are promising building blocks for first-generation molecular machines.

But Isn't Science About Discoveries, Not Tools?

Nobel Prizes are more often awarded for discoveries than for the tools (including instruments and techniques) that made them possible. If the goal is to spur scientific progress, this is a shame. This pattern of reward extends throughout science, leading to a chronic underinvestment in developing new tools. Philip Abelson, an editor of the journal *Science,* points out that the United States suffers from "a lack of support for development of new instrumentation. At one time, we had a virtual monopoly in pioneering advances in instrumentation. Now practically no federal funds are available to universities for the purpose." It's easier and less risky to squeeze one more piece of data out of an existing tool than to pioneer the development of a new one, and it takes less imagination.

But new tools emerge anyway, often from sources in other fields. The study of protein crystals, for example, can benefit from new X-ray sources developed by physicists, and techniques from chemistry can help make new proteins. Because they can't anticipate tools resulting from innovations in other fields, scientists and engineers are often too pessimistic about what can be achieved in their own fields. Nanotechnology will join several fields, and yield tools useful in many others. We should expect surprising results.

What Tools Do Researchers Use to Build Small Devices?

Today's tools for making small-scale structures are of two kinds: molecular-processing tools and bulk-processing tools. For decades, chemists and molecular biologists have been using better and better molecular-processing tools to make and manipulate precise, molecular structures. These tools are of obvious use. Physicists, as we will see, have recently developed tools that can also manipulate molecules. Combined with techniques from chemistry and molecular biology, these physicist's tools promise great advances.

Microtechnologists have applied chip-making techniques to the manufacture of microscopic machines. These technologies—the main approach to miniaturization in recent decades—can play at most a supporting role in the development of nanotechnology. Despite appearances, it seems that microtechnology cannot be refined into nanotechnology.

But Isn't Nanotechnology Just Very Small Microtechnology?

For many years, it was conventional to assume that the road to very small devices led through smaller and smaller devices: a top-down path. On this path, progress is measured by miniaturization: How small a transistor can we build? How small a motor? How thin a line can we draw on the surface of a crystal? Miniaturization focuses on scale and has paid off well, spawning industries ranging from watchmaking to microelectronics.

Researchers at AT&T Bell Labs, the University of California at Berkeley, and other laboratories in the United States have used micromachining (based on microelectronic technologies) to make tiny gears and even electric motors. Micromachining is also being pursued successfully in Japan and Germany. These microgears and micromotors are, however, enormous by nanotechnological standards: a typical device is measured in tens of micrometers, billions of times the volume of comparable nanogears and nanomotors. (In our simulated molecular world, ten microns is the size of a small town.) In size, confusing microtechnology with molecular nanotechnology is like confusing an elephant with a ladybug.

The differences run deeper, though. Microtechnology dumps atoms on surfaces and digs them away again in bulk, with no regard for which atom goes where. Its methods are inherently crude. Molecular nanotechnology, in contrast, positions each atom with care. As Bill DeGrado, a protein chemist at Du Pont, says, "The essence of nanotechnology is that people have worked for years making things smaller and smaller until we're approaching molecular dimensions. At that point, one can't make smaller things except by starting with molecules and building them up into assemblies." The difference is basic: In microtechnology, the challenge is to build smaller; in nanotechnology, the challenge is to build bigger—we can already make small molecules.

(A language warning: in recent years, nanotechnology has indeed been used to mean "very small microtechnology"; for this usage, the answer to the above question is yes, by definition. This use of a new word for a mere extension of an old technology will produce considerable confusion, particularly in light of the widespread use of nanotechnology in the sense found here. Nanolithography, nanoelectronics, nanocomposites, nanofabrication: not all that is *nano-* is molecular, or very relevant to the concerns raised in this book. The terms *molecular nanotechnology* and *molecular manufacturing* are more awkward but avoid this confusion.)

Will Microtechnology Lead to Nanotechnology?

Can bulldozers can be used to make wristwatches? At most, they can help to build factories in which watches are made. Though there could be surprises, the relevance of microtechnology to molecular nanotechnology seems similar. Instead, a bottom-up approach is needed to accomplish engineering goals on the molecular scale.

What Are the Main Tools Used for Molecular Engineering?

Almost by definition, the path to molecular nanotechnology must lead through molecular engineering. Working in different disciplines, driven by different goals, researchers are making progress in this field. Chemists are developing techniques able to build precise molecular structures of sorts never before seen. Biochemists are learning to build structures of familiar kinds, such as proteins, to make new molecular objects.

In a visible sense, most of the tools used by chemists and biochemists are rather unimpressive.

They work on countertops cluttered with dishes, bottles, tubes, and the like, mixing, stirring, heating, and pouring liquids—in biochemistry, the liquid is usually water with a trace of material dissolved in it. Periodically, a bit of liquid is put into a larger machine and a strip of paper comes out with a graph printed on it. As one might guess from this description, research in the molecular sciences is usually much less expensive than research in high-energy physics (with its multibillion-dollar particle accelerators) or research in space (with its multibillion-dollar spacecraft). Chemistry has been called "small science," and not because of the size of the molecules.

Chemists and biochemists advance their field chiefly by developing new molecules that can serve as tools, helping to build or study other molecules. Further advances come from new instrumentation, new ways to examine molecules and determine their structures and behaviors. Yet more advances come from new software tools, new computer-based techniques for predicting how a molecule with a particular structure will behave. Many of these software tools let researchers peer through a screen into simulated molecular worlds much like those toured in the later units.

Of these fields, it is biomolecular science that is most obviously developing tools that can build nanotechnology, because biomolecules already form molecular machines, including devices resembling crude assemblers. This path is easiest to picture, and can surely work, yet there is no guarantee that it will be fastest: research groups following another path may well win. Each of these paths is being pursued worldwide, and on each, progress is accelerating.

Physicists have recently contributed new tools of great promise for molecular engineering. These are the proximal probes, including the scanning tunneling microscope (STM) and the atomic force microscope (AFM). A proximal-probe device places a sharp tip in proximity to a surface and uses it to probe (and sometimes modify) the surface and any molecules that may be stuck to it.

How Does an STM Work?

An STM brings a sharp, electrically conducting needle up to an electrically-conducting surface, almost touching it. The needle and surface are electrically connected (see the left-hand side of Figure I-D.1), so that a current will flow if they touch, like closing a switch. But at just what point do soft, fuzzy atoms "touch"? It turns out that a detectable current flows when just two atoms are in tenuous contact—fuzzy fringes barely overlapping—one on the surface and one on the tip of the needle. By delicately maneuvering the needle around over the surface, keeping the current flowing at a tiny, constant rate, the STM can map shape of the surface with great precision. Indeed, to keep the current constant, the needle has to go up and down as it passes over individual atoms.

The STM was invented by Gerd Binnig and Heinrich Rohrer, research physicists studying surface phenomena at IBM's research labs in Zurich, Switzerland. After working through the 1970s, Rohrer and Binnig submitted their first patent disclosure on an STM in mid-1979. In 1982, they produced images of a silicon surface, showing individual atoms. Ironically, the importance of their work was not immediately recognized: Rohrer and Binnig's first scientific paper on the new tool was rejected for publication on the grounds that it was "not interesting enough." Today, STM conferences draw interested researchers by the hundreds from around the world.

In 1986—quite promptly as these things go—Binnig and Rohrer were awarded a Nobel Prize. The Swedish Academy explained its reasoning: "The scanning tunneling microscope is completely new and we have so far seen only the beginning of its development. It is, however, clear that entirely new fields are opening up for the study of matter." STMs are no longer exotic: Digital Instruments of Santa Barbara, California, sells its system (the Nanoscope) by mail with an atomic-resolution-or-your-money-back guarantee. Within three years of their commercial introduction, hundreds of STMs had been purchased.

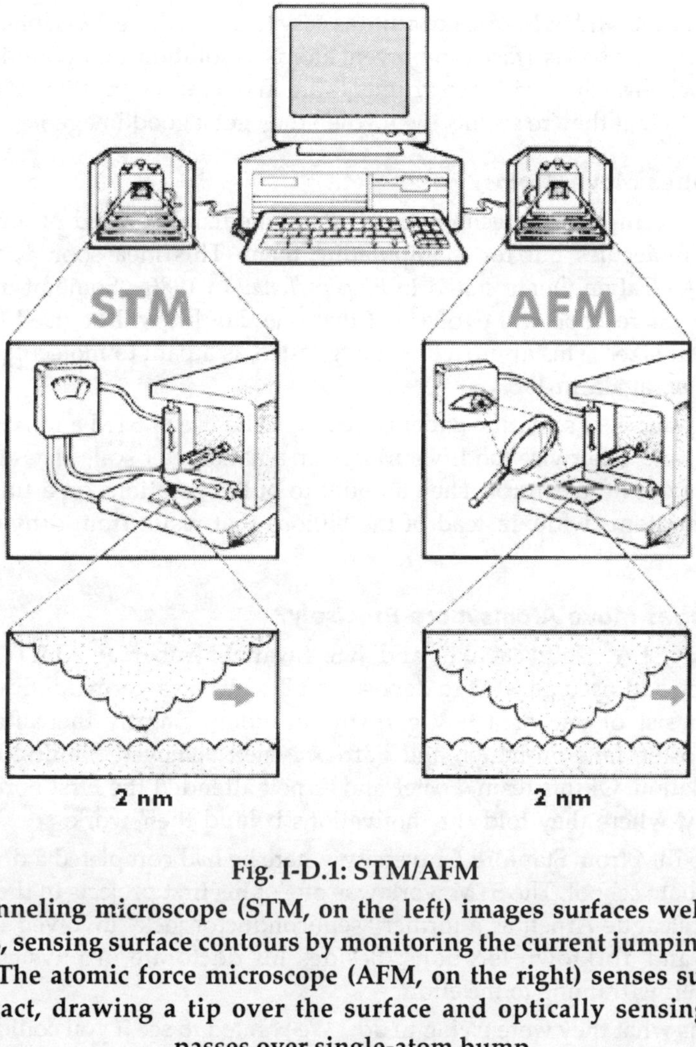

Fig. I-D.1: STM/AFM

The scanning tunneling microscope (STM, on the left) images surfaces well enough to show individual atoms, sensing surface contours by monitoring the current jumping the gap between tip and surface. The atomic force microscope (AFM, on the right) senses surface contours by mechanical contact, drawing a tip over the surface and optically sensing its motion as it passes over single-atom bump.

How Does an AFM Work?

The related atomic force microscope (on the right side of Figure I-D.1) is even simpler in concept: A sharp probe is dragged over the surface, pressed down gently by a straight spring. The instrument senses motions in the spring (usually optically), and the spring moves up and down whenever the tip is dragged over an atom on the surface. The tip "feels" the surface just like a fingertip in the simulated molecular world. The AFM was invented by Binnig, Quate, and Gerber at Stanford University and IBM San Jose in 1985. After the success of the STM, the importance of the AFM was immediately recognized. Among other advantages, it works with nonconducting materials. The next unit will describe how AFM-based devices might be used as molecular manipulators in developing molecular nanotechnology. As this is written, AFMs have just become commercially available.

(Note that that AFMs and STMs are not quite as easy to use as these descriptions might suggest. For example, a bad tip or a bad surface can prevent atomic resolution, and pounding on the table is not recommended when such sensitive instruments are in operation. Further, scientists often have trouble deciding just what they're seeing, even when they get a good image.)

Can Proximal Probes Move Atoms?

To those thinking in terms of nanotechnology, STMs immediately looked promising not only for seeing atoms and molecules but for manipulating them. This idea soon became widespread among physicists. As Calvin Quate stated in *Physics Today* in 1986, "Some of us believe that the scanning tunneling microscope will evolve . . . that one day [it] will be used to write and read patterns of molecular size." This approach was suggested as a path to molecular nanotechnology in *Engines of Creation*, again in 1986.

By now, whole stacks of scientific papers document the use of STM and AFM tips to scratch, melt, erode, indent, and otherwise modify surfaces on a nanometer scale. These operations move atoms around, but with little control. They amount to bulk operations on a tiny scale— *one* fine scratch a few dozen atoms wide, instead of the billions that result from conventional polishing operations.

Can Proximal Probes Move Atoms more Precisely?

In 1987, R.S. Becker, J.A. Golovchenko, and B.S. Swartzentruber at AT&T Bell Laboratories announced that they had used an STM to deposit small blobs on a germanium surface. Each blob was thought to consist of one or a few germanium atoms. Shortly thereafter, IBM Almaden researchers John Foster, Jane Frommer, and Patrick Arnett achieved a milestone in STM-based molecular manipulation. Of this team, Foster and Arnett attended the First Foresight Conference on Nanotechnology, where they told the motivations behind their work.

Foster came to IBM from Stanford University, where he had completed a doctorate in physics and taught at graduate school. The STM work was one of his first projects in the corporate world. He describes his colleague Arnett as a former "semiconductor jock" involved in chip creation at IBM's Burlington and Yorktown locations. Besides his doctorate in physics, Arnett brought mechanical-engineering training to the effort.

Arnett explains what they were trying to do: "We wanted to see if you could do something on an atomic scale, to create a mechanism for storing information and getting it back reliably." The answer was yes. In January 1988, the journal *Nature* carried their letter reporting success in pinning an organic molecule to a particular location on a surface, using an STM to form a chemical bond by applying an electrical pulse through the tip. They found that having created and sensed the feature, they could go back and use another voltage pulse from the tip to change the feature again: enlarging it, partly erasing it, or completely removing it.

IBM quickly saw a commercial use, as explained by Paul M. Horn, acting director of physical sciences at the Thomas J. Watson Research Center: "This means you can create a storage element the size of an atom. Ultimately, the ability to do that could lead to storage that is ten million times more dense than anything we have today." A broader vision was given by another researcher, J.B. Pethica, in the issue of *Nature* in which the work appeared: "The partial erasure reported by Foster et al., implies that molecules may have pieces deliberately removed, and in principle be atomically 'edited,' thereby demonstrating one of the ideals of nanotechnology."

Can Proximal Probes Move Atoms With Complete Precision?

Foster's group succeeded in pinning single molecules to a surface, but they couldn't control the results—the position and orientation—precisely. In April 1990, however, another group at the same laboratory carried the manipulation of atoms even further, bringing a splash of publicity. Admittedly, the story must have been hard to resist: it was accompanied by an STM picture of the name IBM," spelled out with thirty-five precisely placed atoms (Figure I-D.2). The precision here is complete, like the precision of molecular assembly: each atom sits in a dimple on the surface of a nickel crystal; it can rest either in one dimple or in another, but never somewhere between.

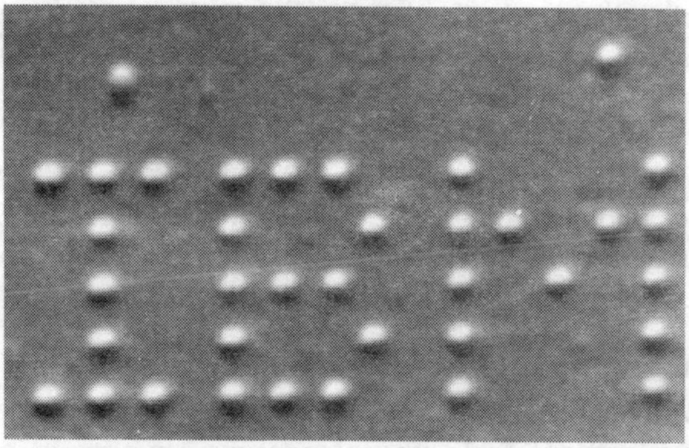

Fig. I-D.2: World Smallest Logo—35 Xenon Atoms.

How Far Can Proximal Probes Take Us?

Proximal probes have advantages as a tool for developing nanotechnology, but also weaknesses. Today, their working tips are rough and irregular, typically even rougher than shown in Figure I-D.1. To make stable bonds form, John Foster's group used a pulse of electricity, but the results proved hard to control. The "IBM" spelled out by Donald Eigler's group was precise, but stable only at temperatures near absolute zero—such patterns vanish at room temperature because they are not based on stable chemical bonds. Building structures that are both stable and precise is still a challenge. To form stable bonds in precise patterns is the next big challenge.

John Foster says, "We're exploring a concept which we call 'molecular herding,' using the STM to 'herd' molecules the way my Shetland sheep dog would herd sheep . . . Our ultimate goal with molecular herding is to make one particular molecule move to another particular one, and then essentially force them together. If you could put two molecules that might be small parts of a nanomachine on the surface, then this kind of herding would allow you to haul one of them up to the other. Instead of requiring random motion of a liquid and specific chemical lock-and-key interactions to give you exactly what you want in bringing two molecules together [as in chemical and biochemical approaches], you could drive that reaction on a local level with the STM. You could use the STM to put things where you want them to be." The next unit will discuss additional ideas for using proximal probes in early nanotechnology.

Proximal-probe instruments may be a big help in building the first generation of nanomachines, but they have a basic limit: Each instrument is huge on a molecular scale, and each could bond only one molecular piece at a time. To make anything large—say, large enough to see with the naked eye—would take an absurdly long time. A device of this sort could add one piece per second, but even a pinhead contains more atoms than the number of seconds since the formation of Earth. Building a Pocket Library this way would be a long-term project.

How Can Such Slow Systems Ever Build Anything Big?

Rabbits and dandelions contain structures put together one molecular piece at a time, yet they grow and reproduce quickly. How? They build in parallel, with many billions of molecular machines working at once. To gain the benefits of such enormous parallelism, researchers can either (1) use proximal probes to build a better, next-generation technology, or (2) use a different approach from the start.

The techniques of chemistry and biomolecular engineering already have enormous parallelism, and already build precise molecular structures. Their methods, however, are less direct than the still hypothetical proximal probe-based molecule-positioners. They use molecular building blocks shaped to fit together spontaneously, in a process of self-assembly.

David Biegelsen, a physicist who works with STMs at the Xerox Palo Alto Research Center, put it this way at the nanotechnology conference: "Clearly, assembly using STMs and other variants will have to be tried. But biological systems are an existence proof that assembly and self-assembly can be done. I don't see why one should try to deviate from something that already exists.

What Are the Main Advantages of Molecular Building Blocks?

A huge technology base for molecular construction already exists. Tools originally developed by biochemists and biotechnologists to deal with molecular machines found in nature can be redirected to make new molecular machines. The expertise built up by chemists in more than a century of steady progress will be crucial in molecular design and construction. Both disciplines routinely handle molecules by the billions and get them to form patterns by self-assembly. Biochemists, in particular, can begin by copying designs from nature.

Molecular building-block strategies could work together with proximal probe strategies, or could replace them, jumping directly to the construction of large numbers of molecular machines. Either way, protein molecules are likely to play a central role, as they do in nature.

How Can Protein Engineering Build Molecular Machines?

Proteins can self assemble into working molecular machines, objects that *do* something, such as cutting and splicing other molecules or making muscles contract. They also join with other molecules to form huge assemblies like the ribosome (about the size of a washing machine, in our simulation view). Ribosomes—programmable machines for manufacturing proteins—are nature's closest approach to a molecular assembler. The genetic-engineering industry is chiefly in the business of reprogramming natural nanomachines, the ribosomes, to make new proteins or to make familiar proteins more cheaply. Designing new proteins is termed protein engineering. Since biomolecules already form such complex devices, it's easy to see that advanced protein engineering could be used to build first-generation nanomachines.

If We Can Make Proteins, Why Aren't We Building Fancy Molecular Machines?

Making proteins is easier than designing them. Protein chemists began by studying proteins found in nature, but have only recently moved on to the problem of engineering new ones. These are called *de novo* proteins, meaning completely new, made from scratch. Designing proteins is difficult because of the way they are constructed. As Bill DeGrado, a protein chemist at Du Pont, explains: "A characteristic of proteins is that their activities depend on their three-dimensional structures. These activities may range from hormonal action to a function in digestion or in metabolism. Whatever their function, it's always essential to have a definite three-dimensional shape or structure." This three-dimensional structure forms when a chain folds to form a compact molecular object. To get a feel for how tough it is to predict the natural folding of a protein chain, picture a straight piece of cord with hundreds of magnets and sticky knots along its length. In this state, it's easy to make and easy to understand. Now pick it up, put it in a glass jar, and shake it for a long time. Could you predict its final shape? Certainly not: it's a tangled mess. One might call this effort at prediction "the sticky-cord-folding problem"; protein chemists call theirs "the protein-folding problem."

Given the correct conditions, a protein chain always folds into one special shape, but that shape is hard to predict from just the straightened structure. Protein designers, though, face the different job of first determining a desired final shape, and then figuring out what linear sequence of amino acids to use to make that shape. Without solving the classic protein-folding problem, they have begun to solve the protein-*design* problem.

What Has Been Accomplished So Far?

Bill DeGrado and his colleagues at Du Pont had one of the first successes: "We've been able to use basic principles to design and build a simple molecule that folds up the way we want it to. This is really the first real example of a designed protein structure, designed from scratch, not by taking an already existing structure and tinkering with it."

Although scientists do the work, the work itself is really a form of engineering, as shown by the title of the field's journal, *Protein Engineering*. Bill DeGrado's description of the process makes this clear: "After you've made it, the next step is to find out whether your protein did what you expected it to do. Did it fold? Did it pass ions across bilayers [such as cell membranes]? Does it have a catalytic function [speeding specific chemical reactions]? And that's tested using the appropriate experiment. More than likely, it won't have done what you wanted it to do, so you have to find out why. Now, a good design has in it a contingency plan for failure and helps you learn from mistakes. Rather than designing a structure that would take a year or more to analyze, you design it so that it can be assayed for given function or structure in a matter of days."

Many groups are pursuing protein design today, including academic researchers like Jane and Dave Richardson at Duke University, Bruce Erickson at the University of North Carolina, and Tom Blundell, Robin Leatherbarrow, and Alan Fersht in Britain. The successes have started to roll in. Japan, however, is unique in having an organization devoted exclusively to such projects: the Protein Engineering Research Institute (PERI) in Osaka. In 1990, PERI announced the successful design and construction of a *de novo* protein several times larger than any built before.

Is There Anything Special About Proteins?

The main advantage of proteins is that they are familiar: a lot is known about them, and many tools exist for working with them. Yet proteins have disadvantages as well. Just because this design work is starting with proteins—soft, squishy molecules that are only marginally suitable for nanotechnology—doesn't mean it will stay within those limits. De Grado points out "The fundamental goal of our work in *de novo* design is to be able to take the next step and get entirely away from protein systems." An early example is the work of Wallace Carothers of Du Pont, who used a *de novo* approach to studying the nature of proteins: Rather than trying to cut up proteins, he tried to build up things starting with amino acids and other similar monomers. In 1935, he succeeded in making nylon.

DeGrado explains "There is a deep philosophical belief at Du Pont in the ability of people to make molecules *de novo* that will do useful things. And there is a fair degree of commitment from the management that following that path will lead to products: not directly, and not always predictably, but they know that they need to support the basic science.

"I think ultimately we have a better chance at doing some really exciting things by *de novo* design, because our repertory should be much greater than that of nature. Think about the ability to fly: One could breed better carrier pigeons or one could design airplanes." The biology community, however, leans more toward ornithology than toward aerospace engineering. DeGrado's experience is that "a lot of biologists feel that if you aren't working with the real thing (natural proteins), you aren't studying biology, so they don't totally accept what we're doing. On the other hand, they recognize it as good chemistry."

Where is Protein Engineering Headed?

Like the IBM physicists, protein designers are moved by a vision of molecular engineering. In 1989, Bill DeGrado predicted, "I think we'll be able to make catalysts or enzymelike molecules, possibly ones that catalyze reactions not catalyzed in nature." Catalysts are molecular machines that speed up chemical reactions: they form a shape for the two reacting molecules to fit into and thereby help the reaction move faster, up to a million reactions per second. New ones, for reactions that now go slowly, will give enormous cost savings to the chemical industry.

This prediction was borne out just a few months later, when Denver researchers John Stewart, Karl Hahn, and Wieslaw Klis announced their new enzyme, designed from scratch over a period of two years and built successfully on the first try. It's a catalyst, making some reactions go about 100,000 times faster. Nobel Prize-winning biochemist Bruce Merrifield believes that "if others can reproduce and expand on this work, it will be one of the most important achievements in biology or chemistry."

DeGrado also has longer term plans for protein design, beyond making new catalysts: "It will allow us to think about designing molecular devices in the next five to ten years. It should be possible ultimately to specify a particular design and build it. Then you'll have, say, proteinlike molecules that self-assemble into complex molecular objects, which can serve as machinery. But there's a limit to how small you can make devices. You'll shrink things down so far and then you won't be able to go any further, because you've reached molecular dimensions."

Mark Pearson shows that management at Du Pont also has this vision. Regarding the prospects for nanotechnology and assemblers, he remarked, "You know, it'll take money and effort and good ideas for sure. But to my way of thinking, there is no absolute fundamental limitation to preclude us from doing this kind of thing." He didn't say his company plans to develop nanotechnology, but such plans aren't really necessary. Du Pont is already on the nanotechnology path, for other—shorter-term, commercial—reasons. Like IBM, if they do decide to move quickly, they have the resources and forward-looking people needed to succeed.

Who Else Builds Molecular Objects?

Chemists, most of whom do not work on proteins, are the traditional experts in building molecular objects. As a group they've been building molecules for over a century, with ever increasing ability and confidence. Their methods are all indirect: They work with billions of atoms at a time—massive parallelism—but without control of the positions of their workpieces. The molecules typically tumble randomly in a liquid or gas, like pieces of a puzzle that may or may not fit together correctly when shaken together in a box. With clever design and planning, most pieces will join properly.

Chemists mix molecules on a huge scale (in our simulation view, a test tube holds a churning molecular swarm with the volume of an inland sea), yet they still achieve precise molecular transformations. Given that they work so indirectly, their achievements are astounding. This is, in part, the result of the enormous amount of work poured into the field for many decades. Thousands of chemists are working on molecular construction in the United States alone; add to that the chemists in Europe, in Japan, and in the rest of the world, and you have a huge community of researchers making great strides. Though it publishes only a one-paragraph summary of each research report, a guide to the chemical literature '*Chemical Abstracts*' covers several library walls and grows by many feet of shelf space every year.

How Can Mixing Chemicals Build Molecular Objects?

An engineer would say that chemists (at least those specializing in synthesis) are doing construction work, and would be amazed that they can accomplish anything without being able to grab parts and put them in place. Chemists, in effect, work with their hands tied behind their backs. Molecular manufacturing can be termed "positional chemistry" or "positional synthesis," and will give chemists the ability to put molecules where they want them in three-dimensional space. Rather than trying to design puzzle pieces that will stick together properly by themselves when shaken together in a box, chemists will then be able to treat molecules more like bricks to be stacked. The basic principles of chemistry will be the same, but strategies for construction will become far simpler.

Without positional control, chemists face a problem something like this: picture a giant glass barrel full of tiny battery-powered drills, buzzing away in all directions, vibrating around in the barrel. Your goal is to take a piece of wood and put a hole in just one specific spot. If you simply throw it in the barrel, it will be drilled haphazardly in many places. To control the process, you must protect all the places you don't want drilled—perhaps by gluing protective pieces of metal over most of the wood surface. This problem—how to protect one part of a molecule while altering another part—has forced chemists to develop ever-cleverer ploys to build larger and larger molecules.

If Chemists Can Make Molecules, Why Aren't They Building Fancy Molecular Machines?

Chemists can achieve great things, but have focused much of their effort on duplicating molecules found in nature and then making minor variants. As an example, take palytoxin, a molecule found in a Hawaiian coral. It was so difficult to make in the lab that it has been called "the Mount Everest of synthetic chemistry," and its synthesis was hailed as a triumph. Other efforts are poured into making small molecules with unusual bonding, or molecules of remarkable symmetry, like "cubane" and "dodecahedrane" (shaped like the Platonic solids they are named after).

Chemists, at least in the United States, regard themselves as natural scientists even when their life's work is the construction of molecules by artificial means. Ordinarily, people who build things are called engineers. And indeed, at the University of Tokyo the Department of Synthetic Chemistry is part of the Faculty of Engineering; its chemists are designing molecular switches for storing computer data. Engineering achievements will require work directed at engineering goals.

How Could Chemists Move Toward Building Molecular Machines?

Molecular engineers working toward nanotechnology need a set of molecular building blocks for making large, complex structures. Systematic building-block construction was pioneered by Bruce Merrifield, winner of the 1984 Nobel Prize in Chemistry. His approach, known as "solid phase synthesis," or simply "the Merrifield method," is used to synthesize the long chains of amino acids that form proteins. In the Merrifield method, cycles of chemical reaction each add one molecular building block to the end of a chain anchored to a solid support. This happens in parallel to each of trillions of identical chains, building up trillions of molecular objects with a particular sequence of building blocks. Chemists routinely use the Merrifield method to make molecules larger than palytoxin, and related techniques are used for making DNA in so-called gene machines: an ad from an Alabama company reads, "Custom DNA—Purified and Delivered in 48 hours."

While it's hard to predict how a natural protein chain will fold—they weren't designed to fold predictably—chemists could make building blocks that are larger, more diverse, and more inclined to fold up in a single, obvious, stable pattern. With a set of building blocks like these, and the Merrifield method to string them together, molecular engineers could design and build molecular machines with greater ease.

How do Researchers Design What They Can't See?

To make a new molecule, both its structure and the procedure to make it must be designed. Compared to gigantic science projects like the Superconducting Supercollider and the Hubble Space Telescope, working with molecules can be done on a shoestring budget. Still, the costs of trying many different procedures add up. To help predict in advance what will work and what won't, designers turn to models.

You may have played with molecular models in chemistry class: colored plastic balls and sticks that fit together like Tinker Toys. Each color represents a different kind of atom: carbon, hydrogen, and so on. Even simple plastic models can give you a feel for how many bonds each kind of atom makes, how long the bonds are, and at what angles they are made. A more sophisticated form of model uses only spheres and partial spheres, without sticks. These colorful, bumpy shapes are called CPK models, and are widely used by professional chemists. Nobel laureate Donald Cram remarks that "We have spent hundreds of hours building CPK models of potential complexes and grading them for desirability as research targets." His research, like that of fellow Nobelists Charles J. Pedersen and Jean-Marie Lehn, has focused on designing and making medium-sized molecules that self assemble.

Although physical models can't give a good description of how molecules bend and move, computer-based molecules can. Computer-based modeling is already playing a key role in molecular engineering. As John Walker (a founder and leader of Autodesk) has remarked, "Unlike all of the industrial revolutions that preceded it, molecular engineering requires, as an essential component, the ability to design, model, and simulate molecular structures using computers."

This has not gone unnoticed in the business community. John Walker's remark was part of a talk on nanotechnology given at Autodesk, a leader in computer-aided design and one of the five largest software firms in the United States. Soon after this talk, the company made its first major investment in the computer-aided design of molecules.

How does Molecular Design Compare to more Familiar Kinds of Engineering?

Manufacturers and architects know that designs for new products and buildings are best done on a computer, by computer-aided design (CAD). The new molecular design software can be called **molecular CAD**, and in its forefront are researchers such as Jay Ponder of the Yale University Department of Molecular Biophysics and Biochemistry. Ponder explains that "There's a strong link between what molecular designers are doing and what architects do. Michael Ward of Du Pont is designing a set of building blocks for a Tinker Toy set so that you can build larger structures. That's exactly what we're doing with molecular modeling techniques.

"All the design and mechanical engineering principles that apply to building a skyscraper or a bridge apply to molecular architecture as well. If you're building a bridge, you're going to model it and see how many trucks can be on the bridge at the same time without it collapsing, what kind of forces you're going to apply to it, whether it can stand up to an earthquake.

"And the same process goes on in molecular design: You're designing pieces and then analyzing the stresses and forces and how they will change and perturb the structure. It's exactly the same as designing and building a building, or analyzing the stresses on any macroscale structure. I think it's important to get people to think in those terms.

"The molecular designer has to be creative in the same way that an architect has to be creative in designing a building. When people are looking at the interior of a protein structure and trying to redesign it to create a space that will have a particular function, such as binding to particular molecules, that's like designing a room to use as a dining room—one that will fit certain sizes of tables and certain numbers of guests. It's the same thing in both cases: You have to design a space for a function."

Ponder combines chemistry and computer science with an overall engineering approach: "I'm kind of a hybrid. I spend about half my time doing experiments and about half my time writing computer programs and doing computational work. In the laboratory, I create or design molecules to test some of the computational ideas. So I'm at the interface." The engineering perspective helps in thinking about where molecular research can lead: "Even though with nanotechnology we're at the nanometer scale, the structures are still big enough that an awful lot of things are classical. Again, it's really like building bridges—very small bridges. And so there are many almost standard mechanical-engineering techniques for architecture and building structures, such as stress analysis, that apply."

Doesn't Engineering Require more Teamwork than Science does?

Getting to nanotechnology will require the work of experts in differing fields: chemists, who are learning how to make molecular machines; computer scientists, who are building the needed design tools; and perhaps STM and AFM experts, who can provide tools for molecular positioning. To make progress, however, these experts must do more than just work, they must work together. Because nanotechnology is inherently interdisciplinary, countries that draw hard lines between their academic disciplines, as the United States does, will find that their researchers have difficulty communicating and cooperating.

In chemistry today, a half-dozen researchers aided by a few tens of students and technicians is considered a large team. In aerospace engineering, enormous tasks like reaching the Moon or building a new airliner are broken down into tasks that are within the reach of small teams. All these small teams work together, forming a large team that may consist of thousands of engineers aided by many thousands of technicians. If chemistry is to move in the direction of molecular systems engineering, chemists will need to take at least a few steps in this direction.

In engineering, everyone knows that designing a rocket will require skills from many disciplines. Some engineers know structures, others know pumps, combustion, electronics, software, aerodynamics, control theory, and so on and so forth down a long list of disciplines. Engineering managers know how to bring different disciplines together to build systems.

In academic science, interdisciplinary work is productive and praised, but is relatively rare. Scientists don't need to cooperate to have their results fit together: they are all describing different parts of the same thing—nature—so in the long run, their results tend to come together into a single picture. Engineering, however, is different. Because it is more creative (it actually *creates* complex things), it demands more attention to teamwork. If the finished parts are going to work together, they must be developed by groups that share a common picture of what each part must accomplish. Engineers in different disciplines are forced to communicate; the challenge of management and team-building is to make that communication happen. This will apply to engineering molecular systems as much as it does to engineering computers, cars, aircraft, or factories.

Jay Ponder suggests that it's a question of perspective. "It's all a matter of what's perceived to be important by the different groups that have to come together to make this work: the chemists doing their bit and the computational people doing their bit. People have to come together and see the big picture. There are people who try to bridge the gaps, but they are rare compared to the people who just work in their own specialty." Progress toward nanotechnology will continue, and as it does, researchers trained as chemists, physicists, and the like will learn to talk to one another to solve new problems. They will either learn to think like engineers and work in teams, or they will be eclipsed by colleagues who do.

Are these Problems Preventing Advances?

With all these problems, the advance toward nanotechnology steadily continues. Industry must gain ever-better control of matter to stay competitive in the world marketplace. The STM, protein engineering, and much of chemistry are driven by commercial imperatives. Focused efforts would yield faster advances, yet even without a clear focus, advances in this direction have an air of inevitability. As Bill DeGrado observes, "We really do have the tools. Experience has shown that when you have the analytic and synthetic tools to do things, in the end science goes ahead and does them—because they are doable." Jay Ponder agrees: "Over the next few years, you're going to see slow evolutionary advances coming from people tinkering with molecular structures and figuring out their principles. People are going to work on a particular problem because they see some application for it or because they got grant funding for it. And in the process of doing something like improving a laundry detergent's ability to clean protein stains, Proctor and Gamble is going to help work out the principles for how to increase molecular stability, and to design spaces inside the molecules."

How Long It Will Take to Develop Molecular Nanotechnology?

Molecular nanotechnology will emerge step by step. Major milestones, such as the engineering of proteins and the positioning of individual atoms, have already been passed. To get a sense of the likely pace of developments, we need to look at how various trends fit together.

Computer-based molecular-modeling tools are spawning computer-aided design tools. These will grow more capable. The underlying technology base—computer hardware—has for decades been improving in price and performance on a steeply rising curve, which is generally expected to continue for many years. These advances are quite independent of progress in molecular engineering, but they make molecular engineering easier, speeding advances. Computer models of molecular machines are beginning to appear, and these will whet the appetites of researchers.

Progress in engineering molecular machines, whether using proximal probes or self-assembly, will eventually achieve striking successes; the objectives of research in Japan will begin to draw serious attention; understanding of the long-term promise of molecular engineering will become more widespread. Some combination of these developments will eventually lead to a serious, public appraisal of what these technologies can achieve—and then the world of opinion, funding, and research fashion will change. Before, advances will be steady but haphazard; afterward, advances will be driven with the energy that flows into major commercial, military, and medical research programs, because nanotechnology will be recognized as furthering major commercial, military, and medical goals. The timing of subsequent events depends largely on when this threshold of serious attention is reached.

In making time estimates, people are prone to assume that a large change must take a long time. Most do, but not all. Pocket calculators had a dramatic effect on the slide-rule industry: they replaced it. The speed of this change caught the slide rule moguls by surprise, but the pace of progress in electronics didn't slow down merely to suit their expectations.

One can argue that nanotechnology will be developed fast: many countries and companies will be competing to get there first. They will be driven onward both by the immense expected benefits—in many areas, including medicine and the environment—as well as by potential military applications. That is a powerful combination of motives, and competition is a powerful accelerator.

A counterargument, though, suggests that development will be slow: anyone who has done anything of significance in the real world of technology—doing a scientific experiment, writing a computer program, bringing a new product to market—knows that these goals take longer than expected. Indeed, Hofstadter's Law states that projects take longer than expected, even when Hofstadter's Law is taken into account. This principle is a good guide for the short term, and for a single project.

The situation differs, though, when many different approaches are being explored by many different groups over a period of years. Most projects may take longer than expected, but with many teams trying many approaches, one approach may prove faster than expected. The winner of a race is always faster than the average runner. John Walker notes, "The remarkable thing about molecular engineering is that it looks like there are many different ways to get there and, at the moment, rapid progress is being made along every path—all at the same time."

Also, technology development is like a race run over an unmapped course. When the first runners reach the top of a hill, they may see a shortcut. A trailing runner may decide to crash off into the bushes, and stumble across a bicycle and a paved road. The progress of technology is seldom predictable because progress often reveals new directions.

How close we are to goal depends on whether technological advances are a constant pace of accelerating. In this diagram, the dashed line represents the current level of technology, and the large dot in the upper right represents a goal such as nanotechnology. With a straight-line advance, it's easier to estimate how far away a goal is. With an accelerating advance, a goal can be reached with little warning.

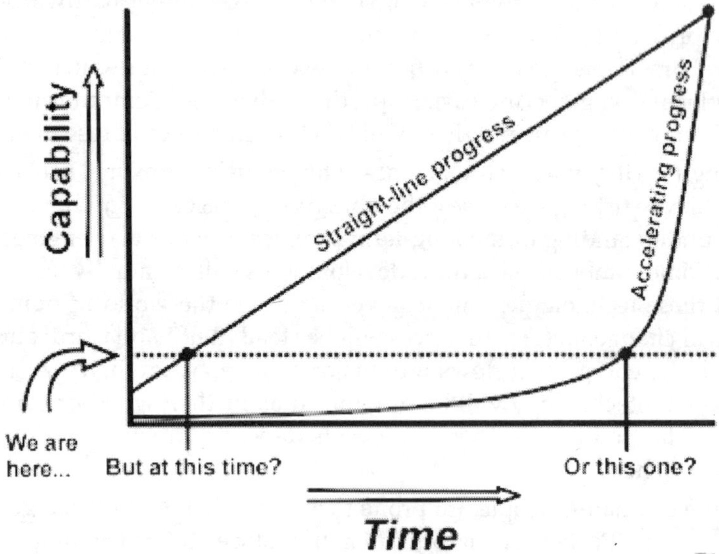

Fig. I-D.3: Graph of Linear Vs. Accelerating Growth of Technology

So how can we estimate a date for the arrival of nanotechnology? It's safest to take a cautious approach: When anticipating benefits, assume it's far off; when preparing for potential problems, assume it's right around the corner. The old folk saying applies: Hope for the best, prepare for the worst. Any dates assigned to "far off" and "right around the corner" can be no better than educated guesses—molecular behavior can be calculated, but not technology timetables of this sort. With those caveats, we would estimate that general-purpose molecular assemblers will likely be developed in the early decades of the twenty-first century, perhaps in the first.

John Walker, whose technological foresight has led Autodesk from start-up to a dominant role in its industry, points out that not long ago, "Many visionaries intimately familiar with the developments of silicon technology still forecast it would take between twenty and fifty years before molecular engineering became a reality. This is well beyond the planning horizon of most companies. But recently, everything has begun to change." Based on the new developments, Walker places his bet: "Current progress suggests the revolution may happen within this decade, perhaps starting within five years.

Unit I–E

The Threshold of Nanotechnology

In the Unit I-D, we looked at the state of current research, but from there to the nanotechnology of even the Pocket Library scenario is a leap. How will this gap be crossed?

In this unit, we outline how emerging technologies can lead to nanotechnology. The actual path to nanotechnology—the one that history books will record—could emerge from any one of the research directions in physics, biochemistry, and chemistry recounted in the last unit, or (more likely) from a combination of them. The availability of so many good options builds confidence that the goal can be reached, even while it decreases confidence that some particular path will be fastest. To see how advances might cross the gap from present technology to early nanotechnology, let's follow one path out of the many possible.

Bridging the Gap

One way to bridge the gap would through the development of an AFM-based molecular manipulator capable of doing primitive molecular manufacturing. This device would combine a simple molecular device—a molecular gripper—with an AFM positioning mechanism. An AFM can move its tip with precision; a molecular manipulator would add a gripper to the tip to hold a molecular tool. A molecular manipulator of this kind would guide chemical reactions by positioning molecules, like a slow, simple, but enormous assembler. (In our standard simulation view, where a molecular assembler arm fits in a room, the AFM apparatus of a molecular manipulator would be the size of a moon.) Despite its limits, an AFM molecular manipulator will be a striking advance.

How might this advance occur? Since we're choosing one path out of many possible, we may as well include more details and tell a story.

Scenario: Developing a Molecular Manipulator

Several years ago, researchers at the University of Brobdingnag began work on developing a molecular manipulator. To reach this goal, a team of a dozen physicists, chemists, and protein researchers banded together (some working full time, some part time) and began the creative teamwork needed to solve the basic problems.

First they needed to attach a gripper to an AFM tip. As grippers, they chose fragments of antibody molecules, the selectively sticky proteins that the immune system uses to bind and identify germs. If they could get the "back" of the molecule stuck onto a tip, then the "front" could bind and hold molecular tools. (The advantage of antibody fragments was this: freedom of tool choice. Since the late 1980s, researchers had been able to generate antibodies able to bind almost any preselected molecule-or molecular tool.) They tried half a dozen methods before finding one that worked reliably, with results like those shown in Figure I-E.1. A graduate student got her PhD, and the AFM tip got its gripper.

In parallel, the U. Brob AFM researchers worked on placing tips in a precise location and then holding them there with atomic accuracy for seconds at a time. This proved straightforward. They used techniques developed elsewhere during the early 1990s, adding only modest refinements.

They now had their gripper and a way of putting it where they wanted it, but they needed a set of tools. The gripper was like the chuck of a drill, waiting to have different bits fitted into its tool-holder slot. So as the final step, the synthetic chemists on the team made a dozen different molecular tools, all identical at one end but different at the other. The similar parts all bound to the same antibody tool-holder, slotting neatly into position. The different parts were all chemically reactive in different ways. Like the molecular tools in the hall of assembler arms, each of these tools could use a chemical reaction to transfer some atoms to a molecular object under construction.

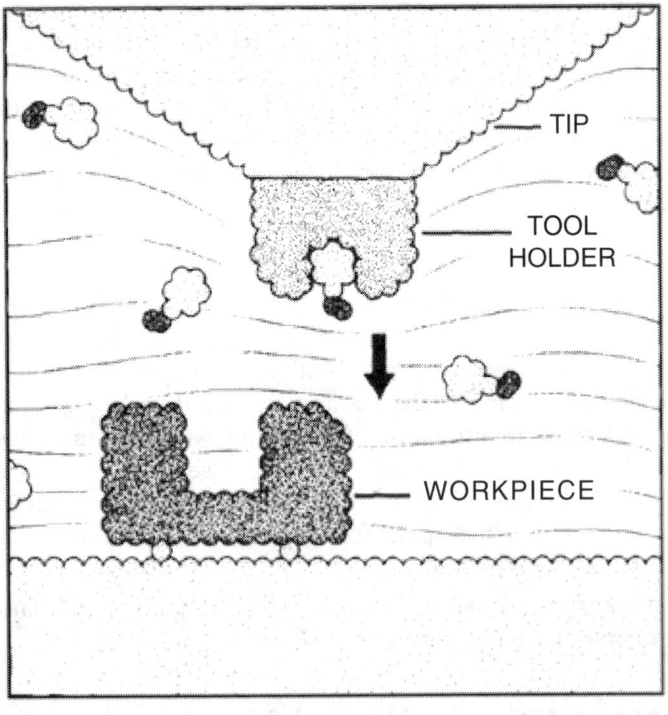

Fig. I-E.1: Molecular Manipulator

A molecular manipulator (AFM tip and tool holder, above) would bind and position reactive molecular tools to build up a workpiece, molecule by molecule.

Developing the molecular tool kit was the toughest part of the project; it took about as much work as had gone into duplicating the palytoxin molecule back in the 1980s. None of the tasks in the project demanded the solution of a deep scientific puzzle, and none demanded the solution of a notoriously difficult engineering problem. Each task had many possible solutions, the problem was to find a compatible set of solutions and apply them. After a few years, the solutions came together and the U. Brob research team began building new molecules by molecular manipulation. Now many teams are doing likewise.

Building with Molecular Grippers and Tools

To build something with the U. Brob team's AFM-based molecular manipulator system, you use it as follows: First, choose a surface to build on and place it under the tip in a pool of liquid. Then dunk the AFM tip into the liquid, bringing it down to the surface, and back it off a little. Construction can now begin as soon as a tool is loaded into the gripper.

Tubes and pumps can flow different liquids over the surface and past the gripper, carrying different tool molecules. If you want to do something with a tool of Type A, you wash in the proper liquid, and a Type A molecule promptly sticks the to the gripper as shown in Figure 5.1. Once it is in the gripper, you can use the AFM mechanism to move it around and put it where you want it. Move it up to the surface at a convenient spot, wait a few seconds, and it reacts, forming a bond and leaving a molecular fragment attached to the spot you chose. To add a different fragment, you can use a tool of Type B: you back up the tip, flow in a fresh liquid carrying the new tools, and in a moment a tool of the new type is bound in place and ready to apply, either on or alongside the first spot. Step by step, you build up a precise molecular structure.

Each step takes only seconds. Molecular tools pop into the gripper in a fraction of a second, and used tools pop off at the same rate. Once the tip has positioned a molecule, it reacts quickly, about a million times faster than unwanted reactions at other sites. In this way, the molecular manipulator gives good control of where reactions will occur (though it is not as reliable as an advanced assembler would be). It is fairly fast by a chemist's standards—per cycle—but still a million times slower than an advanced assembler. It can perform a variety of steps, but isn't as flexible and capable as an advanced assembler. In short, it is hardly the last word in nanotechnology, yet is a great advance over what has gone before.

Products

With its ability to accelerate desired reactions by a factor of a million or so, the U. Brob team's molecular manipulator can perform 10,000 to 100,000 steps with good reliability. Back in the 1980s, chemists making protein molecules struggled to perform just one hundred steps. The U. Brob research team (and its many imitators) can now build structures that are stronger and easier to design than proteins: not floppy, folded chains, but rugged objects held together by a sturdy network of bonds. Though not as strong and dense as diamond, these structures are like bits of a tough engineering plastic. A specially adapted computer-aided design system makes it easy to design molecular objects made from these materials.

Yet the AFM-based molecular manipulator has one grave disadvantage: It does chemistry one molecule at a time, and it ties up a machine as expensive as a car for hours or days to produce that one large molecule. Some molecules, though, are valuable enough to be worth building even one at a time. These draw prompt attention.

A single molecule isn't much use as a dye, a drug, or a floor wax, but it can have substantial value if it provides useful information. The U. Brob team quickly publishes a pile of scientific papers based on experiments with single molecules: they build a molecule, probe it, report the results, and build another. Some of these results show chemists elsewhere in the multibillion-dollar chemical industry how to design new catalysts, molecules that can help make other molecules more cheaply, cleanly, and efficiently. This information is worth a lot.

Three new products of special interest are among the first to be made. The first—molecular electronics—begins with experiments conducted by a research group at a computer chip company. They use their molecular manipulator to build single molecules and probe them, gradually learning how to build the parts needed for molecular electronic computers. These new computers don't immediately become practical, because the costs are too high for making such large molecules with AFM-based technology. Yet some companies begin to produce simpler molecular electronic devices for use in sensors and specialized high-speed signal processing. A specialty industry is born and begins to expand.

The second product is a gene reader, a complex molecular device built on the surface of a chip. The biologists who built the reader combined proteins borrowed from cells with special-purpose molecular machines designed from scratch. The result was a molecular system that binds DNA molecules and pulls them past a read-head-like tape through a tape recorder. The device works as fast as some naturally occurring molecular machines that read DNA, with one key advantage: it outputs its data electronically. At that speed, a single device can read a human genome in about a year. Though still too expensive for a doctor's office, these readers are promptly in great demand from research laboratories. Another small industry is born.

The third product is far more important, in the long run: replacement tips for molecular manipulators, grippers, and tools that are better than the originals. With these new, more versatile devices, researchers are now building more ambitious products and tools.

More Scenario: The Next Step to Nanotechnology

While the physicist-led team at U. Brob was finishing its work on the AFM-based molecular manipulator, a chemist-led team at the University of Lilliput was working furiously. They saw the U. Brob desktop machine as too large and its expected products as too expensive. Even back in the 1980s, David Biegelsen of the Xerox Palo Alto Research Center had noted, "The main drawback I see to using a hybrid protoassembler [AFM-based molecular manipulator] is that it would take a long time to build just one unit. Building requires a series of atom-by-atom construction steps. It would be better to build in parallel from the very beginning, making many trillions of these molecules all at the same time. I think there is tremendous power in parallel assembly. Maybe another field, chemistry or biology, offers a better way to do it." The chemists at U. Lill aimed to develop that better way, building first simple and then more and more complex molecular machines. The eventual result was a primitive molecular assembler able to build molecular objects by the trillions.

Chemist's Tools

How did the chemists achieve this? During the years when the U. Brob team was developing the molecular manipulator, researchers working in protein science and synthetic chemistry had made better and better systems of molecular building blocks. Chemists were well prepared for doing this: by the late 1980s, it had become possible to build stable structures the size of medium-sized protein molecules, and work had begun to focus on making these molecules perform useful work by binding and modifying other molecules. Chemists learned to use these sophisticated catalysts-early molecular devices-to make their own work easier by helping in the manufacture of still more large molecules.

Another traditional chemist's tool was software for doing computer-aided design. The early software designed by Jay Ponder and Frederic Richards of Yale University ultimately led to semi-automatic tools for designing molecules of a particular shape and function. Chemists then could easily design molecules that would self-assemble into larger structures, several tens of nanometers across.

Molecular Construction Machines

These advances in software and chemical synthesis let the U. Lill team tackle the task of building a primitive version of a molecular assembler. Although they couldn't build anything as complex as a nanocomputer or as stiff as diamond, they didn't need to. Their design used sliding molecular rods to position a molecular gripper much like the gripper used at U. Brob, again using the surrounding liquid to control which tool the gripper held. Instead of an AFM's electronic controls, they used the surrounding liquid to control the position of the rods as well. In a neutral solution, the rods would withdraw; in an acid solution, they would extend. How far they moved depended on what other molecules were around to lodge in special pockets and block the motion.

Their primitive assemblers built much the same sorts of products that the U. Brob molecular manipulator did; the tools were similar, and speed and accuracy were about the same. Yet there was one dramatic advantage: About 1,000,000,000,000,000,000,000 U. Lill assemblers could fit in the space occupied by one U. Brob manipulator, and it was easy to produce a mere 1,000,000,000,000,000 times as much product at the same cost.

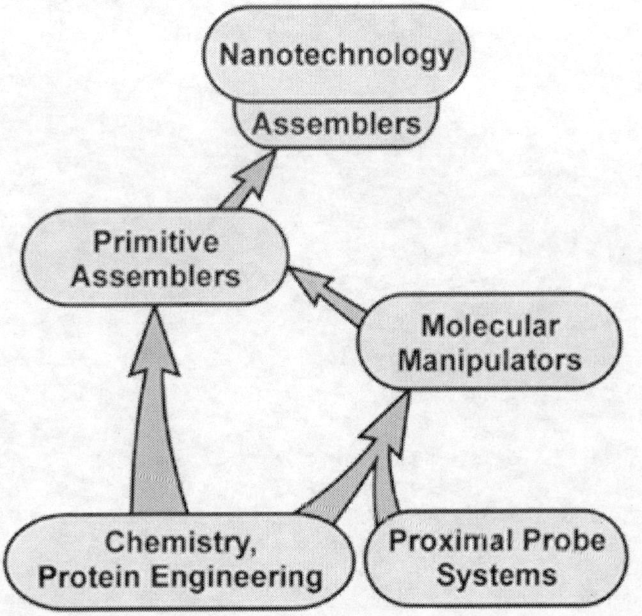

Fig. I-E.2: Paths to Nanotechnology Nanotechnology Development Flow Chart.

With the first, primitive assemblers, construction was slow because each step required new liquid baths and several seconds of soaking and waiting, and a typical product might take thousands of steps. Nonetheless, the U. Lill team made a lot of money licensing their technology to researchers trying to commercialize products they had first researched with the U. Brob machine. After starting an independent company (Nanofabricators, Inc.), they poured their research efforts into building better machines. Within a few years, they had assemblers with multiple grippers, each loaded with a different kind of tool; flashes of colored light would flip molecules from state to state (they copied these molecules from the pigments of the retina of the eye); flipping molecules would change tools and change rod positions. Soaking and waiting become a thing of the past, and soon they were pouring out parts that, when mixed with liquid and added to dishes with special blank chips would build up the dense memory layers that made possible the Pocket Library.

That was when things started moving fast. The semiconductor industry went the way of the vacuum tube industry. Money and talent poured into the new technology. Molecular CAD tools got better, assemblers made it easy to build what was designed, and fast production and testing made molecular engineering as easy as playing with software. Assemblers got better, faster, and cheaper. Researchers used assemblers to build nanocomputers, and nanocomputers to control better, faster assemblers. Using tools to build better tools is an ancient story. Within a decade, almost anything could be made by molecular manufacturing, and was.

Will developments in the late pre-breakthrough days be as just described? Certainly not: the technical approaches will differ, and the U.S. academic research setting implied by the scenario could easily be replaced by academic, commercial, governmental, or military research settings in any of the advanced nations. What do seem realistic are the implied requirements for effort, technology, and time, as well as the basic capabilities of different devices. We are approaching a threshold of capability beyond which further advances will become easy and fast.

Unit I–F

Working with Nanotechnology

The word *manufacturing* comes from the Latin *manufactus*, meaning "handmade." Today, the term brings to mind huge, noisy machines stamping out products and spewing waste. Giving up manufactured products isn't popular or practical—almost everything we use today is manufactured. If all machine-made products were to suddenly vanish, most people in the United States would find themselves naked and outdoors, with very little around them. Expanding manufacturing is an objective of nearly every nation on Earth.

We can't give up manufacturing, but we can replace today's technologies with something radically different. Molecular manufacturing can help us get what we seem to want: high-quality products made at low cost with little environmental impact.

What follows is an exploration of the possible—a look at the devices that could be built once precise molecular control is achieved, and a look at how people might run a manufacturing business based on nanomachines. Try not to think of these sketches as hard-and-fast predictions of precisely how things will be done, but instead as descriptions of capabilities—the sorts of things that can be done once nanotechnology is well in hand. Doubtless there will be better ways to do things than the ways we describe. As usual, references to the 1980s and before are historically accurate; in the rest, the science isn't fiction.

Scenario: Desert Rose Industries

Desert Rose Industries is a diversified wholesale manufacturer of enough furniture, computers, toys, and recreation equipment to have made any twentieth-century captain of industry proud. But if you assembled all Desert Rose employees in front of corporate headquarters, you'd see Carl and Maria Santos standing beside a building the size of a four-bedroom house. This industrial giant is a typical mom-and-pop business, helped along by a network of telecommuters who handle sales and customer support from homes scattered across North America

Their friends chide Carl and Maria as "old-fashioned traditionalists" and tease Maria about abandoning Carl in the factory while she travels to Europe, Asia, South America, and Africa for new business. In the molecular-manufacturing business, familiar personal skills and virtues—honesty, accuracy, good communication—are as important as before. Maria likes to work with the customers. Aided by her S.B. in molecular manufacturing from MIT and her MFA in design, she patiently helps nervous new designers through their first manufacturing experience, and with unflagging courtesy and good humor, handles rush orders, last-second changes, and special orders. Maria's good design ideas and caring personality won them a reputation for being responsive to customer needs. Carl, precise and careful, built their name for accurate manufacturing and delivery on schedule.

Except for Carl's habit of playing Gershwin at full volume with the windows open, the only sounds at the Desert Rose site are the birds along the banks of the stream that winds across the canyon floor; no clanking machinery here. Maria's parents built Desert Rose Industries out here on an old smelter many miles away from human neighbors. They regraded the land and cleaned up the wastes. Maria adapted a molecular processor to convert heavy-metal contaminants back into stable minerals, and shipped them off to help refill the hole they had originally come from, an old open-pit mine. The desert has mostly healed now, and a few tough trees are spreading along the stream again.

New customers coming up the road for a firsthand look at the manufacturing operations get the full tour: a lunch/meeting room, Maria's office, the manufacturing plant, and the warehouse space for parts and products out back. "The plant" is the largest room, and Carl's pride. Twelve manufacturing ponds and their cooling systems—vats ranging in size from a kitchen sink to a small swimming pool—are where Desert Rose uses nanocomputers and assemblers to do their building work. A plumbers' nightmare of piping runs between the ponds and a triple row of containers with labels like CARBON FEEDSTOCK, PREPARED PLATINUM, SIZE-4 STRUCTURAL FIBERS, and PREFAB MOTORS. Carl keeps a good stock of parts and raw materials on hand, with more in the underground warehouse. Sure, some rare things almost never get used, but having them ready to go is one of Carl's secrets for delivering on time and building precisely to specification. Over on a table are Carl's music system and the computers—descendants of the IBM PCs and Macintoshes of the 1980s—that are used to run the manufacturing process. In a space the size of a large living room, Carl and Maria have all the raw materials and all the production equipment—nanocomputers and assemblers—they need for building almost anything.

Occasionally, Carl and Maria need the services of specialized tools, such as disassemblers, that might exist only in labs. A disassembler works like an archeologist, painstakingly excavating the structure of a molecule, removing atom after atom, in order to record and analyze the molecular structure. Because they work so slowly, noting the position of each molecule, disassemblers aren't used for recycling operations—it would be expensive and pointless to record all this unwanted data. But as tools for analyzing the unknown, they're hard to beat.

Maria found this out when a customer sent her an order for tropically scented furniture and fixtures for his restaurant, but instead of including the software instructions for building the perfume, Maria found a plastic bag full of resinous brown gook with a note saying, "I got this stuff in the tropics. Please make the fabric smell like this." Maria (after sniffing the gook and deciding it smelled surprisingly tropically good) shipped the sample to the lab for chemical analysis by disassembler. The lab sent back software with the molecular description and instructions for building the same scent into the furniture.

Carl usually schedules production very tightly: in every tank, assemblers are building products; every computer is directing work. But this morning, listening to the tone of Maria's voice wafting in from the front office, Carl changes his plans: something important is about to happen. He postpones building orders for video wallpaper and commemorative diamond baseballs, and holds three pools and a computer ready. Minutes later, Maria hurries in, her voice tight and anxious. "Carl, that earthquake down south—they need help. Amanda from the Red Cross is sending the software right now."

To build a product, Desert Rose needs design instructions—computer software—for the assemblers. Carl and Maria have their own software library, but usually they buy or rent what they need, or the customers send their own designs.

The software that Amanda sends contains the specifications to manufacture the emergency equipment: a set of instructions to be run on a standard desktop computer. Within minutes, two copies of the Red Cross software arrive electronically. Before starting the build, Carl meticulously checks to make sure that the master copy and backup copy agree and weren't damaged in transit. If the instructions are complete and correct and properly signed with the Red Cross data stamp, then the desktop computer will communicate these building instructions directly to millions of small computers acting as on-the-job foremen directing the work: nanocomputers.

Nanocomputers

While the first, primitive assemblers were controlled by changing what molecules are in the solution around the device, getting the speed and accuracy wanted for large-scale manufacturing takes real computation. Carl's setup uses a combination of special-purpose molecule processors and general-purpose assemblers, all controlled and orchestrated by nanocomputers.

Computers back in the 1990s used microelectronics. They worked by moving electrical charge back and forth through conducting paths—wires, in effect—using it to block and unblock the flow of charge in other paths. With nanotechnology, computers are built from molecular electronics. Like the computers of the 1990s, they use electronic signals to weave the patterns of digital logic. Being made of molecular components, though, they are built on a much smaller scale than 1990s computers, and work much faster and more efficiently. On the scale of our simulated molecular world, 1990s computer chips are like landscapes, while nanocomputers are like individual buildings. Carl's desktop PC contains over a trillion nanocomputers, enough to out-compute all the microelectronic computers of the twentieth century put together.

Back in the dark ages of the 1980s, an exploratory engineer proposed that nanocomputers could be mechanical, using sliding rods instead of moving electrons as shown in Figure 6.1. These molecular mechanical computers were much easier to design than molecular electronic computers would have been. They were a big help in getting some idea of what nanotechnology could do.

Even back then, it was pretty obvious that mechanical computers would be slower than electronic computers. Carl's molecular electronic PC would have been no great surprise, though nobody knew just how to design one. When nanotechnology actually arrived and people started competing to build the best possible computers, molecular electronics won the technology race. Still, mechanical nanocomputers could have done all the nanocomputing jobs at Desert Rose: ordinary, everyday molecular manufacturing just doesn't demand the last word in computer performance.

For Carl, the millions of nanocomputers in the milky waters of his building ponds are just extensions of machines on his desk, machines there to help him run his business and deliver products to his customers—or, in the case of the Red Cross emergency, to help provide time-critical emergency supplies. By reserving those three separate ponds, Carl can either build three different kinds of equipment for the Red Cross or use all the ponds to mass-produce the first thing on the Red Cross list: emergency shelters for ten thousand people. The software is ready, the plumbing is fine, the drums of building materials are all topped up, the Special Mix for this job is loaded: the build is ready to start. "Okay," Carl tells the computer, "build Red Cross tents." Computer talks to nanocomputers. In all three pools, nanocomputers talk to assemblers. The build begins.

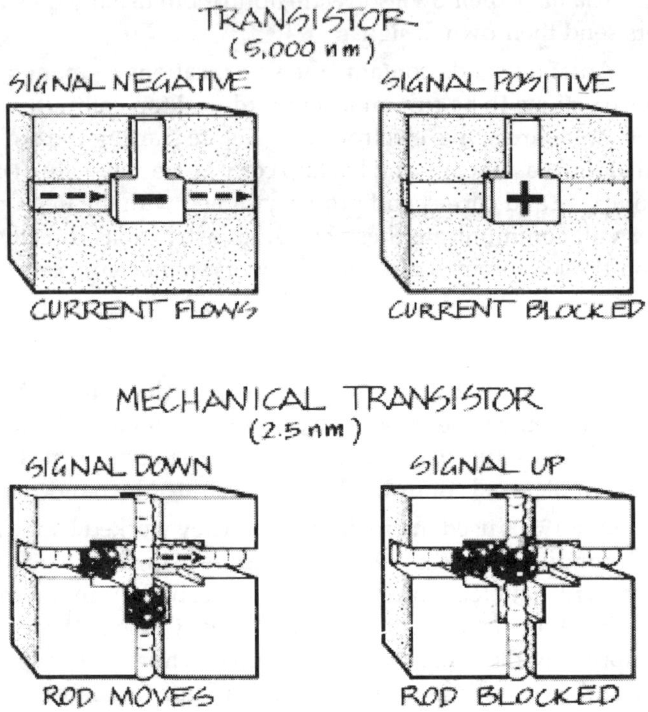

Fig. I-F.1: Mechanical Transistor

An electronic transistor (above) lets current flow when a negative electric charge is applied and blocks current when a positive charge is applied. The mechanical "transistor" (below) lets the horizontal rod move when the vertical rod is down, and blocks the horizontal rod when the vertical rod is up. Either device can be used to build logic gates and computers.

Assembling Products

Some of the building done at Desert Rose Industries uses assemblers much like the ones we saw in the first hall of the plant tour, back in the simulated molecular world of the Silicon Valley Faire. As seen in simulation, they are big, slow, computer-controlled things moving molecular tools. With the right instructions and machinery to keep them supplied with molecular tools, these general-purpose assemblers can build almost anything. They're slow, though, and take a lot of energy to run. Some of the building uses special-purpose assembly systems in the molecule-processing style, like the systems in the basement we saw in the tour of a simulated molecular factory. The special-purpose systems are all moving belts and rollers, but no arms. This is faster and more efficient, but for quantity orders, cooling requirements limit the speed.

It's faster to use larger, prefabricated building blocks. Desert Rose uses these for most of their work, and especially for rush orders like the one Carl just set up. Their underground warehouse has room-sized bins containing upward of a thousand tons of the most popular building blocks, things like structural fibers. They're made at plants on the West Coast and shipped here by subway for ready use. Other kinds are made on site using the special-purpose assemblers. Carl's main room has several cabinet-sized boxes hooked up to the plumbing, each taking in raw materials, running them through this sort of specialized molecular machinery, and pumping out a milky syrup of product. One syrup contains motors, another one contains computers, and another is full of microscopic plug-in light sources. All go into tanks for later use.

Now they're being used. The mix for the Red Cross tent job is mostly structural fiber stronger than the old bulletproof-vest materials. Other building blocks also go in, including motors, computers, and dozens of little struts, angle brackets, and doohickies. The mix would look like someone had stirred together the parts from a dozen toy sets, if the parts were big enough to see. In fact, though, the largest parts would be no more than blurry dots, if you saw one under a normal optical microscope.

The mix also contains block-assemblers, floating free like everything else. These machines are big, about like an office building in our simulation view with the standard settings. Each has several jointed arms, a computer, and several plugs and sockets. These do the actual construction work.

To begin the build, pumps pour the mix into a manufacturing pond. The constant tumbling motions of microscopic things in liquids would be too disorganized for building anything so large as a tent, so the block-assemblers start grabbing their neighbors. Within moments, they have linked up to form a framework spread through the liquid. Now that they are plugged together, they divide up jobs, and get to work. Instructions pour in from Carl's desktop computer.

The block-assemblers use sticky grippers to pull specific kinds of building blocks out of the liquid. They use their arms to plug them together. For a permanent job, they would be using blocks that bond together chemically and permanently. For these temporary tents, though, the Red Cross design uses a set of standard blocks that are put together with amazingly ordinary fasteners: these blocks have snaps, plugs, and screws, though of course the parts are atomically perfect and the threads on the screws are single helical rows of atoms. The resulting joints weaken the tent's structure somewhat, but who cares? The basic materials are almost a hundred times stronger than steel, so there is strength to waste if it makes manufacturing more convenient.

Fiber segments snap together to make fabrics. Some segments contain motors and computers, linked by fibers that contain power and data cables. Struts snap together with more motors and computers to make the tent's main structures. Special surfaces are made of special building blocks. From the human perspective, each tent is a lightweight structure that contains most of the conveniences and comforts of an apartment: cooking facilities, a bathroom, beds, windows, air conditioning, specially modified to meet the environmental demands of the quake-stricken country. From a builder's perspective, especially from a nanomachine's point of view, the tent is just structure slapped together from a few hundred kinds of prefab parts.

In a matter of seconds, each block-assembler has put together a few thousand parts, and its section of the tent is done. In fact, the whole thing is done: many trillions of hands make light work. A crane swings out over the pond and starts plucking out tent packages as fresh mix flows in.

Maria's concern has drawn her back to the plant to see how the build is going. "It's coming along," Carl reassures her. "Look, the first batch of tents is out." In the warehouse, the first pallet is already stacked with five layers of dove-gray "suitcases": tents dried and packed for transport.

Carl grabs a tent by the handle and lugs it out the door. He pushes a tab on the corner labeled "Open," and it takes over a minute to unfold to a structure a half-dozen paces on a side. The tent is big, and light enough to blow away if it didn't cling to the ground so tightly. Maria and Carl tour the tent, testing the appliances, checking the construction of furniture: everything is extremely lightweight compared to the bulk-manufactured goods of the 1990s, tough but almost hollow.

Like the other structures, the walls and floors are full of tiny motors and struts controlled by simple computers like the ones used in twentieth-century cars, televisions, and pinball machines. They can unfold and refold. They can also flex to produce sound like a high-quality speaker, or to absorb sound to silence outdoors racket. The whole three-room setup is small and efficient, looking like a cross between a boat cabin and a Japanese business hotel room. Outside, though, it is little more than a box. Maria shakes her head, knowing full well what architects can do these days when they try to make a building really fit its site. Oh well, she thinks, These won't be used for long.

"Well, that looks pretty good to me," says Carl with satisfaction. "And I think we'll be finished in another hour."

Maria is relieved. "I'm glad you had those pools freed up so fast."

By three o'clock, they've shipped three thousand emergency shelters, sending them by subway. Within half an hour, tents are being set up at the disaster site.

Behind the Scenes and Afterward

Desert Rose Industries and other manufacturers can make almost anything quickly and at low cost. That includes the tunneling machines and other equipment that made the subway system they use for shipping. Digging a tunnel from coast to coast now costs less than digging a single block under New York City used to. It wasn't expensive to get a deep-transit terminal installed in their basement. Just as the tents aren't mere bundles of canvas, these subways aren't slow things full of screeching, jolting metal boxes. They're magnetically levitated to reach aircraft speeds—as experimental Japanese trains were in the late 1980s—making it easy for Carl and Maria to give their customers quick service. There's still a road leading to the plant, but nobody's driven a truck over it for years.

They only take in materials that they will eventually ship out in products, so there's nothing left over, and no wastes to dump. One corner of the plant is full of recycling equipment. There are always some obsolete parts to get rid of, or things that have been damaged and need to be reworked. These get broken down into simpler molecules and put back together again to make new parts.

The gunk in the manufacturing ponds is water mixed with particles much finer than silt. The particles—fasteners, computers, and the rest—stay in suspension because they are wrapped in molecular jackets that keep them there. This uses the same principle as detergent molecules, which coat particles of oily dirt to float them away.

Though it wouldn't be nutritious or appetizing, you could drink the tent mix and be no worse for it. To your body, the parts and their jackets, and even the nanomachines, would be like so many bits of grit and sawdust. (Grandma would have called it roughage.)

Carl and Maria get their power from solar cells in the road, which is the only reason they bothered having it paved. In back of their plant stands what looks like a fat smokestack. All it produces, though, is an updraft of clean, warm air. The darkly paved road, baking in the New Mexico sun, is cooler than you might expect: it soaks up solar energy and makes electricity, instead of just heat. Once the power is used, it turns back into heat, which has to go somewhere. So the heat rises from their cooling tower instead of the road, and the energy does useful work on the way.

Some products, like rocket engines, are made more slowly and in a single piece. This makes them stronger and more permanent. The tents, though, don't need to be superstrong and are just for temporary use. A few days after the tents go up, the earthquake victims start to move out into new housing (permanent, better-looking, and very earthquake resistant). The tents get folded and shipped off for recycling.

Recycling things built this way is simple and efficient: nanomachines just unscrew and unsnap the connectors and sort the parts into bins again. The shipments Desert Rose gets are mostly recycled to begin with. There's no special labeling for recycled materials, because the molecular parts are the same either way.

For convenience (and to keep the plant small), Carl and Maria get most of their parts prefabricated, even though they can make almost anything. They can even make more production equipment. In one of their manufacturing ponds, they can put together a new cabinet full of special-purpose assemblers. They do this when they want to make a new type of part in-house. Like parts, the part-assemblers are made by special-purpose assemblers. Carl can even make big vats in medium-size vats, unfolding them like tents.

If Desert Rose Industries needed to double capacity, Carl and Maria could do it in just a few days. They did this once for a special order of stadium sections. Maria got Carl to recycle the new building before its shadow hurt their cactus garden.

Factory Factories

In the Desert Rose Industries scenario, manufacturing has become cheap, fast, clean, and efficient. Using fast, precise machines to handle matter in molecular pieces makes it easy for nanotechnology to be fast, clean, and efficient. But for it to be cheap, the manufacturing equipment has to be cheap.

The Desert Rose scenario shows how this can work. Molecular-manufacturing equipment can be used to make all the parts needed to build more molecular manufacturing equipment. It can even build the machines needed to put the parts together. This resembles an idea developed by NASA for a self-expanding manufacturing complex on the Moon, but made faster and simpler using molecular machines and parts.

Replicators

In the early days of nanotechnology, there won't be as many different kinds of machines as there are at Desert Rose. One way to build a lot of molecular manufacturing equipment in a reasonable time would be to make a machine that can be used to make a copy of itself, starting with special but simple chemicals. A machine able to do this is called a "replicator." With a replicator and a pot full of the right fuel and raw materials, you could start with one machine, then have two, four, eight, and so on.

This doubling process soon makes enough machines to be useful. The replicators—each including a computer to control it and a general-purpose assembler to build things—could then be used to make something else, like the tons of specialized machines needed to set up a Desert Rose manufacturing plant. At that point, the replicators could be discarded in favor of those more efficient machines.

Replicators are worth a closer look, though, because they show how quickly molecular manufacturing systems can be used to build more manufacturing equipment. Figure I-F.2 shows a design described in Stanford University course CS 404 in the spring of 1988. If we were in one of our standard simulation views, the submicroscopic device at the top of the picture would be like a huge tank, three stories tall when lying on its side. Most of its interior is taken up by a tape memory system that tells how to move the arm to build all the parts of the replicator, except the tape itself. The tape gets made by a special tape-copying machine. At the right-hand end of the replicator are pores for bringing in fuel and raw-material molecules, and machinery for processing them. In the middle are computer-controlled arms, like the ones we saw on the plant trip. These do most of the actual construction.

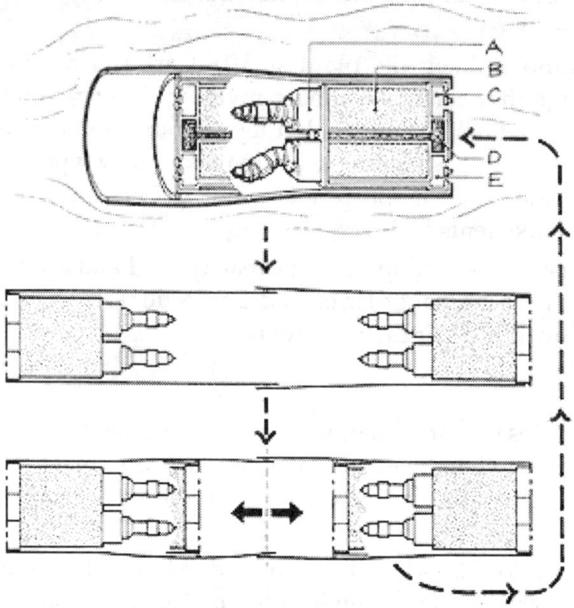

Fig. I-F.2: Replicator

A replicator would be able to build copies of itself when supplied with fuel and raw materials. In the diagram, (A) contains a nanocomputer, (B) a library of stored instructions, (C) contains machinery that takes in fuel and produces electric power, (D) is a motor, and (E) contains machinery that prepares raw materials for use. (All volumes follow calculations presented in a class at Stanford.) The lower diagrams illustrate steps in a replication cycle, showing how the working space is kept isolated from the external liquid, which provides the needed fuel and raw-material molecules. Replicators of this sort are useful as thought experiments to show how nanomachines can product more nanomachines, but specialized manufacturing equipment would be more efficient in practice.

The steps in the cycle—using a copy to block the tube, beginning a fresh copy, then releasing the old one—illustrate one way for a machine to build a copy of itself while floating in a liquid, yet doing all its construction work inside, in vacuum. (It's easier to design for vacuum, and this is exploratory-engineering work, so easier design is better design.) Calculations suggest that the whole construction cycle can be completed in less than a quarter hour, since the replicator contains about a billion atoms, and each arm can handle about a million atoms per second. At that rate, one device can double and double again to make trillions in about ten hours.

Each replicator just sits in a chemical bath, soaking up what it needs and making more replicators. Eventually, either the special chemicals run out or other chemicals are added to signal them to do something else. At that point, they can be reprogrammed to produce anything else you please, so long as it can be extruded from the front. The products can be long, and can unfold or be pieced together to make larger objects, so the size of these initial replicators—smaller than a bacterium—would be only a temporary limitation.

General Assemblers

From the molecular manipulators and primitive assemblers described in the last unit, the most likely path to nanotechnology leads to assemblers with more and more general capabilities. Still, efficiency favors special-purpose machines, and the Desert Rose scenario didn't make much use of general assemblers. Why bother making general-purpose assemblers in the first place?

To see the answer, turn the question around and ask, Why *not* build such a tool? There is nothing outstandingly difficult about a general assembler, as molecular machinery goes. It will just be a device with good, flexible positional control and a system to feed it a variety of molecular tools. This is a useful, basic capability. General-purpose assemblers could always be replaced by a lot of specialized devices, but to build those specialized devices in the first place, it makes sense to come up with a more flexible, general-purpose system that can just be reprogrammed.

So, general purpose machines are likely to find use in making short production runs of more specialized devices. Ralph Merkle, a computers and security expert at Xerox Palo Alto Research Center, sees this as paralleling the way manufacturing works today: "General purpose devices could do many tasks, but they'll do them inefficiently. For any given task, there will be one or a few best ways of doing it, and one or a few special purpose devices that are finely tuned to do that one task. Nails aren't made by a general-purpose machine shop, they're made by nail-making machines. Making nails with a general-purpose machine shop would be more expensive, more difficult, and more time-consuming. Likewise, in the future we won't see a proliferation of general-purpose self-replicating systems, we'll see specialization for almost every task."

What Will These Capabilities Make Possible?

We've surveyed a lot of devices: assemblers of various flavors, nanocomputers, disassemblers, replicators, and others. What's important about these is not the exact distinctions between them, but the capabilities that they will give and the effects they will have on human lives. Again, we are suspending discussion of potential misapplications until later.

If we tease apart the implications of what we've seen in the Desert Rose scenario, we can analyze some of the key impacts of molecular manufacturing in industry, science, and medicine.

Technology and Industry

At its base, nanotechnology is about molecular manufacturing, and manufacturing is the basis of much of today's industry. This is why Desert Rose made a good starting point for describing the possibilities of a nanotechnological world. From an industrial perspective, it makes sense to think of nanotechnology in terms of products and production.

New Products: Today, we handle matter crudely, but nanotechnology will bring thorough control of the structure of matter, the ability to build objects to atom-by-atom specifications. This means being able to make almost anything. By comparison, even today's range of products will feel very limited. Nanotechnology will make possible a huge range of new products, a range we can't envision today. Still, to get a feel for what is possible, we can look at some easily imagined applications.

Reliable Products: Today, products often fail, but for failures to occur—for a wing to fall off an airplane, or a bearing to wear out—a lot of atoms have to be out of place. In the future, we can do better. There are two basic reasons for this: better materials and better quality control, both achieved by molecular manufacturing. By using materials tens of times stronger than steel, as Desert Rose did, it will be easy to make things that are very strong, with a huge safety margin. By building things with atom-by-atom control, flaws can be made very rare and extremely small—nonexistent, by present standards.

With nanotechnology, we can design in big safety margins and then manufacture the design with near-perfection. The result will be products that are tough and reliable. (There will still be room for bad designs, and for people who wish to take risks in machines that balance on the edge of disaster.)

Intelligent Products: Today, we make most things from big chunks of metal, wood, plastic, and the like, or from tangles of fibers. Objects made with molecular manufacturing can contain trillions of microscopic motors and computers, forming parts that work together to do something useful. A climber's rope can be made of fibers that slide around and reweave to eliminate frayed spots. Tents can be made of parts that slide and lock to turn a package into a building. Walls and furniture can be made to repair themselves, instead of passively deteriorating.

On a mundane level, this sort of flexibility will increase reliability and durability. Beyond this, it will make possible new products with abilities we never imagined we needed so badly. And beyond even this, it will open new possibilities for art.

Inexpensive Production: Today, production requires a lot of labour, either for making things or for building and maintaining machines that make things. Labor is expensive, and expensive machines make automation expensive, too. In the Desert Rose scenario, we got a glimpse of how molecular manufacturing can make production far less expensive than it is today. This is perhaps the most surprising conclusion about nanotechnology, so we'll take a closer look at it in the next unit.

Clean Production: Today, our manufacturing processes handle matter sloppily, producing pollution. One step puts stuff where it shouldn't be; the next washes it off the product and into the water supply. Our transportation system worsens the problem as unreliable trucks and tankers spill noxious chemicals over the land and sea. Everything is expensive, so companies skimp on even the half-effective pollution controls that we know how to build.

Nanotechnology will mean greater control of matter, making it easy to avoid pollution. This means that a little public pressure will go a long way toward a cleaner environment. Likewise, it will make it easy to increase efficiency and reduce resource requirements. Products, like the Red Cross tents at Desert Rose, can be made of snap-together, easily recyclable parts. Sophisticated products could even be made from biodegradable materials. Nanotechnology will make it easy to attack the causes of pollution at their technological root.

Nanotechnology will have great applications in the field of industry, much as transistors had great applications in the field of vacuum tube electronics, and democracy had great applications in the field of monarchy. It will not so much advance twentieth-century industry as *replace* it—not all at once, but during a thin slice of historical time.

Science

Chemistry: Today, chemists work with huge number of molecules and study them using clever, indirect techniques. Making a new molecule can be a major project, and studying it can be another. Molecular manufacturing will help chemists make what they want to study, and it will help them make the tools they need to study it. Nanoinstruments will be used to prod, measure, and modify molecules in a host of ways, studying their structures, behaviors, and interactions.

Materials: Today, materials scientists make new superconductors, semiconductors, and structural materials by mixing and crushing and baking and freezing, and so forth. They dream of far more structures than they can make, and they stumble across more things than they plan. With molecular manufacturing, materials science can be much more systematic and thorough. New ideas can be tested because new materials can be built according to plan (rather than playing around, groping for a recipe).

This need not rule out unexpected discoveries, since experiments—even blind searches—will go much faster. A few tons of raw materials would be enough to make a billion samples, each a cubic micron in size. In all of history so far, materials scientists have never tested so many materials. With nanoinstruments and nanocomputers, they could. One laboratory could then do more than all of today's materials scientists put together.

Biology: Today, biologists use a host of molecular devices borrowed from biology to study biology. Many of these can be viewed as molecular machines. Nanotechnology will greatly advance biology by providing better molecular devices, better nanoinstruments. Some cells have already been mapped in amazing molecular detail, but biology still has far to go. With nanoinstruments (including molecule-by-molecule disassemblers), biologists will at last be able to map cells completely and study their interactions in detail. It will become easy not only to find molecules in cells, but to learn what they do. This will help in understanding disease and the molecular requirements for health, enormously advancing medicine.

Computation: Today, computers range from a million to a billion times faster than an old desktop adding machine, and the results have been revolutionary for science. Every year, more questions can be answered by calculations based on known principles of physics. The advent of nanocomputers—even slow, miserable, mechanical nanocomputers—will give us practical machines with a trillion times the power of today's computers (essentially by letting us package a trillion computers in a small space, without gobbling too much money or energy.) The consequences will again be revolutionary.

Physics: The known principles of physics are adequate for understanding molecules, materials, and cells, but not for understanding phenomena on a scale that would still be submicroscopic if atoms were the size of marbles. Nanotechnology can't help here directly, but it can provide manufacturing facilities that will make huge particle accelerators economical, where today they strain national budgets.

More generally, nanotechnology will help science wherever precision and fine details are important. Science frequently proceeds by trying small variations in almost identical experiments, comparing the results. This will be easier when molecular manufacturing can make two objects that are identical, molecule by molecule. In some areas, today's techniques are not only crude, but destructive. Archaeological sites are unique records of the human past, but today's techniques throw away most information during the dig, by accident. Future archaeologists, able to sift soil not speck by speck but molecule by molecule, will be grateful indeed to those archaeologists who today leave some ground undisturbed.

Medicine

Of all the areas where the ability to manufacture new tools is important, medicine is perhaps the greatest. The human body is intricate, and that intricacy extends beyond the range of human vision, beyond microscopic imaging, down to the molecular scale. "Molecular medicine" is an increasingly popular term today, but medicine today has only the simplest of molecular tools. As biology uses nanoinstruments to learn about disease and health, we will learn the physical requirements for restoring and maintaining health. And with this knowledge will come the tools needed to satisfy those requirements—tools ranging from improved pharmaceuticals to devices able to repair cells and tissues through molecular surgery.

Advanced medicine will be among the most complex and difficult applications of nanotechnology. It will require great knowledge, but nanoinstruments will help gather this knowledge. It will pose great engineering challenges, but computers of trillionfold greater power will help meet those challenges. It will solve medical problems on which we spend billions of dollars today, in hopes of modest improvements.

Today, modern medicine often means an expensive way to prolong misery. Will nanomedicine be more of the same? Any reader over the age of, say, thirty knows how things start to go wrong: an ache here, a wrinkle there, the loss of an ability. Over the decades, the physical quality of life declines faster and faster—the limits of what the body can do become stricter—until the limits are those of a hospital bed. The healing abilities we have when young seem to fade away. Modern medical practice expends the bulk of its effort on such things as intensive care units, dragging out the last few years of life without restoring health.

Truly advanced medicine will be able to restore and supplement the youthful ability to heal. Its cost will depend on the cost of producing things more intricate than any we have seen before, the cost of producing computers, sensors, and the like by the trillions. To understand the prospects for medicine, like those for science and industry, we need to take a closer look at the cost of molecular manufacturing.

Unit I–G

The Spiral of Capability

In earlier units, we have stepped forward and backward through time. The last step was a big one, leaping from small laboratory devices to the high-capacity industrial facility of the Desert Rose scenario. Our narrative crossed this gap in a single leap, but the world won't. To understand how nanotechnology might unfold, it makes sense to look at some of its easier and more difficult applications. The result won't be a timetable, or even a series of milestones, but it should give a better picture of what we can expect as nanotechnology develops from simple, crude, costly beginnings to a state of greater sophistication and lower cost.

Improving Quality

Molecular manufacturing will make better products possible. We're likely to see some early applications in at least two areas: stronger materials and faster computers. Strong materials are simple, and will be hard to pass up. Computers are more complex, but the payoff will be enormous.

Computers

The computer industry has been under steady pressure to make computer chips ever smaller. As sizes have shrunk, costs have fallen while efficiency and capabilities have increased. The pressure to continue this process pushes in the direction of nanotechnology; it may even be one of the major motivations behind developing the technology.

John Walker, a founder of Autodesk, explains: "Even technologies with enormous potential can lie dormant unless there are significant payoffs along the way to reward those who pioneer them. That's one of the reasons integrated circuits developed so rapidly; each advance found an immediate market willing to apply it and enrich the innovator that created it.

"Does molecular engineering have this kind of payoff? I think it does. Remembering that we may be less than ten years away from 'hitting the wall' as far as scaling our existing electronics goes, a great deal of research is presently going on in the area of molecular and quantum electronics. The payoff is easy to calculate: You can build devices one thousand times faster, more energy-efficient, and cheaper than those we're currently using—at least one hundred times better than exotic materials being considered to replace silicon when it reaches its limits."

Federico Capasso, head of the Quantum Phenomena and Device Research Department at AT&T Bell Labs, agrees that electronics researchers will keep pushing for smaller devices once silicon's potential has been reached. He explains that "at some point we will reach difficulties: some people say at a hundred fifty nanometers, others think it's beyond that. What will happen then? It's hard to think that the electronics industry will say, 'Stop here. We'll stop evolving because we can't shrink the device.' From an economic point of view, in order to survive, an industry has to innovate continuously."

The computer industry's push toward devices of molecular size has an air of inevitability. Today's researchers struggle to build molecular electronics using bulk techniques, with no products yet in sight; with molecular manipulators, they will finally have the tools they need for fast and accurate experimentation. Once successful designs are developed, packaged, and tested, the pressure will be on to learn to make them in quantity at low cost. The competitive pressures will be fierce, because advanced molecular electronics will be orders of magnitude better than today's integrated circuits, ultimately enabling the construction of computers with trillionfold greater capability.

Strong, Lightweight Structures

At the opposite extreme from molecular electronics—complex and at first worth billions of dollars per gram—are structural materials: worth only dollars per kilogram in most applications, but much simpler in structure. Once molecular manufacturing becomes inexpensive, structural materials will be important products.

These materials play a central role in almost everything around us, from cars and aircraft to furniture and houses. All of these objects get their size, shape, and strength from a structural skeleton of some sort. This makes structural materials a natural place to begin in understanding how nanotechnology can improve products.

Cars today are mostly made of steel, aircraft of aluminum, and buildings and furniture largely of steel and wood. These materials have a certain "strength-to-weight ratio" (more properly, a strength-to-density ratio). To make cars stronger, they'd have to be heavier; to make them lighter, they'd have to be weaker. Clever design can change this relationship a little, but to change it a lot requires a change of materials.

Making something heavy is easy: just leave a hollow space, then fill it with water, sand, or lead shot. Making something light and strong is harder, but often important. Automakers try to make cars lightweight, aircraft manufacturers try harder, and with spacecraft manufacturers it is an obsession. Reducing mass saves materials and energy.

The strongest materials in use today are mostly made of carbon. Kevlar, used in racing sails and bulletproof vests, is made of carbon-rich molecular fibers. Expensive graphite composites, used in tennis rackets and jet aircraft, are made using pure-carbon fibers. Perfect fibers of carbon—both graphite and diamond—would be even better, but can't be made with today's technology. Once molecular manufacturing gets rolling, though, such materials will be commonplace and inexpensive.

What will these materials be like? To picture them, a good place to start is wood. The structure of wood can vary from extremely light and porous, like balsa wood, to denser structures like oak. Wood is made by molecular machinery in plants from carbon-rich polymers, mostly cellulose. Molecular manufacturing will be able to make materials like these, but with a strength-to-weight ratio about a hundred times that of mediocre steel, and tens of times better than the best steel. Instead of being made of cellulose, these materials will be made of carbon in forms like diamond.

Diamond is emphasized here not because it is shiny and expensive, but because it is strong and potentially cheap. Diamond is just carbon with properly arranged atoms. Companies are already learning to make it from natural gas at low pressure. Molecular manufacturing will be able to make complex objects of the stuff, built lighter than balsa wood but stronger than steel.

Products made of such materials could be startling by our present standards. Objects could be made that are identical in size and shape to those we make today, but simultaneously stronger and 90 per cent lighter. This is something to keep in mind next time you're lugging a heavy object around. (If something needs weight to hold it in place, it would be more convenient to add this ballast when the thing is in its proper location than to build in the extra weight permanently.)

Better structural materials will make aircraft lighter, stronger, and more efficient, but will have the greatest effect on spacecraft. Today, spacecraft can barely reach orbit with both a safety margin and a cargo. To get there at all, they have to drop off parts like boosters and tanks along the way, shedding weight. With strong materials, this will change: as in the space-travel-for-business scenario in later units, spacecraft will become more like aircraft are today. They will be rugged and reliable, and strong enough and light enough to reach space in one piece.

Quickening Development

In some areas of high technology—spaceflight has been a notorious example—it takes years, even decades, to try a new idea. This makes progress slow to a crawl. In other areas—software has been a shining example—new ideas can be tested in minutes or hours. Since the space shuttle design was frozen, personal computer software has come into existence and gone through several generations of commercial development, each with many cycles of building and testing.

Fast, Inexpensive Testing

Even in the days of the first operational molecular manipulators, experimentation is likely to be reasonably fast. Individual chemical steps can take seconds or less. Complex molecular objects could be built in a matter of hours. This will let new ideas be put into practice almost as fast as they can be designed.

Later assemblers will be even faster. At a millionth of a second per step, they will approach the speed of computers. And, as nanotechnology matures, experimenters will have more and more molecular instruments available to help them find out whether their devices work or not. Fast construction and fast testing will encourage fast progress.

At this point, the cost of materials and equipment for experiments will be trivial. No one today can afford to build Moon rockets on a hobby budget, but they can afford to build software, and many useful programs have been the result. There is no economic reason why nanomachines couldn't eventually be built with a hobby-size budget, though there are reasons—to be discussed in later units—for wanting to place limits on what can be built.

Early Simplicity

Finally, established technologies are always pushing up against some limit; the easy opportunities have generally been exploited. In many fields, the limits are those of the properties of the materials used and the cost and precision of manufacturing. This is true for computers, for spacecraft, for cars, blenders, and shoes. For software, the limits are those of computer capacity and of sheer complexity (which is to say, of human intelligence). After molecular manufacturing develops certain basic abilities, a whole set of limits will fall, and a whole range of developments will

	Space	*Computers*	*Nanotechnology*
Precursor science and technologies	Physics Sounding rockets	Mathematics Electronics	Theoretical chemistry Chemical synthesis
Crucial advance	Teams combine and improve technologies	Teams combine and improve technologies	Teams combine and improve technologies
Threshold capability	First satellite	First computer	First assembler
Early practical applications	Weather, spy and communication satellites	Scientific calculations Payroll calculations	Molecular sensors Molecular computing
Breakthrough capability	Routine, inexpensive spaceflight	Powerful mass-market desktop computers	Powerful inexpensive molecular manufacturing
Further projected developments	Lunar base, Mars exploration	Widespread electronic publishing	New medical abilities New, inexpensive products
More advanced developments	Mining, development, settlement of solar system	Major automation of engineering design	Help with computer goals Environmental cleanup
Yet more advanced developments	Interstellar flight and settlement feasible	Trillionfold computer power	Help with computer goals General tissue repair

become possible. Limits set by materials properties, and by the cost and precision of manufacturing, will be pushed way back. Competition, easy opportunities, and fast, low-cost experimentation should combine to yield an explosion of new products.

This does not mean immediately, and it does not apply to all imaginable nanotechnologies. Some technologies are imaginable and clearly feasible, yet dauntingly complex. Still, the above considerations suggest that a wide range of advances could happen at a brisk pace. The main bottleneck might seem to be a shortage of knowledgeable designers—hardly anyone knows both chemistry and mechanical design—but improving computer simulations will help. These simulations will let engineers tinker with molecular-machinery designs, absorbing knowledge of chemical rules without learning chemistry in the usual sense.

Climbing Complexity

Making familiar products from improved materials will increase their safety, performance, and usefulness. It will also present the simplest engineering task. A greater change, though, will result from unfamiliar products made possible by new manufacturing methods. In talking about unfamiliar products, a hard-to-answer question arises: What will people want?

Products are typically made because their recipients want them. In our discussions here, if we describe something that people won't want, then it probably won't get built, and if it does get built, it will soon disappear. (The exceptions—fraud, coercion, persistent mistakes—are important, but in other contexts.) To anchor our discussion, it makes sense to look not at totally new products, but instead at new features for old products, or new ways to provide old services. This approach won't cover more than a fraction of what is possible, but will start from something sensible and provide a springboard for the imagination.

As usual, we are describing possibilities, not making predictions. The possibilities focused on here arise from more complex applications of molecular manufacturing—nanotechnological products that contain nanomachines when they are finished. Earlier, we discussed strong materials. Now, we discuss some smart materials.

Smart Materials

The goal of making materials and objects smart isn't new: researchers are already struggling to build structures that can sense internal and environmental conditions and adapt themselves appropriately. There is even a *Journal of Intelligent Material Systems and Structures.* By using materials that can adapt their shapes, sometimes hooked up to sensors and computers, engineers are starting to make objects they call "smart." These are the early ancestors of the smart materials that molecular manufacturing will make possible.

Today, we are used to having machines with a few visible moving parts. In cars, the wheels go around, the windshield wipers go back and forth, the antenna may go up and down, the seat belts, mirrors, and steering wheel may be motor-driven. Electric motors are fairly small, fairly inexpensive, and fairly reliable, so they are fairly common. The result is machines that are fairly smart and flexible, in a clumsy, expensive way.

In the Desert Rose scenario, we saw "tents" being assembled from trillions of submicroscopically small parts, including motors, computers, fibers, and struts. To the naked eye, materials made from these parts could seem as smooth and uniform as a piece of plastic, or as richly textured as wood or cloth—it is all a matter of the arrangement and appearance of the submicroscopic parts. These motors and other parts cost less than a trillionth of a dollar apiece. They can be quite reliable, and good design can make systems work smoothly even if 10 percent of a trillion motors burn out. Likewise for motor-controlling computers and the rest. The resulting machines can be very smart and flexible, compared to those of today, and inexpensive, too.

When materials can be full of motors and controllers, whole chunks of material can be made flexible and controllable. The applications should be broad.

Scenario: Smart Paint

Surfaces surround us, and human-made surfaces—walls, roofs, and pavement—cover huge areas that matter to people. How can smart materials make a difference here?

The revolution in technology has come and gone, and you want to repaint your walls. Breathing toxic solvents and polluting water by washing brushes have passed into history, because paint has been replaced with smarter stuff. The mid-twentieth century had seen considerable progress in paints, especially the development of liquids that weren't quite liquid—they would spread with a brush, but didn't (stupidly) run and drip under their own weight. This was an improvement, but the new material, "paperpaint," is even more cooperative.

Paperpaint comes in a box with a special trowel and pen. The paperpaint itself is a dry block that feels a lot like a block of wood. Following the instructions, you use the pen to draw a line around the edge of the area you want to paint, putting an X in the middle to show where you want the paint to go on; the line is made of nontoxic disappearing ink, so you can slop it around without staining anything. Using the trowel, you slice off a hunk of paperpaint—which is easy, because it parts like soft butter to the trowel, even though it behaves like a solid to everything else. Very high IQ stuff, that.

Next, you press the hunk against the X and start smoothing it out with the trowel. Each stroke spreads a wide swath of paperpaint, much wider than the trowel, but always staying within the inked line. A few swipes spreads it precisely to the edges, whereupon it smooths out into a uniform layer. Why doesn't it just spread itself? Experience showed that customers didn't mind the effort of making a few swipes and preferred the added control.

The paperpaint consists of a huge number of nanomachines with little wheels for rolling over one another and little sticky pads for clinging to surfaces. Each has a simple, stupid computer on board. Each can signal its neighbors. The whole mass of them clings together like an ordinary solid, but they can slip and slide in a controlled way when signaled. When you smooth the trowel over them, this contact tells them to get moving and spread out. When they hit the line, this tells them to stop. If they don't hit a line, they go a few handbreadths, then stop anyway until you trowel them again. When they encounter a line on all sides, word gets around, and they jostle around to form a smooth, uniform layer. Any that get scraped off are just so much loose dust, but they stick together quite well.

This paint-stuff doesn't get anything wet, doesn't stain, and clings to surfaces just tightly enough to keep it from peeling off accidentally. If some experimentally minded child starts digging with a stick, makes a tear, and peels some off, it can be smoothed back again and will rejoin as good as new. The child may eat a piece, but careful regulation and testing has ensured that this is no worse than eating plain paper, and safer than eating a colorful Sunday newspaper page.

Many refinements are possible. Swipes and pats of the trowel could make areas thicken or thin, or bridge small holes (no more Spackling!). With sufficiently smart paperpaint, and some way to indicate what it should do, you can have your choice of textures. Any good design will be washable, and a better design would shed dirt automatically using microscopic brushes.

Removal, of course, is easy: either you rip and peel (no scraping needed), or find that trowel, set the dial on the handle to "strip," and poke the surface a few times. Either way, you end up with a lump ready to pitch into the recycling bin and the same old wall you started with, bared to sight again.

Power Paint

Perhaps no product will ever be made exactly like the smart paint just described. It would be disappointing if something better couldn't be made by the time smart paint is technologically possible. Still, paperpaint gives a feel for some of the features to expect in the new smart products, features such as increased flexibility and better control. Without loading yet more capability into our paint (though there is no reason why one *couldn't*), let's take a look at some other smart properties one might want in a surface.

External walls, roofs, and paving surfaces are exposed to sunlight, and sunlight carries energy. A proven ability of molecular machinery is the conversion of sunlight to stored energy: plants do it every day. Even now, we can make solar cells that convert sunlight into electricity at efficiencies of 30 percent or so. Molecular manufacturing could not only make solar cells much cheaper, but could also make them tiny enough to be incorporated into the mobile building blocks of a smart paint.

To be efficient, this paint would have to be dark—that is, would have to absorb a lot of light. Black would be best, but even light colors could generate some power, and efficiency isn't everything. Once the paint was applied, its building blocks would plug together to pool their electrical power and deliver it through some standard plug. A thicker, tougher form of this sort of material could be used to resurface pavement, generate power, and transmit it over large distances. Since smart solar-cell pavement could be designed for improved traction and a similar roofing material could be designed for amazing leak-resistance, the stuff should be popular.

On a sunny day, an area just a few paces on a side would generate a kilowatt of electrical power. With good batteries (and enough repaved roads and solar-cell roofing), present demands for electrical power could be met with no coal burning, no oil imports, no nuclear power, no hydroelectric dams, and no land taken over for solar power generation plants.

Pretty Paint, Acoustic Paint

The glow of fireflies and deep sea fish shows that molecular devices can convert stored chemical energy into light. All sorts of common devices show that electricity can be converted to light. With molecular manufacturing, this conversion can be done in thin films, with control over the brightness and color of each microscopic spot. This could be used for diffuse lighting—ceiling paperpaint that glows. With more elaborate control, this would yield the marvel (horror?) of video wallpaper.

With today's technology, we are used to displays that glow. With molecular manufacturing, it will be equally easy to make displays that just change color, like a printed page with mobile ink. Chameleons and flatfish change color by moving colored particles around, and nanomachines could do likewise. On a more molecular level, they could use tunable dyes. Live lobsters are a dark grayish green, but when cooked turn bright red. Much of this change results from the "retuning" of a dye molecule that is bound in a protein in the live lobster but released by heat. This basically mechanical change alters its color; the same principle can be used in nanomachines, but reversibly.

How a surface appears depends on how it reflects or emits light. Nanomachines and nanoelectronics will be able to control this within wide limits. They will be able to do likewise for sound, by controlling how a surface moves. In a stereo system, a speaker is a movable surface, and nanomachines are great for making things move as desired. Making a surface emit high-quality sound will be easy. Almost as easy will be surfaces that actively flex to absorb sound, so that the barking dog across the street seems to fade away.

Smart Cloth

Looking further at the human environment we find a lot of cloth and related materials, such as carpeting and shoes. The textile industry was at the cutting edge of the first industrial revolution, and the next industrial revolution will have its effects on textiles.

With nanotechnology, even the finest textile fibers could have sensors, computers, and motors in their core at little extra cost. Fabrics could include sensors able to detect light, heat, pressure, moisture, stress, and wear, networks of simple computers to integrate this data, and motors and other nanomechanisms to respond to it. Ordinary, everyday things like fabric and padding could be made responsive to a person's needs—changing shape, color, texture, fit, and so forth—with the weather and a person's posture or situation. This process could be slow, or it could be fast enough to respond to a gesture. One result would be genuine one-size-fits-all clothing (give or take child sizes), perfectly tailored off the rack, warm in winter, cool and dry in summer; in short, nanotechnology could provide what advertisers have only promised. Even bogus advertising gives a clue to human desires.

Throughout history, the human race has pursued the quest for comfortable shoes. With fully adjustable materials, the seemingly impossible goal of having shoes that both look good and feel good should finally be achieved. Shoes could keep your feet dry, and warm except in the Arctic, cool except in the tropics, and as comfortable as they can be with a person stepping on them.

Smart Furniture

Adaptive structures will be useful in furniture. Today, we have furniture that adapts to the human body, but it does so in an awkward and incomplete manner. It adapts because people grab cushions and move them around. Or a chair adapts because it is a hinged contraption that grudgingly bends and extends in a few places to suit a small range of preferred positions. Occasionally, one sees furniture that allegedly gives a massage, but in fact only vibrates.

These limitations are consequences of the expense, bulkiness, clumsiness, and unreliability of such things as moving parts, motors, sensors, and computers today. With molecular manufacturing, it will be easy to make furniture from smart materials that can adapt to an individual human body, and to a person's changing position, to consistently give comfortable support. Smart cushions could also do a better job of responding to hints in the form of pats, tugs, and punches. As for massage—a piece of furniture, no matter how advanced, is not the same as a masseuse. Still, a typical massage setting on a smart chair would not mean today's "vibrate medium vigorously," but something closer to "five minutes of shiatsu."

And So Forth . . .

This tour through of the potential of smart matter has shown how we could get walls that look and sound as we wish, clothing, shoes, and furniture of greater comfort, and clean solar power. As one might expect, this just scratches the surface.

If you care to think of further applications, here are some ground rules: Components made by molecular manufacturing can be many tens of times stronger than steel, but materials made by plugging many components together will be weaker. For these, strengths in the range of cotton candy to steel seem achievable. The components will be sensitive to heat, and at high temperatures they will break down or burn. Many materials will be able to survive the temperature of boiling water, but only specialized designs would be oven-safe. Color, texture, and (usually) sound should be controllable. Surfaces can be smooth and tightly sealed (this takes some cleverness). Motions can be fairly fast. Power has to come from somewhere; good sources include electricity, stored chemical energy, and light. If nanomachines or smart materials are dunked in liquids, chemical energy can come from dissolved molecules; if they are in the open, energy can come from light; if they are sitting in one place, they can be plugged into a socket; if they are moving around in the dark, they can run on batteries for a while, then run down and quit. Within these limits, much can be accomplished.

"Smart" is a relative term. Unless you want to assume that people learn a lot more about intelligence and programming, it is best to assume that these materials will follow simple rules, like those followed by parts of drawings on computer screens. In these drawings, a picture of a rectangle can be commanded to sprout handles at its corners; pulling a handle stretches or shrinks the rectangle without distorting its right-angle corners. An object made of smart matter could do likewise in the real world: a box could be stretched to a different size, then made rigid again; a door in a smart-material wall could have its position unlocked, its frame moved a pace to the left, and then be returned to normal use.

There seems little reason to make bits of smart matter independent, self-replicating, or toxic. With care, smart matter should be safer than what it replaces because it will be better controlled. Spray paint gets all over things and contains noxious solvents; the paperpaint described above doesn't. This will be a characteristic difference, if we exercise our usual vigilance to encourage the production of things that are safe and environmentally sound.

Falling Costs

It may be fun to discuss wondrous new products, but they won't make much difference in the world if they are too expensive. Besides, many people today don't have decent food, clothes, and a roof over their heads, to say nothing of fancy "nanostuff."

Costs matter. There is more to life than material goods, but without material goods life is miserable and narrow. If goods are expensive, people strive for them; if goods are abundant, people can turn their attention elsewhere. Some of us like to think that we are above a concern for material goods, but this seems more common in the wealthy countries. Lowering manufacturing costs is a mundane concern, but so are feeding people, housing them, and building sewage systems to keep them from dying of cholera and hepatitis. For all these reasons, finding ways to bring down production costs is a worthy goal.

For the poor, for the environment, and for the freeing of human potential, costs matter deeply. Let's take a closer look at the costs of molecular manufacturing.

Can Falling Costs be Realistic?

Inflation produces the illusion that costs rise, when the real story is that the value of money is falling. In the short term, real costs usually don't change very quickly, and this can produce the illusion that costs are stable facts of nature, like the law of gravity or the laws of thermodynamics.

In the real world, though, most costs have been falling by a crucial measure: the amount of human labour needed to make things. People can afford more and more, because their labour, supplemented by machines, can produce more and more. This change is dramatic measured on a scale of centuries, and equally dramatic across the gulf between Third World and developed countries. The rise from Third World to First World standards of living has raised income (dropped the cost of labour time) by more than a factor of ten. What can molecular manufacturing do?

Larger cost reductions have happened, most dramatically in computers. The cost of a computer of a given ability has fallen by roughly a factor of 10 every seven years since the 1940s. In total, this is a factor of a million. If automotive technologies had done likewise, a luxury car would now cost less than one cent. (Personal computer systems still cost hundreds of dollars both because they are far more powerful than the giant machines of the 1940s and because the cost of buying any useful computer system includes much more than just the cost of a bare computer chip.)

Costs: A First Estimate

Some costs apply to a kind of product, regardless of how many copies of it are made: these include design costs, technology licensing costs, regulatory approval costs, and the like. Other costs apply to each unit of a product: these include the costs of labour, energy, raw materials, production equipment, production sites, insurance, and waste disposal. The per-kind costs can become very low if production runs are large. If these costs stay high, it will be because people prefer new products for their new benefits, despite the cost—hardly cause for complaint.

The more basic and easier to analyze costs are per-unit costs. A picture to keep in mind here is of Desert Rose Industries, where molecular machinery does most of the work, and where products are made from parts that are ultimately made from simple chemical substances. Let's consider some cost components.

Energy: Manufacturing at the molecular scale need not use a lot of energy. Plants build billions of tons of highly patterned material every year using available solar energy. Molecular manufacturing can be efficient, in the sense that the energy needed to build a block of product should be comparable to the energy released in burning an equivalent mass of wood or coal. If this energy were supplied as electricity at today's costs, the energy cost of manufacturing would be something like a dollar per kilogram. We'll return to the cost of energy later.

Raw Materials: Molecular manufacturing won't need exotic materials as inputs. Plain bulk chemicals will suffice, and this means materials no more exotic than the fuels and feedstocks that are, for now, derived from petroleum and biomass—gasoline, methanol, ammonia, and hydrogen. These typically cost tens of cents per kilogram. If bizarre compounds are used, they can be made internally. Rare elements could be avoided, but might be useful in trace amounts. The total quantity of raw materials consumed will be smaller than in conventional manufacturing processes because less will be wasted.

Capital Equipment and Maintenance: As we saw in the Desert Rose scenario, molecular manufacturing can be used to build all of the equipment needed for molecular manufacturing. It seems that this equipment—everything from large vats to submicroscopic special-purpose assemblers—can be reasonably durable, lasting for months or years before being recycled and replaced. If the equipment were to cost dollars per kilogram, and produce many thousands of kilograms of product in its life, the cost of the equipment would add little to the cost of the product.

Waste Disposal: Today's manufacturing waste is dumped into the air, water, and landfills. There need be no such waste with molecular manufacturing. Excess materials of the kind now spewed into the environment could instead be completely recycled internally, or could emerge from the manufacturing process in pure form, ready for use in some other process. In an advanced process, the only wastes would be leftover atoms resulting from a bad mix of raw materials. Most of these leftover atoms would be ordinary minerals and simple gases like oxygen, the main "waste" from the molecular machinery of plants. Molecular manufacturing produces no new elements—if arsenic comes out, arsenic must have gone in, and the process isn't to blame for its existence. Any intrinsically toxic materials of this sort can at least be put in the safest form we can devise for disposal. One option would be to chemically bond it into a stable mineral and put it back where it came from.

Labour: Once a plant is operating, it should require little human labour (what people do with their time will change, unless factories are kept running as bizarre hobbies). Desert Rose Industries was run by two people, yet was described as producing large quantities of varied goods. The basic molecular-scale operations of manufacturing have to be automated, since they are too small for people to work on. The other operations are fairly simple and can be aided by equipment for handling materials and information.

Space: Even a manufacturing plant based on nanotechnology takes up room. It would, however, be more compact than familiar manufacturing plants, and could be built in some out-of-the-way place with inexpensive land. These costs should be small by today's standards.

Insurance: This cost will depend on the state of the law, but some comparisons can be made. Improved sensors and alarms could be made integral parts of products; these should lower fire and theft premiums. Product liability costs should be reduced by safer, more reliable products (we'll discuss the question of product safety further in later units). Employee injury rates will be reduced by having less labour input. Still, the legal system in the United States has shown a disturbing tendency to block every new risk, however small, even when this forces people to keep suffering old risks, which are sometimes huge. (The supply of lifesaving vaccines has been threatened in just this way.) When this happens, we kill anonymous people in the name of safety. If this behavior raises insurance premiums in a perverse way, it could discourage a shift to safer manufacturing technologies. Since such costs can grow or shrink independent of the real world of engineering and human welfare, they are beyond our ability to estimate.

Sales, Distribution, Training: These costs will depend on the product: Is it as common as potatoes, and as simple to use? Or is it rare and complex, so that determining what you need, where to get it, and how to use it are the main problems? These service costs are real but can be distinguished from costs of the thing itself.

To summarize, molecular manufacturing should eventually lead to lower costs. The initial expense of developing the technology and specific products will be substantial, but the cost of production can be low. Energy costs (at present prices) and materials costs (ditto) would be significant, but not enormous. They were quoted on a per-kilogram basis, but nanotechnological products, being made of superior materials, will often weigh only a fraction of what familiar products do. (Ballast, were it needed, will be dirt-cheap.) Equipment costs, land costs, waste-disposal costs, and labor costs can be low by the very nature of the technology.

Costs of design, regulation, and insurance will depend strongly on human tastes and are beyond predicting. Basic products, like clothing and housing, can become inexpensive unless we do something to keep them costly. As the cost of improved safety falls, there will be less reason to accept unsafe products. Molecular manufacturing uses processes as controlled and efficient as the molecular processes in plants. Its products could be as inexpensive as potatoes. This may sound to good to be true (and there *are* downsides, as we'll discuss), but why shouldn't it be true? Shouldn't we expect large changes to come with the replacement of modern technology?

A Cycle of Falling Costs

The above estimate made a conservative assumption about future costs: that energy and materials will cost then what they do now, before molecular manufacturing has become available. They won't, because lower costs lead to lower costs.

Let's say that making one kilogram of product by molecular manufacturing requires one dollar for a kilogram of raw materials and four dollars for a generous forty kilowatt-hours of energy. These are typical present-day prices for materials and electrical energy. Assume, for the moment, that other costs are small. One of the resulting five-dollar-per-kilogram products can be solar cell paint suitable for applying to paved roads. A layer of paint a few millionths of a meter thick would cost about five cents per square meter to produce, and would generate enough energy to make another square meter of paint in less than a week, even allowing for nighttime and moderate cloud cover. The so-called energy payback time would thus be short.

Let's assume that this smart paint costs as much to spread and hook up as it does to make, and that we demand that it pay for itself in a single month, so we charge ten cents per square meter per month. At that rate, the cost of solar energy from resurfaced roads would be roughly $0.004 per kilowatt hour—less than a twentieth the energy cost assumed in the initial production-cost estimate. By itself, this makes the cost of production fall to a fraction of what it was before. Most of that remaining fraction consists of the cost of materials.

But the products of nanotechnology will mostly be made of carbon (if present expectations are any guide), and carbon dioxide is too abundant in the atmosphere these days. With energy so cheap, the atmosphere can be used as source of carbon (and of hydrogen, nitrogen, and oxygen). The price of carbon would be a few cents per kilogram—roughly a twentieth the original price assumed for raw materials.

But now, both energy and raw materials are a twentieth the original price, and so the products become cheaper, including the energy-producing products and the raw-material—producing (atmosphere-cleaning) products....

The above scenario is simple, but it seems realistic in its basic outlines: lower costs can lead to lower costs. How far this process can go is hard to estimate precisely, but it could go far indeed.

Power Too Cheap To Meter?

This argument will remind some readers of an old claim—that nuclear energy would lead to "power too cheap to meter." This assertion, attributed to the early nuclear era, has passed into folklore as a warning to be skeptical of technologists promising free goodies. Does the warning apply here?

Anyone claiming that something is free doesn't really understand economics. Using something always has a cost equal to the most valuable alternative use for the thing. Choosing one alternative sacrifices another, and that sacrifice is the cost. As economist Phillip K. Salin says, "There's no such thing as a free opportunity," since opportunities always cost (at least) time and attention. Nanotechnology will not mean free goodies.

But, one might argue, nuclear power hasn't even been inexpensive. If technologists could be so wrong back then, why believe a similar argument today? We are happy to report that the arguments aren't similar: any argument for "nuclear power too cheap to meter" had to be absurd even given the knowledge at the time, and our argument isn't.

Nuclear reactors boil water to make steam to turn turbines to turn generators to drive electrical power through power lines to transformers to local power lines to houses, factories, and so forth. The wildest optimist could never have claimed that nuclear power was a free source of anything more than heat, and a realist would have added in the cost of the reactor equipment, fuel, waste disposal, hazards, and the rest. Even our wild optimist would have had to include the cost of building the boiler, the turbines, the generators, the power lines, and the transformers, and the cost of maintenance on all these. These costs were known to be a major part of the cost of power, so free heat wouldn't have meant free power. Thus, the claim was absurd the day it was made—not merely in hindsight.

In the early 1960s, Alvin Weinberg, head of the Oak Ridge National Laboratory, was a strong advocate of nuclear power, and argued that it would provide "cheap energy." He was optimistic, but did his sums. First, he assumed that nuclear-power plants could be built a little more cheaply than coal-fired power plants of the same size. Then he assumed that the cost of fuel, waste disposal, operations, and maintenance for nuclear plants would be not much more than the cost of operations and maintenance alone for coal plants. Then he assumed that they might last for more than thirty years. Finally, he assumed that they would be publicly operated, tax free at low interest (which merely moves costs elsewhere) and that after thirty years the cost of the equipment would be written off (which is an accounting fiction). With all of that, he derived a power cost that "might be" as low as one half the cost of the cheapest coal-fired plant he mentions. He was clearly an optimist, but he didn't come close to arguing for power too cheap to meter.

Low but Not Zero Costs

People have cried "Wolf!" before about new technologies leading to overwhelming abundance. It was said of nuclear power, and of steam power before it, and perhaps of water wheels, the horse, the plough, and the chipped rock. Molecular manufacturing is different because it is a new way to make almost anything, including more of the equipment needed to do the manufacturing. There has never been anything quite like this before.

The basic argument for low cost production is this: Molecular manufacturing will be able to make almost anything with little labour, land, or maintenance, with high productivity, and with modest requirements for materials and energy. Its products will themselves be extremely productive, as energy producers, as materials collectors, and as manufacturing equipment. There has never been a technology with this combination of characteristics, so historical analogies must be used with care. Perhaps the best analogy is this: Molecular manufacturing will do for matter processing what the computer has done for information processing.

There will always be limiting costs, because resources—whether energy, matter, or design skill—always have some alternative use. Costs will not fall to zero, but it seems that they could fall very low indeed.

Unit I–H

Nanotechnology: Fulfilling the Basics

The hungry, the homeless, and the hunted have little time or energy to devote to human relations or personal development. Food, shelter, and security are not everything, but they are basic. Material abundance is perhaps the best known way to build a contempt for material things and a concern for what lies beyond. In that spirit, let us take a further look at providing heaps of basic material wealth where today there is poverty.

The idea of bringing everyone in the world up to a decent standard of living looks utopian today. The world's poor are numerous and the wealthy are few, and yet the Earth's resources are already strained by our crude industrial and agricultural technologies. For the 1970s and 1980s, with a growing awareness of the environmental impact of human population and pollution, many people have begun to wrestle with the specter of declining wealth. Few have allowed themselves to consider what it might be like to live in a world with far greater material wealth because it has seemed impossible. Any discussion of such things will inevitably have a whiff of the 1950s or 1960s about it: Gee whiz, we can have supercars and Better Living Through (a substitute for conventional) Chemistry!

In the long run, unless population growth is limited, it will be impossible to maintain a decent standard of living for everyone. This is a basic fact, and to ignore it would be to destroy our future. Yet within sight is a time in which the world's poorest can be raised to a material standard of living that would be the envy of the world's richest today. The key is efficient, low-cost production of high-quality goods. Whether this will be used to achieve the goals we describe is more than just a question of technology.

Here, as in the next two units, we continue to focus on how the new technologies can serve positive goals. There is a lot to say, and it needs to be said, in part because positive goals can in some measure displace negative goals. We ask patience of those readers bothered by what may seem an optimistic tone, and ask that they imagine the authors to share their fears that powerful technologies will be abused, that positive goals may end in ruin, that a material paradise may yet harbor human misery.

Third World Nanotechnology

Where wealth is concerned, the least developed countries present the hardest case. Can a capability as advanced as nanotechnology, based on molecular machinery, be of use in the Third World? The answer must be yes. Agriculture is the backbone of Third World economies today, and agriculture is based on the naturally occurring molecular machines in wheat, rice, yams, and the like.

The Third World is short on equipment and skills. (It often has governmental problems as well, but that is another story.) Molecular manufacturing can make equipment inexpensive enough for the poor to buy or for aid agencies to give away. This includes equipment for making more equipment, so dependency could be reduced. As for skills, basic molecular manufacturing will require little labour of any kind, and a little skill will go a long way. As the technology advances, more and more of the products can be easy-to-use smart materials.

Molecular manufacturing will enable the poorest countries to bypass the difficult and dirty process of the industrial revolution. It can make products that are less expensive and easier to use than yams or rice or goats or water buffalo. And with products like cheap supercomputers with huge databases of writing and animation viewed through 3-D color displays, it can even help spread knowledge.

Nanotechnology's role in helping the poorest nations won't be on the minds of the first developers—they'll be in government and commercial labs in the wealthiest nations, pursuing problems of concern to people there. History, though, is full of unintended consequences, and some are for the better.

Construction and Housing

Building large objects is basic to solving problems of housing and transportation. Smart materials can help.

Today, buildings are expensive to construct, expensive to replace, and expensive to make fireproof, tornado-proof, earthquake-proof, and so forth. Making buildings tall is expensive; making walls soundproof is expensive; building underground is expensive. Efforts to relieve city congestion often founder on the high cost of building subways, which can amount to hundreds of millions of dollars per mile.

Building codes and politics permitting, nanotechnology will make possible revolutions in the construction of buildings. Superior materials will make it easy to construct tall (or deep) buildings to free up land, and strong buildings that can ride out the greatest earthquake without harm. Buildings can be made so energy-efficient and so good at using the solar energy falling on them that most are net energy producers. What is more, smart materials can make it easy to build and modify complex structures, such as walls full of windows, wiring, plumbing, data networks, and the like. For a concrete example that shows the principle, let's picture what smart pipes could be like.

Let's say that you want to install a fold-down sink in the corner of your bedroom. The new materials make fold-down sinks practical, and in a house made of advanced smart materials, just sticking one on the wall would be enough—the plumbing would rearrange itself. But this is an old, pre-breakthrough house, so the sink is a retrofit. To do this home-handiwork project, you buy several boxes full of inexpensive tubing, T-joints, valves, and fixtures in a variety of sizes, all as light as wood veneer and feeling like soft rubber.

The biggest practical problem will be to make a hole from an existing water pipe and drainpipe to where you want the sink. Molecular manufacturing can provide excellent power tools to make the holes, and smart paint and plaster to cover them again, but the details depend on how your house is built.

The smart plumbing system does help, of course. If you want to run the drain line through the attic, built-in pumps will make sure that the water flows properly. The flexibility of the pipes makes it much easier to run them around curves and corners. Low-cost power makes it practical for the sink to have a flow-through water heater, so you only need to run a cold-water pipe to have both hot and cold water. All the parts go together as easily as a child's blocks, and seem about as flimsy and likely to leak. When you turn it on, though, the microscopic components of the pipes lock together and become as strong as steel. And smart plumbing doesn't leak.

If your house were made of smart materials, like most of the housing in the Third World these days, life would have been easier. Using a special trowel, wall structures would be reworked like soft clay, doing their structural job all the while. Setting up a plumbing system from scratch with this stuff is easy, and hard to do wrong. Drinking water pipes won't connect to wastewater pipes, so drinking water can't be accidentally contaminated. Drains won't clog, because they can clean themselves better than a rotary steel blade ever could. If you run enough pipes from everything to everything else, built-in pumps will make sure that water flows in the right direction with adequate pressure.

Smart plumbing is one example of a general pattern. Molecular manufacturing can eventually make complex products at low cost, and those complex products can be simpler to use than anything we have today, freeing our attention for other concerns. Buildings can become easy to make and easy to change. The basic conveniences of the modern world, and more, can be carried to the ends of the earth and installed by the people there to suit their tastes.

Food

Worldwide food production has been outpacing population growth, yet hunger continues. In recent years, famine has often had political roots, as in Ethiopia where the rulers aim to starve opponents into submission. Such problems are beyond a simple technological solution. To avoid getting headaches, we'll also ignore the politics of farm price-support programs, which raise food prices while people are going hungry. All we can suggest here is a way to provide fresh food at lower cost with reduced environmental impact.

For decades, futurists have predicted the coming of synthetic foods. Some sort of molecular-manufacturing process could doubtless make such things with the usual low costs, but this doesn't sound appetizing, so we'll ignore the idea.

Most agriculture today is inefficient—an environmental disaster. Modern agriculture is famed for wasting water and polluting it with synthetic fertilizers, and for spreading herbicides and pesticides over the landscape. Yet the greatest environmental impact of agriculture is its sheer consumption of land. In the American East, ancient forests disappeared under the ax, in part to supply wood, in part to clear land. The prairies of the West disappeared under the plow. Around the world, this trend continues. The technology of the ax, the fire, and the plow is chiefly responsible for the destruction of rain forests today. A growing population will tend to turn every productive ecosystem into some sort of farmland or grazing land, if we let it.

No technological fix can solve the long-term problem of population growth. Nonetheless, we can roll back the problem of the loss of land, yet increase food supplies. One approach is intensive greenhouse agriculture.

Every kind of plant has its optimum growing conditions, and those conditions are far different from those found in most farmland during most of the year. Plants growing outdoors face insect pests, unless doused with pesticide, and low levels of nutrients, unless doused with fertilizer. In greenhouses patrolled by "nanoflyswatters" able to eliminate invading insects, plants would be protected from pests and could be provided with nutrients without contaminating groundwater or runoff. Most plants prefer higher humidity than most climates provide. Most plants prefer higher, more uniform temperatures than are typically found outdoors. What is more, plants thrive in high levels of carbon dioxide. Only greenhouses can provide pest protection, ample nutrients, humidity, warmth, and carbon dioxide all together and without reengineering the Earth.

Taken together, these factors make a huge difference in agricultural productivity. Experiments with intensive greenhouse agriculture, performed by the Environmental Research Lab in Arizona, show that an area of 250 square meters—about the size of a tennis court—can raise enough food for a person, year in and year out. With molecular manufacturing to make inexpensive, reliable equipment, the intensive labour of intensive agriculture can be automated. With technology like the deployable "tents" and smart materials we have described, greenhouse construction can be inexpensive. Following the standard argument, with equipment costs, labour costs, materials costs, and so forth, all expected to be low, greenhouse-grown foods can be inexpensive.

What does this mean for the environment? It means that the human race could feed itself with ordinary, naturally grown, pesticide-free foods while returning more than 90 per cent of today's agricultural land to wilds. With a generous five hundred square meters per person, the U.S. population would require only 3 per cent of present U.S. farm acreage, freeing 97 per cent for other uses, or for a gradual return to wilderness. When farmers are able to grow high-quality foodstuffs inexpensively, in a fraction of the room that they require today, they will find more demand for their land to be tended as a park or wilderness than as a cornfield. Farm journals can be expected to carry articles advising on techniques for rapid and esthetic restoration of forest and grassland, and on how best to accommodate the desires of the discriminating nature lover and conservationist. Even "unpopular" land will tend to become popular with people seeking solitude.

The economics of assembler-based manufacturing will remove the incentive to make greenhouses cheap, ugly, and boxy; the only reason to build that way today is the high cost of building anything at all. And while today's greenhouses suffer from viral and fungal infestations, these could be eradicated from plants in the same way they would be from the human body, as will be described later. A problem faced by today's greenhouses—overheating—could be dealt with by using heat exchangers, thereby conserving the carefully balanced inside atmosphere. Finally, if it should turn out that a little bit of bad weather improves the taste of tomatoes, that, too, could be provided, since there would be no reason to be fanatical about sheer efficiency.

Communications

Today, telecommunications systems have sharply limited capacity and are expensive to expand. Molecular manufacturing will drop the price of the "boxes" in telecommunications systems—things such as switching systems, computers, telephones, and even the fabled videophone. Cables made of smart materials can make these devices easy to install and easy to connect together.

Regulatory agencies willing, you might someday be able to buy inexpensive spools of material resembling kite string, and other spools of material resembling tape, then use them to join a world data network. Either kind of strand can configure its core into a good-quality optical fiber, with special provisions for going around bends. When rubbed together, pieces of string will fuse together, or fuse to a piece of tape. Pieces of tape do likewise. To hook up to the network, you run string or tape from your telephone or other data terminal to the nearest point that is already connected. If you live deep in a tropical rain forest, run a string to the village satellite link.

These data-cable materials include amplifiers, nanocomputers, switching nodes, and the rest, and they come loaded with software that "knows" how to act to transmit data reliably. If you're worried that a line may break, run three in different directions. Even one line could carry far more data than all the channels in a television cable put together.

Transportation

Getting around quickly requires vehicles and somewhere for them to travel. The old 1950s vision of private helicopters would be technically possible with inexpensive, high-quality manufacturing, cheap energy, and a bit of improvement in autopilots and air-traffic control—but will people really tolerate that much junk roaring across the sky? Fortunately, there is an alternative both to this and to building ever more roads.

Going Underground

Near the surface of the Earth, there is as much room underground as there is above it. This is usually ignored, because the room is full of dirt, rock, pressurized water, and the like. Digging is expensive. Digging long, deep tunnels is even more expensive. This expense, however, is mostly in the cost of equipment, materials, and energy. Tunneling machines are in wide use today, and molecular manufacturing can make them more efficient and less expensive. The energy to operate them will be no great problem, and smart materials can line tunnels as fast as they are dug, with little or no labour. Nanotechnology will open the low frontier.

With a little care, the environmental impact of a deep tunnel can be trivial. Instead of solid rock far below the surface, there is rock with a sealed tunnel running through it. Nothing nearby need be disturbed.

Tunnels avoid both the aesthetic impact of a sky full of noisy aircraft and the environmental impact of paving strips of landscape. This will make them less expensive than roads, and they can, if desired, be more common than roads in the developed world today. They will even permit faster transportation.

Taking the Subway

Japan and Germany are actively developing magnetic trains, like those in the Desert Rose Scenario. These avoid the limitations of steel wheels on steel rails by using magnetic forces to "fly" the train along a special track. Magnetic trains can reach aircraft speeds at ground level. On long runs through evacuated tunnels, they can reach spacecraft speeds, traveling global distances in an hour or so (less, if passengers are willing to tolerate substantial acceleration).

Systems like this can give "taking the subway" a new meaning. Local transportation would be at fast automotive speeds, but long-distance transportation would be faster than the Concorde. With superconducting electrical systems, fast subways would be more energy efficient than today's slow mass transit.

Getting Your Car

For decades, people have proposed replacing automobiles with some form of mass-transportation system, and it seems that cost revolutions (including inexpensive tunneling) may finally make this practical. Before junking the car, though, it's worth seeing how it might be improved.

Molecular manufacturing can make almost anything better. Automobiles can be made stronger and safer, lighter, higher performance, and higher efficiency, while getting excellent mileage and burning clean, inexpensive fuels, perhaps in fuel cells powering quiet electric motors. Using aerodynamic forces to hold the car to the road, there's no reason why a comfortable passenger car shouldn't be able to deliver uncomfortable, drag-racer acceleration.

To imagine a cheap car built with molecular manufacturing, first imagine loading it with all the attractive features that you've ever heard proposed. This includes everything from today's self-adjusting seats and mirrors, excellent sound systems, and specially tuned steering and suspension systems, through automated navigation displays, emergency braking, and reliable super-duper airbags. Now, instead of just having the position of the seats, mirrors, and so forth adjust to a driver, as some cars do today, our smart-material car can also adjust its size, shape, and color, facing owners with choices such as, "What should our car look like for this occasion?"

Those seeking an image of solid conservatism and wealth won't drive such cheap cars; they will risk their necks in a certified antique car, made from the traditional steel, paint, and rubber. If environmental regulations permit it, the car might even have a genuine gasoline-burning engine. The latter can no doubt be cleaned up by fancy nanotechnology-based emission-control systems.

Opening the Space Frontier

Our transportion system today effectively ends in the upper atmosphere. Travel beyond still takes the form of "historic missions." There is no reason for this situation to continue for long, once molecular manufacturing becomes well established.

The cost of spaceflight is high because spacecraft are huge, fragile things, made in such small numbers that they're almost hand-crafted. Molecular manufacturing will replace today's delicate monsters with rugged, mass-produced vehicles (which, with greater efficiency, needn't be so large). The vehicles will cost little, but the energy? Today, the energy cost of a ticket to orbit in an efficient vehicle would be less than one hundred dollars. Low cost vehicles and energy will drop the total cost to a fraction of this.

We will know that spaceflight has become inexpensive when people see the Earth as just a small part of the world, and understand in their bones that space resources make continued exploitation of Earth's resources unnecessary. In the long run, efficient, clean, low-cost manufacturing can transform the way human beings affect the Earth by their presence. Even stay-at-home humans will be better able to heal the damage they have done.

Unit I–I

Nanotechnology: Aiding the Environment

The 1970s saw a revolution in Western attitudes toward the natural environment. Concern with pollution, deforestation, and species extinction exploded. With the rise of these concerns came an ambivalent attitude toward technology and the wealth it was producing: some said that human beings are destructive to the environment in direct proportion to their power. This immediately suggested that technology and higher living standards were bad, being inherently destructive. "Wealth" came to imply environmental destruction.

The revolution in attitudes toward the environment has changed the idea of wealth. Our national statistics may not reflect it—not every last citizen or politician may agree—but the concept that genuine wealth includes not just houses and refrigerators, factories and machines, cars and roads, but also fields and forests, owls and wolves, clean air, clean water, and wilderness has taken deep root in minds and in politics. "The wealth of nature" has come to include nature as a value in itself, not merely as potential lumber, ore, and farmland.

As a consequence, greater wealth has begun to mean cleaner wealth, greener wealth. Richer countries can afford more expensive, more efficient equipment—scrubbers on smokestacks, catalytic converters on cars—and so they can produce goods with less environmental impact. This trend gives at best a hint of the future.

Lester Milbrath, director of the Research Program in Environment and Society at the State University of New York at Buffalo, observes, "Nanotechnologies have the potential to produce plentiful consumer goods with much lower throughput of materials and much less production of waste, thus reducing carbon dioxide buildup and reducing global warming. They also have the potential to reduce waste, especially hazardous waste, converting it to natural materials which do not threaten life." James Lovelock states that "The future could be good if we regain a sense of purpose and embrace the new industries based on information and nanotechnology. These add enormous value to molecular-sized pieces of matter, and need not be a threat to the environment as were the heavy polluting industries of the past."

Making It Easier to Be Clean

Should we boast of high technology while industry still can't produce without polluting? Pollution is a sign of low technology, of inadequate control of how matter is handled. Inferior goods and hazardous wastes are two sides of one problem.

With processes based on molecular manufacturing, industries will produce superior goods, and by virtue of the same advance in control, will have no need of burning, oiling, washing with solvents and acids, and flushing noxious chemicals down their drains. Molecular-manufacturing processes will rearrange atoms in controlled ways, and can neatly package any unwanted atoms for recycling or return to their source. This intrinsic cleanliness inspired environmentalist Terence McKenna, writing in the *Whole Earth Review*, to call nanotechnology "the most radical of the green visions."

This green vision will not be fulfilled automatically, but only with effort. Any powerful technology can be used for good or ill, and nanotechnology is no exception. Today, we see scattered progress in environmental cleanup and restoration, some slowing of ecological destruction, because of organized political pressure buoyed by a groundswell of public concern. Yet for all its force, this pressure is spread desperately thin, fighting enormous resistance rooted in economic forces.

But if these economic forces vanish, the opposition will crumble. Often, the key to success in battle is to give one's opponents an attractive alternative to fighting. The most powerful cry of the antigreen opposition has been that clearing and polluting the land offer the only path to wealth, the only escape from poverty. Now we can see a clean, efficient, and unobtrusive alternative: green wealth, compatible with natural wealth.

Ending Chemical Pollution, Cutting Resource Consumption

We've already seen how molecular manufacturing can provide clean solar energy without paving over desert ecosystems, and how clean energy and common materials can be turned into abundant, efficient goods, also cleanly. With care, sources of chemical pollution—even of excess carbon dioxide—can, step by step, be eliminated. This includes the pollutants responsible for acid rain, as well as ozone-destroying gases, greenhouse gases, oil spills, and toxic wastes.

In each case, the story is about the same. Acid rain mostly results from burning dirty fuels containing sulfur, and from burning cleaner fuels in a dirty way, producing nitrogen oxides. We've seen how molecular manufacturing can make solar cells cheap enough and rugged enough to use as road surfaces. With green wealth, we can make clean fuels from solar energy, air, and water; consuming these fuels in clean nanomechanical systems would just return to the air exactly the materials taken from it, along with a little water vapor. Fuels are made, fuels are consumed, and the cycle produces no net pollution. With cheap solar fuels, coal and petroleum can be replaced, ignored, left in the ground. When petroleum is obsolete, oil spills will vanish.

The greenhouse gas of greatest concern is carbon dioxide, and its main source is the burning of fossil fuels. The above steps would end this. The release of other gases, such as the chlorofluorocarbons(CFCs) used in foaming plastics, is often a side effect of primitive manufacturing processes: *foaming plastic* will hardly be a popular activity in an era of molecular manufacturing. These materials can be replaced or controlled—and they include the gases most responsible for ozone depletion.

The chief threats to the ozone layer are those same CFCs, used as refrigerants and solvents. Molecular manufacturing will use solvents sparingly (mostly water), and can recycle them without dumping any. CFC refrigerants can be replaced even with current technology, at a cost; with nanotechnology, that cost will be negligible.

Toxic wastes generally consist of harmless atoms arranged into noxious molecules; the same is true of sewage. With inexpensive energy and equipment able to work at the molecular level, these wastes can be converted into harmless forms. Many need never be produced in the first place. Other toxic wastes contain toxic elements, such as lead, mercury, arsenic, and cadmium. These elements come from the ground, and are best returned to the location and condition in which they were found. With nanotechnology, moreover, there will be little reason to dig them up in the first place. Nanotechnology will be able to break materials down to simple molecules and build them back up again. Need it be said that this will permit complete recycling?

It is fair to say that eliminating these sources of pollution would be a major improvement. There doesn't seem to be much more to say, aside from the usual caveats: "Not immediately," "Not all at once," and "Not on a predictable schedule." No one wants to make and dump wastes; they want something else, and get wastes as by-products. With a better way to get what people want, dumping wastes can be stopped.

People will also be able to get what they want while reducing their resource consumption. As materials grow stronger, they can be used more sparingly. As machines grow more perfect—in their motors, bearings, insulation, computers—they will grow more efficient. Materials will be needed to make things, and energy will be needed to run them, but in smaller amounts. What is more, nanotechnology will be the ultimate recycling technology. Objects can be made extremely durable, decreasing the need for recycling; alternatively, objects can be made genuinely biodegradable, designed at the molecular level to decompose after use, leaving humus and mineral grit; alternatively, they can be made of microscopic snap-together pieces, making objects as recyclable as structures built and rebuilt out of a child's blocks; finally, even objects not designed for recycling can be taken apart into simple molecules and recycled regardless. Each approach has different advantages and costs, and each makes current garbage problems go away.

Cleaning Up the Twentieth Century Mess

Still, even after twentieth-century industry is history, its toxic residues remain. Cleaning up waste dumps with today's technology has proved so expensive and ineffective that many in the field have all but given up hope of really solving the problem. What can be done with post-breakthrough technologies?

Cleansing Soil and Water

Nanotechnology can help with the cleanup of these pollutants. Living organisms clean the environment, when they can, by using molecular machinery to break down toxic materials. Systems built with nanotechnology will be able to do likewise, and to deal with compounds that aren't biodegradable.

Alan Liss is director of research for Ecological Engineering Associates, a company that uses knowledge of how natural ecosystems function to address environmental problems such as wastewater treatment. He explains how cleanup could work: "The more we learn about the ecosystem, the more we find that functions are managed by particular organisms or groups of organisms. Nanotech 'managers' might be able to step in when the natural managers are not available, thereby having a particular ecological activity occur that otherwise wouldn't have happened. A nanotech manager might be used for remediation in a situation where toxicants have destroyed some key members of a particular ecosystem—some managerial microbes, for example. Once the needed activities are reinitiated, the living survivors of the stressed ecosystem can jump in and continue the ecosystem recovery effort."

Fig. I–I.1: Environmental Cleanup

By changing the way materials and products are made, molecular manufacturing will free up land formerly used for industrial plants. Toxic materials could be removed from contaminated soil using solar power as the energy source, and the cleanup device and any collected residues could later be carted away.

To see how nanomachines could be used to clean up pollution, imagine a device made of smart materials and roughly resembling a tree, once it has been delivered and unfolded. Above ground are solar-collecting panels; below ground, a branching system of rootlike tubes reaches a certain distance into the soil. By extending into a toxic waste dump, these rootlike structures could soak up toxic chemicals, using energy from the solar collectors to convert them into harmless compounds. Rootlike structures extending down into the water table could do the same cleanup job in polluted aquifers.

Cleansing the Atmosphere

Most atmospheric pollutants are quickly washed out by rain (turning them into soil- and water-pollution problems), but some air pollutants are longer lasting. Among these are the chlorine compounds attacking the ozone layer that protects the Earth from excessive ultraviolet radiation. Since 1975, observers have recorded growing holes in the ozone layer: at the South Pole, the hole can reach as far as the tips of South America, Africa, and Australia. Loss of this protection subjects people to an increased risk of skin cancer and has unknown effects on ecosystems. The new technology base will be able to stop the increase in ozone-destroying compounds, but the effects would linger for years. How might this problem be reversed more rapidly?

Thus far, we've talked about nanotechnology in the laboratory, in manufacturing plants, and in products for direct human use. Molecular manufacturing can also make products that will perform some useful temporary function when tossed out into the environment. Getting rid of ozone-destroying pollutants high in the stratosphere is one example. There may be simpler approaches, without the sophistication of nanotechnology, but here is one that would work to cleanse the stratosphere of chlorine: Make huge numbers of balloons, each the size of a grain of pollen and light enough to float up into the ozone layer. In each, place a small solar-power plant, a molecular-processing plant, and a microscopic grain of sodium. The processing plant collects chlorine-containing compounds and separates out the chlorine. Combining this with the sodium makes sodium chloride-ordinary salt. When the sodium is gone, the balloon collapses and falls. Eventually, a grain of salt and a biodegradable speck fall to Earth, usually at sea. The stratosphere is soon clean.

A larger problem (with a ground-based solution) is climatic change caused by rising carbon dioxide (CO_2) levels. Global warming, expected by most climatologists and probably under way today, is caused by changes in the composition of Earth's atmosphere. The sun shines on the Earth, warming it. The Earth radiates heat back into space, cooling. The rate at which it cools depends on how transparent the atmosphere is to the radiation of heat. The tendency of the atmosphere to hold heat, to block thermal radiation from escaping into space, causes what is called the "greenhouse effect." Several gases contribute to this, but CO_2 presents the most massive problem. Fossil fuels and deforestation both contribute. Before the new technology base arrives, something like 300 billion tons of excess CO_2 will likely have been added to the atmosphere.

Small greenhouses can help reverse the global greenhouse effect. By permitting more efficient agriculture, molecular manufacturing can free land for reforestation, helping to repair the devastation wrought by hungry people. Growing forests absorb CO_2.

If reforestation is not fast enough, inexpensive solar energy can be applied to remove CO_2 directly, producing oxygen and glossy graphite pebbles. Painting the world's roads with solar cells would yield about four trillion watts of power, enough to remove CO_2 at a rate of 10 billion tons per year. Temporarily planting one-tenth of U.S. farm acreage with a solar cell "crop" would provide enough energy to remove 300 billion tons in five years; winds would distribute the benefits worldwide. The twentieth century insult to Earth's atmosphere can be reversed by less than a decade of twenty-first century repair work. Ecosystems damaged in the meantime are another matter.

Orbital Waste

The space near Earth is being polluted with small orbiting projectiles, some as small as a pin. Most of the debris is floating fragments of discarded rocket stages, but it also includes gloves and cameras dropped by astronauts. This is not a problem for life on Earth, but it is a problem as life begins its historic spread beyond Earth—the first great expansion since the greening of the continents, long ago.

Orbiting objects travel much faster than rifle bullets, and energy increases as the square of speed. Small fragments of debris in space can do tremendous damage to a spacecraft, and worse—their impact on an spacecraft can blast loose yet more debris. Each fragment is potentially deadly to a spacefaring human crossing its path. Today, the tiny fraction of space that is near Earth is increasingly cluttered.

This litter needs to be picked up. With molecular manufacturing, it will be possible to build small spacecraft able to maneuver from orbit to orbit in space, picking up one piece of debris after another. Small spacecraft are needed, since it makes no sense to send a shuttle after a scrap of metal the size of a postage stamp. With these devices, we can clean the skies and keep them hospitable to life.

Nuclear Waste

We've spoken of waste that just needs molecular changes to make it harmless, and toxic elements that came from the ground, but nuclear technology has created a third kind of waste. It has converted the slow, mild radioactivity of uranium into the fast, intense radioactivity of newly created nuclei, the products of fission and neutron bombardment. No molecular change can make them harmless, and these materials did not come from the ground. The products of molecular manufacturing could help with conventional approaches to dealing with nuclear waste, helping to store it in the most stable, reliable forms possible—but there is a more radical solution.

Even before the era of the nuclear reactor and the nuclear bomb, experimenters made artificially radioactive elements by accelerating particles and slamming them into nonradioactive targets. These particles traveled fast enough to penetrate the interior of an atom and reach the nucleus, joining it or breaking it apart.

The entire Earth is made of fallout from nuclear reactions in ancient stars. Its radioactivity is low because so much time has passed—many half-lives, for most radioactive nuclei. "Kicking" these stable nuclei changes them, often into a radioactive state. But kicking a radioactive nucleus has a certain chance of turning it into a stable one, destroying the radioactivity. By kicking, sorting, and kicking again, an atom-smashing machine could take in electrical power and radioactive waste, and output nothing but stable, nonradioactive elements, identical to those common in nature. Don't recommend this to your congressman—it would be far too expensive, today—but it will some day be practical to destroy the radioactivity of the twentieth-century's leftover nuclear waste.

Nanotechnology cannot do this directly, because molecular machines work with molecules, not nuclei. But indirectly, by making energy and equipment inexpensive, molecular manufacturing can give us the means for a clean, permanent solution to the problem of wastes left over from the nuclear era.

A Wealth of Garbage

Shortages often spur environmental damage. Faced with a food shortage, herdsmen can graze grasslands down to bare dirt. Faced with an energy shortage, industrial countries can approve destructive projects. The growth of population and the consumption of resources by twentieth-century industry have placed growing pressures on Earth's ability to support us in the manner to which we have become accustomed.

The resource problem will look quite different in the twenty-first century, with a new technology base. Today, we cut trees and mine iron for our structures. We pump oil and mine coal for our energy. Even cement is born in the flames of burning fossil fuels. Almost everything we build, almost every move we make, consumes something ripped from the Earth. This need not continue.

Our civilization uses materials for many things, but mainly to make things with a certain size, shape, and strength. These structural uses include everything from fibers in clothing to paving in roads, and most of the mass of furniture, walls, cars, spacecraft, computers—indeed, most of the mass of almost every product we build and use. The best structural materials use carbon, in forms like diamond and graphite. With elements from air and water, carbon makes up the polymers of wool and polyester, and of wood and nylon. A twenty-first-century civilization could mine the atmosphere for carbon, extracting over 300 billion tons before lowering the CO_2 concentration back to its natural, pre-industrial level. For a population of 10 billion, this would be enough to give every family a large house with lightweight but steel-strong walls, with 95 percent left over. Atmospheric garbage is an ample source of structural materials, with no need to cut trees or dig iron ore.

Plants show that carbon can be used to build solar collectors. Laboratory work shows that carbon compounds can be better conductors than copper. A whole power system could be built without even touching the rich resources of metal buried in garbage dumps.

Carbon can make windows, of plastic or diamond. Carbon can make things colorful with organic dyes. Carbon can be used to build nanocomputers, and will be the chief component of high-performance nanomachines of all kinds. The other components in all these materials are hydrogen, nitrogen, and oxygen, all found in air and water. Other elements are useful, but seldom necessary. Traces would often be ample.

With a new technology base making recycling easy, there need be no steady depletion of Earth's resources, just to keep a civilization running. The sketch just made shows that recycling just one form of garbage—excess atmospheric CO_2—can provide most needs. Even 10 billion wealthy people would not need to strip the Earth of resources. They could make do with what we've already dug up and thrown away, and they wouldn't even need all of that.

In short, a twenty-first-century civilization with a population of 10 billion could maintain a high standard of living using nothing but waste from twentieth-century industry, supplemented with modest amounts of air, water, and sunlight. This won't necessarily happen, yet the very fact that it is possible gives a better sense of what the new technology base can mean for the relationship between humanity, resources, and the Earth.

Green Products

In *The Green Consumer*, Elkington, Hailes, and Makower define a green product as one that:

- Is not dangerous to the health of people or animals.
- Does not cause damage to the environment during manufacture, use, or disposal.
- Does not consume a disproportionate amount of energy and other resources during manufacture, use, or disposal.
- Does not cause unnecessary waste, due either to excessive packaging or to a short useful life.
- Does not involve the unnecessary use of or cruelty to animals.
- Does not use materials derived from threatened species or environments.
- Ideally, does not trade price, quality, nutrition, or convenience for environmental quality.

With its ability to make almost anything at low cost—including products designed for extreme safety, durability, efficiency—without mining, logging, harming animals or environments, or producing toxic wastes, molecular manufacturing will make possible greener products than any yet seen in a store. Nanotechnology can replace dirty wealth with green wealth.

Environmental Restoration

A central problem in environmental restoration is reversing environmental encroachment. We tend to see land as being gobbled up by housing, because the land where we live generally is. Farming, though, consumes more land, and the variant of farming called "forestry" consumes still more. By rolling back our requirement for farmland, and for wood and paper, nanotechnology can change the balance of forces behind environmental encroachment. This should make it more practical, politically and economically, for people to move toward environmental restoration.

Restoring the environment means returning land to what it was—removing what has been added and, where possible, replacing what has been lost. We've seen how this can be done, in part, by removing pollutants and some of the pressures for ploughing and paving. A more difficult problem, though, is restoring the ecological balance where the changes have been biological. Much of Earth's biological diversity has been a result of biological isolation, of islands, seas, mountains, and continents. This isolation has been breached, and reversing the resulting problems is one of the greatest challenges in healing the biosphere.

Imported Species

Human meddling with life in the biosphere has caused enormous ecological disruptions. This hasn't involved genetic engineering—by twisting organisms to better serve human purposes, genetic engineering usually leaves them less able to serve their own purposes, less able to survive and reproduce in the wild. The great disruptions have come from a different source: from globe-traveling human beings taking aggressive, well-adapted species from one part of the planet to another, landing them on a distant island or continent to invade an ecosystem with no evolved defenses. This has happened again and again.

Australia is a classic case. It had been isolated long enough to evolve its own peculiar species quite unfamiliar elsewhere: kangaroos, koalas, duck-billed platypuses. When humans arrived, they brought new species. Whoever brought the first rabbits could not have guessed that they, of all creatures, would be so destructive. They soon overran the continent, destroying crops and grazing lands, unchecked by natural competitors or predators. They were joined by invaders from the plant kingdom: the prickly pear, and others.

The Americas have suffered invasions, too: tumbleweed, a bane of the rancher and farmer, is a relatively recent import from Central Asia. Since 1956, Africanized bees have been spreading from Brazil and moving north—but what they displace, in America, are European bees. Africa, in turn, is being invaded by the American screw-worm fly, an insect with larvae that enter an animal's wounds, including the umbilical wound of a newborn, and eat it alive. The story goes on and on.

People have sometimes tried, with a measure of success, to fight fire with fire: to bring in parasitic species and diseases to attack the imported species and keep its growth within some reasonable bounds. Australia's problem with prickly pear was tackled using an insect from Argentina; the rabbits were cut back—with mixed results—using a viral disease called myxomatosis: "rabbit pox."

Ecosystem Protectors

In many parts of the world, native species have been driven to extinction by rats, pigs, and other imported species, and others are endangered and fighting for their lives. Biological controls—fighting fire with fire—have advantages: organisms are small, selective, and inexpensive. These advantages will eventually be shared by devices made using molecular manufacturing, which avoid the disadvantages of importing and releasing yet more uncontrollable, breeding, spreading species. Alan Liss spoke of using nanotechnological devices to help restore ecosystems at a chemical level. A similar idea can be applied at a biological level.

The challenge—and it is huge—would be to develop insect-size or even microbe-size devices that could serve as selective, mobile, mechanical flyswatters or weed pullers. These could do what biological controls do, but would be unable to replicate and spread. Let's call devices of this sort "ecosystem protectors." They could keep aggressive imported species out, saving native species from extinction.

To a human being or an ordinary organism, an ecosystem protector would seem like just one more of the many billions of different kinds of bugs and microbes in the ecosystem—small things going about their own business, with no tendency to bite. They might be detectable, but only if you sorted through a lot of dirt and looked at it through a microscope, because they wouldn't be very common. They would have just one purpose: to notice when they bumped into a member of an imported species on the "not welcome here" list, and then either to eliminate it or to ensure, at least, that it couldn't reproduce.

Natural organisms are often very finicky about which species they attack. These ecosystem protectors could be equally finicky about which species they approach, and then, before attacking, could do a DNA analysis to be sure. It would be simplest (especially in the beginning while we're still learning) to limit each kind of defender to monitoring only one imported species.

Each unit of a particular kind of ecosystem-defender device would be identical, built with precision by a special-purpose molecular-manufacturing setup. Each would last for a certain time, then break down. Each kind can be tested in a terrarium, then a greenhouse, then a trial outdoors ecosystem, keeping an eye on their effects at each stage until one gains the confidence for larger scale use. "Larger scale" could still be quite limited, if they aren't designed to travel very far. This built-in obsolescence limits both how long each device can operate and how far it can move: getting control of the structure of matter includes making nanomachines work where they're wanted and not work elsewhere.

The agricultural industry today manufactures and distributes many thousands of tons of poisonous chemicals to be sprayed on the land, typically in an attempt to eliminate one or a few species of insect. Ecosystem protectors could also be used to protect these agricultural monocultures, field by field, with far less harm to the environment than today's methods. They could likewise be used in the special ecosystems of intensive greenhouse agriculture.

Unlike chemicals sprayed into the environment, these ecosystem protectors would be precisely limited in time, space, and effect. They neither contaminate the groundwater nor poison bees and ladybugs. In order to weed out imported organisms and bring an ecosystem back to its natural balance, ecosystem protectors would not have to be very common—only common enough for a typical imported organism to encounter one once in a lifetime, before reproducing.

Even so, as the ecosystem protectors wear out and stop working, they would present a small-scale problem of solid-waste disposal. With the exercise of some clever design, all the machinery of ecosystem protectors might be made of reasonably durable yet biodegradable materials or (at worst) materials no more harmful than bits of grit and humus in the soil. So their remains would be like the shells of diatoms, or bits of lignin from wood, or like peculiar particles of clay or sand.

Alternatively, we might develop other mobile nanomachines to find and collect or break down their remains. This strategy starts to look like setting up a parallel ecosystem of mobile machines, a process that could be extended to supplement the natural cleansing processes of nature in many ways. Each step in this direction will require caution, but not paranoia: there need be no toxic chemicals here, no new creatures to spread and run wild. Missteps will have the great virtue of being reversible. If we decide that we don't like the effects of some particular variety of ecosystem protector or cleanup machine, we could simply stop manufacturing that kind. We could even retrieve those that had already been made and dispersed in the environment, since their exact number is known, along with which patch of ground each is patrolling.

If the making and monitoring of ecosystem protectors seems a lot of trouble to go to just to weed out nonnative species, consider this example of the environmental destruction such species can cause. Sometime before World War II, a South African species of fire ant was accidentally imported into the United States. Today, infested areas can have up to five hundred of these ants per square foot. The National Audubon Society—a strong opponent of irresponsible use of pesticides—had to resort to spraying its refuge islands near Corpus Christi when they found these ants destroying over half the hatchlings of the brown pelican, an endangered species.

In Texas, it's been shown that the new ants are killing off native ant species—reducing biodiversity. The USDA's Sanford Porter states that due to them, "Texas may be in the midst of a genuine biological revolution." The ants are heading west, and have established a beachhead in California. Without ecosystem protectors or something much like them, ecologies around the world will continue to be threatened by unnatural invasions. Our species opened the new invasion routes, and it's our responsibility to protect native species made newly vulnerable by them.

Mending the Land

Today, most people are far from the land, tied up in turning the wheels of 20th century industry. In the years to come, those wheels will be replaced by molecular systems that do most of their turning by themselves. The pressure to destroy the land will be less. Time available to help heal the land will be greater. Surely more energy will flow in this direction.

To mend ruined landscapes will require skill and effort. Ecosystem defenders can do flyswatting and weedpulling jobs no humans ever could, but there will also be jobs of shaping, planting, and nurturing. The land has been torn by machines guided by hasty hands, almost overnight. It can gradually be restored by patient hands, whether bare, gloved, or guiding machines able to reshape a ravaged mountain without turning the soil.

The green wealth that can be brought by nanotechnology has raised high hopes among some environmentalists. Again writing in *Whole Earth Review*, Terence McKenna suggests it "would tend to promote . . . a sense of the unity and balance of nature and of our own human position within that dynamic and evolving balance." Perhaps people will learn to value nature more deeply when they can see it more clearly, with eyes unclouded by grief and guilt.

Unit I–J

Nanomedicine

Our bodies are filled with intricate, active molecular structures. When those structures are damaged, health suffers. Modern medicine can affect the workings of the body in many ways, but from a molecular viewpoint it remains crude indeed. Molecular manufacturing can construct a range of medical instruments and devices with far greater abilities. The body is an enormously complex world of molecules. With nanotechnology to help, we can learn to repair it.

The Molecular Body

To understand what nanotechnology can do for medicine, we need a picture of the body from a molecular perspective. The human body can be seen as a workyard, construction site, and battleground for molecular machines. It works remarkably well, using systems so complex that medical science still doesn't understand many of them. Failures, though, are all too common.

The Body as Workyard

Molecular machines do the daily work of the body. When we chew and swallow, muscles drive our motions. Muscle fibers contain bundles of molecular fibers that shorten by sliding past one another.

In the stomach and intestines, the molecular machines we call digestive enzymes break down the complex molecules in foods, forming smaller molecules for use as fuel or as building blocks. Molecular devices in the lining of the digestive tract carry useful molecules to the bloodstream.

Meanwhile, in the lungs, molecular storage devices called hemoglobin molecules pick up oxygen. Driven by molecular fibers, the heart pumps blood laden with fuel and oxygen to cells. In the muscles, fuel and oxygen drive contraction based on sliding molecular fibers. In the brain, they drive the molecular pumps that charge nerve cells for action. In the liver, they drive molecular machines that build and break down a whole host of molecules. And so the story continues through all the work of the body.

Yet each of these functions sometimes fails, whether through damage or inborn defect.

The Body as Construction Site

In growing, healing, and renewing tissue, the body is a construction site. Cells take building materials from the bloodstream. Molecular machinery programmed by the cell's genes uses these materials to build biological structures: to lay down bone and collagen, to build whole new cells, to renew skin, and to heal wounds.

With the exception of tooth fillings and other artificial implants, everything in the human body is constructed by molecular machines. These molecular machines build molecules, including more molecular machines. They clear away structures that are old or out of place, sometimes using machinery like digestive enzymes to take structures apart.

During tissue construction, whole cells move about, amoebalike: extending part of themselves forward, attaching, pulling their material along, and letting go of the former attachment site behind them. Individual cells contain a dynamic pattern of molecules made of components that can break down but can also be replaced. Some molecular machines in the cell specialize in digesting molecules that show signs of damage, allowing them to be replaced by fresh molecules made according to genetic instructions. Components inside cells form their complex patterns by self-assembly, that is, by sticking to the proper partners.

Failures in construction increase as we age. Teeth wear and crack and aren't replaced; hair follicles stop working; skin sags and wrinkles. The eye's shape becomes more rigid, ruining close vision. Younger bodies can knit together broken bones quickly, making them stronger than before, but osteoporosis can make older bones so fragile that they break under minor stress.

Sometimes construction is botched from the beginning due to a missing or defective genetic code. In hemophilia, bleeding fails to stop due to the lack of blood clotting factor. Construction of muscle tissue is disrupted in 1 in 3,300 male births by muscular dystrophy, in which muscles are gradually replaced by scar tissue and fat; the molecule "dystrophin" is missing. Sickle cell anemia results from abnormal hemoglobin molecules.

Paraplegics and quadriplegics know that some parts of the body don't heal well. The spinal cord is an extreme—and extremely serious—case, but scarring and improper regrowth of tissues result from many accidents. If tissues always regrew properly, injury would do no permanent physical damage.

The Body as Battlefield

Assaults from outside the body turn it into a battlefield where the aggressors sometimes get the upper hand. From parasitic worms to protozoa to fungi to bacteria to viruses, organisms of many kinds have learned to live by entering the body and using their molecular machinery to build more of themselves from the body's building blocks. To meet this onslaught, the body musters the defenses of the immune system—an armada of its own molecular machines. Your body's own amoebalike white blood cells patrol the bloodstream and move out into tissues, threading their way between other cells, searching for invaders.

How can the immune system distinguish the hundreds of kinds of cells that should be in the body from the invading cells and viruses that shouldn't? This has been the central question of the complex science of immunology. The answer, as yet only partially understood, involves a complex interplay of molecules that recognize other molecules by sticking to them in a selective fashion. These include free-floating antibodies—which are a bit like bumbling guided missiles—and similar molecules that are bound to the surface of white blood cells and other cells of the immune system, enabling them to recognize foreign surfaces on contact.

This system makes life possible, defending our bodies from the fate of meat left at room temperature. Still, it lets us down in two basic ways.

First, the immune system does not respond to all invaders, or responds inadequately. Malaria, tuberculosis, herpes, and AIDS all have their strategies for evading destruction. Cancer is a special case in which the invaders are altered cells of the body itself, sometimes successfully masquerading as healthy cells and escaping detection.

Second, the immune system sometimes overresponds, attacking cells that should be left alone. Certain kinds of arthritis, as well as lupus and rheumatic fever, are caused by this mistake. Between attacking when it shouldn't and not attacking when it should, the immune system often fails, causing suffering and death.

Medicine Today

When the body's working, building, and battling goes awry, we turn to medicine for diagnosis and treatment. Today's methods, though, have obvious shortcomings.

Crude Methods

Diagnostic procedures vary widely, from asking a patient questions, through looking at X-ray shadows, through exploratory surgery and the microscopic and chemical analysis of materials from the body. Doctors can diagnose many ills, but others remain mysteries. Even a diagnosis does not imply understanding: doctors could diagnose infections before they knew about germs, and today can diagnose many syndromes with unknown causes. After years of experimentation and untold loss of life, they can even treat what they don't understand—a drug may help, though no one knows why.

Leaving aside such therapies as heating, massaging, irradiating, and so forth, the two main forms of treatment are surgery and drugs. From a molecular perspective, neither is sophisticated.

Surgery is a direct, manual approach to fixing the body, now practiced by highly trained specialists. Surgeons sew together torn tissues and skin to enable healing, cut out cancer, clear out clogged arteries, and even install pacemakers and replacement organs. It's direct, but it can be dangerous: anesthetics, infections, organ rejection, and missed cancer cells can all cause failure. Surgeons lack fine-scale control. The body works by means of molecular machines, most working inside cells. Surgeons can see neither molecules nor cells, and can repair neither.

Drug therapies affect the body at the molecular level. Some therapies—like insulin for diabetics—provide materials the body lacks. Most—like antibiotics for infections—introduce materials no human body produces. A drug consists of small molecules; in our simulated molecular world, many would fit in the palm of your hand. These molecules are dumped into the body (sometimes directed to a particular region by a needle or the like), where they mix and wander through blood and tissue. They typically bump into other molecules of all sorts in all places, but only stick to and affect molecules of certain kinds.

Antibiotics like penicillin are selective poisons. They stick to molecular machines in bacteria and jam them, thus fighting infection. Viruses are a harder case because they are simpler and have fewer vulnerable molecular machines. Worms, fungi, and protozoa are also difficult, because their molecular machines are more like those found in the human body, and hence harder to jam selectively. Cancer is the most difficult of all. Cancerous growths consist of human cells, and attempts to poison the cancer cells typically poison the rest of the patient as well.

Other drug molecules bind to molecules in the human body and modify their behavior. Some decrease the secretion of stomach acid, others stimulate the kidneys, many affect the molecular dynamics of the brain. Designing drug molecules to bind to specific targets is a growth industry today, and provides one of the many short-term payoffs that is spurring developments in molecular engineering.

Limited Abilities

Current medicine is limited both by its understanding and by its tools. In many ways, it is still more an art than a science. Mark Pearson of Du Pont points out, "In some areas, medicine has become much more scientific, and in others not much at all. We're still short of what I would consider a reasonable scientific level. Many people don't realize that we just don't know fundamentally how things work. It's like having an automobile, and hoping that by taking things apart, we'll understand something of how they operate. We know there's an engine in the front and we know it's under the hood, we have an idea that it's big and heavy, but we don't really see the rings that allow pistons to slide in the block. We don't even understand that controlled explosions are responsible for providing the energy that drives the machine."

Better tools could provide both better knowledge and better ways to apply that knowledge for healing. Today's surgery can rearrange blood vessels, but is far too coarse to rearrange or repair cells. Today's drug therapies can target some specific molecules, but only some, and only on the basis of type. Doctors today can't affect molecules in one cell while leaving identical molecules in a neighboring cell untouched because medicine today cannot apply surgical control to the molecular level.

Nanotechnology in Medicine

Developments in nanotechnology will result in improved medical sensors. As protein chemist Bill DeGrado notes, "Probably the first use you may see would be in diagnostics: being able to take a tiny amount of blood from somebody, just a pinprick, and diagnose for a hundred different things. Biological systems are already able to do that, and I think we should be able to design molecules or assemblies of molecules that mimic the biological system."

In the longer term, though, the story of nanotechnology in medicine will be the story of extending surgical control to the molecular level. The easiest applications will be aids to the immune system, which selectively attack invaders outside tissues. More difficult applications will require that medical nanomachines mimic white blood cells by entering tissues to interact with their cells. Further applications will involve the complexities of molecular-level surgery on individual cells.

As we look at how to solve various problems, you'll notice that some that look difficult today will become easy, while others that might seem easier turn out to be more difficult. The seeming difficulty of treating disorders is always changing: once polio was frequent and incurable, today it is easily prevented. Syphilis once caused steady physical decline leading to insanity and death; now it is cured with a shot.

Athlete's foot has never been seen as a great scourge, yet it remains hard to cure. Likewise with the common cold. This pattern will continue: deadly diseases may be easily dealt with, while minor ills remain incurable, or vice versa. As we will see, a mature nanotechnology-based medicine will be able to deal with almost any physical problem, but the order of difficulty may be surprising. Nature cares nothing for our sense of appropriateness. Horribleness and difficulty just aren't the same thing.

Working Outside Tissues

One approach to nanomedicine would make use of microscopic mobile devices built using molecular-manufacturing equipment. These would resemble the ecosystem protectors and mobile cleanup machines discussed in the last unit. Like them, they would either be biodegradable, self-collecting, or collected by something else once they were done working. Like them, they would be more difficult to develop than simple, fixed-location nanomachines, yet clearly feasible and useful. Development will start with the simpler applications, so let's begin by looking at what can be done without entering living tissues.

The skin is the body's largest organ, and its exposed position subjects it to a lot of abuse. This exposed position, though, also makes it easier to treat. Among the earlier applications of molecular manufacturing may be those popular, quasimedical products, cosmetics. A cream packed with nanomachines could do a better and more selective job of cleaning than any product can today. It could remove the right amount of dead skin, remove excess oils, add missing oils, apply the right amounts of natural moisturizing compounds, and even achieve the elusive goal of "deep pore cleaning" by actually reaching down into pores and cleaning them out. The cream could be a smart material with smooth-on, peel-off convenience.

The mouth, teeth, and gums are amazingly troublesome. Today, daily dental care is an endless cycle of brushing and flossing, of losing ground to tooth decay and gum disease as slowly as possible. A mouthwash full of smart nanomachines could do all that brushing and flossing do and more, and with far less effort—making it more likely to be used.

This mouthwash would identify and destroy pathogenic bacteria while allowing the harmless flora of the mouth to flourish in a healthy ecosystem. Further, the devices would identify particles of food, plaque, or tartar, and lift them from teeth to be rinsed away. Being suspended in liquid and able to swim about, devices would be able to reach surfaces beyond reach of toothbrush bristles or the fibers of floss. As short-lifetime medical nanodevices, they could be built to last only a few minutes in the body before falling apart into materials of the sort found in foods (such as fiber). With this sort of daily dental care from an early age, tooth decay and gum disease would likely never arise. If under way, they would be greatly lessened.

Going beyond this superficial treatment would involve moving among and modifying cells. Let's consider what can be done with this treatment inside the body, but outside the body's tissues. The bloodstream carries everything from nutrients to immune-system cells, with chemical signals and infectious organisms besides.

Here, it is useful to think in terms of medical nanomachines that resemble small submarines, like the ones in Figure 10.1. Each of these is large enough to carry a nanocomputer as powerful as a mid-1980s mainframe, along with a huge database (a billion bytes), a complete set of instruments for identifying biological surfaces, and tools for clobbering viruses, bacteria, and other invaders. Immune cells, as we've seen, travel through the bloodstream checking surfaces for foreignness and—when working properly—attacking and eliminating what should not be there. These *immune machines* would do both more and less. With their onboard sensors and computers, they will be able to react to the same molecular signals that the immune system does, but with greater discrimination. Before being sent into the body on their search-and-destroy mission, they could be programmed with a set of characteristics that lets them clearly distinguish their targets from everything else. The body's immune system can respond only to invading organisms that had been encountered by that individual's body. Immune machines, however, could be programmed to respond to anything that had been encountered by world medicine.

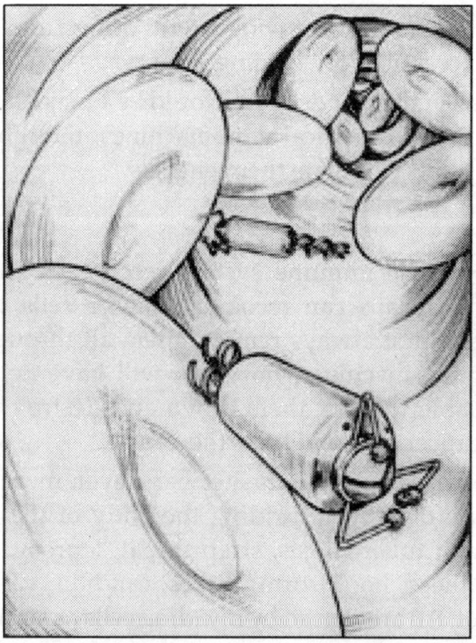

Fig. I–J.1: Immune Machines

Medical nanodevices could augment the immune system by finding and disabling unwanted bacteria and viruses. The immune device in the foreground has found a virus; the other has touched a red blood cell.

Immune machines can be designed for use in the bloodstream or the digestive tract (the mouthwash described above used these abilities in hunting down harmful bacteria). They could float and circulate, as antibiotics do, while searching for intruders to neutralize. To escape being engulfed by white blood cells making their own patrols, immune machines could display standard molecules on their surface-molecules the body knows and trusts already—like a fellow police officer wearing a familiar uniform.

When an invader is identified, it can be punctured, letting its contents spill out and ending its effectiveness. If the contents were known to be hazardous by themselves, then the immune machine could hold on to it long enough to dismantle it more completely.

How will these devices know when it's time to depart? If the physician in charge is sure the task will be finished within, say, one day, the devices prescribed could be of a type designed to fall apart after twenty-four hours. If the treatment time needed is variable, the physician could monitor progress and stop action at the appropriate time by sending a specific molecule—aspirin perhaps, or something even safer—as a signal to stop work. The inactivated devices would then be cleared out along with other waste eliminated from the body.

Working within Tissues

In most parts of the body, the finest blood vessels, capillaries, pass within a few cell diameters of every point. Certain white blood cells can leave these vessels to move among the neighboring cells. Immune machines and similar devices, being even smaller, could do likewise. In some tissues, this will be easy, in some harder, but with careful design and testing, essentially any point of the body should become accessible for healing repairs.

Merely fighting organisms in the bloodstream would be a major advance, cutting their numbers and inhibiting their spread. Roving medical nanomachines, though, will be able to hunt down invaders throughout the body and eliminate them entirely.

Eliminating Invaders

Cancers are a prime example. The immune system recognizes and eliminates most potential cancers, but some get by. Physicians can recognize cancer cells by their appearance and by molecular markers, but they cannot always remove them all through surgery, and often cannot find a selective poison. Immune machines, however, will have no difficulty identifying cancer cells, and will ultimately be able to track them down and destroy them wherever they may be growing. Destroying every cancer cell will cure the cancer.

Bacteria, protozoa, worms, and other parasites have even more obvious molecular markers. Once identified, they could be destroyed, ridding the body of the disease they cause. Immune machines thus could deal with tuberculosis, strep throat, leprosy, malaria, amoebic dysentery, sleeping sickness, river blindness, hookworm, flukes, candida, valley fever, antibiotic-resistant bacteria, and even athlete's foot. All are caused by invading cells or larger organisms (such as worms). Health officials estimate that parasitic diseases, common in the Third World, affect more than one billion people. For many of these diseases, no satisfactory drug treatment exists. All can eventually be eliminated as threats to human health by a sufficiently advanced form of nanomedicine.

Herding Cells

Destroying invaders will be helpful, but injuries and structural problems pose other problems. Truly advanced medicine will be able to build up and restructure tissues. Here, medical nanodevices can stimulate and guide the body's own construction and repair mechanisms to restore healthy tissue.

What is healthy tissue? It consists of normal cells in normal patterns in a normal matrix all organized in a normal relationship to the surrounding tissues. Surgeons today (with their huge, crude tools) can fix some problems at the tissue level. A wound disrupts the healthy relationship between two different pieces of tissue, and surgical glues and sutures can partly remedy this problem by holding the tissues in a position that promotes healing. Likewise, coronary artery bypass surgery brings about a more healthy overall configuration of tissues—one that provides working plumbing to supply blood to the heart muscle. Surgeons cut and stitch, but then they must rely on the tissue to heal its wounds as best it can.

Healing establishes healthy relationships on a finer scale. Cells must divide, grow, migrate, and fill gaps. They must reorganize to form properly connected networks of fine blood vessels. And cells must lay down materials to form the structural, intercellular matrix—collagen to provide the proper shape and toughness, or mineral grains to provide rigidity, as in bone. Often, they lay down unwanted scar tissue instead, blocking proper healing.

With enough knowledge of how these processes work (and nanoinstruments can help gather that knowledge) and with good enough software to guide the process—a more difficult challenge—medical nanomachines will be able to guide this healing process. The problem here is to guide the motion and behavior of a mob of active, living cells—a process that can be termed **cell herding**.

Cells respond to a host of signals from their environment: to chemicals in the surrounding fluids, to signal molecules on neighboring cells, and to mechanical forces applied to them. Cell-herding devices would use these signals to spur cell division where it is needed and to discourage it where it is not. They would nudge cells to encourage them to migrate in appropriate directions, or would simply pick them up, move them along, and deliver them where needed, encouraging them to nestle into a proper relationship with their neighbors. Finally, they would stimulate cells to surround themselves with the proper intercellular-matrix materials. Or—like the owner of a small dog who, on a cold day, wraps the beast in a wool jacket—they would directly build the proper surrounding structures for the cell in its new location.

In this way, cooperating teams of cell-herding devices could guide the healing or restructuring of tissues, ensuring that their cells form healthy patterns and a healthy matrix and that those tissues have a healthy relationship to their surroundings. Where necessary, cells could even be adjusted internally, as we will discuss later.

Rebuilding Tissues

Again, skin provides easy examples and may be a natural place to start in practice. People often want hair where they have bare skin, and bare skin where they have hair. Cell herding machines could move or destroy hair follicle cells to eliminate an unwanted hair, or grow more of the needed cells and arrange them into a working follicle where a hair is desired. By adjusting the size of the follicle and the properties of some of the cells, hairs could be made coarser, or finer, or straighter, or curlier. All these changes would involve no pain, toxic chemicals, or stench. Cell-herding devices could move down into the living layers of skin, removing unwanted cells, stimulating the growth of new cells, narrowing unnaturally prominent blood vessels, insuring good circulation by guiding the growth of any needed normal blood vessels, and moving cells and fibers around so as to eliminate even deep wrinkles.

At the opposite end of the spectrum, cell herding will revolutionize treatment of life-threatening conditions. For example, the most common cause of heart disease is reduced or interrupted supply of blood to the heart muscle. In pumping oxygenated blood to the rest of the body, the heart diverts a portion for its own use though the coronary arteries. When these blood vessels become constricted, we speak of coronary-artery disease. When they are blocked, causing heart muscle tissue to die, we speak of someone "having a coronary," another term for heart attack.

Devices working in the bloodstream could nibble away at atherosclerotic deposits, widening the affected blood vessels. Cell herding devices could restore artery walls and artery linings to health, by ensuring that the right cells and supporting structures are in the right places. This would prevent most heart attacks.

But what if a heart attack has already destroyed muscle tissue, leaving the patient with a scarred, damaged, and poorly functioning heart? Once again, cell-herding devices could accomplish repairs, working their way into the scar tissue and removing it bit by bit, replacing it with fresh muscle fiber. If need be, this new fiber can be grown by applying a series of internal molecular stimuli to selected heart muscle cells to "remind" them of the instructions for growth that they used decades earlier during embryonic development.

Cell-herding capabilities should also be able to deal with the various forms of arthritis. Where this is due to attacks from the body's own immune system, the cells producing the damaging antibodies can be identified and eliminated. Then a cell-herding system would work inside the joint where it would remove diseased tissues, calcified spurs, and so forth, then rework patterns of cells and intercellular material to form a healthy, smoothly working, and pain-free joint. Clearly, learning to repair hearts and learning to repair joints will have some basic technologies in common, but much of the research and development will have to be devoted to specific tissues and specific circumstances. A similar process—but again, specially adapted to the circumstances at hand—could be used to strengthen and reshape bone, correcting osteoporosis.

In dentistry, this sort of process could be used to fill cavities, not with amalgam, but with natural dentin and enamel. Reversing the ravages of periodontal disease will someday be straightforward, with nanomedical devices to clean pockets, join tissues, and guide regrowth. Even missing teeth could be regrown, with enough control over cell behavior.

Working on Cells

Moving through tissues without leaving a trail of disruption will require devices able to manipulate and direct the motions of cells, and to repair them. Much remains to bc learned—and will be easy to learn with nanoscale tools—but today's knowledge of cells is enough for a start on the problem of how to do surgery on cells.

Cell biology is a booming field, even today. Cells can be made to live and grow in laboratory cultures if they are placed in a liquid with suitable nutrients, oxygen, and the rest. Even with today's crude techniques, much has been learned about how cells respond to different chemicals, to different neighbors, and even to being poked and cut with needles. Conducting a rough sort of surgery on individual cells has been routine for many years in scientific laboratories.

Today, researchers can inject new DNA into cells using a tiny needle; small punctures in a cell membrane automatically reseal. But both these techniques use tools that on a cellular scale are large and clumsy—like doing surgery with an ax or a wrecking ball, instead of a scalpel. Nano-scale tools will enable medical procedures involving delicate surgery on individual cells.

Eliminating Viruses by Cell Surgery

Some viral diseases will respond to treatments that destroy viruses in the nose and throat, or in the bloodstream. The flu and common cold are examples. Many others would be greatly improved by this, but not eliminated. All viruses work by injecting their genes into a cell and taking over its molecular machinery, using it to produce more viruses. This is part of what makes viral illnesses so hard to treat—most of the action is performed by the body's own molecular machines, which can't be interfered with on a wholesale basis. When the immune system deals with a viral illness, it both attacks free virus particles before they enter cells, and attacks infected cells before they can churn out too many more virus particles.

Some viruses, though, insert their genes among the genes of the cell, and lay low. The cell can seem entirely normal to the immune system, for months or years, until the viral genes are triggered into action and begin the infective process anew. This pattern is responsible for the persistence of herpes infections, and for the slow, deadly progress of AIDS.

These viruses can be eliminated by molecular-level cellular surgery. The required devices could be small enough to fit entirely within the cell, if need be. Greg Fahy, who heads the Organ Cryopreservation Project at the American Red Cross's Jerome Holland Transplantation Laboratory, writes, "Calculations imply that molecular sensors, molecular computers, and molecular effectors can be combined into a device small enough to fit easily inside a single cell and powerful enough to repair molecular and structural defects (or to degrade foreign structures such as viruses and bacteria) as rapidly as they accumulate. . . There is no reason such systems cannot be built and function as designed."

Equally well, a cell surgery device located outside a cell could reach through the membrane with long probes. At the ends of the probes would be tools and sensors along with, perhaps, a small auxiliary computer. These would be able to reach through multiple membranes, unpackage and uncoil DNA, read it, repackage it, and recoil it, "proofreading" the DNA by comparing the sequences in one cell to the sequences of other cells.

On reading the genetic sequence spelling out the message of the AIDS virus, a molecular surgery machine could be programmed to respond like an immune machine, destroying the cell. But it would seem to make more sense simply to cut out the AIDS virus genes themselves, and reconnect the ends as they were before infection. By doing this, and killing any viruses found in the cell, the procedure would restore the cell to health.

Molecular Repairs

Cells are made of billions of molecules, each built by molecular machines. These molecules self-assemble to form larger structures, many in dynamic patterns, perpetually disintegrating and reforming. Cell-surgery devices will be able to make molecules of sorts that may be lacking, while destroying molecules that are damaged or present in excess. They will be able not only to remove viral genes, but to repair chemical and radiation-caused damage to the cell's own genes. Advanced cell surgery devices would be able to repair cells almost regardless of their initial state of damage.

By activating and inactivating a cell's genes, they will be able to stimulate cell division and guide what types of cells are formed. This will be a great aid to cell herding and to healing tissues.

As surgeons today rely on the spontaneous, self-organizing ability of cells and tissues to join and heal the parts they manipulate, so cell-surgery devices will rely on the spontaneous self-organizing capabilities of molecules to join and "heal" the parts they put together. Healing of a surgical wound involves sweeping up dead cells, growing new cells, and a slow and genuinely painful process of tissue reorganization. In contrast, the joining of molecules is almost instantaneous and occurs on a scale far below that of the most sensitive pain receptor. "Healing" will not begin after the repair devices have done their work, as it does in conventional surgery: rather, when they complete their work, the tissue will have been healed.

Healing Body and Limb

The ability to herd cells and to perform molecular repairs and cell surgery will open new vistas for medicine. These abilities apply on a small scale, but their effects can be large scale.

Correcting Chemistry

In many diseases, the body as a whole suffers from misregulation of the signaling molecules that travel through its fluids. Many are rare: Cushing's disease, Grave's disease, Paget's disease, Addison's disease, Conn's syndrome, Prader-Labhart-Willi syndrome. Others are common: millions of older women suffer from osteoporosis, the weakening of bones that can accompany lowered estrogen levels.

Diabetes kills frequently enough to rank in the top ten causes of death in the United States; the number of individuals known to have it doubles every fifteen years. It is the leading cause of blindness in the United States, with other complications including kidney damage, cataracts, and cardiovascular damage. Today's molecular medicine tries to solve these troubles by supplying missing molecules: diabetics inject additional insulin. While helpful, this doesn't cure the disease or eliminate all symptoms. In an era of molecular surgery, physicians could choose instead to repair the defective organ, so it can regulate its own chemicals again, and to readjust the metabolic properties of other cells in the body to match. This would be a true healing, far better than today's partial fix.

Only now are researchers making progress on another frequent problem of metabolic regulation: obesity. Once this was thought to have one simple cause (consuming excess calories) and one main result (greater roundness than favored by today's aesthetics), but both assumptions proved wrong. Obesity is a serious medical problem, increasing the risk of diabetes mellitus, osteoarthritis, degenerative diseases of the heart, arteries, and kidneys, and shortening life expectancy. And the supposed cause, simple overeating, has been shown to be incorrect—something dieters had always suspected, as they watched thinner colleagues gorge and yet gain no weight.

The ability to lay in stores of fat was a great benefit to people once upon a time, when food supplies were irregular, nomadism and marauding bands made food storage difficult and risky, and starvation was a common cause of death. Our bodies are still adapted to that world, and regulate fat reserves accordingly. This is why dieting often has perverse effects. The body, when starved, responds by attempting to build up greater reserves of fat at its next opportunity. The main effect of exercise in weight reduction isn't to burn up calories, but to signal the body to adapt itself for efficient mobility.

Obesity therefore seems to be a matter of chemical signals within the body, signals to store fat for famine or to become lean for motion. Nanomedicine will be able to regulate these signals in the bloodstream, and to adjust how individual cells respond to them in the body. The latter would even make possible the elusive "spot reduction program" to reshape the distribution of body fat.

Here, as with many potential applications of nanotechnology, the problem may be solved by other means first. Some problems, though, will almost surely require nanomedicine.

New Organs and Limbs

So far we've seen how medical nanotechnology would be used in the simpler applications outside tissues—such as in the blood—then inside tissues, and finally inside cells. Consider how these abilities will fit together for victims of automobile and motorcycle accidents.

Nanomanufactured medical devices will be of dramatic value to those who have suffered massive trauma. Take the case of a patient with a crushed or severed spinal cord high in the back or in the neck. The latest research gives hope that when such patients are treated promptly after the injury, paralysis may be at least partially avoidable, sometimes. But those whose injuries weren't treated—including virtually all of today's patients—remain paralyzed. While research continues on a variety of techniques for attempting to aid a spontaneous healing process, prospects for reversing this sort of damage using conventional medicine remain bleak.

With the techniques discussed above, it will become possible to remove scar tissue and to guide cell growth so as to produce healthy arrangements of the cells on a microscopic scale. With the right molecular-scale poking and prodding of the cell nucleus, even nerve cells of the sorts found in the brain and spinal cord can be induced to divide. Where nerve cells have been destroyed, there need be no shortage of replacements. These technologies will eventually enable medicine to heal damaged spinal cords, reversing paralysis.

The ability to guide cell growth and division and to direct the organization of tissues will be sufficient to regrow entire organs and limbs, not merely to repair what has been damaged. This will enable medicine to restore physical health despite the most grievous injuries.

If this seems hard to believe, recall that medical advances have shocked the world before now. To those in the past, the idea of cutting people open with knives painlessly would have seemed miraculous, but surgical anesthesia is now routine. Likewise with bacterial infections and antibiotics, with the eradication of smallpox, and the vaccine for polio: Each tamed a deadly terror, and each is now half-forgotten history. Our gut sense of what seems likely has little to do with what can and cannot be done by medical technology. It has more to do with our habitual fears, including the fear of vain hopes. Yet what amazes one generation seems obvious and even boring to the next. The first baby born after each breakthrough grows up wondering what all the excitement was about.

Besides, nano-scale medicine won't be a cure-all. Consider a fifty-year-old mentally retarded man, with a mind like a two-year-old's, or a woman with a brain tumor that has spread to the point that her personality has changed: How could they be "healed"? No healing of tissues could replace a missed lifetime of adult experience, nor can it replace lost information from a severely damaged brain. The best physicians could do would be to bring the patients to some physically healthy condition. One can wish for more, but sometimes it won't be possible.

First Aid

Throughout the centuries, medicine has been constrained to maintain functioning tissues, since once tissues stop functioning, they can't heal themselves. With molecular surgery to carry out the healing directly, medical priorities change drastically—function is no longer absolutely necessary. In fact, a physician able to use molecular surgery would prefer to operate on nonfunctioning, structurally stable tissue than on tissue that has been allowed to continue malfunctioning until its structure was lost.

Brain tumors are an example: They destroy the brain's structure, and with it the patient's skills, memories, and personality. Physicians in the future should be able to immediately interrupt this process, to stop the functioning of the brain to stabilize the patient for treatment.

Techniques available today can stop tissue function while preserving tissue structure. Greg Fahy, in his work on organ preservation at the American Red Cross, is developing a technique for vitrifying animal kidneys—making them into a low-temperature, crystal-free glass—with the goal of maintaining their structure such that, when brought back to room temperature, they can be transplanted. Some kidneys have been cooled to -30°C, warmed back up, and then functioned after transplantation.

A variety of other procedures can also stabilize tissues on a long-term basis. These procedures enable many cells—but not whole tissues—to survive and recover without help; advanced molecular repair and cell surgery will presumably tip the balance, enabling cells, tissues, and organs to recover and heal. When applied to stabilizing a whole patient, such a condition can be called **biostasis**. A patient in biostasis can be kept there indefinitely until the required medical help arrives. So in the future, the question "Can this patient be restored to health?" will be answered "Yes, if the patient's brain is intact, and with it the patient's mind."

Sandra Lee Adamson of the National Space Society has her eyes on distant goals. Some have proposed that travel to the stars would take generations, preventing anyone on Earth from ever making the trip. But she notes that biostasis will "give hope to some fearless adventurers who will risk suspension and subsequent reanimation so they can see the stars for themselves."

Plague Insurance

Medical nanotechnologies promise to extend healthy life, but if history is any guide, they may also avert sudden massive death. The word **plague** is rarely heard today, except in relation to AIDS; it calls up visions of the Black Death of the Middle Ages, when one third of Europe died in 1346-50. A virulent influenza struck in 1918, half lost in the news of the First World War: how many of us realize that it killed at least 20 million? People often act as though plagues were gone for good, as if sanitation and antibiotics had vanquished them. But as doctors are forever telling their patients, antibiotics kill bacteria, but are useless for viruses. The flu, the common cold, herpes, and AIDS—none has a really effective treatment, because all are caused by viruses. In some African countries, as much as 10 per cent of the population is estimated to be infected with the AIDS-causing HIV virus. Without a cure soon, the steep rise in deaths from AIDS still lies in the future. AIDS stands as a grim reminder that the great plagues of history are not behind us.

The Threat

New diseases continue to appear today as they have throughout history. Today's population, far larger than that of any previous century, provides a huge, fertile territory for their spread.

Today's transportation systems can spread viruses from continent to continent in a single day. When ships sailed or churned their way across the seas, an infected passenger was likely to show full-blown disease before arrival, permitting quarantine. But few diseases can be guaranteed to show themselves in the hours of a single aircraft flight.

So far as is known, every species of organism, from bacterium to whale, is afflicted with viruses. Animal viruses sometimes "jump the species gap" to infect other animals, or people. Most scientists believe that the ancestors of the AIDS virus could, until recently, infect only certain African monkeys. Then these viruses made the interspecies jump. A similar jump occurred in the 1960s when scientists in West Germany, working with cells from monkeys in Uganda, suddenly fell ill. Dozens were infected, and several died of a disease that caused both blood clots and bleeding, caused by what is now named the Marburg virus. What if the Marburg virus had spread with a sneeze, like influenza or the common cold?

We think of human plagues as a health problem, but when they hit our fellow species, we tend to see them from an environmental perspective. In the late 1980s, over half the harbor-seal population in large parts of the North Sea suddenly died, leading many at first to blame pollution. The cause, though, appears to be a distemper virus that made the jump from dogs. Biologists worry that the virus could infect seal species around the world, since distemper virus can spread by aerosols—that is, by coughing—and seals live in close physical contact. So far its mortality rate has been 60 to 70 per cent.

What of AIDS itself: Could it change and give rise to a form able to spread, say, as colds do? Nobel Laureate Howard M. Temin has said, "I think that we can very confidently say that this can't happen." Nobel Laureate Joshua Lederberg, president of Rockefeller University in New York City, replied, "I don't share your confidence about what can and cannot happen." He points out that "there is no reason a great plague could not happen again. . . . We live in evolutionary competition with microbes—bacteria and viruses. There is no guarantee that we will be the survivors."

Our Inadequate Abilities

Bacterial diseases are mostly controllable today. Sanitation limits the ways in which plague can spread. These measures are just good enough to lull us into imagining the problem is solved.

Viruses are common, viruses mutate; some spread through the air, and some are deadly. Plagues show that fast-spreading diseases can be deadly, and effective antiviral drugs are still rare.

The only really effective treatments for viral diseases are preventive, not curative. They work either by preventing exposure, or by exposing the body beforehand to dead or harmless or fragmentary forms of the virus, to prepare the immune system for future exposure. As the long struggle for an AIDS vaccine shows, one cannot count on modern medicine to identify a new virus and produce an effective vaccine within a single month or year or even a single decade. But influenza epidemics spread fast, and Marburg II or AIDS II or something entirely new and deadly may do the same.

Doing Better

The deaths from the next great plague could have begun in a village last week, or could begin next year, or a year before we learn to deal with new viral illnesses promptly and effectively. With luck, the plague will wait until a year after.

Immune machines could be set to kill a new virus as soon as it is identified. The instruments nanotechnology brings will make viral identification easy. Some day, the means will be in place to defend human life against viral catastrophe.

From eliminating viruses to repairing individual cells, improving our control of the molecular world will improve health care. Immune machines working in the bloodstream seem about as complex as some engineering projects human beings have already completed—projects like large satellites. Other medical nanotechnologies seem to be of a higher order of complexity.

On Solving Hard Problems

Somewhere in the progression from relatively simple immune devices to molecular surgery, we've crossed the fuzzy line between systems that teams of clever biomedical engineers could design in a reasonable length of time and ones that might take decades or prove impossibly complex. Designing a nanomachine capable of entering a cell, reading its DNA, finding and removing a deadly viral DNA sequence, and then restoring the cell to normal would be a monumental job. Such tasks are advanced applications of nanotechnology, far beyond mere computers, manufacturing equipment, and half-witted "smart materials."

To succeed within a reasonable number of years, we may need to automate much of the engineering process, including software engineering. Today's best expert systems are nowhere near sophisticated enough. The software must be able to apply physical principles, engineering rules, and fast computation to generate and test new designs. Call it automated engineering.

Automated engineering will prove useful in advanced nanomedicine because of the sheer number of small problems to be solved. The human body contains hundreds of kinds of cells forming a huge number of tissues and organs. Taken as a whole (and ignoring the immune system), the body contains hundreds of thousands of different kinds of molecules. Performing complex molecular repairs on a damaged cell might require solving millions of separate, repetitive problems. The molecular machinery in cell surgery devices will need to be controlled by complex software, and it would be best to be able to delegate the task of writing that software to an automated system. Until then, or until a lot of more conventional design work gets done, nanomedicine will have to focus on simpler problems.

Aging

Where does aging fit in the spectrum of difficulty? The deterioration that comes with aging is increasingly recognized as a form of disease, one that weakens the body and makes it susceptible to a host of other diseases. Aging, in this view, is as natural as smallpox and bubonic plague, and more surely fatal. Unlike bubonic plague, however, aging results from internal malfunctions in the molecular machinery of the body, and a medical condition with so many different symptoms could be complex.

Surprisingly, substantial progress is being made with present techniques, without even a rudimentary ability to perform cell surgery in a medical context. Some researchers believe that aging is primarily the result of a fairly small number of regulatory processes, and many of these have already been shown to be alterable. If so, aging may be tackled successfully before even simple cell repair is available. But the human aging process is not well enough understood to enable a confident projection of this; for example, the number of regulatory processes is not yet known. A thorough solution may well require advanced nanotechnology-based medicine, but a thorough solution seems possible. The result would not be immortality, just much longer, healthier lives for those who want them.

Restoring Species

A challenging problem related to medicine (and to biostasis) is that of species restoration. Today, researchers are carefully preserving samples from species now becoming extinct. In some cases, all they have are tissue samples. For other species, they've been able to save germ cells in the hope that they will be able to implant fertilized eggs into related species and thus bring the (nearly?) extinct species back.

Each cell typically contains the organism's complete genetic information, but what can be done with this? Many researchers today collect samples for preservation thinking only of the implantation scenario: one that they know has already been made to work. Other researchers are taking a broader view: the Center for Genetic Resources and Heritage at the University of Queensland is a leader in the effort. Daryl Edmondson, coordinator of the gene library, explains that the center is unique because it will "actively collect data. Most other libraries simply collate their own collections." Director John Mattick describes it as a "genetic Louvre" and points out that if genes from today's endangered species aren't preserved, "subsequent generations will see we had the technology to keep [DNA] software and will ask why we didn't do it." With this information and the sorts of molecular repair and cell-surgery capabilities we have discussed, lost species can someday be returned to active life again as habitats are restored.

One such center isn't enough: the Queensland center focuses on Australian species (naturally enough) and has limited funds. Besides, anything so precious as the genetic information of an endangered species should be stored in many separate locations for safety. We need to take out an insurance policy on Earth's genetic diversity with a broader network of genetic libraries, concentrating special attention on gathering biological samples from the fast-disappearing rain forests. Scientific study can wait: the urgency of the situation calls for a vacuum-cleaner approach.

Unit I–K

Limits and Downsides

The discussions of potential economic, medical, and environmental benefits may have given the false impression that nanotechnology will create a wondrous utopia in which all human problems are solved and we all live happily ever after. This is even more mistaken than the idea that new technologies always cause more problems than they solve. Many of the main constraints and difficulties faced by people are based not on technology or its lack, but instead by the very nature of the world we live in and the essence of our humanness.

Increasing affluence based on molecular manufacturing won't end economic problems any more than past increases in affluence have. Wilderness can still be destroyed; people can be oppressed; financial markets can be unstable; trade wars can be waged; inflation can soar; individuals, companies, and nations can go into debt; bureaucracy can stifle innovation; tax levels can become crippling; wars and terrorism can rage. None of these will automatically be stopped by advanced technology.

What is more, the potential benefits of new technologies aren't automatic. Nanotechnology could be used to restore the environment, to spread wealth, and to cure most illness. But will it? This depends on human action, working within the limits set by the real world.

This unit first describes some of the limits to what nanotechnology can accomplish, and then some of the adverse side effects of its basically good applications. The next will discuss the problem of accidents, which seems manageable, and then the far greater problem of potential abuse of new capabilities.

Some Limits of Nanotechnology

The world imposes limits on what we can do. Technology in general (and nanotechnology in particular) can provide padding for us as we throw ourselves against these hard, sharp limitations, and can sometimes help us slip past old limits through previously unknown gaps. Eventually, though, we will encounter new limits. In the end, solid constraints will limit human action no matter how much we juggle atoms and molecules, or the bits and bytes of information. Let's look at some of these, starting with the most abstract and long term—the most definite and hardest to avoid—and moving toward the more personal and near term.

Information Loss

Many problems differ fundamentally from the material problems of limited matter and energy: they involve information. Some of the most precious stores of information in the world today are the genetic codes of the biosphere.

This information, different for virtually every individual organism, is the product of millions of events that we are incapable of modeling or recreating. When this information is lost, it is lost forever. When the atoms encoding this information are thoroughly scattered, there seems to be no way to retrieve it.

With any species, most genetic information is shared in common, found in all members of that species. But the variations in genetic code between individuals are important, both to the individuals themselves and to the health and prospects of the species as a whole. Consider the northern white rhino, whose numbers have dropped to an estimated thirty-two animals, or the California condor, of which only forty remain, all in captivity. Even if biologists succeed in reestablishing these species—eight condors were hatched in 1989—much of the diversity of their genetic information has been lost. Worse yet are extinctions of species for which no tissue samples were saved. The future may see some amazing recoveries: dry skin and bones may yield a complete set of genes when sifted by molecular machinery, and even current techniques have been used to recover genes from an ancient leaf, almost 20 million years old. Our eyes and instruments cannot yet tell us how much information from the past remains, but we do know that genetic information is being lost every day, and once lost, it is irretrievable.

Physical Limits and Nonsense

People have often been wrong about physical limits, confusing the limits of their technology with the limits of the possible. As a result, learned men first dismissed the idea of heavier-than-air flight, and then dismissed the idea of flying to the Moon. Yet physical limits are real, and all technology—past, present, and future—will stay within those limits. There is even reason to suspect that some of those limits are where the learned now believe them to be.

Nanotechnology will make it possible to push closer to the real limits set by natural law, but it will not change those laws or the limits they set. It will not affect the law of gravity, the gravitational constant, the speed of light, the charge of the electron, the radius of the hydrogen atom, the value of Planck's constant, the effects of the uncertainty principle, the principle of least action, the mass of the proton, the laws of thermodynamics, or the boiling point of water. Nanotechnology won't make energy or matter from nothing.

It seems a good bet that no one will build a faster than light spacecraft, or an antigravity machine, or a cable twice as strong as diamond. There are limits. Science today may be wrong about some limits, but scientific knowledge is practically defined to be our best information about how the world works, so it isn't wise to bet against it.

There will be claims that nanotechnology will be able to do things that it can't, or that capabilities are around the corner when they aren't. Sometimes these will be innocent errors, sometimes they will be culpably stupid errors, and sometimes they will be what amounts to fraud. Among the problems that nanotechnology cannot solve is that of misguided claims, by people calling themselves "scientists," "engineers," or "businesspeople," that they have a big technical breakthrough worth a fortune. Every interesting new technology, particularly in its early days, is a chaotic mix of competent workers and charlatans. For every Thomas Edison inventing useful products such as light bulbs or the precursor of movie projectors, there were people promoting electric hairbrushes to cure baldness, and electric shoes, electric belts, electric hats—the list goes on—that authoritatively claimed cures for infertility, overweight, underweight, and all the ills and discomforts of mankind. Today, we laugh at the credulity of our forefathers who bought these gadgets; we shouldn't, unless we laugh at our own times as well.

Population

Natural law imposes limits, but so does the nature of human beings. These will continue as long as people do.

Reproduction is a deeply ingrained instinct enforced by the march of time, which ruthlessly discards the genetic material of all who neglect it. Many would argue that the Earth is already overpopulated. While nanotechnology could enable the current population, and even a greatly increased one, to live more lightly on the Earth, there will still be limits to Earth's capacity.

The norms of human life are shaped by ancient patterns: high rates of infant or childhood mortality have been facts of life for millennia, and having many, many children has been a way to ensure that one or two will survive to work on the farm, and to care for you in your old age. Large families naturally become traditional. When modern medicine and reliable food supplies change those conditions—as they have, in cultural terms, virtually overnight—behavior does not shift as quickly. The result is the Third World population boom. In Western countries, where there has been time for behavior to adapt, a huge family is the exception.

It might seem that our problem is solved. Molecular manufacturing can make everyone wealthy, and wealthy populations today have stable or shrinking populations. The Earth can support more people with advanced technologies, and these will also open up the vast room and resources of the world beyond Earth. Would that this were true.

If 99 per cent of the people in a population respond to wealth by reducing childbearing, the population will indeed stabilize or shrink, for a while. But populations are not uniform. What of the 1 per cent, say, who are members of a minority with different values? If that minority has a growth rate of 5 per cent per year, then in ninety-five years they will be the majority, and in one thousand years their population will have grown by a factor of 1,500,000,000,000,000,000,000, if resource limits or genocide haven't intervened. Note that the Hutterites of North America, a reasonably wealthy religious group viewing fertility control as a sin and high fertility as a blessing, have managed an average of ten children per woman. Given enough time, exponential growth of even the smallest population can consume all the resources in reach.

The right to reproduce is often regarded as basic, as illustrated by the outrage at reports of forced abortion in the People's Republic of China. The Hutterites and many others regard it as part of their freedom of religion. But what happens when parents have more children than they can support—does redistribution solve the problem? If reproduction is not forcibly suppressed, and if resources are forcibly and repeatedly redistributed so that each human being has a roughly equal share, then each person's share will steadily shrink. Even given the most optimistic assumptions regarding available resources, with a policy of resource redistribution and unlimited reproduction, the amount per person would eventually be insufficient to sustain life. This policy must be avoided, because if it is followed, it will kill everyone.

As soon as we grant that any entity is entitled to certain rights—whether that entity be a human child, an animal, or some future artificial intelligence—the question arises of who is responsible for providing resources to support it when it can't do so for itself. The above argument indicates that a policy of coercion by some central power to compel the entire population to support an exponentially exploding population of these individuals would lead directly to disaster. Ultimately, this responsibility must rest with the entities' initiator: the designer of the artificial intelligence, the owner of the pet, the parents of the child. No new technology can magically remove the limits imposed by natural law, and thereby lift the burden of human responsibility.

Solutions Cause Problems

Every time a technology solves a problem, it creates new problems. This doesn't mean that the change is neutral, or for the worse, of course. The Salk and Sabin vaccines for polio virtually destroyed the iron-lung industry, and the pocket calculator virtually destroyed the slide-rule industry, but these advances were worth the price of some economic adjustment.

Molecular manufacturing and nanotechnology will bring far greater changes, placing far greater strains on our ability to adapt. We shouldn't be surprised when basically beneficial applications make someone miserable. Our lives are largely centered around problems. If we can solve many of these problems, the centers of our lives will shift, creating fresh problems. This section sketches some of the issues of change and adaptation more to raise questions than to offer solutions.

Change Causes Problems

Molecular manufacturing offers the possibility of drastic change, a change in the means of production more fundamental than the introduction of industry, or of agriculture. Our economic and social structures have evolved around assumptions that will no longer be valid.

How will we handle the changes in the way we work and live? Nanotechnology will have wide-ranging impact in many areas, including economic, industrial, and social patterns. What do historical patterns in similar circumstances tell us about the future?

Any powerful technology with broad applications revolutionizes lives, and nanotechnology will be no exception. Depending on one's point of view, this may sound exciting or it may sound disturbing, but it most certainly does not sound comfortable.

In comparison to many projections of the twenty-first century, though, nanotechnology may lead to comparatively comfortable change. The changes most often projected—for a future not including nanotechnology—have been ecological disaster, resource shortages, economic collapse, and a slide back into misery. The rise of nanotechnology will offer an alternative—green wealth—but that alternative will bring great changes from the patterns of recent decades.

Times of rapid technological change are disconcerting. For most of humanity's existence, people lived in a stable pattern. They learned to live as their parents had lived—by hunting and gathering, later by farming—and changes were small and gradual. A knowledge of the past was a reliable guide to the future.

Sudden changes, when they did occur, were apt to be ruinous: invasions or natural disasters. These sudden changes were fought or repaired or survived as best one could. Making major changes by choice was rare, and radical innovations were generally for the worse: the old ways at least ensured the ancestors' survival, the new might not. This made cultures conservative.

It is only natural that there be efforts to resist change, but before undertaking such an effort, it makes sense to examine the record of what works and what doesn't. The only examples of successful change fighters have been communities that have created and maintained barricades to isolate themselves from the outside world socially, culturally, and technologically. For the two centuries before 1854, Japan turned its back on the outside world, following a deliberate policy of seclusion. The leaders of Albania restricted contacts for many years; only recently have they started to open up.

Isolation attempts have worked better on a smaller scale, when participation is voluntary rather than decreed by government. Today, within the Hawaiian island chain, the tiny, privately owned island of Niihau, sixteen miles long and six miles wide, is deliberately kept as a preserve of the nineteenth-century Hawaiian lifestyle. Over two hundred full-blooded Hawaiians there speak the Hawaiian language and use no telephones, plumbing, television, and no electricity (except in the school). The Amish of Pennsylvania have no surrounding ocean to help maintain their isolation, but rely instead on tight social, religious, and technological rules aimed at keeping external technology and culture out, and themselves grouped in; those who leave the fold are excluded.

On a national scale, attempts to take only one part of the package—whether social or technological—haven't done well at all. For decades, the Soviet Union and the Eastern bloc nations welcomed Western technology but attempted tight restrictions on the passage of people, ideas, and goods. Yet illegal music, thoughts, literature, and other knowledge still crept in—as they do into the Islamic countries.

Fighting technological change in society at large has had little success, where that change gave some large group what it wanted. The most famous fighters of technological change—the Luddites—were unsuccessful. They smashed "automated" textile machinery that was replacing old hand looms during the early industrial revolution in England, but people wanted affordable clothing, and smashing equipment in one place just moved the business elsewhere. Change has sometimes been postponed, as when a later group, under the banner of "Captain Swing," smashed hundreds of threshing machines in a wide area of southern England in 1830. They succeeded in keeping the old, labour-intensive ways of harvesting for over a generation.

In previous centuries, when the world was less tightly connected by international trade, communications, and transportation, delays of years and even decades could be enforced through violence or legal maneuvers such as tariffs, trade barriers, regulations, or outright banning. Attempting to stop or postpone change is less successful today, when technology moves internationally almost as easily as people do—and human travel is so easy that 25 million people cross the Atlantic each year. Change fighters find that the problems they create mount with time. Products made using the old, high-cost techniques are uncompetitive. There is no way to bring back the "old jobs": they no longer make sense. But old habits die hard, and these same responses to the prospect of technological change continue today—ignoring it, denying it, and opposing it. Societies that have fought change, as Britain did, have fallen behind in a cloud of coal smoke.

Why did the Luddites respond violently? Perhaps their response can be attributed to three factors: First, the change in their lives was sudden and radical; second, it affected a large group of people at one time, in one area; and third, in a world unprepared for rapid technological change, there was no safety net to catch the unemployed. While local economies might have been able to absorb a trickle of hungry laid-off workers, they lacked the size and diversity needed to offer other employment options quickly to large numbers of unemployed.

In the twentieth century, however, societies have of necessity become somewhat better adapted to change. This has been a matter of necessity, because sluggish communities soon fall behind. In the ancient days of peasant stability, there was no need for institutions like Consumer Reports to study and rate new products, or regulators like the Environmental Protection Agency to watch over new hazards. We developed the needs, and we developed the institutions. These mechanisms represent important adaptions, not so much to the technologies of the twentieth century, but to the increasing change in technology during the twentieth century. There is great room for improvement, but they can perhaps provide a basis for adapting to the next century as well.

Even with the best of institutions to cushion shocks and discourage abuse, there will be problems. The very act of solving problems of production—of increasing wealth—will create problems of economic change.

Clean, Decentralized Production Causes Problems

Over centuries, the trend has seemed to be toward centralization, beginning with the rise of factories and industrial towns. What drove these developments was the high cost of machinery and plant operations, the need to be near power sources, the impracticality of transportation among many small, dispersed sites, and the need for face-to-face communication.

Beginning with the first industrial revolution, factories employed large numbers of people in one place, leading to overcrowding and making local economies dependent on one industry and sometimes on a single company. Costly equipment necessitated central locations for textile production, rather than the cottage industries where a lone woman could earn a livelihood carding wool and creating thread on a spinning wheel (providing the origin of the term *spinster*). By the 1930s, the belief in the virtures of centralization and central planning—the supposed efficiencies and economies of scale—led to nationwide or continentwide experiments in centralization. But over the last decade, these large-scale experiments have been dismantled, from Britain's privatization of nationalized utilities to the beginning of a return to the market system in Eastern European countries.

Because the old limits on transportation, energy sources, and communication have fallen, business is now decentralizing. Between 1981 and 1986, the Forbes 500 companies cut their employees by 1.8 million. But during those same years, total civilian jobs went up by 9.2 million. Start-up companies created 14 million jobs; small companies created another 4.5 million. Telecommuting is booming, as are new businesses, independent professionals, and cottage industries.

We've also seen the resurgence of small, but highly diverse stores: gourmet food shops, specialty ethnic shops, tea and coffee purveyors, organic and health food stores, bakeries, yogurt shops, gourmet ice-cream stores, convenience stores offering twenty-four-hour access, shops selling packaged food plus snacks. These stores epitomize something fundamental: At some point, what we want is not a standard good at an ever cheaper price, but special things customized to meet our own individual tastes or needs.

The trend for advanced technologies seems to be leading away from centralization. Will nanotechnology counter or accelerate this trend? By reducing the cost of equipment, by reducing the need for large numbers of people to work on one product, and bringing greater ability to produce the customized goods that people want, nanotechnology will probably continue the twentieth-century trend toward decentralization. The results, though, will be disruptive to existing businesses.

The computer industry perhaps provides a clue to what might happen as costs are lowered by nanotechnology. The computer-software industry is characterized by the garage-shop start-up. When your equipment is cheap—inexpensive PCs built around low-cost chips—and you can make a product by throwing in some ingenuity and human labour, it's possible to start a new industry on a shoestring.

In 1900, when cars were simple, there were many car manufacturers. By the 1980s, if you weren't an industrial giant like General Motors or Ford, Honda or Nissan, you had to be John De Lorean to even get a shot at acquiring the capital to play in the business. If molecular manufacturing can slash the capital costs for producing cars or other plant-intensive equipment, we will see the equivalent of garage-shop businesses springing up to offer new products, and hiring workers away from the industrial giants of today just as the personal computer has destroyed the dominance of the mainframe.

The American dream is to be an entrepreneur, and the technological trends of the twentieth century point in that direction. Nanotechnology probably continues it.

In one area, however, the late twentieth-century trend has been toward uniformity. The nations of Western Europe are in the process of uniting under one set of economic rules, and parts of Eastern Europe are anxious to join them. More and more supranational and transnational organizations knit the world together. The growth of trade has motivated economic integration.

Molecular manufacturing will work against this trend as well, permitting radical decentralization in economic terms. This will help groups that wish to step aside from the stream of change, enabling them to be more independent of the turbulent outside world, picking and choosing what technologies they use. But it will also help groups that wish to free themselves from the constraints of the international community. Economic sanctions will have little force against countries that need no imports or exports to maintain a high standard of living. And export restrictions will likewise do little to hamper a military buildup.

By weakening the ties of trade, molecular manufacturing threatens to weaken the glue that holds nations together. We need that glue, though, to deal with the arms control issues raised by molecular manufacturing itself. This problem, caused by the potential for decentralization, may loom large in the coming years.

Even Wealth and Leisure Cause Problems

Lester Milbrath, professor of sociology and political science, observes, "Nanotechnologies will create the problem of how to meaningfully and sustainably occupy the time of people who need not perform much work in order to have a sufficiency of life's goods. Our society has never faced this problem before, and it is not clear what social restructuring will be required to have a good society in those circumstances. We face much deep social learning."

The world has had little experience with what anthropologists call "abundance economies." The native American tribes of the Pacific Northwest were one of those rarities. Ruth Benedict, in her classic book *Patterns of Culture*, wrote, "Their civilization was built upon an ample supply of goods, inexhaustible, and obtained without excessive expenditure of labour." The Kwakiutls became famous for their "potlatches": contests in which they sought to shame their rivals by heaping more gifts upon them than they could ever return. The potlatches would often be a year in preparation, last for days, and occasionally involve destruction of entire buildings. It was certainly a colorful form of keeping up with the Joneses.

What will motivate us, once we have achieved an abundance economy? What will we regard as worthwhile goals to pursue? Increased knowledge, new art, improved philosophy, eliminating human and planetary ills? Will we find ourselves creating a better, wiser world, or sunk in boredom and jaded now that we have all and want nothing? If boredom gets out of hand, the lively spectacle of wealthy donors seeking to outdo each other to endow the arts, aid the poor, and do other good deeds for the sake of prestige would be welcome.

What will happen as life spans continue to lengthen and the time needed to make a living decreases? Even today, there are people who, when confronted with the prospect of a significantly longer life span, exclaim that they couldn't imagine what they would do with all that time. This response can be hard to understand, when it would take a thousand years to walk all the world's roads, more thousands of years to read all the world's books, and another ten thousand years to have a dinner conversation with each of the world's people—but tastes differ, and even a few decades of bad television might make anyone long for the peace of the grave.

Changing Employment Causes Problems

A major concern, and certainly the single area of greatest upheaval, is employment (which may become hard to distinguish from leisure). Once, people had little choice of employment. To keep a full belly, most had to work at the only job available: peasant farming. Eventually, people will have a complete choice of employment: they will be able to keep a full belly and a wealthy lifestyle while doing whatever they please. Today, we are about halfway between those extremes. In advanced economies, many different jobs are deemed useful enough that other people will offer an adequate income in exchange for the result. Some people can make a living doing something they enjoy—is this work, or leisure?

The impact of nanotechnology on patterns of employment will depend on when it arrives. Current demographics show a shrinking supply of young people entering the work force. Agriculture, the assembly line, and entry level service jobs are experiencing a labour shortage, and no relief is in sight. If these trends continue, nanotechnology may show up in the midst of a shortage of labour. If it arrives late enough, it may compete with industries that are already nearing full automation; "job displacement" may mean replacing an industrial robot with a nanomachine.

Employment patterns have shifted radically in the past. One hundred and fifty years ago, the United States was an agricultural nation—69 per cent of all people worked the land and a growing percentage worked in industry doing things like building steam locomotives for Baldwin Locomotives Works or tanning leather for the giant Central Leather monopoly. By the early twentieth century, agriculture was waning in numbers but increasing in productivity; most people worked in industry, and the tiny information and service sector was beginning to grow. Today, the picture has reversed: 69 per cent of employed Americans work in information or service jobs, only 28 per cent work in industrial production, and 3 per cent in agriculture. This tiny fraction feeds the other 97 per cent of Americans, exports hugely to other countries, and receives subsidies and price support payments to stop them from growing even more food. Manufacturing, even without nanotechnology, seems to be heading toward a similar condition.

With an ever-declining percentage of our population working in manufacturing, we have as everyday products things that were once available only to kings and the high nobility. Yet owning multiple suits of clothes, having personal portraits of ourselves and family members, having music upon our command, having a personal bedroom, and having a coach awaiting our need—these are now regarded as being among the bare necessities of life. It may be possible to adjust to even greater wealth with even less required labour, but the adjustment will surely cause problems. In a world in which nanotechnology reduces the need for workers in agriculture and manufacturing still further, the question will be asked, "What jobs are left for people to do once food, clothing, and shelter are very inexpensive?"

Again, the twentieth century provides some guidelines. As technology has reduced costs by efficiently producing many units of an identical item, people have begun to demand customization to meet individual needs or preferences. As a result, there are ever more jobs in producing custom goods. Today, semi-custom goods that try to help us meet our needs or express our taste abound:

designer linens, ready-to-wear fashions, cosmetics, cars, trucks, recreational vehicles, furniture, carpeting, shoes, televisions, toys, sports equipment, washing machines, microwave ovens, food processors, bread bakers, pasta makers, home computers, telephones, answering machines—are all available in large and ever-changing variety.

Just as varied is the fabulous wealth and diversity of information produced in the twentieth century. Information products are a large factor in the economy: Americans buy 2.5 billion books, 6 billion magazines, and 20 billion newspapers each year. In recent years, new magazines have been invented and launched at the rate of one every business day of the year. A visit to a well-stocked magazine rack shows only a hint of the wealth of highly specialized publications, each one focused on a specialized interest or attitude: hotdog skiing, low-fat gourmet cooking, travel in Arizona, a magazine for people with a home office and a computer, and finely tuned magazines on health, leisure, psychology, science, politics, movie stars and rock stars, music, hunting, fishing, games, art, fashion, beauty, antiques, computers, cars, guns, wrestling.

Motion pictures, which started as a flock of independent production companies and then consolidated into the great studios of the 1930s, have since followed the decentralization and diversification trends of recent years. Now an expanding range of film entertainment comes via network TV, cable channels, private networks, videotapes, music videos. Independent producers are aided by the technology innovations of cable, direct broadcast satellites, videotape technology, laser disks, videocameras.

The arts have burgeoned, with the general public as the new patron of the arts. Any artist or art form that could find and satisfy a market boomed in the twentieth century. Not just the traditional arts of actors, writers, musicians, and painters, but all forms of "domestic" artistry have grown to unprecedented levels: landscape and interior design, fashion design, cosmetics, hairstyling, architecture, bridal consulting.

Providing for these demands are some of the "service and information" jobs created in the late twentieth century. "Service" jobs include many ways of helping other people: from nursing to computer repairs to sales. In "information" jobs, projected to have the fastest percentage growth over the next decade, people find, evaluate, analyze, and create information. A magazine columnist or TV news producer obviously has an "information" job. But so do programmers, paralegals, lawyers, accountants, financial analysts, credit counselors, psychologists, librarians, managers, engineers, biologists, travel agents, and teachers.

"Increasingly," states *Forbes* magazine, "people are no longer labourers; they are educated professionals who carry their most important work tools in their heads. Dismissing them from their jobs, cutting them off from their places of employment may hurt them emotionally and financially. But it doesn't separate them from their vocation in the same way that pushing a farmer off his freshly seeded land does. For centuries workers were more dependent on a particular physical setting than they are now. Modern occupations generally give their practitioners more independence—and greater mobility—than did those of yesteryear."

These human skills that people carry with them will continue to be valued: managing complexity, providing creativity, customizing things for other people, helping people deal with problems, providing old services in new contexts, teaching, entertaining, and making decisions. A reasonable guess would be that many of the service and information industries of the twentieth century will continue to evolve and exist in a world with nanotechnology. What is harder to imagine would be what new industries will come into being once we have new capabilities and lower costs.

Along with the old economic law of supply and demand is another governing factor: price elasticity effects. People's desire for something is "elastic": it expands or contracts when the cost of something valuable goes down or up. If the price of a flight to Europe is five hundred dollars, more people will take a European vacation than if the price is five thousand dollars. When you had to hire a highly trained mathematician to do equations, calculation was slow and expensive. People didn't do much of it unless they absolutely had to. Today, computers make calculation cheap and automatic. So now businesses do sophisticated financial modeling, chemists design protein molecules, students calculate orbital trajectories for spaceships, children play video games, moviemakers do ever more amazing special effects, and the cartoon—virtually extinct because of high labour costs—has returned to movie theaters, all because computers permit cheap calculation. Nanotechnology will offer new, affordable capabilities to these and other people. Today, it's as hard to predict what new industries will be invented as it would have been for the creators of the ENIAC computer to have predicted cheap, handheld game computers for children.

So rather than producing drastic unemployment, nanotechnology seems likely to continue the trend already seen today, away from jobs that can be automated and into jobs where the human perspective is vital. But the true possibilities are, as always in the modern world, beyond predicting.

Change Disrupts Plans

Major shifts in demographics always cause disruptions. Even when we know they are coming, we never prepare for them.

Our plans are based on expectations of what will happen. If things don't go as expected, we find that we have "malinvested." Houston real estate was valuable and looked to become even more so when times were good for the oil business there; when the fortunes of the oil business changed, Houston real estate was found to have been overbuilt, overpriced, and many millions of dollars were lost.

Lengthening life spans push people toward taking a longer-term perspective, but rapid rates of change force a shorter-term perspective in investments. Turbulence in technology and in governmental monetary policy have already shortened time horizons. Businesspeople once routinely built plants with a thirty-year useful life. Today, the rate of change is too fast, and uncertainty regarding inflation and potential changes in tax laws is too great for such investments to make sense. Faster change will shrink time horizons further.

Governments have taken on themselves the burden of looking a lifetime ahead, and the Social Security Administration is in for some rough times. When Otto von Bismarck, Germany's Iron Chancellor, came up with the notion of a guaranteed old age pension, it was a cynically clever and low-cost way to gain popular goodwill. So few people lived to age sixty-five that the amounts paid out in pensions were a pittance. After watching the German experiment for a handful of years, other governments began following suit. None of them expected a world like ours where a baby girl born in the United States today has an average life expectancy of 78.4 years—double that of Bismarck's time—and even this estimate is based on the faulty assumption that her medical care will be no better than her great-grandmother's was.

At present, the Social Security Administration has two models: one they call "positive" and one they call "negative." In the "positive" model, people work like dogs until old age, retire, and promptly die—presumably before they've had a chance to collect substantial social security or medical benefits. In the "negative" model, people retire early, develop illnesses that require medical intervention, and then live a long time making doctor visits and hospital stays during those years. Plans based on these models deserve to be disrupted. A better, more realistic scenario would have people living and able to support themselves for a long time, with illnesses that can be handled easily and inexpensively. Present social security benefits are enough to provide a certain standard of living—food, housing, transportation, and so forth. In a future of great material wealth, these benefits will be easy to provide, and present projections of economic woe resulting from an aging population will seem quaint.

Coping with Change

Back in the seventies, author Alvin Toffler brought out a book called *Future Shock*, describing how disturbing rapid change is for people. The book was a best-seller, but how much actual future shock has been seen in the past decade? Most people seem to have come through the last two decades pretty much all right, not in a state of shock at all. Rather than being shocked by technology, they are instead annoyed about pollution and traffic.

Does this mean Toffler was wrong in predicting future shock? It's true that technology has been advancing rapidly in many areas over the past twenty years. But consider the average person's home life: How much of this rapid technological advance has shown up there? A great deal, yet most of it is hidden, unlike the earlier part of the century, where obvious change was the norm. Electric lights and appliances, automobiles, telephones, airplanes, radio, and television affected almost everyone's private life. One person's life could span the time from horse-and-buggy travel to watching the Moon landings on television.

In contrast, the past twenty years have seen new technologies move more quietly into the home. The VCR and microwave oven don't seem nearly as revolutionary as earlier inventions. Telephone answering machines are useful but haven't caused major changes in lifestyles. Fax machines are handy, but they're much like having very fast mail, and as this is written, fax machines aren't yet in most homes. So it's not surprising that the average person has felt little future shock lately. New medicines taken as pills—which may be radically improved—look just like the earlier pills. The computerized bills that come in the mail aren't any more exciting to pay than the old human-prepared bills.

This situation is unlikely to last. How much longer can technology advance so rapidly in so many fields without major effects on our lifestyles? There's been a respite from future shock in the last three decades; people have had a chance to catch their breath. When nanotechnology arrives, will future shock arrive with it?

Some segments of society today are already getting practice in dealing with rapid technological advance. Those getting the most vigorous workout are in the computer field, where a machine two years old is regarded as obsolete, and software must be updated every few months to keep abreast of the new developments.

But has this terrific rate of progress been dizzying or overwhelming? Not for the consumer—on the contrary, computers have become easier to use. In the 1960s, the New Math that was introduced into American grade schools and junior high schools included extensive study of arithmetic using numbers written in something other than the familiar base 10. This was to prepare the "Adults of Tomorrow" for "The Computer Age" in which we would all be writing assembly language computer programs in binary (base 2) code. But customers now purchase software rather than write it themselves—they need never deal with computer languages at all, much less a primitive assembly language. The rapid increase of computer speed has helped make computers easier to use.

This progression has occurred many times before: Cars started off with external hand cranks, then advanced to starters you could yank from the comfort of the driver's seat; now starters perform invisibly when you turn the key in the ignition. This pattern will surely continue. First, some people will adapt to the technology, but in the long run the technology will adapt to us. The more flexible and powerful the technology, the more easily it will adapt.

Seen from a distance, seemingly trivial patterns of adaptation form part of a larger process that has marked the last century: the Western world has begun to invent mechanisms to handle a world of persistent change. Our mechanisms are by no means perfect or painless, as any unemployed person can testify. Employment agencies and headhunters for job seekers; unemployment and severance packages to ease job transitions; on-the-job training, continuing education, retraining, specialized seminars to update professional skills, professional associations, networking, community resources centers, government training programs, and volunteer agencies are just a few of the inventions dealing with change and transition. Consumer information services, regulatory agencies, and environmental organizations are others. The most effective will endure. More options will continue to be invented.

Unit I–L

Safety, Accidents and Abuse

Some truisms: Almost any technology is subject to use, misuse, abuse, and accidents. The more powerful a technology is when properly used, the worse it is likely to be when abused. Any powerful technology in human hands can be the subject of accidents. Nanotechnology and molecular manufacturing will be no exception. Indeed, if molecular manufacturing replaces modern industry, and if its nanotechnological products replace most modern technologies, then most future accidents will have to involve nanotechnology.

Another truism: In a diverse, competitive world, any reasonably-inexpensive technology with enormous commercial, medical, and military applications will almost surely be developed and used. It is hard to envision a scenario (short of the collapse of civilization) in which nanotechnology will not make its appearance; it seems inevitable. If so, then its problems, however tough, must be dealt with.

Like trucks, aircraft, biotechnology, rockets, computers, boots, and warm clothes, nanotechnology has the potential for both peaceful and aggressive uses. In peaceful uses (by definition), harm to people occurs either by accident or as an unintended consequence. In aggressive uses, harm is deliberate. In a peaceful context, the proper question to ask is **can fallible people of goodwill, pursuing normal human purposes, use nanotechnology in a way that reduces risk and harm to others?** In an aggressive, military context, the proper question to ask is **can we somehow keep the peace?** Our answer to the first will be a clear yes, and to the second, an apprehensive maybe.

Throughout this discussion, we assume that most people will be alert in matters concerning their personal safety, and that some will be alert in matters concerning world safety. During the 1970s, people awakening to the new large-scale, long-term problems of technology often felt isolated and powerless. They naturally felt that technology was out of their control, in the hands of shortsighted and irresponsible groups. Today, there are still battles to be fought, but the tide has turned. When a concern arises regarding a new, obvious technology, it is now much easier to get a hearing in the media, in the courts, and in the political arena. Improving these mechanisms for social vigilance and the political control of technology is an important challenge. Current mechanisms are imperfect, but they can still give a big push in the right directions.

Though we assume alertness, alertness can be a scarce resource. The total amount of concern and energy available for focusing on long-term problems is so limited that it must be used carefully, not squandered on problems that are trivial or illusory. Part of our aim in this unit is to help sort out the issues raised by nanotechnology so that attention can be focused on problems that must be solved, but might not be.

The next few sections deal with accidents of conventional sorts, where safety benefits are obvious. Later sections discuss more novel problems, some tough enough that we have no good answers.

Safety in Ordinary Activities

As countries have grown richer, their people have lived longer despite pollution and automobile accidents. Greater wealth means safer roads, safer cars, safer homes, and safer workplaces. Throughout history, new technologies have brought new risks, including risks of death, injury, and harm to the environment, but prudent people have only accepted new technologies when they offered an improved mix of risks and benefits. Despite occasional dramatic mistakes, the historical record says that people have succeeded in choosing technologies that reduce their personal risks. This must be so, or we wouldn't be living longer.

Molecular manufacturing and its products should continue this trend, not as an automatic consequence, but as a result of continued vigilance, of people exercising care in picking and choosing which technologies they allow into their daily lives. Nanotechnology will give better control of production and products, and better control usually means greater safety. Nanotechnology will increase wealth, and safety is a form of wealth that people value. Public debate, product testing, and safety regulations are standard parts of this process.

Home Safety

In common home accidents, a dangerous product is wrongly applied, spilled, or consumed. Homes today are full of corrosive and toxic materials, for cleaning drains, dissolving stains, poisoning insects, and so forth. All too often, children drink them and die. With advanced technology, none of these tasks will require such harsh, crude chemicals. Cleaning could be performed by selective nanomachines instead of corrosive chemicals; insects could be controlled by devices like ecosystem protectors that know the difference between a cockroach and a person or a ladybug. There will doubtless be room for deadly accidents, but with care and hard work, it should be possible to ensure that nanotechnologies for the home are safer than what they replace, saving many lives.

It is, of course, possible to imagine safety nightmares: nanotechnology could be used to make products far more destructive than anything we've seen because it could be used to extend almost any ability further than we've seen. Such products presumably won't be commonplace: even today, nerve gas would make a potent pesticide, but it isn't sold for home use. Thinking realistically about hazards requires common sense.

Industrial Safety

We've already seen how post-breakthrough technologies can eliminate oil spills by eliminating oil consumption. A similar story could be told of almost any class of industrial accident today. But what about accidents—spills and the like—with the new technologies? Rather than trying to paint a picture of a future technology, of how it could fail and what the responses could be, it seems better to try a thought experiment. What could be done to deal with oil spills, if oil were still in use? This will show how nanotechnologies can be used to cope with accidents:

If there were a spill and oil on the shore, advanced nanomechanisms could do an excellent job of separating oil from sand, removing oil from rocks, and cleaning crude oil from feathers on birds and the feathery legs of barnacles. Oil contamination is a pollution problem, and nanotechnology will be a great aid in cleaning up pollution.

But why should the oil reach the shore? Economical production would make it easy to stockpile cleanup equipment near all the major shipping routes, along with fleets of helicopters to deliver it at the first distress call from a tanker. Oil cleanup equipment built with nanotechnology could surely do an excellent job of scooping oil from the water before it could reach the shore.

But why should the oil leave the tanker? Economical production of strong materials could make seamless hulls of fibrous materials far tougher than steel, with double, triple, or quadruple layers. Smart materials could even make punctures self-sealing. Hulls like this could be run into rocks at highway speeds without spilling oil.

But why should anyone be shipping crude oil across the sea? Even if oil were still being pumped (despite inexpensive solar energy and solar-derived fuels), efficient molecule processing systems could refine it into pure, fluid fuels at the wellhead, and inexpensive tunneling machines could provide routes for deeply buried pipelines.

Any one of these advances would shrink or eliminate today's problem with oil spills, and all of them are feasible. This example suggests a general pattern. If nanotechnology can provide so many different ways to avoid or deal with an oil spill—one of the largest and most environmentally destructive accidents caused by today's industry—it can probably do likewise for industrial accidents in general.

The most direct approach is the most basic: the elimination of anything resembling today's bulk industrial plants and processes. The shift from messy drilling activities and huge tankers to small-scale distributed systems based on solar cells is characteristic of the style in which nanotechnology can be used. The chemical industry today typically relies on plants full of large, pressurized tanks of chemicals. Not surprisingly, these occasionally spill, explode, or burn. With nanotechnology, chemical plants will be unnecessary because molecules can be transformed in smaller numbers, as needed and where needed, with no need for high temperatures, high pressures, or big tanks. This will not only avoid polluting by-products, but reduce the risk of accidents.

Medical Safety

Medicine can be safer too. Drugs often have side effects that can do permanent damage or kill. Nanomedicine will offer alternatives to flooding the body with a possibly toxic chemical. Often, one wants to affect just one target: just the stomach, or perhaps just the ulcer. An antibiotic or antiviral treatment should fight specific bacteria or viruses and not damage anything else. When medicine achieves the sophistication of immune machines and cell-surgery devices, this will become possible.

But what about medical accidents and side effects? Molecular manufacturing will make possible superior sensors to tell medical researchers of the effects of a new treatment, thereby improving testing. Better sensors will also help in monitoring any negative effects of a treatment on an individual patient. With care, only a few cells would be damaged and only small concentrations of toxic by-products would be produced before this was noticed and the treatment corrected.

The resources of nanotechnology-based medicine would then be available for dealing with the problem. With biostasis techniques available, even the worst medically induced illnesses could be put on hold while a treatment was developed. In short, serious medical mistakes could be made far rarer, and most mistakes could be corrected.

The conclusion that follows from these examples of oil spills, chemical plants, and the effects of medical treatments is straightforward. Today our comparative poverty and our comparative technological incompetence press us in the direction of building and using relatively dangerous and destructive devices, systems, and techniques. With greater wealth and technological competence, we will have the option of accomplishing what we do today (and more) with less risk and less environmental destruction: in short, being able to do more, and do it better.

With better-controlled technologies, and with an ample measure of foresight and concern, we will even be able to do a better job of recovering from mistakes. It won't happen automatically, but with normal care we can arrange for our future accidents to be smaller and less frequent than those in our past.

Extraordinary Accidents

The previous section discussed ordinary accidents that would occur during the use of nanotechnology by generally responsible, yet fallible, human beings. Nanotechnology also raises the specter, however, of what have been termed "extraordinary accidents": accidents involving runaway self-replicating machines. One can imagine building a device about the size of a bacterium but tougher and more nearly omnivorous. Such runaways might blow like pollen and reproduce like bacteria, eating any of a wide range of organic materials: an ecological disaster of unprecedented magnitude—indeed, one that could destroy the biosphere as we know it. This may be worth worrying about, but can this happen by accident?

How to Prepare a Big Mistake

The so-called "Star Trek scenario" (named after an episode of **Star Trek: The Next Generation** that featured runaway "nanites") is perhaps the most commonly imagined problem. In this scenario, someone first invests considerable engineering effort in designing and building devices almost exactly like the one just described: bacterial-sized, omnivorous, able to survive in a wide range of natural environments, able to build copies of themselves, and made with just a few built-in safeguards—perhaps a clock that shuts them off after a time, perhaps something else. Then, accidentally, the clock fails, or one of these dangerous replicators builds a copy with a defective clock, and away we go with an unprecedented ecological disaster.

This would be an extraordinary accident indeed. Note well, though, that this accident scenario starts with someone building a highly capable device that is almost disastrously dangerous, but held in check by a few safeguards. This would be like wiring your house with dynamite and relying on a safety-catch to protect the trigger: a subsequent explosion could be called an accident, but the problem isn't with the safety mechanism, it's with the dynamite installation.

Do we need to build nanotechnological dynamite? It's worth considering just how little practical incentive there is for anything even resembling the dangerous replicator just described. (Note that our topic here is accidents; deliberate acts of aggression are another matter.)

How to Avoid it?

With our present technology, which is simpler to build—a car that runs on gasoline, or one that forages for fuels in the forest? A foraging car would be very hard to design, cost more to manufacture, and have more parts to break down. The situation is similar with nanotechnology.

Ralph Merkle of Xerox Palo Alto Research Center discussed this issue at the First Foresight Conference on Nanotechnology. He explains, "It's both uneconomical and more difficult to design a self-replicating system that manufactures every part it needs from naturally occurring compounds. Bacteria do this, but in the process they have to synthesize all twenty amino acids and many other compounds, using elaborate enzyme systems tailored specifically for the purpose. For bacteria facing a hostile world, the ability to adapt and respond to a changing environment is worth almost any cost, for lacking this ability they would be wiped out."

"But in a factory setting, where adequate supplies of all the needed parts are provided, the ability to synthesize parts from scratch is not only unneeded, it consumes extra time and energy, and produces excess waste. Even if we could design artificial self-replicating systems as flexible as existing natural ones, an inflexible and rigid system is better adapted to the controlled factory setting in which it will find itself than a more complex, more adaptable, less efficient design."

What is more, the Desert Rose Industries scenario showed how an expandable factory setup could operate with no self-replicating machines at all: molecular manufacturing doesn't require them. If they are used for some purpose, they will most likely resemble automobiles in their finicky requirements. A self-replicating molecular machine built for industrial purposes (and made as simple as possible) would float in a container of specially selected chemicals. As with the automobile, the best chemicals to use will probably be chemicals not commonly found in nature, and it would be easy to make that a design rule: **never make a replicator that can use an abundant natural compound as fuel.**

If we follow this rule, the idea of a replicator "escaping" and replicating in the wild will be as absurd as the notion of an automobile going feral and refueling itself from tree sap. Whether for replicators or cars, to design a machine that could operate in the wild would not be a matter of a flick of the draftsman's pen, but an intense, sustained research-and-development effort focused on that objective. Crashes and explosions occur in machinery by accident, but complex new capabilities don't.

A simple psychological error frequently occurs when someone first hears about nanotechnology, and hears mention of "molecular machines," and "replicators," and "nanocomputers," and "nanomachines that operate in nature." The error is this: the person makes a single new mental pigeonhole for "nanotechnology," throws everything into it, and stirs. After some mental fermentation, the result is the mythical nanomachine that does everything: it's a replicator, it's a supercomputer, it's a Land-Rover, it slices, it dices—and on reflection, this imaginary nanomachine sounds uncontrolled and dangerous. With enough effort, a do-it-all nanomachine could perhaps be built, but it sounds difficult and there's no good reason to try.

There are advantages to making systems of molecular machinery that can use inexpensive, abundant chemicals, and devices that can operate in nature, but these machines needn't be replicators. A facility like Desert Rose might be designed to use little but electric power from solar panels and molecules from the air, but a setup like this isn't going to slip away. Nanomachines built for cleaning up pollutants and other outdoor tasks could be manufactured in facilities run like Desert Rose and then spread or installed where they're needed.

Extraordinary accidents deserve attention, but with a little care they can be completely avoided. The incentive to build anything resembling a Star Trek-scenario replicator is negligible, even from a military perspective. Any effort toward building such a thing should be seen not as a use of nanotechnology, but as an abuse. Other abuses seem more likely, however, and are quite bad enough.

The Chief Danger: Abuse

The chief danger of nanotechnology isn't accidents, but abuse. The safety benefits of nanotechnology, when used with normal care, will free some of our attention to grapple with this far more difficult problem. As Lester Milbrath observes, "Nanotechnologies have such great power that they could be used for evil or environmentally destructive purposes as easily as they could be used for good and environmentally nourishing purposes. This great danger will require a level of political control far beyond that which most nations know how to exercise. We have a prodigious social learning task that we must face."

Thus far, we've focused on how increased abilities can serve constructive ends. Not surprisingly, the potential consequences—with the huge exception of social and economic disruption—are overwhelmingly positive. Inherently clean, well-controlled, inexpensive, superior technologies, when applied with care, can yield far better results than inherently dirty, messy, costly, inferior technologies. This should come as no surprise, but it is only half of the story. The other half is the application of those same superior technologies to destructive ends.

Readers feeling that all this may be too good to be true can breathe a sigh of relief. This problem looks tough.

Cooperative Controls

Molecular manufacturing will lead to more powerful technologies, but our current, crude technology already has world-smashing potential. We have lived with that potential for decades now. In the coming years, we will need to strengthen institutions for maintaining peaceful security.

If most of the political power in the world, and with it most of the police and military power, sees that the course of self interest lies in peace and stability, then solutions seem possible. (The prospect of an arms race in nanotechnology is terrifying and to be avoided at almost any cost. As of this writing, the end of the Cold War offers a better hope of avoiding this nightmare.) James C. Bennett, a high-tech entrepreneur and public policy commentator affiliated with the Center for Constitutional Issues in Technology, explains the goal: "Advanced technologies, particularly as far-ranging a capability as nanotechnology, will create a strong demand for their regulation. The challenge will be to create sufficient controls to prevent the power-hungry from abusing the technologies, without either smothering development or creating an overbearing international regime."

In the coming decades, preventing major abuse of nanotechnology will take the form of regulation, arms control, and antiterrorist activities. In the field of arms control, nanotechnology should present strong motivation for international cooperation and for intimate mutual inspection in the form of joint research programs.

The sheer productive capabilities of molecular manufacturing will make it possible to move from a working weapons prototype to mass production in a matter of days. In a more exotic vein, dangerous nanomachines could be developed, including programmable "germs" (replicating or nonreplicating) for germ warfare. Either development could bring war. With peace looking so profitable and an arms race looking so dangerous, arms control through cooperative development should look attractive. This does not make it easy, or likely.

Terrorism is not an immediate concern. We have lived with nuclear weapons and nerve gas for decades now, and nerve gas, at least, is not difficult to make. As of this writing, no city has been obliterated by terrorists using these means, and no terrorist has even made a credible threat of this sort. The citizens of Hiroshima and Nagasaki, like the Kurds in Iraq, fell victim to nuclear and chemical weapons wielded by governments, not small groups. So long as nanotechnology is technologically more challenging than the simple chemistry of nerve gas, nanoterrorism should not be a primary concern.

To keep dangerous nanotechnologies unavailable, however, will require regulation. If anyone were free to build anything using molecular manufacturing, then someday as the technology base improves and designs become available for more and more nanodevices, someone, somewhere—if only out of sheer spite—would figure out how to combine those nanodevices to make a dangerous replicator and turn it loose. There will almost surely be warning signs, however: In the natural course of events, causes attract protesters before stone-throwers, and produce letter bombs before car bombs. Abuse of nanotechnology is likely to be visible long before it is devastating, and this at least gives some time to try to respond.

Regulatory Tactics

Abuse of this sort can be delayed, perhaps for a long time, by proper regulation. The goal here isn't to make regulations so tight that people will have to violate them to get anything done. This would encourage holdouts, underground work, and disrespect for the law. Instead, the goal is to draw boundaries loosely enough to cause little difficulty for legitimate work, while making dangerous activities very difficult indeed. This is a delicate balance to strike: those fearful of risk naturally try to apply crude and oppressive regulations, and companies naturally try to loosen and avoid regulation entirely. Nonetheless, the problem must be solved, and this seems the best direction to explore.

In one approach, nanomachines could be divided into two classes: **experimental devices** and **approved products**. Approved products could be made widely available through special-purpose molecular manufacturing systems. Thus, once an experimental device had passed regulatory inspection, it could become inexpensive and abundant. In this way, popular demands for a product could be satisfied without anyone needing to break safety rules.

Approved products could include devices like (but superior to) the full range of modern consumer products, ranging from personal supercomputers with 3-D color displays, through smart construction materials, to running shoes with truly amazing features. The main cost of such goods might be the royalty to the designer. In *Engines of Creation* (the first book to examine this topic), this strategy for producing and distributing approved products is called a "limited assembler system."

Note that both approved products and the limited assemblers that build them would lack the ability to make copies of themselves, to self-replicate. Ralph Merkle sees this ability as the one to keep an eye on: "Self-replicating systems can and should be appropriately regulated. There seems no need, however, to have any more than normal concerns for devices which cannot replicate. While we might, as with any device, need laws to ensure their appropriate use, they pose no extraordinary problems." For most products, normal medical, commercial, and environmental standards would apply; the regulatory bureaucracies are already in place.

There are great advantages to permitting nearly free experimentation in a new technology, allowing creative people to try ideas without seeking prior approval from a cumbersome committee. Surprisingly, this, too, seems compatible with safety.

In the world of nanotechnology, one cubic micron is a large space, with room enough for millions of components. For many purposes, a few cubic microns would amount to a large laboratory space. To a device on a micron scale, a centimeter is an enormous distance. Surrounding a micron-scale device with a centimeter-thick wall would be like surrounding a human being with a wall kilometers thick, and just as hard to penetrate. Further, a micron-scale device can be incinerated in an instant by something as small as a spark of static electricity. Based on observations like these, **Engines of Creation** outlined the idea of a sealed assembler lab, in which a researcher could build anything, even something deliberately designed to be dangerous, and yet be unable to get anything out of the microscopic sealed laboratory except for information.

With a good communications network, a researcher or product developer in Texas could equally easily perform experiments in a remote Maine laboratory run with the security and secrecy of a Swiss bank. A lab would have a responsibility to its customers to keep proprietary work confidential, and a responsibility to regulatory authorities to ensure that nothing but information leaves the laboratory. Researchers could then perform any small scale experiments they wish. Only approved products, of course, would be built outside the sealed laboratories. While this may not be the best pattern of regulation possible, it does show one way in which freedom of experimentation could be combined with strict regulation of use. By providing a clear separation between legitimate and illegitimate activity, it would help with the difficult problem of identifying and preventing research aimed at damaging ends.

A sensible policy will have to balance the risk of private abuse of technology against the risk of government abuse of technology and regulation. Low-cost manufacturing can make surveillance equipment less expensive. Increased surveillance can reduce some risks in society, but the watchers themselves often aren't very well watched. Placing bounds on surveillance is a challenge for today's citizens as well as tomorrow's, and lessons learned in the past can be applied in the future.

In the long run, it seems wise to assume that someone, somewhere, somehow, will escape the bounds of regulation and arms control and apply molecular-manufacturing capabilities to making novel weapons. If by then we have had several decades of peaceful, responsible, creative development of nanotechnology (or perhaps a few years of help from smart machines), then we may have developed both ecosystem protectors and sophisticated immune machines for medicine. There is good reason to think that distributed technologies of this sort could be adapted and extended to deal with the problem of protecting against novel nanoweaponry. Failure to do so could mean disaster. Nonetheless, building protective systems of this sort will be by far the greatest challenge of any we have discussed. The chief purpose of regulatory tactics like those we have described must be to buy time for those peaceful developments, to maximize the chances that this challenge can be met before time runs out.

(Any critic declaring this to be an optimistic book hereby stands charged with having failed to read and understand the above paragraph.)

Guide It, or Stop It?

Potential accidents richly deserve the attention they will get, and we have confidence that this attention will suffice to make nanotechnology a force for improved human and environmental safety. Abuse is the greater danger, and harder to deal with. When considering a proposed policy, the first question should be, "How will this affect the long-term likelihood of serious abuse?"

Guiding Means Making Many Choices

Guiding a technology is a complex task. It means grappling with myriad decisions regarding which applications are beneficial and which harmful in such complex areas as medicine, the economy, and the environment. It means making such happy choices as which of several good approaches to apply in cleaning up toxic-waste dumps and reversing the greenhouse effect. It also means making more difficult choices in planning ecosystem restoration and environmental modification.

These problems will confront us with a range of choices better than we have today, yet choices that throw values into conflict. Which is a better use for a particular piece of land—the slow restoration of a wilderness ecosystem, or development as a recreational park? Either may be far better than pavement, strip mines, and dumps, but the choices will be controversial.

Likewise, in medicine, we will have a choice of developing many different ways to cure cancers, many different ways to cure heart disease, many different ways to cure AIDS. But the technologies that can be used to rebuild damaged heart muscle could be extended to rebuild muscle and connective tissue structures elsewhere in the body, without the harmful side effects of steroid drugs. The range of choices open to people will be enormous, and again will be cause for great debate.

When a new medical technology is discussed today, a frequent comment is, "This procedure raises ethical questions." This is often taken as a signal to delay its use, neglecting such ethical questions as "Is withholding this lifesaving treatment while we ponder akin to murder?" When a choice raises ethical questions or throws values into conflict, it is time to make an ethical decision or to step aside and let others choose for themselves. Deciding to avoid whatever raised the question is itself a decision—and often ethically indefensible. New technologies will face us with uncomfortable decisions, but so does life itself.

Setting up rules for nanotechnology development will be challenging: finding ways to maximize research freedom while preventing serious abuse and making this stick worldwide is a social challenge of the first rank. Beyond this are decisions regarding rules for its application, and the challenge of maximizing freedom of choice and action while preventing serious abuse, again worldwide.

To guide nanotechnology means grappling with a set of decisions that could ultimately remake much of the world—for the better if we are reasonably wise, or for the worse if we are too blundering and incautious. To avoid this responsibility (if we could) would be tempting, yet given the environmental and human stakes, it would, perhaps, be a wrong of historic proportions.

Trying to Stop Means Losing Control

The simplest imaginable approach to "guiding" nanotechnology would be to stop it. The easiest trip to plan is the trip that goes nowhere.

This would have a certain appeal, if it were possible. Because of its enormous potential for abuse, nanotechnology has the potential of doing great harm. If we believe that human beings and human institutions are too incompetent to deal with nanotechnology—that they are too likely to turn it to aggressive military use, or too likely to make it freely available to madmen—then the option of stopping the development of nanotechnology may seem attractive indeed. But the ethical question that must guide human actions is not "Would it be better to stop?", but "Would attempts to stop make things better?"

One option is to push forward, emphasizing the need for caution but also the potential for good applications. The promise of medical, economic, and environmental applications, joined with the threat posed by a new arms race, provides a powerful motive for international cooperation. With positive goals and an inclusive stance, international cooperation is a promising strategy; it could provide a basis for guiding the development and application of nanotechnology.

Another option would be to emphasize the downside, to focus debate on potential abuses in support of a campaign to halt development. In following this strategy, an activist group would want to downplay the civilian applications of nanotechnology and emphasize its military applications. Horror stories of potential abuse (including abuses that regulation could easily prevent) would help to make the technology seem strange and dangerous.

This strategy might succeed in suppressing civilian research in many countries, though probably not all. Unfortunately, it would also guarantee funding for classified military research programs in laboratories around the world, even in the most morally honest countries, because of their then-inevitable fear of the consequences if someone else developed nanotechnology first. In a hostile public atmosphere, research would be pushed into secret programs, and in secrecy the prospects for broad international cooperation would disappear. Attempts to stop nanotechnology for fear of a new, unstable arms race become self-fulfilling prophecies. Afterward, the advocates of this view could then say, "We warned you!" as the world slid toward a war they themselves had helped to prepare.

Attempting to stop technological development is a simple but dangerous idea. The greater its success, the greater the polarization it would cause between technology advocates and technology critics. A moderate success would push research out of the public universities and into corporate and military research labs. A greater success would push research out of the corporate laboratories and into heavily classified programs. A truly amazing success would end most of these, leaving the only remaining military programs in the hands of those states with thoroughly repressive governments or alien ideologies. This, presumably, is not how one would prefer nanotechnology to be developed.

The only genuine success would be a total success, and this would mean banning research not only in the United States, and Germany, and France, and the rest of Western Europe, and Japan, and the Soviet Union, and the People's Republic of China, and Taiwan, but in Korea, South Africa, Iran, Iraq, Israel, Brazil, Argentina, Vietnam, and the part of Colombia controlled by the Medellín Cartel. Later, as computers improve, as chemistry advances, as more and more proximal probe microscopes are built by high school students, total success would require banning kids from tinkering in suburban garages in Pittsburgh.

Competitive pressures are pushing technology toward thorough control of matter, and we have seen that this goal can be reached by many different paths. Preventing one area of research would not prevent the advance, nor would stopping work in one country. When the United States delays drug development through strong regulation by the FDA, drug companies simply switch research overseas, or non-U.S. companies pull ahead. Orbital-launch capability and nuclear weapons capability are other examples. Very seldom has one country given these abilities to another, yet at least eight nations are able to launch satellites to orbit independently, at least seven have detonated nuclear devices, and another two are suspected to be within easy reach of nuclear capability. India and Israel have built bombs and launched satellites, though neither is considered a great power or a leading force in world technology.

Where nanotechnology is concerned, many countries are capable of doing the required research, and more will be in the future. South Korea has both the needed educational levels and the ambition; visitors from the People's Republic of China ask about nanotechnology. A decision at the top directing the resources of a nation could get results almost anywhere. The United States is only gradually being shaken from its illusion that it rules the world of technology. This illusion is a poor basis for decisions and action.

Responsible Action

For all practical purposes, nanotechnology seems inevitable. With work, it can be made beneficial, but only if we exercise ordinary care in avoiding accidents and extraordinary care in preventing abuse.

It's hard to get people to take future technologies seriously. Present-day problems dominate discussions, and ideas about future possibilities take effort to judge. Because of this inertia, broad international regulation of nanotechnology won't be possible until nanotechnology already exists, until people begin to see its results. And then, for regulation to be most effective, researchers and governments in many countries will need to cooperate and be on speaking terms with the technology's critics.

What, then, is the socially responsible course of action, the approach most likely to avoid serious abuse of nanotechnology and most likely to deliver some of its potential benefits? It is, we believe, to point out potential dangers and abuses and how they can be avoided, but also to emphasize the civilian applications in medicine, the environment, and the economy. It is these benefits that provide grounds for advocating open civilian development programs, and for international cooperation that can provide a basis for effective international guidance.

To guide nanotechnology will not be simple. We will be confronted with a range of choices greater than we have faced before in history. It is only by grappling with those choices that we will be able to affect them for the better.

Unit I–M

Nanotechnology: Policy and Prospects

Although exploratory engineering research can show certain technological possibilities, gaining this knowledge can have a paradoxical effect on our feeling of knowledge, on our sense of how much we know about the future. It gives us more information, but it can reveal a range of possibilities so vast that we feel as if we know less than we did before.

The prospect of nanotechnology and molecular manufacturing has this paradoxical effect. It makes certain scenarios—such as a mid-twenty-first-century world of poverty, or choking on pollution caused by massive accumulations of twentieth century-style industry—seem very unlikely indeed. This is useful information in trying to understand our real situation and trying to make sensible plans for the future. And yet the range of new possibilities opened up is broader than we could have imagined before. On the negative side, one can imagine building engines of destruction capable of devastating the world as thoroughly as a nuclear war. On the positive side, one can imagine futures of stable peace with levels of health, wealth, and environmental quality beyond any historical precedent and beyond present expectations.

Within this spectrum of possibilities (and off to its sides) is a range of futures we can't even imagine. Our actions, day by day, are taking us into one of those futures. Not to some future of our present plans or dreams or nightmares, but to a real future, one that will grow from the intended and unintended consequences of our actions, one that we and our descendants will actually have to live in.

Scenarios are useful tools for thinking about the future. They don't represent predictions of what will happen, but instead they present pictures of worlds that one can imagine happening. By looking at these pictures and seeing how they fit together, we can try to get some idea of which events are more likely and which are less likely, and to get some idea of how the choices we make today may affect the shape of things to come.

Scenario 0: Ordinary Expectations

Nanotechnology will have little direct effect on the world until it is well developed, many years from now. The expectation of nanotechnology, however, is influencing how people think and act today. Yet even this expectation is still in the early stages of development and will likely have little effect on world affairs for years to come. In sketching scenarios, it seems sensible to begin with the standard worldview, at least for the next few years, and then to look at how nanotechnology and the expectation of nanotechnology might later begin interacting with large-scale developments.

As this is being written, old projections of East European, Middle Eastern, and world affairs have recently been upended, and expectations are fairly muddy. Still, one can identify the broad outlines of a conventional-wisdom view of expected events in the coming years and decades:

Technology doesn't change much in the next five years, or indeed in the next fifty. Computer power continues to grow rapidly, but with few important effects. The great challenges of technology are environmental: dealing with greenhouse gases and acid rain and the problems of toxic waste.

In parallel, more and more nations climb the ladder of technological capability to such thresholds as the ability to launch satellites, build nuclear weapons, and manufacture computer chips. With the worldwide flow of technical information and the worldwide emphasis on technological development, more and more second-rank countries follow close on the heels of the technological leaders.

Consumer electronics continues to improve, but this leads to a better-entertained population rather than a better-informed one. Exciting announcements like high-temperature superconductors and low-temperature fusion continue to appear, but after hearing cries of "Wolf!" and seeing only puppy dogs and fairy tales, most people discount news of purported breakthroughs.

Even in the thirty-to-fifty-year time frame, most newspaper stories and respected analysts assume there will be little technological change. Fifty-year projections of carbon-dioxide accumulation in the atmosphere assume that most energy will continue to come from fossil fuels. Thirty-year projections of economic crisis due to an aging population and a shrinking work force assume that economic productivity doesn't change greatly.

In terms of productivity and wealth, the United States continues to lose ground relative to the booming economies of Eastern Asia: to Japan, South Korea, Taiwan, and Singapore. In political terms, the Ordinary Expectations scenario is less clear, but expectations seem to run something like this: The breakup of the Eastern bloc and the collapse of communism as a "progressive" ideal lead to a freer and more democratic world. In Eastern Europe, and perhaps in Central Asia, independent countries emerge, each with an industrial base and a population having substantial education in science and technology.

The relative decline of the United States economically and of the erstwhile Soviet Union militarily loosen some of the ties that today bind the world's democracies to one another. The decreased threat of Soviet military power weakens alliances. As NATO loosens, and as the nations of Europe integrate their economic and political lives, gaps between the United States and Europe grow.

In this environment, protectionist pressures increase. An economic crash grows more likely. A shift from friendly relationships to peaceful hostility becomes an ominous possibility. The rise of multiple, nearly equal centers of economic and technological capability provides incentives for greater integration and cooperation, but also motives for great competition and secrecy.

In the long term, however, limited resources and the costs both of pollution and of pollution controls bring economic growth to a halt in an increasingly impoverished world. Population growth during this time has slowed, but creates great economic and environmental pressures. Resource conflicts escalate into war. The climate has changed irreversibly, the old forests are nearly gone, and extinction has swept a majority of species into nothingness.

Variations on the first five to ten years of the Ordinary Expectations scenario can provide a backdrop for scenarios covering the rise of nanotechnology in, perhaps, the next ten to twenty years:

Scenario 1: Pollyanna Triumphant

We are living in a world like that of the Ordinary Expectations scenario where, after years of anticipation, primitive but fairly capable assemblers have recently been developed. For the first time, the media, the public, and policymakers take the prospect of nanotechnology seriously.

It looks very good to them. Technical work has shown that nanotechnology, once developed, can be used in a clean, controlled way, and that it can ultimately displace polluting industries while greatly increasing wealth per capita. The anticipated health benefits are enormous, and after years of a growing death toll from AIDS—only partially stemmed by advances in molecular medicine—the public has become very sensitive to the regular reports of human infection by exotic primate viruses from Africa. Concern about the stability of Earth's climate and ecosystems has grown as forests have shrunk and weather patterns have changed.

The prospect of breaking out of this cycle is appealing. It is clear that nanotechnology is no danger when in the hands of people of goodwill, and a relatively peaceful decade has allowed many people to forget the existence of other motives.

And so, with miraculously undivided popular support drawn from a grand coalition of environmentalists seeking to replace existing industry, industrialists seeking a more productive technology, health advocates seeking better health care, low-income groups seeking greater wealth and so on and so forth, companies and governments plunge into nanotechnology with both feet and without reservation.

Development proceeds at a breakneck pace, and everyone who wants to participate in this great venture is welcome. Primitive assemblers are used to build better assemblers, which are used to build yet better assemblers, in laboratories and hobby shops around the world.

Products begin to pour forth. The economy is thrown into turmoil. Military equipment also begins to pour forth, and tensions begin to build. A military research group with more cleverness than sense builds a monster replicator, it eats everything, and we all die.

This scenario is absurd, at least in part because published warnings already exist. Since the 1960s, uncritical applause for new technologies has been limited to the now-defunct controlled presses of Eastern Europe (and similar places), and even there the resulting environmental disaster has become a matter for public debate, criticism, and correction.

In the expanding free world of today, the benefits, costs, and dangers of any great new technology will be thoroughly examined, expounded upon, and lied about from many different directions. We may or may not manage to make wise choices as a result. But one thing seems sure: Pollyanna will not triumph, because Pollyanna doesn't have the facts on her side.

Scenario 2: Chicken Little Rules the Roost

Again, we are in the world of the Ordinary Expectations scenario, and primitive assemblers have recently been developed. Again, the prospect of nanotechnology is being taken seriously for the first time—but it is somehow portrayed as being just more of the same, but worse. Environmentalists view it not as an alternative to the polluting industries of the twentieth century, but as an extension of human power, and hence of the human ability to do harm. Horror stories of technology gone mad are spun to support this view.

Arms control groups are justifiably alarmed by nanotechnology and emphasize its military applications. Groups seeking arms control via disarmament—and believing in unilateral strategies—work to prevent the development of nanotechnology everywhere they can, that is, everywhere within their political reach. To maximize their political leverage, they portray it as an almost purely military technology of immense and malign power.

Special interest groups in industry see molecular manufacturing as a threat to their business and join the lobbying efforts to prevent it from happening. Unions, neglecting the prospect of greater wealth and leisure for their members, focus instead on possible disruptions in established jobs. They, too, oppose the development of the new technology. As a result, we hear not about how nanotechnology could be used in health care, environmental cleanup, and the manufacture of improved products, but about the insidious threat of tiny, uncontrollable military monster machines that will smash our industry.

After a few years of hearing this, public opinion in the industrial democracies is firmly "against the development of nanotechnology," but this is more a slogan than an enforceable policy. Laws are nevertheless passed to suppress it, and the focus of public debate returns to the old themes of poverty and disease and the newer themes of climatic change and environmental destruction. Solutions seem as distant as ever. No right-thinking person would have anything to do with nanotechnology, so only wrong-thinking people do.

But the initial debate hadn't become serious until assemblers were developed, and research had gone still further before the laws were passed. By then, nanotechnology was just around the corner.

Developing nanotechnology is primarily a matter of tools, just as was developing nuclear weapons. Decades earlier, nuclear-weapons capability had spread from one to two countries in forty-nine months, and to another three in the next fifteen years, despite the requirement for large quantities of exotic materials in each device. By the 1980s, there was already a huge international trade in chemical compounds, and many thousands of chemists who knew how to combine them to make new molecular objects, working not only in university research labs, in corporate research labs, and in civilian and military government research labs, but—as the black market in designer drugs shows—secretly, in criminal research labs.

There is nothing large-scale or exotic about synthetic chemistry or about precise positioning of molecules. And in our scenario, primitive assemblers have already been developed and techniques for constructing them published (as is standard practice) in the open scientific literature.

And so the attempts to suppress the development of nanotechnology succeed only in suppressing the open development of nanotechnology. But governments cannot be sure that other governments are not developing it in secret, and they have now heard so much about its military potential that this is impossible to ignore. Around the world, governments quietly set up secret research programs: some in democracies, others in the remaining authoritarian states.

There are even underground efforts. Once a primitive assembler or even an AFM-based molecular manipulator is in hand, the remaining challenges are chiefly those of design. In the 1990s, personal computers had become powerful enough to use for designing molecules. In the years since then, computer power has continued its exponential explosion. Peculiar elements of the technoculture join with—pick one: radical anarchists, radical reds, radical greens, or radical racists—in a project aimed at bringing down "the corrupt world order" of governments, of companies, of religions, of human beings, or of nonwhite/nonbrown people. With responsible groups out of the technology race, they see a real chance of finding the leverage needed to change the world.

And so years pass in comparative quiet, with occasional rumors of activity or exposure of a project. Then, from an unexpected direction beyond the reach of democratic control, destructive change breaks loose upon an unprepared world. The sky falls, and Chicken Little is vindicated.

With luck, we will find that this scenario is also absurd. Public debate in the coming years will surely present a more balanced picture of the opportunities and dangers posed by the development of nanotechnology. Thoughtful people with conflicting views will become deeply involved. The impracticability of attempting to suppress technologies of this sort will likely become clear enough to give us a chance of keeping development in the open, in relatively responsible hands.

Scenario 3: International Technorivalry

A variant of the Ordinary Expectations scenario has played out for a number of years now. And after years of continuing turbulence, the net result is this: Japanese economic power has grown, with other East Asian economies beginning to close the gap. Their greater investment in long range civilian R&D, with a focus since the late 1980s on engineering molecular systems, has enabled them to take the lead on the path to nanotechnology.

European economic integration and German unification, combined with the pressure of economic and technological competition from the United States and Japan, have turned Europe inward to some extent. Although cultural ties with the United States keep U.S.-European relations on a basically warm basis, hostility between Europe and Japan—already marked in the 1980s—has grown. Europe had long enjoyed great strength in chemistry and basic science, and in the last two decades of the 20th century had led the United States in organizing efforts on molecular electronics. This has placed them in a strong position with respect to nanotechnology, behind Japan but ahead of the United States.

The United States remains an enormously productive economy, but the cumulative effects of an educational system that neglects learning and corporations that emphasize quarterly results have made themselves felt. After decades of emphasizing the short term, people now find themselves living in the long term they had neglected. The reaction to U.S. relative economic decline has not been investment and renewal, but rhetoric and hostility directed toward "foreigners," particularly the Japanese.

It is thus an isolated and somewhat defensive Japan that builds the first molecular manipulator and recognizes its long-term potential. The technology is developed in a government-funded research laboratory with cooperation from major Japanese corporations. As the result of increasing tensions, foreign researchers—those still welcome in Japan—were not invited to participate in this particular effort.

A series of committee meetings formalizes a tacit decision made earlier in choosing researchers, and the specifics of this new development are treated as proprietary. Impressive results are announced, stirring pride in Japanese research, but the specifics of the methods involved are kept quiet.

This scarcely delays the diffusion of the basic technology. After the first demonstration, even the most myopic funding agencies support projects with the same goal. A European project had already been started in a French laboratory: it soon succeeds in building an assembler based on somewhat different principles. European researchers follow the Japanese precedent by keeping the details of their techniques as a loosely held secret, in the name of European competitiveness. The United States follows suit in an effort funded by the Department of Defense.

Public life goes on much as before, dominated by the antics of entertainers and politicians, and by tales of the fate of the environment or the Social Security system in a fantasy-future of extrapolated twentieth-century technology. But more and more, in policy circles and in the media, there is serious discussion of nanotechnology and molecular manufacturing—what they mean and what to do about them.

In Japan, second-generation assemblers have begun to turn out small quantities of increasingly sophisticated molecular devices. These are prototypes of commercially useful products: sensors, molecular electronic devices, and scientific instruments; some are immediately useful even at a price of a hundred dollars per molecule. There are plans on the drawing boards for molecular assemblers that could make these devices at prices of less than one-trillionth of a dollar. There are long-term plans (viewed with hope and anticipation) for full-fledged molecular manufacturing able to make almost anything at low cost from common materials.

This is exciting. It promises to at last free Japan from its decades-old dependence on foreign trade, foreign food, foreign raw materials, and foreign politics. By making spaceflight inexpensive and routine, it promises to open the universe to a people cooped up on a crowded archipelago. Investment soars.

Europe leads America but lags behind Japan and looks on Japanese progress with hostility. Europeans, too, share dreams of a powerful technology, and begin a race for the lead. The United States trails, but its huge resources and software expertise help it pick up speed as it joins the race. Other efforts also begin, and though they advance steadily, they cannot keep pace with the great power blocs.

On all sides, the obvious military potential of molecular manufacturing fires military interest, then research and development in both publicly announced and secret programs. Strategists play nanotechnology war games in their minds, in their journals, and on their computers. They come away shaken. The more they look, the more strategies they find that would enable a technologically superior power to make a safe, preemptive move—lethal or nonlethal—against all its opponents. Defenses seem possible in principle, but not in time.

Yet it becomes obvious that molecular manufacturing can provide defenses against lesser technologies. Even the great, mythical leak-proof missile shield looks practical when the defenders have vastly superior technology and a thousandfold cost advantage building military equipment.

No great power seems particularly hostile. By then, all have formally or informally been in a peaceful alliance for many years. Yet there are still memories of war, and the bonds of alliance and military cooperation are weakened by the lack of a common enemy and the growth of economic rivalry. And so squabbles over trade in obsolescing twentieth-century technologies poison cooperation in developing and managing the fresh technologies of the twenty-first century.

There are a thousand reasons to pursue military research and development in these technologies, and nationalistic economic competition helps keep that work secret on a nationalistic basis. Military planners must concern themselves not so much with intentions as with capabilities.

And so a technology developed in an atmosphere of commercial rivalry and secrecy matures in an atmosphere of military rivalry and secrecy. Advanced nanotechnologies arrive in the world not as advances in medicine, or in environmental restoration, or as a basis for new wealth, but as military systems developed in the midst of an accelerating multilateral arms race, with the quiet goal of preemptive use. Negotiations and development run neck and neck, and then . . . .

Scenario 4: Enough Coherence

Again our world is a variant of the Ordinary Expectations scenario, but the international environment is in a healthier condition. Despite trade friction, global economic integration has continued. Europe, the United States, and Japan all have a large stake in each other's well-being, and they recognize it. International military cooperation has continued, in part as a conscious counterweight to conflicts over trade. International cooperation in research has grown, spurred in part by the Japanese desire for closer international ties. The end of the Cold War has made secret military research programs less commonplace.

It is in this environment that primitive assemblers are developed, and it doesn't make a great difference who gets there first. As is standard in basic research, groups publish their results in the open literature and compete to impress their colleagues at home and abroad with the brilliance of their achievements.

The arrival of the first assemblers spurs serious debate on nanotechnology and its consequences, and that debate is reasonably open and balanced. It covers military, medical, and environmental consequences, with a major emphasis on how clean, efficient manufacturing can raise the level of wealth and spread it worldwide.

Military analysts consider the impact of molecular manufacturing and its potential products, and concerns are grave, so they undertake classified research programs. But—as usual—secrecy slows communication among researchers: those in the classified programs fall behind their more open colleagues, whose informal information-swapping runs far ahead of the published journals.

Some forces push toward rivalry; others push toward cooperation. A healthy pattern emerges: those decision makers who take nanotechnology most seriously are precisely those who see the least reason for future international conflict among democratic nations. They no longer anticipate growing conflict over dwindling resources, inequalities of wealth, and global atmospheric pollution. They see what nanotechnology can do for these problems, without anyone taking anything from anyone else. And so, on all sides, those who take nanotechnology most seriously are those most inclined to look for cooperative solutions to the problems it poses. There are exceptions, but the tide of opinion is against them, and their ideas do not dominate policy.

The public debate on nanotechnology grows, and it ranges far and wide. Enthusiasts suggest many wondrous applications for nanotechnology. Some are soon dismissed as being impossible or just plain undesirable. Some are workable improvements on the horrid technologies of the twentieth century; these are developed and applied almost as soon as they become technically possible. The rest are harder to evaluate, but in the course of years of hard work and careful study some of these are developed and adopted, and others are rejected.

At first, some people proposed that nanotechnology be stopped, but they never proposed a credible way to do it. Realists observing the worldwide technological ferment look for other options to deal with the dangers.

The world's industrial democracies, taken together, hold the decisive lead. They have developed mechanisms for coordinating and controlling technologies with military potential by regulating technology transfer and trade. These mechanisms have been developed, exercised, and honed through decades of Cold War experience not only with nuclear and missile technologies, but with a host of high technology products and devices. These mechanisms aren't perfect, but they are useful.

As concerns about international instability mount, the industrial democracies work to improve their teamwork: they reinforce the tradition of free trade and cooperation within the club, and strengthen regulations that block the flow of critical technologies to the world's remaining dictators.

As a result of these developments, nanotechnology matures in an atmosphere dominated more by economic cooperation than by military competition. The focus of policy is solidly on civilian applications, with due attention to potential military threats. Trust is reinforced by the automatic "mutual inspection" that is a natural part of cooperative research and development.

Hard decisions remain, and the shouting and the arguments grow louder throughout the world's media. But where the problem is clear, and survival or world well-being are at stake, necessary decisions are made and there is enough international coherence to implement them.

Years pass and technologies mature. Health improves, wealth rises, and the biosphere begins to heal. Despite the turbulence and anguish of change—and despite voices saying, "It was better in the old days," at least for them, and despite real losses—many people of goodwill can look at the world, contemplate the whole, and affirm that this change is, on the whole, a change for the better.

Prospects

Today's knowledge about molecules and matter is enough to give a partial picture of what molecular machines and molecular manufacturing will make possible. Even this partial picture shows possibilities that make old views of the 21st century thoroughly obsolete.

Science and technology are advancing toward molecular manufacturing along many fronts, in chemistry, physics, biology, and computer science. Motives for continuing range from the medical to the military to the scientific. Research in these directions is already worldwide, and just beginning to focus on the objective of nanotechnology.

Already, it is easy to describe how known devices and principles can be combined to build a primitive device able to guide molecular assembly. Actually doing it will not be so easy—laboratory research never is—but it will be done, and in not too many years.

The first, slow assemblers will lead to products that include better assemblers. Machines able to put molecules together to make molecular machines will lead to a spiral of falling costs and improving quality, ultimately yielding many results that people fervently want: a cleaner environment, an escape from poverty, health care that heals. These benefits will bring disturbing changes and unsettling choices, as new abilities always do. The pace of change may well accelerate, straining the institutions we have evolved to cope with turbulent times.

Molecular-manufacturing capabilities will lend themselves to abuse, and in particular, to the construction of weapons by those seeking power. To minimize the risk of such abuse, we need to develop broad-based international cooperation and regulation. Domestically, this focus seems the best way to avoid polarization between those concerned with solving old problems and those concerned with avoiding new ones. Internationally, it seems the best way to avoid a sickening slide into a new arms race.

As shown by the four scenarios just sketched, public opinion will shape public policy, helping to determine whether these technologies are used for good or for ill.

We cannot predict the future, and we cannot predict the consequences of our actions. Nonetheless, what we do will make a difference, and we can begin by trying to avoid every major blunder we can identify. Beyond this, we can try to understand our situation, weigh our basic values, and choose our actions with whatever wisdom we can muster. The choices we make in the coming years will shape a future that stretches beyond our imagining, a future full of danger, yet full of promise. It has always been so.

Unit I–N

Nanotechnology Era on the Anvil

The human race is approaching the great historical transition to thorough, inexpensive control of the structure of matter, with all that implies for medicine, the environment, and our way of life. What happens before and during that transition will shape its direction, and with it the future.

Is this worth getting excited about? Look at some of the concerns that bring people together for action:

- Poverty
- Weapons systems
- Deforestation
- Toxic waste
- Social security
- Housing
- Global warming
- Endangered species
- Freedom
- Jobs
- Nuclear power
- Life extension
- Space development
- Acid rain
- AIDS, Alzheimer's disease, heart disease, lung disease, cancer ...

Each of these issues mobilizes great effort. Each will be utterly transformed by nanotechnology and its applications. For many of these issues, nanotechnology offers tools that can be used to achieve what people have been striving to accomplish. For many of these same issues, the abuse of nanotechnology could obliterate everything that has been achieved.

A good companion to the precept "Think globally, act locally" is "Think of the future, act in the present." If everyone were to abandon short-term problems and today's popular causes, the results would be disastrous. But there is no danger of that. The more likely danger is the opposite. The world is heading straight for a disruptive transition with everything at stake, yet 99.9 per cent of human effort and attention is going into either short-term concerns or long-term strategies based on a fantasy-future of lumbering 20th century technology.

What is to be done? For people more concerned with feeling good than with doing good, the answer is simple: Go for the warm feeling that comes from adding one more bit of support to an already-popular cause. The gratification is immediate, even if the contribution is small. For people more concerned with doing good—who can feel good only if they live up to their potential—the answer is less simple: To do the most good, find an important cause that is not already buoyed up by a cheering multitude, a project where one person's contribution almost automatically makes a big difference.

There is, today, an obvious choice for where to look. The potential benefits and drawbacks of nanotechnology generate a thousand areas for research, discussion, education, entrepreneuring, lobbying, development, regulation, and the rest—for preparation and for action. A person's contributions can range from career commitment to verbal support. Both can make a difference in where the world ends up.

Opinion Matters

What people do depends on what they believe. The path to a world prepared to handle nanotechnology begins with the recognition that nanotechnology is a real prospect.

What would be the response to a new idea as broad as nanotechnology, if it were true? Since it doesn't fall into any existing technical specialty, it wouldn't be anyone's job to provide an official, authoritative evaluation. Advanced molecular manufacturing can't be worked on in the lab today, so it wouldn't matter to scientists playing the standard careers-and-funding game. Still, some scientists and engineers would become interested, think about it, and lend support. **Science News**, covering the first major conference on the subject, would announce that "Sooner or later, the Age of Nanotechnology will arrive." This is, in fact, what has happened.

But what if the idea were false? Some curious scientist or engineer would soon point out a fatal error in the idea. Since the sweeping implications of nanotechnology make many people uncomfortable, a good counterargument would spread fast, and would soon be on the lips of everyone who would prefer to dismiss the whole thing.

No such counterargument has been heard. The most likely reason is that nanotechnology is a sound idea. Reactions have been changing from "That's ridiculous" to "That's obvious." The basic recognition of the issue is almost in place.

When nanotechnology emerges from the world of ideas to the world of physical reality, we will need to be prepared. But what does this require? To understand what needs to be done today, it is best to begin with the long term and then work back to the present.

Where We Will Need to Be

When the world is in the process of assimilating molecular manufacturing, years from now, it would be best if people were ready and if the world situation favored peaceful, cooperative applications. Balanced international progress would be better than dominance by any nation. Cooperative development would be better than technological rivalry. A focus on civilian goals would be better than a focus on military goals. A well-informed public supporting sound policies would be better than a startled public supporting half-baked schemes.

All these goals will be best served if politicians aren't forced to act like idiots—that is, if the state of public opinion permits them to make the right decisions, and perhaps even makes bad decisions politically costly. The basic objective is straightforward: a world in which as many people as possible have a basic understanding of what is happening, a picture of how it can lead to a better future, and a broad understanding of what to do (and not to do) to reach that future. The outlines of a positive scenario would then look something like this:

Environmental groups and agencies have thought through the issues raised by nanotechnology, and know what applications they want to promote and what abuses they want to prevent. Likewise, medical associations, associations of retired persons, and the Social Security Administration have thought through the issues raised by dramatically improved medical care and economic productivity, and are ready with policy recommendations. Business groups have done likewise with economic issues, and business watchdog groups are ready to expose policies that merely serve special interests. Labor groups have considered the impact of a deep, global economic restructuring on the jobs and income of their members, and have proposals for cushioning the shock without holding down productivity. Religious leaders have considered the moral dimensions of many applications, and are ready with advice. Military analysts and arms control analysts have done the painstaking work of thinking through strategic scenarios, and have developed an agreed-on core of policies for maintaining stability. International committees and agencies have made the new technologies a focus of discussion and planning, and backed by a healthy climate of opinion, they make international cooperation work.

Overall, supported by a framework of sensible public opinion and sensible politics, the complex process of adapting to change is working rather well. In field after field, group after group has put in the hard work needed to come up with policies that would advance their real interests without wrecking someone else's interests. This is possible more often than most would have expected, because molecular manufacturing makes possible so many positive-sum choices. There are still big battles, but there is also a large core of agreement.

In this time of transition, some people are actively involved in developing and guiding the technologies, but most people act as citizens, consumers, workers, friends, and family members. They shape what happens in the broader world by their votes, contributions, and purchases. They shape what happens in their families and communities by what they say, what they do, and especially by the educational investments they have made or supported. By their choices, they determine what nanotechnology means for daily life.

How We Can Get There

A world like this will require years of preparation. What can people do over the coming years to help this sort of world emerge, to improve the prospects for a peaceful and beneficial transition to new technologies? For the time being, the main task is to spread information.

People within existing organizations can nudge them toward evaluating nanotechnology and molecular manufacturing. A good start is to introduce others in the organization to the concepts, and talk through some of their implications. Follow-up activities will depend on the group, its resources, and its purposes.

For the time being, drafting of new regulations, lobbying of Congress, and the like all seem premature. Getting nanotechnology into the planning process, though, seems overdue. We invite existing organizations with concerns regarding medicine, the economy, the environment, and other issues of public policy to put nanotechnology on their agendas, and to join in debating and ultimately implementing sensible policies.

Some groups are doing relevant research work. Many could bias their choice of projects to favor goals in the direction of molecular systems engineering. For nanotechnology to be taken really seriously, some research group will have to build a reasonably capable molecular manipulator or a primitive assembler. This will require an interdiscipinary team, years of work, and a total cost unlikely to exceed one tenth that of a single flight of the U.S. Space Shuttle.

Other researchers can help by providing further theoretical studies of what advanced molecular manufacturing and nanotechnology will make possible. These studies can help groups know what to anticipate in their planning.

Some scientists and engineers will want to steer their careers into the field of nanotechnology. More students will want to study a combination of physics, chemistry, and engineering that will prepare them to contribute.

We encourage people of common sense and goodwill to become involved in developing nanotechnology. For those who have—or can gain—the necessary technical backgrounds, becoming involved with its development is an excellent way to influence how it is used. For better or for worse, technical experts in a field have a disproportionate influence over related policies.

During these years, there will be a growing need for grass-roots organizations aimed at public education and building a base for political action. Having a few thousand people ready to write five letters to Congress in some crucial year could make the difference between a world that works and a world destroyed by the long-term effects of a shortsighted bill.

What happens will depend on what people do, and what people do will depend on what they believe. The world is overwhelmingly shaped by the state of opinion: people's opinions about what will and won't happen, what will and won't work, what will and won't prove profitable or beneficial for themselves, for their families, for their businesses, for their communities, for the world. This state of opinion—as expressed in what people say to each other, and whether their actions conform to their words—shapes decisions day to day. During these years, it will matter greatly what people are saying to one another about the future, and how to make it work.

Getting Started

With help from new technologies, we can renew the world—not make it perfect, not eliminate conflict, not achieve every imaginable dream, yet clear away many afflictions, both ancient and modern. With good preparation, we can perhaps even avoid creating too many new afflictions to take their place.

Who is responsible for trying to bring this about? Those who want to fight poverty, to earn their share of the benefits to come, to join in a great adventure, to meet people who care about the future, to save species, to heal the Earth, to heal the sick, to be at the cutting edge, to build international cooperation, to learn about technology, to fight dangers, to change the world—not necessarily all together, or all at once.

Unit I–O

Space Applications of Molecular Nanotechnology

Laboratories throughout the world are rapidly gaining atomically precise control over matter. As this control extends to an ever wider variety of materials, processes and devices, opportunities for applications relevant to NASA's missions will be created. This document surveys a number of future molecular nanotechnology capabilities of aerospace interest. Computer applications, launch vehicle improvements, and active materials appear to be of particular interest. We also list a number of applications for each of NASA's enterprises. If advanced molecular nanotechnology can be developed, almost all of NASA's endeavors will be radically improved. In particular, a sufficiently advanced molecular nanotechnology can arguably bring large scale space colonization within our grasp.

Introduction

This unit describes potential aerospace applications of molecular nanotechnology, defined as the thorough three-dimensional structural control of materials, processes and devices at the atomic scale. The inspiration for molecular nanotechnology comes from Richard P. Feynman's 1959 visionary talk at Caltech in which he said, "The problems of chemistry and biology can be greatly helped if our ability to see what we are doing, and to do things on an atomic level, is ultimately developed—a development which I think cannot be avoided." Indeed, scanning probe microscopes (SPMs) have already given us this ability in limited domains. Synthetic chemistry, biotechnology, "laser tweezers" and other developments are also bringing atomic precision to our endeavors.

An expanded version of Drexler's MIT PhD thesis, examines one vision of molecular nanotechnology in considerable technical detail. It proposes the development of programmable molecular assembler/replicators. These are atomically precise machines that can make and break chemical bonds using mechanosynthesis to produce a wide variety of products under software control, including copies of themselves. Interestingly, living cells exhibit many properties of assembler/replicators. Cells make a wide variety of products, including copies of themselves, and can be programmed with DNA. Replication is one approach to building large systems, such as human rated launch vehicles, from molecular machines manipulating matter one or a few atoms at a time. Note that biological replication is responsible for systems as large as redwood trees and whales.

Another approach to nanotechnology is supramolecular self-assembly, where molecular systems are designed to attract each other in a particular orientation to form larger systems. Hollow spheres large enough to be visible in a standard light microscope have been created this way using self-assembling lipids. There are many other examples and this field is rapidly advancing. Biological systems can do most of what molecular nanotechnology strives to accomplish — atomically precise products, active materials, reproduction, etc. However, biological systems are extremely complex and molecular nanotechnology seeks simpler systems to understand, control and manufacture. Also, biological systems usually work at fairly mild temperature and pressure conditions in solution — conditions that are not found in most aerospace environments.

Today, extremely precise atomic and molecular manipulation is common in many laboratories around the world and our abilities are rapidly approaching Feynman's dream. The implications for aerospace development are profound and ubiquitous. A number of applications are mentioned here and a few are described in some detail with references. From this sample of applications it should be clear that although molecular nanotechnology is a long term, high risk project, the payoff is potentially enormous — vastly superior computers, aerospace transportation, sensors and other technologies; technologies that may enable large scale space exploration and colonization.

This unit is organized into two sections. In the first, we examine three technologies—computers, aerospace transportation, and active materials—useful to nearly all NASA missions. In the second, we investigate some potential molecular nanotechnology payoffs for each area identified in NASA's strategic plan. Some of these applications are under investigation by nanotechnology researchers at NASA Ames. Some of the applications described below have relatively near-term potential and working prototypes may be realized within three to five years. This is certainly not true in other cases. Indeed, many of the possible applications of nanotechnology that we describe here are, at the present time, rather speculative and futuristic. However, each of these ideas have been examined at least cursorily by competent scientists, and as far as we know all of them are within the bounds of known physical laws. We are not suggesting that their achievement will be easy, cheap or near-term. Some may take decades to realize; some other ideas may be scrapped in the coming years as insuperable barriers are identified. But we feel that they are worth mentioning here as illustrations of the potential future impact of nanotechnology.

Technology

Computer Technology

The applicability of manufacturing at an ever smaller scale is nowhere more self-evident than in computer technology. Indeed, Moore's law (an observation not a physical law) says that computer chip feature size decreases exponentially with time, a trend that predicts atomically precise computers by about 2010-2015. This capability is being approached from many directions. Here we will concentrate on those under development by NASA Ames and her partners.

Carbon Nanotube SPM Tips

Carbon nanotubes can be viewed as rolled up sheets of graphite from 0.7 to many nanometers in diameter. The smaller tubes are single molecules. Dai placed carbon nanotubes on an SPM tip thus extending our ability to manipulate a single molecule with sub-angstrom accuracy. Not only are the tips atomically precise, but they should have approximately the same chemistry as C_{60}, and thus be functionalizable with a wide variety of molecular fragments. Functionalizing carbon nanotube tips will allow mechanical manipulation of many molecular systems on various surfaces with sub-angstrom accuracy.

One particularly intriguing possibility along this line is to utilize a carbon nanotube SPM tip to engrave patterns on a silicon surface. It should be possible to create features a few nanometers across. These would be perhaps 100 times finer than the current state of the art in commercial semiconductor photolithography. Further, in contrast to approaches such as electron microscope lithography for which the speed of operation now appears to be an insuperable obstacle for industrial production, nanotube SPM-based lithography can be accelerated by utilizing an array with thousands of SPM tips simultaneously engraving different parts of a silicon surface. Also, nanotube SPM lithography could provide a practical means to explore various futuristic electronic device technology ideas, such as quantum cellular automata, which require exceedingly small feature sizes. Needless to say, if these ideas pan out, they could literally revolutionize computer device technology, paving the way for systems that are many times more powerful and more compact than any available today.

For the near term, it should be noted that the semiconductor industry is a major market for SPM products. These are used to examine production equipment. High performance carbon nanotube tips should be of substantial value. NASA Ames is collaborating with Dr. Dai, now at Stanford, to develop these tips.

Data Storage on Molecular Tape

It is possible to store data on long chain molecules (for example, DNA) and it may be possible to read these data with carbon nanotube tipped SPMs. Existing DNA synthesis techniques can be used to write data. If the different DNA base pairs can be distinguished with a carbon nanotube tipped SPM, then the data can be read non-destructively (current techniques allow a destructive read). However, the difference between base pairs is not great. If the base pairs cannot be distinguished, techniques for attaching modified enzymes to specific base pair sequences could be used. Certain enzymes (DNA (cytosine-5) methyltransferases) attach themselves onto a specific sequence of base pairs with a covalent bond. The enzyme then performs its operation and breaks the bond. Smith modified the enzyme such that the initial covalent bond was formed but the subsequent operation was disrupted. The result is that DNA synthesized with the target base pair sequences at the desired location can force precise placement of the enzymes. The presence of an enzyme could represent 1 and its absence 0. Enzymes are sufficiently large that distinguishing their presence should be straightforward. If the DNA/enzyme approach proves impossible, a wide variety of other polymer systems could be examined.

Data Storage on Diamond

Bauschlicher computationally studied storing data in a pattern of fluorine and hydrogen atoms on the (111) diamond surface. If write-once data could be stored this way, 10^{15} bytes/cm^2 is theoretically possible. By comparison, the new DVD write-once disks now coming on the market hold about 10^8 bytes/cm^2. Bauschlicher compared the interaction of different probe molecules with a one dimensional model of the diamond surface. This study found some molecules whose interaction energies with H and F are sufficiently different that the force differential should be detectable by an SPM. These studies were extended to include a two dimensional model of the diamond surface and two other systems besides F/H. Other surfaces, such as Si, and other probes, such as those including transition metal atoms, have also been investigated.

Among the better probes was C_5H_5N (pyridine). Quantum calculations suggest that pyridine is stable when attached to C_{60} in the orientation necessary for sensing the difference between hydrogen and fluorine. Half of C_{60} can form the end cap of a (9,0) or (5,5) carbon nanotube, and carbon nanotubes have been attached to an SPM tip. Thus, it might be possible using today's technology to build a system to read the diamond memory surface.

Avouris has shown that individual hydrogen atoms can be removed from a silicon surface. If this could be accomplished in a gas that donates fluorine to vacancies on a diamond surface, the data storage system could be built. Thummel computationally investigated methods for adding a fluorine at the radical sites where a hydrogen atom had been removed from a diamond surface.

Carbon Nanotube Electronic Components

As mentioned before, carbon nanotubes can be described as rolled up sheets of graphite. Different tubes can have different helical windings depending on how the graphite sheet is connected to itself. Theory suggests that single-walled carbon nanotubes can have metallic or semiconductor properties depending on the helical winding of the tube. Chico, Han, Menon, and others have computationally examined the properties of some of hypothetical devices that might be made by connecting tubes with different electrical properties. Such devices are only few nanometers across — 100 times smaller than current computer chip features.

Molecular Electronic Components

Several authors, including Tour have described methods to produce conjugated macromolecules of precise length and composition. This technique was used to produce molecular electronic devices in mole quantities. The resultant single molecular wires were tested experimentally and found to be conducting. The three and four terminal devices have been examined computationally and look promising. The features of these components are approximately 3 angstroms wide, about 750 times smaller than current silicon technology can produce.

Helical Logic

Helical logic is a theoretical proposal for a future computing technology using the presence or absence of individual electrons (or holes) to encode 1s and 0s. The electrons are constrained to move along helical paths, driven by a rotating electric field in which the entire circuit is immersed. The electric field remains roughly orthogonal to the major axis of the helix and confines each charge carrier to a fraction of a turn of a single helical loop, moving it like water in an Archimedean screw. Each loop could in principle hold an independent carrier, permitting high information density. One computationally universal logic operation involves two helices, one of which splits into two "descendant" helices. At the point of divergence, differences in the electrostatic potential resulting from the presence or absence of a carrier in the adjacent helix controls the direction taken by a carrier in the splitting helix. The reverse of this sequence can be used to merge two initially distinct helical paths into a single outgoing helical path without forcing a dissipative transition. Because these operations are both logically and thermodynamically reversible, energy dissipation can be reduced to extremely low levels. ... It is important to note that this proposal permits a single electron to switch another single electron, and does not require that many electrons be used to switch one electron. The energy dissipated per logic operation can likely be reduced to less than 10^{-27} joules at a temperature of 1 Kelvin and a speed of 10 gigahertz, though further analysis is required to confirm this. Irreversible operations, when required, can be easily implemented and should have a dissipation approaching the fundamental limit of $\ln 2 \times kT$.

Rod Logic

One study not conducted by Ames or partners is particularly worth mentioning since it places a loose lower bound on the computational capabilities of molecular nanotechnology. Drexler designed a number of computer components using small diamondoid rods with knobs that allow or prevent movement to accomplish computation. While this tiny mechanical Babbage Machine is probably not an optimal computational engine, its calculated performance for a desktop computer is 10^{18} MIPS—about a million times more powerful than the largest supercomputer that exists today. Note that with very fast computation energy use and heat dissipation become a severe problem. One approach to addressing this issue is reversible logic.

Aerospace Transportation

Launch Vehicles

Drexler proposed a nanotechnology based on diamond and investigated its potential properties. In particular, he examined applications for materials with a strength similar to that of diamond (69 times strength/mass of titanium). This would require a very mature nanotechnology constructing systems by placing atoms on diamond surfaces one or a few at a time in parallel. Assuming diamondoid materials, McKendree predicted the performance of several existing single-stage-to-orbit (SSTO) vehicle designs. The predicted payload to dry mass ratio for these vehicles using titanium as a structural material varied from < 0 (the vehicle won't work) to 36 per cent, i.e., the vehicle weighs substantially more than the payload. With hypothetical diamondoid materials the ratios varied from 243 per cent to 653 per cent, i.e., the payload weighs far more than the vehicle. Using a very simple cost model ($1000 per vehicle kilogram) sometimes used in the aerospace industry, he estimated the cost per kilogram launched to low-Earth-orbit for diamondoid structured vehicles should be $153-412. This would meet NASA's 2020 launch to orbit cost goals. Estimated costs for titanium structured vehicles varied from $16,000-59,000/kg. Although this cost model is probably adequate for comparison, the absolute costs are suspect.

Drexler used a more speculative methodology to estimate that a four passenger SSTO weighing three tons including fuel could be built using a mature nanotechnology. Using McKendree's cost model, such a vehicle would cost about $60,000 to purchase — the cost of today's high-end luxury automobiles.

These studies assumed a fairly advanced nanotechnology capable of building diamondoid materials. In the nearer term, it may be possible to develop excellent structural materials using carbon nanotubes. Carbon nanotubes have a Young's modulus of approximately one terapascal—comparable to diamond.

Space Elevator

Issacs and Pearson proposed a space elevator—a cable extending from the Earth's surface into space with a center of mass at geosynchronous altitude. If such a system could be built, it should be mechanically stable and vehicles could ascend and descend along the cable at almost any reasonable speed using electric power (actually generating power on the way down). The first incredibly difficult problem with building a space elevator is strength of materials. Maximum stress is at geosynchronous altitude so the cable must be thickest there and taper exponentially as it approaches Earth. Any potential material may be characterized by the taper factor—the ratio between the cable's radius at geosynchronous altitude and at the Earth's surface. For steel the taper factor is tens of thousands—clearly impossible. For diamond, the taper factor is 21.9 including a safety factor. Diamond is, however, brittle. Carbon nanotubes have a strength in tension similar to diamond, but bundles of these nanometer-scale radius tubes shouldn't propagate cracks nearly as well as the diamond tetrahedral lattice. Thus, if the considerable problems of developing a molecular nanotechnology capable of making nearly perfect carbon nanotube systems approximately 70,000 kilometers long can be overcome, the first serious problem of a transportation system capable of truly large scale transfers of mass to orbit can be solved. The next immense problem with space elevators is safety—how to avoid dropping thousands of kilometers of cable on Earth if the cable breaks. Active materials may help by monitoring and repairing small flaws in the cable and/or detecting a major failure and disassembling the cable into small elements.

Interplanetary Transportation

Drexler calculates that lightsails made of 20 nm aluminum in tension should achieve an outward acceleration of ~14 km/s per day at Earth orbit with no payload and minimal structural overhead. For comparison, the delta V from low Earth to geosynchronous orbit is 3.8 km/s. Lightsails generate thrust by reflecting sunlight. Tension is achieved by rotating the sail. The direction of thrust is normal to the sail and away from the Sun. By directing thrust along or against the velocity vector, orbits can be lowered or raised. This form of transportation requires no reaction mass and generates thrust continuously, although the instantaneous acceleration is small so sails cannot operate in an atmosphere and must be large for even moderate payloads.

Active Materials

Today, the smallest feature size in production systems is about 250 nanometers — the smallest feature size in computer chips. Since atoms are an angstrom or so across and carbon nanotubes have a diameter as small as 0.7 nanometers, atomically precise molecular machines can be smaller than current MEMS devices by two to three orders of magnitude in each dimension, or six to nine orders of magnitude smaller in volume (and mass). For example, the size of the kinesin motor, which transports material in cells, is 12 nm. Han computationally demonstrated that molecular gears fashioned from single-walled carbon nanotubes with benzyne teeth should operate well at 50-100 gigahertz. These gears are about two nanometers across. Han computationally demonstrated cooling the gears with an inert atmosphere. Srivastava simulated powering the gears using alternating electric fields generated by a single simulated laser. In this case, charges were added to opposite sides of the tube to form a dipole.

To make active materials, a material might be filled with nano-scale sensors, computers, and actuators so the material can probe its environment, compute a response, and act. Although this document is concerned with relatively simple artificial systems, living tissue may be thought of as an active material. Living tissue is filled with protein machines which gives living tissue properties (adaptability, growth, self-repair, etc.) unimaginable in conventional materials.

Swarms

Active materials can theoretically be made entirely of machines. These are sometimes called swarms since they consist of large numbers of identical simple machines that grasp and release each other and exchange power and information to achieve complex goals. Swarms change shape and exert force on their environment under software control. Although some physical prototypes have been built, at least one patent issued, and many simulations run, swarm potential capabilities are not well analyzed or understood. We briefly discuss some concepts here.

Michael proposes brick-shaped machines of various sizes that slide past each other to assume a variety of shapes. He has generated a large number of videos showing computer simulations of simple motions. Although his web site contains rather extravagant claims, this work has received a U.K. patent.

Yim built a small swarm with macroscopic (size in inches) components called polypod, built a simulator of polypod, and programmed it to move in various ways to study locomotion. There are two brick shaped components in polypod, one of which has two prismatic joints linked by a revolute joint. The second component is a cubic connector with no mechanical motion. Polypod is programmed by tables for each member of the swarm. Each member is programmed to move at various speeds in each degree of freedom for certain amounts of time. The swarm components are implicitly synchronized so there is no clock signal.

Hall proposes a swarm with 10 micron dodecahedral components each with 12 arms that can move in and out, rotate a little, and grab and release each other. This concept is called the "utility fog." Hall estimates that the utility fog would have a density of 0.2, tensile strength of 1000 psi in action and 100,000 psi in a passive mode, and have a maximum shear rate of 100 km/second/meter.

Bishop proposes a swarm consisting of 100 nanometer brick-shaped components that slide past each other to change shape.

Globus proposes a swarm with two kinds of components — edges and nodes. The terms "node" and "edge" are chosen to correspond to those in graph theory. The roughly spherical nodes are capable of attaching to five edges (for a tetrahedral geometry with one free edge per node) and rotating each edge in pitch and yaw. The rod-like edges are capable of changing length, rotating around their long axis, and attaching/detaching to/from nodes. See figure.

Component design, power distribution and control software are significant challenges for swarm development. Consider that with 10 micron components a cubic meter of swarm would contain about 10^{15} devices, each with an internal computer communicating with its neighbors to accomplish a global task.

NASA Missions

NASA's mission is divided into five enterprises: Mission to Planet Earth, Aeronautics, Human Exploration and Development of Space, Space Science, and Space Technology. We will examine some potential nanotechnology applications in each area.

Mission to Planet Earth

EOS Data System

The Earth Observing System (EOS) will use satellites and other systems to gather data on the Earth's environment. The EOS data system will need to process and archive >terabyte per day for the indefinite future. Simply storing this quantity of data is a significant challenge—each day's data would fill about 1,000 DVD disks. With projected nanocomputer processing speeds of 10^{18} MIPS, a million calculations on each byte of one day's data would take one second on the desktop.

Smart Dust

Given a mature nanotechnology, it should be possible to build sensors in balloon-borne systems approximately the size of bacteria. With replication based manufacturing, these should be quite inexpensive. If the serious communication and control problems can be solved, one can imagine spreading billions of tiny lighter-than-air vehicles into the atmosphere to measure wind currents and atmospheric composition. A similar approach might be taken in the oceans—note that the oceans are full of floating microscopic living organisms that can sense and react to their environment. Smart dust might sense the environment, note the location via a GPS-like system, and store that information until close enough to a data-collection point to transfer the data to the outside world.

Aeronautics and Space Transportation Technology

The strength of materials and computational capabilities previously discussed for space transportation should also allow much more advanced aircraft. Stronger, lighter materials can obviously make aircraft with greater lift and range. More powerful computers are invaluable in the design stage and of great utility in advanced avionics.

Active Surfaces for Aeronautic Control

MEMS technology has been used to replace traditional large control structures on aircraft with large numbers of small MEMS controlled surfaces. This control system was used to operate a model airplane in a windtunnel. Nanotechnology should allow even finer control—finer control than exhibited by birds, some of which can hover in a light breeze with very little wing motion. Nanotechnology should also enable extremely small aircraft.

Complex Shapes

A reasonably advanced nanotechnology should be able to make simple atomically precise materials under software control. If the control is at the atomic level, then the full range of shapes possible with a given material should be achievable. Aircraft construction requires complex shapes to accommodate aerodynamic requirements. With molecular nanotechnology, strong complex-shaped components might be manufactured by general purpose machines under software control.

Payload Handling

The aeronautics mission is responsible for launch vehicle development. Payload handling is an important function. Very efficient payload handling might be accomplished by a very advanced swarm. The sequence begins by placing each payload on a single large swarm located next to the shuttle orbiter. The swarm forms itself around the payloads and then moves them into the payload bay, arranging the payloads to optimize the center of gravity and other considerations. The swarm holds the payload in place during launch and may even damp out some launch vibrations. On orbit, satellites can be launched from the payload bay by having the swarm give them a gentle push. The swarm can then be left in orbit, perhaps at a space station, and used for orbital operations.

This scenario requires a very advanced swarm that can operate in an atmosphere and on orbit in a vacuum. Besides the many and obvious difficulties of developing a swarm for a single environment, this provides additional challenges. Note that a simpler swarm might be used for aircraft payload handling.

Vehicle Checkout

Aerospace vehicles often require complex checkout procedures to insure safety and reliability. This is particularly true of reusable launch vehicles. A very advanced swarm with some special purpose appendages might be placed on a vehicle. It might then spread out over the vehicle and into all crevices to examine the state of the vehicle in great detail.

Human Exploration and Development of Space

Nanotechnology-enabled Earth-to-orbit transportation has the greatest potential to revolutionize human access to space by dropping the current $10,000 per pound cost of launch, but this was discussed above. Other less dramatic technologies include:

High Strength and Reliability Materials

Space structures with a long design life (such as space station modules) need high-reliability materials that do not degrade. Active materials might help. The machines monitor structural integrity at the sub-micrometer scale. When a portion of the material becomes defective, it could be disassembled and then correctly reassembled. It should be noted that bone works somewhat along these lines. It is constantly being removed and added by specialized cells.

On Demand Spares and Tools

To effect timely repairs, space stations require a large store of spare parts and tools that are rarely used. A mature nanotechnology might create a "matter compiler," a machine that converts raw materials into a wide variety of products under software control. Contemporary examples of very limited matter compilers are numerically controlled machines and polypeptide sequencers. With a substantially more capable nanotechnology-based matter compiler, a space station crew could simply make spare parts and tools as needed. The programs could be stored on-board or on the ground. New tools invented on Earth could be transferred as software to the station for manufacture. Once used, unneeded tools and broken parts could be ionized in a solar furnace, transferred using controlled magnetic fields, and the constituent atoms stored for later manufacture into new products.

Waste Recycling

An advanced nanotechnology might be able to build filters that dynamically modify themselves to attract the contaminant molecules detected by the air and water quality sensors. Once attached to the filter, the filter could in principle move the offending molecules to a molecular laboratory for modifications to useful or at least inert products. A swarm might implement such an active filter if it was able to dynamically manufacture proteins that could bind contaminant molecules. The protein and bound contaminant might then be manipulated by the swarm for transportation.

With a sufficiently advanced nanotechnology it might even be possible to directly generate food by non-biological means. Then agriculture waste in a self-sufficient space colony could be converted directly to useful nutrition. Making this food attractive will be a major challenge.

Sleeping through RCS Firings

Sleeping crew members in the shuttle experience considerable pain and sleep disruption when the reaction control system fires and they collide with the cabin walls. If crew members were connected to the walls by a swarm, the swarm could absorb most or all of the force before the crew member struck the wall. The swarm could then gradually return the crew member to center (without the oscillations associated with bungee cords) in preparation for the next firing.

Spacecraft Docking

For resupply, spacecraft docking is a frequent necessity in space station operations. When two spacecraft are within a few meters of each other, a swarm could extend from each, meet in the middle, and form a stable connection before gradually drawing the spacecraft together.

Zero and Partial G Astronaut Training

A swarm could support space-suited astronauts in simulated partial-g environments by holding them up appropriately. The swarm moves in response to the astronaut's motion providing the appropriate simulation of partial or 0 gravities. Tools and other objects are also manipulated by the swarm to simulate non-standard gravity.

Smart Space Suits

Active nanotechnology materials (see active materials) might enable construction of a skin-tight space suit covering the entire body except the head (which is in a more conventional helmet). The material senses the astronaut's motions and changes shape to accommodate it. This should eliminate or substantially reduce the limitations current systems place on astronaut range of motion.

Small Asteroid Retrieval

In situ resource utilization is undoubtedly necessary for large scale colonization of the solar system. Asteroids are particularly promising for orbital use since many are in near Earth orbits. Moving asteroids into low Earth orbit for utilization poses a safety problem should the asteroid get out of control and enter the atmosphere. Very small asteroids can cause significant destruction. The 1908 Tunguska explosion, which was calculated to be a 60 meter diameter stony asteroid, leveled 2,200 km^2 of forest. Hills calculated that 4 meter diameter iron asteroids are near the threshold for ground damage. Both these calculations assumed high collision speeds. At a density of 7.7 g/cm^3 Babadzhanov, a 3 meter diameter asteroid should have a mass of about 110 tons. Rabinowitz estimates that there are about one billion ten meter diameter near Earth asteroids and there should be far more smaller objects.

For colonization applications one would ideally provide the same radiation protection available on Earth. Each square meter on Earth is protected by about 10 tons of atmosphere. Therefore, structures orbiting below the van Allen belts would like 10 tons/$meter^2$ surface area shielding mass. This would dominate the mass requirements of any system and require one small asteroid for each 11 $meter^2$ of colony exterior surface area. A 10,000 person cylindrical space colony such as Lewis One with a diameter of almost 500 meters and a length of nearly 2000 meters would require a minimum of about 90,000 retrieval missions to provide the shielding mass. The large number of missions required suggests that a fully automated, replicating nanotechnology may be essential to build large low Earth orbit colonies from small asteroids.

A nanotechnology swarm along with an atomically precise lightsail is a promising small asteroid retrieval system. Lightsail propulsion insures that no mass will be lost as reaction mass. The swarm can control the lightsail by shifting mass. When a target asteroid is found, the swarm spreads out over the surface to form a bag. The interface to the sail must be active to account for the rotation of the asteroid—which is unlikely to have an axis-of-rotation in the proper direction to apply thrust for the return to Earth orbit. The active interface is simply swarm elements that transfer between each other to allow the sail to stay in the proper orientation. Of course, there are many other possibilities for nanotechnology based retrieval vehicles.

Extraterrestrial Materials Utilization

Extraterrestrial materials brought into orbit could be fed into a high-temperature solar furnace and partially ionized. Magnetic fields might then be used to separate the nuclei. These are fed in appropriate quantities to a matter-compiler to build the products desired.

Medical Applications

Several authors, including Freitas have speculated that a sufficiently advanced nanotechnology could examine and repair cells at the molecular level. Should this capability become available—presumably driven by terrestrial applications—the small size and advanced capabilities of such systems could be of great utility on long duration space flights and on self-sufficient colonies.

Terraforming

Self-replicating systems permit efforts of great scope to be pursued economically. Adjusting the environment on another planet to suit the tastes of humans is one such undertaking. Heating and cooling can be achieved by (among many other methods) using space-based mirrors. Chemical modifications of the planetary surface and atmosphere can be achieved in relatively short periods by the use of self-replicating systems that absorb sunlight and raw materials, and convert them into the desired products. Much as plants changed the environment of the earth to what we see today, so self-replicating molecular manufacturing systems might more rapidly convert the environments of other planets.

Suspended Animation

As interstellar trips might last many years, the ability to conserve supplies by maintaining some crew members in a suspended state would be useful. An extremely advanced nanotechnology might use molecular manipulations of each cell to provide (a) better methods of slowing or suspending the metabolic activity of crew members and (b) better methods of restoring metabolic activity to a normal state when the destination is reached.

Space Science

Space Telescopes

Molecular manufacturing should enable the creation of very precise mirrors. Unlike lightsail applications, telescope mirrors require a very precise and somewhat complicated shape. A swarm with special purpose appendages capable of bonding to the mirror might be able to achieve and maintain the desired shape.

Virtual Sample Return

A very advanced nanotechnology would be capable of imaging and then removing the surface atoms of an extra-terrestrial sample. By removing successive surface layers the location of each atom in the sample might be recorded, destroying the sample in the process. This data could then be sent to Earth. Besides requiring a very advanced nanotechnology, there is a more fundamental—but not necessarily fatal—problem: as the outside layer of atoms is removed the next layer may rearrange itself so the sample is not necessarily perfectly recorded.

Meteorological Data

As described earlier in the EOS section, smart dust could be distributed into the atmosphere of another planet to characterize it in great detail.

Space Technology

Solar Power—Low Earth orbit spacecraft generally depend on solar cells and batteries for power. For energy collection, molecular manufacturing can be used to make solar photovoltaic cells at least as efficient as those made in the laboratory today. Efficiencies can therefore be > 30 per cent. In space applications, a reflective optical concentrator need consist of little more than a curved aluminum shell < 100 nanometers thick (photovoltaic cells operate with higher efficiency at high optical power densities). A metal fin with a thickness of 100 nanometers and a conduction path length of 100 microns can radiate thermal energy at a power density as high as 1000 W/m^2 with a temperature differential from base to tip of < 1K.

Accordingly, solar collectors can consist of arrays of photovoltaic cells several microns in thickness and diameter, each at the focus of a mirror of ~100 micron diameter, the back surface of which serves as a ~100 micron diameter radiator. If the mean thickness of this system is ~1 micron, the mass is ~10^{-3} kg/m^2 and the power per unit mass, at Earth's distance from the Sun, where the solar constant is ~1.4 kW/m^2, is > 10^5 W/kg."

By comparison, the U.S. built Photovoltaic Panel Module solar cells currently used on the Mir Space Station and planned for use on the International Space Station generate about 118 W/kg.

Power Storage

Fuel Cells—A critical component in hydrogen/oxygen fuel cells is the PEM (Proton Exchange Membrane). This membrane must: (a) permit the passage of protons while, (b) blocking everything else. Present membranes do a rather poor job. One group at Ames is designing and computationally testing PEMs to study possible energy mechanisms in early life. While these studies are not meant to design optimal membranes for fuel cell use, the basic knowledge and approach may be of value. Another proposal is to design a diamond membrane a few nanometers thick with "proton pores." The pores might be lined with fluorine, oxygen and nitrogen to create a region with a high proton affinity. In addition, a positionally controlled platinum might be held at the mouth of the pore to verify that H_2 can be catalytically split into $H+$ and $e-$, and that the barrier for migration of the $H+$ into the pore is modest in size. Nanotechnology must provide precise control over the manufacturing process of the diamondoid PEM since the pores must be made very precisely.

Hydrogen Storage

Studies of H_2 absorption and packing in carbon nanotubes and nanoropes are in progress at NASA Ames and elsewhere. Nanotubes provide large pore sizes and nanoropes have different pore sizes depending on interstitial and other locations. Dillon estimated that the single walled nanotubes in their sample contained 5 to 10 per cent by weight of H_2. The nanotubes were about 0.1 to 0.2 per cent by weight of the total sample. Computational studies at Ames suggest that to store 7-10 per cent H_2 in single walled nanotubes at room temperature the H_2s must be stored inside the tubes, not merely adsorbed on the walls. This work suggests that carbon nanotubes might be developed into an excellent H_2 storage medium within 3-5 years.

Oxygen Storage

Calculations with oxygen suggest that a diamondoid sphere ~0.1 microns in diameter should easily hold oxygen at ~1,000 atmospheres. While higher pressures are feasible, they offer declining returns. At higher pressures, the pressure-volume relationship becomes severely non-linear and the density approaches a limiting value. Other gases might also be stored if diamondoid spheres can be built, but the analysis has not been done.

Fly Wheels

High strength light-weight materials will allow greater efficiency of energy storage as angular momentum.

Nano Electromechanical Sensors

Many kinds of ultraminiature electromechanical devices have utility on a miniaturized space craft. It has been shown that manipulating carbon nanotubes changes their electrical properties. This might be exploited to build nanometer scale strain devices. This may be achievable within 3-5 years, and simulations along these lines are in progress.

Similar results have been achieved experimentally with C_{60}. The electrical properties of a C_{60} molecule were changed by applying pressure to the molecule with an SPM tip.

Miniature Spacecraft

Smaller, lighter spacecraft are cheaper to launch (current costs are about $10,000/lb) and generally cheaper to build. Diamondoid structural materials can radically reduce structural mass, miniaturized electronics can shrink the avionics and reduce power consumption, and atomically precise materials and components should shrink most other subsystems.

Thermal Protection

Thermal protection is crucial for atmospheric reentry and other tasks. The carbon nanotubes under investigation at NASA Ames and elsewhere may play a significant role. Most production processes for carbon nanotubes create a tangled mat of nanotubes that has a very low mass-to-volume ratio. Like graphite, the tubes should withstand high temperatures but the tangled mat should prevent them from ablating. This may lead to high temperature applications.

Conclusion

Many of the applications discussed here are speculative to say the least. However, they do not appear to violate the laws of physics. Something similar to these applications at these performance levels should be feasible if we can gain complete control of the three-dimensional structure of materials, processes and devices at the atomic scale.

How to gain such control is a major, unresolved issue. However, it is clear that computation will play a major role regardless of which approach—positional control with replication, self-assembly, or some other means—is ultimately successful. Computation has already played a major role in many advances in chemistry, SPM manipulation, and biochemistry. As we design and fabricate more complex atomically precise structures, modeling and computer aided design will inevitably play a critical role. Not only is computation critical to all paths to nanotechnology, but for the most part the same or similar computational chemistry software and expertise supports all roads to molecular nanotechnology. Thus, even if NASA's computational molecular nanotechnology efforts should pursue an unproductive path, the expertise and capabilities can be quickly refocused on more promising avenues as they become apparent.

As nanotechnology progresses we may expect applications to become feasible at a slowly increasing rate. However, if and when a general purpose programmable assembler/replicator can be built and operated, we may expect an explosion of applications. From this point, building new devices will become a matter of developing the software to instruct the assembler/replicators. Development of a practical swarm is another potential turning point. Once an operational swarm that can grow and divide has been built, a large number of applications become software projects. It is also important to note that the software for swarms and assembler/replicators can be developed using simulators—even before operational devices are available.

Nanotechnology advocates and detractors are often preoccupied with the question "When?" There are three interrelated answers to this question:

1. Nobody knows. There are far too many variables and unknowns. Beware of those who have excessive confidence in any date.
2. The time-to-nanotechnology will be measured in decades, not years. While a few applications will become feasible in the next few years, programmable assembler/replicators and swarms will be extremely difficult to develop.
3. The time-to-nanotechnology is very sensitive to the level of effort expended. Resources allocated to developing nanotechnology are likely to be richly rewarded, particularly in the long term.

Unit I–P

Introduction to Fullerenes

Introduction

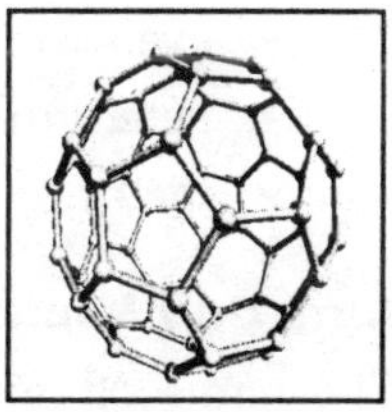

In 1985 the carbon structure family was completed by a new modification of pure carbon. With the discovery of the fullerenes by Curl, Smalley and Kroto and additionally with the possibility for the production of fullerenes in large amouts shown by Krätschmer and Huffman in 1990 the scientific competition in experimental and theoretical research was started.

Fullerenes are polyeders build up by n three times coordinated carbonatoms with 12 pentagons and (n) hexagons, were the minimum for n equal 20 is. Fullerenes fullfill the EULER's theorem, were a polyeder build up from pentagons and hexagons has to contain exact 12 pentagons, to build a closed structure. Following this rule, the dodekaeder with 20 carbonatoms is the smallest possible fullerene.

Actual the smallest fullerene is the C_{60}, because important for the stability of the structure is, that no pentagons are side by side. This is discribed by the Isolated Pentagon Rule (IPR). If two pentagons are annalled, the tension of the binding is increasing and the structure is not anymore energetic stable.

Fullerenes are also found in the natural environment, but only in the ppm-range. Some places are Shunga/Russia, New Zealand) and Sudbury/Canada.

Chemical and Physical Properties

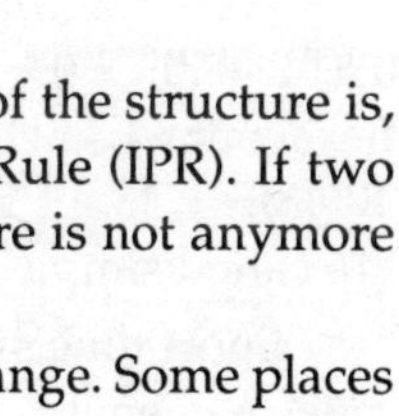

The carbon atoms in fullerenes are sp^m (m = 2...3). The free valence elctrons on the cage building a strong localized π-electron system. This π-electron system influences chemical reactions of the fullerenes. In chemical reactions fullerenes are not reacting aromatic ("superbenzene"), they show aliphatic behavior.

Good solvents for fullerenes are CS_2, o-dichlorobenzene, toluene and xylene. Fullerenes are unsoluble in water and stabil at air. Thin layers of fullerenes are coloured from yellow to yellow-green. Because of $\pi\pi^*$-electron transitions solutions of fullerenes are coloured:

- C_{60} purple/violetred

- C_{70} brick-red
- C_{76} light yellow-green
- C_{2v}-C_{78} maroon, D_3-C_{78} golden
- C_{84} brown and
- C_{86} olive-green

By heating fullerenes without air up to 1500℃ they transform to graphite.

Species of Fullerenes

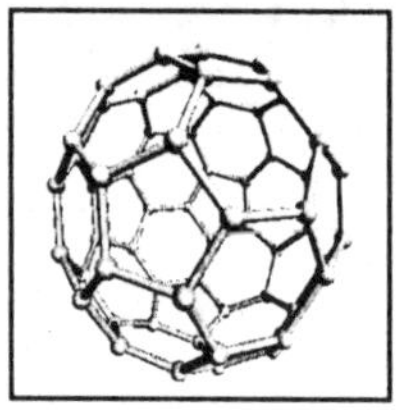

Following species of fullerenes are known:

- fullerenes
- fullerites
- fullerides
- endohedral fullerenes
- exohedral fullerenes
- heterofullerenes
- metcars

Term and Nomenclature

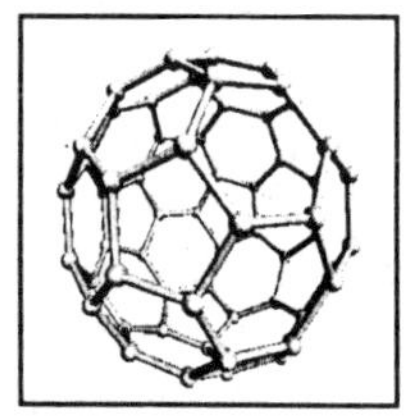

The name of this new class of structures is in memory to the American inventor, architect and philosopher Richard Buckminster Fuller (1895-1993). As architect he designed constructions, build-up from pentagons ans hexagons, for instance the American pavilion on the Expo '67 in Montréal.

By using the IUPAC-nomenclature the name of C_{60} is the following:

Hentriacontacyclo [29.29.0.0.^{2,14}.0^{3,12}. 0^{4,59}.0^{5,10}.0^{6,58}. 0^{7,55}.0^{8,53}.0^{9,21}. 0^{11,20}.0^{13,18}.0^{15,30}. 0^{16,28}.0^{17,25}.0^{19,24}.0^{22,52}. 0$^{23.50}$.0^{26,49}.0^{27,47}. 0^{29,45}.0^{32,44}.0^{33,60}. 0^{34,57}.0^{35,43}.0^{36,56}. 0^{37,41}.0^{38,54}.0^{39,51}.0^{40,48}.0^{42,46}] hexaconta-1,3,5(10),6,8,11,13(18), 14,16,19,21,23,25,27,29(45),30,32(44),33,35(43), 36, 38(54), 39(51), 40(48), 41, 46, 49, 52, 55, 57, 59-triacontaene. Because of the difficuties in using this name the structures are called simple "fullerenes".

Short time after the invention of the fullerenes the researchers realized, that they can not only make a chemitry on the outer site of the fullerenes, no, also on the inner site it should be possible. For coding the inclusion structures the symbol "@" was introduced. For exampie: atom "X" encaged in a fullerene C_{60} - X@C_{60}. This "endohedral" structur differs very strong in their behavior from the "exohedral" structur XC_{60}, where the atom X is bonded at the outsite of the cage.

Unit I–Q

Geometry of Fullerenes

Among all elements, C is the basis of entire life. The whole branch of chemistry—the organic chemistry—is devoted to the study of C-C bonds and different molecules originating from them. Carbon is only 4-valent element able to produce long homoatomic stable chains or different 4-regular nets. The other 4-valent candidate for this could be only Si, with its reach chemistry beginning to develop. After diamond and graphite—the hexagonal plane hollow shell, in 1985 was first synthesized by H.W. Kroto, R.F. Curl and R.E. Smalley the spherical closed pentagonal/hexagonal homoatomic shell: the fullerene C_{60}. Except from this, it possesses some another remarkable properties: the rotational symmetry of order 5, from the geometrical reasons (according to Barlow "crystallographic restriction theorem") forbidden in crystallographic space or plane symmetry groups, and highest possible icosahedral point-group symmetry. After C_{60}, different fullerenes (e.g. C_{70}. C_{76}. C_{78}. C_{82}. C_{84} etc.) are synthesized, opening also a new field for research of different potentially possible fullerene structures from the geometry, graph theory or topology point of view.

General Fullerenes, Graphs, Symmetry and Isomers

From the tetravalence of C result four possible vertex situations, that could be denoted as 31, 22, 211 and 1111 (Fig. I–Q.1a). The situation 31 could be obtained by adding two C atoms between any two others connected by a single bond, and situation 22 by adding a C atom between any two others connected by a double bond (Fig. I–Q.1b). Therefore, we could restrict our consideration to the remaining two non-trivial cases: 211 and 1111. Working in opposite sense, we could always delete 31 or 22 vertices, and obtain a reduced 4-regular graph, where in each vertex occurs at most one double bond (digon), that could be denoted by colored (bold) edge (Fig. I–Q.1a). First, we could consider all 4-regular graphs on a sphere, from which non-trivial in the sense of derivation are only reduced ones. In the knot theory, 4-regular graphs on a sphere with all vertices of the type 1111 are known as "basic polyhedra", and that with at least one vertex with a digon as "generating knots or links". From the chemical reasons, the vertices of the type 1111 are only theoretically acceptable. If all the vertices of such 4-regular graph are of the type 211, such graph we will be called a general fullerene. Every general fullerene could be derived from a basic polyhedron by "vertex bifurcation", this means, by replacing its vertices by digons, where for their position we have always two possibilities (Fig. I–Q.1c). To every general fullerene corresponds (up to isomorphism) an edge-colored 3-regular graph (with bold edges denoting digons). This way, we have two complementary ways for the derivation of general fullerenes: vertex bifurcation method applied to basic polyhedra, or edge-coloring method applied to 3-regular graphs, where in each vertex there is exactly one colored edge. For every general fullerene we could define its geometrical structure (i.e., the positions of C atoms) described by a non-colored 3-regular graph, and its chemical structure (i.e. positions of C atoms and their double bonds) described by the corresponding edge-colored 3-regular graph. In the same sense, for every general fullerene we could distinguish two possible symmetry groups: a symmetry group G corresponding to the geometrical structure and its subgroup G' corresponding to the chemical structure. In the same sense, we will distinguish geometrical and chemical isomers. For example, for C_{60}, $G=G'=I_h=S_5$ of order 120, but for C_{80} with the same G, G' is always a proper subgroup of G, and its chemical symmetry is lower than the geometrical. Hence, after C_{60}, the first fullerene with $G=G'=I_h=S_5$ will be C_{180}, then C_{240}, etc.

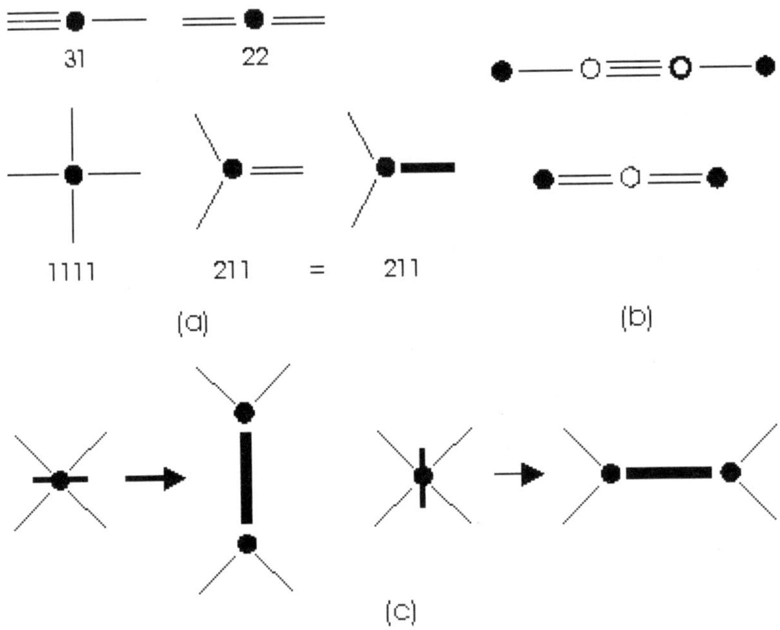

Fig. I–Q.1

Working with general fullerenes without any restriction for the number of edges of their faces, the first basic polyhedron from which we could derive them (after the trivial 1*) will be the regular octahedron or 6*, from which we obtain 7 general fullerenes. From the basic polyhedron 8* with v=8 we derive 30, and from the basic polyhedron 9* we obtain 4 general fullerenes. All the basic polyhedra with v<13 and their Schlegel diagrams are given by Fig. I–Q.2.

5/6 Fullerenes

Among general fullerenes we could distinguish the class consisting of 5/6 fullerenes with pentagonal or hexagonal faces. If n_5 is the number of pentagons, and n_6 the number of hexagons, from the relationship $3v=2e$ and Euler theorem directly follows that n_5=12, so the first 5/6 fullerene will be

C_{20} with $n_6=0$ – the regular dodecahedron, giving possibility for two non-isomorphic edge-colorings, resulting in two chemically different isomers of the same geometrical dodecahedral form (Fig. I-Q.3). The first basic polyhedra generating 5/6 fullerenes will be that with $v=10$ vertices. For $v=10$, there are three basic polyhedrons, but only 10^* and 10^{**} could generate 5/6 fullerenes, each only one of them (Fig. I–Q.4a). On the other hand, they generate, respectively, 78 and 288 general fullerenes. This way, we have two mutually dual methods for the derivation of fullerenes: (a) édge-coloring of a 3-regular graph, with one colored edge in each vertex; (b) introduction of a digon in every vertex of 4-regular graph, giving possibility for a double check of the results obtained. Their duality is illustrated by the example of two C_{20} chemical isomers derived, both of the same geometrical dodecahedral form with $G= [3,5]=I_h =S_5$ of the order 120, but the first with $G'= D_{5d}=[2^+,10]= D_5xC_2$ of the order 20, and the other with $G'=[2,2]^+=D_2$ of the order 4 (Fig. I–Q.3, I–Q.4a). In this case, the symmetry of chemical isomers derived by the vertex bifurcation is preserved from their generating basic polyhedra (Fig. I–Q.4a).

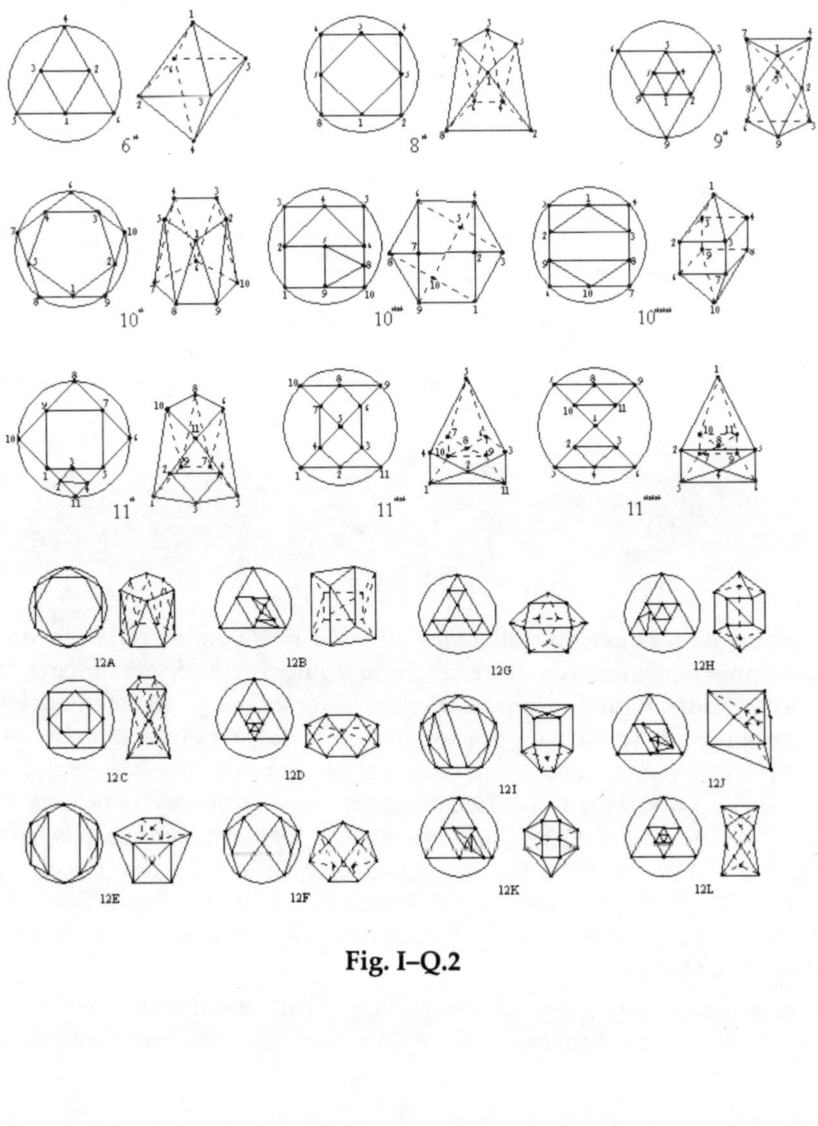

Fig. I–Q.2

Fig. I–Q.3

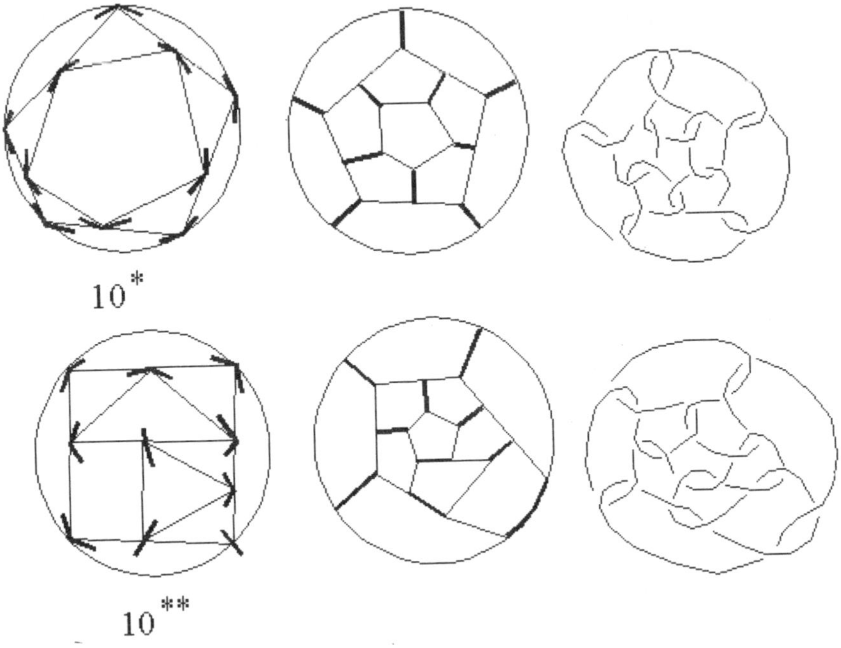

Fig. I–Q.4

For the enumeration of general fullerenes we used Polya enumeration theorem, applied to basic polyhedra knowing their automorphism groups, but with the restriction to 5/6 fullerenes its application is not possible. With the same restriction, the other derivation method: edge-coloring of 3-regular graphs is also not suitable for the application of Polya enumeration theorem, because of the condition that in every vertex only one edge must be colored. The basic polyhedra with n<13 vertices are derived by T.P. Kirkman, and used in the works by J. Conway (only for n<12), A. Caudron and S.V. Jablan (for n<13) (Fig. I–Q.2). The 3-connected 4-regular planar graphs (corresponding to basic polyhedra) are enumerated by H.J.Broersma, A.J.W.Duijvestijn and F. Göbel (n<16) and by B.M.Dillencourt (n<13), but given only as numerical results without any data about individual graphs. The 3-regular graphs with n<13 vertices and their edge-colorings producing 4-regular graphs are discussed by A.Yu.Vesnin.

Proceeding in the same way, it is possible to prove that 5/6 fullerenes with 22 atoms not exist at all, and that they are seven 5/6 fullerenes C_{24} of the same geometrical form with $G=D_{6d}=[2^+,12]=D_{12}$

(Fig. I–Q.5). To distinguish different chemical isomers, sometimes even knowing their chemical symmetry group G' will be not sufficient. For their exact recognition we could use some results from the knot theory: the polynomial invariant of knot and link projections. Every 4-regular graph could be transformed into the projection of an alternating knot or link (and *vice versa*), and the correspondence between such alternating knot or link diagrams and 4-regular graphs is 1-1 (up to enantiomorphism) (Fig. I–Q.4b).

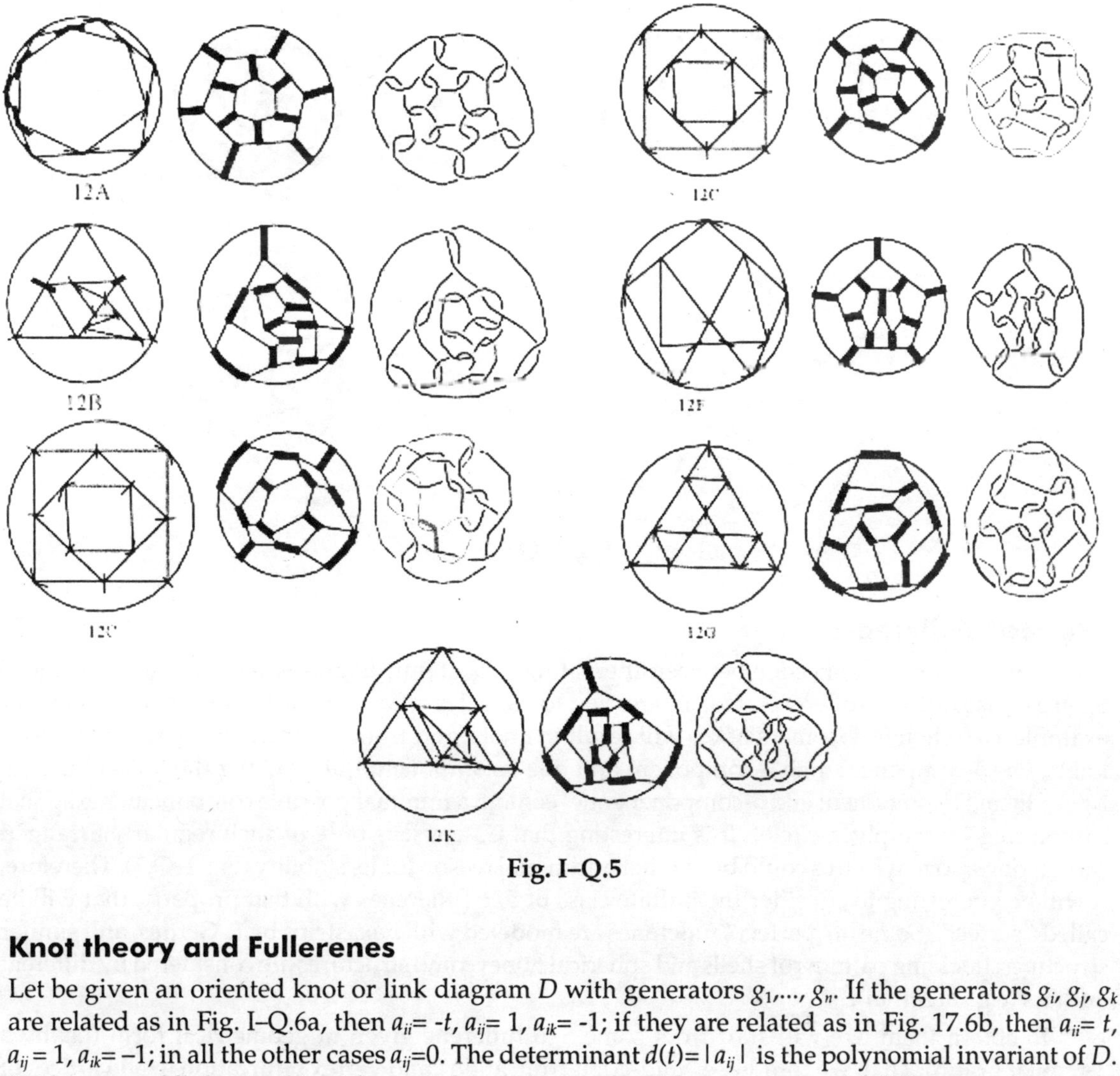

Fig. I–Q.5

Knot theory and Fullerenes

Let be given an oriented knot or link diagram D with generators $g_1,\ldots, g_n$. If the generators g_i, g_j, g_k are related as in Fig. I–Q.6a, then $a_{ii}= -t$, $a_{ij}= 1$, $a_{ik}= -1$; if they are related as in Fig. 17.6b, then $a_{ii}= t$, $a_{ij} = 1$, $a_{ik}= -1$; in all the other cases a_{ij}=0. The determinant $d(t)= | a_{ij} |$ is the polynomial invariant of D. In the case of links, in this polynomial invariant we use for generators of different components different variables (up to a permutation of variables). For example, let us show that two isomers of C_{20} are different (Fig. I–Q.3, I–Q.4, I–Q.6c). After converting their chemical Schlegel diagrams into the alternating knot diagrams, denoting their generators, and calculating the corresponding determinants we obtain, respectively, $D_1=t^{20} -10t^{18}+45t^{16}-120t^{14} +200t^{12}-197 t^{10} +105t^{8}-40t^{6}+25t^{4} -10t^{2}$ and $D_2=t^{20} -10t^{18}+45t^{16}-120t^{14} +208t^{12}-250t^{10} +217t^{8}-130t^{6}+49t^{4} -10t^{2}$, proving their difference. On the other hand, for three-component alternating links corresponding to the diagrams (Fig. 17.6d), using the same multivariable invariant for link projections, we conclude that they both represent the isomorphic Schlegel diagrams of the same fullerene C_{24}.

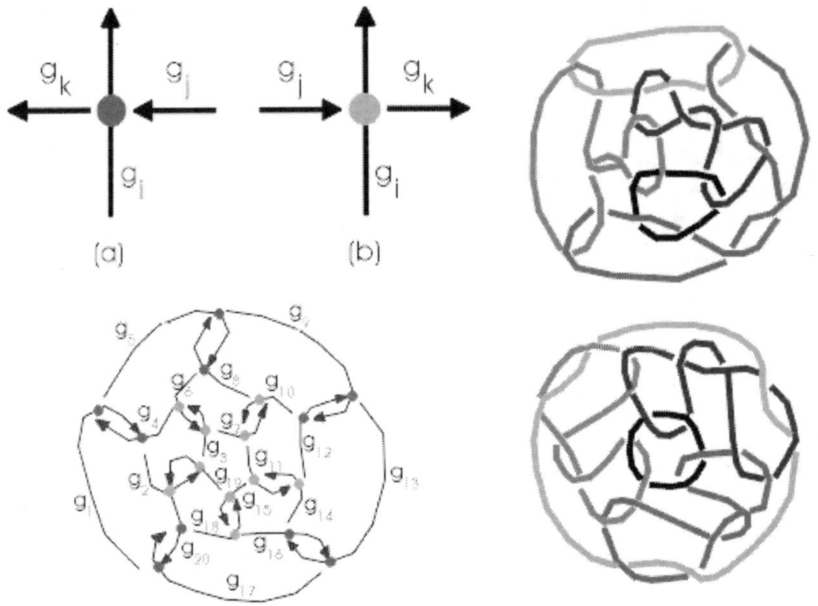

Fig. I–Q.6

"Perfect" Fullerenes

Using the mentioned connection between alternating knot or link diagrams and 4-regular (chemical) Schlegel diagrams of fullerenes, it is interesting to consider all of them after such conversion. For example, two chemical isomers of C_{20} will result in knots, and from 7 isomers of C_{24} we obtain four knots, one 3-component, one 4-component and one 5-component link. Among the links obtained, two of them (3-component and 5-component one) contain a minimal possible component: hexagonal carbon ring (or simply, a circle). It is interesting that C_{60} consists only of such regularly arranged carbon rings, so maybe this could be another additional reason for its stability (Fig. I–Q.7). Therefore, it will be interesting to consider the infinite class of 5/6 fullerenes with that property, that will be called "perfect". Some of "perfect" fullerenes are modeled with hexastrips by P. Gerdes, and similar structures: buckling patterns of shells and spherical honeycomb structures are considered by different authors (e.g. T. Tarnai).

To obtain them, we will start from some 5/6 fullerene given in geometrical form (i.e. by a 3-regular graph). Than we could use "mid-edge-truncation" and vertex bifurcation in all vertices of the triangular faces obtained that way, transforming them into hexagons with alternating digonal edges. Let is given some fullerene (e.g. C_{20}) in its geometrical form (i.e. as 3-regular graph). By connecting the midpoints of all adjacent edges we obtain from it the 3/5 fullerene covered by connected triangular net and pentagonal faces preserved from C_{20}. After that, in all the vertices of the truncated polyhedron we introduce digons, to transform all triangles into hexagonal faces. This way, from C_{20} we derived C_{60} (in its chemical form) (Fig. I–Q.7). The mid-edge-truncation we could apply to any 5/6 (geometrical) fullerene, to obtain new "perfect" (chemical) fullerene, formed by

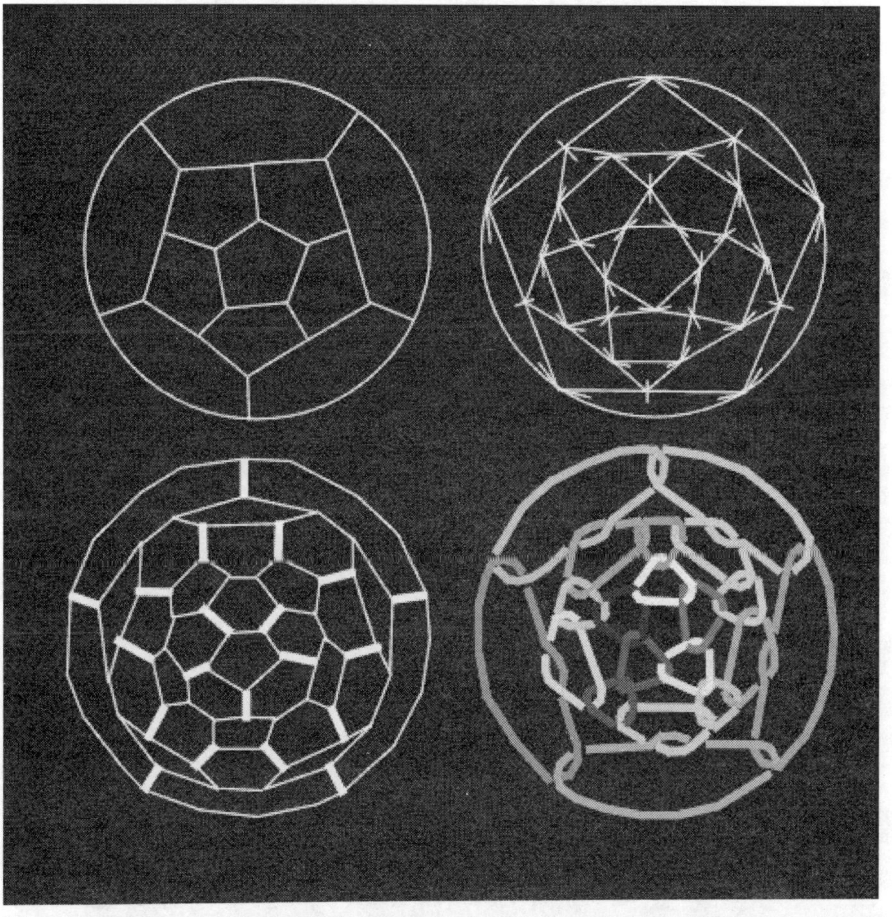

Fig. I–Q.7

carbon rings. This way, from a 5/6 fullerene with v vertices we always may derive new "perfect" 5/6 fullerene with $3v$ vertices (Fig. I–Q.8). Moreover, the symmetry of new fullerene is preserved from its generating fullerene. According to the theorem by Grünbaum & Motzkin, for every non-negative n_6 unequal to 1, there exists 3-valent convex polyhedron having n_6 hexagonal faces. Hence, from the infinite class of 3-regular 5/6 polyhedra with $v=20+n_6$ vertices, we obtain the infinite class of "perfect" fullerenes with $v=60+3n_6$ vertices. The "perfect" fullerenes satisfy two important chemical conditions: (a) the isolated pentagon rule (IPR); (b) hollow pentagon rule (HPR). The IPR rule means that there are no adjacent pentagons, and HPR means that all the pentagons are "holes", i.e. that every pentagon could have only external double bonds. The first 5/6 fullerene satisfying IPR is C_{60}, and it also satisfies HPR. The IPR is well known as the stability criterion: all fullerenes of lower order (less than 60) are unstable, because they don't satisfy IPR. On the other hand, C_{70} satisfies IPR, but cannot satisfy HPR (Fig. I–Q.9).

The same situation is with C_{80}, possessing the same icosahedral geometrical symmetry as C_{60}, but not able to preserve it after edge-coloring, because HPR cannot be satisfied (Fig. I–Q.9). This is

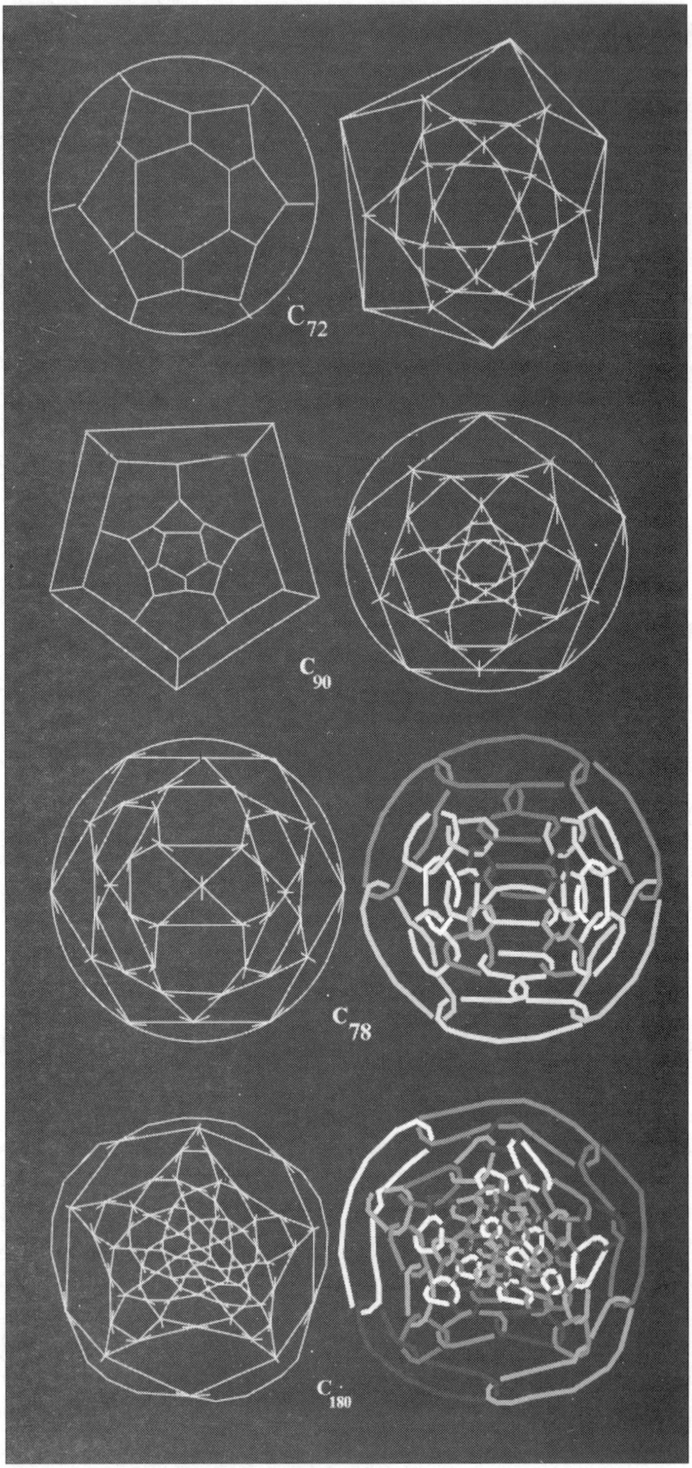

Fig. I–Q.8

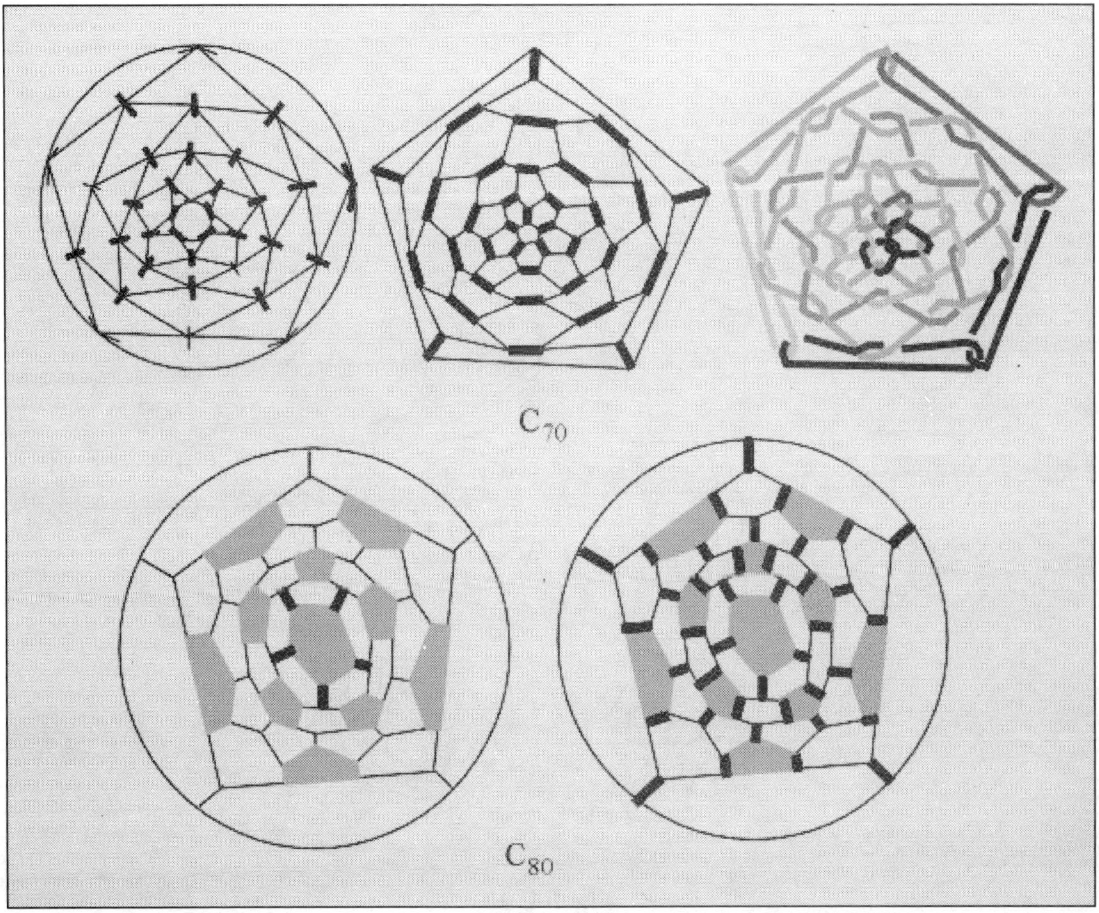

Fig. I–Q.9

the reason that only "perfect" fullerenes, with $G=G'=[3,5]=I_h=S_5$, satisfying both IPR and HPR will be C_{60}, C_{180}, C_{240}, etc. We need also to notice that for n_6= 0,2,3 we have always one 3-regular 5/6 polyhedron (i.e. geometrical form of C_{20}, C_{24}, C_{26}), but for some larger values (e.g. n_6= 4,5,7,9) there are several geometrical isomers of the generating fullerene, and consequently, the same number of "perfect" fullerenes derived from them (Fig. I–Q.10). Hence, considering the fullerene isomers, we could distinguish "geometrical isomers", this means, different geometrical forms of some fullerene treated as 3-regular 5/6 polyhedron, and "chemical isomers"—different arrangements of double bonds, obtained from the same 3-regular graph by its edge-coloring.

Nanotubes, Conical and Biconical Fullerenes and their Symmetry

For denoting different categories of symmetry groups, we will use Bohm symbols. In a symbol $G_{nst...}$ the first subscript n represents the maximal dimension of space in which the transformations of the symmetry group act, while the following subscripts st.. represent the maximal dimensions of subspaces remaining invariant under the action of transformations of the symmetry group, that are properly included in each other. With regard to their symmetry, general fullerenes belong to the

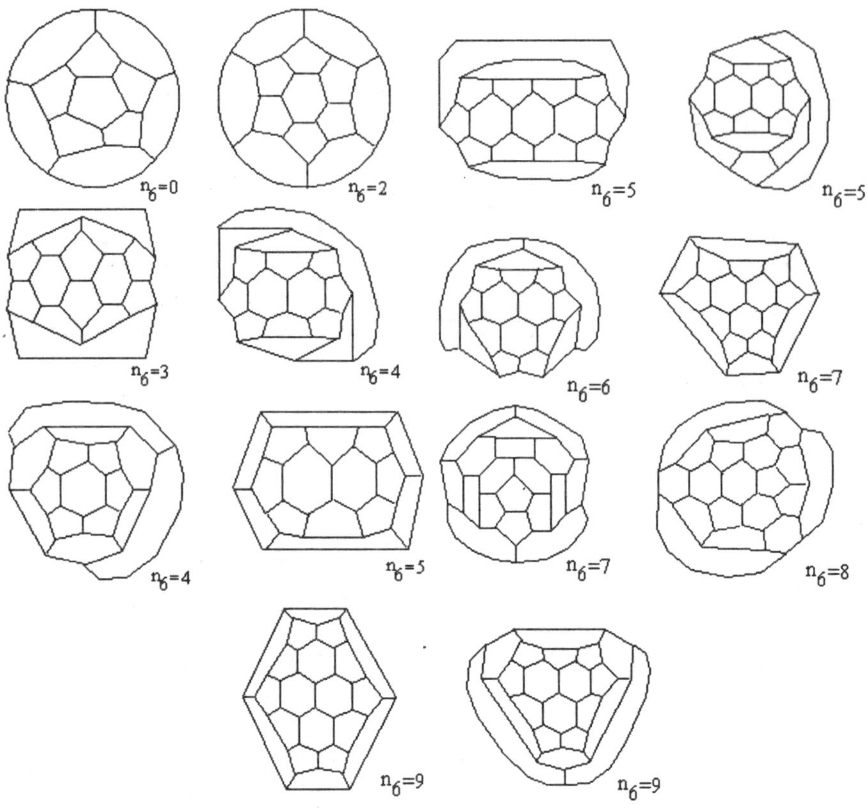

Fig. I–Q.10

category of point groups G_{30}. The category G_{30} consists of seven polyhedral symmetry groups without invariant planes or lines: [3,3] or T_d, $[3,3]^+$ or T, [3,4] or O_h, $[3,4]^+$ or O, $[3^+,4]$ or T_h, [3,5] or I_h, $[3,5]^+$ or I, and from seven infinite classes of point symmetry groups with the invariant plane (and the line perpendicular to it in the invariant point): $[q]$ or C_{qv}, $[q]^+$ or C_q, $[2^+,2q^+]$ or S_{2q}, $[2,q^+]$ or C_{qh}, $[2,q]^+$ or D_q, $[2^+,2q]$ or D_{qd}, $[2,q]$ or D_{qh}, belonging to the subcategory G_{320}. For the groups of the subcategory G_{320}, in the case of rotations of order $q>2$, the invariant line (i.e. the rotation axis) may contain 0,1 or 2 vertices of a general fullerene. According to this, among all general fullerenes with a geometrical symmetry group G belonging to G_{320}, from the topological point of view we could distinguish, respectively, cylindrical fullerenes (nanotubes), conical and biconical ones.

We could simply conclude that for polyhedral 5/6 fullerenes G could be only [3,3] (T_d), $[3,3]^+$ (T), [3,5] (I_h), $[3,5]^+$ (I), because of their topological structure (n_5=12), incompatible with the octahedral symmetry group [3,4] (O_h) or its polyhedral subgroups. In the case of nanotubes (or cylindrical fullerenes) we have infinite classes of 5/6 fullerenes with the geometrical symmetry group $[2,q]$ (D_{qh}) and $[2^+,2q]$ (D_{qd}), and the same chemical symmetry. The first infinite first class of cylindrical nanotubes with $G=G'=D_{5h}$ we obtain from a cylindrical 3/4/5 4-regular graph with two pentagonal bases, 10 triangular and $5(2k+1)$ quadrilateral faces (k=0,1,2,...) and with the same symmetry group

(Fig. I–Q.11). By the vertex bifurcation preserving its symmetry, we obtain the infinite class of nanotubes C_{30}, C_{50}, C_{70},..., with C_{70} as the first of them satisfying IPR. Certainly, the geometrical structure of C_{70} admits different edge colorings (i.e. chemical isomers). Starting from any two of them (Fig. I–Q.12) by "collapsing" (the inverse of "vertex bifurcation", i.e. by deleting digons) we could obtain different generating 4-regular graphs. This example of two different C_{70} isomers, with the same geometrical structure, and with the same G and G', shows that for the exact recognition of fullerene isomers we need to know more than their geometrical and chemical symmetry. In the same way, from 4-regular graphs with two hexagonal bases, 12 triangular and $6(2k+1)$ quadrilateral faces (k=0,1,2,..) we obtain the infinite class of fullerenes C_{36}, C_{60} C_{84},...with the symmetry group $G=G'=D_{6h}$ (Fig. I–Q.13).

The next symmetry groups $[2^+,2q]$ (D_{qd}) with q=5,6 we obtain in the same way, from 4-regular graphs with q-gonal bases, $2q$ triangular and $2kq$ quadrilateral faces (k=1,2,...for q=5; k=0,1,2,...for q=6) (Fig. I–Q.14). As the limiting case, for q=5 and k=0, we obtain C_{20} with the icosahedral symmetry group G, but with $G'=D_{5d}$, that could be used as the "brick" for the complete class of nanotubes C_{40}, C_{60},C_{80},...with $G= D_{5d}$, where all of them could be obtained from C_{20} by "gluing" the pentagonal bases (Fig. I–Q.15). In the same way, fullerene C_{24} obtained for q=6 and k=0 could be used as the building block for the nanotubes C_{48}, C_{72},C_{96},...The geometrical structure of nanotube class with $G=D_{qd}$ (q=5,6) permits the edge coloring preserving the symmetry, so there always exist their isomers with $G=G'$.

If the 3-rotation axis contains the opposite vertices of a fullerene, we have biconical fullerenes (e.g. C_{26}, C_{56}) with $G=D_{3h}$, $G=D_{3d}$, respectively (Fig. I–Q.16). Certainly, after the edge coloring, their symmetry must be disturbed, and for them G' is always a proper subgroup of G. For example, for C_{26} (Fig. I–Q.16), $G=D_{3h}$, $G'=C_{2v}$.

Proceeding in the same way, it is possible to find or construct fullerene representatives of other symmetry groups from the category G_{320}: biconical C_{32} with $G=D_3$, biconical C_{38} or conical C_{34} with $G=C_{3v}$, conical C_{46} with $G=C_3$, or the infinite class of cylindrical fullerenes C_{42}, C_{48}, C_{54}...with $G=D_3$ (Fig. I–Q.17). In general, after edge coloring of their 3-regular graphs, symmetry could not be preserved in all conical or biconical fullerenes mentioned, so their geometrical symmetry is always higher than the chemical.

Fullerenes on other surfaces

Different regular homoatomic C plane nets are discussed by T.Balaban. They could be derived in the same way as the general fullerenes: by introducing digons in the vertices of 4-regular graphs or by an edge-coloring of a 3-regular graph, resulting in 4-regular one. For example, we could start from the regular tessellation {4,4}, Archimedean tiling (3,6,3,6) or 2-uniform tiling ($3,4^2,6;3,6,3,6$) that all are 4-regular, and introduce digons in their vertices, or from regular tiling {6,3} that is 3-regular and color its edges (Fig. I–Q.18). From 3-regular tilings we could also derive "perfect" plane nets in the same way as before.

From such finite {6,3} we could simply derive the corresponding "perfect" hexagonal fullerene on torus. The proposed approach could be extended also to the double, triple, etc. torus with g=2,3,.... Similar transformations of C nets from one surface to the other (e.g. from plane to cylinder, and then to torus) maybe could explain the formation of certain fullerenes and their growing process.

Accepting that the faces could be also heptagons or octagons, from the relationship $2e=3v$ and Euler formula, follows that $n_5-n_7-2n_8= 12(1-g)$. For a sphere without octagons, $n_5-n_7=12$, and for a

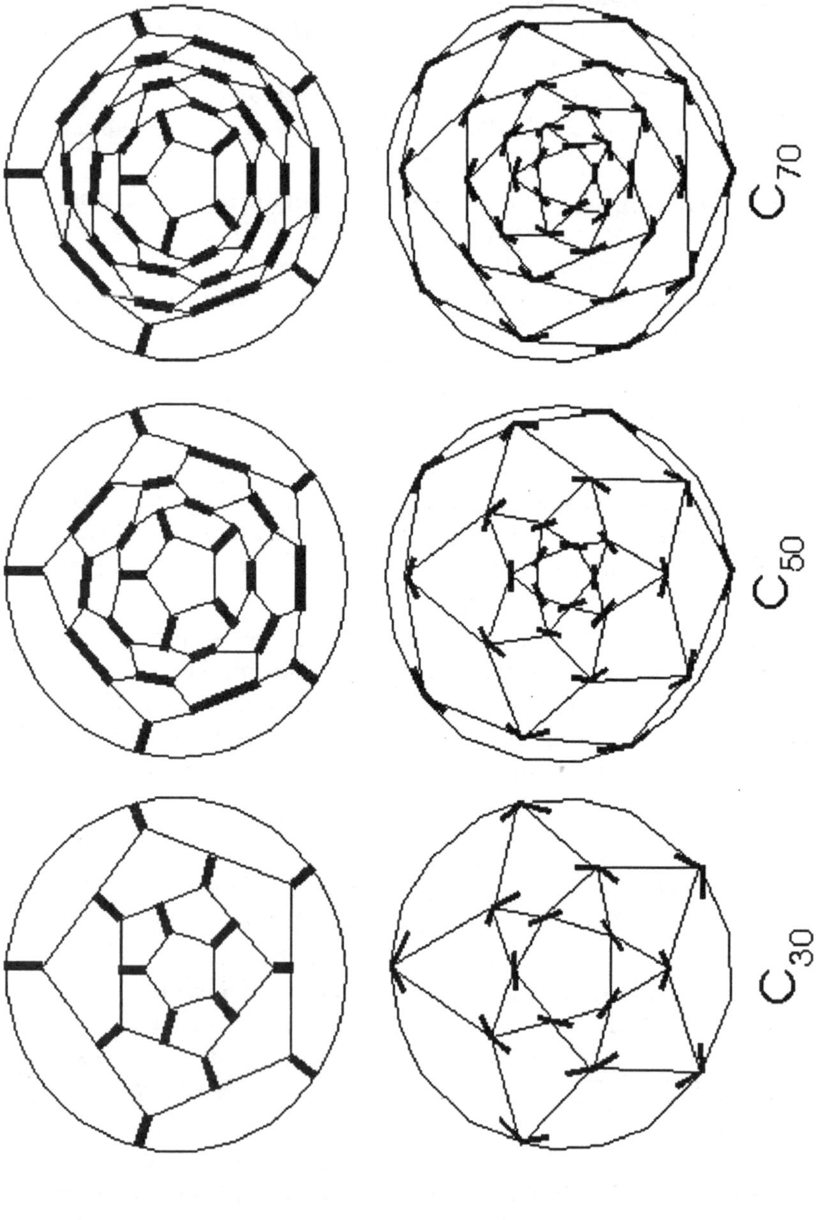

Fig. I–Q.11

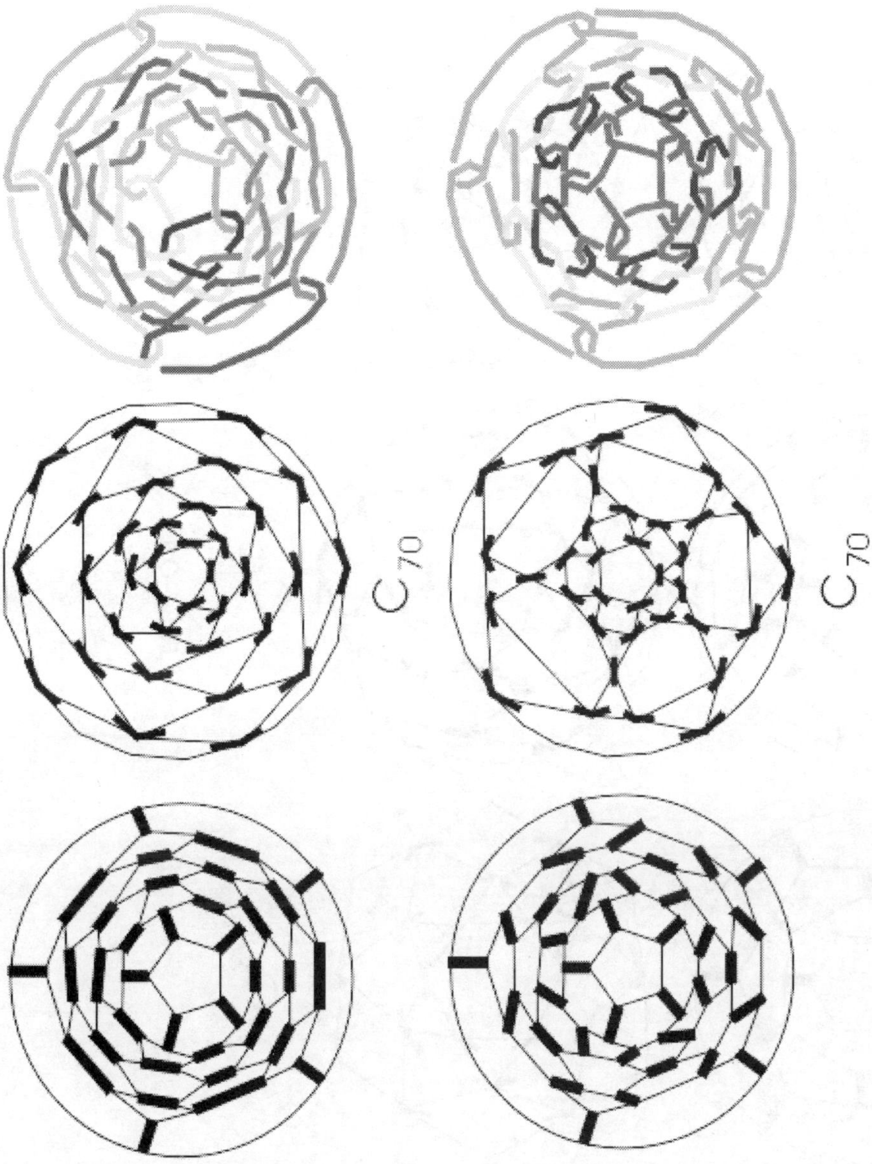

Fig. I–Q.12

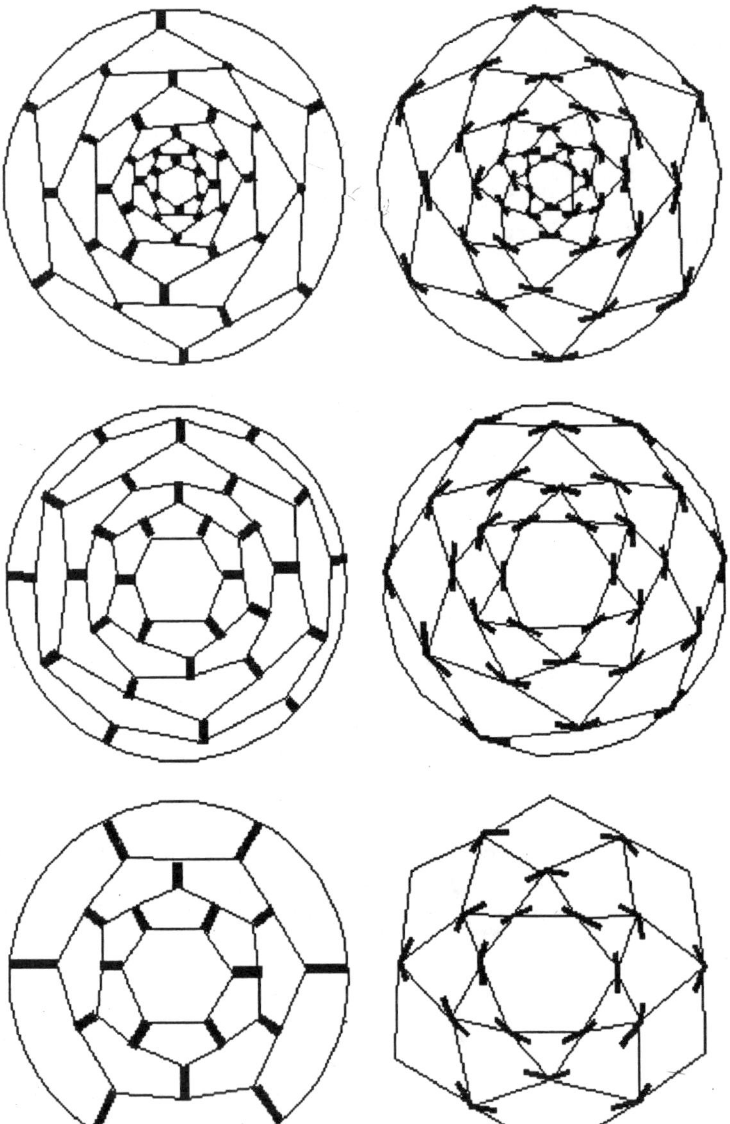

C_{84} C_{60} C_{36}

Fig. I–Q.13

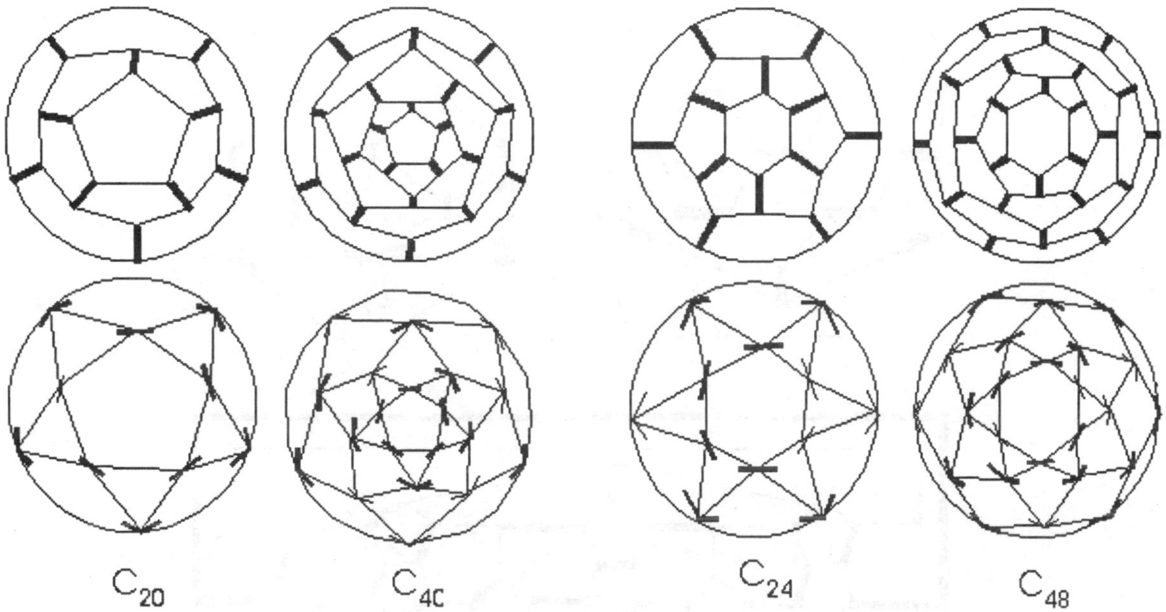

Fig. I–Q.14

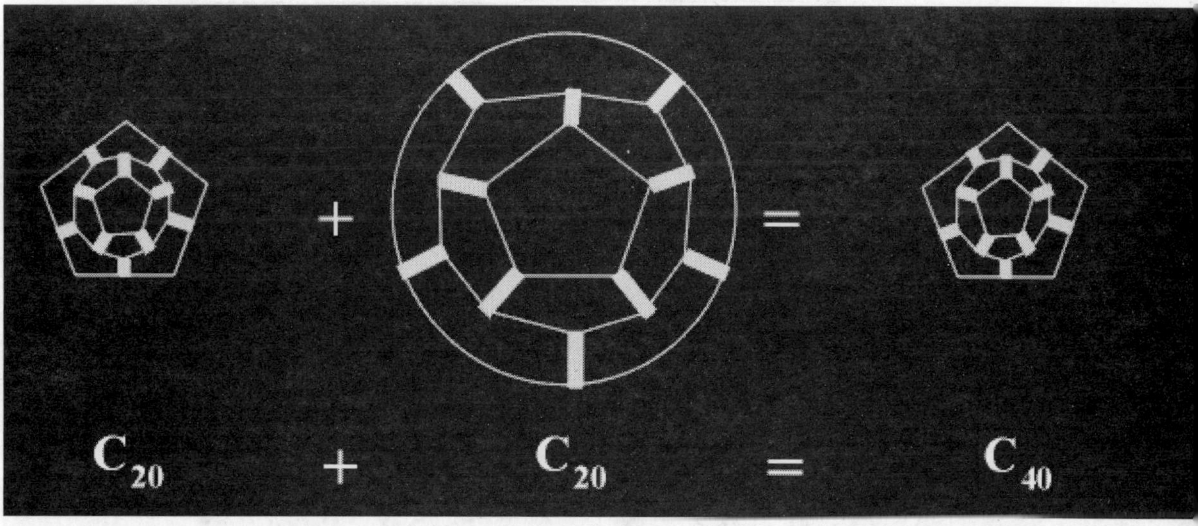

Fig. 17.15

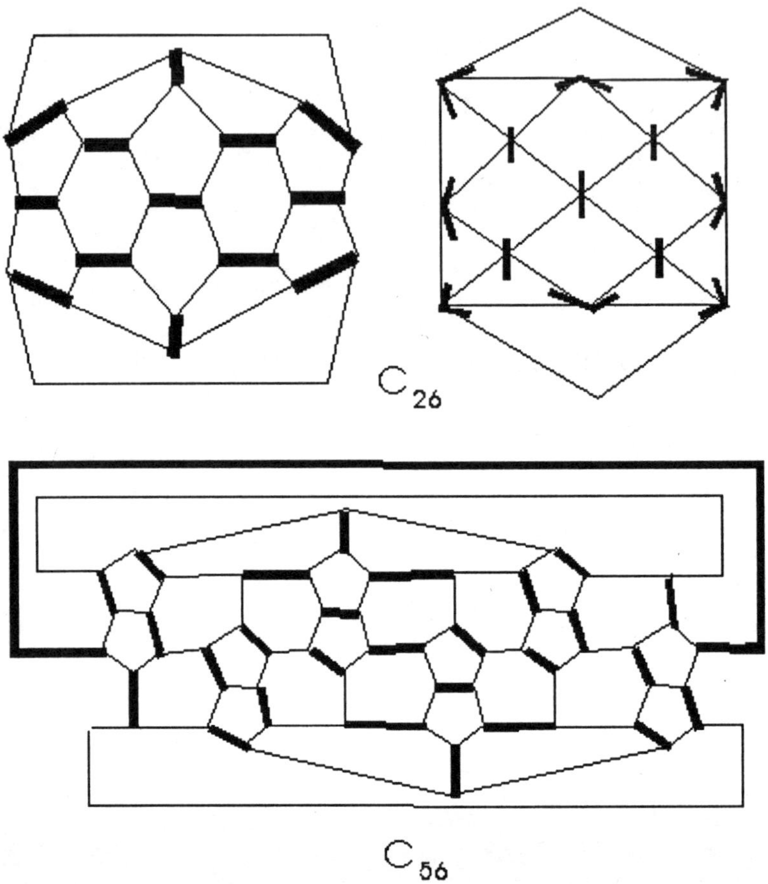

Fig. I–Q.16

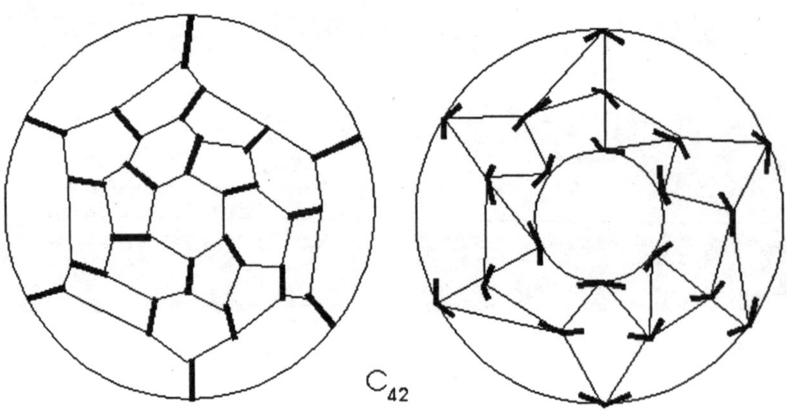

Fig. I–Q.17

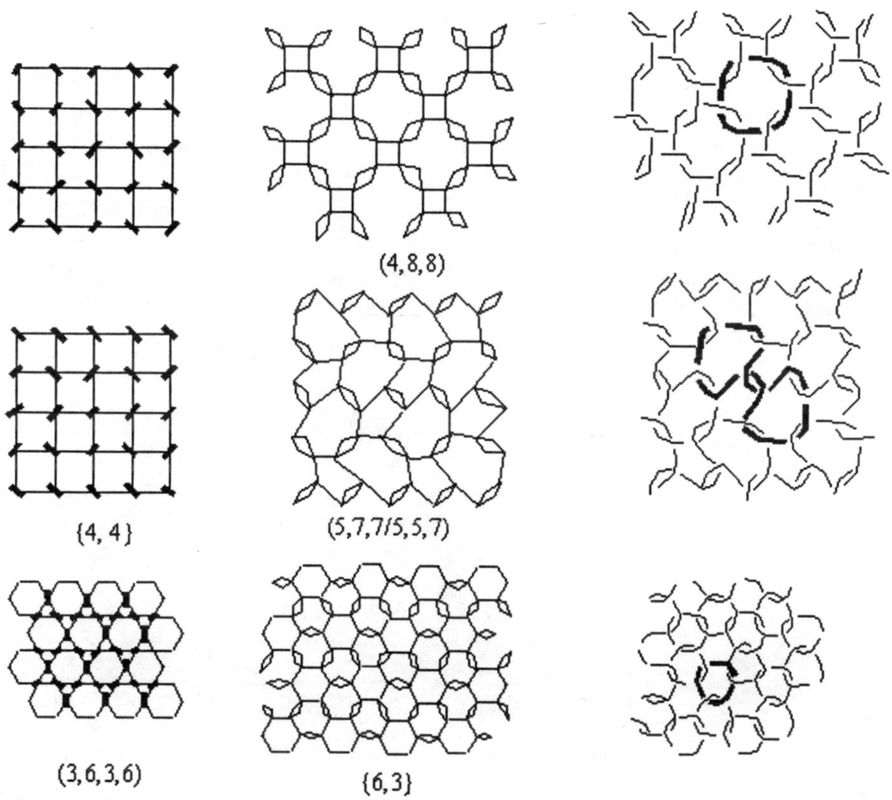

Fig. I–Q.18

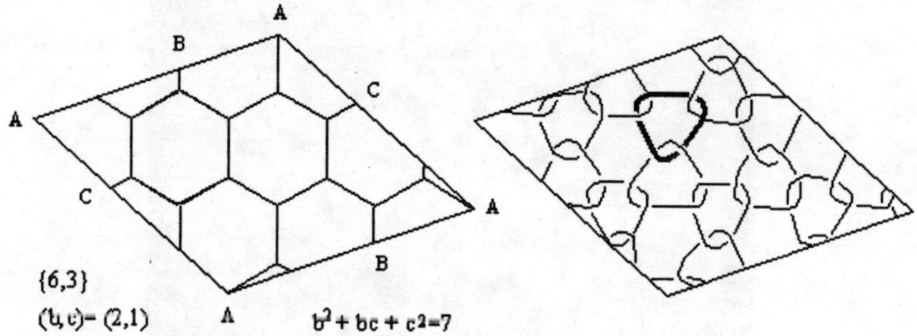

Fig. I–Q.19

torus without octagons $n_5=n_7$. To obtain such general fullerenes with a higher degree of symmetry, we could start from different vertex-transitive structures (e.g. uniform polyhedra, stellated regular and semi-regular polyhedra or infinite polyhedra). For example, different uniform 4-valent polyhedra of the type $(3,q,3,q)$ $(q=7,8,9,10,12,18)$ could be used for the derivation of the corresponding "perfect" fullerenes with q-gonal holes on a double torus ($g=2$) (Fig. I–Q.20, $q=8$). For this, we use the regular vertex-bifurcation of triangular faces, transforming all of them into hexagons. In the same way, the uniform tessellations of the type $(4,q,4,q)$ $(q=5,6,8,12)$ or $(5,10,5,10)$ of a double torus may result in different finite general fullerenes. The interesting classes of infinite general fullerenes with non-euclidean plane symmetry groups could be derived from the tessellations of hyperbolic plane H^2. For example, from the uniform tessellation $(3,7,3,7)$ we derive the infinite perfect 6/7 fullerene in H^2 with heptagonal holes (Fig. I–Q.21).

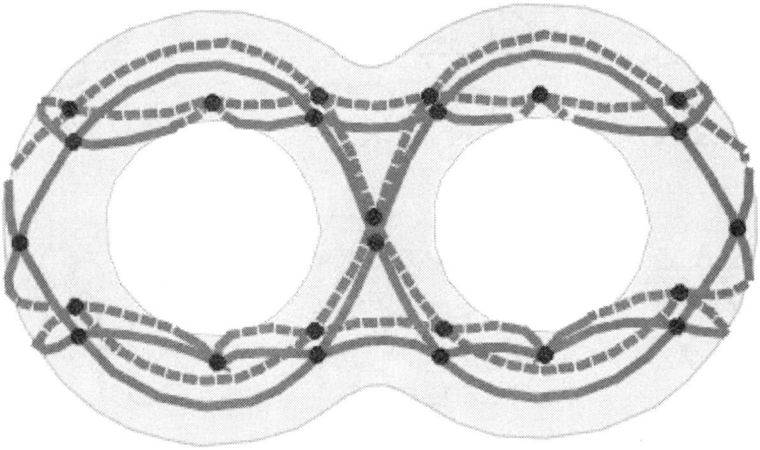

Fig. I–Q.20

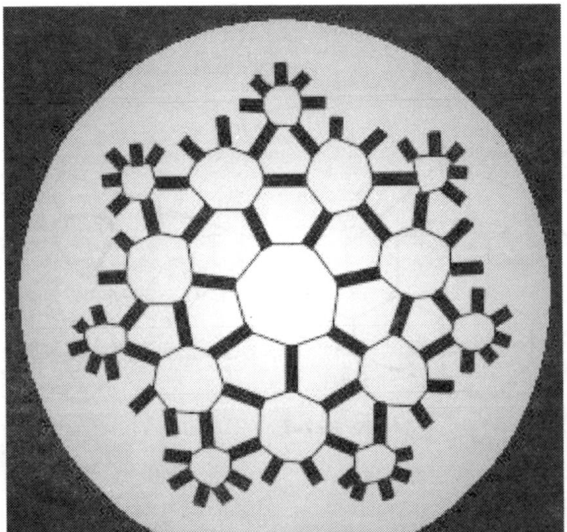

Fig. I–Q.21

Unit I–R

Nanotubes

What is so Special about Carbon Nanotubes?

Carbon nanotubes are one of the most commonly mentioned building blocks of nanotechnology. With one hundred times the tensile strength of steel, thermal conductivity better than all but the purest diamond, and electrical conductivity similar to copper, but with the ability to carry much higher currents, they seem to be a wonder material. However, when we hear of some companies planning to produce hundreds of tons per year, while others seem to have extreme difficulty in producing grams, there is clearly more to this material than meets the eye.

In fact nanotubes come in a variety of flavors: long, short, single-walled, multi-walled, open, closed, with different types of spiral structure, etc. Each type has specific production costs and applications. Some have been produced in large quantities for years while others are only now being produced commercially with decent purity and in quantities greater than a few grams. In this brief white paper we hope to resolve some of the confusion surrounding what may be one of the most significant new materials since plastics.

Carbon Nanotubes and Related Structures

The term nanotube is normally used to refer to the carbon nanotube, which has received enormous attention from researchers over the last few years and promises, along with close relatives such as the nanohorn, a host of interesting applications. There are many other types of nanotube, from various inorganic kinds, such as those made from boron nitride, to organic ones, such as those made from self-assembling cyclic peptides (protein components) or from naturally-occurring heat shock proteins (extracted from bacteria that thrive in extreme environments). However, carbon nanotubes excite the most interest, promise the greatest variety of applications, and currently appear to have by far the highest commercial potential. Only carbon nanotubes will be covered in this part.

Carbon nanotubes are often referred to in the press, including the scientific press, as if they were one consistent item. They are in fact a hugely varied range of structures, with similarly huge variations in properties and ease of production. Adding to the confusion is the existence of long, thin, and often hollow, carbon fibers that have been called carbon nanotubes but have a quite different make-up from that of the nanotubes that scientists generally refer to. To distinguish these we will refer to them as carbon nanofibers.

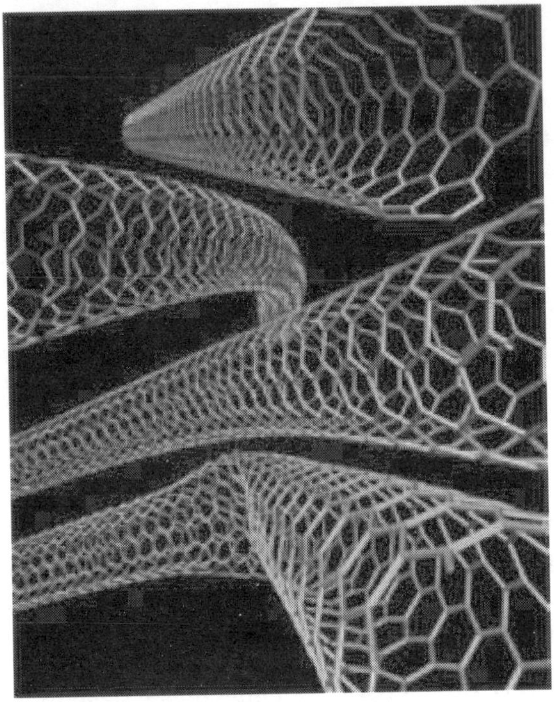

Fig. I–R.1. Bent nanotubes.

Introduction to Carbon Nanotubes

Carbon nanotubes were 'discovered' in 1991 by Sumio Iijima of NEC and are effectively long, thin cylinders of graphite, which you will be familiar with as the material in a pencil or as the basis of some lubricants. Graphite is made up of layers of carbon atoms arranged in a hexagonal lattice, like chicken wire (see figure I–R.2). Though the chicken wire structure itself is very strong, the layers themselves are not chemically bonded to each other but held together by weak forces called Van der Waals. It is the sliding across each other of these layers that gives graphite its lubricating qualities and makes the mark on a piece of paper as you draw your pencil over it.

Now imagine taking one of these sheets of chicken wire and rolling it up into a cylinder and joining the loose wire ends. The result is a tube that was once described by Richard Smalley (who shared the Nobel Prize for the discovery of a related form of carbon called buckminsterfullerene) as "in one direction . . . the strongest damn thing you'll ever make in the universe". In addition to their remarkable strength, which is usually quoted as 100 times that of steel at one-sixth of the weight (this is tensile strength—the ability to withstand a stretching force without breaking), carbon nanotubes have shown a surprising array of other properties. They can conduct heat as efficiently as most diamond (only diamond grown by deposition from a vapour is better), conduct electricity as efficiently as copper, yet also be semiconducting (like the materials that make up the chips in our computers). They can produce streams of electrons very efficiently (field emission), which can be used to create light in displays for televisions or computers, or even in domestic lighting, and they can enhance the fluorescence of materials they are close to. Their electrical properties can be made to change in the presence of certain substances or as a result of mechanical stress. Nanotubes within nanotubes can act like miniature springs and they can even be stuffed with other materials. Nanotubes and their variants hold promise for storing fuels such as hydrogen or methanol for use in fuel cells and they make good supports for catalysts.

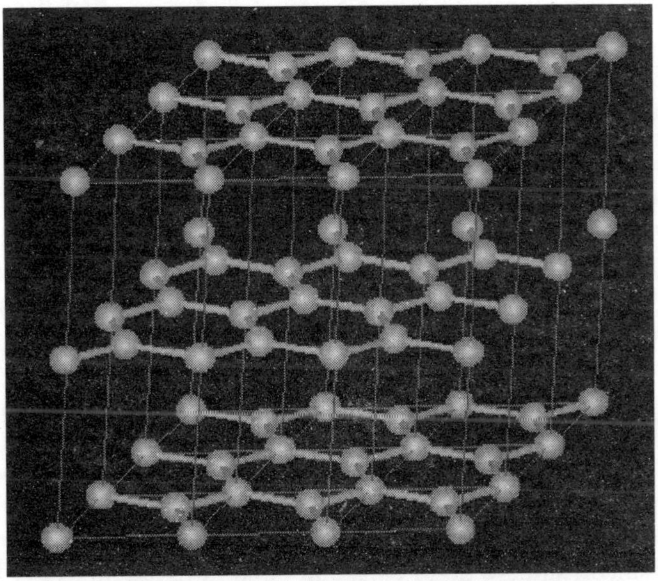

Fig. I–R.2: Layer structure of graphite.

But let's look at some of the different types of nanotubes, and nanotube pretenders.

One of the major classifications of carbon nanotubes is into single-walled varieties (SWNTs), which have a single cylindrical wall, and multi-walled varieties (MWNTs), which have cylinders within cylinders (see figure I–R.3).

The lengths of both types vary greatly, depending on the way they are made, and are generally microscopic rather than nanoscopic, i.e. greater than 100 nanometers (a nanometer is a millionth of

Fig. I–R.3. Computational image of single-and multi-walled nanotubes

a millimeter). The aspect ratio (length divided by diameter) is typically greater than 100 and can be up to 10,000, but recently even this was made to look small. In May 2002 SWNT strands were made in which the SWNTs were claimed to be as long as 20 cm. Even more recently, the same group has made strands of SWNTs as long as 160 cm, but the precise make-up of these strands has not yet been made clear. A group in China has also found, purely by accident, that packs of relatively short carbon nanotubes can be drawn out into a bundle of fibers, making a thread only 0.2 millimeters in diameter but up to 30 centimeters long. The joins between the nanotubes in this thread represent a weakness but heating the thread has been found to increase the strength significantly, presumably through some sort of fusing of the individual tubes.

Single-Walled Carbon Nanotubes (SWNTs)

These are the stars of the nanotube world, and somewhat reclusive ones at that, being much harder to make than the multi-walled variety. The oft-quoted amazing properties generally refer to SWNTs. As previously described, they are basically tubes of graphite and are normally capped at the ends (see figure I–R.4), although the caps can be removed. The caps are made by mixing in some pentagons with the hexagons and are the reason that nanotubes are considered close cousins of buckminsterfullerene (see figure I–R.4), a roughly spherical molecule made of sixty carbon atoms, that looks like a soccer ball and is named after the architect Buckminster Fuller (the word fullerene is used to refer to the variety of such molecular cages, some with more carbon atoms than buckminsterfullerene, and some with fewer).

The theoretical minimum diameter of a carbon nanotube is around 0.4 nanometers, which is about as long as two silicon atoms side by side, and nanotubes this size have been made. Average diameters tend to be around the 1.2 nanometer mark, depending on the process used to create them.

SWNTs are more pliable than their multi-walled counterparts and can be twisted, flattened and bent into small circles or around sharp bends without breaking.

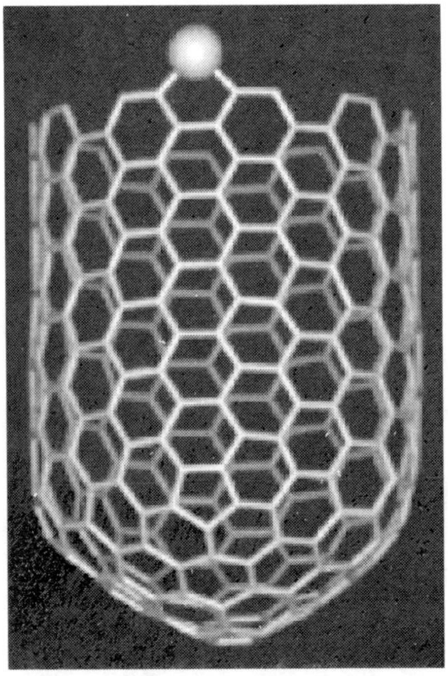

Fig. I–R.4: Simulated structure of a carbon nanotube.

Discussions of the electrical behavior of carbon nanotubes usually relate to experiments on the single-walled variety. As we have said, they can be conducting, like metal (such nanotubes are often referred to as metallic nanotubes), or semiconducting, which means that the flow of current through them can be stepped up or down by varying an electrical field. The latter property has given rise to dreams of using nanotubes to make extremely dense electronic circuitry and the last year has seen major advances in creating basic electronic structures from nanotubes in the lab, from transistors up to simple logic elements. The gulf between these experiments and commercial nanotube electronics is, however, vast.

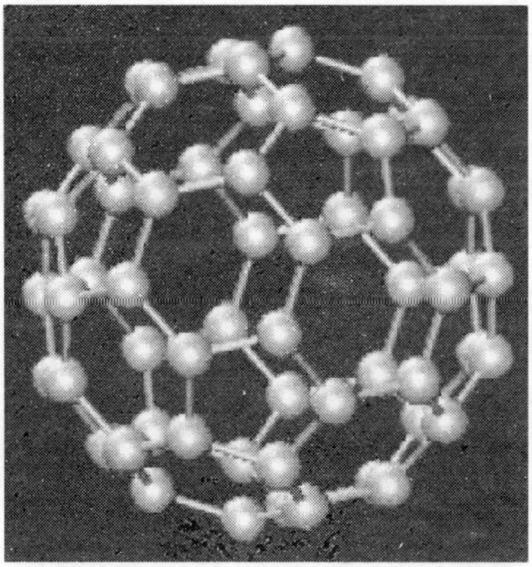

Fig. I–R.5: Buckminsterfullerene.

There are various ways of producing SWNTs, which are briefly discussed later. The detailed mechanisms responsible for nanotube growth are still not fully understood and computer modeling is playing an increasing role in fathoming the complexities. The ambition of SWNT producers is to gain greater control over their diameters, lengths, and other properties, such as chirality (explained below).

The volumes of SWNTs produced are currently small and the quality and purity are variable. Carbon Nanotechnologies Inc. of Houston, Texas, are currently ramping up production to a half a kilogram a day, which is actually huge in comparison to amounts of SWNTs that have been made historically. Various companies pursuing specific nanotube applications produce their own material in house.

Chirality

Imagine again the chicken wire that we roll up to make the nanotube. In fact, imagine a chicken wire fence out of which will be cut a rectangle to roll into a tube. You could cut the rectangle with the sides vertical or at various angles. Additionally, when joining the sides together, you can raise or lower one side. In some cases it will not be possible to make a tube such that the loose ends match and hexagons are formed, but in other cases it will, and these represent the possible permutations of SWNTs (see figure I–R.6). The possibilities are two forms in which a pattern circles around the diameter of the tube, often called zigzag and armchair (not the most intuitive of names, unfortunately, but they are now widely used), and a variety of forms in which the hexagons spiral up or down the tube with varying steepness, these being the chiral forms. There is theoretically an infinite variety of the latter, if you allow for infinite diameters of nanotubes.

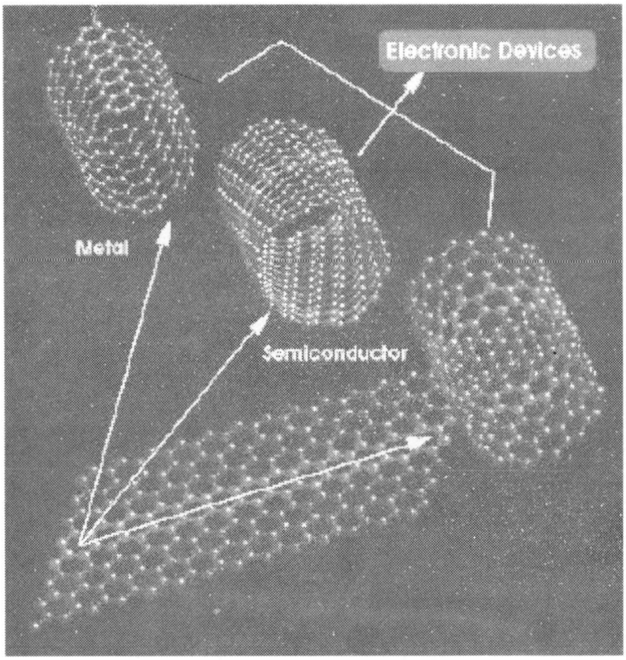

Fig. I–R.6: Schematic representation of rolling graphite to create a carbon nanotube.

Which of these forms a nanotube takes is the major determinant of its electrical properties, i.e. whether the tube is semiconducting or conducting. For a long time, the fact that all known production methods created a mix of types has been considered one of the hurdles to be overcome if the electronic properties are to be exploited. Claims have now been made that it is possible to produce only the semiconducting kind (specifically, the zigzag form). Additionally, there are approaches that can yield only semiconducting nanotubes from a mix of semiconducting and conducting ones. One such approach relies on vaporizing the conducting nanotubes with a strong electric current, leaving only the semiconducting kind behind. A more recent approach is simply to leave the mix of nanotubes lying around for a while—the metallic ones are oxidized and become semiconducting (the process can, of course, be speeded up). In the early part of 2001 there were also reports that nanotubes had been induced to form crystals, each of which contained just one type of nanotube. This would have been a nice separation method but the silence on this approach since then suggests that these results have not been duplicated.

Multi-Walled Carbon Nanotubes (MWNTs)

Multi-walled carbon nanotubes are basically like Russian dolls made out of SWNTs—concentric cylindrical graphitic tubes. In these more complex structures, the different SWNTs that form the

MWNT may have quite different structures (length and chirality). MWNTs are typically 100 times longer than they are wide and have outer diameters mostly in the tens of nanometers.

Although it is easier to produce significant quantities of MWNTs than SWNTs, their structures are less well understood than single-wall nanotubes because of their greater complexity and variety. Multitudes of exotic shapes and arrangements, often with imaginative names such as bamboo-trunks, sea urchins, necklaces or coils, have also been observed under different processing conditions. The variety of forms may be interesting but also has a negative side—MWNTs always (so far) have more defects than SWNTs and these diminish their desirable properties.

Many of the nanotube applications now being considered or put into practice involve multi-walled nanotubes, because they are easier to produce in large quantities at a reasonable price and have been available in decent amounts for much longer than SWNTs. In fact one of the major manufacturers of MWNTs at the moment, Hyperion Catalysis, does not even sell the nanotubes directly but only pre-mixed with polymers for composites applications. The tubes involved typically have 8 to 15 walls and are around 10 nanometres wide and 10 micrometers long. Other companies are moving into this space, notably formidable players like Mitsui, with plans to produce similar types of MWNT in hundreds of tons a year, a quantity that is greater, but not hugely so, than the current production of Hyperion Catalysis. This is an indication that even these less impressive and exotic nanotubes hold promise of representing a sizable market in the near future.

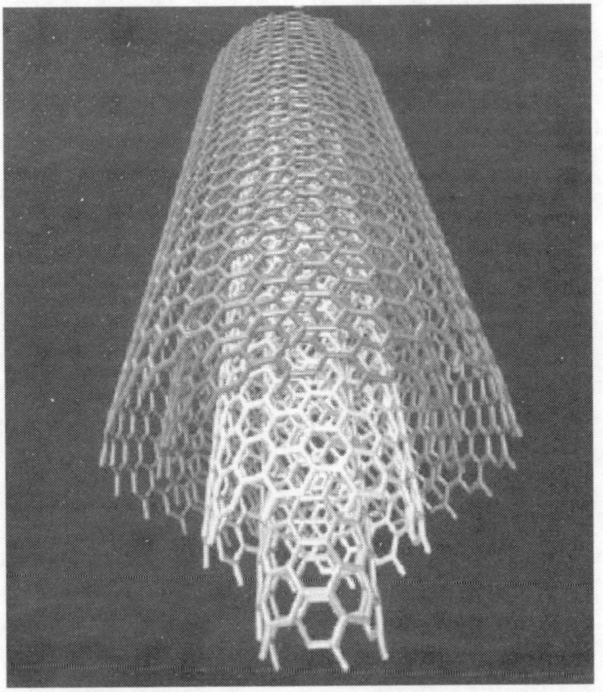

Fig. I–R.7: Representation of a multi-walled carbon nanotube.

Nanohorns

These are single-walled carbon cones with structures similar to those of nanotube caps that have been produced by high temperature treatment of fullerene soot. Sumio Iijima's group at NEC has demonstrated that nanohorns have good adsorptive and catalytic properties (i.e. desired substances stick to them and they enhance chemical reactions), and the company is working on using them in a new generation of fuel cells for personal electronics.

Nanofibers

We use this term to refer to hollow and solid carbon fibers with lengths on the order of a few microns and widths varying from some tens of nanometers to around 200 nanometers. These materials have occasionally been referred to as nanotubes. However, they do not have the cylindrical chicken wire structure of SWNTs and MWNTs but instead consist of a mixture of forms of carbon, from layers of graphite stacked at various angles to amorphous carbon (lacking any large-scale regular structure). Because of this variable structure they do not exhibit the strength of pure nanotubes but can still be quite strong and possess other useful properties. The US company Applied Sciences Inc. is already producing tons of such material a year and there are several major producers in the Far East.

Carbon Nanotube Production Processes

Production processes for carbon nanotubes, crudely described, vary from blasting carbon with an electrical arc or a laser to growing them from a vapor, either en masse (usually in tangled bundles) or on nanoparticles, sometimes in predetermined positions. These processes vary considerably with respect to the type of nanotube produced, quality, purity and scalability. Carbon nanotubes are usually created with the aid of a metal catalyst and this ends up as a contaminant with respect to many potential applications, especially in electronics. IBM have very recently, however, grown nanotubes on silicon structures without a metal catalyst.

Scalability of production processes is an essential commercial consideration—some of the approaches use equipment that simply cannot be made bigger and the only way to increase production is to make more pieces of equipment, which will not produce the economies of scale required to bring down costs significantly.

Nanotube Applications

By now we hope you have an idea of the different types of nanotubes and how their qualities and ease of production vary. Reports of one company producing tons of the material, alongside reports of researchers not being able to get enough to get meaningful results from their research, should no longer be confusing. It would, of course, be useful if commentators got into the habit of making clear the type of material they are talking about.

Understanding these differences is essential for understanding the commercial potential of the various applications of nanotubes and related structures that already exist or have been proposed. The variety of these is vast, and the commercialization timelines involved vary from now to ten years from now or more. Some of the potential markets are enormous. We will leave you with a taste of the possibilities.

The materials markets are already seeing applications for composites based on multi-walled carbon nanotubes and nanofibers. In many ways this is an old market—that of carbon fibers, which have been around commercially for a couple of decades. The benefits of the new materials in these markets are the same as those of carbon fibers, just better; the main properties to be considered being strength and conductivity. Carbon fibers are quite large, typically about a tenth of a millimeter in diameter, and blacken the material to which they are added. MWNTs can offer the same improvements in strength to a polymer composite without the blackening and often with a smaller amount of added material (called the filler). The greater aspect ratio (i.e. length compared to diameter) of the newer materials can make plastics conducting with a smaller filler load, one significant application being electrostatic painting of composites in products such as car parts. Additionally, the surface of the composite is smoother, which benefits more refined structures such as platens for computer disk drives.

When thinking about structural applications such as these, it should be remembered that, in general, as the fibers get smaller so the number of defects decreases, in a progression towards the perfection of the single-walled nanotube. The inverse progression is seen in terms of ease of manufacturing—the more perfect, and thus more structurally valuable, the material, the harder it is to produce in quantity at a good price. This relationship is not written in stone—there is no reason that near-perfect SWNTs should not be producible cheaply and in large quantities. When this happens, and it might not be too far off, the improvements seen in the strength-to-weight ratios of composite materials could soar, impacting a wide variety of industries from sports equipment to furniture, from the construction industry to kitchenware, and from automobiles to airplanes and spacecraft (the aerospace industry is probably the one that stands to reap the greatest rewards). In fact a carbon nanotube composite has recently been reported that is six times stronger than conventional carbon fiber composites.

This is an appropriate point at which to introduce a note of caution, and an area of research worth keeping an eye on. Just because the perfect nanotube is 100 times stronger than steel at a sixth of the weight doesn't mean you are going to be able to achieve those properties in a bulk material containing them. You may remember that the chicken wire arrangement that makes up the layers in graphite does not stick at all well to other materials, which is why graphite is used in lubricants and pencils. The same holds true of nanotubes—they are quite insular in nature, preferring not to interact with other materials. To capitalize on their strength in a composite they need to latch on to the surrounding polymer, which they are not inclined to do (blending a filler in a polymer is difficult even without these issues—it took a decade to perfect this for the new nanoclay polymers now hitting the markets).

One way of making a nanotube interact with something else, such as a surrounding polymer, is to modify it chemically. This is called functionalization and is being explored not just for composite applications but also for a variety of others, such as biosensors. For structural applications, the problem is that functionalization can reduce the valuable qualities of the nanotubes that you are trying to capitalize on. This is an issue that should not be underestimated.

Of course, in theory you don't need to mix the nanotubes with another material. If you want to make super-strong cables, for example, the best solution would be to use bundles of sufficiently long nanotubes with no other material added. For this reason, one of the dreams of nanotube production is to be able to spin them, like thread, to indefinite lengths. Such a technology would have applications from textiles (the US military is in fact investigating the use of nanotubes for bullet-proof vests) to the 'space elevator'. The space elevator concept, which sounds like something straight out of science fiction (it was, in fact, popularized by Arthur C. Clarke) involves anchoring one end of a huge cable to the earth and another to an object in space. The taut cable so produced could then support an elevator that would take passengers and cargo into orbit for a fraction of the cost of the rockets used today. Sounds too far out? It has, in fact, been established by NASA as feasible in principle, given a material as strong as SWNTs. The engineering challenges, though, are awesome, so don't expect that 'top floor' button to be taking you into orbit any time soon.

The materials market is a big one, and there are others, which we'll come to, but smaller ones exist too. Nanotubes are already being shipped on the tip of atomic force microscope probes to enhance atomic-resolution imaging. Nanotube-based chemical and biosensors should be on the market soon (they face stiff competition from other areas of nanotechnology). The thermal conductivity of nanotubes shows promise in applications from cooling integrated circuits to aerospace materials. Electron beam lithography, which is a method of producing nanoscale patterns in materials, may become considerably cheaper thanks to the field emission properties of carbon nanotubes. Recent developments show promise of the first significant change in X-ray technology in a century. Entering more speculative territory, nanotubes may one day be used as nanoscale needles that can inject substances into, or sample substances from, individual cells, or they could be used as appendages for miniature machines (the tubes in multi-walled nanotubes slide over each other like graphite, but have preferred locations that they tend to spring back to).

Big markets, apart from materials, in which nanotubes may make an impact, include flat panel displays (near-term commercialization is promised here), lighting, fuel cells and electronics. This last is one of the most talked-about areas but one of the farthest from commercialization, with one exception, this being the promise of huge computer memories (more than a thousand times greater in capacity than what you probably have in your machine now) that could, in theory, put a lot of the $40 billion magnetic disk industry out of business. Companies like to make grand claims, however, and in this area there is not just the technological hurdle to face but the even more daunting economic one, a challenge made harder by a host of competing technologies.

Despite an inevitable element of hype, the versatility of nanotubes does suggest that they might one day rank as one of the most important materials ever discovered. In years to come they could find their way into myriad materials and devices around us and quite probably make some of the leaders in this game quite rich.

Unit I–S

Applications of Carbon Nanotubes (Buckytubes)

Background

The special nature of carbon combines with the molecular perfection of buckytubes (single wall carbon nanotubes) to endow them with exceptionally high material properties such as electrical and thermal conductivity, strength, stiffness, and toughness. No other element in the periodic table bonds to itself in an extended network with the strength of the carbon-carbon bond. The delocalised pi-electron donated by each atom is free to move about the entire structure, rather than stay home with its donor atom, giving rise to the first molecule with metallic-type electrical conductivity. The high-frequency carbon-carbon bond vibrations provide an intrinsic thermal conductivity higher than even diamond.

In most materials, however, the actual observed material properties—strength, electrical conductivity, etc.—are degraded very substantially by the occurrence of defects in their structure. For example, high strength steel typically fails at about 1 per cent of its theoretical breaking strength. Buckytubes, however, achieve values very close to their theoretical limits because of their perfection of structure—their molecular perfection. This aspect is part of the unique story of buckytubes.

Buckytubes are an example of true nanotechnology: only a nanometer in diameter, but molecules that can be manipulated chemically and physically. They open incredible applications in materials, electronics, chemical processing and energy management.

Applications

Buckytubes have extraordinary electrical conductivity, heat conductivity and mechanical properties. They are probably the best electron field-emitter possible. They are polymers of pure carbon and can be reacted and manipulated using the tremendously rich chemistry of carbon. This provides opportunity to modify the structure and to optimise solubility and dispersion.

Very significantly, buckytubes are molecularly perfect, which means that they are free of property-degrading flaws in the nanotube structure. Their material properties can therefore approach closely the very high levels intrinsic to them.

These extraordinary characteristics give buckytubes potential in numerous applications.

Field Emission

Buckytubes are the best known field emitters of any material. This is understandable, given their high electrical conductivity, and the unbeatable sharpness of their tip (the sharper the tip, the more concentrated will be an electric field, leading to field emission; this is the same reason lightening rods are sharp). The sharpness of the tip also means that they emit at especially low voltage, an important fact for building electrical devices that utilize this feature. Buckytubes can carry an astonishingly high current density, possibly as high as 1013 A/cm^2. Furthermore, the current is extremely stable.

An immediate application of this behaviour receiving considerable interest is in field-emission flat-panel displays. Instead of a single electron gun, as in a traditional cathode ray tube display, here there is a separate electron gun (or many) for each pixel in the display. The high current density, low turn-on and operating voltage, and steady, long-lived behaviour make buckytubes attract field emitters to enable this application.

Other applications utilising the field-emission characteristics of buckytubes include: general cold-cathode lighting sources, lightning arrestors, and electron microscope sources.

Conductive Plastics

Much of the history of plastics over the last half century has been as a replacement for metal. For structural applications, plastics have made tremendous headway, but not where electrical conductivity is required, plastics being famously good electrical insulators.

This deficiency is overcome by loading plastics up with conductive fillers, such as carbon black and graphite fibres (the larger ones used to make golf clubs and tennis racquets). The loading required to provide the necessary conductivity is typically high, however, resulting in heavy parts, and more importantly, plastic parts whose structural properties are highly degraded.

It is well established that the higher aspect ratio of filler, the lower loading required to achieve a given level of conductivity. Buckytubes are ideal in this sense, since they have the highest aspect ratio of any carbon fibre. In addition, their natural tendency to form ropes provides inherently very long conductive pathways even at ultra-low loadings.

Applications that exploit this behaviour of buckytubes include EMI/RFI shielding composites and coatings for enclosures, gaskets, and other uses; electrostatic dissipation (ESD), and antistatic materials and (even transparent!) coatings; and radar-absorbing materials.

Energy Storage

Buckytubes have the intrinsic characteristics desired in material used as electrodes in batteries and capacitors, two technologies of rapidly increasing importance. Buckytubes have a tremendously high surface area (~1000 m^2/g), good electrical conductivity, and very importantly, their linear geometry makes their surface highly accessible to the electrolyte.

Research has shown that buckytubes have the highest reversible capacity of any carbon material for use in lithium-ion batteries. In addition, buckytubes are outstanding materials for supercapacitor electrodes and are now being marketed.

Buckytubes also have applications in a variety of fuel cell components. They have a number of properties including high surface area and thermal conductivity that make them useful as electrode catalyst supports in PEM fuel cells. They may also be used in gas diffusion layers as well as current collectors because of their high electrical conductivity. Buckytubes' high strength and toughness to weight characteristics may also prove valuable as part of composite components in fuel cells that are deployed in transport applications where durability is extremely important.

Conductive Adhesives and Connectors

The same issues that make buckytubes attractive as conductive fillers for use in shielding, ESD materials, etc., make them attractive for electronics materials, such as adhesives and other connectors (e.g., solders).

Molecular Electronics

The idea of building electronic circuits out of the essential building blocks of materials—molecules—has seen a revival the past five years, and is a key component of nanotechnology. In any electronic circuit, but particularly as dimensions shrink to the nanoscale, the interconnections between switches and other active devices become increasingly important.

Their geometry, electrical conductivity, and ability to be precisely derived, make buckytubes the ideal candidates for the connections in molecular electronics. In addition, they have been demonstrated as switches themselves.

Thermal Materials

The record-setting anisotropic thermal conductivity of buckytubes is enabling applications where heat needs to move from one place to another. Such an application is electronics, particularly advanced computing, where uncooled chips now routinely reach over 100°C.

CNI's technology for creating aligned structures and ribbons of buckytubes is a step toward realising incredibly efficient heat conduits. In addition, composites with buckytubes have been shown to dramatically increase the bulk thermal conductivity at small loadings.

Structural Composites

The world-record properties of buckytubes are not limited to electrical and thermal conductivities, but also include mechanical properties, such as stiffness, toughness, and strength. These properties lead to a wealth of applications exploiting them, including advanced composites requiring high values in one or more of these properties.

Fibres and Fabrics

Fibres spun of pure buckytubes have been demonstrated and are undergoing rapid development, along with buckytube composite fibres. Such super strong fibres will have applications including body and vehicle armour, transmission line cables, woven fabrics and textiles.

Catalyst Supports

Buckytubes have an intrinsically high surface area; in fact, every atom is not just on a surface—each atom is on two surfaces, the inside and outside! Combined with the ability to attach essentially any chemical species to their sidewalls provides an opportunity for unique catalyst supports. Their electrical conductivity may also be exploited in the search for new catalysts and catalytic behaviour.

Biomedical Applications

The exploration of buckytubes in biomedical applications is just underway, but has significant potential. Cells have been shown to grow on buckytubes, so they appear to have no toxic effect. The cells also do not adhere to the buckytubes, potentially giving rise to applications such as coatings for prosthetics and anti-fouling coatings for ships.

The ability to chemically modify the sidewalls of buckytubes also leads to biomedical applications such as vascular stents, and neuron growth and regeneration.

Other Applications

There is a wealth of other potential applications for buckytubes, such as solar collection; nanoporous filters; catalyst supports; and coatings of all sorts. There are almost certainly many unanticipated applications for this remarkable material that will come to light in the years ahead and which may prove to be the most important and valuable of all.

Unit I–T

Semiconductor Superlattices

A brief review of semiconductor superlattices is presented—their definition, basics of theory, types of materials and methods used for their preparation, the most important and unusual electrical and optical properties, which have already found their practical applications.

What are semiconductor superlattices

When alternating layers of two semiconductors with different band gaps are deposited and thickness of these layers lies between 1-10 nm (which corresponds approximately to 10–100 atomic layers), what we obtain is semiconductor superlattice. The mostly known example of such a structure is GaAs/AlAs system depicted schematically in Fig. 20.2 together with its corresponding potential profile. Differences in band gaps of materials lead to discontinuities in valence and conduction band and a set of square potential wells (low band gap semiconductor) separated by potential barriers (high band gap semiconductor) forms. Let us consider one of these wells from the quantum mechanical point of view. If we assume that an electron mean free path within a well is substantially longer than well width and the barrier width is infinite, what we obtain is a classical problem of a particle confined in a square well of finite depth. Without going to the details, it is sufficient to state that the discrete electron energy levels appear in the well and an increase, D, of the electron energy from the ground state to its first bound state can be calculated as the lowest order solution of:

$$\cos (kL/2) = k/Q$$

where,

$$k=(2*\pi*(2m^* \Delta)_{1/2}/h \quad Q=(2*\pi*(2m^*V)_{1/2} /h$$

and: V—well depth; L—well width; m*—electron effective mass; h—Planck constant.

Unfortunately, in the real-life systems we cannot obtain perfect separation of adjacent wells and this results in coupling of their wave functions. Thus, instead of discrete levels, this yields us subbands. Fig. I–T.1 shows an example of theoretical analysis of the GaAs/AlAs system shown in Fig. I–T.1, where an equal width, a, of well and barrier layer is assumed. Two main features common for other superlattice structures are easily seen when a is decreased. First, an energy distance between ground state and subsequent bound states increases, which happens because the well width is getting smaller. Second, the discrete levels split into the subbands as a result of thinner barrier layers and poorer separation of neighbouring wells.

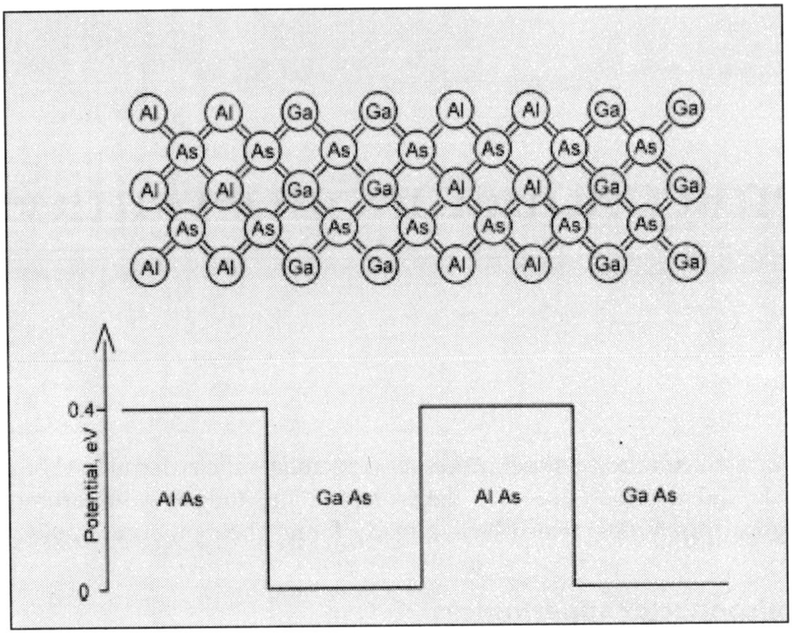

Fig. I–T.1: Schematic illustration of a GaAs/AlAs alternating structure, including the corresponding potential profile.

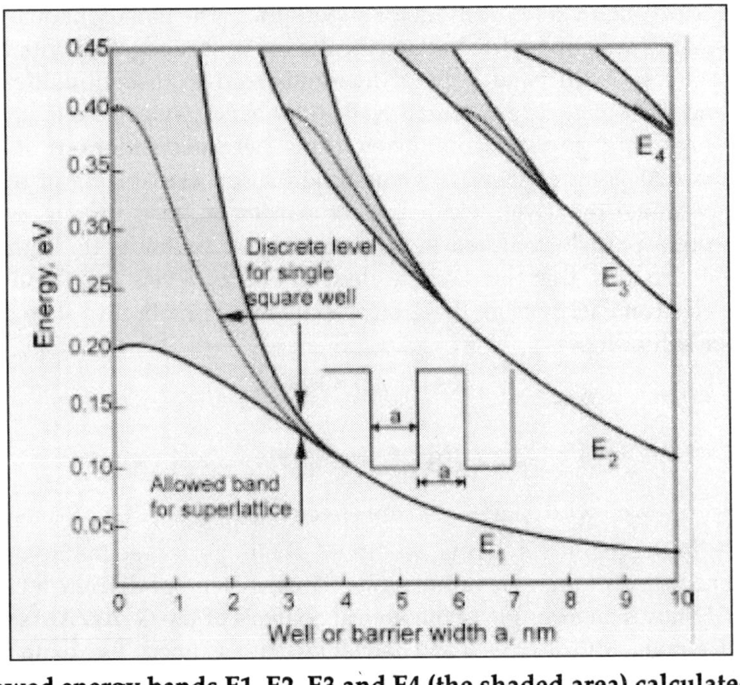

Fig. I–T.2: Allowed energy bands E1, E2, E3 and E4 (the shaded area) calculated as a function of well or barrier width *a*, including the potential profile of superlattice.

At this point, it is also important to notice, that the energy discontinuity, i.e. well depth, cannot be simply derived from such material constants as an energy gap, the conduction and valence band structure etc. in each semiconductor because of the lack of knowledge of the common energy reference. This parameter must be either calculated or measured.

Crystalline or Amorphous Superlattices

Historically, the crystalline superlattices where the first to produce, starting from pioneering work of Esaki and Tsu (1970). What is crucial in fabrication of such structures is nearly perfect match in their lattice constants. Otherwise the density of defects becomes so large that it results in strong electron scattering and obscures quantum phenomena. This is why a molecular beam epitaxy technique is utilised in their deposition and the number of materials, which can be combined to create superlattices, is restricted. The best known and studied are GaAs/AlAs and GaAs/AsxGa1-xAs, InAs/GaSb are also worth mentioning.

If interface defects cause such difficulties, why not to approach the problem from the opposite side—let us build the superlattice from the materials, where lattice constants no longer exist, i.e. amorphous semiconductors. This idea, apparently irrational, proved absolutely sensible in 1983, when Abeles and Tiedje reported in their paper fabrication of a-Si1-xNx:H/a-Si:H and a-Ge:H/a-Si:H systems. They stated that the strict requirements concerning lattice matching in crystalline superlattices can be relaxed due to amorphous structure by its very nature and a hydrogen content which plays an important role by passivating coordination defects. Since then, many other materials were combined to produce superlattices : a-Si:H/a-C:H, a-Si:H/a-SiN:H, a-Si:H/a-SiNx, a-Si:H/a-SiOx, a-Si:H/a-Ge:H, CdSe/SeTe, CdSe/Se, a-Si3N4:H/a-Si:H, a-C:H superlattices where the band gap is tailored by changing deposition conditions. An impressive, comparing to the crystalline case, number of types of amorphous superlattices results from the fact that in this case we have much more freedom in choosing components of our systems as we are not limited by the lattice matching condition.

It is also important to emphasise that the techniques utilised to produce amorphous films - plasma deposition, thermal evaporation, sputtering—are relatively cheaper than molecular beam epitaxy and easier to implement in industrial applications. On the other hand, in case of amorphous multilayers it is more difficult to obtain abrupt interfaces between subsequent layers. Different methods are employed : pulsed application of microwave plasma during continuos r.f. glow discharge in Si2H6/N2, by which N2 is decomposed and incorporated into the growing film, nitridation of a-Si:H to attain a-Si3N4:H, lowering the growth rate in order to avoid fabrication of transition layers when the deposition conditions change, quick exchange of source gases. However, sometimes smooth transition from one layer to another is intentionally introduced, which may lead to sinusoidal potential profile, instead of a step-like, presented in Fig. I–T.1.

Electrical properties of superlattices

For the sake of convenience, electrical properties, particularly conduction phenomena, may be considered in terms of what is happening when we study currents flowing in-plane (parallel to the layers) or perpendicular to them.

As it has been stated above, quantum confinement causes shift of the first bound state allowed for charge carriers. This must of course result in the increase of an activation energy of conductivity, E_a. In general, the smaller well width, the higher E_a, thus in-plane electrical conductivity decreases as compared to the unlayered material. These effects are presented in a number of research papers, typical behaviour for diamond-like carbon films (DLC) superlattices, is shown in Figs. I–T.3 and I–T.4.

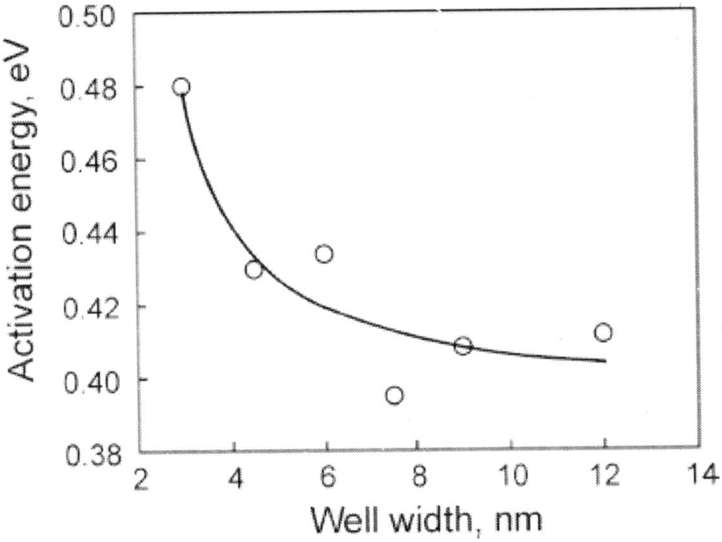

Fig. I–T.3: Activation energy versus well width for the DLC superlattices.

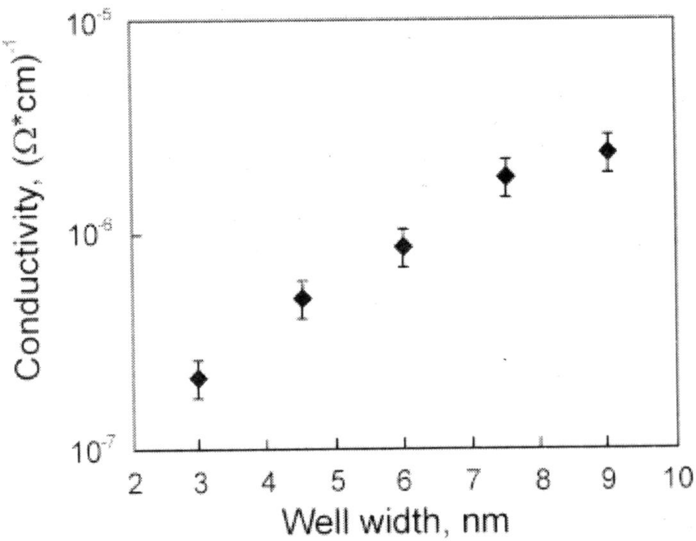

Fig. I–T.4: In-plane conductivity as a function of well width for the DLC superlattices.

However, the opposite situation may occur as well. This is because in amorphous semiconductors, unlike crystalline ones, there is a finite density of states within a mobility gap. When two such materials are put into contact, charge carriers may be transferred through an interface, building up accumulation and depletion layers, and as a result, the conductivity in one of the materials increases to such a degree that obscures quantum effects. A completely different picture appears when conductivity in the direction perpendicular to the layers is considered. This is explained in Fig. I–T.5, which shows superlattice potential profile when external electric field is applied, and corresponding field, F-current, j(F), characteristics.

If there is no electric field (Fig. I–T.5a), the current does not flow of course. When the small field is applied, a tunneling current can flow through the barriers. As the field is increased, an alignment between the energy subbands in adjacent wells is getting worse so observed maximum (Fig. I–T.5b) is a compromise between a driving force (electric field) and subbands mismatch. When the field is increased further, the subbands misalignment is so high, that charge transport becomes very inefficient and the current—field characteristics exhibits minimum (Fig. I–T.5c). If the field becomes higher, we can have the situation, that the lower subband in one quantum well is aligned with the upper one in the adjacent well (Fig. I–T.5d). Transport is efficient again. The most important feature of this characteristics is a region of negative differential resistance where an increase of the electric field leads to a decrease of the electric current. More detailed theoretical considerations of this matter can be found elsewhere. This kind of behaviour has been observed in both crystalline and amorphous multilayers; however, in the crystalline case it is more distinct and resembling model behaviour from Fig. I–T.5.

An example of characteristics for DLC superlattice is shown in Fig. I–T.6, where current-voltage profile together with its first and second differentials is presented. Bumps observed on the current-voltage curve are typical of resonant tunneling. The start of the negative differential resistance region at 3.1 V corresponds to the loss of subbands alignment. It should be noted that the conductivity in the direction perpendicular to the superlattice layers is by its nature very similar to what Esaki found in *p-n* germanium junctions and what is now applied in construction of tunnel diodes.

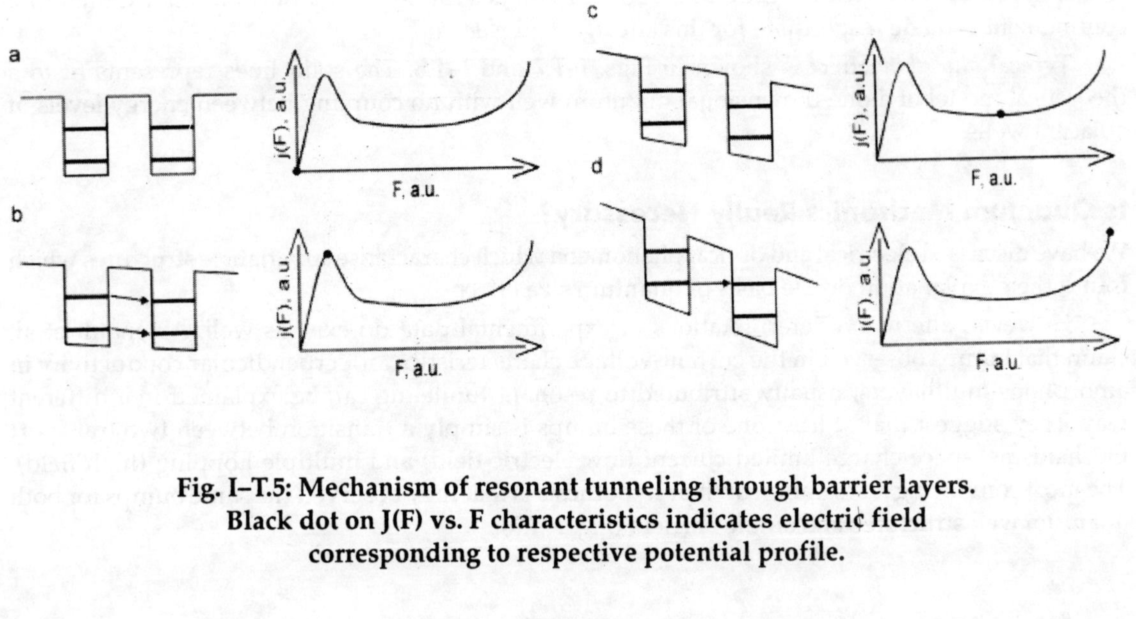

Fig. I–T.5: Mechanism of resonant tunneling through barrier layers. Black dot on J(F) vs. F characteristics indicates electric field corresponding to respective potential profile.

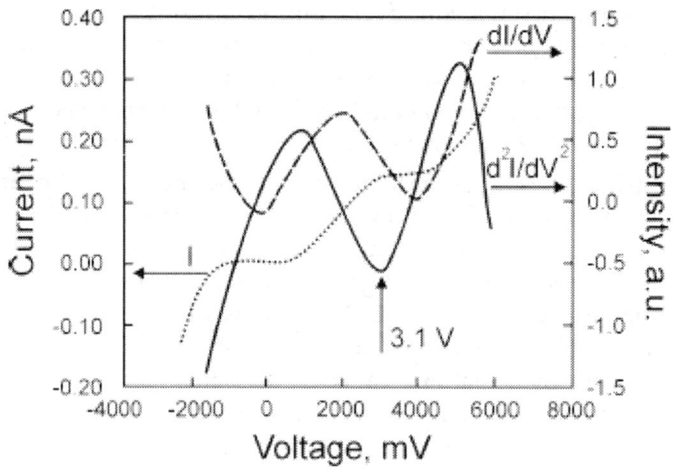

Fig. I–T.6: The current-voltage curve together with its differentials showing negative differential resistance, typical of resonant tunneling.

Optical Properties of the Superlattices

Amorphous semiconductors, unlike crystalline ones, do not exhibit sharp limit of an optical absorption. This arises from the presence of localised electronic states within the mobility gap which are involved in absorption phenomena. However, when the absorption is high enough, it can be described by Tauc formula

$$(\alpha^*E)1/2 = B1/2\ ^*(E - Eopt),$$

where α – absorption coefficient, E – energy of incident light quanta, B – constant.

E_{opt} is a parameter characterizing electronic structure of the amorphous material and called an optical gap. It has been reported in many researches that with the diminishing of the quantum well width Eopt and B increases. Again, like in the case of activation energy of conductivity, quantum confinement is made responsible for this effect, called *blue shift*.

Typical *blue shift* effect is shown in Figs. I–T.7 and I–T.8. The solid lines represents fit to a theoretical model of a one-dimensional quantum well with no coupling between energy levels of adjacent wells.

Is Quantum Mechanics Really Necessary?

We have discussed electrical and optical phenomena which characterise superlattice structures which found their explanation on the basis of quantum size effect.

However, alternative interpretations of experimental data do exist as well. Arsenault et al. claim that bumps observed in the current-voltage characteristics for perpendicular conductivity in amorphous multilayers, usually attributed to resonant tunneling can be explained in a different way. They suggest that at least one of these bumps is simply a transition between two transport mechanisms: space charge limited current (low electric field) and multiple hopping (high field). The most convincing justification of their hypothesis is that they observed the same bumps for both quantum well structures and single barriers (Fig. I–T.9).

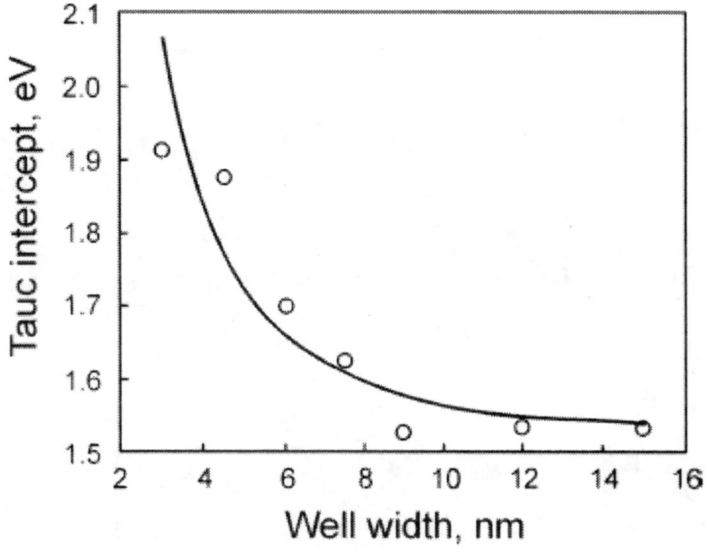

Fig. I–T.7: The variation of the optical Tauc gap as a function of the well width for a DLC superlattice.

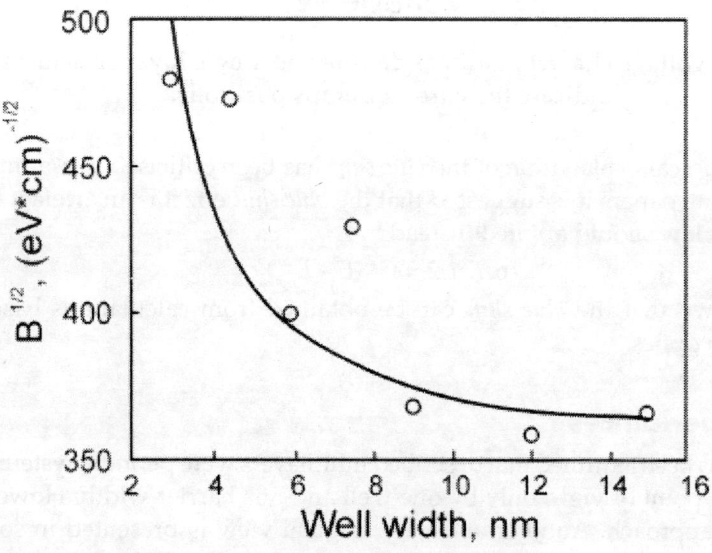

Fig. I–T.8: The variation of the Tauc constant $B^{1/2}$ as a function of the well width for a DLC superlattice.

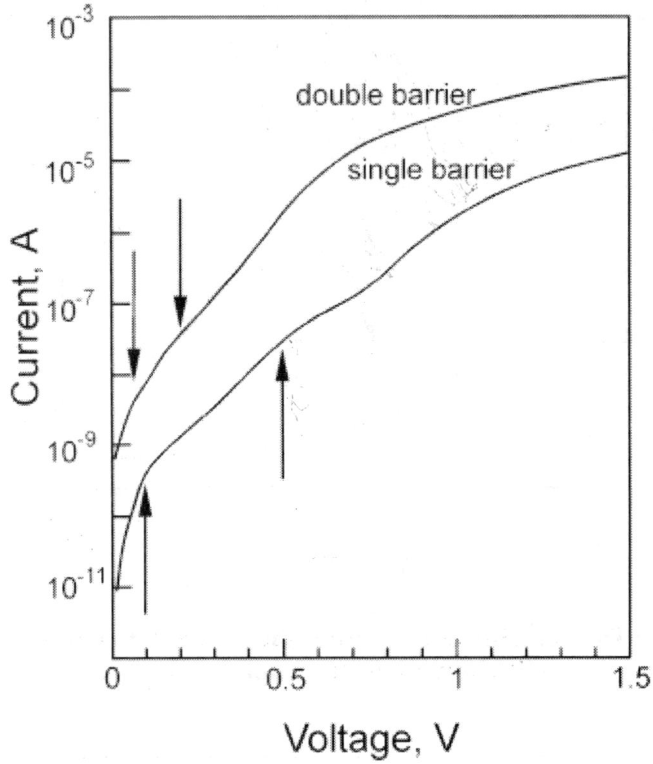

Fig. I–T.9: Current-voltage characteristics of double and single layer structures. The arrows indicate the current bumps positions.

Quantum mechanical explanation of the *blue shift* has been criticised in a number of research papers as well. In some papers it is suggested that the *blue shift* effect is an artefact of the Tauc law and Cody absorption law should applied instead :

$$(\alpha/E)1/2 = C^* (E - E_{opt})$$

It has been shown that the *blue shift* can be obtained from calculations based on classical multilayer thin—film optics.

Non-periodic Superlattices

In our discussion it has been assumed that described multilayers were periodic system, characterized, from the geometrical point of view, only by one well and one barrier width. However it does not have to be the only approach. Another and very original view is presented in some researches. Superlattices of these researches consist of the layers of a-Ge and a-Si deposited using dual-target magnetron sputtering. The most interesting thing is their topology generated by Cantor sets. The modulation algorithm according to the so called "Cantor middle-third" set together with corresponding multilayer structure is presented in Fig. I–T.10.

$$D = in2/in3$$

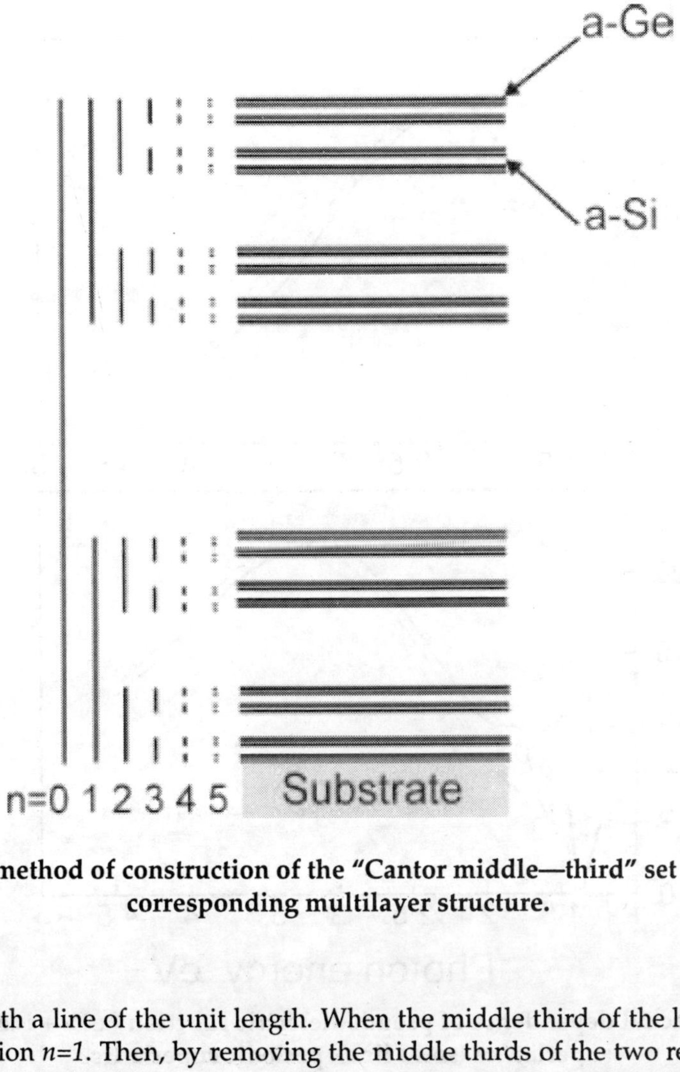

Fig. I–T.10: A method of construction of the "Cantor middle—third" set together with corresponding multilayer structure.

Let us start with a line of the unit length. When the middle third of the line is removed, we obtain first generation *n=1*. Then, by removing the middle thirds of the two remaining parts, the second generation is obtained. Repeating such a procedure gives us self-similar set with a fractal dimension:

Figure I–T.11 shows us, where such a treatment of superlattice matter leads. Graphs present spectral dependence of real, <e1>, and imaginary, <e2> parts of pseudodielectric function for the Cantor superlattice (solid line), periodic superlattice (dashed line) and alloy film (dotted line). Although samples had the same nominal thickness and a-Si/a-Ge volume ratio, differences in the dispersion profiles are clearly visible.

D = in2/in3

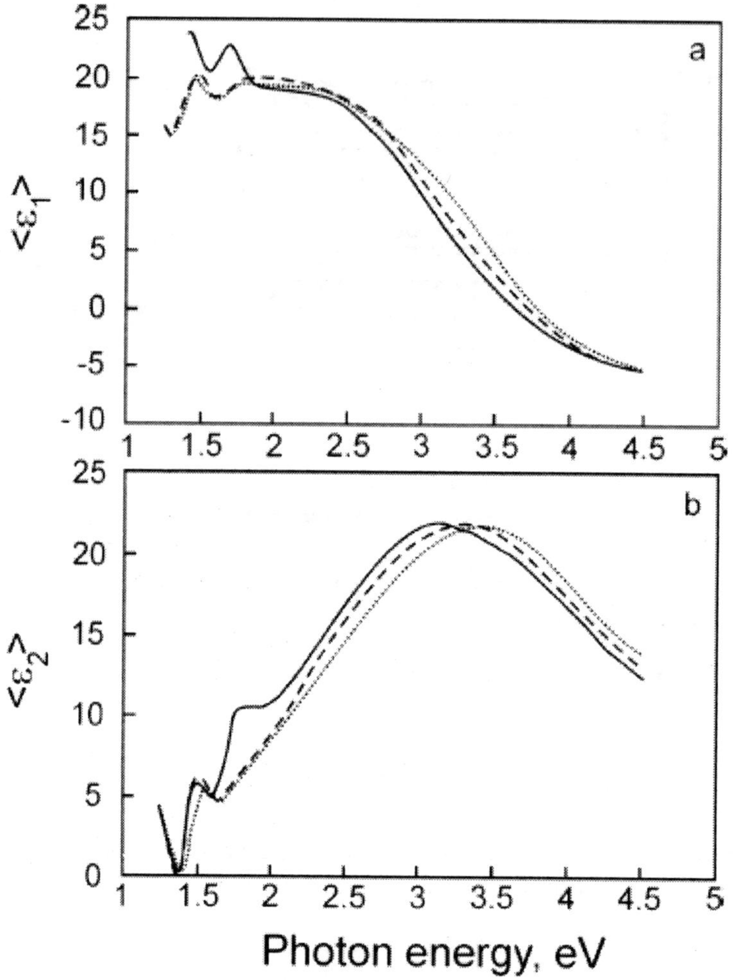

Fig. I–T.11: Spectral dependence of pseudodielectric function <e1> + i<e2> for Cantor superlattice (solid line), periodic superlattice (dashed line) and alloy film (dotted line).

SECTION B

NANOSTRUCTURED MATERIALS

Unit II–A

Nanostructured Materials—An Introduction

What are Nanostructured Materials

- A broad class of materials, with microstructures modulated in zero to three dimensions on length scales less than 100 nm.
- materials with atoms arranged in nanosized clusters, which become the constituent grains or building blocks of the material.
- any material with at least one dimension in the 1-100m range.

Conventional materials have grains sizes ranging from microns to several millimeters and contain several billion atoms each. Nanometer sized grains contain only about 900 atoms each. As the grain size decreases, there is a significant increase in the volume fraction of grain boundaries or interfaces. This characteristic strongly influences the chemical and physical properties of the material. For example, nanostructured ceramics are tougher and stronger than the coarser grained ceramics. Nanophase metals exhibit significant increases in yield strength and elastic modulus. It has also been shown that other properties(electrical, optical, magnetic...) are influenced by the fine grained structure of these materials. Using a variety of synthesis methods, it is possible to produce nanostructured materials in the following forms: thin films, coatings, powders and as a bulk material. There is also considerable interest in the generation of carbon nanostructures, which are related to the famous Buckyball. In addition, the use of nano-sized materials as fillers for composite materials is generating interest. specifically in the case of polymer nanocomposites.

Classes of Nanostructured Materials

There are several different types of nanostructured materials. These range from zero dimensional atom clusters to three dimensional equiaxed grain structure. Each class has at least one dimension in the nanometer range, as shown in Figure 1.1. Atom clusters and filaments are defined as zero modulation dimensionality and can have any aspect ratio from 1 to ∞2.

Any multilayered material with layer thickness in the nanometer range is classified as one-dimensionally modulated. Layers in the nanometer thickness range consisting of ultrafine grains (nanometer range diameter) are two-dimensionally modulated. This class includes coatings, buried layers and thin films. The last class is that consisting of three dimensionally modulated microstructures or nanophase materials.

Synthesis

The methods employed to produce nanostructured materials are numerous, with each method having advantages and disadvantages depending on the desired properties or application. Atom clusters are typically synthesized via a vapor condensation route which is essentially evaporation of a solid metal followed by rapid condensation to form nano-sized clusters. These resulting powders are essentially agglomerates of the nano-sized atom clusters. The powders can be then be used as fillers for composite materials or consolidated into bulk material. The cluster assembly method can be used to produce ceramic or metal nanostructured powders. The main advantage of this method is the extremely low contamination levels and the ability to control the final cluster size by manipulation of the process parameters such as temperature, gas environment and evaporation rate.

Several different methods have been developed using the vapor condensation concept. These include inert gas condensation, chemical vapor condensation, laser ablation, electron beam deposition, arc evaporation, etc.

Chemical Synthesis

Both metals and ceramics can be produced using a variety of chemical approaches such as sol-gel or thermal decomposition. These methods provide large quantities of nano-sized agglomerates at a low cost. Chemical processes also allow effective control of the stoichiometry of the final product. However, the precursor chemicals required may leave residual contamination on the particle surfaces, which leads to difficulties in compaction and sintering. Additionally, powders produced through wet chemical techniques often have difficulties with agglomeration.

Other methods which implement chemical reactions include Examples: A common method of producing nanostructured powders is through mechanical deformation, i.e. milling or shock deformation. These processes produce nanostructured materials through severe mechanical deformation of a coarse grained precursor material. The nanometer sized grains nucleate within the dislocation cell structures located in the shear bands. The final grain size is a function of the amount of energy input during the milling as well as time, temperature during milling and milling atmosphere. The precursor materials can be crystalline or amorphous materials in the form of powders or tapes. Through mechanical milling, it is possible to produce nanostructured powders of systems which are otherwise immicsible. A major disadvantage of this method is the possibility of contamination from the milling media due to the large forces and energies involved in the milling process.

Other synthesis methods based on **mechanical deformation**: friciton/sliding wear. Examples: Three dimensional nanostructured materials are also synthesized through thermal crystallization of an amorphous material. By controlling the nucleation and growth during annealing of an amorphous material, one can produce a bulk material with average grain sizes of less than 20 nm without the need for consolidation and sintering steps. However, this process is limited to material composition which are readily available in the form of a metallic glass, which has an amorphous microstructure.

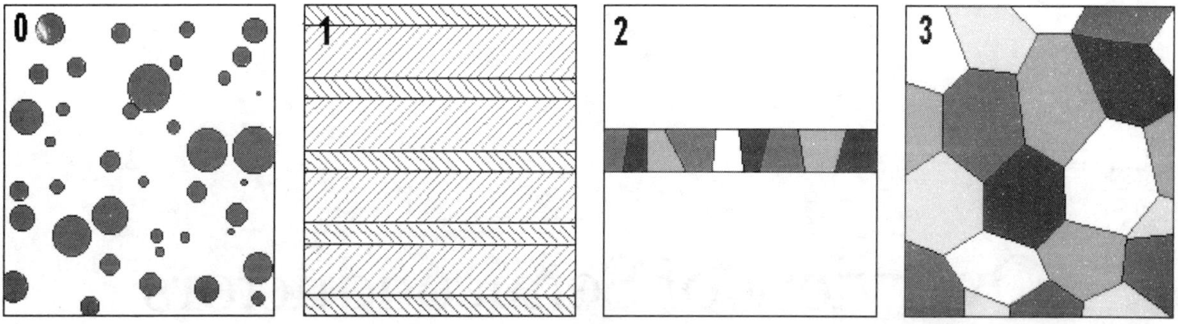

Fig. II–A.1: Each class of nanostructured material having at least one dimension in the nanometer range.

Microstructure

The atomic structure of nanostructured materials is unlike that seen in glasses or crystals because of the large volume fraction (greater than 50 per cent) of grain boundaries or interfaces. Coarse-grained materials have less than 3 per cent of all atoms associated with grain boundaries or interfaces. However, in materials with nanosized grains between 5 per cent and 50 per cent of all atoms are associated with grain boundaries or interfaces. The grain boundary structure in these nanostructured materials is similar to that of the conventional coarse grained materials.

Properties

The material properties of the nanostructured materials show remarkable improvement or deviation from the properties exhibited by the coarser grained material. These unique properties are attributed to the significant increase in grain boundary area due to the small grain size. In terms of the mechanical properties, nanostructured metals have shown increases in hardness values (ultimate and yield strengths). Specifically, pure nanophase metals have shown a clear increase in hardness with decreasing grain size, following the well known Hall-Petch equation. Nanostructured ceramics have exhibited superplastic properties at low temperatures. This is significant as ceramics are conventionally brittle materials. Further deviations from trends observed in conventional materials are observed in creep of nanostructured metals.

Unit II–B

Overview of Semiconductors

What are Semiconductors?

Semiconductors are:

- Solid crystalline materials whose electrical conductivities are intermediate between that of conductors and insulators, and which are usually strongly temperature-dependent.
- A semiconductor is a material that is an insulator at very low temperature, but which has a sizable electrical conductivity at room temperature. The distinction between a semiconductor and an insulator is not very well-defined, but roughly, a semiconductor is an insulator with a band gap small enough that its conduction band is appreciably thermally populated at room temperature.

Before looking at how some of the semiconductor devices work it is necessary to have a basic understanding of what conductors, insulators and naturally what semiconductors are.

Conductors and Non-Conductors

In terms of electricity there are two main classes of material: namely: conductors and non-conductors (or insulators). From their names it can be gathered that conductors will conduct electricity freely, whereas non-conductors act as insulators preventing the flow of an electric current.

An electric current is made up of the flow of electrons. This means that for a current to flow, the electrons must be able to move freely within the material. In some materials electrons are moving freely around the lattice, not particularly attached to a given electron. At any instance electrons are moving freely but randomly. By placing a potential difference across the conductor the electrons can be made to drift in one direction and this constitutes an electric current. Metals are all conductors of electricity, and a number of other substances also conduct it to varying degrees.

Other substances do not have electrons moving freely around the lattice. Electrons are firmly held within their molecules and cannot escape easily. Accordingly when a potential is placed across the substance very few electrons will move and very little or no current will flow. These substances are called non-conductors or insulators. They include most plastics, ceramics and many naturally occurring substances like wood.

Fundamental Semiconductor Physics

In the parlance of solid-state physics, semiconductors (and insulators) are defined as solids in which at 0°K (and without excitations) the uppermost band of occupied electron energy states is completely full. It is well-known from solid-state physics that electrical conduction in solids occurs only via electrons in partially-filled bands, so conduction in pure semiconductors occurs only when electrons have been excited—thermally, optically, etc.—into higher unfilled bands.

At room temperature, a proportion (generally very small, but not negligible) of electrons in a semiconductor have been thermally excited from the "valence band," the band filled at 0°K, to the "conduction band," the next higher band. The ease with which electrons can be excited from the valence band to the conduction band depends on the energy gap between the bands, and it is the size of this energy bandgap that serves as an arbitrary dividing line between semiconductors and insulators. Semiconductors generally have bandgaps of approximately 1 electron-volt, while insulators have bandgaps several times greater.

When electrons are excited from the valence band to the conduction band in a semiconductor, both bands contribute to conduction, because electrical conduction can occur in any partially-filled energy band. The current-carrying electrons in the conduction band are known as "free electrons," though often they are simply called "electrons" if context allows this usage to be clear. The free energy-states in the valence band are known as "holes." It can be shown that holes behave very much like positively-charged counterparts of electrons, and they are usually treated as if they are real charged particles.

Doping of Semiconductors

One of the main reasons that semiconductors are useful in electronics is that their electronic properties can be greatly altered in a controllable way by adding small amounts of impurities. These impurities, called dopants, add extra electrons or holes. A semiconductor with extra electrons is called an n-type semiconductor, while a semiconductor with extra holes is called a p-type semiconductor.

The most common n-type dopants for silicon are phosphorus and arsenic. Notice that these two elements are in Group V of the periodic table, and silicon is in Group IV. When silicon is doped with arsenic or phosphorus atoms, these dopant atoms replace silicon atoms in the semiconductor crystal, but since they have one more outer-shell electron than silicon they tend to contribute this electron to the conduction band. By far the most common p-type dopants for silicon is the Group III element boron, which lacks an outer-shell electron compared with silicon and thus tends to contribute a hole to the valence band.

Heavily doping a semiconductor can increase its conductivity by a factor greater than a billion. In modern integrated circuits, for instance, heavily-doped polycrystalline silicon is often used as a replacement for metals.

Intrinsic and Extrinsic Semiconductors

An intrinsic semiconductor is one which is pure enough that impurities do not appreciably affect its electrical behavior. In this case, all carriers are created by thermally or optically exciting electrons from the full valence band into the empty conduction band. Thus, equal numbers of electrons and holes are present in an intrinsic semiconductor. Electrons and holes flow in opposite directions in an electric field, though they contribute to current in the same direction since they are oppositely charged. Hole current and electron current are not necessarily equal in an intrinsic semiconductor, however, because electrons and holes have different effective masses (crystalline analogues to free inertial masses).

The concentration of carriers is strongly dependent on the temperature. At low temperatures, the valence band is completely full, making the material an insulator (see electrical conduction for more information). Increasing the temperature leads to an increase in the number of carriers and a corresponding increase in conductivity. This principle is used in thermistors. This behavior contrasts sharply with that of most metals, which tend to become less conductive at higher temperatures due to increased phonon scattering.

An extrinsic semiconductor is one that has been doped with impurities to modify the number and type of free charge carriers.

N-type Doping

The purpose of n-type doping is to produce an abundance of carrier electrons in the material. To help understand how n-type doping is accomplished, consider the case of silicon (Si). Si atoms have four valence electrons, each of which is covalently bonded with one of four adjacent Si atoms. If an atom with five valence electrons, such as those from group VA of the periodic table [e.g. phosphorus (P), arsenic (As), or antimony (Sb)], is incorporated into the crystal lattice in place of a Si atom, then that atom will have four covalent bonds and one unbonded electron. This extra electron is only weakly bound to the atom and can easily be excited into the conduction band. At normal temperatures, virtually all such electrons are excited into the conduction band. Since excitation of these electrons does not result in the formation of a hole, the number of electrons in such a material far exceeds the number of holes. In this case the electrons are the majority carriers and the holes are the minority carriers. Because the five-electron atoms have an extra electron to "donate", they are called donor atoms.

P-type Doping

The purpose of p-type doping is to create an abundance of holes. In the case of silicon a trivalent atom, such as boron, is substituted into the crystal lattice. The result is that an electron is missing from one of the four possible covalent bonds. Thus the atom can accept an electron from the valence band to complete the fourth bond, resulting in the formation of a hole. Such dopants are called acceptors. When a sufficiently large number of acceptors are added, the holes greatly outnumber the excited electrons. Thus, the holes are the majority carriers, while electrons are the minority carriers in p-type materials. Blue diamonds (Type IIb), which contain boron impurities, are an example of a naturally occurring p-type semiconductor.

P-N Junctions

A p-n junction may be created by doping adjacent regions of a semiconductor with p-type and n-type dopants. If a positive bias voltage is placed on the p-type side, the dominant positive carriers (holes) are pushed toward the junction. At the same time, the dominant negative carriers (electrons) in the n-type material are attracted toward the junction. Since there is an abundance of carriers at the junction, current can flow through the junction from a power supply, such as a battery. However, if the bias is reversed, the holes and electrons are pulled away from the junction, leaving a region of relatively non-conducting silicon which inhibits current flow. The p-n junction is the basis of an electronic device called a diode, which allows electric current to flow in only one direction. Similarly, a third region can be doped n-type or p-type to form a three-terminal device, such as the bipolar junction transistor (which can be either p-n-p or n-p-n).

Purity and Perfection of Semiconductor Materials

Semiconductors with predictable, reliable electronic properties are difficult to mass-produce because of the required chemical purity, and the perfection of the crystal structure, which are needed to make devices. Because the presence of impurities in very small proportions can have such big effects on the properties of the material, the level of chemical purity needed is extremely high. Techniques for achieving such high purity include zone refining, in which part of a solid crystal is melted. Impurities tend to concentrate in the melted region, leaving the solid material more pure. A high degree of crystalline perfection is also required, since faults in crystal structure such as dislocations, twins, and stacking faults, create energy levels in the band gap, interfering with the electronic properties of the material. Faults like these are a major cause of defective devices in production processes. The larger the crystal, the harder it is to achieve the necessary purity and perfection; current mass production processes use six-inch diameter crystals which are grown as cylinders and sliced into wafers.

Unit II–C

Semiconductor Devices

Semiconductor devices are electronic components that exploit the electronic properties of semiconductor materials, principally silicon, germanium and gallium arsenide. Semiconductor devices have replaced thermionic devices in most applications. They utilize electronic conduction in the solid state, as opposed to the vacuum state or gaseous state. Semiconductor devices are available as discrete units (such as those at Radio Shack) or can be integrated along with a large number—often millions—of similar devices onto a single chip, called an integrated circuit (IC).

Semiconductor Device Fundamentals

If a semiconductor is pure and if it is unexcited by an input like an electric field, it allows very little current to pass through it, and it is practically an insulator. The main reason that semiconductors are so useful is that the conductivity of semiconductors can be manipulated by addition of impurities (doping), by introduction of an electric field, by exposure to light, or by other means. For example, CCDs, the primary unit of digital cameras, rely on the fact that semiconductor conductivity increases with exposure to light. Transistor operation, which will be discussed below, depends on the fact that semiconductor conductivity can be increased by the presence of an electric field.

Current conduction in a semiconductor occurs via "free electrons" and "holes." Holes aren't real particles; in a sense that requires some knowledge of semiconductor physics to understand, a hole is the absence of an electron. Nevertheless, this absence, or hole, can be treated as a positively-charged counterpart to the negatively-charged electron. Indeed, the precise meaning of "free electrons" also requires a background in semiconductor physics to understand. For descriptive ease, "free electrons" are often simply denoted "electrons," but it should be understood that the majority of electrons in a solid, which aren't free, do not contribute to conductivity.

If a semiconductor crystal is perfectly pure, with no impurities, and it is held at a temperature near absolute zero with no excitations (e.g. electric fields or light), it will contain no free electrons and no holes, and thus will be a perfect insulator. At room temperature, thermal excitations produce some free electrons and holes in pairs, but most semiconductors at room temperature are insulators for practical purposes.

Doping a semiconductor, like silicon, with impurity atoms, like boron and phosphorus creates unequal numbers of free electrons and holes. High levels of doping can make a semiconductor a good conductor. When a doped semiconductor contains excess holes it is called "p-type," and when it contains excess free electrons it is known as "n-type." The semiconducting material in devices is almost always carefully doped for engineering purposes. In fact, junctions between n-type and p-type semiconductors, called p-n junctions, are the fundamental elements of many semiconductor devices, such as the p-n diode and the bipolar junction transistor.

An electric field can also create a unequal number of free electrons and holes in a semiconductor. This is the basis for "field effect transistors" like the MOSFET. Exposure to light generally creates electron/hole pairs in a semiconductor, which change its conductivity and allows the light to be sensed.

Transistors

The most important semiconductor device in use today is the MOSFET, a type of transistor whose operation depends on a "gate," from which originates an electric field that controls the conductivity of a "channel." MOSFETs, as well as other transistor types, are used as the building blocks of logic gates. Their role in a microprocessor is somewhat analogous to that of neurons in the brain. In digital circuits like microprocessors, transistors act as on-off switches; in the MOSFET, for instance, the input gate voltage determines whether the switch is on or off.

Transistors also serve very crucial operations in analog circuits. Transistors used for analog purposes do not act as on-off switches; rather, they respond to a continuous range of inputs with a continuous range of outputs. For example, transistors are the workhorses of amplifier circuits, which produce an amplified, but otherwise identical (ideally), version of the input. For years the bipolar junction transistor, or BJT, was the transistor of choice for analog circuits. However, the MOSFET has much more desirable properties for digital circuits, and since it is difficult to integrate BJTs and MOSFETs onto a single chip, MOSFETs are now commonly used for both analog and digital purposes.

More Semiconductor Devices

While the transistor is the most important semiconductor device, there are literally dozens of other families of semiconductor devices. Some of the great many semiconductor devices in current usage are as follows:

Two-terminal devices:

- Semiconductor diode
- Zener diode
- Light-emitting diode
- PIN diode
- Schottky diode
- Avalanche diode
- Laser diode
- DIAC

Three-terminal devices:

- Bipolar transistor

- Field effect transistor
- Darlington transistor
- Thyristor
- Triac

Four-terminal devices:

- Hall effect sensor (magnetic field sensor)

Discrete Semiconductor Devices

The type designators of semiconductor devices are often manufacturer specific. Nevertheless, there have been attempts at creating standards for type codes, and a subset of devices follow those. For discrete devices, for example, there are three standards: JEDEC JESD370B in USA, Pro Electron in Europe and JIS in Japan.

Semiconductor device fabrication

Semiconductor device fabrication is a multiple-step sequence of predominantly chemical processing steps to create chips used in everyday electrical and electronic devices. Silicon is the most common semiconductor used today, although gallium arsenide, germanium, and many other materials are used in special applications.

To create the standard silicon-based chips that are used in most computers, raw silicon dioxide, which is the most common part of sand or glass, is heated in the presence of hydrogen to produce pure silicon and water which is extracted. The silicon is then melted and formed into ingots, that are sliced into wafers approximately 0.5mm thick and ranging from a few centimeters to tens of centimeters in diameter.

Once the wafers are prepared, a large number of process steps are necessary to produce the desired semiconductor integrated circuits. In general the steps can be grouped into four areas: Front End Processing, Back End Processing, Test and Packaging.

Front End Processing refers to the most crucial steps in the fabrication. In this stage the actual semiconductor devices or transistors are created. A typical front end process includes the following: preparation of the wafer surface, patterning and subsequent implantation of dopants to obtain the desired electrical properties, growth or deposition of a gate dielectric, and growth or deposition of insulating materials to isolate neighboring devices.

Once the various semiconductor devices have been created they must be interconnected to form the desired electrical circuits. This "Back End Processing" involves depositing various layers of metal and insulating material in the desired pattern. Typically the metal layers consist of aluminum or more recently copper. The insulating material was traditionally a form of SiO_2 or a silicate glass, but recently new low-K materials are being used. The various metal layers are interconnected by etching holes, called "vias" in the insulating material and depositing tungsten in them.

Once the Back End Processing has been completed, the semiconductor devices are subjected to a variety of electrical tests to determine if they function properly.

Finally, the wafer is cut into individual die, which are then packaged in ceramic or plastic packages with pins or other connectors to the outside world.

Wafer Fabrication

- Wet cleans
- Photolithography
- Ion implantation (in which dopants are embedded in the wafer creating regions of increased (or decreased) conductivity)
- Dry Etching
- Wet Etching
- Plasma ashing
- Thermal treatments
  - — Rapid thermal anneal
  - — Furnace anneals
  - — Oxidation
- Chemical vapor deposition (CVD)
- Physical vapor deposition (PVD)
- Molecular beam epitaxy (MBE)
- Electroplating
- Chemical mechanical polish (CMP)
- Wafer testing (where the electrical performance is verified)
- Wafer backgrinding (to reduce the thickness of the wafer so the resulting chip can be put into a thin device like a smartcard or PCMCIA card.)

Die Preparation

- Wafer mounting
- Die cutting

IC Packaging

- Die attaching

IC Bonding

- Wire bonding
- Flip chip
- Tab bonding

IC Encapsulation

- Electronic design automation
- GDSII
- OASIS

Unit II–D

Semiconductor Devices: Transistor

The transistor is a solid state semiconductor device used for amplification and switching, and has three terminals. A small current or voltage applied to one terminal controls the current through the other two, hence the term transistor; a voltage- or current-controlled resistor. It is the key component in all modern electronics. In digital circuits, transistors are used as very fast electrical switches, and arrangements of transistors can function as logic gates, RAM-type memory and other devices. In analog circuits, transistors are essentially used as amplifiers.

Transistor was also the common name in the sixties for a transistor radio, a pocket-sized portable radio that utilized transistors (rather than vacuum tubes) as its active electronics. This is still one of the dictionary definitions of transistor.

Importance

The transistor is considered by many to be one of the greatest discoveries or inventions in modern history, ranking with banking and the printing press. Key to the importance of the transistor in modern society is its ability to be produced in huge numbers using simple techniques, resulting in vanishingly small prices. Computer "chips" consist of millions of transistors and sell for dollars, with per-transistor costs in the thousandths-of-pennies.

The low cost has meant that the transistor has become an almost universal tool for non-mechanical tasks. Whereas a common device, say a refrigerator, would have used a mechanical device for control, today it is often less expensive to simply use a few million transistors and the appropriate computer program to carry out the same task through "brute force". Today transistors have replaced almost all electromechanical devices, most simple feedback systems, and appear in huge numbers in everything from computers to cars.

Hand-in-hand with low cost has been the increasing move to "digitizing" all information. With transistorized computers offering the ability to quickly find (and sort) digital information, more and more effort was put into making all information digital. Today almost all media in modern society is delivered in digital form, converted and presented by computers. Common "analog" forms of information such as television or newspapers spend the vast majority of their time as digital information, being converted to analog only for a small portion of the time.

Invention

The transistor was invented at Bell Laboratories in December 1947 (first demonstrated on December 23) by John Bardeen, Walter Houser Brattain, and William Bradford Shockley, who were awarded the Nobel Prize in physics in 1956. Ironically, they had set out to manufacture a field-effect transistor (FET) predicted by Julius Edgar Lilienfeld as early as 1925 but eventually discovered current amplification in the point-contact transistor that subsequently evolved to become the bipolar junction transistor (BJT).

How Does a Transistor Work?

A transistor is a three-terminal device. In a bipolar junction transistor (BJT), an electrical *current* is fed into the base (B) and modulates the current flow between the other two terminals known as the emitter (E) and collector (C). In field-effect transistors (FET)s, the three terminals are called gate (G), source (S) and drain (D) respectively, and it is the voltage applied to the gate terminal that modulates the current between source and drain.

Bipolar Junction Transistor (BJT)

The schematic symbols for pnp- and npn-type BJTs.

Conceptually, one can understand a bipolar junction transistor as two diodes placed back to back, connected so they share either their positive or their negative terminals. The forward-biased emitter-base junction allows charge carriers to easily flow out of the emitter. The base is made thin enough so that most of the injected carriers will reach the collector rather than recombining in the base. Since small changes in the base current affect the collector current significantly, the transistor can be used in an electronic amplifier. The ratio of the collector current to the base current, called current gain or β, is on the order of 100 for most types of BJTs. That is, one milliampere of base current usually induces a collector current of about a hundred milliamperes. BJTs prevail in all sorts of amplifiers from audio to radio frequency applications and are also popular as electronic switching devices.

Bipolar transistors can be fabricated to match with like devices much better than FETs, making them useful for high precision analog circuit design. This makes them well suited as components in an op-amps and discrete transistor amplifiers, where combined in precisely-matched pairs they form input structures called current mirrors. Bipolar devices can also be used to measure temperature and to compute analog logarithms as well.

Field-Effect Transistor (FET)

The schematic symbols for p- and n-channel MOSFETs. The symbols to the right include an extra terminal for the transistor body (allowing for a seldom-used channel bias) whereas in those to the left the body is implicitly connected to the source.

The most common variety of field-effect transistors, the enhancement-mode MOSFET (metal-oxide semiconductor field-effect transistor) consists of a unipolar conduction channel and a metal gate separated from the main conduction channel by a thin layer of (SiO_2) glass. This is why an alternative name for the FET is 'unipolar transistor.' When a potential difference (of the proper polarity) is impressed across gate and source, charge carriers are introduced to the channel, making it conductive. The amount of this current can be modulated, or (nearly) completely turned off, by varying the gate potential.

Because the gate is insulated, no DC current flows to or from the gate electrode. This lack of a gate current (as compared to the BJT's base current), and the ability of the MOSFET to act like a switch, allows particularly efficient digital circuits to be created, with very low power consumption at low frequencies. The power consumption increases markedly with frequency, because the capacitive loading of the FET control terminal takes more energy to slew at higher frequencies, in direct proportion to the frequency. Hence, MOSFETs have become the dominant technology used in computing hardware such as microprocessors and memory devices such as RAM. Bipolar transistors are more rugged and hence more useful for low-impedance loads and inductively reactive (e.g. motor) loads.

Power MOSFETs become less conductive with increasing temperature and can therefore be applied in shunt, to increase current capacity, unlike the bipolar transistor, which has a negative temperature coefficient of resistance, and is therefore prone to thermal runaway. The downside of this is that, while the power FET can protect itself from overheating by diminishing the current through it, high temperatures need to be avoided by using a larger heat sink than for an equivalent bipolar device. Macroscopic FET power transistors are actually composed of many little transistors. They are stacked (on-chip) to increase breakdown potential and paralleled to reduce R_{on}, i.e. allowing for more current, bussing the gates to provide a single control (gate) terminal.

The depletion mode FET is a little different. It uses a back-biased diode for the control terminal, which presents a capacitive load to the driving circuit in normal operation. With the gate tied to the source, a DFET is fully on. Changing the potential of a DFET (pulling an N-channel gate downward, for example) will turn it off, i.e. 'deplete' the channel (drain-source) of charge carriers. MOSFETs, formerly called IGFETs (for Insulated Gate Field-Effect Transistor) can be depletion-mode, enhancement-mode, or mixed-mode, but are almost always enhancement mode in modern commercial practice. This means that, with the source and gate tied together (thus equipotential) the channel will be off (high impedance or non-conducting). The n-channel device (reverse for P-channel), like in the DFET, is turned on by raising the potential of the gate. Typically, the gate on a MOSFET will withstand +-20V, relative to the source terminal. If one were to raise the gate potential of an n-channel device without limiting the current to a few milliamps, one would destroy the gate diode, like any other small diode. Why do we typically think of n-channel devices as the default? In silicon devices, the ones that use majority carriers that are electrons, rather than holes, are slightly faster and can carry more current than their P-type counterparts. The opposite is true in GaAs devices.

Substituting the p-n-junction with a Schottky junction gives a MESFET (Metal Semiconductor Field Effect Transistor), used for GaAs and other III-V semiconductor materials. Using bandgap engineering in a ternary semiconductor, like AlGaAs gives a HEMT (High Electron Mobility Transistor).

The FET is simpler in concept than the bipolar transistor and can be constructed from a wide range of materials.

The most common use of MOSFET transistors today is the CMOS (complementary metallic oxide semiconductor) integrated circuit which is the basis for most digital electronic devices. These use a totem-pole arrangement where one transistor (either the pull-up or the pull-down) is on while the other is off. Hence, there is no DC drain, except during the transition from one state to the other, which is very short. As mentioned, the gates are capacitive, and the charging and discharging of the gates each time a transistor switches states is the primary cause of power drain.

The C in CMOS stands for 'complementary.' The pull-up is a P-channel device (using holes for the mobile carrier of charge) and the pull-down is N-channel (electron carriers). This allows busing of the control terminals, but limits the speed of the circuit to that of the slower P device (in silicon devices). The bipolar solutions to push-pull include 'cascode' using a current source for the load. Circuits that utilize both unipolar and bipolar transistors are called Bi-Fet. A recent development is called 'vertical P.' Formerly, BiFet chip users had to settle for relatively poor (horizontal) P-type FET devices. This is no longer the case and allows for quieter and faster analog circuits.

A clever variant of the FET is the dual-gate device. This allows for two opportunities to turn the device off, as opposed to the dual-base (bipolar) transistor which presents two opportunities to turn the device on.

FETs can switch signals of either polarity, if their amplitude is significantly less than the gate swing, as the devices (especially the parasitic diode-free DFET) are basically symmetrical. This means that FETs are the most suitable type for analog multiplexing. With this concept, one can construct a solid-state mixing board, for example.

The power MOSFET has a 'parasitic diode' (back-biased) normally shunting the conduction channel that has half the current capacity of the conduction channel. Sometimes this is useful in driving dual-coil magnetic circuits (for spike protection), but in other cases it causes problems.

The high impedance of the FET gate makes it rather vulnerable to electrostatic damage, though this is not usually a problem after the device has been installed.

A more recent device for power control is the insulated-gate bipolar transistor, or IGBT. This has a control structure akin to a MOSFET coupled with a bipolar-like main conduction channel. These have become quite popular.

Advantages of Transistors over Thermionic Valves

Before the transistor, the thermionic valve, or vacuum tube, was the main component of an amplifier. The key advantages that have allowed transistors to replace their valve predecessors in almost all applications are:

- smaller size, despite continuing miniaturisation of tubes;
- simpler manufacture and, hence;
- lower cost;
- lower operating potentials;
- absence of a heated filament and, as a consequence;
- lower power dissipation;
- higher reliability and greater endurance, in most contexts (although some valves are more resistant to nuclear electromagnetic pulses (NEMP); and
- there is no opportunity to (directly) achieve complementary symmetry with valves. Transformers can be used to work around this, at least in the usual narrow-band case.

Use in Audio Amplifiers

Some audio amplifiers still use valves, their enthusiasts claiming that their sound is superior.

In particular, some argue that the larger numbers of electrons in a valve behave with greater statistical accuracy, although this ignores the facts that tubes generally have a high-impedance control terminal, and that discrete-transistor (as opposed to, say, op-amp) circuits can also be designed to use large currents. (1 milliamp of current carries about 6.24 million billions of electrons per second.)

Others detect a distinctive "warmth" to the tone. The "warmth" is actually distortion caused by the valves, but some audiophiles find a certain amount of "fuzziness" pleasing. This is 'soft-saturation' and occurs when the valves are overdriven, causing poorly designed tube amplifiers to sound better than poorly designed transistor amplifiers.

Tubes are also preferred in guitar amplifiers which are designed to be overdriven, because they have a different non-linear transfer characteristic than transistors, and create a different, more pleasing spectrum of harmonic distortion or "fuzz". Digital signal processing (DSP) can be used to achieve similar effects in the digital domain.

It is possible to mix transistors and valves in the same circuit.

Simple transistor amplifiers use emitter degeneration to achieve negative feedback, which gives a relatively predictable gain compared to the gain of the transistor itself, which varies widely.

Use in Computers

The "second generation" of computers through the late 1950s and 1960s featured boards filled with individual transistors, and magnetic cores. Subsequently, transistors, other components (capacitors, but not high-value inductors or transformers), and the necessary wiring were integrated into a single, mass-manufactured component: the integrated circuit. In modern digital electronics, single transistors are very rare, though they remain common in power and analog applications. Recently, inroads have been made in the integration of analog circuits, also, with the advent of 'programmable analog' circuits. DSP is a technique that can (among other things) be used with A/D and D/A converters to simulate analog circuits. Linear integrated circuits got a bad reputation early on because of the difficulty of creating (high-quaity) PNP transistors, but are much better now.

How semiconductors work?

Operationally, semiconductors and vacuum tubes are similar. Think about a container (borosilicate glass or platinum, say) filled with distilled water. Drop in a pair of resistance probes and one would find that this arrangement does not conduct electricity, because there is no charge carrier, as long as the sense potential is kept below the electrolysis point—i.e. breakdown. Add a pinch of (any) salt, and conduction begins, because mobile carriers (ions) have been added. Increasing the salt concentration helps with conduction, but not very much. A dry lump of salt is non-conductive, because the charge carriers are immobile.

In a valve, on the other hand, the charge carriers (electrons) are generated by thermionic emission from a cathode heated by a resistance wire.

A pure silicon lattice is an insulator, but when e.g. arsenic atoms are added, in a concentration small enough not to disturb the regularity of the lattice, they donate free electrons and allow conduction. This is because we have introduced a mobile carrier of charge, creating what is known as an n-type material. The lattice seems like a vacuum to the charge carrier. Alternatively, we can introduce a boron impurity, which has only three electrons to silicon's four, and we introduce another kind of (virtual) carrier of charge called a 'hole.' This conducts also, yielding p-type material. Valves, in contrast, do not contain positive charge carriers.

Introduce a slightly higher doping concentration, and we have an even more (slightly) conductive material, so long as the lattice remains intact. The part of the bipolar transistor designated as the collector has a lower doping concentration (the ratio directly determines gain) than the emitter. Back-biasing of a diode, or turning off a transistor, is analogous to somehow removing the salt from the water, or freezing it. Note that in all cases the charge carriers repel one another, evening out their distribution in their respective matrices, in the absence of an outside electrostatic force. The grid in a tube deflects the charge carriers rather than extinguishing them. This structure is metallic, rather than semi-metallic as in an transistor, so it can withstand higher ambient temperatures, and dissipate local sources of heat (counter emf from say, droping an inductive load) better, because of its lower specific heat.

Do not think you can make a bipolar transistor out of two diodes. You need to have them on the same crystal, otherwise you don't have epitaxy.

Light-sensitivity

Bipolar transistors can be turned on with light as well as electricity. Devices designed for this purpose are called phototransistors, but these are often standard transistors in a transparent package.

History

All transistors rely on the ability of certain materials, known as semiconductors, to change their electrical resistance under the control of an electric field. In bipolar transistors, the semiconductor is formed into structures called *p-n junctions* that allow electricity to flow in only one direction through them—that is, they are a conductor when voltage is applied in one direction, and an insulator when it is applied in the other direction.

1900s

Semiconductors had been used in the electronics field for some time before the invention of the transistor. Around the turn of the 20th century they were quite common as detectors in radios, used in a device called a "cat's whisker". These detectors were somewhat troublesome, however, requiring the operator to move a small tungsten filament (the whisker) around the surface of a crystal until it suddenly started working. Then, over a period of a few hours or days, the crystal would slowly stop working and the process would have to be repeated. At the time their operation was completely mysterious. After the introduction of the more reliable and amplified vacuum tube based radios, the cat's whisker systems quickly disappeared. The "cat's whisker" is an example of a special type of diode still popular today, called a Schottky diode.

World War II

In WWII, radar research quickly pushed the frequencies of the radio receivers inside them into the area where traditional tube based radio receivers no longer worked well. On a whim, Russell Ohl of Bell Laboratories decided to try a cat's whisker. After hunting one down at a used radio store in Manhattan, he found that it worked much better than tube-based systems.

Ohl investigated why the cat's whisker functioned so well. He spent most of 1939 trying to grow more pure versions of the crystals. He soon found that with higher quality crystals the "oddness" went away, but so did their ability to operate as a radio detector. One day he found one of his purest crystals nevertheless worked well, and interestingly, it had a clearly visible crack near the middle. However, as he moved about the room trying to test it, the detector would mysteriously work, and then stop again. After some study he found that the behaviour was controlled by the light in the room – more light, more conductance. He invited several other people to see this crystal, and Brattain immediately realized there was some sort of junction at the crack.

Further research cleared up the remaining mystery. The crystal had cracked because either side contained very slightly different amounts of the impurities Ohl could not remove—about 0.2 per cent. One side of the crystal had impurities that added extra electrons (the carriers of electrical current) and made it a conductor. The other had impurities that wanted to bind to these electrons, making it an insulator. When the two were placed side by side the electrons could be pushed out of the side with extra electrons (soon to be known as the emitter) and replaced by new ones being provided (say from a battery) where they would flow into the insulating portion and be collected by the filament (the collector). However, when the voltage was reversed the electrons being pushed into the collector would quickly fill up the "holes", and conduction would stop almost instantly. This junction of the two crystals (or parts of one crystal) created a solid-state diode, and the concept soon became known as semiconduction. 'Anode' and 'Cathode' are the terms used to denote the two terminals of a diode. The mechanism of action when the diode is off has to do with the separation of charge carriers around the junction. This is called a 'depletion region.'

Development

Armed with the knowledge of how these new diodes worked, a crash effort started to learn how to build them on demand. Teams at Purdue University, Bell Labs, MIT, and the University of Chicago all joined forces to build better crystals. Within a year germanium production had been perfected to the point where military-grade diodes were being used in most radar sets.

The key to the development of the transistor was the further understanding of the process of the electron mobility in a semiconductor. It was realized that if there was some way to control the flow of the electrons from the emitter to the collector, one could build an amplifier. For instance, if you placed contacts on either side of a single type of crystal the current would not flow through it. However, if a third contact could then "inject" electrons or holes into the material, the current would flow.

Actually doing this appeared to be very difficult. If the crystal were of any reasonable size, the amount of electrons (or holes) supplied would have to be very large – making it less than useful as an amplifier because it would require a large current to start with. That said, the whole idea of the crystal diode was that the crystal itself could provide the electrons over a very small distance. The key appeared to be to place the input and output contacts very close together on the surface of the crystal.

Brattain started working on building such a device, and tantalizing hints of amplification continued to appear as the team worked on the problem. One day the system would work and the next it wouldn't. In one instance a non-working system started working when placed in water. The two eventually developed a new branch of quantum mechanics known as surface physics to account for the behaviour.

Essentially, the electrons in any one piece of the crystal would migrate about due to nearby charges. Electrons in the emitters, or the "holes" in the collectors, would cluster at the surface of the crystal where they could find their opposite charge "floating around" in the air (or water). Yet they could be pushed away from the surface from any other location with the application of a small amount of charge. So instead of needing a large supply of electrons, a very small number in the right place would do the trick.

Their understanding solved the problem of needing a very small control area to some degree. Instead of needing two separate semiconductors connected by a common, but tiny, region, a single larger surface would serve. The emitter and collector would both be placed very close together on one side, with the control lead on the other. When current was applied to the control lead, the electrons or holes would be pushed out, right across the entire block of semiconductor, and collect on the far surface. As long as the emitter and collector were very close together, this should allow enough electrons or holes between them to allow conduction to start.

First Transistor

They made many attempts to build such a system with various tools, but generally failed. Setups where the contacts were close enough were invariably as fragile as the original cat's whisker detectors had been, and would work briefly, if at all.

Eventually they had a practical breakthrough. A piece of gold foil was glued to the edge of a plastic wedge, and then the foil was sliced with a razor at the tip of the triangle. The result was two very closely spaced contacts of gold. When the plastic was pushed down onto the surface of a crystal and voltage applied to the other side (on the base of the crystal), current started to flow from one contact to the other as the base voltage pushed the electrons away from the base towards the other side near the contacts. The point-contact transistor had been invented, a primitive variation of the BJT.

While the device was constructed a week earlier, Brattain's notes describe the first demonstration to higher-ups at Bell Labs on the afternoon of December 23, 1947, often given as the birthdate of the transistor. The PNP point-contact germanium transistor operated as a speech amplifier with a power gain of 18 in that trial.

Shockley was upset about the device being credited to Brattain and Bardeen, who he felt had built it "behind his back" to take the glory. Matters became worse when Bell Labs lawyers found that some of Shockley's own writings on the transistor were close enough to those of an earlier patent that they thought it best that his name be left off the patent application.

Shockley was incensed, and decided to demonstrate who was the brains of the operation. Only a few months later he invented an entirely new type of transistor one day while sitting in his hotel room waiting to give a speech. This new form, the layer transistor, was considerably more robust than the fragile point-contact system, and would go on to be used for the vast majority of all transistors into the 1960s.

With the fragility problems solved, a remaining problem was purity. Making germanium of the required purity was proving to be a serious problem, and limited the number of transistors that actually worked from a given batch of material. One then-small company, Texas Instruments, decided that the solution to this problem was to use silicon rather than germanium, which should be easier to work with. They were right, and germanium disappeared from almost all transistors within only two years of silicon being introduced in the early 1950s.

Now everything was in place, and within a few years, transistor-based products, most notably radios, were appearing on the market. A major improvement in yield came when a chemist advised the fabs to use distilled water rather than tap water: calcium ions were the cause of the problem. 'Zone melting,' a technique using a moving band of molten material through the crystal, makes this whole endeavour possible.

Origin of Name

John R. Pierce coined the name "transistor" in 1949. It was originally thought that the transistor could be usefully considered to be the electronic dual of a vacuum tube. The property equivalent to the transconductance of a tube would have been "transresistance" and the device would then have been a "transresistor," or "transistor" for short. In practice the transistor was not close enough to being a vacuum-tube dual for the concept to have any quantitative usefulness, and the concept of "transresistance" lives on only in the name "transistor."

Early Consumer and Hobbyist Applications

The transistor radio was not the first "mainstream" application of transistors. Even by the 1940s, ordinary consumer radios were rather sophisticated pieces of electronics, using several tubes, and based on Armstrong's brilliantly ingenious superheterodyne technology. To meet consumer expectations, it was necessary for a transistor radio to use similar circuitry. It was not easy in the early days to get transistors to operate reliably as amplifiers and oscillators in the RF range—even the 540-1700 "kilocycle" AM broadcast band. Miniaturized versions of many necessary components such as IF transformers and multiganged tuning capacitors were not available.

The first major consumer application of transistors was the hearing aid, which required only audio amplification, and represented a market where miniaturization was important and low price was not a requirement. Raytheon, which had developed miniaturized and ruggedized vacuum tubes for the military, introduced the first transistorized hearing aids.

Raytheon also introduced the first transistor, the CK722, that was widely available through ordinary commercial channels. Electronics hobbyists of the fifties have a warm place in their heart for the CK722, essentially the only transistor available for almost a decade, and innumerable homebrew projects were designed around it. The CK722's available to hobbyists were, essentially, those that had failed QC for more demanding applications. Germanium based, with low gain and high emitter-to-collector leakage, and high variation in characteristics from unit to unit, designing practical circuits with these components was quite a challenge.

Nowadays, the most visibly popular transistor, at least in the USA, is the 2N3904, a small NPN component that can be found in hundreds, if not thousands, of types of electronic devices. The 2N3906 (its PNP counterpart) and the 2N2222 (NPN) are also very popular. In Europe the most popular general purpose transistors are the BC547 and BC546 (NPN) and the BC557 and BC556 (PNP). All cost only a few cents in quantity, but even single units can be had quite cheaply, making them popular with hobbyists as well.

Unit II–E

Semiconductor Devices: Diode

A **diode** functions as the electronic version of a one way valve. By restricting the direction of movement of charge carriers, it allows an electric current to flow in one direction, but blocks it in the opposite direction.

Applications

Radio Demodulation

The first use for the diode was the demodulation of amplitude modulated (AM) radio broadcasts. The history of this discovery is treated in depth in the radio article. In summary, an AM signal consists of alternating positive and negative peaks of current, whose amplitude or 'envelope' is proportional to the original audio signal, but whose average value is zero. The diode rectifies the AM signal (i.e. it eliminates the negative peaks), leaving a signal whose average amplitude is the desired audio signal. The average value is extracted using a simple filter and fed into a transducer (originally a crystal earpiece, now more likely to be a loudspeaker), which generates sound.

Logic Gates

Diodes can be used to construct logic gates: logical and and logical or.

Power Conversion

A diode is called a half wave rectifier when it is used to convert alternating current electricity into direct current, by removing the negative portion of the current.

A special arrangement of four diodes that will transform an alternating current into a direct current, using both positive and negative excursions of a single phase alternating current, is known as a diode bridge, single-phase bridge rectifier, or simply a full wave rectifier.

With a split (center-tapped) alternating current supply it is possible to obtain full wave rectification with only two diodes. Often diodes come in pairs, as double diodes in the same housing.

When it is desired to rectify three phase power, one could rectify each of the three phases with the arrangement of four diodes used in single phase, which would require a total of 12 diodes. However, due to redundancy, only six diodes are needed to make a three phase full wave rectifier.

Most devices that generate alternating current (such devices are called alternators) generate three phase alternating current.

Disassembled automobile alternator, showing the six diodes that comprise a full-wave three phase bridge rectifier.

For example, an automobile alternator has six diodes inside it to function as a full wave rectifier for battery charge applications. Many of the small wind turbines use three double diodes bolted to the same heatsink.

Over-Voltage Protection

Diodes are frequently used to conduct dangerously high voltages away from sensitive devices, most commonly by being reverse-biased (non-conducting) under normal circumstances, and becoming forward-biased (conducting) when the voltage rises above its normal value. For example, diodes are used in stepper motor and relay circuits to de-energize coils rapidly without the damaging voltage spikes that would otherwise occur. Many integrated circuits also incorporate diodes on the connection pins to prevent external voltages from damaging their sensitive transistors. Specialized diodes are used to protect from over-voltages at higher power.

Diode technology

The first diodes were vacuum tube devices (also known as thermionic valves), arrangements of electrodes surrounded by a vacuum within a glass envelope, similar in appearance to incandescent light bulbs. The arrangement of a filament and plate as a diode was invented in 1904 by John Ambrose Fleming, scientific adviser to the Marconi company, based on an observation by Thomas Edison. Like light bulbs, vacuum tube diodes have a filament through which current is passed, heating the filament. In its heated state it can now emit electrons into the vacuum. These electrons are electrostatically drawn to a positively charged outer metal plate called the anode, or just the "plate". Electrons do not flow from the plate back toward the filament, even if the charge on the plate is made negative, because the plate is not heated.

Although vacuum tube diodes are still used for a few specialized applications, most modern diodes are based on semiconductor p-n junctions. In a p-n diode, conventional current can flow from the p-doped side (the anode) to the n-doped side (the cathode), but not in the opposite direction. When the diode is reverse-biased, the charge carriers are pulled away from the center of the device, creating a depletion region.

Diode Types

There are several types of semiconductor junction diodes:

Normal (p-n) Diodes

These operate as described above. Usually made of doped silicon or, more rarely, germanium.

'Gold Doped' Diodes

The gold causes 'minority carrier suppression.' This lowers the effective capacitance of the diode, allowing it to operate at signal frequencies. A typical example is the 1N914. Germanium and Schottky diodes are also fast like this, as are bipolar transistors 'degenerated' to act as diodes. Power supply diodes are made with the expectation of working at a maximum of 2.5 × 400 Hz (sometimes called 'French power' by Americans), and so are not useful above a kilohertz.

Zener Diodes

Diodes that can be made to conduct backwards. This effect, called Zener Breakdown, occurs at a precisely defined voltage, allowing the diode to be used as a precision voltage reference. Some devices labelled as high-voltage Zener diodes are actually avalanche diodes (see below). Two (equivalent) Zeners in series and in reverse order, in the same package, constitute a transient absorber (or *Transorb*, a registered trademark). They are named for Dr. Clarence Melvin Zener of Southern Illinois University, inventor of the device.

Avalanche Diodes

Diodes that conduct in the reverse direction when the reverse bias voltage exceeds the breakdown voltage. These are electrically very similar to Zener diodes, and are often mistakenly called Zener diodes, but break down by a different mechanism, the Avalanche Effect. This occurs when the reverse electric field across the p-n junction causes a wave of ionization, reminiscent of an avalanche, leading to a large current. Avalanche diodes are designed to break down at a well-defined reverse voltage without being destroyed. The difference between the avalanche diode (which has a reverse breakdown above about 6.2 V) and the Zener is that the channel length of the former exceeds the 'mean free path' of the electrons, so there are collisions between them on the way out. The only practical difference is that the two types have temperature coefficients of opposite polarities. Practical voltage reference circuits feature Zener and switching diodes connected in series and opposite directions to balance the temperature coefficient to near zero.

Transient Voltage Suppression (TVS) Diodes

These are avalanche diodes designed specifically to protect other semiconductor devices from electrostatic discharges. Their p-n junctions have a much larger cross-sectional area than those of a normal diode, allowing them to conduct large currents to ground without sustaining damage.

Light-emitting Diodes (LEDs)

As the electrons cross the junction they emit photons. In most diodes, these are reabsorbed, and are at frequencies that can not be seen (usually infrared). However, with the right materials and geometry, the light becomes visible. The forward potential of these diodes define their color. Thus different materials (extrinsic semiconductors) must be used. 1.2 V corresponds to red, 2.4 to violet. Now, even soft UV diodes are available. The first LED's were red and yellow, and higher-frequency diodes have been developed over time. Polishing the device with parallel faces, so as to form a resonant cavity, yields a 'laser diode.' All LEDs are monochromatic; 'white' LED's are actually combinations of three LED's of a different color, or a blue LED with a yellow scintillator coating. The lower the frequency of emission, the greater the efficiency. So to normalize output when using LED's of different colors, increase current in the higher frequency models. This effect is complicated, somewhat, by the fact that the human eye is most sensitive in the blue-green.

Photodiodes

These have wide, transparent junctions. Photons can push electrons over the junction, causing a current to flow. Photo diodes can be used as solar cells. And in photometry. If a photon doesn't have enough energy, it isn't going to turn the photo-diode on very much. LED's can be used as low-efficiency photodiodes in signal applications. Sometimes a LED is paired with a photodiode or phototransistor in the same package. This device is called an "opto isolator." Unlike a transformer, this scheme allows for DC coupling. These are used to protect hospital patients from shock. Patients with IV's in their bodies are particularly susceptable, sometimes succumbing to 'carpet shock.' They are also used to isolate low-current control or signal circuitry from "dirty" power supply circuits or higher-current motor and machine circuits.

Schottky Diodes

These have a very low forward voltage drop, usually 0.15 to 0.45 V, which makes them useful in battery-powered and low-voltage circuits. Also in mixer circuits for RF.

Snap Diodes

These can provide very fast voltage transitions.

Esaki or Tunnel Diodes

These have a region of operation showing negative resistance caused by quantum tunneling, thus allowing amplification of signals and very simple bistable circuits.

Gunn Diodes

These are similar to tunnel diodes in that they are made of materials such as GaAs or InP that exhibit a region of negative differential resistance. With appropriate biasing, dipole domains form and travel across the diode, allowing high frequency microwave oscillators to be built.

There are other types of diodes, which all share the basic function of allowing electrical current to flow in only one direction, but with different methods of construction.

Point Contact Diode

This works the same as the junction semiconductor diodes described above, but its construction is simpler. A block of n-type semiconductor is built, and a conducting sharp-point contact made with some group-3 metal is placed in contact with the semiconductor. Some metal migrates into the semiconductor to make a small region of p-type semiconductor near the contact. The long-popular 1N34 germanium version is still used in radio receivers as a detector and occasionally in specialized analog electronics.

Tube or Valve Diode

This is the simplest kind of vacuum tube device (referred to as a valve in the UK). Electrons will move from a heated metal surface (cathode) treated with a mixture of barium and strontium oxides into a vacuum (thermionic emission). After leaving the cathode, they can be attracted to positively charged cool surface (anode). However, electrons are not easily released from a cold untreated surface when the voltage polarity is reversed and hence any flow is a very small current. For much of the 20th century they were used in analog signal applications, and as rectifiers in power supplies. Tube diodes were nearly obsolete by 2001, except as rectifiers in tube guitar and hi-fi amplifiers and in a few specialized high-voltage applications.

Gas Discharge Diode

There are two electrodes, not touching, in some kind of gas. One electrode is very sharp. The other has a smoothly curved finish. If a strong negative potential is applied to the sharp electrode, the electric field near the sharp edge or point is enough to cause an electrical discharge in the gas, and a current flows. If the reverse potential is applied, the electrical field strength around the smooth electrode is not enough to start a discharge. (The discharge can only start easily at the negative end because electrons are much more mobile than positive ions.) These are sometimes used for high-voltage high-current rectification in power supply applications.

Varicap or Varactor Diodes

Used as voltage-controlled capacitors. These were important in PLL (phase-locked loop) and FLL (frequency-locked loop) circuits, allowing tuning circuits, such as those in television receivers, to lock quickly, replacing older designs that took a long time to warm up and lock. A PLL is faster than a FLL, but prone to integer harmonic locking (if one attempts to lock to a broadband signal). They also enabled tunable oscillators in early discrete tuning of radios, where a cheap and stable, but fixed-frequency, crystal oscillator provided the reference frequency for a voltage-controlled oscillator.

Other uses for semiconductor diodes include sensing temperature, and computing analog logarithms.

Unit II–F

Semiconductor Devices: Microprocessor and Thermistor

Microprocessor

A **microprocessor** (abbreviated as **µP** or **uP**) is an electronic computer central processing unit (CPU) made from miniaturized transistors and other circuit elements on a single semiconductor integrated circuit (IC) (aka microchip or just chip).

Before the advent of microprocessors, electronic CPUs were made from discrete (separate) TTL integrated circuits; before that, individual transistors; and before that, from vacuum tubes. There have even been designs for simple computing machines based on mechanical parts such as gears, shafts, levers, Tinkertoys, etc. Leonardo DaVinci made one such design, although none were possible to construct using the manufacturing techniques of the time.

The evolution of microprocessors have been known to follow Moore's Law when it comes to steadily increasing performance over the years.

Perspective History

The world's first commercial microprocessor was the 4-bit 4004, released on November 15, 1971, invented by Marcian Ted Hoff. The 4004 was later followed by the 8008. These processors are the precursors to the very successful Intel 8080, Zilog Z80, and derivative Intel 8-bit processors. The competing Motorola 6800 architecture was cloned and improved in the MOS Technology 6502, rivaling the Z80 in popularity during the 1980s.

The 8080 is the ancestor of the 16-bit Intel 8086, the first member of the x86 family which powers most modern PC type computers. Intel introduced the 8086 as a cost effective way of porting software from the 8080 lines, and succeeded in winning much business on that premise. Following up their 8086 and 8088, Intel released the 32-bit 80386 in 1985, cementing Intel's PC dominance with its processor family's backwards compatibility.

Examples of other 16-bit microprocessor families include the Motorola 68000, used in early Apple Macintosh computers; and the Zilog Z8000.

In 2003, about 44 billion US$ worth of microprocessors were manufactured and sold. Although about half of that money was spent on CPUs used in desktop or laptop personal computers, those count for only about 0.2 per cent of all CPUs sold.

About 55 per cent of all CPUs sold in the world are 8-bit microcontrollers.

Less than 10 per cent of all the CPUs sold in the world are 32-bit or more. Of all the 32-bit CPUs sold, about 2 per cent are used in desktop or laptop personal computers.

Microprocessors and Architectures

- AMD K5, K6, K6-2, K6-III, Duron, Athlon
- AMD Opteron
- ARM family, StrongARM
- Atmel AVR architecture (purely microcontrollers)
- CDP1802 (RCA COSMAC)
- DEC Alpha
- Intel 4004, 4040
- Intel 8080, 8085, Zilog Z80
- Intel 8086, 8088, 80186, 80188, 80286, 80386, 80486
- Intel Pentium, Pentium Pro, Celeron, Centrino, Pentium II, Pentium III, Xeon, Pentium 4, Pentium M
- Intel Itanium
- Intel i860, i960
- MIPS architecture
- Motorola 6800, MOS Technology 6502, Motorola 6809, WDC 65816
- Motorola 68000 family, ColdFire
- Motorola 88000, IBM POWER (parents of the PowerPC family)
- NSC 320xx
- OpenCores OpenRISC architecture
- PA-RISC family (HP)
- PowerPC family, G3, G4, G5
- National Semiconductor SC/MP (scamp)
- Signetics 2650
- SPARC, UltraSPARC, UltraSPARC II–IV
- SuperH family
- Transmeta's Crusoe and Efficeon
- INMOS Transputer

Thermistor

A **thermistor** is a type of resistor used to measure temperature changes, relying on the change in its resistance with changing temperature.

If we assume that the relationship between resistance and temperature is linear (i.e. we make a first-order approximation), then we can say that:

$$\Delta R = k\,\Delta T$$

where

ΔR = change in resistance

ΔT = change in temperature

k = first-order temperature coefficient of resistance.

Thermistors can be classified into two types depending on the sign of k. If k is positive, the resistance increases with increasing temperature, and the device is called a positive temperature coefficient (PTC) thermistor, or *posistor*. If k is negative, the resistance decreases with increasing temperature, and the device is called a negative temperature coefficient (NTC) thermistor. Resistors that are not thermistors are designed to have the smallest possible k, so that their resistance remains almost constant over a wide temperature range.

Steinhart Hart Equation

In practice, the linear approximation (above) works only over a small temperature range. For accurate temperature measurements, the resistance/temperature curve of the device must be described in more detail. The Steinhart-Hart equation is a widely used third-order approximation:

$$T = \frac{1}{a+b}\ln R + c\,(\ln R^3)$$

where a, b and c are called the Steinhart-Hart parameters, and must be specified for each device. T is the temperature in kelvin and R is the resistance in ohms. To give resistance as a function of temperature, the above can be rearranged into:

$$R = \mathrm{e}\left(\beta - \frac{\alpha}{2}\right)^{1/3} - \left(\beta + \frac{\alpha}{2}\right)^{1/3}$$

where

$$\alpha = \frac{\mathrm{a} - \dfrac{1}{\mathrm{T}}}{\mathrm{c}}$$

and

$$\beta = \sqrt{\left(\frac{\beta}{3\mathrm{c}}\right)^3 + \frac{\alpha^2}{4}}$$

Conduction Model

Many NTC thermistors are made from a thin coil of semiconducting material such as a sintered metal oxide. They work because raising the temperature of a semiconductor increases the number of electrons able to move about and carry charge—it promotes them into the conducting band. The more charge carriers that are available, the more current a material can conduct. This is described in the formula:

$$I = n \cdot A \cdot V \cdot e$$

where

I = electric current (Ampere)

n = number of charge carriers

A = area of the material (m^2)

V = voltage (volt)

e = charge of an electron (e = 1,602 · 10^{-19} C (coulomb))

The current is measured using an ammeter. Over large changes in temperature, callibration is necessary. Over small changes in temperature, if the right semiconductor is used, the resistance of the material is linearly proportional to the temperature. There are many different semiconducting thermistors and their range goes from about 0.01 kelvin to 2000 kelvin (approx. 1700°C)

Applications

Thermistors are used in smart camera flash guns which adjust for proper film exposure according to the light reflected.

PTC thermistors can be used as current-limiting devices for circuit protection, as replacements for fuses. Current through the device causes a small amount of resistive heating. If the current is large enough to generate more heat than the device can lose to its surroundings, the device heats up, causing its resistance to increase, and therefore causing even more heating. This creates a self-reinforcing effect that drives the resistance upwards, reducing the current and voltage available to the device.

NTC thermistors can be used as inrush-current limiting devices in power supply circuits. They present a higher resistance initially which prevents large currents from flowing at turn-on, and then heat up and become much lower resistance to allow higher current flow during normal operation. These thermistors are usually much larger than measuring type thermistors, and are purpose designed for this application.

Unit II–G

Semiconductor Materials and Spintronics

Semiconductor Materials

Semiconductor materials are insulators at zero temperature that conduct electricity in a limited way at room temperature. The defining property of a semiconductor material is that it can be doped with impurities that alter its electronic properties in a controllable way.

Because of their application in devices like transistors (and therefore computers) and lasers, the search for new semiconductor materials and the improvement of existing materials is an important field of study in materials science.

The most commonly used semiconductor materials are crystalline inorganic solids. These materials can be classified according to the periodic table groups from which their constituent atoms come.

List of Semiconductor Materials

- Group IV elemental semiconductors
  Diamond (C)
  Silicon (Si)
  Germanium (Ge)

- Group IV compound semiconductors
  Silicon carbide (SiC)
  Silicon germanide (SiGe)

- III-V semiconductors
  Aluminium antimonide (AlSb)
  Aluminium arsenide (AlAs)
  Aluminium nitride (AlN)

Aluminium phosphide (AlP)
Boron nitride (BN)
Gallium antimonide (GaSb)
Gallium arsenide (GaAs)
Gallium nitride (GaN)
Gallium phosphide (GaP)
Indium antimonide (InSb)
Indium arsenide (InAs)
Indium nitride (InN)
Indium phosphide (InP)

- III-V ternary semiconductor alloys
  Aluminium gallium arsenide ($Al_xGa_{1-x}As$)

- II-VI semiconductors
  Cadmium selenide (CdSe)
  Cadmium sulfide (CdS)
  Cadmium telluride (CdTe)
  Zinc oxide (ZnO)
  Zinc selenide (ZnSe)
  Zinc sulfide (ZnS)
  Zinc telluride (ZnTe)

- I-VII semiconductors
  Cuprous chloride (CuCl)

- IV-VI semiconductors
  Lead sulfide (PbS)
  Lead telluride (PbTe)
  Tin sulfide (SnS)

- II-V semiconductors
  Cadmium phosphide (Cd_3P_2)
  Cadmium arsenide (Cd_3As_2)
  Cadmium antimonide (Cd_3Sb_2)
  Zinc arsenide (Zn_3As_2)

- Layered semiconductors
  Lead iodide (PbI_2)

Molybdenum disulfide (MoS_2)
Gallium Selenide (GaSe)

- Miscellaneous oxides
  Copper (I) oxide (Cu_2O)
  Copper (II) oxide (CuO)
- Organic semiconductors
- Magnetic semiconductors

Wide Bandgap Semiconductors

Wide Bandgap Semiconductors have applications in optoelectronic and power devices.

Gallium Nitride (GaN) is of particular interest, as it is used to make blue LEDs and lasers. In the future, high brightness, long life white LEDs may replace incandescent bulbs in many situations. The next generation of DVD players, the Blue ray disk, uses GaN based lasers.

Applications for high power and temperature devices are also interesting. Both Gallium Nitride and Silicon Carbide are robust materials well suited for such applications.

Important Wide Bandgap Semiconductors

- Silicon carbide
- Aluminium Nitride
- Gallium Nitride
- Diamond

Spintronics

Neologism for "spin-based electronics"), also known as magnetoelectronics, is an emergent technology which exploits the quantum propensity of electrons to spin as well as making use of their charge state. The spin itself is manifested as a detectable weak magnetic energy state characterised as "spin up" and "spin down".

Conventional usage of electron state within a semiconductor is a purely binary proposition, where an electron's state represents only 0 or 1, and a range of eight bits can represent every number between 0 and 255, but only one number at a time. Spintronics quantum bits (known as *qubits*) exploit the "spin up" and "spin down" states as superpositions of 0 or 1, possessing the ability to represent every number between 0 and 255 simultaneously.

Spintronics is likely to have radical implications in the field of mass-storage devices; recently (in 2002) IBM scientists announced that they could compress massive amounts of data into a small area, at approximately one trillion bits per square inch.

Less radical spintronic devices would allow one to put a pair of signals through a single wire, by using spin polarised electrons and producing a different signal on spin up electrons to the spin down. This has the effect of doubling the bandwidth of the cable.

In order to make a spintronic device, the primary requirement is to have a system that can generate a current of spin polarised electrons, and a system that is sensitive to the spin polarisation of the electrons. The radical devices would also have a unit in between these that does some processing to the current of electrons, dependent on the spin states.

The simplest method of generating a spin polarised current is to pass the current through a ferromagnetic material. In order for this to work, the ferromagnetic must be a single crystal, so that it filters the electrons in a uniform manner. Similarly, putting a ferromagnetic filter in front of a device such as a transistor will make it a spin sensitive detector. If the two magnetic fields are aligned, then a current is passed, whereas if opposed then the resistance of the whole system is higher. This effect is known as giant magnetoresistance (GMR).

It can be claimed that the most successful spintronic device to date is the spin valve. This is a device, utilising a layered structure of magnetic materials, which displays a huge sensitivity to magnetic fields. When in the presence of a magnetic field, it allows all electrons through, but in the absence of such a field, it only lets electrons with some spins through. This show sensitivity to very weak fields, and has been used as the transducer in computer hard disk drive heads since about 2001, becoming common in 2002.

Unit II–H

Two-terminal, Three-terminal, Four-terminal Semiconductor Devices

While the transistor is the most important semiconductor device, there are literally dozens of other families of semiconductor devices. Some of the great many semiconductor devices in current usage are as follows:

Two-terminal Devices

- Semiconductor diode
- Zener diode
- Light-emitting diode
- PIN diode
- Schottky diode
- Avalanche diode
- Laser diode
- DIAC

Three-terminal Devices

- Bipolar transistor
- Field effect transistor
- Darlington transistor
- Thyristor
- Triac

Four-terminal Devices

- Hall effect sensor (magnetic field sensor)

Two-terminal devices

The semiconductor diodes have been discussed in an earlier unit.

Zener Diode

Techniques have produced devices with voltages lower than 5.6 volts with negligible temperature coefficients, but as higher voltage devices are encountered, the temperature coefficient rises dramatically. A 75 volt diode has 10 times the coefficient of a 12 volt diode.

Light-emitting Diode

A light-emitting diode (LED) is a semiconductor device that emits incoherent monochromatic light when electrically biased in the forward direction. This effect is a form of electroluminescence. The color depends on the semiconducting material used, and can be near-ultraviolet, visible or infrared. Nick Holonyak Jr. (1928-) of the University of Illinois at Urbana-Champaign developed the first practical visible-spectrum LED in 1962.

LED Technology

An LED is a special type of semiconductor diode. Like a normal diode, it consists of a chip of semiconducting material impregnated, or doped, with impurities to create a structure called a *pn junction*. Charge-carriers (electrons and holes) are created by an electric current passing through the junction. When an electron meets a hole, it falls into a lower energy state, and releases energy in the form of a photon as it does so.

The wavelength of the light emitted, and therefore its colour, depends on the bandgap energy of the materials forming the pn junction. A normal diode, typically made of silicon or germanium, emits invisible far-infrared light, but the materials used for an LED have bandgap energies corresponding to near-infrared, visible or near-ultraviolet light.

Unlike incandescent bulbs, which can operate with either AC or DC, LEDs require a DC supply of the correct polarity. When the voltage across the pn junction is in the correct direction, a significant current flows and the device is said to be forward biased. The voltage across the LED in this case is fixed for a given LED and is proportional to the energy of the emitted photons. If the voltage is of the wrong polarity, the device is said to be reverse biased, very little current flows, and no light is emitted.

Because the voltage versus current characteristics of an LED are much like any diode, they can be destroyed by connecting them to a voltage source much higher than their turn on voltage. A good LED driver circuit is either a constant current source or an approximation to a current source made by connecting the LED in series with a current limiting resistor to a voltage source. The voltage drop across a forward biased LED increases as the amount of light emitted increases because of the optical power being radiated. One consequence is that LEDs of the same type can be readily operated in parallel. The turn-on voltage of an LED is a function of the color, a higher forward drop is associated with emitting higher energy (bluer) photons. The reverse voltage that most LEDs can sustain without damage is usually only a few volts. Some LED units contain two diodes, one in each direction and each a different colour (typically red and green), allowing two-colour operation or a range of colors to be created by altering the percentage of time the voltage is in each polarity.

Conventional LEDs are made of inorganic minerals such as:

- aluminium gallium arsenide (AlGaAs)—red and infrared

- gallium arsenide/phosphide (GaAsP)—red, orange and yellow
- gallium nitride (GaN)—green
- gallium phosphide (GaP)—green
- zinc selenide (ZnSe)—blue
- indium gallium nitride (InGaN)—blue
- silicon carbide (SiC)—blue
- diamond (C)—ultraviolet
- silicon (Si)—under development

LED development began with infrared and red devices, and technological advances have made possible the production of devices with ever shorter wavelengths.

Commercially viable blue LEDs based on the wide bandgap semiconductor Gallium nitride (GaN) were invented by Shuji Nakamura while working in Japan at Nichia Corporation in 1993 and became widely available in the late 1990s. They can be added to existing red and green LEDs to produce white light. Most "white" LEDs in production today use a blue LED chip covered by a yellowish phosphor coating usually made of cerium doped Yttrium aluminum garnet (YAG:Ce) crystals which have been powdered and bound in a type of viscous adhesive. The LED chip emits blue light, part of which is converted to yellow by the YAG:Ce. The single crystal form of YAG:Ce is actually considered a Scintillator rather than a phosphor. This mixture of blue and yellow light creates the impression of white - hence the bluish or yellowish tint that these diodes usually exhibit. White LEDs can also be made by coating near ultraviolet (NUV) emitting LEDs with a mixture of high efficiency europium based red and blue emitting phosphors plus green emitting copper and aluminum doped zinc sulfide (ZnS:Cu,Al). This is a method analogous to the way fluorescent lights work. The newest method used to produce white light LEDs uses no phosphors at all and is based on homoepitaxially grown Zinc selenide (ZnSe) on a ZnSe substrate which simultaneously emits blue light from its active region and yellow light from the substrate.

Ultraviolet and blue LEDs are relatively expensive compared to the more common reds, greens, yellows and infrareds and are, thus, less commonly used in commercial applications.

The semiconducting chip is encased in a solid plastic lens, which is much tougher than the glass envelope of a traditional light bulb or tube. The plastic may be coloured, but this is only for cosmetic reasons and does not affect the colour of the light emitted.

Most typical LEDs are designed to operate with no more than 30-60 milliwatts of electrical power. Around 1999, commercial LEDs capable of continuous use at one watt of input power were introduced. These LEDs used much larger semiconductor die sizes to handle the large power input. As well, the semiconductor die were mounted to metal slugs to allow for heat removal from the LED die. In 2002, 5 watt LEDs were available with efficiencies of 18-22 lumens per watt. It is projected that by 2005, 10 watt units will be available with efficiencies of 60 lumens per watt. These devices will produce about as much light as a common 50 watt incandescent bulb, and will facilitate use of LEDs for general illumination needs.

In the last few years (up to 2003) there has been much research into organic LEDs or OLEDs, which are made of semiconducting organic polymers. The best efficiency of an OLED so far is about 10 per cent. These promise to be much cheaper to fabricate than inorganic LEDs, and large arrays of them can be deposited on a screen using simple printing methods to create a colour graphic display.

Organic Light-emitting Diodes (OLEDs)

If the emissive layer material of an LED is an organic compound, it is known as an Organic Light Emitting Diode (OLED). To function as a semiconductor, the organic emissive material must have conjugated pi bonds. The emissive material can be a small organic molecule in a crystalline phase, or a polymer. Polymer materials can be flexible; such LEDs are known as PLEDs or FLEDs.

Compared with regular LEDs, OLEDs are lighter and polymer LEDs can have the added benefit of being flexible. Some possible future applications of OLEDs could be:

- Light sources
- Wall decorations
- Luminous cloth

LED Applications

Here is a list of known applications for LEDs, some of which are further elaborated upon in the following text:

- In general, LEDs are commonly used as information indicators in various types of embedded systems (many of which are listed below).
- Thin, lightweight message displays, e.g. in public information signs (at airports and railway stations, among other places).
- Status indicators, e.g. on/off lights on professional instruments and consumers audio/video equipment as infrared LEDs in remote controls (for TVs, VCRs, etc.).
- Clusters of LEDs are used in traffic signals, replacing ordinary bulbs behind coloured glass.
- Bicycle lighting and indicator lights; also for pedestrians to be seen by car traffic.
- Calculator and measurement instrument displays (seven segment displays), although now mostly replaced by LCDs.
- Red or yellow LEDs are used in indicator and (alpha)numeric displays in environments where night vision must be retained: aircraft cockpits, submarine and ship bridges, astronomy observatories, and in the field, e.g. night time animal watching and military field use.
- Red or yellow LEDs are also used in photographic darkrooms, for providing lighting which does not lead to unwanted exposure of the film.
- Illumination, e.g. flashlights (aka torches, UK), and backlights for LCD screens signaling/emergency beacons and strobes.
- As movement sensors, e.g. in mechanical and optical computer mice and trackballs.
- In LED printers, e.g. high-end colour printers.

LEDs offer benefits in terms of maintenance and safety. The typical working lifetime of a device, including the bulb, is ten years, which is much longer than the lifetimes of most other light sources. Further, LEDs fail by dimming over time, rather than the abrupt burn-out of incandescent bulbs. LEDs give off less heat than incandescent light bulbs and are less fragile than fluorescent lamps. Since an individual device is smaller than a centimeter in length, LED-based light sources used for illumination and outdoor signals are built using clusters of tens of devices.

Incandescent light bulbs for traffic signals and pedestrian crosswalks are gradually being replaced by LED clusters.

Lighting systems using incandescent bulbs are cheap to buy but inefficient, generating from about 16 lumen per watt for a domestic tungsten bulb to 22 lm/W for a halogen bulb. Fluorescent tubes are more efficient, from 50 to 100 lm/W for domestic tubes, allowing large energy savings, but are bulky and fragile and require starter circuits. LEDs are robust and moderately efficient, up to 32 lumen per watt, but are expensive, although their cost is falling. The technologies for LED production are developing rapidly.

Because they are monochromatic, LED lights have great power advantages over white lights where a specific color is required. Unlike the white lights, the LED does not need a filter that absorbs most of the emitted white light. Colored fluorescent lights are made, but they are not widely available. LED lights are inherently colored, and are available in a wide range of colors. One of the most recently introduced colors is the emerald green (bluish green, about 500 nm) that meets the legal requirements for traffic signals and navigation lights.

There are applications that specifically require light that does not contain any blue component. Examples are photographic darkroom safe lights, illumination in laboratories where certain photo-sensitive chemicals are used, and situations where dark adaptation (night vision) must be preserved, such as cockpit and bridge illumination, observatories, etc. Yellow LED lights are a good choice to meet these special requirements because the human eye is most sensitive to yellow light (about 500 lm/watt emitted), because their monochromatic light has no blue component, as well as their other advantages discussed above.

Indoor and Outdoor LED Panels

There are two types of LED panels: conventional, using discrete LEDs, and SMD (Surface Mounted Device) panels. Most outdoor screens and some indoor screens are built around discrete LEDs, also known as individually mounted LEDs. A cluster of red, green, and blue diodes is driven together to form a full-color pixel, usually square in shape. These pixels are spaced evenly apart and are measured from center to center for absolute pixel resolution.

Most indoor screens on the market are built using SMD technology — a trend that is now extending to the outdoor market. An SMD pixel consists of red, green, and blue diodes mounted on a chipset, which is then mounted on the driver PC board. The individual diodes are smaller than a pin and are set very close together. The difference is that minimum viewing distance is reduced by 25 per cent from the discrete diode screen with the same resolution.

Indoor use generally requires a screen that is based on SMD technology and has a minimum brightness of 600 nits (a standard unit of luminance — candelas per square meter). This will usually be more than sufficient for corporate and retail applications, but under high ambient-brightness conditions, you may need more punch to compete. Fashion and auto shows are two examples of high-brightness stage lighting that may require a higher LED brightness. Conversely, when your screen may be in a shot on a television show, the requirement will often be for lower brightness levels with lower color temperatures.

For outdoor use, you need at least 2,000 nits for most situations, whereas higher brightness types of up to 5,000 nits cope even better with direct sunlight on the screen. Until recently, only discrete diode screens could achieve that brightness level. (The brightness of LED panels also can be turned down.)

For specific projects, you need to take into account factors such as sight lines, local authority planning requirements (if the installation is to become semi-permanent), vehicular access (trucks carrying the screen, truck-mounted screens, or cranes), cable runs for power and video (accounting for both distance and health and safety requirements), power, suitability of the ground for the location of the screen (check to make sure there are no pipes, shallow drains, caves, or tunnels that may not be able to support heavy loads), and overhead obstructions.

PIN Diode

A **PIN diode** (Positive-Intrinsic-Negative diode) is a photodiode with a large, neutrally doped intrinsic region sandwiched between p-doped and n-doped semiconducting regions.

A PIN diode exhibits an increase in its electrical conductivity as a function of the intensity, wavelength, and modulation rate of the incident radiation.

Schottky Diode

Walter H. Schottky is a semiconductor diode with a low forward voltage drop. A typical application is discharge-protection for solar cells connected to lead-acid batteries. While standard silicon diodes have a forward voltage drop of about 0.6 volts, Schottky diodes have a drop of only about 0.2 volts.

A Schottky diode uses a metal-semiconductor junction as a Schottky barrier (instead of a semiconductor-semiconductor junction as in conventional diodes). This Schottky barrier results in both very fast switching times and low forward voltage drop.

Avalanche Diode

Silicon diode that is designed to break down and conduct at a specified reverse bias voltage.

A common application is protecting electronic circuits against damaging high voltages. The avalanche diode is connected to the circuit so that it is reverse-biased. In other words, its cathode is positive with respect to its anode. In this configuration, the diode is non-conducting and does not interfere with the circuit. If the voltage increases beyond the design limit, the diode suffers avalanche breakdown, causing the harmful voltage to be conducted to earth.

Avalanche breakdown is due to impact ionization. Under a small reverse bias, the diode is almost non-conducting, although a very small current still flows. When the reverse electric field across the p-n junction is large enough, the energy of the few electrons flowing is enough to ionize atoms in the silicon. The ionization process is self-reinforcing, causing a rapid increase in current (theoretically taking only picoseconds to reach its peak). Avalanche breakdown is not destructive, as long as the diode is not allowed to overheat.

The Zener diode exhibits an apparently similar effect, but its operation is caused by a different mechanism, called Zener breakdown. Both effects are actually present in any such diode, but one may dominate the other.

Laser Diode

A laser diode is a laser where the active medium is a semiconductor p-n junction similar to that found in a light-emitting diode. Laser diodes are sometimes referred to (somewhat redundantly) as injection laser diodes or by the acronyms LD or ILD.

Principle of operation: When a diode is forward biased, holes from the p-region are injected into the n-region, and electrons from the n-region are injected into the p-region. If electrons and holes are present in the same region, they may radiatively recombine—that is, the electron "falls into" the hole and emits a photon with the energy of the bandgap. This is called spontaneous emission, and is the main source of light in a light-emitting diode.

Under suitable conditions, the electron and the hole may coexist in the same area for quite some time (on the order of microseconds) before they recombine. If a photon of exactly the right frequency happens along within this time period, recombination may be stimulated by the photon. This causes another photon of the same frequency to be emitted, with exactly the same direction, polarization and phase as the first photon.

In a laser diode, the semiconductor crystal is fashioned into a shape somewhat like a piece of paper—very thin in one direction and rectangular in the other two. The top of the crystal is n-doped, and the bottom is p-doped, resulting in a large, flat p-n junction. The two ends of the crystal are cleaved so as to form perfectly smooth, parallel edges; two reflective parallel edges are called a Fabry-Perot cavity. Photons emitted in precisely the right direction will be reflected several times from each end face before they are emitted. Each time they pass through the cavity, the light is amplified by stimulated emission. Hence, if there is more amplification than loss, the diode begins to "lase".

Types of Laser Diode

The type of laser diode just described is called a *homojunction* laser diode, for reasons which should soon become clear. Unfortunately, they are extremely inefficient. They require so much power that they can only be operated in short "pulses;" otherwise the semiconductor would melt. Although historically important and easy to explain, such devices are not practical.

Double heterostructure lasers: In these devices, a layer of low bandgap material is sandwiched between two high bandgap layers. One commonly-used pair of materials is GaAs with AlGaAs. Each of the junctions between different bandgap materials is called a *heterostructure,* hence the name "double heterostructure laser" or **DH** laser. The kind of laser diode described in the first part of the article is referred to as a "homojunction" laser, for contrast with these more popular devices.

The advantage of a DH laser is that the region where free electrons and holes exist simultaneously—the "active" region—is confined to the thin middle layer. This means that many more of the electron-hole pairs can contribute to amplification—not so many are left out in the poorly amplifying periphery. In addition, light is reflected from the heterojunction; hence, the light is confined to the region where the amplification takes place.

Quantum well lasers: If the middle layer is made thin enough, it starts acting like a quantum well. This means that in the vertical direction, electron energy is quantised. The difference between quantum well energy levels can be used for the laser action instead of the bandgap. This is very useful since the wavelength of light emitted can be tuned simply by altering the thickness of the layer. The efficiency of a quantum well laser is greater than that of a bulk laser due to a tailoring of the distrubution of electrons and holes that are involved in the stimulated emission (light producing) process.

The problem with these devices is that the thin layer is simply too small to effectively confine the light. To compensate, another two layers are added on, outside the first three. These layers have a lower refractive index than the centre layers, and hence confine the light effectively. Such a design is called a separate confinement heterostructure (SCH) laser diode.

Almost all commercial laser diodes since the 1990s have been SCH quantum well diodes.

VCSELs: VCSELs are semiconductor lasers that emit light perpendicular to the chip surface, contrary to conventional diode lasers that are edge-emitting (in-plane) devices.

DIAC

The **DIAC** is a bidirectional trigger diode that conducts current only after its breakdown voltage has been exceeded momentarily. It stops conducting when the current through it drops below a value characteristic for the device. The behavior is typically the same for both directions of current flow. Most DIACs have a breakdown voltage around 30 V. They are typically used for triggering TRIACs.

Three-terminal Devices

Bipolar Transistor

A Bipolar Junction Transistor, (BJT), is a type of transistor, an amplifying or switching device constructed of doped semiconductor. A bipolar transistor is a sandwich of differently doped sections, either NPN or PNP. The center section is called the *base* of the transistor. By varying the current between the base and one terminal called the *emitter*, one can vary the current flow between the emitter and a third terminal known as the *collector*, causing amplification of the signal at that terminal. BJTs can be thought of as current-controlled resistors. BJTs are usually characterised as current amplifiers.

Conceptually, one can understand a bipolar transistor as two diodes placed back to back. In normal operation, the emitter-base junction is forward biased and the base-collector junction is reverse biased. In an npn-type transistor for example, electrons from the emitter wander (or "diffuse") into the base. These electrons in the base are in the minority - there are plenty of holes to recombine with. The base is always made very thin such that most of the electrons diffuse over to the collector before they recombine with holes. The collector-base junction is reverse biased to prevent the flow of holes, but electrons meet a more friendly reception - they are swept into the collector by the electric field around the junction. The proportion of electrons able to run the base "gauntlet" and make it to the collector is very sensitive to the current passing through the base. Hence, a small change of the base current can translate to a large change in electron flow between emitter and collector. For example the ratio of these currents ($I_c \div I_b$ usually called β) in some bipolar transistors is 100 or more.

Cross section of an npn-type BJT as found in integrated circuits (simplified). The base region is the pink layer sandwiched between the green emitter region above and the yellow collector volume below.

Transistors have different regions of operation. In the "linear" regime, collector-emitter current is approximately proportional to the base current but many times larger, making this the ideal mode of operation for current amplification. The BJT enters "saturation" when the base current is increased to a point where the external circuitry prevents the collector current from growing any larger. A residual voltage drop of approximately 0.3V between collector and emitter then remains. A transistor is said to operate in the "cut off" regime when the base-emitter voltage is too small for any significant current to flow. In typical BJTs manufactured from silicon, this is the case below 0.7V or so. BJTs that alternate between cut off and saturation can by viewed as electronic switches.

Field-Effect Transistor

The Field-Effect Transistor (FET) is a family of transistors that rely on an electric field to control the conductivity of a "channel" in a semiconductor material. FETs, like all transistors, can be thought of as voltage-controlled resistors. Most FETs are made using conventional bulk semiconductor processing techniques, using the single-crystal semiconductor wafer as the active region, or channel. The channel region of TFTs (thin-film transistors), on the other hand, is a thin film that is deposited onto a substrate (often glass, since the primary application of TFTs is in liquid crystal displays). For more on TFTs, see thin-film transistor; the remainder of this article deals with the transistors most commonly used in integrated circuits.

The different types of field-effect transistors can be distinguished by the method of isolation between channel and gate:

- The MOSFET (Metal-Oxide-Semiconductor Field-Effect Transistor) utilizes an isolator (typically SiO_2).
- The JFET (Junction Field-Effect Transistor) uses a p-n junction.
- The MESFET (Metal-Semiconductor Field Effect Transistor) uses a Schottky barrier.
- In the HEMT (High Electron Mobility Transistor), also named HFET (heterostructure FET), the fully depleted wide-band-gap material forms the isolation.

The distinguishing feature of the TFT (Thin-film transistor) is the use of amorphous silicon or polycrystalline silicon as channel.

Darlington Transistor

In electronics, the Darlington transistor is a semiconductor device which combines two bipolar transistors in tandem in a single device. This gives it high current gain (commonly written b), and takes up less space than using two discrete transistors in the same configuraton. The total gain of the darlington is the product of the gain of the individual transistors. A typical modern device has a current gain of 1000 or more. It has more phase shift at high frequencies than a single transistor, and hence can become unstable with negative feedback much more easily. The base-emitter voltage is also higher; it is the sum of both base-emitter voltages and for silicon-transistor >1.2 V. This configuration (originally realized as two separate transistors) was invented by Bell Laboratories engineer Sidney Darlington. The idea of putting two or three transistors on a single chip was patented by him, but not the idea of putting an arbitrary number of transistors, which would cover all modern ICs.

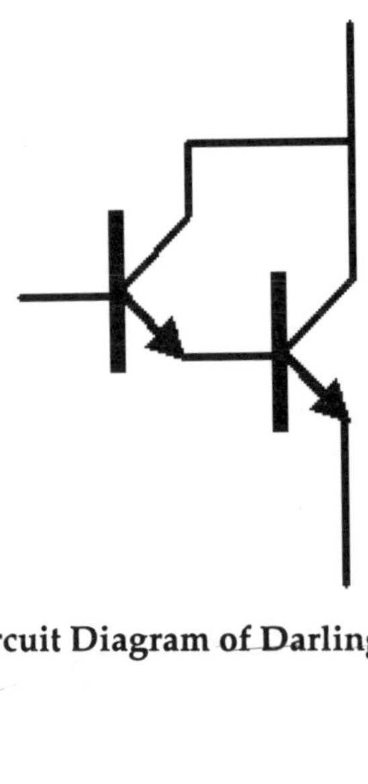

Fig. II–H.1: Circuit Diagram of Darlington Configuration.

Thyristor

The **thyristor** (also called a **silicon-controlled rectifier** or SCR) is a solid-state semiconductor device similar to a diode, with an extra terminal which is used to turn it on. Once turned on, the thyristor will remain on (conducting) as long as there is a significant current flowing through it. If the current falls to zero, the device switches off.

The thyristor is a four-layer semiconducting device, with each layer consisting of an alternately N or P-type material, for example N-P-N-P. The main terminals, labelled anode and cathode, are across the full four layers, and the control terminal, called the gate, is attached to one of the middle layers. The operation of a thyristor can be understood in terms of a pair of tightly coupled transistors, arranged to cause the self-latching action.

Thyristors are mainly used where high currents and voltages are involved, and are often used to control alternating currents, where the change of sign of the current causes the device to automatically switch off. This is known as *synchronous operation* or Zero Cross operation. This principle is used to control the desired loading by adjusting the frequency of the sinusoidal input. The range of frequencies is great because there is no limit to the number of cycles a thyristor can perform, and exhibits no "wear out" modes. This is a frequency domain method of control.

With phase angle control a thyristor is turned on at a specific and adjustable portion of the cycle of the controlling sinusoidal input. Moving the point at which the thyristor is turned on regulates power output. An example of this method of control is a dimmer switch for lights. The turn on point of a thyristor is controlled to occur at a particular point on the sine curve of the AC supply. The thyristor stays on for the remainder of that cycle and the longer the thyristor stays on, the brighter the light. Fine resolution of output is possible with this method and is suitable for fast responding loads such as tungsten filament lamps or temperature variable resistance loads. Phase-angle control is also essential for inductive loads.

The drawback of a thyristor is that, like a diode, it only conducts in one direction. A similar self-latching 5-layer device, called a triac, is able to work in both directions.

An earlier gas filled tube device called a Thyratron provided a similar electronic switching capability, where a small control voltage could switch a large current.

Modern thyristors can switch large amounts of power (up to megawatts). In the realm of very high power applications, they are still the primary choice. However, in low and medium power (from few tens of watts to few tens of killowatts) they have almost been replaced by other devices with superior switching characteristics like MOSFETs or IGBTs. One major problem of thyristor is that is it not a fully controllable switch in the sense that triggering current direction need to be reversed to switch it off. **GTO (Gate Turn-off Thyristor)** is another related device which addresses this problem. In high-frequency applications, thyristors are poor candidates due to large switching times arising out of **bipolar conduction**. MOSFETs, on the other hand, has much faster switching capability because of its **unipolar conduction** (only majority carriers carry the current).

Triac

A **triac** is an electronic component equivalent to two silicon controlled rectifiers joined end to end or back to back with their gates connected together. This results in a two-directional electronic switch which can conduct current in both directions when it is triggered (turned on). It can be triggered by either a positive or a negative voltage. The device turns off again when the current through it drops below a certain threshold value, such as the end of a half-cycle of alternating current. This makes the triac a convenient switch for AC circuits. Applying a trigger pulse at a controllable point in an AC cycle allows one to control the percentage of current that flows through the triac. Lower power triacs are used in lots of applications, especially domestic electrical appliances, such as light dimmers.

Four-terminal Devices

Hall Effect Sensor

The **Hall effect** is a potential difference (voltage) on opposite sides of a thin sheet of conducting or semiconducting material (the Hall element) through which an electric current is flowing, created by a magnetic field applied perpendicular to the Hall element. The ratio of the voltage created to the amount of current is known as the 'Hall resistance', and is a characteristic of the material in the element. This effect was discovered by Dr. Edwin Hall in 1879.

The Hall effect comes about due to the nature of the current flow in the conductor. Current consists of many small charge-carrying "particles" (typically electrons) which see a force due to the magnetic field. Some of these charge elements end up forced to the sides of the conductors, where they create a pool of net charge. This is only notable in larger conductors where the separation between the two sides is large enough.

One important feature of the Hall effect is that it differentiates between positive charges moving in one direction and negative charges moving in the opposite. The Hall effect offered the first real proof that electric currents in metals are carried by moving electrons, not by protons. Interestingly enough, the Hall effect also showed that in some substances (especially semiconductors), it is more appropriate to think of the current as positive "holes" moving rather than negative electrons.

By measuring the Hall voltage across the element, one can determine the strength of the magnetic field applied. So called **Hall effect sensors** are readily available from a number of different manufacturers. The most common types are analog (or Linear) Hall effect sensors, which output a voltage that is proportional to the applied magnetic field, and digital Hall effect sensors, which are often used as magnetically controlled switches — they turn on or off when the applied magnetic field reaches a certain level. These Hall effect switches generally consist of a Hall effect sensor, one or more logic gates and a transistor used to switch the electric current on or off.

Alternately, by applying a known magnetic field (typically from a permanent magnet) one can use the Hall voltage to instead measure the current through the element. This can be particularly useful as it allows one to measure the current in a conductor remotely through induction. This is widely used commercially in "live wire detectors", which allow you to quickly identify which wires are carrying current without plugging into them.

In the presence of large magnetic field strength and low temperature, one can observe the quantum Hall effect, which is the quantization of the Hall resistance.

In ferromagnetic materials, the Hall resistivity also shows an anomalous contribution, known as the **Anomalous Hall Effect**, proportional to the magnetization of the material. Although a well-recognised phenomena, there is still debate about its origins.

Unit II–I

Basics of Superconductors

What are Superconductors?

An element, inter-metallic alloy, or compound that will conduct electricity without resistance below a certain temperature. Resistance is undesirable because it produces losses in the energy flowing through the material.

Once set in motion, electric current will flow forever in a closed loop of superconducting material-making it the closest thing to perpetual motion in nature. Scientists refer to superconductivity as a "microscopic quantum phenomenon."

Superconductivity

Superconductivity is a phenomenon which occurs in certain materials and is characterized by the absence of electrical resistivity. Until recently, this phenomenon had been restricted to metals and alloys with transition temperatures of less than 23K. In 1986, superconductivity was discovered in a ceramic material. This precipitated an onrush of ceramic based superconductors with transition temperatures as high as 120K. The ceramic-based materials are commonly known as high temperature (HTc) superconductors while the metallic and alloy materials are called low temperature (LTc) superconductors. Currently, only low temperature superconductors are of interest to the magnet designer and manufacturer.

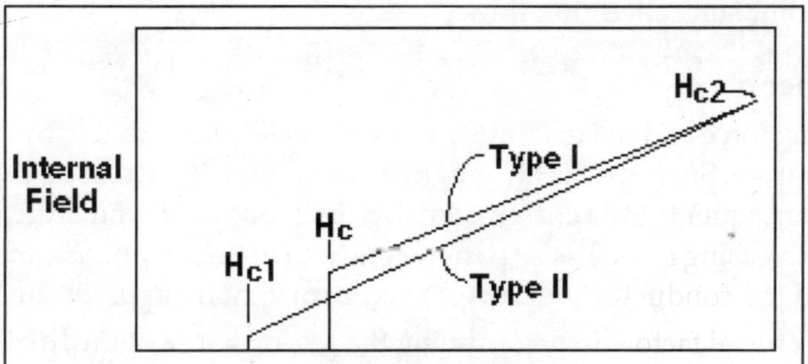

Fig. II–I.1: Superconductivity depicted.

Superconductors are divided into two types depending on their characteristic behavior in the presence of a magnetic field. Type I superconductors are comprised of pure metals, whereas Type II superconductors are comprised primarily of alloys or intermetallic compounds. Both, however, have one common feature: below a critical temperature, T_c, their resistance vanishes.

The critical temperature at which the resistance vanishes in a superconductor is reduced when a magnetic field is applied. The maximum field that can be applied to a superconductor at a particular temperature and still maintain superconductivity is call the critical field, or H_c. This field varies enormously between Type I and Type II superconductors. The maximum critical field (H_c) in any Type I superconductor is about 2000 Gauss (0.2 Tesla), but in Type II materials superconductivity can persist to several hundred thousand Gauss (H_{c2}). At fields greater than H_c in a Type I superconductor and greater than H_{c2} in a Type II superconductor, the conductor reverts to the normal state and regains its normal state resistance.

A Type I superconductor excludes the applied magnetic field from the center of the sample by establishing circulating currents on its surface that counteract the applied field. Type II superconductors, however, permit the field to penetrate through the sample in quantized amounts of flux. These quanta are comprised of circulating vortices of current and the flux contained in the vortices. The total flux in a vortex is 2×10^{-7} Gauss-cm^2. Great numbers of these vortices, or fluxoids as they are frequently called, can exist in a superconductor. For example, at a field intensity of 80 kilogauss (8 Tesla) there are 4×10^{11} fluxoids/cm^2. These fluxoids and their interactions with defects in the superconductor give rise to the high current carrying capabilities of superconducting magnets.

Flux Pinning and Flux Flow

Properties of superconducting materials are altered locally by the presence of defects in the materials. A fluxoid encompassing or adjacent to such a defect in the material has its energy altered and its free motion through the superconductor is inhibited. This phenomenon, known as flux pinning, causes a field gradient in the superconductor and gives rise to a net current in the material. In the absence of defects in a Type II superconductor, no bulk current can be conducted without a transition into the normally conducting resistive state.

Since the pinning force is small, fluxoids can be broken loose from their pinning centers, resulting in a net creep of the flux through a conductor as a function of time. This results in an effective voltage in a Type II superconductor. If the current density is low and the magnetic field is not intense, flux creep is insignificant and the induced voltage and effective resistance of the conductor will be essentially zero. At very high fields and high current densities, fluxoids will migrate rapidly, giving rise to a phenomenon called flux flow.

Conductor Phenomena

The maximum pinning force and the maximum current density sustainable by a conductor increases with reduced temperature. Similarly, as the magnetic induction, B, increases, the sustainable current density decreases. Consequently, the current density, magnetic field, and critical temperature are all interdependent. By increasing any of these parameters to a sufficiently high value, superconductivity can be destroyed and the conductor will revert to a normal, non-superconducting state.

One of the most crucial factors in determining the performance of the final magnet is the design of the conductor. This design affects the ultimate field achieved by the magnet, the rate at which the magnet can be energized and the drift rate in the persistent mode of operation. Several phenomena observed in magnets are caused by the conductor itself.

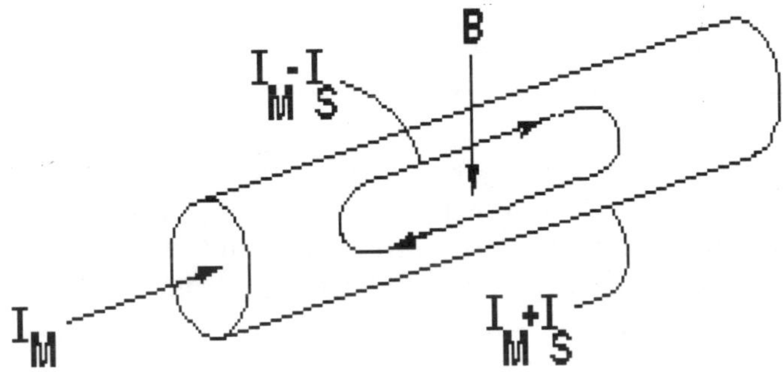

Fig. II–I.2: Single Filament Superconductor.

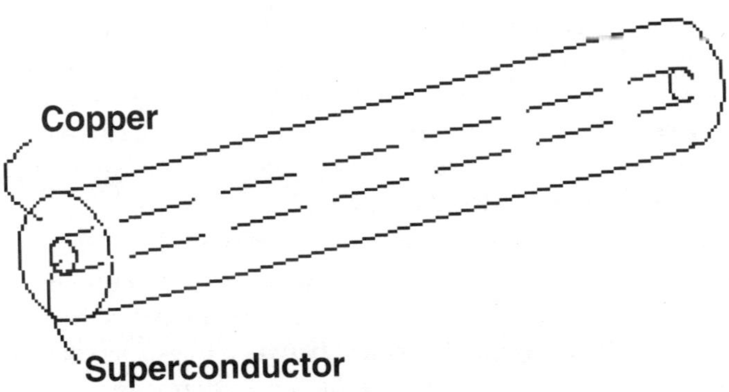

Fig. II–I.3: Copper Shunted Superconductor.

One of the earliest phenomena observed in superconducting magnets wound with single filament conductors was flux jumping. This phenomena arises from current induced in the conductor by the presence of the transverse field generated by the magnet. If a superconductor is placed transverse to the magnetic field, currents are induced in the conductor which shield the bulk of the conductor from the external magnetic field. These circulating currents extend for a finite length along the conductor, flowing in one direction on one side of the conductor and returning on the other side to complete the circuit.

If the heat produced by the penetration of the magnet field into the superconductor cannot escape to the surface rapidly enough, then the temperature increase inside the superconductor triggers a runaway condition known as a flux jump. Consequently, the conductor is driven into the resistive state at fields and currents low in comparison with the critical values. Resistive heat is then dissipated in the small normal zone which increases in temperature causing the normal zone to expand and propagate both along the length of the conductor and transverse to it. This results in the magnet being discharged as the energy in the magnet is dissipated in the resistive portion of the conductor.

To decrease the problem of transition to the normal state, it is common practice to shunt the conductor with a low resistivity normal metal by embedding the superconductor in a copper matrix to form a composite conductor. The copper provides additional heat capacity as well as providing a path for the magnet current while the superconductor is driven normal during a flux jump. If the resistance of the copper is low enough, the temperature of the conductor can remain below the critical temperature at the ambient field, and superconductivity will resume after the currents in the superconductor have decayed.

Embedding the superconductor in a low resistivity metal matrix is effective in reducing the chance of a flux jump which can cause a magnet quench. Magnets constructed with this type of material are dissipative and the heat generated during a flux jump must be conducted to the helium bath. Thus, the magnetic field must be changed slowly to allow time for the heat to be conducted to and dissipated in the liquid helium. Also, the diamagnetic currents in the superconductor contribute to the field generated by the magnet and can reduce its homogeneity.

If, instead of one superconducting filament, many fine filaments of superconductor are used, the heat generated in individual filaments can easily be conducted a short distance to the filament surface thus avoiding flux jumps. Consequently, conductors are made in which many fine filaments of superconductor are coextruded and drawn in a matrix of either copper or aluminum stabilizers. Although this has the desired effect of avoiding flux jumping, circulating currents can again be formed if the conductors are parallel in the highly conductive normal matrix. In this case, the circulating current is between two or more filaments in parallel and the current crosses over through the normally conductive matrix. This gives rise to diamagnetism and unequal distribution of currents in the filaments that limits the rate at which the magnet can be charged.

Problems arising from constructing the superconductor from filaments have been largely circumvented in modern conductors by twisting the filaments in the conductor. This causes the flux from the external magnetic field to be alternated through successive superconducting filaments, thereby reducing the unequal distribution of currents between the superconducting filaments and reducing the diamagnetism of the conductor. This reduction in diamagnetism or hysteresis has two desirable effects. First, it reduces the amount of energy dissipated in the magnet and permits it to be charged more rapidly. Secondly, the reduced diamagnetism causes the current in the magnet and the magnetic field generated by the magnet to be more linearly related. Such conductors are known as intrinsically stabilized conductors.

The History of Superconductors

Superconductors, materials that have no resistance to the flow of electricity, are one of the last great frontiers of scientific discovery. Not only have the limits of superconductivity not yet been reached, but the theories that explain superconductor behavior seem to be constantly under review. In 1911 superconductivity was first observed in mercury by Dutch physicist Heike Kamerlingh Onnes of Leiden University (shown above). When he cooled it to the temperature of liquid helium, 4 degrees Kelvin (-452F, -269C), its resistance suddenly disappeared. The Kelvin scale represents an "absolute" scale of temperature. Thus, it was necessary for Onnes to come within 4 degrees of the coldest temperature that is theoretically attainable to witness the phenomenon of superconductivity. Later, in 1913, he won a Nobel Prize in physics for his research in this area.

Fig. II–I.4: Walter Meissner.

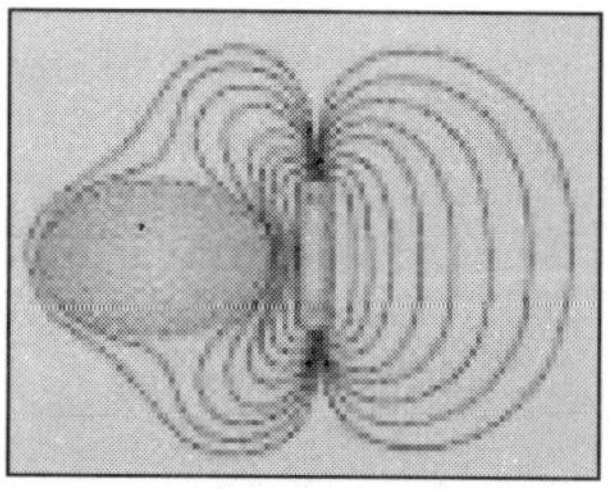

Fig. II–I.5: Meissner effect.

The next great milestone in understanding how matter behaves at extreme cold temperatures occurred in 1933. German researchers Walter Meissner (above) and Robert Ochsenfeld discovered that a superconducting material will repel a magnetic field (below graphic). A magnet moving by a conductor induces currents in the conductor. This is the principle upon which the electric generator operates. But, in a superconductor the induced currents exactly mirror the field that would have otherwise penetrated the superconducting material—causing the magnet to be repulsed. This phenomenon is known as strong diamagnetism and is today often referred to as the "Meissner effect" (an eponym). The Meissner effect is so strong that a magnet can actually be levitated over a superconductive material.

In subsequent decades other superconducting metals, alloys and compounds were discovered. In 1941 niobium-nitride was found to superconduct at 16 K. In 1953 vanadium-silicon displayed superconductive properties at 17.5 K. And, in 1962 scientists at Westinghouse developed the first commercial superconducting wire, an alloy of niobium and titanium (NbTi). High-energy, particle-accelerator electromagnets made of copper-clad niobium-titanium were then developed in the 1960s at the Rutherford-Appleton Laboratory in the UK, and were first employed in a superconducting accelerator at the Fermilab Tevatron in the US in 1987.

Fig. II–I.6: American physicists John Bardeen, Leon Cooper, and John Schrieffer.

The first widely-accepted theoretical understanding of superconductivity was advanced in 1957 by American physicists John Bardeen, Leon Cooper, and John Schrieffer (above). Their *Theories of Superconductivity* became know as the BCS theory—derived from the first letter of each man's last name—and won them a Nobel prize in 1972. The mathematically-complex BCS theory explained superconductivity at temperatures close to absolute zero for elements and simple alloys. However, at higher temperatures and with different superconductor systems, the BCS theory has subsequently become inadequate to fully explain how superconductivity is occurring.

Fig. II–I.7: Brian Josephson.

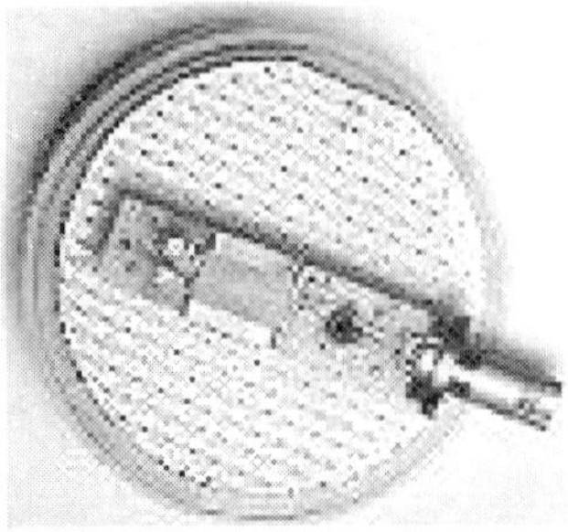

Fig. II–I.8: SQUID.

Another significant theoretical advancement came in 1962 when Brian D. Josephson (above), a graduate student at Cambridge University, predicted that electrical current would flow between 2 superconducting materials—even when they are separated by a non-superconductor or insulator. His prediction was later confirmed and won him a share of the 1973 Nobel Prize in Physics. This tunneling phenomenon is today known as the "Josephson effect" and has been applied to electronic devices such as the SQUID, an instrument capabable of detecting even the weakest magnetic fields. (Below SQUID graphic courtesy Quantum Design.)

The 1980's were a decade of unrivaled discovery in the field of superconductivity. In 1964 Bill Little of Stanford University had suggested the possibility of organic (carbon-based) superconductors. The first of these theoretical superconductors was successfully synthesized in 1980 by Danish researcher Klaus Bechgaard of the University of Copenhagen and 3 French team members. $(TMTSF)_2PF_6$ had to be cooled to an incredibly cold 1.2K transition temperature (known as Tc) and subjected to high pressure to superconduct. But, its mere existence proved the possibility of "designer" molecules - molecules fashioned to perform in a predictable way.

Fig. II–I.9: Alex Müller and Georg Bednorz.

Then, in 1986, a truly breakthrough discovery was made in the field of superconductivity. Alex Müller and Georg Bednorz (above), researchers at the IBM Research Laboratory in Rüschlikon, Switzerland, created a brittle ceramic compound that superconducted at the highest temperature then known: 30 K. What made this discovery so remarkable was that ceramics are normally insulators. They don't conduct electricity well at all. So, researchers had not considered them as possible high-temperature superconductor candidates. The Lanthanum, Barium, Copper and Oxygen compound that Müller and Bednorz synthesized, behaved in a not-as-yet-understood way. The discovery of this first of the superconducting copper-oxides (cuprates) won the 2 men a Nobel Prize the following year. It was later found that tiny amounts of this material were actually superconducting at 58 K, due to a small amount of lead having been added as a calibration standard - making the discovery even more noteworthy.

Müller and Bednorz' discovery triggered a flurry of activity in the field of superconductivity. Researchers around the world began "cooking" up ceramics of every imaginable combination in a quest for higher and higher Tc's. In January of 1987 a research team at the University of Alabama-Huntsville substituted Yttrium for Lanthanum in the Müller and Bednorz molecule and achieved an incredible 92 K Tc. For the first time a material (today referred to as YBCO) had been found that would superconduct at temperatures warmer than liquid nitrogen - a commonly available coolant. Additional milestones have since been achieved using exotic—and often toxic—elements in the base perovskite ceramic. The current class (or "system") of ceramic superconductors with the highest transition temperatures are the mercuric-cuprates. The first synthesis of one of these compounds was achieved in 1993 by Prof. Dr. Ulker Onbasli at the University of Colorado and by the team of A. Schilling, M. Cantoni, J.D. Guo, and H.R. Ott of Zurich, Switzerland. The world record Tc of 138 K is now held by a thallium-doped, mercuric-cuprate comprised of the elements Mercury, Thallium, Barium, Calcium, Copper and Oxygen. The Tc of this ceramic superconductor was confirmed by Dr. Ron Goldfarb at the National Institute of Standards and Technology-Colorado in February of 1994. Under extreme pressure its Tc can be coaxed up even higher - approximately 25 to 30 degrees more at 300,000 atmospheres.

The first company to capitalize on high-temperature superconductors was Illinois Superconductor (today known as ISCO International), formed in 1989. This amalgam of government, private-industry and academic interests introduced a depth sensor for medical equipment that was able to operate at liquid nitrogen temperatures (~ 77K).

While no significant advancements in superconductor Tc's have been achieved in recent years, other discoveries of equal importance have been made. In 1997 researchers discovered that at a temperature very near absolute zero an alloy of gold and indium was both a superconductor and a natural magnet. Conventional wisdom held that a material with such properties could not exist! Since then, over a half-dozen such compounds have been found. Recent years have also seen the discovery of the first high-temperature superconductor that does NOT contain any copper (2000), and the first all-metal perovskite superconductor (2001).

Also in 2001 a material that had been sitting on laboratory shelves for decades was found to be an extraordinary new superconductor. Japanese researchers measured the transition temperature of magnesium diboride at 39 Kelvin—far above the highest Tc of any of the elemental or binary alloy superconductors. While 39 K is still well below the Tc's of the "warm" ceramic superconductors, subsequent refinements in the way MgB_2 is fabricated have opened the door to its use in industrial applications. Laboratory testing has found MgB_2 will outperform NbTi and Nb_3Sn wires in high magnetic field applications like MRI.

Startling—and serendipitous—discoveries like these are forcing scientists to continually re-examine longstanding theories on superconductivity and to consider heretofore-unimagined combinations of elements.

Uses for Superconductors

Magnetic-levitation is an application where superconductors perform extremely well. Transport vehicles such as trains can be made to "float" on strong superconducting magnets, virtually eliminating friction between the train and its tracks. Not only would conventional electromagnets waste much of the electrical energy as heat, they would have to be physically much larger than superconducting magnets. A landmark for the commercial use of MAGLEV technology occurred in 1990 when it gained the status of a nationally-funded project in Japan. The Minister of Transport authorized construction of the Yamanashi Maglev Test Line which opened on April 3, 1997. In December 2003, the MLX01 test vehicle (shown above) attained an incredible speed of 361 mph (581 kph).

Although the technology has now been proven, the wider use of MAGLEV vehicles has been constrained by political and environmental concerns (strong magnetic fields can create a bio-hazard). The world's first MAGLEV train to be adopted into commercial service, a shuttle in Birmingham, England, shut down in 1997 after operating for 11 years. A Sino-German maglev is currently operating over a 30 km course at Pudong International Airport in Shanghai, China. The U.S. plans to have its first (non-superconducting) Maglev train in operation this year on a Virginia college campus. Click this link for a website that lists other uses for MAGLEV.

An area where superconductors can perform a life-saving function is in the field of biomagnetism. Doctors need a non-invasive means of determining what's going on inside the human body. By impinging a strong superconductor-derived magnetic field into the body, hydrogen atoms that exist in the body's water and fat molecules are forced to accept energy from the magnetic field. They then release this energy at a frequency that can be detected and displayed graphically by a computer. Magnetic Resonance Imaging (MRI) was actually discovered in the mid 1940's. But, the first MRI exam on a human being was not performed until July 3, 1977. And, it took almost five hours to produce one image! Today's faster computers process the data in much less time.

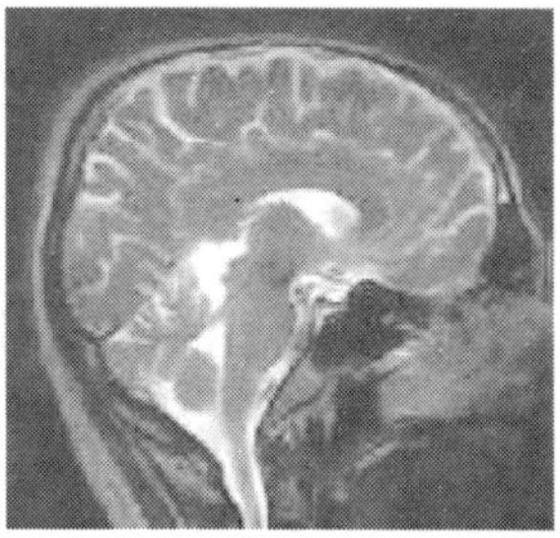

Fig. II–I.10: MRI of a human skull.

The Korean Superconductivity Group within KRISS has carried biomagnetic technology a step further with the development of a double-relaxation oscillation SQUID (Superconducting quantum Interference Device) for use in Magnetoencephalography. SQUID's are capable of sensing a change in a magnetic field over a billion times weaker than the force that moves the needle on a compass (compass: 5e-5T, SQUID: e-14T.). With this technology, the body can be probed to certain depths without the need for the strong magnetic fields associated with MRI's.

Fig. II–I.11: Superconducting Super-Collider project.

Probably the one event, more than any other, that has been responsible for putting "superconductors" into the American lexicon was the Superconducting Super-Collider project planned for construction in Ellis county, Texas. Though Congress cancelled the multi-billion dollar effort in 1993, the concept of such a large, high-energy collider would never have been viable without superconductors. High-energy particle research hinges on being able to accelerate sub-atomic particles to nearly the speed of light. Superconductor magnets make this possible. CERN, a consortium of several European nations, is doing something similar with its Large Hadron Collider (LHC) now under construction along the Franco-Swiss border.

Other related web sites worth visiting include the proton-antiproton collider page at Fermilab. This was the first facility to use superconducting magnets. Get information on the electron-proton collider HERA at the German lab pages of DESY (with English text). Lastly, Brookhaven National Laboratory features a page dedicated to its RHIC heavy-ion collider.

Electric generators made with superconducting wire are far more efficient than conventional generators wound with copper wire. In fact, their efficiency is above 99 per cent and their size about half that of conventional generators. These facts make them very lucrative ventures for power utilities. General Electric has estimated the potential worldwide market for superconducting generators in the next decade at around $20-30 billion dollars. Late in 2002 GE Power Systems received $12.3 million in funding from the U.S. Department of Energy to move high-temperature superconducting generator technology toward full commercialization.

Other commercial power projects in the works that employ superconductor technology include energy storage to enhance power stability. American Superconductor Corp. received an order from Alliant Energy in late March 2000 to install a Distributed Superconducting Magnetic Energy Storage System (D-SMES) in Wisconsin. Just one of these 6 D-SMES units has a power reserve of over 3 million watts, which can be retrieved whenever there is a need to stabilize line voltage during a disturbance in the power grid. AMSC has also installed more than 22 of its D-VAR systems to provide instantaneous reactive power support.

Fig. II–I.12: The General Atomics/Intermagnetics General superconducting Fault Current Controller, employing HTS superconductors.

Recently, power utilities have also begun to use superconductor-based transformers and "fault limiters". The Swiss-Swedish company ABB was the first to connect a superconducting transformer to a utility power network in March of 1997. ABB also recently announced the development of a 6.4MVA (mega-volt-ampere) fault current limiter—the most powerful in the world. This new generation of HTS superconducting fault limiters is being called upon due to their ability to respond in just thousandths of a second to limit tens of thousands of amperes of current. Advanced Ceramics

Limited is another of several companies that makes BSCCO type fault limiters. Intermagnetics General recently completed tests on its largest (15kv class) power-utility-size fault limiter at a Southern California Edison (SCE) substation near Norwalk, California. And, both the US and Japan have plans to replace underground copper power cables with superconducting BSCCO cable-in-conduit cooled with liquid nitrogen. (See fig. II–I.13) By doing this, more current can be routed through existing cable tunnels. In one instance 250 pounds of superconducting wire replaced 18,000 pounds of vintage copper wire, making it over 7000 per cent more space-efficient.

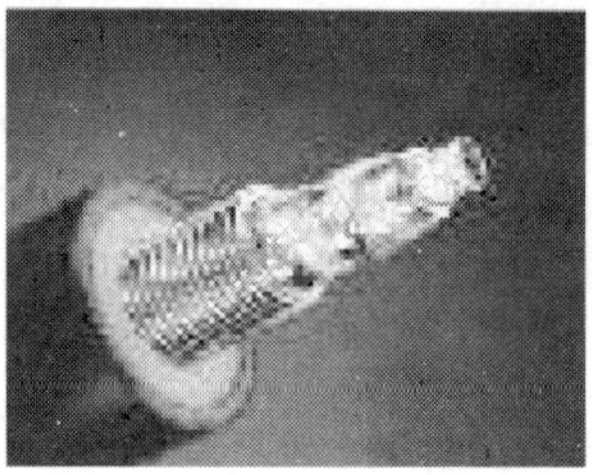

Fig. II–I.13: BSCCO Cable in Conduct cooled with liquid Nitrogen.

An idealized application for superconductors is to employ them in the transmission of commercial power to cities. However, due to the high cost and impracticality of cooling miles of superconducting wire to cryogenic temperatures, this has only happened with short "test runs". In May of 2001 some 150,000 residents of Copenhagen, Denmark, began receiving their electricity through HTS (high-temperature superconducting) material. That cable was only 30 meters long, but proved adequate for testing purposes. In the summer of 2001 Pirelli completed installation of three 400-foot HTS cables for Detroit Edison at the Frisbie Substation capable of delivering 100 million watts of power. This marked the first time commercial power has been delivered to customers of a US power utility through superconducting wire. Intermagnetics General has announced that its IGC-SuperPower subsidiary has joined with BOC and Sumitomo Electric in a $26 million project to install an underground, HTS power cable in Albany, New York, in Niagara Mohawk Power Corporation's power grid. The 350-meter cable, believed to be four times the length of any previously installed HTS cable, will be designed to provide more power and operate at significantly lower loss levels than other HTS installations.

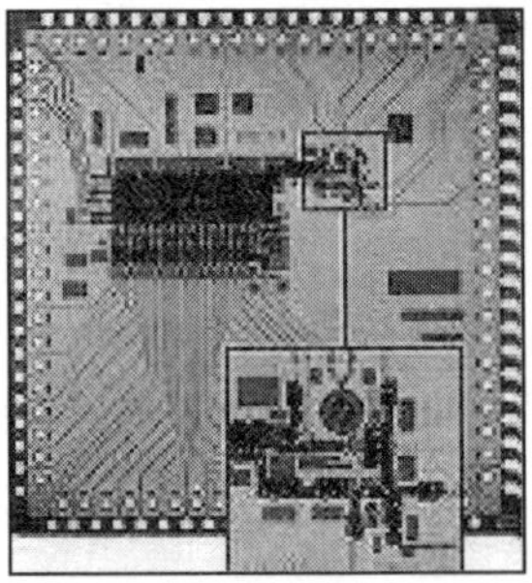

Fig. II–I.14: Hypres Superconducting Microchip, Incorporating 6000 Josephson Junctions.

The National Science Foundation, along with NASA and DARPA and various universities, are currently researching "petaflop" computers. A petaflop is a thousand-trillion floating point operations per second. Today's fastest computing operations have only reached "teraflop" speeds—trillions of operations per second. Currently the fastest is the IBM Blue Gene/L running at 70.7 teraflops per second (multiple CPU's). The fastest single processor is a Lenslet optical DSP running at 8 teraflops. It has been conjectured that devices on the order of 50 nanometers in size along with unconventional switching mechanisms, such as the Josephson junctions associated with superconductors, will be necessary to achieve such blistering speeds. TRW researchers (now Northrop Grumman) have quantified this further by predicting that 100 billion Josephson junctions on 4000 microprocessors will be necessary to reach 32 petabits per second. These Josephson junctions are incorporated into field-effect transistors which then become part of the logic circuits within the processors. Recently it was demonstrated at the Weizmann Institute in Israel that the tiny magnetic fields that penetrate Type 2 superconductors can be used for storing and retrieving digital information. It is, however, not a foregone conclusion that computers of the future will be built around superconducting devices. Competing technologies, such as quantum (DELTT) transistors, high-density molecule-scale processors, and DNA-based processing also have the potential to achieve petaflop benchmarks.

In the electronics industry, ultra-high-performance filters are now being built. Since superconducting wire has near zero resistance, even at high frequencies, many more filter stages can be employed to achive a desired frequency response. This translates into an ability to pass desired frequencies and block undesirable frequencies in high-congestion rf (radio frequency) applications such as cellular telephone systems. ISCO International and Superconductor Technologies are companies currently offering such filters.

Superconductors have also found widespread applications in the military. HTSC SQUIDS are being used by the US NAVY to detect mines and submarines. And, significantly smaller motors are being built for NAVY ships using superconducting wire and "tape". In mid-July, 2001, American Superconductor unveiled a 5000-horsepower motor made with superconducting wire (below).

The military is also looking at using superconductive tape as a means of reducing the length of very low frequency antennas employed on submarines. Normally, the lower the frequency, the longer an antenna must be. However, inserting a coil of wire ahead of the antenna will make it function as if it were much longer. Unfortunately, this loading coil also increases system losses by adding the resistance in the coil's wire. Using superconductive materials can significantly reduce losses in this coil. The Electronic Materials and Devices Research Group at University of Birmingham (UK) is credited with creating the first superconducting microwave antenna. Applications engineers suggest that superconducting carbon nanotubes might be an ideal nano-antenna for high-gigahertz and terahertz frequencies, once a method of achieving zero "on tube" contact resistance is perfected.

Fig. II–I.15: American Superconductor unveiled a 5000-horsepower motor made with superconducting wire.

The most ignominious military use of superconductors may come with the deployment of "E-bombs". These are devices that make use of strong, superconductor-derived magnetic fields to create a fast, high-intensity electro-magnetic pulse (EMP) to disable an enemy's electronic equipment. Such a device saw its first use in wartime in March 2003 when US Forces attacked an Iraqi broadcast facility.

Among emerging technologies are a stabilizing momentum wheel (gyroscope) for earth-orbiting satellites that employs the "flux-pinning" properties of imperfect superconductors to reduce friction to near zero. Superconducting x-ray detectors and ultra-fast, superconducting light detectors are being developed due to their inherent ability to detect extremely weak amounts of energy. Already Scientists at the European Space Agency (ESA) have developed what's being called the S-Cam, an optical camera of phenomenal sensitivity. And, superconductors may even play a role in Internet communications soon. In late February, 2000, Irvine Sensors Corporation received a $1 million contract to research and develop a superconducting digital router for high-speed data communications up to 160 Ghz. Since Internet traffic is increasing exponentially, superconductor technology is being called upon to meet this super need.

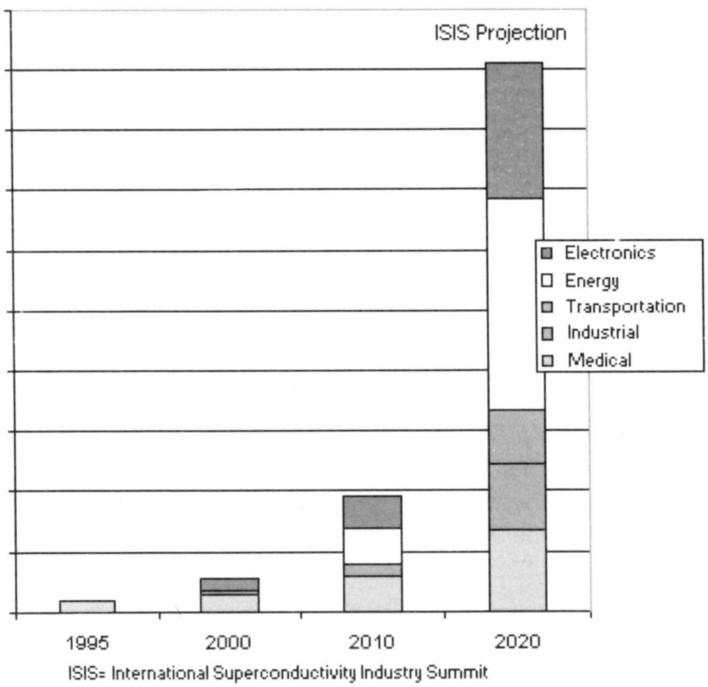

Fig. II–I.16: Depiction of findings of ISIS.

According to June 2002 estimates by the Conectus consortium, the worldwide market for superconductor products is projected to grow to near US $5 billion by the year 2010 and to US $38 billion by 2020. Low-temperature superconductors are expected to continue to play a dominant role in well-established fields such as MRI and scientific research, with high-temperature superconductors enabling the newer industries. The above ISIS graph gives a rough breakdown of the various markets in which superconductors are expected to make a contribution.

All of this is, of course, contingent upon a linear growth rate. Should new superconductors with higher transition temperatures be discovered, growth and development in this exciting field could explode virtually overnight.

Another impetus to the wider use of superconductors is political in nature. The reduction of green-house gas (GHG) emissions has becoming a topical issue due to the Kyoto Protocol which requires the European Union (EU) to reduce its emissions by 8 per cent from 1990 levels by 2012. Physicists in Finland have calculated that the EU could reduce carbon dioxide emissions by up to 53 million tons if high-temperature superconductors were used in power plants.

The future melding of superconductors into our daily lives will also depend to a great degree on advancements in the field of cryogenic cooling. New, high-efficiency magnetocaloric-effect compounds such as gadolinium-silicon-germanium are expected to enter the marketplace soon. Such materials should make possible compact, refrigeration units to facilitate additional HTS applications.

Unit II–J

Types of Superconductors

Type 1: Superconductors and a Periodic Chart Comparison

The Type 1 category of superconductors is mainly comprised of metals and metalloids that show some conductivity at room temperature. They require incredible cold to slow down molecular vibrations sufficiently to facilitate unimpeded electron flow in accordance with what is known as BCS theory. BCS theory suggests that electrons team up in "Cooper pairs" in order to help each other overcome molecular obstacles - much like race cars on a track drafting each other in order to go faster. Scientists call this process phonon-mediated coupling because of the sound packets generated by the flexing of the crystal lattice.

Type 1 superconductors - characterized as the "soft" superconductors—were discovered first and require the coldest temperatures to become superconductive. They exhibit a very sharp transition to a superconducting state (see above graph) and "perfect" diamagnetism—the ability to repel a magnetic field completely. Below is a list of known Type 1 superconductors along with the critical transition temperature (known as Tc) below which each superconducts. The 3rd column gives the lattice structure of the solid that produced the noted Tc. Surprisingly, copper, silver and gold, three of the best metallic conductors, do not rank among the superconductive elements. Why is this?

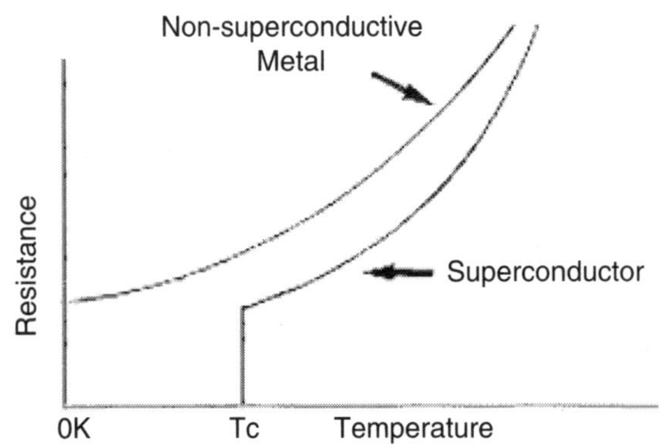

Fig. II–J.1: Type 1 Superconductor.

Lead (Pb)	7.196 K	FCC
Lanthanum (La)	4.88 K	HEX
Tantalum (Ta)	4.47 K	BCC
Mercury (Hg)	4.15 K	RHL
Tin (Sn)	3.72 K	TET
Indium (In)	3.41 K	TET
Palladium (Pd)*	3.3 K	(see note 1)
Chromium (Cr)*	3 K	(see note 1)
Thallium (Tl)	2.38 K	HEX
Rhenium (Re)	1.697 K	HEX
Protactinium (Pa)	1.40 K	TET
Thorium (Th)	1.38 K	FCC
Aluminum (Al)	1.175 K	FCC
Gallium (Ga)	1.083 K	ORC
Molybdenum (Mo)	0.915 K	BCC
Zinc (Zn)	0.85 K	HEX
Osmium (Os)	0.66 K	HEX
Zirconium (Zr)	0.61 K	HEX
Americium (Am)	0.60 K	HEX
Cadmium (Cd)	0.517 K	HEX
Ruthenium (Ru)	0.49 K	HEX
Titanium (Ti)	0.40 K	HEX
Uranium (U)	0.20 K	ORC
Hafnium (Hf)	0.128 K	HEX
Iridium (Ir)	0.1125 K	FCC
Beryllium (Be)	0.023 K (SRM 768)	HEX
Tungsten (W)	0.0154 K	BCC
Platinum (Pt)*	0.0019 K	(see note 1)
Rhodium (Rh)	0.000325 K	FCC

**Note 1*: Tc's given are for bulk (alpha form), except for Palladium, which has been irradiated with He^+ ions, Chromium as a thin film, and Platinum as a compacted powder.

Many additional elements can be coaxed into a superconductive state with the application of high pressure. For example, phosphorus appears to be the Type 1 element with the highest Tc. But, it requires compression pressures of 2.5 Mbar to reach a Tc of 14-22 K. The above list is for elements at normal (ambient) atmospheric pressure. See the periodic table below for all known elemental superconductors (including Niobium, Technetium and Vanadium which are technically Type 2).

Diamagnetic Elements

Bismuth (-280.1 susceptibility/wt.)	Mercury (-33.4)	Zinc (-11.4)
Tellurium (-106.0)	Gold (-28.0)	Beryllium (-9.0)
Antimony (-99.0)	Phosphorous (-26.6/black)	Boron (-6.7)
Molybdenum (-96.5)	Lead (-23.0)	Carbon (-6.0)*
Germanium (-76.8)	Gallium (-21.6)	Copper (-5.5)
Indium (-64.0)	Cadmium (-19.8)	Silicon (-3.9)
Thallium (-50.9/alpha)	Silver (-19.5)	Nitrogen
Tin (-37.0)	Sulfur (-15.0)	Hydrogen
		Noble Gases

*Pyrolytic graphite is the strongest diamagnetic material at room temperature.

EXHIBIT 10.A

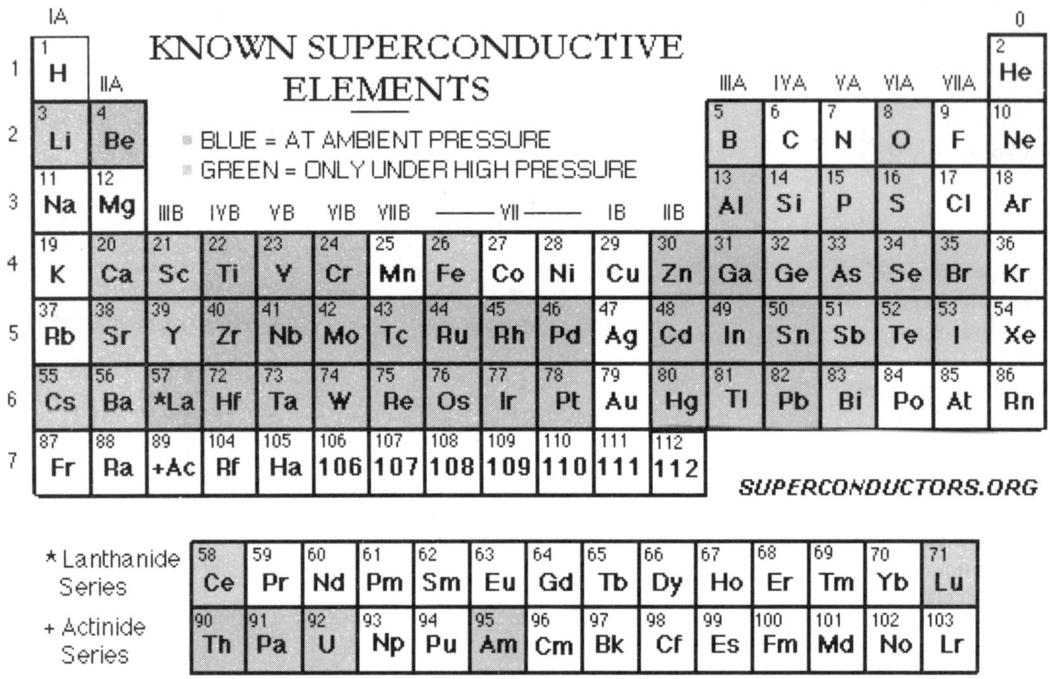

Note: Normally bulk carbon (amorphous, diamond, graphite, white) will not superconduct at any temperature. However, a Tc of 15K has been reported for elemental carbon when the atoms are configured as highly-aligned, single-walled nanotubes. Nanotubes made from other materials, like silicon, boron-nitride or tungsten-disulfide, may also exhibit superconductivity. Since the penetration depth is much larger than the coherence length, nanotubes would be characterized as "Type 2" superconductors.

Type 2: Superconductors

Except for the elements vanadium, technetium and niobium, the Type 2 category of superconductors is comprised of metallic compounds and alloys. The recently-discovered superconducting "perovskites" (metal-oxide ceramics that normally have a ratio of 2 metal atoms to every 3 oxygen atoms) belong to this Type 2 group. They achieve higher Tc's than Type 1 superconductors by a mechanism that is still not completely understood. Conventional wisdom holds that it relates to the planar layering within the crystalline structure (see above graphic). Although, other recent research suggests the holes of hypocharged oxygen in the charge reservoirs are responsible. (Holes are positively-charged vacancies within the lattice.) The superconducting cuprates (copper-oxides) have achieved astonishingly high Tc's when you consider that by 1985 known Tc's had only reached 23 K. To date, the highest Tc attained at ambient pressure has been 138 K. One theory predicts an upper limit of about 200 K for the layered cuprates (Vladimir Kresin, Phys. Reports 288, 347—1997). Others assert *there is no limit*. Either way, it is almost certain that other, more-synergistic compounds still await discovery among the high-temperature superconductors.

Fig. II–J.2: A Type 2 Layered Cuprate.

The first superconducting Type 2 compound, an alloy of lead and bismuth, was fabricated in 1930 by W. de Haas and J. Voogd. But, was not recognized as such until later, after the Meissner effect had been discovered. This new category of superconductors was identified by L.V. Shubnikov at the Kharkov Institute of Science and Technology in the Ukraine in 1936[(1)] when he found two distinct critical magnetic fields (known as H_{c1} and H_{c2}) in $PbTl_2$. The first of the oxide superconductors was created in 1973 by DuPont researcher Art Sleight when $Ba(Pb,Bi)O_3$ was found to have a Tc of 13K. The superconducting oxocuprates followed in 1986.

Type 2 superconductors - also known as the "hard" superconductors - differ from Type 1 in that their transition from a normal to a superconducting state is gradual across a region of "mixed state" behavior. Since a Type 2 will allow *some* penetration by an external magnetic field into its surface, this creates some rather novel mesoscopic phenomena like superconducting "stripes" and "flux-lattice vortices". While there are far too many to list in totality, some of the more interesting Type 2 superconductors are listed below by similarity and with descending Tc's. Where available, the lattice structure of the system is also noted.

$Hg_{0.8}Tl_{0.2}Ba_2Ca_2Cu_3O_{8.33}$	138 K* (record-holder)
$HgBa_2Ca_2Cu_3O_8$	133-135 K
$HgBa_2Ca_3Cu_4O_{10+}$	125-126 K
$HgBa_2Ca_{1-x}Sr_xCu_2O_{6+}$	123-125 K
$HgBa_2CuO_4$+	94-98 K

Lattice: TET

* *Note*: Partial substitution within the oxygen site of Hg-1223 by F or Cl has also reportedly produced a Tc of 138K. However, research by Superconductors.ORG suggests electron doping of the Hg site by both Tl and Pb could produce a Tc even higher than 138K.

$Tl_2Ba_2Ca_2Cu_3O_{10}$	127-128 K
$Tl_{1.6}Hg_{0.4}Ba_2Ca_2Cu_3O_{10+}$	126 K
$TlBa_2Ca_2Cu_3O_{9+}$	123 K

$Tl_{0.5}Pb_{0.5}Sr_2Ca_2Cu_3O_9$	118-120 K
$Tl_2Ba_2CaCu_2O_6$	118 K
$TlBa_2Ca_3Cu_4O_{11}$	112 K
$TlBa_2CaCu_2O_{7+}$	103 K
$Tl_2Ba_2CuO_6$	95 K

Lattice: TET

$Bi_{1.6}Pb_{0.6}Sr_2Ca_2Sb_{0.1}Cu_3O_y$	115 K (thick film on MgO substrate)
$Bi_2Sr_2Ca_2Cu_3O_{10}$***	110 K
$Bi_2Sr_2CaCu_2O_9$***	110 K
$Bi_2Sr_2Ca_{0.8}Y_{0.2}Cu_2O_8$	95-96 K
$Bi_2Sr_2CaCu_2O_8$	91-92K

Lattice: ORTH

*** Though not always listed as a component, a small amount of Lead (x=.2-.26) is often used with Bismuth compounds to help facilitate a higher-Tc crystalline phase.

$Sn_2Ba_2(Ca_{0.5}Tm_{0.5})Cu_3O_x$	~115 K (Superconductors.ORG - 2005)
$Sn_2Ba_2(Sr_{0.5}Y_{0.5})Cu_3O_8$	86 K (Aleksandrov, et al. - 1989)

$Ca_{1-x}Sr_xCuO_2$	110 K (Highest Tc ternary/quaternary compound)
$AuBa_2Ca_3Cu_4O_{11}$	99 K
$AuBa_2(Y, Ca)Cu_2O_7$	82 K
$AuBa_2Ca_2Cu_3O_9$	30 K

Lattice: ORTH

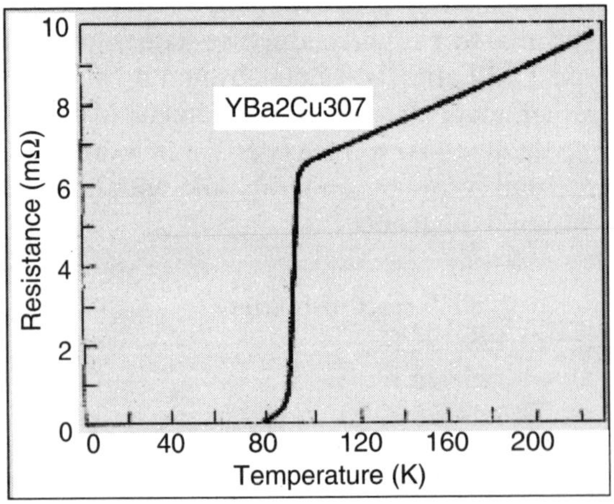

$NdBa_2Cu_3O_7$	96 K
$Y_2Ba_4Cu_7O_{15}$	95 K
$GdBa_2Cu_3O_7$	94 K
$YBa_2Cu_3O_7$	92 K (See above graphic)
$TmBa_2Cu_3O_7$	90 K
$YbBa_2Cu_3O_7$	89 K

Lattice: TET

Comment: "1-2-3" superconductors actually have the 1-2-1-2 structure. Thus, the formula for YBCO could be written $CuBa_2YCu_2O_7$.

$(Ba,Sr)CuO_2$	90 K (infinite layer)
$La_2Ba_2CaCu_5O_{9+}$	79 K (Saurashtra Univ., Rajkot, India,2002)
$(Sr,Ca)_5Cu_4O_{10}$	70 K
$Pb_2Sr_2(Y, Ca)Cu_3O_8$	70 K (Cava, et al - 1989)
$GaSr_2(Y, Ca)Cu_2O_7$	70 K
$(In_{0.3}Pb_{0.7})Sr_2(Ca_{0.8}Y_{0.2})Cu_2O_x$	60 K
$(La,Sr,Ca)_3Cu_2O_6$	58 K
$La_2CaCu_2O_{6+}$	45 K
$(Eu,Ce)_2(Ba,Eu)_2Cu_3O_{10+}$	43 K
$(La_{1.85}Sr_{0.15})CuO_4$	40 K
SrNdCuO****	40 K
$(La,Ba)_2CuO_4$	35-38 K
$(Nd,Sr,Ce)_2CuO_4$	35 K
$Pb_2(Sr,La)_2Cu_2O_6$	32 K
$(La_{1.85}Ba_{.15})CuO_4$	30 K (First HTS ceramic SC discovered, 1986)

**** First ceramic superconductor discovered without a non-superconducting oxide layer.

Comment: All of the above are copper perovskites, even though their metal-to-oxygen ratios are not exactly 2-to-3. The best performers are those compounds that contain one or more of the electron-emitters BaO, SrO or CaO, along with a Period 6 heavy metal like Mercury, Thallium, Bismuth, or Gold.

MgB_2	39 K (Highest Tc Non-Fullerene Alloy)
$Ba_{0.6}K_{0.4}BiO_3$	30 K (First 4th order phase compound)
Nb_3Ge	23.2 K
Nb_3Si	19 K
Nb_3Sn	18.1 K
Nb_3Al	18 K
V_3Si	17.1 K
Ta_3Pb	17 K
V_3Ga	16.8 K
Nb_3Ga	14.5 K
V_3In	13.9 K

Lattice: A15

Comment: Among the binary alloys, these are some of the best performers; combining Group 5B metals in a ratio of 3-to-1 with 4A or 3A elements.

$PuCoGa_5$	18.5 K (First SC transuranic compound)
NbN	16.1 K

Comment: After NbTi (below) this is the most widely used low-temperature superconductor.

$Nb_{0.6}Ti_{0.4}$	9.8 K (First superconductive wire)
$MgCNi_3$	7-8 K (First all-metal perovskite superconductor)
C	15 K (as highly-aligned, single-walled nanotubes)
Nb	9.25 K
Tc	7.80 K
V	5.40 K

Lattice: C=Fullerene, Nb=BCC, Tc=HEX, V=BCC

Comment: These four are the only elemental Type 2 superconductors.

$RuSr_2(Gd,Eu,Sm)Cu_2O_8$	Tc ~58 K (Ruthenium-oxocuprate)
$ErNi_2B_2C$	Tc 10.5 K (Borocarbide)
$YbPd_2Sn$	Tc ~2.5 K (Heusler compound)
UGe_2	Tc ~1K (Heavy fermion)
$URhGe_2$	Tc ~1K (")
$ZrZn_2$	Tc ~1K
$AuIn_3$	Tc 50 *u*K

Comment: The above 7 compounds are all rare ferromagnetic superconductors.

$Sr_{.08}WO_3$	2-4 K (Tungsten-bronze)
$Tl_{.30}WO_3$	2.0-2.14 K (")
$Rb_{.27-.29}WO_3$	1.98 K (")

Atypical Superconductors and the Future

As if ceramic superconductors were not strange enough, even more mysterious superconducting systems have been discovered. One is based on compounds centered around the "*Fullerene*". The fullerene name comes from the late designer-author Buckminster Fuller. Fuller was the inventor of the geodesic dome, a structure with a soccer ball shape. The fullerene—also called a buckminsterfullerene or "buckyball"—exists on a molecular level when 60 carbon atoms join in a closed sphere. When doped with one or more alkali metals the fullerene becomes a "fulleride" and has produced Tc's ranging from 8 K for $Na_2Rb_{0.5}Cs_{0.5}C_{60}$ up to 40 K for Cs_3C_{60}. In 1993 researchers at the State University of New York at Buffalo reported Tc's between 60 K and 70 K for C-60 doped with the interhalogen compound ICl.

Fullerenes, like ceramic superconductors, are a fairly recent discovery. In 1985, professors Robert F. Curl, Jr. and Richard E. Smalley of Rice University in Houston and Professor Sir Harold W. Kroto of the University of Sussex in Brighton, England, accidentally stumbled upon them. The discovery of superconducting alkali metal fullerides came in 1991 when Robert Haddon and Bell Labs announced that K_3C_{60} had been found to superconduct at 18 K.

Larger, non-spherical pure carbon fullerenes that will superconduct have only recently been discovered. In April of 2001, Chinese researchers at Hong Kong University found 1-dimensional superconductivity in single-walled carbon nanotubes at around 15 Kelvin. Silicon-based fullerides like $Na_2Ba_6Si_{46}$ will also superconduct. However, they are structured as infinite networks, rather than discrete molecules. Fullerenes are technically part of a larger family of organic conductors which are described in Fig. II–J.4.

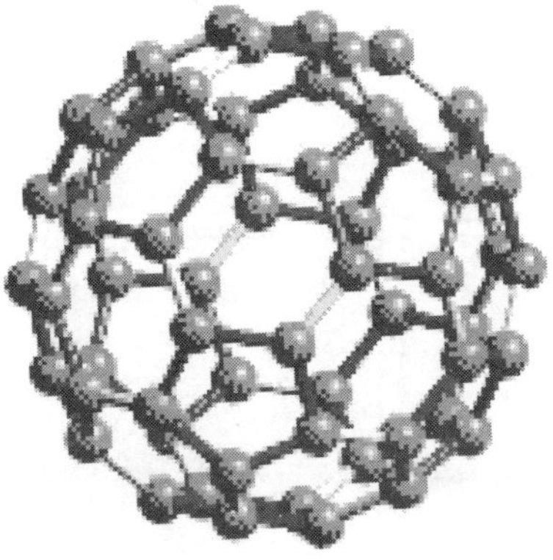

Fig. II–J.3: Carbon fullerenes.

Fig. II–J.4: $(TMTSF)_2PF_6$, The first organic superconductor discovered.

"Organic" superconductors are part of the organic conductor family which includes: molecular salts, polymers and pure carbon systems (including carbon nanotubes and C_{60} compounds). The molecular salts within this family are large organic molecules that exhibit superconductive properties at very low temperatures. For this reason they are often referred to as "molecular" superconductors. Their existence was theorized in 1964 by Bill Little of Stanford University. But the first organic superconductor $(TMTSF)_2PF_6$ was not actually synthesized until 1980 by Danish researcher Klaus Bechgaard of the University of Copenhagen and French team members D. Jerome, A. Mazaud, and M. Ribault. About 50 organic superconductors have since been found with Tc's extending from 0.4 K to near 12 K (at ambient pressure). Since these Tc's are in the range of Type 1 superconductors, engineers have yet to find a practical application for them. However, their rather unusual properties have made them the focus of intense research. These properties include giant magnetoresistance, rapid oscillations, quantum hall effect, and more (similar to the behavior of InAs and InSb). In early 1997, it was, in fact $(TMTSF)_2PF_6$ that a research team at SUNY discovered could resist "quenching" up to a magnetic field strength of 6 tesla. Ordinarily, magnetic fields a fraction as strong will completely kill superconductivity in a material.

Organic superconductors are composed of an electron donor (the planar organic molecule) and an electron acceptor (a non-organic anion). Below are a few more examples of organic superconductors.

$(TMTSF)_2ClO_4$
[tetramethyltetraselenafulvalene + acceptor]

$(BETS)_2GaCl_4$
[bis(ethylenedithio)tetraselenafulvalene + acceptor]

$(BEDO\text{-}TTF)_2ReO_4H_2O$
[bis(ethylenedioxy)tetrathiafulvalene + acceptor]

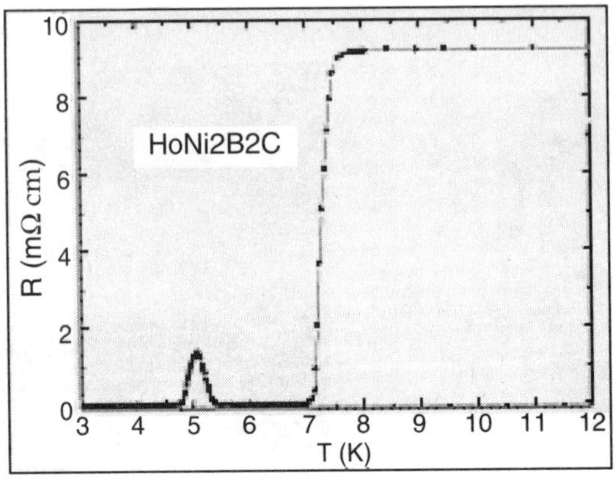

Discovered in 1993 by Bob Cava (currently at Princeton University) and Bell Labs, *"Borocarbides"* are one of the least-understood superconductor systems of all. It has always been assumed that superconductors cannot be formed from ferromagnetic transition metals—like iron, cobalt or nickel. It's the equivalent of trying to mix oil and water. However, in some borocarbides there is a "soap" that acts to bring these adversaries together. The crystallographic sites for the magnetic ions are thought to be isolated from the conduction path. This allows the cooper pairs to detour around the magnetic ions. Further, when combined with an element that has unusual magnetic properties—like holmium—"re-entrant" behavior can also be in evidence in some borocarbides. Below Tc, where it should remain superconductive, there is a discordant temperature at which the material retreats to a "normal", non-superconductive state (see above graphic).

And, not only the borocarbides recede from a superconductive state at extreme low temperatures. In the compounds $HoMo_6S_8$ (Chevrel) and $ErRh_4B_4$ superconductivity suddenly disappears at around 1K. As can be seen from some of the below examples, the first metal site in the molecule is always occupied by a rare earth atom.

YPd_2B_2C	23 K	
$LuNi_2B_2C$	16.6 K	
YNi_2B_2C	15.5 K	
$TmNi_2B_2C$	11 K	(resistance increases below Tc)
$ErNi_2B_2C$	10.5 K	(ferromagnetic)
$HoNi_2B_2C$	7.5 K	(see above graphic)

Note: Other elements that exhibit ferromagnetism have been—by one means or another—integrated into superconducting compounds.

The *"Heavy Fermions"* sound like a family of overweight circus performers. But, they are yet another example of atypical superconductors. Heavy fermions are compounds containing rare-earth elements such as Ce or Yb, or actinide elements such as U. Their (inner shell) conduction electrons often have effective masses (known as quasiparticle masses) several hundred times as great as that of "normal" electrons, resulting in what's known as low "Fermi energy" (Ef). This makes them reluctant superconductors. Yet, at cryogenic temperatures, many of these materials are magnetically ordered, others show strong paramagnetic behavior, and some display superconductivity through a mechanism that quickly runs afoul of BCS theory. Research suggests cooper-pairing in the heavy fermion systems arises from the magnetic interactions of the electron spins (D-wave, P-wave, S-wave), rather than by lattice vibrations.

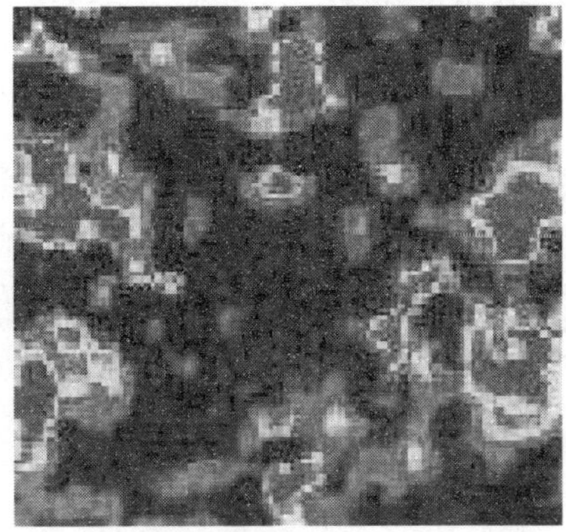

Fig. II–J.5: Flux lattices in UPt3.

The first observation of superconductivity in a heavy fermion system was made by E. Bucher, et al, in 1973 in the compound UBe_{13}; but, at the time was attributed to precipitated uranium filaments. Superconductivity was not actually recognized, per se, in a heavy fermion compound until 1979 when Dr. Frank Steglich of the Max Planck Institute for Chemical Physics in Solids (Dresden, Germany) realized it was a bulk property in $CeCu_2Si_2$.

In April 2003 a heavy-fermion compound unambiguously exhibited the so-called "FFLO" state, where magnetism and superconductivity have a beneficial coexistence. The compound $CeCoIn_5$ confirmed a theoretical model first put forth in 1964 by Fulde, Ferrell, Larkin, and Ovchinnikov (FFLO). Below are some heavy fermion compounds that will superconduct, along with their Tc's. As can be seen, their transition temperatures are in the range of Type 1 superconductors, which severely limits their usefulness.

$CeCoIn_5$	2.3 K	(first confirmed FFLO compound)
UPd_2Al_3	2 K	
Pd_2SnYb	1.79 K	
URu_2Si_2	1.2 K	
UNi_2Al_3	1 K	
Al_3Yb	0.94 K	
UBe_{13}	0.87 K	
$CeCo_2$	0.84 K	
$CePt_3Si$	0.75 K	(first heavy fermion superconductor without a center of symmetry)
UPt_3	0.48 K	(see above graphic)
$CeCu_2Si_2$	.1-.7 K	

Note: UGe_2 and $URhGe_2$ exhibit simultaneous ferromagnetism and superconductivity. Read more about this phenomenon in the below section on Ruthenates.

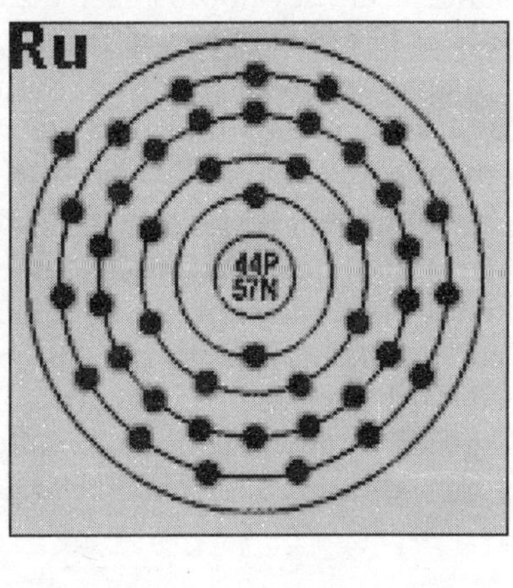

In the mid-1990's, it was discovered that copper-oxygen planes are not the only superconducting facilitators within the layered perovskites. In 1994 physicists at IBM Zurich and at Hiroshima University collaborated to study the atomic planes of ruthenium-oxygen due to their similarity to copper-oxygen planes. Yoshiteru Maeno and colleagues found that the compound Sr_2RuO_4 exhibited superconductivity at 1.5 K. While this is an extremely cold Tc for a superconducting perovskite, it revealed a new area of potential among what are known as *"Ruthenates"*. Shortly after that SrRuO and $SrYRuO_6$ were also found to superconduct at similarly low temperatures.

When the crystalline structure of some of these materials

is broken apart, its surface becomes increasingly ferromagnetic at low temperatures. This phenomenon flies in the face of condensed matter theory. So much so, that researchers have characterized them as an analog to superfluid Helium-3. And, when copper is added to the mix, even stranger things happen. In June 1999 New Zealand researcher Dr. Jefferey Tallon and his German colleague Dr. Christian Bernard discovered a ruthenium-cuprate** whose bulk is both a superconductor and a magnet. Although it was not the first compound discovered that exhibits coexisting ferromagnetism and superconductivity, it's remarkably high Tc of 58 K makes it truly distinct in the world of superconductors. Unlike "normal" superconductors, this compound becomes diamagnetic at about *one-half* Tc. [**$RuSr_2$ (Gd,Eu,Sm) Cu_2O_8 or any parenthetical element partially substituted by Y.]

There are many phase transitions that matter goes through on its way to another state (e.g. ice changing to water requires a sudden increase in heat energy). Among the superconductors this is also the case at Tc, Hc and Jc. However, a superconductor has been discovered that exhibits *no* measurable change in its specific heat (the amount of energy required to increase its temperature by one degree) while going through up to 3 different "critical" magnetic fields. $Ba_{0.6}K_{0.4}BiO_3$ seems to be the first material discovered that enters a "fourth order" phase transition (according to Ehrenfest's phase transition classification scheme) - something that has never before been observed in nature. Roy Goodrich of Louisiana State University and Donovan Hall of the National High Magnetic Field Laboratory announced this discovery in May, 1999. Not only is this an amazing anomaly in the field of superconductivity. It suggests that even higher order phases may exist in nature.

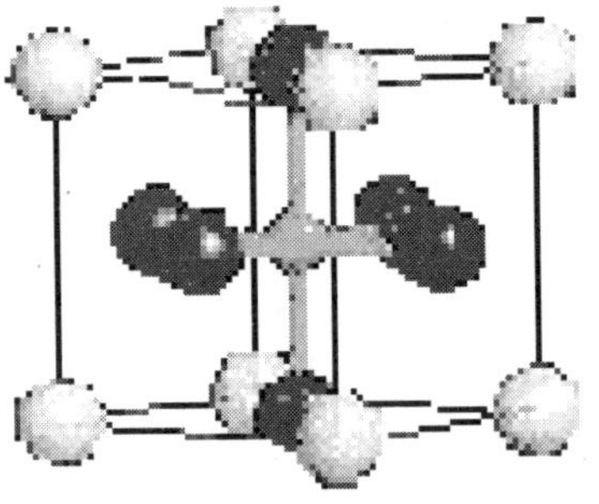

Fig. II–J.6: $Ba_{0.6}K_{0.4}BiO_3$

While no one can predict what future discoveries will be made in the field of superconductivity, some recent developments in the tungsten-bronze system suggest a new vista may be emerging. In July of 1999 researchers Y. Tsabba and S. Reich of the Weizmann Institute in Israel reported possible superconductivity near 91 K in the sodium-doped tungsten-bronze $Na_{0.05}WO_3$. This would be the first known HTS that is not a cuprate. Most tungsten-bronze compounds that are known to superconduct have Tc's below 4K—making this a truly tantalizing find.

Other categories of materials that theory suggests may produce superconductors are the higher silver fluorides and complex fluorides— known as fluoroargentates. Fluoroargentates bear a strong similarity to oxocuprates, compounds that currently have the highest transition temperatures of all known superconductors. In October 2003 researchers Wojciech Grochala, Adrian Porch and Peter P. Edwards reported sudden drops in magnetic susceptibility within a large number of samples of Be-Ag-F. They attribute this to possible spherical regions of superconductivity—with a Tc up to 64 K—couched inside a ferromagnetic host.

With few exceptions (e.g. polysulphur-nitrides), most polymers resist being coaxed into a superconductive state. However, some organic polymers exhibit electrical resistance many orders of magnitude lower than the best metallic conductors. And, they do this at room temperature! These ultraconductors™, materials such as oxidized atactic polypropylene (OAPP), do not have zero resistance. But, their enhanced conductivity at ambient temperatures and pressures may actually allow them to compete with superconductors in certain fields. Polypropylene, for example, is normally an insulator. In 1985, however, researchers at the Russian Academy of Sciences discovered that as an oxidized thin-film, polypropylene can have a conductivity 10^5 to 10^6 higher than the best refined metals. The Meissner effect—the classic criterion for superconductivity—cannot be observed, as the critical transition temperature appears to be above the point at which the polymer breaks down (>700K). However, strong (giant) diamagnetism has been confirmed.

Unit II–K

Nanostructure Science and Technology: Overview

Introduction

Nanostructure science and technology is a broad and interdisciplinary area of research and development activity that has been growing explosively worldwide in the past few years. It has the potential for revolutionizing the ways in which materials and products are created and the range and nature of functionalities that can be accessed. It is already having a significant commercial impact, which will assuredly increase in the future.

A worldwide study of research and development status and trends in nanoparticles, nanostructured materials, and nanodevices (or more concisely, nanostructure science and technology) was carried out by an eight-person panel under the auspices of the World Technology (WTEC) Division of Loyola College. The purpose of the study was to assess the current status and future trends internationally in research and development in the broad and rapidly growing area of nanostructure science and technology. The goals were fourfold:

1. Provide the worldwide science and engineering community with a broadly inclusive and critical view of this field.
2. Identify promising areas for future research and commercial development.
3. Help stimulate development of an interdisciplinary international community of nanostructure researchers.
4. Encourage and identify opportunities for international collaboration.

Findings

There are two overarching findings from this study:

First, it is abundantly clear that we are now able to nanostructure materials for novel performance. That is the essential theme of this field: novel performance through nanostructuring. It represents the beginning of a revolutionary new age in our ability to manipulate materials for the good of humanity. The synthesis and control of materials in nanometer dimensions can access new material properties and device characteristics in unprecedented ways, and work is rapidly expanding worldwide in exploiting the opportunities offered through nanostructuring. Each year sees an ever increasing number of researchers from a wide variety of disciplines enter the field, and each year sees an ever increasing breadth of novel ideas and exciting new opportunities explode on the international nanostructure scene.

Second, there is a very wide range of disciplines contributing to the developments in nanostructure science and technology worldwide. The rapidly increasing level of interdisciplinary activity in nanostructuring is exciting and growing in importance, and the intersections between the various disciplines are where much of the novel activity resides.

The field of nanostructure science and technology has been growing very rapidly in the past few years, since the realization that creating new materials and devices from nanoscale building blocks could access new and improved properties and functionalities. While many aspects of the field existed well before nanostructure science and technology became a definable entity in the past decade, it has only become a coherent field of endeavor through the confluence of three important technological streams:

1. New and improved control of the size and manipulation of nanoscale building blocks.
2. New and improved characterization (spatial resolution, chemical sensitivity, etc.) of materials at the nanoscale.
3. New and improved understanding of the relationships between nanostructure and properties and how these can be engineered.

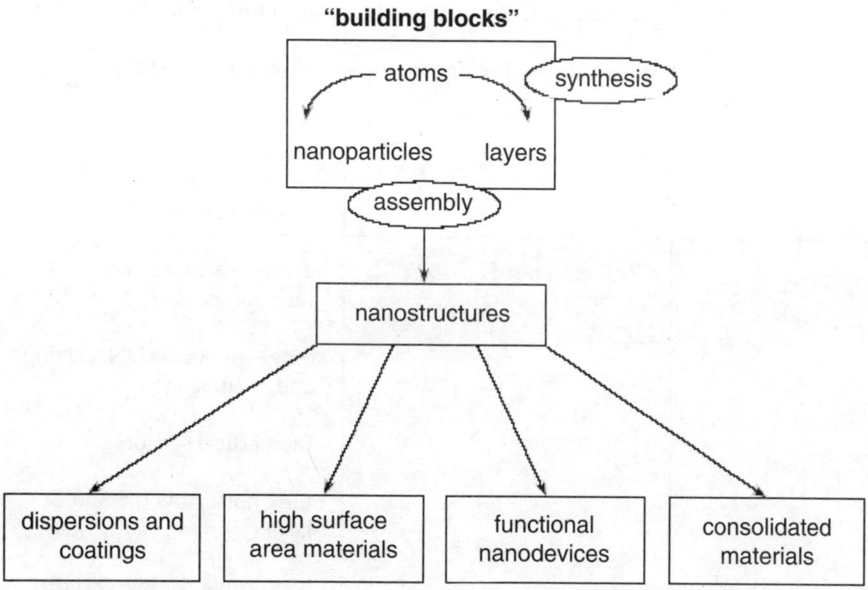

Fig. II–K.1: Organization of nanostructure science and technology

As a result of these developments, a wide range of new opportunities for research and applications in the field of nanotechnology now present themselves. Table II–K.1 indicates some examples of present and potential applications with significant technological impact that were identified in the course of this study. Considerable resources are being expended around the world

Table II–K.1. Technological Impact: Present and Potentia

Technology	Present Impact	Potential Impact
Dispersions and Coatings	Thermal barriers Optical (visible and UV) barriers Imaging enhancement Ink-jet materials Coated abrasive slurries Information-recording layers	Targeted drug delivery/gene therapy Multifunctional nanocoatings
High Surface Area Materials	Molecular sieves Drug delivery	Molecule-specific sensors Large hydrocarbon or bacterial filters
	Tailored catalysts	Energy storage
	Absorption/desorption materials	Grätzel-type solar cells
Consolidated Materials	Low-loss soft magnetic materials	Superplastic forming of ceramics
	High hardness, tough WC/Co cutting tools	Ultrahigh-strength, tough structural materials
	Nanocomposite cements	Magnetic refrigerants
		Nanofilled polymer composites
		Ductile cements
Nanodevices	GMR read heads	Terabit memory and microprocessing
		Single molecule DNA sizing and sequencing
		Biomedical sensors
		Low noise, low threshold lasers
		Nanotubes for high brightness displays
Additional Biological Aspects	Biocatalysis	Bioelectronics Bioinspired prostheses Single-molecule-sensitive biosensors

for research and development aimed at realizing these and a variety of other promising applications. Government funding alone approached half a billion dollars per year towards the end of the 20th century: $128 million in Western Europe; $120 million in Japan; $116 million in the United States; and $70 million altogether in other countries such as China, Canada, Australia, Korea, Taiwan, Singapore, and the countries of the former Soviet Union

Table II–K.2 presents an overall comparison of the current levels of activity among the major regions assessed (Europe, Japan, and the United States) in the various areas of nanotechnology study. These broad areas—synthesis and assembly, biological approaches and applications, dispersions and coatings, high surface area materials, nanodevices, and consolidated materials—constitute the field of nanostructure science and technology.

Table II–K.2. Comparison of Activities in Nanostructure Science and Technology in Europe, Japan, and the United States

Level	*Highest* 1	2	3
Synthesis & Assembly	U.S.	Europe	Japan
Biological Approaches & Applications	U.S./Eur	Japan	
Dispersions and Coatings	U.S./Eur.	Japan	
High Surface	U.S.	Europe	Japan
Nanodevices	Japan	Europe	U.S.
Consolidated Materials	Japan	U.S./Eur	

In the synthesis and assembly area, the United States appears to be ahead, with Europe following and then Japan. In the area of biological approaches and applications, the United States and Europe appear to be rather on a par, with Japan following. In nanoscale dispersions and coatings, the United States and Europe are again at a similar level, with Japan following. For high surface area materials, the United States is clearly ahead of Europe, followed by Japan. On the other hand, in the nanodevices area, Japan seems to be leading quite strongly, with Europe and the United States following. And finally, in the area of consolidated materials, Japan is a clear leader, with the United States and Europe following. These comparisons are, of course, integrals over rather large areas of a huge field and therefore possess all of the inevitable faults of such an integration. At best, they represent only a snapshot of the present, and the picture is admittedly incomplete.

Challenges

We are now at the threshold of a revolution in the ways in which materials and products are created. How this revolution will develop, how great will be the opportunities that nanostructuring can yield, and how rapidly we progress, will depend upon the ways in which a number of challenges are met.

Among the challenges facing us are those concerned with making the necessary advances in enabling technologies in order for rapid progress to continue in this field. We must increase characterization capabilities in visualization and chemical analysis at ever finer size scales. We must be able to manipulate matter at ever finer size scales, and we must eventually use computational approaches in directing this, if we are really going to take full advantage of the available opportunities. Experiment simply cannot do it alone. Theory and modeling are essential. Fortunately, this is an area in which the sizes of the building blocks and their assemblies are small enough that one can, with the ever increasing capabilities in computational sciences, now start doing very serious controlled modeling experiments to guide us in the nanostructuring of matter. Hence, multiscale modeling of nanostructuring and the resulting materials properties across the hierarchy of length scales from atomic, to mesoscopic, to macroscopic is an absolute necessity as we go down the road in the next decades to realizing the tremendous potential of nanostructure science and technology.

Furthermore, we need to understand the critical roles that surfaces and interfaces play in nanostructured materials. Nanoparticles have very high specific surface areas, and thus in their assembled forms there are large areas of interfaces. We need to know in detail not only the structures of these interfaces, but also their local chemistries and the effects of segregation and interaction between the nanoscale building blocks and their surroundings. We need to learn more about the control of nanostructure size and size distribution, composition, and assembly. For some applications, there are very stringent conditions on these parameters; in other applications less so. We must therefore understand the relationships between this stringency and the desired material or device properties.

We also need to be concerned with the thermal, chemical, and structural stability of nanostructured materials and the devices made therefrom, in the face of both the temperature and changing chemistries of the environments in which these nanostructures are asked to function. A nanostructure that is only a nanostructure at the beginning of a process is not of much use to anybody, unless the process is over in a very short time or the process itself is the actual nanostructure advantage. So for many applications, stability is an important consideration, and we must investigate whether natural stability is sufficient or whether we must additionally stabilize against changes that we cannot afford.

To effectively commercialize and utilize the nanostructuring of matter we also need enhancements in statistically driven process control. Achieving reproducibility and scalability of nanoparticle synthesis and consolidation processes in nanostructuring are paramount if successful scale-up is to be effected and if what we do in the laboratory is to contribute to the society that pays for this research. Given a commercial need, the viability of nanostructure production and utilization is wrapped up in the costs of precursors or raw materials, processing costs, and also the costs of dealing with effluent. It is the total integrated cost, in terms of raw materials, synthesis of the building blocks, manufacture of parts from those building blocks, and effluent clean-up costs, that is important and that will ultimately determine commercial viability.

Finally, in order for the field of nanostructure science and technology to truly reach fruition, it is an absolute necessity to create a new breed of researchers who can work across traditional disciplines and think "outside the box." Educating this new breed of researchers, who will either work across disciplines or know how to work with others in the interfaces between disciplines, is vital to the future of nanostructure science and technology. People must start thinking in unconventional ways if we are to take full advantage of the opportunities in this new and revolutionary field.

Unit II–L

Synthesis and Assembly of Nanostructure Materials

Introduction

An important theme for any study on nanostructure materials is the engineering of materials with novel, (i.e. improved) properties through the controlled synthesis and assembly of the material at the nanoscale level. The range of applications is extremely broad, and these will be described in further detail in subsequent units in this report. The corresponding means of synthesis and assembly are similarly wide-ranging. But however multifaceted the synthesis approaches and the ultimate applications, there are common issues and unique defining features of these nanostructured materials.

First, there is the recognition of critical scale lengths that define the material structure and organization, generally in the nanometer range, and that ultimately determine the fundamental macroscopic properties of the material. Research in nanostructured materials is motivated by the belief that ability to control the building blocks or nanostructure of the materials can result in enhanced properties at the macroscale: increased hardness, ductility, magnetic coupling, catalytic enhancement, selective absorption, or higher efficiency electronic or optical behavior.

Synthesis and assembly strategies accommodate precursors from liquid, solid, or gas phase; employ chemical or physical deposition approaches; and similarly rely on either chemical reactivity or physical compaction to integrate nanostructure building blocks within the final material structure. The variety of techniques is shown schematically in Figure II–L.1.

The "bottom-up" approach first forms the nanostructured building blocks and then assembles them into the final material. An example of this approach is the formation of powder components through aerosol techniques and then the compaction of the components into the final material. These techniques have been used extensively in the formation of structural composite materials. One "top-down" approach begins with a suitable starting material and then "sculpts" the functionality from the material. This technique is similar to the approach used by the semiconductor industry in forming devices out of an electronic substrate (silicon), utilizing pattern formation (such as electron beam lithography) and pattern transfer processes (such as reactive ion etching) that have the requisite spatial resolution to achieve creation of structures at the nanoscale. This particular area of nanostructure formation has tremendous scope, is a driving issue for the electronics industry. Another top-down approach is "ball-milling," the formation of nanostructure building blocks through controlled, mechanical attrition of the bulk starting material. Those nano building blocks are then subsequently assembled into a new bulk material.

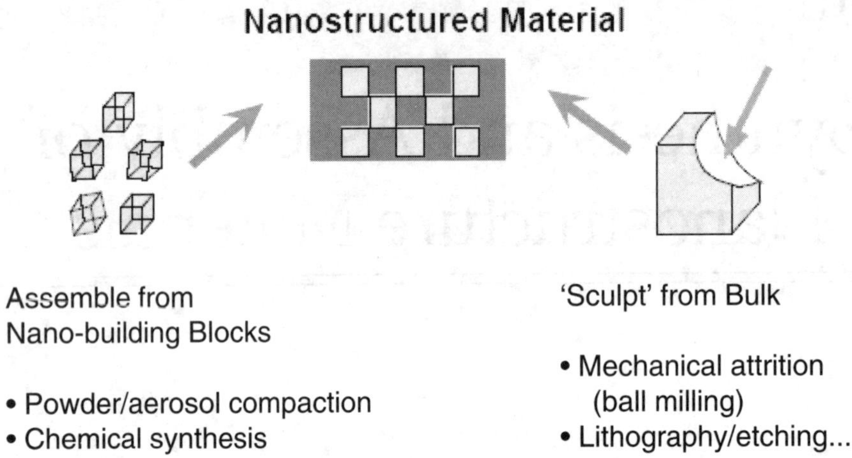

Fig. II–L.1: Schematic of variety of nanostructure synthesis and assembly approaches.

In fact, many current strategies for material synthesis integrate both synthesis and assembly into a single process, such as characterizes chemical synthesis of nanostructured materials. The degree of control required over the sizes of the nanostructure components, and the nature of their distribution and bonding within the fully formed material varies greatly, depending on the ultimate materials application. Achieving selective optical absorption in a material (e.g., UV blocking dispersions) may allow a wide range of sizes of the component nanostructure building blocks, while quantum dot lasers or single electron transistors require a far tighter distribution of size of the nanostructure components. Compaction methods may provide excellent adhesion for nanocomposite materials of improved structural performance (e.g., ductility), but such interfaces may be unsatisfactory for electronic materials. The intention here is not to recapitulate in detail the various synthesis and assembly techniques that have been and are being employed in the fabrication of nanostructured materials. Rather, in attempting to capture the salient features of a new impetus for and interest in a field of nanostructure science and technology, it is more useful to identify the emerging commonalities than the differences among synthesis and assembly approaches.

Critical Issues for Nanostructure Synthesis and Assembly

However broad the range of synthesis approaches, the critical control points fall into two categories:

1. Control of the size and composition of the nanocluster components, whether they are aerosol particles, powders, semiconductor quantum dots, or other nanocomponents.
2. Control of the interfaces and distributions of the nanocomponents within the fully formed materials.

These two aspects of nanostructure formation are inextricably linked; nevertheless, it is important to understand how to exercise separate control over the nucleation of the nanostructure building blocks and the growth (for example, minimizing coagulation or agglomeration) of those components throughout the synthesis and assembly process. This latter issue is related to the importance of the following:

- The chemical, thermal, and temporal stability of such formed nanostructures.
- The ability to scale-up synthesis and assembly strategies for low-cost, large-scale production of nanostructured materials, while at the same time maintaining control of critical feature size and quality of interfaces (economic viability is a compelling issue for any nanostructure technology).

All researchers in this area are addressing these issues.

Common Enabling Technologies

There has been steady technological progress in all fields of nanostructure synthesis and assembly, in no small part because of the more general availability of characterization tools having higher spatial, energy, and time resolution to clearly distinguish and trace the process of nanostructure formation. As transmission electron microscopy and X-ray diffraction techniques helped in an earlier period to relate the improved properties of "age-hardened" aluminum alloys to their nanostructure (Koch 1998), today's technological advances in materials characterization are providing new insights into the role of the nanostructure in determining macroscopic properties. The tightly-coupled iteration between characterizing the nanostructure, understanding the relationship between nanostructure and macroscopic material properties (Figure II–L.2), and improved sophistication and control in determining nanostructure size and placement have accelerated the rate of progress and helped to define the critical components of this "new" field of nanostructure science and technology. Tightly focused (1-2 μm), high brightness synchrotron X-ray sources provide detailed structural information on colloids, polymers, alloys, and other material structures, highlighting the inhomogeneities of the material with suitable spatial resolution.

Another important enabling technology has been the now widely available scanning probe technology, including scanning tunneling microscopy and atomic force microscopy. The power of these techniques has provided impetus for developing even higher performance scanning probe

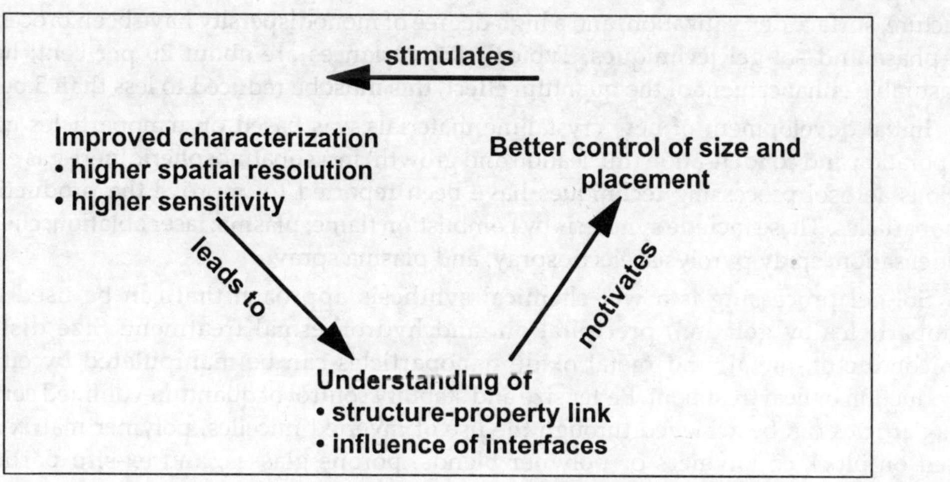

Fig. II–L.2: Interactive cycle of characterization, understanding and enhanced control in the synthesis and assembly of nanostructures.

tips, fabricated through microfabrication techniques. Development of different tip structures in various materials has given rise to an entire family of powerful scanning probe techniques that encompass such a wide range of characterization capabilities that one can envision "a laboratory on a tip". The development of a tip technology also impacts the synthesis and assembly processes themselves: scanning probe technologies have been used as the basis of materials patterning and processing at nanometer scales and have provided information on the mechanical and thermal properties of materials at the nanoscale.

More sophisticated in situ monitoring strategies have provided greater understanding and control in the synthesis of nanostructured building blocks, particularly those formed in vacuum environments. Molecular beam epitaxy (MBE) represents a physical vapor (gas phase) deposition technique where sub-monolayer control can be imposed on the formation of two-dimensional and, more recently, three-dimensional nanostructured materials. A great deal of the understanding and control derives from the ability to carry out sensitive monitoring of the growth process in situ: reflection high energy electron diffraction (RHEED) details the nature of the surface and surface bonding, and oscillations of the RHEED intensity provide information on the growth rate.

The improvements brought about by these advances in technology have been substantial, but perhaps of greater importance for this nascent field of nanostructure science and technology has been the development of strategies and technologies that have been formed across the former disciplines. More reliable means of controlling nanostructure size and placement, with an end view of being able to scale up the production of such materials while maintaining the control over the nanostructure, have given impetus to a common search for novel synthesis and assembly strategies. In that search, it is apparent that the naturally occurring synthesis and assembly of biological materials can provide us with some critical insights.

Nanoparticle Synthesis Strategies

Gas Phase Synthesis and Sol-Gel Processing

Major efforts in nanoparticle synthesis can be grouped into two broad areas: gas phase synthesis and sol-gel processing. Nanoparticles with diameters ranging from 1 to 10 nm with consistent crystal structure, surface derivatization, and a high degree of monodispersity have been processed by both gas-phase and sol-gel techniques. Typical size variances are about 20 per cent; however, for measurable enhancement of the quantum effect, this must be reduced to less than 5 per cent.

Initial development of new crystalline materials was based on nanoparticles generated by evaporation and condensation (nucleation and growth) in a subatmospheric inert-gas environment. Various aerosol processing techniques have been reported to improve the production yield of nanoparticles. These include synthesis by combustion flame; plasma; laser ablation; chemical vapor condensation; spray pyrolysis; electrospray; and plasma spray.

Sol-gel processing is a wet chemical synthesis approach that can be used to generate nanoparticles by gelation, precipitation, and hydrothermal treatment. Size distribution of semiconductor, metal, and metal oxide nanoparticles can be manipulated by either dopant introduction or heat treatment. Better size and stability control of quantum-confined semiconductor nanoparticles can be achieved through the use of inverted micelles, polymer matrix architecture based on block copolymers or polymer blends, porous glasses, and ex-situ particle-capping techniques.

Other Strategies

Additional nanoparticle synthesis techniques include sonochemical processing, cavitation processing, microemulsion processing, and high energy ball milling. In sonochemistry, an acoustic cavitation process can generate a transient localized hot zone with extremely high temperature gradient and pressure. Such sudden changes in temperature and pressure assist the destruction of the sonochemical precursor (e.g., organometallic solution) and the formation of nanoparticles. The technique can be used to produce a large volume of material for industrial applications.

In hydrodynamic cavitation, nanoparticles are generated through creation and release of gas bubbles inside the sol-gel solution. By rapidly pressurizing in a supercritical drying chamber and exposing to cavitational disturbance and high temperature heating, the sol-gel solution is mixed. The erupted hydrodynamic bubbles are responsible for nucleation, growth, and quenching of the nanoparticles. Particle size can be controlled by adjusting the pressure and the solution retention time in the cavitation chamber.

Microemulsions have been used for synthesis of metallic, semiconductor, silica, barium sulfate, magnetic, and superconductor nanoparticles. By controlling the very low interfacial tension (~10-3 mN/m) through the addition of a cosurfactant (e.g., an alcohol of intermediate chain length), these microemulsions are produced spontaneously without the need for significant mechanical agitation. The technique is useful for large-scale production of nanoparticles using relatively simple and inexpensive hardware.

Finally, high energy ball milling, the only top-down approach for nanoparticle synthesis, has been used for the generation of magnetic, catalytic, and structural nanoparticles. The technique, which is already a commercial technology, has been considered dirty because of contamination problems from ball-milling processes. However, the availability of tungsten carbide components and the use of inert atmosphere and/or high vacuum processes have reduced impurities to acceptable levels for many industrial applications. Common drawbacks include the low surface area, the highly polydisperse size distributions, and the partially amorphous state of the as prepared powders.

Other Synthesis Issues

Means to Achieve Monodispersity

One of the most challenging problems in synthesis is the controlled generation of monodispersed nanoparticles with size variance so small that size selection by centrifugal precipitation or mobility classification is not necessary. Among all the synthesis techniques discussed above, gas-phase synthesis is one of the best techniques with respect to size monodispersity, typically achieved by using a combination of rigorous control of nucleation condensation growth and avoidance of coagulation by diffusion and turbulence as well as by the effective collection of nanoparticles and their handling afterwards. The stability of the collected nanoparticle powders against agglomeration, sintering, and compositional changes can be ensured by collecting the nanoparticles in liquid suspension. For semiconducting particles, stabilization of the liquid suspension has been demonstrated by the addition of polar solvent; surfactant molecules have been used to stabilize the liquid suspension of metallic nanoparticles.

Scaleup

Scaleup production is of great interest for nanoparticle synthesis. High energy ball milling, already a commercial high-volume process, as mentioned above, has been instrumental in generating nanoparticles for the preparation of magnetic, structural, and catalytic materials. However, the process produces polydispersed amorphous powder, which requires subsequent partial recrystallization before the powder is consolidated into nanostructured materials. Although gas-phase synthesis is generally a low production rate process (typically in the 100 milligrams per hour range) in research laboratories, higher rates of production (about 20 grams per hour) are being demonstrated at Ångström Laboratory at Uppsala University in Sweden. Even higher production rates (about 1 kg per hour) are now being achieved commercially. For sol-gel processing, the development of continuous processing techniques based on present knowledge of batch processing has yet to be addressed for economical scaleup production of nanoparticles. Other related sol-gel issues concern the cost of precursors and the recycling of solvent. Overall, sol-gel processing is attractive for commercial scale-up production.

Building Nanoparticle Chains or Wires

A recent paradigm shift envisioned for optoelectronics and computational devices involves the assembly of molecular or quantum wires. Large polymeric molecules have been used as nano building blocks for nanoporous molecular sieves, biocompatible materials, optical switching, data processing, and other nonlinear optical components. Chain aggregates of nanoparticles can be considered as polymer-like units with their primary particles composed of a few hundred to a few thousand molecules. Thus, these chain aggregates can be considered "heavy" quantum wires. In fact, nanoparticle chain aggregates have been studied extensively as magnetic materials, as reinforced elastomers, and as additives in concrete. These aggregates have been shown to have chemical and mechanical properties different from those of individual primary particles. Depending on the particle size and its compositional material, the bonding force responsible for holding the aggregates together varies from weak van der Waals force for micrometer particles to strong chemical bonds for nanometer particles to very strong magnetic dipolar bonds for nanosized magnetic particles. The mechanical, optical, and electronic transport properties can be varied by controlling the diameter and the monodispersity of the primary particles, the crystalline structure and morphology, aggregate length, interfacial properties, and material purity. These chain aggregates can be formed by allowing agglomeration of nanoparticles generated by any of the synthesis techniques discussed above, with the exception of high energy ball milling, which generates particles with low surface area and high anisotropic morphologies, both of which are detrimental for the formation of chain aggregates. Depending on the magnetic and electric charging properties of the nanoparticles, an external applied magnetic or electric field can be used to control the fractal dimension of aggregates. For optical applications of chain aggregates, lower fractal dimensions (i.e., relatively straight chain aggregates with few branches) are desirable.

Building Nanometer Fibers or Tubes

Recent advances in the fabrication of nanometer fibers or tubes offer another form of building blocks for nanostructured materials. An effective way to generate nanometer fibers (or tubes) is based on the use of membrane-template techniques. Membranes, with nanochannels generated by fission-fragment tracks or by electrochemical etching of aluminum metal, are used as templates for either chemical or electrochemical deposition of conductive polymers, metal, semiconductor, and other materials for the generation of nanofibers or tubes. Since the nanochannels on membranes are very uniform in size, the diameter and the aspect ratio of the nanofibers (or tubes) synthesized by the membrane template technique can be precisely controlled. This greatly facilitates the interpretation of optical data and the processing of these fibers (or tubes) into 2-D nanostructured materials. Single-crystal semiconductor nanofibers can also be grown catalytically by metalorganic vapor phase epitaxy and laser ablation vapor-liquid-solid techniques. The synthesis of these onedimensional structures with diameters in the range of 3 to 15 nm holds considerable technological promise for optoelectronic device applications, such as the p-n junctions for light emission at Hitachi Central Research Laboratory in Japan.

The advent of carbon-based nanotubes has created yet another way to fabricate nanometer fibers and tubes. These nanotubes have been used as templates for the fabrication of carbide and oxide nanotubes. Synthesis of nanotubes based on BN, BC3 and BC2N have also been reported. These nanotubes potentially possess large third-order optical non-linearity and other unusual properties. Metallic nanofibers synthesized by carbon-nanotube-template techniques are useful in the design of infrared absorption materials. The carbon nanotubes can now be catalytically produced in large quantities and have been used for reinforcement of nanostructural composite materials and concrete.

Biogenic Strategies

The elegant complexity of biological materials represents the achievement of structural order over many length scales, with the full structure developed from the "nested levels of structural hierarchy", in which self-assembled organic materials can form templates or scaffolding for inorganic components. These notions of a multilevel material structure with strong interactions among levels and interplay of perfection and imperfection forming the final material was discussed earlier by Cyril Stanley Smith. Along with characteristic length scales, there are characteristic relaxation times of the material, bringing in the consideration of the temporal stability of the structured materials (Zener 1948).

A more detailed discussion of the role of biological materials as both paradigm and tool for the fabrication of nanostructured materials follows. It is interesting how many of the synthesis and assembly approaches seek out and adapt two key features of biogenic fabrication: that of "self-assembly," and the use of natural nanoscaled templates.

Self-Assembly as a Deliberate Strategy

"Self-assembly" is a term that by now figures prominently in the literature of nanostructured materials and nanofabrication. The term therefore carries a variety of implicit or explicit meanings, and we cite the definition given by Kuhn and Ulman: "Self-assembly is a process in which supermolecular hierarchical organization is established in a complex system of interlocking components." The mechanisms that produce the hierarchical organization are determined by competing molecular interactions (e.g.,interactions between hydrophobic versus hydrophilic components, van der Waals, Coulombic, or hydrogen bonding), resulting in particular microphase separation or surface segregation of the component materials. Thus, the use of a hierarchy of bond strengths and/or chemical specificity can produce a hierarchy of lengths in the final nanostructured material.

As one example of such self-assembled or self-organized materials, McGehee et al. have mixed silica precursors with surfactants that have self-ordered to form various surfactant-water liquid crystals, producing various structures built from walls of amorphous silica, organized about a repetitive arrangement of pores up to a hundred angstroms in diameter. A range of such structures is shown in Figure 12.3. In the "natural" formation of inorganic nanostructures, the addition of organic molecules can strongly influence the resulting structure of inorganic components. Such strategies have been adopted in synthetic formation of nanostructures, such as in the formation of networks of gold clusters.

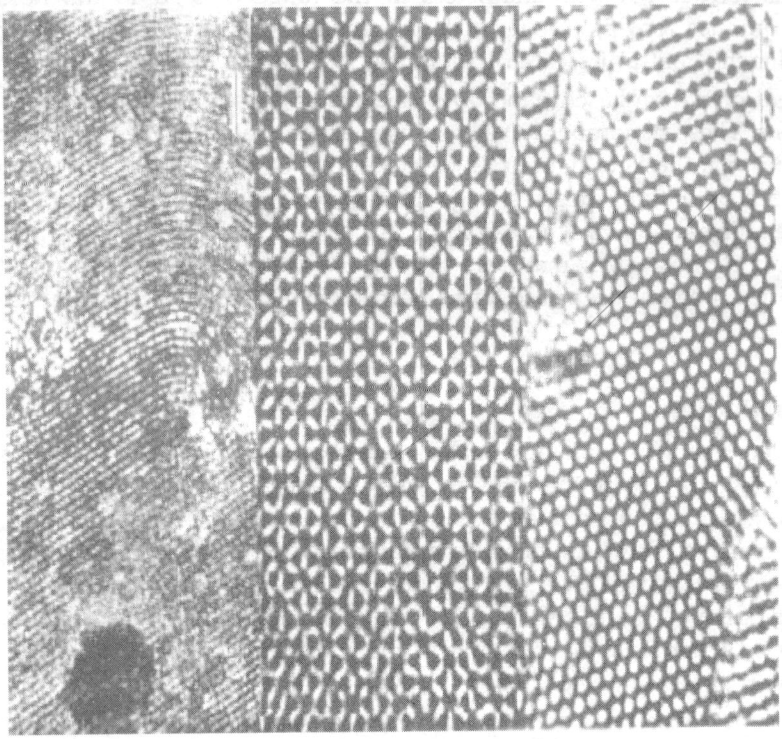

Fig. II–L.3: TEM images of (a) the lamellar morphology, (b) the cubic phase with Ia3d symmetry viewed along its [111] zone axis, and (c) the hexagonal phase viewed along its [001] zone axis of thesilica/surfactant nanostructured composites by co-assembly (bars = 30 nm).

Gold clusters 3.7 nm in diameter, formed in the vapor phase, are encapsulated in organic surfactants, such as dodecanethiol, forming a colloidal suspension. The surfactants prevent the agglomeration of the gold clusters. Addition of small amount of dithiol precipitates out a 3-D cluster network, which can in turn be deposited onto another solid substrate. Figure II–L.4 shows a transmission electron microscope (TEM) image of a cluster array spin-cast onto MoS_2.

The methodology of self-assembly has even been extended to physical vapor deposition processes where it would seem more difficult to control the nucleation and growth of three-dimensional nanostructures. Utilizing the strain inherent in the epitaxial growth of lattice-mismatched materials, and the expected strain-induced transition from two-dimensional (layered) to three-dimensional (islanded) growth, together with careful monitoring of the growth process through RHEED analysis, researchers have been able to form arrays of semiconductor quantum dots, ~ 200-300 Å in diameter, ~ 1011 cm^{-2} in density, and with a size variation of about ±7 per cent. An example of such "self-assembled" semiconductor quantum dots is shown in Figure II–L.5. The achievement of arrays of several billions of quantum dots of these dimensions with such a size variation is beyond the capability of standard high resolution lithographic and pattern transfer processes. Moreover, the controlled formation of critical surfaces and interfaces without the intercession of ion-assisted processing that can introduce potential defects into the materials has produced a rich source of optically and electronically efficient quantum structures. A number of researchers have already incorporated such self-assembled dots into laser structures.

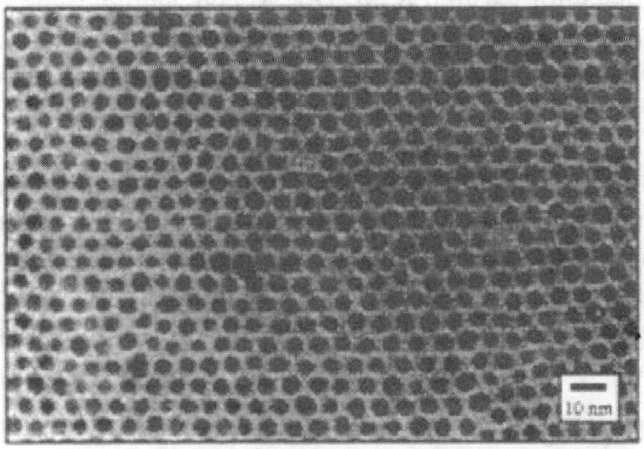

Fig. II–L.4: TEM image of unlinked cluster array of 3.7 nm Au clusters encapsulated by dodecanethiol.

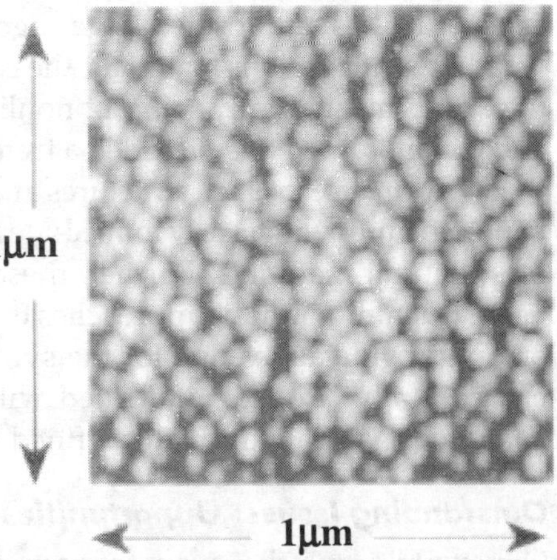

Fig. II–L.5: Array of InAs quantum dot structures grown on GaAs substrates.

Chemical specificity may provide the most robust means of ensuring control of size and placement of nanostructured building blocks, and recent work in the synthesis of compound semiconductor quantum dots from chemical precursors have provided even tighter distributions of size variation (±5 per cent) than those shown in the strain-induced self-assembled dots. The size-dependent and energy-selective optical absorption properties of such chemically synthesized quantum dots are shown in Figure II–L.6.

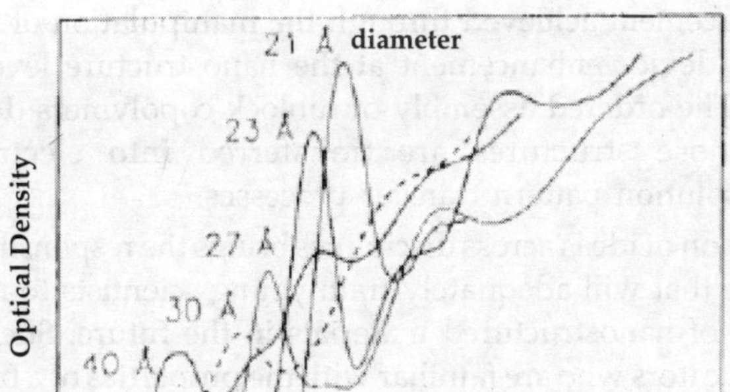

Fig. II–L.6: Variation of optical transparency with diameter of chemically synthesized CdSe nanocrystals.

Natural Templates or "Scaffolds"

Biogenic systems employ natural templates or scaffolding in the construction of nanoscaled materials, where the templates and scaffolding can help set the proper, critical dimensions that are characteristic of the final material. The templates can be formed "artificially," through the use of lithographically defined patterning and processing of the substrate material in order to achieve selective nucleation and growth of the nanostructured materials. However, there are also a host of "natural" templating materials that researchers can avail themselves of. The zeolites have for a long time been utilized as the basis of high surface area materials that enable catalysis. Recent work has taken these natural cage structures and formed large cage structures with varying pore sizes. When zeolite cages are "loaded" with various materials, the controlled proximity enforced by the cage structure can result in various magnetically active or nonlinear optical materials. More recently, the C-based fullerenes and C-nanotubes have provided a tremendously rich basis for nanostructure research. The by now "easy" formation of such structures and ready availability of such materials provide a rich source of nanostructures with nearly immediate applications for electronic devices, storage, and enhancement of structural materials. Further, these materials can serve as the nanostructured precursors of nanostructure synthesis approaches (such as the starting material for cluster or aerosol deposition), or can provide an avenue for the synthesis of templates of varying sizes. Such templates can be electronically "doped" and filled with other materials. Recent work reported the synthesis of "nanorods" of GaN through confined reactions in carbon nanotubes.

Outstanding Issues: Opportunities and Challenges

One could argue that many aspects of the work on nanostructured materials have been long-established efforts, with well-developed techniques that have been brought forth into the manufacturing arena. How do we then explain the current excitement and interest in nanostructured materials as a promising new endeavor? Part of the answer lies in the recognition of common scientific issues and common enabling technologies that link this group of researchers together. Ready availability of ever more sophisticated characterization methods that allow us to visualize and probe materials at the nanoscale has accelerated the pace of activities in this field; at the same time, recognition of the common critical issues of control over nanostructure size and placement motivates sharing of solutions over the boundaries of conventional disciplines. Thus, some of the most exciting findings on nanostructured materials manifest the cross fertilization of techniques and ideas; for example, aerosol particles are being integrated with more traditionally fabricated electronic nanostructures, with placement achieved through the manipulation of STM tips, in order to explore ideas of electronic device enhancement at the nanostructure level. A sample structure is shown in Figure II–L.7. The ordered assembly of diblock copolymers define three-dimensional nanostructures, and those structures are transferred into electronic (semiconductor) substrates through high resolution pattern transfer processes.

Recognizing the need for effective utilization of ideas across disciplines brings the responsibility of establishing the educational infrastructure that will adequately train young scientists to more fully develop the concepts and applications of nanostructured materials in the future. Such an infrastructure will require researchers and educators who are familiar with the properties of a broad range of materials, including polymers, biomolecular materials, metals, ceramics, and semiconductors. It will benefit from an informed perspective on critical applications, and must provide access to a wide range of synthesis and assembly techniques and characterization methods that currently reside separately in the disciplines of physics, chemistry, electrical engineering, biology, etc. Finally, a critical enabler for the future of this field and for its educational infrastructure is further development of computational tools that encompass the full range of atomistic calculations to macroscopic materials properties. This will require a "systems approach" that will span a variety of different computational models, addressing the different critical length scales, starting with solutions of the Schrodinger and Poisson equations, solving interatomic forces, and scaling all the way up to simulations of macroscopic properties and behavior.

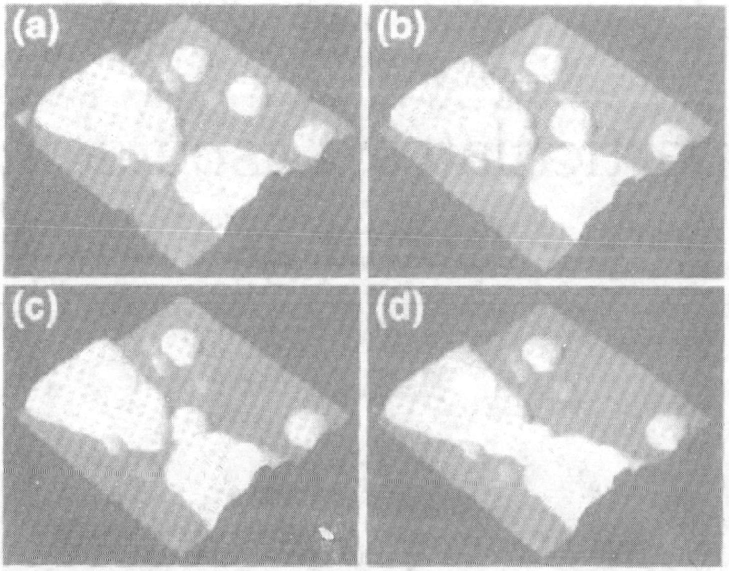

Fig. II–L.7: A sequence of 670 nm by 670 nm AFM images taken during the manipulation of a 50 nm Au particle into the gap between two Au/Ti electrodes.

Increased appreciation of and access to the diverse means of nanoparticle synthesis and assembly that have been developed within many different disciplines, and a common development of enabling tools and technologies, will enhance the pace of accomplishments in this new area of nanoscale synthesis and assembly.

Unit II–M

Nanoparticulates: Dispersions and Coatings

Introduction

The object of the controlled assembly of nanoparticulates is to make materials with new properties and assemble them with practical applications. A unique value of nanoparticulates is their extremely high particle surface area; having many more sites for achieving property enhancements makes them ideal for a wide variety of applications as dispersions and coatings. Dispersive and coating applications of nanoparticles include optical, thermal, and diffusion barriers. Significant work on nanoscale dispersions and coatings is underway worldwide in the areas of ceramics, cosmetics, biosensors, colorants, and abrasion-resistant polymers. Other applications include imaging ink jet materials, electrophotography, pharmaceuticals, flavor enhancers, pesticides, lubricants, and other proprietary applications specific to industry. Still another application is in a new, post-silicon generation of electronic devices that includes nanotubes and fullerenes as constituent units of carbon nanoelectronic devices; here, dispersion takes on a more quantum consideration in which the number of atoms in a cluster is compared to the number of surface atoms to determine its dispersion function. Also in the semiconductor industry, a monolayer or thin film coating of atoms or molecules is deposited on foils, metal sheets, or glass to enhance storage capacity and accelerate responses from the electronic component.

All these applications deal with dispersions or coatings of particles that enhance specific features. Also, ability to manufacture smaller functional systems enhances performance, cost, and efficiency. These considerations power the drive towards miniaturization and precision finishing.

Property Amplification

What's particularly exciting in the study of nanoparticles, nanostructured materials, and nanodevices is their ability to also add value to materials and products through enhancement of specific properties, such as the following:

- **Mechanical Strength**—Nanostructured powders have been produced by plasma processing where the reactor vaporizes coarse metal particles by combustion synthesis where redox reactions take place at elevated temperatures, followed by quenching; and by mechanical alloying with gas atomization. When nanocrystalline powders achieved by these means are compacted and applied as a coating, they lend significant strength and ductility to a variety of conventional materials such as ceramic, composites, and metal alloys.

- **Superconductivity**—Brus describes the nature of the superconductivity effect in detail. It involves a deposition of nanocrystals on substrates, leading to improved optical and electrical properties.
- **Covering power**—Because nanostructuring increases the number of active sites—there are many more atoms per grain boundary—the enhanced surface area leads to a reduced material requirement, which in turn can lower cost.
- **Ability to incorporate high cost materials**—Expensive materials such as colorants and drugs may be effectively dispersed in small and controlled quantities through nanostructuring.
- **Environmental value**—Improvements in environmental impact are achieved by utilizing nanostructure particulates in coatings and thus, eliminating the requirement for toxic solvents. By eliminating hazardous wastes, nanocoatings can both reduce a company's disposal costs and improve its environmental position.

Thus, nanostructured dispersions and coatings can significantly reduce material costs and improve performance and functionality in a large variety of applications.

Enablers

For nanoparticulate dispersions and coatings, certain enablers must be present to achieve success: (1) effective particle preparation, (2) stabilization of the dispersed phase, (3) scaleup and control of the process, and (4) the existence of excellent analytical capabilities. The main methods or issues for each are described below.

1. **Particle preparation**—Wet chemical methods such as liquid phase precipitation or sol-gel methods are of high interest in particle preparation. In the area of hybrid methods, both spray pyrolysis and flame hydrolysis are utilized. Numerous physical methods such as mechanical size reduction are also often employed.
2. **Stabilization of the dispersed phase**—For stabilization of the dispersed phase, it is necessary to understand how particles can be kept as distinct entities. Either a charged stabilized system or a sterically stabilized approach is required. A successful preparation for use as a dispersion should be free from agglomeration in the liquid state so as to maintain particle integrity. The dry-coated format of nanoparticles should also minimize any presence of particle aggregation.
3. **Scaleup and control of the process**—The scaleup and control of nanoparticle processing is well described by Kear. Here, the issue in achieving a high rate of production of powder is the effective pyrolysis of the gas stream containing the precursor. Issues such as scaling and reproducibility from one run of nanoparticle material or dispersions to another are all inherent in the concept of precision engineering or invariant process control.
4. **Analytical capabilities**—Analytical capabilities are absolutely essential for characterizing dispersions and coatings. Particle size determinations, assay analysis, and interfacial

properties are all important. Transmission electron microscopes, atomic force microscopy, nuclear magnetic resonance, and scanning tunneling microscopy are just some of the tools utilized in characterizing nanoparticle dispersions and coatings, particularly at the very small end of the nanoscale.

Applications

Below are some general examples of the vast array of applications of nanoparticulate dispersions and coatings. There are a number of applications in all of the categories below that are already in the public domain; however, most are still highly proprietary, especially drug delivery applications.

Cosmetics—An area of nanoparticle technology that has incredible commercial potential is the cosmetic industry. Here there is a great demonstrated demand, and the technology can be made simple, since properties of color and light fastness are achieved by component mixing in the cosmetic preparation. A survey conducted in the last decade of 20th century indicated a worldwide gross volume of $14-18 billion for toiletries, i.e., traditional hygiene products such as powders, sprays, perfumes, and deodorants. The large markets for sunscreens and skin rejuvenation preparations promise additional revenues.

The diet industry is said to gross $33 billion annually. One way that nanoparticle technology is addressing this market is through introducing nanoparticulate taste enhancers into low-calorie substrates.

Medicine/Pharmacology—In the area of medical applications, finely dispersed pharmaceuticals offer rapid drug delivery and reduced dosages for patients. Dispersions of strong and resilient biocompatible materials suggest opportunities for artificial joints. These generally are ceramic materials containing nanoparticulates.

Overall, much of the demand for nanoparticulate dispersions and coatings comes from the cosmetic and pharmaceutical industries; in particular, liquid dispersion preparations will be widely used to apply topical coatings to the human epidermis because they can be absorbed faster and more completely than conventional coatings.

Microelectromechanical Systems (MEMS)—Although MEMS technologies will support the semiconductor industry in particular, there are many other applications being explored, such as in medicine, ceramics, thin films, metal alloys, and other proprietary applications. In the United States a particular focus is applying sputtering coatings to achieve MEMS technology in concert with these applications.

Printing—In the areas of image capture/image output addressed by ink jet technology, nanoscience can help control the properties of the inks themselves. The production and use of nanoengineered ink products benefits from such complimentary technology as laser-assist delivery of the ink jet droplet to maintain an accurate deposit of the ink on its target. Another application in this field is using nanoscale properties to tailor the inks to achieve ideal absorption and drying times for desired color properties and permanency.

Semiconductors—One form of "bottom up" technology that is receiving considerable attention is thin films for the semiconductor industry. Here single atoms or molecules are deposited by physical vapor deposition, which could be achieved through sputtering, molecular beam epitaxy, or chemical vapor deposition. Sputtering is used on a large scale to coat metal sheets, glass, polymer substrates and other receptive materials in order to produce enhanced electronic properties for information storage and processing speed.

Sensors—Chemical or physical sensors often use nanoparticles because they provide high surface area for detecting the state of chemical reactions, because the quality of detection signals is improved, and because earlier and more accurate determination of leakage reduces waste. Some commercial sensors and actuators composed of thin films are already used for environmental vapor monitoring in reactors, for example.

Other likely applications of nanotechnology involving dispersions and coatings include nanofabricated surface coatings for keeping windows and surfaces clean. Here, the transparent nanocoating on a surface prevents fog and dirt particles from depositing on the substrate. Commercial products (achieved through gas phase condensation) also include aluminum oxide/epoxy dispersions yielding 19 times more wear resistance than conventional products.

Ongoing R&D/Preparation Issues

A wide variety of work on nanostructured dispersions and coatings in various laboratories around the world has been observed .Descriptions follow of a number of research projects and the preparation issues being addressed in the United States and in some of the other laboratories around the world. Table II–M.3 outlines some of the nanoparticle preparation techniques that are currently in use, many of which will be discussed in the paragraphs that follow.

Work in the United States

Significant work on coatings of dispersions is underway in the United States. Some of the highlights of that volume are summarized below. Similar work is ongoing in other countries.

I.A. Aksay describes a ceramic thin film structure that would mirrorimage a self-assembly process in many materials. Here silica precursors when mixed with surfactants yield polymerized templates having structures similar to surfactant-water liquid crystals; what results are highly controlled pores on the 10 to 100 Å scale. Controlling the pore structure and synthesizing the building blocks are two technical challenges facing future work in this area. However, Aksay has shown that he can grow silicate films onto a wide variety of substrates. The layer structure of the bound surfactant molecules is key. Atomic force microscopy can reveal some details on the morphology of these films. The nanostructural patterns obtained in processing ceramics that contain organic/inorganic composites allow self-assembly to take place at lower temperatures.

Table II–M.1: Nanoparticle Preparation for Dispersions/Coatings

Liquid Chemical Methods	*Physical Methods*	*Elevated Temperature Methods*
Sol-Gel	Size reduction by mechanical means (low energy)	Aerosol
Chemical Precipitation Ag-X (Eastman Kodak)		Flame
Colloidal Micelles		Vapor phase condensation

Another example of thin films or coating is the work of Gell, focused on improving both the physical and the mechanical properties of materials. Nanostructured coating can lead to high diffusivity, improved toughness and strength, and better thermal expansion coefficients, with lower density, elastic modulus, and thermal conductivity. In comparisons of nanostructures and conventional materials, use of nanostructured WC/cobalt composites have resulted in as much as a two-fold increase in abrasion resistance and hardness. Deposition is often accomplished by utilizing a sputtering chamber where the nanomaterial is coated on substrates. This can lead to such benefits as resistance to oxidation and cracks in addition to resistance to wear and erosion.

An area that offers exciting possibilities in the area of dispersions is the sol-gel process described by P. Wiltzius. Here, a concentrated dispersion of colloids is chemically converted into a gel body. Drying followed by sintering produces a ceramic or glass product. This process can create nanoparticles, fibers, film, plates, or tubes. All processing is at low temperatures. Lucent Technologies has developed a silica casting process that is reproducible for making tubes of pure silica of one meter in length for use in manufacturing optical fibers. Technical challenges include obtaining pure starting materials, removing refractory particles that lead to breakage in the fiber drawing process, and achieving very tight dimensional tolerances. Colloidal dispersions of this type play a critical role in chemical mechanical polishing. To obtain good yield and high quality, it is necessary to achieve very tight process control.

R. Brotzman at Nanophase Technologies Corporation describes the gas phase condensation process for synthesizing inorganic and metallic powders. Such a process was invented by R. Siegel and his colleagues at Rensselaer Polytechnic Institute in New York. The process involves production of physical vapor from elemental or reacted material followed by sudden condensation and reaction of the vapor into small nanometer particles. The condensation process is rapid and involves dilution to prevent the formation of hard agglomerates and coalescence.

Nanophase Technologies has developed a production system where forced convection flow controls the particle/gas stream and enhances particle transport from the particle growth region so as to generate more metal vapor. The convection flow helps in forming oxides and nitrides from metal crystallites. This process produces commercial quantities of nanosized inorganic powders that have a spherical shape with narrow size distributions. Table II–M.2 shows some properties of these particles, and Figure II–M.1 shows their transparency as a function of particle size.

Table II–M.2: Particle Properties

Particle Properties	TiO_2	ZnO	Fe_2O_3
Refractive Index	2.40	2.01	3.01
Density (gm/cc)	3.95	5.61	5.30
Molecular Weight (μ)	80	81	160
Particle Sizes (see Figure 13.1 below)			

Optical Density (D) = Absorbance (A) = log 10 (lo/l = eCl) C = concentration
l = optical path length e= molar absorption coefficient/2.303.

Nanophase Technologies Corporation has developed a coating process that encapsulates nanoparticles with a surface treatment that ensures individual integrity of the particles in subsequent coating steps. Applications span their use in low dielectric media all the way to water and, if needed, steric stabilizers. Nanophase also has directed efforts in the electronics and industrial catalyst areas. Friedlander describes in good detail aerosol reaction engineering where attention is paid to design of the process and the importance of material properties and process conditions. Key process parameters include time, temperature, and volume concentration. Most commercially produced particles have polydispersity. Important full-scale processes are flame reactors for preparing pigments and powdered materials for optical fibers. Pyrolysis reactors have long been used to prepare carbon blacks.

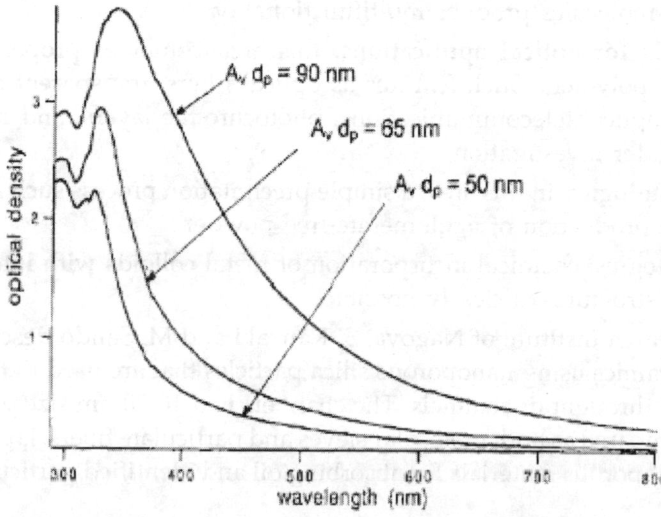

Fig. II–M.1: Transparency as a function of particle size (Nanophase).

Kear and Skandan have discussed two divergent approaches toward fabrication of bulk materials. The first is a powder processing route involving preparations by physical or chemical means followed by condensation and sintering. Materials for cutting tools such as tungsten carbide/cobalt powders were prepared in a controllable way to about 50 nm. Liquid phase sintering completes the formation of the solid dispersion phase. However, a gap in fabrication technology remains that of controlling grain growth during liquid phase sintering.

A second approach to fabrication of bulk materials is spray forming. This procedure avoids contamination and coarsening of the dispersion and its particles when the process has a controlled atmosphere of inert gas at low pressure. In this approach there is need to establish process/product codesign where we understand cause-effect relationships between processing parameters and properties of nanophase spray/deposited materials.

In summary, particle preparation in the United States utilizes a wide variety of methods—both chemical and physical, at both high and low temperatures—all having unique considerations for scale-up and process control. Researchers in other countries are tackling many of these same issues.

Work in Europe and Japan

Cited below is the progress at the Institute for New Materials at Saarbrücken in Germany and Japan's Industrial Research Institute of Nagoya, both of which are exploring direct applications of nanostructured dispersions and coatings.

Drs. Rudiger Nass and Rolf Clasen at the Institute for New Materials at Saarbrücken make specific use of the sol-gel process in which liquid starting materials are utilized at low temperatures for nanoscale metal, ceramic, glass, and semiconductor nanoparticles. The advantages, besides temperature, include the isolation of high purity powders. The institute has focused on four basic areas for spin-off and adaptation towards commercialization:

1. **New functional surfaces with nanomers:** properties such as corrosion protection, wettability, coloration, micropatterned surfaces, porosity, and the ability for selective absorption of molecules produce multifunctionality.
2. **New materials for optical applications:** this area combines properties of lasers and ceramics with polymers. Such features as optical filters, transparent conducting layers, materials for optical telecommunications, photochromic layers, and holographic image storage are under investigation.
3. **Ceramic technologies:** in this area, a simple precipitation process such as sol-gel provides for pilot-scale production of agglomerate-free powder.
4. **Glass technologies:** chemical incorporation of metal colloids with intelligent properties into glasslike structures is clearly possible.

At the Industrial Research Institute of Nagoya, S. Kanzaki and M. Sando described methods for preparing synergy ceramics using nanoporous silica particles that are used to fabricate thin films with one-dimensional throughput channels. The channels had 10-20 nm sizes. High temperature Fe_2SiO_4 oxides were prepared as both molecular sieves and particulate fibers. Japan plans to pursue the preparation of nanoporous materials for absorbing oil and identified particulates.

Alternative Preparation Methods

The quest for low temperature nanoparticle preparation methods has spanned a wide range of systems. One that has been in existence for decades but has not been put into use in other industries is the method of preparing silver halide particles. Eastman Kodak in France, England, and the United States has utilized solution precipitation technology with well-controlled mixing and nucleation control to produce a wide range of grain sizes. "Lippmann"-type grains have a size of about 50 nm. Some of the properties of these fine-grain systems are discussed by Trivelli and Smith.

There are other methods of creating nanoparticles of organic materials such as filter dye applications in photographic films and spectral sensitizing dyes for use in silver halide grains. Ultrafine grinding media are used to almost sandpaper organic crystals to nanoparticle ranges of 20-80 nm. Similar technology by Bishop has been utilized in both pharmaceutical preparations and ink jet applications with good success.

One other exciting area is in polymer science, where dendrimer molecules, often 10 nanometers in diameter, are prepared synthetically. These molecules have been studied in gene therapy, as aids in helping to detect chemical or biological agents in the air, and as a means to deliver therapeutic genes in cancer cells. The preparation of dendritic starburst molecules is described by Salamone. One can imagine applications for coatings in which these molecules with their highly reactive surfaces participate in either classical (sub-nanoscopic chemistry) or novel nanoscopic conversions.

Opportunities and Challenges

There are some outstanding opportunities and challenges that face the nanoscience community. For dispersions and coatings these include four areas:

1. The foremost area of opportunity is controlling the particle preparation process so that

size is reproducible and scaleable. This requires creation of narrow size range particles that can be prepared by processes mentioned in the study such as vapor phase condensation, physical size reduction, or flame and pyrolysis aerosol generators. These same processes must respond to good reproducibility and scaling. In most studies to date, the size of primary particles depends on material properties and the temperature or time history. Two processes, collision and coalescence, occur together, and the processes need to be controlled in order to favorably influence final particle size distribution.

2. The second area of opportunity is process control. In the concept of a process control methodology, the nature of chemical processes makes it imperative to have means of effectively monitoring and initiating change in the process variables of interest. Accordingly, those involved with production of nanoparticles would monitor outputs, make decisions about how to manipulate outputs in order to obtain desired behaviors, and then implement these decisions on the process. Control system configuration will necessarily have a feeding process from the output such that information can be fed back to the controller. It may also have an opportunity to base controller decision-making on information that is being fed forward; decisions could be made before the process is affected by incoming disturbances. Such would apply to process parameters like temperature, flow rate, and pressure.

   In many chemical setups, product quality variables must be considered, and such measurements often take place in the laboratory. An objective analysis is needed of any observed deviation of a process variable from its aim, and an objective decision must be made as to what must be done to minimize any deviation. Here process understanding leads to well behaved reactors in manufacturing that allow for true process verification with data feedback and analysis. This concept of process control requires the application of statistical methodologies for product and process improvements. Current interest in statistical process control (SPC) is due to several factors, in particular, interest in realizing consistent high quality so as to sustain the business and obtain greater market share. Most manufacturing zones are moving toward inline/online sensor technologies coupled with process software and user-friendly computer hardware for factory operators.

   The benefits of process control are many. They include achieving reduced variability and higher quality, safety enhancement, reduction of process upsets, and in many cases, environmental improvements due to achieving mass balance in processes with material in/product out. Poor process design can be inherently overcome through SPC. Reduction in sampling and inspection costs results.

3. A third area of opportunity is the process/product relationship that leads to continuous uniformity—that is, the specification setting by product users that must be available to relate to process control in manufacturing. Here, nanoparticle formulation and the process used to prepare the particles must be linked and interactive.
4. A fourth technical opportunity is to develop process models for various dispersions and coatings that lead to shorter cycle time in manufacturing. Table II–M.3 shows a comparison of enablers and opportunities in Europe, Japan, and the United States.

Table II–M.3: Dispersions and Coatings—Nanotechnology Comparisons Between the United States, Europe, and Japan

LEVEL	*I (Highest)*	2	3	4
Enablers				
Particle Preparation	U.S./Europe	Japan	—	—
Stabilization Scaleup	U.S./Europe	Japan	—	—
Characterization	U.S./Europe	Japan	—	—
Coating	U.S./Japan/Europe	—	—	
Opportunities				
Process Control	—	—	U.S.	Japan/Europe
Process Product	—	—	U.S.	
Co-Design Modeling	—	U.S./Japan		

These four areas of opportunity, then, represent what lies ahead for advancement in nanotechnology with respect to dispersions and coatings. Achieving implementation in the industrial market will require close attention to resolving the challenges in the four areas summarized above. In addition, there remains a large gap between the cost of preparing conventional materials and the cost of preparing nanoparticles. This will remain a challenge for the future if nanomaterials are to be competitive.

Unit II–N

High Surface Area Nanomaterials

The trend to smaller and smaller structures, that is, miniaturization, is well known in the manufacturing and microelectronics industries, as evidenced by the rapid increase in computing power through reduction on chips of the area and volume needed per transistor. In the materials area this same trend towards miniaturization also is occurring, but for different reasons. Smallness in itself is not the goal. Instead, it is the realization, or now possibly even the expectation, that new properties intrinsic to novel structures will enable breakthroughs in a multitude of technologically important areas.

Of particular interest to materials scientists is the fact that nanostructures have higher surface areas than do conventional materials. The impact of nanostructure on the properties of high surface area materials is an area of increasing importance to understanding, creating, and improving materials for diverse applications. High surface areas can be attained either by fabricating small particles or clusters where the surface-to-volume ratio of each particle is high, or by creating materials where the void surface area (pores) is high compared to the amount of bulk support material. Materials such as highly dispersed supported metal catalysts and gas phase clusters fall into the former category, and microporous (nanometer-pored) materials such as zeolites, high surface area inorganic oxides, porous carbons, and amorphous silicas fall into the latter category.

There are many areas of current academic and industrial activity where the use of the nanostructure approach to high surface area materials may have significant impact:

- Microporous materials for energy storage and separations technologies, including nanostructured materials for highly selective adsorption/separation processes such as H_2O, H_2S, or CO_2 removal; high capacity, low volume gas storage of H_2 and CH_4 for fuel cell applications and high selectivity; high permeance gas separations such as O_2 enrichment; and H_2 separation and recovery.
- Thermal barrier materials for use in high temperature engines.
- Understanding certain atmospheric reactions.
- Incorporation into construction industry materials for improved strength or for fault

diagnostics.

- Battery or capacitor elements for new or improved operation.
- Biochemical and pharmaceutical separations.
- Product-specific catalysts for almost every petrochemical process.

In catalysis the key goal is to promote reactions that have high selectivity with high yield. It is anticipated that this goal will be more closely approached through tailoring a catalyst particle via nanoparticle synthesis and assembly so that it performs only specific chemical conversions, performs these at high yield, and does so with greater energy efficiency. In the electronics area one may anticipate manufacture of single electron devices on a grand scale. Manufacture of materials with greatly improved properties in one or more areas such as strength, toughness, or ductility may become commonplace. In separations science new materials with well defined pore sizes and high surface areas are already being fabricated and tested in the laboratory for potential use in energy storage and separations technologies. In addition, many laboratories around the world are actively pursuing the potential to create novel thermal barrier materials, highly selective sensors, and novel construction materials whose bonding and strength depend upon the surface area and morphology of the nanoscale constituents. Many are also engaged in developing molecular replication technologies for rapid scaleup and manufacturing.

The nanoscale revolution in high surface area materials comes about for several reasons. First, since the late 1970s the scientific community has experienced enormous progress in the synthesis, characterization, and basic theoretical and experimental understanding of materials with nanoscale dimensions, i.e. small particles and clusters and their very high surface-tovolume ratios. Second, the properties of such materials have opened a third dimension to the periodic table, that is, the number of atoms, N. N now becomes a critical parameter by which the properties for "small" systems are defined. As a simple example, for metals we have known for decades that the atomic ionization potential (IP) is typically about twice the value of the bulk work function. It is only relatively recently that experiments have shown that the IP (and electron affinity) for clusters containing a specific number N of (metal) atoms varies dramatically and non-monotonically with N for clusters containing less than 100-200 atoms. Other properties such as chemical reactivity, magnetic moment, polarizability, and geometric structure, where they have been investigated, are also found to exhibit a strong dependence on N. Expectations for new materials with properties different from the atom or the bulk material have been realized. The opportunity is now open to precisely tailor new materials through atom-by-atom control of the composition (controlling the types as well as the numbers of atoms) in order to generate the clusters or particles of precision design for use in their own right or as building blocks of larger-scale materials or devices—that is, nanotechnology and fabrication at its ultimate.

Such precision engineering or tailoring of materials is the goal of much of the effort driving nanoscale technology. Scientists and engineers typically have approached the synthesis and fabrication of high surface area nanostructures from one of two directions:

1. The "bottom up" approach in which the nanostructures are built up from individual atoms or molecules. This is the basis of most "cluster science" as well as crystal materials synthesis, usually via chemical means. Both high surface area particles and micro- and mesoporous crystalline materials with high void volume (pore volume) are included in this "bottom up" approach.
2. The "top down" approach in which nanostructures are generated from breaking up bulk

materials. This is the basis for techniques such as mechanical milling, lithography, precision engineering, and similar techniques that are commonly used to fabricate nanoscale material, which in turn can be used directly or as building blocks for macroscopic structures.

A fundamental driving force towards efforts to exploit the nanoscale or nanostructure is based upon two concepts or realizations: (1) The macroscopic bulk behavior with which we are most familiar is significantly different from quantum behavior, and (2) Materials with some aspect of quantum behavior can now be synthesized and studied in the laboratory. Obviously, quantum behavior becomes increasingly important as the controlling parameter gets smaller and smaller. There are numerous examples of quantum behavior showing up in high surface materials: the fact introduced above that clusters are found to exhibit novel (compared to the bulk) electronic, magnetic, chemical, and structural properties; the fact that the diffusivity of molecules through molecular sieving materials cannot be predicted or modeled by hard sphere molecule properties or fixed wall apertures; and the fact that catalysts with one, two, or three spatial dimensions in the nanometer size range exhibit unique (compared to the bulk) catalytic or chemical activity.

Opportunities for Clusters and Nanocrystalline Materials

Clusters are groups of atoms or molecules that display properties different from both the smaller atoms or molecules and the larger bulk materials. Many techniques have been developed to produce clusters, beams of clusters, and clusters in a bottle for use in many different applications including thin film manufacture for advanced electronic or optical devices, production of nanoporous structures as thermal barrier coatings, and fabrication of thin membranes of nanoporous materials for filtration and separation. Figure II–N.1 depicts an apparatus developed at the University of Göteborg to measure cluster reactivity and sticking probability as a function of the number of metal atoms in the cluster.

The unique properties of nanoparticles make them of interest. For example, nanocrystalline materials composed of crystallites in the 1-10 nm size range possess very high surface to volume ratios because of the fine grain size. These materials are characterized by a very high number of low coordination number atoms at edge and corner sites which can provide a large number of catalytically active sites. Such materials exhibit chemical and physical properties characteristic of neither the isolated atoms nor of the bulk material. One of the key issues in applying such materials to industrial problems involves discovery of techniques to stabilize these small crystallites in the shape and size desired. This is an area of active fundamental research, and if successful on industrially interesting scales, is expected to lead to materials with novel properties, specific to the size or number of atoms in the crystallite.

A typical objective of nanoscale catalyst research is to produce a material with exceedingly high selectivity at high yield in the reaction product or product state, i.e. chemicals by design, with the option of altering the product or product state simply by changing the surface functionality, elemental composition, or number of atoms in the catalyst particle. For instance, new catalysts with increasing specificity are now being fabricated in which the stoichiometry may be altered due to size restrictions or in which only one or two spatial dimensions are of nanometer size. Five recent examples where nanocrystalline metallic and ceramic materials have been successfully investigated for catalytic applications are discussed below.

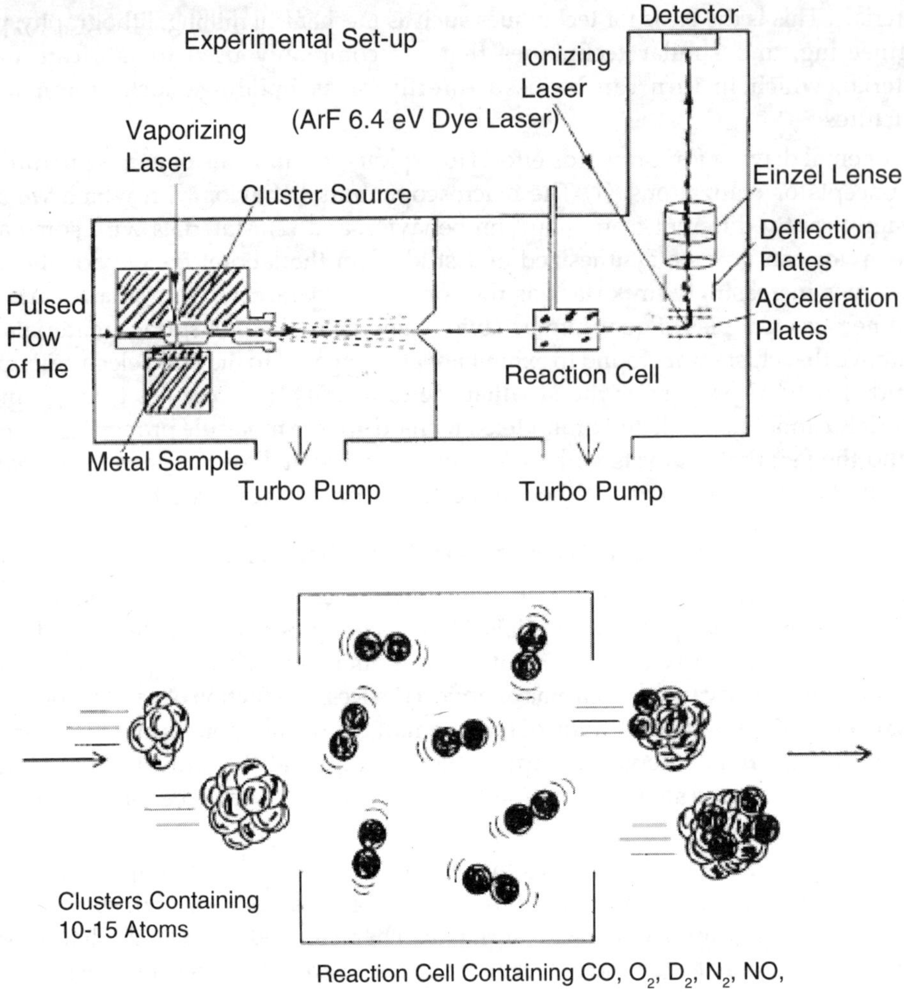

Fig. II–N.1: Schematic drawing of the experimental setup used in Göteborg for studies of chemical reactivity and/or sticking probability of various molecules with the clusters. The production of clusters is via laser vaporization of metal substrates and detection via photoionization time-of-flight mass spectrometry.

Novel Catalytic Properties of Nanostructured Gold Catalysts

A group at Osaka National Research Institute has discovered that nanoscale gold particles display novel catalytic properties. Highly selective low temperature catalytic activity is observed to switch on for gold particles smaller than about 3-5 nanometers in diameter. Accompanying this turn on in catalytic activity is the discovery that these nanoscale gold particles (crystals) also have icosahedral structure and not the bulk fcc structure, again a nanoscale phenomena not available with bulk samples.

In fabricating these novel catalytic materials several issues appear to be crucial. For instance, the Osaka group has shown that (a) the preparation method plays a crucial role for generating aterials with high catalytic activity and selectivity; that (b) the catalytic activity, selectivity and temperature of operation is critically dependent on the choice of catalyst support, and that (c) water (moisture) even in parts per million (ppm) levels dramatically alters the catalytic properties. Figure II–N.2 shows the effect of moisture on the conversion profiles for CO oxidation for nanoscale gold catalysts supported on cobalt oxide. Examples of novel catalytic behavior of nanoscale gold particles include the following:

- CO oxidation at temperatures as low as –70°C, as illustrated in Figure II–N.2 very high selectivity in partial oxidation reactions, such as 100 per cent selectivity at 50°C for oxidation of propylene to propylene oxide as well as near room temperature reduction of nitric oxide with H_2 using alumina-supported gold nanoparticles

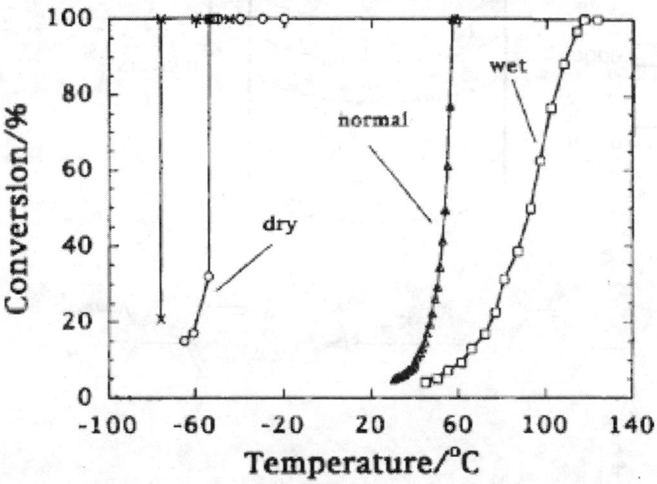

Fig. II–N.2: Effect of moisture on conversion profiles for CO oxidation over Co_3O_4 and Au/Co_3O_4.

The fundamental work on gold catalysts has led to "odor eaters" for the bathroom, based on nanoscale gold catalysts supported on a-Fe_2O_3, a commercialization from Osaka National Research Institute in Japan.

Demonstration of the Importance of Controlling at Least One Dimension in the Nanoscale Size Range for the Industrially Important HDS Reaction

For catalysts based on the layered compound MoS_2, maximum hydrodesulfurization (HDS) activity is obtained only on well-crystallized nanosized materials, while HDS selectivity is determined by the number of layers or "stack height" of the nanocrystalline MoS_2. In the hydrodesulfurization reaction, illustrated in Figure II–N.3, cyclohexylbenzene occurs only on the MoS_2 "rim sites," or those around the edges of the stack, whereas the pathway to biphenyl requires both "rim" and "edge" sites. Thus, the reaction selectivity can be controlled by controlling the aspect ratio of the nanoparticles of MoS_2. Such control of one- and two-dimensional nanostructures for selective chemical advantage is an exciting new area of research. Of course, a major industrial challenge will be to fabricate such nanocrystals in a commercializable form.

Generation of Nanocrystalline, Non-Stoichiometric CeO_2-X by Controlled Post-Oxidation of Cerium Nanoclusters Produced via Inert Gas Condensation Synthesis

The cerium oxide (CeO_2-x) materials have been found to possess a significant concentration of

Ce3+ and oxygen vacancies, even after high temperature (500°C) calcination. Such nanoclusters give rise to a substantial reduction in the temperature of selective SO_2 reduction by CO and exhibit excellent poisoning resistance against H_2O and CO_2 in the feed stream compared to that for conventional high surface area cerium oxide.

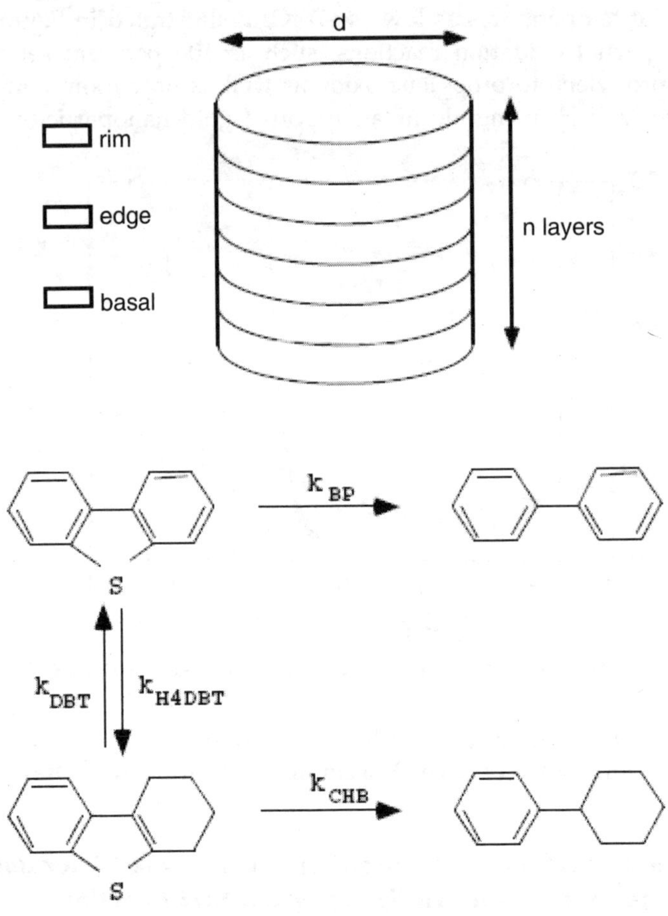

Fig. II–N.3: Hydrodesulfurization reaction. Selective catalysis is controlled by either the edge or rim of MoS_2.

Electrochemical Reduction of Metal Salts

Electrochemical reduction of metal salts is yet another option available to control the size of nanoscale catalyst particles. This has been successfully used to prepare highly dispersed metal colloids and fix the metal clusters to the substrate. Control of the current density during the electrochemical synthesis process allows one to control the size of nanoscale transition metal particles. A combination of scanning tunneling microscopy (STM) and high-resolution transmission electron microscopy (TEM) has allowed surfactant molecules to be visualized on nanostructured palladium clusters.

Energy Storage Enhanced by Catalytic Dissociation of H_2

Materials with higher hydrogen storage per unit volume and weight are considered by many to be an enabling technology for vehicular fuel cell applications. Scientists at Los Alamos National Laboratories have developed an approach that enables materials such as Mg to be used for hydrogen storage. Magnesium is of interest because it can store about 7.7 wt per cent hydrogen, but its adsorption/desorption kinetics are slow, i.e., the rate of charge (hydrogen dissociation and hydride formation) is much slower than in metal hydrides. At Los Alamos, high surface area mixtures of nanoscale Mg and Mg_2Ni particles are produced by mechanical means, ball milling. The addition of Mg_2Ni catalyzes the H_2 dissociation such that the rate of hydrogen adsorption increases to a rate comparable to that of $LaNi_5$. Once a hydride is formed, the hydrogen "spillover" leads to magnesium hydride formation.

Figure II–N.4 illustrates the hydrogen adsorption/desorption characteristics of the mixture of Mg with 23 atomic per cent Ni. As can be seen, a low pressure plateau at about 1500 torr is obtained for this particular sample. Experiments show that the pressure plateau can be tailored through such alloying. Studies with other nanoscale materials, including other catalysts such as FeTi and $LaNi_5$, are presently ongoing to further improve both the capacity and the rate of hydrogen storage.

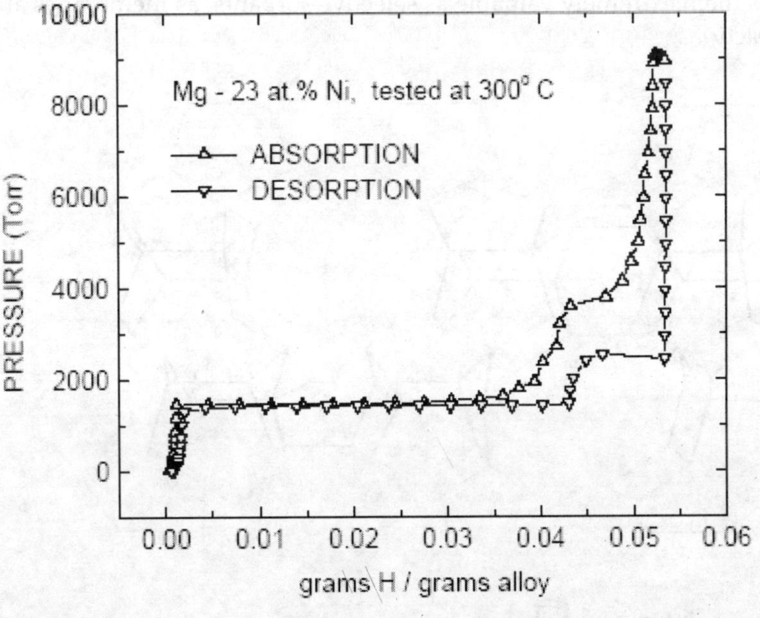

Fig. II–N.4: Hydrogen absorption-desorption characteristics for mixture of Mg and Mg_2Ni prepared by mechanical alloying.

Opportunities in Self-Assembly

In self-assembly large molecular structures are obtained from the organization of a large number of molecules or atoms into a given shape, typically through specific interactions of the molecules among themselves and with a template. The interaction of the different bonding mechanisms is an area of strong fundamental research interest. Only two areas will be highlighted here: zeolites and carbon materials. Both of these materials exhibit characteristics of self-assembly, namely novel and reproducible structures that can be fabricated in industrially significant quantities.

Zeolitic Materials

Aluminosilicates (e.g., zeolites) are crystalline porous nanostructures with long range crystalline order with pore sizes which can be varied from about 4 Å to 15 Å in conventional zeolites. Figure II–N.5 shows a 3-D (e.g., MFI) zeolite cage structure together with a depiction of the straight and ziz-zag channels and a 2-dimensional zeolite with channels only in 2 directions. The vertices in the stick drawing denote position of the O atoms in the crystalline lattice. This particular zeolite has 10 atoms in the zeolite "window." The size of the window is determined by the number of oxygens in the ring. Table 6 gives approximate window dimensions for zeolites as a function of the number of oxygens in the ring.

As can be seen by examination of Table II–N.1, molecules can pass through or be blocked from transport through or into the zeolite depending on the zeolite. For example normal hexane with a kinetic molecular diameter of about 5.1 Å can pass through a 10 ring or larger, whereas cyclohexane with a kinetic molecular diameter of 6.9 Å would be hard pressed to pass through a 10 ring. Thus all other things being equal, a 10-ring zeolite could be used to separate mixtures of normal hexane and cyclohexane. It is this property together with the ability to chemically modify the acidity of zeolitic materials that makes them extremely valuable as selective sorbants, as membranes and for use in selective catalytic reactions.

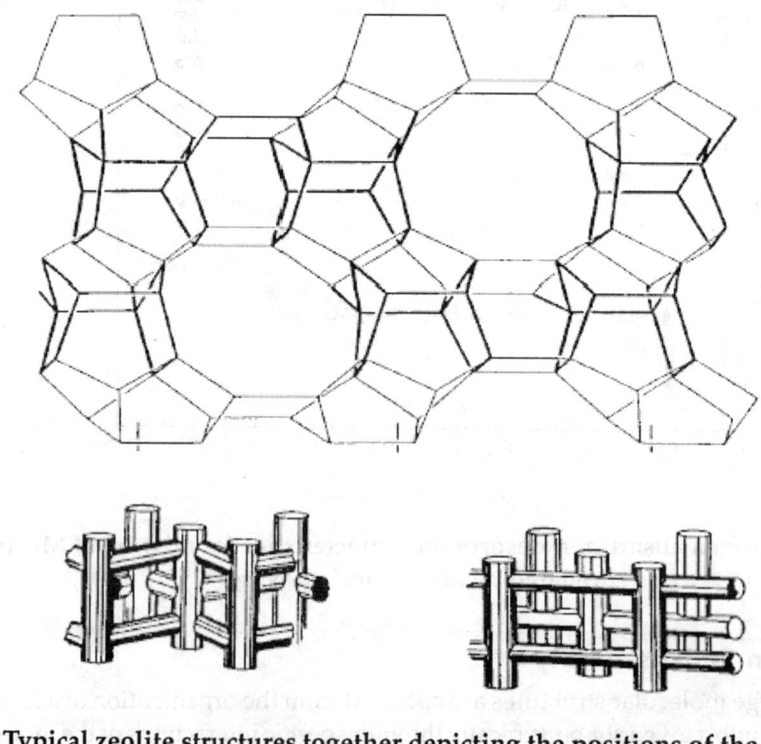

Fig. II–N.5: Typical zeolite structures together depicting the positions of the O atoms (vertices in upper figure) and two different zeolitic structures one (lower left) with a three dimensional structure and (lower right) a zeolite with a two dimensional channel structure.

Table II–N.1: Zeolite Channel "Window" Dimension for Number of Oxygens in Ring

Number of Oxygens in Ring	*Ring Diameter (A)*
4	1.2
5	2.0
6	2.8
8	4.5
10	6.3
12	8.0

In 1992, a new family of aluminosilicates (M41S) with pores sizes between 20 and 100Å in diameter were reported by Mobil researchers. One of particular interest is MCM-41, which consists of hexagonal arrays of uniform 2 to 10 nanometer-sized cylindrical pores. Not only can such materials be synthesized, but novel structures such as "tubules-within-a-tubule" have been fabricated as mesoporous molecular sieves in MCM-41. Of particular interest is the possibility of expanding the so-called "liquid crystal templating" mechanism to non-aluminum dopants within the silicate MCM-41 framework and to derive nonsiliceous MCM-41 type of materials.

Another approach to synthesizing large pore and large single crystals of zeolytic materials is being pioneered by Geoffrey Ozin and his group at the University of Toronto, who have demonstrated that crystals as large as 5 mm can be synthesized. The ability to synthesize such large crystals has important implications for discovery of new sensors (selective chemical adsorbants) and membrane devices (selective transport of molecular species), since large single crystals can now be available to the laboratory researcher to carry out fundamental studies of adsorption and diffusion properties with such materials. These materials are expected to create new opportunities for applications in the fields of separations science, for use directly as molecular sieves or as new molecular sieving sorbant materials; in catalysis, as heterogeneous catalysts; and as supports for other catalytic materials as well as other novel applications. The ability to synthesize zeolitic materials of precise pore size in the range between 4 and 100Å continues to expand the possibilities for research and technological innovation in the catalytic, separations, and sorption technologies.

Carbon Materials

The carbon-based materials of interest from a molecular self-assembly point of view include fullerenes and their relatives, including endohedral fullerenes and metal-coated fullerenes, carbon nanotubes, carbon nanoparticles, and porous carbons. Since 1990 with the discovery of techniques to produce soluble carbon in a bottle, research on and with carbon materials has skyrocketed. Not only can the molecular forms of carbon (the fullerenes and their derivatives) be synthesized, characterized, and studied for applications, but many other new carbon materials such as multi- and single-walled carbon nanotubes can now be produced in macroscopic quantities. Figure II–N.6 illustrates the broad variety of carbon nanotube structures whose properties are now being examined both theoretically and experimentally. A rich literature on these new carbon materials now exists.

Of particular interest for future catalytic applications is the finding that not only can C60 be coated with metal atoms, but that the metal coating can consist of a precise number of metal atoms. For example, $C_{60}Li_{12}$ and $C_{60}Ca_{32}$ have been identified mass spectroscopically. C_{60} has been coated with a variety of different metals, including Li, Ca, Sr, Ba, V, Ta and other transition metals.

Interestingly, addition of more than 3 Ta atoms to C_{60} breaks the C_{60} cage. Replacement of one carbon atom in C_{60} by a transition metal atom such as Co or Ir is being studied for possible catalytic applications. The future technological challenge will be to discover techniques to fabricate large quantities of such materials, so that such catalyst materials can be put in a bottle and not just in molecular beams.

Carbon nanotubes have the interesting property that they are predicted to be either semiconducting or conducting (metallic), depending on the chirality and diameter of the nanotube. Such materials are being studied as conductive additives to plastics and for use in electrochemical applications where the uniformity of the nanotube diameter and length is not overly critical. Another approach is to use the carbon nanotube as a template for a nanotube of an inorganic oxide. Hollow nanotubes of zirconia and yttria-stablilized zirconia have been prepared by coating treated carbon nanotubes with a zirconium compound and then burning out the carbon template. Finally, large scale production of single-walled nanotubes has recently been demonstrated, so one may anticipate a strong upsurge in the characterization and potential usage of single-walled carbon nanotubes in the future.

Porous carbons are of interest as molecular sieve materials, both as sorbants and as membranes, or as nanostraws for filtration. One of the major research objectives is to develop materials or structures with exceedingly high storage capacity per unit volume and weight for gases such as H_2 or CH_4. H_2 or CH_4 could become an economic source of combustion fuel or a means to power fuel cells for ultralow-emission vehicles or for electric power generation. Microporous hollow carbon fibers have exhibited high permeance and high selectivity as hydrogen selective membranes, and development is now underway to scale up these membranes to commercial levels. Carbon fiber materials produced via catalytic decomposition of hydrocarbon vapors have also recently been reported to exhibit exceptionally high hydrogen adsorption capacity. More mundane uses of nanotubes are as nanometer reinforcing rods in polymers or even in concrete. Incorporation of conducting carbon nanotubes in construction materials such as concrete or structural plastics opens opportunities for real time monitoring of material integrity and quality.

Microporous and Dense Ultrathin Films

Research and development of microporous thin films for use as molecular sieving membranes using inorganic crystalline materials such as zeolites or porous silica is another area of active research around the world. For molecular sieving membranes, one critical challenge rests on discovering ways to create large scale, thin, nearly defect-free membranes. One example is the fabrication of mesoporous conducting thin films grown from liquid crystal mixtures. Transmission electron microscopy (TEM) reveals an ordered array of 2.5 nm diameter cylindrical holes in a 300 nm thick Pt film. The hole diameter can be varied either by changing the chain length of the surfactant molecule or by adding an alkane to the plating solution. It is interesting that this technique produces a continuous thin film with nanoscale porosity in an electrically conducting material.

Dense ultrathin films such as single monolayer films would be of significant importance in the semiconductor industry. Thin films of specialized coatings for corrosion, thermal, and/or chemical stability should be valuable for the chemical and aerospace industries. Novel chemical sensors may be anticipated through use of ultrathin films composed of specialized clusters. Typical techniques for production of thin films are physical vapor deposition, chemical vapor deposition, and Langmuir-Blodgett processes.

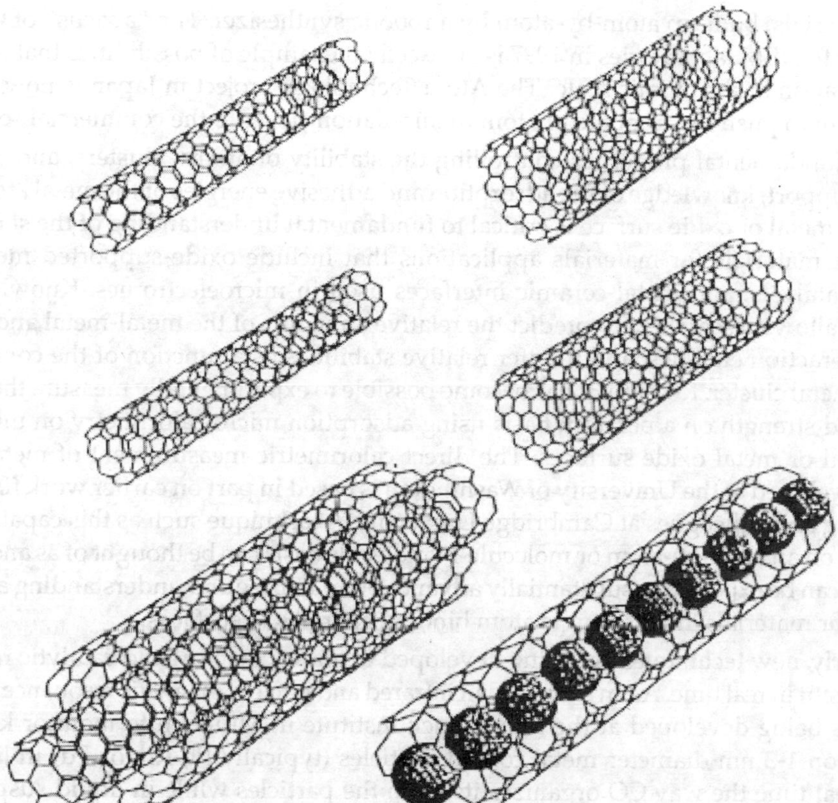

Fig. II–N.6: Examples of carbon nanotube structures, including multi-walled and metal-atomfilled nanotubes.

Opportunities in Characterization and Manipulation at the Nanoscale

Over the last two decades, the development and improvement of new techniques to fabricate and characterize nanoscale materials have fueled much of the enormous growth in nanoscale science and technology, not only by making nanoscale materials relatively easily available for scientific study and characterization, but also in some instances, opening the door for large-scale industrial use. For example, atomic force microscopy and scanning tunneling microscopy are two techniques that have become major workhorses for characterization of nanoscale materials. The strong upsurge in interest and funding in nanoscale materials must to some (large) degree be credited to the recent development of these two techniques. Combining Xray structure, high resolution TEM and low energy, high resolution scanning electron microscopy (SEM), researchers now have the means to physically characterize even the smallest structures in ways impossible just a few years ago. Not only can a nanostructure be precisely examined, but its electronic character can also be mapped out.

Using the scanning probe devices, scientists can both image individual atoms and molecules and also manipulate and arrange them one at a time. This atomic manipulation to build structures is just in its infancy, but it does allow one to imagine a route to the ultimate goal of atomically tailored materials, built up atom-by-atom by a robotic synthesizer. The "abacus" of C_{60} molecules produced at the IBM laboratories in 1997 is an excellent example of possibilities that may lie ahead for manipulation at the atomic scale. The Atom Technology project in Japan is now in its second term program to push the frontiers of atom manipulation closer to the commercial sector.

Of all fundamental properties controlling the stability of atoms, clusters, and particles on a surface or support, knowledge of the adsorption and adhesive energies of the metal atom or particle on the solid metal or oxide surface is critical to fundamental understanding of the stability of high surface area materials for materials applications that include oxide-supported metal catalysts, bimetallic catalysts, and metal-ceramic interfaces used in microelectronics. Knowledge of such parameters allows researchers to predict the relative strengths of the metal-metal and metal atom-support interaction energies, and to infer relative stabilities as a function of the composition and size of the metal cluster. Recently it has become possible to experimentally measure the metal atom-surface bond strength on a peratom basis using adsorption micro-calorimetry on ultrathin single crystal metal or metal oxide surfaces. The direct calorimetric measurement of metal adsorption energies developed at the University of Washington is based in part on earlier work first developed by D.A. King and colleagues at Cambridge University. A technique such as this capable of probing interactions on an atom-by-atom or molecule-bymolecule basis can be thought of as another "atomic probe" that can be expected to substantially advance our database and understanding at the ultimate nanoscale for materials, that is, single atom binding energies to surfaces.

Similarly, new techniques are being developed to allow chemical and catalytic reactions to be followed in situ in real time. As an example, an infrared and nuclear magnetic resonance spectroscopic technique is being developed at the Max Planck Institute in Mülheim to monitor kinetics of CO adsorption on 1-3 nm diameter metal colloid particles (typically Pt, Rh, or Pd) in liquids and to follow in real time the way CO organizes itself on the particles while in liquid suspension. Such techniques will allow one to begin to understand the metal particle properties in solution and thus infer what might occur in real reaction mixtures. Extension of such techniques to real catalytic reactions in solution for catalyst particles of various sizes and composition is likely in the not too distant future.

Summary

The areas where nanoscale high surface area materials may have the greatest future impact are difficult to predict, but some signs point to the possibility of substantial advancement in the areas of adsorption/separations, particularly in gas sorption and separations and in novel chemical catalysis using nanoscale catalyst particles.

At least two major challenges must be faced before utilization and generation of high surface area nanoscale materials becomes a commonplace reality. First is critical dimensional control of the nanoscale structure over long times and varying conditions. In nanoscale catalyst materials the critical chemical selectivity is likely to be intimately associated with the local environment around what is termed the "active" site. This suggests that the size, type, and geometry of the atoms making up the active site will play a critical role in defining the conditions under which this active site will be able to carry out its designed function. Ability to fabricate materials with "exactly" the same structure and composition at each active site has been and will continue to be a major challenge to materials and catalytic scientists.

A second challenge involves thermal and chemical stability control of the fabricated nanostructure. It is generally accepted that the smaller the nanostructure (active site) the more likely it is that the structure may move, aggregate, be poisoned, decompose, or change its shape, composition or morphology upon exposure to thermal and/or chemical cycling. Identifying windows of stable operation in which the specific structure or material will be able to retain the desired (and designed) behavior is critical for commercial applications. On the other hand, the driving force is the fact that nanostructured materials typically exhibit unique properties that are expected to open windows of opportunity previously inaccessible with existing materials.

It is important to recognize that nanoscale science and technology is not a "stand alone" field of endeavor, but rather is more of a "generic" area that is expected to have a critical impact and overlap in many areas of science and technology. The fields that fall under the "nanoscale" umbrella are many and diverse, illustrating that nanoscale science and technology is a collection of many different disciplines and areas of expertise. Such science and technology offers both an opportunity and a challenge to the scientific and technological community. Researchers in some areas of science included under the broad umbrella of nanoscale science and technology in this report do not normally consider (or in some cases, want) their scientific efforts to be labeled as "nanoscale." For this unit, nanoscale science and technology broadly encompasses the science and technology that falls between that involving individual atoms or molecules at the one extreme and that involving "bulk" materials at the other extreme.

In summary, it is important to recognize that the use of nanostructuring or nanostructures to generate, fabricate or assemble high surface area materials is at an embryonic stage. The effect of the nanostructure and our ability to measure it will be increasingly important for future progress and development of materials for the marketplace. That said, it is apparent that so-called "mature" technologies such as catalysis, coatings, separations, etc., are already being impacted. Thus, one may eagerly anticipate exciting new advances in many diverse technological areas growing from our increasing understanding of nanostructuring and nanostructured materials.

Unit II–O

Functional Nanoscale Devices

Introduction

The recent emergence of fabrication tools and techniques capable of constructing structures with dimensions ranging from 0.1 to 50 nm (see Fig. II–O.1) has opened up numerous possibilities for investigating new devices in a size domain heretofore inaccessible to experimental researchers. A review of the research in the United States, Japan, Taiwan, and Europe has revealed that there is considerable nanoscience and technology activity in university, industrial, and government laboratories around the world. The insight gained suggests areas of strength and areas of possible improvement in the field.

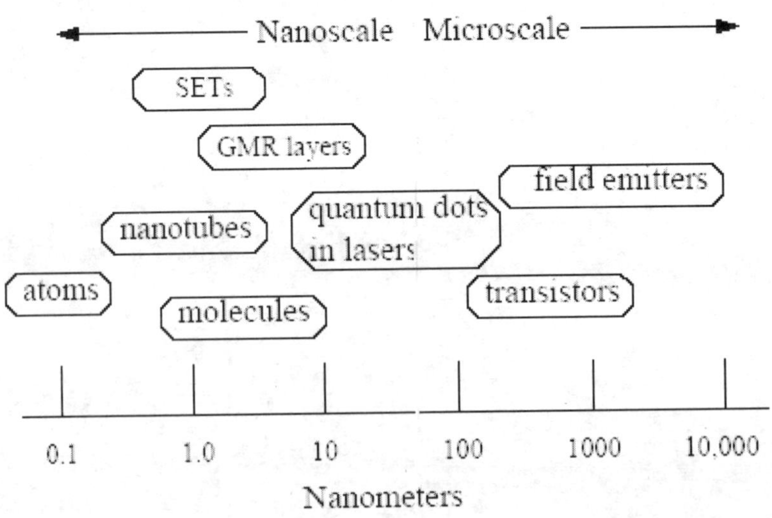

Fig. II–O.1: Functional device scales.

There is intense study around the world to determine the exact point in dimensional scaling where it becomes either physically unfeasible or financially impractical to continue the trend towards reducing the size while increasing the complexity of silicon chips. In some of the same laboratories where research activities on Si are decreasing, research activities on single electron devices (SEDs) are increasing. Although there are myriad questions involving electrical contacts, interconnections, reliability, and the like, one of the fundamental issues in the miniaturization/complexity debate concerns the Si MOSFET itself when the gate length is reduced to less than 50 nm. Does it behave like a long gate device or does the output conductance increase to impractical levels due to short-channel effects?

While the signature current-voltage (I-V) characteristics provide a common basis for comparison of device performance, there are significant variations in the fabrication methods and device structures being considered by the different labs in the countries the panel surveyed that have significant SED activity. The range of research in the surveyed laboratories spans electrical measurements from millikelvin to room temperature and from discrete electronic elements to integrated single electron transistors (SETs). Materials that are used to form the active single electron element range from charge clusters that are shaped by electric fields in a 2-D electron gas to metallic colloids to single oligomers. Progress in the field is hindered by architectures based on conventional circuit approaches that fail to take sufficient advantage of the unique properties of single-charge electronics to achieve significant impact in future high density applications. Most research in SED technology is fundamental and is distributed among universities funded by government agencies. A smaller body of application directed research exists in industrial laboratories; these are mostly in Japan.

The field of magnetics has experienced increasing attention since giant magnetoresistance (GMR) in multilayered structures was discovered in 1988. In these structures ferromagnetic layers are quantum mechanically coupled across a 1-3 nm nonmagnetic metallic layer. GMR structures are under intense study for applications in hard disk heads, random access memory (RAM), and sensors. Several laboratories are investigating the physics of the transition of these layers, which are quantum mechanically confined in one dimension, to layered filaments in which there are one- and two-dimensional confinements. There are numerous experimental process approaches under consideration in fabricating GMR structures, including the following:

- Magnetron or ion beam sputter deposition.
- Epitaxy for layered structures.
- Rubber stamping of nanoscale wire-like patterns.
- Electroplating into nanoscale pores in polymer membranes.

In RAM applications, a high ratio of magnetoresistance combined with a small coercive switching field is key to density, speed, and low power. These features are also achieved in magnetic tunnel junctions in which the ferromagnetic layers are quantum mechanically coupled through a thin dielectric layer. Although research in nanoscale magnetics is underway internationally, most of the activities on the practical applications mentioned above are in the United States.

Optical devices have already benefited from incorporation of nanostructured materials: commercially available semiconductor lasers incorporate active regions comprised of quantum wells, the presence of which modifies the electronic density of states and the localization of electrons and holes, resulting in more efficient laser operation. Extrapolating from these results, even greater improvements are predicted for lasers utilizing either quantum wire or quantum dot active layers. Recent advances in the "self-assembled" formation of quantum dot structures have stimulated progress in the fabrication and characterization of quantum dot lasers in Japan, Europe, and the

United States.

In late 1991, the first synthesis and characterization of carbon nanotubes were reported. The novel material contained a wide variety of multiwalled nanotubes (MWNT) containing 2 to 50 concentric cylindrical grapheme sheets with a diameter of a few nm and a length of up to 1 µm. The material was produced at the negative electrode of an arc discharge and appeared to be mixed with a large amount of other forms of carbon. This initial work led many groups throughout the world to produce and purify nanotubes. The theoretical study of their electronic structure followed in the next year. Soon it became clear that nanotubes have unique electronic and mechanical properties that are expected to lead to ground-breaking industrial applications.

Single-Charge Electronics

Even though the study of single electron charging effects with granular metallic systems dates back to the 1950s, it was the research of Likharev and coworkers almost 10 years ago that laid much of the groundwork for understanding single-charge transport in nanoscale tunnel junctions. The concept was developed of a Coulomb gap that can be exploited to control the transfer of single charges. Since then, many research groups have made use of the Coulomb blockade effect to develop SED technology. Figure II–O.2 through 50 show some of the myriad approaches to developing SEDs and representative laboratories pursuing the various SED concepts. Some of the more recent results are discussed below.

The group at Hitachi Europe uses a side-gated constriction in a delta doped GaAs structure to fabricate a magnetic tunnel junction device in which a series of small islands separated by tunnel barriers are formed (see, for example, Fig. II–O.2). At ~2 K, the Coulomb gap voltage oscillates as a function of the side-gate voltage. Using the MTJ device as a building block, both memory and logic (inverter, NOR) functions have been demonstrated. The fabrication procedure makes use of standard semiconductor processing techniques and does not rely on lithography to define the nanoscale islands, since these are created by disorder in the delta-doped layer.

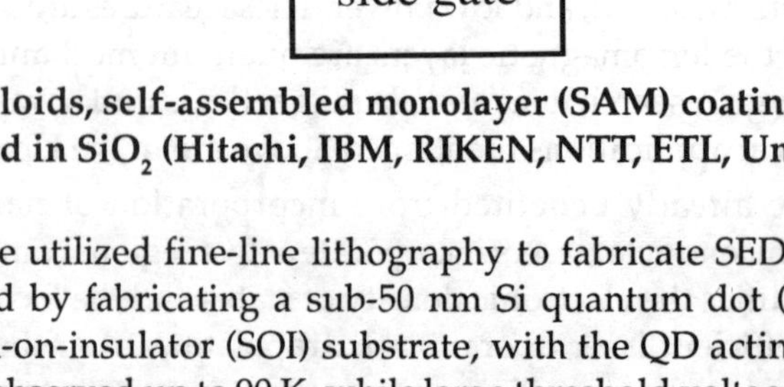

Fig. II–O.2: Metal colloids, self-assembled monolayer (SAM) coatings, polysilicon, quantum dots embedded in SiO_2 (Hitachi, IBM, RIKEN, NTT, ETL, University of Lund).

Other groups have utilized fine-line lithography to fabricate SEDs. At IBM, a flash memory SED was demonstrated by fabricating a sub-50 nm Si quantum dot (QD) on top of a MOSFET channel using a silicon-on-insulator (SOI) substrate, with the QD acting as a floating gate. Single electron charging was observed up to 90 K, while large threshold voltage shifts of up to 0.75 V were measured at 290 K. The University of Minnesota and Fujitsu have also reported similar structures. To overcome the lithography limitation on the QD size, the Toshiba group used a Si edge quantum wire approach. An inversion layer was formed at the 15 nm high Si sidewall of a SOI structure by growing a gate oxide and depositing a poly-Si gate there. Conductance oscillations were clearly seen at 4.2 K in this edge channel MOSFET (Fig. II–O.3). More recently, the Toshiba group has reverted back to a more planar device configuration, with a 50 nm wide Si quantum wire defined by e-beam lithography and oxidation of the surrounding SiO_2.

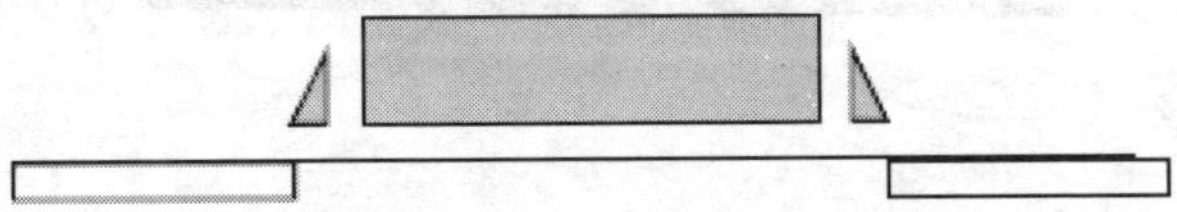

Fig. II–O.3: Sidewall extensions of MOSFET gate (Toshiba).

One method to form semiconductor QDs without depending on fine-line lithography is to make use of the self-organizing nature inherent in the Stranski-Krastanow thin film growth mode. In the initial stages of the heteroepitaxial growth of lattice-mismatched materials, strain-induced coherent relaxation occurs and dislocation-free islands are formed that are in the tens of nanometers range in size. There has been considerable research in these self-organized quantum dots (SOQDs) in the past few years, though much of the work has been of a fundamental nature. More recently, the University of Tokyo has proposed the embedding of InAs SOQDs in AlGaAs/GaAs heterojunction field effect transistors (HFETs) to form a flash-memory SED.

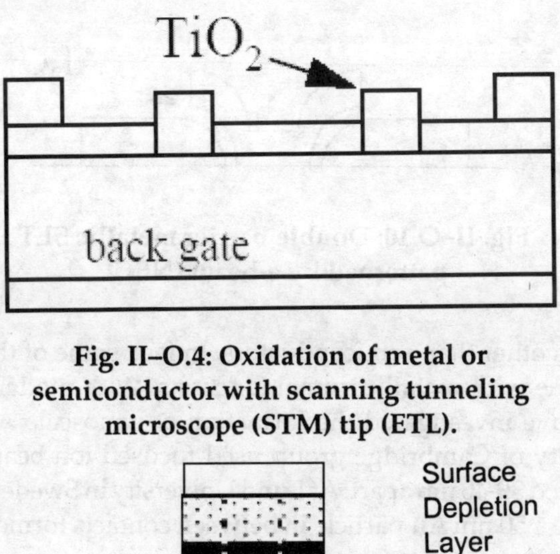

Fig. II–O.4: Oxidation of metal or semiconductor with scanning tunneling microscope (STM) tip (ETL).

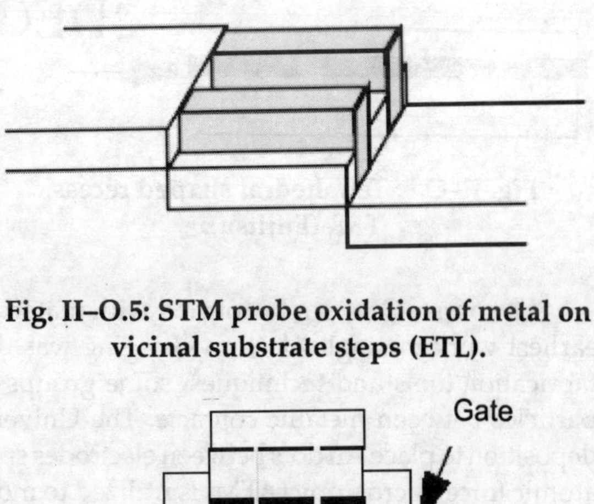

Fig. II–O.5: STM probe oxidation of metal on vicinal substrate steps (ETL).

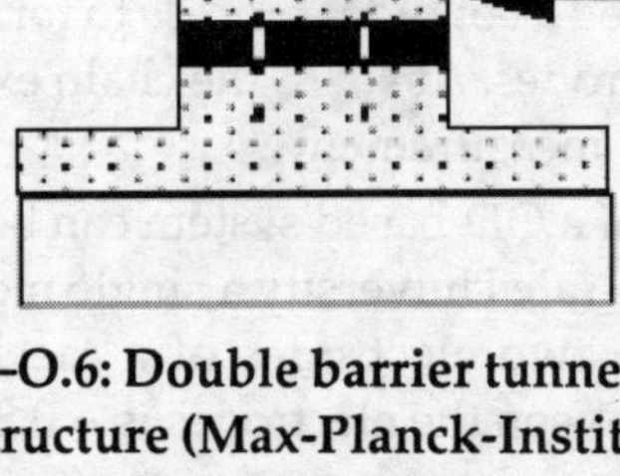

Fig. II–O.6: Double barrier tunnel diode structure (Max-Planck-Institut, Stuttgart; NTT).

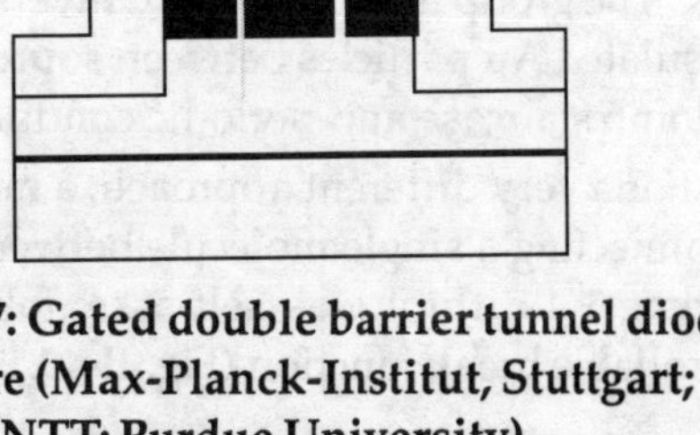

Fig. II–O.7: Gated double barrier tunnel diode structure (Max-Planck-Institut, Stuttgart; NTT; Purdue University).

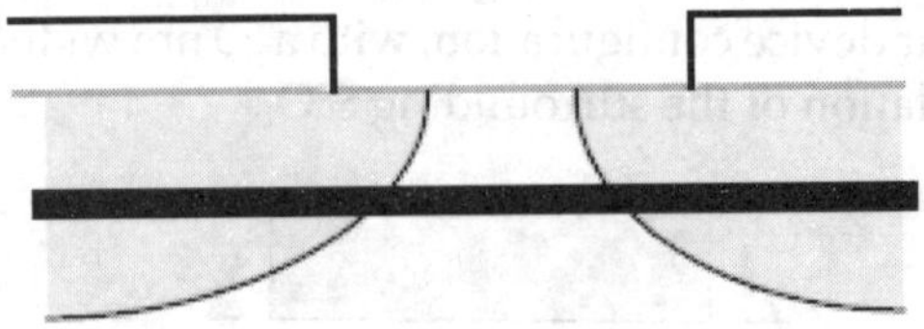

Fig. II–O.8: Depletion layer control of 2DEG area (Hitachi, University of Glasgow, University of Tokyo).

This concept was further demonstrated by the Sony group, which reported observing threshold voltage shifts at 300 K. Fujitsu has also proposed the use of InGaAs QDs in a similar manner. The one difference in this case is that the QDs are formed at the bottom of tetrahedral-shaped recesses formed by substrate patterning. Since SOQDs form in a somewhat random manner on a planar surface, this approach provides positioning control.

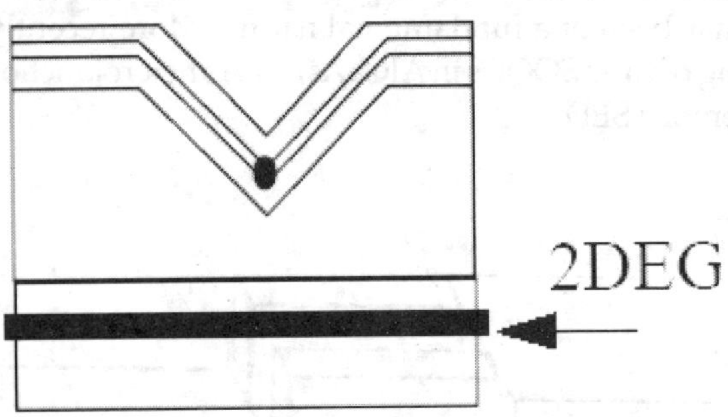

Fig. II–O.9: Tetrahedral shaped recess, TSR (Fujitsu).

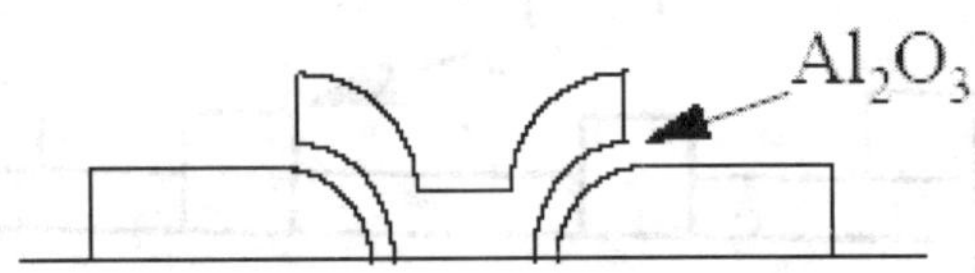

Fig. II–O.10: Double barrier metallic SET patterned by e-beam (NEC).

Of course, QDs can be formed with materials other than semiconductors. In fact, some of the earliest work in single electron charging was done with metallic tunnel junctions. With modern fabrication tools and techniques, some groups have investigated the formation of nanoscale Au particles between metallic contacts. The University of Cambridge group used focused ion beam deposition to place Au dots between electrodes spaced 30-40 nm apart. At Lund University in Sweden, atomic force microscopy (AFM) is utilized to move a 50 nm Au particle in between contacts formed by e-beam lithography. Researchers observed conductance plateaus stable for several minutes at 300 K. The group at Cambridge University/Hitachi Europe used a colloidal process to form a chain of insulated Au particles between source, drain, and gate electrodes. At 4.2 K, the chain exhibited a Coulomb staircase and periodic conduction oscillations in I-V measurements.

For a very different approach, a molecular embodiment of a QD-based system can be realized by connecting a single molecule between metallic contacts. At Yale University a single molecule of benzene-1,4-dithiol was self-assembled from solution onto two electrodes of a mechanically controllable break junction (Fig. II–O.11, left). The spacing between the electrodes is ~ 0.8 nm, and

I-V measurements at room temperature showed a gap ~ 0.7 V wide, which is attributed to a Coulomb gap. The Delft University of Technology in the Netherlands is also working on transport through oligomers (Fig. II–O.11, right).

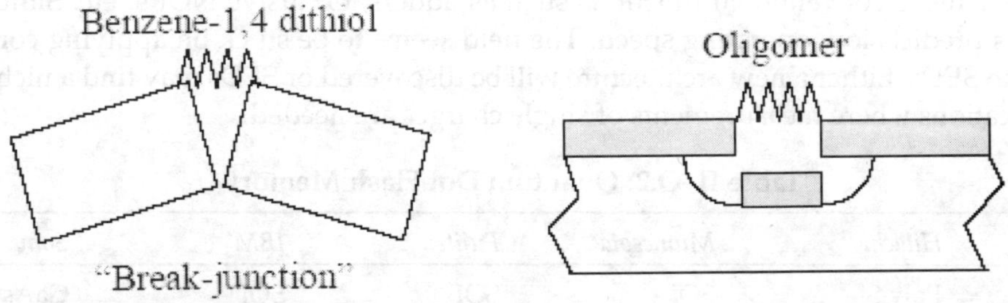

Fig. II–O.11: A single molecule connecting metallic contacts (Yale University, University of South Carolina, Delft University, Karlsruhe University).

Concerning the architecture in which SEDs are utilized, a number of approaches have been proposed. One of the more novel ideas is that of quantum cellular automata (QCA), based on some earlier work at Texas Instruments and developed at the University of Notre Dame. The basic QCA cell is made up of a group of capacitively coupled QDs. Each cell holds two electrons, resulting in two polarization configurations that can represent the logic "0" and "1" states, and each cell interacts via Coulombic forces with neighboring cells. An array of cells can then be used to transmit binary information, which eliminates the need for physical interconnects between devices and represents a paradigm shift for ultralarge-scale integration (ULSI). Basic Boolean operations (AND, OR, etc.) can be implemented using QCA, and more complex functions have been simulated. Most recently, the Notre Dame group has demonstrated a nonlinear, bistable response of a QCA cell, albeit at a very low temperature of less than 20 mK.

The QCA approach is not without its challenges (and critics). Circuit fabrication will be difficult because stringent control in QD positioning is required. Others have pointed out that bistability is only a necessary but not sufficient condition for the operation of Boolean logic circuits, because isolation is needed between the input and output, while background charge fluctuations will hamper logic implementation. Thus, there is also considerable research in the use of SEDs with more conventional architectures. The group at Hitachi Europe uses its MTJ devices in binary decision diagram logic that is commonly used in large-scale integrated (LSI) circuits. The Toshiba group is combining its SED with a MOSFET to compensate for the lack of gain in the former, as are, presumably, other groups working on QDbased flash memory SEDs. Tables II–O.1 and II–O.2 summarize some of these approaches.

Table II–O.1: SET Architectures

Flash Memory	*Digital Logic*	*Cellular Automata*	*Neural Networks*
Hitachi	Hitachi (binary decision diagram logic)	Notre Dame Hitachi	Delft U Technology
IBM	Hakkaido University		
Toshiba			

In summary, while significant progress has been made in nanofabrication techniques, the field of single charge electronics is still limited in scope by the lack of a suitable architecture that fully utilizes the unique aspects of single electron charging. Current approaches require many SED elements to achieve conventional functions such as adders, exclusive NORs, etc. Simulations of such circuits predict slow operating speed. The field seems to be stuck on applying conventional electronics to SEDs. Either a new architecture will be discovered or SEDs may find a niche home in those applications where measurements of single charges are needed.

Table II–O.2: Quantum Dot Flash Memory

	Hitachi	*Minnesota*	*Fujitsu*	*IBM*	*Sony*
Material	Poly-Si	SOI	SOI	SOI	GaAs
QD Material	Poly-Si	Poly-Si	Poly-Si	Poly-Si	InAs
QD Fab Method	E-beam/ etch	E-beam/ etch	E-beam/etch	E-beam/etch	Epitaxial self-assembly
QD Size (nm)	10 (estun.)	7×7×2 h	30×20×25 h	30×20×8 h	25×4 h
$2V^{th}$ (V)	0.5 – 1.0	0.055	0.1	0.75	0.45
Write/Erase	15 V/10 V	>4V	4 V	3 V	>1 V
Retention	1-24 h	10 sec.		>1 wk	10 sec.
Circuits	128 Mb LSI				

For the fabrication of nanoscale electronic devices, the self-organizing technique appears to be the most promising. The field of self-organized semiconductor QDs is quite active, but aside from optical emitters, very few practical electronic functions have been proposed. Of those, the single electron flash memory is attracting attention, but there has been no serious proposal as to how the device could operate under normal integrated circuit performance conditions and reliability specifications. Using the SED as a floating gate to a MOSFET has the same kinds of problems as applying conventional approaches to SED architectures; until someone comes up with a better idea, the future of these approaches remains to be determined.

Nanomagnetics

The discovery in 1988 of GMR in structures of alternating magnetic and nonmagnetic thin layers was the accumulation of several decades of intensive research in thin film magnetism and improvements in epitaxial growth techniques developed mainly in semiconductor materials. Not surprisingly, the first GMR structure was fabricated using molecular beam epitaxy. The high quality magnetic and nonmagnetic metallic films provide electrons with a mean free path exceeding 100 Å; on the other hand, the epitaxial growth allows for each constituent layer of the structure to be as thin as a few atomic layers. The greatly enhanced spin-dependent scattering in these multilayered structures provides magnetoresistance changes as high as 50 per cent. Table II–O.3 shows some of the institutions involved in GMR research and development.

Table II–O.3: Giant Magnetoresistance Activities

Memory	*MRAM*	*HD Heads*	*Structures / Physics / Materials*
• Fuitsu	❑	❑	• IMEC
• Hitachi	❑	❑	• L 'Ecole Polytechnique
• Honeywe	❑		• Lausanne
• IBM	❑	❑	• IBM Zurich
• Motorola	❑		• Tohoku University
• Matsushita	❑	❑	• Nagoya University
• Mitsubishi	❑	❑	• NIST
• Philips		❑	• US Santa Barbara
• Samsung	❑		• US San Diego
• Toshiba	❑	❑	• Camegie Mellon
• Seagate		❑	• Princeton
• Siemens	❑		
• Sony		❑	*Based on Publication, Patents or visit

Two subsequent major developments from IBM enabled the application of GMR materials to hard disk heads, RAM, and sensors. The first development was the demonstration of equally good or better GMR materials using high throughput and production-worthy magnetron sputtering systems. The other development was the invention of magnetically soft spin-valve structures, which allow low field and low power operation.

Industrial R&D efforts on GMR materials initially focused on high density read heads. The major U.S. players are IBM, Seagate, Quantum, ReadRite, and Applied Magnetics. In Japan, all the semiconductor companies are involved, in addition to strong magnetic media powerhouses such as TDK and Yamaha. Korea's Samsung is also actively involved in the GMR race. In Europe, Thomson CSF, Philips, and Siemens seem to have fallen behind. All in all, IBM is in a commanding position to reap the benefits of the GMR phenomenon. In November 1997, it announced the volume production of the first generation of GMR read heads.

In 1995, a different class of high magnetoresistive materials was discovered in which the nonmagnetic layer separating the two ferromagnetic layers is made with an ultrathin insulating material, such as an aluminum oxide layer < 20 Å thick. With the switching of magnetization of the two magnetic layers between parallel and antiparallel states, the differences in the tunneling coefficient of the junction and thus the magnetoresistance ratio have been demonstrated to be more than 25 per cent. A distinctive feature of this MTJ class of materials is its high impedance (> 100 kW-μm^2), which allows for large signal outputs.

The gradual improvement of GMR and MTJ materials have made them attractive for nonvolatile magnetic random access memory (MRAM) applications. The potential to make MRAM a high density, high speed, and low power, general purpose memory prompted the Defense Advanced Reseat Projects Agency to fund three MRAM consortia beginning in 1995, led by IBM, Motorola, and Honeywell, respectively. Other companies engaged in MRAM research are Hewlett-Packard, Matsushita, NEC, Fujitsu, Toshiba, Hitachi, and Siemens.

The key for a competitive MRAM technology is the fabrication of deep submicron-dimension magnetic cells. The further development of lithography tools utilizing e-beam and deep ultraviolet radiation is essential. Magnetic storage elements as small as 0.25 μm have been demonstrated by both Motorola and IBM. Among the steps of MRAM fabrication that are not yet compatible with semiconductor processing is the ion milling of the magnetic cells. The possibility of dry etching the magnetic materials has, nevertheless, been demonstrated. Figures II–O.12 through II–O.15 summarize the major directions in memory R&D. Arrows denote ferromagnetic layers.

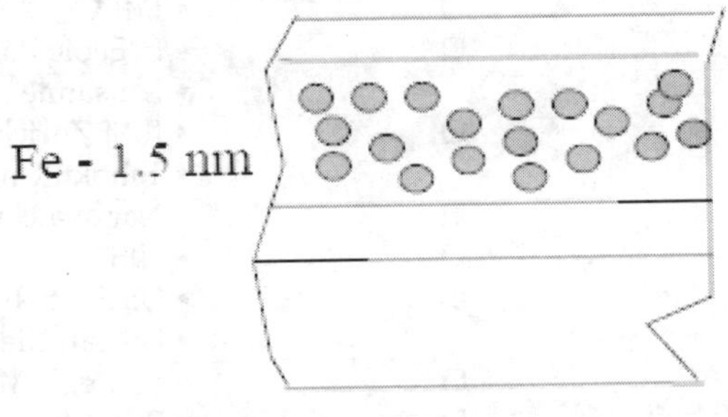

Fig. II–O.12: Granular GMR—Co, Fe (Nagoya University, Tohoku University, CNRS Thomson, UCSB, UCSD).

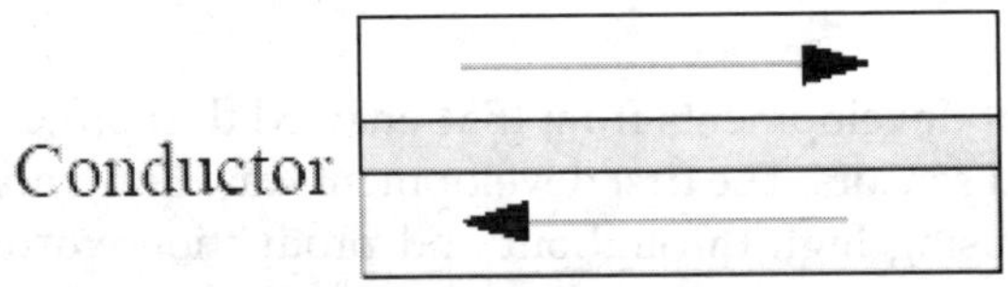

Fig. II–O.13: Current in plane (Matsushita, Fujitsu, Mitsubishi, Toshiba, Hitachi, Thomson, Philips, Siemens, IBM, Univ. Regensburg, IMEC, Nagoya University, Tohoku University, NIST).

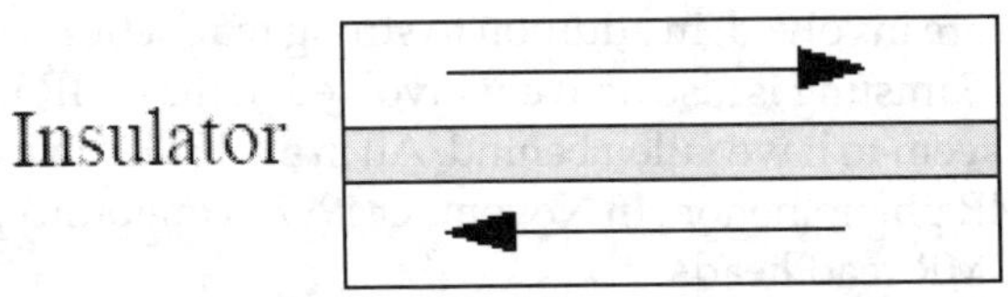

Fig. II–O.14: Magnetic tunnel junction (IBM, MIT, HP, Tohoku University).

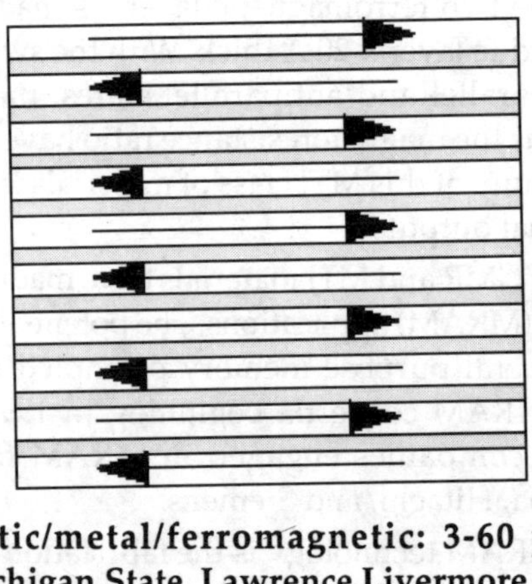

Fig. II–O.15: Ferromagnetic/metal/ferromagnetic: 3-60 periods free-standing (NRL, CNRSThomson, Philips, Michigan State, Lawrence Livermore Labs); plated into pores (L'École Polytechnique Fédérale de Lausanne, Johns Hopkins University, Université Catholique Louven).

The ability to fabricate submicron magnetic elements has opened a very rich and fascinating area of micromagnetics research. Characterization techniques having nanoscale resolution have been utilized and improved to measure and image the complex magnetization patterns in order to understand the magnetization switching characteristics. Examples of such techniques include the following:

- Superconducting quantum interference device (SQUID) magnetometry.
- Magnetic force microscopy (MFM).
- Scanning electron microscopy with polarization analysis (SEMPA).
- Magnetic near-field microscopy.
- Electron holography.

Another essential tool is micromagnetics modeling, which is used to predict complex magnetic domain configurations in patterned magnetic elements and to generate transient pictures that demonstrate the process of forming complex domain configurations.

By combining MFM, SQUID magnetometry, SEMPA, and micromagnetics modeling, researchers at Motorola have conducted a systematic study of the switching characteristics of single-layer and multilayer submicron magnetic structures. Three different phases of the magnetization phase diagram have been identified with regards to material composition, dimension, shape, thickness, and other structural parameters: (1) the quasi-single domain phase can be well described by the coherent rotation model; (2) the end-domain phase is dominated by the nonuniform regions of magnetization at the two ends of the element—the magnetization switching process can either be rotational or through domain wall nucleation and propagation; and (3) the trapped magnetization vortice (TMV) phase, which is characterized by the presence of magnetization vortices. Nucleation from TMV sites requires lower reversal fields than coherent rotation, but a high field is needed to drive out TMVs in the element. When the driving field is not high enough, TMVs can cause unusually large fluctuations in the switching field.

As the size of magnetic elements scales below 20 nm, a superparamagnetic phase emerges in which the room temperature thermal energy overcomes the magnetostatic energy well of the element, resulting in zero hysteresis. In other words, although the element itself is a single-domain ferromagnet, the ability of an individual magnetic "dot" to store magnetization orientation information is lost when its dimension is below a threshold. On the other hand, suitably prepared alloys of immiscible ferromagnetic and nonmagnetic metals that contain single-domain ferromagnetic grains in a nonmagnetic matrix have been shown to exhibit GMR characteristics. The moments of the magnetic grains are aligned at high fields and random at the coercive field, leading to GMR characteristics. In such "granular" metals, GMR has been reported for sputtered alloy films of CoCu, FeCu, NiFe/Ag, and CoAg. GMR values as high as 55 per cent at 4.2 K and 20 per cent at room temperature have been observed. The granular films require magnetic fields of the order of 10 kOe to achieve such a change in electrical resistance.

A very exciting consequence of ultra scaled magnetic particles is quantum tunneling of the magnetization direction of a collection of spins. There is no simple Schrödinger equation that describes this process, since it is not an elementary particle that is tunneling but a collective coordinate. Below its "blocking temperature," at which thermally assisted hopping between magnetic orientations becomes frozen out, magnetic particles of TbCeFe at sizes around 15 nm have been observed to behave independent of temperature and with no freeze-out magnetic relaxation. Because of the coherent tunneling of the magnetization orientation between the symmetric double-well potential, a resonance line in the magnetic susceptibility and noise spectra has been observed at temperatures below 200 mK in zero applied magnetic field. This work has stimulated a number of theoretical investigations into the effects of dissipation and the feasibility of producing quantum effects in larger magnetic structures.

Another interesting type of nanomagnetic structure is nanometer ferromagnetic wires fabricated using conventional nanolithography, nanoimprint lithography, AFM/MFM direct writing, groove deposition, and electrodeposition into pores of template polymer membranes. Such nanowires of either single layer or multilayers may provide new approaches to very small magnetoresistive sensors, ultrahigh-density hard disks, and other extensions of conventional applications. Another intriguing possibility is the suggestion to use heterostructure nanowires to investigate single electron tunneling.

Recently, molecular magnetism has received much attention with the development of a variety of synthesis techniques largely adapted from biology and chemistry. Natural and artificial ferritin proteins are examples of systems obtained using these methods. The ability to add one magnetic ion at a time has resulted in nanoscale magnets precisely defined by atomic weight. Ferritin consists of a segmented protein shell in the shape of a hollow sphere, with an outer diameter of 12.5 nm and an inner diameter of 7.5 nm. In vivo, the inner space is normally filled with a crystal of an iron oxide that is antiferromagnetic below 240 K. The empty protein shells can also serve as vessels for the synthesis of ferrimagnetic magnetite and maghemite. Thus, there exists a system in which its size as well as the nature of its magnetic interactions can be varied.

Another example of molecular magnetism is a cobalt-iron-cyanide-based Prussian blue analog (Sato et al. 1996a). In the ground state the Fe+2 and Co+3 ions are low-spin and diamagnetic, and there is no interaction between them. Red light excitation transfers one electron from an iron site to a cobalt site, resulting in high-spin Fe+3 and Co+2 ions and magnetic interactions between them. The application of a blue light causes a transition back to the initial state and switches off the Fe+3-Co+2 interactions. In addition to this kind of photochemically controllable magnets, electrochemically controllable magnets have also been reported.

Quantum Dot Lasers

Semiconductor lasers are key components in a host of widely used technological products, including compact disk players and laser printers, and they will play critical roles in optical communication schemes. The basis of laser operation depends on the creation of nonequilibrium populations of electrons and holes, and coupling of electrons and holes to an optical field, which will stimulate radiative emission. Calculations carried out in the early 1970s by C. Henry predicted the advantages of using quantum wells as the active layer in such lasers: the carrier confinement and nature of the electronic density of states should result in more efficient devices operating at lower threshold currents than lasers with "bulk" active layers. In addition, the use of a quantum well, with discrete transition energy levels dependent on the quantum well dimensions (thickness), provides a means of "tuning" the resulting wavelength of the material. The critical feature size—in this case, the thickness of the quantum well—depends on the desired spacing between energy levels. For energy levels of greater than a few tens of millielectron volts (meV, to be compared with room temperature thermal energy of 25 meV), the critical dimension is approximately a few hundred angstroms. Although the first quantum well laser, demonstrated in 1975, was many times less efficient than a conventional laser, the situation was reversed by 1981 through the use of new materials growth capabilities (molecular beam epitaxy), and optimization of the heterostructure laser design.

Even greater benefits have been predicted for lasers with quantum dot active layers. Arakawa and Sakaki predicted in the early 1980s that quantum dot lasers should exhibit performance that is less temperature dependent than existing semiconductor lasers, and that will in particular not degrade at elevated temperatures. Other benefits of quantum dot active layers include further reduction in threshold currents and an increase in differential gain—that is, more efficient laser operation. Figures II–O.15 and II–O.16 illustrate some of the key concepts in the laser operation. Stimulated recombination of electron-hole pairs takes place in the GaAs quantum well region, where the confinement of carriers and of the optical mode enhance the interaction between carriers and radiation (Fig. II–O.15). In particular, note the change in the electronic density of states, as a function of the "dimensionality" of the active layer, shown in Figure II–O.16. The population inversion (creation of electrons and holes) necessary for lasing occurs more efficiently as the active layer material is scaled down from bulk (3-dimensional) to quantum dots (0-dimensional). However, the advantages in operation depend not only on the absolute size of the nanostructures in the active region, but also on the uniformity of size. A broad distribution of sizes "smears" the density of states, producing behavior similar to that of bulk material.

Thus, the challenge in realizing quantum dot lasers with operation superior to that shown by quantum well lasers is that of forming high quality, uniform quantum dots in the active layer. Initially, the most widely followed approach to forming quantum dots was through electron beam lithography of suitably small featured patterns (~300 Å) and subsequent dryetch transfer of dots into the substrate material. The problem that plagued these quantum dot arrays was their exceedingly low optical efficiency: high surface-to-volume ratios of these nanostructures and associated high surface recombination rates, together with damage introduced during the fabrication itself, precluded the successful formation of a quantum dot laser.

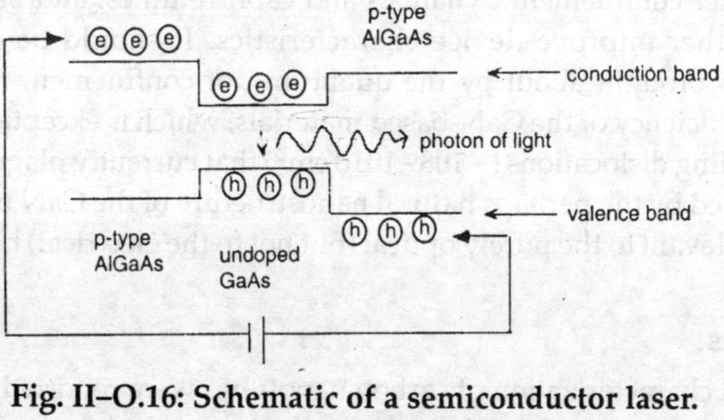

Fig. II–O.16: Schematic of a semiconductor laser.

With the demonstration of the high optical efficiency self-assembled formation of quantum dots formed without need of external processing and having the natural overgrowth of cladding material (which addressed issues of surface recombination), there ensued a marked increase in quantum dot laser research. The first demonstration of a quantum dot laser with high threshold density was reported by Ledentsov and colleagues in 1994. Bimberg et al. achieved improved operation by increasing the density of the quantum dot structures, stacking successive, strain-aligned rows of quantum dots and therefore achieving vertical as well as lateral coupling of the quantum dots. In addition to utilizing their quantum size effects in edge-emitting lasers, self-assembled quantum dots have also been incorporated within vertical cavity surface-emitting lasers. Table II-O.4 gives a partial summary of the work and achievements in quantum dot lasers.

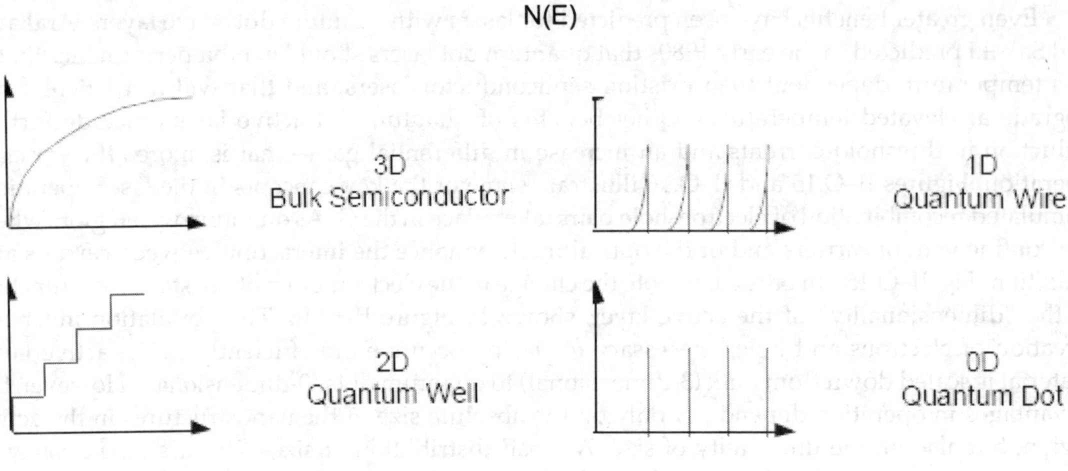

Fig. II–O.17: Density of electronic states as a function of structure size.

As with the demonstration of the advantages of the quantum well laser that preceded it, the full promise of the quantum dot laser must await advances in the understanding of the materials growth and optimization of the laser structure. Although the self-assembled dots have provided an enormous stimulus to work in this field, there remain a number of critical issues involving their growth and formation: greater uniformity of size, controllable achievement of higher quantum dot density, and closer dot-todot interaction range will further improve laser performance. Better understanding of carrier confinement dynamics and capture times, and better evaluation of loss mechanisms, will further improve device characteristics. It should be noted that the spatial localization of carriers brought about by the quantum dot confinement may play a role in the "anomalous" optical efficiency of the GaN-based materials, which is exceptional in light of the high concentration of threading dislocations (~ 108 - 1010 cm^{-2}) that currently plague this material system. The localization imposed by the perhaps natural nanostructure of the GaN materials may make the dislocation largely irrelevant to the purely optical (but not to the electrical) behavior of the material.

Carbon Nanotubes

The first synthesis and characterization of carbon nanotubes were reported by Iijima from NEC in late 1991. The initial theoretical study of their electronic structure was soon followed with the work by Dresselhaus and coworkers at MIT. Since then, the fabrication of nanotubes has been improved by several groups, and methods other than arc discharge have been explored. The main issues are to separate the nanotubes from other forms of carbon also produced in the fabrication process and to increase the yield of single-walled nanotubes (SWNT) for potential applications. Following Iijima's work, macroscopic quantities of MWNT were produced with an improved arc discharge method by Ebbesen and coworkers at NEC (Tsukuba).

It was not until 1995 that Smalley and colleagues at Rice University showed that SWNT can be efficiently produced by laser ablation of a graphite rod. In the following year, that same group produced what is considered to be among the best SWNT material generated so far; over 70 per cent of the volume of material was nanotubes bundled together into crystalline ropes of metallic character.

Also in 1996, a group from the Chinese Academy of Science used chemical vapor deposition to produce a 50 mm thick film of nanotubes that were highly aligned perpendicular to the surface. Progress in recent years leads one to predict that it will indeed be possible to produce high quality carbon nanotubes in macroscopic quantities needed for many of the applications outlined below.

Nanotube bundles form a low density material and are expected to have high stiffness and axial strength as a result of their seamless cylindrical graphitic structure. It is therefore predicted that they can be used to fabricate a material with better mechanical properties than the present carbon fiber materials. Information about the mechanical properties of nanotubes has been gathered recently by a study of the thermal vibrations of a single SWNT attached to a substrate. Ebbesen's group at the Princeton NEC Research Institute found that nanotubes have an exceptionally high Young's modulus (~ 2 x 109 Pa). In order to reach a better understanding of the mechanical properties and intrinsic limitations of nanotubes, Bernholc's group from North Carolina State University theoretically studied the behavior of nanotubes beyond the linear Hooke's law and the nature of the defects leading to dislocations and fractures.

Table II–O.4: Summary of Quantum Dot Laser Results

QD composition & size	*Threshold (ka/cm²)*	*Operating T (K)*	*Wavelength (μm)*
InAs 7 nm	1 0.1	300 77	0.9 0.95
InGaAs 30 mm	76	77	1.26
$In_{0.5}Ga_{0.5}As$ 20 mm	0.8	85	0.92
InP 25 mm	25	300	0.7
$In_{0.3}Ga_{0.7}As$ 1.2	0.5 1	300	1.2
$In_{0.4}Ga_{0.6}As$ 12 mm	0.65	300	1
$In_{0.5}Ga_{0.5}As$ 10 layers	0.06	300	1

Nanotubes are highly polarizable nanoscale straws, a property that confers on them the capacity to ingest inorganic elements by nanocapillarity. As a result, it has been conjectured that they could be used as minute molds to shape nanometer-sized quantum wires and as miniature test tubes. Ajayan and coworkers at NEC (Tsukuba) have first shown that lead can be introduced into carbon nanotubes. The efficiency of their process is low, and prior removal of the caps from the ends of the nanotubes is expected to improve the situation. More information about the mechanism of NT filling was obtained by Pascard and coworkers from the École Polytechnique in France by studying the propensity to form nanowires for 15 encapsulated metal elements. Finally, external decoration of nanotubes with metal atoms has been demonstrated and is predicted to have applications in catalysis. Table II–O.5 summarizes the primary methods of nanotube fabrication and the institutions engaged in specific methods of nanotube fabrication.

Early theoretical studies already showed that the electronic properties of nanotubes strongly depend on their diameter and their chirality leading to metallic or semiconducting structures. It was conjectured that these properties can be used to construct nanoscale electronic devices. While theoretical studies were promptly published, it was only in 1996 that Ebbesen and coworkers at the Princeton NEC Research Institute presented reliable four-point probe conductivity measurements on MWNT, confirming the theoretical predictions. In 1997, two groups, one at Lawrence Berkeley National Laboratory and the second at Delft University in the Netherlands showed that conductivity through nanotubes is controlled by low dimensional effects such as resonant tunneling and single-electron charging effects. Hall effect measurements at the École Polytechnique Fédérale de Lausanne (EPFL) in Switzerland have shown that hole transport is predominant in electronic conductance. Despite these and other very recent and encouraging efforts such as those studying the mean free path of carriers in nanotubes, the conduction mechanism is still only partially understood.

Table II–O.5: Nanotube Fabrication Methods

Method	*Institution*
Laser ablation—SWNT	Rice University
Arc discharge—SWNT	University of Montpellier University of Kentucky
Arc discharge—MWNT	NEC
Chemical vapor deposition—aligned MWNT	Beijing

Metallic nanotubes are strongly polarizable in an electric field and thereby lead to field enhancement at their extremity, the strength of which depends on the ratio of the diameter to the length and can be extremely large for routinely produced nanotubes. For this reason and possibly others related to quantum confinement effects, nanotubes are expected to form outstanding field-emitting materials. In 1995, the Rice University group showed that nanotubes emit electrons very efficiently when immersed in an electric field and irradiated by a laser to remove their cap. They attribute their observation to the unraveling of an atomic wire of carbon atoms. Efficient field emission was also obtained from carefully aligned nanotubes by de Heer and coworkers at EPFL, whereas Collins and coworkers from the University of California at Berkeley have used randomly oriented nanotubes with similar results. Efficient field-emitting material is highly desirable for the production of field-emission displays and microwave tubes.

Recently, a group from Mie University in Japan has built a cathode ray tube (CRT) using nanotube field emitters. In this work, the layers of nanotubes were cut out from the soot produced in an arc discharge chamber. This fabrication method is presently not compatible with industrial production requirements, and more progress must be made before this effort can be translated into an industrial product. Table II–O.6 summarizes the electrical and field emission properties of nanotubes, with the representative institutions pursuing these studies.

Table II–O.6: Electrical and Field Emission Properties of Nanotubes

Results	*Institution*
Theory and experiment: nanotubes can be metallic or Semiconducting	MIT, NEC
Conductivity show low dimensional	LBNL Delft University
Field emission from unraveled carbon chains at the end of nanotubes	Rice university
Field emission from aligned nanotubes attached to scanning probe tip	EPEL
CRTs fabncated with nanotubes as field emitters	Mie University

In summary, macroscopic amounts of good quality nanotubes can presently be fabricated by several groups around the world, and the theoretical understanding of the electronic structure and related properties of nanotubes has reached a very good level. However, despite the fact that many potential applications are mentioned in the literature over and over again, only the outstanding field emission properties of nanotubes have achieved realization in practical devices. One of the main obstacles simply remains the controlled manipulation of nanoscale objects. In this respect it seems that the generation of self-aligned structures is a path to explore further.

Unit II–P

Bulk Behavior of Nanostructured Materials

Introduction

Bulk nanostructured materials are defined as bulk solids with nanoscale or partly nanoscale microstructures. This category of nanostructured materials has historical roots going back many decades but has a relatively recent focus due to new discoveries of unique properties of some nanoscale materials.

Early in the century, when "microstructures" were revealed primarily with the optical microscope, it was recognized that refined microstructures, for example, small grain sizes, often provided attractive properties such as increased strength and toughness in structural materials. A classic example of property enhancement due to a refined microstructure—with features too small to resolve with the optical microscope—was age hardening of aluminum alloys. The phenomenon, discovered by Alfred Wilm in 1906, was essentially explained by Merica, Waltenberg, and Scott in 1919, and the microstructural features responsible were first inferred by the X-ray studies of Guinier and Preston in 1938. With the advent of transmission electron microscopy (TEM) and sophisticated X-ray diffraction methods it is now known that the fine precipitates responsible for age hardening, in Al-4 per cent Cu alloys, for example, are clusters of Cu atoms—Guinier-Preston (GP) zones—and the metastable partially coherent q′ precipitate. Maximum hardness is observed with a mixture of GPII (or q″) (coarsened GP zones) and q′ with the dimensions of the q′ plates, typically about 10 nm in thickness by 100 nm in diameter. Therefore, the important microstructural feature of age hardened aluminum alloys is nanoscale. There are a number of other examples of nanoscale microstructures providing optimized properties. The critical current density JC of commercial superconducting Nb3Sn is controlled by grain size and is inversely proportional to grain size, with grain sizes of 50-80 nm providing high values of JC.

The field of nanocrystalline (or nanostructured, or nanophase) materials as a major identifiable activity in modern materials science results to a large degree from the work in the 1980s of Gleiter and coworkers, who synthesized ultrafine-grained materials by the in situ consolidation of nanoscale atomic clusters. The ultrasmall size (< 100 nm) of the grains in these nanocrystalline materials can result in dramatically improved—or different—properties from conventional grain-size (> 1 μm) polycrystalline or single crystal materials of the same chemical composition. This is the stimulus for the tremendous appeal of these materials. While there are a number of bulk properties that may be dramatically changed when the microstructure is nanoscale, this unit focuses on those for which the recent work with nanostructured materials has been most extensive. These are: (1) the mechanical properties of nanostructured materials for a variety of potential structural applications, and (2) ferromagnetic materials with nanoscale microstructures for potential applications as soft magnetic materials and permanent magnet materials, and for other special applications such as information storage, magnetoresistance spin valves, and magnetic nanocomposite refrigerants. Other bulk applications such as hydrogen storage are discussed briefly.

Mechanical Behavior: Structural Nanostructured Materials

The great interest in the mechanical behavior of nanostructured materials originates from the unique mechanical properties first observed and/or predicted for the materials prepared by the gas condensation method. Among these early observations/predictions were the following:

- Lower elastic moduli than for conventional grain size materials—by as much as 30-50 per cent.
- Very high hardness and strength—hardness values for nanocrystalline pure metals (~ 10 nm grain size) are 2 to 7 times higher than those of larger grained (>1 mm) metals.
- A negative Hall-Petch slope, i.e., decreasing hardness with decreasing grain size in the nanoscale grain size regime diffusional deformation mechanisms.

While some of these early observations have been verified by subsequent studies, some have been found to be due to high porosity in the early bulk samples or to other artifacts introduced by the processing procedures.

Elastic Properties

Early measurements of the elastic constants on nanocrystalline (nc) materials prepared by the inert gas condensation method gave values, for example for Young's Modulus, *E*, that were significantly lower than values for conventional grain size materials. While various reasons were given for the lower values of *E*, it was suggested by Krstic and coworkers that the presence of extrinsic defects—pores and cracks, for example—was responsible for the low values of *E* in nc materials compacted from powders. This conclusion was based on the observation that nc NiP produced by electroplating with negligible porosity levels had an *E* value comparable to fully dense conventional grain size Ni. Subsequent work on porosity-free materials has supported these conclusions, and it is now believed that the intrinsic elastic moduli of nanostructured materials are essentially the same as those for conventional grain size materials until the grain size becomes very small, e.g., < 5 nm, such that the number of atoms associated with the grain boundaries and triple junctions becomes very large. This is illustrated in Figure II–P.1 for nanocrystalline Fe prepared by mechanical attrition and measured by a nano-indentation technique. Thus, for most nanostructured materials (grain size > 10 nm), the elastic moduli are not unique properties and not a "negative."

Hardness and Strength

Hardness and strength of conventional grain size materials (grain diameter, d > 1 μm) is a function of grain size. For ductile polycrystalline materials the empirical Hall-Petch equation has been found to express the grain-size dependence of flow stress at any plastic strain out to ductile fracture. In terms of yield stress, this expression is $\sigma_o = \sigma_i + kd^{-1/2}$, where σ_o = yield stress, σ_i = friction stress opposing dislocation motion, k = constant, and d = grain diameter. Similar results are obtained for hardness, with $H_o = H_i + kd^{-1/2}$. To explain these empirical observations, several models have been proposed, which involve either dislocation pileups at grain boundaries or grain boundary dislocation networks as dislocation sources. In all cases the Hall-Petch effect is due to dislocation motion/ generation in materials that exhibit plastic deformation.

Most of the mechanical property data on nc materials have pertained to hardness, although some tensile test data are becoming available. Several recent reviews have summarized the mechanical behavior of these materials. It is clear that as grain size is reduced through the nanoscale regime (< 100 nm), hardness typically increases with decreasing grain size and can be factors of 2 to 7 times harder for pure nc metals (10 nm grain size) than for large-grained (> 1 μm) metals.

The experimental results of hardness measurements, summarized previously, show different behavior for dependence on grain size at the smallest nc grains (< 20 nm), including: (a) a positive slope ("normal" Hall-Petch behavior), (b) essentially no dependence (~ zero slope), and (c) in some cases, a negative slope. Most data that exhibit the negative Hall-Petch effect at the smallest grain sizes have resulted from nc samples that have been annealed to increase their grain size. It is suggested that thermally treating nanophase samples in the as-produced condition may result in such changes in structure as densification, stress relief, phase transformations, or grain boundary structure, all of which may lead to the observed negative Hall-Petch behavior. Only a few cases of negative Hall-Petch behavior have been reported for as-produced nanocrystalline samples with a range of grain sizes. These include electrodeposited nc alloys and devitrified nc alloys. Nanocrystalline thin films with grain sizes £ 6 nm are also observed to exhibit a negative Hall-Petch effect. While it seems likely that in many cases the observed negative Hall-Petch slopes are due to artifacts of the specimen preparation methods, it is also likely that conventional dislocation-based deformation is not operable in nanocrystalline materials at the smallest grain sizes (< ~30 nm). At these grain sizes, theoretically, mobile dislocations are unlikely to occur; nor have they been observed *in situ* TEM deformation experiments. Thus, the hardness, strength, and deformation behavior of nanocrystalline materials is unique and not yet well understood.

Ductility and Toughness

It is well known that grain size has a strong effect on the ductility and toughness of conventional grain size (> 1 mm) materials. For example, the ductile/brittle transition temperature in mild steel can be lowered about 40°C by reducing the grain size by a factor of 5. On a very basic level, mechanical failure, which limits ductility, is an interplay or competition between dislocations and cracks. Nucleation and propagation of cracks can be used as the explanation for the fracture stress dependence on grain size. Grain size refinement can make crack propagation more difficult and therefore, in conventional grain size materials, increase the apparent fracture toughness. However, the large increases in yield stress (hardness) observed in nc materials suggest that fracture stress can be lower than yield stress and therefore result in reduced ductility. The results of ductility measurements on nc metals are mixed and are sensitive to flaws and porosity, surface finish, and method of testing (e.g., tension or compression testing). In tension, for grain sizes < 30 nm, essentially brittle behavior has been observed for pure nanocrystalline metals that exhibit significant ductility when the grain size is conventional. This is illustrated in Figure II–P.2.

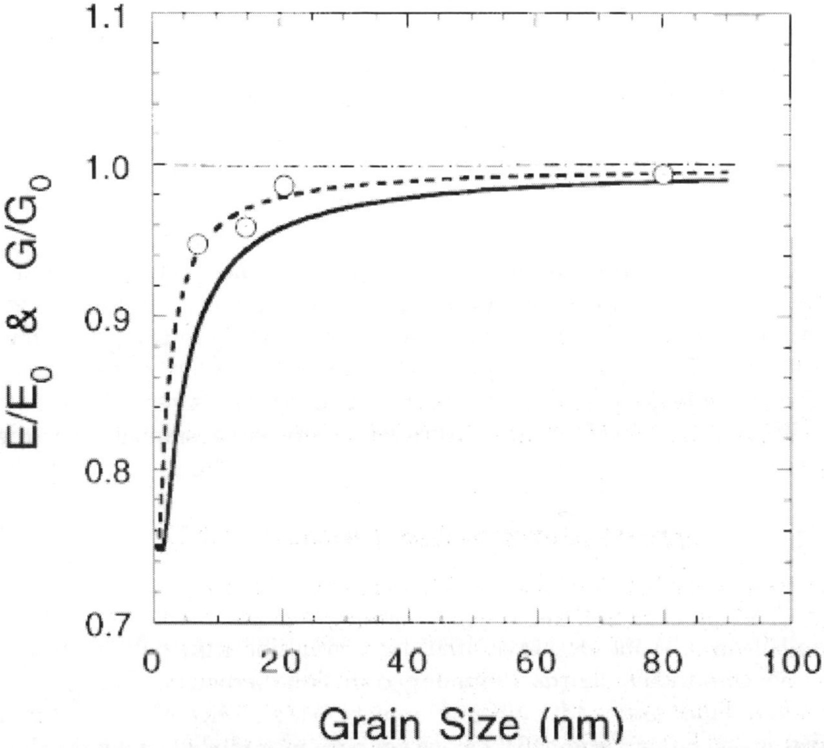

Fig. II–P.1: Ratio of the Young's (E) and shear (G) moduli of nanocrystalline materials to those of conventional grain size materials as a function of grain size. The dashed and solid curves correspond to a grain boundary thickness of 0.5 and 1 nm, respectively.

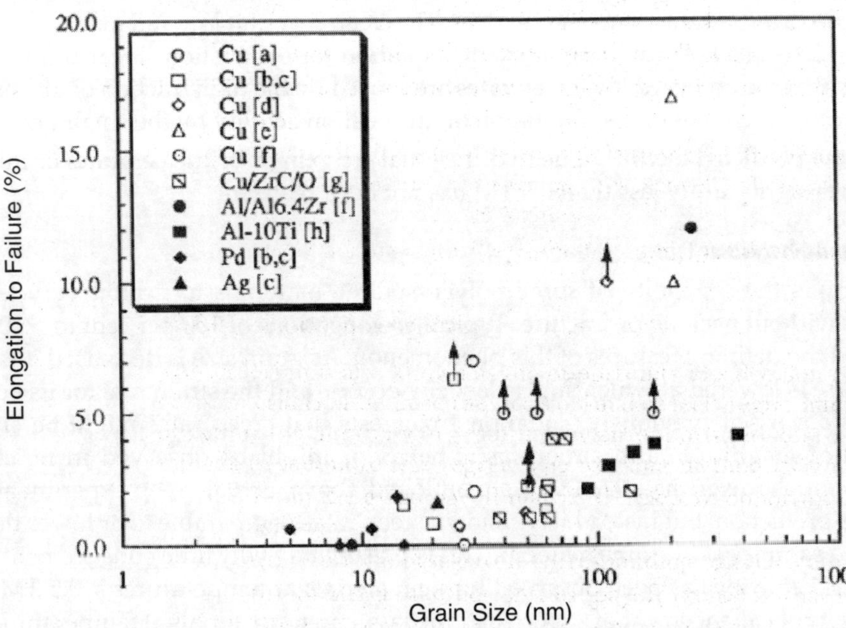

Fig. II–P.2: Elongation to failure in tension vs. grain size for some nanocrystalline metals and alloys.

In some metals, Cu for example, ductile behavior is observed in compression, along with yield strengths about twice those observed in tension. While it is likely that the flaws and porosity present in many nc samples seriously affect the results of mechanical tests and may be partly responsible for the asymmetry of results in compression compared to tension tests, the nature of the deformation process in terms of shear banding (see below) may also be important. The above behavior is presumably due to the inability of usual dislocation generation and motion to occur at these smallest nc grain sizes.

An intriguing suggestion based on early observations of ductile behavior of brittle nc ceramics at low temperatures is that brittle ceramics or intermetallics might exhibit ductility with nc grain structures. Karch and colleagues observed apparent plastic behavior in compression in nc CaF_2 at 80°C and nc TiO_2 at 180°C. These observations were attributed to enhanced diffusional creep providing the plasticity at these temperatures, where conventional grain-size materials would fail in the elastic regime. It was assumed that diffusional creep was responsible for the plasticity; observations were rationalized, with boundary diffusion dominating the behavior such that the strain (creep) rate is defined as

$$\frac{d\varepsilon}{dt} = \frac{B\sigma\Omega\Delta D_b}{d^3 kT}$$

where s is the applied stress, W the atomic volume, d the grain size, k the Boltzmann constant, T the temperature, B a constant, and D_b the grain boundary diffusion coefficients. Going from a grain size of 1 μm to 10 nm should increase de/dt by 106 or more if Db is significantly larger for nc materials. However, these results on nc CaF_2 and nc TiO_2 have not been reproduced, and it is believed that the porous nature of these samples was responsible for the apparent ductile behavior. In addition, the idea of unusually high creep rates at low temperatures has been refuted. Recent creep measurements of nc Cu, Pd, and Al-Zr at moderate temperatures by Sanders et. al. (1997) find creep rates comparable to or lower than corresponding coarse-grain rates. The creep curves at low and moderate homologous temperatures (.24 – .48 TM) could be fit by the equation for exhaustion (logarithmic) creep. One explanation is that the observed low creep rates are caused by the high fraction of low energy grain boundaries in conjunction with the limitation on dislocation activity by the small grain sizes.

In sum, the predicted ductility due to diffusional creep in nc brittle ceramics or intermetallics at temperatures significantly less than 0.5 TM has not been realized.

Superplastic Behaviour

Superplasticity is the capability of some polycrystalline materials to exhibit very large tensile deformations without necking or fracture. Typically, elongations of 100 per cent to > 1000 per cent are considered the defining features of this phenomenon. As grain size is decreased it is found that the temperature is lowered at which superplasticity occurs, and the strain rate for its occurrence is increased. As discussed previously, Equation 1 suggests that creep rates might be enhanced by many orders of magnitude and superplastic behavior might be observed in nc materials at temperatures much lower than 0.5 TM. As mentioned above, actual creep experiments have not borne out this prediction, but instead have shown creep rates comparable to or lower than those in coarse-grained samples of the same material. This is presumably why little enhancement in ductility or superplastic behavior has been observed for nc materials at temperatures < 0.5 TM. However, there is evidence of enhancement of superplastic behavior in nc materials at temperatures > 0.5 TM. Superplasticity has been observed at somewhat lower temperatures and at higher strain rates in nc materials. The evidence for tensile superplasticity is limited and observed typically at temperatures greater than 0.5 TM and in materials that exhibit superplasticity in coarser grain sizes (1–10 μm). For example, Mishra et al. observed superplastic behavior in nc Pb-62 per cent Sn at 0.64 TM and nc Zn–22 per cent Al at 0.52 to 0.60 TM. However, Salishekev et al observed superplastic behavior in submicron—200 nm—Ti and several Ti and Ni base alloys. Here, superplasticity (190 per cent elongation, m = 0.32) was observed in Ti at 0.42 TM. This was at a temperature 50°C lower than for 10 μm grain size Ti. The flow stress for the 200 nm Ti at 550°C was 90 MPa, compared to 120 MPa for 10 μm Ti at 600°C.

Some years back, Mishra and Mukherjee have observed superplastic behavior in Ni3Al with a 50 nm grain size at temperatures of 0.56 to 0.60 TM to strains of 300-600 per cent, but with unusual stress-strain behavior and significant apparent strain-hardening. These new results suggest very different mechanisms may be causing superplastic behavior in these nc materials.

Unique Mechanical Properties of Nanocrystalline Materials

While there are still only limited data on the mechanical behavior—especially tensile properties—of nc materials, some generalizations may be made regarding the deformation mechanisms. It is likely that for the larger end of the nanoscale grain sizes, about 50 - 100 nm, dislocation activity dominates for test temperatures < 0.5 TM. As grain size decreases, dislocation activity apparently decreases. The essential lack of dislocations at grain sizes below 50 nm is presumably the result of the image forces that act on dislocations near surfaces or interfaces. The lack of dislocations in small, confined spaces such as single-crystal whiskers has been known for many years. Creation of new dislocations is also made difficult as the grain size reaches the lower end of the nanoscale (< 10 nm). Stresses needed to activate dislocation sources, such as the Frank-Read source, are inversely proportional to the distance between dislocation pinning points. Since nanoscale grains will limit the distance between such pinning points, the stresses to activate dislocation sources can reach the theoretical shear stress of a dislocation-free crystal at the smallest grain sizes (~ 2 nm). Thus, at the smallest grain sizes we may have new phenomena controlling deformation behavior. It has been suggested that such phenomena may involve grain boundary sliding and/or grain rotation accompanied by short-range diffusion-assisted healing events.

Several examples of deformation by shear banding have been reported for nc materials. Carsley et al have studied nc Fe–10 per cent Cu alloys with grain sizes ranging from 45 to 1,680 nm. In all cases, deformation in compression proceeds by intense localized shear banding. The stress-strain curves exhibited essentially elastic, perfectly plastic behavior; that is, no measurable strain hardening was observed. Shear banding is also the deformation mode observed in amorphous metallic alloys and amorphous polymers. The deformation shear banding in nc Fe–10 per cent Cu was compared to that for metallic glasses, amorphous polymers, and coarse grained polycrystalline metals after significant plasticity and work hardening had taken place. While this suggests a close similarity between deformation in nc materials and amorphous materials, not all tensile data on nc materials exhibit a lack of strain hardening. The Fe–10 per cent Cu samples of Carsley et al. showed shear bands even in their larger grained specimens (i.e., about 1,000 nm).

Theoretical Needs

Central to all of the above discussion is the lack of understanding of the microscopic deformation and fracture mechanisms in nc materials. Clearly, a stronger theoretical effort is needed to guide critical experiments and point the direction for optimizing properties. There has been limited work in this area, especially in Russia, in applying disclination theory to grain rotation, for example. However, a much larger theoretical effort is required. Another alternate deformation mechanism that may be important in nc materials is mechanical twinning. Little theoretical work has been carried out to address this possibility. Another important potential approach to understanding the deformation mechanisms in nc materials is to explore the rich older literature on shear banding—mechanical instabilities that have been observed in, for example, mild steels as grain size decreases; the approach to very low strain hardening (perfectly plastic behavior) is observed. Amorphous materials exhibit many of the phenomenological characteristics of deformation in nc materials, that is, shear banding, asymmetry between tensile and compressive behavior, and perfectly plastic behavior. However, deformation mechanisms are not well understood in amorphous materials either. Recent work on bulk metallic glasses may help clarify this.

Applications

Of the present or near-term applications for nc materials, the hard material WC/Co is an example of several important trends. Nanostructured WC/Co composites have been prepared that can have the following characteristics:

- Nanoscale grain sizes after consolidation of powder.
- Hardness values about twice that of conventional micrograined WC/Co.
- Enhanced wear resistance and cutting performance.

While this material has increased hardness and strength, reports from Stevens Institute of Technology in Hoboken, NJ, and the Royal Institute of Technology in Stockholm, Sweden, point to similar or increased fracture toughness values for nanostructured WC/Co. As noted earlier, single-phase nanostructured materials studied to date have exhibited high strength and hardness but brittle behavior at low homologous temperatures (< 0.5 TM). The results of these studies of WC/Co two-phase nanostructured materials suggest that the combination of high hardness/strength and toughness/ductility may be possible in multiphase nanostructured materials.

Other examples that point to this possibility come from the work of Professor A. Inoue at Tohoku University in Japan. Inoue and coworkers synthesized a variety of multiphase Al, Mg, and Ni-base alloys with nanoscale microstructures. Many of these alloys consist of nanocrystallites in an amorphous matrix. Some Al-rich alloys contain nanoscale quasi-crystalline particles surrounded by crystalline face-centered cubic Al. The fascinating properties of these multiphase nanostructured alloys include extremely high strength coupled with some ductility. Ductility is high in compression, but uniform elongation in tension is limited. Again, this behavior is analogous to that exhibited by ductile amorphous alloys. These results again suggest the possibility for development of nanostructured multiphase composites that combine extremely high hardness and strength with toughness and ductility. Such materials could have many applications as unique structural materials.

Multiphase ceramic nanocomposites are the focus of several efforts. One significant effort is underway at the Research Center for Intermaterials of the Institute of Scientific and Industrial Research at Osaka University in Japan. Professor Koichi Niihara has a large effort studying micro-nano composites such as Al_2O_3/SiC that have enhanced toughness and also a variety of hard matrix/soft dispersion or soft matrix/hard dispersion nanocomposites. The enhanced toughness in some ceramic nanocomposites observed by Prof. Niihara has been verified by parallel studies of Dr. Steve Roberts at University of Oxford in England.

While "low temperature" superplasticity has not been realized in nanostructured materials, enhanced superplastic behavior has been observed at elevated temperatures in terms of a somewhat lower temperature range for superplasticity and, perhaps more significant, a higher strain rate regime. While much needs to be understood about superplastic behavior of nanostructured materials, the possibility for using more conventional strain rates for forming has major industrial implications. This is of particular interest for near net shape forming of ceramics.

Outstanding Issues: Opportunities and Challenges

Perhaps the overriding issue for defining the mechanical behavior of nanostructured materials is the lack of understanding of the mechanisms for plastic deformation and fracture in these materials. An extensive experimental and theoretical research effort should be applied to solve these questions. Only with such understanding can the intelligent design of nanostructured materials with optimum mechanical properties be realized.

Since many of the synthesis routes for nanostructured materials involve using powder or particulate products, compaction of such powders is needed to form bulk parts without coarsening the nanoscale microstructure. While there has been ongoing work on compaction and the thermal stability of nanocrystalline microstructures, more basic work is needed in this critical area. If some unique properties are limited to the finest grain sizes, methods must be found to stabilize the grain size while attaining theoretical density and complete particulate bonding. Furthermore, conventional processing methods are desirable if realistic scaleup and economy are to be realized.

Near-term opportunities for structural applications of nanostructured materials are in the form of coatings prepared by, for example, thermal spray deposition or electrodeposition. There is a significant effort in the United States on thermal spray deposition of nc materials. Several groups in Canada lead in deposition of nc materials by electrodeposition methods. A spin-off company, Nanometals, formed under the auspices of Parteq Research and Development Innovations, has adopted a commercial electrodeposition method to produce nanostructured metals.

Structural applications of nanostructured materials may be viewed as a focused approach to a long standing, well known design of materials with optimum mechanical properties. Use of nanoscale precipitates or dispersoids has been known for decades to improve mechanical behavior. The "new" emphasis on using nanostructured materials comes from two significant factors:

1. The special processing methods to push microstructures to the limits of the nanoscale.
2. The unique properties and phenomena—as yet not well understood—when this limit is approached.

Ferromagnetic Nanostructured Materials

Soft Magnetic Nanocrystalline Alloys

The discovery of nanocrystalline Fe-based soft magnetic materials is less than ten years old. The first class of such materials was the melt-spun Fe-Si-B alloys containing small amounts of Nb and Cu. The Fe-Si-B-Nb-Cu amorphous phase transforms to a body-centered cubic (bcc) Fe-Si solid solution with grain sizes of about 10 nm during annealing at temperatures above the crystallization temperature. The presence of small amounts of Cu helps increase the nucleation rate of the bcc phase while Nb retards the grain growth. These "Finemet" alloys provide low core losses (even lower than amorphous soft magnetic alloys such as Co-Fe-Si-B), exhibit saturation induction of about 1.2 T, and exhibit very good properties at high frequencies, comparable to the best Co-based amorphous alloys. These were first developed in Japan and have stimulated a large amount of research and development worldwide to optimize the magnetic properties. There has been relatively little work in the United States in this area, however.

While many of the soft magnetic properties of Finemet-type nanocrystalline alloys are superior, they exhibit lower saturation inductions than Fe-metalloid amorphous alloys, mainly because of the lower Fe content to attain amorphization and because of the addition of Nb and Cu (or other elements to control the nucleation and growth kinetics). In order to remedy this problem, another class of Fe-based nanocrystalline alloys was developed by Inoue and coworkers at Tohoku University, which is commercialized by Alps Electric Co., Ltd., of Nagaoka, Japan. These "Nanoperm" alloys are based on the Fe-Zr-B system; they contain larger concentrations of Fe (83-89 at per cent) compared to the Finemet alloys (~ 74 at. per cent Fe) and have higher values of saturation induction (~1.6-1.7 T). The Nanoperm nc alloys have very low energy losses at power frequencies (60 Hz), making them potentially interesting for electrical power distribution transformers. The issues of composition modification, processing, and the brittle mechanical behavior of these nanocrystalline/amorphous alloys are discussed by V.R. Ramanan. Figure 16.3 compares the soft magnetic properties of Finemet, Nanoperm, and other materials.

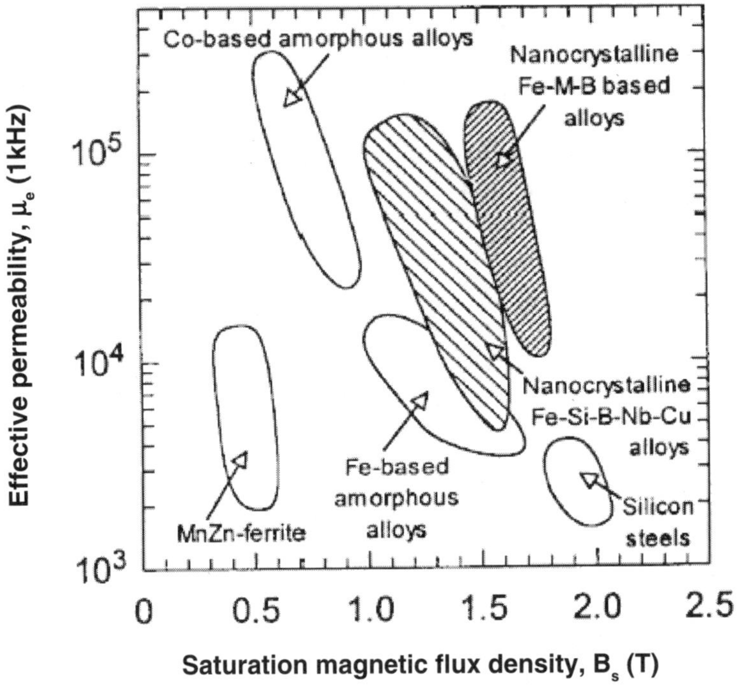

Fig. II–P.3: Effective permeability, μ_e, vs. saturation magnetic flux density, Bs, for soft ferromagnetic materials.

While there has been extensive research on these alloys, particularly in Japan and Europe, most of the development has been carried out in Japan. The Finemet family of alloys is marketed by Hitachi Special Metals. Vacuumschmelze GmbH (Germany) and Impky (France) also market similar alloys. The Nanoperm alloys are being commercialized by Alps Electric Co. (Japan). No extensive research nor any commercialization of these materials has been carried out in the United States.

The small single-domain nanocrystalline Fe particles in the amorphous matrix gives these alloys their unique magnetic behavior, the most dramatic being the lowest energy losses (narrowest B/H hysteresis loop) of any known materials, along with very high permeabilities. These alloys can also exhibit nearly or exactly zero magnetostriction. To date, these materials have been made by crystallization of rapidly solidified amorphous ribbons. Other methods that might provide geometrically desirable products should be explored or developed. Electrodeposition is one such method that requires further work. Electrodeposited nc Fe-Ni soft magnetic alloys are being developed in Canada.

The brittle nature of these materials is a problem for scaleup and transformer manufacture. The brittleness problem must be solved by finding less brittle materials or applying the handling and processing knowledge that exists for embrittled (after annealing) metallic glasses.

Permanent Magnet Materials

The first attempts to produce nanoscale microstructures to enhance the magnetic properties of the Nb-Fe-B permanent magnetic materials used mechanical alloying of blended elemental powders followed by heat treatment. Since the grain structure so obtained does not exhibit any crystallographic texture—and limits the energy product—special processing methods such as die-upsetting were used by Schultz and coworkers to provide the crystallographic anisotropy. While the coercivities of these nanocrystalline alloys are high, the remanent magnetization is decreased.

Recent approaches to increasing the magnetic induction have utilized exchange coupling in magnetically hard and soft phases. The Fe-rich compositions (e.g., $Fe_{90}Nd_7B_3$) result in a mixture of the hard $Fe_{14}Nd_2B$ phase and soft aFe phase. The nanoscale two-phase mixtures of a hard magnetic phase and a soft magnetic phase can exhibit values of remanent magnetization, M_r, significantly greater than the isotropic value of 0.5 Ms. This "remanence enhancement" is associated with exchange coupling between the hard and soft phases, which forces the magnetization vector of the soft phase to be rotated to that of the hard phase. Two important requirements for alloys to exhibit remanence enhancement are a nanocrystalline grain size and a degree of coherence across interphase boundaries sufficient to enable adjacent phases to be exchange-coupled. The significant feature of the exchange coupling is that it allows crystallographically isotropic materials to exhibit remanence values approaching those achieved after full alignment. Such two-phase nanoscale ferromagnetic alloys have been prepared by nonequilibrium methods such as melt-spinning, mechanical alloying, and sputter deposition. Besides the high reduced remanence, the material cost is reduced by reduction in the content of the expensive hard rare earth-containing magnetic phase.

The theoretical understanding of remanence enhancement appears to be developed to a degree enabling prediction of magnet performance; however, this performance, while a significant improvement over single-phase isotropic magnets, does not reach predicted values. Work is required on optimizing the orientation relationships between the hard and soft phases and the interphase properties (coherency) between them.

Research on nanocrystalline hard magnetic alloys has received attention worldwide. The U.S. efforts are summarized in the article by G.C. Hadjipanayis. While less research seems to be carried out in the world on these materials compared to the nanocrystalline soft magnetic alloys, some efforts exist in most countries. Notable programs are those of L. Schultz and coworkers at the Institut für Festköper und Werkstofforschung (IFW) in Dresden and P.G. McCormick and coworkers at the University of Western Australia.

While the very low losses of the nc soft magnetic materials (Finemet or Nanoperm) are dependent on grain size for their properties, the hard magnetic nc alloys with remanence enhancement provide flexibility in processing, especially with powder materials. These remanence-enhanced nc hard magnetic alloys may find many applications as permanent magnet components.

Giant Magnetoresistance (GMR)

The phenomenon of giant magnetoresistance (GMR)—the decrease of electrical resistance of materials when exposed to a magnetic field—was first reported in a number of multilayer ferromagnetic/nonferromagnetic thin film systems. More recently, GMR was observed in equiaxed granular nanocrystalline materials. In particular, GMR systems with low saturation fields offer a wide area for application in magnetoresistive devices. GMR sensors have a higher output than conventional anisotropic magnetoresistive sensors or Hall effect sensors. They can operate at higher magnetic fields than conventional magnetoresistive sensors. In multilayer systems the antiferromagnetic alignment of the ferromagnetic layers in zero field becomes ferromagnetic as the field is applied and causes a decrease in resistance. Granular materials that show GMR consist of small ferromagnetic single-domain particles with randomly oriented magnetic axes in a nonmagnetic matrix. An external field rotates the magnetic axes of all magnetic particles. The rotation towards complete alignment of all magnetic axes again reduces the resistance in a similar way as for multilayers. The GMR in granular systems is isotropic. The explanation for the GMR is spin-dependent scattering of the conduction electrons at the ferromagnetic/nonmagnetic interfaces and, to a lesser extent, within the magnetic grains. The GMR scales inversely with the average particle diameter.

There is worldwide research on the GMR effect. The NIST work described by Shull has provided material with the largest GMR values for the smallest switching fields. Japanese research on GMR includes studies in Prof. Fujimori's group at Tohoku University.

While the theory for GMR of spin-dependent scattering referred to above has been used as an explanation, other explanations taking into account interaction between magnetic regions have been proposed. Combined theoretical and experimental studies should help to clarify the mechanism for this effect.

Other Ferromagnetic Nanocrystalline Materials

Magnetic nanocomposite refrigerants, which have four times the magnetocaloric effects of the best low temperature magnetic refrigerant, were developed by NIST and described by R. Shull. The entropy change at a given (low) temperature for a system of magnetic spins is enhanced when the isolated spins are clustered. Shull et al. have shown that the nanocomposite $Gd_3Ga_5{-}xFexO_{12}$ gives superior magnetocaloric effects, which increase with x up to x = 2.5 and can be extended to higher temperatures than conventional materials.

Magnetostrictive materials such as Terfonol-D ($Tb_{0.3}Dy_{0.7}Fe_2$) have been of scientific and technological interest in recent years. It is suggested by G.C. Hadjipanayis that nanostructured magnetostrictive materials can have improved properties, such as lower saturation fields, with reduced anisotropy and in multilayers with alternate layers of magnetostrictive and soft magnetic materials that are exchange-coupled. Hadjipanayis states that most of the research in this area is carried out in Japan and Europe.

Opportunities and Challenges

Nanocrystalline magnetic materials offer perhaps the nearest-term prospect of significant applications of bulk nanostructured materials. The remanence enhancement in two-phase hard or soft magnetic materials can result in excellent energy products—comparable to those from rapid solidification processing routes—in powder composites. The flexibility in manufacturing this allows should provide many possible permanent magnet applications.

Nanocrystalline soft magnetic materials have the lowest energy losses of any material. While problems such as mechanical brittleness remain, these materials promise to replace existing transformer core materials for power applications. The U.S. research and industrial effort in these materials lags the efforts in Japan.

Other Bulk Applications of Nanostructured Materials

Nanocrystalline Hydrogen Storage Materials

R.B. Schwarz has pointed out that nanostructured materials offer several potential advantages for hydrogen storage materials. Rapid kinetics of absorption or desorption can be aided by refining the microstructure to the nanoscale. For example, nanoscale inclusions of Mg_2Ni in Mg catalyze the decomposition of the molecular hydrogen, increasing the hydrogen absorption or desorption kinetics. Another advantage of the nanoscale microstructure is that the alloy powder does not comminute on repeated charging or discharging with hydrogen. This is not strictly a bulk material, since powder agglomerates or green compacts can be used, thus obviating the need for compaction to theoretical density.

Nanocrystalline Corrosion-Resistant Materials

The limited work to date on corrosion resistance of nanocrystalline materials indicates that no generalizations can be made. Superior localized corrosion resistance in HCl was observed for nanocrystalline 304 stainless steel (Fe–18 per cent Cr–8 per cent Ni) prepared by sputter deposition. This was attributed to the fine grain size and homogeneity of the nc material. However, the average dissolution rate of nc Ni was found to be higher than that for coarse-grained material.

Unit II–Q

Biologically Related Aspects of Nanoparticles, Nanostructured Materials and Nanodevices

Introduction

Biological molecules and systems have a number of attributes that make them highly suitable for nanotechnology applications. For example, proteins fold into precisely defined three-dimensional shapes, and nucleic acids assemble according to well-understood rules. Antibodies are highly specific in recognizing and binding their ligands, and biological assemblies such as molecular motors can perform transport operations.

Although very promising, the bio-related aspects of nanoparticles, nanostructured materials, and nanodevices, are, for the most part, not as well developed as the nonbiological ones. However, a number of workshops, symposia, and books attest to the fact that many of the novel developments in this field are poised for rapid expansion.

This unit first puts into perspective current research directed toward biological synthesis and assembly as it pertains to the building blocks of nanotechnology. It then focuses on the current state of the art in biological aspects of dispersions and coatings, high surface area materials, and functional nanostructures. Finally, it ends with a section on the challenges and opportunities for the biological aspects of nanotechnology.

Nanotechnology Building Blocks

Synthesis

Although it seems at first that Nature has provided a limited number of basic building blocks—amino acids, lipids, and nucleic acids—the chemical diversity of these molecules and the different ways they can be polymerized or assembled provide an enormous range of possible structures. Furthermore, advances in chemical synthesis and biotechnology enable one to combine these building blocks, almost at will, to produce new materials and structures that have not yet been made in Nature. These self-assembled materials often have enhanced properties as well as unique applications.

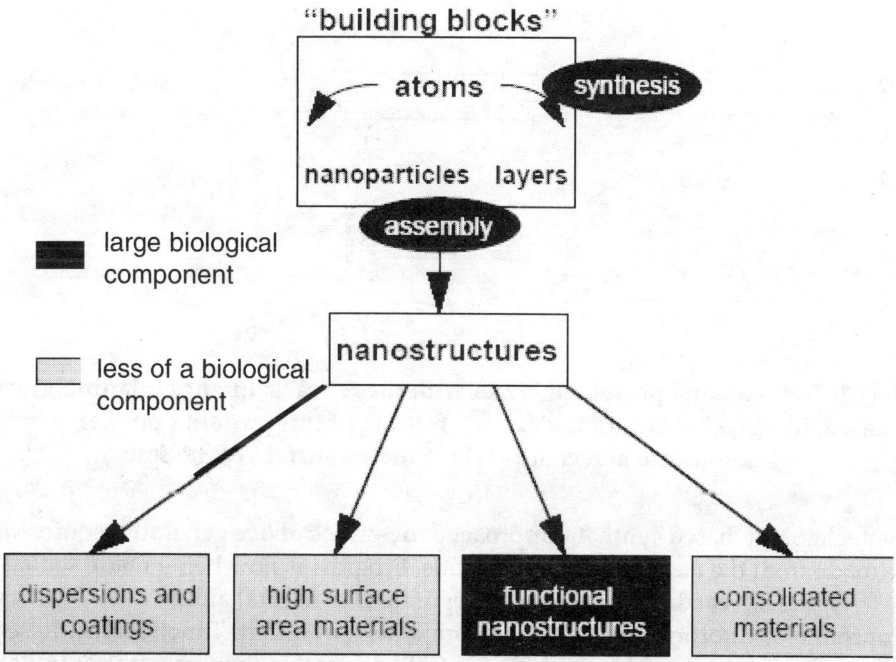

Fig. II–Q.1: Sections with large biological content are indicated.

The selected examples below show ways in which clever synthetic methodologies are being harnessed to provide novel biological building blocks for nanotechnology.

The protein polymers produced by Tirrell and coworkers are examples of this new methodology. In one set of experiments, proteins were designed from first principles to have folds in specific locations and surfacereactive groups in other places (Figure II–Q.2). One of the target sequences was—$((AG)_3EG)_{36}$. The hypothesis was that the AG regions would form hydrogen-bonded networks of beta sheets and that the glutamic acid would provide a functional group for surface modification. Synthetic DNAs coding for these proteins were produced, inserted into an E. coli expression system, and the desired proteins were produced and harvested. These biopolymers formed chain-folded lamellar crystals with the anticipated folds. In addition to serving as a source of totally new materials, this type of research also enables us to test our understanding of amino acid interactions and our ability to predict chain folding.

Biopolymers produced via biotechnology are monodisperse; that is, they have precisely defined and controlled chain lengths; on the other hand, it is virtually impossible to produce a mono disperse synthetic polymer. It has recently been shown that polymers with well-defined chain lengths can have unusual liquid crystalline properties. For example, Yu et al. have shown that bacterial methods for polymer synthesis can be used to produce poly (gamma-benzyl alpha L-glutamate) that exhibits smectic ordering in solution and in films. The distribution in chain length normally found for synthetic polymers makes it unusual to find them in smectic phases. This work is important in that it suggests that we now have a route to new smectic phases whose layer spacings can be controlled on the scale of tens of nanometers.

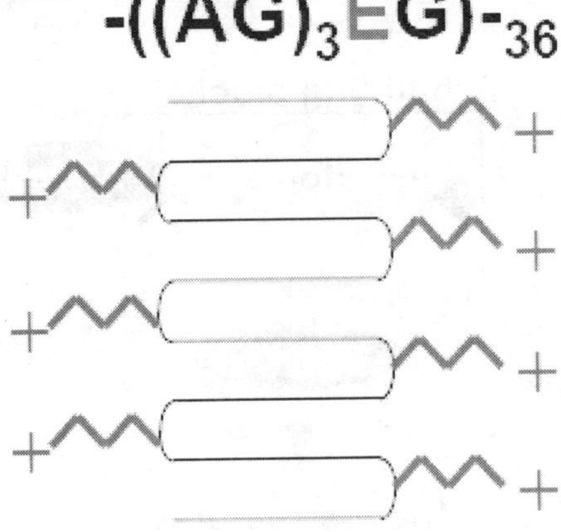

Fig. II–Q.2: Top: a 36-mer protein polymer with the repeat sequence (ulanine-glycine)$_3$ – glutamic acid – glycine. Bottom: idealized folding of this protein polymer, where the glutamic acid sidechains (+) are on the surface of the folds.

The biotechnology-based synthetic approaches described above generally require that the final product be made from the natural or L-amino acids. Progress is now being made so that biological machinery (e.g., *E. coli*), can be co-opted to incorporate non-natural amino acids such as b-alanine or dehydroproline or fluorotyrosine, or ones with alkene or alkyne functionality. Research along these lines opens new avenues for producing controlled-length polymers with controllable surface properties, as well as biosynthetic polymers that demonstrate electrical phenomena like conductivity. Such molecules could be used in nanotechnology applications.

Novel chemical synthesis methods are also being developed to produce "chimeric" molecules that contain organic turn units and hydrogen-bonding networks of amino acids. Another approach includes incorporating all tools of chemistry into the synthesis of proteins, making it possible to produce, for example, mirror-image proteins. These proteins, by virtue of their D-amino acid composition, resist biodegradation and could have important pharmaceutical applications.

Arnold and coworkers are using a totally different approach to produce proteins with enhanced properties such as catalytic activity or binding affinity. Called "directed evolution," this method uses random mutagenesis and multiple generations to produce new proteins with enhanced properties. Directed evolution, which involves DNA shuffling, has been used to obtain esterases with five- to six-fold enhanced activity against *p*-nitrobenzyl esters.

Assembly

The ability of biological molecules to undergo highly controlled and hierarchical assembly makes them ideal for applications in nanotechnology. The self-assembly hierarchy of biological materials begins with monomer molecules (e.g., nucleotides and nucleosides, amino acids, lipids), which form polymers (e.g., DNA, RNA, proteins, polysaccharides), then assemblies (e.g., membranes, organelles), and finally cells, organs, organisms, and even populations. Consequently, biological materials assembly on a very broad range of organizational length scales, and in both hierarchical and nested manners. Research frontiers that exploit the capacity of biomolecules and cellular systems to undergo self-assembly have been identified in two late 20th century National Research Council reports. Examples of self-assembled systems include monolayers and multilayers, biocompatible layers, decorated membranes, organized structures such as microtubules and biomineralization, and the intracellular assembly of CdSe semiconductors and chains of magnetite.

A number of researchers have been exploiting the predictable basepairing of DNA to build molecular-sized, complex, three-dimensional objects. For example, Seeman and coworkers have been investigating these properties of DNA molecules with the goal of forming complex 2-D and 3-D periodic structures with defined topologies. DNA is ideal for building molecular nanotechnology objects, as it offers synthetic control, predictability of interactions, and well-controlled "sticky ends" that assemble in highly specific fashion. Furthermore, the existence of stable branched DNA molecules permits complex and interlocking shapes to be formed. Using such technology, a number of topologies have been prepared, including cubes, truncated octahedra (Figure II–Q.3) and Borromean rings.

Fig. II–Q.3: Idealized truncated octahedron assembled from DNA. This view is down the four-fold axis of the squares. Each edge of the octahedron contains two double-helical turns of DNA.

Other researchers are using the capacity of DNA to self-organize to develop photonic array devices and other molecular photonic components. This approach uses DNA-derived structures and a microelectronic template device that produces controlled electric fields. The electric fields regulate transport, hybridization, and denaturation of oligonucleotides. Because these electric fields direct the assembly and transport of the devices on the template surface, this method offers a versatile way to control assembly.

There is a large body of literature on the self-assembly on monolayers of lipid and lipid-like molecules. Devices using self-assembled monolayers are now available for analyzing the binding of biological molecules, as well as for spatially tailoring the surface activity. The technology to make self-assembled monolayers (SAMs) is now so well developed that it should be possible to use them for complex electronic structures and molecular-scale devices.

Research stemming from the study of SAMs (e.g., alkylthiols and other biomembrane mimics on gold) led to the discovery of "stamping" (Figure II–Q.4). This method, in which an elastomeric stamp is used for rapid pattern transfer, has now been driven to < 50 nanometer scales and extended to nonflat surfaces. It is also called "soft lithography" and offers exciting possibilities for producing devices with unusual shapes or geometries.

Self-assembled organic materials such as proteins and/or lipids can be used to form the scaffolding for the deposition of inorganic material to form ceramics such as hydroxyapatite, calcium carbonate, silicon dioxide, and iron oxide. Although the formation of special ceramics is bio-inspired, the organic material need not be of biological origin. An example is production of template-assisted nanostructured ceramic thin films.

A particularly interesting example of bio-inspired self-assembly has been described in a recent article by Stupp and coworkers. This work, in which organic "rod-coil" molecules were induced to selfassemble, is significant in that the molecules orient themselves and selfassemble over a wide range of length scales, including mushroom-shaped clusters (Figure II–Q.5); sheets of the clusters packed side-by-side; and thick films, where the sheets pack in a head-to-tail fashion. The interplay between hydrophobic and hydrophilic forces is thought to be partially responsible for the controlled assembly.

Dispersions and Coatings

Drug Delivery Systems

Nanoparticles and nanospheres have considerable utility as controlled drug delivery systems. When suitably encapsulated, a pharmaceutical can be delivered to the appropriate site, its concentration can be maintained at proper levels for long periods of time, and it can be prevented from undergoing premature degradation. Nanoparticles (as opposed to micron-sized particles) have the advantage that they are small enough that they can be injected into the circulatory system.

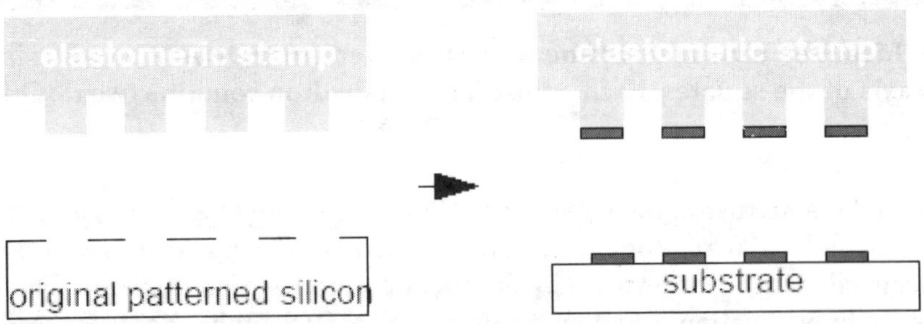

Fig. II–Q.4: An elastomeric stamp (top left) is made from an original master (bottom left). The stamp is dipped into the biological material (top right) and the pattern is transferred to the substrate (bottom right).

Fig. II–Q.5: Mushroom-shaped clusters formed from self-assembly of rod-coil molecules; these clusters can undergo further packing to form sheets.

Highly porous materials are also ideal candidates for controlled drug delivery and for tissue engineering. An example of a controlled drug delivery system comes from the area of microtubules. Phospholipid bilayers can self-assemble into long cylindrical tubes with diameters usually below a micron and lengths up to hundreds of microns. During synthesis, drugs can be entrained in these nanotubes, and the final product can be used as a controlled delivery system. Tubules prepared from phospholipid bilayers are ideal for such applications because they are biocompatible.

Dendrimers can be prepared so they are of discrete size and contain specific functional groups. They can be functionalized and used in biomedicine. Examples include gene transfer agents for gene therapy, made to carry and control the relaxivity of paramagnetic MRI (magnetic resonance imaging) contrast agents, and to deliver drugs on a controlled release basis.

Role of Nanoparticles in Health and Pollution

Although beyond the scope of this review, it is important to keep in mind the potential role of atmospheric nanoparticles in photocatalytic and thermal production of atmospheric pollutants. Atmospheric aerosols in heavily polluted areas have the potential to accelerate ozone formation reactions. Furthermore, because they are respirable, they could represent a health hazard. Two controversial studies (the Harvard University six-city study and the American Cancer Society study) have linked the presence of fine particular matter to premature mortality.

Atmospheric aerosols generally contain two major components: one is composed of amorphous carbon that has fullerene-like particles dispersed in it; the second is inorganic and consists of oxides and sulfides supported on clay minerals. In particular, the iron oxide, manganese oxide, and iron sulfide nanoparticles have band-gaps that could enhance the photocatalytic adsorption of solar radiation. In addition, these materials are acidic and may be coated with water, which would enhance their catalytic ability to crack hydrocarbons and create free radicals. At present this is an underexplored area of research that bears scrutiny.

It is interesting to note that some microorganisms produce and sequester CdSe and CdS nanoparticles in response to high levels of toxic Cd++ in their environment (Brus 1996). A large number of organisms also have the ability to precipitate ferrimagnetic minerals Fe_3O_4 and Fe_3S_4.

High Surface Area Materials

Membranes for Biological Separations

Supported polymeric membranes can be used to remove low molecular weight organics from aqueous solution. They work by filling the pores of microfiltration or ultrafiltration membranes with functional, polymeric, or oligomeric liquids that have an affinity for the compound of interest. These ideas can be extended to microemulsions and used for separations of large biological molecules. For example, non-ionic microemulsions (i.e., oil-water systems) have been used for hemoglobin extraction. These results augur well for the separation of other biological materials. Molecular imprinting can also be used to produce high surface area adsorbents that are enantioselective for amino acids and other biological molecules.

Bacterial Cell Surface Layers as Patterning Elements

Crystalline bacterial cell surface layers (S-layers) are composed of repeating protein units. These layers self-assemble and have a high binding capacity. They have been explored as patterning elements for molecular nanotechnology. For example, they have been used to pattern cadmium sulfide superlattices.

Functional Nanostructures

Molecular Computation

The smallest possible computer would ideally be able to perform computations on a molecular scale. Even though the computation may be carried out on that scale, the issue becomes one of having enough molecules to obtain sufficient signal-to-noise ratio to read out the answer. Consequently, at least at present, these computations must be carried out in bulk or with extremes in temperature. There has been a recent development along this line. Adleman has shown how the rules of DNA selfassembly, coupled with polymerase chain reaction (PCR) amplification of DNA, can lead to a molecular computer of sorts. The system was used to solve the Hamiltonian path problem, a classic and difficult (NP complete) mathematical problem that involves finding a path between vertices of a graph. The starting and ending vertices of each edge of the graph were encoded as the first and second halves of a strand of DNA. A solution to the problem (what is the path between two specific vertices?) was obtained by using PCR primers for the two vertices. Others have extended these ideas and shown that it is possible to make DNA add. Although these are exciting demonstrations, at present this technology has a number of drawbacks, including the labor and time it takes to analyze the results of the computation, and the uncertainties associated with wet chemistry.

Optoelectronic Devices

Biological molecules and assemblies, such as the photochemical reaction center, are capable of capturing light with good quantum efficiency and transforming it into chemical energy. If properly exploited, such assemblies have potential applications as biomolecule information processing units. Bacteriorhodopsin, from the purple membrane bacterium *Halobacterium halobium*, is one such system that has been studied extensively and has been commercialized into optical holographic memories. In the bacterium, the protein bacteriorhodopsin self-assembles into ordered lipid patches. The protein absorbs light and undergoes a cycle involving a complex series of intermediates, resulting in a proton being pumped across the membrane. It was information developed from understanding the basic science behind the way that bacteriorhodopsin works that led to the use of bacteriorhodopsin as a biomolecule information processing unit. To be used for information storage, the protein is placed under nonbiological conditions. It is dispersed and immobilized into a matrix (e.g., collagen or another polymeric substance) and held at liquid nitrogen temperatures. At this low temperature, the protein acts as an optically-driven bistable switch. One form, the light-adapted form of bacteriorhodopsin, absorbs light at 570 nm. When irradiated with green light at 77 K, it switches to a different stable form that absorbs at 630 nm. Using light of different colors enables one to read and write images onto these memories. By subtracting one memory from the other, these memories are especially useful for realtime, rapid comparison of images.

Molecular Motors

The molecular motors found in biology provide for bacterial locomotion, as well as for the active transport and delivery of molecules. For example, the bacterial flagellar motor is about 20 nm in diameter, and is comprised of a complex assembly of more than 10 different proteins. The role of the motor is to rotate the helical flagella of the bacterium so that it is able to swim. Chemical energy (in this case protons or sodium) is transduced into mechanical energy.

Other examples of molecular motors include RNA polymerase, F1-ATPase, myosin, and kinesin. The fuel that powers these motors is ATP (adenosine triphosphate). A number of researchers have proposed schemes by which such motors could be used to deliver molecules, one at a time, for the purpose of the ground-up assembly of nanoscale devices. It is envisioned that the highways could be actin or tubulin, which would need to be immobilized onto a surface. Myosin or kinesin, which naturally travel along these highways, could be used to deliver molecular "packages" to a specific assembly site.

Other Forms of Biological Transport Using Nanoparticles

In the human body, the function of the high density lipoproteins is to transport cholesterol. The ~7.5 nm discoidal lipoprotein assemblies are sandwiched between discs of phospholipids and stacked, poker-chip style. The lipoproteins stabilize the cholesterol particles and assist in their transport. Current research involves manipulating and fusing the assemblies and particles with an atomic force microscope (ATM) tip.

Bioelectronics

There has been considerable research activity in molecular electronics and bioelectronics, particularly in Japan. Although this area of nanotechnology is still not as well developed as others in this report, it bears watching. For example, Shionoya and coworkers at the Institute for Molecular Science of the Okazaki National Research Institutes in Japan have proposed that novel combinations of DNA, metal ligands, DNA templating, and proteins could produce molecular wires; molecular hoops through which DNA could be threaded; and double stranded peptides whose helix pitch could be controlled by an entrained metal (Figure II–Q.6). The active site containing the metal could be induced to go from Cu(I)tetrahedral to CU(II)square planar, perhaps by electrons delivered by a scanning tunneling microscope (STM) tip.

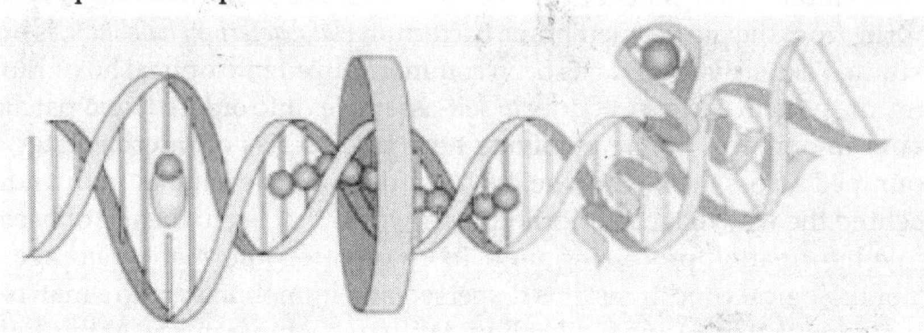

Fig. II–Q.6: Novel combinations of DNA, metal ligands, DNA templating, and proteins are being investigated for molecular wires, inductors, and switches.

Consolidated Materials

Ceramics

Biomolecule/inorganic interactions can be used to produce ceramics with increased toughness. Fundamental studies of biomineralization, in which an organic substance (usually protein or peptide or lipid) interacts with an inorganic phase (e.g., calcium carbonate or hydroxyapatite) have led to the bioinspired synthesis of composite materials.

The structure and porosity of the inorganic phase can be controlled by templating with an organic surfactant, vesicular arrays, or liquid crystalline materials. Micelle-templated synthesis can produce ceramics with 20-100 Å pore dimensions. These tailored pores can be used as catalysts and absorbents, and for gas/liquid separations and thermal and acoustic insulation. Their selectivity makes them very useful for biochemical and pharmaceutical separations. Bioceramics can also be made that are more compatible with teeth and bone.

An interesting example of an organic/inorganic composite is the new packaging material that has been developed to replace the polystyrene "clam-shell" for fast food products. Composed of potato starch and calcium carbonate, this foam combines the advantages of good thermal insulation properties and light weight with biodegradability.

Biological Bar Magnets

It has been well documented that a very large number of organisms have the ability to precipitate ferrimagnetic minerals such as Fe_3O_4 and Fe_3S_4. In addition, linear chains of membrane-bound crystals of magnetite, called magnetosomes, have been found in microorganisms and fish. For example, the Fe_3O_4 domain size in the organism *A. magnetotactum* is about 500 Å, and a chain of 22 of those domains has a magnetic moment of 1.3×10^{-15}. It is not immediately clear how these particles could be exploited for nonbiological applications.

Opportunities and Challenges

This section of the review describes current limits on the biological aspects of nanotechnology. It outlines selected broad areas of research opportunity, and it sets forth challenges for the intermediate and long-term future.

Enabling Technologies

We owe much of the recent progress in the biological aspects of nanotechnology and nanoparticles to important enabling technologies. Those that have had a significant impact on biological measurements include techniques and instrumentation such as optical traps, laser tweezers, and "nano-pokers" to measure femtoNewton forces.

There is also a need to move detection away from the ensemble, to the single molecule scale. This includes driving instrumentation toward singlemolecule spectroscopy; single-spin detection; chemical analysis of nanoliter volumes; nuclear magnetic resonance microcoils; nanoscale electrode arrays and chemical sensing and detecting technology; separations technology; chemical analysis of single cells; new biological transformations; and chemical probes of nanostructures.

Enhanced computational infrastructure, both hardware and algorithm development, will also be required to investigate supra molecular assemblies and to understand interactions that occur over a wide variety of time and length scales. At present, we are fairly well equipped to handle quantum mechanical calculations. At the next level of computational complexity, the molecular dynamics force fields and atomic charges can presently handle up to about a million particles and still retain their accuracy. The next scale of complexity, mesoscale modeling, currently requires the use of pseudo atoms. Much more research is required to improve the accuracy of finite elementcalculations and the modeling of materials applications.

Surface Interactions and the Interface between Biomolecules and Substrates

Much of the progress in the biological aspects of nanotechnology has come from research on surface interactions. It will be important to continue to develop a better understanding of the interface between biomolecules and surfaces. Of particular interest are questions about whether surfaces cause biomolecules to denature, the optimum length of linker groups, and ways to communicate from the biomolecule, through the linker, to the substrate. Another challenge is to produce nanometer ultrathin films that have stable order.

Robustness of Biomolecules and their Interactions in Aqueous Solution

Although much progress has been made on ways to confer additional robustness on biomolecules, it will be necessary to continue to improve the stability and reliability of biomolecular assemblies. Such research may be along the lines of understanding the thermal stability of biomolecules, perhaps by examining extremophiles and molecules produced by directed evolution. It will also be necessary to produce molecules that can work in the absence of water, or to devise ways in which aqueous solutions can be used reliably.

Assemblers or Templating

The hierarchical self-assembly of biomolecules is often touted as one of their key attributes for nanotechnology applications. However, there remains much to be learned about how to make periodic arrays of biomolecular assemblies, how to use biological templating in an efficient manner, how to mimic biological self-assembly with nonbiological molecules, and how to exploit differences between biological and nonbiological self-assembly.

Combinatorial Approaches to Nanotechnology

Combinatorial chemistry has had an enormous impact on drug discovery and development in the past several years. Ability to develop molecularly sensitive screening techniques is one of the limiting factors of combinatorial chemistry. It would be intriguing to discover screening methods that are particularly sensitive for picking up new phenomena exhibited by biological nanoparticles.

Biomimetics and Polymer-Biopolymer Chimeras

The largest impact of biological sciences on nanotechnology may prove to be through the lessons one can learn from biology, rather than from directly employing the biological molecules themselves. To this end, it is important to support research on various aspects of biomimetics, including activities in which synthetic and biologically-based molecules are combined in chimeric fashion.

Technological Comparisons

Table II–Q.1 makes subjective comparisons, weighted by level of effort, impact of research, etc., between the state of biology-related nanotechnology work in Japan, Europe, and the United States.

Table II–Q.1: Comparisons Between Japan, Europe, and the United States in Biologically Related Aspects of Nanotechnology

Level	*1 Highets*	*2*	*3*	*4*
Synthesis	US/Eur/Japan			
Assembly	US/Eur		Japan	
Dispersions and Coatings		US	Eur	Japan
High Surface Area materials		Eur/Japan	US	
Functional Nanostructures		US/Japan		Eur
Bioelectronics		Japan		US/Eur

SECTION C
NANOCOMPOSITES

Unit III–A

Nanocomposites

The definition of nanocomposite material has broadened significantly to encompass a large variety of systems such as one-dimensional, two-dimensional, three-dimensional and amorphous materials, made of distinctly dissimilar components and mixed at the nanometer scale.

The general class of nanocomposite organic or inorganic materials is a fast growing area of research. Significant effort is focused on the ability to obtain control of the nanoscale structures via innovative synthetic approaches. The properties of nanocomposite materials depend not only on the properties of their individual parents but also on their morphology and interfacial characteristics.

This rapidly expanding field is generating many exciting new materials with novel properties. The latter can derive by combining properties from the parent constituents into a single material. There is also the possibility of new properties which are unknown in the parent constituent materials.

The inorganic components can be three-dimensional framework systems such as zeolites, two-dimensional layered materials such as clays, metal oxides, metal phosphates, chalcogenides, and even one-dimensional and zero-dimensional materials such as $(Mo_3Se_3\text{-})_n$ chains and clusters. Experimental work has generally shown that virtually all types and classes of nanocomposite materials lead to new and improved properties when compared to their macrocomposite counterparts. Therefore, nanocomposites promise new applications in many fields such as mechanically reinforced lightweight components, non-linear optics, battery cathodes and ionics, nanowires, sensors and other systems.

The general class of organic/inorganic nanocomposites may also be of relevance to issues of bio-ceramics and biomineralization in which *in-situ* growth and polymerization of biopolymer and inorganic matrix is occurring. Finally, lamellar nanocomposites represent an extreme case of a composite in which interface interactions between the two phases are maximized. Since the remarkable properties of conventional composites are mainly due to interface interactions, the materials dealt with here could provide good model systems in which such interactions can be studied in detail using conventional bulk sample (as opposed to surface) techniques. By judiciously engineering the polymer-host interactions, nanocomposites may be produced with a broad range of properties.

Inorganic layered materials exist in great variety. They possess well defined, ordered intralamellar space potentially accessible by foreign species. This ability enables them to act as matrices or hosts for polymers, yielding interesting hybrid nano-composite materials.

Lamellar nanocomposites can be divided into two distinct classes, intercalated and exfoliated. In the former, the polymer chains alternate with the inorganic layers in a fixed compositional ratio and have a well defined number of polymer layers in the intralamellar space. In exfoliated nano-composites the number of polymer chains between the layers is almost continuously variable and the layers stand >100 Å apart. The intercalated nanocomposites are also more compound-like because of the fixed polymer/layer ratio, and they are interesting for their electronic and charge transport properties. On the other hand, exfoliated nanocomposites are more interesting for their superior mechanical properties.

A subclass of nanocomposites which is the lamellar class of intercalated organic/inorganic nanocomposites and namely those systems that exhibit electronic properties in at least one of the components offers the possibility of obtaining well ordered systems some of which may lead to unusual electrical and mechanical properties. Selected members of this class may be amenable to direct structural characterization by standard crystallographic methods. An important issue in this area is the few structural details that are available, therefore, any system which can be subjected to such analysis is of interest. Nanocomposites also offer the possibility to combine diverse properties which are impossible within a single material, e.g. flexible mechanical properties and superconducting properties. This work is now in its infancy and extensive investigations are underway. Another exciting aspect is the possibility of creating heterostructures composed of different kinds of inorganic layers, which could lead to new multifuntional materials.

Several general synthetic routes have been developed for inserting polymer chains into host structures and many novel nanocomposites have been designed.

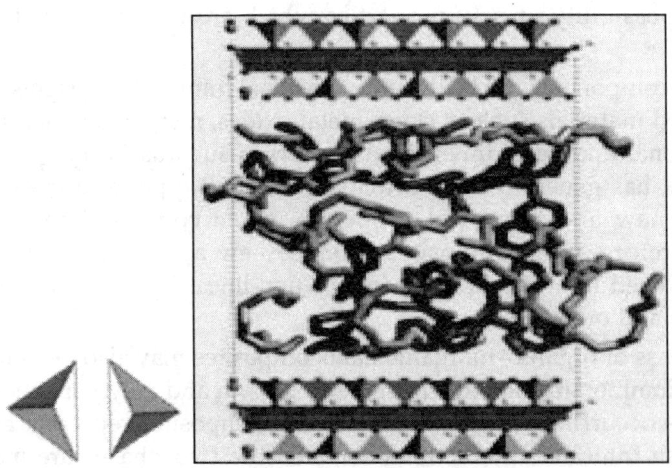

Fig. III–A.1: This computer- generated simulation shows the building blocks of a layered nanocomposite. A copolymer is intercalated between mica-like layers of crystalline silicate. The layers of such a structure are only a nanometer—about three atoms thick.

These include:

(a) *In-situ* intercalative polymerization (ISIP) of a monomer using the host itself as the oxidant. The rationale behind intercalative polymerization is that host matrices with high electron affinity can oxidatively polymerize appropriate monomers in their interior.

(b) Monomer intercalation followed by topotactic intralamellar solid state polymerization. This route creates conjugated polymers inside non-oxidizing hosts.

(c) Direct precipitative encapsulation of polymer chains by colloidally dispersed single layers of a host. This approach gives access to a large variety of nanocomposites with many kinds of polymers and hosts.

Materials with features on the scale of nanometers (10^{-9} meter) often have properties dramatically different from their bulk-scale counterparts. For example, nanocrystalline copper is five times harder than ordinary copper with its micrometer-sized crystalline structure. The development of such materials is currently a research area of great interest.

Important among these nanoscale materials are nanocomposites, in which the constituents are mixed on a nanometer-length scale. They often have properties that are superior to conventional microscale composites and can be synthesized using surprisingly simple and inexpensive techniques.

A nanoscale dispersion of sheet-like inorganic silicate particles in a polymer matrix, for example, is superior to either constituent in such properties as optical clarity, strength, stiffness, thermal stability, reduced permeability, and flame retardancy. The key to the synthesis of nanocomposites is how the silicates are made to disperse in the polymer. Since the organic and inorganic components are typically immiscible (like water and oil), the silicate surface is modified by attaching surfactant molecules (like adding soap to water).

Polymer-based nanocomposites are also being developed for electronics applications such as thin-film capacitors in integrated circuits and solid polymer electrolyes for batteries.

Nature also makes fabulous nanocomposites, and scientists are trying to emulate such processes. The abalone shell, for example, has alternating layers of calcium carbonate and a rubbery biopolymer; it is twice as hard as and a thousand times tougher than its components.

The study of nanocomposite materials requires a multidisciplinary approach, involving novel chemical techniques and an understanding of physics and surface science. It is a field of broad scientific interest with impressive technological promise.

Unit III–B

Nanomaterial Additives

Advantages of Nanosized Additions

Studies have revealed clearly the property advantages that nanomaterial additives can provide in comparison to both their conventional filler counterparts and base polymer. Properties which have been shown to undergo substantial improvements include:

- Mechanical properties e.g. strength, modulus and dimensional stability.
- Decreased permeability to gases, water and hydrocarbons.
- Thermal stability and heat distortion temperature.
- Flame retardancy and reduced smoke emissions.
- Chemical resistance.
- Surface appearance.
- Electrical conductivity.
- Optical clarity in comparison to conventionally filled polymers.

Disadvantages of Nanosized Additions

To date one of the few disadvantages associated with nanoparticle incorporation has concerned toughness and impact performance. Some of the data presented has suggested that nanoclay modification of polymers such as polyamides, could reduce impact performance. Clearly this is an issue which would require consideration for applications where impact loading events are likely. In addition, further research will be necessary to, for example, develop a better understanding of formulation/structure/property relationships, better routes to platelet exfoliation and dispersion etc.

Particle Loadings

In addition it is important to recognise that nanoparticulate/fibrous loading confers significant property improvements with very low loading levels, traditional microparticle additives requiring much higher loading levels to achieve similar performance. This in turn can result in significant weight reductions (of obvious importance for various military and aerospace applications) for similar performance, greater strength for similar structural dimensions and, for barrier applications, increased barrier performance for similar material thickness.

Examples of Mechanical Property Gains due to Nanoparticle Additions

Data provided by Hartmut Fischer of TNO in the Netherlands relating to polyamide – montmorillonite nanocomposites indicates tensile strength improvements of approximately 40 and 20 per cent at temperatures of 23ºC and 120ºC respectively and modulus improvements of 70 per cent and a very impressive 220 per cent at the same temperatures. In addition Heat Distortion Temperature was shown to increase from 65ºC for the unmodified polyamide to 152ºC for the nanoclay-modified material, all the above being achieved with just a 5 per cent loading of montmorillonite clay. Similar mechanical property improvements were presented for polymethyl methacrylate—clay hybrids.

Further data provided by Akkepeddi of Honeywell relating to polyamide-6 polymers confirms these property trends. In addition, the further benefits of short/long glass fibre incorporation, together with nanoclay incorporation, are clearly revealed.

Unit III–C

Nanocomposite Classification System

Nanocomposite materials with an inorganic glass and an organic polymer constitute a relatively new and unique area in material science. The term "ormocers", "ormosils" and "ceramers" are often utilized to describe this class of nanocomposite. By combining at the molecular level inorganic and organic polymeric material a blending of unique physical properties can be achieved. The value in these materials is apparent, from fiber optics to paints these materials may provide the requisite physical properties to achieve the next technological advance.

There are several different ways of synthesizing this class of nanocomposite; therefore a means of classification is necessary. Most developed nomenclature is based on synthetic techniques; Wilkes has a relatively recent and exhaustive categorization. However, we chose to classify these materials upon a simpler system first suggested by Novak. Five categories cover the majority of composites synthesized with more recent techniques being modifications or combinations from this list.

Type I: Organic polymer embedded in an inorganic matrix without covalent bonding between the components.

Type II: Organic polymer embedded in an inorganic matrix with sites of covalent bonding between the components

Type III: Co-formed interpenetrating networks of inorganic and organic polymers without covalent bonds between phases.

Type IV: Co-formed interpenetrating networks of inorganic and organic polymers with covalent bonds between phases.

Type V: Non-shrinking simultaneous polymerization of inorganic and organic polymers.

Before discussing different kinds of nanocomposites it is appropriate to discuss the sol-gel process. Although zirconium, titanium, aluminum and boron oxides have been utilized as the inorganic component, the great majority of nanocomposites incorporate silica from tetraethoxysilane (TEOS). The formation of the inorganic component involves two steps, hydrolysis and condensation as seen in Figure III–C.1. The important insight in the formation of this part of the composite is based upon the relative kinetic rates of each step. For example, if the rate of hydrolysis is fast compared

Condensation

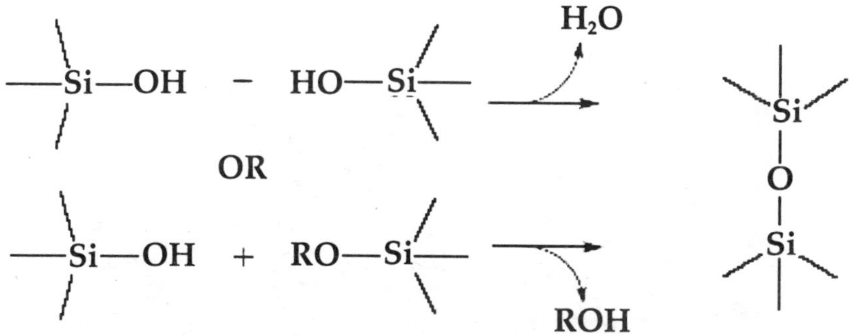

Overall

$$Si\,(OR)_4 - H_2O \xrightarrow[\text{Catalyst}]{\text{Acid, base or salt}} SiO_2 \text{ glass}$$

Fig. III–C.1: Generalized scheme for the hydrolysis and condensation of ceramic species.

to condensation, then simple particles or highly branched silicate matrices are formed. Conversely, if the condensation step is quicker than hydrolysis, then string-like filaments are formed. These changes in morphology of the silicate matrix can be manipulated by the choice of sol-gel catalyst with dramatic effect on the physical properties of the nanocomposite.

Hydrolysis

$$Si(OR)_4 + nH_2O(RO)_4 - nSi(OH)_n + nROH$$

Perhaps the most common and straightforward nanocomposites found in the literature are the Type I composites. Typically, TEOS is used to form an inorganic network around an organic polymer component. The goal in the process is to form a completely interpenetrating network (IPN) of both inorganic and organic phases. Homogeneous nanocomposites with good IPNs are often stronger, more resilient, and optically transparent, whereas heterogeneous composites are often mechanically weaker and opaque. Choice of co-solvents is critical throughout the formation of the inorganic component. As the hydrolysis and/or condensation reaction occurs, changes in the polarity of the solvent mixture can result in undesired phase separation of the inorganic and organic components. Phase separations can produce a range of results from small zones of one component dominance to total inorganic/organic separations. However, the result is that the same sites of instability are created. The timing and physical conditions in which phase separations occur can produce isolated silicate formations within composite formulations as seen in Figure III–C.2.

Type II composites utilize modified organic polymers with sites capable of covalently bonding directly to the forming inorganic phase of the nanocomposite. Most often an organosilane with a reactive pendant group is used to form the covalent bond to the organic polymer component. These pendant groups can be isocyanates, which easily react with polymers containing alcohols or amines.

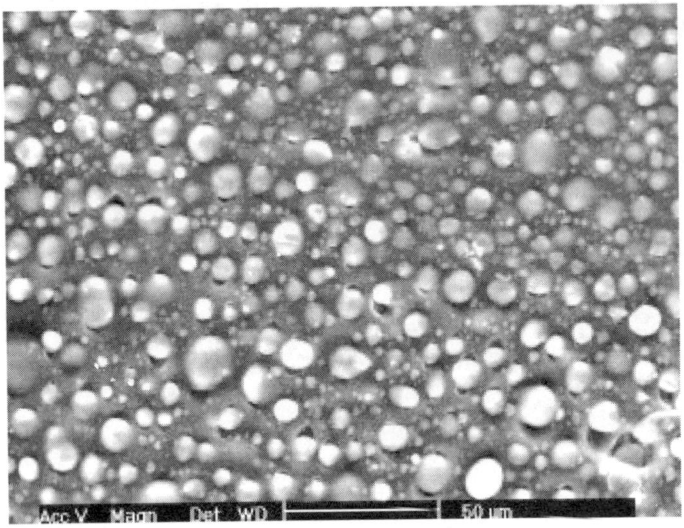

Fig. III–C.2: Formation of isolated ceramic forms in a phase separated composite.

Another option that has been explored is hydrosilation reactions with polymers containing a terminal alkene. Once bound, the silane's remaining pendant ethoxide or chloride groups will be available to bond with such sites on other silane species, forming the inorganic component.

Type V composites are unique examples of these materials because of the physical conditions in which they are formed. Novak and Grubbs developa method of simultaneously polymerizing both inorganic and organic components without shrinkage of the material. The insight was that by replacing the ethoxide found in TEOS with an organic monomer oxide, solvent loss created during the hydrolysis step is eliminated. It is the loss of ethanol (or other solvents) in the sol-gel process, which creates material shrinkage and leads to cracking from capillary pressures. Although extremely innovative, unfortunately a limited number of composites have been formed successfully with this process. This is presumably because of the difficulties in precisely matching the rates of polymerization of the organic component with both the rates of hydrolysis and condensation of the ceramic component to prevent disastrous phase separations.

Experimental

Syntheses

A general synthesis for a base, acid, or salt catalyzed polyphosphazene, polyethylene oxide (PEO), and polyethylene oxide/polypropylene oxide (PPO/PEO) block nanocomposite is as follows: 300 mg of polymer is dissolved into 10 mL of a 50/50 by volume tetrahydrofuran (THF)/ ethanol mixed solvent in a capped vial. To this solution is added TEOS (336 mg). A catalyst is then introduced as an aqueous solution (150 µl) and the mixture is capped and sonicated at 50°C for 30 minutes. The solution is aged from hours to days depending upon the catalyst used in a sealed vial and poured into a teflon mold and loosely covered at room temperature. The nanocomposite self assembles as the volatile solvent slowly escapes during the condensation process.

The synthesis of polyvinyl acetate (PVAc)/silicate nanocomposites requires a different approach from the other nanocomposites. PVAc (300 mg) is dissolved into an 50/50 by volume acetic acid/methanol (10 mL) mixed solvent in a capped vial. To this solution is added TEOS (373 mg). The solution is then sonicated for 5 minutes in a sealed vial at room temperature and poured into a teflon mould and loosely covered at room temperature. The nanocomposite self assembles during the curing process, which typically lasts up to 24 hours. Additional heating at 100°C for 30 minutes aids in removing lingering acetic acid from the nanocomposite.

Mechanical Analyses

Various membrane samples were tested on a Dynamic Mechanical Analyzer (TA Instruments, Model 2980, New Castle, DE) fitted with a tension (film) or penetration (compression) clamp. A simple linear ramp-force protocol was developed to evaluate the membrane strength in tension and compression. Both tension and compression protocols consist of the establishment of 0.03 Newton (N) static force and equilibration to 25°C. A force is then applied to the specimen at a specific rate, 0.5 N per minute for tension, and 1.0 N per minute for compression (18 N maximum) or until the maximum limit of travel is reached (24.7 mm). The membrane specimens are cut with a laboratory fixture consisting of two razor blades mounted on a parallel plate yielding a 6.20 mm by 39.0 mm sample. The specimen's thickness is measured with a caliper (Mitutoyo, Model ID-C112EBS, Japan) in several places along the length to obtain an average membrane thickness.

ESEM Measurements

Various thin films were imaged as an unstressed, swelled, or stretched nanocomposite with a Philips XL30ESEM. A 10 to 20 kV electron beam was utilized. Wet mode analyses employed a 500 μm pressure-limiting aperture (PLA). To image the topography either a gaseous secondary electron detector (wet mode) or secondary electron detector (HiVac) was used.

Polyphosphazene Nanocomposites

The polyphosphazene family of polymers was chosen for the initial investigations into nanocomposites due to their diverse array of physical and solubility properties. The researcher was specifically interested in some heretofore inaccessible phosphazenes that lacked the mechanical stability to be useful for membrane and barrier applications. The polymer poly[*bis*-(2-(2-methoxyethoxy)ethoxy)phosphazene] (MEEP) is such a polymer - one where the chemical properties, such as selective transport of ions are extremely interesting, yet the polymer alone lacks the requisite mechanical integrity to form a stable membrane for evaluation and subsequent use. This water-soluble polyphosphazene is a viscous liquid with two phosphorus bound polyether ligands on an alternating nitrogen phosphorus backbone. In all of the nanocomposites TEOS was employed as the ceramic precursor. The condensation of the ceramic component was catalyzed in a solvent composed of various ratios of water, ethanol and THF. Depending upon the nature of the catalyst employed, differing properties were observed in the resulting nanocomposites. These catalysts are best separated into three categories; acid, base or ionic salt; and each will be discussed separately below.

Acid catalyzed nanocomposites that were examined in this study were found to have the highest tensile strength of any of the catalyst types studied; yet they were also found to be glassy and brittle when compared to salt or base catalyzed analogues. While providing a maximum of physical stabilization for the liquid-like MEEP polymer, this translates directly into extremely poor durability

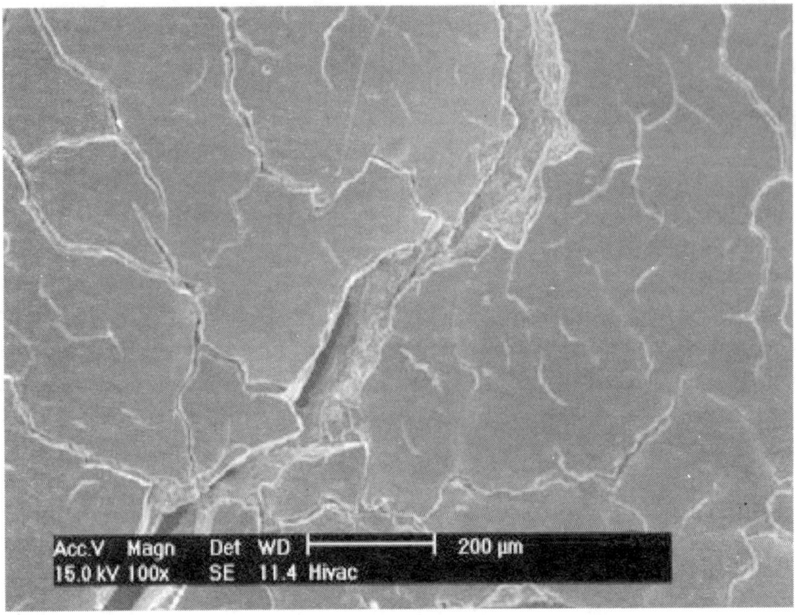

Fig. III–C.3: ESEM of an acid catalyzed nanocomposite after wetting.

when wetted, as exhibited in aqueous swelling tests. Figure III–C.3 is an ESEM image of an aqueous swelled/deswelled membrane, which has fractured from the internal hydrostatic pressure. The dominant reason for this lack of durability when wet is the nature of the ceramic morphology of these acid catalyzed nanocomposites. It is known from the ceramic literature that the general morphology of acid catalyzed sol-gel condensations forms long filament structures instead of star-like or particulate structures. This more rigid morphology imparts a high tensile strength to these nanocomposites, yet leaves them less elastic and highly susceptible to damage upon contact with liquids.

Base catalyzed nanocomposites have the best elastic and swelling characteristics of all the nanocomposites studied. As seen in the acid catalyzed nanocomposites, the morphology of the condensing ceramic component dominates the physical properties of the material. Figure III–C.4 is an ESEM image of a MEEP nanocomposite after being soaked to equilibrium in water. Before swelling, the nanocomposite has a featureless uniform appearance, as it does after drying. During wetting, the hydrophilic polymer component expands as it absorbs the water yet the ceramic is not capable of expansion. This behaviour produces the creases seen in the micrograph. The villae-like features reveal the underlying structure of the ceramic component of the nanocomposite, i.e. areas of higher silicate composition. This is consistent with the morphology expected for a base catalyzed nanocomposite—one that contains interlocking star-like structures. In regions between the ceramic star structures, the polymer is capable of absorbing water and expanding, while regions within the ceramic star structures maintain their structural rigidity. Because a single material contains both of these types of different regions, this renders base catalyzed nanocomposites simultaneously both tough and flexible.

Salt catalyzed nanocomposites exhibit properties that are between those found for acid and base catalyzed nanocomposites. Again, the nature of the ceramic dominates the physical properties of the nanocomposite. There is considerable variation in the properties of this class of nanocomposites

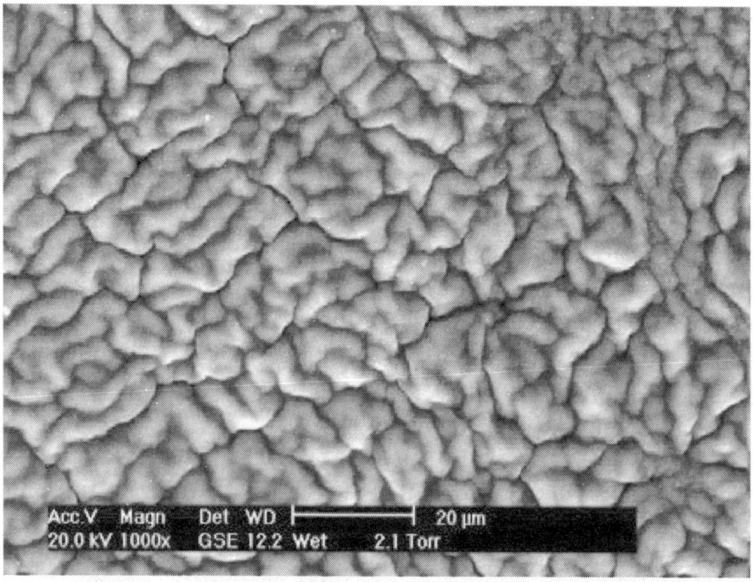

Fig. III–C.4: ESEM of a base catalyzed nanocomposite taken while still wet.

as there are a variety of simple salts that have proven to be effective catalysts. To investigate the nature of ionic salt catalysts in MEEP nanocomposites, a course of fluoride salts with various alkali counter-cations was examined, as it is known that fluoride anion salts are superior ceramic condensation catalysts over other halide salts. As was expected, the mechanical properties of fluoride catalyzed nanocomposites more closely resembled their base catalyzed analogues because the condensation was driven by an anionic catalyst, similar to OH^-.

The initial results indicated salt catalysts that contain large cations like cesium performed better than small ones such as lithium. Intuitively, hard ions such as lithium, the closest of the alkali metal ions to the proton (an extremely effective catalyst) should outperform the softer ionic catalysts, yet exactly the opposite trend is observed. The reason behind the observed trend becomes apparent when one considers that the strongly electronegative fluoride ion has a poorer affinity towards large polarizable cations (like cesium) than harder cations (like lithium). Catalyst availability in these polar, protic organic solvent systems becomes the dominant factor in the total catalytic effectiveness of these systems. There is a linear correlation between the lattice energy of the alkali fluoride salt employed and the final strength of the resulting nanocomposite, with the lowest energy salt yielding the strongest material. This is due to higher availability of the "naked" fluoride anion in systems possessing the lower lattice energies. In these systems, the catalyst ions spend less time during synthesis associated as intimate ion pairs in solution and are consequently more effective in driving the ceramic condensation.

Organic Polymer Nanocomposites

A better understanding of the formation, properties, and morphology of Type I nanocomposites was gained during studies with polyphosphazene polymers. This information was used to try to better understand the properties of nanocomposites employing more conventional organic polymers. Commercially available polymers were used in the preparation of these polymer/silicate materials.

A block polymer consisting of PEO/PPO (M_w = 2K), PEO (2K), polyvinylalcohol (PVA, 30–70K), PVAc (83K and 167K) and poly (acrylonitrile) (PAN, 150K) were the primary organic polymers of interest. Drawing upon previous experience, a 50/50 (w/v) polymer to silicate ratio was used to form these nanocomposites. Acid catalyzed composites were the focus of most of these studies.

Numerous attempts were made with low molecular weight PEO and PEO/PPO block polymers to prepare a nanocomposite with acceptable mechanical integrity. Unfortunately, all attempts resulted in materials that were too brittle to be lifted as a single unit suitable for mechanical analyses. This was surprising in light of the fact that acid catalyzed composites were previously found to yield the strongest nanocomposites. Explanation for this behaviour lies in the low molecular weight of the polymer component. Typical molecular weights for the MEEP polymer were in the 10^6-10^7 range, providing a directing effect for the condensing ceramic. The PEO polymer was sufficiently light that the ceramic exhibits a directing effect on the orientation of the polymer leading to a significantly different ceramic morphology, more base-like in nature.

Polyvinyl alcohol (PVA) and polyvinyl acetate (PVAc) were found to readily form tractable nanocomposites with good mechanical integrity, good flexibility and interesting wetting characteristics. In general, the PVA nanocomposites yielded a significantly higher storage modulus then the PVAc nanocomposites. However, results of a simple destructive force ramp test, at the individual material's glass transition temperature (PVA 85°C, PVAc 30°C), suggests that the PVAc nanocomposite actually is mechanically stronger. Another interesting feature of these two nanocomposites that was noted is that they tend to behave somewhat as a thermoplastic. That is, once heated above the polymer component's glass transition temperature, the material can be easily deformed or bent, and the material will retain its shape once cooled. This is a rare example of where the polymer dictates the physical properties of the entire nanocomposite.

The behaviour of PAN nanocomposites was also explored. However, PAN has limited solubility in solvents commonly used to form this type of nanocomposite due to polarity differences between the polymer (lower polarity) and the solvents or TEOS (higher polarity). Attempts to form PAN nanocomposites through our usual synthetic protocols resulted in an unusual material unlike the other homogeneous nanocomposites previously produced. The polymer and ceramic components retreated into separate phase domains, forming an inhomogeneous composite material. As was expected, the mechanical properties of this material were poor due the poor interfacial characteristics between the polymer and ceramic phases. Additional attempts were made at producing a true homogeneous nanocomposite by using N-methylpyrrolidone as the solvent for the condensation. Though a homogeneous nanocomposite was formed with this method, difficulties in the condensation/curing process resulted in a highly fractured material such that a mechanically acceptable material was never obtained.

Unit III–D

Nanocomposites—Applications

There are a host of applications of nanocomposites. One of the most interesting of these applications is as solid polymer electrolytes (SPE) for lithium batteries. The polyphosphazene MEEP is a well-known SPE with very high room temperature conductivity, however it lacks the mechanical stability to be used in a practical device. Traditional stabilization methods, such as deep UV or electron beam crosslinking methods do improve the physical stability of SPEs, however this crosslinking lowers ionic conductivity—tests performed in the laboratory revealed this to be a factor of 30-45 for MEEP-like phosphazene polymers. This reduction is due to the additional covalent linkages formed during the crosslinking process that inhibit chain segmental motion and ion transfer. Since the nanocomposites formed by the ceramic condensation process do not form bonds to the polymer component, (Type I nanocomposites) mechanical stabilization is achieved without a great loss of ionic conductivity. As illustrated in Figure 4.1, the ceramic component forms a "scaffolding" around the polymer providing the needed stability, while not interfering with the polymer's ability to easily transport ions. It is hoped that these nanocomposites will afford the development of all-solid lithium batteries that will be better, safer, and more environmentally friendly than those available today.

Another set of applications for the nanocomposites is in separations membranes and other selective barriers. Polymers that are used in selective gas separations, and liquid pervaporation separations must usually be mechanically stabilized, typically by crosslinking, before they become practically usable. Since the nanocomposite is able to physically stabilize the polymer component in a manner very different than these traditional methods, a real advantage may be gained through nanocompositing rather than crosslinking.

One final separation application that is being explored is using nanocomposites as subsurface permeable reactive barriers. The nanocomposite matrix is being used to provide a platform for selective chelating agents that remove contaminants from subsurface water. The design employs reproducibly swellable hydrophilic materials (such as the base catalyzed materials discussed above) that allow subsurface water to flow through the barrier, allowing the entrained chelators to selectively sequester the contaminants of concern. This type of *in situ* remediation technology is of particular interest to the U.S. Department of Energy, but is also applicable to other contaminated sites. From auto parts to barrier packaging, the race is on to commercialize nanoclay thermoplastic composites. Just a pinch of these infinitesimally small particles can dramatically raise mechanical, thermal, barrier, and flame-retardant properties.

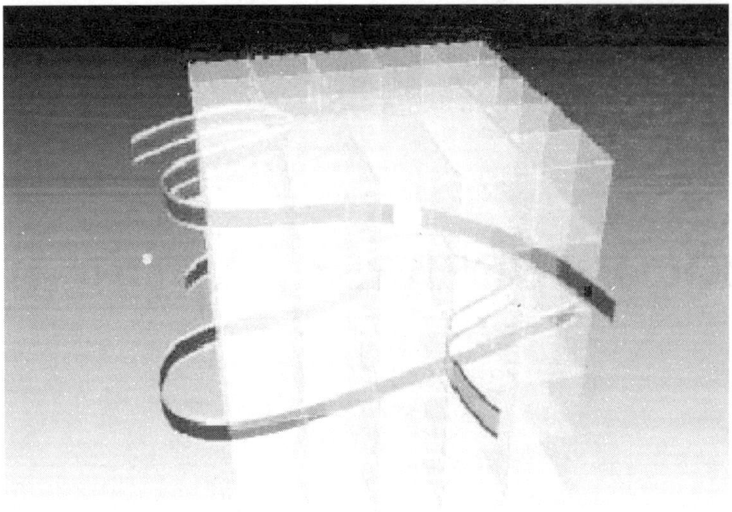

Fig. III–D.1: An illustration of a nanocomposite SPE. The cubes represent the ceramic component, while the ribbons represent the polymer.

Why all the excitement about the new class of mineral-reinforced plastics known as nanocomposites? Because relatively small amounts (2-5 per cent) of nanometer-sized clay particles can provide large improvements in mechanical and thermal properties, as well as gas barrier and flame resistance. They also reduce shrinkage and warpage. And all of these benefits are available without significantly raising the density of the compound or reducing light transmission (nanoclays are in the same size range as visible-light wavelengths).

Most of the current R&D work is focused on automotive parts and packaging. The goals are physical and thermal property enhancements and reduced permeability to gases, moisture, and hydrocarbons. Nylon-based nanocomposites are the first commercial materials to emerge, and there is now a frenzy of activity aimed at nano-reinforcing commodity thermoplastics such as polypropylene and PET.

Two research consortia on nanocomposites have been formed in recent months. The Edison Polymer Innovation Corp. (EPIC) of Akron, Ohio, got one under way in March. Headed by staff scientist Jon Collister, this three-year, $3-million program will coordinate investigations by industry and academic researchers on thermoplastics (PP, PET, PE, and acrylic), advanced composites, coatings, and elastomers.

Meanwhile, the National Institute of Standards and Technology (NIST) in Gaithersburg, Md., has formed a consortium of government and industry scientists to explore nanocomposites' potential for reducing the flammability of thermoplastics.

Where Opportunities are:

The current favorite nanoclay is montmorillonite, a layered alumino-silicate whose individual platelets measure on the order of one micron diameter, giving them an aspect ratio of about 1000:1.

Because it is hydrophilic, montmorillonite is not compatible with most polymers and must be chemically modified to render its surface more hydrophobic. The most popular surface treatments are organic ammonium cations, which can be exchanged for the inorganic cations existing on the silicate surface. The treated clay is then incorporated into the resin matrix either during polymerization or by melt compounding.

As an example of what can result, commercial nylon 6 nanocomposite materials provide tensile modulus, tensile strength, and HDT equivalent to 40 per cent mineral-filled nylon, while specific gravity remains at the level of unfilled nylon 6. With lower-modulus polymers such as polyolefins, nanocomposites have the potential to upgrade physical properties to levels competitive with costlier engineering resins. Nanocomposites already look attractive for molded car parts such as body panels and under-hood components, as well as electrical/electronic parts, power-tool housings, lawnmowers, aircraft interiors, and appliance components.

On the packaging side, nanocomposites can slow transmission of gases and moisture vapor through plastics by creating a "tortuous path" for gas molecules to thread their way among the obstructing platelets. Bottles and food packaging aren't the only areas of interest here. According to EPIC, nanocomposites hold commercial benefits for reducing hydrocarbon emissions from hoses, seals, and auto fuel-system components. Nanocomposites of polypropylene might permit blow molding auto fuel tanks at lower cost than current multi-layer barrier structures.

Market Debut in Nylon

The first nanocomposites on the market are based on nylon 6 and produced in the reactor. Toyota Central R& D Laboratories in Japan pioneered the work on nanocomposites a decade ago. Within the last few years, Toyota has licensed its patented in-reactor nylon nanocomposite technology to companies such as Nanocor and Japan's Ube Industries. Nanocor, in turn, sublicenses the technology to firms that want to make nylon nanocomposites.

Ube was among the first to come out with a commercial nanocomposite—its "nylonclay hybrids" (NCH) of nylon 6 and nylon 6/66 copolymer for film and structural applications. Commercial applications of NCH include nylon 6 barrier film for food packaging and a timing-belt cover for Toyota Motors. Compared with unfilled nylon 6, Ube's NCH is said to have 68 per cent higher tensile modulus and 126 per cent higher flexural modulus. Ube also reports that cast films of NCH-2 grade (2 per cent clay) have half the oxygen permeability of straight nylon 6.

At about the same time, Unitika Co. of Japan introduced a nylon 6 nanocomposite for injection molding, which is marketed and distributed in North America by Toyota Tsusho. Nylon M2350 is produced with Unitika's proprietary technology, which incorporates a synthetic silicate during polymerization.

Nylon M2350 has been used by Mitsubishi Motors for an engine cover on its GDI models, where the nanocomposite is said to offer a 20 per cent weight reduction and excellent surface finish. Lighting covers and knife handles are other applications.

Sometime back, Bayer AG in Germany announced the development of nylon 6 nanocomposites for transparent barrier film packaging. The compounds are made in the reactor using Nanocor's nanoclay. Bayer is currently marketing nanocomposites only in Europe. It has two nylon 6-based nanocomposite grades for cast film: Durethan LPDU 601-1 and LPDU 601-2, which offer different degrees of barrier improvement. Relative to neat nylon 6, they reportedly cut oxygen transmission rate (OTR) by up to 50 per cent, plus offer increased barrier to other gases, chemicals, and flavors/

odors. The nanofilms are also said to be stiffer than unfilled film.

AlliedSignal Corp. has been exploring nylon 6 nanocomposites for potential applications in automotive, consumer, and packaging markets. According to Kris Akkapeddi, team leader in plastics R&D, the company is currently evaluating this technology for some niche applications with specific customers.

Developmental products at AlliedSignal include toughened nylon nanocomposites for injection molding and high-viscosity nanocomposites for blow molding. Compared with unfilled nylon, Akkapeddi says the nanocomposites have 50-80 per cent higher stiffness and up to 126°F higher HDT without much increase in specific gravity.

Compounded Nano-nylons

Nylon nanocomposites produced by melt compounding that reportedly approach the performance of their in-reactor counterparts are now emerging. The trick is to exfoliate (delaminate) the clay particles sufficiently so that the ultimate level of reinforcement is achieved.

Nanocor has introduced a new nanoclay grade, Nanomer I.24TC, which reportedly allows compounders to produce a nylon 6 compound with flexural modulus and HDT about 90 per cent of the levels achieved with in-reactor blends.

RTP Co., a specialty thermoplastics compounder, announced in late April that it has produced what it believed to be the first commercially available compounded nylon 6 nanocomposite. The material is offered in injection molding grades and film grades with 3-5 per cent loadings.

As shown in the table, RTP's injection grade shows 53 per cent higher HDT than unfilled resin—comparable to 20-30 per cent mineral-filled nylon—but with little change in specific gravity. Tensile strength is much better than either unfilled or mineral-filled nylon 6. Reduced flammability is also indicated by a 50 per cent reduction in rate of heat release.

Toyota Central R&D Labs. recently succeeded in producing nylon 6 clay nanocomposites by melt processing in a twin-screw extruder. Dr. Arimitsu Usuki reports that the compounding approach improves productivity and opens the way for making nanocomposites of nylons 12, 66, and 6/66 copolymer.

Showa Denko of Japan recently started commercial production of nylon 66 nanocomposites using a kneading extruder. The materials are said to show improved flame retardancy and rigidity. Two special FR grades contain no halogen or phosphorus. Systemer FE 30600 and 30602 reportedly provide UL 94V-0 and V-2 performance, respectively, at only 1/64-in. thickness. Their flex moduli are 30-80 per cent higher, and HDTs are 54-144°F higher, than those of standard nylon 66. Melt-compounded nylon 6 and 66 nanocomposites are also reported to be forthcoming from BASF AG and from Solutia, respectively. (Solutia markets through Dow Plastics.)

Barriers for PET, EVOH

An agreement announced sometime ago by Eastman Chemical and Nanocor to jointly develop nanocomposites of PET and other packaging materials (e.g., polyethylene) is the first indication that commercialization of such materials may not be far off. Jim Lewis, v.p. of container plastics at Eastman, says his firm has worked with Nanocor since 1995, but several breakthroughs have occurred over the last year and a commercial product introduction is likely to come fairly soon.

Eastman is developing its first PET nanocomposites via the in-reactor approach. Its initial focus is on rigid containers. "Nanocomposites are one of the strategic technology components we are developing to enable emerging markets in food packaging that require superior barrier, including beer bottles and single-serve juice bottles," Lewis says. He hints that these applications are likely to utilize a coinjected multi-layer preform with the nanocomposite barrier layer on the inside.

Active projects on other barrier resins are also under way. Nanocor says it has made significant progress in joint development of EVOH nanocomposites for multi-layer packaging. Nanocor even offers a new nanoclay, Nanomer I.35L, designed for compounding with EVOH. Within the range of 50-80 per cent R.H., it reduces oxygen permeability of EVOH by 66-80 per cent. Besides packaging, EVOH nanocomposites are being investigated as a hydrocarbon barrier for automotive fuel tanks and fuel-system components.

Nanocor has another new nanoclay, Nanomer I.28MC, for compounding with MXD6 aromatic nylon from Mitsubishi Chemical. Applications could include packaging films for moisture- and oxygen-sensitive foods and electronics.

Add muscle to PP & TPO

Automotived-driven development of PP and TPO nanocomposites is one of the hottest areas of current interest. Because PP is a strongly nonpolar resin, it has been especially challenging to come up with an appropriate surface treatment for the clay. Nonetheless, late last year, three development partners—Montell North America, General Motors R& D in Warren, Mich., and Southern Clay—jointly announced the world's first TPO-based nanocomposite.

According to Jim Keeler, advanced program manager at Montell's Automotive Business Group, the partners are testing several TPO nanocomposites for exterior and interior applications. The initial focus is on exterior door and rear quarter panels. Although the nanocomposites are produced by melt compounding, Montell is using its Catalloy in-reactor TPOs in order to simplify the overall compounding process.

According to Keeler, previous PP/nanoclay compounds achieved only 10-15 per cent exfoliation of the layered clay particles, whereas the current work has achieved much greater exfoliation and thus enhanced property development. TPO nanocomposites with specific gravity as low as 0.91 have been produced with better dimensional stability than unfilled TPOs and comparable to talc-filled grades (CLTE of 5×10^{-5} mm/mm/°C).

These developmental TPO nancomposites also have flexural moduli of 150,000 to 250,000 psi. In a 5-mph impact test, the TPO remains ductile down to –30°C. The partners also report that the nanocomposite has improved surface appearance due to the low filler loading.

Work on nanocomposites is also well under way at Ford Motor Co., according to Jeff Helms, director of research and materials science. "We have been focusing on anything polypropylene-based, including TPOs, impact-modified PP, and PP alloys," he says. "We have explored several different methods via compounding with which we can make PP nanocomposites."

Ford's initial target is interior trim—instrument panels, A and B pillars, door panels, and consoles. The chief challenge, Helms says, is to maximize coupling between the silicate and the polymer matrix. He indicates that this could prove simpler with surface-treated synthetic clays. "With natural clays, you may need a surface treatment plus a compatibilizer." He says his group recently has achieved very good results in terms of stiffness/impact balance. The next step is to team up with material suppliers.

Meanwhile, Toyota Central R& D Labs recently produced PP nanocomposites in a twin-screw extruder using 5 per cent clay and maleic anhydride-modified PP oligomer as a compatibilizer. According to Toyota, the "polypropylene clay hybrid" (PPCH) provides higher stiffness than the same loading of talc or glass.

Dow Plastics and automotive processor Magna International, Southfield, Mich., are in their second year of a joint effort to develop nanocomposites of PP and TPO. The five-year project, which is partially funded by the U.S. Commerce Dept.'s Advanced Technology Program, is aimed at developing methods for dispersing the nanoclay in such resins and demonstrating cost-effectiveness of nanocomposites for exterior front and rear car parts.

For auto under-hood and electronics uses, Showa Denko just commercialized the first acetal nanocomposite. It reportedly shows low warpage and few sinks. Compared with unfilled acetal copolymer, it has 40 per cent higher flexural modulus and 45° F higher heat resistance.

Nano Fire Extinguisher

Flame-retardant properties of nanocomposites are of interest on many fronts, as indicated by ongoing work at NIST and its new consortium. Over the last three years, NIST fire researchers demonstrated that the rate of heat release for nylon 6 nanocomposites with 5 per cent clay is 63 per cent lower than the RHR of pure nylon 6. There was also no increase in carbon monoxide or soot generated during combustion. Though the mechanism of nanocomposites' FR action is not known for sure, a durable carbon-silicate char was observed to form very rapidly in the burn tests.

NIST's work last year showed that the reduced flammability of nanocomposites extends to other thermoplastics, including PP and PS. In one study, PP samples were grafted with 0.4 per cent maleic anhydride to make them more compatible with the clay.

Fire researcher Jeffrey Gilman thinks that nanoclay in combination with reduced amounts of conventional FR fillers such as ATH and MgOH at lower use levels are likely to meet FR standards such as 94V-0 and V-2. Gilman notes that Showa-Denko has reported V-0 performance of nylon 6 nanocomposites containing melamine.

Unit III–E

Nanocomposites Using Plastics

Nanosized particles have mega-potential in plastics because just a pinch does so much more than heavy loadings of other additives. Three recent conferences presented almost 200 papers on the feverish pace of 'nano' R&D on boosting plastics' mechanical and barrier properties, flame retardancy, and electrical conductivity.

They are still in their infancy, but if the forecasts are right, nanocomposites could turn out to be the biggest little thing to hit plastics in decades. Polymers reinforced with as little as 2 per cent to 5 per cent of these particles via melt compounding or *in-situ* polymerization exhibit dramatic improvements in thermo-mechanical properties, barrier properties, and flame retardancy. They also can outperform standard fillers and fibers in raising heat resistance, dimensional stability, and electrical conductivity.

Dispersions of nanoscale reinforcements in polymers are already entering the marketplace in automotive and packaging applications, albeit in a low-profile manner and slower than had been anticipated. But that pace is expected to speed up dramatically, as indicated by the enthusiasm of researchers and marketers shown in roughly 200 papers delivered at three technical conferences. These were Nanocomposites 2004 in San Francisco, the SPE Antec 2004 in Chicago, and Nanocomposites 2004 in Brussels, Belgium.

Fig. III–E.1: GM's 2005 Hummer H_2 cargo bed uses about 7 lb of molded-in-colour TPO nanocomposite parts.

A report from market-research firm pegs the total worldwide market for polymer nanocomposites at 24.5 million lb in 2003, valued at $90.8 million. It also projects the market to grow at an average annual rate of 18.4 per cent to reach $211.1 million by 2008. Even if nano developments hit some snags, BCC says some applications will grow faster than 20 per cent per year.

The leading nanoscale fillers in R&D and commercial projects are layered silicate nanoclays and nanotalcs, plus carbon nanotubes and graphite platelets. But other candidates are being actively investigated, such as synthetic clays, polyhedral oligomeric silsesquioxane (POSS), and even natural fibers like flax and hemp.

The Top Contenders

The two types of nanofillers that have been most widely discussed and the first to break into commercial use are nanoclays and carbon nanotubes. Both must be chemically modified with surface treatments in order to achieve the fine dispersion and resin coupling that are required to derive maximum benefit.

Both of these nanofillers have demonstrated improvements in structural, thermal, barrier, and flame-retardant properties of plastics. Carbon nanotubes also enhance electrical conductivity.

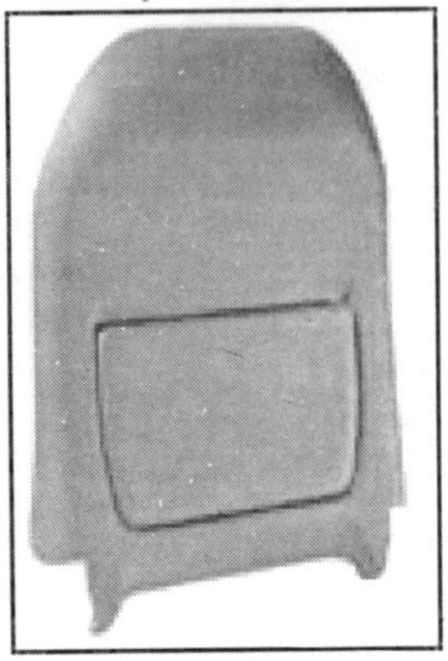

Fig. III–E.2: Noble Polymers' Forte PP nanocomposite was used in the seat backs of the 2004 Acura TL and will be used for the center console of a 2006 light truck.

So far, nanoclays have shown the broadest commercial viability due to their lower cost—$2.25 to $3.25/lb—and their utility in common thermoplastics like PP, TPO, PET, PE, PS, and nylon.

The leading nanoclay is montmorillonite, a layered alumino-silicate whose individual platelets measure around 1 micron diam., giving them an aspect ratio of 1000:1. The two major US producers are Nanocor with its Nanomer line and Southern Clay Products with its Cloisite line. Both companies have formed alliances with suppliers of resins and surfactants, plus compounders and automotive OEMs and packaging firms. While much of their work is proprietary, they have disclosed several commercial successes.

General Motors has taken the lead in putting nanocomposites on the road. GM launched the first commercial auto exterior use of a nanocomposite in the step assist on the 2002 GMC Safari and Chevrolet Astro van. The part also appears on 2003 and 2004 models. More recently, a PP/nanoclay composite appeared on the body side molding of General Motors' highest-volume car, the 2004 Chevrolet Impala. The compound was developed by GM's R&D Center in Warren, Mich., in cooperation with Basell North America and Southern Clay Products.

The latest application is on the 2005 GM Hummer H2 SUT. The vehicle's cargo bed uses about seven pounds of molded-in-colour nanocomposite parts for its center bridge, sail panel, and box-rail protector. The material is Basell's Profax CX-284 reactor TPO with nanoclay.

While nanoclay adds muscle to plastics, carbon nanotubes impart electrical and thermal conductivity. Nanotubes' commercial potential has been limited by their high price tags—reportedly in the range of $100/gram, although they are available in masterbatches for $50/lb and up. Still, nearly every car produced in the U.S. since the late 1990s contains some carbon nanotubes, typically blended into nylon to protect against static electricity in the fuel system. Static-dissipative compounds containing nanotubes are also protecting computer read/write heads.

Carbon nanotubes include both single-and multi-walled structures. The former have a typical outside diameter of 1 to 2 nm while the latter have an OD of 8 to 12 nm. They can range in length from the typical 10 microns to as much as 100 microns and have at least a 1000:1 aspect ratio. Carbon nanotubes have 50 times the tensile strength of stainless steel (100 GPa vs. 2 GPa) and five times the thermal conductivity of copper. When incorporated into a polymer matrix, they have the potential to boost electrical or thermal conductivity by orders of magnitude over the performance possible with traditional fillers such as carbon black or metal powders.

US suppliers of nanotubes include Hyperion Catalysis with its Fibril multi-walled nanotubes and newcomer Zyvex Corp. with its Nanosolve single- or multi-walled tubes. Both suppliers now offer their products in masterbatches that typically contain 15 to 20 per cent nanotubes.

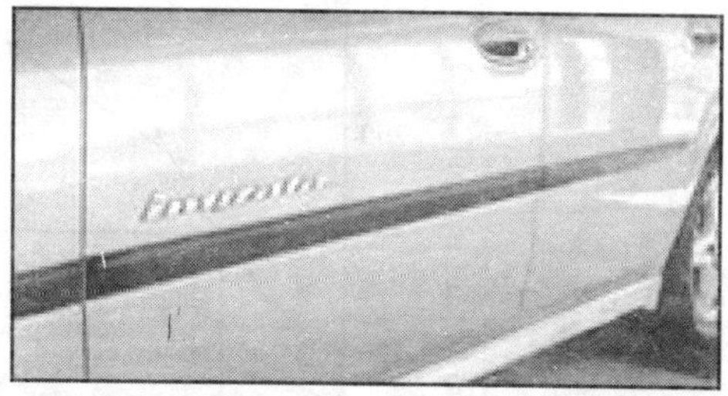

Fig. III–E.3: TPO nanocomposite in the body side molding of GM's highest-volume car, the 2004 Chevrolet Impala, was developed by GM in conjunction with Basell North America and Southern Clay Products.

A different but related category is vapor-grown carbon nanofibers from Pyrograf Products, a spin-off from Applied Sciences. Its Pyrograf III nanofibers reportedly can compete with nanotubes in providing thermal and electrical conductivity and dramatically enhancing mechanical properties and fire resistance (char formation). What's more, nanofibers cost significantly less—around $100 to $150/lb. Evaluations are under way in nylon, PP, and polyurethanes.

Nanoclay Compound Sources

Alliances between Nanocor and two specialty compounders, have resulted in commercial-scale nanocomposite concentrates and compounds for structural and barrier applications.

Noble Polymers new Forte PP nanocomposite compound made its commercial debut in the seat backs of the Honda Acura TL 2004 car. Forte replaced glass-filled PP, which caused processing difficulties, visual defects, and warping. Forte has a low density of 0.928 g/cc, superior mechanical properties, and improved surface quality and recyclability.

Noble reports that the Forte nanocomposite will also be used to produce the center console for a 2006 model light truck. Other applications in the works include office furniture—replacing 20 per cent glass-filled PP—and appliance parts, where Forte reduces weight and possibly material cost.

PolyOne recently introduced the Maxxam LST line of PP homopolymer/nanoclay compounds that boast high stiffness and impact resistance. Through a patent-pending process, PolyOne reports that it has been able to overcome previous problems of incomplete exfoliation and dispersion of the nanoclay, resulting in performance that meets or exceeds many engineering thermoplastics. Lighter weight, aesthetic and processing advantages, and lower cost are also claimed.

PolyOne also offers Nanoblend concentrates of up to 40 per cent nanoclay in homopolymer PP, modified PP, LLDPE, LDPE, HDPE, or an ethylene copolymer. Some grades are tailored specifically for barrier enhancement.

PolyOne reports that applications nearing commercialization include pallets and dunnage, where Maxxam LST compounds are specified as alternatives to engineering resins due to their improved dimensional control, which is critical for robotic assembly. In addition, they boast good impact strength and lighter weight. Maxxam LST is also being considered for consumer disposable applications due to a combination of chemical resistance and stiffness, as well as dramatic cycle-time improvements.

Meanwhile, the Nanoblend concentrates are being considered for auto interior and exterior

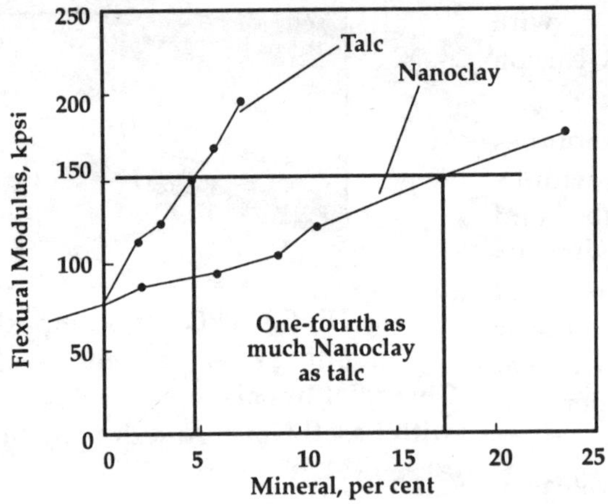

Fig. III–E.4: Modulus vs. Mineral Loading
For optimized flexural modulus in the Chevy Impala TPO body side molding, 75% less nanoclay was required than standard talc.
(Source: Southern Clay Products)

TPO parts. Key drivers are dimensional stability, lighter weight, and stiffness without loss of impact. Nanoblend concentrates are being evaluated in films for enhancing barrier, stiffness, HDT, and controlled release or migration of additives such as biocides and dyes. In blow molded packaging, Nanoblend is being considered for improved barrier and the potential for thinwalling and faster cycles. Thin-walling and faster cycles are also attractions in injection molded containers and totes. Some industry sectors are evaluating the concentrates for improving flame retardancy.

More Muscular TPOs

Papers presented by General Motors and Southern Clay Products discussed numerous enhancements to automotive TPOs obtained with nanoclays. Those advances did not come easily: Early processing problems caused by clay agglomeration were ultimately resolved by optimizing the clay infeed position at the extruder, the screw design, screw speed, temperature, and pressure. Once processing issues were resolved, nanocomposite TPOs outperformed conventional talc-filled TPOs in consistency of properties, retention of low-temperature ductility, elimination of "tiger striping," reduced paint delamination, and improved knit-line appearance, colourability, grain patterns, scratch and mar resistance, and recyclability. What's more, the lower filler level means 3 per cent to 21 per cent lower density (0.92 vs. 0.96 to 1.13 g/cc). Lighter weight requires less adhesive for attachment, which cuts cost.

Fig. III–E.5. Carbon nanotubes, such as the multi-walled Fibrils from Hyperion Catalysis, have 50 times the tensile strength of stainless steel and five times the thermal conductivity of copper.

Among the many auto exterior, interior, and under-hood applications for which nanocomposites appear suited are fascias, rocker covers, side trim, grilles, hood louvers, instrument panels, seat/IP foams, door inners, pillar covers, vertical and horizontal body and closure panels, engine shrouds, fan shrouds, air intakes, fuel tanks, and fuel lines.

In addition to TPO/nanoclay composites, GM has explored using carbon nanotubes to replace current thermoset structural composites. GM is interested in reducing reinforcement levels in Class A applications by replacing continuous carbon fibers with nanotubes or short nanofibers. nanotubes also have the potential to reduce plastics' coefficient of thermal expansion more effectively.

Better Barriers

Polymer barrier technology is also getting a boost from nanoclays. Mitsubishi Gas Chemical (MGC) and Honeywell Specialty Polymers both are using Nanocor's nanoclays in nylons as barrier layers in multi-layer PET bottles and films for food packaging. MGC's MXD6 nylon nanocomposite, called Imperm N, is used commercially in Europe in multi-layer PET bottles for beer and other alcoholic beverages. It is also being evaluated for small carbonated soft-drink bottles. Other Imperm applications that will debut in the next six months are multi-layer thermoformed containers for deli meats and cheeses and flexible multi-layer films for potato chips and ketchup.

Honeywell has aimed its Aegis nylon 6 nanocomposites initially at PET beer bottles. In late 2003, a version containing an oxygen scavenger made a commercial splash with the introduction of the 1.6-liter Hite Pitcher beer bottle from Hite Brewery Co. in South Korea. Aegis is the barrier layer in this three-layer structure, which is said to provide a 26-week shelf life.

Honeywell is aiming other Aegis nanocomposite grades (without oxygen scavenger) as replacements for EVOH in films and pouches. Such grades reportedly are lower in cost than EVOH, provide a better barrier that allows for lightweighting, and also have better puncture resistance and good clarity. (Because of their size, nanoparticles do not interfere with light transmission.)

The U.S. military and NASA, in conjunction with Triton Systems, Inc., Chelmsford, Mass., are looking into nanoclay as a barrier enhancer for EVOH in long-shelflife packaging. An experimental thermoformed food tray was made from EVOH plus 3 per cent of Southern Clay's Cloisite in a layer sandwiched between two PP layers. It reportedly imparts three- to five-year shelf life without refrigeration, plus good clarity, processability, and recyclability.

Fig. III–E.6. Honeywell's Aegis nylon 6 nanocomposites have been used in high-barrier PET beer bottles and are also being considered as a replacement for EVOH in films and pouches.

Alcoa CSI, Crawfordsville, Ind., is seeking a patent on coextruded barrier liners for plastic bottle caps for beer, juice, or carbonated soft drinks. The liners include a layer of nylon 6/nanoclay composite plus one or two EVA layers with oxygen scavengers. This liner is said to outperform other barrier materials at very high humidity (95 per cent to 96 per cent RH).

LG Chem Ltd. of South Korea has developed high-barrier, monolayer blow molded containers of HDPE with 3 per cent to 5 per cent nanoclay for handling toluene and light hydrocarbon fluids. LG reports that permeation of the hydrocarbon solvents is cut by a factor of 40 to 200 compared with neat HDPE.

Versatile Nanocarbons

Since the early 1990s, automotive fuel-line components such as quick connectors and filters have used inner barrier layers consisting of nylon 12 and carbon nanotubes. Hyperion Catalysis now aims to introduce nanotubes into other resins used in auto fuel systems, such modified nylons and fluoropolymers. A new fluoropolymer/nanotube compound is being used to make O-rings for automotive fuel connectors.

In electronics, polycarbonate and polyetherimide (GE's Ultem) components of computer hard drives have been reinforced with nanotubes to render them conductive and very smooth.

Over the last three years, a major automotive OEM in Europe has been using carbon nanotubes in GE's Noryl GTX nylon/PPO alloy to mold exterior fenders. This conductive nanocomposite allows for electrostatic painting.

Michigan State University's Composite Materials and Structures Center in East Lansing developed a new surface-treated graphite nano-platelet. Graphite has a modulus several times that of clay and also has excellent electrical and thermal properties. When incorporated into an epoxy, it results in superior mechanical properties and excellent electrical conductivity compared with standard carbon fibers and nanosized carbon black. MSU sees potential in ESD protection and EMI shielding. Plastic nano-graphite compounds are expected to sell for up to $5/lb, significantly less than compounds based on nanotubes or vapor-grown carbon fibers.

Carbon nanotubes have more going for them than just conductivity. Researchers at the National Institute of Standards and Technology (NIST), Gaithersburg, Md., report that carbon nanotubes in PP not only enhance the material's strength and properties, but also dramatically change how the molten polymer flows, virtually eliminating die swell.

Nano Flame Retardants

Extensive research at NIST has established nanoclays' effectiveness as flame-retardant synergists. Nanoclay levels of 2 per cent and 5 per cent in nylon 6 reduced the rate of heat release by 32 per cent and 63 per cent, respectively, NIST found.

Specialty compounder Foster Corp. recently demonstrated that higher levels (13.9 per cent) of nanoclay can be added to nylon 12 elastomers to achieve UL 94V-0 rating at 1/8-in. thickness. Used as a char former, the nanoclay allows the typical 50 per cent loading of halogen/antimony oxide flame-retardant system to be cut in half, which significantly reduces detrimental effects on physical properties. The company first introduced nylon 12/nanoclay compounds for tubing and film in 2001.

Germany's Sud-Chemie (U.S. office in Louisville, Ky.) offers modified nanoclays called Nanofil as flame retardants. It recently developed halogen-free EVA/PE wire and cable compounds containing 3 per cent to 5 per cent of new Nanofil SE 3000 plus 52 per cent to 55 per cent alumina trihydrate or magnesium hydroxide (typically used at 65 per cent levels). The result is said to be improved mechanical properties, smoother cable, and higher extrusion speeds.

According to Hyperion Catalysis, two recent studies show that multi-walled carbon nanotubes may act as a flame retardant without use of halogen. In both EVA and maleic-anhydride-modified PP, 2.4 per cent to 4.8 per cent loadings of nanotubes show heat-release rates comparable to or better than those obtained with nanoclays.

Nano-nucleators

Among its many virtues, nanoclay can work as a nucleating agent to control foam cell structure and enhance properties of polymeric foams for applications from insulation to packaging. The University of Toronto's Dept. of Mechanical and Industrial Engineering studied extrusion of chemically foamed LDPE/wood-fiber compounds. Addition of 5 per cent nanoclay to the mix decreased the cell size, increased the cell density and facilitated foam expansion. When burned, the foam showed good char formation. Similar results were obtained in LDPE/nanoclay foam blown with CO_2 gas.

Researchers at Ohio State University's Dept. of Chemical Engineering (Columbus) in USA found that small amounts of nanoclay surface-grafted with PMMA can reduce cell size and increase cell density in microcellular PS foamed with CO_2. Another OSU study showed that smaller cell size and higher density can be achieved with 5 per cent nanoclay in polyurethane foams blown with pentane or water.

Louisiana State University's Mechanical Engineering Dept. (Baton Rouge) reports that 4 per cent to 5 per cent nanoclay increases the flexural strength and elongation of epoxy syntactic foams used as core materials for sandwich composites in structural applications.

Unit III–F

Nanocomposite's Role in Automotive and Barrier Packaging

Nanocomposites are gradually gaining acceptance in the mainstream of global plastics processing. These polymer compounds, containing relatively low loadings (under 6 per cent by weight) of nanometer-sized mineral particles, are beginning to show up in polypropylene and TPO-based automotive exterior claddings (see sidebar), barrier beer bottles, nylon packaging films, polyethylene pipe and wire/cable coatings, and more.

Optimism surrounding these novel materials has increased since they burst into industry consciousness two or three years ago. Exploratory effort has intensified as a growing body of data substantiates the potential of established nylon/clay nanocomposites, emerging polyolefin versions, and a range of other resin matrixes and nanofillers (see Table III–F.1).

Real-world applications are coming more slowly, in part because of the need to validate cost-effectiveness in the face of high price tags for nano-particle ingredients. Some early application-development programs have lapsed for cost reasons. Such casualties include an automotive timing-belt cover based on a nylon 6 nanocomposite from Japan's Unitika and an automotive mirror housing of conductive PPO/nylon alloy from GE Plastics.

Fig. III–F.1: General Motors' 2002 Safari and Astro vans represent the commercial debut of polyolefin nanocomposites, appearing in the TPO step assist.

Yet the promise of nanocomposites is undiminished. They can improve polymers' stiffness, HDT, dimensional stability, gas barrier, electrical conductivity, and flame retardancy. Nanoparticles are so small and their aspect ratio (L/D) so high that properties improve with lower loadings and fewer penalties (such as higher density, brittleness, or loss of clarity) than with conventional reinforcers like talc or glass.

Nanoclays are believed to increase barrier

properties by creating a maze or "tortuous path" that slows the progress of gas molecules through the matrix resin. At the same time, these nanoplatelets are only 1 nm thick, less than the wavelength of light, so they do not impede light's passage.

As shown in Table 6.1, nanocomposites now draw on a wider menu of resin matrices, including PP, TPO, EVA, acetal, polycarbonate, biodegradable polylactic acid (PLA), and inherently conductive polyaniline. The nano-particles most widely used so far in these compounds are clays supplied by Nanocor and Southern Clay Products. But a new generation of emerging nanomaterials—including nanostructured silicas, carbon nano-tubes, and ceramic nanofibers—suggest that impressive gains in nanocomposite performance lie just a few years ahead.

Table III–F.1: Partial Listing of Nanocomposite Suppliers

Supplier & Tradename	*Matrix Resin*	*Nano-Filler*	*Target Market*
Bayer AG (Durethan LPDU)	Nylon 6	Organo-clay	Barrier films
Clariant	PP	Organo-clay	Packaging
Creanova (Vestamid)	Nylon 12	Nanotubes	Electrically conductive
GE Plastics (Noryl GTX)	PPO/Nylon	Nanotube	Automotive painted parts
Honeywell (Aegis)	Nylon 6 Barrier Nylon	Organo-clay Organo-clay	Multi-purpose Bottles & film
Hyperion	PETG, PBT PPS, PC, PP	Nanotube	Electrically conductive
Kabelwerk Eupen of Belgium	EVA	Organo-clay	Wire & cable
Nanocor (Imperm)	Nylon 6 PP Nylon MDX6	Organo-clay Organo-clay Organo-clay	Multi-purpose Molding PET beer bottles
Polymeric Supply	Unsaturated polyester	Organo-clay	Marine, transportation
RTP	Nylon 6, PP	Organo-clay	Multi-purpose, electrically conductive
Showa Denko (Systemer)	Nylon 6 Acetal	Clay, mica Clay, mica	Flame retardance Multi-purpose
Ube (Ecobesta)	Nylon 6, 12 Nylon 6, 66	Organo-clay Organo-clay	Multi-purpose Auto fuel systems
Unitika	Nylon 6	Organo-clay	Multi-purpose
Yantai Haili Ind. & Commerce of China	UHMWPE	Organo-clay	Earthquake-resistant pipe

Source: Bins & Associates, Sheyboygan, Wis.

New Focus on Nano-PP

"Progress has moved beyond nylon 6/clay composites to include products based on PP and PE," declares Nanocor president Peter Maul. Nanocor's data show up to 98 per cent stiffness improvement in PP and up to 52° F higher HDT (Table 6.2). The company says the nanocomposite has virtually the same impact strength as unfilled PP homopolymer.

General Motors recently announced the first-ever automotive production part in an olefinic nanocomposite. It's an exterior step assist for 2002 vans, made of a nanoclay/TPO compound from Basell. William Windscheif, Basell's global business v.p. for advanced polyolefins, calls this application "a small step, but a giant one for nanocomposites," adding that it heralds a broader shift to nano-PP in automotive.

Initially, the auto industry expressed most interest in nylon 6 nanocomposites for use under the hood, where higher HDT and lightweighting were the goals. Ingolf Buethe, senior v.p. for polymer research at BASF in Germany, says nylon nanocomposites show considerable promise in terms of enhanced stiffness, heat resistance, and gloss. But a serious downside for a 5 per cent nanoclay nylon compound tested by BASF was a loss of toughness more pronounced than with standard fillers.

More recently, automotive OEMs and molders have turned their attention to PP and TPO nanocomposites. These polyolefin materials potentially offer engineering-thermoplastic properties at 20 per cent lower density and 50 per cent lower cost per pound.

Consultant Kenneth Sinclair, head of STA Research in Snohomish, Wash., says the auto makers find that combination difficult to ignore. He estimates that about 30 per cent of nano-PP usage by 2004 will be in autos, mostly cannibalizing existing PP applications. But replacement of metals and engineering thermoplastics will follow.

Sinclair says nano-PPs are stiffer and process better than standard PPs, so thin-walling of parts by around 40 per cent is feasible. In turn, thin-walling permits around 25 per cent reduction in cycle times, for 60-80 per cent total savings per part.

Nonetheless, an official at one of GM's domestic competitors insists that car companies will balk at paying a premium for new materials. Nanocor, concluding that automotive OEMs want lighter parts at no extra cost, has focused its attention on non-automotive uses, including pallets, electronics, and appliance housings.

Volvo Corp. of Sweden has studied 5 per cent-nanoclay composites based on a Basell TPO modified with maleic anhydride as a coupling agent. Volvo observed 32 per cent to 50 per cent higher stiffness than 20 per cent talc-filled PP. Impact strength was lower than unmodified TPO but higher than 20 per cent talc-filled TPO. Volvo found that nano-TPO still has 68 per cent lower stiffness than aluminum sheet.

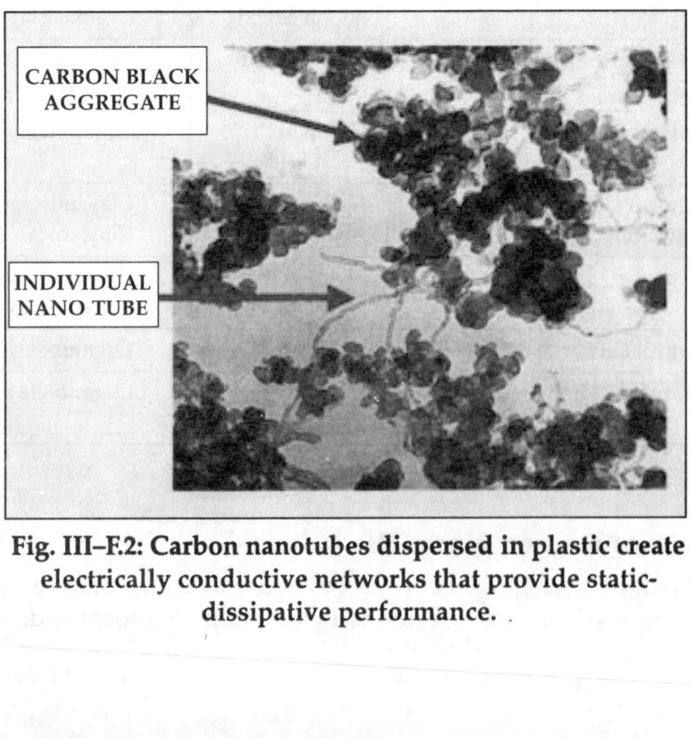

Fig. III–F.2: Carbon nanotubes dispersed in plastic create electrically conductive networks that provide static-dissipative performance.

A long-term goal for Dow Plastics is in-reactor compounding of nano-PP by using nanoclays as the catalyst support for in-situ polymerization of PP homopolymer. Dow's effort is focused on highly loaded (up to 10 per cent clay) nano-PPs for semi-structural automotive uses. Dow sources say preliminary findings show "quite promising" performance of these composites.

Table III–F.2: Effect of 6 per cent Nanoclay on PP Homopolymer Properties

PP Type, MFR	*Flexural Modulus, psi*		*Heat Deflection Temp., F*	
	Unmodified	*Nano-PP*	*Unmodified*	*Nano-PP*
Conventional 4 g/10 min	166,000	296,000	189	241
Conventional 14 g/10 min	173,000	258,000	187	228
Nucleated 35 g/10 min	231,000	335,000	235	250

Source: Nanocor.

Nanocor has developed a 40-50 per cent nanoclay concentrate in PP. One potential use is in heavy-duty electrical enclosures that must meet various fire ratings plus demanding specs for low-temperature toughness and weatherability. Switching to nano-PPs could bring 18 per cent weight savings and permit use of less halogenated FR additive to reach a given UL rating.

In other polyolefins, Kabelwerk Eupen of Belgium says melt blending of nanoclay into EVA shows promise for wire and cable compounds. Calorimeter tests reveal a dramatic decline in heat release at relatively low (3-5 per cent) loadings. Nano-EVAs also exhibit superior mechanical properties, chemical resistance, and thermal stability.

Meanwhile, not all automotive work in nanocomposites involves polyolefins. A role for nanocomposites in polycarbonate automotive glazing is being explored by Exatec of Wixom, Mich., the joint venture of Bayer and GE Plastics that is dedicated to PC auto-glazing development. Exatec marketing director Fritz Stein says nanotechnology is being considered for the exterior coating needed to achieve weatherability and abrasion-resistance without reducing clarity. A Bayer coating containing nano-particles is one of several promising approaches being pursued, Stein reports.

Advances in Packaging

The major application focus for nylon 6 nanocomposites today is in high-barrier packaging. Much of the attention is on PET bottles, where nanocomposites demonstrate improved oxygen and carbon dioxide barrier. Honeywell offers a 2 per cent nanoclay nylon 6 for bottles and has a 4 per cent nanoclay version in development. An early commercial use is a pasteurizable beer bottle that will be introduced in China later this year.

Nanoclays also enhance the oxygen barrier and stiffness of nylon 6 films. That could permit downgauging of packaging of oxygen-sensitive products—pet foods, boil-in bags, vacuum packs, and stand-up pouches. No modification of cast film equipment is needed.

"Nanoclays significantly boost the barrier performance of nylon 6 while retaining most of its existing favorable characteristics," states Lance Altizer, Honeywell's market-development manager.

He notes that nanocomposites retain nylon's toughness, clarity, hot-fill heat resistance, and oil/grease resistance. Honeywell claims that nylon 6 with 2 per cent nanoclay has three times the oxygen barrier of straight nylon 6, and 4 per cent nanoclay confers a six-fold improvement. That makes Honeywell's Aegis NC nano-nylon a candidate for medium-barrier bottles and films—those offering around 0.5 to 1 cc/mil/day O_2 transmission rate (OTR). Data also show a doubling of stiffness, higher HDT, and improved clarity for nano-nylon 6 packaging.

Honeywell has turned its attention to creating nano-nylon materials that can beat the cost of high-barrier plastics or even glass. Its current contender is an active-passive barrier system called Aegis OX, which synergizes nanoclays as the passive barrier and a proprietary, nylon-specific oxygen scavenger as the active agent.

Honeywell claims this one-two punch brings a 100-fold reduction in OTR versus nylon 6, taking oxygen ingress to near-zero levels (Fig. III–F.3). It also addresses a drawback of existing O_2 scavengers: In the Honeywell system, the passive barrier protects the scavenger from premature depletion. Efficiency of the system is also improved by uniform dispersion of the nano-platelets and by ensuring that the scavenger is positioned "to easily find the oxygen," as a Honeywell source puts it.

Aegis materials are being tested by major PET bottle makers. Honeywell says nano-nylon 6 tends to stretch and orient in ways compatible with stretch-blow molding processes. Current barrier requirements for beer bottles (in which Aegis OX would be the core of a three-layer structure) set a maximum limit on oxygen ingress over 120 days, as well as a limit of 10 per cent CO_2 escape in that time. The beer industry appears headed toward a 180-day barrier standard for both gases.

"We feel Aegis OX will compete against any existing barrier system for beer," says Honeywell's Altizer. He claims that Aegis OX meets both 120-day requirements and that the 180-day standard is achievable with processing refinements.

Nanocor has come up with a different high-barrier option. Its Imperm compound supplements the inherent gas barrier (0.35 cc/mil/day) of amorphous MDX6 nylon from Mitsubishi Gas

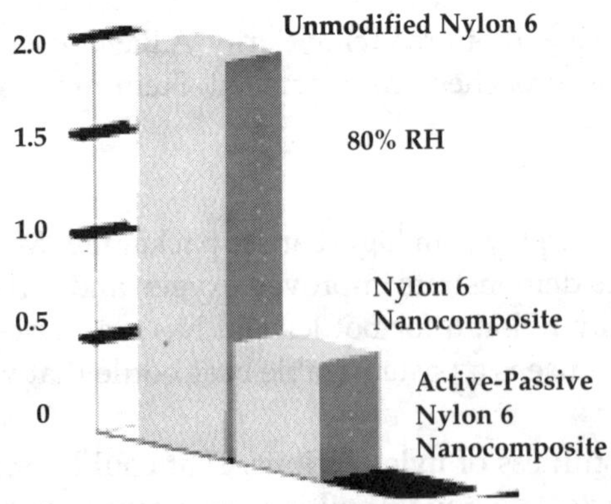

Fig. III–F.3: Oxygen transmission of nylon nanocomposite.
Source: Honeywell

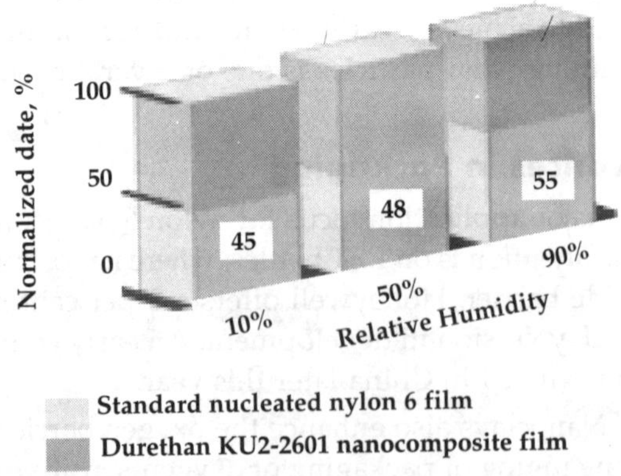

Fig. III–F.4: Oxygen barrier of nanocomposite nylon 6 cast film.
Source: Bayer AG

Chemical with the addition of a nanoclay. Used as the core of a three-layer PET bottle, Imperm is said to have 100-fold lower OTR than that of straight PET. It is being used in a 16-oz, non-pasteurized beer bottle in which the Imperm core (10 per cent of bottle thickness) reportedly ensures a 28.5-week shelf life. Imperm is said to adhere to PET without tie layers. Sufficient clarity is retained to meet requirements for the amber bottle.

Bayer is aiming nylon 6 nanocomposites at cast film for multi-layer packaging, protective films for medical and corrosion-prone items, and more. Bayer's pre-commercial Durethan KU2-2601 compound uses Nanocor's clay to reduce OTR by around 50 per cent versus nucleated nylon 6 (Fig. III–F.4). Stiffness of the nanocomposite is doubled, and its gloss and clarity rival those of a costly high-clarity copolyamide film, Bayer reports. Anti-blocking properties are also improved.

Meanwhile, nanocomposites also limit emissions of gasoline, methanol, and organic solvents. Ube America is developing nanocomposite barriers for automotive fuel systems. It uses up to 5 per cent nanoclay in nylon 6 and 6/66 blends. Nylon 6 with 2 per cent nanoclay is said to be five times more resistant to gasoline permeation than unmodified nylon 6. Ube has developed a coextruded barrier fuel line, trade-named Ecobesta, using nylon 6/66 nanocomposite as the core layer.

Unit III–G

Nanocomposites Improve Package Properties

Nanocomposite technology has been described by some as the next great frontier of material science in packaging. This technology was developed to improve barrier performance pertaining to gases such as oxygen and carbon dioxide. It also enhances the barrier performance to ultraviolet rays, as well as adding strength, stiffness, dimensional stability, and heat resistance. New plastics created with this technology demonstrate an increased shelf life and are less likely to shatter. Once perfected, these plastics will offer these improved characteristics at competitive prices. It will also make them attractive for use in food and beverage packaging and pharmaceutical packaging applications.

In 2001, nearly 100,000 pounds of nanocomposite materials were used in the packaging industry; by 2011, the market in the United States is expected to be around 100 million pounds for rigid and flexible packaging. Polymer nanocomposites consist of resins [either thermoset or thermoplastics] that have fillers added with a least one dimension measured in nanometers. These fillers fall into three categories: nanotubes, nanoscale oxides and metals, and nanoclays.

Nanotube-based nanocomposites are used for electrostatic dissipation applications; nanoscale oxides and metals are used for abrasion-resistant films; and nanoclay-based nanocomposites are used for barrier packaging applications. Nanocomposites have received considerable attention over the past decade because of their potential to achieve property enhancements significantly greater than that attainable using conventional fillers or polymer blends. The use of nanoclay-based nanocomposites in food and beverage packaging applications, and the properties that make them superior to conventional plastics, will be focused on.

Nanocomposites were first referenced in the 1950s and polyamide nanocomposites were reported around 1976. In the early 1980s Toyota's Central Research and Development Labs began working with polymer-layered silicate-clay mineral composites and this is when the technology began to be studied more widely. The clay mineral that is generating the most interest for use in nanocomposites is montmorillonite, generically referred to as nanoclay, and sometimes referred to as bentonite. It is natural clay that is most commonly formed by the in situ alteration of volcanic ash or by the hydrothermal alteration of volcanic rocks. This clay is widely available and relatively inexpensive, thus becoming the most widely used clay in nanocomposite applications.

The first step in the preparation of the nanoclay involves purifying approximately 99 per cent of the montmorillonite. The second step involves surface treatment of the clay. Montmorillonite is hydrophilic and relatively incompatible with most hydrophobic polymers, so it must be chemically modified to make its surface more receptive to dispersion. After the clays are chemically treated, they are dispersed in the polymer. The clays are incorporated into the polymer matrix by one of two approaches: during polymerization or by melt compounding. This is the difficult part of the technology and may limit the use of nanocomposites. The dispersion process requires a custom solution for each polymer used, so developing polymer nanocomposites becomes a capitol intensive research and development project. Very few compounding firms have this kind of capability, and this leaves resin producers or well-funded startup companies to develop these materials

Nanoclay Composites Create a Torturous Path

A nanocomposite behavior is entirely dependent on the size of the clay fibers added. The smaller and thinner the fiber, the more surface area is available to interact with the polymer matrix and an improved plastic. The size of the clay fibers has grown smaller, which means that even tiny amounts of filler (two per cent of total volume) can be added to achieve the same effects. The nanoclays used in plastic composites are extremely small, irregular platelets. They are approximately one nanometer thick and 100 nanometers in diameter. A single human hair is 80,000 nanometers thick, so one can understand how a little goes a long way. Once these tiny flat platelets are dispersed into the plastic, they create a path that gases must follow to move through the material, greatly slowing their transmission.

Nanocomposites Deliver

Today's nanocomposites typically demonstrate unique improvements in materials properties, including rigidity, strength, and barrier characteristics, while maintaining a level of transparency and offering the potential for recyclability.

Polymer-clay promise a veritable explosion of thermoplastic polymer applications because they are lightweight materials that rival metals in stiffness and strength while providing enhanced gas-barrier characteristics, superior dimensional stability, and high heat distortion temperatures—all with low mineral loading and virtually no loss in impact resistance.

Making this technology even more impressive is the fact that the properties improve with few penalties (such as higher density, brittleness, or loss of clarity) than with conventional reinforcements such as talc or glass.

Nanocomposites Find a Niche in Food and Beverage Packaging

Nylon is often touted as one of the most promising resins for nanocomposite additives. In packaging, the major application focus for nylon 6 nanocomposites is in high-barrier packaging. Much of the attention is on PET bottles, where nanocomposites demonstrate an improved oxygen and carbon dioxide barrier.

Honeywell's Aegis Role

Honeywell offers three grades of nanoclay nylon 6 resins; Aegis OX, HFX, and CSD. Aegis OX resin is an oxygen-scavenging nylon formulated for the high oxygen barrier demands of plastic beer bottles. Honeywell has turned its attention to creating nano-nylon materials that can beat the cost of higher-barrier plastics or even glass. The nanoclays act as the passive barrier and the nylon specific oxygen scavenger acts as the active agent

Honeywell claims this one-two punch brings a 100 fold reduction in OTR versus nylon 6, taking oxygen ingress to near-zero levels. Aegis OX demonstrates excellent resistance to delamination, it can be processed easily, has excellent clarity, is recyclable, and cost competitive. Aegis OX is currently used in a three-layer PET bottle where it is the core layer for a 12-ounce premium beer bottle for Anchor Brewing of San Francisco

Honeywell's Aegis HFX Role

Aegis HFX is specifically formulated to address the growing requirement for improved delamination resistance in hot-fill bottle applications. It is a passive-active system that adds more oxygen resistance for foods. Customer trials have demonstrated that Aegis HFX meets shelf life requirements for hot-fill applications including juice, tea, and condiments.

Honeywell's AEGIS CSD Role

For carbonated soft drinks, Aegis CSD provides delamination resistance during production. It also provides a high level of a passive carbon dioxide barrier compared with monolayer PET. At 3.5 per cent of bottle weight, Aegis CSD extends the shelf life of 0.5 liter bottles from 9 to 14 weeks. New data extends the shelf life to 16 weeks.

Honeywell's Specialty Film Roles

Films produced from Honeywell Specialty Polymers' provide excellent barrier properties to oxygen, flavors, and aromas. Each resin provides the extra performance needed for suitable films for a wide range of flexible packaging applications. They also provide toughness, strength, tear, and puncture resistance, and resistance to grease and gas penetration. Applications for this film include processed meat, fresh red meat, poultry, fish, cheese, dried food, and chilled fruit juices.

Nanocor-Mitsubishi's Gas Chemical Company, Incorporated M9 Role

Nanocor has a different high barrier option. Through an alliance with the Mitsubishi Gas Chemical Company, Inc., Nanocor has rights to melt-compound its nanoclay additives with Mitsubishi's MXD6 nylon for use in barrier PET bottles and films under the trade name of "M9".

Mitsubishi Gas Chemical Company, Inc. has supplied MXD6 plastics to the packaging market for 15 years. MXD6 features an excellent barrier for foods and beverages against harmful gases. It is preferred over competitive barrier plastics because it is easy to process and yields packages with high transparency. M9 boosts the CO_2 and O_2 barrier of standard MXD6 respectively by 50 per cent and 75 per cent. The material also retains high clarity and delamination resistance equal to standard MXD6. It is currently being used as the core of a three-layer 16-ounce non-pasteurized PET beer bottle and is said to have 100-fold lower OTR than that of straight PET. It also adheres to PET without tie layers and retains sufficient clarity to meet requirements for the amber bottle.

M9's superior gas barrier provides flexibility in bottle design with a range of layer thickness to meet specific shelf life needs. Layer adhesion is virtually the same as standard N-MXD6, so it can be separated from PET by normal recycling methods. Because M9 is miscible with PET at layer thickness of five percent or lower, recyclers have the option of reprocessing thin-layer bottles in total, eliminating the separation step.

Rancor-Mitsubishi is able to extend the shelf life of a 28 gram weight PET bottle from 14 weeks to 21 weeks. They use the standard N-MXD6 with a 5 per cent barrier layer of M9 due to its high carbon dioxide barrier (see Figure III–G.1). The cut-off is at 90 per cent CO_2 retention.

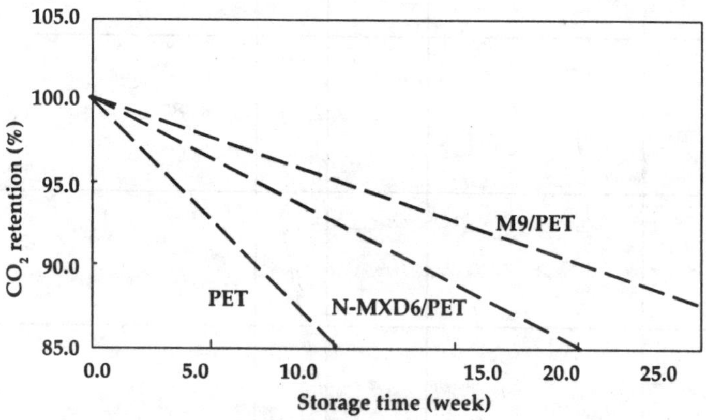

Fig. III–G.1: CO_2 Retention of Multilayer Bottle.

Triton Systems, Incorporated ORMLAS Role

Triton Systems, Inc. also has a class of commodity polymers that have been enhanced with organically modified nanoclays. These materials are sold under the trade name, ORMLAS, and have demonstrated improved barrier performance, impact resistance, as well as flame resistance.

ORMLAS have been incorporated into a long-life food tray that offers high barrier performance and impact resistance. They also include high clarity, ease of manufacture, and recyclability.

Nanoclay Resin Solution

In addition to its alliance with Mitsubishi, Nanocor offers its nanoclay resins to other converters for use in various consumer applications (Figure III–G.2). The dominant property improvement seen is a higher quality shelf life. This improvement can lead to lower weight packages because less material is needed to obtain the same, and if not better, barrier properties. This, in turn, can lead to reduced package cost. Improved shelf life and lower package cost are the reasons why nanotechnology is being pursued in consumer packaging.

Bayer's Durethan Role

Bayer's research team is aiming nylon 6 nanocomposites at cast film. It is referred to as a hybrid system that uses Nanocor's clay to reduce OTR by around 50 per cent versus nucleated nylon. Stiffness of the nanocomposite is doubled, and its gloss and clarity rival those of a costly high-clarity polyamide film. Anti-blocking properties are also improved. The new film material with nanoparticles unites the advantages of EVOH and polyamide 6. The embedded particles prevent gases from penetrating the film and moisture from escaping. Durethan KU 2-2601 is suitable for applications where conventional polyamides are too permeable and EVOH is too expensive. Bayer's Durethan KU 2-2601 makes an ideal candidate as a plastic coating for paperboard juice containers. It can protect highly oxygen-sensitive package contents, such as orange juice, at a much lower cost.

	Improved Shelf Life	*Cost Reduction*	*Hot Strength*	*Door Contract*	*Vitamin Prestntation*	*Flavour Scaling*	*Improved Printability*	*Chemical Resistance*
BOTTLES								
Films	x				x	x		
Soft Drink	x							
Cat Sup	x							
Beer	x					x		
Vegetable Oil	x							
BAGS & PACKS								
Fatty Snacks	x							
Smoked Meat	x			x				
Cheese	x	x		x				
Boil-in-Bags		x	x					
Dog Food	x			x	x			
Electronics		x						
FILMS								
Fresh Meats	x	x						
Vacuum-Pack-Cheese	x	x		x				
CARTONS								
Juice	x	x			x	x		
MISC.								
Soup Lids			x				x	
Stano-up-Pouches		x					x	
Microwave Trays	x		x					
Sausage Casings		x						
Aerosol Can Liners		x						
Bubble Wrap		x						
Antfereeze Containers		x						x
Cigrette Wrappers	x		x					

Fig. III–G.2: Nanocomposite Consumer Packaging Applications versus Property Improvements. (Nanocor, 2004)

Challenges

Real world applications for nanocomposites are coming slowly. There appears to be a reluctance to embrace this new technology due to cost and variability in the quality of some of the products.

John Jones, a market development specialist at Honeywell, believes that manufacturers of these materials have to prove to the market that these new materials can meet their performance expectations.

Another challenge nanocomposite producers face is the production of the nanocomposite itself. Both methods, pre-polymerization and post polymerization, for preparing nanocomposites have drawbacks. Pre-polymerization production can disrupt the polymerization process, which is often critical and requires much developmental time and expense to achieve good yields and controllability, and post polymerization often requires a lot of time to achieve a good dispersion of the nanoparticles in the composite. This then becomes an expensive and low cost-competitive initiative.

Another concern deals with equipment conversion that accepts new material through recalibration; this is a big investment for converters to make. It's a complicated process to go from plastic pellets to a blown bottle. It requires heating and blowing that form to the shape of the bottle. This is expensive equipment, very high-speed equipment, designed for the material that you're going to run. You can't just take another material with different flow characteristics, crystallization rate, and those kinds of things, throw that in there and make it run.

Conclusion: Nanocomposites Greatest Advance in the 21st Century

Today, nanocomposite research is widespread and is conducted by companies and universities across the globe. Plastic suppliers who have already commercialized nanocomposite materials include Basell USA, Bayer, Dow Chemical, Eastman Chemical, Mitsubishi Gas Chemical, Nanocor, Triton Systems, Honeywell, and RTP Co. Most of these efforts are currently focused on either polyolefins or nylons, but, in theory, the clay nanoparticles could be used in any resin family. Optimism surrounding these novel materials has increased since they burst into industry consciousness two or three years ago, and exploratory effort has intensified as a growing body of data substantiates the potential established by nylon/clay nanocomposites, emerging polyolefin versions, and a range of other resin matrixes and nanofillers...the promise of nanocomposites is undiminished.

It appears that the momentum is building. The success of Honeywell, Mitsubishi Gas and Chemical, Bayer, Triton Systems and Nanocor will lead to other successes. As production reaches a sufficient scale, incorporating the clay into polymers will become more cost-effective as well. Equipment recalibration used for conventional plastic resins will also show success.

The popularity of nanocomposites comes from the fact that a little goes a long way. It provides a marked increase in oxygen, carbon dioxide, moisture and odor barrier properties, increased stiffness, strength and heat resistance, and maintains film clarity and impact strength. For the packaging industry, these new materials and their commercial applications are coming into focus.

New Spark in Conductivity

Suppliers of various nano-chemicals, nanofibers, and nanotubes argue that the potential of their technologies exceeds that of current nanoclay materials. But the prices are prohibitively high, and practical impacts probably lie five years ahead.

Static-dissipative applications are emerging as one potentially large market. Applied Sciences has developed a vapor-grown carbon nano-tube made by pyrolysis of coal. Pyrograf-III comes in 100- and 200-nm diam. and has potential as an electrically conductive additive and modifier of plastics' coefficient of thermal expansion.

Max Lake, president of Applied Sciences, says the nanotubes enhance electrical conductivity over a broad resistivity range and boost mechanical properties. Just 0.5 per cent loading provides volume resistivity in the 104 ohm-cm range.

Another class of nanotube is a graphitic carbon type that is designed primarily to enhance electrical conductivity. These nano-tubes or "fibrils" are made from a hydrocarbon gas by Hyperion Catalysis International. They offer surface resistivity of 103 to 105 ohm/sq at 4 per cent to 7 per cent loadings.

"Fibrils are more efficient at building electrical conductivity into plastics than carbon black or PAN carbon fibers," states John Hagerstrom, Hyperion's technical service manager. He says the fibrils' small diameter (10 nm average) and high aspect ratio (1000:1) mean a given level of conductivity is achieved at lower loadings than with conventional carbon particles or fibers. That reportedly means less sacrifice of matrix properties, lower warpage, and better surface smoothness. Fibrils have been used to enhance electrostatic painting of automotive mirrors, and as a static-dissipative additive in semiconductor components and disk drives, where the non-sloughing feature of nanotubes helps retain purity.

Fig. III–G.3: Conductive nanocomposites can assist electrostatic painting of automotive parts.

Hybrid Plastics offers "POSS" nanochemicals, named for their polyhedral oligomeric silsesquioxane molecular building blocks. These molecular silicas are hybrid organic-inorganic materials said to bridge conventional differences between minerals and monomers.

POSS molecules are cage-like structures typically measuring 1.5 nm along each axis. The single molecule particles are truly dispersable and have no affinity for one another. Even loadings of 50 per cent or more by weight reportedly disperse without agglomeration. POSS molecules dissolve in a plastic melt, then recrystallize on cooling into a network that enhances mechanical and thermal properties, as well as flame-retardancy.

These chemical tools allow the creation of superior reinforced, crosslinked, and chemically coupled alloys. A 10 per cent POSS loading elevates PP's flexural modulus by 12 per cent, HDT by 21 per cent, and impact strength by 36 per cent. A 50 per cent POSS loading in crystal PS reportedly has no effect on optical clarity. Hybrid cites optical disks, micro-electronics, and medical products as target niches. At present, POSS costs around $200/lb. A larger plant for making POSS nanochemicals, due on-line in early 2002, could push down the cost of some POSS products to about $15/lb. Hybrid Plastics' sales and marketing arm is Divex, Inc.

Argonide has a technology for producing alumina ceramic nano-whiskers by electro-explosion of metal wire. These NanoCeram whiskers offer potential for reinforcement because of their small size (2 nm diam.) and high aspect ratio (50:1 average). Applications for NanoCeram are being pursued as reinforcements and thermally conductive additives. Current price of pilot-plant materials is $280/lb, but that is likely to fall as usage develops.

Big Step for Nanocomposites in Automotive Exteriors

For the first time, a PP-based nanocomposite has been adopted for use in an automotive production part. A step-assist for 2002 General Motors Safari and Astro vans is molded by Blackhawk Automotive Plastics, Troy, Mich., from a TPO matrix reinforced with 2.5 per cent nanoclay supplied by Southern Clay Products. The part is a low-volume, dealer-installed option. The nano-TPO replaces a 15 per cent talc-filled PP but is molded in the same tool.

This is a beginning. Parties to this effort are investigating other exterior and interior automotive parts.

Nano-TPO delivers stiffness equivalent to the talc-filled PP it replaced, translating into 7-8 per cent weight savings. Further, nano-TPO improves low-temperature impact strength and part surface quality. GM officials termed the project "cost-neutral," explaining that the higher price of Basell's nano-TPO is offset by lightweighting and other cost benefits.

Fig. III–G.4: First automotive production part of a polyolefin nanocomposite is a TPO step-assist for 2002 General Motors Safari and Astro vans.

The nano-TPO processes at a somewhat lower temperature than talc-filled PP, decreasing cycle time, while also decreasing knit-line creation and other surface imperfections. Reports also show molding other prototype exterior claddings, including rocker and sail panels and body side moldings. Newer versions of Basell materials are being investigated, including some containing up to 5 per cent nanoclay and offering stiffness comparable to 30-40 per cent talc-filled PP. It is also noted that the much lower filler levels in nano-TPOs result in smoother surfaces while enhancing scratch resistance.

When GM's R&D unit first announced the nano-TPO project in 1999, emphasis then was on vertical body panels and interior parts. Instead, the first commercial application is a relatively simple, low-volume part. But GM declares that there is no retreat from the applications initially targeted. Work is well under way on a rocker-panel prototype that is 20 per cent lighter than one of talc-filled PP. Interior trim is also being actively explored.

A significant advance in chemistry by GM is critical to success of the van step. It allows more complete exfoliation (layer separation) and better dispersion of nanoclay particles into TPO. GM has granted Basell exclusive rights to use its advances in nano-PP technology. In turn, GM has been guaranteed exclusive rights to use Basell's nano-TPOs in the automotive field.

Unit III–H

More Application of Nanocomposites

Such mechanical property improvements have resulted in major interest in nanocomposite materials in numerous automotive and general/industrial applications. These include potential for utilisation as mirror housings on various vehicle types, door handles, engine covers and intake manifolds and timing belt covers. More general applications currently being considered include usage as impellers and blades for vacuum cleaners, power tool housings, mower hoods and covers for portable electronic equipment such as mobile phones, pagers etc.

Gas Barriers

The gaseous barrier property improvement that can result from incorporation of relatively small quantities of nanoclay materials is shown to be substantial. Data provided from various sources indicates oxygen transmission rates for polyamide-organoclay composites which are usually less than half that of the unmodified polymer. Further data reveals the extent to which both the amount of clay incorporated in the polymer, and the aspect ratio of the filler contributes to overall barrier performance. In particular, aspect ratio is shown to have a major effect, with high ratios (and hence tendencies towards filler incorporation at the nano-level) quite dramatically enhancing gaseous barrier properties. Such excellent barrier characteristics have resulted in considerable interest in nanoclay composites in food packaging applications, both flexible and rigid. Specific examples include packaging for processed meats, cheese, confectionery, cereals and boil-in-the-bag foods, also extrusion-coating applications in association with paperboard for fruit juice and dairy products, together with co-extrusion processes for the manufacture of beer and carbonated drinks bottles. The use of nanocomposite formulations would be expected to enhance considerably the shelf life of many types of food.

Oxygen Barriers

Honeywell have also been active in developing a combined active/passive oxygen barrier system for polyamide-6 materials. Passive barrier characteristics are provided by nanoclay particles incorporated via melt processing techniques whilst the active contribution comes from an oxygen scavenging ingredient (undisclosed). Oxygen transmission results reveal substantial benefits provided by nanoclay incorporation in comparison to the base polymer (rates approximately 15-20 per cent of the bulk polymer value, with further benefits provided by the combined active/ passive system). Akkapeddi suggests that the increased tortuosity provided by the nanoclay particles essentially slows transmission of oxygen through the composite and drives molecules to the active scavenging species resulting in near zero oxygen transmission for a considerable period of time.

Food Packaging

Triton Systems and the US Army are conducting further work on barrier performance in a joint investigation. The requirement here is for a non-refrigerated packaging system capable of maintaining food freshness for three years. Nanoclay polymer composites are currently showing considerable promise for this application.

It is likely that excellent gaseous barrier properties exhibited by nanocomposite polymer systems will result in their substantial use as packaging materials in future years.

A somewhat more esoteric possibility arising from enhanced barrier performance recently suggested has been blown–films for artificial intestines!

Fuel Tanks

The ability of nanoclay incorporation to reduce solvent transmission through polymers such as polyamides has been demonstrated. Data provided by De Bievre and Nakamura of UBE Industries reveals significant reductions in fuel transmission through polyamide–6/66 polymers by incorporation of a nanoclay filler. As a result, considerable interest is now being shown in these materials as both fuel tank and fuel line components for cars. Of further interest for this type of application, the reduced fuel transmission characteristics are accompanied by significant material cost reductions.

Films

The presence of filler incorporation at nano-levels has also been shown to have significant effects on the transparency and haze characteristics of films. In comparison to conventionally filled polymers, nanoclay incorporation has been shown to significantly enhance transparency and reduce haze. With polyamide based composites, this effect has been shown to be due to modifications in the crystallisation behaviour brought about by the nanoclay particles; spherilitic domain dimensions being considerably smaller. Similarly, nanomodified polymers have been shown, when employed to coat polymeric transparency materials, to enhance both toughness and hardness of these materials without interfering with light transmission characteristics. An ability to resist high velocity impact combined with substantially improved abrasion resistance was demonstrated by Haghighat of Triton Systems.

Environmental Protection

Water laden atmospheres have long been regarded as one of the most damaging environments which polymeric materials can encounter. Thus an ability to minimise the extent to which water is absorbed can be a major advantage. Data provided by Beall from Missouri Baptist College indicates the significant extent to which nanoclay incorporation can reduce the extent of water absorption in a polymer. Similar effects have been observed by van Es of DSM with polyamide based nanocomposites. In addition, van Es noted a significant effect of nanoclay aspect ratio on water diffusion characteristics in a polyimide nanocomposite. Specifically, increasing aspect ratio was found to diminish substantially the amount of water absorbed, thus indicating the beneficial effects likely from nanoparticle incorporation in comparison to conventional microparticle loading. Hydrophobic enhancement would clearly promote both improved nanocomposite properties and diminish the extent to which water would be transmitted through to an underlying substrate. Thus, applications in which contact with water or moist environments is likely could clearly benefit from materials incorporating nanoclay particles.

Flammability Reduction

The ability of nanoclay incorporation to reduce the flammability of polymeric materials was a major theme of the paper presented by Gilman of the National Institute of Standards and Technology in the US. In his work Gilman demonstrated the extent to which flammability behaviour could be restricted in polymers such as polypropylene with as little as 2 per cent nanoclay loading. In particular heat release rates, as obtained from cone calorimetry experiments, were found to diminish substantially by nanoclay incorporation. Although conventional microparticle filler incorporation, together with the use of flame retardant and intumescent agents would also minimise flammability behaviour, this is usually accompanied by reductions in various other important properties. With the nanoclay approach, this is usually achieved whilst maintaining or enhancing other properties and characteristics.

Unit III–I

Polymer Nanocomposites

There has been enormous interest in the commercialization of nanocomposites for a variety of applications, and several of these applications will be successful in the near future.

Mineral fillers, metals, and fibers have been added to thermoplastics and thermosets for decades to form composites. Compared to neat resins, these composites have a number of improved properties including tensile strength, heat distortion temperature, and modulus. Thus, for structural applications, composites have become very popular and are sold in billion pound quantities. These filled thermoplastics are sold in even larger volumes than neat thermoplastics. Furthermore, the volume of fillers sold is roughly equal to the volume of thermoplastic resin sold. Clearly, the idea of adding fillers to thermoplastics and thermosets to improve properties, and in some cases decrease costs, has been very successful for many years.

More recently, with advances in synthetic techniques and the ability to readily characterize materials on an atomic scale has lead to interest in nanometer-size materials. Since nanometer-size grains, fibers and plates have dramatically increased surface area compared to their conventional-size materials, the chemistry of these nanosized materials is altered compared to conventional materials.

Polymer nanocomposites combine these two concepts, i.e., composites and nanometer-size materials. Thermoplastics filled with nanometer-size materials have different properties than thermoplastics filled with conventional materials. Some of the properties of nanocomposites, such as increased tensile strength, may be achieved by using higher conventional filler loading at the expense of increased weight and decreased gloss. Other properties of nanocomposites such as clarity or improved barrier properties cannot be duplicated by filled resins at any loading.

Polymer nanocomposites were developed in the late 1980s in both commercial research organizations and academic laboratories. The first company to commercialize these nanocomposites was Toyota, which used nanocomposite parts in one of its popular car models for several years. Following Toyota's lead, a number of other companies also began investigating nanocomposites.

Most commercial interest in nanocomposites has focused on thermoplastics. Thermoplastics can be broken into two groups: less expensive commodity resins and more expensive (and higher performance) engineering resins. One of the goals of nanocomposites was to allow substitution of more expensive engineering resins with a less expensive commodity resin nanocomposite.

Substituting a nanocomposite commodity resin with equivalent performance as a more expensive engineering resin should yield overall cost savings.

By a strict definition of nanocomposites, i.e., any filler submicron in size, there already are significant volumes of nanocomposites being produced. These amount to more than 20 million pounds. However, since these fillers are on the upper end of the nanocomposite size range, most sources have excluded them from consideration. This report covers these products but does not add their volumes into aggregate numbers of nanocomposite markets or production volumes to comply with the established convention.

At this point in time, there has been much less commercial interest in thermoset nanocomposites compared to thermoplastics. This neglect may not continue much longer since thermoset nanocomposites have some distinct advantages over neat thermoset resins. This report covers some possible markets for thermoset nanocomposites.

There are two types of recognized polymer nanocomposites:

- Thermoplastic nanocomposites: these materials are broken into two major categories, i.e., commodity resins and engineering resins. The potential of polymer nanocomposite commodity resin is covered for each resin. Engineering resin nanocomposites are restricted to resins that have been under development.
- Thermoset nanocomposites: thermoset nanocomposites have received less commercial interest in their development than thermoplastic nanocomposites, but these materials may be relatively straightforward to bring into production. Furthermore, thermoset nanocomposites can offer some significant advantages over conventional thermosets.

Processing Polymer Nanocomposites

Polymer nanocomposite (PNC) is a polymer or copolymer having dispersed in it nanoparticles. These may be of different shape (e.g., platelets, fibers, spheroids), but at least one dimension must be ca. 1 to 50 nm (diameter of pencil lead is about 1 mm, or 1,000,000 nm). PNC with all three types of nanoparticles have been prepared (e.g., polycarbonate with carbon nanotubes, polyamide with iron oxide spheres), but only PNC with clay platelets are on the market for the structural (high volume) applications. These PNC's belong to the category of multi-phase systems (MPS, viz. blends, composites, and foams) that consume ca. 95 per cent of plastics production. The key to industrial success of MPS is the desired and stable morphology. Thus, these systems require controlled mixing/compounding, stabilization of the achieved dispersion, orientation of the dispersed phase, and optimization of interactions in the finished product. In spite of differences in the nature of the dispersed phase, the compounding strategies for all MPS, including PNC, are similar.

Mixing

Mixing is the most important operation in polymer processing. Uniformity of the molecular weight and its distribution, entanglement, composition, as well as the stress history and temperature, is the key to good performance. Dispersive (or intensive) and distributive (or extensive) mixing of molten polymers is carried out within the laminar flow regime. The dispersive action is responsible for the breaking of original domains of the minor component into the desired size, while the distributive provides for their homogeneous distribution within the matrix. Microrheology may be used to describe the dispersive process in the shear or extensional flow field, while distributive mixing may be treated in terms of the laminar flow theory. Both models

- microrheology and laminar mixing - indicate superiority of the extensional flow field. The presence of the vorticity component in the shear tensor is responsible for rotation of the flow elements that cause excessive energy consumption, polymer degradation and attrition of anisometric solid particles - shear mixing is inefficient.

Mixing in the extensional flow field may be carried out by means of the convergent-divergent flow geometry. A motionless extensional flow mixer (EFM) and its newer, dynamic version (DEFM) are fully adjustable, general-purpose mixers used for homogenization of reaction products, removal of "fish eyes" from films, alloying, blending or dispersing of nanoparticles[4]. In the search for synergism the industry moves toward progressively complex MPS, e.g., foaming of nanoclay reinforced polymer blends. Here, the difficulty is the presence of several phases with potentially different thermodynamic and rheological character as well as reinforcing and foaming the appropriate blend components. However, once the difficulties are over, the advantages translate into superior performance.

Unit III–J

Clay-containing Polymeric Nanocomposites (CPNC)

The clay-containing polymeric nanocomposites (CPNC) constitute new type of polymeric systems, with a great variety of molecular structures and processing difficulties. The only type of nanoparticles used in commercial production of CPNC is the lamellar, expandable, smectite clay. For the best performance of CPNC the clay platelets (ca. 1 nm thick and 300 nm diameter) must be homogeneously dispersed as individual crystalline lamellae (known as exfoliation). To accomplish this task, a balance between the thermodynamics and mechanical mixing must be achieved. Owing to the large aspect ratio, at a concentration of clay exceeding ca. 1 wt per cent the platelets start to interact, creating 3D structures. In commercial CPNC the clay content is about twice as large, hence there is local ordering. Generation of the most suitable orientation is the key for the optimization of the performance. Three recent developments in the CPNC science and technology are worth noting.

Dispersion in Extensional Flow Field

The use of an elongational flow field greatly accelerates dispersion and orientation, facilitating the processing. Several CPNC types (e.g., with polyethylene terephthalate, PET, polystyrene, PS, or polypropylene, PP, matrix) have been prepared using commercial EFM attached either to single-screw extruder (SSE) or twin-screw extruder (TSE). The results demonstrated significant beneficial role even of the standard EFM (developed for melt homogenization and polymer blending). For example, PET was extruded with organoclay using a TSE and the same machine, but with attached to it EFM. The presence of EFM increased the PNC properties over that from TSE by (data from EFM Inc.):

- Tensile modulus by ca. 67 per cent.
- Tensile strength by ca. 60 per cent.
- Impact strength 80 per cent.

In Union Chemical Laboratory, Industrial Technology Research Institute (UCL/ITRI) in

Taiwan, PP was compounded with organoclay and a compatibilizer (maleated polypropylene, PP-MA)[5]. Three processing methods were used: #1: TSE; #2: TSE + EFM; and #3: SSE + EFM. The method #1 resulted in mixtures with clay dispersed as micron-sized aggregates. By contrast, process #2 produced well-intercalated CPNC with short stacks of organoclay uniformly dispersed in the molten PP. However, the best dispersion was obtained from #3, i.e., using SSE + EFM. This unexpected result was most likely due to higher thermo-mechanical degradation in TSE.

Degradation of Organoclay

The ammonium-ion intercalated organoclay is the main type used for the production of CPNC. However, these materials start to degrade at temperatures above 180°C by the Hofmann elimination mechanism, leaving the clay surface bare. Numerical simulations were carried out to evaluate the effects of independent parameters on the degree of CPNC dispersion. The most dramatic worsening of the thermodynamic conditions for exfoliation was obtained in the presence of bare clay platelets —even as small amount as 1 or 2 per cent of clay surface precludes exfoliation.

New Production Method for CPNC

Kato et al. from the Toyota Central R&D Lab proposed a revolutionary method of CPNC production. The method uses un-treated clay, e.g., sodium montmorillonite (Na-MMT), which becomes exfoliated and the morphology stabilized during the mixing/compounding step. The process uses a long (L/D = 77) TSE, divided by a pair of seal rings into three zones. In the first zone PP + PP-MA + Na-MMT were melt mixed together. In zone #2 ca. 10 per cent water was injected into the melt and the pressure of the water vapor was well controlled. Adsorption of water by clay platelets, followed by reaction with compatibilizer (PP-MA), resulted in exfoliation and stabilization of the dispersion. Finally, in zone #3 devolatilization took place. The resulting CPNC had almost the same properties as the conventionally prepared one, but at a fraction of cost. The method bypasses the need for the costly intercalation of Na-MMT with ammonium ions, avoids the introduction of readily degradable organoclay to molten polymer (which in turn gets degraded by combination of organoclay degradation and air) and at the same time eliminates the need of costly in time and energy pre-drying step.

Evidently, the method is in an early stage of development. Its extension to other polymers (e.g., polyesters or polyamides), optimization of the compatibilizer, processing equipment and process parameters are required.

Future of CPNC

The main advantages of CPNC stem from the improved stiffness (by a factor of 2 at 5 wt per cent loading), reduction of gas and vapor permeability (by a factor of 100 at 10 wt per cent loading), and reduced flammability. Since these enhancements are achieved with only a slight increase of density, the transport, and packaging industries are expect to be the first to profit from this new technology. Up to now two principal obstacles hindered progress of CPNC into the market: the cost and reproducibility of performance. The demonstrated by Kato et al. suitability of Na-MMT reduces the cost of the nano-filler from ca. $7 to $2 per kilogram, and at the same time reduces the probability of thermo-mechanical degradation during processing and forming. Many questions remain the foremost being whether the achieved improvement of technology is sufficient to lower the cost and improve stability to satisfy the industry standards. Evidently, the future of CPNC technology hinges on the intelligent application of the available compounding and processing technologies, adopted for the specific needs of the nano-scale mixing.

Unit III–K

Clay-based Nanocomposites

The subject of hybrids based on layered inorganic compounds such as clays has been studied for a considerable time, but the area is enjoying a resurgence of interest and activity as a result of the exceptional properties which can be realized from such nanocomposites.

Materials variables which can be controlled and which can have a profound influence on the nature and properties of the final nanocomposite include the type of clay, the choice of clay pre-treatment, the selection of polymer component and the way in which the polymer is incorporated into the nanocomposite. The last of these may be dictated by the available processing methods and whether the prospective user is an integrated polymer manufacturer or a specialist processor.

Clays and Clay Modifications

Common clays are naturally occurring minerals and are thus subject to natural variability in their constitution. The purity of the clay can affect final nanocomposite properties. Many types of clay are aluminosilicates, which have a sheet-like (layered) structure, and consist of silica SiO_4 tetrahedra bonded to alumina AlO_6 octahedra in a variety of ways. A 2:1 ratio of the tetrahedra to the octahedra results in smectite clays, the most common of which is montmorillonite. Other metals such as magnesium may replace the aluminium in the crystal structure. Depending on the precise chemical composition of the clay, the sheets bear a charge on the surface and edges, this charge being balanced by counter-ions, which reside in part in the inter-layer spacing of the clay. The thickness of the layers (platelets) is of the order of 1 nm and aspect ratios are high, typically 100-1500. The clay platelets are truly nanoparticulate. In the context of nanocomposites, it is important to note that the molecular weight of the platelets (ca. 1.3×10^8) is considerably greater than that of typical commercial polymers, a feature which is often misrepresented in schematic diagrams of clay-based nanocomposites. In addition, platelets are not totally rigid, but have a degree of flexibility. The clays often have very high surface areas, up to hundreds of m^2 per gram. The clays are also characterized by their ion (e.g. cation) exchange capacities, which can vary widely. One important consequence of the charged nature of the clays is that they are generally highly hydrophilic species and therefore naturally incompatible with a wide range of polymer types. A necessary prerequisite for successful formation of polymer-clay nanocomposites is therefore alteration of the clay polarity to make the clay 'organophilic'. An organophilic clay can be produced from a normally hydrophilic clay by ion exchange with an organic cation such as an alkylammonium ion. For example, in montmorillonite, the sodium ions in the clay can be exchanged for an amino acid such as 12-aminododecanoic acid (ADA):

Na^+-CLAY + HO_2C-R-$NH_3^+Cl^-$ → .HO_2C-R-NH_3^+-CLAY + NaCl

The way in which this is done has a major effect on the formation of particular nanocomposite product forms and this is discussed further below. Although the organic pre-treatment adds to the cost of the clay, the clays are nonetheless relatively cheap feedstocks with minimal limitation on supply. Montmorillonite is the most common type of clay used for nanocomposite formation; however, other types of clay can also be used depending on the precise properties required from the product. These clays include hectorites (magnesiosilicates), which contain very small platelets, and synthetic clays (e.g. hydrotalcite), which can be produced in a very pure form and can carry a positive charge on the platelets, in contrast to the negative charge found in montmorillonites.

Synthetic Processing of Clay-Based Nanocomposites

The synthetic route of choice for making a nanocomposite depends on whether the final material is required in the form of an intercalated or exfoliated hybrid (Figure III–K.1). In the case of an intercalate, the organic component is inserted between the layers of the clay such that the inter-layer spacing is expanded, but the layers still bear a well defined spatial relationship to each other. In an exfoliated structure, the layers of the clay have been completely separated and the individual layers are distributed throughout the organic matrix. A third alternative is dispersion of complete clay particles (tactoids) within the polymer matrix, but this simply represents use of the clay as conventional filler.

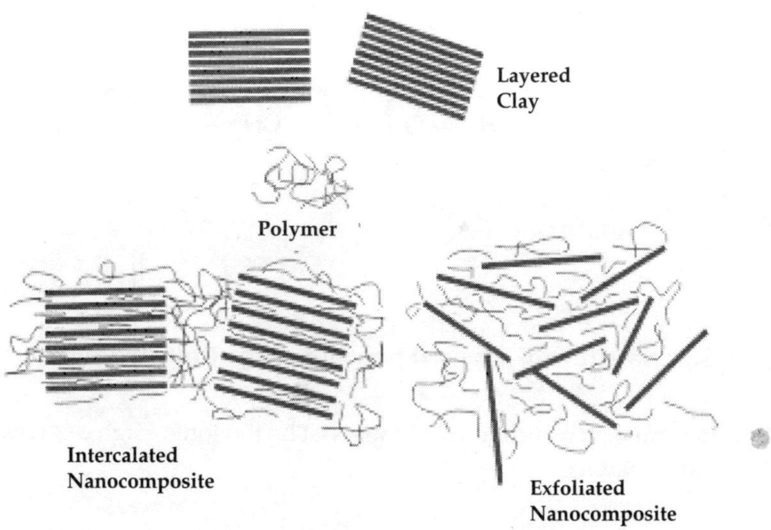

Fig. III–K.1: Formation of intercalated and exfoliated nanocomposites from layered silicates and polymers. (not to scale)

Factors Affecting the Type of Organo-Clay Hybrid Formed

In recent years, there has been extensive study of the factors which control whether a particular organo-clay hybrid can be synthesised as an intercalated or exfoliated structure. Since clay nanocomposites can produce dramatic improvements in a variety of properties, it is important to understand the factors which affect delamination of the clay. These factors include the exchange capacity of the clay, the polarity of the reaction medium and the chemical nature of the interlayer cations (e.g. onium ions). By modifying the surface polarity of the clay, onium ions allow thermodynamically favourable penetration of polymer precursors into the interlayer region. The ability of the onium ion to assist in delamination of the clay depends on its chemical nature such as its polarity. The loading of the onium ion on the clay is also crucial for success and it should be borne in mind that a commercial organo-clay may not have the optimum loading for a particular application. For positively charged clays such as hydrotalcite, the onium salt modification is replaced by use of a cheaper anionic surfactant. Other types of clay modification can be used depending on the choice of polymer, including ion-dipole interactions, use of silane coupling agents and use of block copolymers.

An example of ion-dipole interactions is the intercalation of a small molecule such as dodecylpyrrolidone into the clay. Entropically-driven displacement of the small molecules then provides a route to introducing polymer molecules. Unfavourable interactions of clay edges with polymers can be overcome by use of silane coupling agents to modify the edges. These can be used in conjunction with the onium ion treated organo-clay. An alternative approach to compatibilising clays with polymers has been introduced by TNO (Netherlands), based on use of block or graft copolymers where one component of the copolymer is compatible with the clay and the other with the polymer matrix. This is similar in concept to compatibilisation of polymer blends. A typical block copolymer would consist of a clay-compatible hydrophilic block and a polymer-compatible hydrophobic block (Fig. III–K.2).

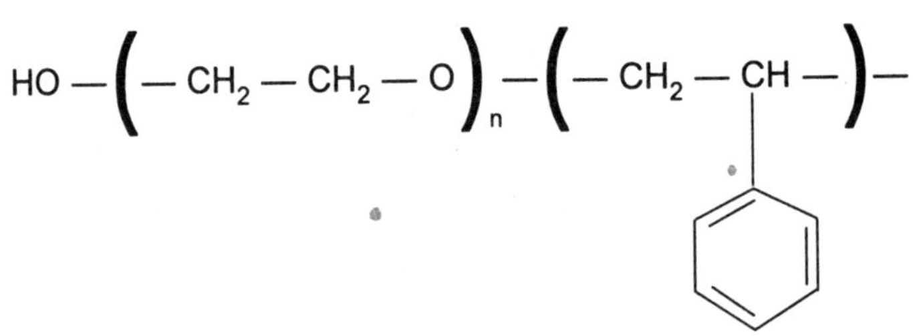

Fig. III–K.2: Structure of a typical polymer-compatible hydrophobic block.

The block length must be controlled and must not be too long. High degrees of exfoliation are claimed using this approach.

Polymer Incorporation

The correct selection of modified clay is essential to ensure effective penetration of the polymer or its precursor into the interlayer spacing of the clay and result in the desired exfoliated or intercalated product. Indeed, further development of compatibiliser chemistry is undoubtedly the key to expansion of this nanocomposite technology beyond the systems where success has been achieved to date. Polymer can be incorporated either as the polymeric species itself or via the monomer, which is polymerized in situ to give the corresponding polymer-clay nanocomposite. The second of these is the most successful approach to date, although it probably limits the ultimate applicability of these systems. Polymers can be introduced either by melt blending, for example extrusion, or by solution blending. Melt blending (compounding) depends on shear to help delaminate the clay and can be less effective than in situ polymerization in producing an exfoliated nanocomposite. Both thermo sets and thermoplastics have been incorporated into nanocomposites, including:

+ Nylons
+ Polyolefins, e.g. polypropylene
+ Polystyrene
+ Ethylene-vinyl acetate (EVA) copolymer
+ Epoxy resins
+ Polyurethanes
+ Polyimides
+ Poly(ethylene terephthalate) (PET)

Unit III–L

Polyolefin Nanocomposites

General Description

Polyolefin nanocomposites are useful for a wide variety of applications in packaging, engineering and fire retardant markets. These nanocomposites are near molecular blends of montmorillonite nanoclay and a polyolefin matrix. Concentrate trade name. Here performance properties are outlined for typical polyolefins with explanations of options for creating nanocomposites.

Polyolefin Nanocomposites

Polyolefins represent a wide range of homo and copolymers, principally containing propylene and ethylene. Also included are polyolefin/rubber alloys usually called TPO's and polyolefin/EVA blends. Because propylene and ethylene contents as well as polydispersities vary widely, some Nanomer Concentrate screening is required to determine the optimum match between the matrix and nanoclay surface treatment.

Creating Polyolefin Nanocomposites

Polyolefin nanocomposites are produced via extrusion compounding using Concentrates. Refer to data for concentrate specifications and grade selection. Film or bottle grade products require twin screw compounders or intensive mixers. Most engineering and fire retardant applications can be accommodated using single screw technology. A specially designed screw, called a NanoMixer™, must be used. Compounding guidelines are available on Tech Data Sheets P-803 through P-806. The figure III–L.1 illustrates the relationship between Nanomer powders, concentrates and nanocomposites.

Nanocomposite Performance

Nanocomposite performance depends on the degree of nanoclay de-agglomeration and dispersion within the polyolefin matrix. The process of de-agglomeration/dispersion is commonly referred to as exfoliation. Three factors influence exfoliation: the nanoclay surface treatment, nanocomposite

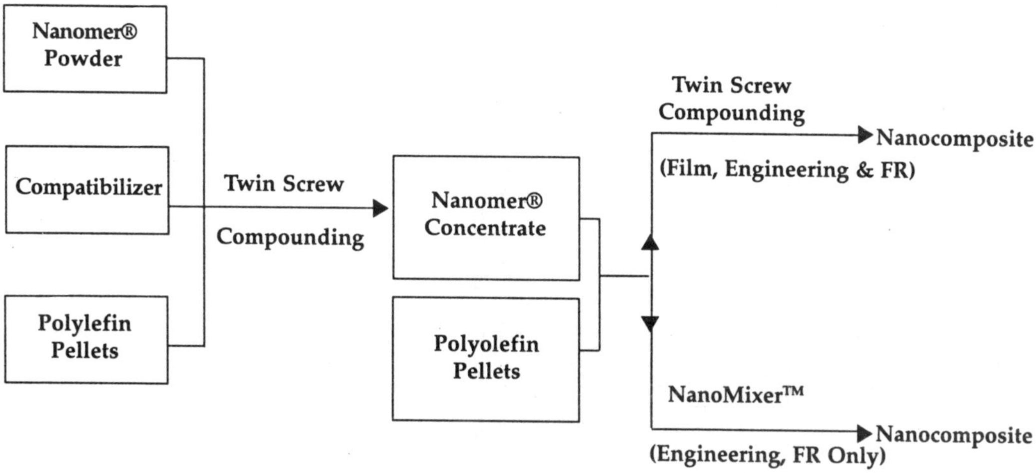

Fig. III–L.1: Relationship between Nanomer powders, concentrates and nanocomposites.

loading, and the amount of dispersive energy used.

The optimum nanoclay surface treatment is readily determined through screening, as described earlier. Nanomer loading will not hinder exfoliation within the 2-8 per cent wt/wt range and is optimized at 6 per cent wt/wt. The proper choice of compounding equipment and use of the guidelines assists in achieving the dispersive energy requirement. However, another factor impacting dispersive energy is matrix resin melt flow. Resin melt flow influences the level of shear resulting from dispersive energy input. The lower the melt flow, the greater the shear on Nanomer particles. It follows that exfoliation is maximized in lower melt flow resins, where shear is the highest.

Mechanical Properties

Polyolefin nanocomposites exhibit mechanical properties similar to highly filled systems, but with the added advantage of low density. The density (and resulting part weight) is just slightly higher than unfilled resin. Nanocompositing also provides a high degree of reinforcement without significant loss of impact. In most cases notched Izod impact remains the same as neat (unfilled) resin. Table III–L.1 provides typical mechanical performance data for a variety of polyolefins and melt flows. The Concentrate addition level is 12 per cent wt/wt in all cases. This corresponds to a final Nanomer loading of 6 per cent wt/wt.

Polyolefin nanocomposites have specific gravities and melt flows nearly the same as neat resin, so they process like neat resin. Part output rates will be significantly higher than conventionally filled composites and machine wear will also improve noticeably.

Nanocomposites can be used in conjunction with conventional fillers, i.e. glass fiber or talc, to create synergistic materials with customized property sets. For example a 6 per cent wt/wt nanocomposite containing 10-12 per cent wt/wt glass fiber will deliver a property set similar to 30 per cent wt/wt straight glass fiber, but lower specific gravity and part weight.

Fire Resistant Properties

Polyolefin nanocomposites are potent char formers. In fire retardant (FR) applications they function by insulating the resin substrate from the air-surface interface, starving the flame by reducing the

Table III–L.1: Typical mechanical performance data for a variety of polyolefins and melt flows

Resin	*Melt Flow (gms/10min)*	*Condition*	*Tensile Strength (Mpa)*	*Elongation @ Break (%)*	*Flexural Nodulus (Mpa)*	*Notched Izod ft-lb/In)*	*HDT[2] (°C)*
Homopolymer	4	Neat	32	280	1150	0.7	86
Polypropylene		Nanocomposite[1]	38	20	2040	0.7	115
Homopolymer	35	Neat	35.2	105	1590	0.4	113
Polypropylene		Nanocomposite[1]	39.9	11	2310	0.4	121
Copolymer	6.5	Neat	26.3	420	840	0.7	71
Polypropylene		Nanocomposite[1]	30.3	93	1350	1	81
HDPE	4	Neat	23.8	460	680	3	71
		Nanocomposite[1]	26.5	200	1170	1.1	77
TPD	12	Neat	19.5	320	780	9.8	71
		Nanocomposite[1]	21.8	42	1230	9.8	85

1. 12% wt/wt concentrate addition (6% wt/wt Nanomer loading)
2. 66 psi

Note: Specific gravity for the nanocomposites is 0.93 compared to 0.90 for neat resins. Reinforcement is linear across the effective loading range (2-8% wt/wt) Nanomer. The magnitude of increase per percent loading decreases above the range due to nanoclay packing. Thermal improvement is curve-linear in the effective range because nanoclay surface treatment exerts some plasticizing effect on the nanocomposite. Table 4 graphs reinforcement for homopolymer polypropylene.

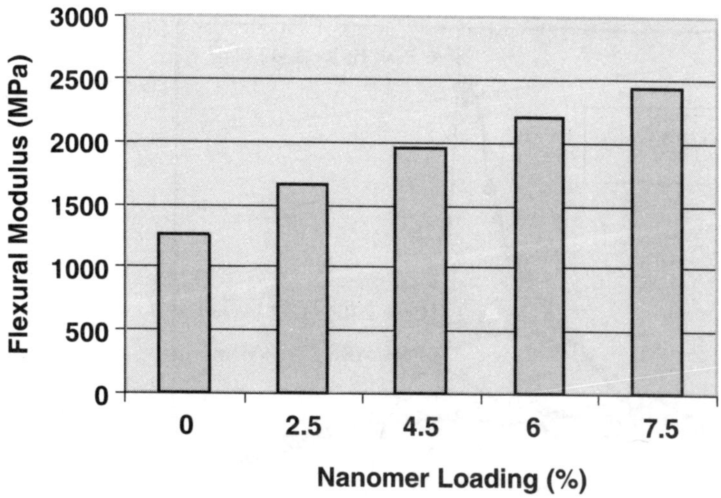

Fig. III–L.2: Graph of reinforcement for homopolymer polypropylene.

fuel needed to continue the burning process. The result is a dramatically lower peak heat release rate (HRR).

Nanocomposites alone do not conform to a UL V0 rating. But in conjunction with reduced levels of conventional FR additives they act synergistically to achieve V0. Often traditional FR additives can be reduced by 50 per cent wt/wt. With less FR additive dilution and the enhanced reinforcing effect of nanocompositing, mechanical properties are restored, usually at lower cost than would be the case without nanocompositing. Another benefit of nanocomposite/FR additive synergy is lower density and part weight, since a lot of high density additive has been removed from the formulation.

Gas Barrier Enhancement

As polyolefins take an increased role in packaging applications, gas barrier becomes an important requirement. Nanocomposites improve oxygen transmission rates (OTR) between 1.2-2X. Carbon dioxide transmission rates (CO2TR) improve 1.4-1.8X. There is also some improvement in water vapor transmission (WVTR), around 10-15 per cent. Improved gas barrier follows the tortuous path model and is valid for cast, blown and molded products. Tables III–L.2A and III–L.2B present performance data for a variety of polyolefins.

Predictably, nanocomposite films possess greater tensile strength and modulus, although elongation-atbreak is reduced. Simultaneous improvement in tensile properties and gas barrier provides the formulator with opportunities to downguage films and containers. Tear strengths in both the machine and transverse directions remain the same as neat polyolefin films of the same mil thickness. Tear strength of downguaged nanocomposite film drops in proportion to the reduction in film thickness.

Clay-Nylon Nanocomposites

The earliest example of the in situ polymerization method was work by Toyota on synthesis of clay-nylon nanocomposites and this remains probably the most studied system, including work by

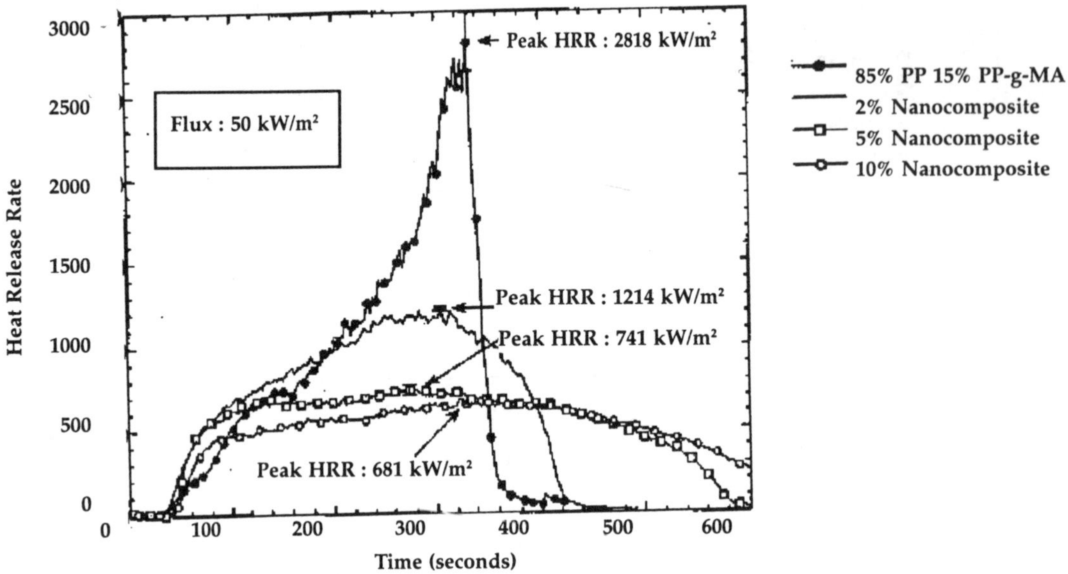

Fig. 12.3: HRR for Polypropylene Nanocomposites
Gilman, Jeffery W., Kashiwagi, Takashi, Morgan, Alexander B., Harris, Richard H. Jr., Brassell, Lori, Vanlandingham, Mark, Jackson, Catheryn L., National Institute of Standards and Testing (NIST), "Flammability of Polymer Clay Nanocomposites Consortium: Year One Annual Report".

Bayer and Ube. In a typical synthesis ADA-modified clay is dispersed in the monomer caprolactam, which is polymerized to form the nylon-6-clay hybrid as an exfoliated composite. Complete exfoliation may be preceded by intercalation of the monomer in the clay. Generally, low concentration of clay (a few per cent) are incorporated in these nanocomposites, partly because this is often sufficient to modify the desired properties significantly, but also because higher levels of clay can adversely increase the system viscosity leading to poor processability, although the viscosity increase is shear rate dependent. Other nylons and copolyamides (e.g. nylon-6/6,6) have also been incorporated in clay nanocomposites. Functionality such as hydroxyl groups can be introduced into the onium salt modifiers to improve compatibility with the nylon via hydrogen bonding—this can lead to an enhancement of desirable nanocomposite properties.

Similar modification of the chemistry may be required for successful exfoliation with other types of polymer system. In the case of ethylene-vinyl alcohol (EVOH) copolymers, use of hydroxylated quaternary ammonium ions improves compatibility between the clay and the EVOH by introducing favourable hydroxyl group interactions. In polypropylene (PP) nanocomposites, maleic anhydride grafted PP is used as a compatibiliser. Polymerization initiators can be anchored to the clay platelet surface and this approach has been extended to living free radical polymerization of styrene, where an initiating species bonded to the TEMPO free radical is attached to the surface of the clay platelets via ion exchange.

Tables III–L.2A and III–L.2B: Performance data for a variety of polyolefins Film/Bottle Strength Properties

Resin	*Melt Flow (gms/10min)*	*Product Form*	*Condition*	*Tensile Strength (Mpa)*	*Tensile Modulus (Mpa)*	*Elongation @ Break (%)*
Homopolymer olypropylene	1.3	Cast Film	Neat Nanocomposite[1]	36.4 33.4	1230 1710	660 480
Copolymer Polypropylene	2	Cast Film	Neat Nanocomposite[1]	38.7 31.7	910 1500	650 480
		Stretched Film	Neat Nanocomposite[1]	N/A N/A	N/A N/A	N/A N/A
		Cast Film	Neat Nanocomposite[1]	N/A N/A	N/A N/A	N/A N/A
HDPE	4	Cast Film	Neat Nanocomposite[1]	22.0 22.5	700 854	800 690
TPO	0.9	Cast Film	Neat Nanocomposite[1]	23.4 27.6	56 178	730 780
TPE	N/A	Injection Molded Liner	Neat Nanocomposite[1]	N/A N/A	N/A N/A	N/A N/A

1. 12% wt/wt concentrate addition (6% wt/wt Nanomer loading)

Table 12.2B: Gas Barrier Properties

Resin	*Melt Flow (gms/10min)*	*Product Form*	*Condition*	*Tensile Strength (Mpa)*	*Tensile Modulus (Mpa)*	*Elongation @ Break (%)*
Homopolymer olypropylene	1.3	Cast Film	Neat	36.4	1230	660
			Nanocomposite[1]	33.4	1710	480
Copolymer Polypropylene	2	Cast Film	Neat	38.7	910	650
			Nanocomposite[1]	31.7	1500	480
		Stretched Film	Neat	N/A	N/A	N/A
			Nanocomposite[1]	N/A	N/A	N/A
		Cast Film	Neat	N/A	N/A	N/A
			Nanocomposite[1]	N/A	N/A	N/A
HDPE	4	Cast Film	Neat	22.0	700	800
			Nanocomposite[1]	22.5	854	690
TPO	0.9	Cast Film	Neat	23.4	56	730
			Nanocomposite[1]	27.6	178	780
TPE	N/A	Injection Molded Liner	Neat	N/A	N/A	N/A
			Nanocomposite[1]	N/A	N/A	N/A

1. 12% wt/wt concentrate addition (6% wt/wt Nanomer loading)

High Temperature Thermoplastics in Clay-Based Nanocomposites

For preparation of nanocomposites from high temperature engineering thermoplastics, a major limitation in the use of conventional onium ion modified clays is the thermal instability of the alkylammonium species during processing. The block copolymer route developed by TNO (vide supra) offers one potential solution to this problem. Imidazolium salts such as that depicted in Figure III–L.4 are also more thermally stable than the ammonium salts.

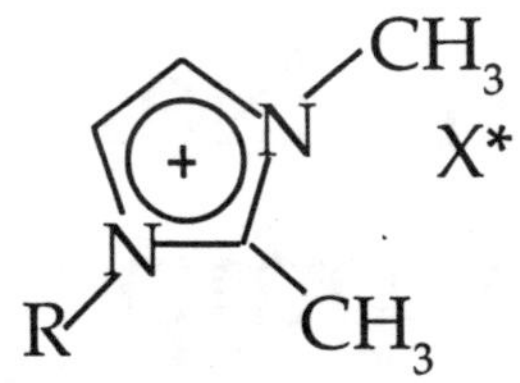

Fig. III–L.4: Structure of an imidazolium salts.

A further alternative is the use of phosphonium salts in place of ammonium salts—this can lead to an increase in the degradation temperature of the organo-clay from 200-300°C to >300°C. Wholly synthetic organo-clays clays, on the other hand, can exhibit thermal stability to beyond 400°C. By using these last two approaches, Triton Systems has succeeded in producing nanocomposites from high temperature resins such as polyetherimide (PEI). Forming the nanocomposite provides a potential route to producing 'speciality' engineering resin performance from cheaper 'standard' engineering resins.

The Future

Future goals include the extension of this nanotechnology to additional types of polymer system, where the development of new compatibilisation strategies is likely to be a prerequisite. Production of PVC-based systems is still some way off and challenges remain to be solved in PET nanocomposites. Additional reinforcement of clay nanocomposites by glass fibre is currently being investigated. There is also interest in the development of electrically conducting clay nanocomposites.

Unit III–M

Polymer-Silicate Nanocomposites via Melt Compounding

Polymer nanocomposites, particularly polymer-silicate nanocomposites, represent an emerging class of specialty plastic materials. The melt-processing route would be a powerful approach in the manufacture of these compounds. Techniques such as transmission electron microscopy and x-ray diffraction were used to determine that nylon-6 nanocomposites could be produced via melt compounding. As a result of the "nano"-sized dispersion, reinforcement efficiency is realized at low loading levels (3-5 per cent) which is a small cost in terms of specific gravity. Dramatic enhancements are observed in heat distortion temperature and stiffness. Besides mechanical properties, nanocomposites also have much lower gas permeability than the neat resin. Other potential benefits include good surface appearance, dimensional stability, and lower heat release.

Nanocomposites, particularly polymer-silicate nanocomposites, are materials that have been given much attention recently and represent an emerging class of specialty plastic compounds. One of the reasons for this attention is that due to the "nano"-scale dispersion, which results in high aspect ratio and surface area, the reinforcement efficiency of polymer-silicate nanocomposites can be significantly better than conventional mineral fillers. The potential benefits of polymersilicate nanocomposites include higher heat resistance, increased flexural modulus, lower gas permeation rate, easy flow, dimensional stability, good surface appearance, and lower heat release rate. Typically, the type of silicate used in polymersilicate nanocomposites is montmorillonite which is a 2:1 layered smectite clay. This clay has a platy structure with individual platelets having thicknesses of one nanometer and surface lengths on the order of 100 to 1000 nanometers. Montmorillonite is hydrophillic which makes proper exfoliation and dispersion into conventional polymers difficult. Thus, montmorillonite is usually modified through substitution of sodium ions with organic onium ions. The result is an organo-clay complex. The intercalation procedure expands the gallery and improves the compatibility of the complex with the polymer so that individual platelets can be more easily separated. Please refer to Figure III–M.1 for possible morphologies of polymers and organo-clay complexes.There are several potential approaches for the production of the polymer-silicate nanocomposite. One procedure involves the synthesis of the compound in the polymerization stage with the complex *in-situ* with the liquid monomer. Also, nanocomposites could be prepared from solvents in which both polymer and complex are dissolved.

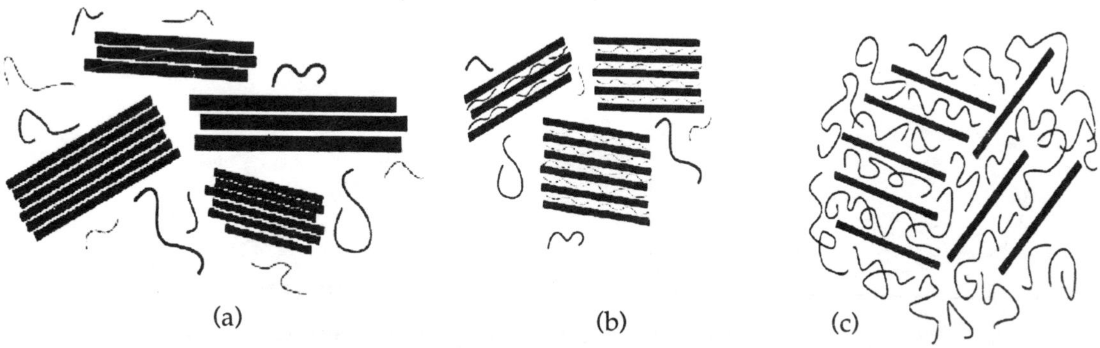

Fig. III–M.1: Structures of possible morphologies developed when processing organo-silicate complexes with polymers. (a) immiscible, typical of conventional mica-type clays, (b) intercalated, and (c) exfoliated.

Another method involves the preparation of polymer-silicate nanocomposites via melt processing. In this process, the application of shear during compounding assists exfoliation and dispersion. The melt-compounding route is a powerful approach in the production of polymersilicate nanocomposites since existing technologies and equipment could be utilized and easily scaled to commercial quantities. Additionally, this approach allows specialty materials such as those with custom colours, static dissipation, or wear resistance to be readily manufactured.

Experimental

Polymer-silicate compounds were prepared by compounding various organo-clay complexes at different loading levels with commercially available nylon-6 using a twin-screw extruder. Neat nylon-6 and the compounds were dried at 80°C under vacuum prior to compounding, molding, and rheometry.

The extent of intercalation, exfoliation, and dispersion of the complexes in the compounds was determined. This was accomplished through wideangle x-ray diffraction (XRD) for preliminary determinations. Transmission electron microscopy (TEM) was used to confirm the morphology development established by XRD. Test specimens were prepared by injection molding. The mechanical and electrical properties of the specimens were evaluated according to ASTM test methods: D-638 for tensile, D-790 flexural, D-256 for Izod impact, D-648 for deflection temperature, D-955 for shrinkage, and D-570 for water absorption. The barrier capability of these materials was explored by measuring oxygen transmission rates (OTR) with a Mocan apparatus. OTR measurements were performed at constant temperature and humidity for 200 hours on specimens thermoformed from injection molded plaques. Compound melt flow behavior was analyzed with a capillary rheometer. The effect of shear rate at 235°C and melt flow stability at 235°C and 1000 sec^{-1} were determined.

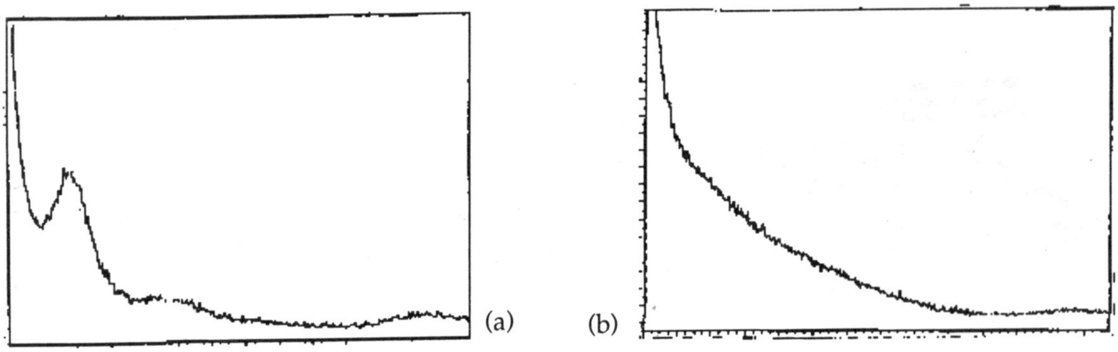

Fig. III–M.2: X-ray Diffraction patterns of a nylon-6 compound containing two different type of complexes.

Results and Discussion

Suitable morphology development in a nanocomposite via the compounding route consists of wetout, intercalation, exfoliation, and dispersion. Therefore, the selection of an appropriate complex for the matrix polymer plays a critical role in determining the potential degree of exfoliation and dispersion in a polymer-silicate compound. One procedure that can be used to assess the extent of intercalation and exfoliation is to analyze x-ray diffraction patterns. Figure III–M.2 are XRD patterns for nylon-6 compounds containing two different types of complexes. Figure III–M.2(a) is for complex #1 and shows a very pronounced peaks. Additionally, the primary d-spacing of the peak is very similar to the complex prior to melt processing. This would indicate that intercalation and exfoliation of complex #1 was negligible. Figure III–M.2(b), on the otherhand, pronounced peaks in the XRD pattern are not discernable for complex #2. This would indicate that complex #2 is exfoliated and dispersed and would be considered a good complex for nylon-6 nanocomposites.

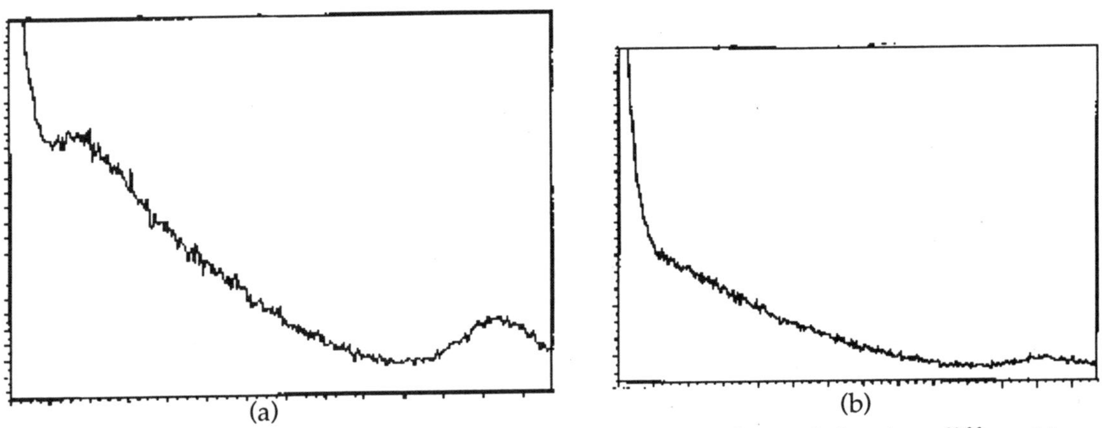

Fig. III–M.3: X-ray Diffraction patterns of a nylon-6 compound containing two different type of complexes and two different processing set-ups.

Along with a suitable selection of complex for the desired matrix polymer, it is also expected that the processing technique is highly crucial to the production of nanocomposite compounds.

Similar to complex selection, XRD patterns should be able to determine which processing techniques are more suitable to nanocomposite development as seen in Figures (a) and (b) of Figure III–M.3. For instance, a peak is observed for the material in Figure (a) of Figure III–M.3 while no peak is seen for the compound in Figure (b) of Figure III–M.3. This indicates that the process setup used for th material in Figure (b) of Figure 13.3 produced an improved nanocomposite.

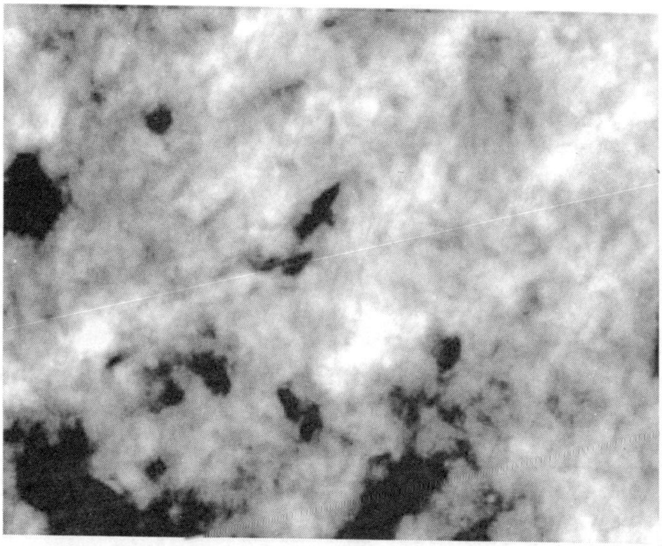

Fig. III–M.4: TEM of a nylon-6 compound with organo-clay complex #2 and process setup #1.

As we have seen, XRD patterns appear to be good indicators for nanocomposite development. In order to confirm this, TEMs were obtained for the compounds as shown in Figures (a) and (b) of Figure III–M.3. In Figure III–M.4, large aggregates are visible and the platelets are not well dispersed. This agrees with Figure (a) of Figure III–M.3, which indicated poor exfoliation by having observable peaks present in the XRD pattern. Conversely, individual 'edge-on' platelets are easily observed and appear to be well dispersed in Figure III–M.5. This agrees with Figure (b) of Figure III–M.3, which indicated good exfoliation. Finally, not only does this show that XRD patterns can be used to evaluate nanocomposite development but also that nylon-6 nanocomposites are indeed possible via melt compounding. The properties of a typical nylon-6 nanocomposite are given in Table III-M.1 and compared to the unfilled matrix nylon-6. At almost no cost in specific gravity, significant improvements in certain properties are realized. For instance, flexural modulus is increased by almost 50 per cent and deflection temperature raised by over 35°C. Also, tensile strength is increased by over a third. For comparison purposes, the properties of typical mineral and glass fiber filled nylon-6 compounds are given. Reinforcement efficiencies of nanocomposites compared to conventional mineral fillers on a weight basis are shown in Figures III–M.6 and III–M.7. The flexural modulus as a function of loading level is shown in Figure III–M.6 while the deflection temperature as a function of loading level is given in Figure III–M.7. As a result of the high surface area and aspect ratio upon exfoliation, nancomposites are able to provide more efficient reinforcement than conventional mineral fillers. For instance, at a 5 wt per cent loading level, a nylon-6 nanocomposite has a stiffness equivalent to a 20 wt per cent conventional mineral-filled nylon-6. The same nylon-6 nanocomposite also has a deflection temperature better than a 30 wt per cent mineral-filled nylon-

6.

Fig. III–M.5: TEM of a nylon-6 compound with organo-clay complex #2 and process setup #2.

Table III–M.1: Properties of Nylon 6 compound

	Nanocomposite	*Unfilled*	*25 per cent Mineral*	*10 per cent Glass Fiber*
Specific Gravity	1.14	1.13	133	1.20
Tensil Strength (MPa)	92	71	71	93
Flexural Modulus (GPa)	4.3	2.9	4.5	4.3
Notched Izod (J/m)	34	37	33	36
Deflection Temperature @ 1.82 MPa (°C)	100	65	80	185

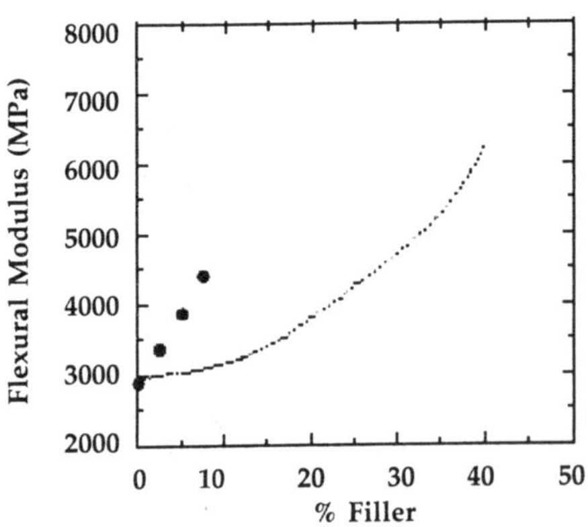

Fig. III–M.6: Flexural modulus of nylon-6 composites (_) as a function of loading level compared to a conventional mineral filler (L).

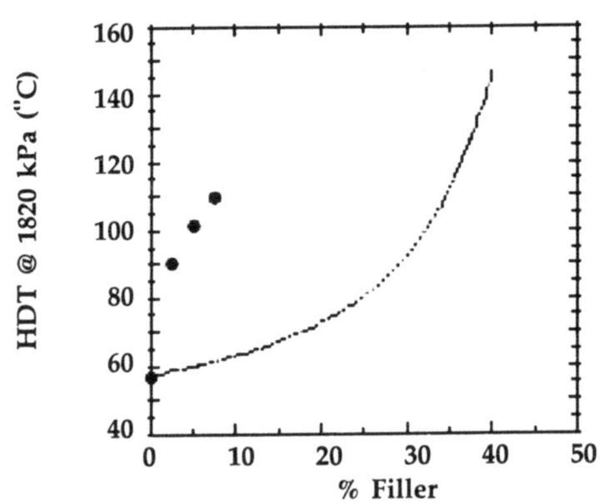

Fig. III–M.7: Deflection temperature of nylon-6 composites (_) as a function of loading level compared to a conventional mineral filler (L).

Table III–M.2: Barrier Properties of Potential Monolayer Materials

	OTR (cc-mil/100in²/day)
Nylon-6 Nanocomposite	0.5 - 1.0
Typical Nylon-6/66	1.5 - 6.5
PET	5.0
PET Blends	1.0 - 2.5
Amorphous Nylon	0.15 - 2.0
EVOH	0.01 - 1.0

Besides mechanical and heat resistance property improvements, nanocomposite compounds can exhibit dramatically enhanced barrier properties as represented by lower oxygen transmission rates. From Table III–M.2, nylon-6 nanocomposites can achieve an OTR almost four times lower than the unfilled nylon-6. The reduction in OTR is typically attributed to the increase in effective diffusion length as solutes must travel a tortuous path around well-dispersed platelets of high aspect ratio. It is also important to point out that as a result of the nanometer length scale that excellent transparency is retained in sheets and films formed from nylon-6 nanocomposites.

There has been some concern about the ability to process nanocomposites due to the high surface area of the nanoclay upon exfoliation. For instance, the effect of other high surface area additives such as high structure carbon black is to dramatically increase the viscosity. In Figure III-M.8, the viscosity of a typical nylon-6 nanocomposite is compared to the viscosity of the neat polymer as a function of shear rate. Contrary to the concerns, the viscosity is observed to be lower than the neat resin at the shear rates of most interest to potential processors. As a result, nylon-6 nanocomposites may be processed as easily as the neat resin. There are several potential explanations for the shear thinning at higher shear rates. One possible reason could be due to orientation of the platelets upon shear. Another reason could be that the polymer has been degraded during compounding with the nanoclay. However, degradation was not obvious in the nylon-6 nanocomposite samples. In Figure III–M.9, the melt flow stability of a nylon-6 nanocomposite was determined. The viscosity was found to not significantly decrease with time. Additionally, the viscosity is equivalent to the viscosity measured under similar conditions during prior testing of the effect of shear rate. Therefore, the degradation if present, would not be expected to continue on further processing.

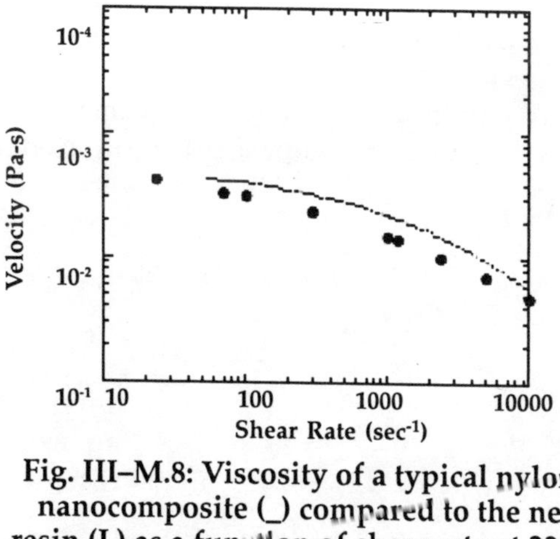

Fig. III–M.8: Viscosity of a typical nylon-6 nanocomposite (_) compared to the neat resin (L) as a function of shear rate at 235°C.

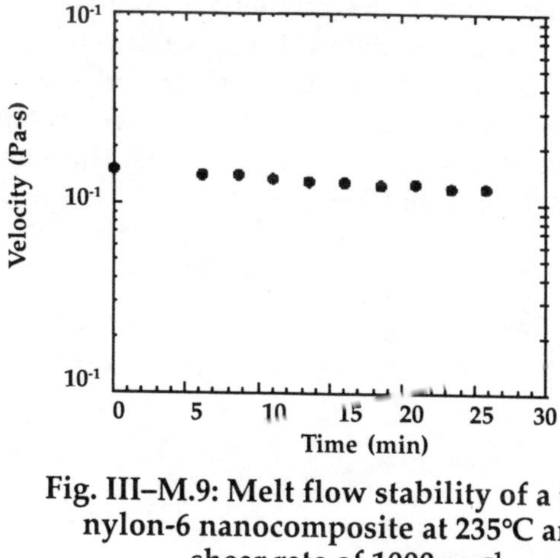

Fig. III–M.9: Melt flow stability of a typical nylon-6 nanocomposite at 235°C and a shear rate of 1000 sec⁻¹.

The dimensional stability of nylon-6 nanocomposites is compared in Table III–M.3 to the unfilled resin and other typical reinforced compounds. As seen, the water absorption of the nanocomposite is dramatically reduced by as much as 55 per cent. This is comparable to a filled compound which contains much higher content. This is due to the barrier properties since resin replacement cannot entirely account for all of the change as is the case for conventional mineral and glass compounds. Also observed in table is that the shrinkage of nylon-6 nanocomposites is more similar to the base resin due to the lower filler content. Additionally, the shrinkage in the transverse direction does not differ significantly from the flow direction. This gives a warpage factor much closer to one as shown in Table 8 than a mineral or glass compound. Thus, the nylon-6 nanocomposites would not be expected to warp much more than the unfilled nylon-6.

Table III–M.3: Nylon-6 Dimensional Stability Comparison

	Unfilled	*nanocomposite*	*40 per cent Mineral*	*30 per cent Glass Fiber*
Water Absorption (per cent)	1.8	0.99	0.98	1.1
Shrinkage Comparison (flow/$flow_{unfiled}$	1.0	0.93	0.49	0.30
Warpage Factor (shrinkage of flow/transverse)	0.91	0.82	0.75	0.56

Conclusions

Polymer-clay nanocomposites have been developed via the compounding route in nylon-6. XRD and TEM analysis confirmed that good exfoliation and dispersion was attained. Notable property improvements were observed in tensile strength, flexural modulus, heat deflection temperature, and oxygen transmission rate. Proposed application areas include those requiring structural properties. For structural applications, nylon-6 nanocomposites exhibit equivalent or better properties than typical 20-30 wt per cent mineral filled compounds depending on the property being investigated. It is important to note that these gains come at almost no cost in specific gravity from the unfilled nylon-6. However, the properties of nylon-6 nanocomposites have yet to approach those of glass fiber reinforced compounds. Even so nylon-6 nanocomposites are excellent candidates when improved heat deflection temperature, stiffness, or strength is required especially when glass fiber reinforcement cannot be considered either due to specific gravity, mold shrinkage/warpage, or wear. Another potential application area is barrier. A significant four-fold improvement was observed in the oxygen transmission rate. This may allow nylon-6 nanocomposites to be utilized in certain packaging and film applications.

Unit III–N

Dendrimer Nanocomposites in Medicine

Dendrimer based nanocomposites are novel organic-inorganic hybrid nanoparticles synthesized by dispersing very small inorganic domains in nanoscopic size polymeric networks. Due to the molecular level mixing of their components they display unique material properties in addition to properties represented by their individual components. Dendrimers are often used as templates in making these particles because they can be made uniform, and using these well-controlled and uniform templates leads to well controlled and uniform composite particles. Apart from obvious materials science applications, such as catalysis and photonic materials, dendrimer nanocomposites (DNC) have a great potential in biomedical applications due to their controlled composition, predetermined size, shape and optional surface functionalities. They may be used as drug delivery vehicles to deliver bioactive guests such as silver or for an incorporated radioactive isotope. Radioactive dendrimer nanocomposites can be delivered directly to the tumor by injecting nanoparticle solutions directly into the tumor microvasculature. If a short-lived isotope is used, a procedure like that may serve as an alternate method to direct irradiation. It has also been envisioned that if the surface of such a nanoparticle had been equipped by appropriate targeting moieties the nanoscopic device could find its way selectively to cancerous cells anywhere in the body.

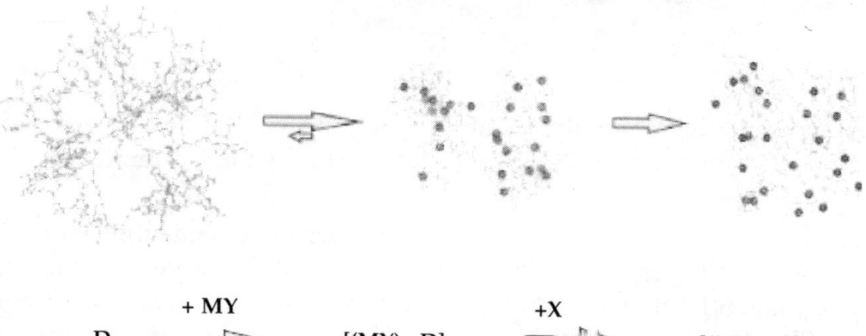

Fig. III–N.1: Dendrimer Complex Nanocomposite.

Poss-Based Nanocomposites

A relatively new type of hybrid is based on the silsesquioxane cage structure, which can be regarded as one of the smallest forms of silica or even as 'molecular silica'. This approach has been developed by Hybrid Plastics (USA) and is called polyhedral oligomeric silsesquioxane (POSS) nanotechnology. The particle diameter of the POSS molecules is in the range 0.7-30 Å, on average much smaller that the diameter of, for example, colloidal silica particles. Polymers containing the POSS molecules are therefore true nanocomposites. The physical form of the materials varies from liquids through waxes to crystalline solids. The cost of the POSS chemicals is projected to fall between that of fumed silica and the silicones. The POSS structures can be functionalised with a wide variety of groups, leading to a range of POSS monomers. The nature of the functional group determines compatibility with the polymer matrix.

The R group can be simple alkyl, cycloalkyl or aryl, or reactive/ polymerisable groups such as acrylic, α-olefin, styrene, epoxide, carboxylic acid, isocyanate, amine, alcohol and silane. Using such functionalisation, the POSS structures can be either copolymerised with a range of monomers, or grafted onto polymer chains. A variety of phase morphologies is possible, ranging from phase separated to phase inverted to single phase, truly molecular level reinforcement. The POSS compounds can be readily incorporated into polymers using standard processing methods. Interestingly, and in contrast to clay nanocomposites, incorporation of POSS can lead to a reduction in the polymer melt viscosity, which should represent a distinct advantage for many applications. A major effect of POSS incorporation, e.g. in polypropylene, is retention of modulus above the glass transition temperature (Tg) of the unreinforced polymer. Increasing the concentration of the POSS structures in copolymers can have a dramatic effect in increasing the Tg of the polymer. POSS nanocomposites occupy the performance region indicated as 'Hybrids'. Flammability behaviour of the polymer is also claimed to be improved by incorporating POSS.

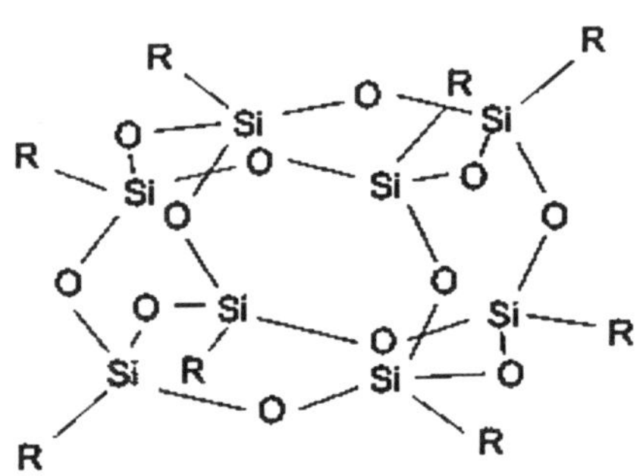

Fig. 14.2: Representation of a polyhedral oligomeric silsesquioxane or POSS.

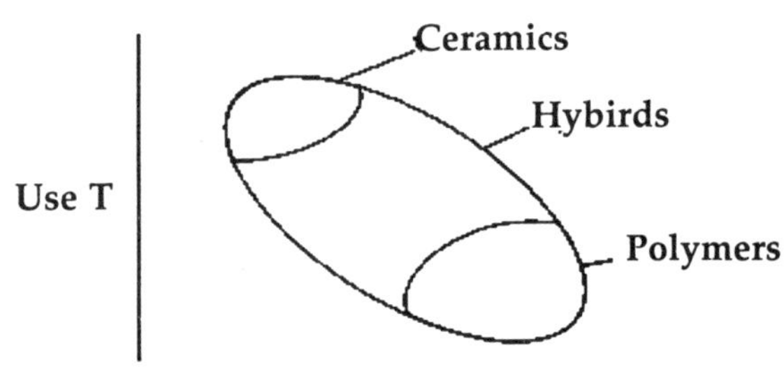

Fig. 14.3: Relative property performance of polymers, ceramics and hybrids.

POSS silanols can be used as coupling agents for fillers and offer a number of advantages over conventional silane coupling agents, including moisture stability, low volatility, no requirement for water addition and no formation of volatile by-products. Surface coverage by POSS cages is also very high. Use of POSS surface treatment can enhance the reinforcing effect of the filler.

POSS products are attracting interest in a wide range of materials markets, including aerospace, sporting goods, medical applications and electronics.

Unit III–O

Nanoparticles

Nanoscience has taken scientists around the world by storm. It claims to revolutionize the world we live in with radical breakthroughs in areas such as materials and manufacturing, electronics, medicine and healthcare, environment and energy, chemical and pharmaceutical, biotechnology and agriculture, computation and information technology.

What are Nanoparticles?

Nanoparticles are small clusters of atoms about 1 to 100 nanometers long. 'Nano' derives from the Greek word "*nanos*", which means dwarf or extremely small. It can be used as a prefix for any unit like a second or a liter to mean a billionth of that unit. A nanosecond is a billionth of a second. A nanoliter is a billionth of a liter. And therefore a nanometer is a billionth of a meter or 10^{-9} m.

How Small is Nanometer?

- If a baseball is the size of Earth, a nanoparticle would be the size of an apple.
- We can also compare it with things in the natural world.
- The smallest synthetic nanotechnology made so far are the individual components of an Intel Pentium III microprocessor and they span about 200 nanometers.

What Makes Nanoparticles Different?

Nanoparticles are larger than individual atoms and molecules but are smaller than bulk solid. Hence, they obey neither absolute quantum chemistry nor laws of classical physics and have properties that differ markedly from those expected.

There are two major phenomenons that are responsible for these differences. First is the high dispersity of nanocrystalline systems. As the size of a crystal is reduced, the number of atoms at the surface of the crystal compared to the number of atoms in the crystal itself, increases. For example, a 4 nm diameter CdS nanoparticle has about 1500 atoms, of which about a third are on the surface. Properties, which are usually determined by the molecular structure of the bulk lattice, now become increasingly dominated by the defect structure of the surface.

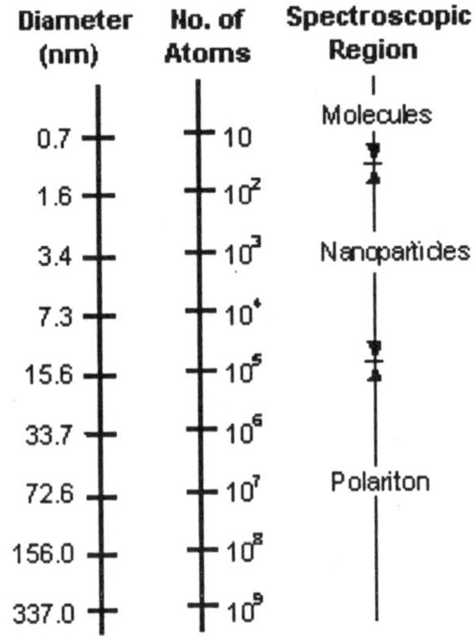

Fig. III–N.1: The diagram shows a schematic representation of the different sizes.

The second phenomenon occurs noticeably only in metals and semiconductors. It is called size quantisation and arises because the size of a nanoparticle is comparable to the de Broglie wavelength of its charge carriers (*i.e.* electrons and holes). Due to the spatial confinement of the charge carriers, the edge of the valance and conduction bands split into discrete, quantized, electronic levels. These electronic levels are similar to those in atoms and molecules.

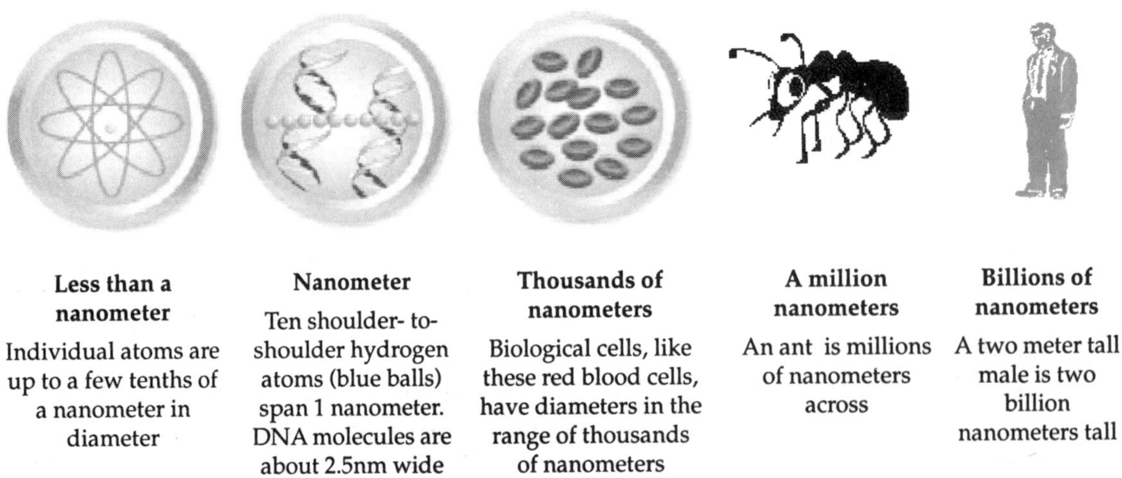

Fig. III–N.2: Size of nanoparticles.

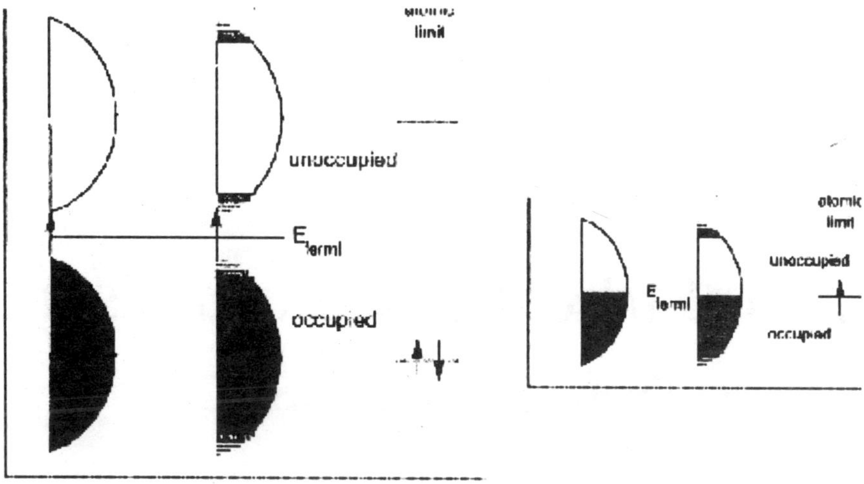

Fig. III–N.3: In diagrams the conduction and valence bands in the solid, nanoparticle and atom state of a semiconductor (left) and a metal (above) are compared. It can be seen here again that nanoparticles represent a state of matter in the transition state between bulk solids and individual atoms.

The spacing of the electronic levels and the bandgap increases with decreasing particle size. This is because the electron hole pairs are now much closer together and the Coulombic interaction between them can no longer be neglected giving an overall higher kinetic energy.

This increase in bandgap can be observed experimentally by the blue-shift in the absorption spectrum or sometimes even visually by the colour of the samples. A larger bandgap means that more energy is required to excite an electron from the valance band to the conduction band and hence light of a higher frequency and lower wavelength would be absorbed.

Fig. III–N.4: By changing the particle size of the same material, and hence the energy bandgap, different coloured samples can be obtained.

The increase in molar absorption coefficient shows that the oscillator strength also increases as particle size decreases. This is due to the strong overlapping of the wave functions of confined charge carriers. Experiments carried out in the past have found the absorption spectrum to be potential dependent, with decreasing wavelengths as applied potential decreases. This change is fully reversible before the threshold potential.

Unit III–P

Nanoparticles, Nanotubes and Nanocomposites

The first generation of nanofillers is emerging from R&D into commercial application in composites.

Federal funding for nanotechnology research and development in USA has increased six-fold, from $116 million (USD) in 1997 to an estimated $961 million in 2004. In the U.S. alone, the FY 2005 budget request that President George W. Bush has sent to Congress calls for a total National Nanotechnology Initiative (NNI) budget of $982 million, which represents a 2 per cent increase. With all this research, what has come out of the R&D pipeline ready for commercial application? As it turns out, plenty.

The following are new and, surprisingly, not-so-new applications of two types of nanotechnology: ceramic nanoparticles and carbon nanotubes used as highly functional fillers/additives in polymer products of interest to composites manufacturers. Introduction of these materials can result in the improvement of the polymer's mechanical properties in molded parts and coatings on a wide variety of fronts, including increased part strength, ductility and dimensional stability as well as resistance to damage by abrasion and exposure to caustic chemicals and/or extreme heat.

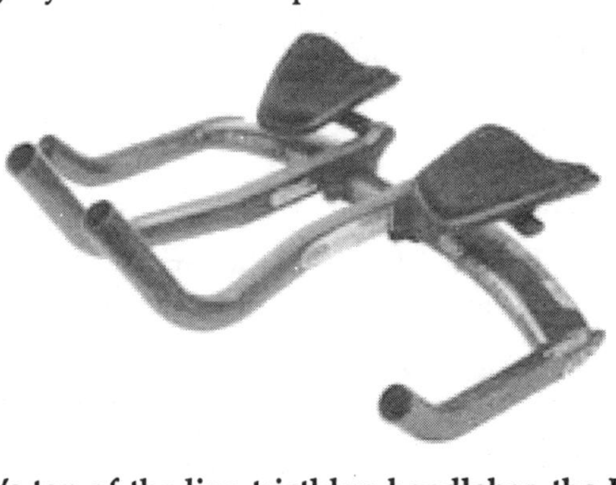

Fig. III–P.1: Easton Bicycle's top-of-the-line triathlon handlebar, the DeltaForce with Aeroforce Clip-ons, incorporates nanotubes in the bar and the clip-ons for a total weight of only 535g/18.9 oz.

Nanotechnology involves the control and manipulation of materials at the nanoscale—particle sizes from 1 to 100 nanometers (nm), where 1 nm equals 1 *billionth* of a meter—to create new materials and structures that have novel properties due to their small size. Conventional multifunctional materials, such as fiber-reinforced or filled polymers, require a trade-off between various mechanical and physical properties, cost and processability. However, combinations of properties derived at the nanoscale provide opportunities to circumvent these traditional trade-offs. The transition from microparticles to nanoparticles leads to a number of changes in physical properties. One of the major factors is the increase in the ratio of surface area to volume. As the surface area of a particle increases, the portion of its constituent atoms at or near the surface increases exponentially, creating more sites for bonding, catalysis or reaction with surrounding materials, resulting in improved properties such as increased strength or chemical or heat resistance. Additionally, the fact that nanoparticles have dimensions below the critical wavelength of light can render them transparent, which has implications for part aesthetics.

Nanoparticles: Silicates and Metal Oxides

One of the most common groups of the new generation of commercial nanotechnologies is ceramics. Ceramics can be categorized into two subgroups: silicates (or silicon oxide), generally in the form of nanoscale flakes of clay, and metal oxide ceramics such as titanium, zinc, aluminum and iron oxides.

Nanoclays—The most common type is montmorillonite (MMT), a layered aluminosilicate in the smectite family of clays. MMT is a component of bentonite clay, formed when volcanic ash settled into ancient brine seas, and now mined in Wyoming, western Europe, China, Australia, Iran and elsewhere worldwide. Unlike clay minerals such as talc and mica that have been used as fillers for years, MMT can be delaminated and dispersed into individual layers only one nanometer thick by about 70 nm to 150 nm across. The result is a radical increase in the surface area-to-volume ratio: With a surface area of 750 m^2/g, 20g of MMT platelets could cover a football field!

Southern Clay Products Inc. (Austin, Texas), a manufacturer of functional clay-based additives, began MMT product development in the early 1990s. The company's Cloisite additives, commercially available since 1999, consist of organically modified, nanometer-sized MMT platelets.

Unseparated MMT particles are often referred to as tactoids. When tactoids are partially separated in a polymer, they are referred to as being intercalated. When thoroughly separated, individual platelets are said to be exfoliated. To form a nanocomposite, the clay must first be modified with an appropriate organic surface treatment to enable complete dispersion into and miscibility with the polymer matrix. Then the tactoids must be exfoliated into platelets either through the shear stresses exerted during melt compounding or via chemical processes during polymerization of the matrix. As Karl Kamena, Southern Clay's commercial manager of Cloisite nanoclays, describes it, "The incorporation of single-digit percentages of high surface-area nanoclays into host polymer systems improves the performance of the specific polymer matrix without significant deterioration of properties." The numerous benefits that Cloisite can confer on the host polymer include increased mechanical properties, such as modulus, melt strength and scratch resistance, as well as increased heat resistance, dimensional stability and flame retardation. The extent of enhancement is related directly to the degree of exfoliation. Currently, about 40 per cent of the MMT platelets can be fully exfoliated, with the remainder forming tactoids of 2 to 10 platelets. Southern Clay continues to improve exfoliation methods, anticipating that, eventually, remaining tactoids will contain as few as one to three platelets.

Fig. III–P.2: The composite cargo bed trim, center bridge, sail panels and box rail protectors of this GM *Hummer* use a molded-in-colour, nanoclay-filled TPO.

In 1998, Southern Clay began working with General Motors (GM, Detroit, Mich.) and resin manufacturer Basell (Hoofddorp, The Netherlands) to develop a nanoclay-reinforced thermoplastic olefin (TPO) that would outperform the talc-filled TPO then in use. An 8-micron Cloisite clay, each particle of which contains one million platelets, was selected as the filler. Work carried out jointly by the companies allowed the new material to be formulated to different shrinkage rates and to match colour with conventionally filled molded-in-colour parts. Early on, the 8-micron particles had a tendency during the extrusion stage (when the clay was mixed into the TPO) to aggregate into 200-micron clumps rather than exfoliate, causing surface defects and painting problems in the injection molded parts. The team then proved that this issue could be resolved by making adjustments to the extrusion temperature, the extruder screw rotation speed, the material feed rate and other variables during the nanocomposite manufacturing process. GM first introduced a 3 per cent nanoclay-filled TPO on the step assists for the 2002 GMC *Safari* and Chevrolet *Astro* vans, which had been using a 20 per cent talc-filled TPO. The same material moved into a higher volume application in the body side trim moldings for the 2004 and 2005 Chevrolet *Impala*, molded at Blackhawk Automotive Plastics (Troy, Mich.). Will Rodgers, GM staff research scientist, says GM saw weight savings of 3 to 21 per cent on these redesigned parts. With increased confidence in the material, GM now is using 7 lb of molded-in-colour nanocomposite parts to make this year's *Hummer H2 SUT* cargo bed trim (the center bridge, the sail panels and the box rail protectors), all molded by SportRack Automotive (Sterling Heights, Mich.). These are the first parts that GM has designed as nanocomposites from conception, choosing the material primarily because of its low coefficient of thermal expansion (CTE) and improved appearance. For example, by varying the surface finish of the tool, three levels of gloss were molded into the box rail protectors to match adjoining parts, something that was never possible with talc-filled TPO. GM also reports better scratch/mar performance, increased cohesive strength between the paint and the part, greater ductility at low temperatures, lower flammability, a larger injection molding process window, sharper feature lines, improved knit line appearance and the elimination of tiger striping, (Tiger striping is caused when a filled material groups itself into lines of higher and lower fill weight or viscosity as it flows into a mold, giving the part a striped appearance.) Although the MMT-filled TPO is about an order of magnitude more expensive than the talc-filled material (about $3/lb vs. $0.35/lb), the improvements in part performance and appearance apparently justify the cost.

Metal oxides—Scratch resistance is a highly sought quality in coatings, such as automotive finishes. However, use of silica or clay fillers in quantities sufficient to improve scratch resistance tends to make the coating hazy and reduce its viscosity. Alumina particles are harder than silica but also lead to haziness. This is particularly troublesome in clear coats. Nanophase Technologies Corp. (Romeoville, Ill.) and BYK-Chemie (Wesel, Germany) have teamed to address this problem with two new additives marketed under the tradenames Nanobyk-3600 and Nanobyk-3601. Both additives contain Nanophase's 37-nm diameter Nanodur aluminum oxide particles in concentrations of 30 to 50 per cent by weight, and are intended for use with aqueous (Nanobyk-3600) and nonaqueous UV-curable coatings (Nanobyk 3601). Nanophase Technologies produces its aluminum oxide nanoparticles via physical vapor synthesis: Arc energy is applied to a metallic precursor to generate a vapor at a high temperature. A reactant gas is introduced to the vapor and then the mixture is cooled at a controlled rate so that it condenses and forms nanoparticles, which then are coated with a thin polymeric shell that ensures their compatibility with various polymers and fluids. When the additives are blended into resins and coatings at very low 1.5 to 6 per cent concentrations, scratch resistance reportedly goes up dramatically. In Nanophase tests, particle loadings of 1 per cent resulted in five- to ten-fold increases in scratch resistance, depending on the resin formulation tested, with less than 0.5 percent haze. (At greater than 1 per cent haze, or about a 2 per cent loading of Nanodur alumina particles, a resin is no longer considered "clear.")

Nanotubes: Conductivity and Mechanical Performance

Exploration of carbon nanotubes is relatively new. Single wall nanotubes (SWNTs) are long, thin cylinders made of carbon atoms arranged in a graphitic lattice structure that may or may not have closed ends. Multiwall nanotubes (MWNTs) are concentric cylinders of carbon in a similar graphite structure, held together by weak intermolecular-level forces. SWNTs and MWNTs have length-to-diameter ratios that can range from 100:1 to 10,000:1. Their large surface-area-to-volume ratios mean that carbon nanotubes (CNTs) affect matrix properties far beyond what would be predicted from rule-of-mixtures calculations. Nanotubes are currently being commercialized in applications where electrical conductivity and increased mechanical performance are primary design drivers.

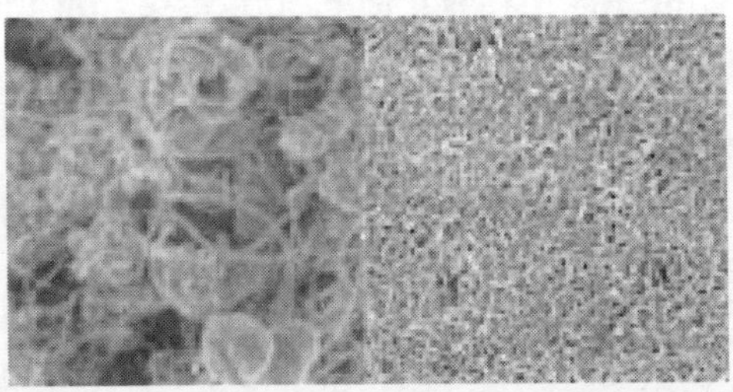

Fig. III–P.3: Nanotube dispersion comparison: The scanning electron micrographs (SEMs) illustrate untreated multi-wall nanotubes vs. Zyvex functionalized tubes. The pictures were taken with 50Kx magnification at 300 nm resolution (samples were prepared with equal concentrations).

Electrical properties. SWNTs can possess electrical conductivity of 1 billion amps/cm^2, Young's modulus of more than 1 TeraPascal and a maximum strain of 10 per cent. MWNTs have lesser properties but are easier to produce, though they also are less well understood, due to their greater complexity and variety. Orders of magnitude less costly than SWNTs, MWNTs are more attractive for commercial applications. While they have been used for the last 10 years in thermoplastics as a direct replacement for carbon black in electrical applications, recent developments are enabling their use in thermosets, and significant efforts are underway to exploit their mechanical properties as well. Carbon black and carbon nanotubes are both pure graphitic carbon and so have the same electrical properties. Carbon black particles, however, are spherical, and the rod-like CNTs have a higher aspect ratio, which means far fewer CNT particles are needed to form a conductive "bridge" through a nonconductive polymer. Typical carbon black loading in polymers is 7 to 18 percent (depending on the polymer and the level of conductivity needed), which will make the polymer brittle. CNT's achieve the same conductivity in 1 to 3 percent loadings. Moreover, the CNT's conductivity-versus-loading curve is more gradual and more easily tailored to a specific level of conductivity than carbon black's curve.

In 1984, Hyperion Catalysis (Cambridge, Mass.) was the first company to file a patent on a nanotube manufacturing method. Hyperion's MWNTs have 8 to 12 walls and are about 10 nm wide by 10 micrometers long, an aspect ratio of 1,000:1. They are made through a process which Hyperion's sales and marketing manager Pat Collins describes as "continuous, catalyzed vapor phase decomposition and recombination, using light hydrocarbon feedstock." When an atomized metallic catalyst comes in contact with a hydrocarbon gas in a reactor under appropriate time, pressure, and temperature conditions, nanotubes will grow on the catalyst. Each nanotube has a tiny remnant of catalyst inside it. The MWNTs, as formed, tend to clump together in aggregates, attracted to each other by intermolecular Van der Waals forces. The high shear forces which are necessary to disperse the CNTs into the resin are obtained by using a high-viscosity resin and a twin-screw extruder typically used to compound thermoplastics. The disentangled and dispersed nanotube/resin mixture is formulated in concentrations of up to 20 per cent. A wide variety of nanotube-filled thermoplastics can be purchased, either ready to use with low CNT loadings or as a master batch with higher nanotube concentrations, which then can be let down (diluted) to the appropriate usage level. The ready-to-use resins are offered in antistatic, electrostatic dissipative and conductive grades, with volume resistivities ranging from 1010 ohm/cm for antistatic grades to 1 ohm/cm for conductive grades (as compared to 1.68 x 10-6 ohm/cm for copper). Because the nanotubes are in such low concentrations and are so small, their addition has little effect on the base resin's properties, allowing the material to be processed on the injection molder's existing equipment under conditions similar to those used to process the base resin. Hyperion's continuous process runs 24 hours a day, seven days a week, producing tons of pure (no byproducts, such as amorphous carbon) nanotubes per week.

Automotive fuel systems have to be designed for electrostatic discharge. The turbulent flow of fast-moving fuel creates a static charge which must be dissipated. If charge is allowed to build up it can cause pinholes to develop in the fuel system and may damage electronic sensors. On rare occasions, sparks may even discharge during refueling. The fuel system, which includes the filler neck, inlet check valves, fuel module (pump/sender unit) inside the tank, the fuel lines and the fuel filter, must all, therefore, be conductive. (The Society of Automotive Engineers (SAE) will soon issue its newly revised *Recommended Practice, SAE J1645*, which will outline how all materials that touch fast-moving fuel must be grounded.)

Fuel systems static dissipation is a rapidly growing segment of Hyperion's business, according to Collins. BASF (Ludwigshafen, Germany) has already commercialized a conductive polyoxymethylene (POM) with Hyperion nanotubes, specifically for automotive fuel systems. CNT-reinforced nylon 6, manufactured by Degussa (Parsippany, N.J.), is used for the tubing in the flexible portion of the fuel line in 70 percent of cars made in North America. O-rings in the system also must be conductive or they will build unwanted capacitance. Precix (New Bedford, Mass.) recently introduced its F98 series of O-rings, made by incorporating Hyperion's multiwall nanotubes into Precix's own proprietary flouroelastomer resin system. Compared to the company's carbon black-filled O-rings, the F98 compound is said to have three to five times greater resistance to fuel vapor permeation, with significantly reduced fuel swell (the expansion of part dimensions due to absorption of the fuel) as well as a durometer of 79 versus 88, allowing for much lower insertion pressure when assembling parts.

Hyperion also supplies various conductive thermoplastics to manufacturers such as GE Plastics (Pittsfield, Mass.) for use as injection molding material for electrostatically painted automotive body parts, such as mirror housings, door handles and fenders. According to a *CT* web search, the Renault *Clio* and *Mégane* currently sport fenders made from a nanotube-filled polyamide/polyphenylene ether (PA/PPE) blend. Since CNT loadings are typically 2 to 3 percent rather than the 12 to 15 per cent necessary with carbon black, the plastics do not become brittle, thus negating the need for impact modifiers which would produce the undesirable side effect of reducing the plastic's glass transition temperature (Tg). The CNT-reinforced parts need no primer and can be painted at the same temperature and on the same electrostatic line used with metal parts, reducing cost and improving colour matching. Because the CNT concentrations are low, the surface finish also is smoother than plastics with other types of conductive fillers.

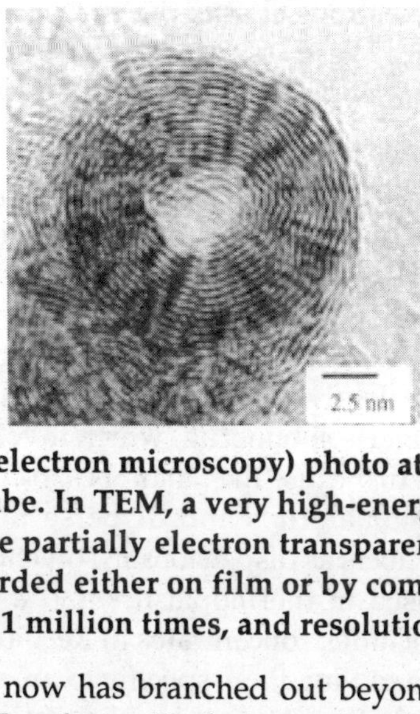

Fig. III–P.4: TEM (transmission electron microscopy) photo at 700Kx magnification of the end-on view of a multiwall nanotube. In TEM, a very high-energy electron beam is placed on a sample that is thin enough to be partially electron transparent, and the electron "shadow" of the sample is viewed and recorded either on film or by computer. TEM imaging allows for magnifications up to 1 million times, and resolution of better than 2 nm.

Collins adds that Hyperion now has branched out beyond thermoplastics into thermosets, with nanotube-filled versions of Resolution Performance Products' (Ennis, Texas) Epon 1009 and Epon 828 resins. Epon 1009 is a high molecular weight, solid pellet form of bisphenol-A, which can be compounded into liquid epoxies or vinyl esters. Hyperion sells the Epon 828 nanocomposite in paste form, which is made by grinding the Epon 1009 nanocomposite pellets into a powder and blending the powder into the 828 with a multishaft double planetary mixer in a liquid dispersion process. Collins also notes that Hyperion will introduce easier-to-use liquid master batches of other nanotube-filled epoxies in the near future, primarily for electrical applications, such as conductive epoxies, potting compounds and connectors. Although the nanotubes may add $5/lb to $6/lb to the final polymer mix, Collins says there are many applications where the cost is justified either through cost savings in the final product manufacturing process or because the necessary properties cannot be obtained any other way.

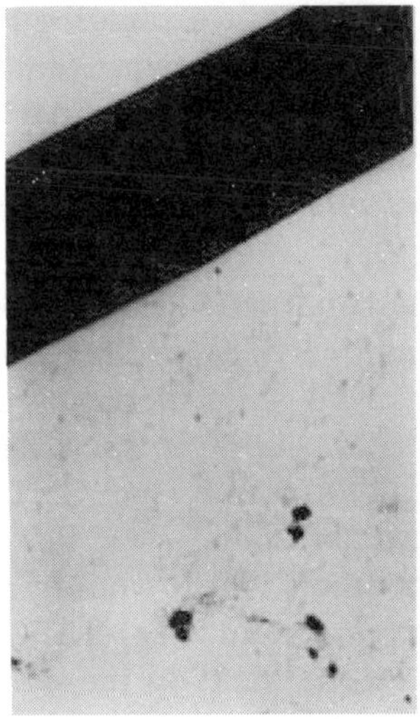

Fig. III–P.5: An SEM (scanning electron microscopy) image compares a single carbon fiber to dispersed nanotubes (the small specks) and agglomerated nanotubes (larger specks). SEM uses a fine beam of electrons to scan the surface of a specimen to generate electrons from its surface. These secondary electrons are then collected by a detector and processed to form images.

Mechanical properties. Zyvex Corp. (Richardson, Texas) approaches the market from a different angle, seeking to use specially modified carbon nanotubes to increase the mechanical performance in sporting goods, aerospace and defense applications. Zyvex does not make its own nanotubes, but instead uses SWNTs and MWNTs from other suppliers, and certifies them for quality and batch-to-batch consistency using a series of standard analytical tests. Zyvex uses a patented process to functionalize the surface of the nanomaterials with a non-covalently bonded "Kentera" molecule which non-covalently bonds the nanomaterial surface via pi-pi bonding and also adheres well to the host matrix. Pi-pi bonding, which involves a slightly stronger molecular force than the intermolecular forces that cause the nanotubes to clump together, enables the Kentera molecule to slip between the nanomaterials and disperse them. Zyvex offers its Nanosolve functionalized nanotubes and nanofibers as dispersions in solvents or water additives, or in solvent-free concentrates, which can be used in thermoplastics and a subset of thermosets comprising polyurethanes and epoxies. This includes concentrates in Resolution Performance Products' Epon 826, 828 and 862 at customized weight loadings specific to an application, whether it is thermal, electrical or mechanical. Marni Rutkofsky, Nanosolve chemist and technical sales representative, is overseeing third-party ASTM physical and mechanical testing on several materials. She says preliminary results at less than 5 per cent loading of MWNTs in thermoplastic polycarbonate show increases of 80, 50 and 150 per cent, respectively, in tensile strength, elongation and Young's modulus. Results at similar loading in polyurethane show a four-fold increase in thermal conductivity. Data on epoxy is expected later this year.

One beneficiary of Zyvex research is Easton Sports (Van Nuys, Calif.) a leading manufacturer of bicycling, hockey, baseball, softball and archery equipment. Three years ago, Easton embarked on an ambitious project to explore the advantages that nanotubes could bring to bicycle components bladder molded at its production facility in Dongguan City, Guangdong Province, China. Easton drum winds its own prepreg and planned to incorporate CNTs into the resin before introducing the resin to the fiber. The challenge was to find a nanotube chemistry that would remain well dispersed in the resin during the course of Easton's manufacturing process and yield improved properties at a reasonable cost. Zyvex and Easton tried several iterations of the resin filled with the CNT/Kentera chemistry before developing a formulation that met the challenge.

"We were looking more for strength increase than weight decrease," says John Harrington, VP of Easton's Bicycle Div. "Our products are already some of the lightest on the market, and we want to be conservative with weight reductions when we are using a new material." However, Easton has shaved 10g to 20g from most of the CNT-filled prepreg components, while achieving an average 15 per cent increase in product flexural strength. Road, mountain, and triathlon handlebars, forks, seatposts and OEM seatstays and chainstays are already on the market. The CNT Technology logo distinguishes components which have incorporated nanotubes, and most of the rest of Easton's bicycle components will follow suit. Retail prices on the logoed components have gone up an average of only 10 per cent—an increase that Harrington says consumers are willing to pay for the increase in performance.

While nanosized fillers aren't an overnight revolution, it is clear that there is significant activity on many fronts, with increasing use on all. Beyond the applications highlighted here, nanofillers have been incorporated into in golf clubs and golf balls, tennis rackets and a number of other products, with many more to come. The main barriers to nanomaterials use now seem to be development cost and the end-user's ability to obtain proper dispersion in and compatibility with resins. However, all of the nanosuppliers interviewed above are focused on making materials less expensive and more accessible by scaling up production, standardizing products, issuing test data and offering them in user-friendly forms.

Unit III–Q

More Applications of Nanoparticles

You must be wondering by now after all the 'over-the-head' theory about how different nanoparticles are from bulk and atomic material, what this really mean to us average everyday individuals. A lot!!

Due to the two unique phenomenon which occurs in nanoparticles, their properties (electrical, optical, chemical, mechanical, magnetic, etc.) can be selectively controlled by engineering the size, morphology, and composition of the particles. These new substances will have enhanced or entirely different properties from their parent materials.

Nanoparticles can make metals stronger and harder, give ceramics enhanced ductility and formability, enable normally insulating materials to conduct heat or electricity, and make protective coatings transparent. Hence, industries can re-engineer many existing products and design novel new products/processes that function at unprecedented levels.

Use the menu bar on the top to read about more about some examples of the possible applications.

Solar Cells

Solar cells are photovoltaic devices, which are devices that can convert light into electricity. They are much in demand as they offer an inexhaustible and environmentally benign energy source. Most solar cells are made of amorphous silicon. The problem with this is that the silicon must be of a very high purity and have a near perfect crystal structure. This makes it very expensive to produce. The efficiency of such a cell is also very small, typically converting only 13-18 per cent of sunlight to electricity. However, low efficiency wouldn't matter if huge arrays of cells could be produced cheaply. After all, nature's solar cells, chloroplasts in plants are less than 1 per cent efficient.

Most solar powered devices rely on the same principle: a photon of sunlight boosts an electron in the material into a mobile state so that it can be used to generate electricity. The problem with this simple mechanism is that the electrons are negatively charged and will leave a positive charge. These opposite charges attract one another and therefore will tend to recombine, squandering the absorbed energy as heat or as re-emitted light.

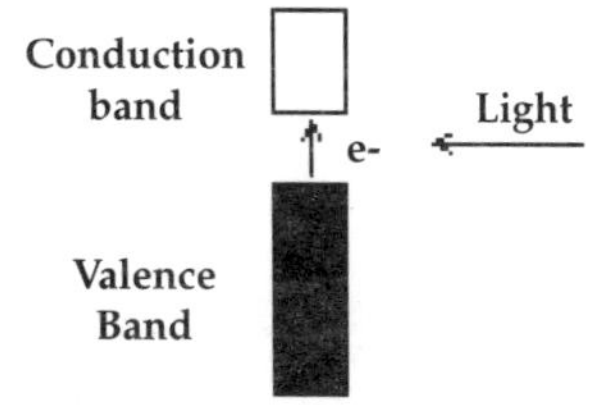

Fig. III–Q.1: Solar cells in the field. Fig. III–Q.2: Electrons jump from the valence bands into the conduction band, where they are mobile.

Silicon solar cells use an electric field to push the negatively charged electrons and positive charges apart. While chloroplasts adopt a more subtle approach of separate charges by making a distinction between the units that generate the electron and those that transport it away.

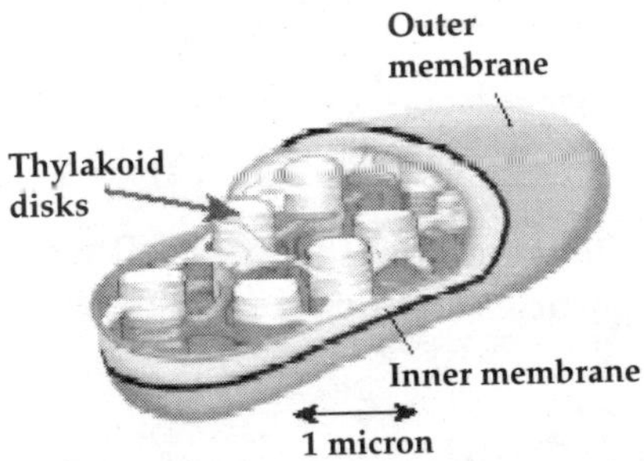

Fig. III–Q.3: Natural nanotechnology: Chloroplast contain nanoscale molecular machinery (including pigment molecules like chlorophyll) arranged inside stacked structures, called thylakoid disks, that conevert light and carbon dioxide into biochemical energy.

Deciding to copy nature's trick, Michael Gratzel and Brian O'Regan at the Swiss Federal Institute of Technology began research and produced the Grätzel cell in 1991.

The Grätzel cell uses intensely coloured organic dye molecules to capture light energy to inject an electron from the dye into a semiconductor such as titanium dioxide (TiO_2). This remarkably efficient charge separation reaction initiates current flow and the output of electrical energy by the cell.

How does Nanoscience Help?

The ideal material used in the cell must have a high surface area for light absorption and charge separation. Nanoparticles, having a comparable surface area to volume ratio, provides for just that. Titanium dioxide nanoparticles are used to make nanoporous thin film supported upon a glass substrate. The material obtained has optical transparency, excellent stability and good electrical conductivity.

The benefit of these novel photoelectrical solar cells is that they can be fabricated from cheap, low purity materials by simple and low cost procedures. Contrary to expectation, some of the new devices also have strikingly high conversion efficiency. The size-tunable bandgaps of the semiconductor nanoparticles, due to size quantisation, also means more efficient solar cells can be produced for photovoltaics (electricity production) and water splitting (hydrogen production) processes.

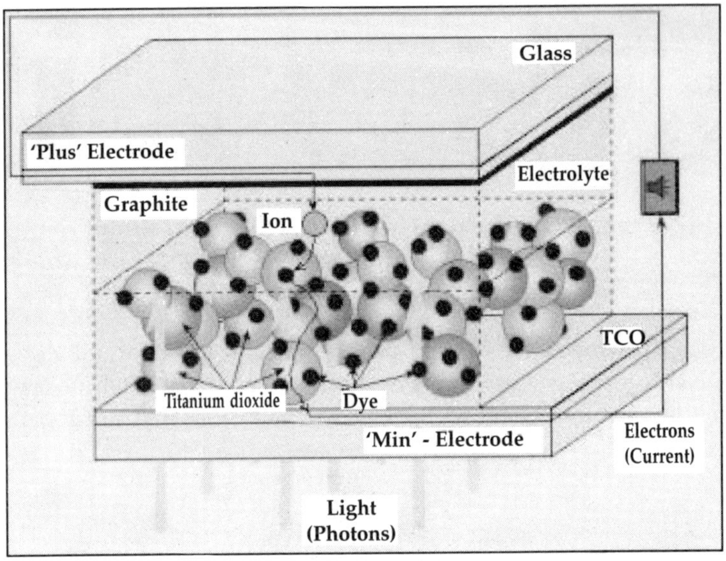

Fig. III–Q.4: Schematic diagram showing how a Grätzel cell work. The dye molecules attached onto the semiconductor particles are sandwiched between two glass electrodes.

Electrochromic Devices

Electrochromism is defined as a reversible colour change of a material caused by the application of an electrical current or potential. It has potential applications in making smart windows for the modulation of incoming light, as well as displays, a field in which competition with liquid crystal displays (LCD) is possible.

In the Past

There used to be only two types of electrochromic devices. In the first type, the electrodes are made of conducting glass and covered with an organic or inorganic polymer. The two materials would usually display complementary electrochromism and hence produce the same colour change when one is oxidized while the other is reduced. This type of construction is used in electrochromic windows. It is bistable, which means that once the colour change has occurred, the state of the device remains even in absence of applied voltage. The limitation of this type of electrochromic devicesis the slowness of the colour change, due to thelow migration rate of the counterions in the bulk polymer. It is also difficult to obtain strong colour changes or bright colours.

In the second type, two complementary electrochromic molecules are dissolved in a solvent. This type of system is very simple to build, reacts very fast and can produce dark or bright colours. The drawback of this type is that an electrical current is needed to maintain the coloured state because the two types of coloured molecules diffuse through the system and react with each other to restore the bleached states. It therefore cannot be used for large area devices or for battery-powered displays.

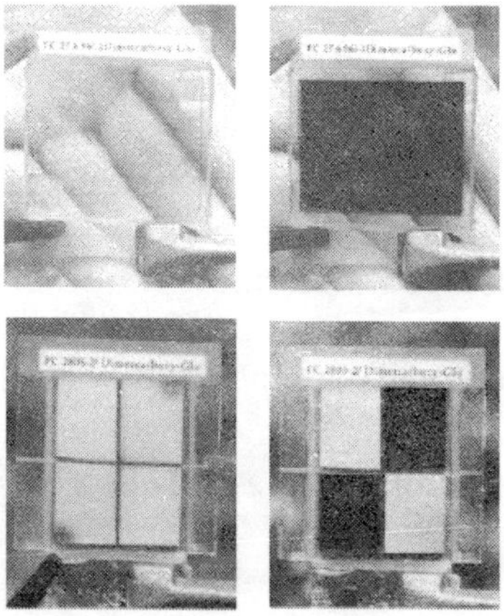

Fig. III–Q.5: An electrochromic window while the bottom pictures show an electrochromic display.

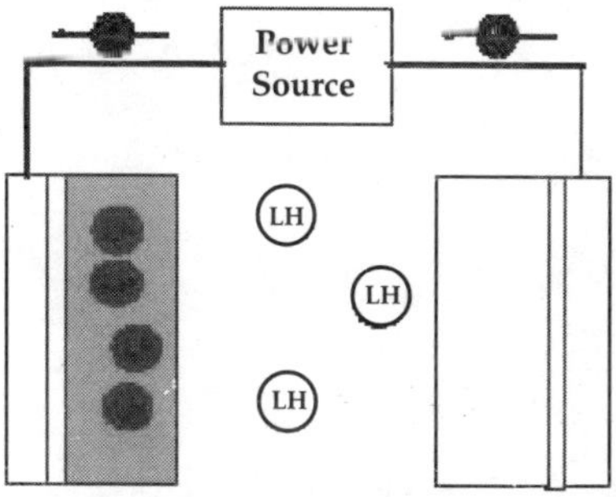

Fig. III–Q.6: The inorganic/electrode type. The power source produces an electric current which reduces the polymer giving it a blue colouration.

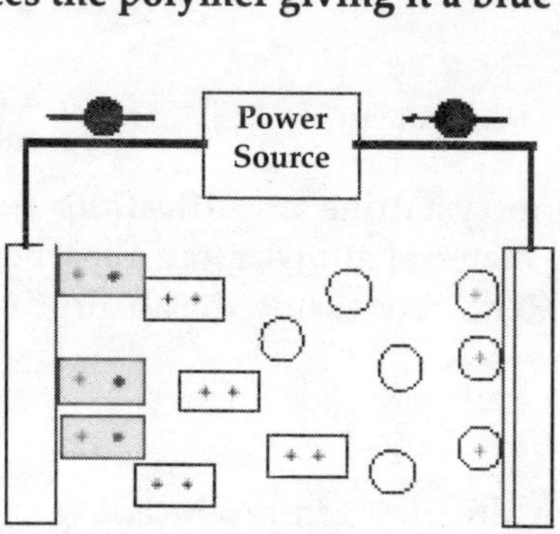

Fig. III–Q.7: The molecular/solution type. The particles can move rapidly this time giving high switching speeds.

Problem and Solution

We therefore need a system which is both bistable and changes colour rapidly. This can be done by attaching a suitable molecule that is colourless in the oxidized state and coloured in the reduced state onto the surface of a colourless semiconductor on conducting glass. When a sufficiently negative potential is applied, electrons are injected from the conducting glass into the conduction band of the semiconductor and reduce the adsorbed molecules. The reverse process occurs when a positive potential is applied and the molecules get bleached. The advantage of this system is that combines immobility of the electrochromic material with the rapidity and colouration efficiency of molecular systems.

However, a single molecular monolayer does not absorb a perceptible amount of light. This is where nanoparticles come in!

Nanocrystalline Amplification

By using nanocrystalline semiconductor films, it was possible to amplify the light absorption of the molecular monolayer to obtain visible colour changes. The nanocrystalline layer are highly porous and usually the real surface of a 5 μm thick layer is about 500 times the projected area. Therefore, as light passes through the layer, it crosses several hundreds of monolayers of coloured molecules giving a strong absorption.

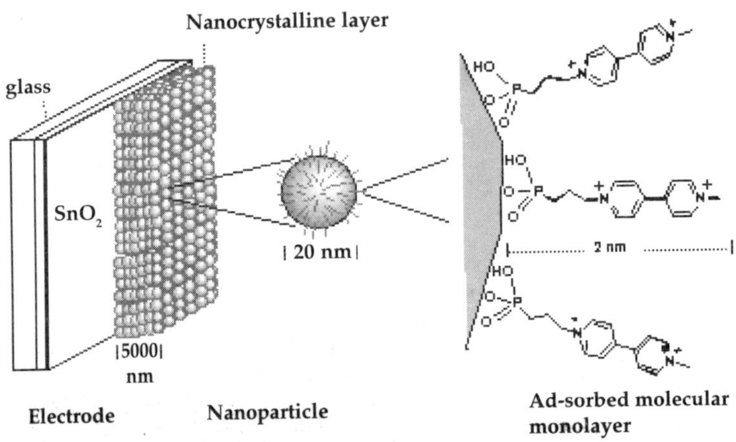

Fig. III–Q.8: Principle of the nanocrystalline amplification. The highly porous nanocrystalline layer of the semiconducting material allows more absorbed molecular monolayer to be present. Hence increasing the absorption of light.

Product

Some commercial products have already been produced!

The most successful is the automatic dimming rear-view mirror sense ambient light and glare from behind, and reduce annoying glare from bright lights.

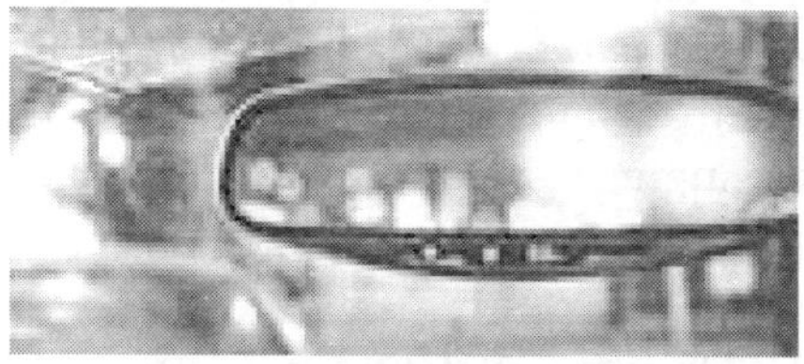

Fig. III–Q.9: Automatic dimming rear-view mirror.

While the company NTera has produced paper quality displays. These displays are bi-stable, not requiring much power keep to a pixel coloured, and reflective, not needing power for backlighting, and therefore can operate with very little power. Coupled with the fast switching, (*i.e.* changing of colour) due to amplification with nanoparticles, this technology could in most flat panel displays.

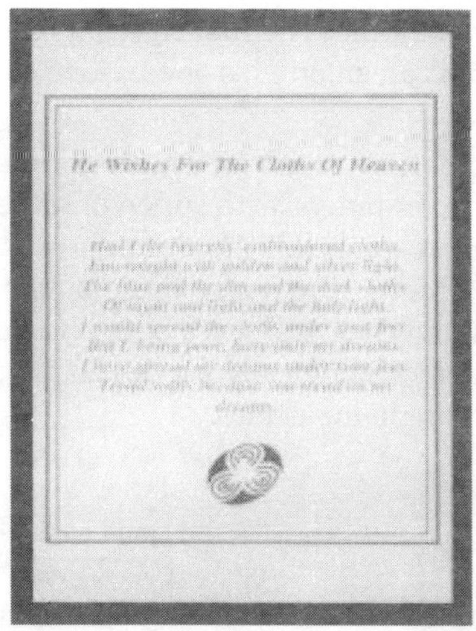

Fig. III–Q.10: The picture above and below show examples of the paper quality displays from NTera.

Electronics

The invention of the transistor in 1948 by John Bardeen and William Shockley, triggered a new era in electronics. The trend since then has been to create smaller and smaller product using fewer chips of greater complexity and smaller 'feature' sizes. The development of integrated circuits and storage devices have continued to progress at an exponential rate. At present it takes two or three years for each successive halving of component size.

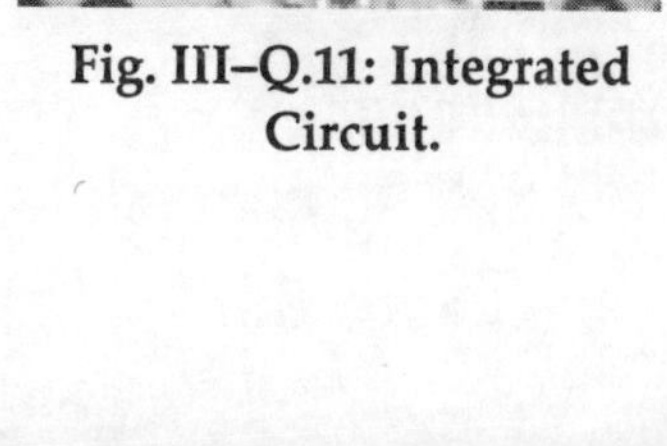

Fig. III–Q.11: Integrated Circuit.

However, the technologies currently in use have fundamental limits below, which the devices no longer function in a predictable manner. For instance, oxide layers used in Complimentary Metal Oxide Semiconductors (CMOS) devices are becoming so thin that leakage currents are conducted quantum mechanically by electron tunneling. Therefore, replacement technology to advance miniaturisation even further down, eventually to the dimensions of single atoms and molecules, is keenly sought.

In 1985, Dmitri Averin and Konstantin Likharev, proposed the idea of a new three-terminal device called a single electron tunneling (SET) transistor. Two years later Theodore Fulton and Gerald Dolan at Bell Labs in the US fabricated such a device and demonstrated how it operates. Since then much research has gone into this area and better single electron devices has been fabricated.

These devices usual have a molecule or atom as a functional unit and operate based on the quantum effect. Due to the small dimensions of the device (a couple of nanometers and hence is a nanoparticle), there are repulsive Coulombic interactions between the electrons and energy is required to add an extra electron. This energy cost can be tuned to zero by applying a voltage to the gate electrode. The electrostatic potential at this gate will be such that an extra electron can hop from the source onto the molecule. However, due to Coulomb repulsion, a second, extra, electron cannot hop on at the same time. The first electron must leave the molecule, moving into the drain before the next electron can enter. This one-by-one electron motion, is known as the single electron quantum tunneling.

The creation an ultrafast, single electron transistor would lead to the development of "quantum" computers with supercomputer powers and the size of a thumbtack. However, there is still a long road ahead before atomic or molecular transistors can be assembled into viable, dense, fast logic-circuits. Currently these single electron devices are still no competition for silicon transistors. But they will serve both scientifically, for studying electron motion through nanoscale objects, and technologically, for developing chemical techniques with which to fabricate electronic devices on single molecules.

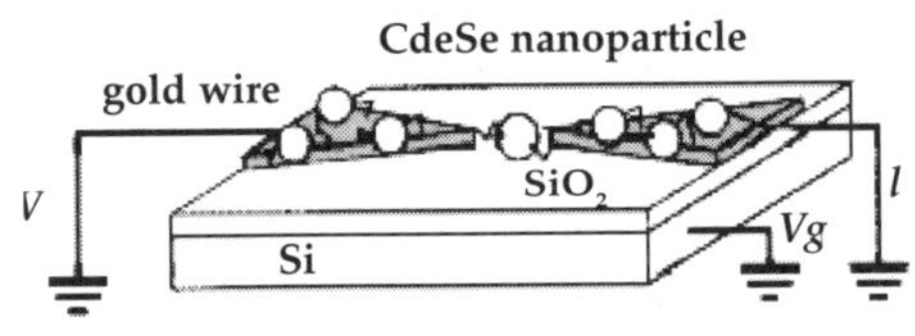

Fig. III–Q.12: Schematic diagram of a single electron transistor in which a CdSe nanoparticles is placed in between two gold wires.

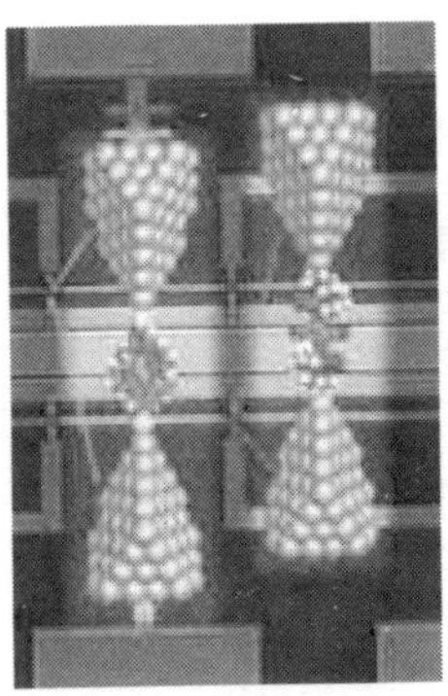

Fig. III–Q.13: Another representation of the single electron transistors. Nanoart by Felice Frankel.

Unit III–R

Nanoart

Nanoart comes in two flavors:

- Works created by artists envisioning our nanotechnological future
- Creations and designs produced by nanotechnologists

Let's look at some intriguing examples of nanoart . . .

Nanoguitar

In 1997, researchers at Cornell University created the world's smallest guitar. Carved out of crystalline silicon it measures only 10 micrometers long, about the size of a single cell, and has six strings, each about 50 nanometers wide.

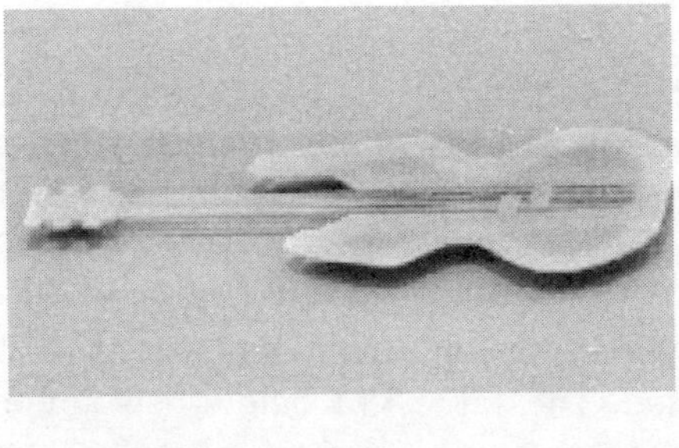

Fig. III–R.1: Nanoguitar—Photo by D. Carr and H. Craighead, Copyright © Cornell University

Using an atomic force microscope (AFM) the nanoguitar can be played, but it cannot be heard as its strings resonate at inaudible frequencies.Its creation demonstrates a new technology that could have a variety of uses in fiber optics, displays, sensors and electronics leading to a new generation of microelectromechanical devices.

Nanoplotter

In 1998, writing and printing techniques were miniaturized by a research team led by Chad Mirkin and Charles and Emma Morrison of Northwestern's Center for Nanotechnology when they created the world's smallest pen, utilized by an automated four-colour nanoplotter.

The nanolithography had previously allowed us to draw tiny lines with only a single ink or type of molecule. Now, with the nanoplotter, multiple inks, or different kinds of molecules can be placed side by side with such accuracy that the chemical purity of each line can be retained.

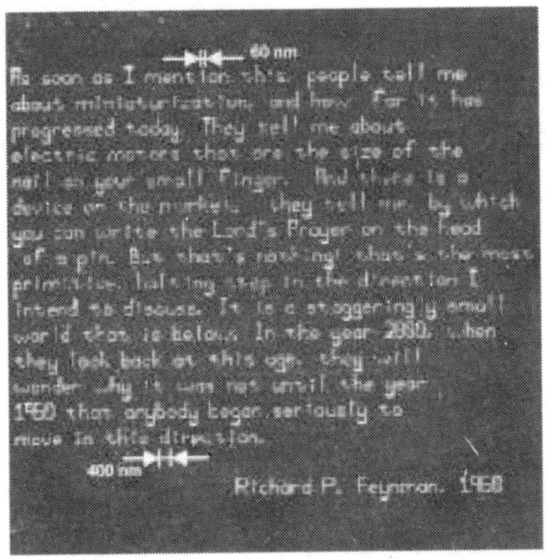

Fig. III–R.2: A paragraph from a famous speech by Nobel Prize-winning physicist Richard P. Feynman illustrates an 'overwriting' capability of the automated nano-plotter. The words were typed into the system's computer and then automatically output onto a gold substrate. In this example, the letters are written with an acid 'ink' and the lines are 60 nm wide and one molecule thick. They are surrounded by a one-molecule- thick layer of octadecanethiol that was laid down using a raster scanning method. It took only 10 minutes to generate this paragraph. Picture was taken by researchers in the Northwestern University.

By 2001, the Northwestern University researchers Seunghun Hong and Mirkinin had invented a nanoplotter that can draw multiple copies of a nanometre-sized pattern: at assorted places on a surface, simultaneously. Their eight-pen nanoplotter could be adapted to use a thousand pens!

Other Examples

The examples so far has focused only on small dimensions of nanoparticles. There are other properties that can be exploited. The Table 18.1 below lists some of these properties and their possible future applications.

Table III–R.1: Properties and Possible Future Applications of Nanoparticles

Property	*Applications*
Single magnetic domain	Magnetic recording, Data encryption
Small mean free path of electrons in a solid	Special conductors.
High & selective optical absorption of metal particles	Colors, filters, solar absorbers, photovoltaics, photographic material, phototropic material, Molecular Filters
Formation of ultra fine pores due to superfine agglomeration of particles	Molecular Filters
Uniform mixture of different kinds of superfine particles	R&D of New Materials
Grain size too small for stable dislocation	High strength and hardness of metallic materials
Surface/ Interface Large specific surface area	Catalysis, sensors
Large surface area, small heat capacity	Heat-exchange materials, Combustion Catalysts
Lower sintering temperature	Sintering accelerators
Superplastic behavior of ceramics	Ductile ceramics
Cluster coating and metallization	Special resistors, temperature sensors
Multi-shell particles	Chemical activity of catalysts

Unit III–S

Nanocrystals

Semiconductor nanocrystals (also called nanocrystals or quantum dots) can be described as a state of matter intermediate between molecular and bulk crystalline. Their small size and high optical activity make them a very interesting material for study in disciplines ranging from opto-electronics (integrating optical and electronic signals on a single substrate, i.e. crystalline Si) to fluorescence microscopy (as a bright, photostable and minimally perturbative fluorescent marker). This work will concentrate on experiments designed to optically characterize silicon nanocrystals as fluorescent beacons in solution since in biology fluorescent labels are invariably used to image structures or track/probe particles/biomolecules in the liquid state. Images of single immobilized Si nanocrystals on quartz substrates will also be presented. Just as bulk semiconductors are characterized by the structure of their conduction and valence bands as being direct or indirect, so are semiconductor nanocrystals. Direct semiconductor nanocrystals (e.g. CdSe, CdS) are characterized by having the minimum transition energy to promote an electron from the valence band to the conduction band without a change in the electron momentum (the energy of this energy separation is known as the band-gap). For indirect semiconductors however excitation at the band gap energy must be accompanied by a change in the electron's momentum (supplied by a phonon).

While both types of semiconductors are efficient absorbers of light, it has only been the direct semiconductors which have been emissive enough to be used as an optical source (e.g. in semiconductor GaAs quantum well lasers, Saleh and Teich, 1991) or as fluorescent labels (CdSe) (Chan and Nie, 1998, Bruchez et al, 1998). The discovery of porous silicon as an efficient luminescent material (Canham, 1990) has created great excitement as a prospective candidate for opto-electronic circuit integration. [Integrating direct semiconductor structures on silicon has been unsuccessful due to lattice mismatch].

Though the waited for advances have not materialized, it has been the starting point for experimental and theoretical investigations on nanocrystalline silicon.

Nanocrystalline Material

Nanocrystalline materials are build from nanoparticles. These particles are also called quantum dots or artificial atoms. These nanoparticles are usually spherical colloids of a semi-conducting material with a diameter of nanometers. Because these colloids are so small new material properties exist which are not present in bulk materials.

An important property is that the energy difference between the energy bands, the valence and conducting bands in bulk materials, increases and separate energy levels can be formed. These changes in the energy levels and the small size of the particles result in a new optical and electrical properties which can be used in the devices which were mentioned before.

The Structure and Properties of Nanocrystalline Materials

Due to their novel properties and varied potential applications, nanocrystalline materials, with typical grain sizes <100 nm, are attracting increasing attention from researchers all over the world. These materials are traditionally classified as one-dimensional (1-D), or layered structures, two-dimensional (2-D), or rod- or wire-like, and three-dimensional (3-D), or equiaxed, with the 1-D layered and 2-D wire-like structures being extensively investigated for coating and electronic applications and the 3-D equiaxed structures exploited for potential structural or bulk applications. Because of the small grain size of these materials, and consequently the large volume fraction of atoms in or near the grain boundaries, these **materials** exhibit properties that are often superior, and sometimes completely new, in comparison to those of conventional coarse-grained **materials**. However, it should be realized that a full understanding of the structure, thermal stability, and properties of these novel materials is not yet available. A number of issues and concerns still need to be solved.

Synthesis/Processing

Nanocrystalline materials can be synthesized by a number of techniques with the starting material existing either in the vapor, liquid, or solid state. Inert-gas condensation, in which the evaporated material is quenched onto a cold substrate, was the first technique used to synthesize nanocrystalline metals and alloys. Subsequently, plasma processing and other physical and chemical vapor deposition techniques have also been used. Electrodeposition and rapid solidification processing use the liquid state as the starting material.

Mechanical alloying, friction stir welding, severe plastic deformation, spark erosion, sliding wear, and repeated cold rolling produce nanocrystalline materials in the solid state. While some of these techniques are commercially in use for large-scale production purposes (e.g., inert-gas condensation, electrode-position, and mechanical alloying), many of the other techniques are essentially used for laboratory investigations. The choice of technique for the synthesis of nanocrystalline materials is determined by the ease of the process, its economic viability, its scalability, and the purity of the end product desired. Most of the techniques referred to previously produce the nanocrystalline material in the form of powder. The application of nanocrystalline materials requires that these powders are consolidated to full density with minimal or no porosity and this is not a trivial matter. Full densification with complete inter-particle bonding requires exposure of the powder to high temperatures and pressures for extended periods of time, which is likely to coarsen the micro structural features. Retention of the material in the nanocrystalline state, however, warrants that the powder is not taken to high temperatures for long periods of time; this may not result in complete densification. Thus, successful consolidation to full density requires innovative methods of consolidation. However, since nanocrystalline materials have a large fraction of the atoms in the grain boundaries, the (effective) diffusivity of nanocrystalline materials is much higher than that of a coarse-grained material of the same composition. This will facilitate achieving full consolidation of nanocrystalline materials at temperatures 300°C to 400°C lower than those required for the coarse-grained materials. Successful consolidation of nanocrystalline powders has been achieved by electro-

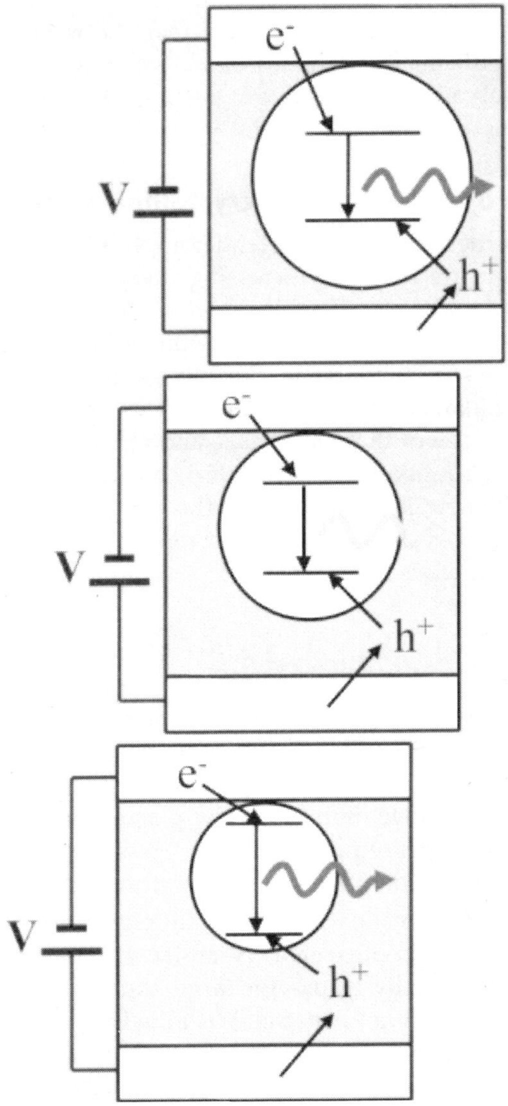

Fig. III–S.1: Light from nanoparticles

Via the reverse proces as in a solar cell, electrons (e^-) and holes (h^+) can be inserted in the conductance and valence band of a semiconductor colloid with means of a powersupply (V) like a battery. The electron and the hole will recombine and produce a photon. The particle will emit light with an energy corresponding to the energy difference between the onductance and valence band energy levels. Light can be generated covering the entire vissible spectrum by tuning the size and chemical composition of the semiconducting particles.

discharge compaction, plasma-activated sintering, shock (explosive) consolidation, hot isostatic pressing (HIP), Ceracon processing, hydrostatic extrusion, strained powder rolling, and sinter forging. Consolidation is not required if the powder could be used as-is (e.g., in slurries, as catalysts, or for coating purposes).

Unit III–T

Nanoclusters

A nanocluster is a fragment of solid comprising somewhere between a few atoms to a few tens of thousands of atoms. Nanoclusters are therefore a novel state of matter with properties that are neither those of a bulk crystal nor those of individual atoms and molecules. Over the past 10 years huge advances have been made both in the synthesis of size-tunable, monodisperse (i.e. size-selected) nanoclusters of various chemical compositions and in the development of techniques for their assembly into nanostructured solids (facilitating the synthesis of what have been termed "designer materials").

Fig. III–T.1: Nanocluster.

Clusters with sizes in the order of several nanometers can have physical properties with are very different from the bulk material. In just the same way are the (electronic) properties of a cluster very different from those of the individual atoms which make up the cluster. This makes the nanoclusters a very attractive research area in the solid state physics. One of the most characterizing properties of these clusters is the very large fraction of surface atoms. This makes the cluster very sensitive to changes at the border of the clusters. Influences from the outside of the clusters are therefore having a great impact on the properties of these clusters. The study of these influences at the border is the main topic of our study. In our research we try to investigate in which way the vibrational state of a cluster is influenced by the surrounding matrix. At the same time we want to get a detailed picture of how these clusters are being formed from a (over)saturated solution, formed by ion implantation. We focus on two different cluster systems. On one hand we study Sn clusters in SiO_2, which is a soft metal in a hard matrix and on the other hand we study Co clusters in Ag which is a hard metal in a soft matrix. These two systems are so to say each others counterpart in the dynamical behavior we expect.

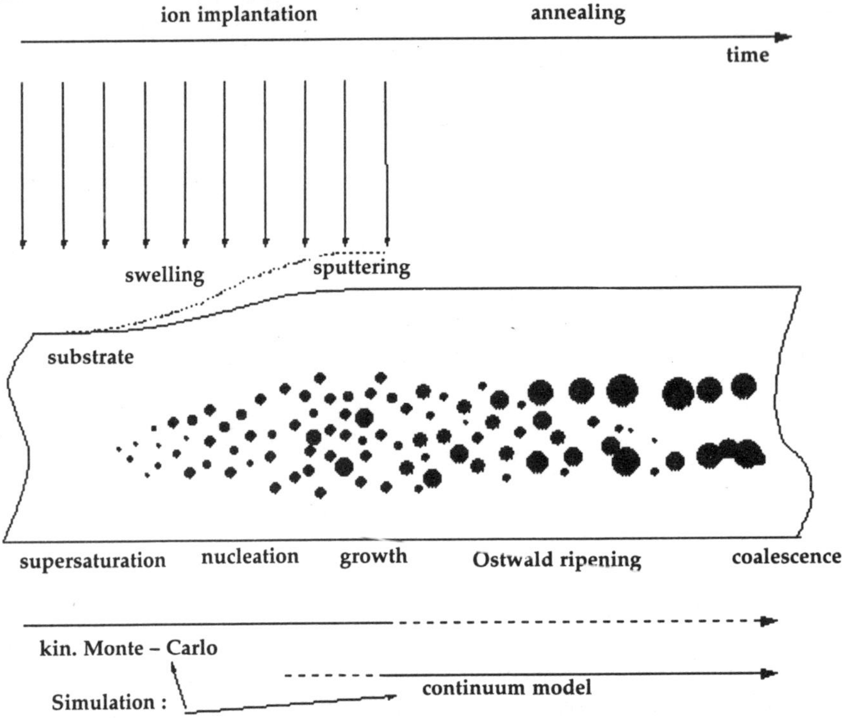

Fig. III–T.2: Formation of nanoclusters.

How Nanoclusters Work?

Nanoclusters have a very high zeta potential (negative electrical charge) and are small enough to easily penetrate any cell membrane in the body. Zeta potential represents a basic law of nature. It plays a vital role in the health of all forms of plant and animal life. It is the force that maintains the discreteness of the trillions of circulating cells that nourish an organism. If zeta potential is low, toxins can not be suspended for elimination and nutrients can not be suspended for transportation to the cells. The whole system becomes clogged. Nanoclusters, with their high zeta potential, may increase the absorption of nutrients and the elimination of toxins by helping to maintain the proper polar attraction of negatively charged nutrients to the positively charged intracellular fluid. When these two electrical charges are in the appropriate balance, nutritional absorption and cellular metabolism are at optimum levels for maximum energy and an overall sense of well being.

Silica Nanoclusters

The properties of silica clusters at temperatures from 1500 to 2800 K are investigated using classical molecular dynamics simulations for particles containing up to 1152 atoms. It was found that atoms in the cluster were arranged in a shell-like structure at low temperatures, and that the radial density profile peaked near the outer edge of the particle. Smaller clusters have much higher pressures, with the magnitudes corresponding quite well to those obtained from the Laplace–Young equation, when evaluated in a self-consistent manner using our derived surface tension. The computed surface tension did not show any significant size-dependent behavior in contrast to the decreasing surface tension observed for Lennard–Jones liquid clusters. Finally the computed diffusion coefficients in the liquid state are seen to be larger than bulk computed diffusivities.

One of the most import methods of synthesizing nanoparticles is through vapor phase nucleation and growth. Several methods are used, including combustion, plasmas, thermal reactors, and evaporation-condensation. The evolution of an aerosol undergoing a gas-to-particle growth process is described by a master equation, the "aerosol general dynamic equation" (GDE). The GDE is a population balance equation for the behavior in time and space of the particle size distribution function, which can include all the driving forces for particle growth (nucleation, surface growth, coagulation/coalescence, transport, etc.). Less conventionally, but of interest for nanoparticle synthesis research, the GDE can be extended to track the degree of coalescence of particle aggregates. One of the biggest challenges in the implementation of the GDE is knowledge of the rates of these individual processes. For nanoparticle synthesis, one of the most fundamental and important kinetic properties of interest in nanoparticle growth is the sintering or coalescence rate between particles during vapor-phase growth. A knowledge of these rates and their dependence on process parameters ultimately impacts the ability to control primary particle and agglomerate growth, which are of critical importance to nanoscale particles whose properties depend strongly on size, morphology, and crystal structure. Previous experimental and theoretical studies of sintering in flame reactors demonstrated good agreement for the case of titania particle growth, 7–9 based on a characteristic coalescence time obtained from a solid-state diffusion model, in which the characteristic time is written as

$$\tau_f = 3k_BT\nu_p/64\pi D\sigma\nu_0,$$

where T is the gas temperature, vp is the particle volume, D is the diffusion coefficient usually reported as an Arrhenious σ_i is the surface tension, and v_0 is the molecular volume for diffusion. In the case of silicon dioxide (silica), however, the primary particle size is typically underpredicted based on a viscous flow coalescence time written as

$$\tau_f = 2\eta R_p/\sigma,$$

where η is the viscosity, and R_p is the radius of the particle. Ehrman *et al.* discussed the possible causes of the enhanced rate of silica particle sintering observed in an experiment, including the presence of impurities such as alkali metals which could result in a lower viscosity of silica. On the other hand, in theoretical considerations kinetic coefficients such as the diffusion coefficient and the viscosity, as well as the surface tension, were taken from the properties of bulk material. However, we know that what makes the nanoparticles so interesting is that their fundamental properties are intimately related to the physical size of the component. For an example, a 10-nm-diameter iron particle has almost 20 per cent of its atoms on the surface. This high fraction of surface atoms alters properties such as the melting point and presumably the surface tension14 and accounts for the high reactivity of nanoparticles both toward each other and also toward other materials and gases. One of the interesting aspects of nanoparticles properties is the variation of the internal pressure with particle size. Laplace's equation

$$\Delta P = 2\sigma/R_p,$$

tells us that the pressure difference between a particle and the surrounding environment ΔP goes as the surface tension s, divided by the particle radius. This implies that as the particle becomes smaller, the pressure will approach infinity as the radius becomes vanishingly small. Of course, this assumes that the surface tension is size independent, but in fact, at some point, the surface tension will tend toward zero. Nevertheless, we should expect that the pressure of a particle can be many hundreds of atmospheres, and may impact the chemical, phase and crystalline structure of particles. For an example, the pressure inside a 10-nm-diameter TiO_2 particle $\sigma = 0.5$ J/m^2 is on the order of 2000 atmospheres. Such elevated pressures may have a significant impact on properties such as diffusion coefficients and the viscosity, which in turn affect the rates of particle sintering· Given these constraints, data obtained at large grain sizes when extrapolated to smaller dimensions may lead to considerable errors. Experimental studies are certainly desirable for investigating these characteristics; however, experiments aimed at determining the fundamental transport and thermodynamic properties of fine aerosols are not easily obtained, because of the ensemble nature of most studies.

Here we study the nanoscale silica clusters of different sizes (N=72, 288, 576, and 1152) and at different temperatures (1500–2800 K) using a classical moleculardynamics (MD) simulation to determine the structural properties, internal pressure, diffusion coefficients, and surface tension. We will particularly focus on the particle size dependence of these properties, and its influence on coalescence/sintering modeling.

Model System

The clusters consist of N classical atoms which interact through a pair model potential developed by Tsuneyuki *et al.*,which takes into account the steric repulsion due to atomic sizes, screened Coulomb interaction from charge transfer, charge-dipole interactions due to the large electronic polarizability of the anions, and dipole-dipole interactions to mimic the short- and intermediate-order interaction. The interaction potential has an analytical form of the Born-Huggins-Mayer type

$$V_{ij} = q_i q_j / r_{ij} + a_{ij} \exp(-b_{ij} r_{ij}) - c_{ij} / r_{ij}^{\,6},$$

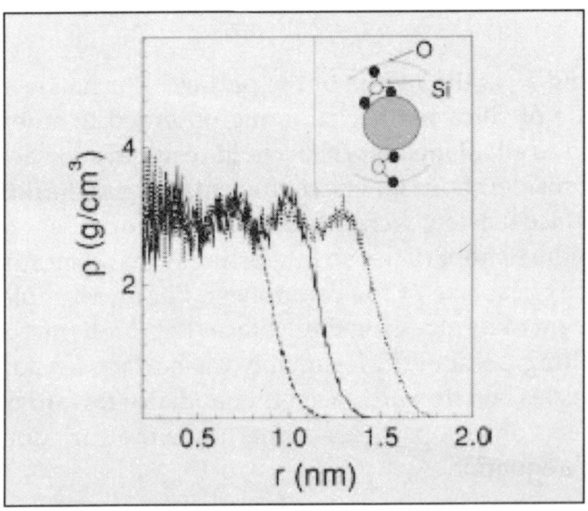

Fig. III–T.3: The radial density distribution for clusters with 288 atoms (dashed line), 576 atoms (solid line) and 1156 atoms ~dotted line! at T52080 K. The sketch of surface structure is shown in the inset.

where q_i is the charge of the i-type atom, r_{ij} is the interatomic distance, and a_{ij}, b_{ij}, and c_{ij} are the parameters taken. For higher temperature the potential was modified by adding a Lennard-Jones term, in order to avoid the occasional overriding of the O-O repulsion barrier at a shorter interatomic distance at a higher temperature. The modified interatomic interaction is referred to as the TTAMm potential with the form $V_{ij}^* = V_{ij} + 4\varepsilon_{ij}\,[(\sigma_{ij}/r_{ij})^{18} - (\sigma_{ij}/r_{ij})^{6}]$, where ε_{ij} and σ_{ij} are the usual Lennard–Jones parameters. The equations of motion for each atom was solved using the standard velocity version of the Verlet algorithm. In our calculations we constructed a silica cluster with an appropriate number of atoms, and placed it in the center of a spherical cavity of radius R^*, which comprised the calculational cell. The center of the calculation cell (r50) coincides with the center of mass of the cluster. The radius of the cell was kept at 2 nm for all clusters. In order to conserve mass at higher temperatures, we force any cluster fragment which escapes the cluster due to evaporation, and reaches the boundary of the cell, to be elastically reflected. All clusters were initially heated to a high temperature liquid state at T=4000 K. Several clusters were then prepared at a variety of temperatures between 1500 and 2800 K by slow numerical cooling. Clusters were allowed to anneal for 35000 time steps (with a time step $\delta t = 1 - 4$ fs), followed by a test at constant energy to evaluate if the system had reached an equilibrium state. Production runs for cluster properties were performed at a fixed temperature, and typically required about $1.5\text{x}10^5$ time steps to obtain a reasonable statistical averaging of cluster characteristics. In order to verify the equilibration procedure for each prepared cluster, we gradually decreased the temperature from 4000 to 1500 K, followed by a reversal of this procedure to the desired target temperature. A comparison of cluster properties such as the radial density distribution obtained by cooling and then reverse heating, shows good agreement. The total angular momentum of cluster motion is always reset to zero by the transformation of velocities in order to evaluate transport properties such as diffusion coefficients. It is not conducive to analyze the cluster behavior at higher temperatures T>3000 K, where a significant dissociation of fragments takes place during the time interval of the calculation ($t \propto 0.15$ ns). This is because this interatomic potential cannot realistically describe the vaporization processes of silica. For instance, in the calculations we observe the vaporization of SiO_2 molecules, while it is well known that silica vaporizes by decomposition to SiO and O_2.

Results

Particle Density

At lower temperature T = 1500 K in a solid "glass" state, all clusters exhibit an oscillating density distribution. With increasing temperature there seems to be a tendency to dampen out the radial density gradient, though some peaks still remain. The radial dependence of the mass densities of different-sized clusters is shown in Fig. 20.4 at a temperature T=2080 K. Note that the interpeak space is about the mean Si-Si interatom distance which is 0.3175 nm. To explore the effect of density, in Fig. 20.5 we show the radial mass density distribution for a 576-atom cluster at three temperatures. It is clear that the radius of the SiO_2 nanoparticles and, consequently, the averaged density, are effectively insensitive to temperature over quite a large temperature range (1500–2800 K). This is in accordance with the fact that bulk silica has an extremely low thermal expansion coefficient. Over this temperature interval a cluster with N=288 atoms has a radius R=0.92 nm and an average density of p=53.2 g/cm^3. The 576-atom cluster has a radius R=1.16 nm and a density p=52.9 g/cu cm and the cluster of N=1152 atoms has R=1.5 nm and p=2.8 g/cu cm. The radius of a cluster is defined as the distance where the density drops to 0.5 of its interior value. Note that smaller clusters have a higher density, but the values are in close agreement with that calculated for bulk silica in the coesite phase, with a density of 2.9 g/cu cm. The density profiles presented in Figs. III–T.3 and III-T.4 show that near the surface the clusters have a maximum in the density, just before the rapid decrease at the cluster edge. A similar behavior was observed by Roder *el al* using a BKS interaction potential. They obtained a density of 2.3 g/cu cm. This implies that clusters of different sizes have similar shell structures. Moreover, the cluster surface has the same structure and width regardless of size. The rapid density decrease extends over a distance which is more than a Si-O bond length equal to 0.162 nm. A careful temporal observation makes it clearer that the cluster can be separated into "core"and surface components. On the surface both silicon and oxygen have dangling bonds, with the oxygen atom tending toward the surface (see the inset in Fig. III–T.3). These fragments residing on the surface tend to be highly mobile, so that when averages of the density are taken we see a smooth monotonic decrease in the density profile at the edge of the cluster corresponding to about 0.2–0.3 nm. In general, then, we observe an enhanced density for smaller clusters, small thermal expansion, and an oscillating density profile.

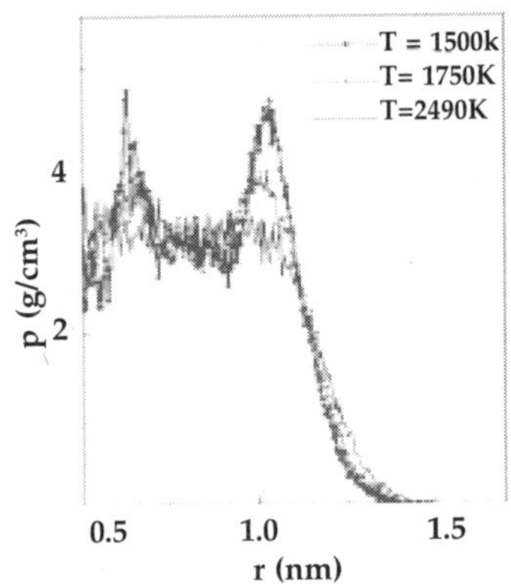

Fig. III–T.4: Radial density distribution for the cluster with 576 atoms at different temperatures.

Structure

In order to obtain further insight into this oscillating radial density behavior, we consider the structure of the clusters more closely. Shown in Fig. III–T.5 is a cross-section slice for two different clusters (N=288 and 576 atoms) at T=500 K. The slice is taken through the center of a cluster of thickness 0.328 nm (which is slightly larger than two of nominal Si-O bond lengths) What is being shown are Si and O atom positions following averaging over 75 ps [Figs. III–T.5 (a) and III–T.5 (b)] and the trajectories for O atoms recorded every 5 ps [Figs. III–T.5 (c) and III–T.5 (d).

The slice map of smaller cluster looks like a network of atoms organized in rings [Fig. 20.5 (a)]. The larger cluster has apparently a spherically symmetric shell structure [Fig. 20.5 (b)], and the external shell incorporates both O and Si atoms. The defects in the smaller cluster are denoted by atom trajectories that wander. For an instance, in Figs. III–T.5 (a) and III–T.5 (c) the arrow labeled by 1 shows a vacancy for a silicon atom. The arrow labeled by 2 is another example of a defect and the enhanced mobility of atoms containing dangling bonds. The distribution of oxygen and silicon atoms as a function of radial position is presented in Fig. III–T.6 for the cluster prepared with 576 atoms (192 Si atoms and 384 O atoms). It clearly indicates that oxygen is preferentially found on the surface layer of the cluster under all conditions. Note that the change of temperature has a pronounced

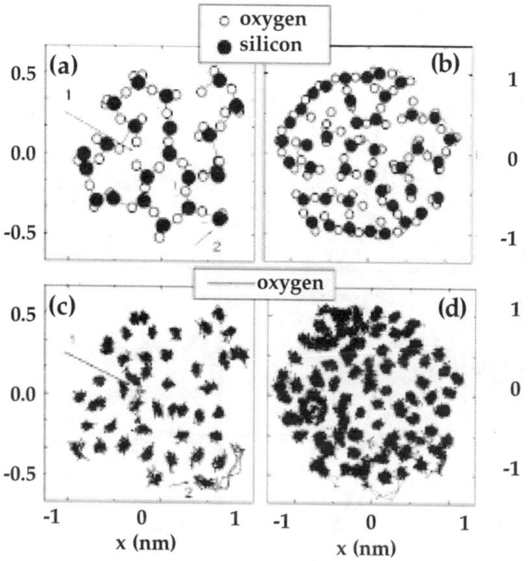

Fig. III–T.5: Averaged oxygen atom positions ~open circles! and silicon positions (solid circles) for clusters with 288 atoms (a) and 576 atoms (b). The trajectories of oxygen atoms for the clusters with 288 atoms (c) and 576 atoms. Shown is a slice of the clusters at T=1500 K.

effect on the internal arrangement of the particles. In particular, at *T*=1600 K the peak in the Si concentration corresponds to a minimum in the O atom density. With increasing temperature to *T*=1680 K, we find that a thermally induced structural transition takes place. The second Si peak labeled by 1 in Figs. III–T.6 (a) and III–T.6 (b) shifts, and the second peak in O atoms distribution labeled by 2 in Fig. III–T.6 (a) disappears and indicates an altogether different shell structure. In order to understand the change in the cluster structure we analyze the angular distribution, varying the temperature. In Fig. III–T.7 the O-Si-O and Si-O-Si angular distributions are shown for a cluster with 576 atoms. At a temperature of *T*=1600 K, the Si-O-Si angle distribution has two maxima which are at 97.7° and 142°. The former denotes the presence of four member rings, and the latter refers to six member rings. As the temperature increases to *T*=1680 K, the Si-O-Si angular distribution has only one maximum at 142°. The observation of the angular distribution change clarifies the structural rearrangement in the shell structure presented in Figs. III–T.6 (a) and III–T.6 (b). Note that the O-Si-O angular distribution only slightly broadens with increasing temperature as expected, and retains a monomodal character. The results agree well with the angular distributions in the amorphous and molten bulk SiO_2 studied by Vashishta *et al.* who used a three-body potential. For example, at *T*=2500 K the O-Si-O and Si-O-Si angular distributions have maxima at 110.1° and 141° for the cluster with 576 atoms, and at 109.5° and at 142.5°, respectively, for bulk material. The angular distributions are, however, much broader for the clusters compared to the bulk because of the spherical surface and shell structure. A similar broadening was observed by Roder *et al.*

The partial pair correlation functions were calculated for the O-O, Si-O, and Si-Si bonds to understand the enhanced density of smaller particles and mechanism of cluster expansion at higher temperature. It turns out that the first and second peaks of the O-O and Si-O pair correlation functions have the same positions for all clusters. In contrast to this fact the different clusters have various Si-

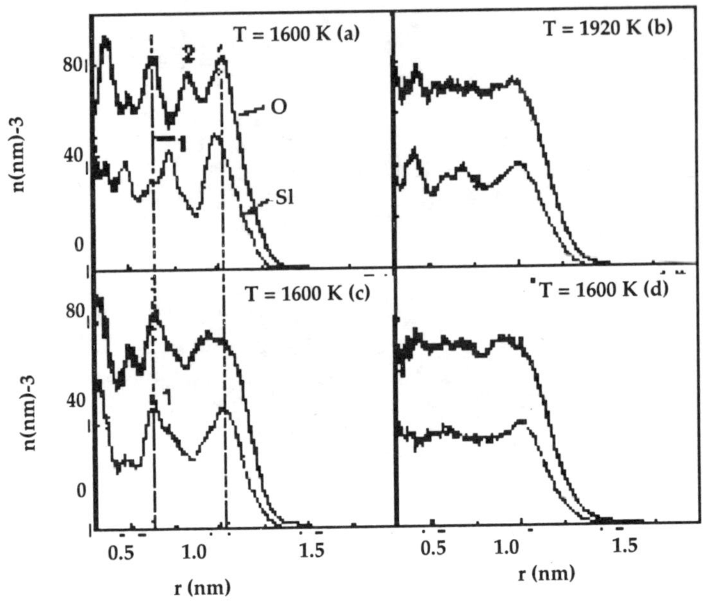

Fig. III–T.6: Averaged radial distribution of oxygen atoms (solid lines) and silicon atoms (dotted lines) in a cluster with 576 atoms at temperatures T=1600 K (a) 1680 K (b) 1920 K (c) and 2480 K (d).

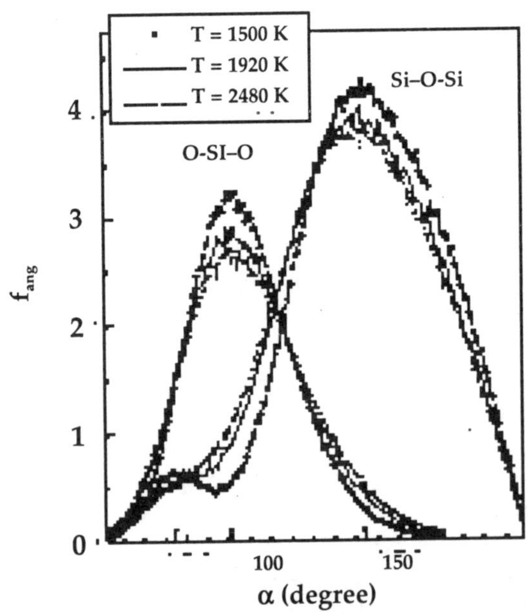

Fig. III–T.7: Angle distribution for the Si-O-Si and O-Si-O bonds at different temperatures for the 576-atom cluster.

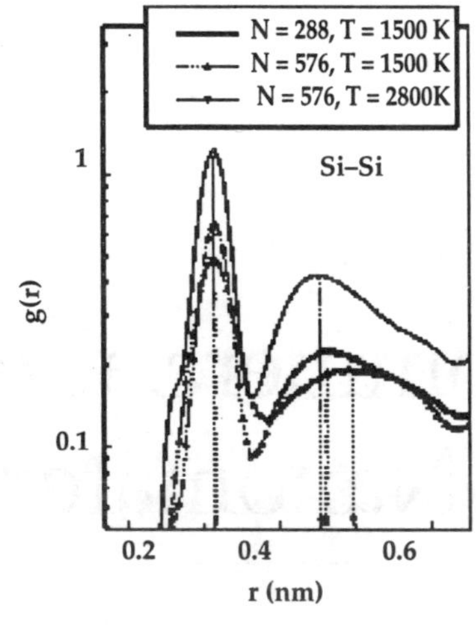

Fig. III–T.8: Si-Si pair correlation function for clusters with 288 atoms at *T*=1500 K (solid line), with 576 atoms at *T*=1500 K (triangles) and with 576 atoms at *T*=2800 K (circles).

Si pair correlation functions. One can see in Fig. III–T.8 that the smaller cluster (N=288 atoms) has a closer Si-Si packing than the largest one (N=576 atoms at T=1500 K. With increasing temperature the peaks of the O-O and Si-O pair correlation function become broader, and their positions remain independent of temperature, whereas the Si-Si pair correlation function is sensitive to the temperature. Its first peak at r = 0.317 nm does not respond to temperature, but a second peak shifts to the larger coordinate (see Fig. III–T.8). Thus, we can conclude that the variation of the density with the cluster size as well as the expansion of a cluster with temperature are explained by the change in the Si-Si atom packing due to the strong repulsive nature of the Si-Si interaction.

Unit III–U

Current Commercial Applications of Nanomaterials

This quote from "The Next Big Thing Is Really Small" by Jack Uldrich and Deb Newberry sums it up nicely: This is not to say that nanotechnology is a far-off, fuzzy, futuristic technology. It is not. It has already established a beachhead in the economy. The clothing industry is starting to feel the effects of nanotech. Eddie Bauer, for example, is currently using embedded nanoparticles to create stain-repellent khakis. This seemingly simple innovation will impact not only khaki-wearers, but dry cleaners, who will find their business declining; detergent makers, who will find less of their product moving off the shelf; and stain-removal makers, who will experience a sharp decrease in customers. This modest, fairly low-tech application of nanotechnology is just the small tip of a vast iceberg—an iceberg that threatens to sink even the "unsinkable" companies."

Nanocomposites

Examples

Researchers at Pacific Northwest National Laboratory have developed a coating process to make sponge-like silica latch onto toxic metals in water. Self-Assembled Monolayers on Mesoporous Supports easily captures such metals as lead and mercury, which are then recovered for reuse or contained in-place forever. PNNL One example of a SAMMS nanocomposite (**S**elf-Assembled Monolayers on Mesoporous Supports). An hexagonally close-packed cluster of tubular pores (end view) is shown in the foreground. A single pore, in this case coated with a mercaptopropylsiloxy monolayer, is shown in the background. A model of one surfactant molcule is also shown.

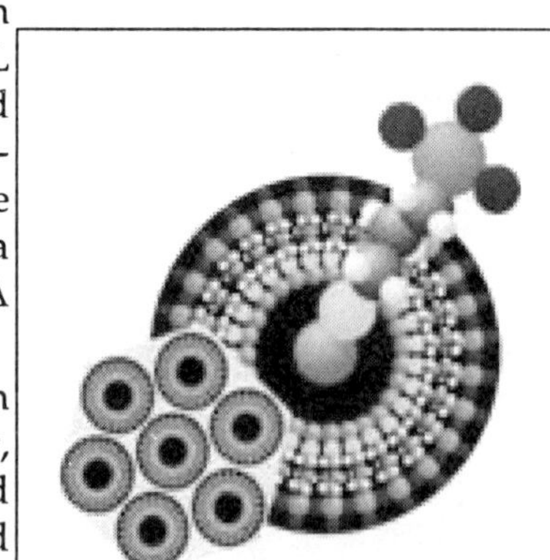

A plastic nanocomposite is being used for "step assists" in the GM Safari and Astro Vans. It is scratch-resistant, light-weight, and rust-proof, and generates improvements in strength and reductions in weight, which lead to fuel savings and increased longevity. And in 2001, Toyota started using nanocomposites in a bumper that makes it 60 per cent lighter and twice as resistant to denting and scratching.

Impact: Will likely be used on other GM and Toyota models soon, and in other areas of their vehicles, as well as the other auto manufactures, lowering weight, increasing milage, and creating longer-lasting autos. Likely to impact repair shops (fewer repairs needed) and auto insurance companies (fewer claims). Will also likely soon be seen everywhere weight, weather-proofing, durability, and strength are important factors. Expect NASA, the ESA, and other space-faring organizations to take a serious look, soon, which will eventually result in lower lift costs, which will result in more material being lifted into space.

Electron micrograph of typical silicon nanocomposite cross section showing uniform distribution of conductive carbon nanotube network. Photo courtesy of U.S. Air Force.

Nanocrystals

Examples

Nanocrystals absorb then re-emit the light in a different colour—the size of the nanocrystal (in the Angstrom scale) determines the colour.

Six different quantum dot solutions are shown, excited with a long-wave UV lamp.

Quantum dots are molecular-scale optical beacons. Qdot nanocrystals behave like molecular LEDs (light emitting diodes) by "lighting up" biological binding events with a broad palette of applied colours.

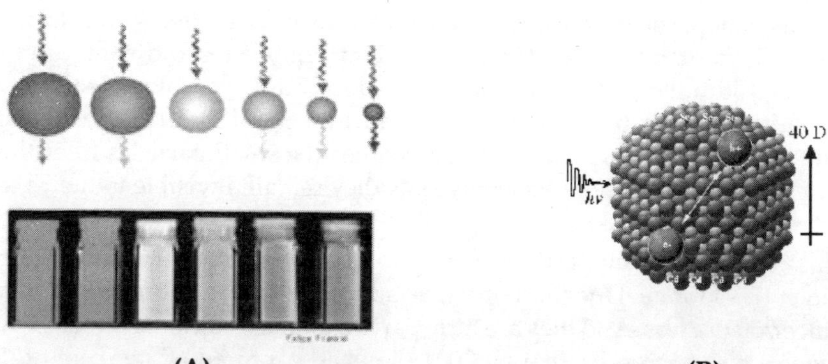

(A) (B)

Fig. III–U.1: (A) Metal nanocrystals might be incorporated into car bumpers, making the parts stronger, or into aluminum, making it more wear resistant. Metal nanocrystals might be used to produce bearings that last longer than their conventional counterparts, new types of sensors and components for computers and electronic hardware.

(B) Nanocrystals of various metals have been shown to be 100 percent, 200 percent and even as much as 300 percent harder than the same materials in bulk form. Because wear resistance often is dictated by the hardness of a metal, parts made from nanocrystals might last significantly longer than conventional parts.

Smith & Nephew markets an antimicrobial dressing covered with nanocrystalline silver (A patented Technology of NUCRYST Pharmaceuticals). The nanocrystalline coating of silver rapidly kills a broad spectrum of bacteria in as little as 30 minutes.

Nanocrystals are an ideal light harvester in photovoltaic devices. (They) absorb sunlight more strongly than dye molecules or bulk semiconductor material, therefore high optical densities can be achieved while maintaining the requirement of thin films. Perfectly crystalline CdSe nanocrystals are also an artificial reaction center, separating the electron hole pair on a femtosecond timescale. Fluorescent nanocrystals have several advantages over organic dye molecules as fluorescent markers in biology. They are incredibly bright and do not photodegrade. Drug-conjugated nanocrystals attach to the protein in an extracellular fashion, enabling movies of protein trafficking. (They) also form the basis of a high-throughput fluorescence assay for drug discovery.

Nanoparticles

Stain-repellent Eddie Bauer Nano-Care khakis, with surface fibers of 10 to 100 nanometers, uses a process that coats each fiber of fabric with "nano-whiskers. Developed by Nano-Tex, a Burlington Industries subsidiary. Dockers also makes khakis, a dress shirt and even a tie treated with what they call "Stain Defender", another example of the same nanoscale cloth treatment.

Impact: Dry cleaners, detergent and stain-removal makers, carpet and furniture makers, window covering makers.

BASF's annual sales of aqueous polymer dispersion products amount to around $1.65 billion. All of them contain polymer particles ranging from ten to several hundred nanometers in size. Polymer dispersions are found in exterior paints, coatings and adhesives, or are used in the finishing of paper, textiles and leather. Nanotechnology also has applications in the food sector. Many vitamins and their precursors, such as carotinoids, are insoluble in water. However, when skillfully produced and formulated as nanoparticles, these substances can easily be mixed with cold water, and their bioavailability in the human body also increases. Many lemonades and fruit juices contain these specially formulated additives, which often also provide an attractive colour. In the cosmetics sector, BASF has for several years been among the leading suppliers of UV absorbers based on nanoparticulate zinc oxide. Incorporated in sun creams, the small particles filter the high-energy radiation out of sunlight. Because of their tiny size, they remain invisible to the naked eye and so the cream is transparent on the skin.

Sunscreens are utilizing nanoparticles that are extremely effective at absorbing light, especially in the ultra-violet (UV) range. Due to the particle size, they spread more easily, cover better, and save money since you use less. And they are transparent, unlike traditional screens which are white. These sunscreens are so successful that by 2001 they had captured 60 per cent of the Australian sunscreen market.

Impact: Makers of sunscreen have to convert to using nanoparticles. And other product manufactures, like packaging makers, will find ways to incorporate them into packages to reduces UV exposure and subsequent spoilage. The $480B packaging and $300B plastics industries will be directly effected.

Using aluminum nanoparticles, Argonide has created rocket propellants that burn at double the rate. They also produce copper nanoparticles that are incorporated into automotive lubricant to reduce engine wear.

AngstroMedica has produced a nanoparticulate-based synthetic bone. Human bone is made of a calcium and phosphate composite called Hydroxyapatite. By manipulation calcium and phosphate at the molecular level, we have created a patented material that is identical in structure and composition to natural bone. This novel synthetic bone can be used in areas where natural bone is damaged or removed, such as in the in the treatment of fractures and soft tissue injuries.

SECTION D

NANOELECTRONICS

Unit IV–A

Nanoelectronics: Introduction

What is Nanoelectronics?

Nanoelectronics refers to electronics at the sub-micron scale. Many chip components in production already feature sizes at the nanoscale. However, nanoelectronics also encompass what is known as molecular electronics, which uses individual molecules in electronics. Other advances in nanotechnology with applications in electronics include carbon nanotubes that can be used in both chips and displays, and nanomaterials that can be used in films that make smaller, more flexible displays and improved hard disks.

Nanoelectronics in Japan and Korea tends to focus on next generation semiconductor devices and single electron devices (SED).

Who's Who in Nanoelectronics?

Below is a list of individuals who have been instrumental in advancing the state of the art in the field of nanoelectronics and who are among the world's experts on this next-generation technology. The list is organized according to areas of expertise.

The Big Picture

- **Dr. Jane Alexander**—Dr. Alexander is the initiator of the DARPA ULTRA Program for nanoelectronics research.
- **Larry Cooper**—Larry Cooper is a long-time advocate and program manager for a broad range of U.S. Navy programs in advanced electronics.
- **Dr. Gerry Iafrate**—Dr. Iafrate is a visionary program manager and an active research investigator who is widely acknowledged as having been one of the initiators of the U.S. research effort in nanoelectronics during the early 1980s.
- **Lt. Col. Gernot Pomrenke**—Lt. Col. Pomrenke is the Program Manager of the the DARPA ULTRA Program for nanoelectronics research.
- **Dr. William Tolles**—Dr. Tolles is the author of two major reports on nanotechnology, including one on Nanotechnology in Europe.

Design and Architecture for Nanocomputers

Craig Lent

Wolfgang Porod

Gary Bernstein

Craig Lent, Wolfgang Porod and Gary Bernstein are on the faculty at the University of Notre Dame, where they are closely associated at the Notre Dame Microelectronics Lab. They are performing among the most important and comprehensive work in the field of pure nanoelectronics. Lent and Porod have proposed and simulated successfully a wireless design for an electronic nanocomputer based upon quantum cellular automata and a two-state nanoelectronic device they term a "quantum dot cell". Bernstein is attempting to fabricate such devices and the logic based upon them. Their quantum cellular automata are a very important new idea that is stimulating important new research into potential designs and architectures for electronic nanocomputers.

- ***Terry Fountain***

Professor Fountain performs research on computer architecture at the University College London, UK. He is a source of many innovative ideas about how to organize a densely integrated computer system with nanoscale components.

- ***Brosl Hasslacher and Ronnie Mainieri***

Drs. Hasslacher and Mainieri are on the staff at the Los Alamos National Laboratory. They are leading experts on the limits of nanoscale, low-power electronic devices and the difficult technical issues that must be overcome to build a nanocomputer. They have performed extensive theoretical work and calculations in this area.

- ***Rolf Landauer***

Dr. Landauer works for IBM where he is a well-known expert on the theory and physics of computation, especially in its nanoscale limits.

- ***Konstantin Likharev***

Professor Likharev is on the faculty at the State University of New York at Stony Brook. In collaboration with Alexander Korotkov, he helped originate a cellular automaton proposal designed particularly to alleviate some of the weaknesses in the Lent-Porod quantum dot cell approach (which also has been improved). Likharev also is one of the world's preeminent experts on single-electron devices.

- ***John Storrs-Hall***

Dr. Storrs-Hall is a senior scientist in the Laboratory for Computer Science Research (LCSR) at Rutgers University. He is an expert on the issue of reversibility in nanocomputers. He is also a noted proponent of alternative, nanomechanical approaches for the design and fabrication of nanocomputers.

- ***Tomasso Toffoli***

Professor Toffoli, of Boston University, is an expert on cellular automaton architectures for nanocomputers and on quantum computation.

Development of Hybrid Nanoelectronic—Microelectronic Devices and Solid-State Nanoelectronic Devices

- ***Gary Frazier and Alan Seabaugh***

Drs. Frazier and Seabaugh are on staff at Texas Instruments Corporation (TI). Gary Frazier is the Manager of the Texas Instruments Nanoelectronics Research and Development Group, while Alan Seabaugh is the Technical Leader. TI is among the leaders in the field of nanoelectronics, in large part due to the efforts of Drs. Frazier and Seabaugh, who are involved in many different aspects of the field. Their work in hybrid nanoelectronics—microelectronics technology, as well as their efforts in silicon nanoelectronics research, and in the modeling of nanoelectronic devices are of particular importance. This work could lead to the introduction of the first mass-manufactured nanoelectronics products.

- ***Federico Capasso***

Capasso works at AT&T where he performed some of the earliest work on "resonance tunneling devices," the class of quantum effect devices that are viewed as among the more likely candidates for employment as a nanometer-scale transistor.

- ***Robert Bate***

Bob Bate now is a senior investigator at Texas A&M. He is one of the godfathers of nanoelectronics and an early visionary about the potential of this area. He is also the author of a number of fundamental papers on nanoelectronics. In the early 1980s, Bate started and helped to build the Nanoelectronics Group at Texas Instruments prior to his retirement from TI several years ago.

- ***Tom McGill***

Professor McGill is on the faculty at the California Institute of Technology, in Pasadena, CA, where he has performed innovative fabrication and experimentation with quantum effect devices.

- ***Herb Goronkin***

Dr. Goronkin is a senior scientist at the Motorola Corporation. A leading investigator in the area of solid-state quantum effect devices, he is an expert and proponent for using methods of chemical self-assembly for the fabrication of next-generation nanoelectronic devices.

- ***Professor Yasuaki Masumoto***

Director of the Single Quantum Dot Project in Japan's ERATO Initiative and Professor at the University of Tsukuba in Tsukuba, Japan.

- ***Professor James Meindl***

Professor Meindl is on the faculty of the Georgia Institute of Technology, where he has done comprehensive work on the constraints that apply in scaling down solid-state devices (i.e., transistors) to nanometer dimensions. He is noted for his extremely clear, comprehensive lectures on this subject, based upon years of research that are summarized in a number of fundamental articles.

- ***Barry Gilbert***

Gilbert is on the Staff of the Mayo Clinic, where he leads the Mayo Special Purpose Processor Development Group. His research on design and fabrication of solid-state quantum effect devices for signal processing is widely recognized as being among the most advanced in the nanoelectronics field.

Molecular Nanoelectronics

- ***Mark Reed***

Professor Reed is Chairman of the Electrical Engineering Department at Yale University and Director of the Yale Center for Microelectronic Materials and Structures. An acknowledged expert on quantum dots, with much experience fabricating and experimenting with solid-state "dots", Prof. Reed concluded some years ago that they must get much smaller in order to operate at room temperature. Thus, Professor Reed now is performing very sensitive experiments on prototype molecular-scale electronic devices, along with Professor James Tour of the University of South Carolina. They are hoping to produce a molecular-scale quantum dot.

- ***James Tour***

Tour is a Professor in the Department of Chemistry and Biochemistry at the University of South Carolina, where he is performing pioneering research on molecular electronics by synthesizing the molecules on which Mark Reed of Yale is performing experiments. Recently, Professor Tour and his collaborators successfully fabricated and tested the first functioning molecular wire.

- ***Ari Aviram***

A long-time proponent of molecular-scale electronic devices, Dr. Aviram is a senior scientist at the IBM Watson Laboratory. He is the author of a series of very influential papers on the subject, dating back to the 1974 classic paper that he co-authored with Professor Mark Ratner. A more recent talk by Dr. Aviram served as the catalyst for the chemical nanofabrication research of Professor James Tour, who has successfully fabricated a conducting molecular wire.

- ***Mark Ratner***

Professor Ratner, now at Northwestern University in Evanston, Illinois, is one of the originators of the idea of molecular electronics, back in the 1970s. Along with Ari Aviram (see above), Professor Ratner published the seminal technical paper on this topic. He continues to publish innovative research results on the dynamics and modeling of molecular devices, such as those that are being experimented upon by James Tour and Mark Reed.

- ***Yasuo Wada***

Wada is a senior member of the research technical staff at Hitachi Corp. in Saitama, Japan. He is an expert on nanoprobes. He also has originated an innovative approach for molecular-scale electrical switches.

Nanofabrication and Self-Assembly of Electronic Devices

- ***Noel MacDonald***

Noel MacDonald is Professor in the School of Electrical Engineering at Cornell University where his conducting research in the Cornell University Nanofabrication Facility Professor MacDonald's research focuses on fabricating an array of scanning tunneling electron microscopes (STMs) on a chip using the techniques for fabricating micro-electromechanical systems (MEMS). This work is a groundbreaking engineering achievement and it is potentially very important for a variety of applications in nanofabrication, dense molecular memories, and the exploration of chemical reactions and surface properties. For now, he is focused primarily considering applications of STM arrays to dense memories.

- ***Calvin Quate***

Quate is Leland T. Edwards Professor of Electrical Engineering and Applied Physics at Stanford University and the inventor of Atomic Force Microscopy (AFM), a primary tool for the manipulation and investigation of nanostructures. More recently, has been focusing his research on using the techniques for fabricating microelectromechanical systems (MEMS) to build an array of a large number of AFM tips on a single silicon "chip." The comments made above for Noel MacDonald apply here, except Professor Quate is focused on using his arrays for nanofabrication.

- ***Phaedron Avouris***

Avouris is a member of the research staff at IBM. He is an expert on nanoprobes for nanofabrication.

- ***George Whitesides***

George Whitesides is Mallinckrodt Professor of Chemistry at Harvard University. Professor Whitesides is the one of the world's leading investigators in the area of chemical self-assembly for nanofabrication. He has pioneered in the technique of applying self-assembling monolayers (SAMs) for this purpose. Research in the Whitesides Group covers a wide range of disciplines, including chemistry, biology, physics, and electrical engineering. The work of Professor Whitesides and his collaborators on SAMs and contact printing for nanoelectronic circuitry is an important step toward the fabrication of nanocomputers.

- ***Steven Chou***

Professor Chou is leading a wide-ranging research program in the Nanostructure Laboratory

(NSL) in the Department of Electrical Engineering at the University of Minnesota where, among his other achievements, he has pioneered highly refined techniques for nanoscale "printing." He has applied this technique to fabricate an ultra-dense magnetic information storage structure. He is working on at least a half dozen projects in VLSI and ULSI (Ultra-Large-Scale Intergration).

- ***Tom Mallouck***

Professor Mallouck is on the faculty of the Chemistry Department at Penn State University, where he has led experimental efforts that resulted in the chemical self-assembly of a single-electron transistor.

- ***Harold Craighead***

Professor Craighead works at Cornell University Nanofabrication Facility and is the head of the Cornell Nanotechnology Lab and an expert on self-assembly.

- ***James Harris***

Professor of Electrical Engineering at Stanford University and leader of the Harris Research Group there. His group specializes in nanofabrication by Molecular Beam Epitaxy (MBE), as well as in studies of the dynamics of quantum dots.

Modeling and Simulation of Nanoelectronic Devices

- ***James Ellenbogen***

As Lead Scientist of the Nanosystems Group at The MITRE Corporation, Dr. Ellenbogen has been performing research to develop faster and more accurate electron-density-based methods for modeling nanoelectronic devices.

- ***Kevin Empson***

Program Manager of the U.S. Government's Nanoelectronic Modeling (NEMO) Project.

- ***David K. Ferry***

Dave Ferry is Professor of Electrical Engineering and Computer Science at Arizona State University and Director of the Nanostructures Research Group at ASU. He is an authority on the physics and modeling of solid-state nanoelectronic devices, having written two text books and many technical articles on the subject.

- ***William Frensley***

Bill Frensley is Professor and Head of the Department of Electrical Engineering at the University of Texas, Dallas. He has used his expertise and experience in the simulation of solid-state devices and quantum-effect nanoelectronic devices to develop deep insight into which device designs can work and which probably won't, and why. Also, he has developed a clear, unified picture of device designs, their evolution and relationships.

- ***Harold Grubin***

As President of SRA, Inc., Hal Grubin is an active participant in a diverse range of modeling and simulation activities at the forefront of solid-state nanoelectronics research.

- ***Gerhard Klimeck***

A software engineer in the Nanoelectronics Research Group at the Texas Instruments Corporation in Dallas, TX, Gerhard is the driving force behind the innovative design of the software for the Nanoelectronic Modeling (NEMO) software.

- ***Roger Lake***

A scientist working in the Nanoelectronics Research Group at the Texas Instruments Corporation in Dallas, TX, Roger Lake is responsible for the design and development of the unusually fast and accurate algorithms that have been employed in the Nanoelectronic Modeling (NEMO) simulation software for modeling solid-state, nanometer-scale quantum-effect electronic devices.

Unit IV–B

Nanoscale Electronics

At first there were only a few of them, but recently, their numbers have multiplied wildly. Newspaper headlines, magazine articles, journal papers, even television commercials now are loaded with those big "nano" words: nanometer, nanoscale, nanosecond, and nanotechnology, to name a few. And it seems that every week some organization is announcing yet another "nanoconference."

Why all the excitement? Because in recent years, as scientists have begun to get a handle on controlling matter at the nanometer scale, they've recognized that these skills can be used to make new materials with unique and useful properties—perhaps leading to a range of commercial applications in sensing, electronics, and other areas. Impressive laboratory demonstrations of nanoscale dexterity have been widely reported in the scientific and popular press, thereby drawing the attention of more scientists, companies, and technology investors.

Nanoelectronics—a developing field in which circuitry is composed of nanometer-sized electronic components—is one topic that's attracting a lot of interest in scientific and nonscientific corners alike. Researchers are driven to explore the field because further miniaturizing today's already small electronic circuits will lead to faster, more sophisticated, and more portable devices. Yet it's widely believed that in a decade or so, silicon-based circuitry will have been shrunk as small as is physically possible. And so the search is on for alternative materials from which nanoscale circuits can be constructed.

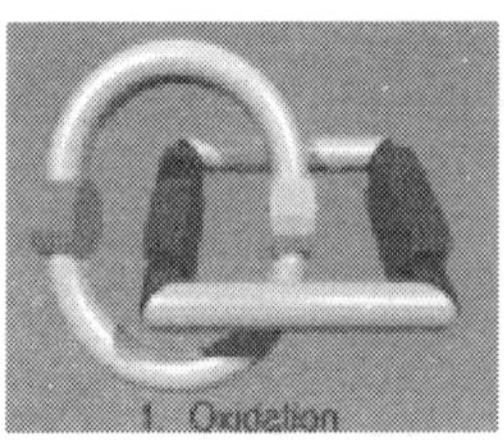

Switching of a catenane

Switching of a rotaxane in a device setting

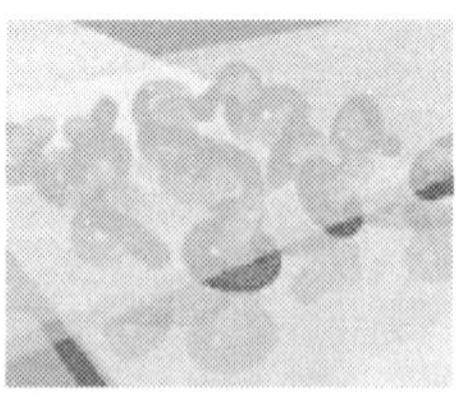

Construction of a crossbar device

Fig. IV–B.1: Some examples of nanoscales devices.

Carbon nanotubes and nanowires of materials such as gallium arsenide are being studied for just that purpose by several research groups. Other investigators, whose work is often labeled "molecular electronics," focus on the electrical behavior of individual molecules or small numbers of molecules. Scientists who work in these fields aim to design and build new types of computers in which the nanosized entities lie at the heart of logic circuits. Skeptics say that the proposed applications are exaggerated and that reports are full of "hype." But researchers around the globe are exploring and discovering nanometer-scale phenomena, and they are rapidly turning out large numbers of scientific papers.

"These are exciting and heady times," says Mark A. Ratner, surveying the combined topics of molecular and nanoelectronics. "There are wonderful experiments being reported these days," he says. "The excitement is justified." A nearly 30-year-old paper by the Northwestern University chemistry professor is often cited as the beginning of modern molecular electronics. In that theoretical study, Ratner and Arieh Aviram, who was at that time a graduate student, proposed that under certain conditions, individual organic molecules could function as p-n junction diodes. These simple electronic devices typically are formed at the interface between positive charge-carrying (p-type) and negative charge-carrying (n-type) semiconductors and can be used as rectifiers to convert AC current to DC current.

Ratner notes that few advances were made in molecular electronics until scanning probe microscopes became available in the late 1980s. The new tools gave the field a much-needed boost by making it possible to measure a whole range of phenomena, including the type Ratner had described in the 1970s. He adds that progress in other areas, for example, in methods for preparing self-assembled monolayers of molecules, was also instrumental in jump-starting nanoscale electronics. "Now the advances are coming thick and fast," he says.

"Thick and fast" aptly describes the flurry of research being carried out on carbon nanotubes. According to Phaedon Avouris, manager of nanoscale science and technology at IBM's T.J. Watson Research Center, Yorktown Heights, N.Y., the strawlike all-carbon structures with nanometer-sized diameters are endowed with unique properties that make them excellent candidates for nanoelectronic applications.

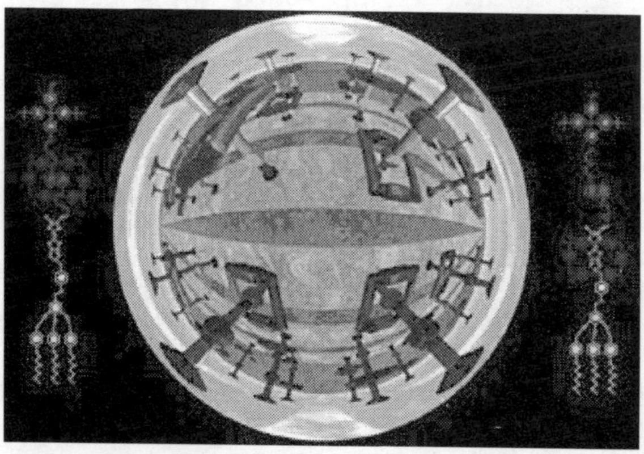

Fig. IV-B.2: TINY WORLD Molecules that can be switched between conducting and nonconducting states—such as the UCLA-synthesized [2]rotaxanes shown here—lie at the heart of some of today's lab demonstration circuits. In one state (right structure and upper right schematic), a cationic cyclophane ring (blue) encircles a tetrathiafulvalene (TTF) unit (green). Oxidizing the TTF unit causes electrostatic forces to drive the cyclophane ring to a dioxynaphthalene unit (red), switching the molecule to the other state (left).

Nanotubes can be metals or semiconductors, depending on their chirality, Avouris notes. And because of their strong chemical bonds and satisfied valencies, the materials boast high thermal, mechanical, and chemical stability. In addition, carbon nanotubes can be efficient conductors as a result of their tiny diameters, long lengths, and defect-free structures that make them ideal one-dimensional systems.

Some years ago, Avouris' research group and, independently, scientists working with Cees Dekker, a professor of applied physics at Delft University of Technology, in the Netherlands, demonstrated that field-effect transistors (FETs) could be fashioned from carbon nanotubes. FETs are a type of switch in which a semiconducting channel bridges two electrodes designated "source" and "drain." Current flow between these electrodes is controlled by a third electrode known as a "gate." By applying a voltage to the gate, the semiconductor's state can be changed—reversibly—from insulating to conducting, thereby switching the transistor on or off. In conventional FETs, the bridging channel is made of silicon. In nanotube devices, the channel is a single carbon nanotube.

Today's sophisticated silicon-based integrated circuits owe much of their success to advances in FET technology. Manufacturers pack tens of millions of the tiny transistors into the postage-stamp-sized computer chips used in modern microprocessors. For that reason, the IBM research group and others have been working on improving the properties of nanotube FETs. Much of the work focuses on understanding and controlling electrical conduction through the tiny tubes.

One problem that plagues researchers looking to fashion circuit components from nanotubes is separating metallic tubes from the ones that are semiconducting. Common synthesis procedures produce spaghetti-like mixtures of nanotube ropes that are unusable for semiconductor applications because they contain both types of tubes.

But last year, Avouris and coworkers came up with a way to sort through the tangle. After fashioning electrodes around nanotube bundles via lithography methods, the team applied certain voltages to the gate electrodes to switch off the semiconducting tubes, converting them to insulators. Then, by applying high voltage to the circuit, the IBM group oxidized the metallic tubes, causing them to break down without affecting the semiconducting tubes. The group used the method to prepare FET arrays from single-walled nanotube ropes and to peel apart multiwalled nanotubes shell-by-shell.

Just a few months later, the IBM group showed that elementary computing circuits known as logic gates—which typically are constructed from combinations of FETs—can be fashioned from a single nanotube bundle. Shortly thereafter, Delft's Dekker reported similar advances in constructing multi-FET nanotube logic circuits. Complex combinations of AND, OR, and NOT logic gates make up the core of digital processors.

To make NOT gates using nanotubes, Avouris and coworkers needed to come up with a procedure for preparing n-type nanotubes. NOT gates require both types of transistors. But the synthesis techniques available at that time, produced only p-type tubes. Hunting for a solution, Avouris and IBM coworkers Vincent Derycke, Richard Martel, and Joerg Appenzeller found that p-type nanotubes can be converted to n-type simply by annealing (heating) in vacuum. Armed with the new preparation trick, the team prepared separate n- and p-type transistors and fabricated a NOT gate.

Then they went a step further and fabricated a NOT gate using just one nanotube. For that feat, the team selectively converted a short length of a nanotube to n-type by doping it with potassium through a tiny window in a protective polymer coating while leaving the unexposed portion p-type. The single tube with p- and n-type segments was then used to construct an intramolecular NOT gate.

Further advances have come from transistor design modifications. Earlier this year, for example, the IBM group reported that their so-called top-gated nanotube FETs outperform state-of-the-art silicon FETs in terms of switching rate and the amount of current they can carry per width of conductor. One key difference between the latest design and earlier designs is that the silicon wafers that support the FETs do not function as gates. Instead, the gate is fabricated above the nanotube—allowing all FETs in contact with the silicon wafer to be switched independently. In addition, the new FET design benefits from switching voltages that are an order of magnitude lower than those needed to switch older FETs.

Avouris

Just recently, the IBM team developed a catalyst-free procedure for preparing single-walled carbon nanotubes. Conventional methods for preparing single-walled tubes, such as discharge and vapor deposition techniques, rely on particles of nickel and cobalt or other catalytic transition metals. A problem with those procedures is that the metal particles left behind in the products disturb the nanotubes' electrical properties, forcing scientists to take complex purification measures.

But now Avouris and coworkers have shown that single-walled nanotubes can be prepared from SiC in the absence of metal catalysts at temperatures above 1500°C. The method produces nanotubes that are 1.2 to 1.6 nm wide and are aligned along the face of the support. Other researchers have reported techniques for growing vertically aligned products, which they dub nanotube forests. But nanotubes that grow horizontally are amenable to standard vapor deposition methods and may be connected in parallel to lower their electrical resistance.

Despite the recent advances, Avouris asserts that nanotube FETs remain "far from optimized." Improvements could be made by using thinner insulators with higher dielectric constants, he suggests. "But what's really needed is a better understanding of the mechanism of electrical switching in nanotube FETs."

Carbon nanotubes aren't the only game in town. Nanowires made of semiconductors such as silicon, gallium arsenide, and indium phosphide are being investigated as candidate materials for nanoelectronics. The field is advancing rapidly as researchers are making fast progress in synthesis, device fabrication, and testing.

Although Harvard University chemistry professor Charles M. Lieber has published widely on carbon nanotubes, semiconducting nanowires figure prominently into his research program. The tiny wires are versatile building blocks that can be used in a bottom-up approach to constructing nanoelectronic circuits, Lieber notes, because the size, structure, and functional properties of nanowires can be controlled readily via synthesis procedures.

In 1998, Lieber and coworkers described a vapor-liquid-solid synthesis method in which laser light is used to ablate nanometer-sized metal clusters that serve as nucleation centers and catalysts for nanowire growth. The Harvard researchers used the technique to prepare uniform single-crystalline nanowires of silicon and germanium with diameters as small as 3 nm and lengths up to 30 mm. Since that time, the procedure has been used to prepare nanowires with even smaller diameters, and it's been extended to a wide variety of materials. Examples include III-V semiconductors such as GaAs and II-VI semiconductors such as ZnSe.

Recently, a number of research groups boosted the complexity of materials that can be prepared by the cluster-nucleation method. The teams prepared modulated structures—nanowires composed of dissimilar segments. The two-tone materials are made by turning the supplies of reactants on and off during synthesis with pulsed lasers or by other methods. These "heterostructured" products open the door to new sophisticated applications, such as terahertz-frequency photon emitters.

Lieber, graduate student Mark S. Gudiksen, and coworkers used modulation methods to prepare nanowires with 21 alternating segments of GaAs and GaP. And they prepared Si and InP nanowires with modulated doping, such that the wires were endowed with alternating p- and n-type regions. Using similar methods, Peidong Yang, an assistant chemistry professor at the University of California, Berkeley, prepared Si-SiGe nanowires. And Lars Samuelson of Lund University, in Sweden, synthesized InAs-InP nanowires.

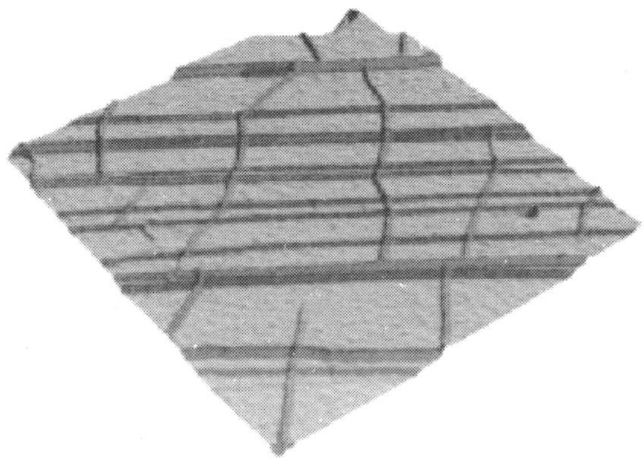

Fig. IV-B.3: STOP METALING a catalyst-free synthesis procedure developed at IBM produces aligned single-walled carbon nanotubes (vertical lines) from SiC without residual metal impurities.

To evaluate the usefulness of the tiny wires in nanoscale electronics, the Harvard researchers fashioned the wires into test circuits that were fabricated using lithography methods and then studied the circuits' electrical properties. According to Lieber, Si nanowire FETs, for example, perform remarkably well. The devices exhibit large on/off ratios ($>10^5$)—a requirement for effective switches—and charge-carrier mobilities that exceed those measured in conventional silicon devices.

In addition to constructing FETs from nanowires, Lieber's team has built and tested photodetectors, biological and chemical sensors, and more complex devices, including light-emitting diodes (LEDs) and logic gates. Most of the papers in which the results are reported were published in just the past 18 months.

In the chemical sensing study, Lieber, graduate students Yi Cui and Qingqiao Wei, and Harvard assistant chemistry professor Hongkun Park found that by functionalizing boron-doped Si nanowires, they could prepare efficient sensors. Modifying the wires with an amino-silane compound, for example, produced pH sensors that are linear over a large dynamic range. And they observed that biotin-modified wires can be used to detect picomolar levels of streptavidin. In addition, the Harvard chemists found that the wires could be used to detect proteins reversibly in real time.

By crossing an n-type and a p-type InP nanowire and then attaching electrodes to the wires lithographically, Lieber and coworkers made a nanometer-sized junction—at the point of overlap—that behaved as a current rectifier and an LED.

Based on the study, Lieber says that more complex devices can be constructed from semiconducting nanowires simply by crossing them and forming tic-tac-toe patterns. But taking advantage of the crossed-nanowire idea requires developing a method for preparing the patterns efficiently and reproducibly.

By combining surface patterning techniques and microfluidic alignment methods, the Harvard group developed the needed procedure. Their idea is to flow nanowires from a suspension into tiny parallel channels in a mold, in a layer-by-layer fashion, while rotating the mold orientation—and hence, nanowire flow direction—between layers.

With the crossed-nanowire method for making p-n junctions firmly in hand, Lieber and coworkers were well on their way toward making computational logic circuits. Near the end of 2001, the group published a study in Science showing that their procedures could be used to make arrays of nanowire FETs configured as AND, OR, and NOR logic gates. Lieber stresses that in crossed-nanowire structures, such as the Harvard logic gates, the width and length of the FET channels are controlled by synthesis and assembly—not lithography—unlike today's microelectronic components.

These fabrication concepts make it possible "to envision a straightforward path to nanometer-scale electronics of the future," Lieber says. His vision, which is supported by experiments conducted by his group in the past several months, includes a nanoscale logic device known as a decoder that can address individual elements in crossed-wire arrays.

Using a handful of molecules—or even just one of them—to make nanoscale switches or other devices is an idea that's gathered momentum in recent times. In 1999, researchers working with chemistry professors James R. Heath and J. Fraser Stoddart of the University of California, Los Angeles, demonstrated that simple circuits—fuses—could be made from a layer of [2]rotaxane molecules. Then a team headed by Rice University chemistry professor James M. Tour and Yale University electrical engineer Mark A. Reed reported success in making reversible switches using benzene thiol molecules.

Other groups have also been investigating the possibilities of single-molecule systems. For example, a team at UC Berkeley including Park (now at Harvard), Paul L. McEuen (now at Cornell), and A. Paul Alivisatos, trapped a single C_{60} molecule in a tiny gap between a pair of electrodes to study electronic conduction mechanisms. And just recently, a single molecule of a transition metal-organic complex—containing either one Co atom or two V atoms—was turned into a single-molecule transistor in independent studies by Park and McEuen.

Before molecular devices can be wired up and tested, organic chemists need to synthesize the molecules that drive them. In the UCLA collaboration, Stoddart's group prepares the compounds and Heath's group turns them into circuit components. To make the molecules switchable, Stoddart's group designs them so that one segment of a molecule can move (rotate or translate) relative to another segment.

Stoddart notes that, contrary to remarks made by many researchers, the molecules are not difficult to prepare nowadays. But it took some years to build up the necessary expertise in molecular recognition concepts, p-donor/p-acceptor interactions, and other subjects. He adds that his group has synthesized "hundreds, if not thousands" of catenane and rotaxane analogs since 1989.

In 2000, the UCLA team demonstrated a reusable switch based on a [2]catenane—a molecule with a pair of interlocked rings. The switching state of the redox-active molecule was shown to change reversibly, depending on an applied potential. Electrostatic repulsion drives rotation of one ring relative to the other between open and closed states.

Fig. IV-B.4: MAGNITUDE, a series of micrographs in which each image is magnified roughly 10 times more than the previous one (left to right) reveals details of a new HP Labs molecule-based 64-bit prototype memory and logic chip. Using a newly patented procedure, researchers patterned a silicon wafer with 625 memory devices (one shown at top right). At the center of each memory circuit, an 8 3 8 array of crossed metal nanowire electrodes (bottom row) sandwiches roughly 1,000 rotaxane molecules forming 64 bits of nonvolatile memory.

Looking to improve device performance by tailoring the molecules, Jan O. Jeppesen, a visiting graduate student in Stoddart's lab, and postdoctoral associate Hsian-Rong Tseng prepared various rotaxane analogs of the catenane that brought success in 2000. The series of amphiphilic bistable [2]rotaxanes are dumbbell-shaped molecules capped with large stoppers that are threaded through a cyclophane ring. As with the [2]catenanes, the ring's position relative to the other piece of the molecule is controlled by redox interactions.

Heath and coworkers constructed circuits with the new molecules using a layer-by-layer deposition method to fabricate tic-tac-toe structures with a few thousand molecules situated at each intersection. The team used this fabrication method to prepare 64-bit random-access memory circuits. And Hewlett-Packard Labs announced earlier this month that R. Stanley Williams, Yong Chen, and coworkers have also succeeded in using the UCLA rotaxane molecules to fabricate a 64-bit memory circuit.

Only a few years ago, many scientists sat through molecular electronics presentations thinking that the far-fetched nanocomputing goals would never be reached. But the flurry of demonstration results in just the past two years appears to have decreased the number of nonbelievers. Yet some

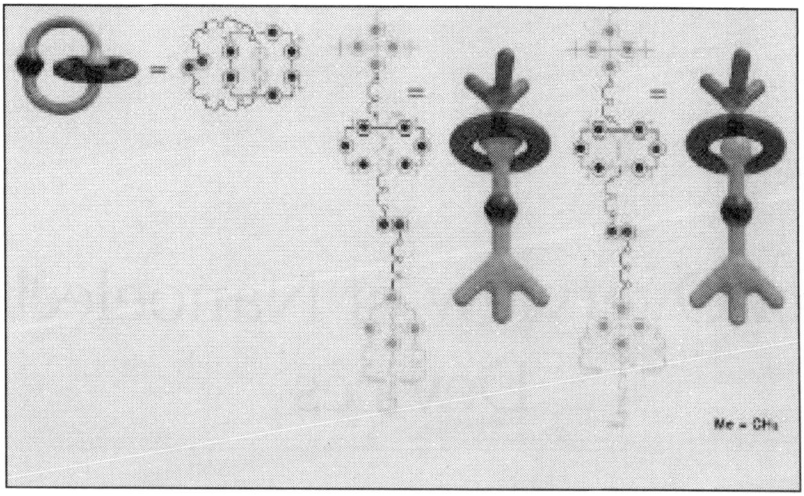

Fig. IV–B.5: Modular Molecules Ucla chemists use synthesis techniques to prepare multipart molecules with customized structures and properties for nanoelectronic devices. [2]Catenancs (left) and [2]rotaxanes (center and right) can be switched on and off depending upon redox state, which determines where in the molecule a cyclophane ring (blue) resides.

naysayers remain. Just recently, a California professor was furious about a C&EN story on molecular electronics and wrote to the editor, exclaiming, "In reality, there is no such topic!"

The professor has a few surprises in store. In reality, molecular and nanoelectronics are bustling fields. Large numbers of commercial products may not be available in the near future, but as the vibrant research activity reveals, they're coming.

Unit IV–C

A Brief Overview of Nanoelectronic Devices

In this unit we discuss about nanometer-scale, quantum effect alternatives to micron-scale, and bulk effect transistors in digital circuits. The status of R&D and recent important advances are reviewed briefly.

For the past forty years, electronic computers have grown more powerful as their basic sub-unit, the transistor, has shrunk. However, the laws of quantum mechanics, plus the limitations of materials and fabrication techniques soon are likely to inhibit further reduction in the minimum size of today's bulk effect semiconductor transistors. Researchers have projected that as the overall size of the bulk-effect, semiconductor transistor is aggressively miniaturized to approximately 0.1 micron (i.e., 100 nanometers) and beyond, the devices may no longer function as well. Thus, in order to continue the miniaturization of integrated circuits well into the next century, it is likely that present day, micron-scale or microelectronic device designs will be replaced with new designs for devices that take advantage of the quantum mechanical effects that dominate on the much smaller, nanometer scale. (Note that 1 nanometer, abbreviated nm, is one billionth of a meter or 10 atomic diameters.)

Here we briefly survey such quantum effect, nanometer-scale alternatives to micron-scale, bulk-effect transistors for use in electronic digital circuits.

The topic of nanoelectronic devices subdivides into two broad areas, as follows:

- Solid-state quantum effect nanoelectronic devices.
- Molecular electronic devices.

The comments above are summarized and amplified in Figure IV-C.1, as well as in the section below.

Solid-State Quantum-Effect Nanoelectronic Devices

Solid-state quantum-effect nanoelectronic devices are the nearer-term alternative for continuing to increase the density and speed of information processing. This technology, which changes the

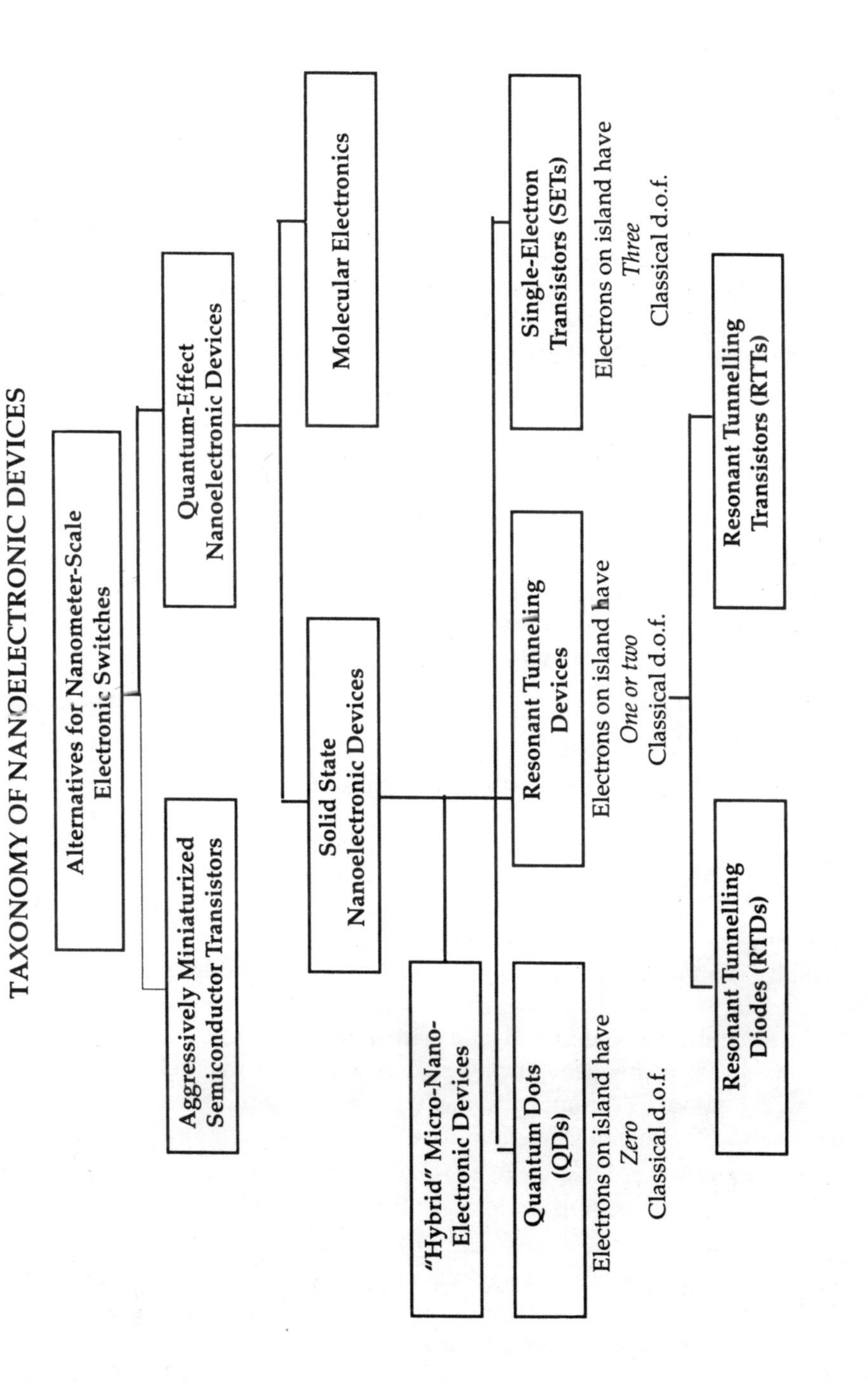

Fig. IV–C.1: Taxonomy of Nanoelectronic Devices

operating principles, but not the solid fabrication medium for integrated circuits, might extend Moore's Law of electronics miniaturization to produce devices down to 25 nm in length, beyond the domain of the bulk effect transistor. Present-day field effect transistors (FETs) on commercial integrated circuits are approximately 1 micron or 1000 nm across, and even the most optimistic proponents of aggressively miniaturized bulk effect FETs predict that they will cease to function effectively when their gate lengths dip below 25 nm, which corresponds to an overall device length of approximately 100 nm.

By contrast, quantum effect switching devices tend to function better when they are made smaller. Solid-state quantum-effect nanoelectronic devices might be made as small as approximately 12 to 25 nanometers across and still function effectively if they could be fabricated reliably and uniformly on that scale. Presently, mass fabrication on this small scale does present a set of formidable challenges, no matter what the device design. However, a number of solid-state nanoelectronic switching devices have been fabricated and demonstrated, and prototype solid-state nanoelectronic processors already are operational, as well.

Moreover, the drastic decrease in size that could result from the wide-scale introduction of such nanoelectronic devices might produce ultrafast, low power integrated circuits with as many as 100 billion or even 1 trillion switching devices on a single CPU chip, as well as Terabyte nonevanescent memories on a chip. Such great advances would require, however, the reliable, uniform mass fabrication of features only 5 to 10 nanometers wide in solids. This is at least twenty-five times as small as is achievable using UV lithography, the present industrial method of choice. More modest, near term goals do not require such extreme precision, but take advantage of other very desirable properties of solid-state quantum-effect devices, such as low power consumption and high speed.

Types of Solid-state Nanoelectronic Devices

As is indicated in Figure IV-C.1, there are 3 basic categories of solid-state nanoelectronic devices:

(1) Quantum Dots (or "artificial atoms")

(2) Resonant Tunneling Devices

(3) Single Electron Transistors (SETs)

Quantum Dots—A small "island" composed of semiconductor or metal in which electrons may be confined is the essential structural feature that all these quantum-effect, solid-state nanoelectronic devices have in common. The island in a nanoelectronic device assumes a role analogous to that of the channel that forms beneath the gate in a familiar microelectronic field-effect transistor when it is switched on. The composition, shape, and size of the island gives the different types of solid state nanoelectronic devices their distinct properties.

Controlling these factors permits the designer of the device to employ quantum effects in different ways to control the passage of electrons on to and off of the island.

Resonant-tunneling devices—Here, we focus primarily on explaining the operation of resonant tunneling devices, because they employ quantum effects in their simplest form. Presently, these devices usually are fabricated from layers of two different III/V semiconductor alloys, such as the pair GaAs and AlAs. The simplest type of resonant tunneling device is the resonant tunneling diode (RTD). As depicted in Figure IV-C.2(a), a resonant-tunneling diode is made by placing two insulating barriers in a semiconductor, creating between them an island or potential well where electrons can reside. Resonant tunneling diodes are made with center islands approximately 10 nanometers in width. Whenever electrons are confined between two such closely spaced barriers, quantum mechanics restricts their energies to one of a finite number of discrete "quantized" levels. This energy quantization is the basis for the operation of the resonant-tunneling diode. The only way for electrons to pass through the device is to "tunnel," quantum mechanically, through the two barriers. The probability that the electrons can tunnel is dependent on the energy of the incoming electrons compared to the energy levels on the island of the device. As illustrated in Figure IV-C.2(b), if the energy of the incoming electrons differs from the energy levels allowed inside the potential well on the island, then current does not flow. The device is switched "off."

However, when the energy of the incoming electrons aligns with that of one of the internal energy levels, as shown in Figure IV-C2(c), the energy of the electrons outside the well is said to be "in resonance" with the allowed energy inside the well. Then, current flows through the device—i.e., the device is switched "on."

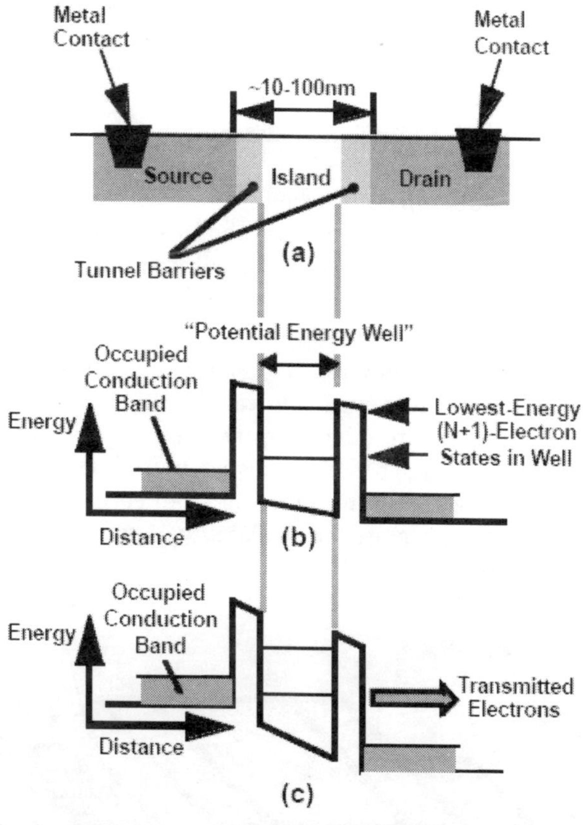

Fig. IV–C.2: Operation of Solid-state Resonant Tunneling Diode.

By adding a small gate electrode over the island of an RTD one may construct a somewhat more complex resonant tunneling device called a resonant tunneling transistor (RTT). In this three-terminal configuration, a small gate voltage can control a large current across the device. Because a very small voltage to the gate can result in a relatively large current and voltage across the device, amplification or "gain" is achieved. Thus, an RTT can perform as both switch and amplifier, just like the conventional bulk effect transistor.

Unlike conventional bulk effect transistors which usually have only two switching states, "on" and "off," resonant tunneling devices like RTDs and RTTs can have several switching states. This occurs because the quantum well on the island may exhibit several possible energy levels.

Thus, as the voltage bias on an RTD is increased from zero, where it is "off," the device initially switches "on" when the first energy level comes in resonance with the incident electrons. Then, it switches off again as the bias is increased further, past resonance. However, if there is a second energy level in the quantum well, then the device can switch on again as the bias voltage of the RTD (or the gate voltage of an RTT) is increased still further. This multistate switching behavior permits each device to "count higher" and represent more logic states.

"Hybrid" Microelectronic—Nanoelectronic Resonant Tunneling Transistors—Purely nanometer-scale RTTs tend to be difficult to fabricate with sufficient uniformity in large quantities because of their relatively complex structure, small size, and the sensitivity of the effects they employ. However, there are also nanoelectronic devices which are "hybrids" of solid-state quantum-effect devices combined with micron-scale transistors. One such hybrid transistor-like device, the RTD-FET, is constructed by building tiny, nanometer-scale quantum-effect RTDs into the drain (or source) of a bulk effect micron-scale FET. A schematic of such a device is shown in Figure IV–C.3.

Such a hybrid RTT can exhibit multistate switching behavior of the type described in the preceding section for purely nanometer-scale RTTs. For this reason, the hybrid device can represent more logic states than a pure bulk effect, microelectronic FET with the same "footprint" on the integrated circuit. Thus, the density of the logic can be increased using the multistate switching characteristics of hybrid micro-nanoelectronic RTTs, without appreciably increasing the density of the devices on an integrated circuit. Also, these hybrid RTTs share the advantages of low power and high speed exhibited by the purely nanoelectronic RTTs.

Most important, fabricating circuits with this relatively large, hybrid type of three-terminal RTT is easier than fabricating circuits with the much tinier, complex structures for purely nanoelectronic RTTs. For this reason a number of groups are experimenting with such devices. In recent months, robust prototype circuits based upon such hybrid devices have been demonstrated. Hybrid logic can be viewed as an important, practical engineering step on an evolutionary path toward nanoelectronics. It should accelerate the availability of quantum-effect devices in integrated circuits with very dense functionality.

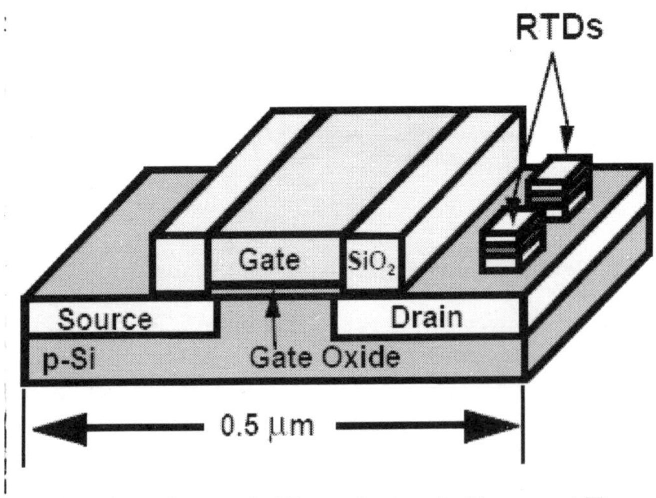

Fig. IV–C.3: "Hybrid" Microelectronic-Nanoelectronic Resonant Tunneling Transistors.

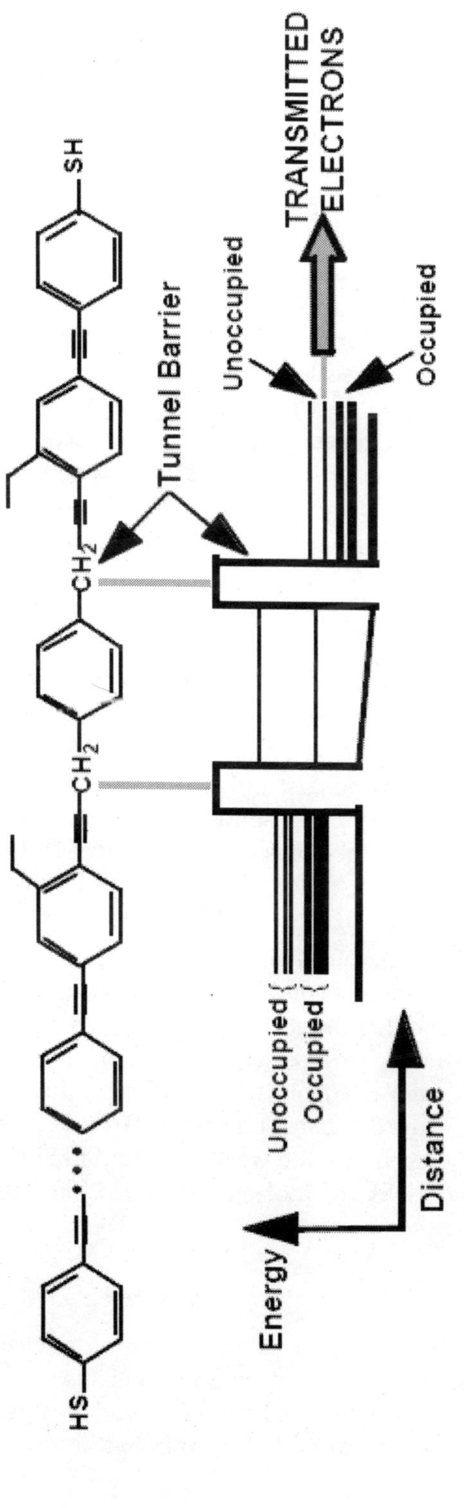

Fig. IV–C.4: Molecular Electronic Resonant Tunnelling Diode.

Molecular Electronic Devices

Ultimately, however, it will be desirable to build ultra dense, low-power nanoelectronic circuits made from purely nanometer-scale switching devices and wires. Molecular electronics—using individual covalently bonded molecules to act as wires and switching devices—is the longer-term alternative for achieving this increase in density and for continuing Moore's Law down to the nanometer scale. Individual molecular switching devices could be as small as 1.5 nanometers across, with densities of approximately 1012 devices per sq. cm. This decrease in size could result in Terabyte memories on a chip and in excess of one trillion switching-devices on a single CPU chip. A primary advantage of molecular electronics is that molecules are natural nanometer-scale structures that can be made absolutely identical in vast quantities (approximately 10^{23} at one time). Also, synthetic organic chemistry—the chemistry of carbon-based compounds—offers more options for designing and fabricating these intrinsically nanometer-scale devices than the technology presently available for producing solid-state chips with nanometer scale features. It has been shown during the past few years that small organic molecules can conduct and control electricity, that large organic bio-molecules are conductive, and that individual molecules of the newly discovered molecular forms of pure carbon called buckminsterfullerenes ("buckyballs" and "buckytubes") have interesting and useful electrical properties.

Great progress is being made in understanding and harnessing the electrical properties of individual organic molecules. There has been particular experimental progress in demonstrating that individual molecules can conduct electrical current to function as wires and that they also can act as switches to turn this current on and off. No logic circuits using these molecular switching devices have been fabricated yet. However, designs for such circuits have been proposed recently

To some extent, the research and development of molecular electronic switches is recapitulating that for the corresponding solid-state devices, except on a much smaller scale. Thus, it is not surprising, that molecular electronic switching proposals and prototypes may be categorized into molecular analogs of solid-state diodes and analogs for solid-state three-terminal FETs. A molecular analog of the solid state RTD that is discussed above is shown in Figure IV-C.4. This molecular RTD first was fabricated by Tour and then demonstrated by a group led by Reed.

In the molecular structure depicted at the top of Figure IV-C.4, the hexagonally shaped components are known as benzene "rings." Chains of these rings, such as appear at the right and left ends of the molecule in the figure have been shown to function as conductive "molecular wires". The "CH_2" groups (termed "methylene" groups), on the other hand, act as insulators or "barriers" to electron flow. They can trap electrons on the single benzene ring between them, which forms an island. This molecular sandwich-like structure, with the island in the middle, produces a potential energy well like that sketched at the bottom of Figure 3.4. Except that it is 10 to 100 times smaller, the molecular potential well in Figure 3.1 plainly resembles the well illustrated in Figure IV-C.2 for the solid state RTD. This similarity is the reason why the molecule sketched in Figure IV-C.1 behaves like an RTD.

Conclusion: Challenges

Despite enormous recent progress in the fabrication and demonstration of nanoelectronic devices, many challenges remain. For solid state nanoelectronics, one of the most important challenges is to be able to produce reliably and uniformly in silicon the characteristic nanometer-scale features required for nanoelectronics: nanometer-scale islands, barriers, and "heterojunctions" between islands and barriers. Up to now, solid state nanoelectronic devices largely have been fabricated in III/V semiconductor compounds such as gallium arsenide and aluminum arsenide. It is believed, however, that silicon nanoelectronics would go a long way toward permitting the inexpensive mass manufacture of nanoelectronic devices.

For molecular electronics, one of the great challenges is to develop two-terminal and three-terminal devices than can be incorporated in circuits. Then, it must be demonstrated that such molecular switches can be used to make reliable logic, such as has been proposed recently. Using either the solid-state or the molecular electronic approach, in order to fulfill the promise of nanoelectronics it will be necessary to refine greatly both the fabrication techniques and the architectural concepts to permit the useful assembly of small, low-power computing structures that contain trillions of nanoelectronic switching devices.

Unit IV–D

Quantum Dots: An Introduction

Overview

Quantum dots are nanometer-scale "boxes" for selectively holding or releasing electrons. Over the past decade or so they have been transformed from laboratory curiosities to the building blocks for a future computer industry.

Quantum dots are small devices that contain a tiny droplet of free electrons. They are fabricated in semiconductor materials and have typical dimensions between nanometres to a few microns. The size and shape of these structures and therefore the number of electrons they contain, can be precisely controlled; a quantum dot can have anything from a single electron to a collection of several thousands.

Atomic physics progressed rapidly at the beginning of the last century, thanks, in large part, to optical spectroscopy.

Quantization and spin were discovered through optical studies, as were other fundamental atomic properties. With the advent of the laser, physicists learned how to manipulate atomic wave functions by applying coherent optical fields. More discoveries followed. Now, at the beginning of the new century, optical techniques are being used to explore a new scientific frontier: the atom-like entities known as quantum dots (QDs).

Measuring 1-100 nm across, QDs are semiconductor structures in which the electron wave function is confined in all three dimensions by the potential energy barriers that form the QD's boundaries. A QD's electronic response, like that of a single atom, is manifest in its discrete energy spectrum, which appears when electron hole pairs are excited. Although the wave function of a QD electron, and its corresponding hole, extends over many thousands of lattice atoms, the pair—termed an 'exciton'—behaves in a quantized and coherent fashion. The coherence is relatively easy to detect and control optically—for two reasons. First, the superposition of the ground and excited states dephases more slowly in QDs than in higher-dimensional semiconductor structures. Second, QDs have large dipole moments (50-100 times larger than those of atoms). Thanks to these advantages, it is possible to probe and manipulate the wave function of a single QD.

QDs possess another attractive property. Their size, shape, and composition can all be tailored to create a variety of desired properties. These "artificial atoms" can, in turn, be positioned and assembled into complexes that serve as new materials. Researchers who work on QDs anticipate that a host of complex, customized QD-based materials will become available.

The physics of quantum dots shows many parallels with the behaviour of naturally occurring quantum systems in atomic and nuclear physics. As in an atom, the energy levels in a quantum dot become quantized due to the confinement of electrons. Unlike atoms however, quantum dots can be easily connected to electrodes and are therefore excellent tools to study atomic-like properties. There is a wealth of interesting phenomena that have been measured in quantum dot structures over the past decade. This page shows a few examples from our group. The next paragraph first discusses briefly the parallels between atoms and quantum dots.

Artificial Atoms

The three-dimensional (3D) spherically symmetric potential around atoms yields degeneracies known as shells, 1s, 2s, 3s, 3p... Each shell can hold a specific number of electrons. The electronic configuration is particularly stable when these shells are completely filled wih electrons, occurring at 'magic' atomic numbers 2, 10, 18, 36,... In a similar way, the symmetry of a two-dimensional (2D), disk-shaped quantum dot leads to a shell structure with magic numbers 2, 6, 12, 20,... The lower degree of symmetry in 2D results in a different sequence of magic numbers than in 3D.

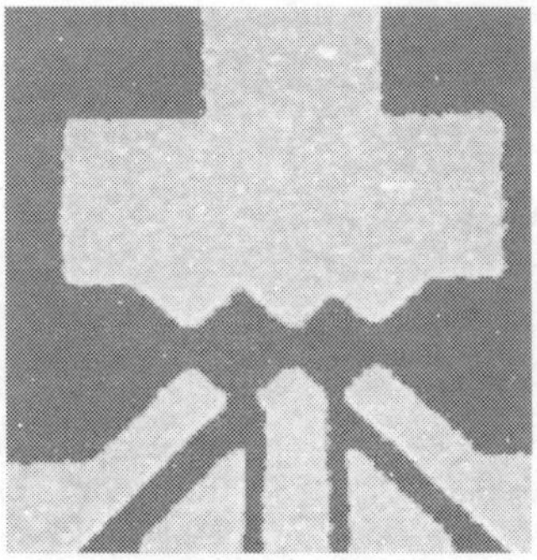

Fig. IV–D.1

By measuring electron transport through quantum dots, a periodic table of artificial 2D elements can be obtained. For this purpose, dots are connected via potential barriers to source and drain contacts. If the barriers are thick enough enough, the number of electrons on the dot, N, is a well defined integer. This number changes when electrons tunnel to and from the dot. However, due to Coulomb repulsion between electrons, the energy of a dot containing N+1 electrons is larger than when it contains N electrons. Extra energy is therefore needed to add an electron to the dot. Consequently, no current can flow which is known as the Coulomb blockade.

The blockade can be lifted by means of a third electrode closeby, known as the gate contact.

A negative voltage applied to this gate is used to supply the extra energy and thereby change the number of free electrons on the dot. This makes it possible to record the current flow between source and drain as the number of electrons on the dot, and hence its energy, is varied. The Coulomb blockade leads to a series of sharp peaks in the measured current (see figure below). At any given peak, the number of electrons on the dot alternates between N and N+1. Between the peaks, the current is zero and N remains constant. The distance between consecutive peaks is proportional to the so-called addition energy, which is the difference in energy between dots with N+1 and N electrons. The magic numbers can be identified because significantly higher voltages are needed to add the 2nd, 6th and 12th electron.

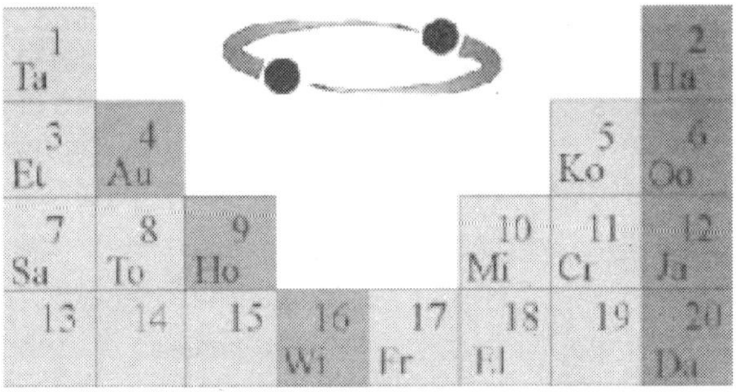

Fig. IV–D.2: 2D Artificial Atoms.

Quantum dots are 2D analogies for real atoms. But since they have much larger dimensions they are suitable for experiments that can not be carried out in atomic physics. It is especially interesting to observe the effect of a magnetic fieldd, B, on the atom-like properties. A magnetic flux-quantum in an atom requires typically a B-field as high as 10^6 T, whereas for dots this is of the order 1 T, which is experimentally accessible.

Quantum Dot Cells and Wireless Electronic Logic

A quantum dot is a box that holds a discrete number of electrons. This number can be changed by adjusting electric fields in the neighborhood of the dot, for example by applying a voltage to a nearby metal gate. In the following, I will refer to quantum dots that have either zero or one electron in them. Since quantum dots are fabricated in solids, not in vacuum, of course there are many electrons in them. However, almost all of these are tightly bound to atoms in the solid. The few electrons that I will speak of are extra ones beyond those that are tightly bound. These extra electrons could roam free in a solid were they not confined in a quantum dot.

Lent's idea begins with a cell containing four quantum dots, arranged at the corners of a square:

```
0 0
0 0.
```

We arrange that two electrons reside in each cell. They may hop (tunnel) from dot to dot within the cell to find a favorable, low-energy configuration. Of course, because electrons are charged, they repel each other, so the two electrons will not both choose to sit in the same dot within the 4-dot cell. In fact, they will choose to occupy two opposite corners of the cell, since this gives them their maximum possible separation. The electrons would be equally happy in the following two configurations:

```
* 0 0 *
and
0 *, * 0,
```

with electrons sitting in either pair of opposite corners.

This is nice, because it gives us two possible "polarization" states for a cell, which we can associate with a binary zero and one. We can use a voltage on an external gate to force a cell into one of its two a priori equally good alternative states. This is a signal, analogous to a high or low voltage at a point within a conventional integrated circuit. A conventional integrated circuit uses two tools to transmit a voltage signal over long distances: an aluminum or silicon wire (to carry signals from place to place) and a transistor used as an amplifier (to refresh weakened signals so that they may travel farther without corruption). In the quantum dot circuits proposed by Lent, both of these tasks are accomplished simply by arranging as many cells as necessary in a row. If two cells are placed next to each other, and the polarization of first is fixed, the second will match its own polarization to that of the first:

```
* 0 * 0

0 * 0 *.
```

This, extended to a row of many cells, is the task normally carried out by a wire. In fact, even if the polarization of the first dot is between zero and one (the electrons are not fully localized on one diagonal), as more dots are added to the chain, the polarization of later dots will rapidly approach a perfect zero or one. This is the task of amplification or refreshing signals, the trademark of digital circuitry.

Although we will not illustrate them, one can also build a corner, fan-out, or inverter (some of the building blocks of digital circuits). "Wires" carrying signals can even cross each other in the plane. So far, we could make a simple translation from objects in standard integrated circuits to objects in quantum dot circuits. However, the cornerstone of digital circuitry, logic gates, are not so simply translated into this new architecture. A standard logic gate has two inputs and an output determined by the inputs. For example, an AND gate puts out a high output only if both input 1 _and_ input 2 are high. The natural basic gate in Lent's scheme is a 3 input majority logic gate (i.e. the inputs "vote" and the more popular polarization wins and is represented on the output):

```
* 0
In
1
0 *
* 0 * 0 * 0

In Mid Out
2
0 * 0 * 0 *
0 *

In
3
* 0
```

Here, inputs 1 and 2 vote together and hence determine the middle cell, which in turn determines the output.

One can build any ordinary logic using these gates (most clumsily, one can make an OR gate by fixing input 3 to have polarization 1, an AND gate by forcing it to have polarization 0). More impressively, if the center-to-center distance between dots within a cell in 20 nm (extremely aggressive though not impossible with today's electron-beam lithography), a single bit full adder will fit in an area of 1 micron^2. By contrast, the semiconductor industry projects that in 2010 the maximum packing density for transistors will be approximately 1 micron^2. Forty transistors would be needed to build an adder! This dramatic difference in density is explained by two facts:

1. Much of the space in a conventional integrated circuit is taken up by wiring, which is unnecessary here.
2. The minimum feature size required to pack 1 transistor/micron^2 would be 70 nm, and relative alignment between features produced in different lithographic steps generally needs to be good to 10% of this minimum feature size, or 7 nm. Only one stage of quantum dot manufacture need involve small features, so no sensitive alignment is required. This means that the minimum feature size could probably be made smaller than 70 nm even abandoning electron beams in favor of optical techniques favored by industry for their high throughput. In addition, the diameter of a single cell is only a few times the minimum feature size, as compared to 1 micron/70 nm which is approximately 14 times.

In short, for given lithographic capabilities, quantum dot cells can be packed far more densely than can transistors.

Switching

Lent's quantum dot cells can be switched from one state to another adiabatically (reversibly). This is accomplished by lowering the barriers between dots within each cell, and removing the global inputs to the circuit, hence relaxing the polarization of each cell, then applying new global inputs and raising the barriers again. This produces very little heat, an important quality in densely integrated circuits (a DEC Alpha already has a fan mounted on top of it, continually blowing cold air on it so that it doesn't melt).

Readout

Readout could be accomplished by passing a current through a very narrow constriction near a corner of a cell. An electron in that corner of the cell could shut off current. This was demonstrated by Field et al.

Drawbacks

This whole discussion has assumed that when new boundary conditions (inputs) are imposed, the system rapidly and reliably finds its ground (or lowest energy) state. This would not be true if the ambient temperature were comparable to the difference in energy between a "correct" and an "incorrect" polarization of a cell. For the quoted 20 nm dot-to-dot distance, this energy associated with an error in one cell's polarization would be 1 millielectron volt, or one twentieth of thermal energies (kT) at room temperature, so this is a major worry. Even if the error energy could be made many times larger, which could be accomplished by making the individual dots smaller and closer together, entropy would limit the size of an array of cells to about 20 thousand at error energy = 10 kT. Finally, no one knows how long an array of cells would take to reach its ground state, though Lent estimates 30 picoseconds per cell in a "wire".

Stray charges within 1 cell's length of a "wire" would lock a nearby cell permanently into one polarization state, stymieing computation. Charges are always sprinkled liberally on surfaces of semiconductors, so this would seem to be a major barrier to dense integration using this architecture.

Perhaps most seriously, an attendee at the talk pointed out that the ground state of a large array may not be the obvious one determined by propagating signals "correctly" from inputs to outputs (though Lent has showed by modeling that his adder would not suffer from this problem). An example of the problem is the following:

```
1??????????
?
?
1?????????????????out
? ?
? ?
0??????????
?
?
0???
```

(a set of 2 linked majority logic gates), where the 1s and 0s are inputs in different polarizations, the ?s are cells, and "out" is the output. The stable configuration should logically develop as follows:

```
11111111111
1
1
1111111111???????out
1 ?
1 ?
000????????
0
0
0000
11111111111
1
1
1111111111????????out
1 0
1 0
00000000000
0
0
0000
```

```
11111111111
1
1
11111111111111111out
1 0
1 0
00000000000
0
0
0000
```

giving an output of 1. However, what costs energy is not really a logical error but rather a 1 next to a 0 (a "kink"). Another configuration with the same number of kinks, and so a very similar energy, is

```
11111111111
1
1
11000000000000000out
0 0
0 0
00000000000
0
0
0000
```

which gives an output of 0 instead. It may be possible to find a more definitive counter example, though.

The Production of Quantum Dots by Etching

What length scales are required to produce a quantum dot? In a quantum well the wavelike properties of the electrons only become apparent for layer widths of about a hundredth of a micron. To produce a quantum dot we need to achieve similar length scales along the other two dimensions. Using present techniques it is relatively easy to reduce the lateral dimensions of the islands to about a tenth of a micron. With great effort further reductions can be achieved, but producing a structure one hundredth of a micron across presents an immense technical challenge. Fortunately nature is on our side. Suppose that we etch two parallel channels through the layered materials to leave a ridge approximately one-tenth of a micron across. In the etching process the atoms on the vertical surfaces of the ridge are disturbed so that the properties of the material are changed considerably. In fact, surfaces in general have quite different properties to bulk material. Let us consider an example using a crystal of silicon. We know that each silicon atom tends to bond with four neighboring atoms to obtain a full complement of electrons in its outer shell. However, atoms at a surface cannot always find another four atoms with which to bond. This means that surface atoms have a tendency to hold on to any nearby free electrons in order to obtain a full shell. This process alone tends to remove conduction electrons from the surface layers. In addition, the accumulation of negative charges trapped by these surface atoms creates an electric field which repels other electrons from this region. As a result we find that the only free electrons in our tiny ridge-like structure are confined to an even smaller channel in the center. It is rather like a coaxial cable in which the central copper core is surrounded by an insulating layer. For a ridge which is a tenth of a micron across, the electrons are confined to a region which is only about a hundredth of a micron wide. This is just the right length scale required to observe the wavelike nature of the electron. So although at first sight the dimensions appear to be far too large, we can observe lateral quantum confinement in these structures. In this case we have created a quantum wire, since the electrons are still free to travel along the length of the ridge. From this stage the construction of a quantum dot is relatively straightforward. If we etch another set of similarly spaced grooves at right angles to the first set, then the surface effects in these directions are sufficient to cause the electrons to be trapped in a tiny box. Using these techniques it is possible to pattern a relatively large area to produce a regular grid of quantum dots.

Finite Fermion Systems

Low-dimensional nanometer-sized systems have defined a new research area in condensed-matter physics within the last 20 years. Modern semiconductor processing techniques allowed the artificial creation of quantum confinement of only a few electrons. Such finite fermion systems have much in common with atoms, yet they are man-made structures, designed and fabricated in the laboratory. Usually they are called "quantum dots", referring to their quantum confinement in all three spatial dimensions. A common way to fabricate quantum dots is to restrict the two-dimensional electron gas in a semiconductor heterostructure laterally by electrostatic gates, or vertically by etching techniques. This creates a bowl-like potential in which the conduction electrons are trapped. In addition to the many possible technological applications, what makes the study of these "artificial atoms" or "designer atoms" interesting are the far-reaching analogies to systems that exist in nature and have defined paradigms of many-body physics: atoms, nuclei, and, more recently, metallic clusters or trapped atomic gases. Quantum dots added another such paradigm. Their properties can be changed in a controlled way by electrostatic gates, changes in the dot geometry, or applied magnetic fields. Their technological realization gave access to quantum effects in finite low- dimensional systems that were largely unexplored.

After the initial success in the fabrication and control of mesoscopic semiconductor structures, which are typically about one hundred nanometers in size and confine several hundred electrons, many groups focused on the further miniaturization of such devices. A breakthrough to the "atomic" regime was achieved with the experimental discovery of shell structure in fluctuations of the charging energy spectra of small, vertical quantum dots: the borderline between the physics of bulk condensed matter and few-body quantum systems was crossed. Much of the many-body physics that was developed for the understanding of atoms or nuclei could be applied. In turn, measurements on artificial atoms yielded a wealth of data from which a fundamental insight into the many-body physics of low-dimensional finite fermion systems was obtained. With further progress in experimental techniques, artificial atoms will continue to be a rich source of information on many-body physics and undoubtedly will hold a few surprises."

Unit IV–E

Quantum Dots: More Details

Quantum Dots and Semiconductors

The "Kondo effect" is a large anomalous increase in the resistance of certain dilute alloys of magnetic materials in nonmagnetic hosts as the temperature is lowered. In general, the Kondo effect occurs when an impurity atom with an unpaired electron is placed in a metal, producing an interaction of localized electrons with delocalized electrons. The "Kondo temperature" is the temperature at which the Kondo effect predominates.

In general, in this context, the term "Coulomb blockade" refers to an effective blockade of quantum mechanical tunneling produced by specific energy barrier constraints.

Kondo Resonance in a Single Molecule Transistor

When an individual molecule nanocrystal nanotube, or lithographically defined quantum dot is attached to metallic electrodes via tunnel barriers, electron transport is dominated by single-electron charging and energy-level quantization. As the coupling to the electrodes increases, higher-order tunneling and correlated electron motion give rise to new phenomena, including the Kondo resonance. To date, all of the studies of Kondo phenomena in quantum dots have been performed on systems where precise control over the spin degrees of freedom is difficult. Molecules incorporating transition-metal atoms provide powerful new systems in this regard, because the spin and orbital degrees of freedom can be controlled through well-defined chemistry.

Devices were prepared by an extension of the methods previously used in constructing single-C60 and single-nanocrystal transistors. Using electron-beam lithography, a narrow gold bridge was fabricated on an aluminum pad with a 3-nm oxide layer serving as a gate electrode. The electromigration-induced break-junction technique was then used to create two closely spaced gold electrodes. Scanning electron microscope imaging and tunnel current measurements reveal that the narrowest gap between the two electrodes is consistently 1 nm.

The observation of the Kondo effect in single-molecule transistors, where an individual divanadium molecule serves as a spin impurity. The authors find that the Kondo resonance can be tuned reversibly using the gate voltage to alter the charge and spin state of the molecule. The resonance persists at temperatures up to 30 K and when the energy separation between the molecular state and the Fermi level of the metal exceeds 100 meV. The authors suggest the present study demonstrates that molecules can provide a Kondo system where critical parameters of Kondo physics, such as the spin and orbital degrees of freedom, are defined by chemical synthesis. With the recent advances of synthetic methodology, the preparation of molecular clusters possessing adjustable magnetic properties is becoming feasible. Future investigations of such species are expected to provide detailed insight into electron transport through a molecular system where the spin and orbital degeneracies are precisely controlled.

Quantum dots are small electrically conducting regions, typically less than 1 micron in diameter, that contain from one to a few thousand electrons. Because of the small volume, the electron energies within the dot are quantized, and the behavior of the quantum dot is intermediate between that of an atom and that of a classical macroscopic object. Such intermediate systems are called "mesoscopic" systems, and in the past several years great attention has been devoted to the physics of such systems, since they apparently can provide insights into quantum systems in general. The electronic states in quantum dots can be probed by transport when a small *tunnel coupling is allowed between the dot and nearby source and drain leads.

Cronenwett et al report the realization of a tunable Kondo effect in small quantum dots, with the capability of switching a dot from a Kondo system to non-Kondo system as the number of electrons on the dot is changed from odd to even. The Kondo temperature can be tuned by means of a gate voltage as a single particle energy state nears the Fermi energy. Measurements of the temperature and magnetic field dependence of a Coulomb-blockaded dot show good agreement with prediction of both equilibrium and nonequilibrium Kondo effects.

Localization-Delocalization in Quantum Dots

N.B. Zhitenev *et. al.* report a study of the electron localization- delocalization transition in quantum dots. The problem of electron localization has remained a prime focus of experiment and theoretical research over the past 40 years. Single electron capacitance spectroscopy precisely measures the energies required to add individual electrons to a quantum dot. The spatial extent of electronic wave functions was probed by investigating the dependence of these energies on changes in the dot confining potential. For low electron densities, electrons occupy distinct spatial sites localized within the dot. At higher densities, the electrons become delocalized, and all wave functions are spread over the full dot area. Near the delocalization transition, the last remaining localized states exist at the perimeter of the dot. Unexpectedly, these electrons appear to bind with electrons in the dot center.

Nanomechanical Oscillations in a Single-C60 Transistor

The motion of electrons through quantum dots is strongly modified by single electron charging and the quantization of energy levels. Much effort has been directed towards extending studies of electron transport to chemical nanostructures, including molecules, nanocrystals, and nanotubes. The authors report the fabrication of single molecule transistors based on individual C60 molecules connected to gold electrodes. The authors performed transport measurements that provide evidence for a coupling between the center-of-mass motion of the C60 molecules and single electron hopping—a conduction mechanism that has not been observed previously in quantum dot studies. The coupling is manifest as quantized nano mechanical oscillations of the C60 molecule against the gold surface, with a frequency of about 1.2 THz. This value is in good agreement with a simple theoretical estimate based on van der Waals and electrostatic interactions between C60 molecules and gold electrodes.

Single-C60 transistors were prepared by depositing a dilute toluene solution of C60 onto a pair of connected gold electrodes fabricated using electron-beam lithography. A break-junction technique was then used to create a gap between these electrodes by the process of electromigration. The typical lateral size of the fabricated electrodes was of the order of 100 nm at the point of the gap formation, and the height of the electrodes was 15 nm. Scanning electron microscope images of fabricated electrodes reveal that the gap between two electrodes is not uniform and that the narrowest gap is formed only between small protrusions (10 nm) of two gold electrodes. Current voltage measurements of these electrodes at cryogenic temperatures without deposited C60 molecules show that the size of the gap is consistently about 1 nm. In a significant fraction of the C60 devices, the conductance of the junction after initial breaking is substantially enhanced compared to devices with no C60 deposited, indicating that C60 molecules reside in the junction. The entire structure was defined on a SiO_2 insulating layer on top of a degenerately doped silicon wafer that serves as a gate electrode that modulates the electrostatic potential of C60.

It was suggested by the researchers that the transport measurements presented demonstrate that single electron-tunneling events can be used both to excite and probe the motion of a molecule: indeed, the single-C60 transistor behaves as a high-frequency nanomechanical oscillator. Furthermore, the oscillations of the C60 molecule must be treated in a quantized fashion, showing that this is a true quantum "mechanical" system. The authors expect that the coupling between the quantized electronic and mechanical degrees of freedom will be generically important in electron transport through nanomolecular systems.

A Single Electron Transistor made from a Cadmium Selenide Nanocrystal

The techniques of colloidal chemistry permit the routine creation of semiconductor nanocrystals whose dimensions are much smaller than those that can be realized using lithographic techniques. The sizes of such nanocrystals can be varied systematically to study quantum size effects or to make novel electronic or optical materials with tailored properties. Preliminary studies of both the electrical and optical properties of individual nanocrystals have been performed recently. These studies show clearly that a single excess charge on a nanocrystal can markedly influence its properties.

The researchers present measurements of electrical transport in a single electron transistor made from a colloidal nanocrystal of cadmium selenide. This device structure enables the number of charge carriers on the nanocrystal to be tuned directly, and so permits the measurement of the energy required for adding successive charge carriers. Such measurements are invaluable in understanding the energy-level spectra of small electronic systems, as has been shown by similar studies of lithographically patterned quantum dots and small metallic grains.

The researchers suggest their work represents a new type of spectroscopy for single nanocrystals. Unlike optical measurements, where electron hole pairs are created, these measurements probe the energy for adding a single type of charge carrier. The authors hope that these results will stimulate new calculations of the hole addition energies for a CdSe nanocrystal that include the effects of exchange, correlations, and screening by the metallic electrodes. Future measurements will investigate how the ground-state and excited-state properties vary with the size, shape and composition of nanocrystals as well as with the composition of the leads.

Identification of Atomic-like Electronic States in Indium Arsenide Nanocrystal Quantum Dots

Semiconductor quantum dots, due to their small size, mark the transition between molecular and solid-state regimes, and are often described as "artificial atoms". This analogy originates from the early work on quantum confinement effects in semiconductor nanocrystals, where the electronic wavefunctions are predicted to exhibit atomic-like symmetries. Spectroscopic studies of quantum dots have demonstrated discrete energy level structures and narrow transition linewidths, but the symmetry of the discrete states could be inferred only indirectly.

The researchers report they use cryogenic scanning tunneling spectroscopy to identify directly atomic-like electronic states with s and p character in a series of indium arsenide nanocrystals. These states are manifest in tunneling current voltage measurements as two- and six-fold single electron-charging multiplets respectively, and they follow an atom-like Aufbau principle of sequential energy level occupation.

The researchers suggest that the good correlation between optical transitions and the spacing of levels observed in the tunneling spectra indicates that the charging did not have a significant effect on the QD level structure. The tunneling IV curves are thus well described as an algebraic sum of single electron charging energies and the QD level spacings.

Quantum Dots and Photon Devices

Indistinguishable Photons from a Single-Photon Device

Single-photon sources have recently been demonstrated using a variety of devices, including molecules, mesoscopic quantum wells, color centres, trapped ions and semiconductor quantum dots. Compared with a *Poisson*-distributed source of the same intensity, these sources rarely emit two or more photons in the same pulse. Numerous applications for single-photon sources have been proposed in the field of quantum information, but mostly including linear-optical quantum computation—also require consecutive photons to have identical wave packets. For a source based on a single quantum emitter, the emitter must therefore be excited in a rapid or deterministic way, and interact little with its surrounding environment.

When identical single photons enter a 50-50 beam splitter from opposite sides, quantum mechanics predicts that both photons must leave in the same direction, if their wave packets overlap perfectly. This two-photon interference effect originates from the Bose Einstein statistics of photons. This bunching effect was first observed using pairs of highly correlated photons produced by parametric down coversion, but it should also occur with single, independently generated photons. Most proposed applications for single-photon sources in the field of quantum information (with the notable exception of quantum cryptography) involve two-photon interference. Such applications include quantum teleportation, post-selective production of polarization-entangled photons, and linear-optics quantum computation. It is therefore important to demonstrate that consecutive photons emitted by a single-photon source are identical and exhibit mutual two-photon interference effects.

The researchers report a test of the indistinguishability of photons emitted by a semiconductor quantum dot in a microcavity through a Hong Ou Mandel type two-photon interference experiment. The authors find that consecutive photons are largely indistinguishable, with a mean wave-packet overlap as large as 0.81, making this source useful in a variety of experiments in quantum optics and quantum information.

Quantum Dots and Electroluminescence

Electroluminescence has taken a new turn: S. Coe et al reports the fabrication of high-efficiency organic light-emitting diodes (LEDs) in which the light-emitting centers are cadmium-selenium (CdSe) nanocrystals, or quantum dots. Organic light-emitting diodes (LEDs) bring with them the advantages of robust fabrication technique and high performance, which, when coupled with the excellent luminescent properties of nanocrystals, offer exciting prospects for real, workable devices. Mobile phones with small colour displays using organic LEDs are already commercially available. The images are generated through fluorescence, as electrons make transitions between orbital states of pi-conjugated organic molecules (the pi-bond arises from the overlap of the 2p orbitals of electrons in carbon atoms). As well as having high quantum efficiency for electron-to-photon conversion, pi-conjugated molecules in organic LEDs have the advantage of colour tunability, so that they can be used to build full-color displays of red green blue (RGB) emitters.

But there is also a drawback: the emission spectra of pi-conjugated molecules are very broad, typically spanning 50 to 100 nm (the "full width at half-maximum", or FWHM). This range of molecular fluorescence is caused by the vibrational and rotational motion of atoms inside the pi-conjugated molecules. So with an organic LED, it is difficult to get, for example, pure red light emitted with high quantum efficiency. Nevertheless, sharp RGB emission has been demonstrated from LEDs based on some specific materials, such as europium chelates, aggregated structures of cyanine dyes or layered inorganic organic perovskite compounds. But these LEDs have not achieved the emission efficiency or device durability needed for practical display applications.

Some inorganic nanocrystals emit visible light with sharp emission spectra that are less than 30 nm FWHM. The nanocrystals are in effect quantum dots and confine charge so well within their small volume that a high quantum efficiency, exceeding 50 per cent, is possible. It might be expected, then, that quantum dots incorporated in organic LEDs would make excellent emission centres 4. In fact, electroluminescence has been observed by simply mixing inorganic nanocrystals with -conjugated polymers, but the emission efficiency was far lower than that of conventional polymer LEDs. Coe et al have fabricated an organic LED with a single layer of CdSe quantum dots sandwiched between organic thin films. Remarkably, the efficiency of their device is about 25 times higher than that achieved so far with quantum-dot LEDs. Thinking ahead to the design of device architectures, there are two noteworthy aspects here: the structure of this quantum-dot LED is already close to that for an optimum device and the process of fabricating a layer of quantum dots is simple.

Optical Activity of Quantum Dots

An electron in a quantum dot can be described by a quantum wave function that is similar to that used for an electron in a single atom, although the energy of the electron in the quantum dot is spread in a coordinated way (spread "coherently") over the lattice of atomic nuclei. The electronic wave functions of quantum dots are often labeled with atomic notation, but quantum dots are very much solid-state nanostructures that can be tailored into different shapes. Recent studies describe the optical behavior of individual quantum dots and quantum rings, and such behavior is of considerable interest because quantum dots that emit light are expected to form the basis of a new generation of lasers.

In these optical studies of quantum dots, the semiconductor dots and rings are made from indium arsenide embedded in gallium arsenide, and the structures were grown using techniques developed within the past decade that allow much smaller nanostructures to be created than were previously possible. In these new experiments, electrons are introduced one by one into individual quantum dots while the optical emission of the dots is measured with great precision. These studies provide new perspectives on the internal quantum mechanical workings of quantum dots; the ultimate goal is to create useful electronic and optical nanomaterials that have been quantum mechanically engineered by tailoring the shape, size, composition, and position of various quantum dots.

Concerning the physics of quantum dots, adding even a single electron to such a system requires a significant amount of extra energy because of the repulsion between the negatively charged electrons as they are forced into a smaller volume. One result of this, called the "Coulomb blockade", is to make possible a greater laboratory control of the number of electrons in a quantum dot, i.e., researchers can tune the number of electrons by manipulating input energy.

In general, optical excitation of a semiconductor leads to the creation of a quasi-particle known as an "exciton"-a negatively charged electron bound together with a positively charged "hole". In contrast to the Coulomb blockade resulting from electrical injection of electrons into a quantum dot, such dots remain neutrally charged following optical excitation, and the quantum dot exciton has been studied in detail by measuring the light emitted when the hole and electron recombine.

The researcher concludes: "Quantum dots have great flexibility because their properties can be artificially engineered, but this comes at a price. Nature has given us atoms; scientists must make quantum dots. Further advances in this exciting field of science and technology will depend heavily on the creativity of physicists, chemists, and materials scientists who make these tiny structures."

Quantum Dots and Computers

Quantum Dots and Quantum Information Processing

Physicists are actively seeking to transfer the weird phenomena of the quantum world from their laboratories to the real world. In particular, the prospect of quantum information processing has attracted considerable attention for its potential to improve the speed and reliability of data handling. Information would be encoded in "quantum bits", and the search is on for a physical system that could form a reliable, controllable quantum bit.

In contrast to classical bits, which can be in either state 0 or state 1, quantum bits exist as a combination (a linear superposition) of two quantum logic states, represented as $|0\rangle$ and $|1\rangle$. In a quantum computer, the quantum bits first have to be controlled individually in order to initialize the quantum register in which information is stored. Then a controllable interaction between the quantum bits must be established so that the quantum states become entangled. It is this "entanglement" that is the key to a quantum computer's power: in effect, a rigid coupling is introduced between the quantum bits; which can then no longer be considered individually but are affected simultaneously by a calculational operation. It is thanks to this capacity for parallel processing that a quantum computer should be able to perform calculations much faster than a classical computer.

The original proposals for quantum computers were based on atomic systems(1), such as atoms held in traps, where the quantum bit is formed by two energy levels between which an atomic electron can make transitions. Now that semiconductor quantum dots have been synthesized, it opens up the possibility of mimicking these approaches in a solid-state environment.

Quantum dots, tiny clusters of semiconductor material, are often called "artificial atoms", because the charge carriers in these systems (electrons or holes) can only occupy a restricted set of energy levels, just like the electrons in an atom. In fact, quantum dots offer a variety of two level systems, based on charge or spin (or both). One such two level system is a coupled electron hole pair-an exciton. The absence (equivalent to the state |0>) and presence (state |1>) of an exciton in a semiconductor quantum dot could represent the two levels of a quantum bit.

Coherent Properties of a Two Level System Based on a Quantum Dot Photodiode

Present-day information technology is based mainly on incoherent processes in conventional semiconductor devices. To realize concepts for future quantum information technologies, which are based on coherent phenomena, a new type of "hardware" is required. Semiconductor quantum dots are promising candidates for the basic device units for quantum information processing. One approach is to exploit optical excitations (excitons) in quantum dots. It has already been demonstrated that coherent manipulation between two excitonic energy levels—via so called Rabi oscillations can be achieved in single quantum dots by applying electromagnetic fields.

The researchers report they make use of this effect by placing an InGaAs quantum dot in a photodiode, which essentially connects it to an electric circuit. The researchers demonstrate that coherent optical excitations in the quantum dot two level system can be converted into deterministic photocurrents. The researchers find that this device can function as an optically triggered single-electron turnstile.

A coherently driven single-QD photodiode is an optically triggered single-electron source, which offers, just like the related single electron turnstile device, single electrons (or holes) on demand. Both devices have the potential to deliver frequency-controlled currents according to $I = fe$. With circularly polarized excitation, a single-QD photodiode has the potential to provide deterministic streams of spin-polarized charges. More subtle points concerning the absolute accuracy of this device, like the influence of incomplete tunneling or possible avalanche multiplication, will certainly be the subject of future work. By means of single-QD photodiodes, it becomes possible to transfer optical excitations of single quantum systems into deterministic electric currents. This optoelectronic functionality provides a link between coherent optical excitations and electric currents on the single electron level.

The researchers believe that this will allow the electric readout of excitonic quantum gates, and could have a substantial impact on quantum information technology.

Allowed and Forbidden Transitions in Artificial Hydrogen and Helium Atoms

The strength of radiative transitions in atoms is governed by selection rules that depend on the occupation of atomic orbitals with electrons. Experiments have shown similar electron occupation of the quantized energy levels in semiconductor quantum dots, often described as artificial atoms. But unlike real atoms, the confinement potential of quantum dots is anisotropic, and the electrons can easily couple with phonons of the material.

The researchers report electrical pump-and-probe experiments that probe the allowed and "forbidden" transitions between energy levels under phonon emission in quantum dots with one or two electrons (artificial hydrogen and helium atoms). The forbidden transitions are in fact allowed by higher-order processes where electrons flip their spin. The authors find that the relaxation time is about 200 microseconds for forbidden transitions, 4 to 5 orders of magnitude longer than for allowed transitions. This indicates that the spin degree of freedom is well separated from the orbital degree of freedom, and that the total spin in the quantum dots is an excellent quantum number. The authors suggest this is an encouraging result for potential applications of quantum dots as basic entities for spin-based quantum information storage.

Quantum Dots and Nanotubes

Band Gap Modulation of Carbon Nanotubes by Encapsulated Metallofullerenes

The technical and economic difficulties in further miniaturizing silicon-based transistors with the present fabrication technologies have motivated a strong effort to develop alternative electronic devices, based, for example, on single molecules. Recently, carbon nanotubes have been successfully used for nanometer-sized devices such as diodes, transistors, and random access memory cells. Such nanotube devices are usually very long compared to silicon-based transistors.

Since the discovery of carbon nanotube a decade ago, researchers have synthesized a variety of forms of carbon nanotubes. Recently, it has been shown experimentally that fullerenes or endohedral etallofullerenes such as Gd encapsulated inside C82 can be inserted into single-wall nanotubes (SWNTs), forming a pea-pod-like structure. When the diameter of the inserted fullerene is smaller than that of the SWNT minus a certain length (0.7 nm, roughly twice the van der Waals bond length), the fullerenes are inserted exothermally. When the diameter of a fullerene slightly exceeds the length described above, the fullerene can be inserted endothermally, but the resultant SWNT is elastically strained. A theoretical study has predicted that the electronic structure is severely modified, including the positions of the van Hove singularities (VHS), when an SWNT is uniaxially strained. Combining these ideas, one can perform "local bandgap engineering" at the site where a fullerene is endothermally inserted.

The researchers report a method for dividing a semiconductor nanotube into multiple quantum dots with lengths of about 10 nm by inserting Gd-C82 endohedral fullerenes. The spatial modulation of the nanotube electronic bandgap is observed with a low-temperature scanning tunneling microscope. The researchers find that a bandgap of 0.5 eV is narrowed down to 0.1 eV at sites where endohedral metallofullerenes are inserted. This change in bandgap can be explained by local elastic strain and charge transfer at metallofullerene sites. The researchers suggest this technique for fabricating an array of quantum dots could be used for nano-electronics8 and nano-optoelectronics.

Scanned Probe Imaging of Single Electron Charge States in Nanotube Quantum Dots

Single electron charging phenomena are ubiquitous in atoms, molecules, and small electronic devices, and their effects are central to an understanding of the physics and technology of nanoscale systems. Single electron effects arise because the number of electrons residing on a small, quasi-isolated, conducting island is quantized. Adding an additional charge to such a quantum dot costs an electrostatic energy on the order of U = e2/C, where C is the capacitance of the dot and e is the electronic charge. This charging energy suppresses charge transport when U >> kBT, where kBT is the thermal energy, leading to the Coulomb blockade of charge motion on and off the dot.

Although Coulomb blockade phenomena have been studied extensively with transport measurements, such measurements lack the spatial discrimination necessary to probe the interior of a dot or to probe complex multidot systems. An alternative approach is to detect single-charge motion using scanned probe techniques, such as scanned capacitance microscopy, scanned single electron transistors, and atomic force microsopy (AFM). The first two of these have excellent charge sensitivity but are technically very difficult; moreover, they are not easily able to image the topography of the device under study. AFM-based techniques, on the other hand, can be used both to image the sample and to interact with it in a variety of ways. For example, electrostatic force microscopy (EFM), which measures the electrostatic force between a sample and a metallized AFM tip, has been used to detect the motion of single charges during contact electrification of insulating surfaces and to image the potential profile in carbon nanotubes. In addition, scanned gate microsopy (SGM), in which the AFM tip is used to perturb the conducting properties of a sample, has been used to image electron trajectories and scattering centers in two-dimensional electron gases and barriers in carbon nanotubes.

In summary: The researchers report that an atomic force microscope was used to study single electron motion in nanotube quantum dots. By applying a voltage to the microscope tip, the number of electrons occupying the quantum dot could be changed, causing Coulomb oscillations in the nanotube conductance. Spatial maps of these oscillations were used to locate individual dots and to study the electrostatic coupling between the dot and the tip. The electrostatic forces associated with single electrons hopping on and off the quantum dot were also measured. These forces changed the amplitude, frequency, and quality factor of the cantilever oscillation, demonstrating how single-electron motion can interact with a mechanical oscillator.

Kondo Physics in Carbon Nanotubes

The "Kondo effect", first explained by the theoretical physicist Jun Kondo, is observed when magnetic ions occur as dilute impurities in nonmagnetic crystals. If the host crystal is a metal (e.g., copper), the magnetic impurities (e.g., iron) contribute to the electrical resistivity, with the conduction electrons scattered by the magnetic impurity. The scattering is strong at low temperatures, and increases slightly as temperature decreases, an anomalous resistivity-temperature relation. The effect essentially results from the interactions between the localized spins of the magnetic impurities and the free electrons involved in conduction. There is a characteristic temperature, called the "Kondo temperature", which depends on the impurity and on the host crystal, the resistivity increasing at low temperature, starting near the Kondo temperature. The Kondo temperature of the system iron-in-copper is 24 degrees kelvin. Nearly all transition metal atoms are found as magnetic impurities in copper or gold, and each system has a different Kondo temperature, which varies from 1000 degrees kelvin to a fraction of 1 degree kelvin.

Fullerenes are large molecules composed entirely of carbon, with the chemical formula C(n), where n is any even number from 20 to over 100. They apparently have the structure of a hollow spheroidal cage with a surface network of carbon atoms connected in hexagonal and pentagonal rings. Carbon nanotubes are similar to fullerenes, except their shape is tubular. They were first discovered by Sumio Iijima (NEC Laboratories, JP) in 1991, they come in both multi wall and single wall versions, with single wall nanotubes having a diameter of approximately 1 nanometer and multi wall versions having diameters of the order of 10 to 30 nanometers. There have been rapid developments in understanding the chemistry and physics of carbon nanotubes, and there is much excitement in both the materials science and electronics communities concerning possible applications of these unique structures.

In general, the term "quantum dot" refers to an artificial atom. As realized in the laboratory, quantum dots are small electrically conducting regions, typically less than 1 micron in diameter, that contain from one to a few thousand electrons. Because of the small volume, the electron energies within the dot are quantized, and the behavior of the quantum dot is intermediate between that of an atom and that of a classical macroscopic object. Electron transport in conductors is usually well described by so-called "Fermi-liquid theory", which assumes that the energy states of the electrons near the *Fermi level are not qualitatively altered by Coulomb interactions. In one-dimensional systems, however, even weak Coulomb interactions cause strong perturbations, and the resulting system, known as a "Luttinger liquid" (after Joaquim Luttinger, who did major work on such systems in the 1950s) is predicted to be distinctly different from its 2- and 3-dimensional counterparts. Single walled carbon nanotubes are model systems for the study of 1-dimensional electron transport, and according to theoretical predictions, such systems should be very sensitive to Coulomb repulsion and become insulators at low temperatures. However, researchers have demonstrated that it is possible to create superconducting junctions when carbon nanotubes are connected to superconducting contacts, a phenomenon called "proximity-induced superconductivity". In the case of nanotubes, the junction properties are apparently related to its highly 1-dimensional character, and such junctions may have potential applications as thermal detectors of weak infrared radiation (bolometers).

In this context, the term "ballistic" (as opposed to the term "diffusive") refers to the passage of electrons through a conductor whose length is less than the mean free path of electrons in the conductor, with the result that most of the electrons pass through the conductor without scattering.

The researchers point out that the connection of electrical leads to wire-like molecules is a logical step in the development of molecular electronics, but such a system also allows studies of fundamental physics. For example, metallic carbon nanotubes are essentially quantum wires that have been found to act as 1-dimensional quantum dots, Luttinger liquids, proximity-induced superconductors, and ballistic and diffusive 1-dimensional metals.

The researchers report that electrically-contacted single-walled carbon nanotubes can serve as powerful probes of Kondo physics, demonstrating the universality of the Kondo effect. Arising in the prototypical case from the interaction between a localized impurity magnetic moment and delocalized electrons in a metallic host, the Kondo effect has been used to explain enhanced low-temperature scattering from magnetic impurities in metals, and also occurs in transport through semiconductor quantum dots. The far greater tunability of quantum dots (in this case, nanotubes) compared with atomic impurities makes new classes of Kondo-like effects accessible to study. The nanotube devices used by the researchers differ from previous systems in which Kondo effects have been observed in that they are 1-dimensional quantum dots with 3-dimensional metal (gold) reservoirs. This allows observations of Kondo effects for very large electron numbers in the quantum dot, numbers estimated as of the order of tens of thousands.

Quantum Dots and Chemical Systems

Probing and Controlling the Bonds of an Artificial Molecule

Coupled quantum dots in semiconductor materials are often referred to as artificial molecules, because these electron boxes allow us to precisely control the number of charge carriers and their mutual interaction. For future quantum information processing, entangled quantum states within such mesoscopic systems need to be implemented in electronic circuits. This requires precise control of the bonds of the artificial molecule in situ. In this context, coupled dots can be used for investigating the probing mechanism of quantum physics; that is, how to attach leads to real molecules. Generally the interaction of quantum systems with their environment induces decoherence and the decay of quantum superpositions. Key questions are how accurately the coupling of two artificial atoms can be adjusted and how coherent coupling can be monitored noninvasively without destroying the molecular quantum states.

It has already been shown in transport and microwave spectroscopy that the transition from ionic to covalent binding in quantum dot molecules can be traced. The detection mechanism for those techniques relies on the transmission of a single electron through the coupled dot system. In contrast, the researchers have used a co-tunneling mechanism, involving two electrons tunneling, simultaneously, to probe the coupling between the dots. Even though the probed quantum states are only weakly coupled to adjacent reservoirs, the dynamics resemble the Kondo effect, in which hybridization of electron states with source and drain contacts is desired.

In summary: The researchers demonstrate how molecular quantum states of coupled semiconductor quantum dots are coherently probed and manipulated in transport experiments. The applied method probes quantum states by the virtual co-tunneling of two electrons and hence resolves the sequences of molecular states simultaneously. This result is achieved by weakly probing the quantum system through parallel contacts to its constituting quantum dots. The overlap of the dots' wave functions and, in turn, the splitting of molecular states are adjusted by the direct influence of coupling electrodes.

Electrochemistry and Electrogenerated Chemiluminescence from Silicon Nanocrystal Quantum Dots

In a bulk semiconductor, electrons and holes move freely throughout the crystal. However, in a nanocrystal, confinement of the electrons and holes leads to a variety of optical and electronic consequences, including size-dependent molecular-like optical properties, greater electron/hole overlap for enhanced photoluminescence (PL) efficiencies, and discrete single-electron/hole charging. Because of their enormous surface area-to-volume ratios, nanocrystals (NCs) are highly susceptible to heterogeneous redox chemistry with the surrounding environment. Depending on the semiconductor and the surface chemistry, this chemical reactivity can lead to either fatal chemical degradation or new useful properties, such as reversible photocatalytic and electrochromic properties and redox reactivity.

Semiconductor NCs have been prepared with narrow size distributions, controlled surface chemistry, and internal bulk crystal structure, and adsorbed capping ligands are often used to control size and prevent irreversible aggregation. Although the electrochemical properties of monolayer-protected metallic NCs have been well documented, reports concerning the electrochemical properties of semiconductor NCs remain scarce. Difficulties include the limited potential window available in aqueous solutions, the limited solubility of many NCs in nonaqueous solvents, and the need for highly pure, isolated, monodisperse dots. Compound semiconductor NCs, such as CdS, are also chemically unstable upon electron transfer.

Reversible electrochemical injection of discrete numbers of electrons into sterically stabilized silicon nanocrystals (NCs) (approximately 2 to 4 nanometers in diameter) was observed by the authors by differential pulse voltammetry (DPV) in N,N′-dimethylformamide and acetonitrile. The electrochemical gap between the onset of electron injection and hole injection —related to the highest occupied and lowest unoccupied molecular orbitals—grew with decreasing nanocrystal size, and the DPV peak potentials above the onset for electron injection roughly correspond to expected Coulomb blockade or quantized double-layer charging energies. Electron transfer reactions between positively and negatively charged nanocrystals (or between charged nanocrystals and molecular redox-active coreactants) occurred that led to electron and hole annihilation, producing visible light. The electrogenerated chemiluminescence spectra exhibited a peak maximum at 640 nanometers, a significant red shift from the photoluminescence maximum (420 nanometers) of the same silicon NC solution. The authors suggest these results demonstrate that the chemical stability of silicon NCs could enable their use as redox-active macromolecular species with the combined optical and charging properties of semiconductor quantum dots.

Microwave Spectroscopy of a Quantum Dot Molecule

Quantum dots are small conductive regions in a semiconductor, the regions containing a variable number of electrons (from 1 to 1000) that occupy well-defined, discrete quantum states—for which reason they are often referred to as "artificial atoms". Connecting quantum dots to current and voltage contacts allows the discrete energy spectra of the system to be probed by charge-transport measurements. Two quantum dots can be connected to form an "artificial molecule", and depending on the strength of the inter-dot coupling (which supports quantum-mechanical tunneling of electrons between the dots), the two dots can form "ionic" or "covalent" bonds. In the ionic bond case, the electrons are localized on individual dots, and in the covalent bond case, the electrons are delocalized over both dots.

T.H. Oosterkamp et al report a transition from ionic bonding to covalent bonding in a quantum dot "artificial molecule" probed by microwave excitations. The authors suggest their results demonstrate controllable *quantum coherence in single electron devices, an essential requirement for practical applications of quantum dot circuitry in the construction of quantum computers.

Incorporation of Quantum Dots in Colloids

Nanoscale metal and semiconductor particles are of current interest because they mark a material transition range between quantum and bulk properties. With decreasing particle size, bulk properties are lost as the continuum of electronic states becomes discrete (the quantum size effect) and as the fraction of surface atoms becomes large. The electronic and magnetic properties of metallic nanoparticles and nanoclusters show new characteristics that can be utilized in novel applications in areas that range from nonlinear optical switching and catalysis to high-density information storage.

Numerous methods have been developed to synthesize metal nanoparticles. A major difficulty with scale-up of these methods is that the metal colloid stability is often controlled by electrostatic interactions across the Debye double layer and sterically through adsorption of steric stabilizing agents such as polymers and surfactants. As a consequence, such metal colloids are extremely sensitive to their environment.

One way to improve the stability of metal nanoparticles is to coat them with silica, which is very resistant to coagulation, even at high volume fractions. It has been reported that particles of noble metals such as Ag and Au can be coated with silica shells, but these procedures usually involve a multi-step process, and only single metal particle cores could be coated.

The researchers report the development of a new method to fabricate nanocomposite silicon dioxide spheres (approximately 100 nanometers in diameter) containing homogeneously dispersed Ag quantum dots (2 to 5 nanometers in diameter). The inclusion morphology is controlled through the timing of the photochemical reduction of silver ions during hydrolysis of tetraethoxysilane in a microemulsion.

Quantum Dots and Biochemical Systems

In Vivo Imaging of Quantum Dots Encapsulated in Phospholipid Micelles

Nanometer-scale semiconductor crystallites (known as nanocrystals or quantum dots) could dramatically improve the use of fluorescent markers in biological imaging. Because these colloidal particles act as robust broadly tunable nanoemitters that can be excited by a single light source, they could provide distinct advantages over current in vitro and in vivo markers (e.g., organic dyes and fluorescent proteins).

However, before nanocrystals can be widely used as biolabels, they must maintain three properties under aqueous biological conditions: efficient fluorescence, colloidal stability, and low nonspecific adsorption. Unfortunately, despite recent advances, these conditions have not been simultaneously satisfied, limiting the development of in vivo applications of nonaggregated (or individual) semiconductor nanocrystals.

The main challenge is that the quantum dots (QDs), as synthesized, have hydrophobic organic ligands coating their surface. To make the QDs water soluble, these organophilic surface species are generally exchanged with more-polar species, and both monolayer and multilayer ligand shells have been pursued. Although the monolayer method is reproducible, rapid, and produces QDs with a regular, well-oriented, thin coating, their colloidal stability is poor. In contrast, the multilayer method yields QDs that are stable in vitro, but the coating process is long and the coating is difficult to control. A more serious concern is that both approaches still produce QDs that tend to aggregate and adsorb nonspecifically. To resolve this problem, researchers have explored two additional coatings. First, the outer ligand shell of the QD has been overcoated with proteins adsorbed through hydrophobic or ionic interactions. Other layers can then be added to allow conjugation with specific biomolecules. Indeed, this method has provided new reagents for fluoroimmunoassays. Second, the outer ligand shell has been overcoated with surfactants or polymers to prevent nonspecific adsorption of biomolecules while still permitting bioconjugation.

For example, silica-coated QDs have been further modified with small monomers of poly(ethylene glycol) to reduce nonspecific adsorption. Despite these efforts, nonspecific adsorption and aggregation still occur when QDs are used in biological environments. Studies of cellular uptakes of QDs report large aggregate formation inside the cell. The same aggregation problems are reported when QDs are used for fluorescence in situ hybridization, or as markers for molecular recognition on cell surfaces. Consequently, the use of QDs in biological applications is still limited and primarily confined to in vitro studies.

In summary: Fluorescent semiconductor nanocrystals (quantum dots) have the potential to revolutionize biological imaging, but their use has been limited by difficulties in obtaining nanocrystals that are biocompatible. To address this problem, the researchers encapsulated individual nanocrystals in phospholipids block-copolymer micelles and demonstrated both in vitro and in vivo imaging. When conjugated to DNA, the nanocrystal-micelles acted as in vitro fluorescent probes to hybridize to specific complementary sequences. Moreover, when injected into Xenopus embryos, the nanocrystal-micelles were stable, nontoxic (<5 ž109 nanocrystals per cell), cell autonomous, and slow to photobleach. Nanocrystal fluorescence could be followed to the tadpole stage, allowing lineage-tracing experiments in embryogenesis.

Ordering of Quantum Dots using Genetically Engineered Viruses

Building ordered and defect-free two- and three-dimensional structures on the nanometer scale is essential for the construction of next-generation optical, electronic, and magnetic materials and devices. Traditional assembly approaches have been based on hydrogen bonding, coulombic interactions, and van der Waals forces. Although a bacterial synthetic method was reported to make monodisperse modified polypeptides, it has been difficult to tune the layer spacing and structure of conventional synthetic polymers because of their polydisperse chain lengths. Efforts have been directed toward the use of soft materials to organize inorganic materials at the nanoscale.

Protein cages have been used as templates to synthesize nanoscale materials in capsids. DNA recognition linkers have been successfully used to construct specific gold nanocrystal structures. ZnS and CdS have been nucleated in a lyotropic liquid crystalline medium to make nanowires and nanocrystal superlattice structures by a surfactant assembly pathway. However, these methods have limitations with respect to length scale and type of inorganic material.

Monodisperse biomaterials that have an anisotropic shape are promising as components of well-ordered structures. Liquid crystalline structures of wild-type viruses (Fd, M13, and TMV) have been tunable by controlling the solution concentrations, the solution ionic strength, and the external magnetic fields applied to the solutions. The authors recently demonstrated that engineered viruses can recognize specific semiconductor surfaces through the method of selection by combinatorial phage display. These specific recognition properties of the virus can be used to organize inorganic nanocrystals, forming ordered arrays over the length scale defined by liquid crystal formation.

In summary: The researchers report that a liquid crystal system was used for the fabrication of a highly ordered composite material from genetically engineered M13 bacteriophage and zinc sulfide (ZnS) nanocrystals. The bacteriophage, which formed the basis of the self-ordering system, were selected to have a specific recognition moiety for ZnS crystal surfaces. The bacteriophage were coupled with ZnS solution precursors and spontaneously evolved a self-supporting hybrid film material that was ordered at the nanoscale and at the micrometer scale into approximately 72-micron domains, which were continuous over a centimeter length scale. In addition, suspensions were prepared in which the lyotropic liquid crystalline phase behavior of the hybrid material was controlled by solvent concentration and by the use of a magnetic field.

Unit IV–F

Resonant Tunneling Devices (RTDs)

Introduction

Here we talk about the future nanoelectronic devices using quantum effects employed in semiconductor devices for fabrication of computer devices and high frequency oscillators. The underlying principles used for defining the perpendicular transport of electrons in low dimensional structures up to room temperature and its current and voltage characteristics are explained using energy band diagrams.

MOSFETS with its special characteristic features serves two purposes. First case its acts as an excellent conductor device even with its scaling properties and even with small changes in its bias voltage results in large changes in its conductivity, hence its acts as an amplifier device.

Though it functions as a two state device there some factors need to be considered for its implementation. The obstacles associated with MOSFET are high electric field, heat power dissipation, vanishing bulk properties, shrinkage of depletion layers and unevenness of the depletion layer. Corrections have to be applied for these devices. Once the devices approach the nanoelectric properties, the bulk properties of semiconductor devices can be replaced by quantum physics which depend on the electron properties, energy quantization effects and tunneling come into effect. When quantum effects are applied to semiconductor devices the consideration of doped materials no longer considered as major issue. These devices are of great importance for high speed electronics. Here we discuss the quantum properties, general used terms such a quantum dot, tunneling effect, island, subband and electron behavior in quantized energy levels to deal with the resonant tunneling devices and the application of the quantum properties in various devices such as Resonant Tunneling Diodes (RTDs), Resonant Tunneling Transistors (RTTs). Here we explain the theory related to tunneling and the mathematical derivation for transmission probability of the particle in the double barrier junction.

Solid State Quantum Effects

Electron Fundamentals in Semiconductors

The de Broglie Wavelength: The mass of an electron is semiconductor is less than the free electron mass which is denoted by m*, thus the debroglie wavelength of electron is greater than that of free electron

_0.

Where is the wavelength of the electron _ = h/p = h/(2m* E)1/2 = _o (m/m*) ½

Momentum of electron is p, E is the energy of the electron.

The zero dimension or quantum dot is the condition in which the quantization occurs in all three direction and the electron cannot move freely in any direction.

Lx ≈ Ly ≈ Lz ≈ _

The one dimensional or quantum wire case in which quantisation takes place in two directions, the lectrons move freely only along the length of the wire.

Where Lx ≈ Ly ≈ _ « Lz

The two dimensional or quantum is a case in which the quantization of the electron motion occurs in one direction while in the other two directions electron motion is free.

where _ ≈ Lx « Ly, Lz

The three-dimensional or bulk case in which the electron spectrum quanization is not important at all and an electron behaves like a free particle characterized by the effective mass m*.

Where _ « Lx, Ly, Lz.

The thermal energy for the electrons E = KB

T where T = 300K is the ambient temperature and KB Boltzmans constant its value is 1.38602 * 10^{-23} JK –1.

The Debroglie wavelength for typical semiconductors with m* in the range of (0.01 – 1) nm is the order of 730 – 73 A.

As the temperature decreases to 3K, the debroglie wavelength increases by 1 order of magnitude. This wavelength becomes comparable with the sizes of semiconductor structures and devices fabricated by modern nanofabrication technology. Thus understanding of quantum phenomena is essential to understand the physics of nanostructures.

Resonant Tunneling Devices

An island confining electrons with one or two classical degrees of freedom. These solid state nanoelectronic devices are fabricated by using semiconductor materials. The V group element in the periodic table Si/Ge are generally used, but in current nanoelectronic devices III–IV group elements in periodic table such as GaAs Gallium Arsenide and AlGaAs Aluminium Gallium arsenide are used because of its excellent mobility of electrons and easier to fabricate defect free junctions than it is for IV group elements S/Ge. It's a quantum heterostructure exhibiting quantum perpendicular transport of electrons up to room temperature. An example for such a property is resonant electron transmission through a double barrier heterostructure. If more than one material is used in a device is referred as heterostructure. A heterojunction is formed when two different materials are joined together at a junction. The pairs of semiconductor materials that form the heterostructures are GaAs/AlGaAs, GaInAs/InP, GaInAs/AlInAs, GaSb/AlSb, GaN/AlN, InN/GaN etc. The total energy U ı__ for adding an electron to the island when a resonant tunneling device is subject to a voltage bias between the input and the output produces a current Vs voltage plot. The study of these devices is based on the concept of quantisation and tunneling properties.

Island, Potential well and Quantization Effects

A Quantum Well is potential structure which spatially confines electrons. Molecular Beam Epitaxy method is for growing the hetrostructures provide one dimensional confinement along the superlattice axis. The basic theory of Quantum Mechanics illustrates that an electron subjected to potential confinement has its energy quantized and a discrete energy spectrum is expected for the electron system. The quantum wells with the smallest dimension of the island in the quantum devices ranges from approximately 5 to 100nm. This island is a layer or region different from the surrounding material; otherwise edges of the island create electric fields from small electrodes patterned in the shape of the desired island. The island is surrounded by two narrow walls called tunneling barriers or potential barriers of other material or an insulating defect zone in the substrate. The island surrounded by the potential barrier impedes the flow of electrons in and out of the island region. But within the island mobile electrons tend to form puddles which are smaller than the dimensions of the island. Thus fabrication of the device becomes complicated for defining confinement of the electrons in the island.

There are two basic concepts for defining the confinement of the electron in quantum mechanical effects for nanometer scale islands between closely spaced potential barriers. Firstly, quantum mechanics restricts electrons energy to one of the finite number of one electron energy levels. The smaller the distance between the barriers the smaller the island, the more widely spaced energy levels for the electrons in the potential well between the barriers. __ is the energy space between the energy levels in the potential well. If the potential barriers are thin approximately 5 to 10 nm or less depends on the height of the barrier, electrons occupying the energy levels lower than the height barrier has a finite probability of "Tunneling "through the barrier to get on or off the island. But for the electron of a given energy to tunnel through a barrier empty state must exist on one other side of barrier. Energy quantization and tunneling strongly influence the flow of electrons through a nano-electronic device. The Quantum well diagram shown below in Fig. IV-F.1 describes the features of the tunneling device.

Quantum Theory

The familiar Schrodinger wave equation for the electron in the free space is given by

$$-h^2/2m + \_\ \nabla^2 =$$
$$-h^2/2m\ (\partial^2\_/x^2 + \partial^2\_/y^2 + \partial^2\_/z^2\ )$$

where '*m*' is the mass of free electron. The wave function of the electron or particle _ k (r) = 1/ (2_)3 (e^{ikr}) are phase waves.

The wave vector $k = (kx, ky, kz)$

Its energy given by

$= \_^2k^2/2m = \_2/2m\ (kx^2 + ky^2 + ky^2)$

Vector components of "K" are the quantum numbers for the free motion of the electron, for each degree of freedom.

For a K space, the number of states in a volume dk, $dk = dkx, dky, dkz$.

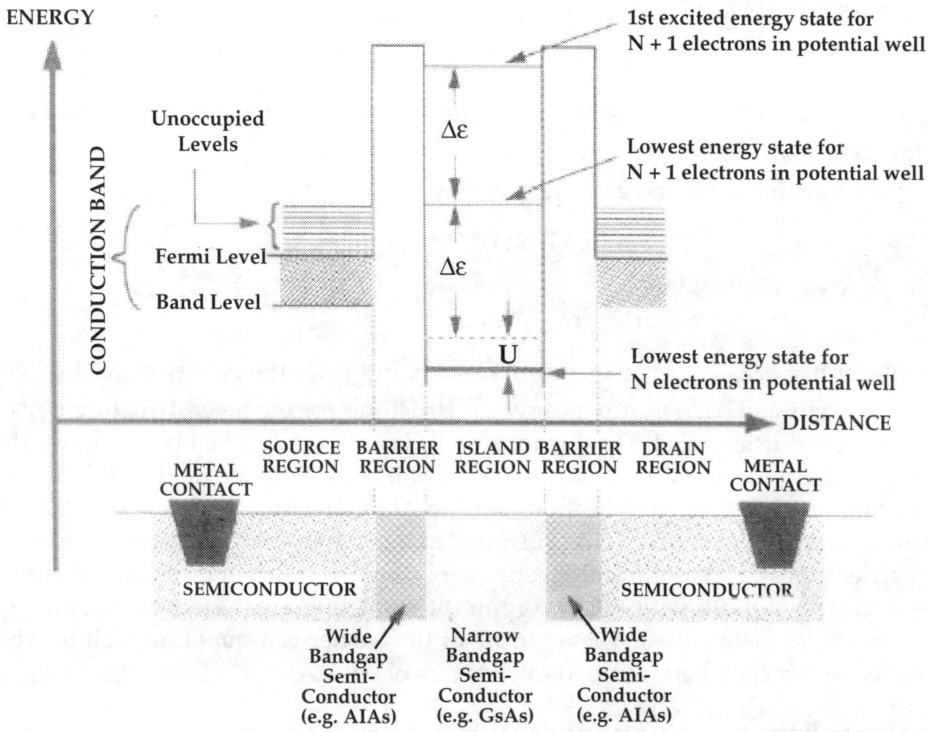

Fig. IV–F.1: Quantum Well.

Tunneling Transport

Wave Mechanics

Electron is a free particle with potential $V(x) = 0$.

The familiar time dependent Schrodinger equation is independent of potential given in (2) for a free particle. Electron as wave packet moves with group velocity and phase velocity and the wavelets inside the wave move with a phase velocity.

$$(-h^2/2m)\ d^2\ \_/dx^2 = E\ \_\ (x)$$

E is the energy of the particle. Electron as a wave packet moves with the group velocity and the wavelets inside the wave move with a phase velocity Vph.

Bound Particles in a Quantum Well

The movement of the electron is restricted in a finite region of space rather than to propagate in space is defined as" Quantum well or a particle in a box". It's an infinitely deep square well and has zero potential energy in the region $0 < x < a$, and infinitely high potential barrier preventing to flow above the region. The width 'a 'is the full width and is sometimes studied as half width. The Schrodinger equation of motion of particle is same for the particle inside the well as there is no potential energy.

The three solutions for:

$$E = h^2K^2/2m$$

Are Exp(ikx), exp(–ikx), Sinkx or Coskx E is +ve for k is imaginary and waves are real exponential. E is –ve for Hyperbolic function.

The wave functions and energies are given by

$_{-n}$ (x) = A_n Sin (n_x)/a (4)

_n = _o (kn)

= $\_^2K^2$ n/2m = $\_^2n^2$ $\_^2$/2m a^2

The integer n = 1, 2, 3. ..... is a quantum number indicates the energy states in the well.

Quantum wells and barriers are the layered structures for the heterostructure devices for the electrons in the conduction band. The quantum well GaAs is sandwiched between two thick layers of AlGaAs and the well exists for conduction and valence bonds for the heterojunction devices. The conduction band has its complications for the Alx Ga1-x Asx with x > 0.45. The tunneling barrier is provided by a layer of AlGaAs surrounded by GaAs. Viewing classically electron cannot cross the barrier unless it has sufficient kinetic energy, but in quantum mechanics the transmission coefficient increases rapidly with the energy in the tunneling area. Electron selects high energy states to cross the tunneling area. The opposite of the barrier is a quantum well in which a thin layer of GaAs sandwiched between two thick layers of AlGaAs.

Resonant Tunneling

Quantum well with thin barriers where an electron can "leak out" arising to "Quasibound or Resonant State" instead of three bound state. This double barrier structure provides the active region of Resonant -Tunneling Diode. Let us consider electrons that impinge on the barrier from the left of energy "E". Their probability of crossing the barrier is given by the transmission coefficient T. The transmission coefficient is defined as the product of transmission coefficients of the two individual barriers (i.e Left and right barriers TL and TR) respectively. T ~ TL * TR, TL, TR are small for typical barriers. For energies near resonance T value increases to higher than TL and TR. Resonant tunneling diodes are studied using this concept at extremely high frequencies.

From the following figure IV–F.2 it is inferred that for a wave to propagate in three regions for which potential energy is constant in each region separated by two barriers. The T-matrices are multiplied to define the transmission amplitudes. T(d) matrix is defined from the origin T(0) in three steps.

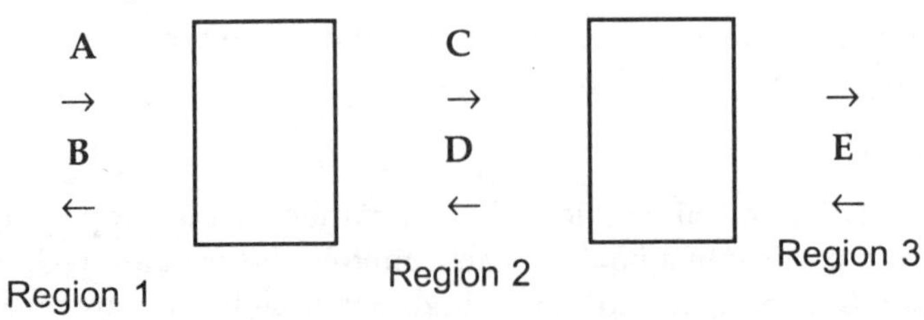

Fig. IV–F.2: Wave in three regions defines transmission coefficients.

Step I: Lets imagine the electron is moved from "d" to the origin given by Z′ = Z–d. The incoming wave on the left, exp(ik1 Z), becomes exp(ik1 Z′) exp(i k1d); the amplitude of the outgoing wave from left is multiplied by exp(-i k1d).

Step II: The T-matrix at the origin is calculated given by T(0) to calculate the amplitude at the right.

Step III: The wave or object is restored i.e. reflected back to its position 'd' Z = Z + d which introduces phase factors and the wave numbers k1 to k2.

T_d matrix for the wave or object at the origin is given by

$$T_d = \begin{pmatrix} e^{-ik1d} & 0 \\ 0 & e^{-ik1d} \end{pmatrix} T(0) \begin{pmatrix} e^{-ik2d} & 0 \\ 0 & e^{-ik2d} \end{pmatrix}$$

The barrier on the left with transmission and reflection amplitudes is given by matrix

$$TL = \begin{pmatrix} 1/t_L^* & r_L/t_L \\ r_L^*/t_L^* & 1/t_L \end{pmatrix}$$

Similarly defining TR for the right barrier with reflection and transmission amplitudes is

$$TR = \begin{pmatrix} 1/t_R^* & -r_R^*/t_R^* \\ -r_R^*/t_R^* & 1/t_R \end{pmatrix}$$

The quantum well shifts the left barrier by –1/2* a and the right barrier to1/2 *a from 'T'matrix

$$T = \begin{pmatrix} e^{-ika/2} & 0 \\ 0 & e^{-ika/2} \end{pmatrix} \begin{pmatrix} 1/t_R^* & -r_R^*/t_R^* \\ -r_R^*/t_R^* & 1/t_R \end{pmatrix} \begin{pmatrix} e^{ika/2} & 0 \\ 0 & e^{-ika/2} \end{pmatrix}$$

$$X \begin{pmatrix} e^{ika/2} & 0 \\ 0 & e^{-ika/2} \end{pmatrix} \begin{pmatrix} 1/t_L^* & -r_L^*/t_L^* \\ -r_L^*/t^* & 1/t_L \end{pmatrix} \begin{pmatrix} e^{-ika/2} & 0 \\ 0 & e^{ika/2} \end{pmatrix}$$

$$= \begin{pmatrix} (1 - r_L^* r_R e^{-2ika})/t_L^* t_R^* & r_L e^{ika} - r_R^* e^{-ika})/t_L^* t_R^* \\ r_L^* e^{-ika} - r_R e^{ika})/t_L^* t_R^* & 1 - r_L r_R e^{2ika})/t_L t_R \end{pmatrix}$$

The transmission amplitude coefficient becomes

$$t = t_L t_R \; / \; (1 - r_L r_R e^{2ika})$$

The complex reflection amplitudes is written in polar form

$r_L = |r_L| \exp(i\_L)$

The flux transmission coefficient becomes

$$T = |t|^2$$

$$= \frac{T_L T_R}{(1+R_L R_R - 2(R_L R_R)^{1/2} \text{Cos}(2ka+\_+\_R))}$$

$$= \frac{T_L T_R}{(1-(R_L R_R)^{1/2})^2 + 4(R_L R_R)^{1/2} \sin^2 1/2\_)}$$

where the phase _ = 2Ka + _L + _R Resonance is due to the change in phase of the wave between the barriers 2Ka, 'T' has peaks when sine in the denominator vanishes _ = 2n_

$$T = T_{pk}$$

$$= \frac{T_L T_R}{(1-R_L R_R)^{1/2})^2}$$

$$T \approx \frac{4\, T_L T_R}{(T_L+T_R)^2}$$

The transmission coefficient is function of energy given as where the resonance is centered on Epk. This equation is called Lorentzian approximation.

$$T(E) \approx [T_{pk}\, [1 + ((E - E_{pk})/1/2\_)^2]^{-1}.$$

How does Resonant Tunneling take Place through Quantum Wells?

For our study of resonant tunneling devices we consider a resonant tunneling diode is shown below. The sequential layer structure of the RT Diode is shown below in Fig IV–F.3. The thickness of the barriers, wells and spacers can be varied as desired. The layer structure of GaAs/AlGaAs with two tunnel barriers of AlGaAs sandwiched by a quantum well of GaAs. The quantum well is thin enough for the bottom of conduction band to reach the quantum well in order to form 1 or more 2D energy levels in the quantum level. The barriers AlGaAs are thin enough electrons are capable of crossing the barrier through the quantum well and current flows between the heavily doped contacts. The well consist of numerous quantized states these form the quasi-bound states, as there is minimum but finite probability with which the electrons can tunnel through the well. This tunneling is key aspect for the finite lifetime of the electrons on these levels leading to some broadening of the quantum wells. The resonant tunneling diode is a system with two contacts containing three dimensional electrons and quantum well containing two dimensional electrons (if any). These subsystems are weakly coupled through tunneling. The Energy Band diagram of this RTD device can be explained under three voltage bias conditions.

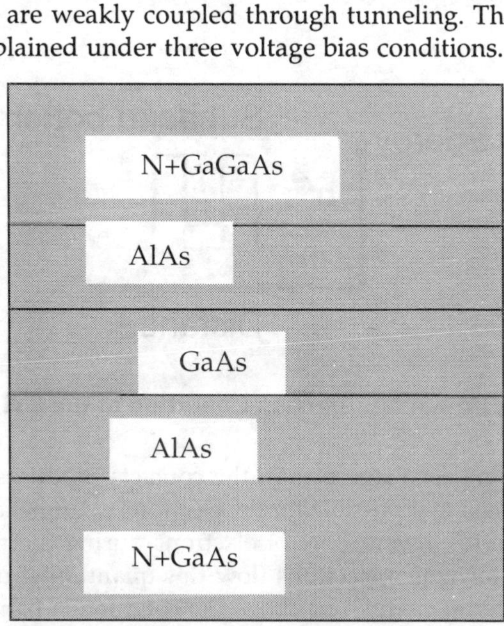

Fig. IV–F.3: Layer Design for double barrier Resonant tunneling structures.

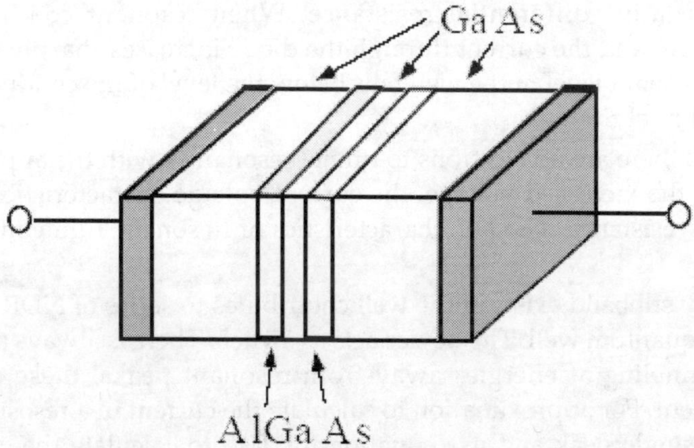

Fig. IV–F.4: Resonant Tunneling Diode.

From The energy band diagram of Resonant Tunneling Diode shown below in Fig. IV–F.6 following conditions can be studied

Condition 1: The equilibrium condition exists for the structure when no voltage is applied, but in the well there exists a single level of energy e1 which is lower of the bottom of the two-dimensional sub-band.

The parameters of the diode are chosen in such a way even in the nonbiased state energy e1 exists, which lies above the Fermi–energy level. The equilibrium condition for RTD is shown below in Fig. IV–F.5.

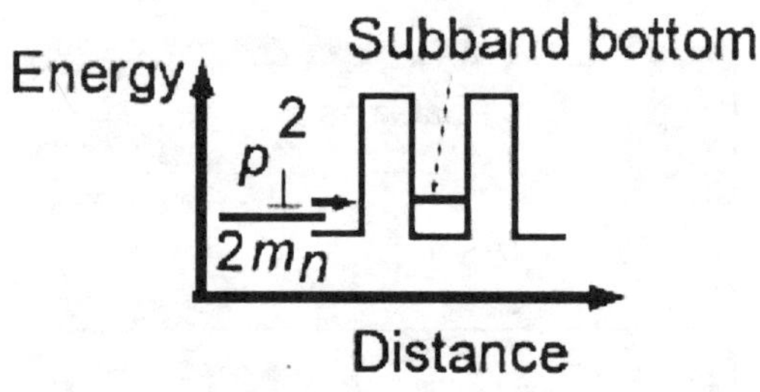

Fig IV–F.5: Equilibrium Condition of the RTD.

Condition 2: By applying a voltage bias in the contacts produces a downward shift of the level in the quantum level making it possible for the electrons to tunnel through the double barrier. The structure is designed in to prevent flow of electrons during thermal agitation process over the barriers. The condition for which electrons flow this quantum structure is when the quasi – bound level lies below the fermi energy but above the conduction band bottoms of the contacts. For this case the electrons from the emitter (left contact) whose kinetic energy of the perpendicular motion is given by E_ = _2 k2/2 m* coincides with the e1 are transmitted through the structure with finite probability. This is called resonant tunneling process which forms an important case study exhibiting negative differential resistance. When resonant condition is reached the transmission coefficient and the current through the diode increases sharply as the voltage bias is increased, and the resonant level in the well falls below the level of the conduction band and there are no electrons to tunnel through.

Condition 3: As there are no electrons to tunnel resonantly with the applied bias, the current decreases despite of the increased voltage, the current voltage characteristics is identified by the negative differential resistance. The I–V characteristics of Resonant Tunneling Diode is shown in Fig. IV–F.7.

If more than one subband exists in the well contributes to series of NDR peaks as it passes to each subband in the quantum well. The device acts as switch. There is always a background current due to electrons tunneling at energies away from resonant peaks, these peaks dominate the transmission coefficient. For approximation to calculate the current in a resosnant tunneling diode we consider only a single peak and use equation for T(E) to calculate the current. The effect of bias on a resonant tunneling structure across the diode is small, Epk is above sea of the incoming electrons and current flows. As the applied bias increases this decreases the energy of the resonant level, so the sea of electrons on the left can pass through it and large current flows. In three dimensions the current increases linearly as Epeak approaches the bottom of the electrons on the left as as shown in Energy band diagram of Fig IV–F.6(c). A further increase of the bias decreases the energy of the resonant level down and no longer available to electrons and the current decreases abruptly. The result is a current–voltage characteristics that shows Differential Conductivity is used for the application of amplifier or oscillator.

$I \approx 2e / h \int T(E)\, dE$

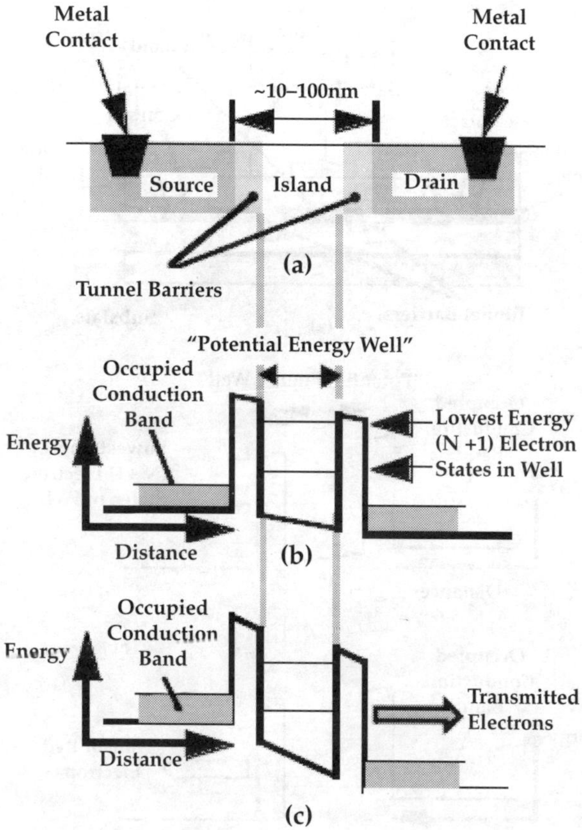

Fig. IV–F.6: Energy Band diagram of Resonant Tunneling Diode.

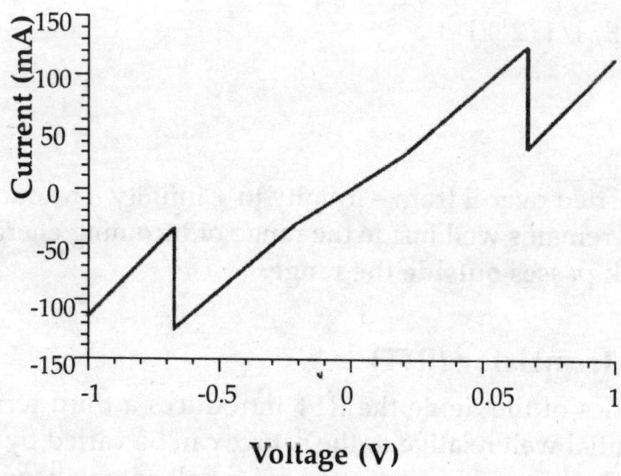

After Capasso et. al. (1990)

Fig IV–F.7: I-V characteristics of Resonant Tunneling Diode.

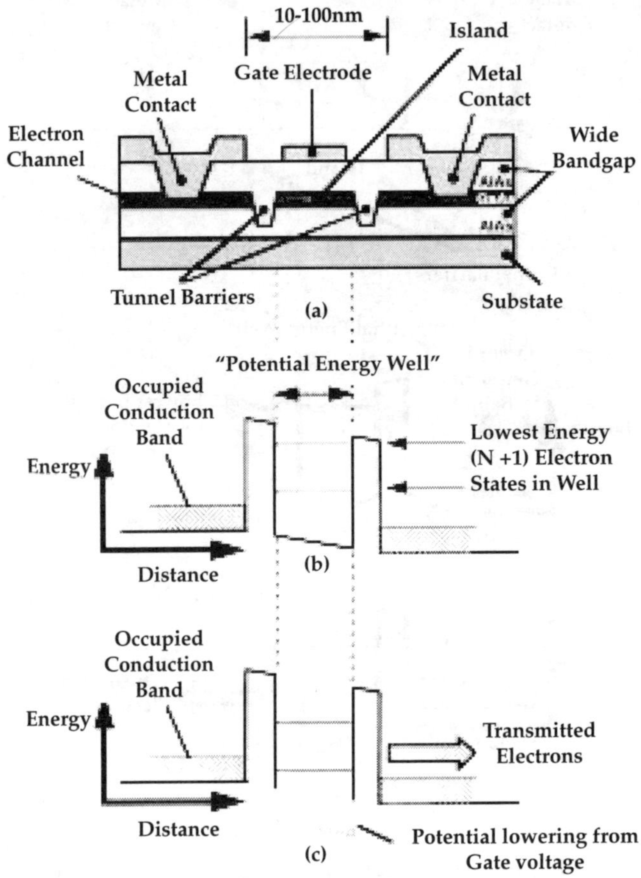

Fig. IV–F.8: RTT device.

$$\approx 2e / h \int [1+((E - E_{pk})/1/2\_)^2]^{-1} d$$

$$I = \frac{2e \_/2T_{pk}}{h}$$

The integration is carried over E from – infinity to + infinity. The current remains constant at this value while the peak remains well inside the range of incoming energies and falls to zer over a width of about _ as **Epk passes outside the range.**

Resonant Tunneling Transistor (RTT)

Based on the characteristics of the diode the RTT introduces a third terminal called GATE. The energy levels in the potential well relative to the source can be varied by applying voltage on the third (gate) terminal rather than voltage on source. A small gate voltage controls the flow of the current through the device. The barriers in the RTT create potential around the well or island and prevent charge from flowing through the device, even it is under a voltage bias. Potential well is configured in the same way as RTD, until the gate voltage is charged by the voltage bias the energy levels of the well are lowered for all the states and bring them into resonance with the mobile electrons in the conduction band in the source, so that an electron current can be transmitted through the device. The schematic diagram of RTT is shown below in Fig. IV–F.9 (a) Lateral type of RTT constructed by Rand et al. (b) Energy band diagram with potential well and barriers obstruct the flow of current (c) The energy levels in the well are lowered by the applied gate voltage.

Comparison on Tunneling Devices

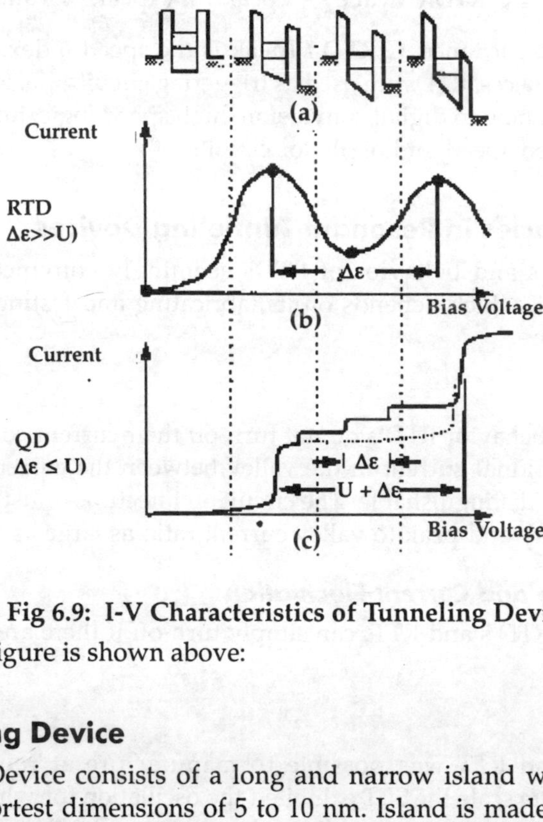

Fig 6.9: I-V Characteristics of Tunneling Devices.

The comparison figure is shown above:

Resonant Tunneling Device

Resonant Tunneling Device consists of a long and narrow island which is a quantum wire or quantum dot with shortest dimensions of 5 to 10 nm. Island is made of semiconductor material with number of mobile electrons. The short dimensions of the island makes $\Delta\varepsilon$ large, large dimensions keep U small, i.e $\Delta\varepsilon \gg U$. The energies of electrons in the quantum well are defined by $\Delta\varepsilon$ as "U" is negligible term in calculating the total energy $U + \Delta\varepsilon$ for the addition of an electron in the island. Thus when a resonant tunneling device is subject to a voltage bias between source and drain produces a current and voltage plot. The distance between current peaks is determined by $\Delta\varepsilon$ and little effect of charging energy.

Advantages of Resonant Tunnelling Devices

The RTDs and RTTs employ quantization effects and tunneling property which is used for best amplification and multiple on and off states associated with multiple discrete quantum levels inside the potential well on a very small well or narrow islands. If there is a second energy level in the quantum well, then the device can switch on again as the bias voltage of the RTD (or the gate voltage of an RTT) is increased still further. This multi-state switching behavior permits each device to "count higher" and represent more logic states. The RTD is expected to operate up to terahertz (1000 GHz). Brown and coworkers fabricated microwave oscillators operating around 712 GHz. The maximum power generated for a microwave oscillator is given by:

$$1/8\ (\text{Vvalley} - \text{Vpeak})\ (\text{Ipeak} - \text{I valley})$$

Diamond et al proposed the switching characteristics of RTD given by:

$$t = \text{C RTD}\ (\text{Vvalley} - \text{Vpeak})\ /\ (\text{Ipeak} - \text{I valley})$$

where C RTD is the RTD capacitance, **C RTD / Ipeak** is the speed index. Switching times is few picoseconds with current devices. RTDs are used as triggering circuits, pulse forming and sampling, low-power memory cell, Analog to digital conversion, high speed logic, fuzzy logic which reduces circuit complexity, improved speed and oscillator circuits.

Drawbacks and Obstacles in Resonant Tunneling Devices

Though the characteristics and behavior of RTDs seemingly convinced but there are some challenges that are focused, which depends on its fabricating and testing of these nanoelctronic devices. They drawbacks are

Valley Current

Multistate quantum effect behavior RTD's do not turn-off their current completely when they are offresonance. There is a residual current in the valley between the current peaks and leading to the on and off states barely distinguishable. The circuit architectures must be designed keeping in mind its potential sensitivity and peak to valley current ratio as large as possible.

Sensitivity to i/p Voltage and Current Fluctuation

Launder pointed out that RTD's and RTTs can simply turn-off if there are any fluctuations in the input voltage.

Cryogenic Operation

The operation of RTD and RTT was possible to manufacture at room temperature but at temperatures below absolute scale the SET exhibited the oscillation for which the random thermal motion often provide electrons with small additional energy to get to the island, this could increase the relevant energy separation U for states in the island, resulting switching off instead of switching on. This limit is only possible in molecular electrons.

Materials III–V semiconductors

Device fabrication is concentrated on III–V semiconductors because of its excellent purity and easy manufacturing. But these materials contaminated easily when they are exposed to mobile charge carriers leads to instability of the device, also they cannot withstand electric field . While silicon, SiO_2 is amorphous crystalline is the best insulating material can acts as barrier as little as 5nm can be grown uniformly and prevents electrons to cross the barrier. The new trend is to make use of Si/Ge in the heterostructure devices particularly sandwiches of Si/Ge.

Background Charge Problems

The random background charges tend to accumulate in semiconductor for a small operating quantum- effect or single electron device. This can be overcome by the use of Silicon-On-Insulator (SOI) technology. SOI could be used in molecular electronics for its relatively non-conductive, nonpolarizable organic compound.

Extreme Sensitivity

The exponential sensitivity of the tunneling current to width of potential barrier device operability depends on the nanoscale barriers which are made with extreme precision and uniform width.

Extreme Difficulty of Making Barriers and Islands

Sufficient reliability and precision, predictable behavior of large number of devices is very difficult to achieve on a several-nanometer scale in solids. Even though it is possible in hetero-structure Devices such as RTDFETs accuracy is achieved in one of the three dimensions.

Conclusion

Rapid scaling on microelectronic devices, employing quantum effects in nanoscale and using high resolution techniques for fabrication devices has put some challenges that to be worked on for its efficient functions. For solid state nanoelectronics, one of the most important challenges is to be able to produce reliably and uniformly in silicon the characteristic nanometer-scale features required for nanoelectronics: nanometer-scale islands, barriers, and "heterojunctions" between islands and barriers. Up to now, solid state nanoelectronic devices largely have been fabricated in III/V semiconductor compounds such as gallium arsenide and aluminum arsenide. It is believed, however, that silicon nanoelectronics would go a long way toward permitting the inexpensive mass manufacture of nanoelectronic devices.

Unit IV–G

Resonant Tunneling Device Logic Circuits

Current State of NanoScale Circuit Design

During the past decades the scaling of electronic devices and the development of data-processing systems such as microprocessors, memories and more recently complete systems on chip are the driving forces behind the progress in semiconductor industry. Today, the combination of sub-μm lithography and heterostructure growth techniques with atomic layer precision enables the fabrication of devices so small that quantum mechanical effects are observable.

Among solid-state electronics the most promising devices are resonant tunneling and single electron devices. Energy quantization due to coulombic repulsion of single charges and quantum confinement are their underlying physical phenomena to control the charge transport through this nanoscale structures. Both device categories have in common a current voltage characteristics with several, in the case of single electron devices periodical, negative-differential resistance (NDR) regions. Handling these NDR characteristics and exploiting them in a logic family or memory cell is a widely accepted approach to nanoscale circuits, especially as a method to improve the computational functionality of basic circuit components.

From the technological point of view resonant tunneling devices are presently the most mature type of quantum effect device because their interfacing with conventional transistors has reached an advanced level and even in silicon room temperature operation of NDR-devices is achievable. Besides the high-speed operation in the GHz regime the combination of the NDR with electronic amplification is attractive to reduce circuit complexity, that is the number of devices and the wiring. This has motivated the development of different three terminal devices. Significant examples are resonant hot electron transistors, gated resonant tunneling diodes and two dimensional electron gas tunneling field effect transistors. Here a monolithic integration of a resonant tunneling diode (RTD) and a heterostructure field effect transistor (HFET) is used. This approach has proven to be large-scale integration capable on the InP material system and is the basis for different resonant tunneling device circuits investigated by industrial as well as academic research groups.

In addition to the technologically oriented research a further prerequisite for nano-scale integration is the investigation of suitable logic families, architectures and the development of a design framework. Recapitulating some milestones, the first approach to build a logic circuit with RTDs is a 1-bit full adder based on XNOR gates proposed by Capasso in 1989. By taking advantage from the multistate behavior of the RTD the circuit complexity of the adder is reduced. Other applications following this idea are different kinds of multiple-valued logic gates as well as resonant tunneling diode memory cells and analog-digital mixed circuits.

For large scale integration of nano-scale devices an essential aspect is the question if these functionally integrated circuits could be combined with existing VLSI algorithms for bit-level computations. Independent of the specific technology and the operating principles of the different categories of nano-scale devices, following aspects have to be considered:

- The use of quantum effects to reduce the logic depth of a circuit.
- A flexible and intuitive methodology to transfer a give logic functionality into a circuit design. The impact of fabrication tolerances and physics related parameter variations on the circuit functionality and error rate.
- The capability to fabricate logic and memory circuit components with one technology.
- A regular layout with a small number of different circuit modules to enable the a library based design.
- Local interconnections on the circuit and the system level to solve the wiring problem.
- The impact of increasing clocking frequencies on the circuit design indicated by a continuing decrease of gate delays per clock cycle (Fig. IV–G.1) which limits the number of logic circuit stages between two synchronizing registers.

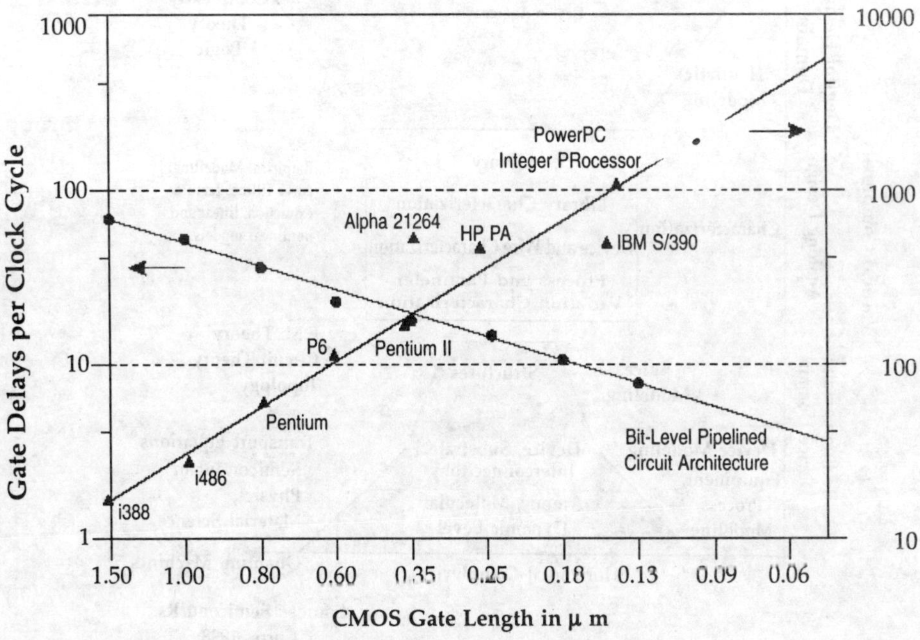

Fig. IV–G.1: Impact of increasing clocking frequencies on the gate delays per clock cycle for scaled CMOS microprocessors.

- ➢ Additional circuit overhead due to clocking and temporal data storage in latches.
- ➢ The implementation of concurrent VLSI algorithms to achieve a low latency and a high data throughput for basic low-level operations such as addition, multiplication, and accumulation.
- ➢ The enormous design complexity ranging from device and technology to the system level. In this context sever problems arouse from the pure increase of the number of gates, devices, and wires (horizontal complexity) as well as from themore complicated modeling of these components (vertical complexity) (Fig. IV–G.2).

Even today most of these principles are a substantial part of modern CMOS-VLSI design. Thus, a potential impact of solid-state nano-scale devices on future information processing will strongly depend on these aspects and on an interdisciplinary approach of device physics, semiconductor technology, and circuit design.

RTD-HFET Boolean Logic

Monostable-Bistable Logic Transition Element

For clocked logic families various high speed circuits based on the monostable-bistable transition

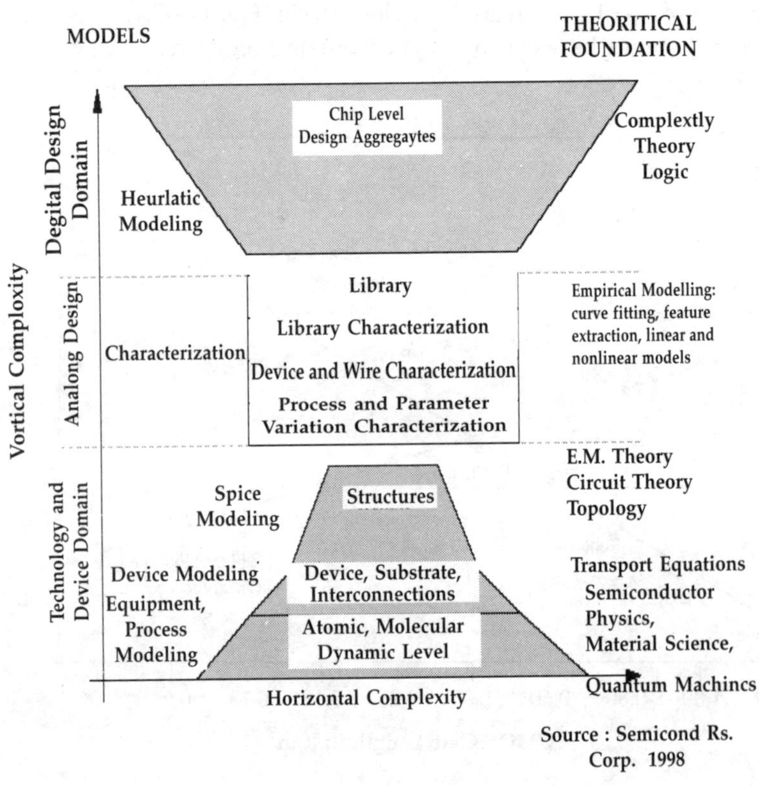

Fig. IV–G.2: Vertical and horizontal design complexity of future integrated circuits.

logic element (MOBILE) have been proposed and operated at clocking frequencies of several GHz. The MOBILE is a rising edge triggered, current controlled gate. It consists of two RTDs which are switched into a monostable or bistable state depending on an oscillating bias voltage VCLK.

In the sense of a pulsed power supply the bias voltage also serves as clock (Fig. IV–G.3) to synchronize the gates. The dynamic behavior of a MOBILE circuit is expressed by a time-variant first order differential equation:

$$C_L \frac{dV_{out}}{dt} = I_{D2}\,(V_{CLK}\,(t) - V_{OUT}) - I_{D1}\,(V_{OUT}) = \Delta I$$

$$V_{CLK}\,(t) = V_{CLK}\,(t + T_{CLK})$$

Here, C_L is the total load capacitance including the RTD-capacitance and the gate source capacitances of subsequent stages. The oscillating bias voltage V_{CLK} is a trapezoid waveform with clock period and an amplitude larger than twice the RTD peak voltage: $V_{CLK}^{max} > 2Vp$. The specific logic functionality of a MOBILE is determined by embedding an input stage which causes the input current _I. This input stage is composed of RTD/Schottky diodes or HFETs. In this work an RTD-HFET was used and is added in parallel to the driver RTD D1 so that the input current is $\Delta I = -I\,(T_1)$ and an inverter function is obtained. Due to the pulsed power supply V_{CLK} (t) the output of the gate is either monostable, in a metastable transition state, or bistable:

$$V_{OUT} = \begin{cases} \text{monostable for } V_{CLK}\,(t) < 2\,V_P \\ \text{metastable for } V_{CLK}\,(t) \approx 2\,V_P \\ \text{bistable for } V_{CLK}\,(t) > 2\,V_p \end{cases}$$

The bistable configuration occurs if V_{CLK} exceeds twice the peak voltage and results in two self stabilizing digital output states. In Fig. IV–G.3 the two logic states V_L and V_H appear at a bias voltage of $V_{CLK} > 0.55$ V where the central equilibrium point of the circuit becomes unstable.

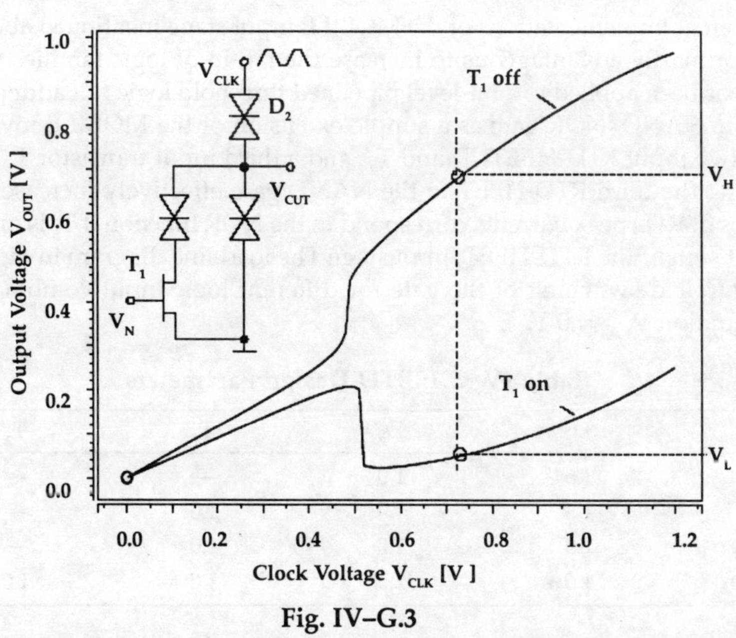

Fig. IV–G.3

The basic idea of implementing logic gates by means of such a monostable-bistable transition in a symmetric configuration of two tunneling diodes has been investigated first by Goto in 1960. Since field effect transistors were not available at that time to isolate inputs and outputs these circuits have not played an important role in microelectronics. Today, their comeback in nanoelectronics is primarily due to the technological progress in the field of quantum effect devices and themore profound understanding of nonlinear dynamic phenomena.

Optimization of RTD Device Parameters

In the previous section it was shown that the monostable-bistable behavior of MOBILE circuits strongly depends on the RTD peak voltage. Accordingly it is also to expect that the shape of the resonance and the exponentially increasing thermionic current will influence the equilibrium points of the circuit. Fig. IV–G.4 shows the stable (solid lines) and unstable equilibrium points (dashed lines) of three MOBILEs composed of RTDs with different I-V characteristics. As it is depicted in Fig. IV-G.4b a reduced peak to valley ratio significantly decreases the output voltage swing.

In contrast to this the RTD in Fig. IV–G.4c has small peak voltage of $V_P = 0.2$ V, an non-symmetric resonance peak and a large valley region due to the slowly increasing thermionic current. As a result a distinct tristable region appears for clock voltages of $0.7\text{ V} < V_{CLK} < 1.25$ V. In digital logic circuits especially the central stable state in the tristable region is a potential error source if the amplitude of the clock voltage exceeds 0.7 V and the output remains at $V_{OUT} = V_{CLK}/2 = 2$ instead of switching to the logic high level. For the first RTD in Fig. IV–G.4a this tristable region is less pronounced. These investigations illustrate that the use of RTDs in MOBILE circuits requires a careful optimization of the device parameters and a precise control of the double barrier heterostructure. Since the RTD depicted in Fig. IV–G.4a has the large bistable region this device has been used as basis for more complex Boolean logic and threshold logic gates.

Programmable NAND/NOR Gate with RTD-HFET Input Stage

Among the different implementations of the MOBILE input stage mentioned above, the RTD-HFET input stage is primarily advantageous to increase the fan-in of logic families withmultiple input terminals and has been applied to a bit-level pipelined threshold logic full adder. Fig. IV–G.5 shows a programmable NAND/NOR gate as a simple extension of the MOBILE inverter. The NAND/ NOR gate has two input RTD-HFETs T_1 and T_{22} and a third input transistor T_3 to modify the logic function. Adding the third RTD-HFET to the NAND gate effectively increases the input current so that the ratios of RTD peak currents correspond to the NOR function if T_3 is on. The input current _I is the current sum of the RTD-HFET input stage The load line diagram in Fig. IV–G.6 illustrates the current controlled switching of the gate for different logic input combinations (X; Y) in the NAND configuration $V_{PRO} = 0$ V.

Table IV–G.1: RTD Design Parameters λ

Logic Gate	D_1	D_2	T_1	T_2	T_3
NOT 1.0	1.5	1.0	—	—	
NOR 1.0	1.5	1.0	1.0	—	
NAND 1.0	2.5	1.0	1.0	—	
NAND/NOR	1.0	2.5	1.0	1.0	1.0

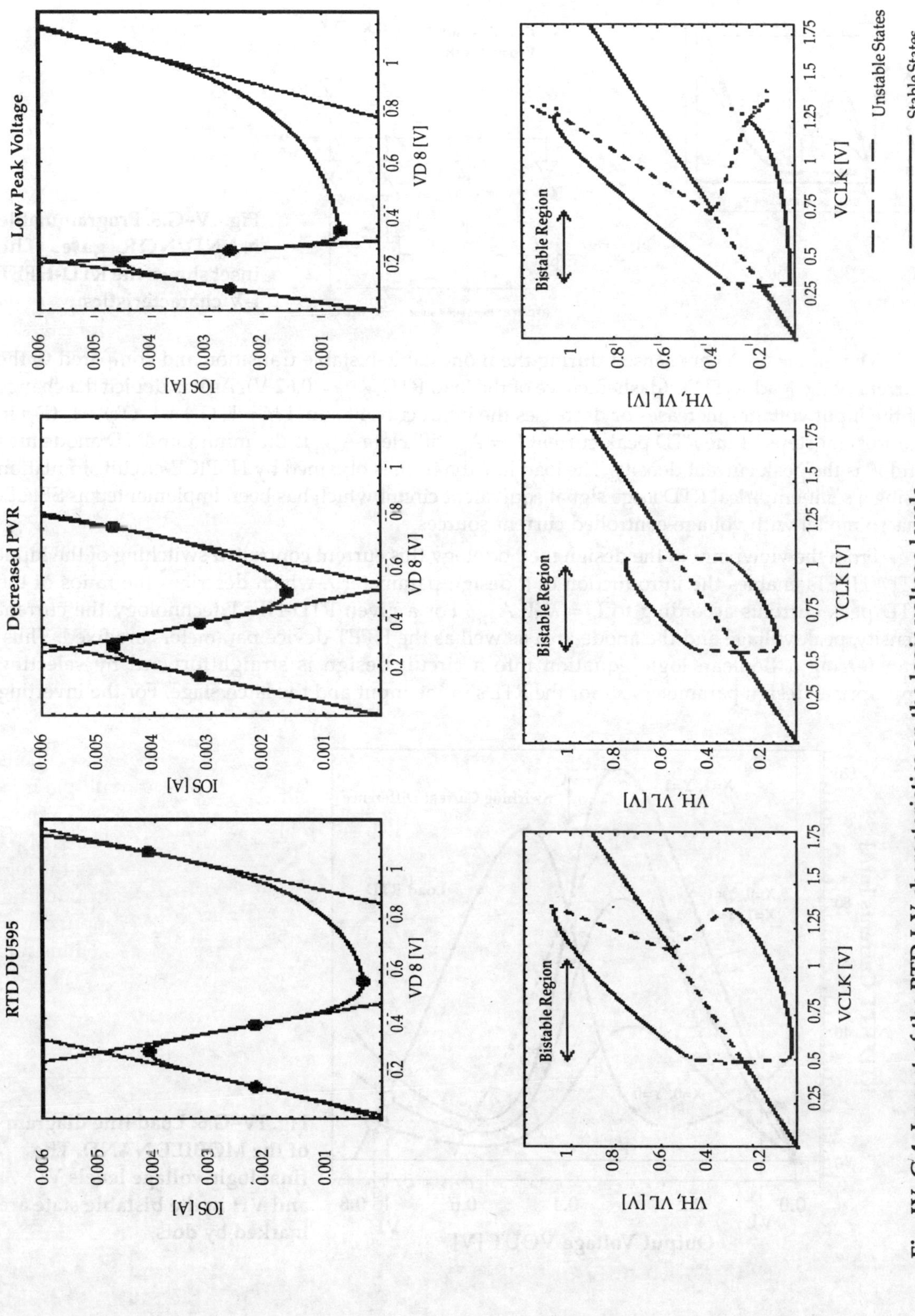

Fig. IV–G.4: Impact of the RTD I-V characteristics on the logic voltage levels.

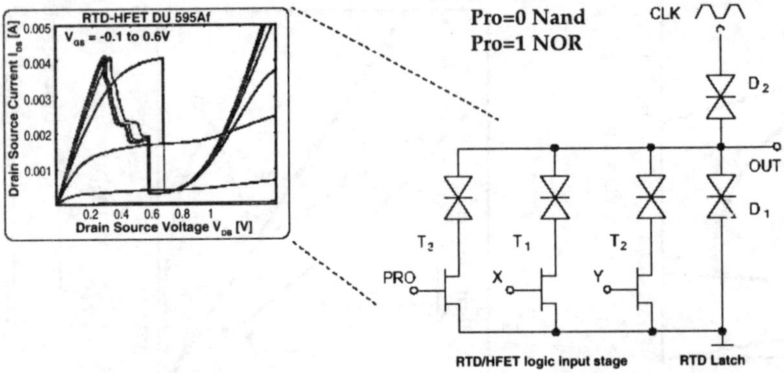

Fig. IV–G.5: Programmable NAND/NOR gate. The inset shows the RTD-HFET I-V characteristics.

The inputs (X, Y) are sensed during the monostable-bistable transition and compared to the current of the load RTD D_2 (dashed curve of the load RTD $V_{CLK} = 0.62$ V). As it is depicted a change of the input voltages increases or decreases the input current sum $I_P^{tot} = I_P(T_1) + I_P(T_2) + I_P(D_1)$ in equidistant steps of the RTD peak current $I_P = A_{RTD}$ jP. Here A_{RTD} is the minimum RTD anode area and jP is the peak current density. The load line diagram is obtained by HSPICE-circuit simulation using a semi-empirical RTD large signal equivalent circuit which has been implemented as SPICE-macro model with voltage-controlled current sources.

From the viewpoint of the design methodology this current controlled switching of the input RTD/HFETs enables the introduction of a design parameter λ which describes the ratios of the RTD peak currents according to $I_P = \lambda$ jP A_{RTD}. For a given RTD-HFET technology the current density, peak voltage and the anode area as well as the HFET device parameters are fixed. Thus, transferring a Boolean logic equation into a circuit design is straightforward by selecting appropriate design parameters λ_ for the RTDs in the input and the latch stage. For the inverting

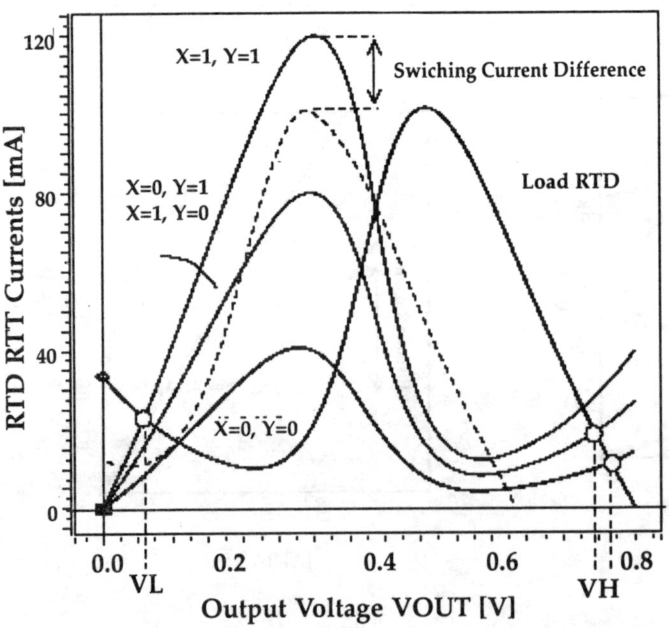

Fig. IV–G.6: Load line diagram of the MOBILE NAND. The final logic voltage levels VL and VH in the bistable state are marked by dots.

logic family this is illustrated in Table IV–G.1. Here, the only difference between the NAND and NOR function is the increased anode area of load RTD D_2. Compared to previous RTD logic families with different devices in the input stage and the RTD latch this is a simple methodology to scale the circuits.

RTD-HFET Technology and Experimental Results

The RTD-HFET is amonolithically integrated, 3-dimensional InP based heterostructure where the double barrier quantum well of the RTD is grown on the HFET drain contact layer. The resulting I-Voutput characteristics exhibits a gate controllable NDR-like behavior if the HFET channel current I_{DS} is larger than the RTD peak current I_P (inset Fig. IV–G.7, W=28.0 μm, L=1.0 μm, A_{RTD} = 10 μm²).

In this case the RTD acts as current limiting device and the peak current is independent of the gate voltage if a logic high level of $V_H > 0.3$ V is applied. For gate voltages in the region of the threshold voltage the device shows the usual amplifying HFET characteristics.

The RTD structure is composed of an n-InGaAs cathode layer, an i-InGaAs spacer and an InGaAs quantum well which is enclose by two AlAs-barriers. At the boundary between the RTD an HFET structure an InAlAs etch stop is included. The HFET has alloyed Ni/Ge/Au contacts. Between the InP substrate and the 20 nm InGaAs channel a 115nm InAlAs buffer is inserted. The Schottky barrier of the HFET is made of a Pt/Ti/Pt/Au gate, a 22nm InAlAs/AlAs barrier, and a 4nm InAlAs spacer. Since the logic swing of MOBILE circuits is $\Delta V_{OUT} = V_H - V_L$ the layer structure has been optimized to obtain a low peak voltage of V_P = 0.27 V so that gate currents due to HFET Schottky gate ($V_H < V_{GS,max} \approx 0.8$ V) are avoided for logic high input voltages $V_H \approx 0.7$ V.

Recent investigations of the homogeneity have shown that a RTD peak voltage and current variation of 16 mV and 0.41 μA over a quarter 2" wafer are achievable without any optimization. These values are comparable to standard deviations reached with other technologies.

A technological advantage of the RTD-HFET input stage are the relaxed precision requirements of the HFET parameters. In the original version of this RTD-based LTG proposed by Yamamoto et al. at NTT the input terminals are single HFETs with different transistor widths. Here, the HFET acts as a switch. Consequently, the gates are more robust against device parameter variations than in the original MOBILE where theHFET input stage has to provide an exactly defined amount of current at a distinct bias voltage.

Based on this RTD-HFET technology and the design methodology discussed above a NAND/NOR gate with 3.0 μm² mode area and j_p = 9 kAcm^{-2} has been fabricated and tested (Figs. IV–G.7 and IV–G.8). The RTD HFET has a gate length of L = 1.0 μm and a gate width of W = 3 μm. Fig. IV–G.8(b) demonstrates that the input and output voltage levels are compatible. For a clock voltage of V_{CLK} = 0.69 V the logic swing ΔV_{OUT} = 0.65 V is sufficiently large to switch the RTD-HFETs in a following circuit stage. The voltage of the third input is V_{PRO} = 0 V (NAND function).

MOBILE Circuits for Scaled Devices

For laterally scaled gates with sub-_m gate lengths and RTD anode areas of A_{RTD} = 1.0 μm the circuit configuration is modified. First, the load RTD D_2 is substituted by an depletion type RTD-HFET. The clock V_{CLK} is now applied to the HFET gate and therefore a constant supply voltage of $V_{DD} = 0.8 \text{ V} > 2V_P$ is used instead of the difficult pulsed power supply scheme of the original MOBILE.

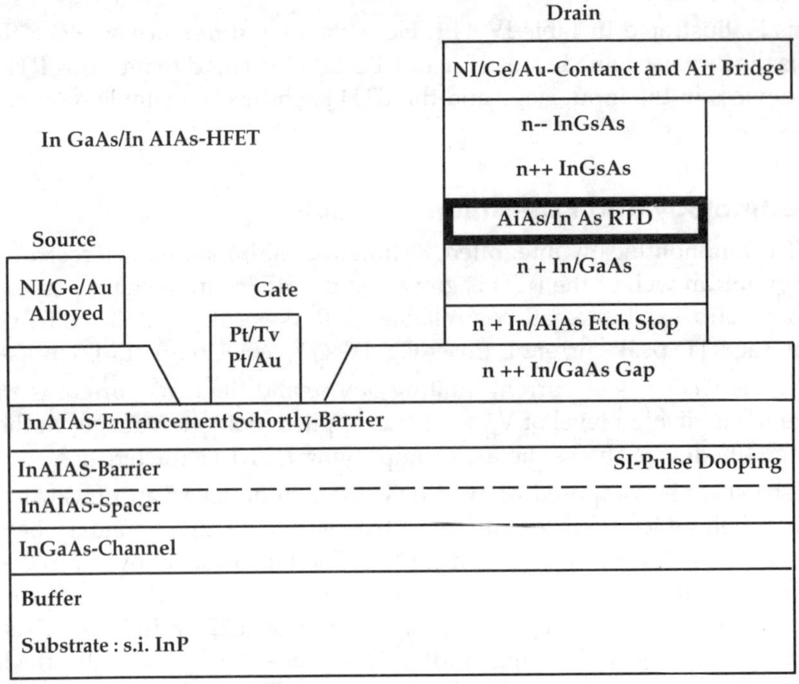

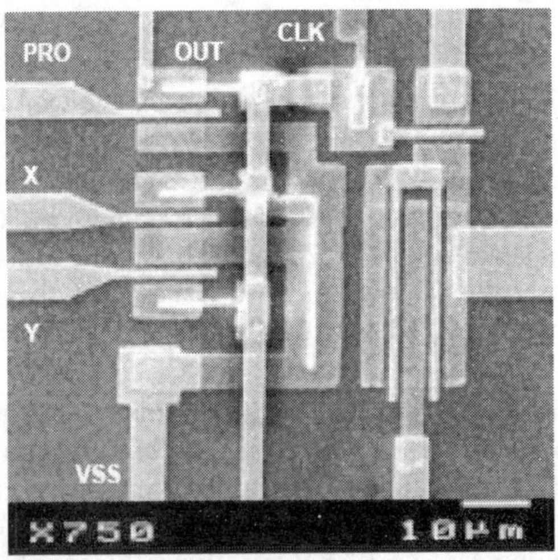

Fig. IV–G.7: Layer structure of an InP-based RTD-HFET combining an InGaAs/InAlAs HFET and an AlAs/InAs RTD and SEM picture of a NAND/NOR MOBILE with three self-aligned RTD-HFET inputs.

Fig. IV–G.9 shows the monostable-bistable transition of an inverter for different supply voltages from V_{DD} = 0.6 . . . 1.0 V. Compared to the RTD-HFET inverter with pulsed power supply here the voltage levels in the bistable region are more robust against variations of the clock voltage. Furthermore, by modifying the supply voltage the output voltage swing can be optimized independent of the clock signal amplitude. A further improvement is achieved by inserting a pull down HFET in the output branch which is controlled by the inverse clock V_{CLK}. The pull down

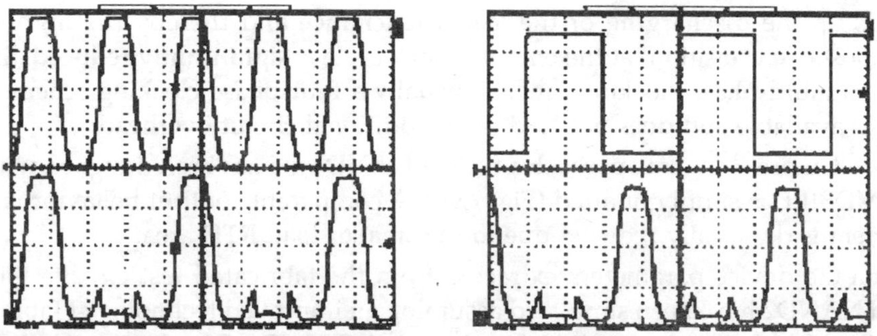

Fig. IV–G.8: Switching behaviour (a) and input/output voltage level compatibility (b) of the MOBILE NAND.

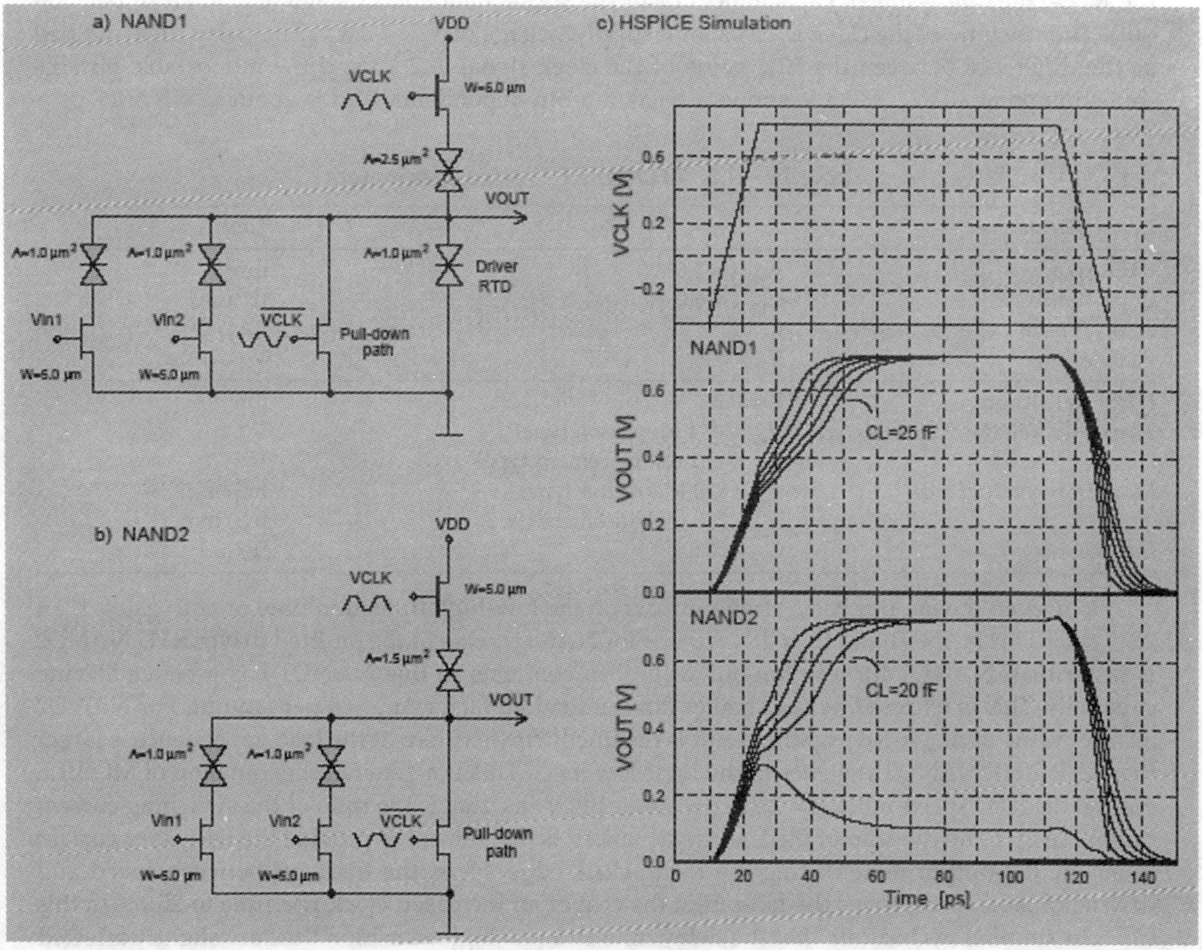

Fig. IV–G.9: Novel MOBILE NAND gates with active pull down device and depletion type clock HFETs (a, b) and driving capability for the critical logic combinations ($X = 1, Y = 0$) and ($X = 1, Y = 1$) (c) The load capacitance is varied from $C_L = 5 \ldots 25$ fF in steps of 5 fF.

HFET speeds up the discharging of the load capacitance and thereby the high-lowtransition (bistable-monostable transition) at the end of a clock cycle is significantly reduced. Since MOBILE gates require a multi-phase clocking scheme, usually a four phase clocking scheme, this inverse clock signal is available and does not lead to an additional circuit overhead.

In a second, more stringent approach a further NAND gate has been designed where the driver RTD of the MOBILE is simply omitted (Fig. 7.9b). A comparable output behavior is obtained but the total current is drastically reduced due to the smaller load RTD area.

Based on the device parameters extracted from the fabricated test circuits the novel gates NAND1 and NAND2 have been simulated assuming an improved technology (Table IV–G.2) with 200 nm gate length enhancement and depletion mode HFETs. Fig. 7.9c shows a HSPICE simulation of these scaled MOBILE gates. For both gates the driving capability at the critical logic input combination ($V_{in1} = 0$, $V_{in2} = 1$) and ($V_{in1} = 1$, $V_{in2} = 0$) has been investigated for load capacitances of $C_L = 5.0 \ldots 25$ fF, corresponding to fan-out values up to 5. The load capacitance C_L is the sum of the RTD-HFET gate-source capacitances of in the subsequent circuit stage ($C_{GS} = 4.8$ fF per fan-out). The rise time of the clock is 15 ps for a supply voltage of $V_{DD} = 0.8$ V. The delay time, defined as the difference between the 50% value of the clock signal (0:2 V) and the monostable-bistable transition point at $V_{OUT} = 0.3$ V, shows a weak fan-out dependency and is about $t_D = 3 \ldots 7$ ps.

Table IV–G.2: RTD-HFET Device Parameters

Device Parameters		*Unit*
RTD Area	$A_{RTD} = 1.0$	μm^2
Capacitance	$C_{RTD} = 6.0$	$fF/\mu m^2$
Peak Current	$I_P = 800$	μA
Peak Voltage	$V_P = 0.27$	V
HFET gate length	$L = 0.2$	μm
Threshold Voltage	$V_{t0} = -0.4$ (depletion type)	V
	$V_{t0} = -0.1$ (enhancement type)	V
Max. Transconductance	$g_m = 850$ (depletion type)	mS/mm
	$g_m = 780$ (enhancement type)	mS/mm
Capacitance	$C_{GS} = 4.8$	$fF/\mu m^2$

However, the fan-out has a strong impact on the low-high transition time of both gates: $t_{LH} = 23 \ldots 40$ ps for NAND1 and $t_{LH} = 19 \ldots 36$ ps for NAND2. Due to the omitted driver RTD NAND2 is faster than NAND1 for low fan-out values. In contrasts to this NAND1 has a better driving capability. This is reflected by the smaller fan-out loading factor $\Delta t_{LH} = 4$ per fan-out. For NAND2 a larger value of $\Delta t_{LH} = 5.8$ ps per fan-out is obtained. Furthermore, if the load capacitance is larger than $C_L = 20$ fF NAND2 switches to the logic low level. This is a general phenomenon of MOBILE gates if the RTD speed index $S = I_P/2C_L \approx 10 \ldots 100$ V/ns, that is the ratio of the switching current difference $I_P/2$ and the sum of the load capacitances, is insufficient to provide the necessary current to charge the output node during the rising clock edge. Here, the trade-off between speed and driving capability improves the fan-out at the cost of an increased clock rise time to 25 ps. In this case our simulations indicate that the delay time and low high transition time are slightly affected only.

Due to the pull-down HFET the high-low transition time t_{HL} is comparable to the clock fall time and less fan-out dependent than the low-high transition. Therefore, total clock periods of T =

140 . . . 150 ps and multi GHz operation will be possible with a sub-_m scaled RTD-HFET technology. Concerning the power dissipation NAND2 performs better because the static power dissipation as well as the switching current during the monostable-bistable transition are depending on the area of the load RTD (Table IV–G.3).

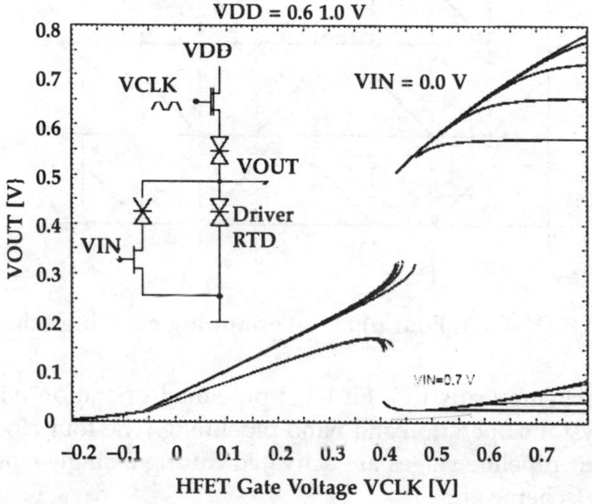

Fig. IV–G.10: Logic voltage levels of an inverter with clock HFET for different supply voltages V_{DD},

RTD-HFET Circuit Design

Clocking Scheme and Bit-Level Pipelined Operation

Before the design of larger threshold logic circuits is presented we will discuss the question of an appropriate clocking scheme for MOBILE-type gates. To operate cascaded threshold gates in circuits composed of several stages it has to be ensured that the evaluation of one stage only starts after the previous stage has finished the computation. A possible solution is a four phase overlapping clocking scheme with a phase delay between the clocks of two adjacent stages (Fig. IV–G.10).

Table IV–G.3: Comparison of NAND1 and NAND2

Circuit Parameters	*NAND1*	*NAND2*	*Units*
Delay time t_D	3 . . . 7	3 . . . 7	ps
Low high transition time t_{LH}	23 . . . 47	19 . . . 36	ps
Fan-out loading factor Δt_{LH}	4.0	5.8	ps/Fan out
Maximum load capacitance C_L^{max}	40	20	fF
Static power dissipation $P_{ST\,AT}$	0.3	0.2	mW
Switching power dissipation P_{SW}	1.5	0.85	mW
Dynamic power dissipation $P_{DY\,N}$ (C_L = ?0 fF)	0.15	0.14	mW

Since MOBILE circuits are self-latching as soon as they have reached the bistable state this

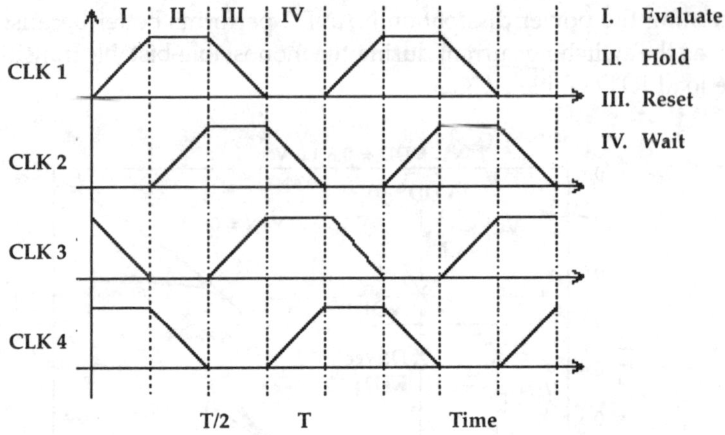

Fig. IV–G.11: Four phase overlapping clocking scheme.

four phase clocking scheme leads to a bit-level pipelined operation, also known as fine-grain pipelining, bit-level systolic operation and nano-pipelining. The four clock phases have an equal length T/4 so that four pipeline stages are activated during a single clock cycle T. The different phases of the clocking scheme are:

➢ Phase 1: $0 \leq t < T/4$: *Evaluation*

During the slowly rising edge of the clock the gate evaluates the output (transition from monostability to bistability) and charges the load capacitance to the logic high voltage V_H or logic low voltage V_L, respectively.

➢ Phase II: $T/4 \leq t < T/2$: Hold

After the evaluation phase has finished the gate holds the result as long as the supply voltage remains constant at $V_{CLK} = 3V_P$. During this hold phase the output is valid and the subsequent circuit stage starts with the evaluation phase.

➢ Phase III: $T/2 \leq t < 3/4T$: *Reset*

In the reset phase, that is the falling edge of the clock, the load capacitance is discharged and the gate returns to the initial, monostable state. Since this reset operation is a bistable monostable transition the switching process is symmetric for equal rise and fall times.

➢ Phase IV: $3/4T \leq t < T$: *Wait*

The clock cycle is finished by an inactive phase. During this phase the inputs of the gate are changed according to the results of the preceding pipeline stage.

As it is depicted in Fig. IV–G.11 the clock signals of the four phase clocking scheme have slowly rising and falling clock edges ($t_{rise} = t_{fall} = T/4$) which are frequently used in adiabatic circuit architectures. From the perspective of the power dissipation a result of such an adiabatic operation is a reduction of the dynamic switching energy by slowly charging and discharging the load capacitances. First estimations for MOBILE-type circuits consisting of laterally scaled resonant tunneling devices with minimum feature sizes below 200 nm have shown that the dynamic power dissipation is the major source of power dissipation. Compared with a non-adiabatic operation (t_{rise}, $t_{fall} << T/4$) the symmetric adiabatic clocking scheme decreases the total power dissipation by about one order of magnitude.

RTD-HFET Threshold Logic

To extend the computational capabilities of RTD logic gates the current controlled switching of the MOBILE in connection with the RTD-HFET input stage allows a functional implementation of linear threshold gates (LTGs). The characteristic feature of LTGs is the parallel processing of multiple inputs. A LTG calculates the weighted sum χ of the digital inputs x_k, $k = 1, \ldots, N$. The weighted sum is converted into a digital output y by comparing χ with a given threshold value θ (Fig. IV–G.12). If the weights w_k and the threshold value θ are selected properly the LTG computes any linear separable Boolean function of the N inputs. The output y of a threshold gate is given by

$$y(\chi) = \text{sign}(\chi - \theta) = \begin{cases} 1 \text{ if } \chi \geq \theta \\ 0 \text{ if } \chi > \theta \end{cases}$$

$$\chi = \sum_{k=1}^{N} w_k \,.\, x_k,\ x_k = \{0, 1\},\ w_k = \{0, \pm 1, \ldots, \pm w_{max}\},$$

$$\theta = \{0, \pm 1, \ldots, \pm w_{max}\}$$

Compared to a Boolean logic gate, a threshold gate combines an internal multiple-valued computation of the weighted sum with digital encoded input and output states. Actually, this capability of processing multiple input signals enables it to design circuits with bit-level parallelism and reduced complexity. Thus, our approach shows a certain similarity to multiple valued logic, but differs in regard to the digital input and output logic states. Although not explicitly mentioned the programmable NAND/NOR gate of the previous subsections is basically a LTG with two negative weighted inputs and a modifiable threshold value.

The circuit in Fig. IV–G.13 shows an RTD-based threshold gate composed of two serially connected RTDs and four RTD-HFETs with two positive weighted inputs (x_1, x_2) and two negative weighted inputs (x_3, x_4). The threshold value is implemented by modifying the RTD anode area of the driver RTD. According to the current controlled switching the weighted sum of a LTG is then given by the total input current at the output node according Kirchhoff's current law. If the threshold gate has M_p positive weighted and M_n negative weighted inputs, the total input current at the monostable-bistable transition point is:

$$\Delta I = \sum_{k=1}^{M_p} w_k I_P (V_{GSK}) - \sum_{k=1}^{M_n} w_k I_P (V_{GSK})$$

Here, $I_P (V_{GSK})$ is the peak current of an RTD with minimum area A_{min} (weight $w_{min} = 1$). The weight factors w_k express the area ratios between the RTDs with weighted inputs and the RTD with minimum area: $A_w = w \,.\, A_{min}$. Thus, according to the design methodology of the NAND/NOR gate, the weight factors w_k are equivalent to the RTD design parameters: $|w_k| = \lambda_k$. The inputs x_k of the threshold gate are set by the gate source voltage V_{GSK} and load line diagrams similar to

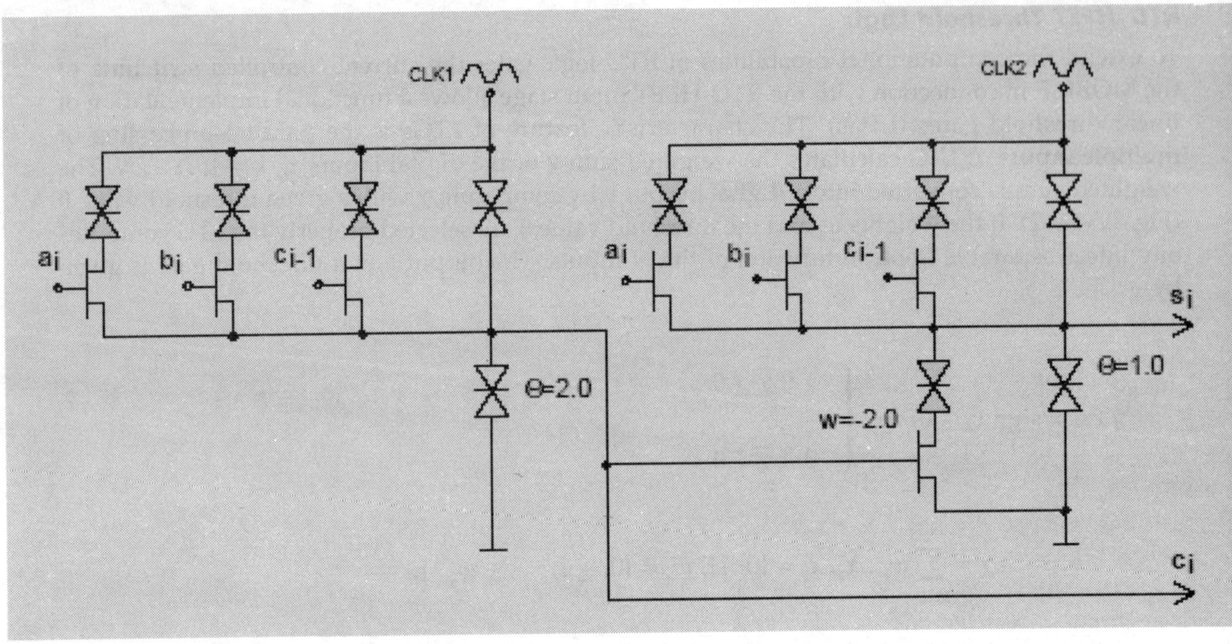

Fig. IV–G.12: Linear threshold gate $y = \text{sign}\,(x_1 + x_2 - x_3 - x_4 - \Theta)$ and MOBILE circuit.

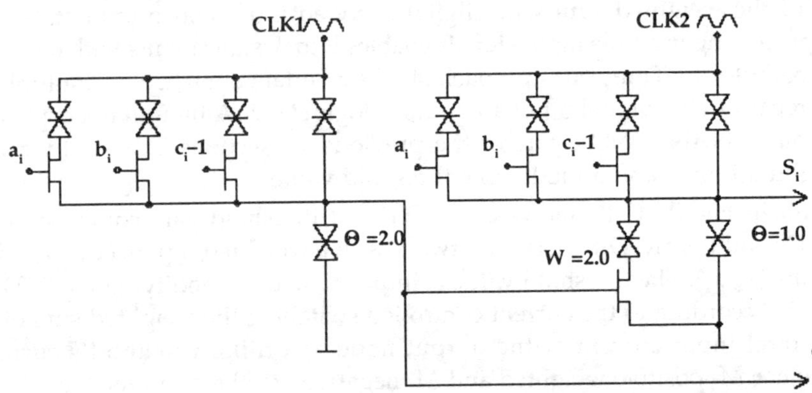

Fig. IV–G.13: Full adder composed of two threshold gates

Fig. IV–G.13 can be derived. In the proposed circuit the metastable transition is an efficient way to implement the required comparison function because the sign of the input current is equivalent to the sign of the weighted sum. Since the threshold value Θ is the peak current difference of the driver and load RTD $I\Theta = \Theta I_p$ the internal weighted sum (i.e. the input current) is finally converted into a digital output:

$$V_{OUT} = \begin{cases} V_H \text{ for } \Delta I > I_\Theta \\ V_L \text{ for } \Delta I < I_\Theta \end{cases}$$

Threshold Logic Adder Circuits

Addition is the most frequently used operation in general purpose and application specific processors. Therefore, the design of an efficient adder is essential for every emerging circuit technology and is the motivation to investigate two threshold logic full adder circuits as basic components of a n-bit ripple carry adder (RCA). As a further improvement a pipelined threshold logic version of a parallel n-bit carry look ahead adder (CLA) is presented using a logarithmic depth tree structure. In connection with the bit-level pipelined operation of the RTD-HFET threshold logic gates our approach is interesting for future digital signal processing applications where data throughput is of great relevance. Due to the inherent bistability of RTD-HFET threshold logic gates, an advantage of the proposed bit-level pipelined adders is that the area and latency increasing methodology of introducing D-latches between two different logic stages in pipelined arithmetic circuits can be avoided.

Non-Pipelined Full Adder—As mentioned before, the main objective in developing threshold logic based RTD-HFET gates is to reduce the complexity of basic circuit components. To obtain a full adder function three digital signals the two operand bits (a_i, b_i) of position i and the carry c_{i-1} of the previous bit position i - 1 are added to compute the sum s_i and the carry c_i In the case of threshold logic four different cases depending on the input sum $\chi = a_i + b_i + c_{i-1}$ are considered:

- $s_i = 0$ and $c_i = 0$ if $a_i + b_i + c_{i-1} = 0$;
- $s_i = 1$ and $c_i = 0$ if $a_i + b_i + c_{i-1} = 1$;
- $s_i = 0$ and $c_i = 1$ if $a_i + b_i + c_{i-1} = 2$;
- $s_i = 1$ and $c_i = 1$ if $a_i + b_i + c_{i-1} = 3$;

This is achieved by connecting two RTD-HFET threshold gates in a serial way (Fig. IV–G.13) so that the first gate computes the carry

$$c_i\,(\chi) = a_i b_i + c_{i-1}\,(a_i{+}b_i) = \text{sign}\,(a_i + b_i + c_{i-1} - 2)$$

and the second gate performs the computation of the sum

$$s_i\,(\chi) = \text{sign}\,(a_i + b_i + c_{i-1} - 2c_i - 1)$$

Here, the carry c_i is also used as an internal carry flag in the second gate to compute the sum bit. Since two threshold gates are the most compact way to implement a full adder, this design demonstrates that threshold logic offers the opportunity to reduce the logic depth and the number of gates. However, taking into consideration the operation of RTD-HFET gates using the four phase clocking scheme, a disadvantage of this first adder design is the multiple use of the input operands (a_i, b_i, c_{i-1}) in the second stage. Due to the overlapping clocking scheme the second stage starts the evaluation after a time delay of $T/4$. This causes an additional circuit overhead because the three operands have to be stored in a register or in three D-latches until the computation of the sum bit has finished. Consequently, this adder design is not the best choice for a bit-level pipelined operation and one of the most characteristic features of MOBILE circuits is not efficiently exploited.

Bit-level Pipelined Addition—To operate a threshold logic full adder in a bit-level pipelined fashion the first design has to be modified in such a way that the input operands (a_i, b_i, c_{i-1}) are not required to compute the sum bit si in the subsequent stage. An alternative adder design is shown in Fig. IV–G.14 where the complete adder cell comprises four threshold gates which are arranged in two circuit stages. The underlying idea of this second threshold logic algorithm is to exploit the periodical relationship between the sum bit *si* and the operand sum $\chi = a_i + b_i + c_{i-1}$. The first stage of the full adder contains three gates having the threshold values $\Theta_i = 1,2,3$ and classifies the operand sum according to the intervals shown in Fig. IV–G.14(b). Finally, the second stage detects whether the operand sum lies in one of the two high intervals $\chi = [1, 2]$ and $\chi = [3]$ and or not. In terms of threshold logic equations the alternative full adder is described by:

$$x_i (\chi) = \text{sign} (a_i + b_i + c_{i-1} - 1)$$
$$c_i (\chi) = \text{sign} (a_i + b_i + c_{i-1} - 2)$$
$$y_i (\chi) = \text{sign} (a_i + b_i + c_{i-1} - 3)$$

and

$$s_i (\chi) = \text{sign} (x_i - c_i + y_i - 1)$$
$$= 1 \text{ if } a_i + b_i + c_{i-1} \in \{1,3\}$$

Since the inputs of the second circuit stage are the intermediate results (x_i; y_i) and the new carry ci, the input operands of the first stage (a_i; b_i; c_{i-1}) are not required in the second clock cycle. Therefore, at the cost of two additional gates the modified full adder can be fully pipelined and is ideally suited for a pipelined n-bit adder.

The most simple hardware algorithm for parallel addition of two n-bit numbers is a ripple carry adder (RCA) composed of pipelined full adder cells. The carry bits are propagating in diagonal direction so that the logic depth comprises $d = n + 1$ pipeline stages (Fig. IV–G.14). This carry-propagation path is the critical path of the ripple carry adder. Since the four phase clocking scheme leads to four gate delays per clock cycle the overall delay (latency) of an n-bit ripple carry adder is $T_{add} = (n+1)\, T/4$. Apart from n RTD-HFET full adders the RCA also contains two triangular arrays of delay elements which are typical for bit-level pipelined circuits to consider the propagation or the carries along the diagonal direction. This ensures that the operands a_i and b_i are arriving simultaneously with the carry c_{i-1} at the corresponding full adder. The delay elements are implemented as RTD-HFET threshold gate with a single input similar to the RTD-HFET inverter. Also shown in Fig. IV–G.14 are the input registers where the n-bit operands ($a_0, \ldots, a_{n-1}$) and ($b_0, \ldots, b_{n-1}$) are stored before they are loaded into the first pipeline stage. The final sum bits ($s_0, \ldots, s_n$) are stored in the output register. Both, the input and output registers are updated synchronously with the full adder array at each clock cycle and thus providing a high data throughput.

Conclusion

Above was a comprehensive analysis of the design and application of RTD/HFET-based circuit architectures. Considering the present state of technology the impact of these quantum effect devices is investigated for different aspects of the circuit design, such as signal timing, clocking scheme and design methodology.

After investigating the impact to the RTD I-V characteristics on the circuit behavior a high speed logic family for three terminal resonant tunneling devices with drastically reduced design parameter sensitivity (RTD area only) is presented. Experimental data prove the functionality at

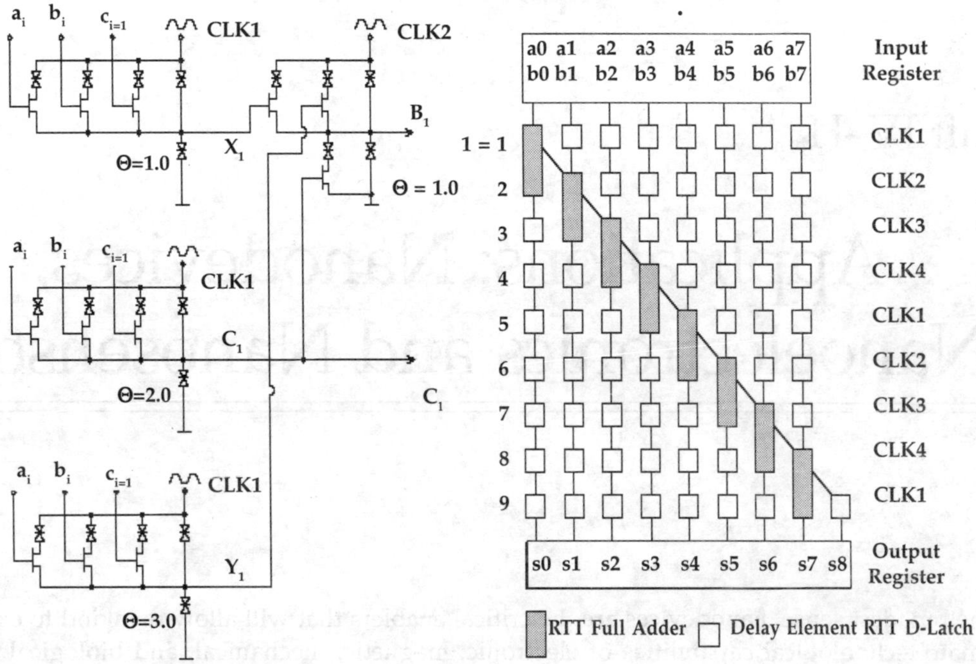

Fig. IV–G.14: Full adder optimized for bit-level pipelined operation and ripple carry adder.

ultra-low voltages V_{DD} = 0.8 V, V_H = 0.7 V, V_L = 0.1 V). HSPICE simulations of two novel scaled NAND gates based on experimental results indicate multi-GHz operation with a fan-out up to 5, a gate delay below 10 ps, and a clock rise time of 15 ps. The total power dissipation of these gates is below 2 mW and 1:2 mW, respectively. This motivates a further lateral scaling of resonant tunneling devices, that is sub-_m² RTD anode areas, to achieve power delay products below 1 fJ for MOBILE based logic gates. It has been shown that the increased functionality of monolithic integrated nanoelectronic devices can be transfered to the circuit level to reduce the complexity of the designs by using threshold logic gates. The advantages of a threshold logic design have been illustrated by the fast carry computation in two full adder designs and a pipelined n_bit ripple carry adder. The proposed circuits demonstrate that the use of threshold logic as a novel design methodology for resonant tunneling devices offers a significant performance improvement for basic arithmetic components.

Unit IV–H

Applications: Nanodevices, Nanoelectronics and Nanosensors

In the broadest sense, nanodevices are the critical enablers that will allow mankind to exploit the ultimate technological capabilities of electronic, magnetic, mechanical, and biological systems. While the best examples of nanodevices at present are clearly associated with the information technology industry, the potential for such devices is much broader. Nanodevices will ultimately have an enormous impact on our ability to enhance energy conversion, control pollution, produce food, and improve human health and longevity.

Current Scientific and Technological Advancements

In the past decade, our ability to manipulate matter from the top down, combined with advances and in some cases unexpected discoveries in the synthesis and assembly of nanometer-scale structures, has resulted in advances in a number of areas. Particularly striking examples include the following:

Current Scientific Advances: The unexpected discovery and subsequently more controlled preparation of carbon nanotubes and the use of proximal probe and lithographic schemes to fabricate individual electronic devices from these materials:

- The ability in only the last one or two years to begin to place carefully engineered individual molecules onto appropriate electrical contacts and measure transport through the molecules.
- The explosion in the availability of proximal probe techniques and their use to manipulate matter and thereby fabricate nanostructures.
- The development of chemical synthetic methods to prepare nanocrystals, and methods to further assemble these nanocrystals into a variety of larger organized structures.
- The introduction of biomolecules and supermolecular structures into the field of nanodevices.
- The isolation of biological motors, and their incorporation into nonbiological environments.

Current Technological Advances: A number of examples of devices in the microelectronics and telecommunications industries rely on nanometer-scale phenomena for their operation. These devices are, in a sense, "one-dimensional" nanotechnologies, because they are micrometer-scale objects that have thin film layers with thicknesses in the nanometer range. These kinds of systems are widely referred to in the physics and electronics literature as two-dimensional systems, because they have two classical or "normal" dimensions and one quantum or nanoscale dimension. In this scheme, nanowires are referred to as one-dimensional objects and quantum dots as zero-dimensional. In this document, and at the risk of introducing some confusion, we have chosen to categorize nanodevices by their main feature nanodimensions rather than by their large-scale dimensions. Thus, two dimensional systems such as two-dimensional electron gases and quantum wells in our notation are one-dimensional nanotechnologies, nanowires are two-dimensional nanotechnologies, and quantum dots are three-dimensional nanotechnologies. Examples include high electron mobility transistors, heterojunction bipolar transistors, resonant tunneling diodes, and quantum well optoelectronic devices such as lasers and detectors.

The most recent success story in this category is that of giant magnetoresistance (GMR) structures. These structures can act as extremely sensitive magnetic field sensors. GMR structures used for this purpose consist of layers of magnetic and nonmagnetic metal films. The critical layers in this structure have thicknesses in the nanometer range. The transport of spin-polarized electrons that occurs between the magnetic layers on the nanometer length scale is responsible for the ability of the structure to sense magnetic fields such as the magnetic bits stored on computer disks. GMR structures are currently revolutionizing the hard disk drive magnetic storage industry worth $30-40 billion/year. Our ability to control materials in one dimension to build nanometer-scale structures with atomic scale precision comes from a decade of basic and applied research on thin film growth, surfaces, and interfaces.

The extension from one nanodimension to two or three is not straightforward, but the payoffs can be enormous. Breakthroughs in attempting to produce three-dimensional nanodevices include the following:

- Demonstration of Coulomb blockade, quantum effect, and single electron memory and logic elements operating at room temperature.
- Integration of scanning probe tips into sizeable arrays for lithographic and mechanical information storage applications.
- Fabrication of photonic band-gap structures.
- Integration of nanoparticles into sensitive gas sensors.

Goals for the Next 5-10 Years: Barriers and Solutions

In order to exploit nanometer-scale phenomena in devices, we must have a better understanding of the electronic, magnetic, and photonic interactions that occur on and are unique to this size scale. This will be achieved through experiment, theory, and modeling over the next decade. In addition, new methods to image and analyze devices and device components will be developed. These might include three-dimensional electron microscopies and improved atomic scale spectroscopic techniques.

Over the same time period, we believe that it will become possible to integrate semiconductor, magnetic, and photonic nanodevices as well as molecular nanodevices into functional circuits and chips.

The techniques now being developed in biotechnology will merge with those from nanoelectronics and nanodevices. Nanodevices will have biological components. Biological systems will be probed, measured, and controlled efficiently with nanoelectronic devices and nanoprobes and sensors.

There will be significant progress in nanomechanical and nanobiomechanical systems, which will exhibit properties that are fundamentally different from their macroscopic counterparts.

There are important applications for instruments that will fly into space: nanocomponents are needed to achieve overall instrument sizes in the micron or millimeter range. Some of the same issues apply to battlefield sensors for situational awareness.

Finally, a significant goal is the development of nanometer-scale objects that manipulate and perform work on other nanometer-scale objects, efficiently and economically achieving the same things we currently rely on scanning tunneling microscopy (STM) or atomic force microscopy (AFM) to carry out. A first step towards this goal might be the integration of nanometer-scale control electronics onto micromachines.

Paradigm Shifts

In the information technology arena, nanodevices will both enable and require fundamentally new information processing architectures. Early examples of possible architectural paradigm shifts are quantum computation, quantum dot cellular automata, molecular electronics, and computation using DNA strands. Such architectures will fundamentally change the types of information technology problems that can be attacked. Effective implementation of these types of architecture will require nanodevices.

Other paradigm shifts include the emergence of quantized magnetic disks; single photonic systems that will allow efficient optical communication; nanomechanical systems; a broad class of structures and devices that merge biological and non-biological objects into interacting systems; and use of nanocomponents in the shrinking conventional circuit architectures.

Research on nanodevices using nanoscale wiring and molecular logic, as well as new principles for devices such as spin electronics, have made significant inroads in the past year or two.

Scientific and Technological Infrastructure

The exploration and fabrication of nanodevices requires access to sophisticated and sometimes expensive tools. More and better access to such equipment as well as rapid prototyping facilities is needed. Of equal importance is the recognition that success in nanodevices will draw upon expertise from a broad range of traditional disciplines. Therefore, it is imperative that programs be established that facilitate and strengthen cross-fertilization among diverse disciplines and that allow rapid adoption of new methods across field boundaries.

R&D Investment and Implementation Strategies

Nanodevices are in some ways the most complicated nanotechnological systems. They require the understanding of fundamental phenomena, the synthesis of appropriate materials, the use of those materials to fabricate functioning devices, and the integration of these devices into working systems. For this reason, success will require a substantial funding level over a long period of time. There is strong sentiment for single investigator funding as well as for structured support of interdisciplinary teams.

Conclusions and Priorities

Priorities in Research and Development

- Development of new systems and architectures for given functions.
- Study of interfaces and integration of nanostructures into devices and systems.
- Multiscale, multiphenomena modeling and simulation of complex systems.

Priorities in Modes of Support

- Establishment of consortia or centers of excellence for the research priorities identified above, by using vertical and multidisciplinary integration from basic research to prototype development.
- Encouragement of system integration at the nanoscale in research and education.

Examples of Current Achievements and Paradigm Shifts

Organic Nanostructures: The Electrical Conductivity of a Single Molecule—By combining chemical self-assembly with a mechanical device that allows them to break a thin gold wire with nanometer scale control, researchers have succeeded in creating a "wire" consisting of a single molecule that can connect two gold leads (Fig. IV–H.1). Using this structure, they have been able to begin to measure and study the electrical conductivity of a single molecule.

Molecular Electronics—If the reduction in size of electronic devices continues at its present exponential pace, the size of entire devices will approach that of molecules within a few decades. However, well before this happens, both the physics upon which electronic devices are based and the manufacturing procedures used to produce them will have to change dramatically.

This is because current electronics are based primarily on classical mechanics, but at the scale of molecules, electrons are quantum mechanical objects. Also, the cost of building the factories for fabricating electronic devices, or fabs, is increasing at a rate that is much larger than the market for electronics; therefore, much less expensive manufacturing process will need to be invented.

Thus, an extremely important area of research is molecular electronics, for which molecules that are quantum electronic devices are designed and synthesized using the batch processes of chemistry and then assembled into useful circuits through the processes of self-organization and self-alignment. If molecular electronics achieves the ultimate goal of using individual molecules as switches and carbon nanotubes as the wires in circuits, we can anticipate nonvolatile memories with one million times the bit area density of today's DRAMs and power efficiency one billion times better than conventional CMOS circuitry. Such memories would be so large and power-efficient that they could change the way in which computation is performed from using processors to calculate on the fly to simply looking up the answer in huge tables.

A major limitation of any such process is that chemically fabricated and assembled systems will necessarily contain defective components and connections. By describing a silicon-based computer that was designed to operate perfectly in the presence of huge numbers of manufacturing defects, researchers from Hewlett-Packard (HP) and the University of California–Los Angeles

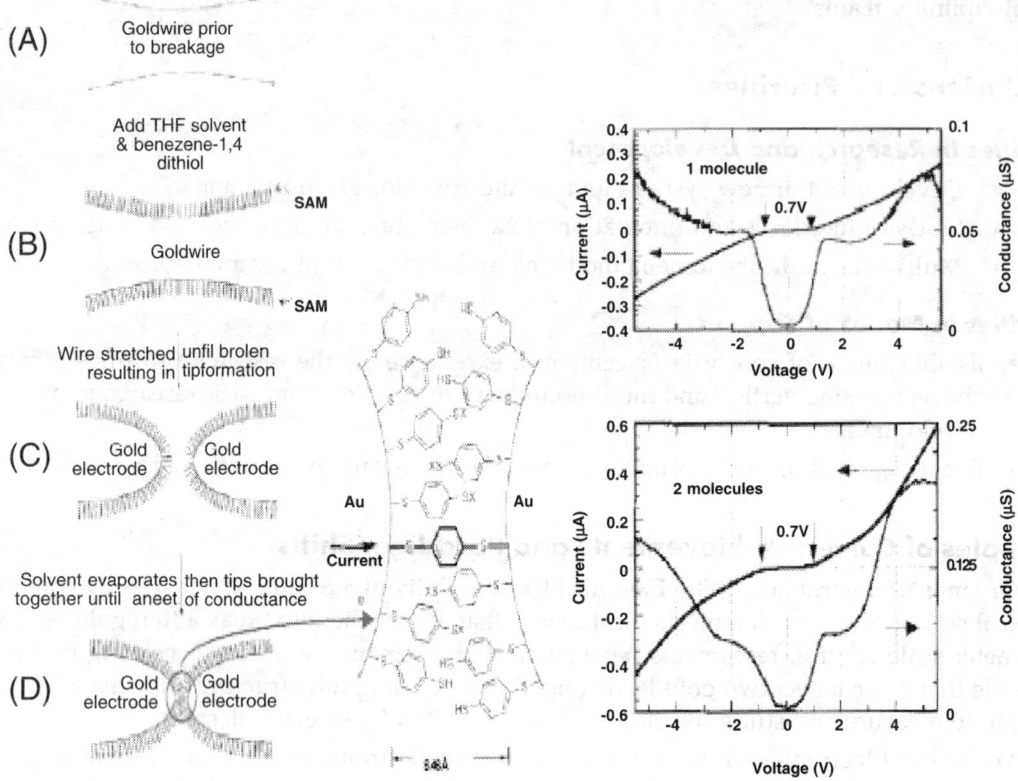

Fig. IV–H.1: Organic nanostructures: on left, showing self-assembly of benzene-1,4-dithiol onto Au electrodes; on right, showing room-temperature I-V measurements suggesting presence of a Coulomb gap.

(UCLA) presented an architectural solution to the problem of defects in molecular electronics, as described in Figure IV–H.2, and thus demonstrated in principle that manufacture by chemical assembly is feasible.

In 1999, researchers from HP Labs and UCLA experimentally demonstrated the most crucial aspect for such a system, an electronically addressable molecular switch that operates in a totally "dry" environment. As illustrated in Figure IV–H.3, logic gates were fabricated from an array of configurable molecular switches, each consisting of a monolayer of electrochemically active rotaxane molecules sandwiched between metal electrodes.

Figure IV–H.4 Illustrates the operation of the switches. In the "closed" state, current flow is dominated by resonant tunneling through the electronic states of the molecules. The switches are irreversibly opened by applying an oxidizing voltage across the device. In this case, since the memory of the molecules is not volatile, only one set of wires is needed to set and read the state of the molecules, and in principle, one molecule can replace seven transistors in a conventional silicon circuit. In the demonstration, several devices were configured together to produce AND

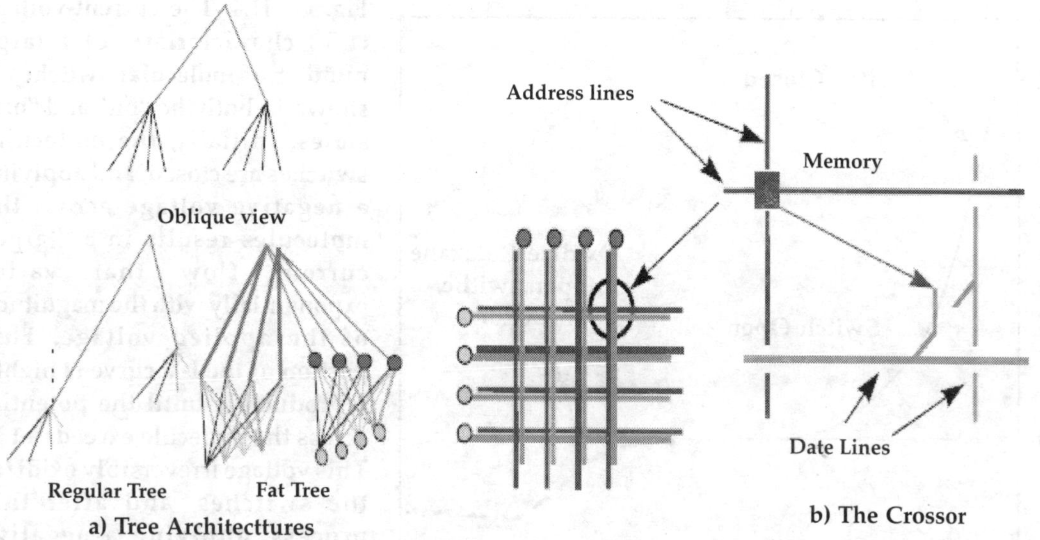

Fig. IV–H.2: The logical design of a defect-tolerant circuit: (a) shows a "fat tree" architecture in which every member of a logical level of the tree hierarchy can communicate with every member at the next level; in the case of a defective component, this structure enables one to route around and avoid the defect; (b) shows how this architecture is implemented using cross bars, which are very regular structures and look like something that can be built chemically. The complexity required for a computer is programmed into the cross bars by setting the switches to connect certain elements of the tree together. Using silicon circuitry, two completely separate sets of wires (address and data lines) are required for the cross bars, and seven transistors are required for each switch, since a continual application of electrical power is required to hold the sense of the switches.

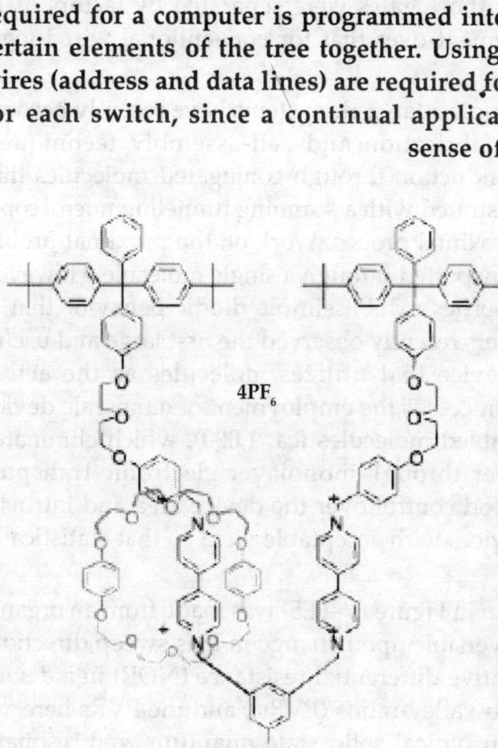

Fig. IV–H.3. The atomic structure of one of the molecular switches used in the devices described above, which is known as a rotaxane. This molecule conducts via resonant tunneling through unoccupied molecular orbitals when it is in its reduced chemical state (switch closed), but it is a tunneling barrier in its oxidized state (switch open). The switch can be closed electronically in a solid-state circuit by applying the appropriate voltage across the molecule.

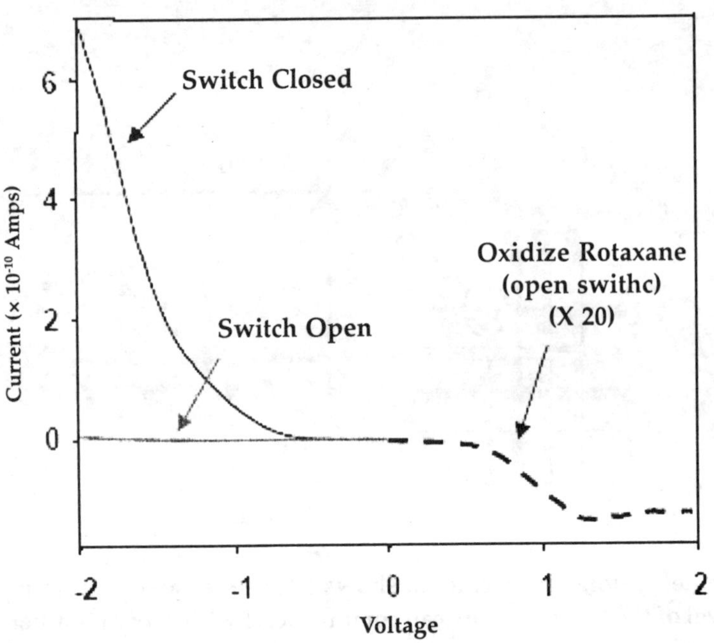

Fig. IV–H.4. The current-voltage (I-V) characteristic of a large number of molecular switches is shown in both the "on" and "off" states. Initially, the molecular switches are closed, and applying a negative voltage across the molecules results in a "large" current flow that varies exponentially with the magnitude of the applied voltage. This portion of the I-V curve is highly reproducible until the potential across the molecule exceeds +1 V. This voltage irreversibly oxidizes the switches, and after this process, applying a negative voltage results repeatedly in a "small" current, demonstrating that the switch is open.

and OR logic gates. The high/low current levels of those gates were separated by factors of 15 and 30, respectively, which is a significant enhancement over that for conventional wired-logic gates.

Molecular Logic: Electron transport studies in molecular-scale systems have recently become possible with the utilization of advanced microfabrication and self-assembly techniques. Investigations are now possible of the electronic conduction through conjugated molecules that are end-bound onto surfaces; these have been demonstrated with a scanning tunneling microscope, with micromachined silicon nanopores, and with proximal probes. Work on the proximal probes demonstrated that 0.1 microamp of current can be transported through a single molecule. However, in all of the past embodiments, the electronic properties exhibit simple diodic behavior that is unsuitable for potential circuit applications. Researchers recently observed the first large and useful reversible switching behavior in an electronic device that utilizes molecules as the active component. The essential feature of the fabrication process is the employment of nanoscale device area that gives rise to a small number of self-assembled molecules (ca. 1,000), which eliminates pinhole and other defect mechanisms that hamper through-monolayer electronic transport measurements. This technique has demonstrated good control over the device area and intrinsic contact stability and produces a large number of devices with acceptable yield so that statistically significant results can be produced.

The active electronic component (synthesis shown in Figure IV–H.5) was made from an organic compound upon exposure to Au. The I-V is fully reversible upon change in bias sweep direction. This is the first observation of robust and large negative differential resistance (NDR) in a device where molecules form the active region with peak-to-valley-ratios (PVRs); and the PVRs here are >1000:1. The performance exceeds that observed in typical solid state quantum well resonant tunneling heterostructures. Therefore, in addition to the obvious size advantages for scaling, the

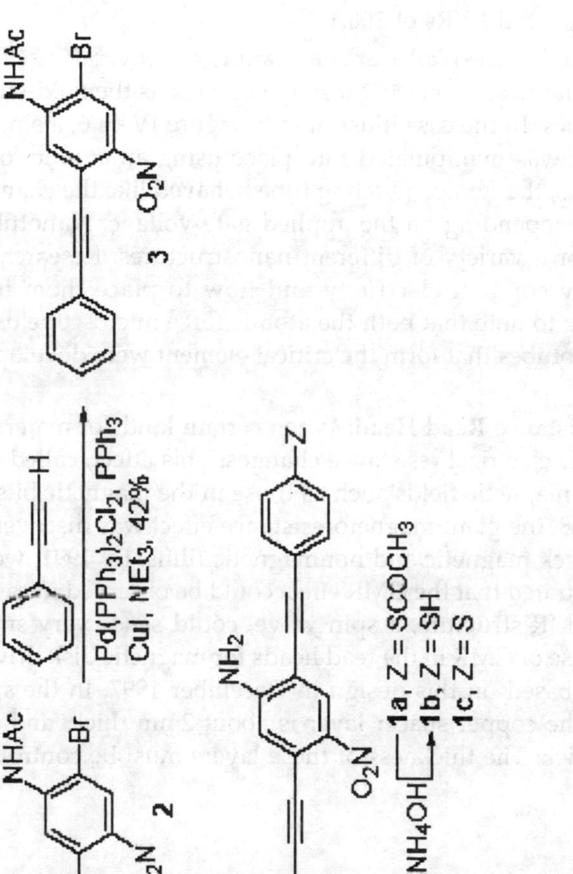

Fig. IV–H.5: Schematic of the synthesis of the active molecular compound and its precursors (1a-c).

intrinsic device characteristics (i.e., valley current shutoff) may be superior to solid state embodiments; present silicon devices rarely exceed PVRs of 100:1.

A Field-Effect Transistor Made from a Single-Wall Carbon Nanotube: Several research groups around the world have succeeded in fabricating electrical switches such as the field-effect transistor from single-walled carbon nanotubes. In the case illustrated in Figure IV–H.6, a single-walled carbon nanotube 1.6 nm in diameter was manipulated into place using an atomic force microscope. Once placed on the metal contacts, the semiconducting tube behaved like the channel in a field-effect transistor, turning on or off depending on the applied gate voltage. Nanotubes hold great promise as electronic elements for a variety of different nanostructures. Researchers are just beginning to understand how they conduct electricity and how to place them into appropriate device structures. It is interesting to note that both the atomic force microscope used to fabricate this structure and the carbon nanotubes that form the critical element were developed only in the past decade.

A Commercial IBM Giant Magnetoresistance Read Head: When certain kinds of materials systems are exposed to a magnetic field, their electrical resistance changes. This effect, called the magnetoresistive effect, is useful for sensing magnetic fields such as those in the magnetic bits of data stored on a computer hard drive. In 1988, the giant magnetoresistance effect was discovered in specially prepared layers of nanometer-thick magnetic and nonmagnetic films. By 1991, work at the IBM Almaden Research Center demonstrated that the GMR effect could be observed in easily made samples and that a special kind of GMR structure, a spin valve, could sense very small magnetic fields. This opened the door to the use of GMR in the read heads for magnetic disk drives. IBM first announced a commercial product based on this design in December 1997. In the spin valve GMR head shown in Figure IV–H.7, the copper spacer layer is about 2 nm thick, and the cobalt GMR pinned layer is about 2.5 nm thick. The thickness of these layers must be controlled with atomic precision.

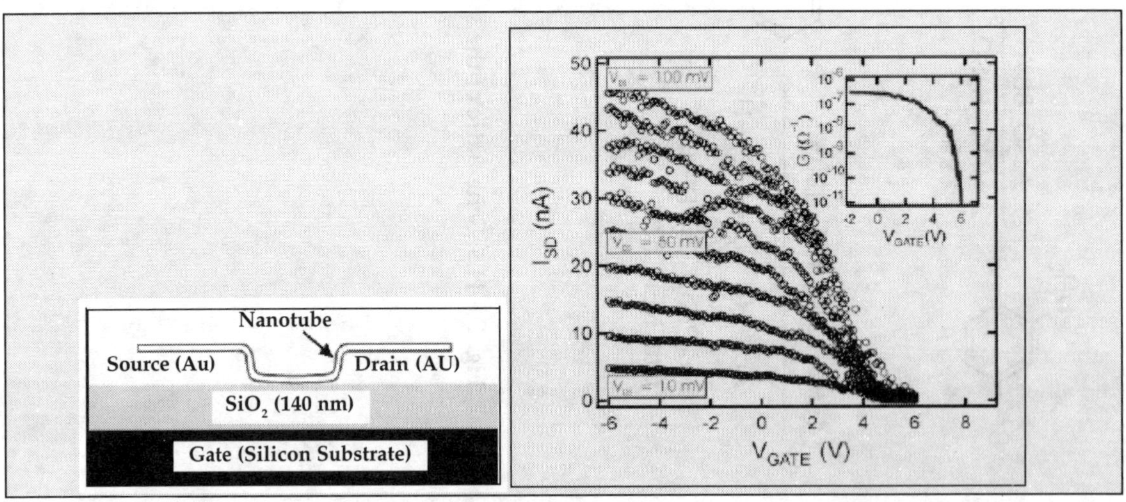

Fig. IV–H.6: Field-effect transistor based on a single 1.6 nm diameter carbon nanotube.

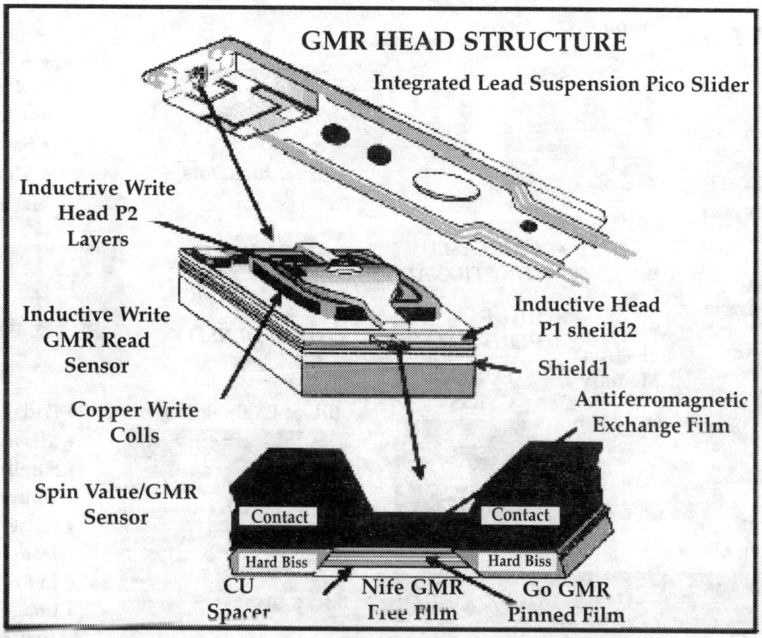

Fig. IV–H.7: Commercial IBM giant magnetoresistance read head.

Nanoelectronic Devices

Nanoelectronics offers a broad set of opportunities by focusing on quantum devices and addressing their potential for high performance through increases in density (factors of 5 to 100), speed (factors of 10 to 100), and reduced power (factors of more than 50) (see Figure IV–H.8). Resonant tunneling devices are being explored with demonstrated successes in multivalued logic and various logic circuits and memory circuits. SET logic and memory concepts are being explored with focus on memory applications. Molecular electronics and self-assembly approaches have shown a path towards manufacturing alternatives and device options for regimes beyond traditional scaling. Spin devices in the form of nanomagnetics using the magnetoresistive effect in magnetic multilayers have demonstrated their use for nonvolatile, radiation-hard memory. Quantum cellular automata and coupled quantum dot technology are being explored and their potential assessed for transistorless computing. By exploring Si-based heterojunctions, bandgap engineering, vertical device structures, and quantum devices, inroads are being made into extending CMOS capabilities. Potential applications are in digital radar, electronic support measures (ESM) receivers, ATM data stream processing, wide bandwidth communications, digital image processing, waveform generation, and the broad area of analog to digital (A/D) applications. Demonstrations have shown the efficacy of resonant tunneling devices in various network environments. The long-term vision for nanoelectronics sees the use of quantum devices in other high performance systems especially in telecommunications for signal processors and electronics for A/D converters in detectors.

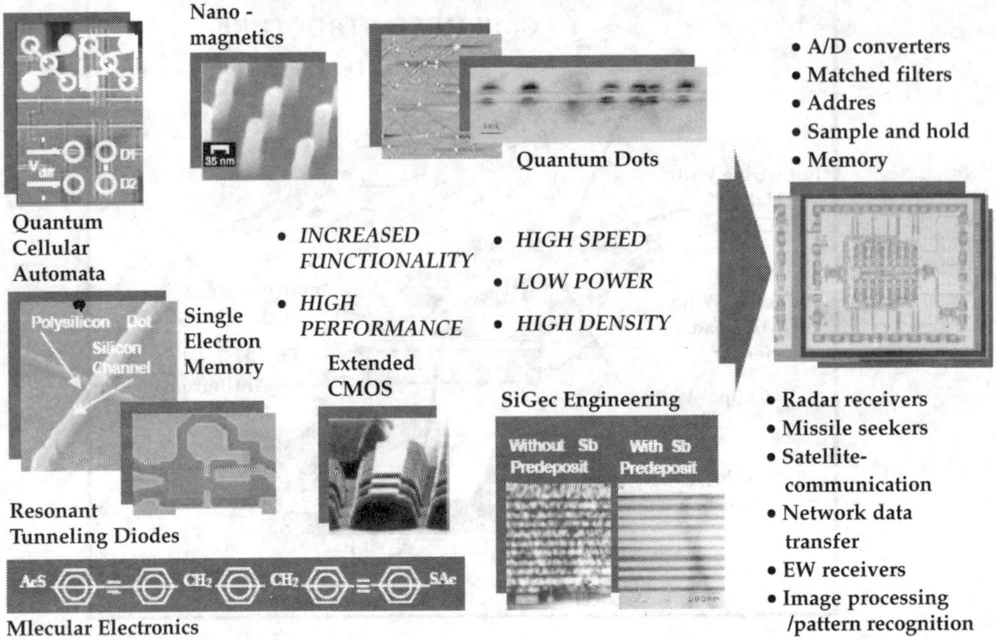

Fig. IV–H.8. Nanoelectronics: device and architecture options for high-performance electronics.

Nanodevices and Breakthroughs in Space Exploration

After more than three decades of exploring space, the National Aeronautics and Space Administration (NASA) has completed an initial reconnaissance of our solar system. The next missions will involve sending spacecraft to destinations that are much more difficult to travel to, like the Sun or Pluto. Also, spacecraft will be required to perform more difficult tasks, such as landing on a celestial body, collecting a sample of its material, and returning it to Earth. To carry out such technically challenging missions at an affordable cost, NASA has created the Deep Space Systems Technology Program, known as X2000. Every two to three years starting in the year 2000, the program will develop and deliver advanced spacecraft systems and body structures to missions bound for different areas of the solar system and beyond. In order to achieve reduction in the size of spacecraft, the avionics systems of the spacecraft are being reduced in size with each delivery of X2000, in part by means of integrating nanotechnology with microtechnology. Figure 8.9 attempts to chart the forecasts of the mass, volume, and power of future avionics systems of spacecraft. The leftmost column shows the Mars Pathfinder spacecraft, which represents the current state of the art.

The first delivery for X2000 is an integrated avionics system that subsumes the functionalities of command and data handling, attitude control, power management and distribution, and science payload interface. Advanced packaging technologies as well as advanced design automation techniques are used to define a highly integrated, modular, building-block architecture for highly reliable and long-term survivable deep-space planetary missions. Advanced low-power techniques and architectures will drastically reduce overall power consumption compared to currently available flight hardware. "System On A Chip" (SOAC) will prototype single chip and multichip module solutions that lead towards an avionics system on a chip. This chip will integrate the avionics system that is being developed for the X2000 avionics deliverable. That is, the chip will include power management, sensor technology, and telecommunications modules, together with CPU and storage technology. To accomplish this, nanotechnology will be needed to miniaturize and integrate the different subsystems.

The goal for the year 2020 is to establish and maintain a world-class program to research revolutionary computing technologies (RCT) that will not only take us beyond the limits of semiconductor technology scaling but also will enable the vision of a "thinking spacecraft." A thinking spacecraft would be a totally autonomous, highly integrated, extremely capable spacecraft that operates at ultralow power. To achieve this goal, without a doubt, we need to employ nanoscience. In spite of the phenomenal advances in digital computing in recent years and those expected in the near future, even future supercomputers cannot compete with biological systems in performing certain ill-defined tasks such as pattern recognition, sensor fusion, fault-tolerant control, and adaptation to the environment. Biological systems address these types of problems with extreme ease and very low power. The forth column from the left in Figure IV–H.9 (Fifth Delivery) depicts two different technologies based on nanoscience that may have a great impact on the capabilities of the spacecraft by the year 2020:

➢ *Quantum computing*, that is, a joint venture between computer science and quantum physics. Although, the concept of a quantum computer is simple, its realization is not. Two issues motivate quantum computing:
  - Quantum mechanical concepts must be applied to solve intractable (NP-complete) computing problems.
  - From a computer miniaturization point of view, the size limit of a bit of information is important. Recently, this issue has attracted increased attention, due to the current development of nanotechnology and the design problems of semiconductor and metal devices that are approaching the quantum size limit. Consequently, the idea of quantum computing, in which the elements that carry the information are atoms, has attracted the attention of many scientists.

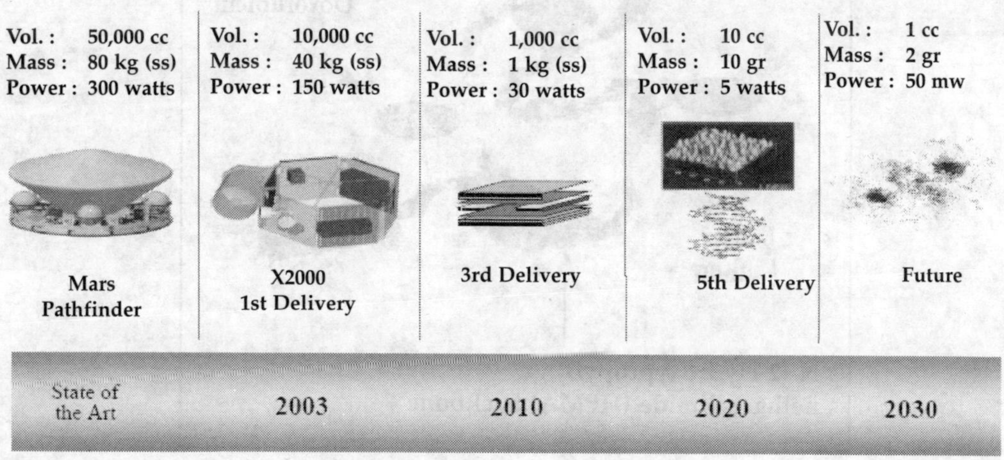

Fig. IV–H.9: Avionics roadmap.

- *Biomimetics*, that is, systems or technologies inspired by architectures, functions, mechanisms, and principles found in biological systems, for example:
  - One gram of DNA could possibly store all the data in the Library of Congress.
  - The human brain contains about 1014 interconnects and operates at 1016 operations per second, using ultra low power and imprecise computing elements.
  - Humans are endowed with an immune system that provides recovery from illness—a "self-repair system".

As devices become smaller, lighter, and consume less power, NASA will be able to design and fly space probes on missions that are not currently possible.

A Biological Nanodevice for Drug Delivery: The nanobiological anticancer agent PK1 (Figure IV–H.10) exploits the enhanced permeability and retention (EPR) effect associated with disease tissues with low integrity vasculature in order to deliver cytotoxin (doxorubicin) to tumors. The synthetic backbone of PKI (N-2-(hydroxypropyl) methacrylimide or HPMA) gives the complex a size (diameter in the mid-nanometer range) that makes it unable to extravasate efficiently into healthy tissues with normal vasculature. Tumor vasculature is abnormally permeable, allowing preferential accumulation of PK1 in tumor tissue. HPMA-bound doxorubicin is non-toxic, limiting toxicity to healthy tissue, and active doxorubicin is released from the complex preferentially in tumor tissues. The labile peptidic linker tethering doxorubicin to HPMA was selected because it is the substrate for a protease known to be over-expressed in the target tumor types. PK1 increases the tolerated doxorubicin dose by more than an order of magnitude by virtue of EPR-based targeting and its engineered tumor-preferred doxorubicin release properties. It is in human clinical trial in Europe. Contemplated embellishments to this and similar polymer therapeutics include use of monodisperse nanopolymers (dendritic polymers) to enhance control of EPR properties, incorporation of protein docking domains that recognize tumor associated antigens to tether the complex following its delivery to the tumor, and incorporation of additional antitumor agents thought to have synergistic effects with cytotoxins, that is, angiostatic agents, among others.

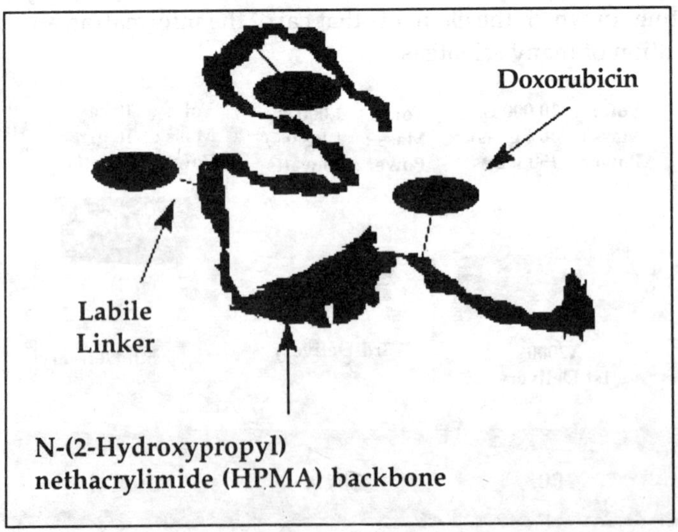

Fig. IV–H.10: The nanobiological anticancer agent PK1.

Nanotechnology on a Chip: A New Paradigm for Total Chemical Analysis Systems, the ability to make chemical and biological information much cheaper and easier to obtain is expected to fundamentally change healthcare, food safety, law enforcement, national security, and many other areas of direct interest to the American public. The vision of broadly available chemical analysis is fueling an international effort to develop "mChemLabTM-on-a-chip" technology. Micro-total analysis systems (m-TAS, as they are often referred to) are distinguished from simple sensors because they conduct a complete analysis; a raw mixture of chemicals goes in and an answer comes out. Sandia National Laboratories is developing a hand-held mChemLabTM demonstrator that will analyze for air-born chemical warfare agents and liquid-based explosives agents. The mChemLabTM development project brings together an interdisciplinary team of about 50 staff members from throughout the laboratory in areas of expertise including microfabrication, chemical sensing, microfluidics, and information sciences. Although nanotechnology plays an important role in current mTAS efforts, most mTAS approaches use miniaturized versions of conventional architecture and components to achieve system tasks. Small valves, pipes, pumps, separation columns, etc. are patterned after their macroscopic counterparts. Even though we are finding that these miniaturized components can work as well as (and sometimes better than) their macroscopic analogs, they simply will not allow for the vision of chemical laboratories in a grain of sand. Nanotechnology will enable a completely new architecture, or nano-TAS. The ability to build materials with switchable molecular functions could provide completely new approaches to valves, pumps, chemical separations, and detection. For example, fluid streams could be directed by controlling surface energy without the need for a predetermined architecture of physical channels. Switchable molecular membranes and the like could replace mechanical valves. By eliminating the need for complex fluidic networks and micro-scale components used in current m-TAS efforts, nano-TAS is a fundamentally new approach to allow greater function in much smaller, lower power total chemical analysis systems.

The Development of Useful Nanotech Robotic Systems: It is potentially feasible to manufacture nanorobots that are capable of sophisticated symmetric behaviors, either through independent function or by assembling themselves into collective units. Imagine, for example, high-resolution video screens that can repair themselves simply by having a microscopic robot at each screen element. These "pixelbots" would be capable of producing light, but also be smart enough to remove themselves from the video array should they ever fail. Other pixelbots would sense the vacancy left by any defective device and reorganize themselves to fill the hole. Another example is the incorporation of autonomous "nurse" robots into the human body that are chemically benign but are capable enough to remove cancer cells at the source. Having the ability to discriminate between healthy and cancerous cells, the "nurse-bots" would function independently to continuously heal tissue in ways beyond the current ability of the human body. The same notion can be implemented in self-optimizing silicon memories or processors, where a blown transistor would mean one just had to wait for the computer to heal itself.

One key to establishing such capabilities is research into autonomous self-assembled systems. Researchers at Los Alamos National Laboratory are exploring these systems by creating very inexpensive macro-scale robots. These sense and adapt to their environment, including assimilating other robots to execute such tasks as searching for and marking the location of unexploded land mines (Figure IV-II.11). The capabilities of these intelligent cellular systems are readily scaled, providing untapped possibilities for large numbers of inexpensive nano-machines to become microscopic building blocks for heretofore unimaginable functions—a form of "nano-Lego" for the new millennium with novel, untapped market potential.

Fig. IV–H.11: Models for nanoscale: Three-inch-diameter self-assembled robots mark the spot where an unexploded mine rests under the surface. Such robots are cheap, solar-powered, and have no processor to make application or miniaturization difficult.

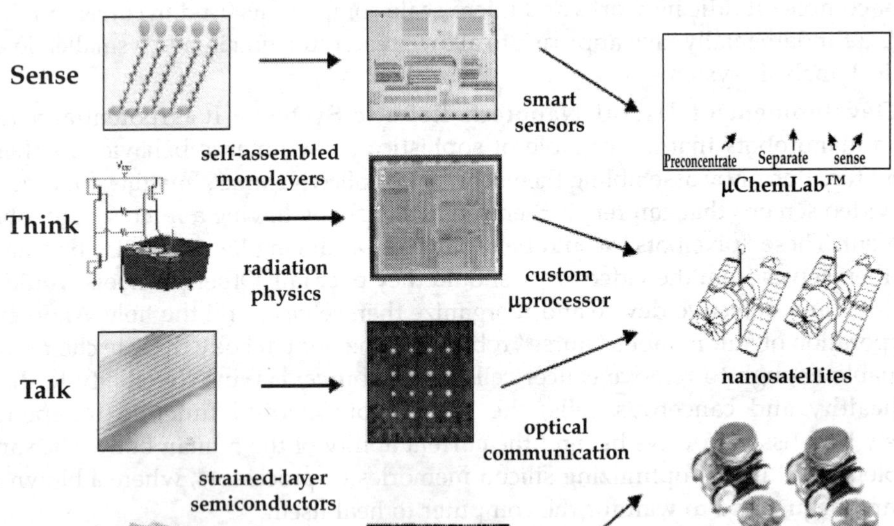

Fig. IV–H.12. The control of mechanical, electrical, optical, and chemical properties at the nanoscale will enable significant improvements in integrated microsystems.

Integrated Nanotechnology in Microsystems: Advances in nanotechnology will have a profound effect on the future of integrated microsystems. The integration of microelectronic, microelectromechanical, optical, and chemical microsensors into "systems on a chip" is an area that may involve mechanical, optical, and/or chemical functions as well. As illustrated in Figure IV–H.12, these advances will make possible miniaturized systems that sense, think, talk (communicate), and act. However, these microscale systems will only become a reality if enabled by the control of performance at the nanoscale. Thus, for example, advances in microelectromechanical systems (MEMS) and photonics shown in the figure depend on discoveries in nanoscience and nanoscale fabrication.

Unit IV–I

Nano-Optics: Introduction

What is Nano-Optics

Today we encounter a strong trend towards nanoscience and nanotechnology. This trend is motivated by the fact that as we move to smaller and smaller scales the underlying physical laws change from macroscopic to microscopic. The exploitation of quantum effects for technological applications is the most obvious driving force behind the current miniaturization. The recent rapid advances are due in large part to our newly acquired ability to measure and manipulate individual structures on the nanoscale (scanning probe techniques, optical tweezers, and high-resolution electron microscopes).

The increasing trend towards nanoscience and nanotechnology makes it necessary to address the key issues of optics on the nanometer scale. Since the diffraction limit does not allow us to focus light to dimensions smaller than roughly half a wavelength, traditionally it was not possible to interact selectively with nanoscale features. In recent years several new approaches have been put forth to 'shrink' the diffraction limit (confocal microscopy) or to even overcome it (near-field microscopy). For example, with our tip-enhancement technique we are able to do Raman spectrocopy and multiphoton fluorescence imaging with a spatial resolution of less than 20nm. To date, this is the highest optical resolution of a spectroscopic measurement.

The reason for the hard effort to advance the field of optics to the nanometer scale is the fact that the energy of light lies in the range of electronic and vibrational transitions in matter. Therefore, the interaction of light with matter renders unique information about the structural and dynamical properties of matter. These unique spectroscopic capabilities are of great importance for the study of biological and solid-state nanostructures. We are applying near-field optical techniques to probe complex semiconductor nanostructures as well as individual protein molecules. We plan to explore the possibility to optically interact with semiconductor nanostructures on length scales smaller than the extent of their quantum wavefunctions. Probing and manipulating these wavefunctions might open up various exciting applications such as data storage and optical switching based on quantum logic.

Nano-optics addresses the broad spectrum of optics on the nanometer scale covering technology and basic science. On the technological side, we find topics like nanolithography and high density optical data storage. On the basic sciences end, we might mention atom-photon interactions in the optical near-field and their potential applications for atom trapping and manipulation experiments. Compared with free propagating light the optical near-field is enriched by so-called virtual photons. These are the same sort of particles responsible for molecular binding (van der Waals/Casimir forces) and are therefore promising for selective probing of atomic structures. The consideration of virtual photons in the field of quantum optics will enlarge the range of fundamental experiments and applications.

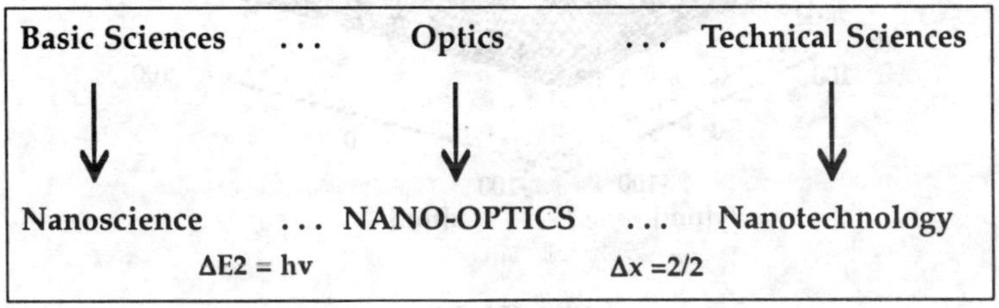

Fig. IV–I.1

Nano-optics deals with the controlled manipulation of evanescent optical fields and their exploitation in investigations of nanometer scale structures. The basic idea is to detect the evanescent field distribution, which is significant in the near-field region that extends to only a few tens of nanometers from the sample surface. The detection of the evanescent waves is typically done with a nanoprobe. The probe acts as a local scattering center transforming some of the evanescent field components into homogeneous (propagating) waves that can be detected in the far zone. By scanning the nanoprobe over the sample surface, the evanescent field distribution can be detected with a resolution limited mainly by the size of the nanoprobe and the scanning height. In practice, the resolution achieved in scanning near-field optical microscopy (SNOM) is on the order of a few tens of nanometers, i.e., the technique offers an order of magnitude improvement to the best conventional far-zone measurements.

Our research has concentrated mainly on three specific areas. The main efforts have been focused on developing the shear-force feedback system of our SNOM apparatus, on the near-field investigation of a broad-area semiconductor laser, and on the theoretical modeling of the near-field-matter interaction.

Insight into the origin of the high spatial resolution in a SNOM measurement may be gained by investigating the contributions of the homogeneous and evanescent electromagnetic waves separately in some selected geometries. The simplest model for the sample-probe system in SNOM can be obtained by considering the probe tip and the sample structure as being composed of a collection of point-dipoles. Of fundamental importance in this approach is to understand how the electromagnetic field of a single dipole is decomposed into its evanescent and homogeneous parts. Knowing this one can then proceed to generalize these results to systems containing several point-dipoles by including the couplings between the dipoles in a self-consistent calculation.

The point-dipole model has been widely employed as an aid in discussing the properties of near-field microscopy. Caution should, however, be exercised when interpreting the results. The point dipole does not offer an adequate model for describing mesoscopic structures in near-field investigations, since strong evanescent field gradients exist close to the sources. By applying the point-dipole model, however, we can investigate how the nanoprobe affects the near field and how the resolution depends on the scanning height of the probe over the sample in the different

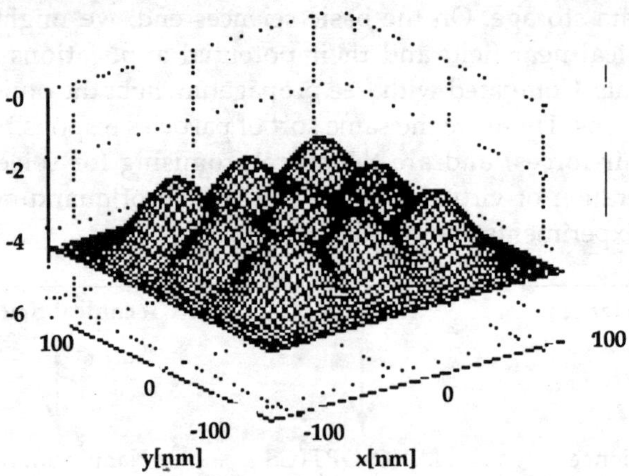

Fig. IV–I.2

operation modes of SNOM. Furthermore, we can study the effects of different polarization directions of the nanoprobe and the external polarizing field on the resolution of a SNOM measurement. In particular, we can see the central role of the evanescent field components in breaking the classical diffraction barrier.

As an example, in Fig. IV–I.2 is illustrated the homogeneous, evanescent and total field distribution at the probe site when a z-polarized nanoprobe is scanned at the height of 40 nm over the sample. The probe dipole locally illuminates the sample and the same probe is also used to collect the scattered light. As a result of the local interaction between the probe and the sample dipoles the scanned homogeneous field distribution now reveals the same information about the sub-wavelength scale structures of the sample as the evanescent field, although in lower magnitude.

Nanooptics, i.e. optical spectroscopy with a spatial resolution of <100 nm, opens up fascinating insights into the physical processes occuring on a nanometer scale. In our group, we apply this technique to investigate *single* semiconductor quantum dots, sometimes refered to as 'artificial atoms' because of their δ-like density of states.

Unit IV–J

Wafer-level Nano-optics

"Embossing" is an ancient technical term in English traceable to the period 1350 to 1400, and used by Chaucer in its present sense. The technical meaning, creating a raised design or pattern by pressing a die having the raised negative image of the design against the material to be patterned, has not changed during the ensuing 650 years. What has changed are the applications, materials used, and the scale of the design. Today's embossed patterns may have features measured in nanometers, bringing new capabilities to integrated optical systems.

Nanoembossing, often referred to as nano-imprint lithography (NIL), can rapidly pattern relatively large area substrates with feature sizes below 100 nm. Creating patterns at this scale previously required far slower and more costly techniques such as electron-beam writing. NIL, used in combination with related technologies, can create an entire wafer of integrated optical structures with nanometer features in minutes. Figure IV–J.1 shows an array of replicated lenses formed on the surface of a 4-in. silicon wafer.

Wafer-level NIL promises reductions in the size, cost and complexity of integrated optics, while improving performance. Integrating embossed components on a wafer offers major size reductions. Producing truly "monolithic" optical systems could reduce centimeter-sized systems to millimeter-sized die. Optical performance is improved by locating components physically closer together, reducing the optical path length and the attendant losses. For example, optical elements such as lenses can be fabricated directly onto optical sources, such as lasers. Figure IV–J.2 shows diffractive lenses formed directly on a VCSEL laser wafer.

Surprising to the non-expert, the small physical size of the nano-structures also creates optical elements that differ significantly in some characteristics from normal macro-optics. Optical elements with feature sizes smaller than the wavelength of the incident light, sometimes referred to as sub-wavelength elements, interact differently with the light than optical elements with larger feature sizes. More specifically, light interacts in different ways with periodic structures having spacing smaller than its wavelength. The altered interactions create a variety of devices with characteristics that differ from similar but larger structures.

The sub-wavelength optical elements offer some new, desirable characteristics. Sub-wavelength grating structures, for example, may have a wider acceptance angle than conventional gratings. Wider acceptance angles relieve manufacturing tolerances and may permit automated assembly, further reducing costs. Sub-wavelength gratings also perform more consistently over a broader range of wavelengths than conventional gratings.

Fig. IV–J.1: Replicated lens array on a silicon wafer. (CSEM photo).

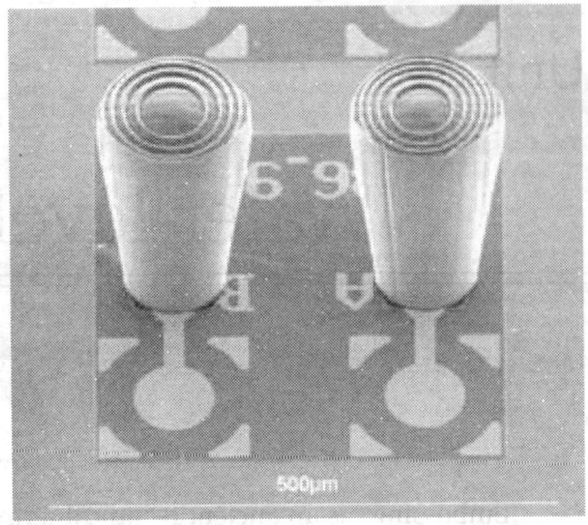

Fig. IV–J.2: Portion of a diffractive lens array on a VCSEL wafer, 200x.

NIL has demonstrated a growing variety of nano-optical components, including gratings, filters, anti-reflective coatings, refractive and diffractive components, fresnel lenses, waveguides, and lens arrays. With all of these potential products and proven benefits, why has nano-optics languished so long in the laboratories? Because formidable difficulties must be surmounted to replicate these tiny dimensions in volume manufacturing processes.

Nanoembossing Processes

To understand the manufacturing challenges that are now being overcome, look at the basic processing techniques that have been developed. NIL in its various forms has five processing steps:

- Design and fabricate a master stamp with features patterned by high-resolution techniques.
- Impress the master stamp on a layer of suitably prepared material.
- Maintain the master stamp pressure while the material first flows to duplicate the stamp, and then solidifies to preserve the duplicate pattern.
- Separate the patterned replica from the master stamp, without damaging nanoscale features.
- Post-process the replica further for its desired optical function.

The starting point for NIL is the creation of a master stamp or mold. The ultimate resolution capability of the embossing depends solely on the master stamp. The design and fabrication of the master stamp is critical to minimum feature size, uniformity, fluid flow, and ease of separation. Designing a stamp to allow quick and complete fluid flow of the softened material is challenging. Once the design is satisfactory, it must be transferred to the master. Fine-line approaches, such as electron beam, are used to create nanoscale features on the master stamp. Electron beams have fine-line capability, but direct e-beam writing has a working area only a few millimeters square, and is relatively slow. Stamps used for hot embossing or for injection molding are commonly made from nickel or steel. Stamps for UV embossing must be of UV-transparent materials.

While the completed master stamp could be directly used to produce many replicas, wear or deterioration through repeated use would require creating a new master stamp. In practice, to preserve its pristine quality, the completed master stamp generally is cloned, using the replication techniques described here. Cloned sub-masters are used to produce the nanoscale features in volume production, either simultaneously over a whole substrate, or repetitively, at chip size, onto a substrate.

The substrates may be silicon, quartz, glass, or other materials. Replication layers are generally coatings of heat-curable or UV curable polymers, with carefully controlled coating uniformity and thickness. While polymers are common materials for embossing, they may not offer the dimensional precision or the long-term stability under challenging environmental conditions demanded of some system components. This need has been filled by the development of special materials, such as of sol-gel organic/inorganic composite materials. These composites can be UV cured or thermally cured like polymers, and after curing display optical and mechanical properties close to those of glass.

Embossing and curing the pattern on the substrate varies with different processes. Once the coating is cured, the stamp must be separated from the cured material, without damaging the replicated pattern. As feature size becomes smaller, clean separations are more difficult. A special coating may be applied to the stamp before molding, to facilitate separation.

The embossed wafer may be post-processed by etching, deposition, or by repeating further lithographic processes. The cured patterned material can be the etch resist, for reactive ion etching of the underlying substrate. Alternatively, the patterned material may be used as a resist for metal deposition and lift-off. Post-processing might also include additional lithography, further depositions, and the addition of antireflective or protective coatings.

Hot Embossing

Hot embossing of wafers uses the cloned master die as the wafer stamp. Wafers are first coated with a thermoplastic polymer, that is, a polymer that softens when heated. The wafer is positioned in a controlled-atmosphere stamping chamber. After heating the wafer, the stamp is brought down with pre-determined pressure and temperature to soften the polymer and simultaneously form the image on the whole wafer. After allowing time for polymer flow during heating, the stamp and wafer are cooled to harden the image. To maintain the pattern, pressure must continue to be applied until both stamp and substrate are cooled below the thermoplastic softening point.

The hot-embossed wafer may be used for its embossed features, or it may be post-processed as a resist for etching, material deposition, and lift-off patterning. For resist use, curing may be followed by a light etch to remove the remaining traces of materials from the clear areas.

Step and Stamp

An alternative hot embossing approach is step-and-repeat stamping, using a smaller than wafer size master stamp. Step-and-stamp imitates the action of step-and-repeat optical steppers used in semiconductor fabrication. The smaller master stamp is simpler to design, produce, and control. It is also less costly than a full wafer stamp. The step-and-repeat approach mitigates the dimensional, planarity, flow, and separation anxieties inherent in wafer embossing, by literally reducing the problem. The tradeoffs include the added time required for multiple repetitions of the stamping process, the need to confine heating to the stamp area, and the cost and complexity of a precision step-and-repeat apparatus.

UV Embossing

Cold embossing uses UV light to polymerize and solidify materials after forming by the stamp. The starting materials are thin films of UV-curable materials. Typical materials may be monomers, or inorganic-organic hybrid polymers, which are polymerized by UV to polymers with desirable optical properties. The films may be deposited onto silicon, gallium arsenide, or similar wafers. The UV-transmissive patterned master template is pressed against the film, imprinting the pattern, which is UV cured before pressure is released.

UV embossing allows accurately aligning and overlaying multiple levels of features. UV embossing has a clear advantage over hot embossing in allowing multi-level applications, since the first level is polymerized after exposure, making it irrevocably formed and dimensionally stable, regardless of subsequent layer applications and exposures.

UV embossing is a low pressure, room temperature process, with little shrinkage of the pattern. It also allows partial UV casting, by limiting the illumination to maintain clear areas on the wafer. Cold embossing may be done on one or both sides of the wafer. Double-sided embossing allows micromolding. The laser lens array shown in Figure IV–J.1 was formed by cold embossing.

As with hot embossing, UV embossing may be used with a single wafer size stamp, or with a smaller stamp in a "step and flash" procedure similar to the "step and stamp" described above.

Injection Molding

Injection molding of optical components offers throughput and cost advantages for high-volume production. Injection molding can produce 3-D objects and integrated multicomponent systems. To move beyond conventional injection molding and capture nanofeatures, electroformed metal inserts are nanopatterned and included as part of the master mold.

Double-sided injection molding at wafer scale has been used to create optical structures on both sides of a supporting substrate. These structures may be the same on both sides of the substrate, or they may be different; for example, combining a diffractive lens on one side with a refractive lens on the other. The two sides may be processed separately, or simultaneously. In either case, precise alignment must be made between the features through the intervening substrate.

Injection molding has been used to produce optical components in hundreds of thousands. However, the high tooling cost can only be justified for these large quantities. Injection molding draws upon a wider range of materials than direct embossing. Acrylics or polycarbonates may be used for all-plastic systems in low-cost, high-volume injection molding, producing inexpensive commercial products in the hundreds of thousands.

Comparison of NIL Techniques

Table IV–J.1 provides a summary comparison of the three embossing processes. Hot embossing, at wafer level or by step-and-stamp, lends itself to single side patterning on thermoplastic materials. Process temperatures are typically 90°C above the glassivation temperature. Throughput is limited by the time required to heat, hold, and cool the die and substrate. Step and stamp avoids the high cost and tolerances of a wafer-size stamp, at a penalty in multiplying the time delays.

Table IV–J.1: Comparison of embossing production processes

Process	*Cure method*	*Process temp*	*Process time*	*No. of sides*	*No. of layers*	*Other*
Hot:						
Wafer	Thermo-	High	Medium	Single	Single	Roll to roll
Step/Stamp	plastic		Slower			possible
UV:						
Wafer			Medium			Requires UV
	UV cure	Room		Double	Multi-	
Step/flash			Slower		layer	transparency
Injection	Heat cure	Varies	Faster	Double	3-D	Tooling cost

Hot embossing has some well established and perhaps more profitable commercial applications. Music CDs are commonly replicated in volume by embossing. Roll-to-roll hot embossing has been used to add interesting color effects to wrapping paper, and to produce material with difficult to counterfeit optical patterns for security documents, such as identification cards.

Cold (UV) embossing may be preferred for critically dimensioned materials. It can be a room temperature process, eliminating the heating and cooling times and thermal expansions/contractions of hot embossing. It allows single or repeated multiple patterns to be layered on one or both sides of a substrate with accurate alignments. Both sides of the substrate may be embossed with identical or different patterns. For applications with more critical planarity, dimensional tolerances, or environmental requirements, cold embossing of photosensitive organic-inorganic sol-gel materials gives results close to those of glass.

Injection molding produces 3-D molding, allowing entire systems to be molded as a unit. It is not a room temperature process; processing time and temperature depend on the materials used. These may be inexpensive commercial plastics, or highly accurate specialty materials. Patterning can be single- or double-sided, in 3-D, although not generally multilayered as in cold embossing. The high tooling cost makes injection molding practical for high-volume (e.g. 100,000 units and up) products.

Manufacturing Progress

A mixture of techniques and equipment from the semiconductor and micropackaging industries has been adapted to nano-optics fabrication. Much of the development work reported above used one-of-a-kind laboratory equipment, built for that purpose, or modified equipment, originally developed for other purposes. One report referred to a modified semiconductor mask aligner. Another recent paper reported demonstrating hot embossing using a standard flip chip aligner-bonder. One nano-optics manufacturer has developed customized robotic embossing systems specifically for volume production. Only recently have major equipment manufacturers announced production equipment intended to meet the challenges of nanometer optics. Dedicated aligner-steppers with sub-20-nm embossing resolution are now available.

Product applications based upon NIL include wavelength monitors, beam splitters, pattern generators, polarization filters, switches, and many others. Figure IV–J.3 shows a nano-optic pattern generator for forming a cross-hair pattern from a single laser beam. Nano-optics allows generating essentially an arbitrary pattern of dots in this manner.

Also commercially available now are replicated components that are high performance building blocks for fabricating optical assemblies. These offer manufacturing advantages in comprising a common family of similar products, allowing standardization of manufacturing operations and equipment. Figure IV–J.4 shows a typical component size.

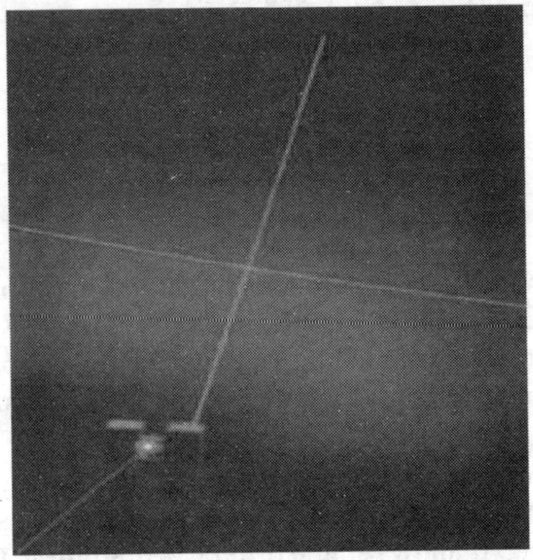

Fig. IV–J.3: Replicated pattern generator forming cross-hair pattern from a single laser beam.

Fig. IV–J.4: Nano-optical building blocks, 1.4mm square, 0.5mm thick.

Embossing has come a long way since Chaucer. After decades of laboratory development, nanoembossing promises to open new optical frontiers with truly monolithic production systems. The growing commercialization of nano-optic systems will subject them to the scrutiny and judgment of the marketplace. Success in surmounting the manufacturing challenges will determine whether nanoembossed optical systems remain a technology niche, or become a broad new technology opportunity.

Unit IV—K

Near-Field Optical Microscopy

Background and Motivation

Near field scanning optical microscopy (NSOM) was developed in the mid 1980s as a means to break the diffraction limit on spatial resolution attainable with optical measurement. Traditional NSOM utilizes a tapered, metal-coated optical fiber with a small aperture as either an excitation source or collection device. The spatial resolution attainable with this technique is roughly defined by the size of the aperture itself. Unfortunately, rough metallic coatings on glass fibers, coupled with the finite skin depth of metals at optical frequencies, place a limit on the spatial resolution attainable with fiber probes. Typical spatial resolution for the fiber approach to NSOM is ~50-100 nm, though careful probe and sample preparation have yielded ~30 nm resolution.

Research groups have long been interested in mapping out the spatial distribution of the proteins in photosynthetic membranes. These proteins are approximately 10 nm in diameter, and electron microscopy indicates that they are closley packed in their native configuration. The density of this arrangement precludes optical mapping with either far-field measurement or traditional aperture NSOM.

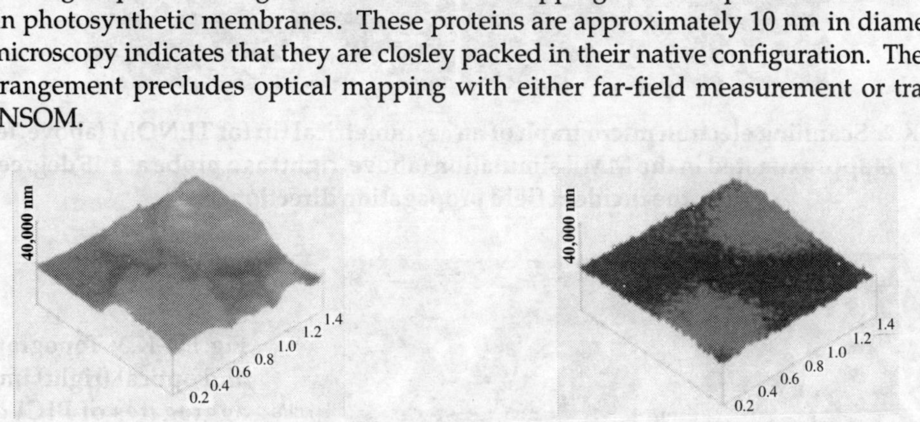

Fig. IV–K.1: Aperture NSOM image of thylakoid membrane fragments from chlamydamonis. Left image is topography, right image is near field fluorescence signal. The spatial resolution in the optical image is insufficient to identify individual protein units in the membrane.

Apertureless NSOM

In the mid-1990's, a number of groups began working on what would prove to be the solution to the fundamental resolution limitations in aperture NSOM. The idea was to spatially map an optical signal from the surface of a sample by scattering that signal with a solid probe. The choice of probe ranged from tungsten STM probes to metallized AFM tips to silicon probes. While optical image interpretation was complicated by topographic coupling, the spatial resolution in the apertureless technique was improved to ~10's of nm.

In 1999 the Xie group began work on a type of apertureless NSOM that takes advantage of the phenomenon of field enhancement by noble metal probes. With the appropriate combination of excitation polarization and probe geometry, a gold or silver probe is capable of strongly enhancing the magnitude of the incident illumination. This strong, localized field can be used as an excitation source for a spectroscopic transition of interest, such as fluorescence or Raman scattering. By locally exciting a spectroscopic transition in the sample and detecting the Stokes shifted light in the far field, the interpretation of apertureless NSOM images is greatly simplified.

They utilize the nonlinear response of the sample by exciting two-photon fluorescence in the sample. This tip enhanced nonlinear optical microscopy (TENOM) amplifies the contrast. The enhancements achievable with specifically fabricated gold probes are sufficient to yield near-field two photon fluorescence signal above the background generated in the diffraction-limited far-field region. Furthermore, TENOM has improved spatial resolution to the sub-20 nm regime. This high resolution and strong signal generation should be capable of, for example, mapping the spatial distribution of PSI and PSII in the thylakoid membrane.

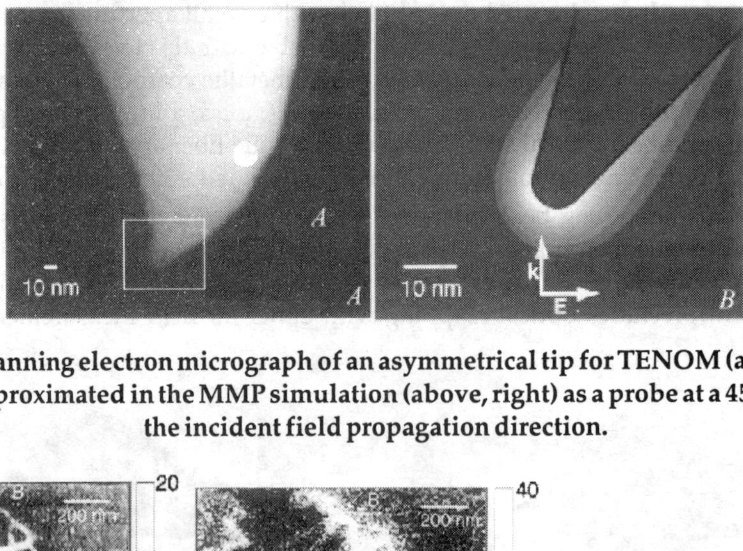

Fig. IV–K.2: Scanning electron micrograph of an asymmetrical tip for TENOM (above, left). This geometry is approximated in the MMP simulation (above, right) as a probe at a 45 degree angle to the incident field propagation direction.

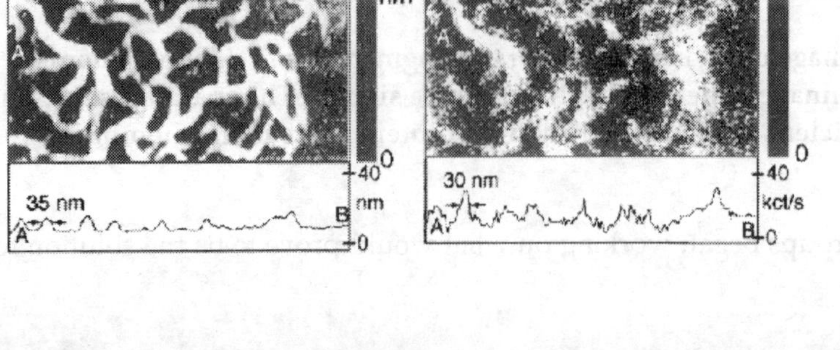

Fig. IV–K.3: Topographic (left) and optical (right) images of J-aggregates of PICI dye. The optical signal is two-photon excited fluorescence. Fast energy transfer along the J-aggregate fiber and away from the tip reduces the quenching of the fluorescence by the tip.

While TENOM has presented a solution to issues of high resolution optical imaging, it does have several intrinsic drawbacks. These drawbacks include the quenching of non-energy-transferred fluorescence, the necessity to carefully tailor the probe geometry to yield high field enhancement, and the practical requirement of a nonlinear signal source.

Current Research

Currently exploration is on to get a variety of approaches to eliminate the few drawbacks of TENOM and to expand the capabilities of apertureless NSOM in general. The first concern has been to address the problem of fluorescence quenching by the metal probe. To this end, an ion- and electron-beam assited deposition (IBAD and EBAD, respectively) of dielectrics has been applied to the apertureless probes. The growth of glass with these techniques allows us to build a spacer layer between the probe and the sample, reducing quenching while still maintaining high field enhancement at the sample. The theoretical treatment of fluorescence quenching by probes of complex geometries is currently being explored through the use of finite difference time domain (FDTD) simulation and Poynting vector integration.

Due to the complex nature of light-metal interactions, electromagnetic simulation is vital in the TENOM and apertureless NSOM research. The FDTD technique is a flexible, powerful method for fully three dimensional simulation of the response of metal probes to incident radiation. The power of the FDTD technique lies largely in its conceptual simplicity. A geometry of interest is defined by a number of cells, called Yee cells. The complex permittivities of each cell are assigned. After a geometry has been defined, an incident optical field is simulated, and the Maxwell's equations are integrated in a leapfrog fashion. The result is a method that can calculate the resultant fields for any arbitrary geometry illuminated with optical sources.

FDTD has been applied to the study of NSOM in the past. Unfortunately, the large computational burden associated with this technique made three dimensional simulations quite difficult. Two dimensional calculations have generated insight into the aperture method of NSOM. In this research a commercial FDTD package (RemCom's XFDTD) was utilized to simulate apertureless probes in three dimensions, working to adapt this microwave-focused package to the complex permittivities intrinsic to the optical regime behavior of real metals. This commercial software, coupled with the recent tremendous improvement in computational power, has provided a general method for simulating complex geometries, such as groups of spheres.

In order to expand apertureless NSOM to linear signal sources, intensity enhancements on the order of 10,000 were required in these experiments. In order to develop a physical intuition for the nature of strong field enhancement, and to rationally design probes for high field enhancement, FDTD was used to model apertureless NSOM. The simulations indicated that probes with finite size on the scale of the excitation light were preferable to the "quasi-infinite" tips used in the past. Furthermore, by properly tuning the size of a right trigonal pyramid to the excitation light, field enhancements sufficient for exciting linear signal sources and yielding signal above the far field background were predicted. The simulated high field enhancement has been experimentally verified.

A focused ion beam instrument was used to make apertureless NSOM probes. This instrument (FEI Strata DB-235) allowed to mill materials with ~10 nm resolution, image with ~5 nm resolution, and deposit glass or platinum with ~20 nm resolution. This powerful tool for nano-optics and near field microscopy applications, including surface enhanced Raman scattering (SERS) and chemical sensors was utilized.

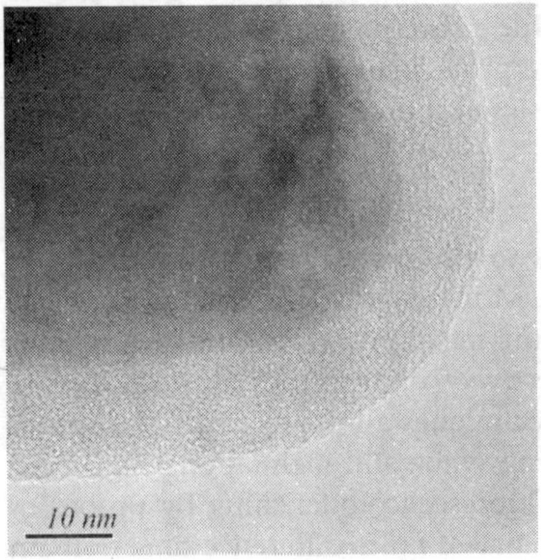

Fig. IV–K.4: TEM image of a SiOx coated metal probe (above). The dielectric coatings on these types of probes should help alleviate the problem of fluorescence quenching by the metal tip.

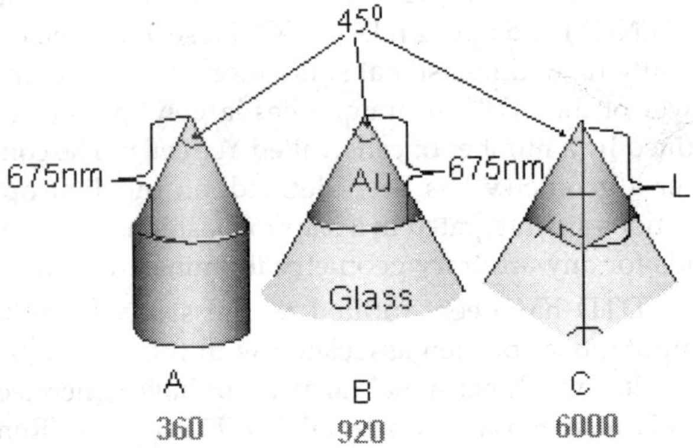

Fig. IV–K.5: Schematic representation of gold probes modelled with FDTD (above). From left to right, the probes are a quasi-infinite tip, a finite cone, and a right trigonal pyramid. The approximate intensity enhancement factors are indicated below the renderings.

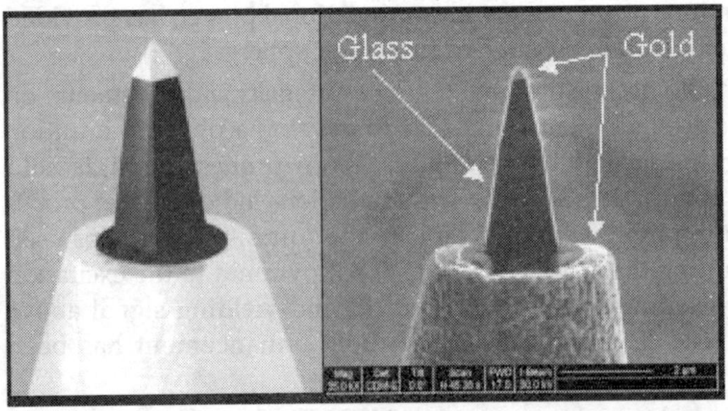

Fig. IV–K.6: An artist's rendering of a gold pyramidal probe (left). FDTD simulation shows that this type of probe should yield large field enhancement. A scanning electron micrograph of a gold pyramidal probe (right.) This probe was made with an FEI focused ion beam (left), as are many of our probes.

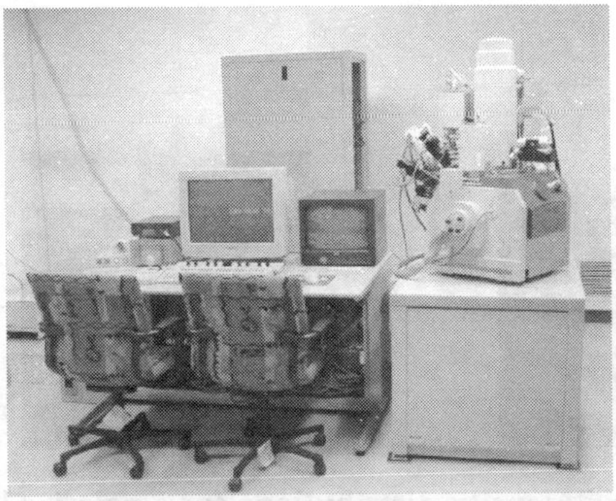

Fig. IV–K.7: FEI focused ion beam allows milling NSOM probes on the nanometer scale.

Electromagnetic coherence theory for optical near fields refers to the electromagnetic field at the immediate vicinity of the sample where the evanescent or non-propagating waves are strong. The near field is a primary example of a highly non-paraxial field and, in general, it is composed of three orthogonal electric field components. The conventional tools of electromagnetic coherence theory are, however, valid mainly for beam-like fields. We have put forward the definitions for the degree of polarization of general three-dimensional fields and for the electromagnetic degree of coherence. We have shown that in the limit of complete coherence, the electromagnetic coherence tensors, both in the space-time and space-frequency domains, behave analogously to the corresponding scalar functions. Furthermore, we have rigorously formulated the electromagnetic theory of coherent modes, and constructed the mode expansion for a class of statistically homogeneous and isotropic electromagnetic fields. Currently, we are investigating the optical coherence in a near-field detection system, and tightly focused electromagnetic fields.

Degree of Polarization in Tightly Focused Fields

We study the state of partial polarization of tightly focused optical fields. In the vicinity of such a focus the electromagnetic field displays a complicated three dimensional structure. For example, the light is fully polarized on concentric rings around the optical axis even when the incident light is unpolarized in the 2D sense. The distribution of the degree of polarization in the image plane of an imaging system of high numerical aperture is shown in Fig. IV–K.8. The rings of full polarization are clearly resolved in the image.

Boundary-Integral Method

We have modeled the electromagnetic field in various nanostructures using a boundary-integral method in two dimensions. In the boundary-integral method the electromagnetic scattering problem is casted into a form of an integral equation involving integrations only over the boundaries of the system. We have applied the method to analyze changes in the degree of polarization in light transmission through a tip aperture of a scanning near-field optical microscope. Furthermore, we have carried out spectral analysis of light transmission through sub-wavelength slits in metal films.

Nanoparticle spectroscopy

We study experimentally the detection and spectroscopy of very small gold nanoparticles. Such particles are an interesting model system to study the transition from individual atoms to the optical

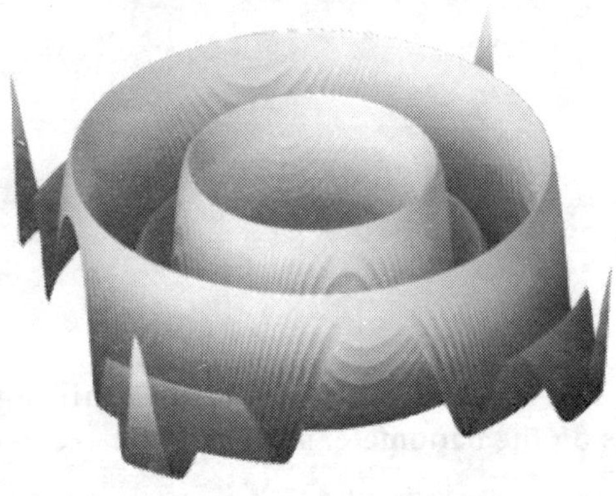

Fig. IV–K.8: . The distribution of the degree of polarization in the focal plane of an imaging system of numerical aperture 0.9. The incident field is taken to be unpolarized in the two-dimensional sense.

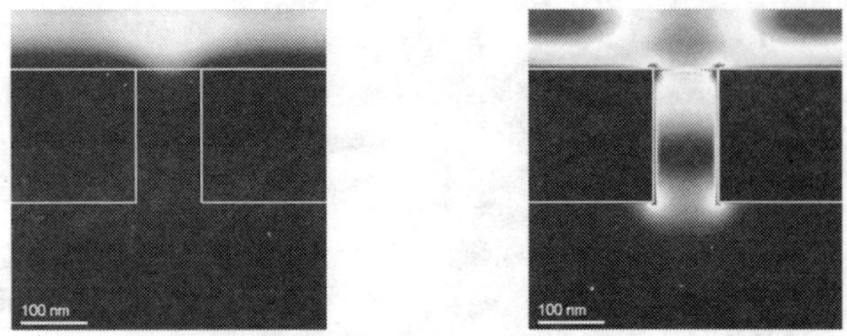

Fig. IV–K.9: Intensity distributions for s-polarized (left) and p-polarized (right) light (λ = 532.5 nm) around a nanoslit structured on aluminum film on a glass substrate.

properties of bulk matter. Nanoparticles also have important applications in chemical sensing and as labels in biology. In our first experiments we used supercontinuum light produced in a photonic crystal fibre to detect sub 10-nm gold particles. Figure IV–K.10 displays a confocal microscope image of 15 nm gold particles. The cross section shows that the supercontinuum light can be focused to almost a diffraction limited spot. Figure IV–K.11 shows the spectrum of the light scattered by an individual particle. The scattering spectrum displays a clear resonance which can be used to distinguish gold nanoparticles from other scatterers in a biological medium. This work was performed in the nano-optics group of professor Vahid Sandoghdar at ETH Zürich. We are currently continuing this work at HUT with experiments on the absorption spectroscopy of individual metallic nanoparticles.

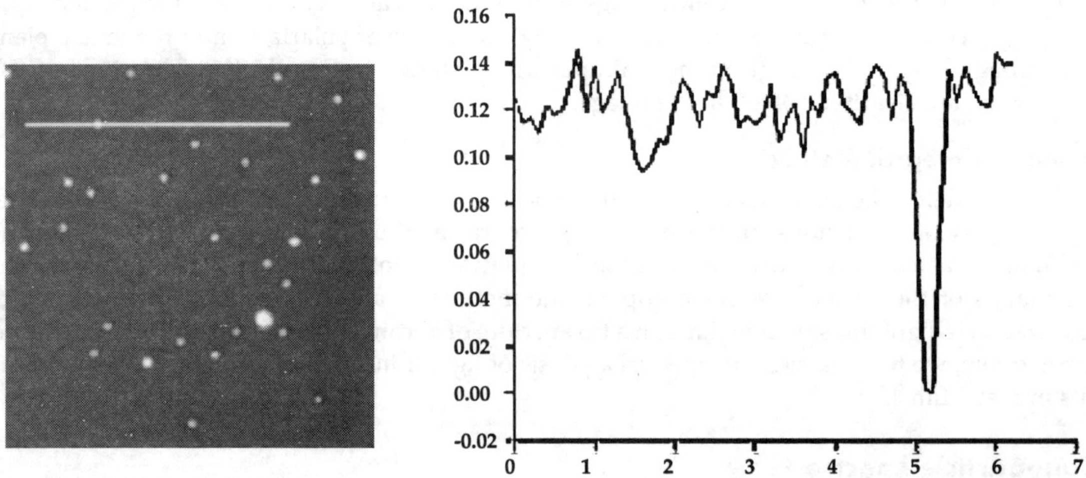

Fig. IV–K.10: A confocal microscope image of 15 nm gold particles. The cross sections demonstrates the excellent focusability of our white-light source.

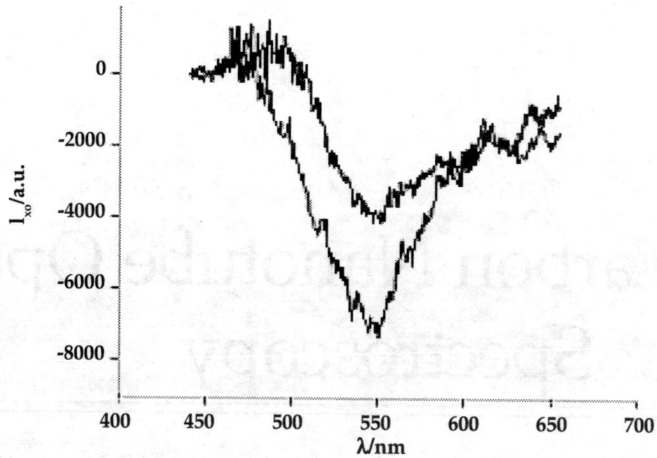

Fig. IV–K.11: The scattering spectrum of two separate 10 nm gold particles. The plasmon resonance is clearly seen in the spectrum.

Unit IV–L

Single Carbon Nanotube Optical Spectroscopy

Carbon Nanotube Optical Properties

Single walled carbon nanotubes (SWNTs) are tubular graphitic molecules that can be microns long and less than one nanometer wide. The combination of the unique properties of the carbon–carbon bond and their interesting molecular structure have given SWNTs unique physical properties and provided the potential for revolutionary applications. SWNT structure is characterized by two integers (n,m) that can be easily related to both its diameter and chirality. SWNTs have an unusual electronic configuration: each SWNT with a distinct (n,m), where n–m divisible by three is metallic, while the rest are semiconducting Further, SWNT electronic properties exhibit a strong structural dependence which helps drive potential applications.

Simple one electron models indicate that the density of states (DOS) of a nanotube is characterized by singularities, with each singularity corresponding to a quantum subband. (Figure IV–L.1) Recent studies of individual SWNT electronic structure using scanning tunneling spectroscopy (STS) agree with these general predictions. Simple one electron models also imply that the dipole-allowed optical transitions occur between mirror image peaks in the DOS (Figure IV–L.1). Absorption spectra of SWNT bundles contain three broad absorption regions corresponding to the first and second allowed transitions for semiconductor SWNTs, and the first allowed transition for metallic SWNTs, respectively. However, these spectra do not exhibit any features identified with a particular nanotube structure, since strong electronic coupling mixes the energy states from the different SWNTs in the bundle. This strong intertube coupling, combined with the large inhomogeneity in nanotube structures present in an ensemble, completely obscures any fine structure in the absorption spectra. Isolation of individual nanotubes is critical for decoupling the optical response of one nanotube from another, and was recently demonstrated by O'Connell et al. for nanotubes manufactured by the high-pressure CO chemical vapor deposition (HiPCo) process (H-SWNTs). As shown in Figure IV–L.2, absorption spectra of an ensemble of isolated SWNTs (Mic-SWNTs) show many more highly resolved features than the corresponding corresponding absorption spectrum from nanotube bundles. These distinct features correspond to dipole-allowed optical transitions from SWNTs with different (n,m) values and/or from higher

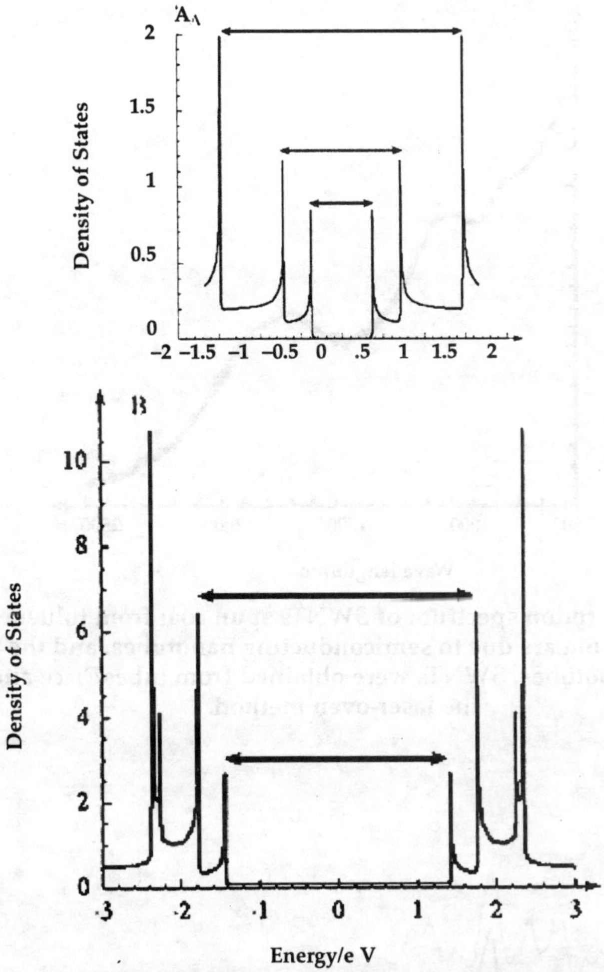

Fig. IV–L.1: Calculated density of states for A) a (10,5) nanotube and B) a (9,0) nanotube within the tight-binding approximation. The absence of electronic states at the Fermi level (0 eV) indicates the (10,5) tube is semiconducting, while the presence of electronic states at the Fermi level indicates the (9,0) nanotube is metallic. The arrows correspond to dipole-allowed optical transitions.

interband transitions for a nanotube with a given (n,m) value. The general isolation procedure can be applied using numerous surfactants, as well as nanotubes produced by various methods such as the laser oven (L-SWNTs), and arc discharge (A-SWNTs) processes (Figure IV–L.2).

Bundled SWNTs do not fluoresce since any metallic tubes present in the bundle provide an efficient nonradiative pathway for photoexcited electrons. However, for HiPCo SWNTs, Mic-SWNT samples show strong, dipole allowed fluorescence at the bandedge (Figure IV–L.3). The measured Stokes shift between the absorbing and emitting lines is quite small (_40 cm_1). SWNTs are poor emitters; the fluorescence quantum yield (QY) of SWNTs is on the order of QY _10_4.

The dependence of the SWNT electronic structure on diameter and chirality [(n,m) values] is fundamentally important and is also of technical interest. Simple one electron models predict that

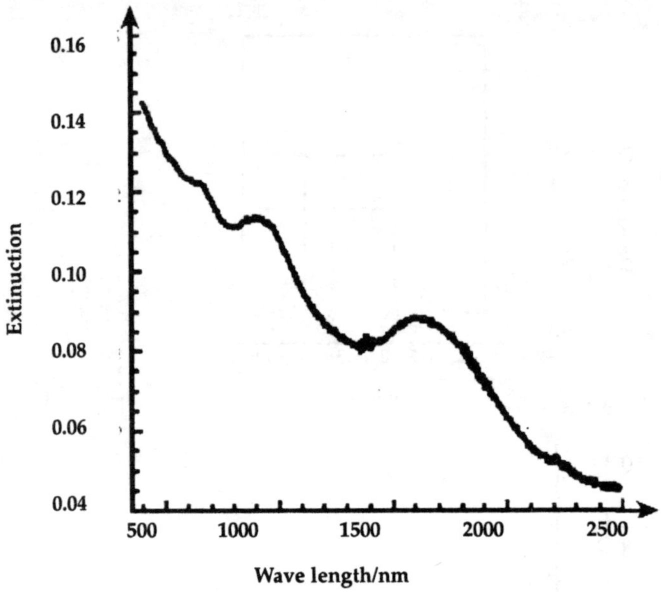

Fig. IV–L.2: Optical absorption spectrum of SWNTs spun coat from toluene onto quartz. The peaks at 1700 and 950 nm are due to semiconducting nanotubes, and the peak at 650 nm is due to metallic nanotubes. SWNTs were obtained from tubes@rice and produced by the laser-oven method.

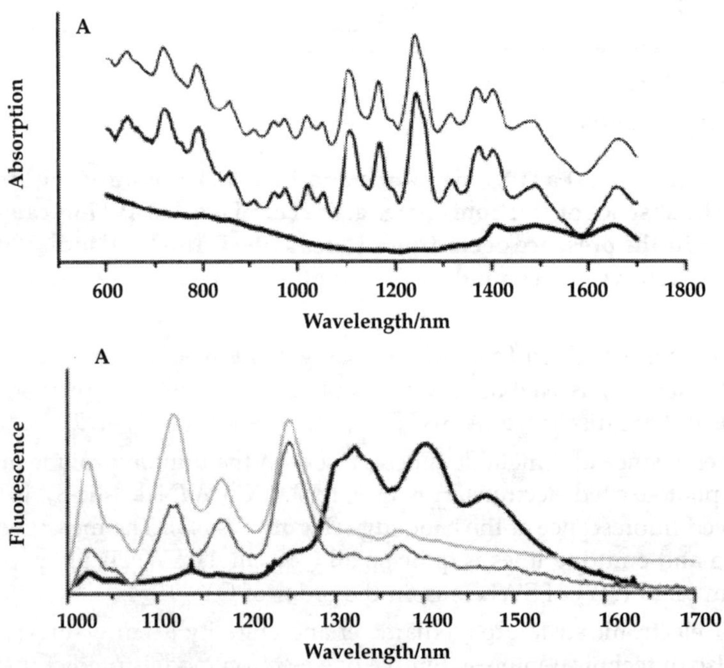

Fig. IV–L.3: Absorption (A) and fluorescence (B) spectra of Mic-SWNT samples isolated in surfactant micelles in D2O. The top, middle and bottom curves correspond to H-SWNTs, A-SWNTs, and L-SWNTs, respectively. Spectra are vertically offset for clarity. Due to detector responsivity fluorescence spectra do not extend past 1650 nm.

the energy of the dipole-allowed optical transitions depends linearly on the nanotube diameter. However, the observation of structurally dependent photoluminescence from SWNTs represented a major breakthrough, as the dependence of electronic transition energies was shown to depend on (n,m) values explicitly and not simply on the SWNT diameter. In particular, the magnitude and sign of deviations from the predicted linear dependence were observed to depend on the value of (n-m) mod 3, suggesting that trigonal warping effects were playing a significant role. Empirical fitting parameters have been introduced modifying the simple one-electron picture to compensate for these effects, thus allowing for a precise prediction of the transition energies for different tube (n,m) structures. However, despite the observation of fluorescence spectral features attributable to specific nanotube (n,m) structures, it is not fully understood why the transition frequencies differ significantly from simple theoretical predictions, although many-particle interactions and/or trigonal warping effects likely play a role in this discrepancy.

Notwithstanding these recent advances in SWNT separation and isolation, the optical spectra shown in Figure IV–L.3 clearly suffer from severe inhomogeneous broadening due to the distribution of different nanotube structures in the ensemble. Thus, early optical studies could only determine an averaged overall behavior that obscured crucial electronic characteristics of the individual nanotube. For example, since spectra from SWNTs with different (n,m) values will overlap, it is difficult to make definite conclusions regarding important emission features such as the spectral line width or details of the band shape, or to observe evidence of coupling to vibrations. Therefore, new approaches are needed to further advance the understanding of SWNT optical properties and accelerate the de velopment of novel SWNT photonic applications, such as nanometer scale integrated electroluminescent devices.

Single Nanotube Fluorescence

Spin-casting the Mic-SWNT suspension onto a cover slip allows for isolation of individual nanotubes by standard confocal microscopy (Figure IV–L.4). A typical fluorescence image from the spun Mic-SWNT sample is shown in Figure IV–L.5, with a spatial resolution of 300 nm. The image was obtained by collecting an optical spectrum at every pixel and then binning regions of the individual spectra where strong fluorescence signals are expected according to the ensemble spectrum (Figure IV–L.5). Since atomic force microscopy (AFM) measurements reveal predominantly short SWNTs (length 200–300 nm) lying on top of residual surfactant patches, distinct tubular features in the Mic-SWNT fluorescence images are typically unresolved. Alternatively, by dispersing SWNTs directly in organic solvents such as dichloroethylene, followed by spin coating, occasionally individual nanotubes that fluoresce can be isolated, as shown in Figure IV–L.6. In Figure IV–L.6 both the fluorescence and Raman images clearly show the characteristic one-dimensional features of the SWNT thus proving that the fluorescence in fact results from the single nanotube.

Fluorescence images from the isolated Mic-SWNTs show distinct, bright spots at different positions for different emission wavelengths. Figure IV–L.7 displays representative spectra detected from one of these bright spots. Each spectrum now exhibits a single emission band with a smooth line shape; all traces of contributions to the fluorescence spectra from the different SWNT (n,m) structures have been eliminated. The fluorescence maximum for all three spectra matches three transitions observed in the ensemble spectrum (Figure IV–L.6). In so much as the ensemble fluorescence spectrum is a weighted superposition of spectra from individual SWNTs, the spectra in Figure IV–L.4 are attributed to individual SWNTs.

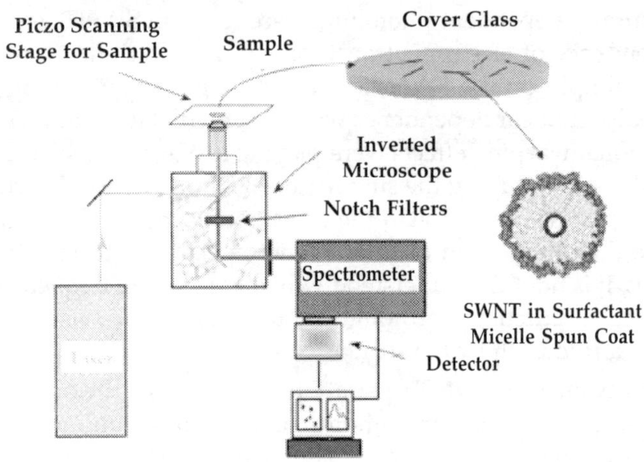

Fig. IV–L.4: Standard confocal microscopy setup for observing fluorescence from individual SWNTs. Excitation was typically at 633 nm (tunable dye laser) with intensities on the order of 40 kWcm^{-2}. Laser excitation at 633 nm ensures that all Raman signals between 633 and 770 nm are spectrally distinct from the fluorescence signals above 850 nm.

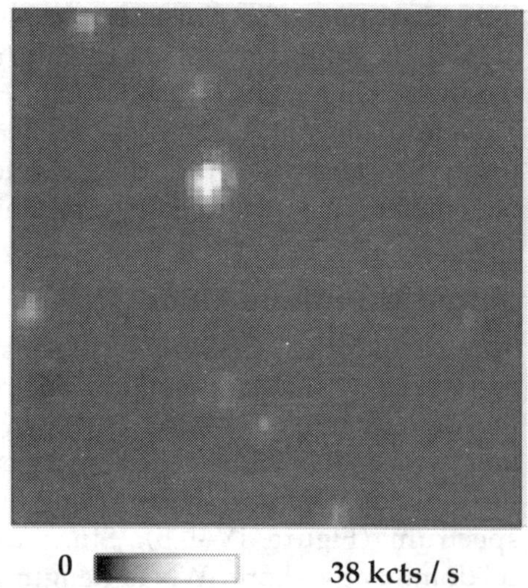

Fig. IV–L.5: Typical confocal fluorescence image (10_10 mm^2) of single spatially isolated SWNTs on glass acquired by raster scanning the sample and simultaneous detection of an optical spectrum at every pixel. The image contrast was obtained by integrating the detected spectrum between 945–975 nm.

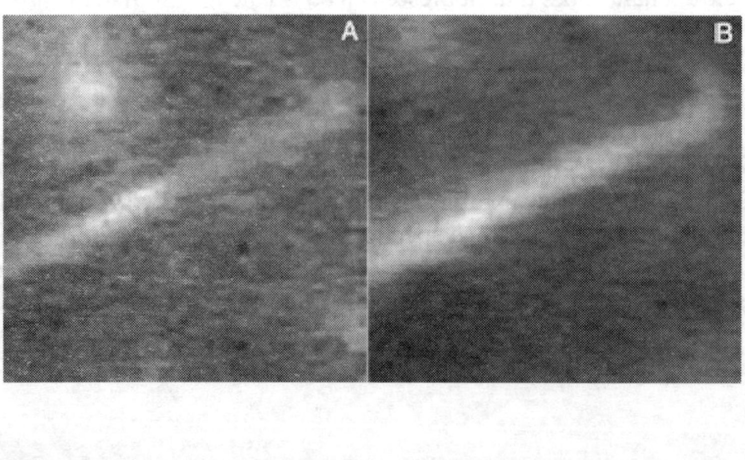

Fig. IV–L.6: Fluorescence image (2.5_2.5 mm^2) of a single spatially isolated SWNT on glass. The image was obtained by filtering the detected optical emission between 925–975 nm. (B) Resonance Raman image from the same area as (A). The image represents a spatially resolved detection of the tangential (Gband) at 1550 cm^{-1}.

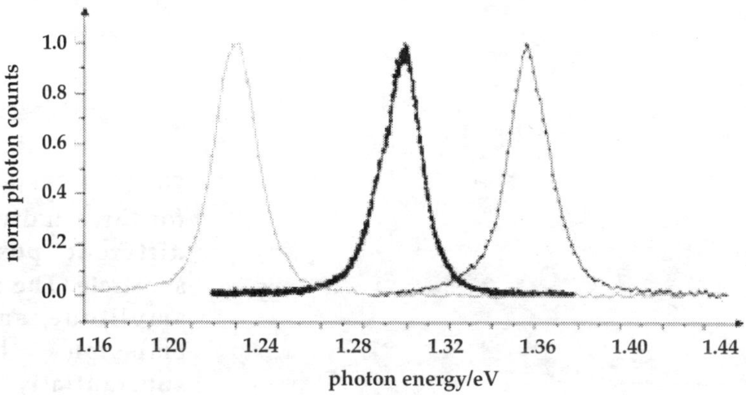

Fig. IV–L.7: Fluorescence spectra detected for three different sample positions. Fluorescence maxima are at wavelengths of 1016, 955, and 914 nm, respectively, from left to right.

Single nanotube spectroscopy clearly eliminates the overlapping spectral features associated with ensemble averaging and thus allows for the potential discovery of crucial SWNT optical and electronic properties. For example, SWNT fluorescence spectra are described by a single, featureless, symmetrical band corresponding to a single Lorentzian line shape function (solid lines in Figure 12.8. The room-temperature-broadened fluorescence line width is approximately 23 meV, in agreement with the approximately 25 meV linewidths for fluorescence from macroscopic samples. Also, single nanotube spectroscopy reveals that the fluorescence spectra from SWNTs with identical (n,m) structural parameters are not the same. Indeed, as shown in Figure IV–L.8 fluorescence spectra from SWNTs with the same structure can be significantly different. This deviation suggests that SWNT fluorescence may come from trap states caused by structural or chemical defects. Alternatively, the deviations in fluorescence energy could arise from perturbations in the electronic band structure due to fluctuations in the local environment, such as from electrostatic surface potential changes or localized charges. Some insight is provided by recent experiments whereby fluorescence from only SWNTs was monitored. As shown in Figure IV–L.7, as the excitation wavelength is tuned, different nanotubes in the optical image move in and out of resonance. Thus, different nanotubes with the same (n,m) structure have different absorption energies (and therefore different electronic properties) arising from an inhomogeneity in their local environments. Assignment of a SWNT (n,m) structure with a particular optical transition relies on correlating that spectral feature with the radial breathing mode (RBM) frequency through resonant Raman scattering. For macroscopic samples this connection is difficult, since uncertainties in parameters used for calculating the relevant (resonant) optical transition energy, and the cluster of different SWNT structures with similar diameters lead to several possibilities for assignments. Thus, complicated fitting procedures were employed to assign optical transition energies to particular (n,m) values. Single nanotube spectroscopy has the advantage in that it allows for a direct assignment of fluorescence energies to specific (n,m) nanotube structures without any ambiguity arising from using complicated fitting procedures (Table IV–L.8). Thus, as a result of the single nanotube spectroscopic measurements, the assignments were directly confirmed.

SWNT Fluorescence Intermittency

The fluorescence from individual semiconductor quantum dots and from small molecules exhibit

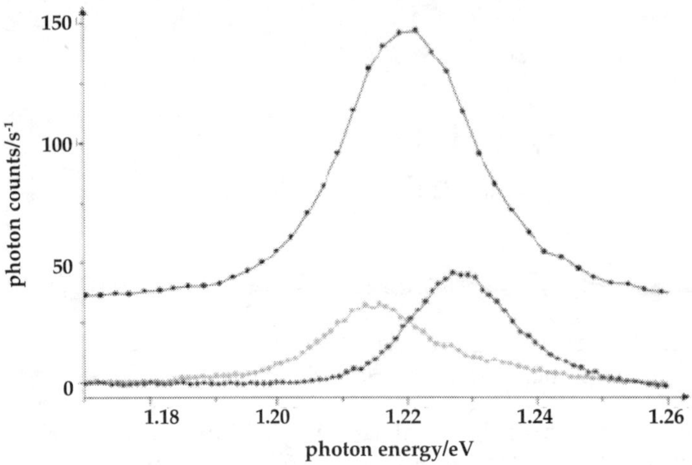

Fig. IV–L.8: Fluorescence spectra for three individual SWNTs at different positions on the substrate. The spectral position, amplitude, and shape of the emission bands differ substantially. The spectra are offset for clarity.

an emission intermittency or on-off "blinking" behavior for all excitation intensities, all temperatures, and on time-scales spanning many orders of magnitude (Figure IV–L.9). In fact, fluorescence intermittency is commonly used as the conclusive indicator of single chromophore optical emission. Surprisingly, the fluorescence intensity from a single SWNT at 300 K recorded using moderate excitation intensities (<70 kWcm^2) is constant over many seconds with no indication of fluctuations on the 40-ms to 100-second time scale (Figure IV–L.9 A). At extremely high excitation powers (>100 kWcm^2), fluorescence intensity fluctuations can be observed, likely due to laser-induced sample heating.

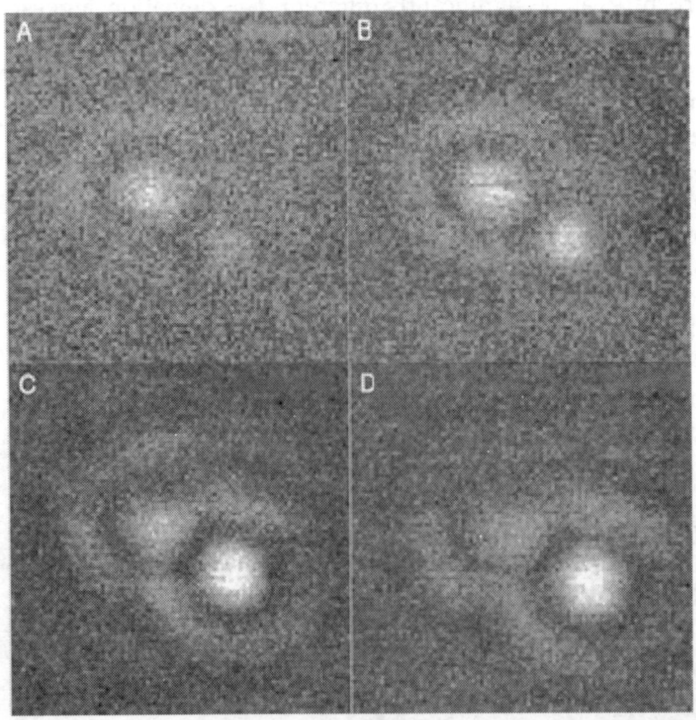

Fig. IV–L.9: Fluorescence images acquired through an 880_20 nm filter that selects fluorescence from only (6,4) nanotubes. The excitation wavelength is A) 600 nm, B) 595 nm, C) 580 nm, D) 575 nm. Scan range is 2.3_2.3 mm.

The use of single-photon sources is extremely desirable for applications in quantum optics, such as quantum cryptography, since the presence of even two photons in a signal pulse is an opportunity for an eavesdropper to steal or manipulate information. Thus, single molecules or single nanoparticles are attractive as single-photon "on demand" sources in quantum optics. However, their unavoidable fluorescence intermittency drastically hampers their potential utility. For example, in quantum cryptography it is difficult for the receiver to determine whether an "off" period is related to the information being transferred, or just the result of a random blinking event. Some stable sources of single photons do exist, such as nitrogen vacancies in diamond nanoparticles. However, these sources have a very broad linewidth typically over 100 nm, thus limiting the bandwidth of any potential devices.

Table IV–L.1: Spectral data and structural assignments for SWNTs. λ_{Em} ($h\nu_{Em}$) is the measured average fluorescence wavelength (energy). ν_{RBM} is the SWNT radial breathing mode frequency

λ_{Em} *[nm]*	$h\nu_{Em}$ *[eV]*	$h\nu_{Em}$ *[eV] literature)*	*Observed* ν_{RBM} *[cm^{-1}] +3 cm^{-1}*	*Predictd* ν_{RBM} *[cm^{-1}]*	*(n,m)Assign-ment*
1023	1.212	1.212	282	281.9	(7.5)
976	1.270	1.272	308	307.4	(6.5)
955	1.298	1.302	296	298.1	(8.3)
915	1.355	1.359	309	307.4	(9.1)
881	1.407	1.420	337	335.2	(9.4)

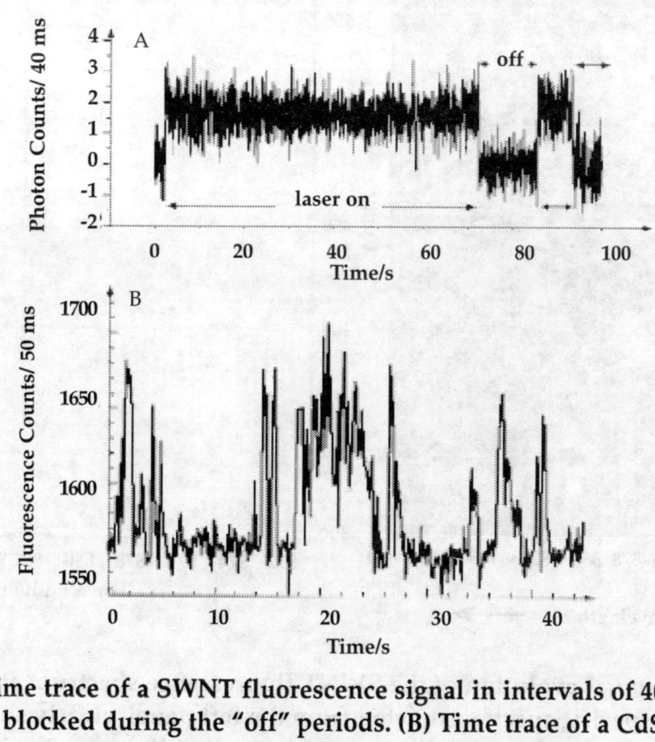

Fig. IV–L.10: (A) Time trace of a SWNT fluorescence signal in intervals of 40ms. The excitation was physically blocked during the "off" periods. (B) Time trace of a CdSe quantum dot fluorescence signal in intervals of 150ms.

SWNTs have many advantages for possible use as single-photon sources for quantum optics applications. On the millisecond to minutes time scale, and at room temperature, SWNT fluorescence is stable showing no signs of intensity or spectral blinking. SWNTs also have an extremely narrow linewidth, and can be pumped electrically which allows for the exciting possibility of future nanometer scale integrated single-photon photonic devices. SWNTs also are resistant to photobleaching, as their fluorescence intensity does not change despite hours of intense photoillumination. Finally, unlike dye molecules or impurities in diamond, SWNTs have an easily size-tunable emission that is in the infrared at technologically relevant wavelengths (1.3–1.5 mm).

Single molecule sources are also highly desirable for applications in biology and biotechnology. For example, the use of single fluorophores to track individual molecules inside a cell is a compelling method for studying intracellular processes. However, photobleaching of organic dye labels severely restricts the single molecule processes that can be studied in cells. Detection of a single analyte as reported through changes in the optical emission of a single fluorophore is also a promising approach to achieving the ultimate in detection sensitivity. On the other hand the fluorescence intermittency of single molecules makes the fluorometric detection of single analytes extremely difficult, since the modulation of fluorescence upon binding of the analyte cannot be distinguished from a random blinking event.

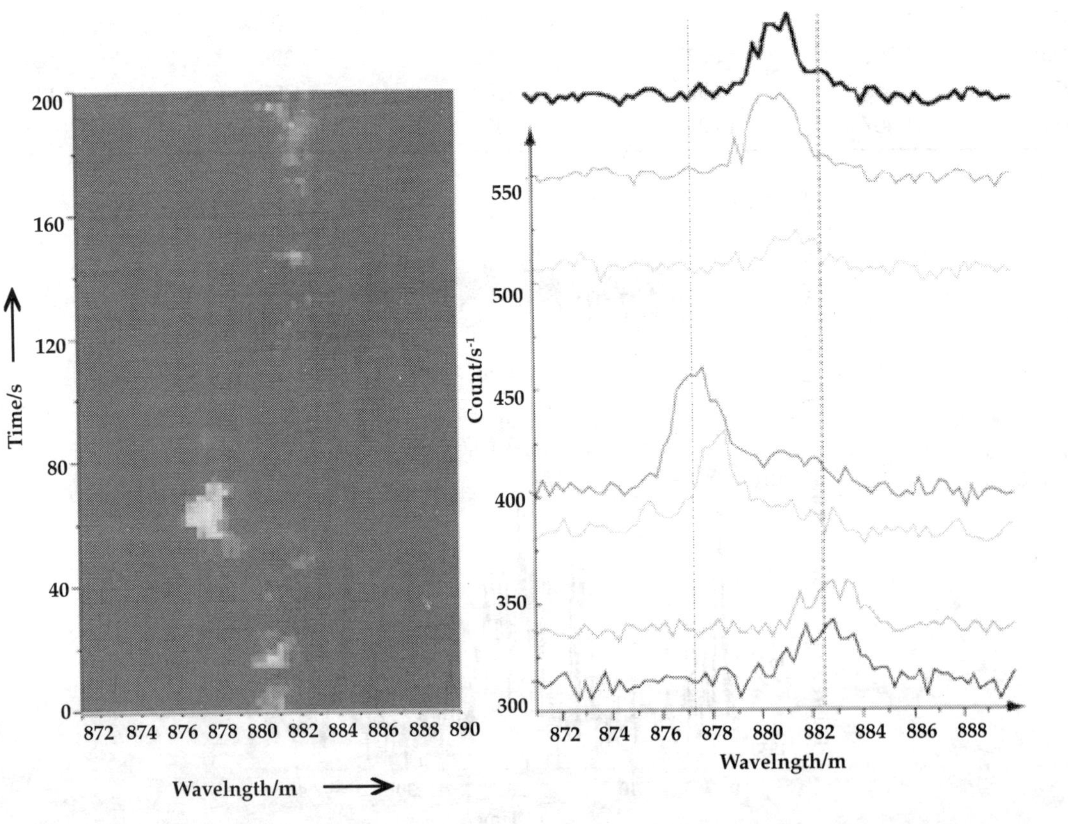

Fig. IV–L.11:. Temporal evolution of the SWNT fluorescence spectrum at T=1.8 K. The right panel represents specific spectra extracted from the left panel, clearly showing evidence of spectral diffusion and intensity blinking.

SWNTs hold great promise for possible use as single-molecule fluorophores. First, SWNT fluorescence does not photobleach at 300 K even after hours of continuous laser excitation. Thus, extremely long-lived biological processes on the singlemolecule level can be followed much more easily. Second, the fact that SWNTs do not exhibit fluorescence blinking allows them to be used as fluorometric reporters of a single biological agent. SWNT photonic devices show strong near-infrared (NIR) electroluminescence, allowing for easy "on-chip" integration of single SWNT sensors with silicon technology. Finally, SWNTs are optically active at wavelengths in the range of 800–1100 nm, which allows for NIR excitation and detection. This NIR optical activity has the tremendous advantage that background fluorescence from the sample can be dramatically reduced allowing for much improved sensitivity on the single fluorophore level.

For some nanoparticles the processes that contribute to fluorescence blinking events slow down as the temperature is lowered. For example, in CdSe semiconductor nanocrystals a higher fraction of nanoparticles are "on" (or bright) at low temperatures versus at room temperature. Interestingly, at low temperatures (T=1.8 K) a significant fraction (about 2/3) of SWNTs show an intermittency in their fluorescence intensity, and also a spectral wandering of their fluorescence energy as shown in Figure IV-L.11. Spectral diffusion cannot be observed at room temperature, since the fluorescence linewidth is much larger than the change in energy during a spectral jump. However, it is currently unclear whether the processes that cause blinking "turn on" at low temperatures, or whether SWNTs blink inherently on timescales much faster than 10 ms. With the time resolution of past experiments, fluctuations in the sub-millisecond regime were not observable.

Conclusions

Combined fluorescence and Raman scattering measurements on the single molecule level open up new ways to characterize SWNTs. Single nanotube studies enable the connection between fluorescence energy, (n,m) structural parameters, and RBM frequency of a particular SWNT with a single measurement, and without any complicated fitting procedures necessitated by working with macroscopic samples. Single nanotube studies also reveal new insights into nanotube optical properties, such as a variation of the fluorescence energies of nanotubes having the same structure. Unlike for single molecules, for SWNTs the photoluminescence unexpectedly does not show any intensity or spectral fluctuations at room temperature. The fact that SWNTs show no emission intensity blinking or bleaching demonstrates that SWNTs have the po tential to provide a stable, single molecule infrared photon source, with an extremely narrow linewidth, which would have a significant impact for future nanometer scale integrated photonic devices and which shows promise for applications in quantum optics and biological sensing.

Unit IV–M

Optical Processing

Nano-Optics Redefine Rules for Optical Processing

A new class of devices is smashing the size, manufacturing, cost and reliability hang-ups that have hampered optical elements to date.

The commercial introduction of wafer-based, nanofabrication techniques now makes it possible to create a new class of optical components called subwavelength optical elements (SOEs). With physical structures far smaller than the wavelength of light, the physics of the interaction of these fine-scale surface structures with light yields new arrangements of optical-processing functions. These arrangements have greater density, more robust performance and greater levels of integration when compared with many existing technologies and could fundamentally alter approaches to optical system design.

However, as with any new technology, the designer must clearly understand the general concepts underlying the technology as well as its potential impact (in this case, on optical component design), as well as implications for the evolution of the technology.

Current Technology Limits

Integrated optical component and module design has achieved a high degree of sophistication in the combination of discrete optical elements. The properties of optical materials and their combinations, clever optics design, and advances in alignment and assembly are applied to increase component and module density, reduce costs and increase reliability. But there are limits. The natural properties of many discrete optical elements are relatively fixed, limiting the flexibility of the optical component designer and reducing transferability from design to design. This is where SOEs come in.

SOEs are the realization of nanotechnology as applied to optical elements. As such, they offer excellent optical properties while being easy to integrate with other optical materials in a broad range of configurations. In addition, they self-integrate, allowing the combination of optical functions to reduce parts count and increase reliability, and to allow increased flexibility in optical component design. A broad range of optical effects can be achieved by the manipulation of nanostructures in the optical beam path.

Operating Principles

The interaction of light with subwavelength grating structures-gratings with one or more dimensions one or more orders of magnitude smaller than the wavelength of the incident light-produces a range of controllable effects. To explain those effects, a simple subwavelength grating structure is shown in Figure IV–M.1. It comprises a subwavelength grating structure on an optical substrate.

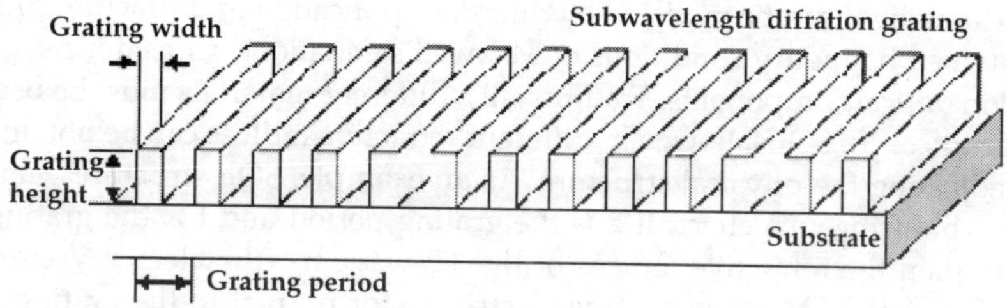

Fig. IV–M.1: Controlling the dimensions of a subwavelength grating determine whether it'll form a polarizer, waveplate or polarization-dependent filter.

When light is normally incident on the structure, the transmitted portion of the light-that which passes through the grating structure-is affected. The effect on the passed light can be varied by adjusting the size, shape and spacing of the grating structure. In particular, for the one-dimensional grating structure in Fig. IV–M.1, appropriate selection of dimensions will result in behavior as a polarizer, waveplate or polarization-dependent filter. More sophisticated effects, such as polarization-independent filters, can be achieved by two-dimensional gratings, as shown in Figure IV–M.2. The general class of these structures is referred to as SOEs.

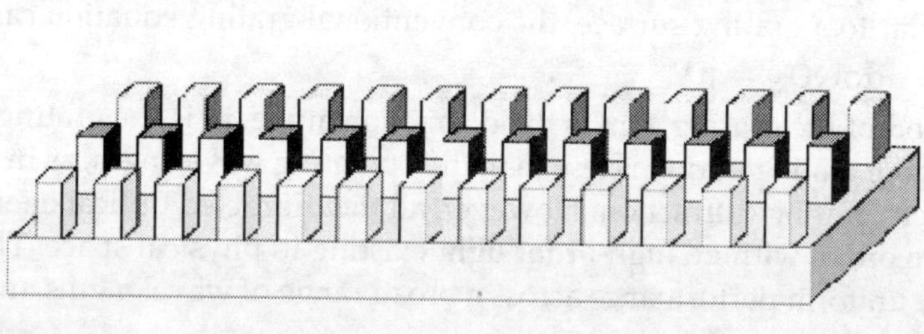

Fig. IV–M.2: Multi-dimensional grating structures enable more sophisticated functionality.

Because the diameter of the incident light beam is often much larger than the structures of the grating, the effect on the transmitted light is really the statistical sum of numerous local interactions between the light and the grating. For example, if a light beam measuring 300 microns in diameter is incident on a two-dimensional grating such as that shown (Figure IV–M.1 again), where the structure size is roughly 100 x 100 nanometers, more than a million nanostructures will be illuminated. Therefore, varying the grating dimensions spatially across the incident beam front allows additional control over optical processing effects.

The physics of the SOE depends on rigorous application of the boundary conditions of Maxwell's equations to describe the interaction of light with the structures. At the wavelengths of light used in telecommunications-980 through 1,800 nm-the structures required to achieve those effects have some dimensions on the order of tens to a few hundred nanometers. At the lower end of the scale, single electron or quantum effects may also be observed.

Whereas the principles of reflection, refraction, diffraction and interference describe the behavior of traditional optical elements, with SOEs the equations describing conventional optical behavior do not fully cover the resultant phenomena because quantum mechanical effects now come into play.

In many applications, subwavelength structures act as a nanoscale diffraction grating whose interaction with incident light can be modeled by rigorous application of diffraction-grating theory and the abovementioned boundary conditions of Maxwell's equations. Consider refraction, a key property exploited in optical components. Traditionally, different materials must be used to obtain different refractive indices. In SOEs, however, different refractive indices can be obtained with the same material by adjusting the physical structure. As an example, SOE structures can be used to create an "artificial" birefringence effect. If a is the grating period and t is the grating width (as per Figure IV–M.3), then the refractive indices of the TE wave n_{te} (the electric vector parallel to the grating grooves) and the TM wave n_{tm} (the electric vector normal to the grating groove) are expressed as:

$$n_{te}^2 = fn_1^2 + (1\text{-}f)n_2^2$$
$$n_{tm}^2 = n_1^2 n_2^2 / (fn_2^2 + (1\text{-}f)\, n_1^2)$$

where n_1 is the dielectric constant of the grating material, n_2 is the dielectric constant of the fill material and f is the grating filling factor and is defined as f = t/a. By choosing the SOE material and adjusting the grating filling factor, a birefringence effect can be achieved that is much larger that those achieved with standard components.

Many SOEs have periodic patterns, and so may be viewed as optical gratings. When incident light is perpendicular to a grating surface, the conventional grating equation can be expressed as:

$$\alpha\sigma\iota\nu\Theta_m = \mu\lambda$$

where a is the period of the grating, Qm is the diffraction angle, m is the grating order and l is the wavelength. When the grating period is less than the operating wavelength, as in SOEs, the incident light is still subject to grating diffraction. However, all the diffracted optical energy of the incident light will enter zero order, with no high-order light existing in physical space. This results in SOEs that have relatively uniform performance across a broad range of wavelengths and wide acceptance angles.

SOEs—Why Now?

With that sort of flexibility, why are SOEs just now becoming available for telecommunications applications? The primary reason has been manufacturability. While these optical effects have been researched for at least 20 years, cost-effective manufacturing of the optical elements has not been available. Building subwavelength grating structures in a research environment has generally required high-energy techniques, such as electron-beam lithography, or extremely precise process control, for instance, to grow nanostructures via "self-alignment." Second, there must be a consistent process for creating a broad range of nanostructure patterns. Many techniques for creating nanostructure patterns are limited in the range of possible patterns.

Nanoimprint lithography overcomes those limits and has four key steps: making a mold inscribed with the complement of the desired nanostructure, impressing this mold into a resist-coated wafer, separating the mold and selectively removing the resist with reactive ion etching to transfer the nanopattern to the target material (Figure IV–M.3).

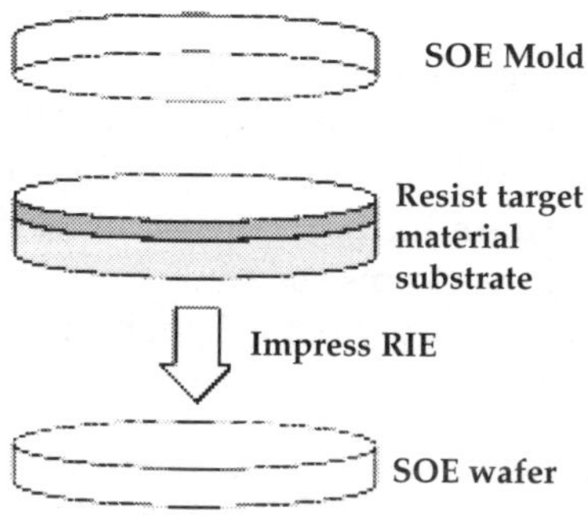

Fig. IV–M.3: Nano-imprint lithography solves the manufacturing issues associatedwith subwavelength optical elements (SOEs).

After that, postimprint processing may be used to add metal layers to enhance performance and to provide a protective overcoat for handling in a standard manufacturing environment. Testing and dicing follow. Because nanoimprint lithography forms the nanostructures of SOEs by a direct, physical process rather than by beams of energy, the wave diffraction, scattering and interference in the resist do not limit fabrication resolution.

Silicon dioxide is commonly used to make a mold patterned with the desired nanostructures. Various techniques, including electron beam lithography, can be used to create the desired negative image of a nanostructure. Because the mold can be replicated and reused, complex multistep and multiprocess methods can be used to create the desired nanostructure. Since those initial tooling steps do not need to be repeated for each production wafer, they are amortized over the full production run of a particular SOE. Using different molds with different nanostructure patterns allows the same manufacturing process to create the full range of SOEs.

Unpackaged for use in free-space applications, the final SOE is an optical chip with a subwavelength grating on one side of a substrate as shown in Figure IV–M.4. The overall thickness of the element is dependent on the thickness of the substrate. The (x,y) dimensions depend on application requirements. Figure IV–M.4 shows SOEs that are 1.4 × 1.4 × 0.5 mm thick.

Fig. IV–M.4: SOE optical chips are sized to meet optical component architecture requirements; here they are 1.4 × 1.4 × 0.5 mm.

Applying SOEs

The practical consequences of using SOEs can be explored by looking at a specific realization—an SOE polarization beam splitter/combiner, or PBS/C (Figure IV–M.5). By appropriate selection of the dimensions of a one-dimensional subwavelength grating structure, one polarization will be transmitted, while the orthogonal polarization will be reflected, as illustrated. At that scale, the subwavelength-diffraction grating exhibits only zero-order diffraction, resulting in a number of useful element characteristics. These include consistent performance across a broad range of wavelengths, the grating structure performs the same for wavelengths from 980 nm to 1,800 nm, and wide variations in the angle of incidence of the light beam that are up to 20 degrees of deviation from normal.

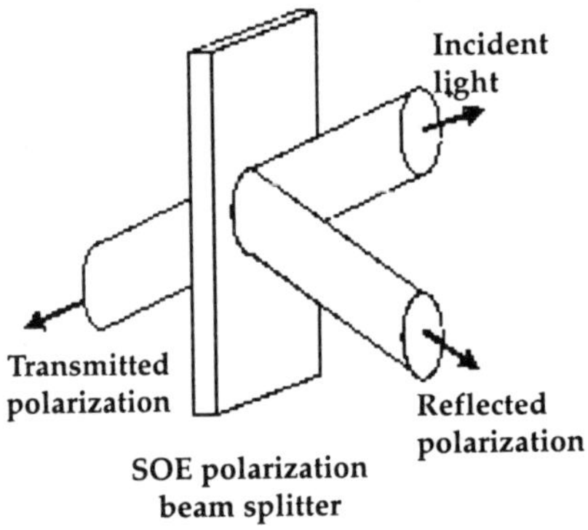

Fig. IV–M.5: An SOE-based polarization beam splitter/combiner achieves 180 degrees of effective separation in a thickness of less than a micron.

Typical performance parameters for an SOE PBS reveal insertion loss of less than 0.13 dB on both the reflected and the transmitted beams. Extinction ratios are greater than 40 dB and 20 dB for the transmitted and reflected beams, respectively. The flexibility of SOEs allows trade-offs around those parameters.

The benefits of a PBS/C SOE can be realized in two areas. First, there are architectural benefits in optical component design. Because of their small form factor, they allow more compact component designs. Reduced insertion losses are also realized because SOEs can be abutted to other elements with minimum free-space separation.

There are differences in the way that the elements handle light, for instance, the polarization beam splitter is a reflective device with 180 degrees of effective separation obtained in less than 1 micron. Because of those differences, the beam path can be laid out to simplify device design and possibly eliminate other optical elements. Further, the reflective nature of the SOE PBS lets it support high-energy applications associated with laser transmitters and fiber amplifiers.

Second, there are advantages for manufacturing. The wide acceptance angle simplifies the alignment process, reducing manufacturing time and cost. It is also possible to use automated "pick-and-place" manufacturing technologies. The SOEs are robust: with proper selection of materials they can be subjected to temperature ranges of –200°C to 400°C. Consequently, they can be subjected to a broad range of manufacturing processes. Finally, reduced size may both simplify packaging and reduce packaging costs.

Applications of the SOE PBS/C have been explored for fiber amplifiers, circulators and isolators, interleavers, optical switches and variable optical attenuators. Here, their size and low power consumption are attractive relative to traditional technologies.

The Future of SOEs

Current devices are the initial applications of SOEs, but the possibilities are expanding in several directions. SOE-based building-block functionality will be introduced both as chips and as packaged devices. Demonstrated SOE functionality spans polarizers, polarization beam splitters/combiners, filters, photodetectors and photonic bandgap devices. Dynamic control for switching, attenuation and tuning is also possible.

Monolithic integration of SOEs can be achieved by stacking SOE layers to create aggregated optical effects. Nanoimprint lithography permits direct layering on of SOEs without resorting to lamination techniques. Combining SOEs with optically active layers allows optical control circuits to be built, resulting in complex optical components "on-a-chip." Multilayer SOE integration has already been demonstrated by combining filters with photodetector arrays to create dynamic optical feedback loops.

Because SOEs are are self-compatible and fabricated via wafer-scale manufacturing with relatively minor differences from device to device, the means exist to implement them.

Unit IV–N

Optoelectronic Space Applications

Nano-Optoelectronic Devices for Potential Use in Space Applications

Background

Diffractive optical elements, optoelectronic transducers and photonic components, which play an important role in optical data communication, can be substantially improved by lateral nanostructures. With the development of lateral optoelectronic nanostructures the way to controllable diffractive optics is paved. For this, elements with specific interference structures are necessary, which act as specific and possibly controllable transmission or reflection filters. Nanostructured optoelectronic components (e.g. quantum well or quantum dot lasers, photonic crystals) offer large market potentials in the future, e.g. for optical data communication or in the range of consumer electronics (for example laser television).

Space Applications that will Use Nanostructured Optoelectronic Components

Nanostructured optoelectronic components offer promising application in space in the fields of optical satellite telecommunications or sensor technology (infrared sensors, high resolution charge-coupled devices, etc). With optical wireless data links (OWL) for intrasatellite communication as well as optical intersatellite links, significantly smaller and lighter devices a higher bandwidth could be realized in comparison to conventional microwave communications. Optical intersatellite links were demonstrated in the frame of the ARTEMIS mission of the ESA. For the data transmission, extreme frequency-stable solid state lasers (Nd:YAG lasers) are used, which are pumped with diode lasers.

How Laser Terminals can be Used in Scientific Applications

The German company, Tesat, is a leading manufacturer of laser terminals for optical intersatellite communication. Such laser terminals are also interesting for scientific applications; for example, as an injection seeder for a satellite-based Doppler-Lidar (ALADDIN), as a satellite-based measuring device for gravitation wave detectors (LISA, with SMART as demonstration mission), or as a frequency normal for a satellite-based FT-spectrometer (POISON).

Benefits of Using Quantum Dot (QD) Lasers

Semiconductor quantum dots, which have been manufactured in high quality by means of self-organization for about the last 5 years, offer a new degree of freedom in selecting the working wavelength of photonic elements. They allow manufacturers to cover almost completely the entire spectral region from the ultraviolet to the far infrared, with a small number of substrate materials. Further advantages of quantum dot lasers are a small energy consumption through low threshold current densities, a high modulation range for high-speed applications as well as an improved temperature stability. Figure IV–N.1 represents the threshold current densities of different types of semiconductor diode lasers.

Predicted Timescale for Commercial Use of Quantum Dot Lasers

Beyond that, an improved radiation hardness has already been proven by quantum dot lasers compared with quantum well lasers. First commercial uses of quantum dot lasers are expected in 2003. Figure IV–N.2 shows the schematic structure of a quantum dot Vertical Cavity Surface Emitting Quantum Dot laser (VCSEL).

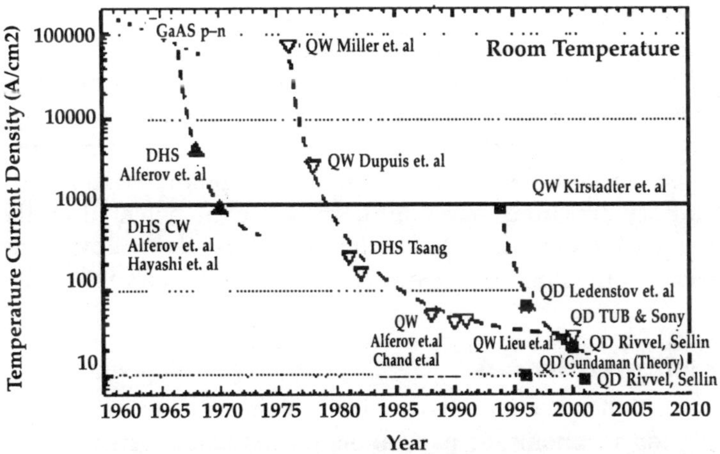

Fig. IV–N.1: Comparison of threshold current densities of different semiconductor lasers.

Fig. IV–N.2: Schematic structure of a QD Vertical Cavity Surface Emitting Laser (VCSEL).

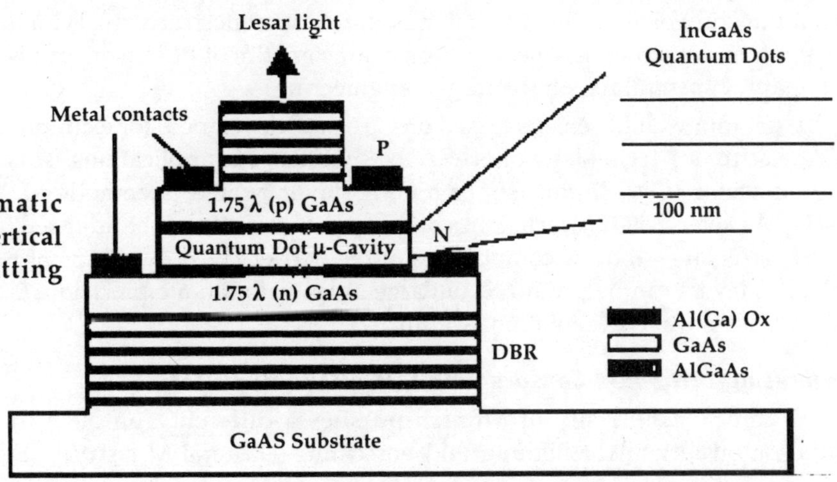

What Quantum Dot Lasers can be Used for in Space Applications

Due to their radiation hardness and the low energy consumption, quantum dot lasers in principle are relevant for space applications, e.g. as pump lasers for solid state lasers, which are needed for different applications. In order to realize the potential of quantum dot lasers in space applications, appropriate measures have to be accomplished by the space industry for the specification, system integration and space qualification.

Photonic Crystals

Photonic crystals are a further example of nano-optoelectronic components with application potential in optical data communication. Photonic crystals exhibit a periodic refractive index and possess an analogy to semiconductors in electronics, a "photonic band gap" for certain frequency in the visible and IR wavelength ranges. The lattice constant of photonic crystals lies in the range of half the wavelength of the light in the medium. For visible light this means that for the production of photonic crystals, a precision within the range of 10 nm is necessary. Two-dimensional structures today can be routinely manufactured with high precision.

Developing Three-Dimensional Photonic Crystals

At present, intensified efforts are made for the development of three-dimensional photonic crystals, e.g. with utilization of lithography and self-organization procedures, in which nanoscale colloids (e.g. from polymers or silicates) arrange spontaneously to a cubic lattice. These lattices are used as templates for lattices from more interesting materials such as metals and metal oxides. Three-dimensional photonic crystals would open up new possibilities in optical data communication (light could be guided and branched to arbitrary directions) and offer in principle the potential for the realization of purely optical circuits (optical computing). Such photonic transistors are however at present still very far from realization. In the long run, photonic crystals will find applications in optical satellite communications.

Potential Space Applications for Infra-red Sensors

IR sensors offer a multiplicity of application potential in space, e.g. for the satellite-based earth observation and atmosphere research, for astronomy, as navigation aid for space systems or for optical data communication. Approaches for the miniaturization and further improvement of infrared sensors are based among other things on the application of two- (quantum well), one- (quantum wire) or zero-dimensional (quantum dot) nanostructures. With the help of quantum well or quantum dot structures the detection characteristics of IR sensors can be adjusted selectively to the relevant spectral region (band gap engineering).

Quantum well IR sensors, based on GaAs are developed, for example, by the Center for Space Microelectronics Technology of NASA for special space applications. This quantum well infrared photodetector (QWIP) consists of a GaAs-layer, which is embedded, sandwich-like, in two $Al_xGa_{1-x}As$ layers. The characteristics of the quantum wells can be adjusted by varying the thickness of the GaAs layer and the composition of the barrier layer. By means of molecular beam epitaxy, nm-thick layers can be produced on large areas with atomic precision. Ga-As QWIP can also be used for long-wave IR radiation > 6 µm.

Optimising Infra-Red Sensors with Si/Ge Quantum Dots

The Technical University of Munich pursues a different approach in the frame of a BMBF (Bundesministerium für Bildung und Forschung—Federal Ministry of Education and Research) joint project on the self organization of Si/Ge islands on silicon. The research activities are focused on controlling the characteristics of epitactical, defect free Si/Ge islands on silicon substrates, which can be produced with the self-organizing, parallel Stranski-Krastanow procedure in the material system Si/SiGe. The objective here is to develop coupled systems with several layers of quantum dots in a homogeneous layer system, which exhibit new functions by charge transfer and electrostatic coupling and can be used as optical detectors particularly in the mid IR (MIR) range. With combined QD/QW structures a 50-fold increased photoresponse can be obtained compared with sole quantum dot structures. Expected advantages of QD MIR (mid-infrared) detectors to be mentioned are the extended spectral range, a high durability and reproducibility, radiation hardness and a low dark current. For the year 2003, the realization of a prototype 2-colour IR detector in co-operation with DaimlerChrysler was done.

Unit IV–O

Integrated Optical Circuits

Creating High Performance, Integrated Optical Circuits

Nano-optics is a valuably novel class of technology that takes advantage of light's unique interaction with subwavelength, nanoscale patterned materials, and nanotechnology enabled fabrication methods to create a broadly applicable optical device and manufacturing platform.

In optical circuits used in consumer electronics, communications, and security applications, nano-optic devices offer significant benefits by virtue of their novel behavior, small size, ease of integration, and lower overall costs. Through these virtues, nano-optics delivers compelling alternatives to conventional bulk optical solutions, and offers functionality and form factors that conventional optics cannot match. The following is a basic review of nano-optics theory and applications.

Function and Structure

The unique optical properties of nanooptic devices derive from appropriately patterning material on the nanoscale with structures whose critical dimensions are sized to be several times less than the wavelength of light at which they operate, thereby creating optical materials with highly useful modified optical functions.

Through a proper combination of materials and structures, nano-optic devices can perform any passive optical function including polarization filtering, phase retardation, spectral filtering, and propagation management (e.g., lenses and beam splitters). Various functions can be designed for both free-space- and waveguide-based applications. The former is far and away the most common in commercial applications.

Nano-optic devices also can be designed to operate over any wavelength range. The basic methodology is applicable to UV, visible, and IR wavelengths with appropriate variations in structural dimensions and materials. Additionally, they can form the core of an optical system. Practical nano-optic applications require a complete optical system comprised of the nano-patterned material and adjacent materials including the optical substrate and thin-film coatings.

The nano-optic structures are defined as a combination of material and physical attributes, as illustrated in Figure IV–O.1. Attributes can be divided in to three types:

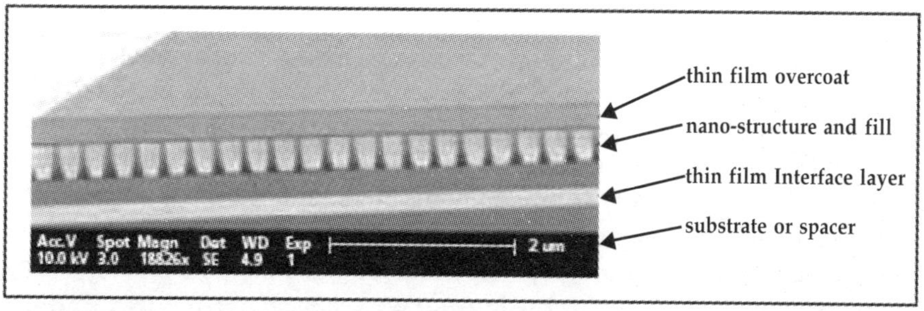

Fig. IV–O.1: A nano-optic system is a combination of a nano-structured grating layer, which can composed of a variety of materials, and bounding thin film layers on an optical substrate. The typical thickness of the nano-structure layer is <500 nm, with a period of 200 nm and finger widths of <100 nm.

- **Structural:** pattern (e.g., linear, mesh, and circular), dimensions (e.g., period, thickness, and duty cycle), and spatial variations (e.g., "chirping", arrays, and multiple layers).
- **Material:** nano-structure material (e.g., dielectrics, metals, and polymers), interstructure fill material, adjacent thin film materials, and substrate material (e.g., glass, plastics, dielectrics, and crystalline material).
- **System:** relationship of the nano-optic structures with the optical beam path and any actuator materials and structures (e.g., liquid crystal, solid state materials, electro-optic polymers, and MEMS).

Subwavelength structure size is a key to nano-optics' optical and physical advantages. Nano-optic elements employ microstructures one or more orders of magnitude smaller than the wavelengths of the incident light—with dimensions typically on the order of 10s to a few 100s of nanometers. At such dimensions, these structures may interact with light both according to the principles of classical optics and also in a realm of comparatively small-scale local interactions; rigorous application of the boundary conditions of Maxwell's equation describes the interactions of light with nano-optic structures. For very small structures, single electron or quantum effects may be observed. As a consequence, many nano-optic devices exhibit unique optical behaviors—such as high polarization discrimination in a micron-thin space, achromatic phase retardance, and single layer combinations of optical functions like polarization and spectral filtering—either by coincidence or intent. These nano-optics provide useful alternatives for optical circuit design; others can simply be designed as drop-in replacements for classical bulk optics.

In addition to unique behaviors, the devices also achieve their optical effects over relatively short distances—nano-optic devices are generally very thin, usually one micron or less in thickness. This has strong positive repercussions for integrated optics since nano-optic structures can often be applied, similar to thin films, as a "coating layer" during a complex manufacturing process. Single electron and quantum effects also enable optical performance to be readily customized. Modifying material properties by adding nano-structures results in a continuous spectrum of optical functionality—essentially this combination creates hybrid, "artificial" optical materials—that allows performance specifications and optical wavelengths to be incrementally and accurately varied to meet specific application requirements.

Manufacturing and Integration

Nano-optic structures can be realized through a variety of manufacturing methods. For commercial usefulness, an optical device is not optimally practical unless it can be manufactured in high volumes, in multiple application specific variations, and readily integrated with other relevant technologies. The first criterion is necessary to achieve economies of scale; the second, to ensure that the relevant market opportunity is broad enough to be interesting; and the third, to facilitate easy integration of nano-optic elements into more complex systems. The available fabrication approaches (self-assembly and nano-lithography) achieve these objectives to varying degrees.

Some materials can be caused to form regular, nanoscale structures under appropriate, controlled conditions—a.k.a. self-assembly. The difficulty with this approach is lack of flexibility in achievable structures and usable materials, which limits the functions that can be realized.

Nanolithography, which is akin to standard semi-conductor manufacturing, generally uses some method to create an image in a polymer resist layer; this image is then used as an etching mask to transfer the nanoscale pattern into the target material. Nanolithography methods include interference lithography, e-beam (electron beam) lithography, and nanopattern replication (see sidebar at bottom for definitions). Overall, each of these methods is optimal for some set of nanostructure patterns, materials, and volume requirements.

As a manufacturing method, nanolithography supports a range of integration modes for optical circuits:

- Optimization of traditional multi-element, multi-technology optical circuit architectures is possible by taking advantage of the unique optical behavior of nano-optics.
- Monolithic "self-integration" can be achieved by layering one atop the other to create aggregate optical effects.
- Spatial integration can be accomplished by organizing varying optical functions into an array structure via nanopattern replication.
- Hybrid integration is achieved by adding a nano-optic layer(s) to functional optical materials.
- Integration of nano-optics with optically active layers, such as liquid crystal or MEMS structures, can create electronically controllable optical devices.

Integrated optical devices and subassemblies based on nano-optic elements realize a range of benefits. Compared to conventional bulk optics, their size and optical properties make alignment in assembly more forgiving, less labor-intensive, and less costly. When physically combined with other nano-optics or other materials in manufacturing to create monolithic integrated optics, this eliminates the complication and cost of multidevice lamination, while improving reliability. Also, arrayed nano-optic devices can be applied in multi-beam or multi-path optical circuits, eliminating the need to individually align discrete optics.

In general, the process by which nano-optic elements are fabricated is flexible and robust, allowing for cost-saving manufacturing process integration in creating hybrid and monolithic optics; again, the resultant optics are often smaller, more robust, more functional, and easier to assemble. Because nano-optic devices are manufactured utilizing wafer-based processes proven in semi-conductor fabrication, this enables flexible sharing of manufacturing capacity and supports high-volume manufacturing throughput. Figure IV–O.2 shows a processed wafer and nano-optic devices.

Fig. IV–O.2: Nano-optic wafers are manufactured using a nanolithography process on 100 or 150 mm diameter wafers. Nano-optic chips are diced from the wafer.

From Concept to Application

Nano-optics draws on a growing "design vocabulary" that is functionally equivalent to the design libraries used in semiconductor applications. Multiple design elements can be used with complex, multi-step, and multi-process methods to create particularly sophisticated nano-patterns and multi-layer devices. The broad range of different nano-structure patterns, materials, and modes of integration means that there are no inherent limits in the functionality and operational wavelength of nanooptic devices. These attributes are solely driven by the needs of the application and the optical circuit.

To date, the commercialization of nano-optics has demonstrated the portability of the manufacturing platform and its products across multiple markets including optical data storage, digital imaging, projection display, and telecommunications. As nano-optic technology continues to propagate, it promises to further expand both in applications and in benefits realized by users of nano-optic based devices.

Nano-lithography Methods

Interference Lithography: Readily creates interference patterns with useful dimensions using UV light sources. The benefit of this method is simplicity; the difficulty is in creating complex shapes and arrays.

E-Beam Lithography: Writes arbitrary, complex patterns using a focused beam of electrons. The benefit is that almost any pattern can be created; a drawback for commercial production is that individual wafer processing times can be fairly long (tens to thousands of hours).

Nano-Pattern Replication: Relies on other methods, e.g., the two above—to create a master plate capturing the inverse of the desired pattern. A printing-like transfer method then patterns the polymer resist. The benefit is high-fidelity reproduction allowing extremely small feature size, complex patterns, effective material independence during repetitive production, and for different structures in the same production line. It is not applicable for limited runs or when only a fixed interference pattern is required.

Optical Disk Drives, Communications and Projection Displays

A nano-optic polarization beam splitter/combiner (PBS/C) provides 180 degrees of effective separation—meaning that one polarization is transmitted and the other is reflected—via a submicron-thick nanoustructural layer. This device's beam separation and thin form allow adjacent components to be closer coupled compared with conventional optics, reducing existing optical path lengths by up to one-half. In turn, shorter optical path lengths reduce beam divergence, improving coupling efficiency and reducing power loss. Many optical circuit designs also incorporate a wave plate adjacent to the PBS/C.

Digital Imaging, Communications and Sensors: Physically, a nano-optic enabled variable optical attenuator (VOA) consists of a crossed pair of polarizing nano-structures sandwiching a liquid crystal cell; applying an electrical field across the liquid crystal will rotate the liquid crystal molecules and control the fraction of light that is passed through the structure. This tunable optic addresses the frequent need to control and block the output power of polarized light sources, such as optical transceivers, in order to dynamically optimize their signals and to facilitate maintenance. Because the nano-optic polarizer is directly integrated into the liquid crystal cell construction, this allows the entire component to be less than 1 mm thick.

SECTION E

NANOROBOTICS

Unit V–A

Introduction to Nanorobotics

What is a Nanorobot?

A nanorobot is a tiny machine designed to perform a specific task or tasks repeatedly and with precision at nanoscale dimensions, that is, dimensions of a few nanometers (nm) or less, where 1 nm = 10^{-9} meter. Nanorobots have potential applications in the assembly and maintenance of sophisticated systems. Nanorobots might function at the atomic or molecular level to build devices, machines, or circuits, a process known as molecular manufacturing. Nanorobots might also produce copies of themselves to replace worn-out units, a process called self-replication.

Nanorobots are of special interest to researchers in the medical industry. This has given rise to the field of nanomedicine. It has been suggested that a fleet of nanorobots might serve as antibodies or antiviral agents in patients with compromised immune systems, or in diseases that do not respond to more conventional measures. There are numerous other potential medical applications, including repair of damaged tissue, unblocking of arteries affected by plaques, and perhaps the construction of complete replacement body organs.

A major advantage of nanorobots is thought to be their durability. In theory, they can remain operational for years, decades, or centuries. Nanoscale systems can also operate much faster than their larger counterparts because displacements are smaller; this allows mechanical and electrical events to occur in less time at a given speed.

Nanorobotics

Nanorobotics is an emerging field that deals with the controlled manipulation of objects with nanometer-scale dimensions. Typically, an atom has a diameter of a few Ångstroms (1 Å = 0.1 nm = 10^{-10} m), a molecule's size is a few nm, and clusters or nanoparticles formed by hundreds or thousands of atoms have sizes of tens of nm. Therefore, Nanorobotics is concerned with interactions with atomic and molecular-sized objects and is sometimes called Molecular Robotics.

Molecular Robotics falls within the purview of nanotechnology, which is the study of phenomena and structures with characteristic dimensions in the nanometer range. The birth of Nanotechnology is usually associated with a talk by Nobel Prize winner Richard Feynman entitled

"There is plenty of room at the bottom. Nanotechnology has the potential for major scientific and practical breakthroughs. Future applications ranging from very fast computers to self-replicating robots are described in Drexler's seminal book. In a less futuristic vein, the following potential applications were suggested by well-known experimental scientists at the Nano conference held in Palo Alto in November 1995:

- Cell probes with dimensions ~ 1/1000 of the cell's size.
- Space applications, e.g. hardware to fly on satellites.
- Computer memory.
- Near field optics, with characteristic dimensions ~ 20 nm.
- X-ray fabrication, systems that use X-ray photons.
- Genome applications, reading and manipulating DNA.
- Nanodevices capable of running on very small batteries.
- Optical antennas.

Nanotechnology is being pursued along two converging directions. From the top down, semiconductor fabrication techniques are producing smaller and smaller structures. For example, the line width of the original pentium chip is 350 nm. Current optical lithography techniques have obvious resolution limitations because of the wavelength of visible light, which is in the order of 500 nm. X-ray and electron-beam lithography will push sizes further down, but with a great increase in complexity and cost of fabrication. These top-down techniques do not seem promising for building nanomachines that require precise positioning of atoms or molecules.

Alternatively, one can proceed from the bottom up, by assembling atoms and molecules into functional components and systems. There are two main approaches for building useful devices from nanoscale components. The first is based on self-assembly, and is a natural evolution of traditional chemistry and bulk processing see e.g. [Gómez-López *et al.* 1996]. The other is based on controlled positioning of nanoscale objects, direct application of forces, electric fields, and so on. The self-assembly approach is being pursued at many laboratories. Despite all the current activity, self-assembly has severe limitations because the structures produced tend to be highly symmetric, and the most versatile self-assembled systems are organic and therefore generally lack robustness. The second approach involves Nanomanipulation, and is being studied by a small number of researchers, who are focusing on techniques based on scanning probe microscopy.

A top-down technique that is closely related to Nanomanipulation involves removing or depositing small amounts of material by using an SPM. This approach falls within what is usually called Nanolithography. SPM-based Nanolithography is akin to machining or to rapid prototyping techniques such as stereolithography. For example, one can remove a row or two of hydrogen atoms on a silicon substrate that has been passivated with hydrogen by moving the tip of an SPM in a straight line over the substrate and applying a suitable voltage. The removed atoms are "lost" to the environment, much like metal chips in a machining operation. Lines with widths in the order of 10 to 100 nm have been written by these techniques.

Nanorobotics research has proceeded along two lines. The first is devoted to the design and computational simulation of robots with nanoscale dimensions. Drexler's nanorobot uses various mechanical components such as nanogears built primarily with carbon atoms in a diamondoid structure. A major issue is how to build these devices, and little experimental progress has been made towards their construction.

The second area of Nanorobotics research involves manipulation of nanoscale objects with macroscopic instruments. Experimental work has been focused on this area, especially through the use of SPMs as robots.

To begin with we focus on the state of the art in the field of nano-robotics by describing various molecular level systems and associated design and control issues. Nanorobots are controllable machines at the nano (10^{-9}) meter or molecular scale that are composed of nanoscale components. With the modern scientific capabilities, it has become possible to attempt the creation of nanorobotic devices and interface them with the macro world for control. There are countless such machines that exist in nature and there is an opportunity to build more of them by mimicking nature. Even if the field of nanorobotics is fundamentally different than that of macro robots due to the differences in scale and material, there are many similarities in design and control techniques that eventually could be projected and applied. There are many applications for nanorobotic systems and its biggest impact would be in the area of medicine.

Nanotechnology can best be defined as a description of activities at the level of atoms and molecules that have applications in the real world. A nanometer is a billionth of a meter, that is, about 1/80,000 of the diameter of a human hair, or 10 times the diameter of a hydrogen atom. The size-related challenge is the ability to measure, manipulate, and assemble matter with features on the scale of 1-100nm. In order to achieve cost-effectiveness in nanotechnology it will be necessary to automate molecular manufacturing. The engineering of molecular products needs to be carried out by robotic devices, which have been termed **nanorobots**. A nanorobot is essentially a controllable machine at the nano meter or molecular scale that is composed of nanoscale components. The field of nanorobotics studies the design, manufacturing, programing and control of the nanoscale robots.

We talk about the state of the art in the emerging field of nanorobotics, its applications and discusses in brief some of the essential properties and dynamical laws which make this field more challenging and unique than its macroscale counterpart.

Nanorobots would constitute any passive or active structure (nanoscale) capable of actuation, sensing, signaling, information processing, intelligence, swarm behavior at nanoscale. These functionalities could be illustrated individually or in combinations by a nanorobot (swarm intelligence and co-operative behavior). So, there could be a whole genre of actuation and sensing or information processing nanorobots having ability to interact and influence matter at the nanoscale. Some of the characteristic abilities that are desirable for a nanorobot to function are:

(i) Swarm Intelligence—decentralization and distributive intelligence.
(ii) Cooperative behavior—emergent and evolutionary behavior.
(iii) Self assembly and replication—assemblage at nanoscale and 'nano maintenance'.
(iv) Nano Information processing and programmability—for programing and controlling nanorobots (autonomous nanorobots).
(v) Nano to macroworld interface architecture—an architecture enabling instant access to the nanorobots and its control and maintenance.

There are many differences between macro and nanoscale robots. However, they occur mainly in the basic laws that govern their dynamics. Macroscaled robots are essentially in the Newtonian mechanics domain whereas the laws governing nanorobots are in the molecular quantum mechanics domain. Furthermore, uncertainty plays a crucial role in nanorobotic systems. The fundamental barrier for dealing with uncertainty at the nanoscale is imposed by the quantum and the statistical mechanics and thermal excitations. For a certain nanosystem at some particular temperature, there are positional uncertainties, which can not be modified or further reduced.

The nanorobots are invisible to naked eye, which makes them hard to manipulate and work with. Techniques like Scanning Electron Microscopy (SEM) and Atomic Force Microscopy (AFM) are being employed to establish a visual and haptic interface to enable us to sense the molecular structure of these nanoscaled devices. Virtual Reality (VR) techniques are currently being explored in nanoscience and biotechnology research as a way to enhance the operator's perception (vision and haptics) by approaching more or less a state of 'full immersion' or 'telepresence'. The development of nanorobots or nanomachine components presents difficult fabrication and control challenges. Such devices will operate in microenvironments whose physical properties differ from those encountered by conventional parts. Since these nanoscale devices have not yet been fabricated, evaluating possible designs and control algorithms requires using theoretical estimates and virtual interfaces/environments. Such interfaces/simulations can operate at various levels of detail to trade-off physical accuracy, computational cost, number of components and the time over which the simulation follows the nanoobject behaviors. They can enable nanoscientists to extend their eyes and hands into the nanoworld and also enable new types of exploration and whole new classes of experiments in the biological and physical sciences. VR simulations can also be used to develop virtual assemblies of nano and bio-nano components into mobile linkages and predict their performance.

Nanorobots with completely artificial components have not been realized yet. The active area of research in this field is focused more on molecular robots, which are thoroughly inspired by nature's way of doing things at nano scale. Mother nature has her own set of molecular machines that have been working for centuries, and have been optimized for performance and design over the ages. As our knowledge and understanding of these numerous machines continues to increase, we now see a possibility of using the natural machines, or creating synthetic ones from scratch, using nature's components. This chapter focuses more on molecular machines and explores various designs and research prevalent in this field. The main goal in the field of molecular machines is to use various biological elements—whose function at the cellular level creates motion, force or a signal—as machine components. These components perform their preprogramed biological function in response to the specific physiochemical stimuli but in an artificial setting. In this way proteins and DNA could act as motors, mechanical joints, transmission elements, or sensors. If all these different components were assembled together in the proper proportion and orientation they would form nano devices with multiple degrees of freedom, able to apply forces and manipulate objects in the nanoscale world. The advantage of using nature's machine components is that they are highly efficient and reliable.

Nanorobotics is a field which calls for collaborative efforts between physicists, chemists, biologists, computer scientists, engineers and other specialists to work towards this common objective. Fig. V–A.1 details the various fields which come under the field of bionanorobotics (this is just a representative figure and not exhaustive in nature). Currently this field is still evolving, but several substantial steps have been taken by great researchers all over the world and are contributing to this ever challenging and exciting field.

The ability to manipulate matter at the nano scale is one core application for which nanorobots could be the technological solution. A lot has been written in the literature about the significance and motivation behind constructing a nanorobot. The applications range from medical to environmental sensing to space and military applications. Molecular construction of complex devices could be possible by nanorobots of the future. From precise drug delivery to repairing cells and fighting tumor cells; nanorobots are expected to revolutionize the medical industry in the future. These applications come under the field of nanomedicine which is a very active area of research in

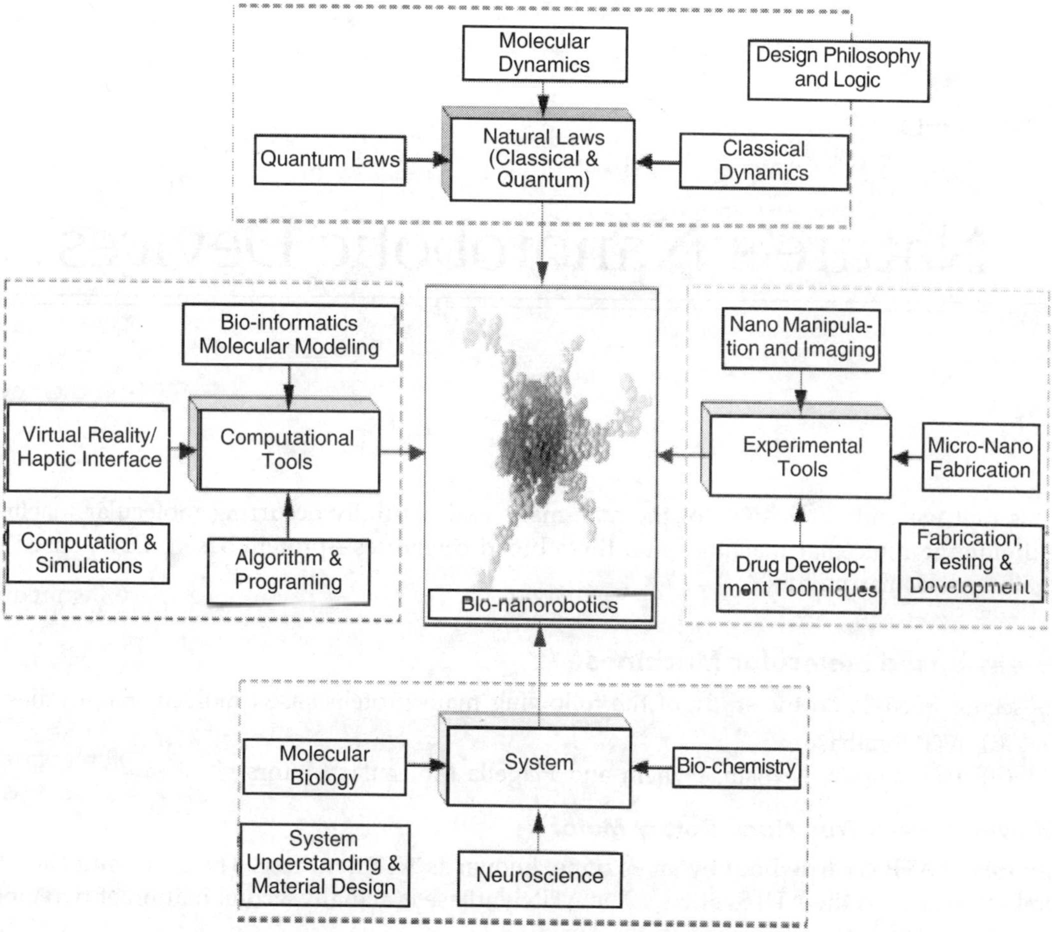

Fig. V–A.1: Bio-nanorobotics—a truly multidisciplinary field.

nanotechnology. These molecular machines hence form the basic enablers of future applications. In the next section, we shall try to understand the principles, theory and utility of the known molecular machines and look into the design and control issues for their creation and modification. A majority of natural molecular machines are protein-based, while the DNA-based molecular machines are mostly synthetic. Nature deploys proteins to perform various cellular tasks—from moving cargo, to catalyzing reactions, while it has kept DNA as an information carrier. It is hence understandable that most of the natural machinery is built from proteins. With the powerful crystallographic techniques available in the modern world, the protein structures are clearer than ever. The ever increasing computing power makes it possible to dynamically model protein folding processes and predict the conformations and structure of lesser known proteins. All this helps unravel the mysteries associated with the molecular machinery and paves the way for the production and application of these miniature machines in various fields including medicine, space exploration, electronics and military.

Unit V–B

Nature's Nanorobotic Devices

In this unit we will detail some of the man-made and naturally occurring molecular machines. We divide the molecular machines into three broad categories—protein-based, DNA-based and chemical molecular motors.

Protein-based Molecular Machines

This section focuses on the study of the following main protein based molecular machines:

(i) ATP Synthase
(ii) The Kinesin, Myosin, Dynein and Flagella Molecular Motors

ATP Synthase—A True Nano Rotary Motor

Synthesis of ATP is carried out by an enzyme, known as ATP Synthase. The inner mitochondrial membrane contains the ATP Synthase. The ATP Synthase is actually a combination of two motors functioning together as described in the Fig. V–B.1.

This enzyme consists of a proton-conducting F_0 unit and a catalytic F_1 unit. The figure also illustrates the subunits in side the two motor components. F_1 constitutes of $\alpha_3\beta_3\gamma\delta\varepsilon$ subunits. F_0 has three different protein molecules, namely, subunit a, b and c. The γ-subunit of F_1 is attached to the c subunit of F_0 and is hence rotated along with it. The $\alpha_3\beta_3$ subunits are fixed to the b-subunit of F_0 and hence do not move. Further the b-subunit is held inside the membrane by a subunit of F_0 (shown in the figure by Walker).

ATP Synthase 'nano' Properties

1. *Reversibility of the ATP Synthase*: There are two directions in ATP Synthase system and these two directions correspond to two different functionalities and behavior. This two-way behavior is because of the reversible nature of the ATP-ADP cycle and the structure of the ATP Synthase. Let us term the forward direction as when the F_0 drives the γ-subunit (because of proton motive force) of F_1 and hence ATP Synthesis takes place. And the backward direction is when hydrolysis of ATP counter-rotated the ã-subunit and hence the F_0 motor and leads to pumping back the protons. Therefore the forward direction is powered by the proton motive force and the backward direction is powered by the ATP hydrolysis. Which particular direction is being followed depends upon the

situation and the environmental factors around the ATP Synthase.

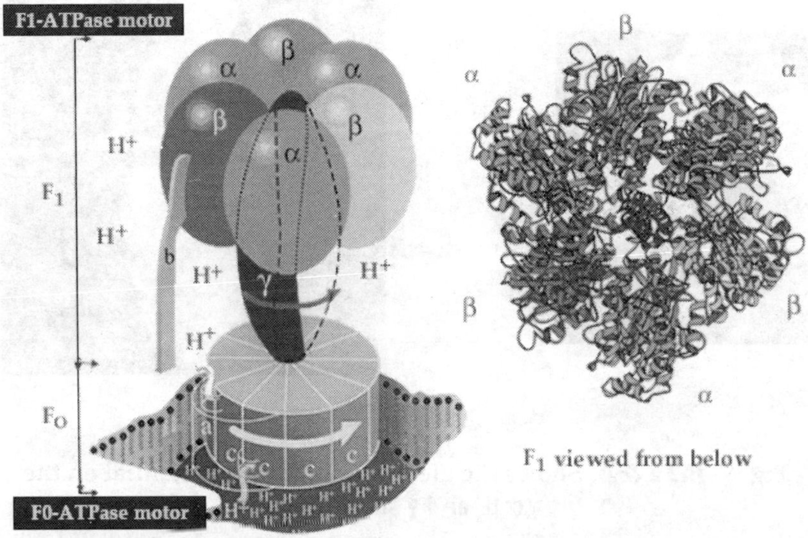

Fig. V–B.1: Shows the basic structure of the ATP Synthase. Shown is the flow of protons from the outer membrane towards the inner through the F_0 motor. This proton motive force is responsible for the synthesis of ATP in F_1. (©Copyright 2004. The Nobel Foundation).

2. ***Coupling of Proton Flow (F_0) and the ATP synthesis and hydrolysis (in F_1):*** Boyer proposed a model which predicted that the F_0 and F_1 motors are connected through the γ-subunit. Further he proposed that this connection was mechanical in nature. The following figure illustrates the electrostatic surface potential on $\alpha_3\beta_3$ subunits. The red colour represents a negatively charged surface and blue a positively charged surface. Shown in the Fig. V–B.2a, a predominately neutral hole in the $\alpha_3\beta_3$ subunit through which the γ-subunit protrudes. Fig. V–B.2b shows a γ-subunit with a significantly negatively charged region around half across its length. The γ-subunit slides through the hole of the $\alpha_3\beta_3$ subunits.

3. ***Boyer's binding Change Mechanism***: Boyer isolated the F_1 part of the ATP Synthase complex. It was found that the alpha and beta subunits alternate in this cylindrical part of the F_1 structure. As per this model each α and β pair forms a catalytic site. The rotation of the γ subunit induces structural conformation in the $\alpha_3\beta_3$ subunits. Although the three catalytic units are identical in their chemistry but they are functionally very different at any given point in time. These conformal changes induce a change in the binding affinities of all the three catalytic sites towards the ATPase reactants (ADP, P_i, ATP etc.). Fig. V–B.3 shows the binding change mechanism as proposed by Boyer:

The three catalytic sites could be in three distinct forms. O form stands for open which implies that it has got very low affinity for reactants; L form stands for loose, which implies that it would bind the reactants loosely but is still catalytically inactive; T form stands for tight, which binds the reactant tightly and is catalytically active.

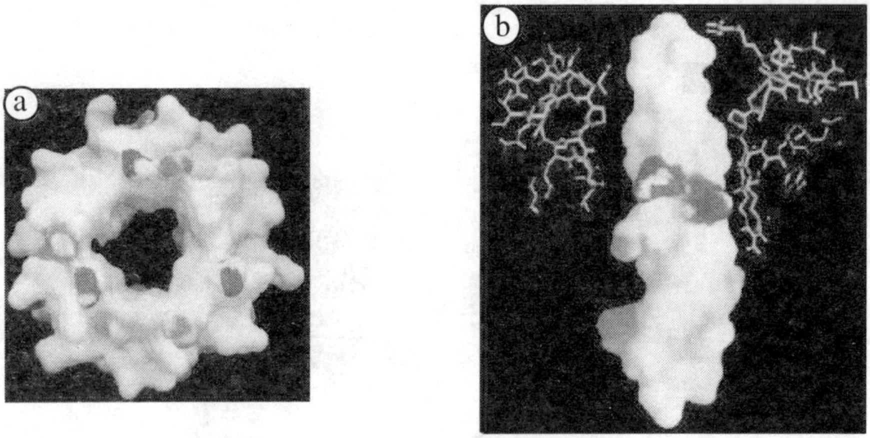

Fig. V–B.2 a & b: Shows the electrostatic surface potential on the $\alpha_3\beta_3$ and γ subunits.

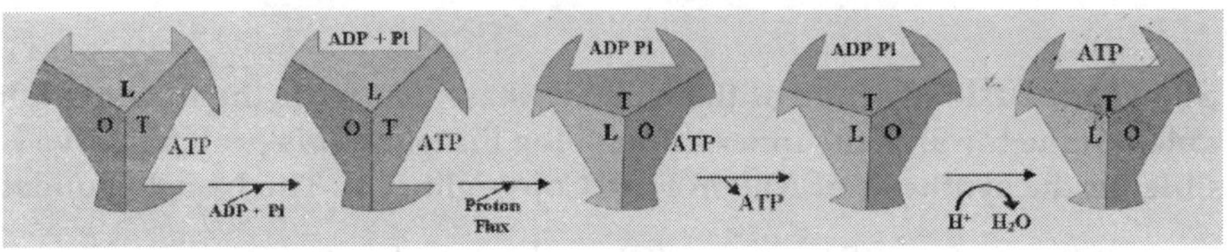

Fig. V–B.3: Boyer's binding change mechanism.

The situation depicted in the Fig. V–B.3 above shows one of the sites being bound to ATP at T. The site for ADP and P_i would probably be L as it is more binding than the O site. Now when the energy is input to the system the sites are changed. T becomes O; O becomes L and L becomes T. Due to these changes the ATP which was previously in T is now released. Further the ADP and P_i are now tightly bound due to the conversion of L site to the T site. This tight binding of ADP and P_i allows them to spontaneously get converted into ATP. Hence, this model proposed that the proton flow from the F_0 is coupled to the site inter conversions in the F_1 unit which triggers synthesis-hydrolysis of ATP. Boyer's theory was supported when Walker and his group solved the structure of the F_1-ATPase motor. The high resolution structure thus obtained gave hints towards many experiments which proved the fact that γ subunit indeed rotated against the alpha and beta subunits.

4. F_1-ATPase a true nano rotary motor: Till today the exact mechanism of the molecular motor characterized by F_1-ATpase has not been fully determined. Research by Kinosita's lab is a step towards this goal and proposes some very conclusive models for the same. The results obtained show not only the various methods through which we can analyze these nano devices, but also predicts many characteristics for these.

What is known till now is that γ subunit rotates inside the alpha-beta hexamer, but whether the rotation is continuous or is random was not known. Kinosita's lab solved this problem by imaging the F_1-ATPase molecule. The objective of their experiment was to determine the uniqueness of the rotary motion and its characteristics. Fig. V–B.4 depicts the experiment that was performed by this group. They attached a micrometer long actin filament to the γ subunit. This actin filament was fluorescently label, so that its fluorescence could be measured under a microscope. Hydrolysis of the ATP (when introduced in the experiment) led to the rotation of the γ subunit and in effect the rotation of the actin filament. As reported by the authors, not all the actin filaments were observed to have rotation. But some percentage of them did rotate and that too in a unique direction and without having much reversibility in the direction. This direct imaging proved that the structure solved by Walker and group was indeed correct and there exists rotary motion between γ subunit and the alpha and beta hexamer.

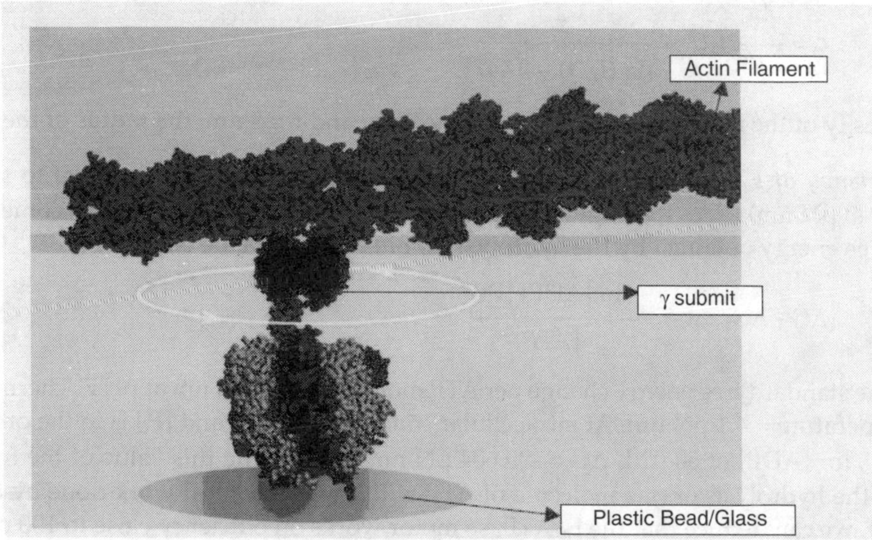

Fig. V–B.4: Experiment performed for Imaging F_1-ATPase.

Fig. V–B.5 shows the imaging of the actin filament that was obtained. The rotation is not restricted to some particular rotational angle but is continuously progressive is a particular direction. As the rotational motion is a continuous one, possibility of it being a twisting motion is ruled out. Hence, there should be no direct linkage between the γ subunit and the alpha-beta hexamer in F_1-ATPase. It was further observed by the authors that the rotation of the actin filament was in certain steps. These step sizes were approximately equal to 120 degrees.

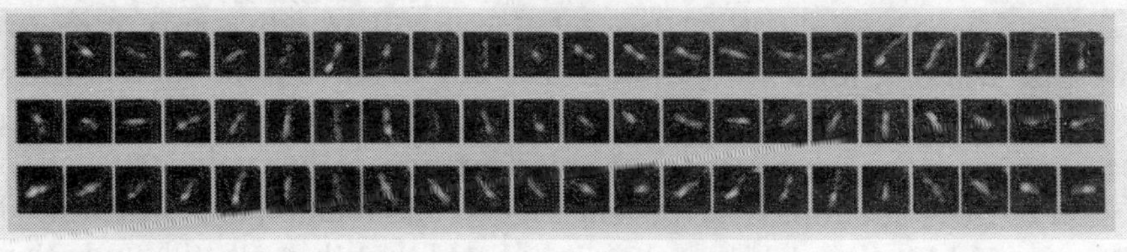

Fig. V–B.5: Images of a rotating actin filament (sequential image at 33ms intervals)

Other observed behavior of the F_1-ATPase motor

Rotational rates: Rotational rates were dependent upon the length of the actin filament. Rotational rates were found to be inversely proportional to the filament length. This could be attributed to the fact that the hydrodynamic friction is proportional to the cube of the length of the filament, therefore, longer length implied that the frictional forces were higher and rotational speeds in effect slower. As reported in the principle article, if the filament were to rotate at the speed of 6 rev per sec and the length of the actin filament were 1 μm, then as per the relation.

$$N = \omega\xi$$

The torque required would be 40 pN nm. Here ξ is the drag coefficient given by the relation:

$$\xi = \frac{4\pi}{3}\eta L^3 \frac{1}{[\ln (L/r) - 0.447]}$$

μ is the viscosity of the medium = 10^{-3} N sec per meter sq and r = 5 nm, the radius of the filament.

- *Efficiency of F_1:* Work done in one step, i.e., 120 degree rotation is equal to the torque generated (=40 pN nm) times the angular displacement for that step (=2pi/3). This comes to about 80pN nm. Free energy obtained by the hydrolysis of one ATP molecule is:

$$\Delta G - \Delta G_0 + k_D T \frac{\ln [\text{ADP}] [\text{Pi}]}{[ATP]}$$

Here, the standard free energy change per ATP molecule = -50 pN nm at pH 7. Thermal energy at room temperature = 4.1 pN nm. At intracellular conditions, [ATP] and [P_i] is in the order of 10^{-3} M. Therefore, for [ADP] of 50 μM, ΔG = -90.604 pN nm. Comparing this value of the free energy obtained by the hydrolysis of one molecule of ATP with the mechanical work done by the motor (=80 pN nm), we can deduce that the F_1-ATPase motor works on efficiency close to 100 per cent!

- *Chemical synthesis of ATP powered by mechanical energy:* In another study conducted by Hiroyasu Itoh, Akira Takahashi, Kengo Adachi, Hiroyuki Noji, Ryohei Yasuda, Masasuke Yoshida & Kazuhiko Kinosita Jr, evidence has been provided to justify the claim that the chemical synthesis of ATP occurs when propelled by mechanical energy. This basic nature of the F_1-ATPase was known for quite some while, but it is first time that it is being experimentally verified. To prove this concept, the F_1 was bound to the glass surface through histidine residues attached at the end of the γ subunits. A magnetic bead coated with streptavidin was attached to the γ subunit (Fig. V–B.6).

Electric magnets were used to rotate this bead attached to the γ subunit. The rotation resulted in appearance of ATP in the medium (which was initially immersed in ADP). Thus, the connection between the syntheses of ATP as a result of the mechanical energy input is established.

The exact mechanism of the F_1-ATPase rotation is still an active area of research today and many groups are working towards finding it. The key to solving the mechanism is solving the transient conformation of the catalytic sites and the γ subunit when rotation is taking place. What is not clear is the correspondence between the chemical reactions at the catalytic sites and their influence on the rotation of the γ subunit. Which event triggers the rotation and which not has still to be exactly determined? Many models have been predicted, but they all still elude the reality of the rotational mechanism.

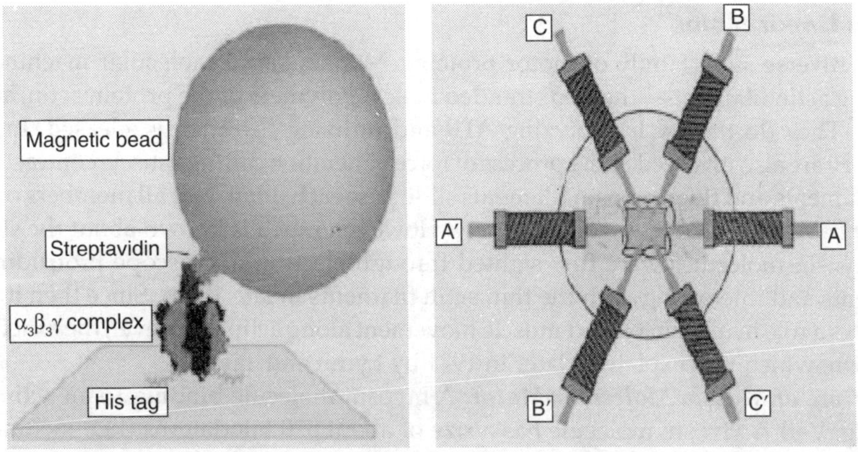

Fig. V–B.6: Magnetic bead is attached to the γ subunit here (left figure) and the arrangement of the magnets (right figure).

The Kinesin, Myosin, Dynein and Flagella Molecular Motors

With modern microscopic tools, we view a cell as a set of many different moving components powered by molecular machines rather than a static environment. Molecular motors that move unidirectionally along protein polymers (actin or microtubules) drive the motions of muscles as well as much smaller intracellular cargoes. In addition to the F_0-F_1-ATPase motors inside the cell, there are linear transport motors present as tiny vehicles known as motor proteins that transport molecular cargoes that also require ATP for functioning. These minute cellular machines exist in three families—the kinesins, the myosins and the dyneins. The cargoes can be organelles, lipids or proteins etc. They play an important role in cell division and motility. There are over 250 kinesin-like proteins, and they are involved in processes as diverse as the movement of chromosomes and the dynamics of cell membranes.

The only part they have in common is the catalytic portion known as the motor domain. They have significant differences in their location within cells, their structural organization, and the movement they generate. Muscle myosin, whose study dates back to 1864, has served as a model system for understanding motility for decades. Kinesin, however, was discovered rather recently using in vitro motility assays in 1985. Conventional Kinesin is a highly processive motor that can take several hundred steps on a microtubule without detaching whereas muscle myosin executes a single "stroke" and then dissociates. A detailed analysis and modeling of these motors has been done.

Kinesin and myosin make up for an interesting comparison. Kinesin is microtubule-based; it binds to and carries cargoes along microtubules whereas myosin is actin-based. The motor domain of kinesin weighs one third the size of that of myosin and one tenth of that of dynein. Before the advent of modern microscopic and analytic techniques, it was believed that these two have little in common. However, the crystal structures available today indicate that they probably originated from a common ancestor.

The Myosin Linear Motor

Myosin is a diverse superfamily of motor proteins. Myosin-based molecular machines transport cargoes along actin filaments—the two stranded helical polymers of the protein actin, about 5-9 nm in diameter. They do this by hydrolyzing ATP and utilizing the energy released. In addition to transport, they are also involved in the process of force generation during muscle contraction, wherein thin actin filaments and thick myosin filaments slide past each other. Not all members of the myosin superfamily have been characterized as of now. However, much is known about the structure and function. Myosin molecules were first sighted through electron microscope protruding out from thick filaments and interacting with the thin actin filaments in late 1950s. Since then it was known that ATP plays a role in myosin related muscle movement along actin. However, the exact mechanism was unknown, which was explained later in 1971 by Lymn and Taylor.

- *Structure of Myosin Molecular Motor:* A myosin molecule binding to an actin polymer is shown in Fig. V–B.7. Myosin molecule has a size of about 520 kilodaltons (kD) including two 220 kD heavy chains and light chains of sizes between 15-22 kD. They can be visualized as two identical globular 'motor heads', also known as motor domains, each comprising of a catalytic domain (actin, nucleotide as well as light chain binding sites) and about 8 nm long lever arms. The heads, also sometimes referred to as S1 regions (subfragment 1) are shown in blue, while the lever arms or the light chains, in yellow. Both these heads are connected via a coiled coil made of two α-helical coils (grey) to the thick base filament. The light chains have considerable sequence similarity with the protein 'calmodulin' and troponin C, and are sometimes referred to as calmodulin-like chains. They act as links to the motor domains and do not play any role in their ATP binding activity but for some exceptions. The motor domain in itself is sufficient for moving actin filaments. Three-dimensional structures of myosin head revealed that it is a pear-shaped domain, about 19 nm long and 5 nm in maximum diameter.
- *Function of Myosin Molecular Motor:* A crossbridge-cycle model for the action of myosin on actin has been widely accepted since 1957. Since the atomic structures of actin monomer and myosin were resolved this model has been refined into a 'lever-arm model' which is now acceptable. Only one motor head is able to connect to the actin filament at a time, the other head remains passive. Initially the catalytic domain in the head has ADP and P_i bound to it and as a result, its binding with actin is weak. With the active motor head docking properly to the actin-binding site, the P_i has to be released. As soon as this happens, the lever arm swings counterclockwise due to a conformational change. This pushes the actin filament down by about 10 nm along its longitudinal axis. The active motor head now releases its bound ADP and another ATP molecule by way of Brownian motion quickly replaces it, making the binding of the head to the actin filament weak again. The myosin motor then dissociates from the actin filament, and a new cycle starts. However, nanomanipulation of single S1 molecules (motor domains) show that myosin can take multiple steps per ATP molecule hydrolyzed, moving in 5.3 nm steps and resulting in displacements of 11 to 30 nm.

The Kinesin Linear Motor

Kinesin and Dynein family of proteins are involved in cellular cargo transport along microtubules as opposed to actin in the case of myosin. Microtubules are 25 nm diameter tubes made of protein tubulin and are present in the cells in an organized manner. Microtubules have polarity; one end being the plus (fast growing) end while the other end is the minus (slow growing) end. Kinesins move from minus end to plus end, while dyneins move from plus end to the minus end of the

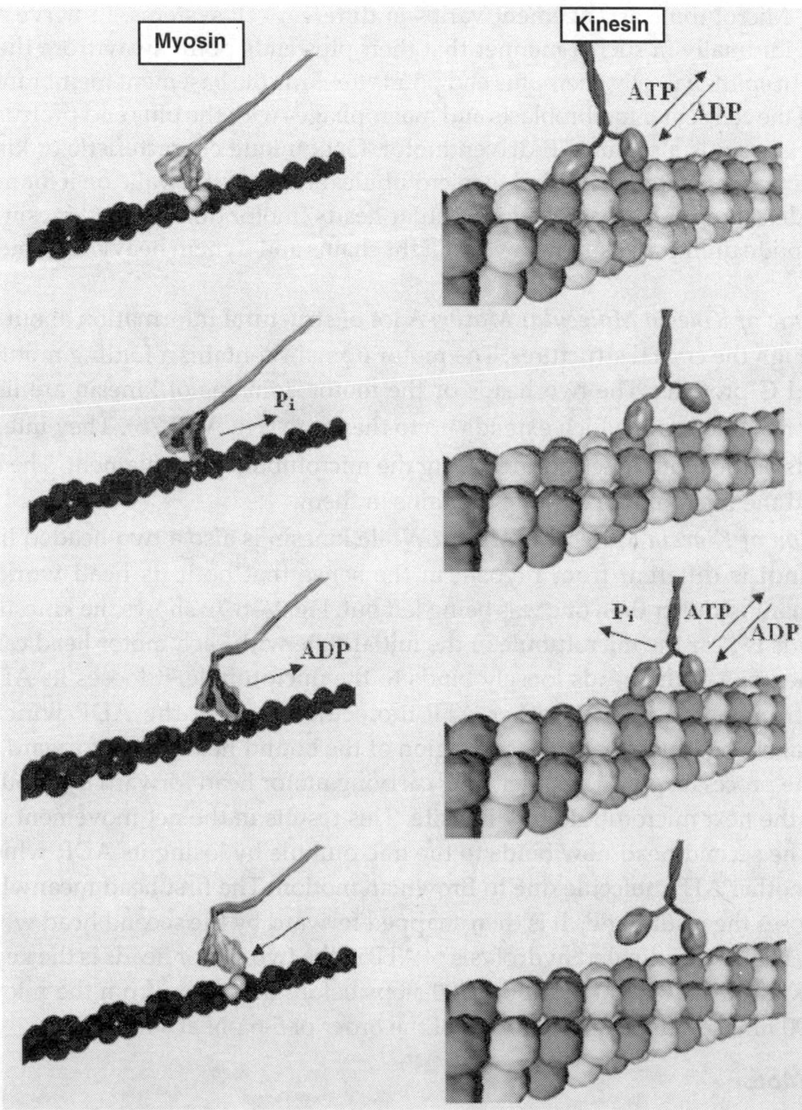

Fig. V–B.7: The kinesin-myosin walks: (a) Myosin motor mechanism. (i) Motor head loosely docking to the actin binding site; (ii) The binding becomes tighter along with the release of P_i; (iii) Lever arm swings to the left with the release of ADP, and; iv) replacement of the lost ADP with a fresh ATP molecule results in dissociation of the head; (b) Kinesin heads working in conjunction. (i) Both ADP-carrying heads come near the microtubule and one of them (black neck) binds; (ii) Loss of bound ADP and addition of fresh ATP in the bound head moves the other (red neck) to the right; (iii) The second head (red) binds to microtubule while losing its ADP, and replacing it with a new ATP molecule while the first head hydrolyses its ATP and loses P_i; (iv) The ADP-carrying black-neck will now be snapped forward, and the cycle will be repeated.

microtubules. Microtubule arrangement varies in different cell systems. In nerve axons, they are arranged longitudinally in such a manner that their plus ends point away from the cell body and into the axon. In epithelial cells, their plus end points towards the basement membrane. They deviate radially out of the cell center in fibroblasts and macrophages with the plus end protruding outwards. Like myosin, kinesin is also an ATP-driven motor. One unique characteristic of kinesin family of proteins is their processivity—they bind to microtubules and literally 'walk' on it for many enzymatic cycles before detaching. Also, each of the globular heads/motor domains of kinesin is made of one single polypeptide unlike myosin (heavy and light chains and dynein heavy, intermediate and light chains).

- ***Structure of Kinesin Molecular Motor:*** A lot of structural information about kinesin is now available through the crystal structures. The motor domain contains a folding motif similar to that of myosin and G proteins. The two heads or the motor domains of kinesin are linked via 'neck linkers' to a long coiled coil, which extends up to the cargo (Fig. V–B.7b). They interact with the α and β-subunits of the tubulin hetrodimer along the microtubule protofilament. The heads have the nucleotide and the microtubule binding domains in them.

- ***Function of Kinesin Molecular Motor:*** While kinesin is also a two-headed linear motor, its modus operandi is different from myosin in the sense that both its head work together in a coordinated manner rather than one was being left out. Fig. V–B.7b shows the kinesin walk. Each of the motor heads is near the microtubule in the initial state with each motor head carrying an ADP molecule. When one of the heads loosely binds to the microtubule, it looses its ADP molecule to facilitate a stronger binding. Another ATP molecule replaces the ADP which facilitates a conformational change such that the neck region of the bound head snaps forward and zips on to the head. In the process it pulls the other ADP carrying motor head forward by about 16 nm so that it can bind to the next microtubule-binding site. This results in the net movement of the cargo by about 8 nm. The second head now binds to the microtubule by losing its ADP, which is promptly replaced by another ATP molecule due to Brownian motion. The first head meanwhile hydrolyses the ATP and loses the resulting P_i. It is then snapped forward by the second head while it carries its ADP forward. Hence coordinated hydrolysis of ATP in the two motor heads is the key to the kinesin processivity. Kinesin is able to take about 100 steps before detaching from the microtubule while moving at 1000 nm/sec and exerting forces of the order of 5-6 pN at stall.

The Dynein Motor

Dynein superfamily of proteins was introduced in 1965. Dyneins exist in two isoforms, the cytoplasmic and the axonemal. Cytoplasmic dyneins are involved in cargo movement, while axonemal dyneins are involved in producing bending motions of cilia and flagella. Fig. V–B.8 shows a typical cytoplasmic dynein molecule.

- ***Structure of Dynein Molecular Motor:*** The structure consists of two heavy chains in the form of globular heads, three intermediate chains and four light intermediate chains. Recent studies have exposed a linker domain connecting the 'stem' region below the heads to the head itself. Also from the top of the heads the microtubule binding domains (the stalk region, not visible in the figure) protrude out. The ends of these stalks have smaller ATP sensitive globular domains which bind to the microtubules. Cytoplasmic dynein is associated with a protein complex known as dynactin, which contains ten subunits. Some of them are shown in the figure as p150, p135, actin related protein 1 (Arp1), actin, dynamitin, capping protein and p62 subunit. These play an important regulatory role in the binding ability of dynein to the microtubules. The heavy chains forming the two globular heads contain the ATPase and microtubule motor domains.

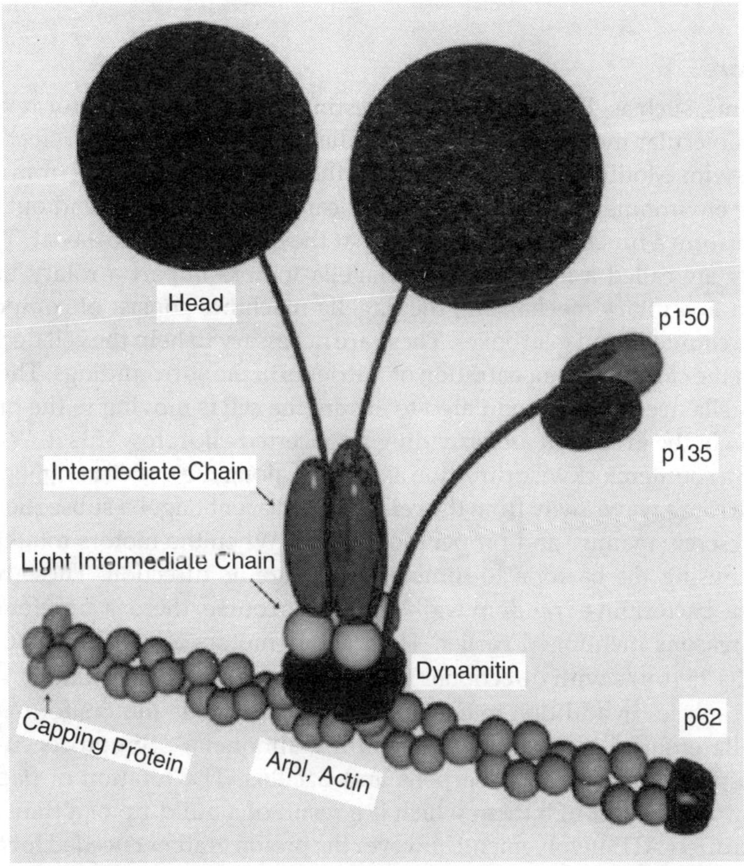

Fig. V–B.8: A dynein molecule. Shown in blue are the globular heads (heavy chains) connected to the indermediate chains (red) and the light chains (light blue). Dynactin complex components p150, p135, dynamitin, p62, capping proteins, Arp1, Actin is also shown.

One striking difference that dynein exhibits compared to kinesins and myosins is that dynein has AAA (ATPases Associated with a variety of cellular Activities) modules, which indicate that its mode of working will be entirely different from kinesins and myosins. This puts dyneins into the AAA superfamily of mechanoenzymes. The dynein heavy chains contain six tandemly linked AAA modules with the head having a ring-like domain organization, typical of AAA superfamily. Four of these are nucleotide binding motifs, named P1-P4, but only P1 (AAA1) is able to hydrolyse ATP.

• ***Function of Dynein Molecular Motor:*** Since dynein is larger and more complex structure as compared to other motor proteins, its mode of operation is not as well known. However, very recently, Burgess et al., have used electron microscopy and image processing to show the structure of a flagellar dynein at the start and end of its power stroke; giving some insight into its possible mode of force generation. When the dynein contains bound ADP and V_i (vandate), it is in the pre-power stroke conformation. The state when it has lost the two, known as the apo-state is the more compact post power stroke state. There is a distinct conformational change involving the stem, linker, head and the stalk that produces about 15 nm of translation onto the microtubule bound to the stalk.

The Flagella Motors

Unicellular organisms, such as, E. coli have an interesting mode of motility (for reviews see. They have a number of molecular motors, about 45 nm in diameter, that drive their 'feet' or the flagella that help the cell to swim. Motility is critical for cells, as they often have to travel from a less favorable to a more favorable environment. The flagella are helical filaments that extend out of the cell into the medium and perform a function analogous to what the oars perform to a boat. The flagella and the motor assembly are called a flagellum. The flagella motors impart a rotary motion into the flagella. In addition to a rotary mechanism, the flagella machines consist of components such as rate meters, particle counters, and gearboxes. These are necessary to help the cell decide which way to go, depending on the change of concentration of nutrients in the surroundings. The rotary motion imparted to the flagella needs to be modulated to ensure the cell is moving in the proper direction as well as all flagella of the given cell are providing a concerted effort towards it. When the motors rotate the flagella in a counterclockwise direction as viewed along the flagella filament from outside, the helical flagella create a wave away from the cell body. Adjacent flagella subsequently intertwine in a propulsive corkscrew manner and propel the bacteria. When the motors rotate clockwise, the flagella fly apart, causing the bacteria to tumble, or change its direction. These reversals occur randomly, giving the bacterium a 'random walk', unless of course, there is a preferential direction of motility due to reasons mentioned earlier. The flagella motors allow the bacteria to move at speeds of as much as 25 μm/s with directional reversals occurring approximately 1 per second. A number of bacterial species in addition to E. coli., depend on flagella motors for motility. Some of these are Salmonella enterica serovar Typhimurium (Salmonella), Streptococcus, Vibrio spp., Caulobacter, Leptospira, Aquaspirrilum serpens and Bacillus. The rotation of flagella motors is stimulated by a flow of ions through them which is a result of a build-up of a transmembrane ion gradient. There is no direct ATP-involvement; however the proton gradient needed for the functioning of flagella motors can be produced by ATPase.

• ***Structure of the Flagella Motors:*** A complete part list of the flagella motors may not be available as of now. Continued efforts dating back to early 1970s have however revealed much of their structure, composition, genetics and function. Newer models of the motor function are still being proposed with an aim to explain observed experimental phenomena. That means that we do not fully understand the functioning of this motor. A typical flagella motor from E. coli. consists of about 20 different proteins, while there are yet more that are involved in the assembly and functioning. There are 14 Flg-type proteins named FlgA to FlgN; 5 Flh-type proteins called FlhA to FlhE; 19 Fli-type proteins named FliA to FliT; MotA and MotB making a total of 40 related proteins. The name groups Flg, Flh, Fli and Flg originate from the names of the corresponding genes. Out of these the main structural proteins are FliC or the filament; FliD (filament cap); FliF or the MS-ring; FliG; FliM and FliN (C-ring); FlgB, FlgC and FlgF (proximal rod); FlgG (distal rod); FlgH (L-ring); FlgI (P-ring); FlgK and FlgL (hook-filament junction); and MotA-MotB (torque generating units). Earlier it was believed that the M and S are two separate rings and M was named after membrane and S after supramembranous. Now they are jointly called the MS-ring as it has been found that they are two domains of the same protein FliF. The C-ring is named after cytoplasmic, while the names of the P and L-rings come from 'peptidoglycan' and 'lipopolysaccharide' respectively. The FlhA,B,

FliH,I,O,P,Q,R constitute the 'transport apparatus'.

The hook and filament part of the flagellum is located outside the cell body, while the motor portion is embedded in the cell membrane with parts (the C-ring and the transport apparatus) that are inside the inner membrane in the cytoplasmic region. MotA and MotB are arranged in a circular array embedded in the inner membrane, with the MS-ring at the center. Connected to the MS-ring is the proximal end of a shaft, to which the P-ring, which is embedded in the peptidoglycan layer, is attached. Moving further outwards, there is the L-ring embedded in the outer cell membrane followed by the distal shaft end that protrudes out of the cell. To this end there is an attachment of the hook and the filament, both of which are polymers of hook-protein and flagellin respectively.

• ***Function of the Flagella Motors:*** The flagellar motors in most cases are powered by protons flowing through the cell membrane (protonmotive force, defined earlier) barring exceptions such as certain marine bacteria, for example, the Vibrio spp., which are driven by Na+ ions. There are about 1200 protons required to rotate the motor by one rotation. A complete explanation of how this proton flow is able to generate torque is not available as of today. From what is known, the stator units of MotA and MotB play an important role in torque generation. They form a MotA/MotB complex which when oriented properly binds to the peptidoglycan and opens proton channels through which protons can flow. It is believed that there are eight such channels per motor. The protonmotive force is a result of the difference of pH in the outside and the inside of the cell. The E. coli cells like to maintain a pH of 7.6-7.8 on their inside, so depending on the pH of the surroundings, the protonmotive force will vary, and hence the speed of rotation of their motors. To test how the speed of rotation depends on the protonmotive force, the motors were powered by external voltage with markers acting as heavy loads attached to them. It was found that the rotation indeed depends directly on the protonmotive force. According to the most widely accepted model, MotA/MotB complex interacts with the rotor via binding sites. The passage of protons through a MotA/MotB complex (stator or torque generator) moves it so that they bind to the next available binding site on the rotor, thereby stretching their linkage. When the linkage recoils, the rotor assembly has to rotate by one step. Hence, whichever complex receives protons from the flux will rotate the rotor, generating torque. The torque-speed dependence of the motor has been studied in detail and indicates the torque range of about 2700 pN-nm to 4600 pN-nm.

DNA-based Molecular Machines

As mentioned earlier, nature chose DNA mainly as an information carrier. There was no mechanical work assigned to it. Energy conversion, trafficking, sensing etc. were the tasks assigned mainly to proteins. Probably for this reason, DNA turns out to be a simpler structure—with only four kinds of nucleotide bases Adenosine, Thiamine, Guanine, and Cytosine (A, T, G & C) attached in a linear fashion that take a double helical conformation when paired with a complementary strand. Such structural simplicity vis-à-vis proteins—made of twenty odd amino acids with complex folding patterns—results in a simpler structure and predictable behavior. There are certain qualities that make DNA an attractive choice for the construction of artificial nano machines. In recent years, DNA has found use in not only mechanochemical, but nano electronic systems as well. A DNA double-helical molecule is about 2 nm in diameter and has 3.4-3.6 nm helical pitch no matter what its base composition is; a structural uniformity not achievable with protein structures if one changes their sequence. Furthermore, double stranded DNA (ds-DNA) has a respectable persistence length of about 50 nm which provides it enough rigidity to be a candidate component of molecular machinery. Single stranded DNA (ss-DNA) is very flexible and cannot be used where rigidity is required; however this flexibility allows its application in machine components like hinges or nanoactuators. Its persistence length is about 1 nm covering up to 3 base pairs at 1M salt concentration.

Other than the above structural features there are two important and exclusive properties that make DNA suitable for molecular level constructions. These are—molecular recognition and self-assembly. The nucleotide bases A and T on two different ss-DNA have affinity towards each other, so do G and C. Effective and stable ds-DNA structures are only formed if the base orders of the individual strands are complementary. Hence, if two complementary single strands of DNA are in a solution, they will eventually recognize each other and hybridize or zip-up forming a ds-DNA. This property of molecular recognition and self-assembly has been exploited in a number of ways to build complex molecular structures. In the mechanical perspective, if the free energy released by hybridization of two complementary DNA strands is used to lift a hypothetical load, a force capacity of 15 pN can be achieved, comparable to that of other molecular machines like kinesin (5 pN).

Dr. Nadrian Seeman and colleagues built the first artificial DNA based structure in the form of a cube in 1991. They then went on to create more complex structures such as knots and Borromean rings. In addition to these individual constructs, two dimensional arrays with the help of the double-crossover (DX) DNA molecule. This DX molecule gave the structural rigidity required to create a dynamic molecular device, the B-Z switch. DNA double helices can be of three types—the A, B or the Z-DNA. The B-DNA is the natural, right-handed helical form of DNA, while the A-DNA is a shrunken, low-humidity form of the B-DNA. Z-DNA can be obtained from certain C-G base repeat sequences occurring in B-DNA that can take a left-handed double helical form. The CG repeated base pair regions can be switched between the left and the right-handed conformations by changing ionic concentration. The switch was designed in such a way that it had three cyclic strands of DNA, two of them wrapped around a central strand that had the CG-repeat region in the middle. On the two free ends of the side strands fluorescent dyes were attached in order to monitor the conformational change. With the change in ionic concentration the central CG-repeat sequence could alternate between the B and the Z modes bidirectionally which was observed through fluorescence resonance energy transfer (FRET) spectroscopy.

The DNA Tweezers

In the year 2000, Dr. Bernard Yurke and colleagues made an artificial DNA based molecular machine that also accepted DNA as a fuel. The machine, called DNA tweezers, consisted of three strands of DNA labeled A, B and C. Strands B and C are partially hybridized on to the central strand A with overhangs on both ends. This conformation of the machine is the open conformation. When an auxiliary fuel strand F is introduced, that is designed to hybridize with both overhang regions, the machine attains a closed conformation. The fuel strand is then removed from the system by the introduction of its exact complement, leaving the system to go back to its original open conformation. This way, a reversible motion is produced, which can be observed by attaching fluorescent tags to the two ends of the strand A. In this case the 5′ end was labeled with the dye TET (tetrachloro-fluorescein phosphoramidite) and the 3′ end was labeled with TAMRA (carboxy-tertamethylrhodamine). Aside from the creation of a completely new molecular machine, this showed a way of selective fueling of such machines. The fuel strands are sequence-specific, so they will work on only those machines towards which they are directed, and not trigger other machines surrounding them. This machine was later improved to form a Three-State Device which had two robust states and one flexible intermediate state. A variation of the tweezers came about as the DNA-scissors.

Inorganic (Chemical) Molecular Machines

In the past two decades, chemists have been able to create, modify and control many different types of molecular machines. Many of these machines carry a striking resemblance with our everyday macro-scale machines such as gears, propellers, shuttles etc. Not only this, all of these machines are easy to synthesize artificially and are generally more robust than the natural molecular machines. Most of these machines are organic compounds of Carbon, Nitrogen and Hydrogen, with the presence of a metal ion being required occasionally. Electrostatic interactions, covalent and hydrogen bonding play essential role in the performance of these machines. Such artificial chemical machines are controllable in various ways—chemically, electrochemically and photochemically (through irradiation by light). Some of them are even controllable by more than one ways, rendering more flexibility and enhancing their utility. A scientist can have more freedom with respect to the design of chemical molecular machines depending on the performance requirements and conditions. Rotaxanes and Catenanes make the basis of many of the molecular machines described in this section. These are families of interlocked organic molecular compounds with a distinctive shape and properties that guide their performance and control.

The Rotaxanes

Rotaxane family of molecular machines is characterized by two parts —a dumb-bell shaped compound with two heavy chemical groups at the ends and a light, cyclic component, called macrocycle, interlocked between the heads as shown in Fig. V–B.9. It has been shown that a reversible switch can be made with a rotaxane setup. For this, one needs to have two chemically active recognition sites in the neck region of the dumb-bell. In this particular example, the thread was made of polyether, marked by recognition sites hydroquinol units and terminated at the ends by large triisoproplylsilyl groups. A tetracationic bead was designed and self-assembled into the system that interacts with the recognition sites.

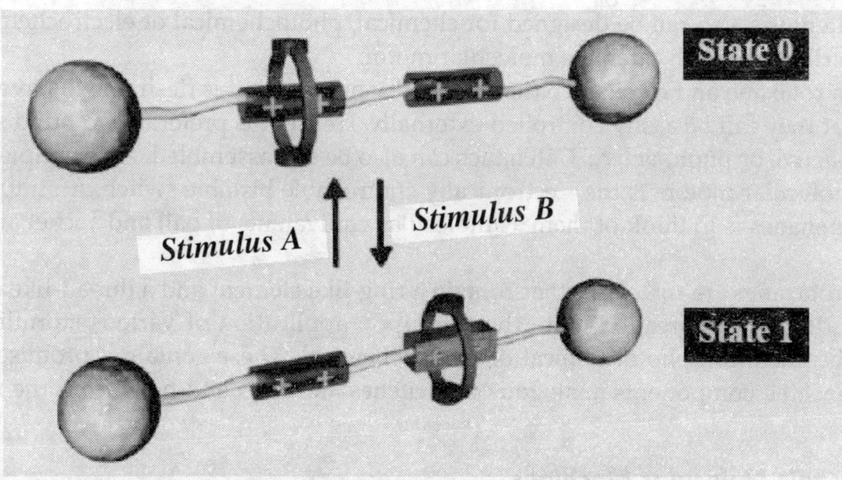

Fig. V–B.9: A typical rotaxane shuttle set-up. The macrocycle encircles the thread-like portion of the dumb-bell with heavy groups at its ends. The thread has two recognition sites which can be altered reversibly so as to make the macrocycle shuttle between the two sites.

The macrocycle has a natural low energy state on the first recognition site, but can be switched among the two sites reversibly upon application of suitable stimuli. Depending on the type of rotaxane setup the stimuli can be chemical, electrochemical or photochemical. The stereo-electronic properties of the recognition sites can be altered protonation or deprotonation, or by oxidation or reduction, thereby changing their affinity towards the macrocycle. In a recent example, light induced acceleration of rotaxane motion was achieved by photoisomerization, while similar controls through alternating current (oscillating electric fields) was shown before.

The Catenanes

The catenanes are also special type of interlocked structures that represent a growing family of molecular machines. They are synthesized by supramolecular assistance to molecular synthesis. The general structure of a catenane is that of two interlocked ring-like components that are non-covalently linked via a mechanical bond, that is, they are held together without any valence forces. Both the macrocyclic components have recognition sites that are atoms of groups that are redox-active or photochemically reactive. It is possible to have both rings with similar recognition sites. In such a scenario, one of the rings may rotate inside the other with the conformations stabilized by noncovalent interactions, but the two states of the inner ring differing by 180° will be undistinguishable (degenerate). For better control and distinguishable molecular conformations, it is desirable to have different recognition sites within the macrocycles. Then they can be controlled independently through their own specific stimuli. The stereo-electronic property of one recognition site within a macrocycle can be varied such that at one point it has more affinity to the sites on the other ring. At this instant, the force balance will guide the rotating macrocycle for a stable conformation that requires that particular site to be inside the other macrocycle. Similarly, with other stimulus, this affinity can be turned off, or even reversed along with the affinity of the second recognition site on the rotating macrocycle increased towards those on the static one. There is a need for computational modeling, simulation and analysis of such molecular machine motion. Like rotaxanes, catenanes also can be designed for chemical, photochemical or electrochemical control. Fig. 2.10 describes one such catenane molecular motor.

For both rotaxane and catenane based molecular machines, it is desirable to have recognition sites such that they can be easily controlled externally. Hence it is preferable to build sites that are either redox-active or photo-active. Catenanes can also be self assembled. An example of catenane assembled molecular motors is the :ectronically controllable bistable switch. An intuitive way of looking at catenanes is to think of them as molecular equivalents of ball and socket and universal joints.

Pseudorotaxanes are structures that contain a ring-like element and a thread-like element that can be 'threaded' or 'dethreaded' onto the ring upon application of various stimuli. Again, the stimuli can be chemical, photochemical or electrochemical. These contain a promise of forming molecular machine components analogous to switches and nuts and bolts from the macroscopic world.

Other Inorganic Molecular Machines

Many other molecular devices have been reported in the past four decades that bear a striking resemblance to macroscopic machinery. Chemical compounds behaving as bevel gears and propellers that were reported in the late 1960s and early 1970s are still being studied today. A molecular propeller can be formed when two bulky rings such as the aryl rings are connected to one central atom, often

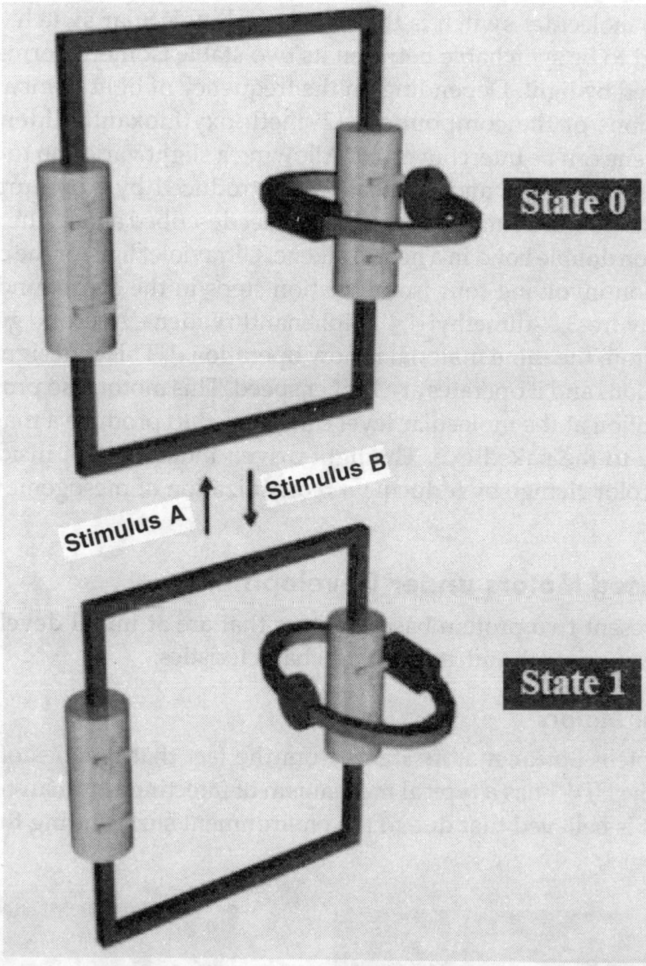

Fig. V–B.10: A non-degenerate catenane. One of the rings (the moving ring) has two different recognition sites in it. Both sites can be turned 'off' or 'on' with different stimuli. When the green site is activated, the force and energy balance results in the first conformation, whereas when the red one is activated, the second conformation results. They can be named states 0 and 1 analogous to binary machine language.

called the focal atom. Clockwise rotation of one such ring induces a counterclockwise rotation of the opposite ring about the bond connecting it to the central atom. It is possible to have a three-propeller system as well. Triptycyl and amide ring systems have been shown to observe a coordinated gear-like rotation. 'Molecular Turnstiles' which are rotating plates inside a macrocycle, have been created. Such rotations however were not controllable. A rotation of a molecular ring about a bond could be controlled by chemical stimuli, and this was shown by Kelly et al., when they demonstrated a molecular brake. A propeller-like rotation of a 9-triptycyl ring system, which was used in gears, this time connected to a 2, 2'-bipyridine unit could be controlled by the addition and subsequent removal of a metal. Thus, free rotations along single bonds can be stopped and released at will.

Another type of molecular switch is the 'chiroptical molecular switch'. Another large cyclic compound was found to be switchable between its two stable isomeric forms P and M' (right and left handed) stimulated by light. Depending on the frequency of light bombarded on it the cis and the trans conformations of the compound 4-(2'-meth-oxythioxanthylidene)]-7-methyl-1,2,3,4-tetrahydrophenanthrene can be interconverted. Allowing a slight variation to this switch, a striking molecular motor driven by light and/or heat was introduced by Koumura and colleagues. As opposed to the rotation around a single bond in the ratchet described above, the rotation was achieved around a carbon-carbon double bond in a helical alkene. Ultraviolet light or the change in temperature could trigger a rotation involving four isomerization steps in the compound (3R,3'R)-(P,P)-trans-1,1',2,2',3,3',4,4'-octahydro-3,3'-dimethyl-4,4',-biphenanthrylidene. A second generation motor along with 8 other motors from the same material is now operational. This redesigned motor has distinct upper and lower portions and it operates at a higher speed. This motor also provides a good example of how controlled motion at the molecular level can be used to produce a macroscopic change in a system that is visible to the naked eye. The light-driven motors when inside liquid crystal (LC) films can produce a color change by inducing a reorganization of mesogenic molecules.

Other Protein Based Motors under Development

In this section we present two protein based motors that are at initial developmental stages and yet possess some very original and interesting characteristics.

Viral Protein Linear Motors

The idea of Viral Protein Linear motors stems from the fact that families of retroviruses like the influenza virus and the HIV-1 has a typical mechanism of infecting a human cell. When such a virus comes near the cell it is believed that due to the environment surrounding the cell it experiences a

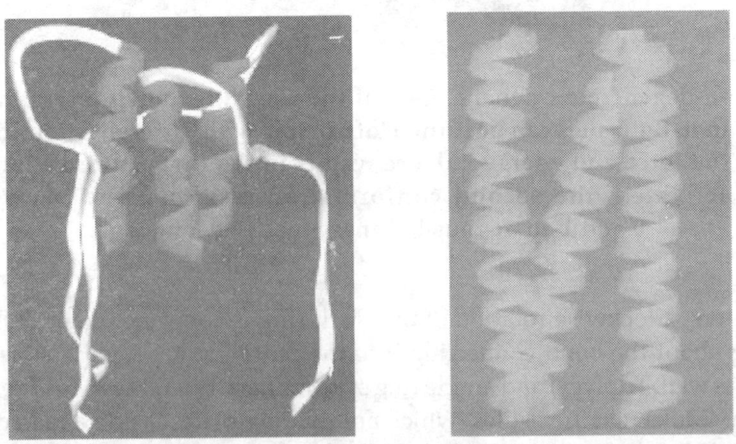

Fig. V–B.11: VPL Motor at (a) neutral (b) acidic pH. (a) Front view of the partially α-helical triple stranded coiled coil. VPL motor is in the closed conformation; b) VPL Motor in the open conformation. The random coil regions (white) are converted into well defined helices and an extension occurs at lower pH.

drop in pH of its surroundings. This is a kind of a signal to the virus that its future host is near. The drop of pH changes the energetics of the outer (envelope glycoprotein) protein of the viral membrane in such a way that there is a distinct conformational change in a part of it. A triple stranded coiled coil domain of the membrane protein changes conformation from a loose random structure to a distinctive α-helical conformation. It is proposed to isolate this domain from the virus and trigger the conformational change by variation of pH in vitro. Once this is realized, attachments can be added to the N or C (or both) terminals of the peptide and a reversible linear motion can be achieved. Fig. V–B.11 shows a triple stranded coiled coil structure at a pH of 7.0; the inverted hairpin-like coils shown in the front view in Fig. V–B.11a and top view in Fig. V–B.11b change conformation into extended helical coils as seen in Fig. V–B.11b.

Synthetic Contractile Polymers

In a recent development, large plant proteins that can change conformation when stimulated by positively charged ions were separated from their natural environment and shown to exert forces in orthogonal directions. Proteins from sieve elements of higher plants that are a part of the microfluidics system of the plant were chosen to build a new protein molecular machine element. These elements change conformations in the presence of Ca^{2+} ions and organize themselves inside the tubes so as to stop the fluid flow in case there is a rupture downstream. This is a natural defense mechanism seen in such plants. The change in conformation is akin to a balloon inflating and extending in its lateral as well as longitudinal directions. These elements, designated as 'forisomes' adhered on to glass tubes were shown to reversibly swell in the presence of Ca^{2+} ions and shrink in their absence, hence performing a pulling or pushing action in both directions. Artificially prepared protein bodies like the forisomes could be a useful molecular machine component in a future molecular assembly, producing forces of the order of micronewtons. Unlike the ATP-dependant motors discussed previously, these machine elements are more robust because they can perform in the absence of their natural environment as well.

Unit V–C

Nanorobotics Design and Control

Design of Nanorobotic Systems

Designing nanorobotic systems deal with vast variety of sciences, from quantum molecular dynamics, to kinematic analysis. The rules applicable to nanorobotics depend upon the nano material we intend to use. Nanomechanical robotic systems deal with science significantly different from the biological or inorganic nanorobotic systems. Eric Drexler, author of the famous book, *Nanosystems: Molecular machinery, manufacturing and Computation*, detailed on many areas of science which influences design of nanomechanical systems. In this review chapter on various laws which govern the designing of nanorobotics, we will concentrate on biological systems. We will consider that the components that details a nanorobot is made of biological components, such as, proteins and DNAs. There doesn't exist any particular guideline or a prescribed manner which details the methodology of designing a bio-nanorobot (bio-nanorobot implies nanorobots made up of biocomponents) up to the date. There are many complexities which are associated with using biocomponents (such as protein folding, presence of aqueous medium), but the advantages of using these are also quite considerable. These biocomponents offer immense variety and functionality at a scale where creating a manmade material with such capabilities would be extremely difficult. These bio components have been perfected by nature through millions of years of evolution and hence these are very accurate and efficient. As noted in the review section on Molecular Machines, F_1-ATPase is known to work at efficiencies which are close to 100 per cent. Such efficiencies, variety and form are not existent in any other form of material found today.

Also, the other significant advantages in using protein-based bio-nano composites is the development and refinement over the last 30 years of tools and techniques that enable researchers to mutate proteins in almost any way imaginable. These mutations can consist of anything from simple amino acid side-chain swapping, to amino acid insertions or deletions, incorporation of non-natural amino acids, and even the combination of unrelated peptide domains into whole new structures. An excellent example of this approach is the engineering of the F_1-ATPase, which is able rotate a nanopropeller in the presence of ATP. A computational algorithm was used to determine the mutations necessary to engineer an allosteric zinc-binding site into the F_1-ATPase using site-directed mutagenesis. The mutant F_1-ATPase was then shown to rotate an actin filament in the presence of ATP with average torque of 34 pN nm. This rotation could be stopped with the addition of zinc, and restored with the addition of a chelator to remove the zinc from the allosteric binding site. This type of approach can be used for the improvement of other protein-based nanocomponents.

Hence, these bio components seem to be a very logical choice for designing nanorobots. Some of the core applications of nanorobots are in the medical field and using biocomponents for these applications seems to be a good choice as they both offer efficiency and variety of functionality. This idea is clearly inspired by nature's construction of nanorobots, bacteria and viruses which could be termed as intelligent organisms capable of movement, sensing and organized control. Hence, our scope would be limited to the usage of these bio components in the construction of bio-nanorobotics. A roadmap is proposed which details the main steps towards the design and development of bio-nanorobots.

The Roadmap

The roadmap for the development of bio-nanorobotic systems for future applications (medical, space and military) is shown in Fig. V–C.1. The roadmap progresses through the following main steps:

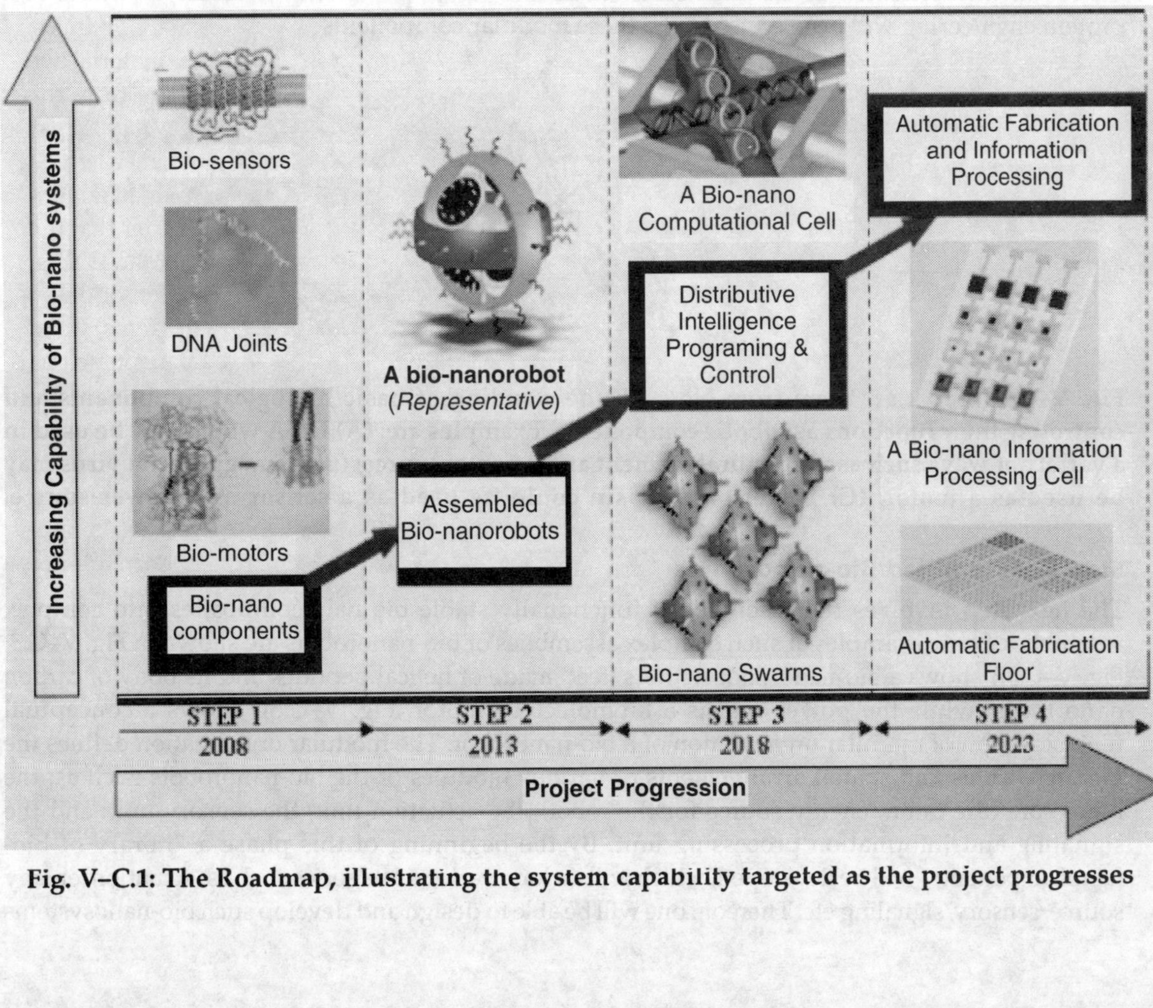

Fig. V–C.1: The Roadmap, illustrating the system capability targeted as the project progresses

Step 1: Bio-nanocomponents

Development of bio-nanocomponents from biological systems is the first step towards the design and development of an advanced bio-nanorobot, which could be used for future applications (see Fig. V–C.2). Since the planned systems and devices will be composed of these components, we must have a sound understanding of how these behave and how could they be controlled. From the simple elements such as structural links to more advanced concepts such as motors, each component must be carefully studied and possibly manipulated to understand the functional limits of each one of them. DNA and carbon nanotubes are being fabricated into various shapes, enabling possibilities of constructing newer and complex devices. These nanostructures are potential candidates for integrating and housing the bio-nanocomponents within them. Proteins such as rhodopsin and bacteriorhodopsin are a few examples of such bio-nanocomponents. Both these proteins are naturally found in biological systems as light sensors. They can essentially be used as solar collectors to gather abundant energy from the sun. This energy could either be harvested (in terms of proton motive force) for later use or could be consumed immediately by other components, such as the ATP Synthase nano rotary motor. The initial work is intended to be on the bio-sensors, such as, heat shock factor. These sensors will form an integral part of the proposed bio-nano assemblies, where these will be integrated within a nano structure and will get activated, as programed, for gathering the required information at the nano scale. Tools and techniques from molecular modeling and protein engineering will be used to design these modular components.

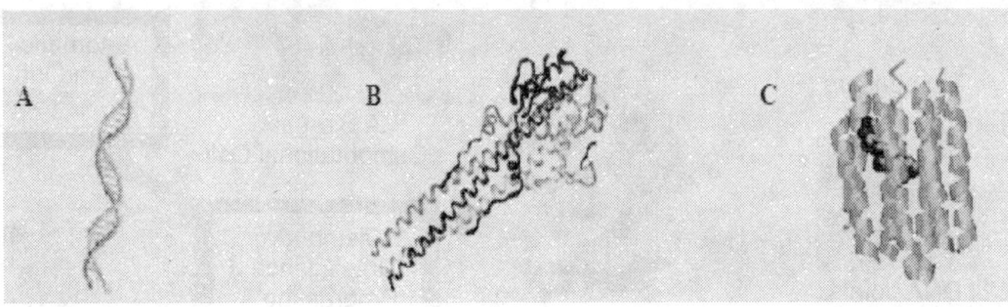

Fig. V–C.2 (Step 1): 5 Years from Now—Understanding of basic biological components and controling their functions as robotic components. Examples are: (A) DNA which may be used in a variety of ways such as a structural element and a power source; (B) Hemagglutinin virus may be used as a motor; (C) Bacteriorhodopsin could be used as a sensor or a power source.

Step 2: Assembled Bio-nanorobots

The next step involves the assembly of functionally stable bio-nanocomposites into complex assemblies. Some examples of such complex assemblies or bio-nanorobots are shown in Fig. V–C.3. Fig. V–C.3A shows a bio-nanorobot with its 'feet' made of helical peptides and its body of carbon nano tubes, while the power unit is a biomolecular motor. Fig. V–C.3B shows a conceptual representation of modular organization of a bio-nanorobot. The modular organization defines the hierarchy rules and spatial arrangements of various modules of the bio-nanorobots such as: the inner core (the brain/energy source for the robot); the actuation unit; the sensory unit; and the signaling and information processing unit. By the beginning of this phase a "library of bio-nanocomponents" will be developed, which will include various categories such as, actuation, energy source, sensory, signaling etc. Thereon, one will be able to design and develop such bio-nanosystems

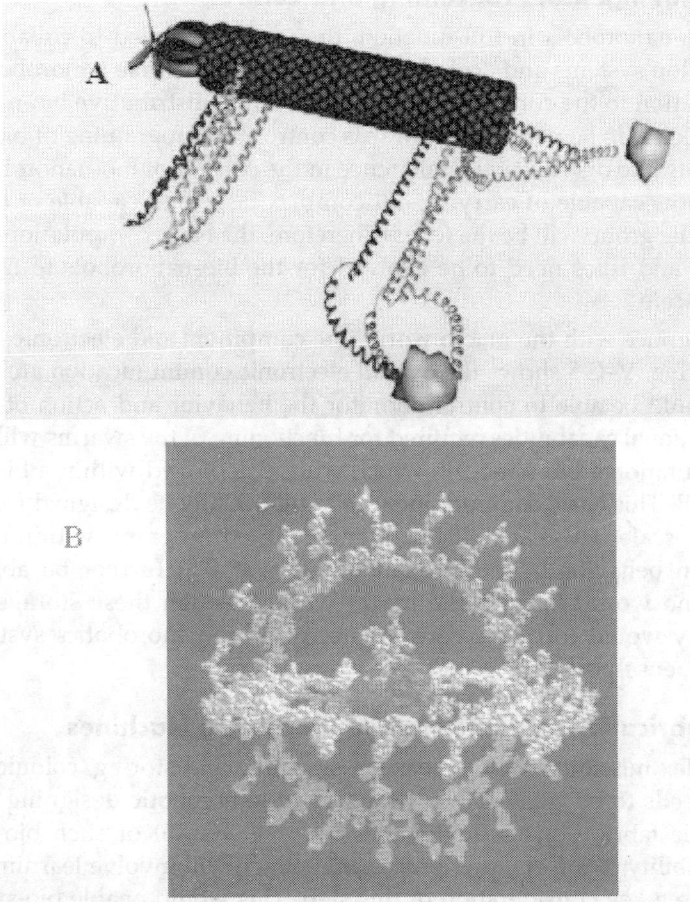

Fig. V–C.3 (Step 2): (A) the bio-nano components will be used to fabricate complex bio robotic systems. A vision of a nanoorganism: carbon nanotubes form the main body; peptide limbs can be used for locomotion and object manipulation and the biomolecular motor located at the head can propel the device in various environments. (B) Modular Organization concept for the bio-nanorobots. Spatial arrangements of the various modules of the robots are shown. A single bio-nanorobot will have actuation, sensory and information processing capabilities.

that will have enhanced mobile characteristics, and will be able to transport themselves as well as other objects to desired locations at nano scale. Furthermore, some bio-nanorobots will have the capability of assembling various bio-components and nanostructures from in situ resources to house fabrication sites and storage areas, while others will just manipulate existing structures by repairing damaged walls or making other renovations. There will also be robots that not only perform physical labor, but also sense the environment and react accordingly. There will be systems that will sense an oxygen deprivation and stimulate other components to generate oxygen, creating an environment with stable homoeostasis.

Step 3: Distributive Intelligence, Programing and Control

With the individual bio-nanorobots in full function, they will now need to collaborate with one another to further develop systems and "colonies" of similar and diverse nanorobots. This design step will lay the foundation to the concept of bio-nano swarms (distributive bio-nanorobots) (see Fig. V–C.4A). Here work has to be performed towards control and programing of bio-nanoswarms. This will evolve concepts like distributive intelligence in the context of bio-nanorobots. Designing swarms of bio-nanorobots capable of carrying out complex tasks and capable of computing and collaborating amongst the group will be the focus. Therefore, the basic computational architectures needs to be developed and rules need to be evolved for the bio-nanorobots to make intelligent decisions at the nano scale.

To establish an interface with the macro world, the computers and electronic hardware have to be designed as well. Fig. V–C.5 shows the overall electronic communication architecture. From a location, humans should be able to control, monitor the behavior and action of these swarms. Also, the basic computational capabilities required for functioning of the swarms will be developed. A representative computational bio-nanocell, which will be deployed within a bio-nanorobot, is shown in Figure V–C.4B. This basic computational cell will initially be designed for data retrieval and storage at the nanoscale. This capability will enable us to program (within certain degrees of freedom) the swarm behavior in the bio-nanorobots. We will further be able to get their sensory data (from nano world) back to the macro world through these storage devices. This programming capability would form the core essence of a bio-nanorobotics system and hence enables them with immense power.

Step 4: Automatic Fabrication and Information Processing Machines

For carrying out complex missions, such as sensing, signaling and storing, colonies of these bio-nanorobotic swarms needs to be created. The next step in nanorobotic designing would see the emergence of automatic fabrication methodologies (see Fig. V–C.6) of such bio-nanorobots in vivo and in vitro. Capability of information processing, which will involve learning and decision making abilities, will be a key consideration of this step. This would enable bio-swarms to have capability of self-evolving based on the environment they will be subjected to. These swarms could be programed to search for alternate energy sources and would have an ability to adapt as per that resource. Energy management, self-repairing, and evolving will be some of the characteristics of these swarms.

Design Philosophy and Architecture for the Bio-nanorobotic Systems

Modular Organization: Modular organization defines the fundamental rule and hierarchy for constructing a bio-nanorobotic system. Such construction is performed through stable integration of the individual 'bio-modules or components', which constitute the bio-nanorobot. For example, if the entity ABCD, defines a bio-nanorobot having some functional specificity (as per the Capability Matrix defined in Table V–C.1) then, A, B, C, and D are said to be the basic bio-modules defining it. The basic construction will be based on the techniques of molecular modeling with emphasis on principles such as energy minimization on the hyper surfaces of the bio-modules; hybrid quantum-mechanical and molecular mechanical methods; empirical force field methods; and maximum entropy production in least time.

Modular organization also enables the bio-nanorobots with capabilities such as, organizing into swarms, a feature, which is extremely desirable for various applications. Fig. V–C.7A & B shows

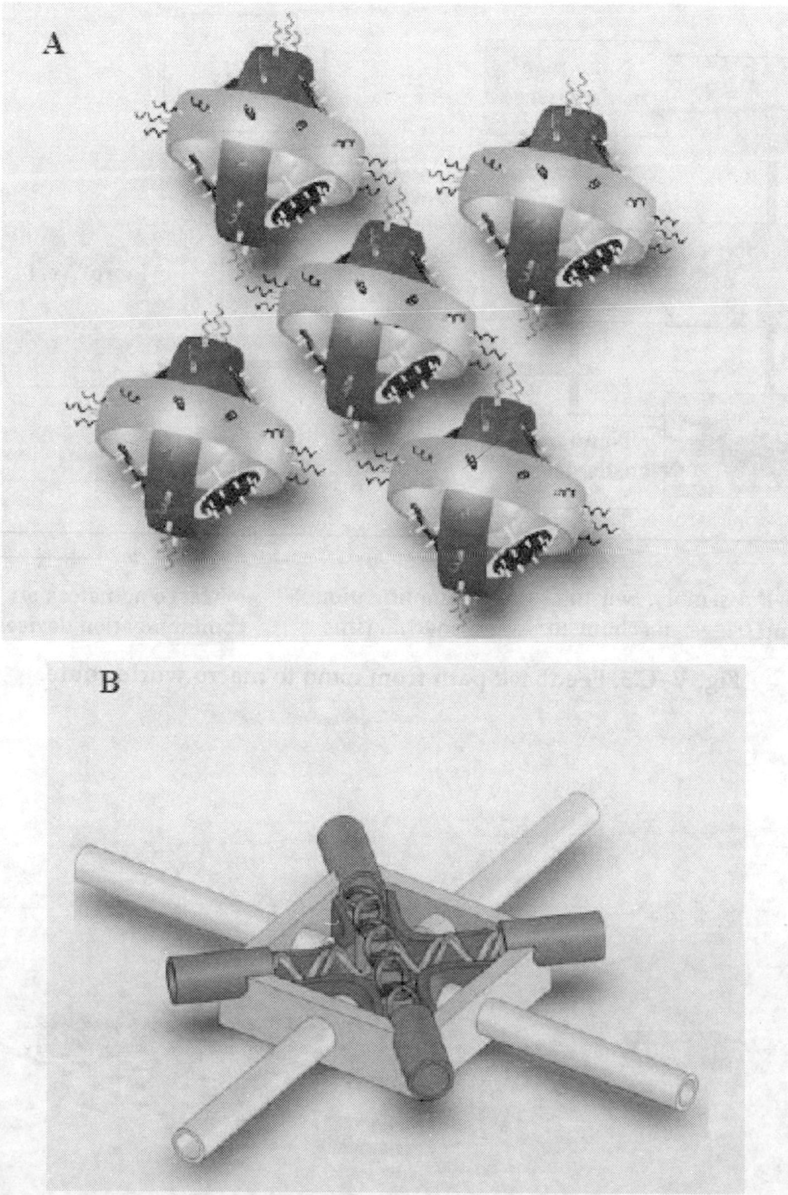

Fig. V–C.4 (Step 3): (A) Basic bio-nanorobot forming a small swarm of five robots. The spatial arrangement of the individual bio-nanorobot will define the arrangement of the swarm. Also, these swarms could be reprogramed to form bindings with various other types of robots. The number of robots making a swarm will be dependent of the resulting capability required by the mission. Also the capability of attaching new robots at run time and replacing the non functional robots will be added. (B) A basic bio-nano computational cell. This will be based on one of the properties of the bio molecules, which is "reversibility".

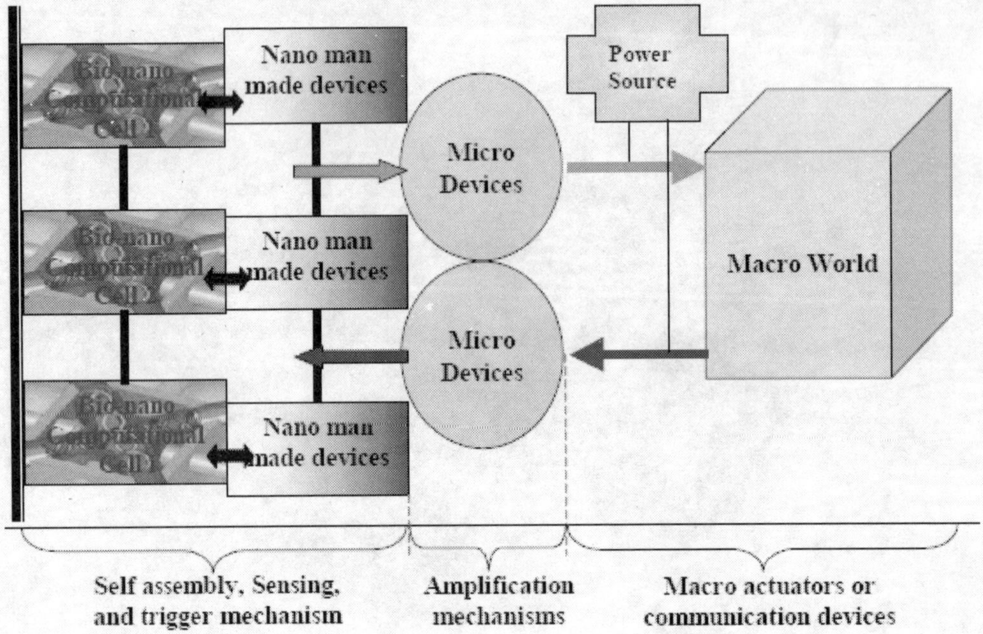

Fig. V–C.5: Feedback path from nano to macro world route

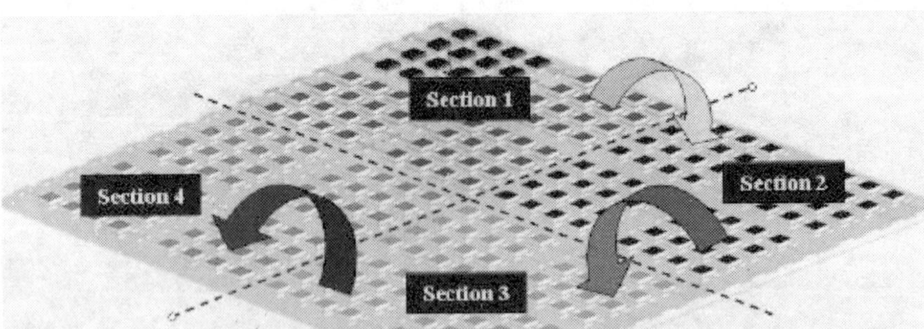

Fig. V–C.6 (Step 4): 20–30 Years—An automatic fabrication floor layout. Different color represents different functions in automatic fabrication mechanisms. The arrows indicate the flow of components on the floor layout. Section 1: Basic stimuli storage—Control expression; Section 2: Bio molecular component manufacturing (actuator/sensor); Section 3: Linking of bio-nano components; Section 4: Fabrication of bio-nanorobots (assemblage of linked bio-nanocomponents).

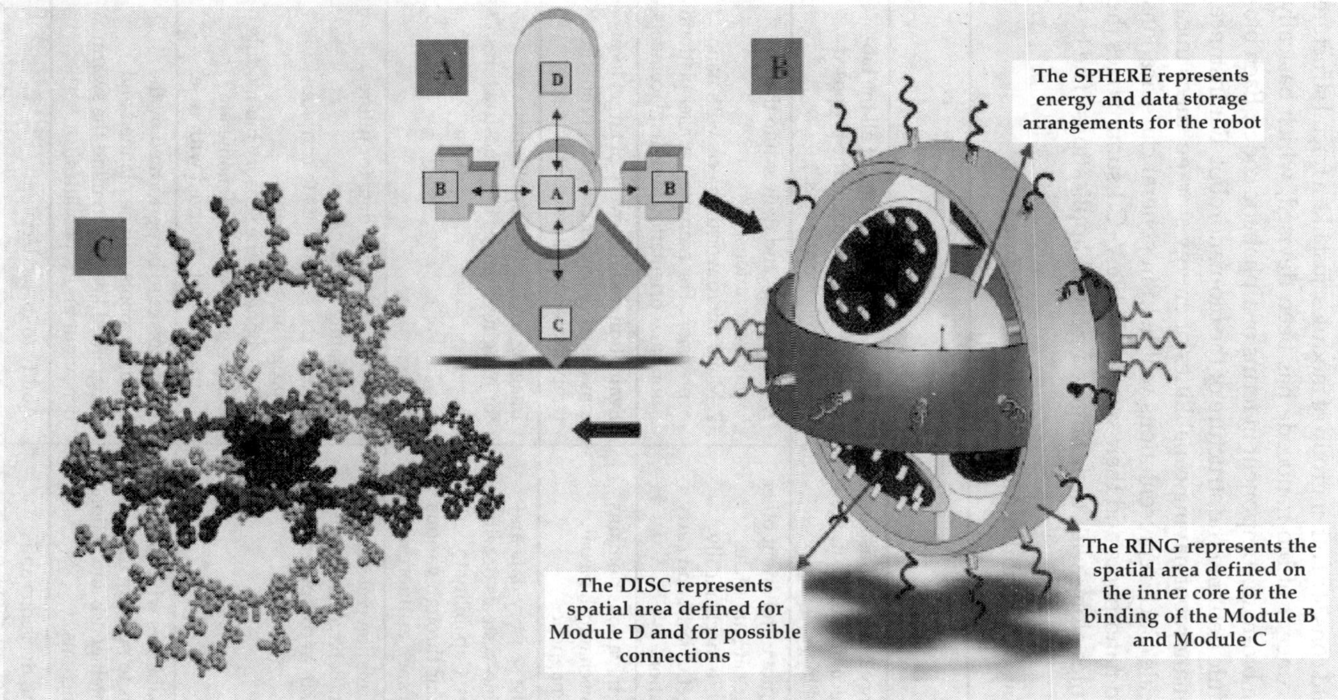

Fig. V–C.7: (A) A Bio-nanorobotic Entity 'ABCD', where A, B, C and D are the various Bio Modules constituting the bio-nanorobot. In our case these bio modules will be set of stable configurations of various proteins and DNAs. (B) A Bio-Nanorobot (representative), as a result of the concept of Modular Organization. All the modules will be integrated in such a way so as to preserve the basic behavior (of self-assembly, self-replication and self organization) of the bio-components at all the hierarchies. The number of modules employed is not limited to four or any number. It's a function of the various capabilities required for a particular mission. (C) A molecular representation of the figure in part B. It shows the red core and green and blue sensory and actuation bio-modules.

the conceptual representation of modular organization. Fig. 3.7C shows a more realistic scenario in which all the modules are defined in some particular spatial arrangements based on their functionality and structure. A particular module could consist of other group of modules, just like a fractal structure (defined as fractal modularity). The concept of bio-nanocode has been devised, which basically describes the unique functionality of a bio-nanocomponent in terms of alphabetic codes. Each bio-nanocode represents a particular module defining the structure of the bio-nanorobot. For instance, a code like E-M-S will describe a bio-nanorobot having capabilities of energy storage, mechanical actuation and signaling at the nanoscale. Such representations will help in general classifications and representative mathematics of bio nanorobots and their swarms. Table V–C.1 summarizes the proposed capabilities of the bio-modules along with their targeted general applications. The bio-

Table V–C.1: Defining the Capability Matrix for the Bio-Modules

Functionality	*Bio Nano Code*	*Capabilities Targeted*	*General Applications*
Energy Storage and Carrier	E	Ability to store energy from various sources such as, Solar, chemical for future usage and for its own working.	Required for the working of all the bio-chemical mechanisms of the proposed bio-nanorobotic systems.
Mechanical	M	Ability to precisely move and orient other molecules or modules at nanosacle. This includes ability to mechanically bind to various target objects, and carry them at desired location.	1. Carry drugs and deliver it to the precise locations. 2. Move micro world objects with nano precision. For example, parallel platforms for nano orientation and displacements.
Sensory	S	Sensing capabilities in various domains such as, chemical, mechanical, visual, auditory electrical, magnetic.	Evaluation and discovery of target locat-ions based on either chemical properties, temperature or others characteristics.
Signaling	G	Ability to amplify the sensory data and communicate with bio-systems or with the micro controllers. Capability to identify their locations through various trigger mechanisms such as fluorescence.	Imaging for Medical applications or for imaging changes in Nanostructures.
Information Storage	F	Ability to store information collected by the sensory element. Behave similar to a read-write mechanism in computer field.	1. Store the sensory data for future signaling or usage. 2. Read the stored data to carry out programed functions. 3. Back bone for the sensory bio-module. 4. Store nano world phenomenon currently not observed with ease.
Swarm Behaviour	W	Exhibit binding capabilities with "similar" bio-nanorobots so as to perform dis-tributive sensing, intelligence and action (energy storage) functions.	All the tasks to be performed by the bio-nanorobots will be planned and programed keeping in mind the swarm behaviour and capabilities.
Bio Nano Intelligence	I	Capability of making decisions and performing intelligent functions	Ability to make decision.
Replication	R	Replicate themselves when required	1. Replicate at the target site and 2. Replication of a particular bio-module as per the demand of the situation.

nanocode EIWR | | M | | S | | FG representing the bio-nanosystem shown in Fig. V–C.7B which could be decoded as shown in Fig. V–C.8. This depicts the "Fractal Modularity" principle of the proposed concept.

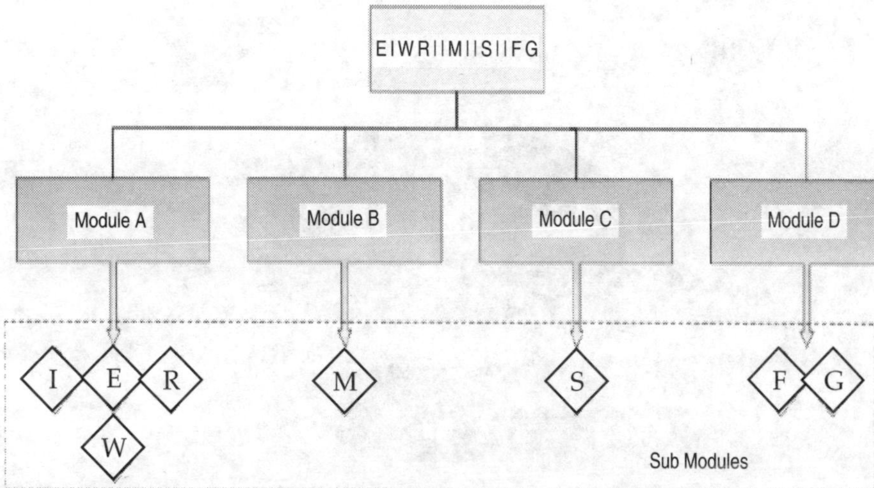

Fig. V–C.8: Showing the bio-nanocode and the fractal modularity principle. The letter symbols have the values specified in Table 1. The "| |" symbol integrates the various bio-modules and collectively represents a higher order module or a bio-nanorobot.

The Universal Template—Bio-nano Stem System: The modular construction concept involves designing a universal template for bio-nano systems, which could be 'programed and grown' into any possible Bio-nano coded system. This concept mimics the embryonic stem cells found in the human beings, that are a kind of primitive human cells which give rise to all other specialized tissues found in a human foetus, and ultimately to all the three trillion cells in an adult human body. Our Bio-nano Stem system will act in a similar way. This universal growth template will be constituted of some basic Bio-nanocodes, which will define the Bio-Nano-stem system. This stem system will be designed in a manner that could enable it to be programed at run-time to any other required bio-module. Fig. V–C.9 shows one such variant of the Bio-nano stem system, having the Bio-nanocode: EIWR | | M | | S | | FG and having enhanced sensory abilities.

Computational and Experimental Methods—Designing Bio-nanorobotic Systems

Computational Methods

Molecular modeling techniques in sync with extensive experimentations would form the basis for designing these bio-nano systems. As per the roadmap, various bio-nano components would have to be designed and then these functional elements would have to be bound into assemblies which could be controlled and programed. Some of the molecular modeling techniques we are emphasizing are described below.

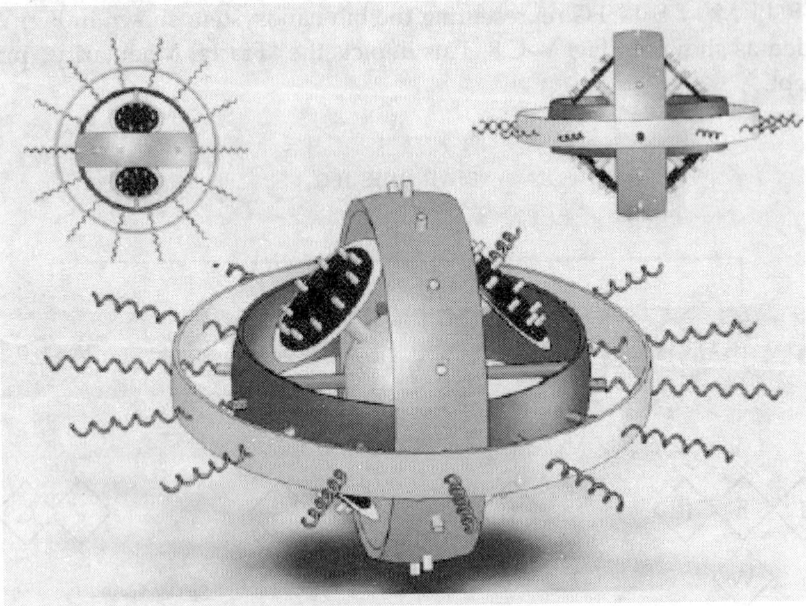

Fig. V–C.9: Shows a variant of the initial bio-nano stem system, fabricated with enhanced bio nano code S, which defines it as a bio-nanorobot having enhanced sensory capabilities. The other features could be either suppressed or enhanced depending upon the requirement at hand. The main advantage of using bio-nano Stem system is that we could at run-time decide which particular type of bio-nanorobots we require for a given situation. The suppression ability of the bio nano Stem systems is due to the property of "Reversibility" of the biocomponents found in living systems.

• ***Empirical force field methods: (Molecular mechanics):*** In this method, motion of the electrons is ignored and energy of the system is calculated based on the position of the nucleus in a particular molecular configuration. There are several approximations which are used in this method the very basic is the Born-Oppenheimer. Modeling of bio-nanocomponents or the assemblies could be done by one of the energy functions:

$$V(r^N) = \sum_{\text{bonds}} \frac{k_i}{2}(l_i - l_{i,0})^2 + \sum_{\text{angles}} \frac{k_i}{2}(\theta_i - \theta_{i,0})^2 + \sum_{\text{torsions}} \frac{v_n}{2}(1+\cos(n\omega - \gamma)) +$$

$$\sum_{i=1}^{N} \sum_{i=i+1}^{N} \left(4\varepsilon_{ij} \left[\left(\frac{\sigma_{ij}}{r_{ij}}\right)^{12} - \left(\frac{\sigma_{ij}}{r_{ij}}\right)^{6} \right] + \frac{q_i q_j}{4\pi\varepsilon_0 r_{ij}} \right)$$

Where $V(r^N)$ denotes the potential energy of N particles having position vector (r).

$$V(r^N) = \sum_{\text{bonds}} \frac{k_i}{2}(l_i - l_{i,0})^2$$ Denotes the energy variations because of bond extension

$\sum_{angles} \frac{k_i}{2}(\theta_i - \theta_{i,0})^2$ Denotes the energy variations due to bond angle bending

$\sum_{torsions} \frac{v_n}{2}(1+\cos(n\omega - \gamma))$ Denotes the energy variations due to bond torsion

$\sum_{i=1}^{N}\sum_{i=i+1}^{N}\left(4\varepsilon_{ij}\left[\left(\frac{\sigma_{ij}}{r_{ij}}\right)^{12} - \left(\frac{\sigma_{ij}}{r_{ij}}\right)^{6}\right] + \frac{q_i q_j}{4\pi\varepsilon_0 r_{ij}}\right)$ Denotes the energy due to non-bonded terms, usually modeled using Coulomb's potential

• ***Energy Minimization Methods:*** The potential energy depends upon the co-ordinates of the molecular configuration considered. In this method, points on the hypersurface (potential energy surface) are calculated for which the function has the minimum value. Such geometries which correspond to the minimum energy are the stable states of the molecular configuration considered. By this method we can also analyze the change is configuration of the system from one minimum state to another. Methods such as the Newton-Raphson and Quasi-Newton are employed to calculate these minima. This method of finding minimum energy points in the molecule is used to prepare for other advanced calculations such as molecular dynamics or Monte-Carlo simulations. It is further used to predict various properties of the system under study.

• ***Molecular Dynamics Simulation Methods:*** To begin predicting the dynamic performance (i.e. energy and force calculation) of a biocomponent (say a peptide) molecular dynamics (MD) is performed. This method utilizes Newton's Law of motion through the successive configuration of the system to determine its dynamics. Another variant of molecular dynamics employed in the industry is the Monte Carlo simulation, which utilizes stochastic approach to generate the required configuration of a system. Simulations are performed based on the calculation of the free energy that is released during the transition from one configuration state to the other (for example using the MD software CHARMM, Chemistry at Harvard Molecular Mechanics). In MD, the feasibility of a particular conformation of a biomolecule is dictated by the energy constraints. Hence, a transition from one given state to another must be energetically favorable, unless there is an external impetus that helps the molecule overcome the energy barrier. When a macromolecule changes conformation, the interactions of its individual atoms with each other—as well as with the solvent—constitute a very complex force system. With one of the aspects of CHARMM, it is possible to model say, a peptide based on its amino acid sequence and allow a transition between two known states of the protein using Targeted Molecular Dynamics (TMD). TMD is used for approximate modeling of processes spanning long time-scales and relatively large displacements. Say to achieve a large conformational change the biomolecule is 'forced' towards a final configuration 'F' from an initial configuration 'I' by applying constraints. The constraint is in the form of a bias in the force field. If we define the 3N position coordinates corresponding to N atoms in the molcculc as:

$$\mathbf{X} = (X_1, X_2, \ldots, X_{3N})^T$$

where 3N are the Cartesian coordinates of the position vectors $r_1, r_2, \ldots, r_N$ of each individual atom, then for each configuration x, its distance, ρ, to the target configuration 'F' is defined as:

$$\rho = |X - X_F| = [\sum (x_{Ii} - x_{Fi})]^2$$

The distance ρ is a purely geometric control parameter here, which will be used to force the macromolecule to undergo the desired transformation. The constraint applied for this is equal to:

$$f(x) = |X - X_F|^2 - \rho^2$$

This results in an additional constraint force:

$$F_e = \lambda \frac{df}{dx} = 2\lambda [X - X_F]$$

where λ is the Lagrange parameter. The TMD algorithm steps are:

(1) Set $\rho = \rho_0 = |x_1 - x_F|$ where 'I' is the initial and 'F' is the final conformation.

(2) Choose initial coordinates $x_i(0) = x_{1i}$ and appropriate initial velocities.

(3) Solve, numerically, the equations of motion with the additional constraint force F_c.

(4) After each time step Δt diminish ρ by $\Delta\rho = (\rho_0 - \rho_f)\,\Delta t/t_s$ where t_s is the total simulation time. At the end of the simulation, the final distance fρ is reached. In this way, a monotonous decrease of ρ forces the system to find a pathway from X_I to a final configuration X_F.

Molecular kinematic simulations are also being used to study the geometric properties and conformational space of the bio molecules (peptides). The kinematic analysis is based on the development of direct and inverse kinematic models and their use towards the workspace analysis of the bio molecules.

This computational study calculates all geometrically feasible conformations of the bio molecule i.e. all conformations that can be achieved without any atom interference. This analysis suggests the geometric paths that could be followed by a bio molecule during the transition from the initial to a final state, while molecular dynamics narrow downs the possibilities by identifying the only energetically feasible paths. The workspace analysis also characterizes the geometrically feasible workspace in terms of dexterity.

• ***Modular Pattern Recognition and Clustering Function***: The instantaneous value of the property A of a molecular system can be written as, $A(p^n(t), r^n(t))$, where $(p^n(t)$ and $(r^n(t)$ represent the n momenta and positions of a molecular system at time t. This instantaneous value would vary with respect to the interactions between the particles. There are two parts to this function (termed as M function). One part (function A), evaluates and hence recognizes the equivalent modules (in term of properties) in the molecular system. The second part of the function (D) forms the cluster (like a bioisosteres, which are atom, molecules or functional groups with similar physical and chemical properties) of various modular patterns according to the characteristic behaviors as identified by function A and also tracks their variations with rest to time.

$$M \equiv \{A(p^n(t), r^n(t), C^n[n]), D(A(p,r,C,t), f(x, y, \ldots), Bn(t))\}$$

Where, A is the first part of the function; $(p^n(t)$ and $(r^n(t)$ represents the n momenta and positions of a molecular system at time t; and $C^n[n]$ is an n dimensional matrix element for the individual components of the molecular system, which would store the categorized values of the modular patterns. D is the second part of the function, and it takes on function A recursively. It also maps these clusters based on a fitness function and stores the time-variant value in sub-function B. Fig. V–C.10, represents the conceptual version of the M-function.

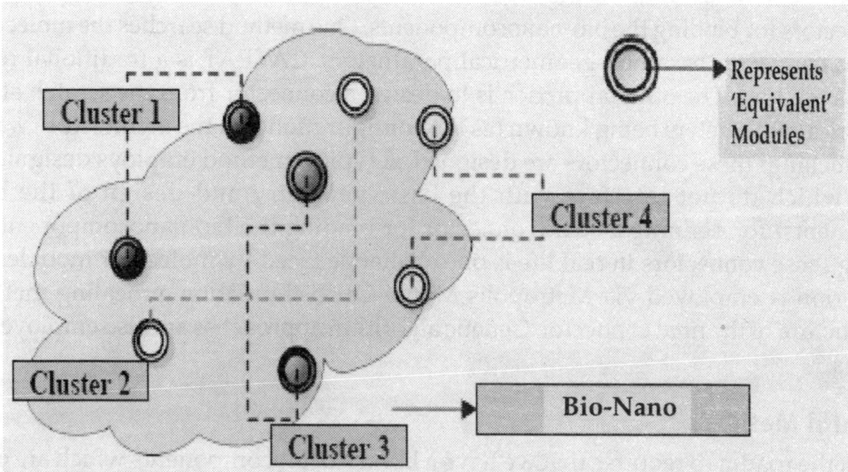

Fig. V–C.10: Representing the Modular Pattern Recognition and Clustering Function (the M—Function).

This function is based on the hypothesis that in a complex molecular system, there are certain parts which have similar properties and behavior as the system goes from one state to the other. By identifying these property patterns we can considerably reduce the simulation and computational time frames and can have better predictability of the system. Having established the design for the bio-nano components, computational studies would focus on determining the overall design and architecture of the bio-nano assemblies. One of the methods used in the industry is of Molecular Docking.

- ***Molecular Docking and Scoring Functions:*** Molecular docking method is very important for the design of nanorobotic systems. This method is utilized two fit two molecules together in 3D space. The method of molecular docking could be used to computationally design the bio-nanoassemblies. The algorithms like DOCK, genetic algorithms and distance geometry are being specifically explored. Many drug design companies are using depth-first systematic conformational search algorithms for docking, and therefore, is a primary candidate for current research. Docking algorithm generates a large number of solutions. Therefore, we require scoring functions to refine the results so produced. Scoring functions in use today approximate the "binding free energy" of the molecules. The free energy of binding can be written as an additive equation of various components reflecting the various contributions to binding. A typical equation would be:

$$\Delta G_{bind} = \Delta G_{solvent} + \Delta G_{cont} + \Delta G_{int} + \Delta G_{rot} + \Delta G_{t/r} + \Delta G_{vib}$$

Where, $\Delta G_{solvent}$ is the solvent effects contribution, $c\Delta G_{conf}$ is due to conformational changes in the biomolecule, ΔG_{int} due to specific molecular interactions, ΔG_{rot} due to restrictive internal rotation of the binding bio-molecules, $\Delta G_{t/r}$ is due to loss in the translational and rotational free energies due to binding, and ΔG_{vib} is due to vibrational mode variations. Once the overall architecture to bind the bio-nanocomponents is carried out through molecular docking, we would need to combine the components into real assemblies. The molecular component binding is the next step in the design scheme.

- ***Connecting bio-nano components in a binding site:*** There are typically two approaches to design connectors for binding the bio-nanocomponents. One method searches the molecular database and selects a connector based on geometrical parameters. CAVEAT is a traditional routine which employs this method. The other approach is to design a connector from the scratch atom by atom. The geometrical parameters being known (as the configuration and the architecture was defined by molecular docking) these connectors are designed. A typical method employs designing molecular templates (which do not interfere with the basic geometry and design of the binding bio-nanocomponents) for defining a basic connector for binding the bio-nanocomponents. Feasibility of designing these connectors in real life is one challenge faced by molecular modelers. Minimum energy criterion is employed via Metropolis Monte-Carlo simulation annealing method to select the basic structure of the final connector. Genetic algorithm approaches are also employed to generate such designs.

Experimental Methods

The step 1 in the roadmap requires that we have a library of biocomponents which are characterized based on their potentials for using them for designing and developing nanorobots. Various molecular techniques could be employed for assessing the usefulness of the biocomponents. For developing a nanorobot, we require the knowledge of the structure of the biocomponents used and the degree of conformational change that each element is capable of undergoing, and the environmental cues (e.g., T, pH) that induce these structural changes. Based on the computational methods detailed in the previous section, the following experimental techniques could be used to obtain this information.

- ***Circular dichroism spectroscopy (CD):*** CD spectroscopy can be utilized to gain insights into the secondary structure of each bio molecules such as peptides. In this technique, polarized far-UV light is used to probe the extent of secondary structure formation in solutions. This technique represents a proven method for estimating the structural composition of the nanorobotic elements during a change in environmental conditions. The effect of temperature on the structure of the nanorobotic elements can be evaluated by cooling the samples to +1°C and then heating to +85°C, with a five minute equilibration at each temperature interval. Intervals could be determined by first conducting CD scans with broad intervals, with subsequent narrowing as the location of the transition is identified. The effect of pH and ionic strength on the transitions could be monitored using separate, equilibrated preparations for each condition.
- ***Förster Resonance Energy Transfer (FRET):*** Another method that can be used to rapidly assess the extent of conformational changes by the nanorobotic elements is FRET. In order to employ this method, the nanorobotic components have to be engineered to have cysteine residues at both ends. One end of the biocomponent then has to be labeled with a donor dye molecule (Alexa Fluor 488, Molecular Probes), and the other end labeled with an acceptor dye molecule (Alexa Fluor 594). The resulting FRET signal can be monitored with a laser (488 nm), whereby the signal changes would depend on the spatial separation of the ends of the biocomponent.
- ***Nuclear Magnetic Resonance (NMR):*** NMR is utilized to determine the exact 3-D structure of the nanorobotic components. Although CD spectroscopy can be used to estimate the changes in the structural composition of the biocomponents, it does not provide the exact 3-D configuration. This information can be obtained through the use NMR spectroscopy and the labeling of the biocomponent with NMR-active atoms. In addition to determining the endpoint structures of the components, NMR can also be used to follow the kinetics of the biomolecule conformational changes, in real time.

• ***Laser-Based Optical Tweezers (LBOT) and Single Molecule Fluorescence:*** LBOT and single molecule fluorescence could be used to estimate the forces exerted by the nanorobotic elements, and the structural changes that occur during the genesis of these forces, after exposure to different stimuli. LBOT can provide force and displacement results with a resolution of sub-picoNewton and approximately one nanometer, respectively. Single molecule fluorescence measurements, based on the previously described FRET method, can simultaneously provide information about the structural state of the molecule at each point of the force-displacement curve.

To correlate force measurements with structural changes, combination of the LBOT method with FRET can be used. The recent development of this technique has enabled the simultaneous assessment of nanoscale structural changes and their associated biomechanical forces.

Nano Manipulation—Virtual Reality-based Design Techniques

For scanning and manipulating matter at the nanoscale, Scanning Tunneling Microscopy (STM), Scanning Electron Microscopy (SEM), Atomic Force Microscopy (AFM) and Scanned-Probe Microscope (SPM) seem to be the common tools. Current work is mainly focused on using atomic force microscopes (AFM) nanoprobes for teleoperated physical interactions and manipulation at the nanoscale. Precise manipulation could help scientists better understand the principles of nanorobots.

To achieve this it is essential to visualize the atom-to-atom interaction in real-time and see the results in a fully immersive 3D environment. Also, to facilitate user input in nanorobotic systems it is essential to develop voice, gesture and touch recognition features in addition to the conventional visualization and manipulation techniques. Virtual Reality (VR) technology is applied here, which not only provides immersive visualization but also gives an added functionality of navigation and interactive manipulation of molecular graphical objects. VR technology comes to our aid by providing the experience of perception and interaction with the nano world through the use of sensors, effectors and interfaces in a simulated environment. These interfaces transform the signals occurring at nanoscale processes into signals at macro-level and vice-versa. The requirement is that the communication with the nano-world must be at high-level and in real-time, preferably in a natural, possibly intuitive 'language'. Considering the nano-specific problems related to task application, tools and the interconnection technologies it leads to many flexible nano manipulation concepts. They can range from pure master/slave tele-operation (through 3-D visual/VR and/or haptic force feedback), over shared autonomy control (where for example some degrees of freedom are tele-operated and other are operating autonomously) to fully autonomous operation.

In order to precisely control and manipulate biomolecules, we need tools that can interact with these objects at the nanoscale in their native environments. Existing bio-nanomanipulation techniques can be classified as non-contact manipulations including laser trapping and electro-rotation, and contact manipulation referred to as mechanical stylus-, AFM- or STM-based nanomanipulation. The rapid expansion of AFM studies in biology/biotechnology results from the fact that AFM techniques offer several unique advantages: first, they require little sample preparation, with native biomolecules usually being imaged directly; second, they can provide a 3-D reconstruction of the sample surface in real space at ultra-high resolution; third, they are less destructive than other techniques (e.g. electron microscopy) commonly employed in biology; and fourth, they can operate in several environments, including air, liquid, and vacuum. Rather than drying the sample, one can image quite successfully with AFM in fluid. The operation of AFM in aqueous solution offers an unprecedented opportunity for imaging biological molecules and cells in their physiological environments and for studying biologically important dynamic processes in real time.

Currently, these bio-nanomanipulations are conducted manually; however, long training, disappointingly low success rates from poor reproducibility in manual operations, and contamination call for the elimination of direct human involvement. Furthermore, there are many sources of spatial uncertainty in AFM manipulation, e.g., tip effects, thermal drift, slow creeping motion, and hysteresis. To improve the bio-nanomanipulation techniques, automatic manipulation must be addressed. Visual tracking of patterns from multiple views is a promising approach, which is currently investigated in autonomous embryo pronuclei DNA injection. Interactive nanomanipulation can be improved by imaging 3D viewpoint in a virtual environment. Construction of a VR space in an off-line operation mode for trajectory planning combined with a real time operation mode for vision tracking of environmental change ensures a complete 'immersed' visual display. Fig. V–C.11 shows an example of 3D bio-micro/nano manipulation system with a 3D VR model of the environment including the biocell, and carrying out the user viewpoint change in the virtual space.

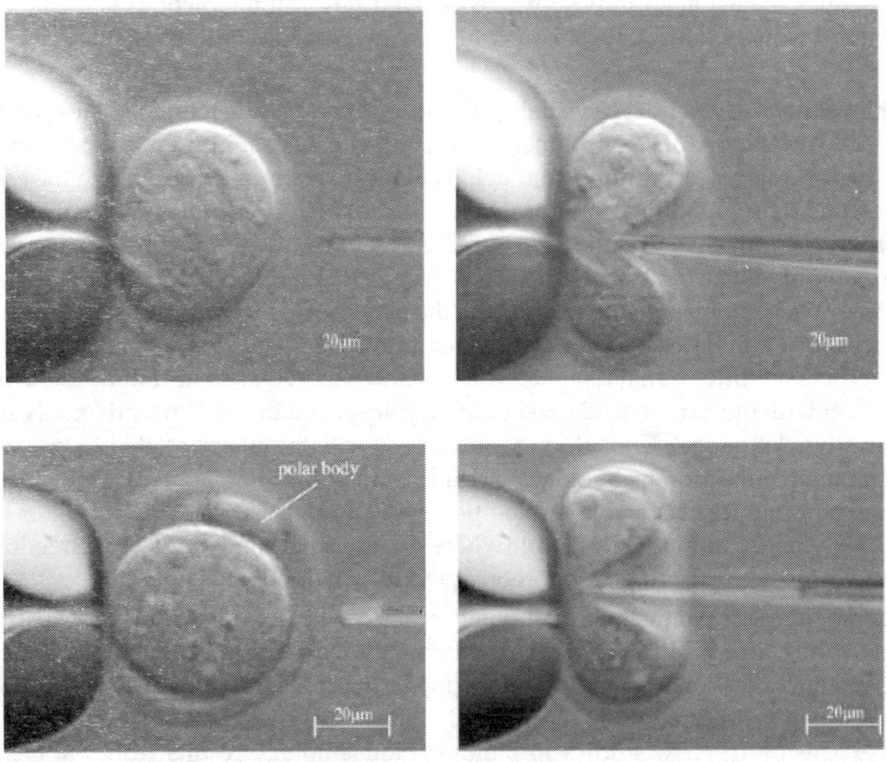

Fig. V–C.11: Haptic force measurement process on the Zona Pellucida of mouse oocytes and embryos.

Molecular Dynamics Simulations in Virtual Environment: Molecular Dynamics (MD) simulations of complex molecular systems require enormous computational power and produces large amount of data in each step. The resulting data includes number of atoms per unit volume, atomic positions; velocity of each atom, force applied on each atom and the energy contents. These results of MD simulations need to be visualized to give the user a more intuitive feel of what is happening. Haptic interaction used in conjunction with VR visualization helps the scientist to control/ monitor the simulation progress and to get feedback from the simulation process as well. Fig. V-C.12 shows the virtual reality visualization of molecular dynamics simulation. Fig. V–C.13 shows a VR representation of MD simulations in CAVE Virtual Environment.

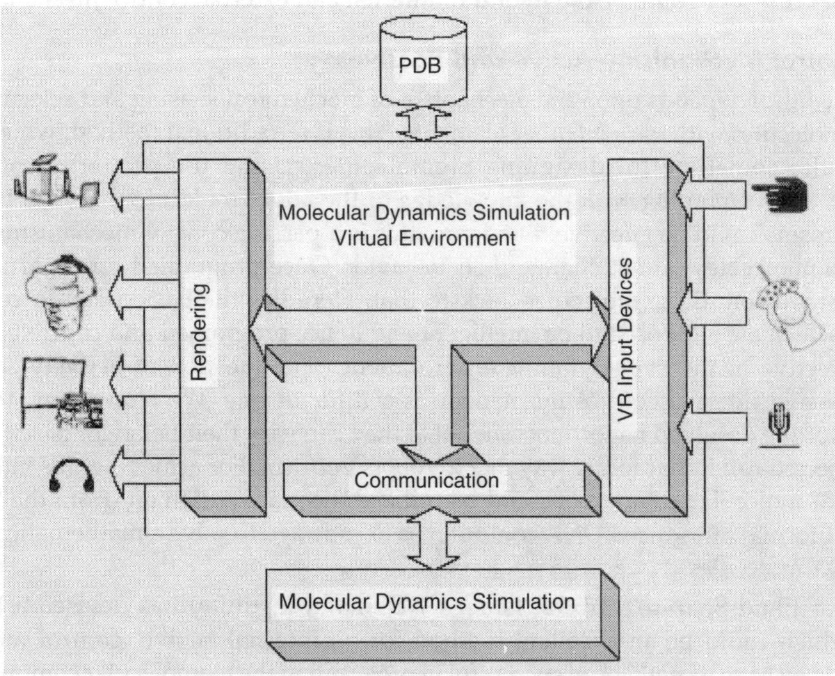

Fig. V–C.12: A block diagram showing virtual environment for Molecular Dynamics simulations.

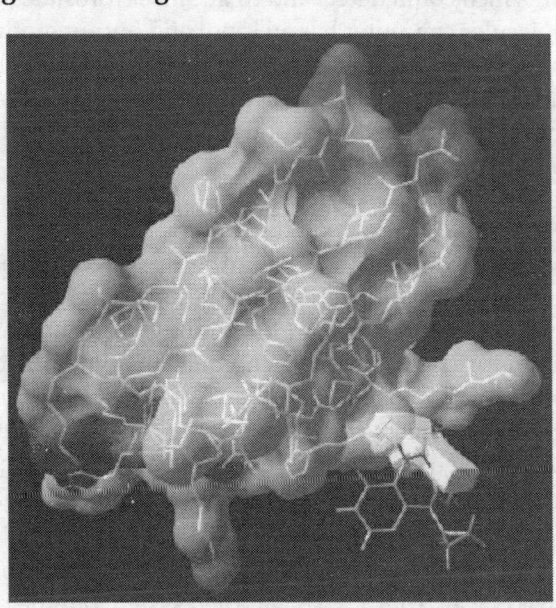

Fig. V–C.13: Molecular Dynamics simulation visualized in the CAVE.

Control of Nanorobotic systems

The control of nanorobotic systems could be classified in two categories:

(i) Internal control mechanisms.

(ii) External control mechanisms.

The other category could be the hybrid of internal and external control mechanisms.

Internal Control Mechanism—Active and Passive

This type of control depends upon the mechanism of biochemical sensing and selective binding of various biomolecules with various other elements. This is a traditional method, which has been in use since quite sometime for designing biomolecules. Using the properties of the various biomolecules and combining with the knowledge of the target molecule that is to be influenced, these mechanisms could be effective. But again, this is a passive control mechanism where at run time these biomolecules cannot change their behavior. Once programed for a particular kind of molecular interaction, these molecules stick to that. Here lies the basic issue in controlling the nanorobots which are supposed to be intelligent and hence programed and controled so that they could be effective in the ever dynamic environment. The question of actively controling the nanorobots using internal control mechanism is a difficult one. We require an 'active' control mechanism for the designed nanorobots such that they can vary their behavior based on situations they are subjected to, similar to the way macro robots perform. For achieving this internal control, the concept of molecular computers could be utilized. Leonard Adleman (from the University of Southern California) introduced DNA computers a decade ago to solve a mathematical problem by utilizing DNA molecules.

Professor Ehud Shapiro's lab (at Israel's Weizmann Institute) has devised a biomolecular computer which could be an excellent method for an internal 'active' control mechanism for nanorobots. They have recently been successful in programing the biomolecular computer to analyze the biological information, which could detect and treat cancer (prostate and a form of lung cancer) in their laboratory. The molecular computer has an input and output module which acting together can diagnose a particular disease and in response produce a drug to cure that disease. It uses novel concept of software (made up of DNAs) and hardware (made up of enzymes) molecular elements. This molecular computer is in generalized form and can be used for any disease which produces a particular pattern of gene expression related to it.

External Control Mechanism

This type of control mechanism employs affecting the dynamics of the nanorobot in its work environment through the application of external potential fields. Researchers are actively looking at using MRI as an external control mechanism for guiding the nano particles. Professor Sylvain Martel's Nanorobotics Laboratory (at École Polytechnique de Montréal, Canada) is actively looking at using MRI system as a mean of propulsion for nanorobots. An MRI system is capable of generating variable magnetic field gradients which can exert force on the nanorobot in the three dimensions and hence control its movement and orientation. Professor Martel's laboratory is exploring effect of such variable magnetic field on a ferromagnetic core that could probably be embedded into the nanorobots.

Other possibilities being explored are in the category of 'hybrid' control mechanisms where the target is located and fixed by an external navigational system but the behavior of the nanorobot is determined locally through an active internal control mechanism. The use of nanosensors and evolutionary agents to determine the nanorobots behavior is suggested by the mentioned reference.

Conclusion

Manipulating matter at molecular scale and influencing their behavior (dynamics and properties) is the biggest challenges for the nanorobotic systems. This field is still in very early stages of development and still a lot has to be figured out before any substantial outcome is produced.

The recent explosion of research in nanotechnology, combined with important discoveries in molecular biology have created a new interest in bio nanorobotic systems. The preliminary goal in this field is to use various biological elements—whose function at the cellular level creates a motion, force or a signal—as nanorobotic components that perform the same function in response to the same biological stimuli but in an artificial setting. In this way proteins and DNA could act as motors, mechanical joints, transmission elements, or sensors. If all these different components were assembled together they can form nanorobots and nanodevices with multiple degrees of freedom, with ability to apply forces and manipulate objects in the nanoscale world, transfer information from the nano to the macroscale world and even travel in a nanoscale environment.

Ability to determine the structure, behavior and properties of the nanocomponents is the first step which requires focused research thrust. Only when the preliminary results on these nanocomponents are achieved, steps towards actually building complex assemblies could be thought of. Still problems like protein (basic bio nano component folding, precise mechanism behind working of the molecular motors like ATP Synthase have still has to be solved. Active control of nanorobots has to be further refined. Hybrid control mechanisms, where in, a molecular computer and external (navigational) control system work in sync to produce the precise results seems very promising. Further, concepts like swarm behavior in context of nanorobotics is still have to be worked out. As it would require colonies of such nanorobots for accomplishing a particular task, concept of co-operative behavior, distributed intelligence has to be evolved.

The future of bio-nanorobots (molecular robots) is bright. We are at the dawn of a new era in which many disciplines will merge including robotics, mechanical, chemical and biomedical engineering, chemistry, biology, physics and mathematics so that fully functional systems will be developed. However, challenges towards such a goal abound. Developing a complete database of different biomolecular machine components and the ability to interface or assemble different machine components are some of the challenges to be faced in the near future.

Unit V–D

SPM and Nanomanipulation

In this unit we describe SPM principles, survey SPM use in nanomanipulation, look at the SPM as a robot, and conclude with a discussion of some of the challenges that face nanorobotics research.

Scanning Probe Microscopes

The Scanning Tunelling Microscope (STM) was invented by Binnig and Rohrer at the IBM Zürich laboratory in the early 1980s, and won them a Nobel Prize four years later. The principles of the instrument can be summarized with the help of Figure V–D.1.

A sharp conducting probe, typically made from tungsten is placed very close to a sample, that must also be a conductor. The tip is biased with respect to the sample as shown in the figure. The tip can be moved towards or away from the sample, i.e., in the $-z$ or $+z$ directions in the figure, by means of piezoelectric actuators. At Ångstrom-scale distances, a quantum-mechanical effect, called **tunneling**, causes electrons to flow across the tip/sample gap, and a current can be detected. To first approximation, the tunneling current depends exponentially on the distance between tip and sample. This current is kept constant by a feedback circuit that controls the piezoelectric actuators. Because of the current/distance relationship, the distance is also kept constant, and the z accuracy is very high because any small z variation causes an exponential error in the current. Additional piezo motors drive the tip in a xy scanning motion. Since the tip/sample gap is kept constant by the feedback, the scanning tip traverses a surface parallel to the sample surface. The result of a scan is a $z(x, y)$ terrain map, with enough resolution to detect atomic-scale features of the sample, as indicated diagrammatically in the figure.

Various instruments analogous to the STM have been built. They exploit physical properties other than the tunneling effect on which the STM is based. The most common of these other instruments is the Atomic Force Microscope (AFM), which is based on interatomic forces. All of these instruments are collectively known as Scanning Probe Microscopes (SPMs). The principles of operation of the AFM are shown in Figure V–D.2. The forces between atoms in the tip and sample cause a deflection of the cantilever that carries the tip. The amount of deflection is measured by means of a laser beam bouncing off the top of the cantilever. (There are other schemes for measuring deflection.) The force depends on the tip/sample gap, and therefore servoing on the force ensures that the distance is kept constant while scanning, as in the STM. The AFM does not require conducting tips and samples, and therefore has wider applicability than the STM.

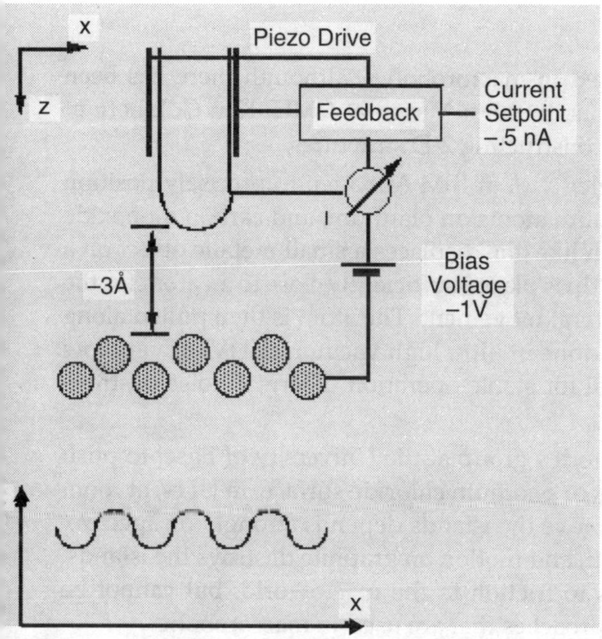

Fig. V–D.1: STM principles.

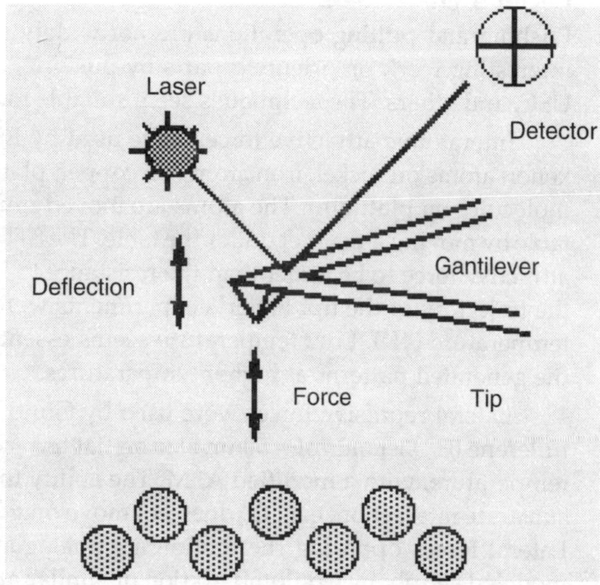

Fig. V–D.2: AFM principles.

If the tip of the AFM is brought very close to the sample, at distances of a few Å, repulsive forces prevent the tip from penetrating the sample. This mode of operation is called contact mode. It provides good resolution but cannot be used with delicate samples, *e.g.* biomaterials, which are damaged by the contact forces. Alternatively, the AFM tip can be placed at distances in the order of several nm or tens of nm, where the interatomic forces between tip and sample are attractive. The tip is vibrated at a frequency near the resonance frequency of the cantilever, in the kHz range. The tip/sample force is equivalent to a change in the spring constant of the cantilever, and causes a change in its resonance frequency. This change can be used as the error signal in the feedback circuit that controls the tip. (There are alternative detection schemes.) This mode of operation is called non-contact mode. It has poorer resolution than contact mode but can be used with delicate samples.

Although the SPM is just about twenty years old, it has had a large scientific impact.

Nanomanipulation with the SPM

Since the early days of the SPM it was known that tip/sample interaction could produce changes in both tip and sample. Often these were undesirable, for example, a blunt probe due to a crash into the sample. But it soon became clear that one could produce new and desirable features on a sample by using the tip in a suitable manner. One of the first demonstrations was done by Becker and co-workers at Bell Labs, who managed to create nanometer-scale germanium structures on a germanium surface by raising the voltage bias of an STM tip. Much of the subsequent work falls under the category of nanolithography and will not be discussed here. In the following subsections we survey nanomanipulation research involving the SPM.

Pushing and Pulling

Pushing and pulling operations are not widely used in macrorobotics, although there has been interesting work on orienting parts by pushing, done by Matt Mason at CMU, Ken Goldberg at USC, and others. The techniques seem suitable for constructing 2-D structures.

Interatomic attractive forces were used by Eigler *et al.* at IBM Almadén to precisely position xenon atoms on nickel, iron atoms on copper, platinum atoms on platinum, and carbon monoxide molecules on platinum. The atoms are moved much like one displaces a small metalic object on a table by moving a magnet under the table. The STM tip is placed sufficiently close to an atom for the attractive force to be larger than the resistance to lateral movement. The atom is then pulled along the trajectory of the tip. Eigler's experiments were done in ultra high vacuum (UHV) at very low temperature (4K). Low temperature seems essential for stable operation. Thermal noise destroys the generated patterns at higher temperatures.

Lateral repulsive forces were used by Güntherodt's group at the University of Basel to push fullerene (C_{60}) islands of ~ 50nm size on flat terraces of a sodium chloride surface, in UHV, at room temperature, with a modified AFM. The ability to move the islands depends strongly on species/substrate interaction, e.g., C_{60} does not move on gold, and motion on graphite destroys the islands. Lateral forces opposing the motion are analogous to friction in the macroworld, but cannot be modeled simply by Coulomb friction or similar approaches that are used in macrorobotics.

Mo at IBM Yorktown rotated pairs of antimonium atoms between two stable orientations 90 degrees apart. This was done in UHV at room temperature, on a silicon substrate, by scanning with an STM tip with a higher voltage than required for imaging. The rotation was reversible, although several scans were sometimes necessary to induce the desired motion.

Samuelson's group at the University of Lund succeeded in pushing galium arsenide (GaAs) nanoparticles of sizes in the order of 30 nm on a GaAs substrate at room temperature in air. The sample is first imaged in non-contact AFM mode. Then the tip is brought close to a nanoparticle, the feedback is turned off and the tip is moved against the nanoparticle. Schaefer *et al.* at Purdue University push gold clusters with an AFM in a nitrogen environment at room temperature. They first image the clusters in non-contact mode, then remove the tip oscillation voltage, and sweep the tip across the particle in contact with the surface and with the feedback disabled.

A similar technique is being used at USC's Laboratory for molecular robotics to push colloidal gold nanoparticles with 15 nm diameters on a mica substrate at room temperature and in ambient air. We image in non-contact mode, then disable the feedback and push by moving the tip in a single line scan, without removing the tip oscillation voltage. Figure 4.3 shows on the right a "USC" pattern written with gold nanoparticles. The z coordinate is encoded as brightness in this figure. On the left is the original random pattern, before manipulation with the AFM. Figure V–D.4 shows the same "USC" pattern displayed in perspective.

Smaller objects have been arranged into prescribed patterns at room temperature by Gimzewski's group at IBM's Zürich laboratory. They push molecules at room temperature in UHV by using an STM. They have succeeded in pushing porphyrin molecules on copper, and more recently they have arranged bucky balls (i.e., C_{60}) in a linear pattern, using an atomic step in the copper substrate as a guide.

C_{60} molecules on silicon also have been pushed with an STM in UHV at room temperature by Maruno *et al.* in Japan, and Beton *et al.* in the U.K. In Maruno's approach the STM tip is brought closer to the surface than in normal imaging mode, and then scan across a rectangular region with the feedback turned off. This causes many probe crashes. In Beton's approach the tip also is brought close to the surface, but the sweep is done with the feedback on and a high value for the tunneling current. Their success rate is in the order of only 1 in 10 trials.

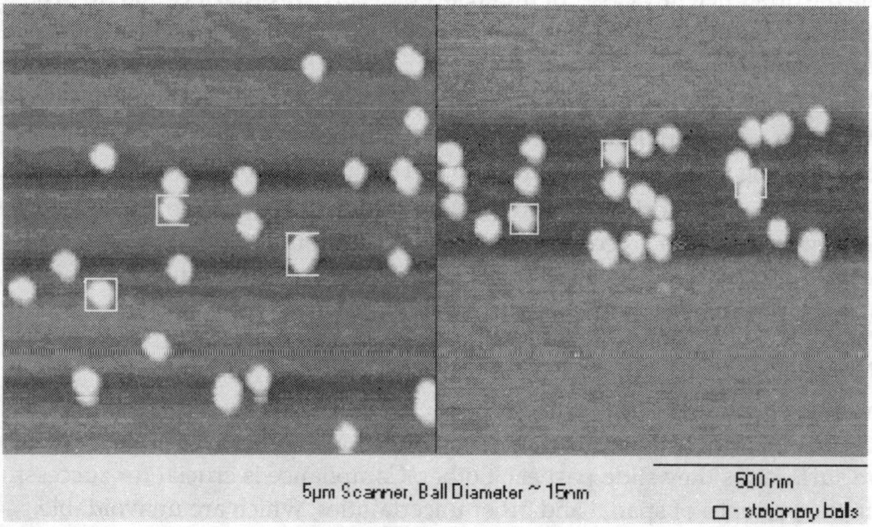

Fig. V–D.3: The initial pattern of 15 nm Au balls (left) and the "USC" pattern obtained by nanomanipulation (right).

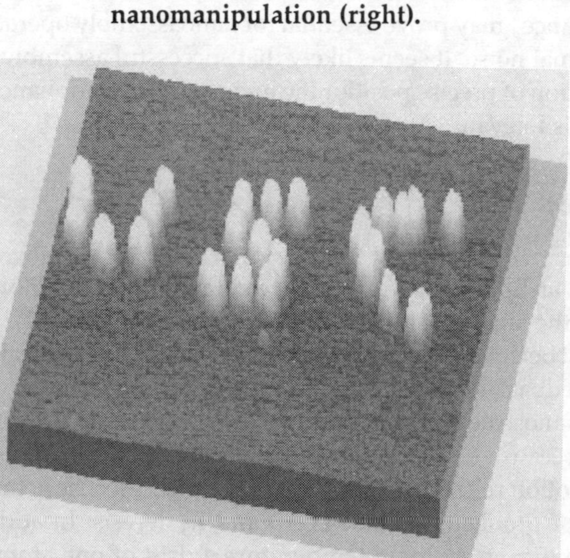

Fig. V–D.4: The "USC" pattern viewed in perspective.

Picking and Placing

Much of industrial macrorobotics is concerned with pick and place operations, which typically do not require very precise positioning or fine control. There are a few examples of experiments in which atoms or molecules are transferred to SPM tips, these are moved, and the atoms transferred back to the surfaces.

Eigler *et al.* succeeded in transferring xenon atoms from platinum or nickel surfaces to an STM tip by moving the tip sufficiently close for the adsorption barriers of surface and tip to be comparable. An atom may leave the surface and become adsorbed to the tip, or vice-versa. Benzene molecules also have been transferred to and from tips.

Eigler's group also has been able to transfer xenon atoms between an STM tip and a nickel surface by applying voltage pulses to the tip. This is attributed to electromigration, caused by the electric current flowing in the tunneling junction. All of Eigler's work has been done in UHV at 4K.

Avouris' group, at IBM Yorktown, and the Aono group in Japan have transferred silicon atoms between a tungsten tip and a silicon surface in UHV at room temperature, by applying voltage pulses to the tip. The mechanism for the transfer is believed to be field-induced evaporation, perhaps aided by chemical phenomena at the tip/surface interface in Avouris work.

Compliant Motion

This is the most sophisticated form of macrorobotic motion. It involves fine motions, as in a peg-in-hole assembly, in which there is accomodation and often force control, for example to ensure contact between two surfaces as they slide past each other. Compliance is crucial for successful assembly operations in the presence of spatial and other uncertainties, which are unavoidable.

The study of the nanoscale analog of compliant motion seems to be virgin territory. We speculate that the analog of compliance is chemical affinity between atoms and molecules. It is suspected that such "chemical compliance" may prove essential for nanoassembly operations at room temperature, in the presence of thermal noise. It seems likely that successful assembly of nanoscale components will require a combination of precise positioning and chemical compliance. Therefore, work on self-assembling structures is relevant.

The SPM as a Robot

Motion

To first approximation, an SPM is a 3 degree-of-freedom robot. It can move in x, y, and z, but cannot orient its tip, which is the analog of a macrorobotic hand. (How could a 6 degree-of-freedom SPM be built, and what could be done with it, are interesting issues.) The vertical displacement is controlled very accurately by a feedback loop involving tunneling current or force (or other quantities for less-common SPMs). But nanoscale x, y motion over small regions (e.g., within a 5 micron square) is primarily open-loop, because of a lack of suitable sensors that can be used in a feedback scheme. Accurate horizontal motion relies on calibration of the piezoelectric actuators, which are known to suffer from a variety of problems such as creep and hysteresis. In addition, thermal drift of the instrument is very significant. At room temperature a drift of one atomic diameter per second is common, which means that manipulation of atomic objects on a surface is not unlike picking parts from a conveyor belt. Thermal drift is negligible if the SPM is operated at very low temperatures, and all the experiments in atomic-precision manipulation to date have been done at 4K. This involves complex technology and is clearly undesirable. For room-temperature nanomanipulation, drift, creep and hysteresis must be taken into account. Ideally, compensation should be automatic. Research is under way at USC on how to move accurately an SPM tip in the presence of all these sources of error.

To complicate matters further, the SPM often must operate in a liquid environment, especially if the objects to be manipulated are biological. Little is known about nanomanipulation in liquids.

Sensing

The SPM functions both as a manipulator and a sensing device. The lack of a direct and independent means of establishing "ground truth" while navigating the tip causes delicate problems. The SPM commands are issued in instrument or robot coordinates, and sensory data also are acquired in robot coordinates, whereas manipulation tasks are expressed in sample or task coordinates. These two coordinate systems do not coincide, and indeed are in relative motion due to drift. Accurate motion in task coordinates may be achieved by a form of visual servoing, by tracking features of the sample and moving relative to them.

A simple example of a practical implementation of these ideas is illustrated in Figure V–D.5. To construct the pattern shown in Figures V–D.3 and V–D.4 we move the tip to the approximate position of a nanoparticle to be pushed, and then search for it through single line scans.

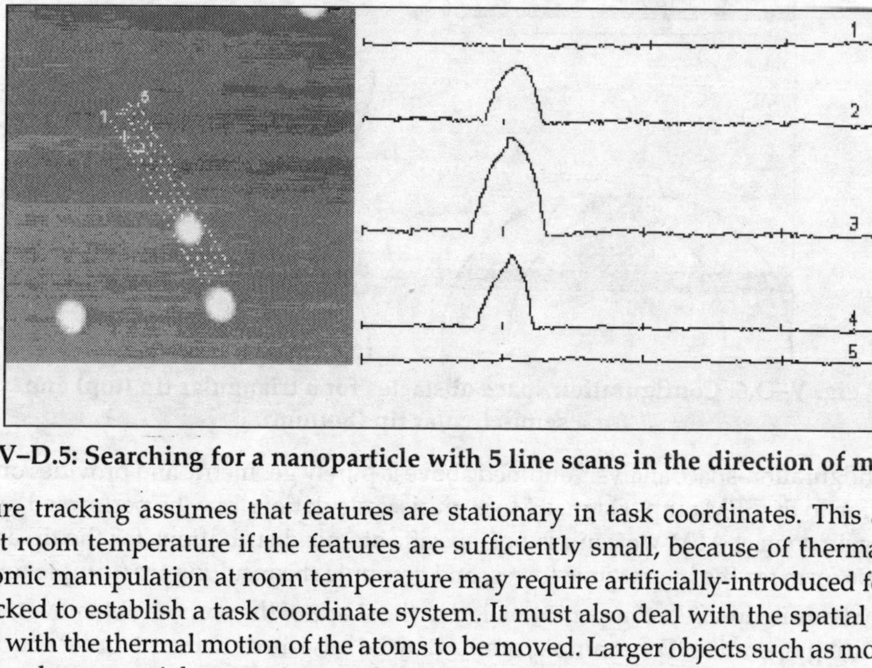

Fig. V–D.5: Searching for a nanoparticle with 5 line scans in the direction of motion.

Feature tracking assumes that features are stationary in task coordinates. This assumption may fail at room temperature if the features are sufficiently small, because of thermal agitation. Hence, atomic manipulation at room temperature may require artificially-introduced features that can be tracked to establish a task coordinate system. It must also deal with the spatial uncertainty associated with the thermal motion of the atoms to be moved. Larger objects such as molecules and clusters have lower spatial uncertainty, and should be easier to handle.

The SPM output signal depends not only on the topography of the sample but also on the shape of the tip, and on other characteristics of the tip/sample interaction. For example, the tunneling current in an STM depends on the electronic wavefunctions of sample and tip. To first approximation one may assume that the tip and sample are in contact. Under this assumption one can use configuration-space techniques to study the motion of the tip]. Figure V–D.6 illustrates the procedure in 2-D. On the top of the figure we consider a tip with a triangular end and a square protrusion in the sample. In the bottom we consider a tip with a semi-circular end and the same protrusion. We choose as reference point for the tip its apex, and reflect the tip about the reference point. The configuration-space obstacle that corresponds to the real-space protrusion obstacle is obtained by sweeping the inverted tip over the protrusion so that the apex remains inside the obstacle. Mathematically, we are calculating the Minkowski sum of the inverted tip and the protrusion. The path of the tip in its motion in contact with the obstacle is the detected topographical signal. As shown in the figure, the sensed topography has been broadened by the dimensions of the tip. Note, however, that the detected height is correct. (Minkowski operations and related mathematical morphology tools were introduced in the SPM literature only recently.) Tip effects are sometimes called "convolution", by analogy with the broadening of an impulse passing through a linear system, and one talks of "deconvolving" the image to remove tip effects.

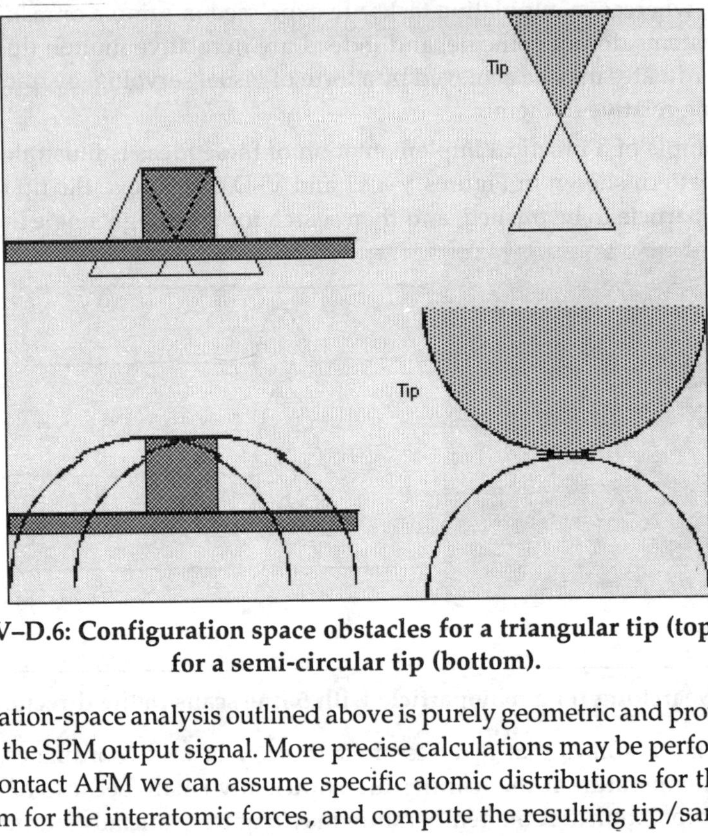

Fig. V–D.6: Configuration space obstacles for a triangular tip (top) and for a semi-circular tip (bottom).

The configuration-space analysis outlined above is purely geometric and provides only a coarse approximation to the SPM output signal. More precise calculations may be performed numerically. For example, in contact AFM we can assume specific atomic distributions for the tip and sample, and a specific form for the interatomic forces, and compute the resulting tip/sample force.

A major issue in tip-effect compensation is that the shape of the probe is not known, and indeed may vary during operation. For example, atoms may be adsorbed on the tip or lost because of the contact with the sample. The most promising approach for dealing with tip effects consists of estimating the tip shape by using it to image known features. If necessary, artificial features may be introduced into the scene for tip estimation purposes. The estimated tip shape can then be used to remove (at least in part) the tip effects from the image by using Minkowski operations. Removal procedures that take into account more sophisticated, non-geometric effects do not appear to be known.

Sensor fusion techniques may be used, at least in principle, for increasing the quality of the sensory data, because it is possible to access several signals during an SPM scan. For example, vertical and lateral force (akin to friction) can be recorded simultaneously in typical AFMs. To our knowledge, sensor fusion has not been attempted in SPM technology.

It is clear that faithful sensory data should facilitate manipulation tasks. What is not clear is whether clever manipulation strategies can compensate for the imperfections of SPM data.

End Effectors

The SPM tip is the primary end effector in nanomanipulation. A plain, sharp tip seems to be adequate for most pushing operations. In some cases it may also suffice for picking and depositing objects, especially in conjunction with electrostatic forces generated by applying a suitable bias to the tip. This requires both a conducting tip and a conducting substrate. Pick-and-place tasks, however, usually require the nanoscale analog of a gripper. Very little is known about molecular grippers.

One can think of a nanogripper as a molecule or cluster of molecules that are attached to a tip and are capable of picking up and transporting other molecules or particles. (Tips with attached molecules are said to be functionalized.) Ideally these grippers should be switchable, so as to pick and release objects on command. Candidates for grippers are certain molecules such as cyclodextrins, which have large cavities that can carry other molecules. Coating a tip with strands of DNA may also permit picking up objects that are coated with complementary strands. In both of these examples, switching the gripper on and off is not a solved problem.

Techniques for changing SPM tips automatically do not exist. Changes must be done manually, and it is very difficult or impossible to maintain sample registration, *i.e.*, to return to the same position on the sample after tip replacement. This implies that one often must image using a tip with an attached gripper. Again, little is known about imaging with such functionalized tips.

Fixtures

In the macroworld fixtures are often necessary to hold mechanical parts during assembly and other manufacturing operations. In micromechanics sacrificial layers are used as scaffolding to fabricate certain microelectromechanical systems (MEMS). At the nanoscale, the analogs of fixtures are substrates that ensure that certain objects remain fixed during manipulation, while others are allowed to move. Substrate selection seems to be highly dependent on the chemistry of the objects being manipulated.

Manipulation Processes

In macrorobotics the physical processes involved are mechanical and relatively well understood. At the nanoscale, the processes are chemical and physical, and still an area of active research. In addition, nanomanipulation takes place in several different environments, such as liquids, air or UHV. The environment has a strong influence on the physics and chemistry of the processes. Nanomanipulation is not restricted to mechanical interactions. For example, light, electrostatic fields, and the pH of a liquid all are candidates for controlled interaction with nanoparticles.

Programing and Planning

High level programing and planning systems are highly desirable, and indeed essential for assembling complex structures. One must begin with relatively low-level programing primitives and build upon them constructs at a higher level of abstraction. High-level commands must be compiled into low-level primitives. This compilation may involve sophisticated computations, for example to ensure collision-free paths in an environment with large spatial uncertainty. What are the relevant high-level manipulation tasks? For example, what is the nanoscale equivalent of a peg-in-hole insertion? In short, we may need to adapt much of what is known about macrorobotics to the nanoworld. It is likely that new concepts will also be needed, because the physics and chemistry of the phenomena and objects are quite different in these two worlds.

Assembly Applications

What hardware primitives are suitable as building blocks? The nanotechnology literature suggests hardware primitives based on DNA structures such as those built in Seeman's lab, proteins, and diamondoid structures. Biomaterials such as DNA and proteins may be too flimsy, whereas diamondoid structures are expected to be very strong. No experiments have yet been reported in which any of these components are successfully assembled into a composite structure.

Which tasks should one attempt first? What should one try to build? Here the options are in the realms of electronics, photonics, mechanics, or biomaterials. On-going research at USC is attempting to build nanowires as assemblies of gold particles, and arrays of nanostructures with photonic properties on patterned semiconductor substrates.

Challenges

Nanorobotics manipulation with SPMs is a promising field that can lead to revolutionary new science and technology. But it is clearly in its infancy.

Typical nanomanipulation experiments reported in the literature involve a team of very skilled, Ph.D.-level researchers working for many hours in a tightly-controlled environment (typically in ultra high vacuum and at low temperature, often 4K) to build a pattern with tens of nanoparticles. It still takes the best groups in the world some 10 hours to assemble a structure with about 50 atoms. This is simply too long-changes will occur in many systems, *e.g.* contamination or oxidation of the components, on a timescale that will constrain the maximum time available for nanomanipulation. Requiring all operations to take place at 4K and in UHV also is not practical for widespread use of nanomanipulation. In short, nanomanipulation today is more of an experimental tour-de-force than a technique that can be routinely used. It is clear that complex tasks cannot be accomplished unless the SPM is commanded at a higher level of abstraction. Compensation for instrument inaccuracies should be automatic, and the user should be relieved from many low-level details.

Building a high level programing system for nanomanipulation is a daunting task. The various component technologies needed for nanomanipulation must be developed and integrated. These technologies include: substrates that serve as nanofixtures or nanoworkbenches on which to place the objects to be manipulated; tips, probes and molecules that serve as grippers or end-effectors; chemical and physical nanoassembly processes; primitive nanoassembly operations that play a role analogous to macroassembly tasks such as peg-in-hole insertion; methods for exploiting self-assembly to combat spatial uncertainty, in a role analogous to mechanical compliance in the macroworld; suitable hardware primitives for building nanostructures; and algorithms and software for sensory interpretation, motion planning, and driving the SPM. This is a tall order, and requires an interdisciplinary approach that combines synergistically the knowledge and talents of roboticists and computer scientists with those of physicists, chemists, materials scientists, and perhaps biologists.

SPM-based assembly methods face a major scale-up challenge. Building complex structures one atom (or even one nanoparticle) at a time is very time consuming. We believe that SPMs will have applications in the exploration of new structures, which may later be mass produced by other means. This is the nanoworld analog of rapid protyping technologies such as stereolithography that are becoming popular at the macroscale.

There are at least two approaches for fighting the serial nature of SPM manipulation. The first involves the use of large arrays of SPMs on a chip. These chips are being developed at Cornell and Stanford. Programing such arrays for coordinated assembly tasks poses interesting problems. The second approach is subtler, and consists of using the SPM to construct structures that are capable of self-replication. The best known such structures involve DNA, but other systems also exist. Self-replication is inherently an exponential process.

In summary, nanomanipulation with SPMs may have a revolutionary impact on science, technology, and the way we live. To fully exploit its potential we will have to develop powerful systems for programing nanorobotic tasks. Much of what is known in macrorobotics is likely to be relevant, but may have to be adapted to the nanoworld, where phenomena and structures are quite different from their macroscopic counterparts. Research at USC and elsewhere is progressing, with promising results.

Nanomanipulation, perhaps coupled with self-assembly, is expected to eventually succeed in building true nanorobots, *i.e.*, devices with overal dimensions in the nanometer range and capable of sensing, "thinking", and acting. Complex tasks are likely to require a group of nanorobots working cooperatively. This raises interesting issues of control, communications, and programing of robot "societies".

Unit V–E

Nanorobots: Medicine of the Future

"Living organisms are naturally-existing, fabulously complex systems of molecular nanotechnology."

—*Dr. Gregory Fahy*

The above statement raises the interesting possibility that machines constructed at the molecular level (nanomachines) may be used to cure the human body of its various ills. This application of nanotechnology to the field of medicine is commonly called as nanomedicine.

Nanorobots as already explained are nanodevices that will be used for the purpose of maintaining and protecting the human body against pathogens. They will have a diameter of about 0.5 to 3 microns and will be constructed out of parts with dimensions in the range of 1 to 100 nanometers. The main element used will be carbon in the form of diamond/fullerene nanocomposites because of the strength and chemical inertness of these forms. Many other light elements such as oxygen and nitrogen can be used for special purposes. To avoid being attacked by the host's immune system, the best choice for the exterior coating is a passive diamond coating. The smoother and more flawless the coating, the less the reaction from the body's immune system. Such devices have been designed in recent years but no working model has been built so far.

The powering of the nanorobots can be done by metabolising local glucose and oxygen for energy. In a clinical environment, another option would be externally supplied acoustic energy. Other sources of energy within the body can also be used to supply the necessary energy for the devices. They will have simple onboard computers capable of performing around 1000 or fewer computations per second. This is because their computing needs are simple. Communication with the device can be achieved by *broadcast-type acoustic signaling*.

A navigational network may be installed in the body, with stationkeeping navigational elements providing high positional accuracy to all passing nanorobots that interrogate them, wanting to know their location. This will enable the physician to keep track of the various devices in the body. These nanorobots will be able to distinguish between different cell types by checking their surface antigens (they are different for each type of cell). This is accomplished by the use of chemotactic sensors keyed to the specific antigens on the target cells.

When the task of the nanorobots is completed, they can be retrieved by allowing them to exfuse themselves via the usual human excretory channels. They can also be removed by active scavenger systems. This feature is design-dependent.

Fields of Application of Nanorobots in Medicine

Some possible applications using nanorobots are as follows:

1. To cure skin diseases, a cream containing nanorobots may be used. It could remove the right amount of dead skin, remove excess oils, add missing oils, apply the right amounts of natural moisturising compounds, and even achieve the elusive goal of 'deep pore cleaning' by actually reaching down into pores and cleaning them out. The cream could be a smart material with smooth-on, peel-off convenience.
2. A mouthwash full of smart nanomachines could identify and destroy pathogenic bacteria while allowing the harmless flora of the mouth to flourish in a healthy ecosystem. Further, the devices would identify particles of food, plaque, or tartar, and lift them from teeth to be rinsed away. Being suspended in liquid and able to swim about, devices would be able to reach surfaces beyond reach of toothbrush bristles or the fibres of floss. As short-lifetime medical nanodevices, they could be built to last only a few minutes in the body before falling apart into materials of the sort found in foods (such as fibre).
3. Medical nanodevices could augment the immune system by finding and disabling unwanted bacteria and viruses. When an invader is identified, it can be punctured, letting its contents spill out and ending its effectiveness. If the contents were known to be hazardous by themselves, then the immune machine could hold on to it long enough to dismantle it more completely.
4. Devices working in the bloodstream could nibble away at arteriosclerotic deposits, widening the affected blood vessels. Cell herding devices could restore artery walls and artery linings to health, by ensuring that the right cells and supporting structures are in the right places. This would prevent most heart attacks.

Chemical Elements for Medical Nanorobot Composition

The typical medical nanodevice will probably be a micron-scale robot assembled from nanoscale parts. These parts could range in size from 1-100 nm (1 nm = 10^{-9} meter), and might be fitted together to make a working machine measuring perhaps 0.5-3 microns (1 micron = 10^{-6} meter) in diameter. Three microns is about the maximum size for bloodborne medical nanorobots, due to the capillary passage requirement.

Carbon will likely be the principal element comprising the bulk of a medical nanorobot, probably in the form of diamond or diamondoid/fullerene nanocomposites—largely because of the tremendous strength and chemical inertness of diamond. Many other light elements such as hydrogen, sulfur, oxygen, nitrogen, fluorine, silicon, etc. will be used for special purposes in nanoscale gears and other components.

From a medical standpoint, it makes sense to regard the nanorobot as having two spaces which should be considered separately—its interior and its exterior. It is true that the nanorobot exterior will be exposed to the diverse chemical brew that makes up our human biochemistry. But the interior of the nanorobot may be a highly controlled environment, possibly a vacuum, into which external liquids cannot normally intrude.

Of course it may often be necessary for a nanorobot to import external fluids in a controlled manner for chemical analysis or other purposes. But the important thing is that the device will be watertight and airtight. Body fluids cannot get into the interior of the device, unless these fluids are purposely pumped in for some specific reason.

Physical Appearance of Humans Injected with Medical Nanorobots

In most cases a human patient who is is undergoing a nanomedical treatment is going to look just like anyone else who is sick. The typical nanomedical treatment (e.g. to combat a bacterial or viral infection) will consist of an injection of perhaps a few cubic centimeters of micron-sized nanorobots suspended in fluid (probably a water/saline suspension). The typical therapeutic dose may include up to 1-10 trillion (1 trillion = 10^{12}) individual nanorobots, although in some cases treatment may only require a few million or a few billion individual devices to be injected. Each nanorobot will be on the order of perhaps 0.5 micron up to perhaps 3 microns in diameter. (The exact size depends on the design, and on exactly what the nanorobots are intended to do.)

The adult human body has a volume of perhaps 100,000 cm^3 and a blood volume of ~5400 cm^3, so adding a mere ~3 cm^3 dose of nanorobots is not particularly invasive. The nanorobots are going to be doing exactly what the doctor tells them to do, and nothing more (barring malfunctions). So the only physical change you will see in the patient is that he or she will very rapidly become well again. Most symptoms such as fever and itching have specific biochemical causes which can also be managed, reduced, and eliminated using the appropriate injected nanorobots. Major rashes or lesions such as those that occur when you have the measles will take a bit longer to reverse, because in this case the broken skin must also be repaired.

Appearance of a Typical Nanorobot

It is impossible to say exactly what a generic nanorobot would look like. Nanorobots intended to travel through the bloodstream to their target will probably be 500-3000 nanometers (1 nanometer = 10^{-9} meter) in characteristic dimension. Nonbloodborne tissue-traversing nanorobots might be as large as 50-100 microns, and alimentary or bronchial-traveling nanorobots may be even larger still. Each species of medical nanorobot will be designed to accomplish a specific task, and many shapes and sizes are possible.

Finally, and perhaps most importantly, no actual working nanorobot has yet been built. Many theoretical designs have been proposed that look good on paper, but these preliminary designs could change significantly after the necessary research, development and testing has been completed.

An Example of a Simple Medical Nanorobot

One very simple nanorobot that I designed a few years ago is the artificial mechanical red cell, which I call a *"respirocyte."* The respirocyte measures about 1 micron in diameter and just floats along in the bloodstream. It is a spherical nanorobot made of 18 billion atoms. These atoms are mostly carbon atoms arranged as diamond in a porous lattice structure inside the spherical shell. The respirocyte is essentially a tiny pressure tank that can be pumped full of up to 9 billion oxygen (O_2) and carbon dioxide (CO_2) molecules. Later on, these gases can be released from the tiny tank in a controlled manner. The gases are stored onboard at pressures up to about 1000 atmospheres. (Respirocytes can be rendered completely nonflammable by constructing the device internally of sapphire, a flameproof material with chemical and mechanical properties otherwise similar to diamond.)

The surface of each respirocyte is 37 per cent covered with 29,160 molecular sorting rotors (Nanosystems) that can load and unload gases into the tanks. There are also gas concentration sensors on the outside of each device. When the nanorobot passes through the lung capillaries, O_2 partial pressure is high and CO_2 partial pressure is low, so the onboard computer tells the sorting rotors to load the tanks with oxygen and to dump the CO_2. When the device later finds itself in the oxygen-starved peripheral tissues, the sensor readings are reversed. That is, CO_2 partial pressure is relatively high and O_2 partial pressure relatively low, so the onboard computer commands the sorting rotors to release O_2 and to absorb CO_2.

Respirocytes mimic the action of the natural hemoglobin-filled red blood cells. But a respirocyte can deliver 236 times more oxygen per unit volume than a natural red cell. This nanorobot is far more efficient than biology, mainly because its diamondoid construction permits a much higher operating pressure. (The operating pressure of the natural red blood cell is the equivalent of only about 0.51 atm, of which only about 0.13 atm is deliverable to tissues.) So the injection of a 5 cm^3 dose of 50 per cent respirocyte aqueous suspension into the bloodstream can exactly replace the entire O_2 and CO_2 carrying capacity of the patient's entire 5,400 cm^3 of blood!

Respirocytes will have pressure sensors to receive acoustic signals from the doctor, who will use an ultrasound-like transmitter device to give the respirocytes commands to modify their behavior while they are still inside the patient's body. For example, the doctor might order all the respirocytes to just stop pumping, and become dormant. Later, the doctor might order them all to turn on again.

What if you added 1 liter of respirocytes into your bloodstream, the maximum that could possibly be safe? You could then hold your breath for nearly 4 hours if sitting quietly at the bottom of a swimming pool. Or if you were sprinting at top speed, you could run for at least 15 minutes before you had to take a breath!

It is clear that very "simple" medical nanodevices can have extremely useful abilities, even when applied in relatively small doses. Other more complex devices will have a broader range of capabilities. Some devices may have mobility—the ability to swim through the blood, or crawl through body tissue or along the walls of arteries. Others will have different shapes, colors, and surface textures, depending on the functions they must perform. They will have different types of robotic manipulators, different sensor arrays, and so forth. Each medical nanorobot will be designed to do a particular job extremely well, and will have a unique shape and behaviour.

Do "Old Nanorobots" Left in the Body Cause Problems when they Fail

Following most simple treatments, nanodoctors of the 21st century will want to remove their therapeutic nanorobots from the patient's body as soon as the nanodevices have finished the job. So there will be little danger of "old nanorobots" breaking down or malfunctioning, or causing something unpleasant to happen to the patient after the original disease or traumatic condition has been treated.

Additionally, nanorobots will be designed with a high level of redundancy to ensure fail-operational and fail-safe performance, further reducing the medical risk.

How can they be Retrieved from the Body

Some nanodevices will be able to exfuse themselves from the body via the usual human excretory channels; others will be designed to allow ready exfusion by medical personnel using apheresis-like processes (commonly called nanapheresis) or active scavenger systems. It is very design dependent. In the case of the respirocytes, the removal procedure is fairly simple:

> "Once a therapeutic purpose is completed, it may be desirable to extract artificial devices from circulation. Onboard water ballast control is extremely useful during respirocyte exfusion from the blood. Blood to be cleared may be passed from the patient to a specialized centrifugation apparatus where acoustic transmitters command respirocytes to establish neutral buoyancy. No other solid blood component can maintain exact neutral buoyancy, hence those other components precipitate outward during gentle centrifugation and are drawn off and added back to filtered plasma on the other side of the apparatus. Meanwhile, after a period of centrifugation, the plasma, containing mostly suspended respirocytes but few other solids, is drawn off through a 1-micron filter, removing the respirocytes. Filtered plasma is recombined with centrifuged solid components and returned undamaged to the patient's body. The rate of separation is further enhanced either by commanding respirocytes to empty all tanks, lowering net density to 66 per cent of blood plasma density, or by commanding respirocytes to blow a 5-micron O_2 gas bubble to which the device may adhere via surface tension, allowing it to rise at 45 mm/hour under normal gravitational acceleration."
>
> (Quoted from Robert A. Freitas Jr., "Exploratory Design in Medical Nanotechnology: A Mechanical Artificial Red Cell".)

Are Medical Nanorobots Attacked by the Immune System, as soon as they are Placed Inside the Human Body

Immune system response is primarily a reaction to a "foreign" surface. Nanorobot size is also an important variable, along with device mobility, surface roughness, surface mobility, and other factors. Yet the problem of nanodevice biocompatibility is in principle no more difficult than the biocompatibility of medical implants generally. In some ways it may even be an easier problem, because many medical nanorobots will have only temporary residence in the body. Even today, application of immunosuppressive agents during the treatment period would allow poorly-engineered non-bioinactive nanorobots to perform their repair work without trouble.

Passive diamond exteriors may turn out to be ideal. Several experimental studies hint that the smoother and more flawless the diamond surface, the less leukocyte activity and the less fibrinogen adsorption you will get. So it seems reasonable to hope that when diamond coatings can be laid down with almost flawless atomic precision, making nanorobot exterior surfaces with near-nanometer smoothness, that these surfaces may have very low bioactivity. Due to the extremely high surface energy of the passivated diamond surface and the strong hydrophobicity of the diamond surface, the diamond exterior is almost completely chemically inert and so opsonization should be minimized.

However, even if flawless diamond surfaces alone do not prove fully bioinactive as hoped, active surface management of the nanorobot exterior can be used to ensure complete nanodevice biocompatibility. Allergic and shock reactions are similarly easily avoided.

How Fast do Medical Nanorobots Replicate Inside the Human Body

This is a very common error. Medical nanorobots need not EVER replicate. In fact, it is unlikely that the FDA (or its future equivalent) would ever approve for general use a medical nanodevice that was capable of *in vivo* replication. Except in the most unusual of circumstances, you would never want anything that could replicate itself to be turned loose inside your body. Replicating bacteria are trouble enough!

Replication is a crucial basic capability for molecular manufacturing. But aside from the most aggressive applications, there is simply no good reason to risk manufacturing "fertile" nanorobots inside the human body, when "mule" nanorobots can be manufactured so cheaply, conveniently, and in such vast numbers outside of the human body. Replicators will almost certainly be very tightly regulated by governments everywhere.

Do Medical Nanorobots Possess a Humanlike Artificial Intelligence

This is another common error. Many medical nanorobots will have very simple computers on board each device. Respirocytes, for example, have only a ~1,000 operations/sec computer on board each device—far less computing power than an old Apple II.

Most cellular repair nanorobots will not need more than 10^6-10^9 operations/sec of onboard computing capacity to do their work. This is a full 4-7 orders of magnitude below (even the potential for) true human-equivalent computing at 10 teraflops (~10^{13} operations/sec). Faster computing capacity is simply not required for most medical nanorobots.

How will a Medical Nanorobot be Powered

One of the earliest proposals by Drexler in *Engines of Creation* was that an *in vivo* medical nanodevice could metabolize local glucose and oxygen for energy. Another possibility is externally supplied acoustic power, which is probably most appropriate in a clinical setting. There are literally dozens of useful power sources that are potentially available in the human body.

Can you Communicate with the Machines as they do their Work

There are many different ways to do this. One of the simplest ways to send broadcast-type messages into the body, to be received by *in vivo* nanorobots, is acoustic messaging. A device similar to an ultrasound probe would encode messages on acoustic carrier waves at frequencies between 1-10 MHz. Thus, the supervising physician can easily send new commands or parameters to nanorobots already at work inside the body. Each nanorobot has its own power supply, computer, and sensorium, thus can receive the physician's messages via acoustic sensors, then compute and implement the appropriate response.

The other half of the process is getting messages back out of the body, from the working nanodevices out to the physician. This can also be done acoustically. However, onboard power requirements for micronscale acoustic wave generators in water dictate a maximum practical transmission range of at most a few hundred microns for each individual nanorobot. Therefore, it is convenient to establish an internal communications network that can collect local messages and pass them along to a central location, which the physician can then monitor using sensitive ultrasound detectors to receive the messages. Such a network can probably be deployed inside a patient in less than an hour, may involve up to 100 billion mobile nanorobotic network nodes, and may release at most 60 watts of waste heat (less than the 100-watt human body basal rate) assuming a (worst case) full 100 per cent network duty cycle.

There are many other techniques that may be used as well—this one is just the easiest to describe.

If Medical Nanorobots are Infused into the Human Body, Intravenously, how will one Track their Location

A navigational network may be installed in the body, with stationkeeping navigational elements providing high positional accuracy to all passing nanorobots that interrogate them, wanting to know their location.

Physical positions can be reported continuously using an *in vivo* communications network. Since the typical therapeutic dose may involve billions or trillions of nanorobots (e.g. up to a few cm^3 of injection), it will usually be impractical to address nanorobots individually, though this is in principle possible for treatments involving only a few million devices, or fewer.

Form of detection system will medical nanorobots use to distinguish between differing cell types

Each cell type has its own unique set of surface antigens. Other cell surface antigens indicate the health status of the cell, the parent organ type, the species of the animal, and even the identity of the individual—a kind of biochemical Social Security Number.

So the short answer to this question is: use chemotactic sensors (crudely analogous to chemical force microscopy), keyed to the specific known antigens on the target cells you are looking for.

How will Chemical Agents (e.g. an anti-cancer drug) be Transported and Delivered to a Target Cell

Once you've identified a group of cells that needs some chemical substance delivered to it, you can simply release the agent from onboard tanks after the nanorobot arrives on the scene. A 1 cm^3 injection of 1-micron nanodevices could probably hold at least 0.5 cm^3 of chemical agent. Virtually all of these billions of nanites (in the 1 cm^3) will be smart enough to show up at the correct group of cells that are targeted for destruction, so delivery efficiency is virtually 100 per cent. Onboard sensors can test for ambient levels of the chemical agent, to prevent overdose.

However, this question is a good example of an "anachronistic" application—one that could be done using medical nanorobots, but in fact would probably never be done that way, because in an era of advanced nanotechnology much more efficient and much less destructive ways would exist to get the same job done. In the above example, bulk delivery of cytotoxins to tissue cells is completely unnecessary if the means exists to reverse the carcinomatous process at the cellular and genetic level.

Is it Possible to Image in vivo Medical Nanomachines using an Imaging Technique such as MRI

Yes, nanodevices could probably be observed at work inside the body using MRI, especially if their diamond components were manufactured using mostly ^{13}C atoms rather than the more common natural ^{12}C isotope of carbon, since ^{13}C has a nonzero nuclear magnetic moment. But in the nanomedical era, such an approach may again be somewhat anachronistic. Here's why.

Applying the classical medical model to a typical nanomedical treatment, the medical nanodevices would first be injected into a human body, and would then go to work—say, in a specific organ or tissue mass. The physician wants to be able to monitor their progress, and make certain that the nanodevices have gotten to the correct target treatment region. So the first instinct of the contemporary physician who is contemplating a prospective nanomedical treatment will be to insist on the ability to directly image the nanorobots. In other words, the doctor wants to be able to scan a section of the body, and actually see the nanodevices congregated neatly around their target, say, a tumor mass, so that he can be absolutely certain that the therapy is proceeding as (and where) planned.

However, if the technology exists to fabricate nanorobots to molecular precision, then this same technology will allow communication and navigational mechanisms to be designed and built into each and every nanorobot, and will also allow communications and navigational networks to be deployed inside the body. Therapeutic nanodevices may be programed to home in on a very precisely-specified set of surface antigens populating the surface of the target tumor mass. As an additional guide, internal reference frame navigation with millimeter (or better) accuracy may be used to direct the nanorobots to the close vicinity of the target treatment volume.

So the correct medical treatment model in the nanomedical era is as follows: Injected and circulating nanorobots would remain absolutely inactive outside of the target volume. Even once inside the target treatment volume, nanorobots would still remain inactive until the precise antigenic signature of the target tissue was chemotactically detected by nanorobot sensors. The nanorobots could further be programed to remain inactive until they received an acoustic command from the physician telling them that they were free to begin the active treatment, perhaps after the physician receives confirmation through the navigational grid that most of the devices have reached their proper destinations. The physician retains complete control throughout the course of the treatment process; a signal ordering all nanorobots to halt may be sent at any time.

Equally important, nanorobots can communicate their positions, operational statuses, and the success or failure of the treatment as the treatment progresses. Bandwidth limitations will require considerable information pooling, but this should not present a problem. In this treatment model, the physician receives continuous reports from the active nanorobots. They tell you their physical coordinates in the body, so you know where they are. They tell you how many cancer cells they have encountered and inactivated (or whatever is the appropriate metric to establish progress for the particular treatment). They will have multiple-redundant systems (like the five consensus computers onboard the space shuttle), establishing a fail-operational or a fail-safe design—upon detecting a critical component failure, the device places itself in shutdown mode in preparation for exfusion.

So in this kind of scenario, it may be quite unnecessary to image the nanodevices directly, because the feedback available to the supervising physician from other means will be far more sophisticated, reliable, useful, and complete.

To Monitor the Progress of the Nanorobots during a Treatment, could you Biopsy the Tissue and the Image the Machines using TEM

Yes, it would be possible to biopsy tissue and then image the embedded nanomachines using transmission electron microscopy. However, the normal presumption will be that your medical nanorobots are working properly. Because of fail-safe design, the nanodevices should only rarely be a part of the problem. On the contrary, they will likely be a major part of the solution.

In the usual biopsy situation, your primary interest is in the condition of the tissue, not the condition of the nanodevices embedded in it. But nanodevices can be used to rapidly examine a given piece of tissue, surveying its biochemistry, biomechanics, and histometric characteristics in great detail. Indeed, in an era of proficient nanomedicine, it should rarely be necessary to remove tissue samples from patients for testing at all. Most testing should be possible *in situ*, with the added benefits of reduced intrusiveness, increased patient comfort, and greater fidelity of results since the target tissue can be examined in its active state in the actual host environment.

Errors during a Nanomedical Procedure

The incompetence or negligence of medical personnel is always a potential concern. However, in the nanomedical era, as today, such occurrences should be infrequent and notorious.

A true glitch will come from some direction that nobody anticipated. Biocompatibility problems are well anticipated, and multiple-redundant onboard computers should ensure safe operation, correct operation, and reprogramability of operational parameters even after the devices have been launched on their mission—especially to permit deactivation if anything goes wrong. Fail-stop protocols may be particularly appropriate in high-risk missions where large numbers of replacement nanorobots are readily available.

Therefore, the most serious problems may devolve from the inherent complexity of a trillion machines independently trying to cooperatively work on a very complex repair problem in a short period of time. One class of malfunction might involve some unexpected emergent machine-machine interaction—the kind of subtle interaction that is unlikely to have been exhaustively tested in full-up systems, in advance.

As a simple example, consider two nanorobot species that are jointly repairing a given block of tissue. If the nanorobot programing allows species A to interpret the repair work of species B as a new tissue flaw that lies within species A's original repair mission parameters, and vice versa, then it would be possible for the two species to become locked in an endless recursive cycle, as each species attempted repeatedly to undo the other's work.

But even in such cases, control over the devices is not lost. The supervising physician, upon observing the fault, would simply shut down one or the other species to allow the work to proceed, or would shut down both species and reprogram them both (while they are still inside the body) to avoid the unwanted emergent behavior. The doctor must always be able to "pull the plug" on the nanomachines. This is one of the most important design constraints, one that will probably become a strict and universal regulatory requirement for all medical nanodevices.

Biggest Benefit for Human Society from Nanomedicine

Nanomedicine will eliminate virtually all common diseases of the 20th century, virtually all medical pain and suffering, and allow the extension of human capabilities—most especially our mental abilities.

Consider that a nanostructured data storage device measuring ~8,000 micron^3, a cubic volume about the size of a single human liver cell and smaller than a typical neuron, could store an amount of information equivalent to the entire Library of Congress. If implanted somewhere in the human brain, together with the appropriate interface mechanisms, such a device could allow extremely rapid access to this information.

A single nanocomputer CPU, also having the volume of just one tiny human cell, could compute at the rate of 10 teraflops (10^{13} floating-point operations per second), approximately equalling (by many estimates) the computational output of the entire human brain. Such a nanocomputer might produce only about 0.001 watt of waste heat, as compared to the ~25 watts of waste heat for the biological brain in which the nanocomputer might be embedded.

But perhaps the most important long-term benefit to human society as a whole could be the dawning of a new era of peace. We could hope that people who are independently well-fed, well-clothed, well-housed, smart, well-educated, healthy and happy will have little motivation to make war. Human beings who have a reasonable prospect of living many "normal" lifetimes will learn patience from experience, and will be extremely unlikely to risk those "many lifetimes" for any but the most compelling of reasons.

Unit V–F

Nanorobots in the Bloodstream—The Promise of Nanomedicine

In just a few decades physicians could be sending tiny machines into our bodies to diagnose and cure disease. These nanodevices will be able to repair tissues, clean blood vessels and airways, transform our physiological capabilities, and even potentially counteract the aging process.

It is the year 2031, and the age of advanced nanomedicine has arrived. A young man arrives at his physician's office with a mild fever, nasal congestion, discomfort and a cough. The physician pulls from her pocket a lightweight, handheld device resembling a pocket calculator. She unsnaps from it a cordless, self-sterilizing, pencil-sized probe and inserts it into the patient's mouth as if it were a tongue depressor. On the tip of the probe are billions of molecular assay receptors, mounted on hundreds of self-guiding retractile stalks. Each receptor is sensitive to the chemical signature of a specific kind of bacterium or virus. "Ahhh," says the patient, and a few seconds later a three-dimensional, color-coded map of his throat appears on the display panel of the device. Beneath the map scroll columns of data, revealing the unique molecular signature of a known and unwelcome bacterial pathogen.

With the diagnosis complete, the infectious microorganism can be exterminated. No need for antihistamines, cough drops and a week-long course of antibiotics. The physician keeps several generic classes of nanorobots in her office for just such a circumstance. Using a desktop appliance in her office, she programs billions of nanorobots to find, recognize and destroy the particular microbial strain. The nanomachines are suspended in a carrier fluid that the patient inhales into his lungs, after which the mobile devices march down the patient's throat, propelled on tiny legs. Following a search pattern, the nanorobots ingest and destroy the harmful bacteria they encounter using mechanical and chemical phagocytosis. The patient feels nothing: nanorobots are the size of bacteria, which constantly crawl on and inside the body without ever being noticed. After several minutes, the physician activates an acoustic homing beacon to guide the nanorobots back into the patient's mouth, where she retrieves them through a collection port on the tip of the homing device. A further survey with the original diagnostic probe reveals no evidence of the pathogen.

This is a wonderful vision of medicine in the future, but how do we get there from here?

Nanotechnology ("nano" from the Greek word for dwarf) involves engineering and manufacturing at the molecular, or nanometer, level. One nanometer is one billionth of a meter, about the width of six bonded carbon atoms. How does one build at this scale?

Scientists at New York University have adopted the self-assembly approach, and have succeeded in producing complementary strands of DNA that can zip themselves into complex structures. They have built cubes, octahedra and other solids made of just a few thousand nucleotides each, by the billions per batch. Meanwhile at Cornell University, researchers have genetically engineered a natural biomotor normally found in the enzyme ATPase to allow it to incorporate nonbiological parts such as a silicon nitride bar 100 nanometers in length, making the first artificial hybrid nanomotor. In a microscopic video presentation, dozens of bars attached to artificial molecular motors in a large precise array could be seen spinning at 200 revolutions per minute, like a field of tiny propellers.

Another way to build at the molecular level is by positional assembly—picking and placing molecular parts exactly where you want them. A device capable of positional assembly would work much like the robot arms that manufacture cars on automobile assembly lines, or like the ribosome that assembles amino acids one by one into proteins in our cells. The US engineering firm Zyvex Corp. is aiming to be the first to create an artificial molecular assembler using positional assembly to manufacture atomically precise structures.

Elsewhere, researchers at Harvard University have constructed the first general-purpose "nanotweezers," using a pair of electrically controlled carbon nanotubes These have successfully grasped 300-nanometer clusters of polystyrene spheres and extracted a single semiconductor wire 20 nanometers wide from a mass of entangled wires. The scientists hope eventually to produce nanotweezers small enough to grab individual large molecules.

Such nanoresearch will eventually lead to applications affecting almost every aspect of modern life, including the home, transport, computers, energy and the environment. Some of the most important and life-changing applications, however, are likely to be seen in medicine. We may see the first nanomedical materials and devices in use within the next few years. Scientists at the University of Michigan's Center for Biologic Nanotechnology are currently pursuing the use of dendrimers as a safer and more effective genetic therapy agent. Dendrimers are branching synthetic molecules that can be grown nanometer by nanometer to reach the desired size. They take on a spherical shape, and have sufficiently large openings and cavities to carry small molecules—dendrimers could therefore be used to sneak DNA into cells without triggering an immune response. At the same center, researchers have reported using dendrimer "nanodecoys" to trap and deactivate influenza virus particles.

Relatively simple nanodevices could soon offer cures for major conditions such as diabetes. Researchers at Ohio State University and the University of Illinois have managed to construct silicon-based microcapsules highly perforated with "nanopores," each as small as 20 nanometers in diameter. These pores are large enough to allow small molecules such as oxygen, glucose and insulin to pass through, but are small enough to impede the passage of much larger immune system molecules such as immunoglobulins. Microcapsules containing islet cells could be implanted beneath the skin of some diabetes patients, temporarily restoring the body's delicate glucose control feedback loop without the need for powerful immuno-suppressants that can leave the patient at serious risk for infection. The same method of encapsulation could be used to treat other enzyme- or hormone-deficiency diseases. Encapsulated neurons could also be implanted in the brain and then electrically stimulated to release neurotransmitters, possibly as part of a future treatment for Alzheimer's or Parkinson's disease.

Artificial "biobots" could be in our bodies within five to ten years. Advances in genetic engineering are likely to allow us to construct an artificial microbe—a basic cellular chassis—to perform certain functions. These biobots could be designed to produce vitamins, hormones, enzymes or cytokines in which the host body was deficient, or they could be programed to selectively absorb and break down poisons and toxins. A new company called EngeneOS, Inc., founded in late 2000, has already announced plans to develop artificial Engineered Genomic Operating Systems using the techniques of molecular biology. These systems will comprise a library of component device modules and proprietary modular components. This will allow the engineering and construction of programmable biobots with novel form and function.

The greatest power of nanomedicine will emerge in the longer term, perhaps 10-20 years from now, when we learn how to design and construct complete artificial nanorobots using strong diamond-like materials, nanometer-scale parts, and onboard subsystems including sensors, motors, manipulators, power plants, and molecular computers.

One example is an artificial mechanical cell called a respirocyte, which could be used to keep a patient's tissues safely oxygenated for up to about four hours (at maximum dosage) if their heart has stopped beating. These cells (side panel, 4) could also enable a healthy person to sprint at top speed for at least 15 minutes without breathing, or to sit underwater at the bottom of a swimming pool for hours. Still entirely theoretical, the respirocyte is a micron-wide spherical nanorobot made of 18 billion atoms precisely arranged in a diamondoid structure to form a tiny 1,000-atmosphere pressure tank. Several billion molecules of oxygen and carbon dioxide can be absorbed into, or released from, this tank using computer-controlled molecular pumps powered by serum glucose and oxygen. External gas concentration sensors would allow respirocytes to mimic the action of the natural hemoglobin-filled red blood cells, with oxygen released and carbon dioxide absorbed in the tissues, and vice versa in the lungs. Each respirocyte would be able to hold 200 times more gas per unit volume than a natural red cell, so a few cubic centimeters injected into the human bloodstream would exactly replace the gas carrying capacity of the patient's entire 5.4 liters of blood.

Other proposed medical nanorobots offer equally astonishing performance improvements over nature. For instance, micronsize artificial mechanical platelets could allow complete hemostasis (control of bleeding by clot formation) in just one second, even for moderately large wounds, a response time 100-1,000 times faster than the natural system. This would be achieved by rapidly unfurling a compactly stowed onboard biodegradable mesh under the control of an onboard computer. These "clottocytes" would be about 10,000 times more effective as clotting agents than an equal volume of natural platelets.

Nanorobotic phagocytes called microbivores could patrol the bloodstream, seeking out and digesting unwanted pathogens including bacteria, viruses or fungi. Each nanorobot could completely destroy one pathogen in just 30 seconds—about 100 times faster than natural leukocytes or macrophages—releasing a harmless effluent of amino acids, mononucleotides, fatty acids and sugars. No matter that a bacterium has acquired multiple drug resistance to antibiotics or to any other traditional treatment. The microbivore will eat it anyway, achieving complete clearance of even the most severe septicemic infections in minutes to hours, as compared to weeks or even months for antibiotic-assisted natural phagocytic defenses, without increasing the risk of sepsis or septic shock. Related nanorobots could be programed to recognize and digest cancer cells, or to clear circulatory obstructions within minutes in order to rescue stroke patients from ischemic damage.

More sophisticated medical nanorobots will be able to intervene at the cellular level, performing surgery within cells. Physician-controlled nanorobots could extract existing chromosomes from a diseased cell and insert newly manufactured ones in their place, a process called chromosome replacement therapy. This would allow a permanent cure of any pre-existing genetic disease, and permit cancerous cells to be reprogramed to a healthy state.

In recent years, many gerontologists have begun to think of aging as a genetic disease that might be cured. Evidence from the Human Genome Project suggests that just a few hundred genes at most may be directly involved in aging. If we can completely understand these few genes and how they work, we may be able to alter them to eliminate this unwanted syndrome. Using cytosurgical nanorobots, corrected genes could be installed in every one of the 10 trillion tissue cells in our bodies. We would then no longer naturally age, and our bodies would again repair themselves as well as they did when we were children.

Although nanotechnology is in its infancy, researchers are steadily making major breakthroughs. If we can learn to harness and precisely control the ability to manipulate molecules, then many aspects of our lives will change forever. In particular, the ability to carry out medical procedures at the molecular level will revolutionize medical practice. The next few decades will be very interesting indeed.

Unit V–G

Nanorobots in Surgery

Introduction

A new era on medicine are expected to happen in the coming years. Due to the advances in the field of nanotechnology, nanodevice manufacturing has been growing gradually. From such achievements in nanotechnology, and recent results in biotechnology and genetics, the first operating biological nanorobots are expected to appear in the coming five years, and more complex diamondoid based nanorobots will become available in about ten years. In terms of time it means a very near better future with significant improvements in medicine. In this work we present a practical approach taken on developing nanorobots for medicine in the sense of using computational nanomechatronics techniques as ancillary tools for investigating manufacturing design, nanosystems integration, sensing and actuation for medicine applications. Thus, the work describes pathways that could enable design testability, but also help scientists and profit corporations in providing the helpful information needed to test and design integrated devises and solutions towards manufacturing biomedical nanorobots.

The use of robots in surgery has provided additional tools for surgeons enabling minimally invasive intervention or even long distance teleoperated surgeries. Indeed we may trust on human creativeness and technical capabilities that can ever be improved in terms of technical achievements. In recent years the medicine has enabled significant wellness for the life quality and longevity of the world population. And for the coming years, we may be prepared to experiment even more benefits, as results from advances that are being pursued step by step in new fields of science, such as nanobiotechnlogy. With the expected miniaturization of devices provided by several works on nanoelectromechanical systems (NEMS), nanomanufacturing has actually become a reality.

Hence, with the NEMS recent advances on building nanodevices, and the development of interdisciplinary works, altogether may be translated in few years through the development of integrated nanomachines, also known as nanorobots. With the use of techniques that are advancing rapidly, such as nano-transducers, and biomolecular computing, nanorobots are expected to be able to operate in a well defined set of behaviors performing pre-programed tasks. Thus, in the coming few years, nanorobots being teleoperated to perform surgery, or even nanorobots continually supervising the human body in order to assist organs that may require some kind of repair, is one of the most expected revolutionary tools for biomedical engineering problems.

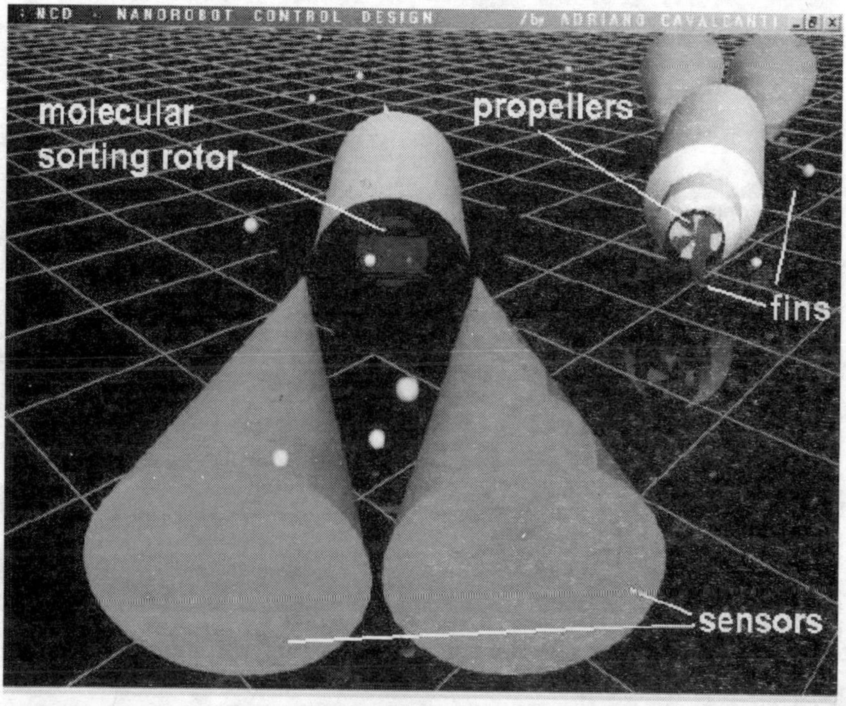

Fig. V–G.1: The depicted blue cones shows the sensors "touching" areas that triggers the nanorobots' behaviors.

The development of nanorobots is an emerging field with many aspects under investigation. Simulation is an essential tool for exploring alternatives in the organization, configuration, motion planning, and control of nanomachines exploring the human body. Nanorobot applications could be focused mainly on two major areas, as follows: nanorobots for surgical interventions, as well as their utilization for patients that need constant monitoring. The nanorobots require specific controls, sensors and actuators, basically in accordance with each kind of biomedical problem.

Advanced simulations can include various levels of detail, giving a trade-off between physical accuracy and the ability to control large numbers of nanorobots over relevant time scales with reasonable computational effort. Another advantage is that simulation can be done in advance of direct experimentation. It is most efficient to develop the control technology in tandem with the fabrication technologies, so that when we are able to build these devices, we will already have a good background in how to control them.

Computational mechatronic approaches were proposed as a suitable way to enable the fast development of nanorobots operating in a fluid environment relevant for medical applications. Unlike the case of larger robots, the dominant forces in this environment arise from viscosity of low Reynolds number fluid flow and Brownian motion and such parameters are been implemented throughout a set of different investigations. Practical and innovative paradigms have been developed based on the Nanorobot Control Design (NCD) simulator that allows fast design testability comparing various control algorithms for nanorobots and their application for different tasks. Also such information generated by the NCD can be useful as parameters for building nanodevices, such as transducers and actuators.

Fig. V–G.2: Rendering schematically the nanorobot sensors' collision detection for chemical signals molecular identification.

Nanorobots for Medicine

In future decades the principal focus in medicine will shift from medical science to medical engineering, where the design of medically-active microscopic machines will be the consequent result of techniques provided from human molecular structural knowledge gained in the 20th and early 21st centuries. For the feasibility of such achievements in nanomedicine, two primary capabilities for fabrication must be fulfilled: fabrication and assembly of nanoscale parts. Through the use of different approaches such as biotechnology, supramolecular chemistry, and scanning probes, both capabilities had been demonstrated to a limited degree as early as 1998. Despite quantum effects which impart a relative uncertainty to electron positions, the quantum probability function of electrons in atoms tends to drop off exponentially with distance outside the atom. Even in most liquids at their boiling points, each molecule is free to move only ~0.07 nm from its average position. Developments in the field of biomolecular computing have demonstrated positively the feasibility of processing logic tasks by biocomputers, a promising first step toward building future nanoprocessors with increasing complexity. There has been progress in building biosensors and nanokinetic devices, which also may be required to enable nanorobotic operations and locomotion. Classical objections related to the feasibility of nanotechnology, such as quantum mechanics, thermal motions and friction, have been considered and resolved and discussions of techniques for manufacturing nanodevices are appearing in the literature with increasing frequency.

Motivation

One important challenge that has become evident as a vital problem in nanotechnology industrial applications is the automation of atomic-scale manipulation. The starting point of nanotechnology to achieve the main goal of building systems at the nanoscale is the development of control automation for molecular machine systems, which could enable the massively parallel manufacture of nanodevice building blocks. Governments all around the world are directing significant resources toward the fast development of nanotechnology. At least 30 countries have initiated activities in this field, and beyond that, with government investments to nanotechnology of US$ 800 million in Japan and US $ 774 million in USA. The U.S. National Science Foundation has launched a program in "Scientific Visualization" in part to harness supercomputers in picturing the nanoworld. A US $ 1 trillion market consisting of devices and systems with some kind of embedded nanotechnology is projected by 2015. More specifically, the firm 'display search' predicts rapid market growth from US $ 84 million today to $ 1.6 billion in 2007. A first series of commercial nanoproducts has been announced as foreseeable by 2007. To reach the goal of building organic electronics, firms are forming collaborations and alliances that bring together new nanoproducts through the joint efforts of companies such as IBM, Motorola, Philips Electronics, Xerox/PARC, Hewlett-Packard, Dow Chemical, Bell Laboratories, and Intel Corp., among others. For such goals, new methodologies and theories to explore the nanoworld are the key technology.

Fig. V–G.3: Textured vein inside view without the red cells. The tumor represented by the pink sphere is located left at the wall. All the nanorobots swim near the wall searching for the abnormal tissues.

Developed Simulations

A useful starting point for achieving the main goal of building nanoscale devices is the development of generalized automation control for molecular machine systems which could enable a manufacturing schedule for positional nanoassembly manipulation. In our work we consider more specialized scheduling problems with a focus on nanomedicine: describing in a detailed fashion the nanorobot control designs and the surrounding virtual workspaces modelling that are required for the main kinematics aspects in the physically-based simulations (Figure V–G.1).

Here the biomolecular assembly manipulation could be automatically performed by smart agents, which are given a set of possible tasks for biomedical engineering problems, embedded in a complex 3D environment. Virtual Reality could be considered as a suitable technique for nanorobot design and for the use of macro and microrobotics concepts given certain theoretical and practical aspects that focus on its domain of application.

The collective nanorobotics approach proposed is one possible method to perform a massively-parallel positional nanoassembly manipulation (Figure V–G.2). We constructed and demonstrated the applicability of multi-robot teams in timely sequenced set of works with practical applications that could enable the establishment of generalized control guidelines for nanorobotics. The use of multirobot teams working cooperatively to achieve a single global task applied to nanotechnology is a field of research that is very new and challenging. Research on collective robotics suggests that we should consider emulating the methods of the social insects to build decentralized and distributed systems that are capable of accomplishing tasks through the interaction of agents with the same structures and pre-programed actions and goals. Thus, a careful decomposition of the main problem task into subtasks with action based on local sensor-based perception could generate multirobot coherent behaviors. Several techniques was applied for such aims, as Neural Networks algorithms, Evolutionary computation, chemical based sensors and actuators (Figure V–G.3), and even temperature time-gradients (Figures V–G.4 and V–G.5), just to quote a few.

Fig. V–G.4: View of the NCD simulator workspace showing the cardiovascular occlusion, red blood cells and nanorobots.

Among other interesting nanorobot applications, we could foresee their use to process specific chemical reactions in the human body as ancillary devices for injured organs. Nanorobots equipped with nanosensors could be used to detect glucose demand in diabetes patients. Moreover, nanorobots could also be applied in chemotherapy to combat cancer through superior chemical dosage administration, and a similar approach could be taken to enable nanorobots to deliver anti-HIV drugs.

Fig. V–G.5: The atherosclerotic lesion was reduced due nanorobots activation. The temperatures in the region turn in expected levels.

Conclusion

Successful nanorobotic systems must be able to respond efficiently in real time to changing aspects of microenvironments not previously examined from a control perspective. Unlike some prior simulators for simple robots, in the present work we have developed nanorobots that are not restricted to a fixed grid nor behave as simple cellular automata with very simple environments. Also by contrast, most CAD approaches provide only animation or visualization tools, while the NCD is a physically-based simulator.

The set of experiments we have carried out include the main physical properties existent in the environment where the nanorobots are being currently projected to be operating in the coming years. Hence, the majority of results from our investigation should be useful as well on analysing integrated manufacturing capabilities. In fact, including aspects of the physical environment in conjunction with graphical visualization provide a feasible approach for automation and control design. Furthermore, the architecture that was developed intended to enable the incorporation and evaluation of several control methods and distinct nanorobot shapes analyses.

The automation, control, and manufacturing of nanorobots is a challenging and very new field. Realizing revolutionary applications of nanorobots to health or environmental problems raises new control challenges. The design and the development of complex nanomechatronic systems with high performance should be addressed via simulation to help pave the way for future applications of nanorobots in biomedical engineering problems.

Unit V–H

Vasculoid—A Personal Nanomedical Appliance to Replace Human Blood

Introduction

The vasculoid is a single, complex, multisegmented nanotechnological medical robotic system capable of duplicating all essential thermal and biochemical transport functions of the blood, including circulation of respiratory gases, glucose, hormones, cytokines, waste products, and cellular components. This nanorobotic system, a very aggressive and physiologically intrusive macroscale nanomedical device comprised of ~500 trillion stored or active individual nanorobots, weighs ~2 kg and consumes from 30-200 watts of power in the basic human model, depending on activity level.

The vasculoid system conforms to the shape of existing blood vessels and serves as a complete replacement for natural blood. Here we present a preliminary theoretical scaling analysis including transport capacity, thermal conduction, control and biocompatibility considerations, along with a hypothetical installation scenario and a description of some useful optional equipment.

We begin with a very preliminary technical examination of an idea originally referred to as "roboblood". The concept stimulated informal interest in a particularly aggressive nanomedical design and elicited considerable online popular discussion on sci.nanotech and elsewhere.

Here we examine a set of appropriate design constraints, scaling issues, and reference designs to investigate whether or not the basic idea of a blood replacement appliance might be feasible, and to determine key limitations of such designs.

The Vasculoid Concept

The idea of the vasculoid originated in the asking of a simple question: Once a mature molecular nanotechnology becomes available, could we replace blood with a single, complex robot? This robot would duplicate all essential thermal and biochemical transport functions of the blood, including circulation of respiratory gases, glucose, hormones, cytokines, waste products, and all necessary cellular components. The device would conform to the shape of existing blood vessels. Ideally, it would replace natural blood so thoroughly that the rest of the body would remain, at least physiochemically, essentially unaffected, but sustained in a cardioplegic state. It is, in effect, a mechanically engineered redesign of the human circulatory system that attempts to integrate itself as an intimate personal appliance with minimal adaptation on the part of the host human body.

A robotic device that replaces and extends the human vascular system is properly called a "vasculoid," a vascular-like machine. But the vasculoid is more than just an artificial vascular system. Rather, it is a member of a class of space- or volume-filling nanomedical augmentation devices whose function applies to the human vascular tree. The device is extremely complex, having ~500 trillion independent cooperating nanorobots. In simplest terms, the vasculoid is a watertight coating of nanomachinery distributed across the luminal surface of the entire human vascular tree. This nanomachinery uses a ciliary array to transport important nutrients and biological cells to the tissues, containerized either in "tankers" (for molecules) or "boxcars" (for cells). The basic device weighs ~2 kg and releases ~30 watts of waste heat at a basal activity level and a maximum of ~200 watts of power at peak (e.g., Olympic sprint) activity level. The power dissipation of the human body ranges from ~100 watts (basal) to ~1600 watts (peak), so the device presents no adverse thermogenic consequences to the user. The appliance is powered by glucose and oxygen, as may be common in medical nanorobotic systems.

The most important basic structural component of the exemplar vasculoid robot (Table V–H.1) is a ~300 m^2 two-dimensional vascular-surface-conforming array of ~150 trillion "sapphiroid" (i.e., using sapphire-like building materials) basic plates. (Thermal conductivity favors sapphire over diamond) These square plates are nanorobots that cover the entire luminal surface of all blood vessels in the body, to one-plate thickness. Each basic plate is an individual self-contained nanorobot ~1 micron thick and ~2 $micron^2$ in surface area, a size small enough to allow adequate clearance even in the narrowest human capillaries. Molecule-conveying "docking bays" comprise ~24 trillion, or 16 per cent, of all vasculoid basic plates. Tankers containing molecules for distribution can dock at these bays and load or unload their cargo. Cell-conveying "cellulocks" are built on "cellulock plates" which span the area of 30 basic plates, or 60 $micron^2$ each. Boxcars containing biological cells for distribution can dock at these cellulocks and load or unload their cargo. With only 32.6 billion cellulock plates in the entire vasculoid design, cellulocks occupy the area of 0.978 trillion basic plates or only 0.65 per cent of the entire vasculoid surface. The remaining ~125 trillion basic plates are reserved for special equipment and other as-yet undefined applications. All nanomachinery within each plate is of modular design, permitting easy replacement and repair by mobile repair nanorobots called vasculocytes.

Adjacent plates abut through flexible but watertight mechanical interfaces on metamorphic bumpers along the entire perimeter of each plate. Each bumper has controllable variable volume, permitting the vasculoid surface: to slightly expand or contract in area, or to flex, either in response to macroscale body movements or in response to vascular surface corrugations or other irregularities to the same degree or better than the natural endothelium. Thus, plated surfaces readily accommodate the natural cyclical volume changes of various organs such as lung, bladder, or spleen. Rigidity of the plate array is also subject to engineering control and to localized real-time control as well, via the bumpers; diamondoid or sapphire plating may be made substantially stiffer than natural endothelium, if desired. The rupture strength of individual plates is ~40,000 atm, taking plate wall thickness t_{wall} ~ 0.1 mm, effective radius R ~ 0.5 mm, and working stress s_w ~10^{10} N/m^2. Bumper actuation power is of order ~0.1 pW, comparable to the power draw of a single plate cilium.

Table V–H.1: Number, Mass and Volume of Exemplar Vasculoid System Components

System Component	*Number of Nanorobots*	*Subsystem Mass*	*Subsystem Volume*
Surface Plates		1.05 kg	
Docking bays	24×10^{12}	—	48.0 cm^3
Cellulocks	0.0326×10^{12}	—	1.956 cm^3
Applications plates	125×10^{12}	—	250.0 cm^3
Container Fleet			
Tankers			
Active	166.2×10^{12}	0.133 kg	166.2 cm^3
Backups in storage	166.2×10^{12}	(0.133 kg)	(166.2 cm^3)
Boxcars			
Active	0.032×10^{12}	0.0768 kg	91.7 cm^3
Backups in storage	0.032×10^{12}	(0.0768 kg)	(91.7 cm^3)
Vasculocytes			
Active	0.200×10^{12}	0.002 kg	0.6 cm^3
Backups in storage	2.000×10^{12}	(0.02 kg)	(6.0 cm^3)
Storage Vesicles	0.5×10^{6}	0.460 kg	500.0 cm^3
Other structure, unspecified	16.3×10^{12}	0.2782 kg	141.544 cm^3
TOTALS	500.0×10^{12}	2.000 kg	1200.0 cm^3

(items in parens are already included in storage vesicles)

Some Benefits of Vasculoid Installation

The advantages of installing a vasculoid are potentially numerous. Many of these benefits theoretically could be provided on a temporary or more limited basis using terabot-quantity doses of considerably less aggressive bloodborne nanomedical devices. However, the vasculoid appliance simultaneously provides all benefits on an essentially permanent and whole-body basis. Additionally, some benefits appear unique to the vasculoid and can be achieved in no other way. Whether the entire package is sufficiently attractive to warrant installation will probably be a matter of personal taste rather than of medical necessity, since a molecular nanotechnology capable of building and deploying a complex vasculoid is likely to offer complete non-vasculoid cures for most circulatory and blood-related disorders that plague humanity today, and biological enhancements may also be available. The choice between an augmentation technology that works alongside a natural system (e.g., respirocytes) and an augmentation technology that entirely replaces a natural system (e.g., vasculoid) may involve significant safety, psychological, and even ethical considerations.

The most important benefits of vasculoid installation may include:

- Exclusion of parasites, bacteria, viruses, and metastasizing cancer cells from the bloodflow, thus limiting the spread of bloodborne disease. Such microorganisms and cells are easily eliminated from the blood using ~cm^3 doses of appropriately programed nanobiotics, but such individual nanorobotic devices might not normally be deployed on a permanent basis. Intracellular pathogens that can infect motile phagocytic cells (e.g., the tuberculosis *Mycobacterium* or the bacterium *Listeria*, both of which can reside inside macrophages) cannot be directly excluded from the tissues when infected cells are transported by the vasculoid. However, cell surface markers will often reveal such infection, so vasculoid systems can check for the presence of such markers and thus deny these cells re-entry to human tissues. For example, the membrane surface of macrophages infected by *Mycobacterium microti* is antigenically different from that of uninfected macrophages; Listeria, derived peptides are found acting as integral membrane proteins in the plasma membrane of infected macrophages, and other *Listeria*-infected antigen-presenting cells display hsp60 on their plasma membranes only when infected. Note that the natural bacterial inhabitants of the human gastrointestinal tract should be relatively unaffected by the installation of a vasculoid appliance, since gut microbes normally do not enter the bloodstream; similar considerations apply to the natural microbial flora of the skin, mouth, nasal passages, and so forth. The vasculoid could also significantly reduce the prospect of death by cancer by refusing either to transport angiogenesis factors or to transport nonvalidated native stem cells (thus preventing metastasis), or by refusing to install itself in any capillaries arising in locations where angiogenesis is disallowed in an adult body.

- Faster and more reliable trafficking of lymphocytes throughout the secondary lymphoid organs, allowing them to survey for targeted antigens in minutes or hours, rather than days (because both white cells and antigenic sources can be efficiently concentrated), thus greatly speeding the natural immune system response to foreign antigen. This lymphocyte function might also be augmented or replaced using individual histomobile medical nanorobots or biorobots. If biorobots are developed first, many vasculoid installations might take place in patients possessing largely artificial immune systems, thus obviating the need for much of the cellular component trafficking.
- Eradication of most serious circulatory-related pathological conditions including all vascular disease (e.g., aortic dissection, vessel blockages, spasms, aneurysms, phlebitis, varicose veins), heart disease, syncope (including orthostatic hypotension) and shock, strokes, and bleeding, due to the elimination of unconstrained metabolite and fluid circulation. Certain other conditions due to localized prevention of blood flow such as bedsores and subclinical paresthesias (e.g., "pins-and-needles" sensation) can also be ameliorated, since stiffened blood vessels will not be nearly so easy to close via external compression. Again, many of these conditions may already have adequate nanomedical treatments by the time the vasculoid can be built, but other conditions might not yet be readily or as conveniently treatable, such as the dangers of large-scale bleeding (both internal and external).
- Reduced susceptibility to chemical, biochemical, and parasitic poisons of all kinds, including allergenic substances in food, air and water, although bloodborne nanotankers or pharmacytes may be able to partially duplicate this function as well. Note that toxins, novel metabolites or other unknown foreign substances may arrive in the tissues via solvents penetrating the dermis (e.g., DMSO) or by other nonvascular routes, and would be removed from the natural blood in due course by the kidney which extracts all small molecules by default, then actively reabsorbs only useful molecules such as glucose, electrolytes, vitamins, and water. By contrast, in the present design the vasculoid transports only recognized substances. Novel undesired substances must therefore be removed by employing a small number of previously unallocated applications plates: (a) as miniature analytical laboratories

which can detect foreign analytes and alter a programmable binding site to bind them, allowing transport to the kidney for disposal; (b) as miniature chemical processing systems, possibly involving well-designed sets of artificial digestive enzymes, which can break down the foreign substance into well-recognized simpler substances; or (c) as analytical laboratories or chemical processing systems previously described which can receive appropriate new instructions directly from the patient, the physician, or other external sources of information. The lymphatic system is unmodified in the basic appliance design and may also serve to remove nonvascular toxins. Processing of such toxins may be deferred to the points where the lymph is reabsorbed into the vasculoid system, using either dedicated application plates or specialized processing stations.

- Faster metabolite transport and distribution, significantly improving physical endurance and stamina, including the ability to breathe at low O_2 partial pressures and the ability to flush out unwanted specific biochemicals from the body (a feature which might be duplicated using bloodborne respirocyte-class devices). The architecture would also permit convenient long-term storage of protein, or amino acid recovery and recycling, which could prove nutritionally useful. As another example, an installed vasculoid could mitigate the effects of weightlessness or acceleration on nutrient transport and somatic fluid distribution by decoupling transport from gravity. The appliance can differentially control the release of water and electrolytes in various regions of the body. In addition, by making the transport function independent of gravity, vasculoid avoids the hydrostatic effects of extreme, varying, or absent gravity vectors. This would eliminate the facial puffiness seen in astronauts and the dependence of fighter pilots on external prostheses to force blood out of the legs during high-G maneuvers.
- Direct, rapid user control of many hormonal- and neurochemical-mediated, and all blood-mediated, physiological responses. It would be difficult (though not impossible) to provide equivalent comprehensive whole-body physiological control using individual micron-scale bloodborne nanorobots alone. Additionally, the appliance can continuously monitor the basic physiochemical status of vascularized tissues to near-micron resolution if necessary, since each micron-size plate is initialized with its physical position relative to the vascular tree, and medically relevant information is communicated back to a central user (or physician) interface. This can permit extremely rapid and specific detection of health problems as will generally be the case in nanomedicine, al though the vasculoid may provide an easier and more thorough method for accomplishing this.
- Voluntary control of capillary conductance and rigidity permitting conscious regulation of thermal energy exchange with the environment and at least limited control of whole-body morphological structure, rigidity (e.g., stiffness, bending modulus, etc.), and volume with ~millisecond response times.
- At least partial protection from various accidents and other physical harm (e.g., insect stings, animal bites, collisions, bullet or shrapnel penetrations, falling from heights). This is perhaps the only specific benefit of the vasculoid appliance that could not be achieved by any less radical means: extreme trauma resistance, especially resistance to exsanguination and cushioning against mechanical shock.

Medically oriented people might properly wonder why anyone would want to discuss replacing a perfectly functional natural fluid transport system with an untested, complex, artificial, dry system with which humans have no experience today. There are several answers to this very good question. First of all, medical skeptics should bear in mind that the vasculoid appliance is clearly a highly sophisticated medical nanosystem. It cannot be built without using a manufacturing system based on a mature molecular nanotechnology. Its use would come only after many decades of previous engineering experience in building, testing, and operating such highly complex systems inside the human body. In the future nanomedicine-rich milieu in which it would be deployed, the vasculoid as a medical intervention may be closer to the typical than to the extreme (as it might appear today).

In the nanomedical era, it will be a matter of personal preference and choice for each patient, in consultation with their physician, whether the aforementioned benefits of the vasculoid appliance are worth the risks. The following chapters are only a preliminary scaling study to investigate whether the machinery involved will fit in the allotted space, perform all required functions fast enough to be useful, operate within an acceptable energy budget even at peak loads, promise a strong likelihood of being safe, reliable, and biocompatible, and so forth.

Unit V–I

Vasculoid: Physiological Materials Transport

Introduction

As outlined below, at peak average male human metabolic rates, ~5400 cm^3 of circulating blood transports up to ~125 cm^3 of essential molecules and up to ~90 cm^3 of essential cellular elements (excluding erythrocytes). Consequently ~96 per cent of blood volume (mostly water) may be rendered superfluous if its salvation and suspension functions can be replaced with a more efficient and highly reliable nanomechanical transport mechanism.

The resulting baseline vasculoid device conveys physiologically useful molecules and cells throughout its interior, and thence to the biological tissues, by containerizing all materials and then using a ciliary transport system to distribute the containers to appropriate destinations. This architecture thus utilizes the fractal branching vascular network that is characteristic of efficient fluid transport systems, although many architectures are possible in this design space.

Molecular Transport

In natural human physiology, the circulating blood moves ~2 × 10^{26} molecules/sec. However, most of these molecules are solvent water not strictly necessary in human metabolism. In the vasculoid, all non-cell materials transport takes place via ciliary transport of tanker vessels at a very conservative mean transport velocity of 1 cm/sec. (For comparison, blood velocity in the natural vasculature ranges from 0.02-0.15 cm/sec in the capillaries up to 117.5 cm/sec pulses in the femoral artery). Tanker vessels are $(1\text{-micron})^3$ mostly-sapphire (possibly chamfered) cubes with 0.75 $micron^3$ useful interior storage volume and ~8 × 10^{-16} kg dry mass. Gases are stored in tankers at 1000 atm pressure; this is highly conservative, as a sapphire pressure vessel should easily withstand ~10^5 atm without bursting.. Liquids and solids are packed at normal macroscopic densities. All tanker mechanical subsystems (other than passive hull) have at least tenfold redundancy, hence are extremely reliable.

The following is an approximate inventory and assessment of the minimum human physiological molecular transport requirement.

Respiratory Gases

The human body consumes 1-20 × 10^{20} molecules/sec of O_2, depending on activity level. Similar numbers of CO_2 molecules are generated as waste products. The human red blood cell mass (~2.4 liters of RBCs) has a total storage capacity of 3.2 × 10^{22} O_2 molecules, but since hemoglobin operates between 70-95 per cent saturation in normal physiological conditions, the active capacity of human blood is only 8.1 × 10^{21} O_2 molecules. Blood CO_2 capacity is roughly the same, and O_2/CO_2 are loaded/unloaded reciprocally at lungs and tissues. Packing density of CO_2 molecules (1.11 × 10^{28} molecules/m^3) is slightly lower than for O_2 molecules (1.26 × 10^{28} molecules/m^3) at 1000 atm, so CO_2 packing density controls respiratory tanker population.

Assuming a mean circulatory circuit of ~1.4 meter and a 1 cm/sec transport speed, continuous transport at the maximum 20 × 10^{20} CO_2 molecules/sec rate requires a circulating fleet of 33.6 trillion respiratory gas tankers. At normal basal metabolic rates, only 1.68 trillion active tankers are required; the remainder are held in reserve to support periods of more intense physical activity. Each tanker can hold 9.48 × 10^9 oxygen molecules or 8.36 × 10^9 molecules of carbon dioxide at 1000 atm pressure. Highly efficient reciprocal regenerative energy recovery is assumed during gas loading and unloading operations and unrecoverable gas compression energy losses should be minimal. Isothermal compression of ~10^{10} O_2 or CO_2 molecules requires—$P_i V_i \ln(V_f/V_i) \sim (10^5 N/m^2)$ (0.75 × 10^{-18} m^3/tanker) ln(1/1000) = 0.5 pJ/tanker. Assuming a need for 2 × 10^{20} molecules/sec for O_2 and CO_2 at basal demand implies ~2 × 10^{10} tanker fills/sec or ~0.01 watt for both sets of gases, or up to ~0.2 watts at peak demand, for compression of gases.

Conservatively taking the CO_2 concentration in the vasculoid-endothelium interface fluid as equal to the normal CO_2 venous plasma concentration of C = 0.45-1.14 × 10^{24} molecules/m^3, then the diffusion limit for loading CO_2 from the tissues to one side of a square plate area of L^2 = 2 micron², with the diffusion coefficient D = 1.9 × 10^{-9} m^2/sec for CO_2 at 310 K, is $J = (4/p^{1/2})$ LDC = 2.7-6.9 × 10^9 molecules/sec; a 10-second fill time allows system operation at about one order of magnitude below the diffusion limit. Similarly, normal alveolar partial pressure of O_2 is >100 mmHg, or C ~ 3.1 × 10^{24} molecules/m^3; taking D = 2.0 × 10^{-9} m^2/sec for O_2 at 310 K, the diffusion limit for loading oxygen at the lungs is J ~ 2 × 10^{10} molecules/sec, which again allows system operation at about one order of magnitude below the diffusion limit assuming a 10-sec tanker fill time.

Tankers are used to transport O_2 and CO_2, rather than the vasculoid interior in part because the theoretical rupture pressure of the natural cylindrical aorta is p_{max} ~ 1 atm (taking aortic wall thickness t_{wall} = 1.5 mm, aortic radius R = 12.5 mm, and working stress $s_w \sim 10^6$ N/m^2 and the rupture pressure of the vasculoid plate coating is p_{max} ~ 8 atm (taking vasculoid plate tube thickness t_{wall} = 1 micron, tube radius R = 12.5 mm, and working stress $s_w \sim 10^{10}$ N/m^2. At peak exertion levels, respiratory gases stored in the tanker fleet at 1000 atm would fill the 4.2 liters of free vasculoid internal volume to 3.2 atm of pressure, too high for reliable safe operation under all foreseeable circumstances. By comparison, the rupture strength of individual tankers exceeds ~40,000 atm (the rupture strength of individual plates).

Water

The human body excretes 6-12 × 10^{20} molecules/sec of urinary water under normal circumstances. The kidney continuously filters ~18 gallons/hour of blood (~7 × 10^{23} H_2O molecules/sec) so much higher rates of urine generation are possible in theory, but it seems unnecessary to accommodate the vasculoid design to such extreme cases because the removal of waste molecules can be accomplished by direct molecular sorting rotor extraction rather than by bulk filtration and solvent repumping—that is, the principle blood-cleansing function of the kidneys is to pump nonaqueous molecules, not bulk water solvent molecules. With the vasculoid in place, only specific waste molecules and minimal quantities of carrier water need be discharged into the glomeruli of the ~10^6 nephrons of the kidney, with the result that gross water flows through the kidney and energy consumption in this biological organ can be significantly reduced—but not reduced so far as to induce urinary stone disease (urinary lithiasis or urolithiasis) or renal cytotoxicity from excessive pH, salinity (e.g., hyperosmotic stress), or toxin concentrations. The precise volume and concentration of urine generated by the biological organ is regulated by hormones that control kidney function, and these hormones may in turn be regulated by the vasculoid appliance.

Additionally, 2-186 × 10^{20} molecules/sec of water are perspired through the skin, depending upon air temperature, humidity, activity level and mental state. Assuming the same transport parameters as before, the maximum requirement of 2.0 × 10^{22} H_2O molecules/sec (~0.6 cm^3/sec) requires 110.5 trillion water tankers to be available, of which only 4.5 trillion will normally be active with the rest held in reserve for periods of peak exertion. Each tanker can hold 2.51 × 10^{10} water molecules in the liquid state.

Glucose

Typically 1.6 × 10^{22} molecules of glucose are present in the human bloodstream. A human on a standard 2000 kcal/day diet metabolizes 3.08 x 10^{19} glucose molecules/sec throughout the body. However, during strenuous exercise this requirement can briefly rise to as much as 33,000 kcal/day, a 4.93 × 10^{20} molecule/sec rate. Assuming the usual ciliary transport parameters, we require 17.6 trillion glucose tankers, providing storage of 6.91 x 10^{22} glucose molecules in the system, four times the number present in normal blood. (Slightly smaller capacity might be realized if the glucose must be stored as a 70 per cent aqueous solution concentrate to facilitate molecular mobility for sorting rotors. Only 1.1 trillion glucose tankers are needed to service basal metabolic needs; the remainder are held in reserve for periods of extreme physical activity.

The average human cell is a ~30 picowatt (basal) device, so each glucose tanker, with its full complement of 3.93 x 10^9 glucose molecules, yields ~400 basal cell-seconds of energy (assuming 65 per cent efficiency for the natural cellular metabolic pathways using glycolysis and tricarboxylic acid cycles). Note that at higher power levels, the delivery rate increases correspondingly, so that even with random delivery at peak demand all cells should receive a sufficient fuel supply. Additionally, the average tissue cell normally contains an internal inventory of ~3 × 10^{10} glucose molecules, an in cyto buffer supply equivalent to ~8 full tanker loads.

Nonprotein Nitrogenous Molecules

Normal bloodstream concentrations of urea, uric acid, creatinine, creatine, ammonium salts, cerebroside, amino acids, etc. imply 1.73-2.49 gm present in the circulation and a maximum of 2.3 trillion tankers for full encapsulation and transport.

Plasma Proteins

All plasma proteins including most notably albumen, globulins (including antibodies), complement, fibrinogen and prothrombin constitute ~7 per cent of adult blood plasma, or 210 gm, which would require 215 trillion tankers to encapsulate. Fortunately, on a transactional basis this figure is a gross overestimate. The RDA (Recommended Daily Allowance) active requirement for protein is ~70 gm/day for a 70 kg male human, an 8.1 × 10^{-7} kg/sec transport demand requiring only 0.12 trillion protein tankers for continuous ciliary transport, reducing mandatory instantaneous system storage capacity to ~0.12 gm of plasma proteins. However, almost the entire RDA protein intake is hydrolyzed to amino acids and dipeptides, so the actual protein transport needs should be far less. With no need to maintain osmotic pressure, there would be almost no reason to transport albumin within the system, although the dosage of protein-bound medication and hormones might need to be corrected. Finally, there is sufficient tanker capacity to increase protein transport capacity at least 100 fold above the RDA requirement if necessary, so the design is not particularly sensitive to this transport requirement.

Note also that the five antibody classes can be efficiently transported using just five types of sorting rotors having binding pockets complementary only to the constant Fc regions without regard to the numerous variable domains these antibodies may encode on their antigen-binding Fab regions, thus vastly reducing the required vasculoid rotor specificities needed to accomplish this task. Complement (immune system) components may be transported with similar efficiency.

The limited free enzyme activities that occur in natural blood can still take place in the fluids at the vasculoid-endothelial interface, with due care taken to avoid inducing localized hyperenzymemia (which is often benign, as in hypertransaminasemia), although higher concentrations may be expected because of the lower total volume of the vasculoid-endothelial interface compared to the natural blood volume.

Lipids

The mass of lipids including fatty acids, cholesterol, triacylglycerides (transported in the plasma lipoproteins as triacylglycerols), and phospholipids present in the human bloodstream totals roughly 35-42 gm, which would require 55-66 trillion tankers to encapsulate. However, the RDA for fat is 23 per cent of caloric intake. (Many people consume several times this amount, but others survive on fat-free diets, so the figure represents a reasonable compromise estimate of the typical human active requirement.) This implies a demand of 6.9×10^{-7} kg/sec for a 2000 kcal/day diet, requiring only 0.15 trillion lipid tankers for continuous ciliary transport and reducing the required vascular storage capacity to ~0.1 gm. As with plasma proteins, there is sufficient tanker capacity to increase lipid transport capacity to at least 100 times the RDA requirement, so once again the design is not particularly sensitive to this transport requirement.

Other Molecules

Human blood plasma contains ~40 gm of salts (e.g., electrolytes) including sodium, potassium, calcium, magnesium, chloride, phosphate, and sulfate; ~1.2 gm lactic acid and other non-nitrogenous waste products; ~0.23 gm of RNA and DNA; ~0.21 gm of enzymes (protein); ~0.15 gm testosterone in the male, ~0.03 gm progesterone in the female, ~0.02 gm of corticosteroids, and a total of ~0.002 gm of more than fifty other major biologically active hormones; ~0.04 gm of vitamins including A, B2, niacin, pantothenic acid, biotin, pteroylglutamic acid, choline, inositol, C, D, and E; ~0.03 gm of essential trace elements including iron, zinc, copper, manganese, iodine, and cobalt; and smaller masses of various cell-signaling (e.g., cytokines, autocrines, neuropeptides) and numerous other specialty proteins. Other nonprotein regulatory molecules affecting numerous signaling pathways must also be locally regulated with precision by the vasculoid appliance. For example, in the present context the simple molecule NO is of particular importance as a neurotransmitter and vasoconstrictor, though it is normally present only in exceedingly small quantities. Other molecules that mediate certain regulatory pathways may become unnecessary to transport after the vasculoid is installed, as for example the renin-angiotensin pathway (since there is no blood pressure to regulate).

The capacity demand totals ~42 gm, which would require ~43 trillion tankers for full encapsulation. However, the RDA for all minerals, electrolytes, vitamins and trace elements is just 6.8-16.1 gm/day for healthy adults, a maximum of 1.87×10^{-7} kg/sec, requiring only 0.03 trillion tankers assuming ciliary transport parameters as described above. (Sodium and potassium ion transport in individual neurons is ~10^6 ions/discharge; assuming ~10^{10} active neurons firing at near-maximum ~100 Hz, whole-brain ion transport rate is ~10^{18} ions/sec or ~7×10^{-8} kg/sec, well within the aforementioned mass flow budget for mineral and electrolyte transport. However, Na^+ and K^+ are very quickly recycled after each discharge, so the true transport requirement is more accurately measured by the small amounts of these electrolytes that are lost in the gut, sweat, and lymph.) Transport of lactic acid, RNA/DNA, enzymes (e.g., those present as a result of normal tissue breakdown such as CPK and transaminases), hormones, cytokines and so forth (~1.8 gm) requires 1.9 trillion additional tankers. In some cases these "other" molecules may be shipped mixed in a single container.

Summary of Molecular Transport

A total of 166.2 trillion tankers are required to transport all-important physiological molecules at maximum human metabolic rates. At basal metabolic rates, only 11.78 trillion tankers will typically be active. If, contrary to our assumption, RDA levels are found to be an inadequate substitute for natural blood storage of plasma proteins, lipids, minerals and vitamins, or if production and demand occur at different times and the blood volume performs a significant storage function, then vasculoid design may need to include up to ~300 cm³ of auxiliary bulk storage or caching to extend the "just in time" inventorying of these vital substances.

Cellular Transport

In addition to conveying molecules, the vasculoid must also transport physiologically important cellular species throughout the body. Cells to be transported (comprised mainly of white cells, platelets, and the few remaining red cells) generally range in size from 2-20 microns. These are shipped in cylindrical mostly-sapphire "boxcars" 100 microns long and 6 microns in diameter. These containers, of which 75 per cent (2120 micron³) is usable storage volume (~2.4×10^{-12} kg dry mass per boxcar), will just fit through average sized capillaries (~8 microns diameter, ~1000 microns in length) and all larger vessels, after accounting for plate thickness, and thus the boxcars have access to the vast majority of human tissues. The largest transportable cells (20-micron monocytes) must be temporarily dehydrated 50 per cent by volume prior to loading; such desiccated cells quickly rehydrate by natural osmosis upon release into the aqueous intercellular environment. Spherical cells 17.5 microns in diameter or smaller need not be dehydrated during loading, as they can fit into the available volume via change of shape. Typical transit time from point of entry to point of debarkation is ~70 sec (i.e., ~half the mean circuit length of 1.4 m, at 1 cm/sec), well within cellular ischemic survival times except possibly for oxygen, which may need slight supplementation during transit.

Onboard computers and sensors embedded in interior boxcar walls allow the detection and interpretation of surface antigens on passenger cells to ensure that all transported cells are non-hostile, and if not, to refuse transport, convey them to a disposal site, or destroy them after entry.

The following is an assessment of the minimum human cell transport requirement for the vasculoid appliance.

Leukocytes and Stem Cells

Under normal physiological conditions, 27-54 billion leukocytes circulate in the human blood volume, although in cases of chronic myelogenous leukemia the total count may increase to 0.13-2.7 trillion white cells. These 54 billion leukocytes consist approximately of 35-41 billion neutrophils (10-12 microns in diameter), 11-14 billion lymphocytes (8 microns), 1.6-4.3 billion monocytes (12-20 microns, average 15 microns), 1.1-2.7 billion eosinophils (12 microns), and 0.3 billion basophils (10 microns). Based on maximum container packing densities for each type of cell, 12-28 billion boxcars are required to encapsulate and transport all normal populations of bloodstream leukocytes. This will frequently include more than one cell per boxcar—multiple cells should be sedated for transport (using reversibly rotored molecular agents) to avoid unwanted interactions such as clonal expansion or induction of apoptosis. Principal origination sites are the bone marrow, the spleen, and the lymph nodes, but leukocytes may be distributed to almost any location in the body. Giant macrophages (e.g. >30 microns) and fibroblasts already present in the tissues (that normally do not enter the bloodstream) cannot be relocated by the appliance, although fresh stem cell precursors of fibroblasts with appropriate activating cytokines can be relocated from their natural source site, as required.

A comparatively small population of circulating stem cells must also be transported by the vasculoid. For example, vasculogenesis is assisted by hematopoietic stem cells (surface marker CD34+) that originate in the bone marrow and enter the peripheral blood in response to colony stimulating factors, with 1-3 per cent expressing the common leukocyte surface antigen CD45. Upon reaching the vascular surface and in the presence of the cytokine IL-3, these cells may differentiate either into leukocyte precursor (CD34+/CD45+) or into endothelial precursor (CD34+/CD45–); with the cytokine VEGF present, the latter cells further differentiate into new endothelial cells. The vasculoid must provide transport for stem cells of these and other types, along with the required cytokines, to ensure vascular repair of damaged endothelium and for other purposes. Fortunately there are only a limited number of such cell types, and the total population of stem cells requiring transport is many orders of magnitude smaller than the population of leukocytes. As a result, stem cell transport should be a comparatively minor—if physiologically essential—task for the previously-specified boxcar fleet.

Platelets

Although substitution of vasculoid transport for liquid blood eliminates much of the conventional health risk due to bleeding, blood vessel breaches that penetrate the endothelium but do not penetrate the vasculoid will still exude large amounts of extracellular serous fluid and other critical substances if the leakage is not promptly staunched. Furthermore, platelets are storehouses for a variety of molecules that affect vascular tone, fibrinolysis (subsequent clot dissolution), and wound healing. These substances are released during the clotting process. Platelets also communicate via chemical messengers with other blood cells (e.g. macrophages and fibroblasts) and with the endothelial cells that coat the interior of all blood vessels. Hence, platelets still have an important role in human physiology even after the vasculoid has been installed.

Approximately 2 trillion platelets (~2 microns in diameter) circulate in human blood. Each boxcar can hold about 500 platelets, giving a requirement of 4 billion additional containers for platelet transport. As with white cells, multiple cells could easily activate and aggregate, hence their function should be heavily suppressed during transport (again, using reversibly rotored molecular agents).

Erythrocytes

In the vasculoid, all respiratory gases are transported in bulk and at high pressure. Boxcars could in principle be used to transport red cells, but RBCs have an effective storage pressure <<1 atm which is far less efficient than the 1000 atm storage pressure of the respiratory gas tankers. Consequently, the ~30 trillion red cells normally present in the human bloodstream are superfluous and may be permanently removed from circulation. Beyond their respiratory function, red cells also are mechanically involved in the clotting process. However, the vascular plug typically contains mostly platelets, plasma fibronectin and factor XIII-crosslinked fibrin, plus small amounts of tenascin, thrombospondin, and SPARK (secreted protein acidic and rich in cysteine) with a relatively minor RBC contribution, so RBCs probably are not essential to the clotting process or are necessary only in very small numbers.

A sufficient oxygen concentration maintained in kidney peritubular cells should reduce erythropoiesis to at most 1 per cent of the normal rate of red cell production , requiring only ~34,700 red cells/sec to be conveyed from the erythroid marrow directly to the liver or spleen for disposal. Alternatively, erythropoiesis can be reduced by directly regulating the amount of erythropoietin that is allowed to reach the bone marrow. With few erythrocytes remaining to be destroyed in the liver, hepatic heme catabolism drops significantly, greatly reducing the production of bilirubin, a yellow bile waste pigment that plays no role in fat digestion in the gut but imparts the brown color to feces; as with hepatitis patients, the stool of the envasculoided user may become chalky white. Ciliary transport of boxcars containing the few remaining red cells over a ~1.4 meter circuit at a conservative mean velocity of 0.1 cm/sec (allowing plenty of time for docking and cell unloading) requires a circulating subfleet of just 34.7 million boxcars (only ~0.1 per cent of the entire boxcar fleet).

Summary of Cellular Transport

A maximum of 32 billion boxcars are required to convey all essential bloodborne cells during normal physiological conditions, although at times as few as 16 billion boxcars may be actively involved in cellular transport. Since a principal benefit of vasculoid installation is a significant decline in susceptibility to microbial invasion, the above figures can probably be further reduced in actual clinical practice.

Ciliary Distribution Subsystem

The interior volume of the vasculoid is lined with a sapphire surface (comprised of watertight adjacent installed plates) from which protrude trillions of mechanical cilia (arranged in patterns designed to maximize transport speed and reliability. A secondary role is to assist the vasculocytes (small mobile nanorobots) in cleaning up after accidental internal spills, component malfunctions, or other pathological events. Even by 1997, prototype MEMS ciliary arrays of up to 1024 cilia had already demonstrated transport speeds up to 200 microns/sec with ~3 micron positional accuracy.

Description of the Cilium

While a highly efficient specialized ciliary mechanism can undoubtedly be designed, for the present study each vasculoid cilium is conservatively assumed to be similar in size, shape, and performance characteristics to the vacuum-sealed robotic manipulator arm described by Drexler. In brief, each cilium is a cylindrical assembly 100 nm long and 30 nm in diameter, having a transverse travel of

100 nm and a lateral speed of 1 cm/sec, consuming 0.1 picowatt during continuous operation at low load. Note that even if the vasculoid interior contained gas at 1 atm, viscous drag would contribute an additional power loss of only ~0.001 pW per cilium (taking viscosity $h_{air} = 1.83 \times 10^{-5}$ kg/m-sec at 293 K).

Mean cilium/cargo contact time is ~10 microseconds. This implies an acoustic frequency >100 KHz, well above the conventional human threshold of hearing; however, ultrasonic hearing up to 108 KHz via otolithic conduction has been reported in humans, so further study is required to ensure restriction of transport motions to sufficiently high frequencies to preclude any perceptible hum directly due to ciliary operations, and possibly involving a variety of noise cancellation, anti-resonance (e.g., out-of-phase operation) and acoustic dampening techniques. Additionally, physical discomfort, intra-articular pain, and a lowering of electrical pain sensation threshold due to ultrasound exposure have also been reported in humans. The possibility of indirect audible or infrasonic vibrations due to beat frequencies, spatial acoustic interference patterns, or various resonances should be investigated further, along with the possibility of undesired vibrational energy losses into the tissues.

Each cilium can apply ~1 nanonewton at the tip. This is sufficient force to accelerate a 1 micron^3 block of water (weight ~0.01 piconewton) at 10^5 G, or a 100-micron long, 6-micron wide water-filled sapphire-shell cylindrical boxcar at ~30 G. The human body is normally subject to accelerations far less than 30 G; with tenfold grasping redundancy, the probability that a container will be shaken loose from the cilia and tumble uncontrolled into the vascular lumen, due to normal macroscale accelerations of the patient's body, is extremely remote. In the rare instance where such detachment occurs, the container will eventually be caught and returned to the flow; kinetic impact energy in such cases is relatively small. Whether such cross-luminal impacts can trigger a detachment avalanche should be investigated using computer simulations.

Changeable cilium tool tips may permit rapid, reversible, "blind" container grappling using van der Waals adhesive forces, which forces may be controlled by varying adhesive pad surface corrugations at the nanometer scale, or by other means. The cilia might also have a reversing function (e.g., for loosening container jams). Molecular dynamics simulations of these processes would be useful.

Ciliary Subsystem Requirements

The principal task of the ciliary distribution subsystem is to reliably transport a maximum of 166.2 trillion 1-micron cubical tankers and a maximum of 32 billion cylindrical 6 mm x 100 mm boxcars through a physical circuit ~1.4 meters in length. Because of their smaller size and greater number, ciliary subsystem design is driven almost entirely by tanker, not boxcar, requirements.

Although a single cilium can apply sufficient force to grapple and manipulate containers of either size, tenfold grappling is employed in the baseline design to achieve firm contact, to avoid physical escape of containers from the cargo traffic stream, to ensure massive redundancy hence high subsystem reliability, and to allow extremely precise steering. If a tanker occasionally encounters a patch of 9-contact cilia, or the even rarer 8-contact patch, steering or grappling capability should not be significantly reduced. Fail-safe grappling modes must be designed for cilium end-effectors.

Each tanker is normally in continuous contact with 10 cilia at all times during transport, so maximum ciliary spacing is 0.32 micron. A minimum of 3000 trillion cilia are needed to achieve 100 per cent service over the entire ~300 m^2 vascular (mostly capillary) surface. A set of 3000 trillion cilia physically occupies ~2.1 m^2 (~7 per cent) of the vascular surface. Assuming natural background radiation damage causes 1.5 × 10^7 fatal hits/kg/sec for nonredundant structures, the mean time to failure of a ~10^{-19} kg cilium is ~10^{12} sec, which implies a mean failure rate of only ~2000 cilia/sec (~20 nanograms/day) throughout the entire vasculoid. Using redundant structures mean failure rate may be considerably reduced.

Under resting metabolic conditions, 11.78 trillion tankers are transported by 117.8 trillion cilia generating waste heat of ~11.8 watts, an increment of only ~12 per cent over the basal human metabolic rate. During periods of maximum exertion and thermoregulatory stress, 166.2 trillion tankers are transported by 1662 trillion cilia generating excess heat of ~166 watts, well below the maximum human metabolic rate of 1600 watts and just slightly exceeding the ~100 watt effective load error that could trigger a response from the human thermoregulatory control system.

Docking Bays and Cellulocks

Once molecules or cells have been transported to the appropriate site, the container in which they reside must dock with the vasculoid surface on the interior side, and pass the delivered materials through this surface to the interstitial side of the vasculoid wall, whereupon the cargo can diffuse or migrate into the tissues as required. Tankers offload their molecular cargo in docking bays whereas boxcars unload their cellular passengers at cellulocks.

Possible undesired long-term chemical interactions of transported nutrients discharged into the fluids of the vasculoid-endothelial interface at above-physiological concentrations must be carefully analyzed. For example, high concentrations of glucose akin to chronic hyperglycemia could enhance protein glycosylation of long-lived proteins (e.g., vascular and myocardial collagen) which undergo continual cross-linking during aging because of the formation of advanced glycosylation end-products (AGEs) , most significantly (for this paper) in endothelial cells , which may be partly responsible for arterial stiffening and higher blood pressure with age (and which drugs like ALT-711, benfotiamine , nitric oxide and other substances may reverse). AGEs can also induce apoptosis and vascular endothelial growth factor overproduction in capillary pericytes.

Docking Bays for Tankers

Physiologically relevant molecules are delivered to specific locations by docking at 2 micron^2 (~1.4-micron square) tanker docking bays embedded at appropriate spatial intervals across the vasculoid surface. After securely connecting to the docking bay structure, and after the tanker's cargo manifest is scanned by the docking bay computer and said cargo is approved for receipt (Section 5), the tanker can be pumped dry by molecular sorting rotors in ~10 seconds, then sealed, remanifested, undocked, and released empty back into the traffic. Empty tankers may be reloaded with molecular cargo by a similar process in a similar time. Note that tankers carrying liquid or gas may discharge their contents into the narrow end of a funnel-shaped manifold leading to the active rotor banks at the wider end, to minimize the required tanker connection aperture. In different operating modes, tankers might be only partially emptied at each stop, or might be partially emptied more rapidly via bulk-flow exhausting rather than molecular rotoring. Docking bays include buffer tanks so that offloaded materials can be metered out to the underlying tissues over time periods longer than 10 seconds.

If the average endothelial cell (of which blood vessel walls are comprised) has a flat luminal area of 300-1200 micron^2, then the ~300 m^2 vascular surface may be comprised of up to ~1 trillion endothelial cells. This vascular surface could theoretically accommodate a maximum of 300 trillion tankers positioned side by side. At basal metabolic rates, only 11.78 trillion tankers are actively on the move. Even during peak exertion, a maximum of 166.2 trillion tankers are being transported, requiring just 55 per cent of the available vasculoid luminal surface for their passage assuming all tankers are traveling in monolayer. Other functionally equivalent modes might include operating all tankers at reduced capacity or reduced duty cycle during periods of basal demand, or reducing the transport velocity. A detailed computer simulation study of optimal traffic patterns near peak capacity would be valuable to perform.

A maximum of 166.2 trillion tankers circulating through an average ~1.4 meter circuit at ~1 cm/sec implies a docking bay cargo unloading requirement of ~1.2 trillion tankers/sec. Docks operate on a 50 per cent duty cycle to allow plenty of time for maintenance and repair, although in a less conservative design the duty cycle could possibly be boosted as high as 99 per cent. Tankers are emptied in 10 sec, so ~24 trillion docking bays are needed for reliable operation at peak loads. Docking bays occupy ~16 per cent of the vasculoid surface and have a mean center-to-center separation of ~3.5 microns. Each endothelial cell is serviced by up to ~24 docking bays.

Normally only a small fraction of the body's tissue cells are directly bathed in blood. Hence, delivery of molecules to each endothelial cell surface must be sufficient to supply the needs of the up to ~30 tissue cells that lie, on average, within ~3 cell widths of the nearest capillary. The maximum whole-body transport requirement of 2.25×10^{22} molecules/sec, consisting principally of respiratory gases, water, and glucose, gives an endothelial surface mass transport rate of 2.25×10^{10} molecules/sec/endothelial cell. This transport rate may be satisfied using a minimum of 45,000 molecular sorting rotors per endothelial cell operating on a 50 per cent duty cycle.

With 24 docking bays per endothelial cell, the requirement is ~0.94 billion molecules/sec per docking bay. Allowing tenfold multiplicity of sorting rotors to ensure high reliability requires 18,750 sorting rotors per docking bay, of which only 67 will be active at the basal rate and up to 9375 will be active during maximum physical exertion; 18,750 sorting rotors measuring 98 nm^2 on the ventral surface (facing the endothelium) cover 1.84 micron^2, essentially coating the entire undercarriage of the 2 micron^2 docking bay facility. This provides a ~7.5×10^8 molecule/sec supply per tissue cell.

At peak metabolic load, maximum molecule offloading rates as constrained by possible effervescence and crystallescence effects appear acceptable in this application for all low-volume molecules and for glucose and CO_2, but unfortunately exceed the effervescence limit for O_2 by a factor of ten, due to the relatively poor aqueous solubility of oxygen. Hence at peak load this gas must be offloaded either ~10 times more slowly than indicated above for a single release site, or at the indicated flow rate from ten discrete release sites all spatially separated by more than the largest possible O_2 effervescence bubble radius.

To simplify rotor system design, there are four classes of docking bays approximately matching the anticipated four principal tanker populations. Of the ~24 docking bays overlying each endothelial cell, ~5 stations are designated to handle tankers carrying respiratory gases, and possess rotors (on the tissue side) designed to reversibly bind only oxygen or carbon dioxide molecules. Another ~14 docking bays receive only water-bearing tankers, and employ rotors specializing in water molecule transport. Another ~3 stations accept only glucose cargoes, and use only glucose sorting rotors. The remaining ~2 docking bays are general-purpose stations equipped with rotors of up to tens of thousands of different types capable of reversibly binding all remaining biologically useful molecules (or classes of molecules) that must be transported through the vasculoid surface.

In addition to gas, water, glucose, and specialty tankers, there is a fifth class called power tankers that support the docking bay oxyglucose metabolism. Power tankers transport stoichiometric parcels of oxygen and glucose, and also remove carbon dioxide byproduct for disposal at the lungs. (Vasculoid power may be generated by glucose engines, glucose fuel cells, or other energy conversion devices that consume oxygen and native glucose; water produced by glucose engines may be vented into the biological tissues) All docking bays can accept power tankers, which supply the docking bay's own internal energy needs. Tankers reload their own oxyglucose engines and fuel tanks as they make their rounds of the various docking bays. Tanker duties are not energy intensive, so tankers refuel at most once a day, at the docking bays.

Rotors conveying molecules from tankers (at high concentration) through the vasculoid integument to the endothelial surface (at low concentration) generate positive mechanical energy, which may be used to largely offset the work of concentration that must be expended to transfer molecules in the reverse direction when loading tankers from the tissue space. Concentration work losses can be reduced almost to zero in an efficient system, with systemwide entropic losses in the compression-expansion cycle as low as 0.01-0.2 watts. Hence the primary energy loss is the drag power per rotor, $\sim 10^{-16}$ watts or ~0.1 zJ/molecule assuming a transfer rate of $\sim 10^{6}$ molecules/rotor-sec. Taking the whole-body transport requirement range as $10\text{-}225 \times 10^{20}$ molecules/sec, the net molecular transport power dissipated per docking bay is thus 0.004-0.09 picowatt depending on physical exertion level, or 0.1-2.25 watts for the entire vasculoid docking bay subsystem. (Rotor systems that do not include efficient energy recovery and must dissipate the work of concentration, perhaps up to ~20 zJ/molecule, may require different appliance architecture)

To hold docking bay subsystem total power requirement for computational tasks to a 10-watt budget, nanocomputers using an appropriate mix of local and distributed computing costing $\sim 6 \times 10^{-17}$ watts/(ops/sec) can provide ~200 billion MIPS systemwide or ~10,000 ops/sec per docking bay, costing ~0.6 pW per docking bay. These rates have been enhanced by employing reversible logic. Hence the total power per docking bay for both mechanical and computational tasks ranges from ~0.604 pW (basal) to ~0.69 pW (peak), or 14.5-16.6 watts for the entire system of 24 trillion docking bays. If power is supplied by a glucose engine achieving mechanochemical power conversion at $\sim 10^{9}$ watts/m^3, a 16.6 watt peak energy budget for all docking bays requires one glucose engine only ~0.0007 $micron^3$ in volume (~0.03 per cent of bay volume), per dock, and a similar volume of fuel tankage.

Cellulocks for Boxcars

Biological cells are delivered to specific locations by docking at cellulock stations embedded at appropriate intervals across the vasculoid surface, often preferentially located near specialized post-capillary blood vessels less than 30 microns in diameter found in lymphoid tissues, commonly known as high endothelial venules (HEVs). Cellulock density requirements also vary on a tissue-by-tissue basis—for instance, intestines, lungs, throat and nasal cavity probably have much higher requirements than heart or brain. After securely docking with their loading face pressed firmly against the vasculoid surface, boxcar doors dilate open to make a 30 $micron^2$ (~6.2-micron diameter circle) aperture. Cellulock doors (constructed of self-cleaning sliding spiral graphene segments in snug facial contact) then dilate to the same diameter, exposing passenger cells to the interstitial fluid. Passenger cells are gently pistoned out through the opening, taking care to minimize extracellular fluid entry. The cycle ends as all doors reseal and the boxcar undocks, resuming its travels.

Biochemical signals of cellular distress (e.g. cytokines such as interleukin-1 or tumor necrosis factor) are received by sensors on the interstitial side of the vasculoid. Upon receipt of this chemical signal, the cilia at that site are programed to call for leukocyte activity there. Cellulock and boxcar interiors may employ a small set of presentation semaphores which expose to the interstitial fluid minute quantities of various chemoattractants and chemorepellents keyed to each major type of cell to be transported. These chemicals are permanently bound to the rotors and are not released from the vasculoid; like semaphores, each rotor position may present differing concentrations of one chemical species, or a selection of different chemical species. White cells of the type desired to be transported can be encouraged to enter or exit, as required, by positioning presentation rotor settings so as to create "chemotactic funnels"—programable patterns of varying concentration gradient to steer motile cells toward, or away from, the cellulock aperture (e.g., haptotaxis). Contact sensors around the aperture detect the presence of a motile cell "requesting entry" to the transport system. Motile cells presenting themselves for entry at the interstitial side of the cellulock are, at minimum, subjected to a rudimentary cell protein coat identification test before they are allowed admittance. Affirmatively recognized harmful foreign pathogens such as bacteria or viruses and infected native motile cells may be transported to a specific disposal site or destroyed, rather than being admitted freely for transport elsewhere in the tissues.

The cellulock mechanism covers 60 micron^2 (~7.7-micron square) of vasculoid surface. If spiral door segments slide past each other at 0.004 m/sec over an average of one-half rotation or ~15.4 microns, dilation (or resealing) of the cellulock door takes 3.85 milliseconds. Sliding friction for 16 nm^2 contact surfaces traveling 20 nm at 0.004 m/sec dissipates ~10^{-24} joules, or ~3.13 joules/m^2-meter. Thus ~30 micron^2 cellulock sliding door surfaces moving ~15.4 microns to open, then again to close, require ~3×10^{-15} joules to complete one dilation cycle. Assuming a maximum rate of 32 billion boxcars per (normal blood circulation time of) ~60 sec to emulate typical physiological delivery rates, the population of vasculoid cellulocks must process a total of 533 million boxcars/sec. If there are therefore also 533 million cellulock dilation cycles/sec, frictional waste heat generated by moving door segments in the entire vasculoid cellulock subsystem is a negligible ~1.5 microwatts. Taking cellulock power as equal to the basal rate for 30 docking bays gives a requirement of ~18 pW/cellulock, or ~0.59 watts for the entire cellulock subsystem. Serum glucose and oxygen to provide cellulock power (a minor volumetric requirement) may be rotored in directly from the interstitial fluid on the external side of the vasculoid surface.

Since the active tanker/boxcar ratio varies from 736 (basal) to 5194 (max) depending on metabolic load, there should be no less than one cellulock per 736 tanker docking bays, or a total of 32.6 billion cellulocks in the vasculoid surface, again allowing a 50 per cent duty cycle for maintenance and repair. Thus there is approximately one cellulock for every boxcar in circulation. Cellulocks occupy ~0.65 per cent of the vasculoid surface and have a mean center-to-center separation of ~96 microns. Each cellulock services an average of ~31 endothelial cells, though the actual spatial distribution of cellulocks is expected to be highly nonuniform (e.g., clustering near HEVs). The average capillary is ~1 mm long and is made up of ~60 endothelial cells, and there are ~19 billion capillaries in the human body, so on average ~1 cellulock resides at either end of every capillary vessel in the body.

The maximum white cell transit rate through injured or diseased tissue is ~1 cell/sec/mm^3. With 32.6 billion cellulocks there are ~163 cellulocks/mm^3 of tissue. Each cellulock has ~163 sec to pass each white cell, even at maximum traffic flows. This should be sufficient, given that cellulock doors open and close in milliseconds, and natural white cell transendothelial diapedesis may occur in as little as ~3 minutes. The creation of unnecessary interior voids where unwanted opportunistic pathogens might flourish should be avoided. It is probably unnecessary to maintain voids between vasculoid and endothelium near cellulocks because: (1) the time allocated to cycle the cellulocks appears adequate to permit WBC diapedesis, and (2) diapedesis can probably be expedited using chemical factors emitted by the vasculoid.

Container Routing and Identification

The ideal method for container routing will be completely automatic, consuming no incremental power or compute cycles. To this end, the outer surfaces of each of the five types of tankers (Gas, Water, Glucose, Power, and other) and the four types of boxcars (Red Cell, White Cell, Platelet, and other) bear a specific repeating interlock pattern, crudely analogous to Braille. These patterns remain in place on the container surface unless the container is affirmatively "reset" to carry a different cargo, in response to varying physiological needs and local container availability.

Cilia surrounding each docking bay or cellulock are outfitted with gripper pads that adhere adequately to all container surfaces, but strongly only to the surface pattern of the container type sought by the docking bay or cellulock with which the cilia are associated. These pad "recognition" types normally remain constant over time, but, like the container surfaces, may be modified as required in special circumstances. Efficient methods for handling other-type tankers that can avoid the need for time-consuming docking and downloading of manifests should be investigated further.

Cilia that have "recognized" a desired container transmit a trigger signal to nearby cilia, activating a collective stereotypical docking sequence. Thus, although containers follow statistically random paths across the vasculoid surface, each docking bay and cellulock automatically receives the cargo it requires from the passing container stream. Docking acquisition is demand-regulated locally.

Potential Transport Bottlenecks

The exemplar vasculoid has been scaled to transport essential molecules and cells at maximum physiological transfer rates. However, all engineering systems involve design tradeoffs and limitations, and the present design invokes a number of potential transport bottlenecks including intestinal water absorption and renal water elimination, glucose absorption and elimination at peak loading, boxcar geometrical clearance in narrow capillaries, and geometrical limitations of tanker monolayer transport.

Water Absorption and Elimination—The ~6 meter-long intestinal tract has a simple cylindrical surface area of ~0.65 m^2, but the ~5 × 10^6 villi of the small intestines (which is most of the length) increase the effective absorbing surface to ~10 m^2. The ileum, the lower portion of the small intestine, absorbs water into the bloodstream at the maximum rate of ~0.07 cm^3/sec. The cylindrical villi, typically measuring ~200 microns wide and ~1000 microns long, are heavily capillarized; assuming ~1000 capillaries/mm^3, each villus has ~30 capillaries presenting a total ~8 × 10^5 micron^2 of absorbing surface—for the entire ileum, that's ~4 × 10^{12} micron^2, a water transport rate of ~6 × 10^8 molecules/sec-micron^2. If maintenance activities are temporarily suspended during peak demand, water tankers loaded with 2.51 × 10^{10} water molecules at 2 micron^2 docking bays can provide the same transport rate of 6 x 10^8 molecules/sec-micron^2 if 49 per cent of the vasculoid surface in the ileum is given over to water transport—potentially crowding out some "applications plates" at this surface. (The requirement for alimentary water can also be greatly lessened by reducing the need for water solvent in urine by reducing the need for water cooling in sweat and possibly by reducing water losses in the alveoli (which could be covered with an improved waterproofing surfactant.)

Excretion of ~0.023 cm^3/sec aqueous waste at the kidneys occurs primarily through the renal glomeruli which have a total surface area of ~1.4 m^2, giving a maximum water export rate of ~5 × 10^8 molecules/sec-$micron^2$. If maintenance activities are temporarily suspended during peak demand, water tankers loaded with 2.51 × 10^{10} water molecules at 2 $micron^2$ docking bays can provide the required transport rate if 40 per cent of the vasculoid surface in the glomeruli is given over to water transport. The ~0.001 gm/sec (~4 × 10^6 molecules/sec-$micron^2$) solute waste stream is conveyed on mixed-cargo tankers carrying ~10^9 solute molecules per tanker, which may be unloaded at mixed-cargo docking bays covering only 8 per cent of the vasculoid surface in the glomeruli.

Glucose Absorption and Elimination—Glucose is absorbed in the small intestine, mostly in the duodenum and the jejunum, at a maximum rate of ~10^{20} glucose molecules/sec, somewhat less than our maximum glucose requirement of 5 × 10^{20} molecule/sec. Given a capillary absorptive area of ~3 × 10^{12} $micron^2$ in the upper small intestine, the natural glucose transport rate is ~3 × 10^7 molecules/sec-$micron^2$. As before, if maintenance activities are temporarily suspended during peak demand, glucose tankers loaded with 3.93 × 10^9 glucose molecules at 2 $micron^2$ docking bays can provide the same transport rate if 17 per cent of the vasculoid surface in the duodenum and jejunum is given over to glucose transport.

A related issue is the vasculoid's response to severe gastrointestinal overloading, as for example the rapid ingestion of a large quantity of pure sugar by the user. Sugar has a tremendous irritating effect on the stomach, with 15-30 gm (60-120 kcal) producing great outpourings of stomach mucous; as little as 0.25 liter of 20 per cent sugar solution (78 gm, 300 kcal) is sufficient to cause vomiting in some patients with chronic gastritis, according to century-old laboratory experiments. A rapid 2000 kcal (1.75 x 10^{24} glucose molecules) ingestion would represent a bulging stomach-full (<1.5 liters) of >22 per cent sugar solution (~526 gm of sugar) which would be severely irritating to tissues and toxic to cells, almost certainly exceeds the natural absorption abilities of the small intestine, and would likely prove emetic. Urination at the physiological maximum ~10^{20} glucose molecule/sec elimination rate unloads a 2000 kcal sugar binge in ~5 hours, a renal export rate of 7.14 × 10^7 molecules/sec-$micron^2$ that can be handled at glucose docking bays covering 36 per cent of the vasculoid surface in the renal glomeruli. Alternatively, some excess sugar can be routed to the liver or other organs for conversion to glycogen or starch.

Boxcar Clearance in Narrow Capillaries—Cell-carrying boxcars 6 microns in diameter cannot squeeze through the narrowest of capillaries, most notably those found in the human retina which may be as small as 4 microns wide. This has two operational implications.

First, boxcars carrying cellular cargo destined for tissues having capillaries <8 microns must onload and offload at cellulocks located outside such tissues, relying on natural transtissue mobility to achieve the desired placement of the transported cells. In this the vasculoid mimics the natural transport process.

Second, boxcars must be capable of being rerouted around narrow capillary bottlenecks to their flow, else the carriers may lodge in the narrow passage and become an obstacle to the free flow of materials as sometimes occurs with leukocyte plugging in certain pathological conditions such as diabetes. Cilia with boxcar-coded gripper pads located at the entrances of difficult passages may be programed to refuse boxcars, and may also utilize reverse ciliary flows to assist this process. Alternatively, boxcars may be designed with flexible metamorphic surfaces (and thus a limited ability to change shape) to allow tight passage when traveling empty. With metamorphic boxcars, cilia can allow boxcars to pass that will empty their contents before reaching a capillary. Another alternative is that the vasculoid could be used directly to induce changes in capillary size.

Tanker Monolayer Transport—If container traffic is strictly limited to a single monolayer at the artery or vein wall, then ciliary transport produces a tanker bottleneck in medium- and large-diameter vessels of the arteriovenous (noncapillary) vasculature. This is most clearly illustrated in the aorta. The circulation time of ~140 sec (1.4 m circuit at v_{tanker} = 1 cm/sec transport speed) implies a system flux of 8.4 x 10^{10} tankers/sec at basal load and 1.2 x 10^{12} tankers/sec at peak load. However, the luminal surface of an aorta of diameter $2R_{aorta}$ = 2.5 cm has room to accommodate only a flux of 2p R_{aorta} v_{tanker}/S_{tanker} = 7.9 × 10^8 tankers/sec per monolayer of coverage, assuming S_{tanker} = 1 mm^2 tankers traveling at v_{tanker} = 1 cm/sec.

The operational implication is that most tankers must be allowed to transit the aorta as a multilayer averaging 0.11 mm deep (~110 tanker layers) or <1 per cent of available luminal diameter at basal load, and up to 1.5 mm deep (~1,500 tanker layers) or ~12 per cent of available luminal diameter at peak load. This bottleneck extends to a lesser degree throughout the arteriovenous vasculature, with the number of required tanker transport layers decreasing rapidly with decreasing vessel diameter. Purely monolayer tanker transport becomes tenable in blood vessels with diameters <200 microns (e.g., arterioles and large venules) at basal load or <20 microns (e.g., metarterioles) at peak load. Tankers can employ one or more simple grapples or temporary coupling mechanisms to join up into progressively lengthening chains (of linked tankers) as they pass blood vessel bifurcations moving upstream, and then to gradually detach and shorten the chains as they move downstream. Even a single cilium can exert sufficient force to pull a fully-loaded 1000-tanker chain at ~100 G, although the tanker at the base of each chain will normally be grasped by up to 10 independent cilia. Nevertheless, chain linkage reliability and the effects of such extended chains on flow viscosity should be studied more thoroughly. Other approaches, including facultative lateral links between chains to improve transport stability, tanker bloc motions, independently ciliated tankers (allowing high flow rates in large vessels with low inter-laminar slip rates), mechanically valved free flow, extensive water and respiratory gas caching to reduce tanker count, and increasing transport speeds up to 1 m/sec in large blood vessels, are potentially useful strategies but will not be explored further here.

Mobile Vasculocytes

An additional small population of mobile legged nanorobots, or "vasculocytes," continuously patrols the vasculoid interior. Vasculocytes are independent nanodevices several microns in size, replete with ambulatory appendages, manipulator arms, repair and assembly tools/materials, onboard computers with mass memories, communications and sensory equipment, and independent power supplies and fuel tankage. A description of a similar system is in; a fuller discussion of these capabilities is in and is beyond the scope of this paper. Like other vasculoid subsystems, to ensure high reliability vasculocytes are designed with the usual tenfold redundancy in most major components.

Vasculocytes in the vasculoid appliance have many important functions, including: (1) general maintenance and repair activities; (2) plugging internal leaks and breaches, cleaning spills, leak scavenging, and maintaining an orderly particle-free internal environment; (3) repairing or replacing malfunctioning ciliary leg mechanisms (designed for modular changeout); (4) clearing of jammed docking bays or cellulocks; (5) reconstruction of the vasculoid sapphire surface plate array after a breach or other catastrophic failure; (6) physical reconfiguration and remodeling of the vasculoid microstructure, as for example to extend additional segments into new capillaries following angiogenesis, to remove vasculoid tendrils from dead capillaries, or to expand or contract existing segments (details deferred to another paper); and (7) installation and removal of vasculoid systems from the human body.

With a vasculocyte mass of ~10 picograms, volume ~3 micron^3, and peak power consumption ~50 picowatts to allow bursts of more intensive computation (at up to nearly ~1 megaflop/sec) or other activity, a 10-watt vasculocyte subsystem energy budget permits the deployment of 200 billion continuously active devices, a total 2 gm mass of nanorobots. Deployment of this number of devices would allow, for example, at least one vasculocyte to visit all 24 trillion docking bays once a day, spending 720 sec visit, enough time to perform ~10^9 manipulatory operations using a 10^6 Hz robotic arm . Alternatively, this would allow 7200 visits/day per docking bay, each visit lasting ~100 msec (~10^5 operations; travel time between adjacent bays at 1 cm/sec is only 0.35 msec) which is long enough to run a simple diagnostic set and to detect plate malfunction within just ~12 seconds (on average) of its initial occurrence. It would also allow one visit per day to each of 3000 trillion cilia, spending ~6 sec/visit (~10^7 operations). This seems sufficient.

Assuming ~0.1 pW per vasculocyte leg (similar to cilium) and 4 legs in motion, simple walking consumes only ~0.4 pW of mechanical power; a 10,000 ops/sec computation budget to control simple walking, costing ~6 × 10^{-17} watts/(ops/sec), adds another ~0.6 pW, giving a total power requirement of just ~1 pW for continuous simple walking. An onboard 1 micron^3 fuel tank with energy storage density of ~2 × 10^{10} J/m^3 for externally-supplied O_2 gives each 1-50 pW vasculocyte a range of 400-20,000 sec between refueling stops, or ~100-5,000 sec if the O_2 is stored internally. The power draw of the entire 200 billion vasculocyte fleet is 0.2-10 watts.

During vasculoid installation, the patient is injected with a population of vasculocytes adequate to construct the initial configuration in a reasonable period of time. After installation, a sufficient number of spare vasculocytes are stored internally in dormant condition in the vesicles until needed for repair, maintenance, or injury response functions. Active vasculocytes have highly redundant subsystems, hence fail infrequently. Upon such infrequent failure, they are replaced with reactivated dormant devices (self-repairing vasculocytes are possible in principle—e.g., modular repair-by-replacement—but would add significantly to overall system complexity, hence are not included in this scaling study.). Nanorobot detritus and internal waste is containerized and ciliated to an appropriate internal holding area located in the vesicles. New vasculocytes, replacement modules, and other necessary spare parts and repair materials may be exchanged into the user as consumables, when required.

Vasculocytes performing general repair and maintenance on tanker docking bays have been allocated up to 10 sec docking cycle for this task a very conservative 50 per cent duty cycle for docking bays. Vasculocyte system subtasks which must be subsumed within this time budget may include: (1) Detection and analysis of system fault; (2) determination of appropriate response; (3) dispatch of appropriate repair mechanism to problem site, including travel time and docking; (4) on-site verification that the problem is as reported, and that the repair plan is correct; (5) deployment of tools and materials; (6) performance of repair work possibly including plate changeout, rotor replacements, sensor recalibration, etc.; (7) verification of correct completion of repairs and that the faulty subsystem is now performing properly; (8) retracting/repacking all tools with final verification of job completion; and (9) undocking and return to transport stream for storage or to receive instructions for the next assignment. The vasculoid configuration, including numbers of components, their locations, their accessibility for repair and their ease of disconnection/reconnection will influence the rapidity of these steps.

Of course, most of the time there will be no fault at a particular docking bay, and running a diagnostic is much quicker than correcting a fault. Since there are only enough vasculocytes to allow up to 720 seconds/day of repair and maintenance activity at each docking bay (see above), and since a 50 per cent duty cycle implies that each docking bay is allocated ~43,200 sec/day for this activity, then it might be possible to significantly increase the docking bay duty cycle up to (86,400—720) / (86,400) ~ 99 per cent in a less conservative design.

Appliance Power Requirement

Power is supplied to the vasculoid appliance via the chemical combination of oxygen and glucose, both of which are plentiful in the human body. Total body power dissipation is ~100 watts at the basal rate and ~1600 watts at peak. As a further point of comparison, the basal human heart power output is $p_{heart} \sim P_{systolic} V_{blood} / t_{circ} = 1.4$ watts (taking pumping pressure $P_{systolic} = 120$ mmHg (1.59×10^4 N/m^2), pumped blood volume $V_{blood} = 5.4$ liters, and circulation time $t_{circ} = 60$ sec), rising to ~8.0 watts when cardiac output reaches ~30 liters/minute at peak exertion, assuming no rise in pressure. Experiments confirm that the basal power output of a 350 gm resting heart is ~3.5 milliwatts/gm or ~1.2 watts, and patients with chronic congestive heart failure can achieve measured peak cardiac outputs no higher than ~2 watts; efficient total artificial hearts (TAH) may draw ~5 watts of power.

The four most significant subsystems requiring a regular energy supply include ciliary transport (11.8-166.2 watts), docking bays (14.5-16.6 watts), vasculocytes (0.2-10 watts), and cellulocks (~0.6 watts), giving a requirements range of 27.1 watts (basal) to 193.4 watts (peak). There are also 125 trillion applications plates which are unused in the basic vasculoid model described in this paper. If each applications plate is allotted a 0.6 pW energy budget comparable to the docking bays, and if all possible applications plates are simultaneously activated,, then the applications subsystem could require an additional 75 watts, boosting the total power draw of the vasculoid appliance to 102.1-268.4 watts, which is still only at the borderline of thermogenic significance.

Unit V–J

Biocompatibility of Vasculoid Systems

Preliminary Thermal Conductivity Analysis

Besides providing systemic materials transport throughout the human body, the water component of blood also serves an important thermoregulatory function. Removing most of the circulating water from the vasculature eliminates one of the means by which the body regulates its gross thermal conductivity. In addition, the extremely high thermal conductivity of diamond imposes the requirement for a vasculoid design using mostly sapphire, rather than diamond, construction materials, even though diamondoid materials have been more extensively discussed in connection with molecular nanotechnology and medical nanorobots and may still be employed in lesser quantities (e.g., as thin coatings) throughout the vasculoid structure.

Aside from blackbody radiation, sweating, and behavioral thermoregulation (including respiratory cooling), the body regulates its temperature and offloads excess heat principally via two mechanisms:

First, there is passive conduction. Heat travels by pure conduction through fat and muscle from the body core out to the periphery. The thermal conductivity of human tissue is $K_t \sim 0.5$ watts/m-K, so for a typical L = 10 cm path length (~half-torso thickness), heat flow $H_f \sim K_t/L = 5$ watts/m^2-K, or ~10 watts/K for a 2 m^2 human body. In a cold room, the mean temperature differential between core and periphery DT ~ 11 K, so $H_f \sim 100$ watts, which is approximately the basal metabolic rate. Experiments confirm that 5-9 watts/m^2-K is the minimum heat flow in very cold conditions (the actual value depending largely upon the thickness of subcutaneous fat layers). In this case, the peripheral capillary blood flow has slowed to a trickle, producing the minimum thermal conductivity of the human body in cold conditions. On the other hand, in a warm room or during heavy exercise, DT ~ 1 K, so $H_f \sim 10$ watts. Thus, paradoxically, at warmer temperatures when the human body is generating considerable surplus heat, the body's passive heat flow is actually very low because of the smaller temperature differential between core and periphery.

Second, heat is transported via the active blood flow. In warm rooms, not only are the peripheral capillary sphincters fully dilated, allowing more blood to flow through the peripheral capillaries relative to the core capillaries, but also the total volume of blood flow may increase. (During heavy exercise, total blood flow volume may rise by a factor of 4 or 5.) Diathermy experiments suggest that the active blood flow mechanism alone may carry off 100-200 watts of heat before core temperature starts to rise. In cold rooms and in the absence of heavy exercise, peripheral capillary sphincters are maximally contracted, thus minimizing blood flow (and hence heat transport) to the periphery.

To summarize: The passive conduction mechanism can throw off ~100 watts of waste heat when the human body is in a cold room but only ~10 watts when the body is in a warm room, while the active conduction mechanism can throw off negligible heat in a cold room but up to 100-200 watts in a warm room. Thus, as the external environment warms up, the human body shifts from passive conduction to active conduction (via increased blood flow and capillary sphincter widening); if the environment becomes hotter still, or the person begins exercising, then sweating eventually comes to dominate both processes.

If we now remove the active blood flow regulatory mechanism, inhibit capillary sphincter expansion, and emplace artificial tubes inside all the blood vessels and capillaries of the body (i.e., install the vasculoid), the sweating process and surface thermal radiation remain unaltered. Thus the principal changes to the human thermoregulatory system are in the passive and active conduction systems.

First, consider the active conduction system. In a working vasculoid operating at the maximum tanker flow rate, we have ~0.133 kg (dry mass) of tankers moving around the body in ~100 sec per circuit. We assume all tankers are filled with water, for a total water mass of 0.125 kg. Diamond has a heat capacity of 519 joules/kg-K; sapphire is 728 joule/kg-K and water is 4218 joule/kg-K at 310 K. Hence, the circulating tanker fleet can transport at most ~6 watts/K of heat from the core to the periphery. Given that DT might be as small as 1 K, the "active conduction system" is essentially disabled by installation of a vasculoid that is constructed either of diamond or sapphire.

Second, consider the passive conduction system. For a natural biological-tissue body, heat flow is $H_f = K_t / L = 5$ watts/m^2-K. For a human body shape comprised entirely of pure diamond ($K_t \sim 2000$ watts/m-K at 310 K) and again taking L = 10 cm, then $H_f = 20{,}000$ watts/m^2-K. For a diamond-envasculoided human body, taking a mass of ~1.7 kg of diamond thoroughly interwoven with 68.3 kg of mostly aqueous biological tissue mass (for a standard 70 kg male body), as a crude estimate the effective heat flow becomes $H_f \sim 500$ watts/m^2-K, or ~100 times more thermally conductive than before. For comparison, a pure metal human form would have $H_f \sim 170$ watts/m^2-K for stainless steel, ~350 watts/m^2-K for lead, ~780 watts/m^2-K for iron, or ~3800 watts/m^2-K for copper. Hence a diamond-envasculoided human body would have passive conduction properties similar to those of solid metal.

This has implications for the maximum DT that can be maintained between core and periphery. Consider a human-shaped tissue-mass with half-thickness L ~ 10 cm and surface area A ~ 2 m^2, sufficiently heated from the inside to cause P ~ 100 watts (human basal rate) to flow via passive conduction from core to periphery, establishing a temperature differential $DT \sim P L / A K_t \sim$ 10 K for natural human tissue with $K_t = 0.5$ watts/m-K. Upon switching to diamondoid envasculoided tissue, mean tissue thermal conductivity would rise to $K_t \sim 50$ watts/m-K and so DT would fall to ~0.1 K. In effect, the entire human body would become isothermal to within 100 millikelvins; even at the peak power output of 1600 watts for the human body, DT rises to just ~1.6 K. Thus, a diamond-envasculoided human body would tend to become isothermal with its surroundings very quickly (although partially offset by subcutaneous fat), a clear hazard to normal human health especially in very hot or very cold environments. The thermal equilibration time is approximately $t_{EQ} \sim L / v_{thermal} \sim 0.1$ millisec, where $v_{thermal} \sim K_t / h_{plate} C_V = 1000$ m/sec for neighboring vasculoid plates in good thermal contact with each other and having thickness $h_{plate} \sim 1$ micron, with $K_t = 2000$ watts/m-K and $C_V = 1.8 \times 10^6$ joules/m^3-K for diamond at 310 K, and taking L = 10 cm as before. This is far shorter than the typical 1-10 sec thermal response time of the purely-biological human vasculature.

Substitution of sapphire for diamond significantly improves thermal performance. The thermal conductivity of synthetic sapphire may be as low as $K_t \sim 2.3$ watts/m-K for sapphire at 310 K, roughly a thousand fold lower than for diamond, when measured in the direction normal to the symmetry or optic axis (c-axis); heat capacity ($C_V = 2.9 \times 10^6$ J/m^3-K) and density (3970 kg/m^3) of sapphire are slightly higher than for diamond. Thus for a sapphire-envasculoided human body, taking a mass of ~1.9 kg of sapphire (at 25 watts/m^2-K for L = 10 cm) thoroughly interwoven with 68.1 kg of mostly aqueous biological tissue mass (at 5 watts/m^2-K), the total heat flow is just 5.5 watts/m^2-K, which differs insignificantly from natural biological tissue. At P = 100 watts, DT falls to 2 K compared to 10 K for natural tissue and 0.1 K for diamond-envasculoided tissue; $t_{EQ} \sim 1$ sec for sapphire vs. 10^{-4} sec for diamond.

Two complications regarding sapphire require additional research. **First**, the thermal conductivity of sapphire may vary significantly with both composition and crystallographic orientation, a fact which might impose additional and unknown constraints on the present design. For instance, one source reports heat flow values interpolated to 310 K of 21 watts/m-K normal to the c-axis and 23 watts/m-K parallel to the c-axis; minor extrapolations of other sources to 310 K (i.e. slightly outside the exact temperature ranges measured experimentally) imply values of 2.0 and 2.3 watts/m-K for heat flows normal to the c-axis. However, all reported values for sapphire are at least two orders of magnitude more insulating than diamond.

Second, much like diamond, the thermal conductivity of sapphire varies with temperature. For example, at ~200 K (near dry ice temperature) sapphire's thermal conductivity rises to 5 watts/m-K. At liquid nitrogen temperature (77 K), K_t soars to ~1000 watts/m-K; the peak is ~6000 watts/m-K at 35 K. (Diamond's conductivity also rises as it cools). At the other temperature extreme, sapphire's thermal conductivity rises to 3.9 watts/m-K by 523 K. Diamond thermal conductivity also varies significantly with isotopic composition (e.g., ^{12}C vs. ^{13}C); it is unknown whether similar opportunities may exist for the engineering of desired levels or patterns of thermal conductivity in isotopically-controlled sapphire.

Finally, we note that much burn damage occurs due to a failure to dissipate heat. Selectively increasing the thermal conductivity of certain parts of the body at certain chosen times is a useful feature that could mitigate the effects of transient local heating. Note that blood is an excellent electrical conductor, and its replacement by an inorganic construct may reduce the susceptibility of the body to electrical burns under ordinary conditions. Pure single-crystal alumina (sapphire), roughly analogous to the core material of the vasculoid plates, is one of the best electrical insulators known. Pure bulk diamond is also an excellent insulator, but hydrogen-terminated diamond surface shows a high p-type surface conductivity, and slight impurities (e.g., p-type (boron) or n-type (phosphorus) or dislocations may also permit conduction, so a diamond-coated sapphire vasculoid plate should be somewhat more electrically conductive than bulk sapphire. The vasculoid is envisioned as a primarily mechanical system and so its electrical characteristics have not yet been investigated. The susceptibility of the appliance to electrostatic charge transfer, electromagnetic interference or electromagnetic pulses (EMP) is unknown.

Biocompatibility of Vasculoid Systems

Mechanical Interaction with Vascular Endothelium

Perhaps the single most important biocompatibility factor involving the vasculoid has been termed "vascular mechanocompatibility" by Freitas. Vasculoid nanorobots in all their forms must be as mechanically biocompatible with the vascular walls as are stents and must not produce destructive mechanical vasculopathies or disrupt the endothelial glycocalyx.

Modulation of Endothelial Phenotype and Function—The luminal surfaces of all blood and lymph vessels consist of a thin monolayer (the intima) comprised of flat, polygonal squamous endothelial cells (EC) covering a much thicker layer (the media) comprised of vascular smooth muscle cells (SMC). Under normal physiological conditions, both layers are subject (and respond) to tangential fluid shear stresses across the endothelial cell surface due to the bulk flow of blood, normal hydrostatic pressure stress acting radially on the vessel wall due to the propagation of the pressure wave, and cyclic stretch or strain due to blood vessel circumferential expansion in vivo, and thus might also be sensitive to similar mechanical stresses that may be applied by stationary or cytoambulatory intravascular nanorobots such as the vasculoid basic plates.

Endothelial cells (EC) are randomly oriented in areas of low shear stress but elongated and aligned in the direction of fluid flow in regions of high shear stress. In vitro endothelial cells previously acclimatized to physiological fluid shear stresses respond to artificial changes in local fluid shear stress only very slowly, and in three stages. In the first stage, EC initially respond to the imposition of stress within 3 hours by enhancing their attachments to the substrate and to neighboring cells; the cells elongate and have more stress fibers, thicker intercellular junctions, and more apical microfilaments. In the second stage, after 6 hours the EC show constrained motility as they realign, losing their dense peripheral bands and relocating more of their microtubule organizing centers and nuclei to the upstream region of the cell. In the third stage, after 12 hours the EC become elongated cells oriented in the new apparent direction of fluid flow; stress fibers are thicker and longer, the height and thickness of intercellular junctions are higher, and the number and height of apical microfilaments are increased. This produces a new cytoskeletal organization that alters how the forces produced by fluid flow act on the cell and how the forces are transmitted to the cell interior and substrate.

Physiological fluid mechanical stimuli (e.g., fluid shear stresses*) are important modulators of regional endothelial phenotype and function. For example, endothelium exposed to fluid shear stress undergoes cell shape change, alignment, and microfilament network remodeling in the direction of flow (though this may be blocked via microtubule disruption using nocodazole). Interestingly, the application of a steady laminar shear stress (a physiological stimulus) upregulates the human prostaglandin transporter (hPGT) gene at the level of transcriptional activation, whereas a comparable level of turbulent shear stress (a nonphysiological stimulus) or low stress (such as would be produced by a vascular surface coated with sessile nanorobots) does not . A few of the many quantitative experimental observations include:

1. shear stresses from 0.02-1.70 N/m^2 produce flow-induced membrane K^+ currents ;
2. physiological shear stresses of 0.35-11.7 N/m^2 stimulates mitogen-activated protein kinase in a 5-min peak response time;

3. 0.04-6 N/m^2 shear stress increases inositol trisphosphate levels in human endothelial cells, with a 10-30 sec peak response time;
4. shear stresses from 0.5-1.8 N/m^2 regulate (in frequency and amplitude) oscillating K^+ currents known as spontaneous transient outward currents or STOC which are observed both in isolated bovine aortic endothelial cells and in intact endothelium; activation of STOC depends on the existence of a Ca^{++} influx and is blocked by 50 microM of Gd or is significantly reduced by 20 microM of ryanodine;
5. shear stress of 1.2 N/m^2 induces transcription factor activation over response times ranging from 0.3-2 hours;
6. Arterial shear stresses of 1.5-2.5 N/m^2 (but not a venous shear stress of 0.4 N/m^2) induce endothelial fibrinolytic protein secretion;
7. shear stress of 2 N/m^2 induces TGF-b1 transcription and production in a ~60 sec initial response time, with a sustaining increase in expression after 2 hours;
8. a shear stress of 2 N/m^2 suppresses ET-1 mRNA on confluent bovine aortic endothelial cell monolayers; these effects of shear may be completely blocked (thus allowing ET-1 to be expressed) using 875 nM of herbimycin to inhibit tyrosine kinases or 10 microM of quin 2-AM to chelate intracellular Ca^{++}, partially inhibited using 3mM of tetraethylammonium (TEA), or attenuated by elevated extracellular K^+ at 70 mM or completely inhibited by K^+ at 135 mM;
9. shear stress of 3 N/m^2 induces Ca^{++} membrane currents in a 30 sec peak response time;
10. shear stress of 6 N/m^2 applied for 12 hours causes endothelial cells to align with their longitudinal axes parallel to flow;
11. membrane hyperpolarization occurs as a function of local shear stress up to 12.0 N/m^2, with an exponential approach to steady state in ~1 minute; the process is fully reversed once the artificial fluid flow stress is removed;
12. critical shear stress of 42 N/m^2 is the disruptive threshold for endothelial cells, inducing cell mobility; and
13. Shearing stresses of 5-100 N/m^2 occur at the contact interface when a leukocyte is adhering to or rolling on the endothelium of a venule.

Endothelial cells thus respond to sustained physiological fluid shear stresses from 0.02-100 N/m^2, spanning the range of normal arterial wall fluid shear stresses of 1.0-2.6 N/m^2 from the aorta through the capillaries and 0.14-0.63 N/m^2 for the venous circulation. By contrast, legged vasculomobile medical nanorobots may apply shear stresses during luminal anchorage or cytoambulation at velocities up to 1 cm/sec of at least 40-200 N/m^2 or higher. (Self-expanding aortic stents forcibly pulled from the vessel require an extraction force of ~400 N/m^2 assuming a 10-cm length, rising to ~1200-3600 N/m^2 for stents anchored with hooks and barbs; varying the radial force applied by stents against the vascular wall has little impact on the required extraction force.) Such shear forces, if imposed unidirectionally by large numbers of closely-packed co-ambulating nanorobots for time periods of $>10^3$ sec, may induce significant changes in shape, orientation, and physiological function in the underlying endothelial cell population. If instead these forces are applied in randomized directions varying over short time periods (<1 hour); by vasculoid nanorobots, then mechanically-induced modulation of endothelial phenotype and function should not occur.

All shear forces must not be eliminated, however. A nanorobot aggregate that shields vascular cells from fluid shear for an extended time may induce those cells to revert to their flow-unstressed phenotype or to undergo apoptosis. Endothelial cells cultured in the absence of shear stress tend to become dedifferentiated. In one study, after blood shear was artificially reduced near a wound lesion for 24 hours the local endothelial cells became less elongated, contained fewer central microfilament bundles, and exhibited a slower repair process. In another study, vein grafts removed from the higher-shear arterial circulation and reimplanted in the lower-shear venous circulation of the same animal showed regression of intimal hyperplasia and medial rethickening in 14 days, apparently due to induction of smooth muscle cell apoptosis by a reduction in pressure or flow forces.

Endothelial cells can also respond to persistent static overstretching in many ways, up to and including apoptosis. For instance, hypertension caused by hydrostatic edema can induce apoptosis in capillary EC.

Additionally, vascular wall cells respond to lateral stretch forces due to cyclical blood vessel expansion in vivo. For example, in one experiment bovine aortic endothelial cells were seeded to confluence on a flexible membrane to which cyclic strain was then applied at 1 Hz (0.5 sec strain, 0.5 sec relaxation) for 0-60 min. After 15 minutes of this cyclic stretching, there was an increase in adenylyl cyclase (AC), cAMP, and protein kinase A (PKA) activity of 1.5-2.2 times at 10 per cent average strain as compared to unstretched cells, but there was no activity increase at 6 per cent strain—evidently, cyclic strain activates the AC signal transduction pathway in endothelial cells by exceeding a strain threshold, thus stimulating the expression of genes containing cAMP-responsive promoter elements. Stretch-activated cation channels in bovine aortic EC are inhibited by $GdCl_3$ at 10 microM. Human umbilical EC subjected to a 3-sec stretch pulse show an intracellular rapid-increase Ca^{++} spike, followed by a (ryanodine-inhibitable) slow-decline, due to Ca^{++} entry into the cell through stretch-activated channels; Mn^{++} also permeates mechanosensitive channels (but not Ca^{++} channels) and enters the intracellular space immediately after an application of mechanical stretch. Cyclic strains of 10 per cent at 1 Hz induce intracellular increases in Ca^{++}, diacylglycerol, inositol trisphosphate and protein kinase C (PKC) in peak response times of 10-35 sec, often sustained for up to ~500 sec, and induce transcription factor activation over response times ranging from 0.25-24 hours. Several endothelial cytokines are induced by cyclic mechanical stretch and cyclic mechanical strain modulates tissue factor activity differently in endothelial cells originating from different tissues.

Similarly, bovine aortic smooth muscle cells (SMCs) seeded on a silastic membrane and subjected to cyclic strains up to 24 per cent enhanced SMC proliferation at any strain level, although SMC under high strain (7-24 per cent) showed more proliferation than SMC at low strain (0-7 per cent) in this experiment. High-strained SMC aligned themselves perpendicular to the strain gradient, whereas low-strained SMC remained aligned randomly; PKA activity and CRE (cAMP response element) binding protein levels increased for highly strained cells, compared to low-strained cells. Other experiments have found that small mechanical strains of 1-4 per cent at 1 Hz applied to human vascular smooth muscle cells can inhibit intracellular PDGF- or TNFa-induced synthesis of matrix metalloproteinase (MMP)-1; that saphenous vein SMC distention by 0.5 atm pressure subsequently elevates cell apoptosis; that cyclic mechanical strain at normal physiological levels decreases the DNA synthesis of vascular smooth muscle cells, holding SMC proliferation to a low level; that 1 Hz, 10 per cent cyclic strain on SMC activate tyrosine phosphorylation and PKC, PKA, and cAMP pathways over response times from 10 sec to 30 min and that vascular SMC exhibited stretch-induced apoptosis when subjected to cyclic 20 per cent elongation stretching at 0.5 Hz for 6 hours. Consequently, medical nanorobot aggregates which shield the vasculature from normal cyclical strains might elicit excess growth of vascular smooth muscle cells, which growth is normally held in check by the rhythmic stretching from the arterial pulse. Intravascular nanorobot aggregates that apply cyclic mechanical strains exceeding a few percent might encourage increased SMC proliferation and activate mechanosensitive and stretch-activated channels in EC, along with cellular realignment and subsequent SMC apoptosis at the highest strain levels.

In 2002 it was unknown whether high frequency (>KHz) cyclic mechanical strains likely to be employed by vasculomobile medical nanorobots would have biological effects similar to or different from those described above for low-frequency cyclic strains, excepting certain specialized mechanoreceptor cells such as the cochlear stereocilia, other hair cells and somatosensory neurons, since most mechanical cell stimulation experiments have been conducted at low frequencies. Unrecognized effects that might be triggered by high-frequency cyclic strains cannot be ruled out. However, given the relative safety of procedures involving intravascular ultrasound with its low complication rate (e.g., only 1.1 per cent, including spasms, vessel dissection and guidewire entrapment using frequencies as high as 10-20 MHz, it seems improbable that KHz or MHz acoustic waves of the intensities that might be employed by medical nanorobots for communication or power supply will damage the luminal vascular walls. Continuous low-power ul trasound exposures exceeding ~10^4 sec are considered safe but there are few studies on the safety of long-duration chronic exposures. (Interestingly, relatively high-intensity intravascular ultrasound has been used to dissolve occlusive platelet-rich thrombi safely and effectively in myocardial infarctions and in restenosed stents.) If necessary, large wave motions and pressure patterns that are characteristic of normal blood flow can in principle be simulated by the vasculoid plate array by a combination of motile ciliary activity and periodic manipulations of interplate bumpers.

It is likely that the vasculoid appliance will need to control smooth muscle cell proliferation in the simplest case releasing specific cytokines into the vasculoid-endothelial space. Such factors may include known SMC proliferation promoters such as thrombin (esp. alpha-thrombin), PDGFs (esp. PDGF-AA), FGF (esp. basic FGF), HBEGF (heparin binding epidermal growth factor), TGFb (transforming growth factor-beta) at low concentrations, angiotensin II, thrombospondin-1 (stretch/tension), and known SMC proliferation inhibitors such as heparin/heparan sulfate, TGFb (transforming growth factor-beta) at high concentrations, nitric oxide, prostaglandins, calcium antagonists, agonists that activate guanylate and adenylate cyclases, inhibitors of angiotensin-converting enzyme, interferon gamma, 18-beta-estradiol, sodium salicylate, and the topoisomerase I inhibitor topotecan. Adult arterial walls contain both differentiated and immature SMCs. R. Bradbury notes that further research is needed regarding how SMCs handle conflict resolution between "divide" and "don't divide" signals that it may be receiving from both internal and external sensors. Given the large number of signals that SMCs currently respond to, it seems highly likely that the vasculoid can "manage" them, along with other secretion products of SMCs that play important roles in the prevention of vascular disease, such as extracellular superoxide dismutase.

Vascular Response to Stenting—Mechanical biocompatibility must also be demonstrated by intravascular nanorobots that are intended to remain in long-term near-contact with blood vessel walls. A related medical analog is the vascular stent—a flexible metal coil or open-mesh tube that is surgically inserted into a narrowed artery, expanded, and pressed into the vascular wall at up to 10-20 atm pressure, in order to ensure long-term local vascular patency by providing a scaffold to hold the artery open. Within 4 days, SMC begin to appear in the intima. After a few months the stent is completely overgrown with new endothelium, forming a neointima, although the media is usually compressed with smooth muscle cell atrophy in all stented regions; stenosis is prevented in vessels 10 mm or greater in diameter but is not precluded in vessels smaller than 6 mm. Histologically, in-stent restenosis appears to derive almost exclusively from neointimal hyperplasia, which appears more abundant following stent implantation than balloon angioplasty and in stents of greater stent length and smaller vessel caliber (or inadequate stent expansion). Restenosis occurs in 22-46 per cent of all stents emplaced within 6-12 months), in some cases requiring the insertion of a second stent into the first, and varies according to the material used. In one experiment, the thickness of the neointimal layer formed over wire-mesh stents placed in canine aortas was 83.9 microns thick for gold, 103.6 microns for stainless steel, 115 microns for Teflon, 209.6 microns for silicone, and 228.6 microns for silver; a copper stent produced severe erosion of the vessel wall, marked thrombus formation, and aortic rupture. Stent surface coating and texture also affects leukocyte-platelet aggregation and platelet activation.

Improved prospects are reported for diamondoid stents. For instance, diamond-coated stents produced by Phytis Corp are inserted at 16 atm pressure (vs. 2-3 atm normally for stents) and yet do not dislodge surface (thrombogenic) antigens and selectins: "The results show that in none of the control systems (systems without stents) [could] a change in glycoprotein expression (i.e. all antigens) ... be detected. With the exception of one measurement also the structural epitopes (CD 41a, CD 42b) show no significant changes during the test period. The reason is most probably that the shearing strength of the system was too weak or the expression of the antigens was too strong. Whereas for the thrombocyte activation, remarkable difference in the expression of CD 62p and CD 63 could be detected. The thrombocyte activity marker CD 62p is reduced in diamond-like coated stents and diamond-like coated stents with heparin compared with uncoated stents."

However, stent devices are far from ideally mechanocompatible with blood vessel walls. For example, stents placed endovascularly in dog aorta for 4-45 weeks and then examined histologically show medial atrophy, intimal hyperplasia (tissue ingrowth), and proliferation of the vasa vasorum (the microvasculature of the aorta) more prominently for covered stents than for bare stents, probably due to hypoxia in the aortic wall. Cellular proliferation is highest when the artery wall is most hypoxic; but vasculoid nanorobot aggregates (e.g., basic plates) that entirely cover the vascular endothelium can precisely regulate oxygenation of the underlying tissue (using data from oxygen sensors in the plate wall contacting the endothelium), thus largely eliminating the possibility of hypoxia. Stentlike vascular-coating structures also may be able to inhibit stenosis due to vascular smooth muscle proliferation, migration, and neointima formation, without inducing apoptosis, by releasing the topoisomerase I SMC-proliferation inhibitor topotecan in a localized 20 min exposure, or by using other similar drugs.

Nanorobotic Destructive Vasculopathies—The physical configurations or activities of medical nanorobotic aggregates could in some circumstances be destructive of vascular tissue. Owens and Clowes point out that the severity of arterial injury is important in determining the ultimate pathophysiologic response and describe a classification system based on the immediate histologic effect of the injury:

Type I injuries involve no significant loss of the vessel's basic cellular architecture, although there may be a slight change in endothelial architecture and associated cellular adhesion. Examples include the fatty streak (an early atherosclerotic lesion), hemodynamic factors and flow disturbances which produce, at most, only a modification of the established cellular architecture.

Type II injuries involve loss of the endothelial layer, perhaps inducing platelets to adhere and begin forming a thrombus at the area of loss, but the internal elastic lamina remains intact and there is little or no damage to the media. Examples include injuries incurred during simple arterial catheterization, endovascular procedures, vein graft preparation, or gentle filament-induced endothelial denudation of the carotid artery in a rat model.

Type III injuries involve transmural damage in which the endothelium is removed, the internal elastic lamina is often disrupted, and a significant portion of the medial cells are killed; platelets deposit and a thrombus forms at the site of endothelial loss, and an inflammatory response (vasculitis) including intimal hyperplasia is initiated within the vessel wall. Examples include spontaneous vascular dissection and various forms of surgical repair or reconstruction such as balloon angioplasty, endarterectomy and atherectomy.

Medical nanorobot device and mission designs should always seek to avoid *Type II* and *Type III* injuries, although in some special circumstances the potential even for *Type III* injuries may be inescapable. Destructive mechanical vasculopathies that might be caused by vasculoid nanorobots may be classified as ulcerative, lacerative, or concussive.

Nanorobotic Ulcerative Vasculopathy—Pressure ulcers are normally caused by a prolonged mechanical pressure against epidermal tissues (e.g., the skin of a person who is lying down; decubitus ulcer, or bedsores), typically at sites over bony or cartilaginous prominences including sacrum, hips, elbows, heels and ankles. The combination of pressure, shearing forces, friction and moisture leads to tissue death due to a lack of adequate blood supply; if untreated, the ulcer progresses from a simple erosion to complete involvement of the dermal deep layers, eventually spreading to the underlying muscle and bone tissue. In rare cases, mechanical frictional stimulation of the skin can precipitate systemic cutaneous necrosis and calciphylaxis, a state of induced tissue sensitivity characterized by calcification of tissues. Stercoral ulcers are caused by the necrosis of intestinal epithelium due to the pressure of impacted feces. Fluid mattresses can greatly reduce pressure ulcers in long-duration surgeries and pressure-relieving surfaces have been investigated for surgical patients, wheelchair users, and for other circumstances) It is generally recommended that the interior surface should employ materials having roughly the same mechanical properties as the enclosed tissue, and the applied interfacial pressures should be reduced to below 1 psi or ~ 50 mmHg.

Similarly, a macroscale nanorobot aggregate such as a vasculoid appliance may cause luminal vascular ulceration by prolonged mechanical pressure against intimal tissues, similar to epithelial pressure ulcers, necrotizing vasculitis, or pressure necrosis. For example, mechanical stretch induces apoptosis in mammalian cardiomyocytes and hypertension caused by hydrostatic edema can induce apoptosis in capillary endothelial cells. Another example is IUD-induced metrorrhagia (nonmenstrual uterine bleeding), wherein the IUD (intra-uterine device) elicits a vascular reaction most pronounced in the endometrium adjacent to the device and includes increased vascularity, degeneration with defect formation, congestion, and poor hemostatic responsiveness to increased vascular permeability and damage, leading to interstitial hemorrhage due to vascular damage from mechanical stress transmitted by the IUD through the endometrium to its vascular network.

Indwelling catheters can rest very snugly against the vascular walls without complication for brief periods. A biological-like interface may reduce ulceration in longer-term missions. In one study, a stented aortic graft was placed endovascularly in the native aorta of male sheep, and a histological examination 6 months later found good incorporation of the graft with no pressure necrosis, although there was a foreign body reaction around the graft and an organized blood clot was noted between the graft and the aortic wall (the inner, but not the outer, aortal surfaces would be expected to have clotting-suppressive properties).

However, long-duration nanoaggregates (such as vasculoid plates) that must maintain close contact with endothelium should employ a mechanically compliant coating having properties similar to extracellular matrix. All such linkages should be not just immunocompatible but also mechanocompatible, possessing equivalent elasticity or mechanical compliance as the underlying tissue to which attachment must be secured. With conventional implants, compliance design may include assessments of circumferential compliance (measurement of changes in vessel diameter over a complete cardiac cycle, including pressure-radius curves, dynamic compliance, and mechanical hysteresis effects), longitudinal compliance (elasticity of selected lengths of the vascular system, including any localized stiffening), tubular compliance (imparity of elasticity between a prosthetic conduit and the native artery, elastic energy reservoiring, and pulsatile energy losses due to interfacial impedance mismatches), and anastomotic compliance (suture line anastomotic compliance mismatch and the para-anastomotic hypercompliance zone, localized regions of excessive mechanical stress, and cyclic stretch effects on replication of vascular SMC and extracellular matrix). A mismatch in mechanical properties between relatively compliant arteries and less-compliant metallic stents and tissue grafts has been thought to influence patency and pseudointimal hyperplasia. Larger more central arteries are more compliant than the distal small caliber arteries, wall shear stress from blood flow differs on either side of a curving vessel and the stress is out of phase with the pulsing circumferential stretch strain, and compliance mismatch between host artery and prosthetic graft may promote subintimal hyperplasia. Analogous compliance issues may be assessed once the static and dynamic stress patterns in the vasculoid transport system are more precisely known.

Post-installation vascular conditioning can be maintained on a permanent basis because the vasculoid appliance exercises precise control over the transport of most metabolic and cellular traffic—e.g., of cholesterol, leukocytes and platelets, the principal participants in the arteriosclerogenic process, or of various circulating pluripotent stem cells and their activating factors.

Nanorobotic Lacerative Vasculopathy—Individual vasculomobile nanorobots or nanorobotic aggregates may occasionally scratch, scrape, or gouge the vascular luminal surface, causing partial or complete loss of local endothelium (*Type II* damage), a form of mechanical vasculitis or capillaritis, particularly during installation of the vasculoid appliance. Since the typical dimensions of nanorobotic vasculoid components approximate the endothelial thickness, transmural *Type III* damage to the media is unlikely. Turnover studies of rat endothelium show that (a) injured endothelium can recover an area one cell wide (~1000 micron^2) in ~3 hours, (b) the natural loss rate is ~0.1 per cent of endothelial cell area per day (~1 micron^2/day) [267], and (c) the steady-state vascular denudation area is ~0.125 micron^2/cell.

Smooth nanorobot hulls, boundary layer effects and low fluid flow velocity throughout most of the vasculature during installation should ensure that major "sandblasting" type erosion is unlikely to occur inside human blood vessels even at the highest nanocrits consistent with continuous flow. Free-floating nanorobots that collide with blood vessel walls (given the no-slip condition at the wall) produce minimal shear forces, on the order of <~0.1 N/m^2—this is less than the 1.0-2.6 N/m^2 shear forces normally encountered in arteries and capillaries due to normal blood flow and the 0.14-0.63 N/m^2 shear forces in veins, but may be sufficient to cause a small biological response from the vascular endothelium. Applying the maximum possible bloodstream velocity of 1 m/sec to the impact-scratch relation, it is clear that particle-wall collisions should produce only harmless submicron nicks even in the most turbulent arteries.

Nevertheless, some caution is warranted because natural endothelial cell wounding of ~1-18 per cent of all cells, possibly erosionally-derived, has been observed in rat aorta. Erosion of cultured fibroblast monolayers (simulating the vascular endothelium) using MHz ultrasound at acoustic pressures of ~10^6 N/m^2 is enhanced by the presence of a microbubble (particulate) contrast agent. Injection of crystalloid cardioplegic solutions into canine hearts at pressures >110 mmHg and at peak flow rates >25 ml/sec also causes a higher incidence of mechanical-physical trauma to the vascular endothelium and the endocardium. In another unusual case, intravenous self-injection by a drug abuser of dissolved tablets containing microcrystalline cellulose as filler material produced numerous microcrystalline cellulose pulmonary emboli, intravascular foreign body granulomas, focal necrosis and edema of the pulmonary parenchyma, and fatal vascular destruction.

Endothelial abrasion alone may not stimulate neointimal thickening but inevitably must involve some endothelial cell loss and other biological responses. For example, mechanically scraping cultured endothelial cells causes growth factor to be released within 5 minutes, not abating for at least 24 hours thereafter, due to plasma membrane disruption. In the case of vascular dissection, a piece of the endothelium peels up, making an intimal flap that defines regions of true and false lumina, and sometimes may induce an intramural hematoma in the aortic wall. Endothelial cells mechanically damaged with a razor blade activate extracellular-signal-regulated kinases within ~300 sec, releasing fibroblast growth factor (FGF-2) which in turn induces intimal hyperplasia. Nanorobots which detect FGF-2 are alerted that mechanical endothelial injury has taken place; by absorbing the cytokine using molecular sorting rotors, the hyperplasia signal may be suppressed by a nanorobot, if desired. However, shear-induced endothelial denudation of healthy canine arterial endothelium appears not to occur at shear stresses up to at least 200 N/m^2. The role of erythrocyte collisions with vascular walls on the detachment rate of endothelial cells is just starting to be seriously investigated.

K. Clements suggests that some provision might be also be made for emergency plate disconnects in certain unusual accident situations, e.g., where the user has irreversibly caught his hand in a machine applying superior force, and now risks not just amputation of the original biological limb, but also faces either (1) forcible extraction of the diamondoid appliance from the remaining biological tissues, or (2) progressive amputation of additional biological tissues because the vasculoid network has become caught in the machine.

Nanorobotic Concussive Vasculopathy—If a patient experiences significant external crushing or concussive forces, resident medical nanorobots that are present in small numbers can simply move out of the way, as described previously by Freitas in connection with the risks of dental grinding. In the case of macroscale intravascular nanoaggregates, there are several additional risks that should be avoided in specific appliance designs.

First, there is the possibility that a sudden mechanical external tissue compression could push macroscale nanorobotic aggregates through the soft tissues, causing deep tissue penetrations, perforations, or other serious mechanical trauma. Similarly, because the vasculoid materials may have higher density than the surrounding biological tissues, very high accelerations could produce effects on those tissues that would be not unlike pushing gelatin through a metal wire strainer. Simple activities such as hand clapping or fistfights are unlikely to produce the high accelerations required for serious damage, but this risk should be quantified in future studies. Possibly relevant analogies in the medical literature include:

(1) Ulnar artery erosion, thromboemboli, digital ischemia and skin necrosis from a glass foreign body in a patient's hand.

(2) Tantalum coil stent damage that was induced or aggravated by intravascular ultrasound inside a coronary artery.
(3) A chronic indwelling catheter that led to erosion and rupture of the anterior wall of the right ventricle, producing a near-exsanguinating hemorrhage.
(4) Cardiac perforation by a subclavian catheter.
(5) Pulmonary artery catheter-induced right ventricular perforation during coronary artery bypass surgery.
(6) An ICD patch that migrated and perforated the right ventricular cavity.
(7) A stent that migrated to an oblique position across the aorta, producing a 7-cm pseudoaneurysm after 3 years.
(8) Catheter-induced pulmonary artery rupture (a well-recognized complication of invasive monitoring) that often leads to fatal hemorrhage.
(9) Femoral artery catheterization trauma producing hematoma, pseudoaneurysms and arteriovenous fistulas of the femoral vessel.
(10) Iatrogenic subclavian artery injury due to central venous catheterization.
(11) Repeated and prolonged vein catheterization that led to subsequent stenosis (presumably due to luminal vascular mechanical damage).
(12) High-pressure injection injury that induced inflammation and foreign body granulomatous reaction, progressing to necrosis.
(13) Mechanical tearing of arteries due to overstretching.
(14) Spontaneous coronary artery dissection (mechanical arterial wall failure).

(Most of these cases pertain to injury from objects much stiffer and larger than vasculoid components, or are due instead to intrinsic vessel dysfunction (as in spontaneous arterial dissection.)

Second, a sudden external tissue compression could force nanorobotic aggregates into physical contact with neighboring nanoaggregates, possibly causing major structural damage or fragmentation of the devices. This risk increases as the nanodevices become more densely packed, especially along the crushing axis. Nanoorgans (as well as looser aggregates) can be crushed if sufficient force or mechanical shock is applied. Again, a few possibly relevant analogies from the medical literature include:

(1) External compression of emplaced stents that produced premature stenosis.
(2) A transabdominal teflon stent that broke intraperitoneally during tuboplasty procedure.
(3) A strongly-beating heart that sheared off a pericardial drainage catheter.
(4) A Hickman catheter that suffered rupture and embolization during normal use.
(5) An indwelling catheter that fractured and a distal remnant embolized to the right ventricular outflow tract and main pulmonary artery, precipitating cardiopulmonary near-collapse.
(6) A catheter embolism that was produced when a catheterized patient engaged in power training exercises, externally crushing the catheter, although no symptoms or complications accompanied this event.
(7) Spontaneous fracture of indwelling venous catheter, leading to vascular leakage.
(8) Other instances of catheter fracture and embolism, including one case that led to cardiac arrest.

(Here, too, many of these cases might differ substantially from the vasculoid, since the catheters were made from a bulk material that depended on its integrity for function while the vasculoid would consist of a large number of semi-independent components capable of restoring functional connections on their own.)

Third, there is a small risk that poor device design, poor mission design, or a loss of control might cause nanoaggregates to operate in a dangerous manner, causing macroscale concussive injury to biological tissues. For example, iatrogenic vascular trauma caused by intra-aortic balloon pumps is well-known. In one case, balloon expansion during angiography ruptured the pulmonary artery. In a canine model, an aortic valve balloon dilation to 5-12 atm pressure produced valve leaflet connective-tissue injury and hemorrhage. In an experiment with canine intestine, excessive lymph pressures (>850-1630 N/m^2) produced artificially in the central lacteals caused fluid to leak out of the villi and caused intestinal epithelial cells to be shed into the intestinal lumen. These risks can be largely avoided or at least minimized by good design and experience with related systems, initially in animal models.

Mechanical Interactions with Glycocalyx—Endothelial cells are surrounded by a well-developed extracellular glycocalyx. If this outer margin is traumatized, receptor sites and fibronectin may be exposed which could then become available for bacterial adhesion. Nanorobots that rely upon absorption of local oxygen and glucose for their power supply or whose missions include extensive small-molecule exchanges with the environment may have ~10^4-10^5 molecular sorting rotors embedded in their exterior surfaces. These spinning sorting rotors are unlikely to cause direct physical damage to cell surfaces for several reasons. First, rotors are atomically smooth and recessed into the housing, reducing physical contact with colliding surfaces and eliminating potential nucleation sites that may trigger thrombogenesis, gas embolus formation, or foaming. Second, only a small fraction of all available sorting rotors may be actively spinning at one time, further reducing the likelihood of physical trauma. And such limited contact, when it occurs, should be relatively benign: Maximum rotor rim velocity of 2.6 mm/sec is less than 1 per cent of mean aortic blood velocity and lies only slightly above maximum capillary flow speed.

Many disease processes are known that involve damage to the glycocalyx, including some bacteria that phagocytose or otherwise destroy the cellular glycocalyx during an infection. Damage to the glycocalyx creates conditions that favor the binding of immune complexes, complement activation, and intravascular coagulation, with loss of gradients between blood and parenchyma; desialylated glycocalyx of endothelium also allows an increased rate of endothelial cell detachment from arterial walls. Could the glycocalyx strands present at all tissue and nontissue cells surfaces get trimmed, even by a recessed sorting rotor? Nanorobot sorting rotor binding sites for small molecules (<20 atoms) involve pockets measuring <2.7 nm in diameter, too small to physically accommodate the 10-20 nm thick plasma membrane or the main body of the glycocalyx projections typically measuring 5-8 nm thick and 100-200 nm long, consisting of glycoproteins comprised of 10,000 atoms or more. While an occasional sugar residue may get clipped, binding sites can be designed for maximum steric incompatibility with glycocalyx glycoproteins and proteoglycans, further minimizing the opportunities for trimming. Note that clipping a covalent C-C, C-O, or C-N bond probably requires a clipping energy >500 zJ/molecule, but sorting rotors designed to pump against pressures of ~30,000 atm can only apply ~100 zJ/molecule (i.e., per binding site) so an accidentally-bound glycocalyx moiety seems more likely to jam the rotor than to be clipped off by the rotor. If this happens, the result may be a glycocalyx-tethered nanorobot, in which case a rotor-dejamming protocol will be required to free the trapped nanorobot.

Natural rates of glycocalyx damage are just starting to be quantified, and many tissue cells replace their glycocalyx or are retired in times ranging from 10^3-10^6 sec. For example: Schistosome parasites can shed some tegument-bound complexes in only ~1200 sec to 3600 sec; plasma membrane turnover rate is ~1800 sec for macrophage and ~5400 sec for fibroblast; cholesterol turnover rate in

RBC membrane is ~7200 sec membrane phospholipid half-life averages ~10,000 sec; neutrophil lifespan in blood is ~11,000 sec; enterocyte glycocalyx is renewed in 14,000-22,000 sec, as vesicles with adhered bacteria are expelled into the lumen of small and large intestine; some schistosome membrane antigen turnover may require from 68,000 to 160,000-430,000 sec; typical protein turnover half-life is ~200,000 sec; cell turnover time is ~86,000 sec in gastric body, ~200,000 sec for duodenal epithelium, ~240,000 sec for ileal epithelium, and ~400,000 sec for gastric fundus; neutrophil lifespan in tissue is ~260,000 sec; glycocalyx turnover in rat uterine epithelial cells is ~430,000 sec; and platelet lifespan is ~860,000 sec).

As the cell coat is a secretion product incorporated into the plasma membrane that undergoes continuous renewal, any trimmed glycocalyx glycoproteins from tissue cells would be rapidly replaced via biosynthesis in the ribosomes of the endoplasmic reticulum, followed by final assembly with the oligosaccharide moiety in the Golgi complex and subsequent export to the plasma membrane. Glycoprotein strands or stray sugar residues released into the extracellular medium as a consequence of such trimming are nonimmunogenic and would be quickly metabolized, although it is possible that nearby parasites could absorb this released material onto their surface, affording themselves some camouflage protection against natural host immune defenses but little protection against vasculoid defenses since parasite antigens should still be visible at cell surfaces.

Interruption of Plasmatic Water and Lymphatic Flows

A nanorobotic aggregate covering a macroscale area of the capillary luminal surface may reduce the normal flow of plasma water and other substances that leaves the circulation via ultrafiltration, unless the necessary water is replaced by the nanosystem. The plasma water flow helps to remove waste products from the extracellular spaces around tissue cells, a function that could be compromised by the shielding presence of the nanoaggregate unless the aggregate replaces this flow with water transported through or around the device, by various means. Consequently, both lymph volume and the gross water flows between tissues may be affected by vasculoidization, given that the vasculoid substitutes encapsulated fluid transport in place of bulk-flow and diffusive fluid transport in the principal global materials distribution system of the human body. More specifically, ~20 liters/day (0.23 cm^3/sec) of plasma water exit the natural circulation via ultrafiltration through leaky capillaries, of which 18 liters/day are reabsorbed after passing through the lymphatic capillaries and back into the venous loops, leaving a net flow of ~2 liters/day to pass onward through the lymphatic system.

The maximum vasculoid water transport rate has been scaled to 0.60 cm^3/sec or ~52 liters/day, which is the most that the water tanker subsystem can handle. Under normal conditions only ~4 per cent of the water tanker fleet is in use, which still allows the transit of at least 0.024 cm^3/sec or ~2.1 liters/day of water which if not re-imported by the vasculoid on the venous side of the capillary bed is forced to enter the lymphatic system, maintaining normal flow and volume. This flow should be sufficient because the direct removal from the intercellular fluids of pure waterborne electrolytes, solutes, and other physiological substances by the vasculoid (through the plates) can create pericellular concentration gradients sufficient to ensure injection and removal of such substances from the vicinity of the bathed cells. This should greatly reduce the gross intercellular fluid flow rate required to maintain proper physiological conditions.

One minor additional concern might arise because lymph is moved primarily by peristaltic one-way valving. Envasculoided tissues adjacent to lymphatic capillaries may have slightly higher mechanical stiffness, hence may transmit somewhat less peristaltic action resulting in reduced lymphatic flow rates. But lymph fluids transport solutes at far below maximum concentration—for example, lymph normally contains ~0.003 gm/cm^3 NaCl, two orders of magnitude less than the ~0.36 gm/cm^3 maximum solubility at 310 K– and bacterial entry will be largely precluded by the appliance, so small reductions in lymph flow rates are probably tolerable. Lymphatic venules do exhibit small-amplitude pulsations on their own, and it may be desirable for other reasons to simulate the stiffness and motion (including motion resulting from blood flow) of pre-vasculoid tissue. Lymph movement also may be replaced by an optional "lymphovasculoid" appliance.

Similar considerations and conclusions apply to other fluid spaces within the body that communicate, directly or indirectly, with the blood, including the choroid plexus and cerebrospinal fluid, as well as the pericardial, pleural, peritoneal, synovial, and intraocular fluid spaces.

Immune System Interactions

CVD (chemical vapor deposition) diamond coatings are said to have "low immunoreactivity" and there were no reports of diamond immunogenicity in the medical literature as of 2001. Indeed, diamond is often used as an experimental control because it is so chemically inert and biologically inactive. Pure sapphire also appears fairly nonimmunogenic, although similar hydrophilic surfaces do adsorb immunoglobulin IgG; single-crystal sapphire has excellent biological inertness and chemical stability. However, both diamond nanoparticles and various soluble aluminum salts (e.g., alum, aluminum hydroxide, and aluminum phosphate) have been shown to serve as adjuvants which enhance vaccines or immune system responses to foreign antigens.

Exposed to water, the polished single-crystal a-alumina (0001) surface elicits a hydration reaction, with a water vapor pressure of ~1 torr sufficient to fully hydroxylate the surface. Alumina is corrosion-resistant because it exists in the highest oxidation state that aluminum metal can possess, and has the potential for microstructural control of the interface (with tissue) without formation of toxic corrosion products. Yet it is also known that a-alumina is very slightly soluble in highly acidic or alkaline aqueous environments. Since Al^{+++} ions can produce "dialysis dementia" and are generally considered toxic, it is of interest to determine whether or not these ions can leach into the body from alumina implants or sapphire nanorobots. Early studies in the 1970s found no movement of known contaminants into the surrounding tissue from sintered alumina implants inserted into the iliac crests (hip bones) and mandibles of rabbits. During the 1980s and 1990s, small increases in blood aluminum concentrations were demonstrated in smelter workers, though the potential exposure level is several orders of magnitude greater for body uptake of more soluble aluminum compounds used as food additives, as antacid medication, or from food packaging materials and cooking utensils. In 1990, Lewandowska-Szumiel and Komender investigated aluminum release from an alumina bioceramic during standardized biocompatibility testing in an animal experiment. Alumina implants introduced into rat femurs and guinea-pig mandibles and then removed 6-8 months later were found to be well tolerated, and no changes in the surfaces of the removed implants were observed under SEM examination. The researchers decided to compare the aluminum content of the femurs of experimental and control rats using atomic absorption spectroscopy, and discovered that the level of aluminum was higher in the bones of the experimental animals. In 1991, Arvidson et al investigated the corrosion resistance of single-crystal sapphire implants with respect to the release of aluminum ions, and found no ions in the test solutions. The next year, Christel reported that alumina exhibited greater bioinertness than all other implant materials currently available for joint replacement, and that no lymphocyte or plasma-cell infiltration into joint implants is observed "because of the absence of soluble component release." However, two studies in the early 1990s found some detectable aluminum ion release, so more research is clearly required on this issue. In any case, a thin veneer of diamond on sapphire should suffice to prevent aluminum ion release in vivo.

An allergic reaction or "hypersensitivity" is an acquired and abnormal immune system response to a substance, called an allergen, that normally does not cause a reaction. An allergy requires an initial exposure to an allergen which produces sensitization to it subsequent contact with the allergen then results in a broad range of inflammatory responses. Sapphire or alumina ceramic is considered nonallergenic—ceramic coatings are used to eliminate metal allergies on implant surfaces and hypersensitivity to oral ceramic is reported only rarely. There are no reports of allergenicity for diamond, sapphire, fullerenes, or other probable diamondoid nanorobot exterior materials and such allergenicity appears unlikely, but experiments should be done to positively confirm this expectation. The immune system could also react to small subcomponents like cilia, especially if they form aggregates with proteins, possibly requiring clonal deletion or tolerization to deal with such issues.

Inflammation

Could the vasculoid surface in contact with the vascular endothelium trigger general inflammation in the human body? One early experiment to determine the inflammatory effects of various implant substances placed subdermally into rat paws found that an injection of 2-10 mg/cm^3 (10-20 micron particles at 10^5-10^6 particles/cm^3) of natural diamond powder suspension caused a slight increase in volume of the treated paw relative to the control paw. However, the edematous effect subsided after 30-60 minutes at both concentrations of injected diamond powder employed. Another experiment at the same laboratory found that intraarticulate injection of diamond powder was not phlogistic (i.e., no erythematous or edematous changes) in rabbit bone joints and produced no inflammation. Diamond particles are traditionally regarded as biologically inert and noninflammatory for neutrophils and are typically used as experimental null controls. CVD diamond and DLC diamond surfaces elicit minimal or no inflammatory response, and atomically smooth diamond may perform even better. Diamond particles are said to have little or no surface charge but unmodified graphene surfaces readily acquire negative charges in aqueous suspension, so experiments are needed to determine if negatively charged fullerenes or other diamondoid substances can contact-activate Hageman factor or kallikrein and trigger an inflammation reaction.

Experiments with sapphire have generally found no serious inflammation in soft tissues or bony tissues, or only mild reactions, though there are a few modest exceptions including a brief acute inflammatory response.

Thrombogenesis

Blood coagulation involves a complex series of reactions in which various proteins are enzymatically activated in a sequential manner, transforming liquid blood into a gel-like clot which is then stabilized to form a thrombus (clot) consisting of platelets, fibrin, and red cells. The series of reactions is classically divided into two pathways—extrinsic and intrinsic—involving more than a dozen factors that converge on a single common final pathway, resulting in clot formation. Since, these factors are carried by the blood, the vasculoid can regulate their local concentration and thus, prevent thrombogenic pathways from proceeding to completion.

Additionally, platelets must undergo adhesion and activation for coagulation to occur. The adhesion of platelets to exposed collagen in injured blood vessels is mediated by a bridging molecule called von Willebrand's factor that is secreted by endothelial cells into plasma, which prevents platelets from detaching under the high shearing stresses developed near vessel walls. The activation of normally quiescent platelets is a complex phenomenon that includes changes in cell shape, increased movement, release of the contents of their granules (containing nucleotidyl phosphates, serotonin, various factors, enzymes and plasma proteins), and aggregation. The most potent activator of platelets in vivo is thrombin, which interacts with a receptor on the platelet plasma membrane, followed by transmembrane signaling and subsequent activation of the cell. Collagen is the other most important platelet activator; ADP can stimulate aggregation but not granule release. In principle, the blood-contacting surfaces of a nanoorgan, or of nanorobots in sufficient bloodstream numbers and concentrations, could activate platelets (and thus either of the two coagulation pathways), but careful choices of materials and of allowable mechanical motions should reduce or eliminate inherent nanodevice thrombogenicity. For example, DLC diamond-coated stents, heart valves and other blood-contacting LVAD surfaces or substrates generally show reduced thrombogenicity and no platelet activation. Sapphire (alumina ceramic) has low thrombogenicity and both platelet adhesion and activation are low. Hemolysis is near-zero for diamond and alumina powders. Additionally, in the vasculoid plateletogenesis and release of platelets could be actively regulated and reduced to minimal levels.

Future experiments must determine if ordinary diamondoid surfaces will have to be supplemented with additional antithrombogenic coatings in order to achieve vasculoid performance objectives. If such coatings are required, one simple possibility is surface-immobilized heparin, a ~15 kD straight-chain anionic (acidic) mucopolysaccharide (glycosaminoglycan) that forms polymers of various lengths. Heparin, first discovered in 1916, is produced naturally by human liver mast cells and basophil leukocytes, and inhibits coagulation primarily by accelerating the interaction between antithrombin and thrombin. Nanorobot exteriors can be "heparinized", and thereby rendered thromboresistant by immobilized heparin on all blood-contacting surfaces at ~monolayer surface concentration (e.g., 7-10 pmol/cm^2). Cellulose membranes coated with 3.6 pmol/cm^2 of endothelial-cell-surface heparan sulphate show complete inhibition of platelet adhesion. If satisfactory passive nonthrombogenic surfaces cannot be found, nanorobots might employ any of several active strategies to prevent iatrogenic coagulation.

Regulation of Angiogenesis and Vasculogenesis

During development and after physical injury, new blood vessels may originate from pre-existing blood vessels by angiogenesis or from endothelial cell precursors (angioblasts) by a process called vasculogenesis; both processes are mediated by paracrine growth factors. In certain circumstances, such as wound healing, post-ovulation capsule repair, and exercise-related capillary formation to support development of new skeletal muscle, angiogenesis is critical in the adult human body and its lack has been associated with chronic renal failure and other pathological conditions. Without the ability to incorporate vasculoid plates into new capillary vessels, masses of new non-envasculoided tissue would accumulate around the original (envasculoided) tissues over time, reducing the effectiveness of the appliance. In principle, vasculoid can support angiogenesis and vasculogenesis by extending itself into the newly-formed space. Watertight multicomponent metamorphic surfaces have been described by Freitas), but special techniques will be required to extend a new vascular branch while maintaining continuous watertightness. An angiogenesis event can be detected by monitoring concentrations of angiopoietin-2, VEGF, and other relevant factors. Conversely, detection of tumor angiogenesis factors could induce a vasculoid-mediated reduction in oxygen supply, increased local delivery of antibodies, refusal to deliver angiogenesis factors, and optional intervention such as delivery of chemotherapeutic drugs or external signaling to call for targeted therapeutic intervention.

It is envisioned that self-repair capabilities via active replacement of damaged components (from onboard inventories of spare parts including spare plates will be an important feature of the complete vasculoid system design. These capabilities may be extended to support angiogenesis. When angiogenesis is not desired (for example, near tumors), local concentrations of angiogenic factors can be regulated by the vasculoid appliance to minimize vessel formation, and the vasculoid can refuse to deliver nutrients to the undesired tissues even if ersatz capillaries form. Attention must also be paid to the removal and disposal (or temporary onboard warehousing) of damaged plates. In rare cases, traumatic events may damage plates (rather than simply causing them to separate temporarily), and radiation damage may cause rare failures. A capillary damaged beyond repair (e.g., in a severe contusion) may be abandoned and dismantled by the body; in such a case the vasculoid should cease delivery to that region and remove the plates.

Damaged endothelial cells can be detected based on sensory data indicating localizing chemical imbalances, and new stem cells or endothelial precursor cells can be transported to the vicinity of the damaged site to help restore the natural vascular integrity. Such restoration in some cases may require repositioning of a few plates using the motive cilia.

Vascular Patency and Relaxation

One important function of fluid flow in blood vessels is to ensure vascular patency. But replacing bulk fluid with a vasculoid appliance is unlikely to cause the smaller vessels to collapse because the buckling force is increased by up to several orders of magnitude by the addition of the plates in the capillaries and related small vessels, even in the absence of bulk fluid. The largest vessels such as the aorta will lose significant luminal support (even if filled with gas), but—in addition to the external muscular and elastic support from the surrounding tunica media and tunica adventitia layers—such vessels also possess an extensive capillary vasculature which, greatly stiffened and at typical capillary densities, should provide adequate replacement support. Intravascular (cross-luminal) spring-loaded diamondoid scaffolding could also be added at need.

Arterioles in the natural vascular system perform extensive regulatory functions. These and related functions (including vasoconstriction and vasodilation) can be controlled, in turn, by the vasculoid (e.g., by monitoring local NO concentrations). The underlying vascular musculature will probably be maintained in the relaxed state if this maximizes tissue stability and simplicity of control, although further research of this question is warranted.

Unit V–K

Control Systems and Computational Requirements and Hypothetical Vasculoid Installation Scenarios

Numerous local, intermediate-scale, and global control systems and protocols will be required to ensure the proper autonomic operation of the vasculoid appliance. Special communications subsystems for local plate configuration maintenance, cargo traffic control, and fast systemic response to large-scale external stimuli must be designed. An interplate packet-switching network for long-range communications, and to support real-time autogenous user control, should be considered. System control might be simplified using local computing centers throughout tissues, directing immune, angiogenetic, and trophic functions. Besides the interplate network, trans-tissue sonic or optical communication in principle could be employed between spatially close but topologically distant portions of the vasculoid network. Precise knowledge of the functions of various cytokines and other substances will enable the vasculoid to respond appropriately to sensor data reporting local extravascular concentrations of signaling molecules.

Complexity of stimulation patterns must mirror physiological conditions as much as possible. Highly responsive and user-friendly graphic user interfaces may be useful to allow the patient to communicate quickly and easily with his or her appliance; several aspects of such interfaces have been described elsewhere, but considerably more work remains to be done. The computational requirements of directed container switching/routing by the ciliary system should be investigated further.

A great number of software and computational architectural issues are as yet unresolved, including a detailed molecular routing logic for each substance to be transported, comprehensive details of nanorobot control protocols, container traffic flow patterns (e.g. flow lensing, lane definition, congestion waves and gridlock, flow viscosity, etc.), and software requirements including software complexity for large-scale control issues, a discussion of which are beyond the scope of this paper. (For example, it might be interesting to apply existing vehicular traffic simulations to transport in fractal structures.) Only since the mid-1990s have multirobot control issues begun to be seriously addressed by the broader research and technical community. Computation for vasculoid accident recovery and repair will be considerably more intensive than computation for maintenance or installation, and must also be deferred to another paper.

However, a few basic observations may be made here regarding vasculoid materials transport computation requirements. Concentrations of water and glucose are normally maintained by the appliance near typical serum levels, using simple feedback loops driven by concentration sensors located on the tissue side of the vasculoid device. Oxygen and carbon dioxide are similarly controlled using sensors capable of measuring partial pressures, though the precise triggering thresholds may be modified in special circumstances (e.g. high altitudes, deep sea diving, hyperbaric chambers). A simple rate-control mechanism driven by sensor data compiled once every second, employing ~1000 distinguishable concentration levels requires ~10 bits/sec per molecular type, or ~40 bits/sec for all four molecules. Most of the time, entire rotor banks are engaged using a single command. Ciliary motions, largely stereotypical, likely require a similarly small computational budget.

Establishing or maintaining appropriate concentrations of mixed-cargo molecules is far more computationally complex. These other molecules include minerals, vitamins, lipids and waste products, but the most numerous and difficult to coordinate are the blood proteins and sterols including control molecules such as insulin, leptin, gastrin, resistin, estrogen, or testosterone; synaptic transmitters and other neural biochemicals such as acetylcholine, adrenalin, dopamine, nitric oxide, or serotonin; cytokines and other signaling molecules such as follicle stimulating hormone, epidermal growth factor, or tumor necrosis factor; and many thousands of specific antibodies that reflect the body's unique learned recognition of specific antigens. (Embedded analytical units scattered throughout the vasculoid surface can test for rising concentrations of hitherto unrecognized nonmetabolizable molecules, and configure a set of programmable binding sites to rotor these molecules out of the tissues as well, though this should be a relatively rare occurrence.) Since the human genome contains ~40,000 genes, most of which may encode several proteins each, we assume there may be on the order of ~100,000 distinct biomolecules that must be recognized, monitored, transported, and regulated at the docking bays. This is likely a generous estimate, given that most of the proteins produced remain intracellular, but our conclusions do not depend sensitively upon the precise figure chosen.

Upon receipt of any tanker, each docking bay scans the tanker manifest before offloading the contents. In the case of gas, water, and glucose tankers, the manifest is extremely brief, since only one or two molecules are involved. This content information can be encoded in as little as 15 bits. A precise molecule count of up to 3×10^{10} molecules per tanker requires an additional 35 bits, but in most cases a precise molecule will not be necessary and a range figure requiring fewer bits should suffice. At the maximum rate of one tanker unloading every 20 seconds, the manifest-reading bit rate is at most ~2.5 bits/sec.

In the case of mixed-cargo tankers, 17 bits uniquely specify the identity of each of 100,000 distinct molecules; each tanker can hold a total mixed-cargo molecule count of ~109, requiring up to 30 bits per item to ensure an exact count. Based on existing human physiology, we posit a specification that each vascular cell must have access to at least 5 per cent of all ~100,000 molecular types during one blood circulation time, ~60 sec. Since each cell is supplied by two mixed-cargo docking bays, then restricting tankers to a maximum of 834 different molecules per load fulfills our specification, requiring in the worst case a 39,198 bit manifest to be read in 20 seconds, a maximum bit rate of ~2,000 bits/sec. Again assuming ~10 bits/sec per molecular type for rate-control of rotor mechanisms, rotor control will require a maximum ~8,340 bits/sec. (One design alternative is to allow stochastic mixing and transport of the low-concentration molecules which would mimic the natural process, reduce the computational needs, and impose only a very small burden on the transport capacity.)

In sum, docking bay process control appears to require ~10,000 bits/sec of computation. Cellulock control requires reading manifests of up to ~1000 cells of at most ~1000 different types, which is ~20 bits/cell type or ~20,000 bits/manifest. Cellulocks unloading one boxcar every 60 seconds will require a bit rate of ~333 bits/sec. Other control tasks will add only fractionally to the ~10,000 ops/sec budget previously estimated for individual docking bays, a capacity that can be used for other purposes between docking events.

Overall System Reliability

A detailed assessment of vasculoid system reliability also lies beyond the scope of this paper, especially given the possibility for prompt and severe damage if the robot is subjected to extreme accelerative loads or unplanned segmentations. However, if we assume normal gravitational loading and ignore catastrophic accidents, we find that the vasculoid promises an annual survival probability well in excess of "six nines" (failure probability $< 10^{-6}$) at least against radiation damage if all major subsystems incorporate tenfold redundancy.

Adopting Drexler's radiation damage model for the first part of this analysis and applying it to a system comprised of N components, with each component comprised of n redundant parts, the probability P that the system remains operational after T years is approximated by: $P = \exp[-N(1-p)n]$, where p = the probability that an individual part remains operational; $p \sim \exp(-10^{15} Dm)$, where m = mass of the part in kg and D = radiation dose in rads ~ 0.5 T for normal background radiation in the terrestrial environment. As a further simplifying assumption for this analysis, we shall presume that all vasculoid modular "parts" are similar in mass to the ~4,000,000-atom robotic manipulator device, after which the vasculoid cilium is patterned. From the above formulae, p ~ 0.999960 for each such cilium-like part, per year, even with no internal redundancy within the part.

The vasculoid ciliary subsystem includes a minimum requirement of N = 300 trillion cilia incorporating a redundancy of n = 10, giving a total of 3000 trillion cilia. From the above, the annual probability of failure of the ciliary subsystem is $< 10^{-29}$; assuming a redundancy of only n = 5, failure probability is still $< 10^{-7}$.

Each vasculoid plate is ~2 micron^3 in volume. Given that there are ~150 trillion plates, with zero redundancy among them, and then to be assured of a complete plate subsystem annual failure rate $< 10^{-6}$, the annual probability of failure per plate must be reduced to $\sim 10^{-20}$. This is achieved for plates comprised of a total of ~88,000 "parts" as described above, with a redundancy of ~6 among such parts. However, in this study we have specified a more generous (i.e., safer) redundancy of 10 among such parts.

Each vasculocyte repair robot is $\sim 10^{-14}$ kg in mass. Given that there are ~200 billion active vasculocytes, and assuming zero redundancy among them, then to be assured of a vasculocyte subsystem annual failure rate $< 10^{-6}$, the annual probability of failure per robot must be reduced to $\sim 10^{-17}$. This is achieved for robots comprised of a total of ~125,000 "parts" as described above, with a redundancy of ~5 among such parts; we have again assumed a more generous redundancy of 10 among such parts.

Non-radiation failure mechanisms have not been examined in the above analysis.

Software failure modes have been ignored, and even plate computers may be complex enough to host intentionally disruptive programs such as computer viruses. Many other more subtle or second-order difficulties also have been ignored here, such as the potentially destabilizing effects of dynamic oscillations and resonances among the various moving components which would require a detailed design before undertaking a comprehensive analysis. For instance, we should try to avoid Per Bak's notion of self-organized criticality when complex systems are pushed too close to their design limits, as in car traffic simulations. Numerous specific failure modes have not yet been exhaustively analyzed, including complete single-plate failures and the seizing or locking of individual cilia while a transfer operation is in progress, e.g., near a capillary entrance where such failures could prove most troublesome. However, the vasculocyte fleet has been scaled to accommodate a reasonable repair mission requirement.

Vasculoid tankers are micron-size compressed-gas containers which may rupture explosively—though this event is extremely improbable, since tankers with rupture strength exceeding 40,000 atm are loaded to only 1000 atm pressure. If this unlikely rupture event occurs in a large vessel such as the aorta, the possible failure modes may resemble those involved in a detachment avalanche, but the low tanker mass should preclude any serious direct infrastructural damage. However, if the explosive event occurs within a smaller vessel such as a capillary, the risk of breach is considerably greater. However, the maximum overpressure that can be explosively applied to the appliance walls is limited to the ~1000 atm compression pressure of the tanker contents, which likely does not exceed the theoretical rupture strength of the vasculoid walls in capillaries which may be crudely estimated as pmax ~ 3300-33,000 atm, taking vasculoid plate tube thickness twall = 1 micron, capillary tube radius R ~ 3 microns and working stress sw ~ 1010 N/m^2 for solid tube walls or maximum interbumper working stress sw ~ 109 N/m^2 for more realistic plated walls jointed with bumper interconnects. Whether such explosive events will occur with sufficient frequency to warrant explicit corrective protocols (beyond conventional vasculoid self-repair activities) deserves further study. A shockwave cascade failure (such as recently occurred at the Super Kamiokande neutrino detector facility, wherein the initial implosion of a single photomultiplier tube (PMT) during refilling of the water tank triggered a runaway cascade, destroying 6,665 of the 11,146 PMTs in a few seconds seems unlikely in the vasculoid but also should be investigated theoretically.

In practice, one should expect that a few contaminant molecules will make their way into the vasculoid, perhaps from local strain that exceeds the adjustment capabilities of the metamorphic bumpers, accidental spills, and so forth. A molecule that is small enough to be mobile at room temperature is small enough to be captured by a sorting rotor.

Larger molecules can be physically handled by vasculocytes. A specialized class of vasculocyte, also measuring several microns in size and moving at ~1 cm/sec, can employ a 0.25-mm-wide chemotactic detector bank on each mobile nanorobot to search for, bind, and remove large molecules—a complete sweep of the entire vasculoid luminal surface once a minute requires the activity of only 1 per cent of the total active vasculocyte fleet. The motion of the tankers creates a net motion of the nitrogen atmosphere that can sweep molecules toward centrally placed filters. Thus a combination of internal sorting rotors, filters, and vasculocyte activity can quickly extract contaminants from the vasculoid volume. Once contaminants are trapped, they may be dealt with in ways suggested previously. The volumetric capacity of the appliance to deal with internal contamination must be scaled according to anticipated applications and event scenarios.

Hypothetical Vasculoid Installation Scenarios

Installation of the vasculoid involves, at the least, complete exsanguination of a sedated patient and an intricate vascular plating operation. Two hypothetical installation scenarios are presented. The first scenario is a detailed description of a procedure that would be feasible using the technology available currently. This is to demonstrate that vasculoid installation can in principle be carried out without violating any well-established medical or physical limits. However, the authors are aware that by the time a vasculoid-class device can be built, medical technology will have advanced significantly. We therefore also briefly sketch out a highly speculative second scenario which, if practicable in some future era, might be considerably more convenient and up to 100 times faster. This second procedure would be unduly aggressive by today's standards but might be feasible and safe, given the supporting technology available in a nanotechnology-rich medical environment.

The principal installation scenario involves complete exsanguination of a sedated and hypothermic patient, replacement of the natural circulatory fluid with various installation fluids, followed by mechanical vascular plating, defluidization, and finally activation of the vasculoid and rewarming of the patient. Installation takes ~6.5 hours from start to finish and requires a peak ~200-watt power draw midway through the procedure. (By comparison, present-day kidney dialysis treatments require 4-12 hours and the equipment also draws a few hundred watts.)

Our discussion first considers the lifespan of cells temporarily denied access to external molecular transport mechanisms. We next describe the details of vasculoid installation including patient preparation (~24 hours), vascular washout (~4 hours), vascular plating (~1 hour), defluidization (~0.3 hour), and initialization and cold start (~1.2 hours). We conclude with a brief discussion of vasculoid removal.

The hypothetical installation protocol described below has been selected for maximum comfort, reversibility, and reliability. The correct performance of the system is verified at every step, providing a plateau of safe operation before moving on to the next step. Since the entire procedure is intended to be fully reversible at each step, any step which fails or does not proceed in a manner acceptable to the installing physician may be quickly abandoned with a retreat to the previous plateau of established safety, after which a decision may be made to try again or to abort the installation.

Patients must be made aware that they are about to undergo a major medical procedure which involves replacing ~8 per cent of their body mass with complex nanomachinery. They must be psychologically prepared to deal with the personal implications of this. From the turn-of-the-century perspective, vasculoid installation appears massively intrusive especially when compared to other superficially related procedures with which we are commonly familiar such as intubation, kidney dialysis, blood transfusion and blood replacement therapies, installation of pacemakers or artificial organs, and coronary angioplasty or fiberoptic endoscopy. However, in a future era when nanomedical systems are widely employed and generally accepted as standard treatment, vasculoid installation may be regarded with considerably less trepidation than it would today.

Cellular Ischemispecific Limits

In classical medicine, "ischemia" refers to a local inadequate blood supply which is usually caused by a mechanical arterial obstruction (e.g. clot), a spasm wherein a blood vessel pinches shut, arterial narrowing (e.g. arteriosclerosis), or cessation of cardiac activity. The principal clinical outcome of local ischemia is a shortage of oxygen. Because of the high continuous biological power density of nerve cells, neurons and the brain succumb most rapidly to oxygen starvation. The precise survival limit of human brain tissue under hypoxic conditions (after which time some decline begins to occur) depends on many factors. Traditionally this has been reported as ~4 minutes (240 sec), although more recently it has been shown that 10 minutes of warm ischemia (even followed by another 10 minutes of trickle-flow CPR) is reversible without neurological deficit if followed by mild hypothermia when normal blood flow is restarted, and even modest advances in technology will likely extend this still further. Much longer periods of cold ischemia are tolerable—for example, survival with complete functional and histologic cerebral recovery has been achieved with brain temperature at 5-10°C during 1 hour of circulatory arrest, and stable spontaneous circulation has restored after water ice submersion of up to 90 minutes.

In nanomedicine, where it is possible to achieve precise control of human physiology at the cellular and molecular levels, ischemia may refer to an interruption in the extracellular mass transport of any vital molecular species within the human body. Failure to adequately import a particular nutrient, or to adequately export a specific harmful waste product, or to properly regulate the transport of a particular signaling molecule, may produce some of the classical symptoms of cellular ischemia. Since the entire bloodflow must be interrupted during the vasculoid installation process (producing whole-body ischemia), it is useful to estimate how long tissue cells can maintain normal metabolism and avoid toxemia after cellular access to the bloodflow is denied. This time period—the ischemispecific limit—varies with each important cytometabolic molecule, as summarized below. (The limit also varies by cell type; a more detailed study would likely reveal specific tissues where ischemic sensitivity is higher than average for a given metabolite.)

Oxygen—The average (20 micron)^3 tissue cell consumes ~107 molecules/sec of O_2 at the basal (resting) rate of ~30 picowatts. The cytosol of such a cell can dissolve up to ~6 × 10^8 O_2 molecules at 310 K. If suddenly cut off from all external supply, the average cell has only enough O_2 in inventory to survive ~60 sec at the basal metabolic rate.

Glucose—Assuming ~50 per cent energy conversion efficiency and ~10^{-3} gm/cm^3 glucose in the cytosol, then ~3 × 10^{10} glucose molecules are available in the average tissue cell. For a power demand of 30 picowatts, ~4800 zJ/glucose molecule, and ~50 per cent energy conversion efficiency, then the basal glucose consumption rate is 1.25 × 10^7 glucose molecules/sec and the basal cellular ischemispecific limit for glucose is ~2400 sec.

Carbon Dioxide—The metabolic "combustion" of one molecule of glucose with six molecules of O_2 produces six molecules of CO_2, hence the average tissue cell produces ~107 molecules/sec of CO_2 at the basal rate. The cytosol of an average cell should normally dissolve up to ~1010 molecules of CO_2 at 310 K (assuming working tissue PCO_2 = 54 mmHg and applying the Henry's law constant for CO_2, an estimate which compares favorably with the measured ~5 millimoles/kg intracellular CO_2 concentration (2.4 × 10^{10} molecules/cell assuming an 8000 micron^3 cell) in various frog tissue cells at 302 K. Hence the cellular ischemispecific limit for CO_2 is ~1000 sec, consistent with maximum physiological cellular and plasma pH levels. It may be possible to dissolve more CO_2 but this will lower pH, leading to toxic acidosis of the cell. (Acidosis has two major components, CO_2 and lactic acid. The absence of oxygen blocks electron transport and stalls the Krebs cycle, halting CO_2 production, so glucose is shunted to lactic acid in anoxia.) Additional amounts of CO_2 may be stored in combination with plasma proteins as carbamate.

Lactate—Under anaerobic conditions, lactic acid is produced during the metabolism of glucose (glycolysis) in most cells, glycogenolysis in muscle and glucolysis; two molecules of lactic acid are produced per molecule of glucose metabolized. Thus at the basal rate the average anaerobic tissue cell can produce at most 2.5 × 10^7 lactate molecules/sec—though this is an overestimate by perhaps a factor of 2 because in most cells the pyruvate (the immediate precursor of lactic acid) formed at the end of glycolysis enters the TCA cycle and is further oxidized by mitochondria. The maximum normal blood concentration of lactate equates to ~1.1×10^{10} molecules/cell, potassium channel openings are induced by 2-20 nM lactate ($0.96\text{-}9.6 \times 10^{10}$ molecules/cell) applied to the cytosol of rabbit ventricular myocytes, and the brain lactate threshold for cerebral ischemic damage is 17 mmol/gm or ~8.1×10^{10} molecules/cell, so the ischemispecific range for lactate is ~380-3800 sec.

Nitrogenous Waste Products—Urea is the principal mammalian waste product due to protein, purine, and pyrimidine nitrogen metabolism, constituting ~85 per cent of human nitrogen excretion. However, urea is formed in the human liver through reactions of the Krebs ornithine cycle, a process not available to tissue cells denied access to the circulation. Ammonia is the chief byproduct of protein and amino acid metabolism in the cell, with one molecule of ammonia produced per molecule of amino acid broken down. The daily RDA for protein represents an upper limit on the breakdown rate of ~5×10^5 amino acid molecules per cell-sec, assuming mean amino acid MW ~ 100 Da, implying a maximum ammonia generation rate of ~5×10^5 molecules/cell-sec. Given the maximum observed concentration of ammonia of ~6×10^8 NH3 molecules/cell in whole blood cells, the ischemispecific limit for nitrogenous wastes is ~1200 sec.

Acetate—Acetic acid (MW = 60 Da) is the simplest possible fatty acid having an even number of carbon atoms. It is most notably produced during the breakdown of acetaldehyde (the second step in the metabolism of alcohol), bacterial fermentation in the gut, and in other circumstances. The typical human intracellular production of acetic acid has been roughly estimated from rat studies as ~8×10^6 molecules/cell-sec. This chemical is highly metabolizable in vivo, but assuming a maximum nontoxic limit of ~10^{-4} gm/cm^3 (vs. ~4×10^{-3} gm/cm^3 for all esterified fatty acids in cells) or ~10^{10} molecules/cell, then the ischemispecific limit for acetic acid is ~1300 sec.

Ketones—Absent circulatory removal and in cases of low glucose levels, the metabolism of fats may produce abnormal amounts of toxic ketones including primarily beta-hydroxybutyric and acetoacetic acids (and their decarboxylation product acetone) which may be present up to ~6×10^8 molecules/cell assuming MW ~ 86 Da for ketones. Given the RDA for lipids of ~7×10^{-7} kg/sec and since one palmitic acid (typical lipid) molecule (MW = 256 Da) would convert to ~3 molecules of acetoacetic acid (MW = 86 Da), the maximum natural ketone generation rate is ~5×10^5 molecules/cell-sec and the minimum ischemispecific limit for ketones is >1200 sec.

Summary of Cellular Ischemispecific Limits—If the average human tissue cell is denied access to all extracellular molecular transport systems, the major byproducts of normal cellular metabolism may build to near-toxic levels from a near-zero initial concentration in ~10^3 sec. Glucose reserves last >10^3 sec, but oxygen runs short in ~10^2 sec without external resupply. These theoretical limits appear crudely consistent with the concept of the "Golden Hour" in traditional trauma care. For instance, ischemia induced by aortic cross-clamping in dogs has been survived for up to 20-60 minutes without inducing paraplegia (though decline of spinal cord electrical function is detected in the ~20-30 minute range for dogs and rabbits). Cardiac ischemia induced by aortic cross-clamping in dogs is survived for 30 minutes during normothermic (37°C) cardioplegia, 45 minutes during mild hypothermic (28-30°C) cardioplegia, and 90 minutes using potassium verapamil during profound hypothermic cardioplegia at 8-10°C. The duration of cold ischemic tolerance is enhanced by the administration of excess insulin and other pretreatments, possibly involving gene products that will be well-known in a future nanomedical era in which vasculoid installation is commonplace. (Interestingly, wood frogs can endure freezing for >2 weeks with no breathing, no heart beat or blood circulation, and with up to 65 per cent of their total body water as ice)

Patient Preparation (~24 hours)

Before the vasculoid may be installed, the patient must be prepared as follows:

1. **Vascular Conditioning**—The day before the installation, the patient receives an injection containing a treatment dosage (~1 cm^3) of vascular repair nanorobots or ~70 billion individual devices. These 8-picogram, 7 $micron^3$ mobile legged artery-walking nanodevices clean out all fatty streaks, plaque deposits, complex atherosclerotic lesions, infections, vascular wall tumors, and parasites, and repair all other vascular lesions as required in less than 24 hours, in a multistep process described elsewhere in detail. As one additional task, these nanorobots inject each of the ~1012 endothelial cells lining the human vascular tree (a) with cytokine blockers to at least partially inhibit the cells' natural surface pressure and shear force responses, and (b) with adhesion-inducing glycoproteins to forestall cell migration during the installation procedure by encouraging firm anchoring. All vascular repair devices are then exfused before vasculoid installation begins; the results of their vascular reconnoiter protocol may be downloaded to an external computer and used to prepare a detailed map of the patient's vascular tree to improve efficiency during plating and plate initialization. The patient may be started on a diuretic (e.g. furosemide) to reduce blood volume a few hours before the installation procedure begins.
2. **Sedation**— On the day of the operation, the patient arrives at the installation facility and is administered a preoperative sedative such as sodium pentobarbital an hour before the procedure is to begin, to encourage drowsiness.
3. **Cannulation**— A standard closed-chest cardiopulmonary bypass (aka. heart-lung machine support) or CPB is employed. Usually this involves a combination of femoral and thoracic vessel cannulation, but in this case a fem-fem bypass (cannulation of the vena cava and aorta via the femoral vein and artery) might suffice. This procedure, by definition, continuously supplies the equivalent of resting cardiac output through the femoral vessels and the blood pump/oxygenator circuit—typically a 50-70 cm^3/sec flow rate during hypothermic CPB, allowing the entire human blood volume can be exchanged about once every few minutes. Fresh oxygenated blood flows retrograde up the femoral artery into the aorta, and oxygen-depleted blood flows retrograde down the vena cava to return to the heart lung machine via the femoral vein. The leg inferior to the cannulation point is supplied by collateral vessels. At this point in the procedure, the patient's own blood continues to circulate through the body. (Note that the above "manual cannulation" scenario is included for illustrative purposes only. In an era when advanced medical nanotechnology is available, self-directing nanocannulators will make it easy to quickly and safely establish flow-regulated channels into any desired artery or vein, up to and including the aorta and vena cava, so much higher flow rates than the conservative figures used here are theoretically available if required.)
4. **Heparinization**—Heparin and streptokinase are injected (a) to prevent clotting and promote lysis of hemostatic fibrin, (b) to break up axial red cell rouleaux that form spontaneously at low blood flow shear rates, and (c) to help ensure patency of the catheters throughout the procedure. The conventional cocktail of drugs given to patients placed on bypass leading up to a cold circulatory arrest procedure is rather complicated and will not be elaborated further here, but there are established procedures in conventional medical practice for cooling patients and diluting their blood for work at very cold temperatures.

5. **Cytoactivity Restraint**—The installing physician next administers: (a) an erythropoietin antagonist to maximally suppress erythrocyte production; (b) agents to temporarily suppress all leukocyte production, reversibly depress leukocyte intracellular metabolism and cytokine sensitivity, and briefly reduce or block antigen reactivity; (c) agents to temporarily suppress platelet production, reversibly depress platelet intracellular metabolism and cytokine sensitivity, and temporarily toughen the platelet cytomembrane to reduce the likelihood of shear damage; (d) minute traces of normal bloodstream hormones, cytokines, and other control biochemicals designed to minimize production of natural secretions such as insulin, adrenalin and testosterone, glucose and cholesterol, sebum and semen, etc., as well as angiogenesis inhibitors to halt the development of new capillaries; and (e) broad-spectrum antibiotics or programmable nanobiotics such as microbivores designed to prevent microbial attack during the brief period of immunological vulnerability. For example, one element of this complex and as yet incompletely specified biochemical cocktail might be dexamethasone, the most powerful anti-inflammatory and immunosuppressive adrenocortical steroid; another element might be anti-CD18, an antibody known to prevent leukocytes from sticking to blood vessel walls in the brain. (Some of these and related restraints may not be absolutely essential, since the processes in question are quite slow compared to the time course of installation.)

The patient is now ready for vascular washout.

Vascular Washout (~4 hours)

Vascular washout (to prepare the patient for vasculoid plating) may be performed as follows:

1. **General Anesthesia**—A general anesthetic such as propofol is administered by the installing physician at sufficient dosage (~90 mg) to establish a condition of surgical anesthesia.
2. **Respirocyte Infusion**—Over a period of 3 hours, the patient's entire blood volume is replaced with a suspension of fully charged respirocytes (1 micron^3 spherical O_2/CO_2 1000-atm pressure vessels) in an isotonic aqueous 0.01 M glucose solution (double the natural serum concentration) also containing most of the cytoactivity restraint substances described earlier and additionally a mixture of appropriate electrolytes and other components commonly found in hypothermal blood substitutes. A 5 per cent (by volume) respirocyte suspension, containing ~10^{11} respirocytes/cm^3, is sufficient to provide oxygen and carbon dioxide transport equivalent to the entire human red cell mass for ~104 sec (almost 3 hours) after the cessation of respiration. The respirocyte fleet generates ~17 watts of waste heat while supplying oxygen at the human basal rate, producing at most a negligible ~0.1 °F rise in core body temperature. Typically ~20 liters of infusant is required to completely clear the vessels of all natural blood components. Exfused blood cells are removed, collected and saved for emergency reperfusion, if required.

   It is true that the reduction of cerebral blood flow by 50 per cent can produce marked disturbances in brain metabolism, and at 20 per cent of normal flow the neurons can depolarize with rapid loss of intracellular potassium into extracellular spaces. But the infusant composition can be adjusted to maintain physiological electrolyte (esp. Na+, K+, and Ca++) concentrations and osmotic balances, and to extract any excitotoxins that might be released—or else respirocyte-class pharmacyte nanorobots can be employed to similar effect. Since blood cells have been removed, leukocyte plugging and other forms of neutrophil-related damage cannot occur. The basal metabolism remains sufficiently active throughout the procedure so that tissue cells suffer minimal ischemic damage. The infusant constitutes an adequate temporary replacement for the natural blood which is being removed. This is based on an established medical technology: It is well-known that the entire blood supply of a human being can be replaced with cell-free blood substitute at a low temperature, with the patient then reperfused with blood and recovered, as was first demonstrated clinically 30 years ago as a treatment modality for hepatic coma.

3. **Cooldown**—Once the entire blood volume has been completely exchanged with the respiro-infusant described in (2) in the unconscious patient on CPB (as in conventional medical settings), control of perfusate temperature will suffice to control body temperature. Core temperature follows the perfusate temperature fast and close, with any perfusate/core temperature difference decaying with a ~10 minute half-life. The heat capacity of the body is so large that the 200 watts of power dissipated by vasculoid installation over an hour or so would heat the patient by only a couple of degrees, which is tolerable. (By comparison, external cooling of a body in circulatory arrest is so inefficient that even placing the patient in an ice water bath would yield surface-to-core temperature differences decaying with a half-life of hours.) The patient's average core temperature is reduced from 310 K (37°C) to 280-290 K (7-17°C) in ~1 hour, a cooldown rate of ~0.5°C/minute. This cooldown schedule is qualitatively similar to hypothermic schedules commonplace in current resuscitation medicine and hypothermic CPB procedures. Mivacurium, rocuronium, or older agents such as vecuronium or metubine (recently withdrawn) may be administered to inhibit shivering and as a reversible nondepolarizing muscle relaxant; the effect of vecuronium bromide may be reversed by acetylcholinesterase inhibitors such as neostigmine or pyridostigmine.

   B. Wowk notes that there are three broad categories of clinical hypothermia used in medicine. First, there is mild hypothermia (a few degrees below normal body temperature). This is very commonly used during cardiopulmonary bypass for open heart surgery, and often occurs as an incidental effect of general anesthesia during other surgeries. Second, there is deep hypothermia (15-20°C). This is used when significant periods of circulatory arrest (up to 1 hour) are required for complex bloodless surgeries such as repair of cerebral aneurisms or congenital defects of the aortic arch. Hemodilution (partial blood substitution) is used during these surgeries, in part to avoid red blood cell agglutination at low temperatures. Third, there is profound (or ultraprofound) hypothermia (0-10°C). Profound hypothermia does not yet have routine clinical use but is the subject of active investigation in relevant animal models with an eye toward human use, and preliminary human clinical results recently have been reported. Total blood substitution, as proposed in the vasculoid installation protocol, is the norm for profound hypothermia. Specialized perfusates such as hypothermosol are necessary to overcome physiological problems with this temperature regime that ordinary plasma can't handle. Dogs have been held for three hours in profound hypothermia during perfusion of a blood substitute, and the record for profound hypothermic perfusion without neurological deficit is said to be ~6 hours; the record for profound hypothermic circulatory arrest (as distinct from continuous perfusion) in dogs at <5°C is ~3 hours.

4. **Cardioplegia**—After the cooldown process has begun, the heart (already in bradycardia

from the muscle relaxant) may be stopped by direct infusion of cold potassium chloride solution (reversible with atropine or digitalis) or other standard cardioplegic solution. However, if the target is ultraprofound hypothermia there is no need to stop the heart with drugs during this process because the heart will stop itself when the temperature falls below 15-20°C.

In ~4 hours, washout, cooldown and cardioplegia are completed. Cytometabolic processes continue, but, at 280 K, with reduced metabolic requirements and hence reduced oxygen demand. This potentially allows infused respirocytes to provide complete respiratory gas maintenance for up to ~105 sec (~1 day) before they would be exhausted after complete cessation of circulatory flow via the catheters, giving a significant safety margin for the next phase of the installation procedure. However, B. Wowk notes other risks of profound hypothermic circulatory arrest beyond mere hypoxia, including the failure of ion pumps (causing growing intra- and extracellular imbalances of calcium ions), the alteration of cell membrane permeability (allowing normally extracellular solutes to slowly leak into cells), and the decoupling of certain cellular metabolic processes (which does not occur during mild hypothermia). The vasculoid installation process may fit within the circulatory arrest times achievable with the standard off-the-shelf deep hypothermic (not profound hypothermic) protocols of conventional medicine. Further study is required to determine the optimum temperature for hypothermic installation.

The anesthetized patient is finally ready for intravenous deployment of vasculoid components.

Vascular Plating (~1 hour)

The original 5 per cent respirocyte suspension is replaced by a new suspension containing 1 per cent fully-charged respirocytes and 10 per cent cargo-bearing vasculocytes (plus cytoactivity inhibitors and glucose), creating a mixture whose viscosity and flow characteristics are roughly equal to normal-hematocrit human blood. At an ~11 per cent nanocrit, partial plug flow might ensue in a few of the smallest capillaries, but complete plug flow (requiring much higher pumping power and pressure than laminar flow) can be avoided.

Each 3 micron^3 vasculocyte grasps a single 2 micron^3 plate, ready for installation, thus the new vasculo-infusant contains ~20 × 109 vasculocytes/cm^3. Installing ~150 × 1012 basic plates thus demands a minimum of 7500 cm^3 of vasculo-infusant, requiring ~3800 sec or ~1 hour assuming a very gentle flow rate of only ~2 cm^3/sec.

Each vasculocyte drifts quietly in the flow until it encounters a vessel wall for the Nth time (N is an integer control variable), which activates it, causing it to attempt to release its cargo in a clear space. (Slowly increasing N during the installation process produces a crudely progressive plating pattern.) If the immediate area is already fully plated, the legged vasculocyte walks across the surface until it reaches a clear area to deposit its cargo. Corner registration of adjacent plates is verified prior to plate release, to ensure a maximally dense tiling pattern. Because of their small size, vasculocytes can enter even the narrowest capillaries and install plates by tiling motions. This process may be loosely regarded as a robotically-guided variant of fluidic self-assembly, a well-known existing commercial process. Special procedures must also be devised to handle encounters with nonendothelial materials (e.g., fibrin strands) or stray cells (e.g., adherent leukocytes) that may be blocking the endothelial surface, or denuded patches of vasculature lacking full endothelial cell coverage, although most of these defects should have been remedied during preoperative vascular conditioning.

Once its cargo plate is in place, the vasculocyte releases back into the flowing fluid, powers down, and is eventually exfused from the body. Approximately 42 billion plates/sec are deposited during this 1-hour process. We generously assume that each 1-50 picowatt vasculocyte requires ~100 sec of active operation up to peak power to find a clear space to deposit its cargo before resuming its fluidborne dormancy—though early arrivers will spend less time searching for gaps in the structure than later arrivers. Thus there are at most ~4 trillion vasculocytes active at any moment and the total power released as waste heat by the vasculo-infusant is under ~200 watts, thermal energy which is easily carried off by the cold infusant. Given that one plate is installed by one vasculocyte carrier that performs ~106 mechanical motions/sec during a ~10 sec install time, this implies ~107 motions/plate and allows an allocation of ~1800 motions per nanometer of plate perimeter during the installation of each plate, which seems sufficient.

The plates are not passive during the installation process. Besides the 20 cilia positioned atop each plate to provide tanker and boxcar mobility as part of the ciliary subsystem, plates also possess "motive cilia" to assist in both installation and repair operations. Each motive cilium is fully retractable when not in use. Complete positional, rotational and translational control during installation (and functional redundancy) requires at least one motive cilium on each of the 4 sides and at least four on the underside to establish a stable tripod while walking. The motive cilia allow limited trans-endothelial cytoambulation by each plate (using adherent tool tips) and fine control of plate/plate jostling motions.

A working cilium consumes 0.1 picowatt during continuous operation at 1 cm/sec roughly the speed of jostling motions involving motions of ~1 per cent plate width per microsecond. Even with all motive cilia operating at once, plate power is under ~1 picowatt; even with all 150 trillion jostling at once, maximum power would be under 150 watts—energy easily obtained from glucose and oxygen provided by the vasculo-infusant medium, and waste heat that is easily dissipated by thermal conduction. However, the motive ciliary power draw of the entire plate population normally should not exceed ~1 watt during installation because once properly positioned in a regular grid pattern (in <1 sec) all local plate jostling ceases and the motive cilia are stowed.

The installation process may be made more efficient if plates are programed to dynamically shift holes (in the plating pattern) upstream. This ensures that gaps are unlikely to persist in the developing vasculoid structure (except in cases of plate malfunction), and also ensures that shifted holes will arrive in a known region of the upper vascular tree where they are most convenient for vasculocytes to quickly find and fill. Out-of-register plate domains that collide during this dynamic repositioning process require a simple interaction protocol to allow them to mutually align correctly (such a protocol has not yet been devised). Once plates are in place, their surface cilia can be adjusted to provide variable drag on the circulating fluid. This can be used as a crude method of directing resources for more efficient construction.

Plates also assist the vasculocytes in system validation. Upon installation, plates briefly activate all subsystems to verify that everything is working properly. With all 18,750 sorting rotors spinning, individual plates momentarily draw up to ~2 picowatts. Malfunctioning plates jettison themselves back into the fluid flow for ultimate removal, or if this is not possible, take other action to bring their condition to the attention of the circulating vasculocyte fleet so that they may be replaced at once.

Following supravascular positioning and subsystem validation, each plate inflates fluidtight metamorphic bumpers along its contact perimeter with its neighbors. This provides at least ~14 per cent linear effective elasticity (1.6 microns vs. 1.4 microns, center-to-center, comparable to the usual stretch requirements of the elastic arteries during normal cardiac pumping) and permits plates to track lateral movements of the underlying tissue while avoiding any relative movement of the opposing surfaces.

Ventral lipophilic anchors dropped into the lipid bilayer cytomembrane of the adjacent endothelial cells of the vascular wall serve as sensors to detect any such relative movement, providing continuous feedback to control bumper inflation—although with cardioplegia such wall movements will be quite small because infusant pressure may be held nearly constant throughout the installation procedure. Neighboring plates lock their bumpers firmly together with reversible fasteners embedded in the bumpers.

Along vessels whose diameters change rapidly as a function of axial position, orthogonal plate alignment along a circumference can be maintained via bumper expansion. Once maximum bumper extension is reached, a further increase in vessel circumference is accommodated in the next row by deflating bumpers in the circumferential direction and inserting one additional plate. Incommensurate offsets are reconciled using multiple ports in the bumper structure. If bumpers can expand by up to 20-30 per cent, then the blood vessel being plated can change in width by 20-30 per cent before such a pattern discontinuity is required, so such discontinuities should be relatively uncommon in most of the plated vasculature. Computer modeling of the dynamic adjustments required to achieve continuous plating at blood vessel bifurcations and in tapered vessels would be useful.

As the natural vascular surface is covered by the growing vasculoid, individual plates that have completed self-testing procedures and are locked in place activate a sufficient number of rotors to begin providing full molecular transport services to the underlying tissues. Any plate malfunctions are corrected by the circulating vasculocytes, via plate changeout. Metabolism in the underlying cellular tissues continues normally, just as it did before these tissues were plated with the vasculoid. Rotors on the ventral plate face (away from the lumen) gain access to nutrients in the vasculo-infusant fluid via gated channels leading to the dorsal plate face (toward the lumen) which is directly exposed to the fluid; cellular waste products escape to the fluid by a similar route. Plate power requirements normally range from ~0.004 picowatt (basal rate) to ~0.09 picowatt (peak rate) plus ~0.6 pW for computation. At 280 K the patient's cells would require only a fraction of the basal rate; after plating is complete and in the absence of significant computation, the entire population of 150 trillion installed plates might generate only ~1 watt.

Specific procedures for cellulock installation have not been examined in detail because these vasculoid components are relatively few in number and require a net installation rate of only ~9 million/sec (compared to 42 billion/sec for basic plates). To the extent cellulock placement can be governed by easily detected geometric factors such as vascular diameter (e.g., near capillary entrances), cellulock placement may occur similarly to plate deposition, but using vasculocytes programed with specific geometric release criteria. Some modest number of cellulocks destined for specific vascular addresses may be installed after plate initialization using a combination of vasculocytes and now-active cargo cilia, by replacing temporary plates with cellulocks.

Nontubular vascular segments including flaps, valves, sphincters, portals, sinuses, plexuses, nodes, and discontinuous microvascular beds such as the red pulp of the spleen constitute only a tiny fraction of the vascular surface but may require special procedures by the vasculocytes.

Endothelial pressure/shear responses may be managed by continuous emission of appropriate blockers, inhibitors or cytokines; the response may be reversibly disabled or eliminated using a designed vector to insert a short segment of revised DNA into the cells' nucleus. The vasculoid also includes an explicit mechanism to inhibit the electrical output of the sinoatrial node, thus maintaining a permanent (but in theory reversible) state of cardioplegia.

After ~1 hour, the structure of the vasculoid is almost complete. All major components have been tested and are in good working order—indeed, the still-submerged vasculoid is already maintaining full cell metabolism in the tissue below. The patient is now ready for defluidization.

Defluidization (~0.3 hour)

At this stage, the 1-micron thick monolayer of nanorobotic plates forms a chemically inert, flexible sapphire liner on the luminal (interior) surface of the entire vascular tree. This liner may be made fluidtight and airtight to at least modest pressures. Vasculo-infusant fluid is purged from the body by introducing ~6 liters of oxygenated pure anhydrous acetone as a dehydrating rinse followed by pressurized dry air at 2 atm through the input catheters and allowing all lavage fluids to exit through the output catheters for ~15 minutes. If necessary, surface cilia may assist in the distribution of fluid or air flow to help ensure that no pockets of fluid remain trapped anywhere in the system, and to ensure that no respirocytes, inoperative or damaged vasculocytes, or other stray particles are left behind.

The upper limit for laminar flow (Reynolds number ~ 2000) of aqueous vasculo-infusant fluid passing through a smooth 9-mm diameter exit catheter occurs at ~1 m/sec.

Assuming a constant laminar outflow velocity of 1 m/sec, flow volume through each of the two cannula is ~60 cm^3/sec and the vasculoid lumen is emptied of all installation and rinse fluids in ~100 sec. (This step is easily reversed by re-introducing an aqueous oxygen-rich or respirocyte-laden nutrient solution.) Acetone viscosity is ~40 per cent that of water, so pumping power may be reduced as the rinse proceeds.

Oxygen needed to sustain human life is slightly restricted for only ~100 sec during the pure acetone rinse, and thereafter is continuously available at the requisite concentration from the dry purge gas. A temporary supply of dry air with 20 per cent O_2 at 2 atm puts 5×10^{22} O_2 molecules into the vasculoid lumen, which when burned with glucose at 50 per cent efficiency in tissue cells produces ~20,000 joules, equivalent to ~2000 sec of power at the reduced metabolic rate of ~10 watts in a human body cooled to 280 K. The plate population consumes only an additional ~1 watt. As long as fresh purge gas continues to circulate, the oxygen requirements of both human and vasculoid can be satisfied indefinitely.

Cellular glucose reserves will last ~10^5 sec at the reduced metabolic rate; subject to cautions noted earlier, cellular waste products may not approach toxic levels for ~10^4 sec. Vasculoid plates can also continue functioning in the temporary absence of glucose. Assuming a power requirement of ~0.6 pW/plate at the basal rate, then a ~1 hour supply of glucose may be stored in a small fuel tank inside each plate, representing only ~9 per cent of the 2 micron^3 plate volume.

Initialization and Cold Start (~1.2 hours)

The final steps of vasculoid installation include:

(1) *Plate Initialization.* To aid in injury response and medical diagnosis, each plate may be assigned a unique position and identity code which can be echoed upon interrogation. With 200 billion operational vasculocytes and 150 trillion plates to initialize, each active vasculocyte must contact and initialize ~750 plates. Traveling at a net velocity of 10 microns/sec across the vasculoid,

a vasculocyte can traverse (and initialize) 750 plates in ~100 sec. The address block is 100 bits in length; 50 bits are sufficient to specify 2^{50} = 1000 trillion unique objects in the system, but additional bits are required to specify each plate's branch level in the fractal vascular tree and 10 bits are reserved for self-correcting parity checks.

(2) *Install Storage Vesicles.* The vesicles are small storage garages containing reserves of mobile and cargo-carrying nanodevices, other auxiliary nanodevices, spare parts, replacement modules and repair materials, critical biochemical consumables, and compressed refuse. Vesicles may also incorporate nanoscale bioprocess plants to manufacture small quantities of artificial biochemicals not normally produced in the body. Each vesicle is ~1 mm^3 in volume, providing enough space to store ~1 billion tankers, ~300 million vasculocytes, or 350,000 boxcars. Vesicles may be attached directly to the vasculoid surface using sapphire struts affixed to reinforced base plates, or may be attached to each other to build convenient three-dimensional configurations (e.g. "bunch of grapes" formations).

As an order-of-magnitude estimate of the total number of vesicles that might be required, the volume required to store a complete replacement for the entire mobile container population would include 166.2 trillion tankers (1 $micron^3$/tanker) or 166.2 cm^3, 32 billion boxcars (2827 $micron^3$/boxcar) or 91.7 cm^3, and 2 trillion backup vasculocytes for tenfold redundancy (3 $micron^3$/vasculocytes) or 6 cm^3, totaling ~264 cm^3 of storage. Including additional space for bioprocess plants, spare parts, consumables and refuse storage may boost total vesicle volume to ~500 cm^3, requiring ~0.5 million vesicles which would fill roughly the combined internal volume of the four cardiac chambers which are no longer needed for pumping blood. However, conservative design principles would suggest that vesicles should be widely distributed throughout the body to maximize system survivability in the event of massive trauma.

Infusion of 500 cm^3 through the two catheters at a modest infusion velocity of 1 cm/sec requires ~200 sec for infusion of all vesicle components and their contents. Each 3 mg (fully loaded) vesicle is installed by a team of ~3000 vasculocytes which can collectively apply ~30,000 nanonewtons of force (using only 10 active legs among the ~100 available on the ventral side of each vasculocyte) resulting in a vesicle acceleration of ~1 g (enough to overcome gravity in a supine patient, and easily increased if needed).

Simultaneously installing all 500,000 vesicles requires the cooperative activity of ~1.5 billion 1-50 picowatt vasculocytes which consumes a total power of 1.5-75 milliwatts during the installation.

(3) *Install* sealed carotid, jugular, or navel access port, as required (~40 sec).

(4) *Activate* ciliary distribution system (~10 sec). This should probably include a brief stereotypical self-cleaning protocol to further ensure that all particles and stray nonessential nanorobots have been removed from the appliance interior.

(5) *Introduce* into the vasculoid interior the entire 257.9 cm^3 operational tanker and boxcar population, requiring ~100 sec at a modest ~1 cm/sec dry infusion velocity.

(6) *Warm* the patient (~1 hour) using heat generation from excess ciliary activation, energy from onboard thermogenerative systems or external heat sources imported via enhanced plate conductivity (or simple electrical or rf heating. Restart normal respiration as soon as possible. (The heart remains permanently inhibited, although the endocardium is coated with vasculoid plates.)

(7) *Reverse* cytoactivity inhibition that was initiated during patient preparation as appropriate.

(8) *Remove* catheters and seal all vascular and dermal breaches.

The vasculoid is now fully operational and self-contained. The patient is warm and breathing.

All essential metabolic and immunological systems have returned to normal function. Healthy vascular conditioning is maintained on a permanent basis (post-installation) because the vasculoid exercises precise control over the transport of cholesterol, leukocytes and platelets, the principal participants in the arteriosclerogenic process. Skin and tissue suppleness may be controlled by adjusting the spring constants of linked bumpers between adjacent plates.

A number of cosmetic issues must be addressed. For example, unpigmented tissues (e.g. the tongue and gums, fingernail dermae, the uvea and lacrimal apparatus of the eye, and brain tissue) will appear a pale, waxy translucent white owing to the complete absence of red cells; pigmented tissues (e.g. dark skin) should be largely unchanged. For more traditionalist users or for other aesthetic reasons, semi-natural coloration might be restored by retaining blue sapphire in venous channels but substituting red ruby (chemically similar to sapphire) in arterial plate materials. However, transmission and reflection properties of these substances may differ markedly, and such colors would likely require at least small numbers of potentially troublesome impurity atoms (10^{-3}-10^{-4} of Fe, Ti, or Cr atoms; to be present. Alternatively, a thin diamondoid-veneered sublayer of organic chromophores of appropriate colors might be employed.

After a brief period of rest, postoperative checkout, and familiarization with user interfaces, the patient is released from the installation facility.

Vasculoid Removal

The procedures outlined above for vasculoid installation are designed to be fully reversible at every step. Thus, vasculoid removal may be accomplished by reversing the installation procedure, except that patient preparation should be performed first. If the patient has worn his or her vascular appliance for a considerable time, heart muscles, venous valves, splenic filtration systems and the erythroid marrow may have significantly atrophied, requiring remedial cellular repair. Homologous formed blood elements such as RBCs must also be premanufactured for postoperative infusion prior to vasculoid removal, and any genetically modified endothelial cells must be reprogramed back to their original state.

Aggressive Installation Scenario

For our highly speculative second scenario, we briefly (and only superficially) describe an aggressive installation procedure that attempts to install the vasculoid into a fully metabolizing normothermic human body about 100 times faster than the hypothermic procedure detailed in the principal scenario. As discussed earlier vascular conditioning and mapping should occur before installation, in part to allow the manufacture of a folded, pre-assembled custom appliance tailored to the patient's unique vasculature which is then presented to the physician for installation. Our aggressive procedure involves three steps: (1) pre-charging the body with respirocyte infusant and removal of bloodborne cells; (2) physical installation of vasculoid plates as continuous rolling sheets, coincident with removal of infusant fluid; and (3) configuration and permanent connection of the appliance, followed by removal of support structures and system activation. It is intended that all steps will be carried out within the ~10^3 sec (~15 min) ischemic time limit imposed by non-respiratory metabolite concentrations.

The aggressive procedure begins by accessing one or more large veins to permit rapid infusion and exfusion. The blood is quickly replaced, as it circulates, with an infusant solution containing electrolytes, glucose, and other essential substances, plus a sufficient nanocrit of respirocyte-class devices to provide respiratory support. The transfusion is speeded by using moderate numbers of "dragnet" style nanorobots as described in other contexts by Freitas—each such nanorobot manipulates an adjustable 1-100 micron diameter net in order to physically gather bloodborne cells that may be trapped in eddies near venous valves, vascular sinuses, and similar spaces. Within a few circulation times, perhaps ~200 sec, the vascular compartment is cleared of all free cells and most macromolecules. Exfused blood is retained until the procedure is complete, to facilitate reversal in the event of a medical emergency. The heart is now stopped. Respirocytes continue to provide oxygen and absorb carbon dioxide; even a modest 1 per cent respirocrit can provide ~25 minutes of resting metabolism without recharging, perhaps slightly less in the most energy-intensive tissues such as the brain. But because fresh infusant can be introduced continuously, the patient can be parked in this state indefinitely if need be.

Next, the vasculoid is introduced into the arteriovenous vasculature directly through the heart via cardiopuncture or cardiocentesis. Cardiocentesis involves the surgical puncture of the heart, most often used today *in utero* on developing fetuses either to extract or to insert fluids into the vasculature (e.g., blood transfusion), or more rarely on adult organisms for related purposes. Cardiocentric installation of the vasculoid is required because the human circulation consists of two independent arterial circuits (pulmonary and systemic) each containing their own capillary beds.

The appliance is installed as a continuously-everting concentric tube, a process called progressive fractal eversion that may be visualized as turning a glove inside out. Basic plates, docking bays and cellulocks are prefastened in the proper configuration to fit the various diameters and branchings of the blood vessels that they will coat. The necessary flexibility of this sheet of plates is provided by sophisticated watertight jointed interplate bumpers that have yet to be designed. Eversion may be powered by compressed gas, ciliary action between opposed plates, or by other appropriate means; the presence of pressurized gas (e.g., pure dry N_2) would also help to minimize the occurrence of leaks during installation. Each docking bay has an appropriate tanker already docked. As soon as the plate makes contact with the endothelial surface, the docking bay begins to work. Since the vasculoid contains a sufficient number of docking bays to accommodate peak metabolic loads, the preattached tankers can supply oxygen, glucose, and other nutrients, and accumulate wastes. Most critically, 5 trillion oxygen tankers are attached to all available respiratory-tanker docking bays, providing ~474 sec (~8 min) of oxygen at the basal rate. This sustainable duration is easily extended to ~1318 sec (~22 min) of basal oxygen supply by redesigning 13.9 trillion of the 24 trillion docking bays that are not needed for nonrespiratory transport to include the capacity for gas transfer during installation, since at the basal rate docking bays require only 67 active sorting rotors.

Installation is initiated cardiocentrically as two primary segments imported via two cardiopuncture entry points through the relatively thin walls of the right atrium and the left atrium, respectively. Installation works outward from each chamber opening, usually in the natural direction of valve motion, moving towards the distal capillary beds. From the right atrial entry point, the unfolding right atrial appliance segment has four "fingers" and simultaneously plates: (1) the superior vena cava; (2) the inferior vena cava; (3) the coronary sinus (which receives cardiac veins from heart tissue); and moves through (and plates) the right atrial and ventricular chambers, thence to plate; (4) the pulmonary artery. The right atrial segment also must plate all of the numerous Thebesian veins; these venules return blood from the myocardium without entering the venous current, and open directly into the right atrium. From the left atrial entry point, the unfolding left atrial appliance segment has five "fingers" and simultaneously plates: (1&2) the two left pulmonary veins (which frequently terminate by a common opening); (3&4) the two right pulmonary veins;

and moves through (and plates) the left atrial and ventricular chambers, thence to plate (5) the aorta.

Additionally, a third vasculoid segment called the portal segment must be inserted through a third abdominal entry point in order most efficiently to plate the portal vein, which lies midway between the hepatic capillary beds and the intestinal capillary beds.

Advancing at 1 cm/sec (2 cm/sec internal speed between the installed surface and the inverted tube), the vasculoid requires only 70 sec to plate the maximum ~70-cm main arterial or venous course. Fluid and respirocytes are withdrawn through the center of the inverted tube. As the vasculoid reaches the capillaries, the progression of the eversion may slow to ~10 micron/sec of travel, requiring another ~100 sec to complete all capillary plating. Human capillaries contain ~14 per cent of blood volume but comprise ~95 per cent of the surface area of the vascular system, so most of the appliance's plates are installed during this final ~100 seconds. If all 150 trillion plates are active throughout this process, the power draw is a physiologically-tolerable 150 watts. Upon first contacting the endothelium, each 2 mm^2 plate has ~0.1414 seconds—time enough for 14,140 localized 100-nm movements at 1 cm/sec (100 KHz)—to adjust its position and bumper configurations, and to clear away any detritus trapped beneath it (e.g., respirocytes, stray cells, etc.), before the next plate is placed. This seems sufficient.

Once the vasculoid has filled the capillaries, the arterial and venous branches are joined, detaching the internal fluid transfer channels running through the vasculoid lumen which are quickly retracted from the interior. The ciliary system can then be activated and tankers will begin to be delivered to and from the capillaries, starting ~200 sec after the first transcardial introduction of the vasculoid mechanism. Interior rinse is unnecessary because the luminal surface of the appliance was installed dry. Vesicle installation is performed post-installation via component injection, and assembled using vasculocytes, at leisure.

Since this advanced model of the vasculoid comes preassembled, it can continually report its status and monitor physiological indications of the patient during installation, via an interplate communication network. In the event of an unforeseen medical catastrophe during installation, the vasculoid may be evacuated using an emergency extraction protocol if the appliance itself has not suffered a major loss of integrity and the transfer channels are still attached. In the emergency extraction protocol, the vasculoid undergoes partial flattening with circumferential fission in the capillaries. Then the arterial and venous sides of the vasculoid are withdrawn as they were inserted, using a reverse-eversion process with progressively increasing velocities possibly reaching ~10 cm/sec in the largest arteries. During such an extraction, only a ~1.4-micron-wide annular ring will be moving at this high speed adjacent to tissue—the installed vasculoid does not move, and the moving vasculoid is surrounded by immobile vasculoid. As a result, installed appliance plate-sheets may be pulled out by applying an extraction force to the fluid transfer channel structures that are still attached to the terminus of each capillary tube. Continuous pressurized return of respirocyte-charged infusant to the natural vasculature through these channels prevents pulling a vacuum behind the retreating tendrils of the appliance, which would cause the collapse of blood vessels as the appliance was removed. Thus, in less than a minute, the patient has returned to the relative safety of the respirocyte-infused parked state from which the installation procedure was begun. In the context of a medical technology capable of manufacturing a vasculoid, the temporary bloodless state would not appear to be a cause for concern.

Unit V–L

Vasculoid Optional Equipment

The following is a selection of a few additional devices, options, or subsystems that may be added to the basic vasculoid installation to achieve enhanced functionality.

External Ports

To facilitate physical communication with the external environment, the vasculoid may be equipped with mechanical external ports which permit the ready attachment of cables or peripheral equipment as an alternative to purely wireless (e.g., radio, optical) communication links. Vasculoid ports include an appropriate macroscopic connector receptacle in hard contact with the sapphire structure. Ports may include connections to the internal plate-to-plate acoustic communications network or to docking bays and cellulocks to permit access to the internal material flow and thus allow inserting or extracting: (1) molecules and cells; (2) fresh vasculocytes and spare parts; (3) bioprovisions including respiratory gases, water and energy supplies; and (4) nanomechanical or other waste material. Extracorporeal user control interfaces could also be connected via an external port, if needed. Convenient and aesthetic locations for external ports include the navel or the nape of the neck.

Respiratory gas and glucose transport requirements could be reduced by importing externally-supplied electrical energy—e.g., wall sockets (110 VAC at 0.9 amps equals the 100 watt human basal rate), backpack generators, solar collectors, nuclear batteries etc.—to be distributed throughout the appliance via wiring in the plates. These sources could be used to power: (1) local recycling of CO_2 and H_2O waste back into O_2 and high-energy carbohydrate fuel (mimicking photosynthesis energized by light), (2) the reconstitution of amino acids and proteins from urea and other nitrogenous wastes (as found in ruminants and in hibernating bears), and (3) other "reversible nutrition" processes, converting the body of the envasculoided user into a more closed-cycle system with reduced dependence on certain material and energy inputs. However, this approach would correspondingly increase user dependence on the chosen external power source and would require new in-plate or in-vesicle chemical processing plants without completely eliminating the need for any existing subsystem, hence may not be worth the added complexity.

Thermal Comfort Subsystem

Vasculoid installation leaves the user's gross thermal mass largely unchanged, as only ~4.4 kg of blood borne water (~9 per cent of the ~50 kg normally present in the human body) is removed and replaced with ~2 kg of sapphire vasculoid with a heat capacity equivalent to ~0.2 kg of water. Total passive aqueous heat capacity drops from 50.4 kcal/K down to 46.3 kcal/K.

Nevertheless, 20th century patients with extensive metallic implants (e.g. pins, plates, bolts and joints) occasionally report brief chilly sensations during periods of cold weather and in other circumstances, due to the high thermal conductivity of metal compared with natural biomaterials or with plastics. Some organs such as the cornea are quite sensitive to sudden temperature changes—as little as a 0.3 K drop over a 0.785 mm^2 area for a duration of 0.9 sec (~700 nanojoules) is detectable by patients. A thermally conductive sapphire or diamondoid-coated sapphire vascular implant could extend these unpleasant sensations throughout the entire body, even for activities as simple as manually grasping a cold object. A targeted thermogenerative subsystem could help to eliminate this effect.

Perspiratory thermoregulation will continue as before. However, as explained earlier the active heat transfer mechanism of blood will be completely disabled. Normal capillary vasoconstriction/vasodilation mechanisms may be at least partially suppressed to avoid unnecessary tensions at the vasculoid-vessel interface. Detection of these responses, or direct temperature measurement, may be used to adjust passive thermal conductivity over a wide range, as follows. Each plate abuts neighboring plates through metamorphic bumpers with optimal thermal contact along stripe-like diamondoid buttons. Within each bumper, we may place an opposed pair of diamondoid pistons, separated by vacuum when they are pulled apart. When pulled apart, heat can be conducted only through the sapphire infrastructure, or vacuum in the plate, or through a surrounding aqueous medium, and is thus very slow, almost the same as normal tissue. But when the two pistons are pressed tightly together against diamondoid bumper buttons that are in good thermal contact, heat conduction largely bypasses the poorly-conducting sapphire regions and flows almost exclusively through the diamondoid contact region. Thus with pistons apart, the envasculoided human body has near-normal thermal conductivity; with pistons in contact, the body's thermal conductivity becomes near-metallic, perhaps as conductive as stainless steel. This transition is subject to user or program control, is switchable in microseconds, and may be directed only to specific volumes or pathways within the body if so desired. A more complex system might employ thermal rectifiers, or, as J.S. Hall suggests, "heat pumps could be placed in the joints to make the whole phenomenon usefully controllable, augmented if necessary with tankers of ice and/or steam."

Internal Caching

Installation of a full vasculoid appliance permanently displaces ~4.2 liters of natural blood volume, freeing up ~1 gallon of internal storage volume. At least some of this volume may be used for containerized temporary caching of consumables including surplus glucose, fats, water, oxygen (e.g. a high-pressure nanolung), minerals or other useful biomaterials, or various bio- and nano-wastes. For example, a 1-liter 1000-atm O_2 solid-walled cache holds ~34 hours of oxygen at the human basal metabolic rate or ~2 hours at the maximum rate.

The free ~4.2-liter volume could also be used to store a wide variety of useful equipment or tools including computers, computer memories, external communications or navigational devices, solar energy accumulators, weapons, spare vasculocytes and other special purpose nanorobots, spare parts, or useful tools. An entire spare vasculoid (~558 cm^3) could even be stored in this volume.*

Breathing in Low-Oxygen Environments

The vasculoid, like the micron-sized respirocytes proposed elsewhere, will offer the ability to breathe at low O_2 partial pressures. The vasculoid gas tanker fleet can hold ~20 minutes of oxygen at the basal metabolic usage rate, or up to ~100 minutes if most of the tanker fleet is diverted to respiratory gas carriage in lieu of other applications.

Vasculoid installed in underwater divers who are breathing pressurized air to modest depths should allow only enough nitrogen into the body to forestall mechanical tissue damage, then rotor it back out again as the diver surfaces to avoid decompression sickness. At 100 per cent saturation the body absorbs ~10^{21} excess N_2 molecules per meter of depth. Each tanker could hold ~7.98×10^9 nitrogen molecules (at 1000 atm), requiring the storage capacity of 0.125 trillion tankers per each 1 meter of decompression. Given a reconfigurable fleet of ~166 trillion tankers, expedited decompressions from ~100 meters are probably achievable using the present design. To establish a greater vasculoid operating depth, in theory an auxiliary 1000-atm storage tank "nanolung" installed in the vasculoid wall could provide ~9.43 cm^3 N_2 storage capacity per each 100 meters of decompressible diving depth. However, nitrogen should not be used to hyperpressurize the tissues due to nitrogen narcosis. It should also be noted that obesity is a risk factor in decompression sickness because nitrogen is 5 times more soluble in fat than in water and because of reduced blood access to adipose tissue, and also compression and decompression are asymmetrical: a nitrogen load acquired in a few minutes may take hours to fully deplete purely by diffusion (and the extraction rate varies markedly by tissue type), so extravasculoid respirocyte-class nanorobots may be required for optimal results. Clearly the vasculoid can achieve a decompression rate comparable to breathing pure oxygen, but without the oxygen toxicity. Helium may be useful for mixed-gas diving, but is prone to leakage some will be lost from the body during use and cannot effectively be replaced from the environment.

Lymphovasculoid and Pathogen Disposal

The lymphatic system extends into all the same tissues as capillaries and is structurally similar to the venule network. However, since lymph vessels constitute a "cul-de-sac" system, not a "circulatory" system, it is unnecessary to install vasculoid systems in the lymphatics in order to achieve most of the positive benefits expected from an arteriovenous vasculoid. With the circulatory vasculoid in place, the workload of the natural lymphatic system may be greatly reduced—protein leakage from capillaries, entry of particulates into the tissues, and the presence of pathogens requiring immune system response all should be greatly reduced.

More properly, lymphovasculoid should be regarded as optional equipment which may permit more precise control of the immune system in general and of lymphocyte traffic in particular. Total adult lymph flow is typically ~2 liters/day, so transport requirements within the lymphovasculoid should not be particularly severe. If implemented, the lymphovasculoid would be emplaced as two distinct installations—a right lymphatic and a thoracic subsystem—following the natural division in the human body. Special lymph-recycling facilities may be required at the locations where the thoracic duct and the right lymphatic duct join the venous tree.

Pathogen disposal sites using an enzyme-based digest and discharge protocol may be located near lymphoid tissues and organs (to permit immune system processing) and near excretory organs such as kidney, liver, gallbladder, and gut (to facilitate removal from the body). Specialization of this function at specific locations appears more efficient than on-site pathogen processing systems which would need to be numerous, widely distributed, and usually idle.

Extravasculoid Devices

It may be useful to deploy sensor probes which can leave the vasculoid and enter the surrounding tissues to detect and monitor remote events (such as angiogenesis, tumors, or pathogens) that otherwise might not be conveniently detected from within the vasculoid. Additionally, large blocks of tissue are only lightly vascularized or have no capillaries whatsoever, as for instance the synovial chambers in skeletal joints and the epidermis. Medical nanorobots capable of migrating through noncellular tissue might be useful both for repair purposes and for maintaining a disease-free condition in these tissues. Extravasculoid devices may also constitute general-purpose mobile cell repair machines with even broader injury response and prevention capabilities.

Active Thermal Damage Suppression

Although the vasculoid itself is essentially fireproof (sapphire and ruby will not burn in oxygen), the overlying human tissue is not. Thus another optional upgrade to the basic vasculoid package is an active damage control subsystem that offers some protection from severe thermal burns. This optional subsystem would include installation of biocompatible sensor-tipped diamondoid or sapphire pores that pass from the epidermis through the dermis to the nearest envasculoided capillaries. The comparative human skin response to sapphire vs. fluid cooling has been studied experimentally.

Upon detection of a potentially harmful thermal event, sacrificial water is pumped from internal reservoirs into channels in the plates, and from there into the pores, eventually emerging as billions of aqueous microstreams. Touching a red hot object stimulates a nearly instantaneous emission of ablative water, producing a protective layer of heated water and steam between the hot object and the skin. Protection of the human hand (~150 cm^2) up to ~1300 K (~decomposition temperature of diamond) for a ~1 second exposure requires a 162 kilowatt/m^2 heat dissipation rate and a ~1 cm^3 sacrificial water reservoir. Of course, this system could easily be overwhelmed if the user is exposed to intense IR radiation, such as inside burning buildings or near large explosions, and in any case the hair is not protected as well as the skin.

Active Resistance to Mechanical Damage

The overall static structure of the basic vasculoid appliance is an array of rigid plates, strongly fastened together, entirely covering a curved two-dimensional surface embedded in a three-dimensional space. The plates conform to ordinary movement, but may adequately resist certain kinds of destructive movement such as separation due to cutting, rapid accelerations or decelerations, and crushing injuries. Although local endothelial cells may be damaged during such sharp movements, an envasculoided tissue should be somewhat more resistant to gross damage.

For example, the critical buckling pressure of a hollow tube of circular cross-section deformed into an elliptical cylinder by external pressure is given by Freitas as $p_{crit} = (E\ h_{tube}^3)/(4\ r_{tube}^3\ (1 - c_{Poisson}^2))$ where E is Young's modulus (~10^6 N/m^2 for vascular tissue and ~10^{11} N/m^2 for sapphire, h_{tube} is tube wall thickness (~1 mm for aorta, ~ 1 mm for capillary and for vasculoid plate-tube, r_{tube} is tube inner radius (~25.0 mm for aorta, ~8 mm for capillary, ~7 mm for vasculoid plate-tube; and $c_{Poisson}$ is the Poisson ratio for the material (~0.3 for vascular tissue, ~0.1 for diamondoid). Using this formula and the stated values, p_{crit} for aorta is ~59 N/m^2 and the vasculoid coating adds only negligible additional resistance to crushing, p_{crit} ~ 0.002 N/m^2. However, for the ubiquitous capillaries p_{crit} ~ 540 N/m^2 but adding the vasculoid coating can dramatically increase crushing resistance up to ~7 x 10^7 N/m^2, or ~700 atm of overpressure. Since the mechanical coupling between plate bumpers is subject to both design and conscious user control the crushing resistance of capillary beds can be autogenously varied over five orders of magnitude. Of course, high overpressures may seriously damage the underlying natural endothelium but this might nevertheless be acceptable over small areas during emergency situations. For instance, resistance to deep incision-slash wounds would be markedly improved. Crushing strength may be increased using thicker plates or stronger bumpers.

Similarly, the bending stiffness of a hollow tube is given by Freitas as $k_{shaft} = 3p\ E\ (R_{tube}{}^4 — r_{tube}{}^4) / (4\ L_{tube}{}^3)$, where the outside tube radius is $R_{tube} = r_{tube} + h_{tube}$ and the tube length is L_{tube} (~400 mm for aorta, ~1 mm for capillary. Using this formula and the stated values, the resistance to bending (stiffness) of the aorta increases 60-fold (to ~230 N/m) when coated with a single layer of vasculoid plates, but the stiffness of capillaries increases nearly 70,000-fold (to ~0.4 N/m) when coated with diamondoid plates. The buckling strength of coated vessels compared to natural vessels may exhibit similar orders-of-magnitude improvement, though in practical systems some of this potential will be lost because interbumper connection rupture strength may be only 10^7-10^9 N/m^2, 1-3 orders of magnitude less than for solid diamondoid materials. The natural musculature may be too weak to move soft tissues that are thoroughly penetrated with ultrastiff capillaries, but selective autogenous control of plate bumper expansion and contraction should make possible any necessary macroscale voluntary movements, or on-demand mechanical rigidification of specific limbs or organs. Bumpers may be driven at ~KHz frequencies, permitting positional adjustments on millisecond timescales. Such motions will require additional energy expenditure.

A complete examination of the acceleration tolerance of the envasculoided whole human body would have to take into account a wide variety of factors including differential density of body parts (e.g., lung, soft tissue, skeleton), magnitude and direction of the stress vector (e.g., positive or negative, axial or transverse, etc.), duration and timing of the stress vector (e.g., acute or chronic, linear or periodic), rate of onset and pulse shape of the acceleration, and the mechanical characteristics both of linked vasculoid components and of the vasculoid-endothelial interface. However, a simple order-of-magnitude estimate may be cautiously ventured, as follows.

A vasculoid appliance of differential density Dr_{vasc} relative to soft tissue and mean thickness h_{vasc} in contact with biological tissues that are subjected to a linear acceleration of a = (a_G g), where g = 9.81 m/sec^2, pushes the vasculoid into the tissues with a pressure force of $P_{vasc} \sim h_{vasc}\ Dr_{vasc}\ g\ a_G$. Because the vasculoid tube has 2 opposed walls, each 1 mm thick, and because the tanker fleet covers 4 per cent-55 per cent of the surface of each of these opposed walls with tankers that are 1 mm thick, then the effective capillary thickness of pushed vasculoid under acceleration ranges from $h_{vasc} \sim$ 2.1-3.1 microns. Tissue density is $r_{tiss} \sim$ 1050 kg/m^3; sapphire tanker density is $r_{tank} \sim$ 993-1738 kg/m^3 if filled with vacuum or water at 310 K, and plate density is $r_{plate} \sim$ 2000 kg/m^3, giving an effective vasculoid density of $r_{vasc} \sim$ 1643-1988 kg/m^3 under various usage conditions. Hence $Dr_{vasc} = r_{vasc} — r_{tiss} \sim$ 593-938 kg/m^3 and so the maximum tolerable G-force for envasculoided tissue is $a_G \sim P_{vasc} / (h_{vasc}\ Dr_{vasc}\ g) = 24.4\ P_{vasc} \sim$ 4880 G, taking $P_{vasc} \sim$ 200 N/m^2 as the tentative limit for acute vascular damage because this has been found experimentally not to cause gross injury or denudation of canine arterial endothelium by shear force.

However, endothelial transcription factor responses (protein expression) can be triggered by shear forces as low as 0.02-2 N/m^2 ($a_G \sim$ 0.5-48.8 G) and normal physiological blood shear forces are 0.14-2.6 N/m^2 ($a_G \sim$ 3.4-63.4 G), so the maximum chronic external acceleration indefinitely tolerable by an envasculoided human body probably may not exceed 50-100 G without producing altered, possibly pathological, cytochemical states. (The operational limit of the current vasculoid design is also ~30-100 G). This still would represent a substantial improvement over the natural acceleration tolerance of the unaided human body—the mean relaxed tolerance is ~3.23 G and loss of brain blood flow occurs at $+G_z > 4.5$ G. Consciousness has been retained for a maximum of 45 seconds at 9 G using G-protective equipment and straining maneuvers developed for the U.S. Air Force, and for 4 minutes at 12 G or 4 seconds at ~16 G using water immersion. Severe impact injury to humans occurs from $+G_z$ accelerations of 30 G exceeding 100 millisec or 100 G exceeding 2 millisec—the recordholding primate is evidently one of Colonel Stapp's chimpanzee test subjects that survived ~1 millisec of 247 G in the $-G_x$ direction on a rocket sled, suffering only "moderate" injuries. Thus, the basic vasculoid may improve maximum human acceleration tolerance by a factor of 10-20, or more, and higher-G-tolerant architectures can probably be devised.

With sufficient capability, the vasculoid structure could allow the non-local distribution of stress—for example, by spreading the force of an impact in a manner similar to a bulletproof vest, also helping to quickly close any vascular breaches that might occur. A common and significant personal injury mechanism is brain-skull impact, and a tough interwoven vascular support scaffolding could help to fix the brain within the skull, potentially reducing the danger of concussion for at least moderate decelerative loads. (Active suppression of rotational impact trajectories may enhance cerebral durability, as demonstrated using high-speed cinematograph films of woodpeckers which regularly survive 6-7 m/sec cranial impact velocities with ~1000 G decelerations.) Positional information and selective sectioning could allow unavoidable partitions of tissue to occur along straight planes, producing a minimum of damage when compared with ordinary tearing injuries. Whether the vasculoid appliance would be a liability in explosive overpressure situations, where a fluid-filled vasculature might not (e.g., lethal effects in humans noted for 40 psi to 50 psi overpressure shockwave; 300 psi (~20 atm) record for successful deep water submarine escape), deserves further study.

Conclusion

The previous chapters presented a preliminary scaling study for a conceptual design of a single, complex, multisegmented nanotechnological robot that appears capable—with numerous caveats, as noted—of duplicating all essential thermal and biochemical transport functions of the blood, including circulation of respiratory gases, glucose, specialty biochemicals, waste products, and all blood borne cellular elements. The vasculoid, a 2-kg ~200-watt intimate personal appliance, conforms to the shape of the existing vasculature and may serve as a complete replacement for natural blood while greatly improving the durability and functionality of the human body.

The device described in these chapters would represent a most extreme intervention using a very advanced medical molecular nanotechnology. It also should be noted that our current knowledge of the biological functions of the circulatory system is incomplete, so the design presented here must be considered provisional at best. But the principal challenge was to advance a plausible argument that a nanomechanical whole-body thermal and biomaterials transport system would violate no known physical, engineering, or medical principles, could presumptively be made adequately safe for the user, and might confer some significant advantages over simpler whole-body systems exclusively employing unlinked populations of individual blood borne and tissue-borne nanorobots.

Ultimately, and from the standpoint of human-guided evolution, the body exists primarily to ensure the survival of the mind—not the replication of the genes, which was the ancient paradigm. It would seem that a somewhat more advanced and compact version of the proposed device could function independently of nearly all noncortical tissue. Thus the vasculoid is most fascinating because it may represent one last outpost of humanity at the final frontier of biological evolution.

Unit V–M

NanoTribology

Introduction

The term tribology is derived from the Greek word *tribo* meaning rubbing and *logy* meaning knowledge. The original applications by the Greeks of tribology were in trying to understand the motion of large stones across the earth's surface. Today tribology has grown to include the methodical study of friction, lubrication, and wear.

Tribology plays a critical role in diverse technological areas. In the advanced technological industries of semiconductor and data storage, tribological studies help optimize polishing processes and lubrication of data storage substrates. In traditional industries such as automotive and aerospace, tribological studies help increase the lifespan of mechanical components.

Many industrial processes require a detailed understanding of tribology at the nanometer scale. The development of lubricants in the automobile industry depends on the adhesion of nanometer layers (mono layers) to a material surface. Assembly of components can depend critically on the adhesion of materials at the nanometer length scale.

There are a number of traditional tools for characterizing friction, lubrication and wear. The most common characterization tool is the tribometer having several configurations such as pin-on-disk, ball on flat, and flat on flat, etc. Generating motions at the nanometer scale is extremely challenging. New characterization techniques are required to understand tribology at the nanometer scale.

The atomic force microscope is now being routinely applied for studying nanoscale tribology. The natural extension of the AFM for tribology applications is derived from the motion of a nanometer-sized stylus in the AFM over a surface. Although traditional tribology testing is not done with an AFM, many new types of applications are possible.

Examples of the application of AFM to tribology include:

- Direct three-dimensional visualization of wear tracks, or scars on a surface.
- Measurement of the thickness of solid and liquid lubricants having nanometer or even monolayer thickness.
- Measurement of frictional forces at the nanometer scale.
- Surface characterization of morphology, texture, and roughness.

- Evaluation of mechanical properties such as hardness and elasticity, and plastic deformation at the nanometer scale.

A major advantage of the AFM for tribological studies is that the AFM can be routinely used on all types of materials. Materials commonly studied include: ceramics, metals, polymers, semiconductors, magnetic, optical, and biomaterials. AFM investigations are usually made in ambient air environment. It is possible to make AFM studies in a vacuum or liquid environment.

Nanoscale Wear Analysis

The effects of wear at the nanometer scale become critical to the optimization and stability of machines as the tolerances in precision machines become smaller and smaller. Traditional microscopes such as the optical and scanning electron microscopes facilitate visualization of wear in 2-dimensions. For example, with the SEM it is possible to get a magnified view of wear tracks in the x-y axis but cross sectioning is required for measuring the depth of wear tracks.

The AFM allows direct 3-dimensional visualization of wear tracks and scars. The images may be displayed in a 2-D projection and a 3-D projection. Direct measure of wear track depth can be easily measured with a line profile derived from the AFM image.

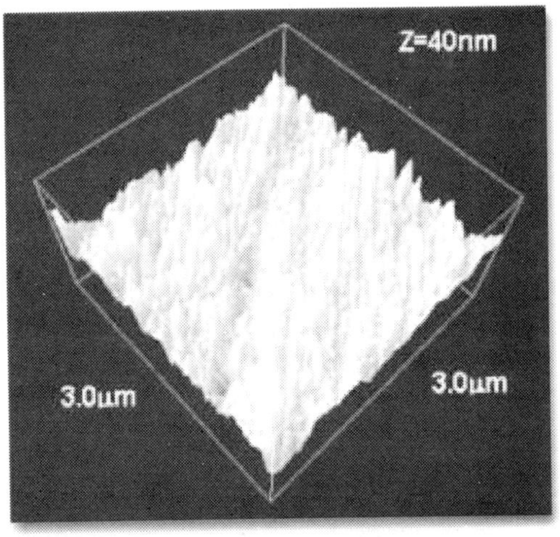

Fig. V–M.1: Three-dimensional atomic force microscope (AFM) image of a polish mark on a piece of steel. The scan range in X and Y is 3 micrometers and the entire Z range is 40 nanometers.

Lubricant Studies

It is well known that layers of lubricants on surfaces that are less than 100 nm can dramatically affect lubrication behavior. Characterization of such films is necessary for developing optimized lubricating films. However, nanometer scale characterization of lubrication films offers a substantial challenge. Optical techniques such ellipsometers can be used for measuring lubrication thickness of large sections, (greater than 10 square micrometers), of a surface. Measurement of the localized (less than 1 micron) film thickness is not possible with the ellipsometer.

The probe is mounted at the end of a cantilever in an AFM making it possible to measure interaction forces between the probe and the surface by monitoring the deflection of the cantilever. A graph, called a force/distance curve, shows the forces on the probe as the distance between the probe and the surface are reduced. The nature of the force/distance curve depends on the force constant of the cantilever, the lubrication density, probe geometry, and the lubrication thickness.

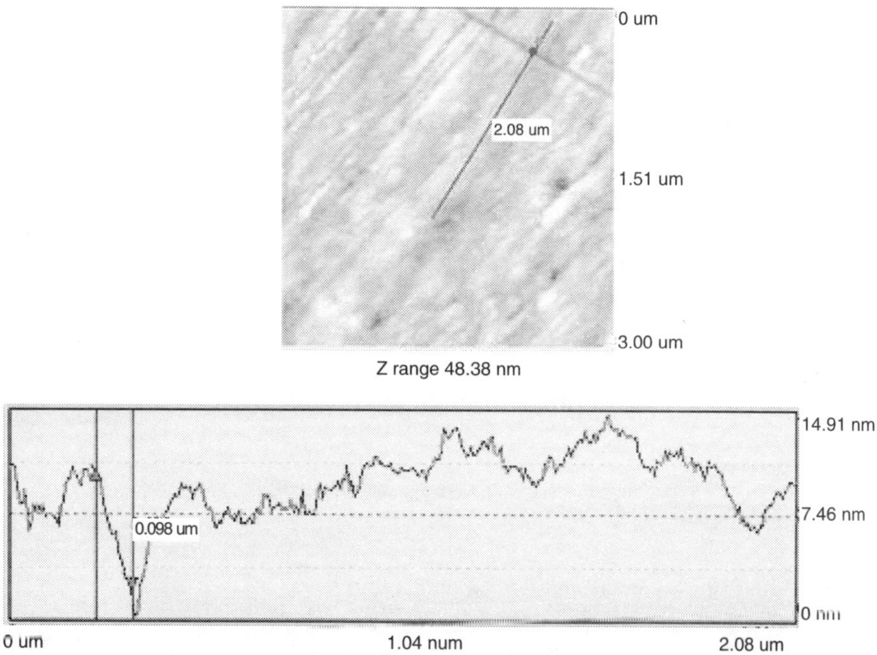

Fig. V–M.2: Metrological study of a scratch mark in the surface of a polished material (stainless steel). Because the AFM directly measures three-dimensional data, the depth of the scratch mark is easily quantified. In this case the mark is 8.6 nm deep. Z= 8.6 nm.

By measuring the changes in force/distance curves in an AFM it is possible to directly ascertain the thickness of lubrication films. Below is an example of a force/distance curve for a surface with no lubrication film compared to one with a lubrication film. The thickness of the film is established from the force/distance curve.

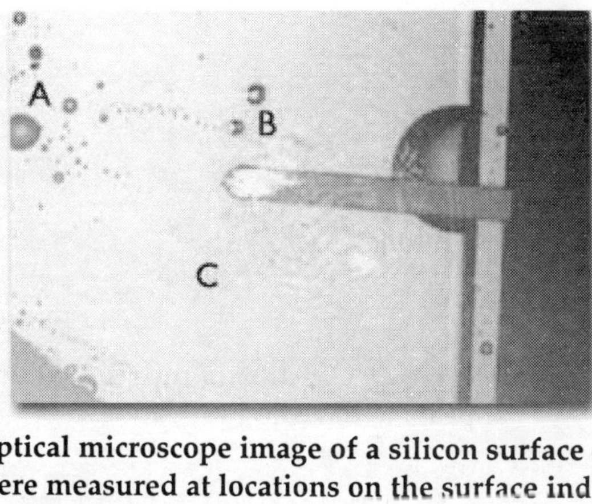

Figure V–M.3: Video optical microscope image of a silicon surface coated with a lubricant. Force/Position curves were measured at locations on the surface indicated with the letters A, B, and C. The red light in the video microscope image is from the laser used for the force sensor in the atomic force microscope.

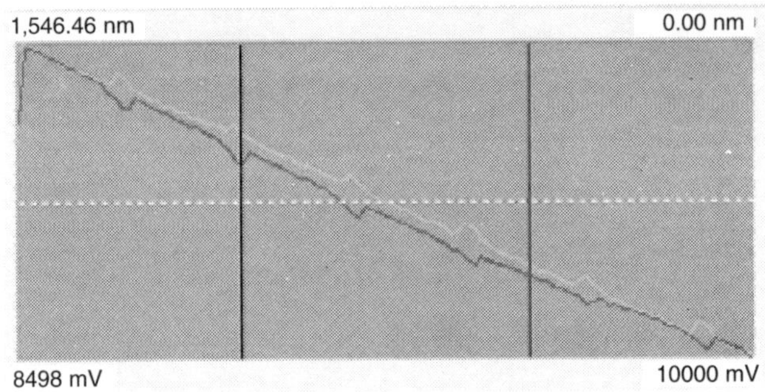

A: Force/Position curve measured on one of the droplets of lubricant. The slope represents the force required to move the AFM probe through the liquid surface.

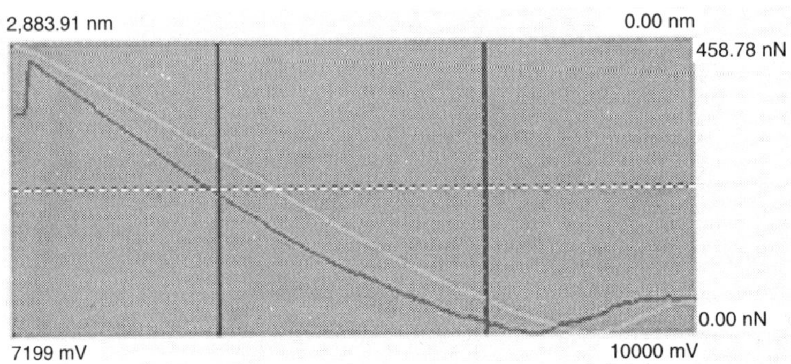

B: Force/Position curve at a place on the surface where there is apparently a thin film of lubricant material.

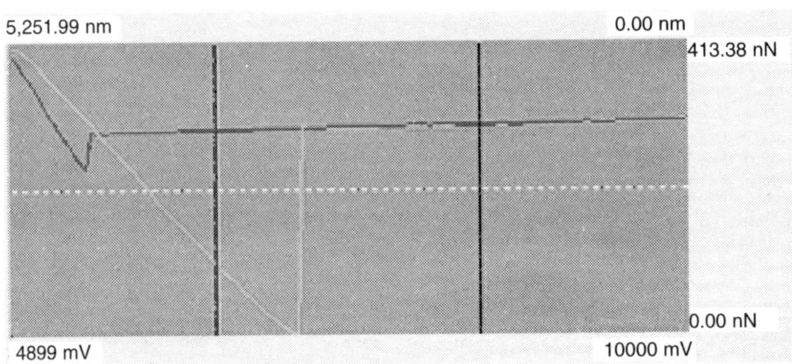

C: Force/Position curve on a section of the silicon surface that appears to have no lubrication.

Frictional Forces

Friction between two surfaces depends on the chemical and mechanical interaction between the surfaces. Changes in chemical composition giving rise to friction are measurable with the AFM. The technique for measuring these forces is called lateral force, or frictional force microscopy.

As the probe moves over a surface in the AFM, changes in the chemical composition of the surface can give rise to torsions of the cantilever on which the probe is mounted. The torsion of the cantilever is then proportional to the friction between the probe and the surface.

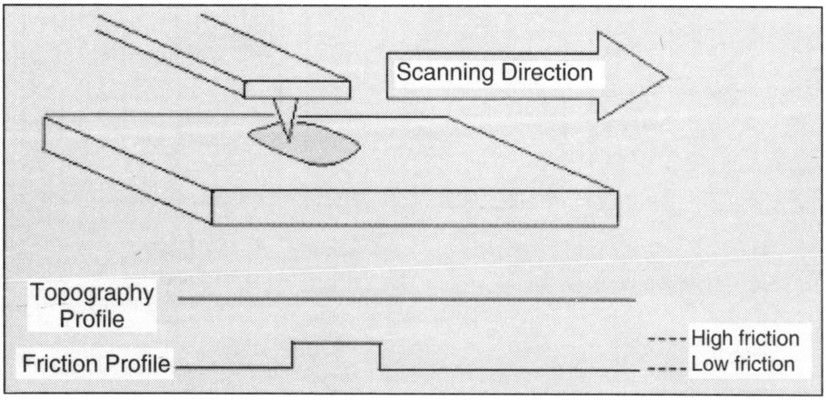

Fig. V–M.4: Illustration of the principle that allows lateral force or frictional fore microscope images to be measured. The cantilever will twist as the probe interacts with the surface while scanning.

In an AFM it is possible to simultaneously measure topography and frictional force images. The topography image is derived from monitoring the vertical forces on the cantilever and the friction image is acquired simultaneously by monitoring the lateral motions of the cantilever. Below is a FFM image of a sample illustrating changes in the friction.

Surface Texture/Morphology/Roughness

The AFM gives extremely high contrast on surfaces that are flat at the nanometer scale. Optical and electron microscopes are not able to resolve surface texture that is easily measured with the AFM. Applications include the visualization of surface topography in 2-D and 3-D perspectives, line roughness measurements, and area roughness measurements. All of the traditional area and surface roughness parameters can be calculated after the AFM image is acquired.

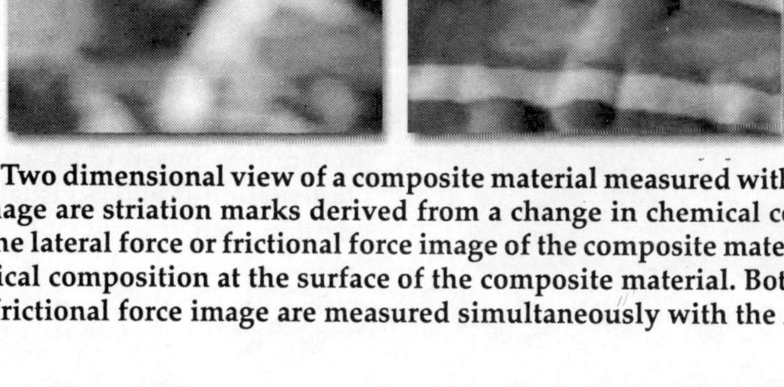

Fig. V–M.5: (left) Two dimensional view of a composite material measured with the AFM. Barely visible in this image are striation marks derived from a change in chemical composition at the surface. (right) The lateral force or frictional force image of the composite material clearly shows changes in chemical composition at the surface of the composite material. Both the topography and frictional force image are measured simultaneously with the AFM.

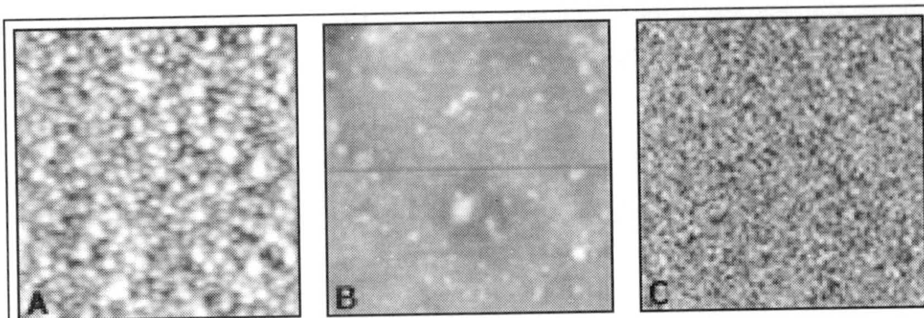

Ra:	0.71 nm	Sa:	0.157 NM
Rq:	0.90 nm	Sq:	0.197 NM
Rsk:	.45 nm	Ssk:	0.0119 NM
Rku:	3.67	Sku:	2.29
Rz:	4.63 NM	SZ:	1.38 NM

Fig. V–M.6: (A) AFM image of a metal bonding pad on a semiconductor device (B) Use of the AFM to measure line roughness of a polymer sample (C) AFM image of polished silicon and the area roughness of the silicon.

Nanoscale Mechanical Properties

Mechanical properties such as hardness, elastic modulus, stiffness and compressibility as well as material behavior such as plastic deformation, and fracture can be studied with the AFM.

It is possible to study nanohardness by directly pressing an AFM probe into a sample's surface; however, it is advantageous to use an instrument that is optimized for nanoindentation. The primary advantage of the nanoindenter over an AFM for nanohardness measurements is that it is easier to get calibrated measurements with the nanoindenter. It is useful to use the AFM to measure the three-dimensional topography of indentations made with a nanoindenter. AFM images allow direct visualization of material deformation or fracture behavior.

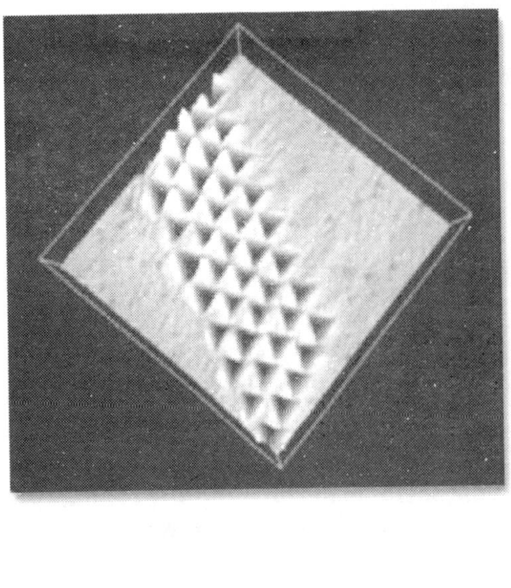

Fig. V–M.7: Three-dimensional view of nano-indents in a material surface. From the AFM image it is possible to see the depth of the indentation as will as visualizing the material deformation.

Using techniques such as pulsed force mode, the stiffness of a sample at a matrix of locations is measurable. From this data it is possible to create a stiffness mapping of a surface. Stiffness maps can only be made on samples where the stiffness of the surface is lower than the stiffness of the cantilever. Such stiffness images are routinely measured on polymer samples.

Adding a fixture to the stage of the AFM makes the study of material behavior such as plastic deformation and fracture possible. The fixture permits creating forces on a sample while AFM images are being taken. A variety of materials may be studied with such a technique.

Unit V–N

Nanotribology: Modeling Atoms when Surfaces Collide

Tribiology (friction, lubrication, and wear) is the science of interacting surfaces in relative motion. The word "tribology" is from the Greek *tribein*, meaning to rub. Although the term may be unfamiliar to many, its meaning is clearly conveyed among some of the oldest written records. An Egyptian painting dating back to 1880 BC depicts workers dragging a sled containing a heavy statue. One worker pours a liquid on the ground just before the runners to make the going easier. Leonardo da Vinci addressed the same problem when he wrote in 1519, "All things and everything whatsoever however thin it be, which interposed in the middle between objects that rub together, lighten the difficulty of this friction." An excellent source for further historical reading is *Dowson's History of Tribology*.

We have first-hand experience with friction any time we walk on ordinary ground and compare the outcome with what happens when we try to walk on ice. Wear of mechanical parts—as every car owner knows—eventually leads to failure and is one of the most costly problems facing industry.

To design and engineer mechanical parts, we need a macroscopic understanding of tribological processes. However, today the tolerances of many components, such as optical devices, computer chips, and ultra-smooth surfaces, are approaching the microscopic (atomic) length scale. In fabricating such precise components, we can greatly benefit from an atomic scale understanding of what happens when surfaces interact. The emerging field that lets us obtain this atomic scale understanding of fundamental processes of surfaces in motion is called molecular tribology or nanotribology. (A nanometer is one billionth or 10Ð9 of a meter; the size of an atom is about 0.3 nm.)

Recent progress in nanotribology is being driven by advances on several fronts. Of primary importance is the advent of probes that allow researchers to examine interacting surfaces at a microscopic level. One such probe is the atomic-force microscope, which has a very small tip with a radius of about 10 nm used to probe a surface. The surface force apparatus has also become a highly useful tool. Other major developments in the last decade are better models of interatomic forces, which, coupled with improved performance of computers, are giving researchers a new window into tip—surface interactions through computer simulations.

The Basics of Molecular Dynamics Modeling

The molecular dynamics method was originally developed in the late 1950s to study the statistical mechanical properties of a collection of atoms at equilibrium. The method is now a well-established and important tool in the fields of physics, chemistry, biology, and engineering. Recent advances in computer hardware and software allow us to simulate relatively large systems using increasingly realistic models of the forces between atoms.

In principle, MD modeling is simple. We start by identifying the position and velocities of all atoms in a system we want to model. Then we calculate the force on every atom from its neighbors and advance the positions of the modeled atoms according to Newton's equations of motion (force = mass × acceleration). During a simulation, we evaluate the response of a material—often copper, silicon, or silica glass—as it is subjected to an external force by following the response of each atom. Examples of external forces include stress fields, moving boundaries, and heat baths.

Despite the simplicity of MD modeling, the number of users and applications today remains limited; in addition, many researchers do not use it to its full potential. The reason for this under use is that this type of modeling can be "overkill" in that it is too expensive and requires too much detail for solving most practical problems. Furthermore, there is a certain art in selecting the appropriate boundary conditions and constructing interatomic-force models.

Consider, for example, Figure V-N.1, which shows the boundary conditions used in many of the simulations of indentation and cutting of a clean metal surface by a diamond-like tool. In this figure, the dashed line forms a box surrounding all the atoms that are represented. This box is called a simulation cell, which is the "window" on the system being modeled. Both the metal surface and the V-shaped diamond tool above it have "handles" to position the surfaces relative to one another. These are the dark circles forming the boundary. In practice, the tool is fixed in the rotational direction and slide the work surface from left to right during a simulation of cutting.

Immediately next to the boundary atoms is a region in which the atoms are constrained to be at room temperature. From an atomic point of view, temperature is a manifestation of the vibrational energy of the atoms. The role of the thermostat atoms in the tribology simulations is to draw away heat generated at the tip of the tool, mimicking a much larger piece of material than is represented in the illustration. The remaining atoms represented by open circles make up the Newtonian region, meaning that these atoms are free from further constraint and move according to Newton's equations. The tribological processes in this region are of interest in the MD simulation.

In many cases, we would like to simulate cutting or scraping over lengths that are far greater than the dimensions of the dozen or so atoms shown in Figure 14.1. However, as mentioned earlier, following the motion of a large number of atoms is prohibitive in terms of computer time and cost. To help overcome this problem, a relatively simple approach is used. Atoms are allowed to leave the simulation cell at the "downstream" end (to the right in Figure 14.1 as the simulated work piece moves under the tool). At the same time, periodically new atoms are inserted from the left at the upstream boundary. This approach restricts the population of atoms simulated at any given time, while allowing to model nanotribological processes in a much larger work piece. The extension of this model from two to three dimensions is straightforward.

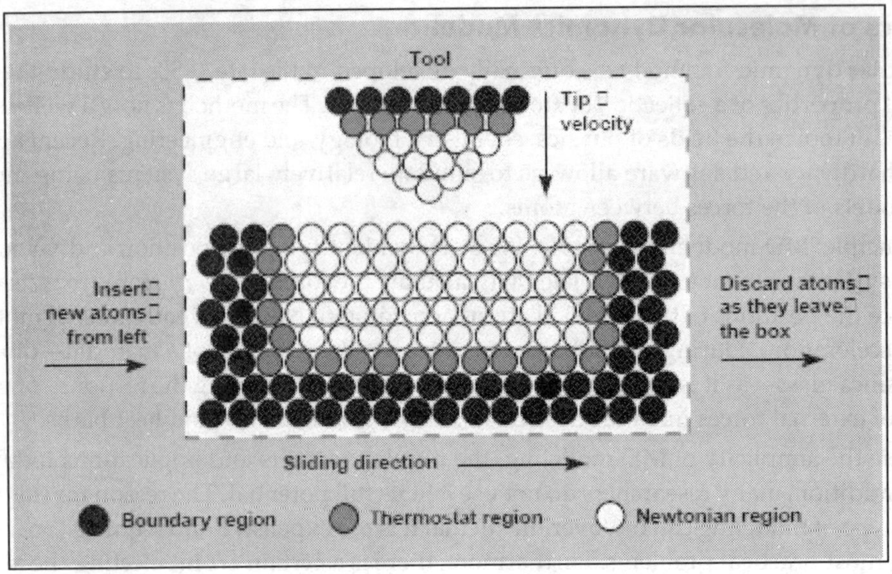

Fig. V–N.1: Geometry of a tool tip and work piece for a molecular dynamics (MD) simulation of indentation or cutting. The box, represented by the dashed line surrounding all the atoms, is the "window" into the system being modeled. Only the atoms in this box are evolved during an MD simulation. The top and bottom layers of boundary atoms (dark circles) are rigid and serve as "handles" to position the tool relative to the surface. The next layer of atoms (gray circles) are maintained at room temperature. Their purpose is to draw away heat produced at the tool tip. The remaining atoms (open circles) are free from constraint; that is, they obey Newton's equations of motion. Deformation processes in this interior region are the ones of interest in the MD simulations.

MD Simulations

Indenting Surfaces

One of the first simulations we refer to here involved the indentation of a metal surface by a blunted, triangular diamond tip moving vertically at a constant velocity ranging from 1 to 1000 m/s. Keep in mind that in this and all the subsequent work talked about here, the objective is to understand—at a more microscopic, atomistic level than was previously possible—what is happening when surfaces in relative motion interact with one another.

Figure V–N.2(a) is a "snapshot" of what happens when the tool indents the first three layers of metal. The interatomic forces in a metal arise from two effects: the interaction between electrons localized on the atoms, and the interaction of atoms with free electrons.

An embedded-atom model of the metal is used that incorporates both of these effects. In this figure, the atoms are shaded by the local value of stress. At first, the surface responds elastically (deforms without permanent loss of size or shape), and we see circular regions of constant stress, called the Hertzian stress field, which is well known from elasticity studies of contact mechanics. When the tool is pushed in further to indent six layers of metal, as in Figure V–N.2(b), the surface begins to flow plastically, and we obtain shear and discontinuity. In a 2D simulation like this one, dislocation edges are clearly visible within the metal along the darker colored slip bands that appear

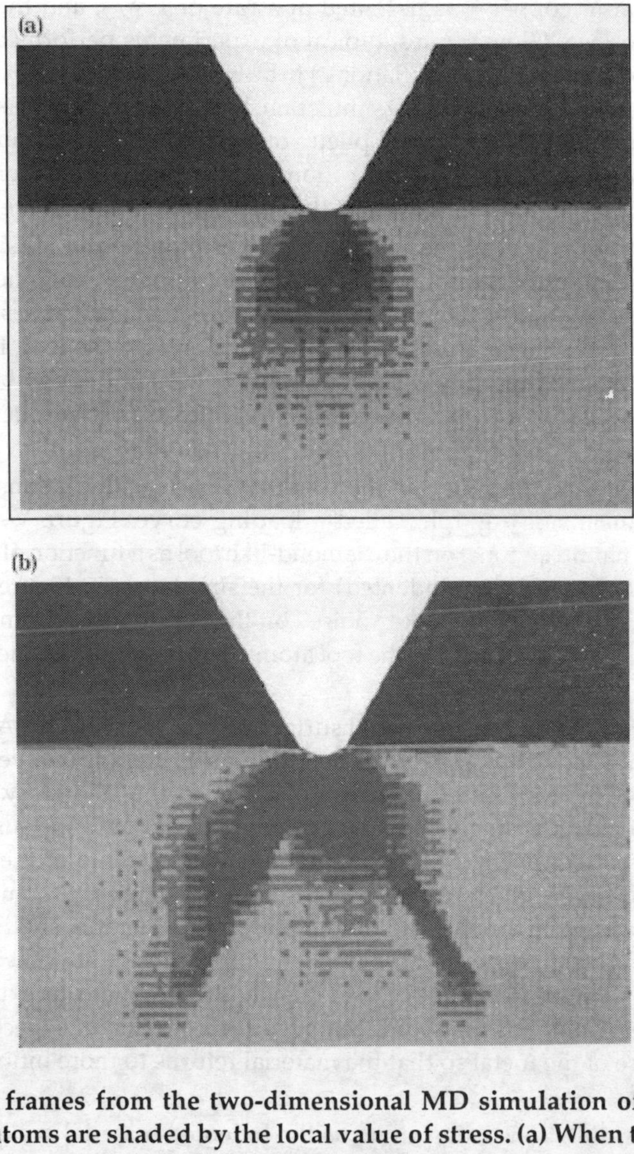

Figure V–N.2. Two frames from the two-dimensional MD simulation of the indentation of a metal surface. The atoms are shaded by the local value of stress. (a) When the tip is pressed three layers into the surface, the response is entirely elastic; that is, there is no permanent change in size or shape of the material. The circular regions of stress are well known from continuum elasticity studies of contact mechanics. (b) When the tip is pressed six layers into the surface, the material yields plastically by creating dislocations along slip bands visible at ±60 degrees from the horizontal surface of the metal. The plastic deformation means that the metal's change in shape remains when the tool tip is removed from the surface.

at 60° angles with the surface. In a close-packed metal such as this, the atoms behave like hard spheres (for example, billiard balls), and slip is analogous to sliding stacks of billiard balls over each other A 3D simulation of the indentation of copper gives a slightly different picture. Figure 14.3(a) is an image obtained after copper was indented at a rate of 1 m/s, and the simulated tool was withdrawn from the surface. (Slower rates, typical of experiments performed in a laboratory, are beyond the current capabilities of MD simulations.) In a 3D simulation, we do not see the distinctive bands of dislocation that are observed in a 2D simulation. Moreover, there is relatively little distortion at the metal surface. Instead, we find a small pileup of atoms on the metal surface, a few atoms in interstitial positions (between regular lattice positions), and a small dislocation loop (observable in Figure V–N.3a as the "step" or ledge of atoms on the surface) extending to the bottom of the crater.

Figure V–N.3(b) helps to explain the difference between the 2D and 3D simulations. Here, the atoms are colored according to the initial layering before the computer experiment. This cross section at the moment of full indentation of the copper surface shows that very little surface distortion has actually occurred. Remarkably few atoms have bulged out around the tool. Instead of distinctive dislocation bands, we see disorder only within a few layers of the sides and bottom of the tool tip. This localized disorder arises from point mechanisms (interstitials, vacancies, and so forth) of plasticity.

There are many other ways to display the results for such a simulation. One of the best and most revealing is by means of a graph, called a loading curve. Figure V–N.4 shows the load (instantaneous force in nanoNewtons) on the diamond-like tool as a function of depth of indentation (shown here as the number of layers indented) for the simulation in Figure V–N.3(a). The very rapid fluctuations in load—which look like "noise" on the curve—arise from the rapid motion of surface atoms repeatedly colliding against the tool atoms. The larger peaks and valleys are far more interesting in terms of what they reveal.

At first, the load rises linearly as the metal surface responds elastically. After indenting about 1.5 layers of copper, the load drops abruptly when the elastic stress is relieved. This abrupt drop, called "critical yielding," is reminiscent of what is observed in the laboratory. The first yielding in the simulation corresponds to a single copper atom popping out onto the surface from under the tool tip and relieving the stress energy. With more indentation, more of these point events take place. After indenting seven layers of copper, the tool is stopped. At this point, the surface accommodates nearly all the tip through elastic and plastic deformation, and there is little pileup of atoms around the tip, as observed in the cross section in Figure V–N.3b. When the direction of the tool is reversed (lifted from the surface), the load quickly drops to zero, as expected. However, the load suddenly rises once again as the tool is removed further. This unexpected rise is caused by annealing at the surface of the metal so that the material returns to more intimate contact with the tool, and the load rises.

The results from simulations using silver as the indented material are quite similar to those just described for copper. However, because silver is softer, the yielding occurs at a smaller load than for copper.

Cutting and Chip Formation

Cutting is a widely used process in fabricating components. To obtain a more basic understanding of what happens when a metal is cut, a simplified geometry is used. We idealize the model slightly by limiting the motion to two dimensions or thin, periodic slabs in three dimensions. This idealization, known as orthogonal cutting, assumes that the material being cut is sufficiently wide that edge effects have essentially no influence on the results.

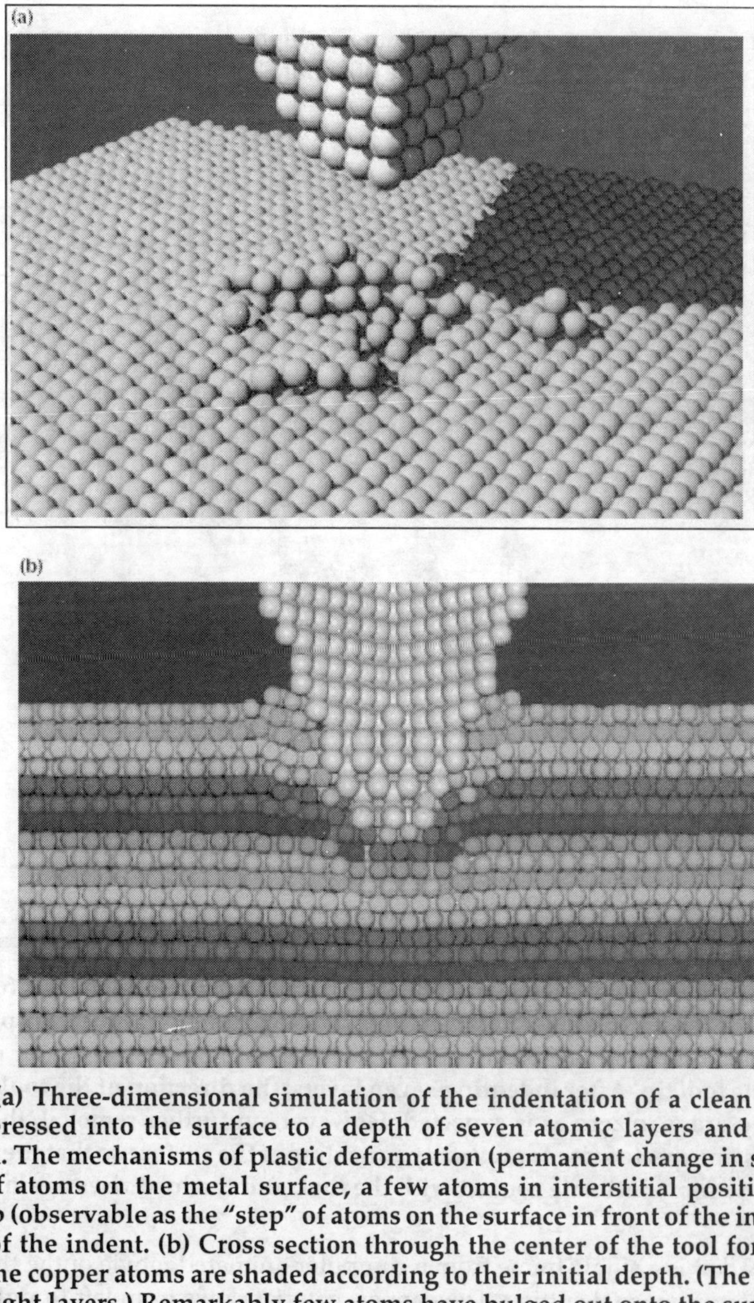

Figure V–N.3. (a) Three-dimensional simulation of the indentation of a clean copper surface. The tool was pressed into the surface to a depth of seven atomic layers and removed to the position shown. The mechanisms of plastic deformation (permanent change in shape) include a small pileup of atoms on the metal surface, a few atoms in interstitial positions, nd a small dislocation loop (observable as the "step" of atoms on the surface in front of the indent) extending to the bottom of the indent. (b) Cross section through the center of the tool for the simulation shown in (a). The copper atoms are shaded according to their initial depth. (The shading pattern repeats every eight layers.) Remarkably few atoms have bulged out onto the surface around the tool. Through elastic and plastic deformation, the surface is able to accommodate nearly the entire volume of the tool.

Orthognal cutting is stimulated in both two and three dimensions. The 2D simulations more accurately model the length scales and time scales that are common in laboratory experiments using single point, diamond-turning machines. In this type of simulation, cutting speeds range from about 10 to 100 m/s. We have also varied the sharpness of the rigid tool tip used for cutting. We have used tool-edge radii of curvature ranging from 1 to 20 nm. The bluntest tip (20 nm) approaches the sharpest real-life tool tip, which has a radius of curvature of about 35 nm.

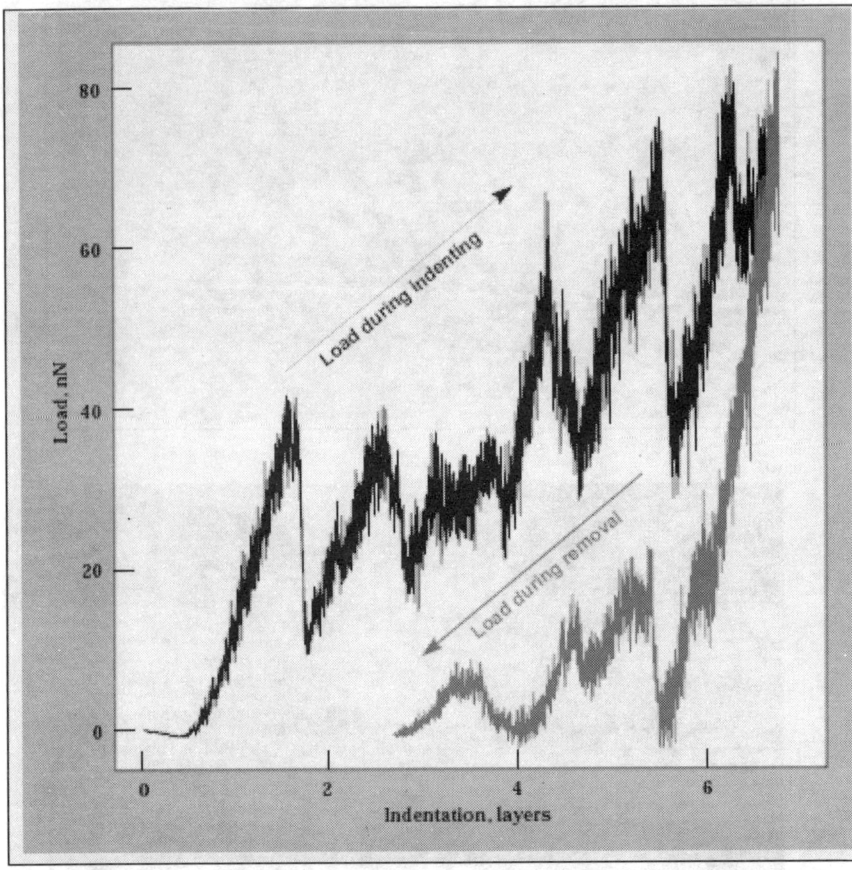

Fig. V–N.4: The load (normal force in nanonewtons) on the diamond tool as a function of depth of indentation for the simulation shown in Figure V–N.3(a). The abrupt drops in the loading curve correspond to individual plastic events, such as an atom popping out onto the surface from beneath the tool tip. After indenting seven layers, the direction of the tool is reversed, and the load quickly drops to zero as is expected. However, on further removal, the load suddenly rises again. This unexpected rise is caused by annealing at the surface so that the material returns to more intimate contact with the tool, and the load rises again.

Figure V–N.5(a) is a single frame from a computer-animated movie of the three-dimensional MD simulation of orthogonal cutting. The copper material flows from left to right at 100 m/s. In this simulation, the carbon atoms in the diamond-like tool are not permitted to move, so they serve as the fixed frame of reference.

Figure V–N.5(b) is a projection in two dimensions of all the atomic positions in Figure V-N.5(a). This type of projection makes it easier to see some important effects. During our simulation of cutting, the system forms a chip. The chip is reminiscent of what is actually observed during real experiments. The chip remains crystalline, but it has an orientation different from that of the surface.

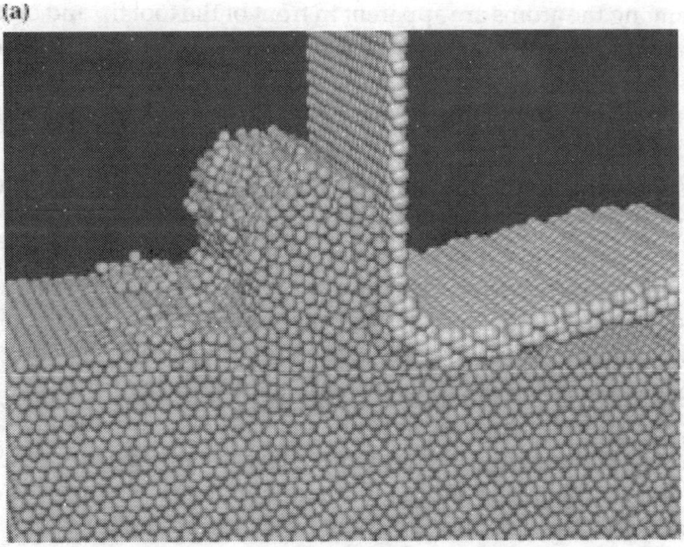

Fig. V–N.5: A 3D simulation of orthogonal cutting of copper (a) after the formation of a chip and (b) with all atomic positions projected into two dimensions. In this MD simulation, the carbon atoms in the tool are not permitted to move, and the copper material is flowing from left to right. The chip remains crystalline, and regions of disorder are seen in front of the tool tip and on the surface in front of the chip. These regions of disorder, clearest in (b), are connected by a zone of intense shear.

Regions of disorder among the atoms are apparent in front of the tool tip and on the surface in front of the chip. Unlike the 3D studies of point indentation, dislocations readily form during orthogonal cutting. Unlike the single point in the indentation simulations, the cutting tool in the orthogonal cutting geometry forms an infinite line, like a knife edge. This geometry provides much more energy for the creation of dislocations.

The calculations show that the cutting force strongly depends on the sharpness of the tool. This relation is also observed experimentally. Dull tools (those with a large tip radius) require larger forces to achieve the same depth of cut as sharp tools. A good measure to quantify the observations is the work (force x distance) performed by the cutting tool divided by the volume of material removed, also known as the specific energy. This energy increases dramatically with decreasing depths of cut. For our shallowest (nanometer scale) cut, the specific energy exceeds the energy required to vaporize the material, though the material remains a solid.

The dependence of specific work on depth of cut has been observed in macroscopic metal cutting for many years and is known as the size effect. However, the macroscopic size effect is much less dramatic than the size effect observed in the simulations or in experiments using single-point, diamond-turning machines. The transition between macroscopic and microscopic behavior occurs at a length scale of a few micrometers, comparable to the average grain size in most metals. This result is interpreted as a change in the mechanisms of deformation, from grain-boundary sliding and motion of existing dislocations at the macroscopic scale to the creation of new dislocations and other point mechanisms of deformation at the microscopic, atomic scale. These dislocation-creation and point mechanisms of deformation consume significantly greater energies, leading to the observed size effect.

In contrast to copper, materials like silicon and iron are not considered machinable with diamond tools. The reason is that the diamond tool wears rapidly, and it becomes difficult to maintain contour accuracy. The increased wear occurs for all materials that form strong chemical bonds (covalent bonds) with carbon. We have investigated and identified some of the underlying processes that give rise to these effects.

In covalent materials like silicon, the nature of interatomic forces is much different from that in metals. The strength of a covalent bond depends on the local environment, and the material forms an open structure. The basic method is the same as that for cutting copper, except that now we evolve not only the silicon surface material but also the carbon atoms in the diamond tool to allow for tool wear. It has been found that both copper and silicon show ductile behavior during cutting; however, the underlying mechanisms that allow this behavior are very different for the two materials. Whereas a copper chip remains crystalline, a silicon chip is transformed into a completely different state.

Figure V–N.6 is a cross-sectional view, just as the one in Figure V–N.5(b), but this time of a silicon-cutting simulation. In this case, the diamond wedge is cutting at a speed of 540 m/s. Although no wear of the tool is apparent yet, a single layer of silicon atoms has coated the diamond tip. Chip formation occurs between this layer of silicon atoms and the crystalline surface. In related simulations of diamond asperities (abrasive tips) scraping a silicon surface (as in grinding processes), it was found that the diamonds wear by forming small clusters of silicon carbide, which remain on the silicon surface.

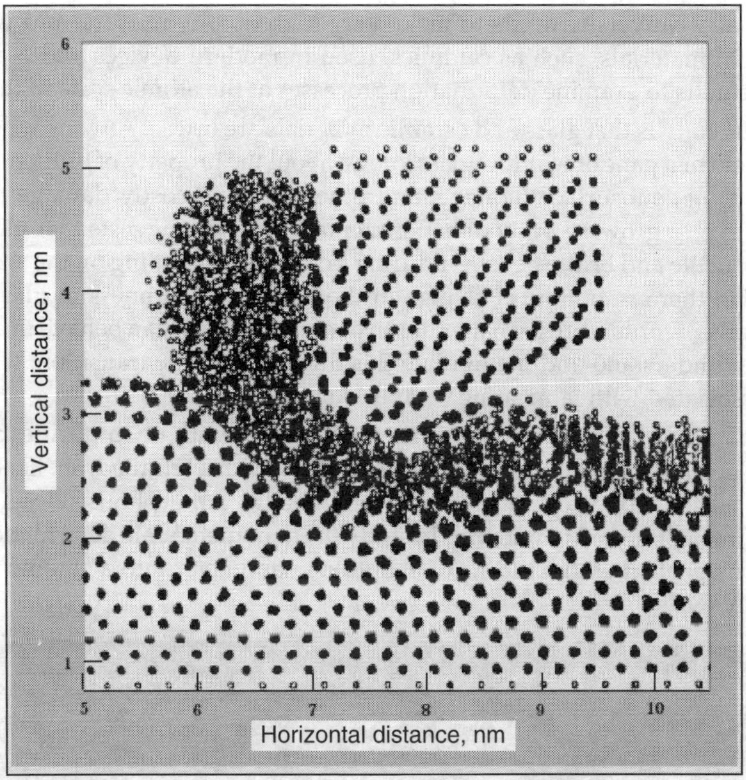

Fig. V–N.6: A 3D simulation of orthogonal cutting of silicon. As in Fig. V–N.5(b), all the atomic positions are projected into two dimensions to make certain features more visible. Unlike copper, silicon is transformed from a crystalline state into an amorphous state during cutting. Both the chip and the first few layers of newly machined surface are amorphous. A thin layer of silicon atoms also coats the diamond tool. The chip forms between the bulk silicon and this thin coating.

In Figure V–N.6, the silicon material in the chip and in the first few layers of newly cut surface appears to be amorphous or possibly to have melted. The temperature in the chip, calculated from the vibrational motion of silicon atoms, is comparable to the melting temperature of bulk silicon. However, the silicon atoms in the chip are not diffusing, which demonstrates that the material is, indeed, in a solid state.

The following idea has been developed to explain chip formation in crystalline silicon. Because of the nature of covalent bonds in crystalline silicon, the energy needed to shear the crystal is enormous. In fact, less energy is required to transform the crystal into an amorphous solid and to shear the amorphous solid than to shear the crystal. The interplay between the energetics of different deformation processes and the dependence on length scale may explain the transition from brittle to ductile behavior that is observed in ceramic materials.

How Cracks Propagate in Glass

The ability to precisely machine ceramic surfaces like silica glass affects many different fabrication technologies and is central to constructing a wide range of sophisticated optics systems.

A laser program at a University needs to make very high-quality mirrors, and we need to know more about special materials, such as ceramics, used in modern devices and equipment. We are applying MD methods to examine deformation processes at the atomic scale in these materials.

The main difficulty is that glass and ceramic materials are brittle. Anyone who has seen a rock impact a windshield or a pane of window glass knows about the property of brittleness. This property can result in cracking, subsurface damage, and other kinds of costly damage during precision machining. However, a growing body of experimental evidence suggests that glass and ceramics can exhibit both ductile and brittle behavior during grinding, depending on the size of the abrasive used. In particular, there is an abrupt change in the surface smoothness of glass with a smaller abrasive size. In other words, with a smaller depth of cut, even glass can behave in a ductile manner. If we could better understand and manipulate this ductile-to-brittle transition, we could improve the economics associated with fabricating ceramic components.

To address this problem, simulation of fused silica glass with no impurities was undertaken. In silica, four oxygen atoms surround each silicon atom to form a tetrahedron. Each oxygen atom is shared with two silicon atoms to form connected tetrahedra. Amorphous silica glass is a random network of these interconnected tetrahedra. To study the nanometerscale deformation of fused silica, a diamond tip was pushed with a radius of about one nanometer into a smooth silica surface, as shown in Figure V–N.7.

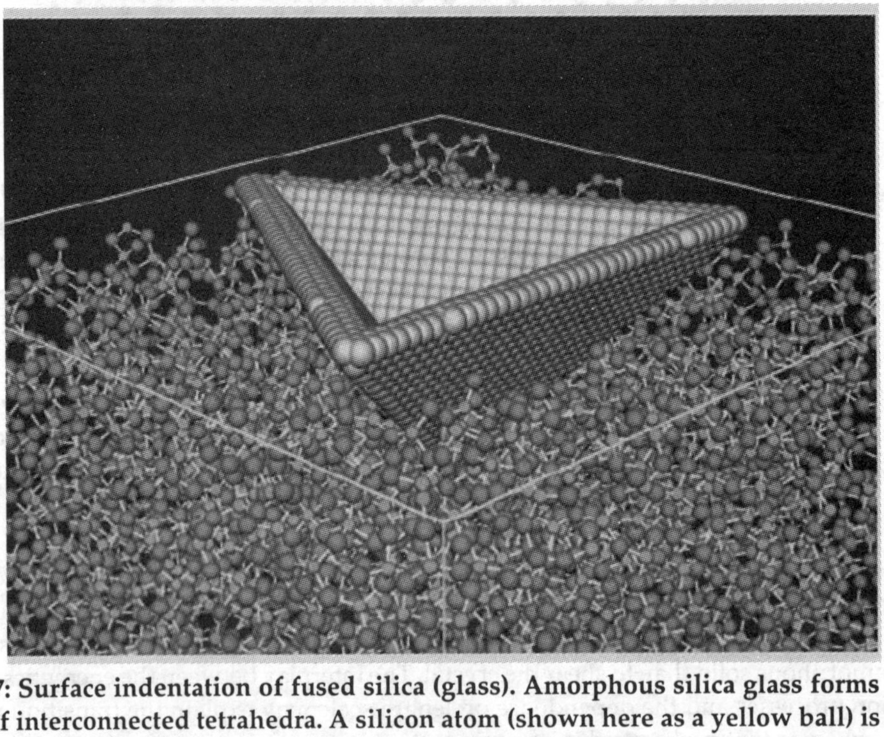

Fig. V–N.7: Surface indentation of fused silica (glass). Amorphous silica glass forms a random network of interconnected tetrahedra. A silicon atom (shown here as a yellow ball) is located at the center of each tetrahedron; an oxygen atom (blue ball) is located at each of the four corners. During our simulated indenting of a silica surface, we do not observe the abrupt drops in the loading curve that were seen in copper. Instead, the response of a silica surface is almost entirely elastic up to 1.25 nm. After 1.25 nm, there is a slight change in slope of the loading curve, indicating the onset of plastic deformation.

It was found that the silica surface responds much more elastically than did either the copper or silicon surfaces. For indentations up to 1.25 nm, it was observed that no plastic deformation occured. At an indentation of 1.25 nm, a significant rearrangement of the silica network occurs directly beneath the tool tip. This rearrangement, an example of which is shown in Figure V–N.8, leads to a decrease in the slope of the loading curve. When the tool is removed, the unloading curve does not follow the loading curve, as it does for elastic indentations up to 1.25 nm. The area between the two curves is a measure of the work performed by the tool in rearranging the silica network. These calculated loading and unloading curves are strikingly similar to those observed experimentally.

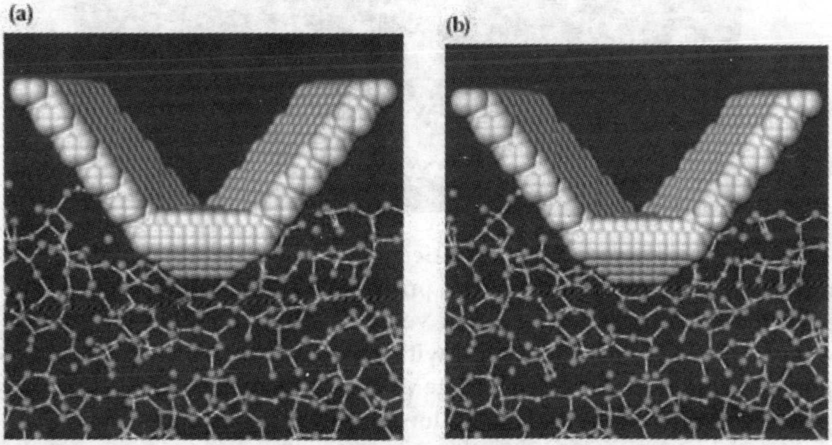

Fig. V–N.8: A cross-sectional plot through the center of the diamond tool for the simulation shown in Figure 50 immediately before (a) and immediately after (b) the onset of plasticity when the material yields. The bond shaded in red in (a) is one of the bonds that breaks when the silica network rearranges.

It is essential to learn more about cracks for a better understanding of deformation in glass. To simulate the propagation of a crack into a silica surface, we put a notch on the top surface of the glass and let the computer simulation pull apart the sample at a constant velocity. The critical stress for crack propagation depends on the geometry and length of a pre-existing crack. Longer cracks, for example, yield at lower applied stress than do shorter cracks. In the simulation shown in Figure V–N.9, it was observed that the crack length elongated by 30 per cent, stopped, and then yielded again at lower stress. This "stick-slip" phenomenon was been observed during crack propagation in fiber reinforced materials. The sticking is determined by the microstructure. In this case, the crack stopped after entering a large void. The void permits additional stress relaxation, and the crack tip halts until sufficient stress builds up again.

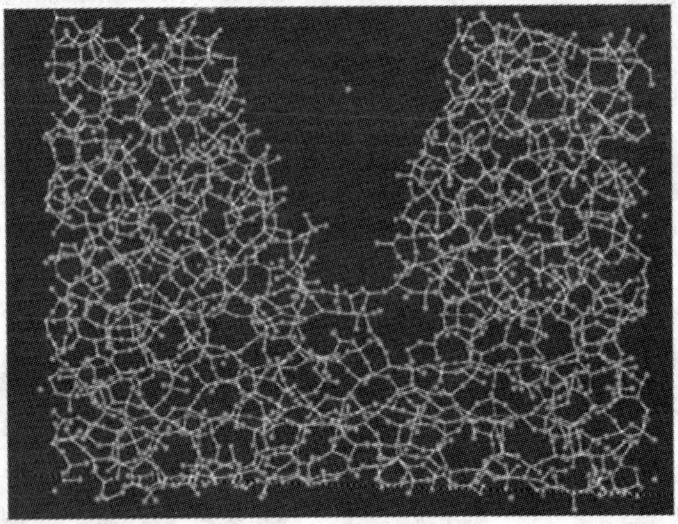

Fig. V–N.9: Large-scale, or macroscopic, plasticity in silica is dominated by the propagation of cracks. We simulated the mechanism of crack propagation by creating a notch on a silica surface and pulling the material apart at a constant velocity. Our calculated critical stress (the stress required to move the crack tip) agrees well with experiments. We also observed a novel stick-slip phenomenon. As the crack tip enters a large void, such as the one just below and to the right of center, the tip momentarily halts as the stress is lowered, then the crack continues to propagate again.

SECTION F

NANOMEDICINE

Unit VI—A

The Prospect of Nanomedicine

A Noble Enterprise

The history of disease is vastly older than that of humankind itself. Indeed, disease and parasitism have been inseparable companions to life since the dawn of life on Earth. Fossilized bacteria similar to those responsible for many infections that afflict people today have been found in geological formations that are 500 million years old. Fossil shells dating from an era almost equally remote show clear evidence of disturbance by injury and parasites. Examination of the skeletons of long-extinct dinosaurs and other great reptiles show that these creatures suffered from fractures, bone tumors, arthritis, osteomyelitis, dental caries and other diseases that still plague us in the 20th century. While available fossil evidence is largely limited to changes observable in bones and teeth, it is probably safe to assume that disease processes were equally prevalent in the soft organs and tissues that have not been geologically preserved, and that the general pattern of disease has not changed in its essentials during the hundreds of millions of years that animal life has existed on this planet.

Since its first appearance on the prehistoric stage millions of years ago, the human body has also been constantly subject to assault and injury, invasion by parasites, extremes of heat and cold, and infections. Early man probably suffered from a number of diseases due to nutritional factors and body chemistry disorders. Diseases that may date back more than 25 million years to the ape ancestors of modern apes and man are amoebic dysentery, malaria, pinworm infections, syphilis, yaws, and yellow fever. Diseases which may have appeared and evolved with man include leprosy and typhoid; certain modern diseases such as cholera, measles, mumps, smallpox, whooping cough, and the common cold require large concentrated populations to support them, thus probably could not have existed in the prehistoric era.

The art of surgery probably must first have originated as a response to immediate crises. The first and most obvious course of action of a wounded man would be to protect the site of injury from the influence of external forces or agents. For this there was, and remains to this day, only one means—the application of a dressing. Many observations were made and many substances tried, and in time a body of experience was accumulated and passed on orally to others for use in similar emergencies, eventually creating a considerable sum of inherited empirical knowledge. It is believed that the art of dressing wounds long constituted the whole of medicine—the use of internal remedies or herbs, and use of the knife or fire, came much later.

Present-day primitive and folk medicines provide additional clues to early medical practice. For example, many early peoples developed effective methods to control bleeding—the use of cobwebs is an ancient folk remedy, as is the application of tourniquets, packing with absorbent materials, the laying on of snow or, at the other temperature extreme, the application of cautery by hot knife or spear.

While able to deal with wounds, early man did not admit the existence of disease from "natural causes". Internal diseases were generally ascribed to malevolent influences exercised by a supernatural entity or a human enemy. As centuries passed, many internal diseases were eventually recognized and simple treatments empirically established. But the rational basis for internal medicine awaited basic physiological knowledge such as the circulation of the blood, finally proven by William Harvey (1578-1657) in 1628, and the discovery and acceptance of the theory of infection by microorganisms in the 1800s. Throughout most of human history, life has been short indeed—Table VI–A.1 shows the average age of death for a sampling of persons buried in Roman times. (Some variation may represent regional differences in burial practices.) Even by the 17th century, more than half of all children never lived past the age of ten, often succumbing to diseases such as cholera, diphtheria, scarlet fever, or whooping cough, or being scarred for life by smallpox. It was a common and widely accepted fact of life that people would die at all ages, although wealthy families could leave town every year during the cholera season.

Table VI–A.1: Average Age of Death of Persons Buried in Roman Times

Burial Location	*Average Age of Death*
City of Rome	29.9 years
Iberia	31.4 years
Britain	32.5 years
Germany	35.0 years
North Africa	46.7 years

Bacteria are among the last remaining "wild animals" on Earth that threaten man. As our instrumentalities continue to progress from the macroscale to the microscale, and finally to the molecular or nanoscale, all of the remaining natural "wild things" that endanger human life and health—whether viruses, bacteria, protozoa, metazoan parasites, or even our own pathological native cells—will be confined and tamed, reconstructed or eliminated. Using smaller tools, we hunt smaller prey. As with sabertooths in the post-Neolithic era, and with wolves in post-medieval times, people of the 21st century will no longer fear or need suffer predation by wild microbes or tumor cells run amok.

Humanity is poised at the brink of completion of one of its greatest and most noble enterprises. Early in the 21st century, our growing abilities to swiftly repair most traumatic physical injuries, eliminate pathogens, and alleviate suffering using molecular tools will begin to coalesce in a new medical paradigm called nanomedicine. Nanomedicine may be broadly defined as the comprehensive monitoring, control, construction, repair, defense, and improvement of all human biological systems, working from the molecular level, using engineered nanodevices and nanostructures.

Here we shall explore in preliminary fashion a few of the technical details, specific requirements, and physical limitations of these engineered nanodevices, and the meaning of nanomedicine within the context of the traditional healing arts.

Current Medical Practice

In order to fully appreciate the changes that nanomedicine will inevitably bring, it is useful first to review the history and development of current medical practice. After summarizing the history of scientific medicine and placing nanomedicine in its proper historical context, we define exactly what is meant by "medicine" and present a new model for "disease". The modern medical treatmentmethodology is examined in light of the changes that will be wrought by nanomedicine, and this is followed by a discussion of several important issues in the evolution of medical bedside practice and the changing view of the human body.

The Evolution of Scientific Medicine

A study of the history of "scientific" or Western medical practice suggests a continuity in certain aspects of the medical paradigm even since ancient times—as for instance the basic precepts of observation and diagnosis, followed by treatment—but reveals changes in attitudes, techniques, and instrumentalities, and a gradual evolution in the epistemology and ontology of medical logic. Figure VI–A.1 presents a simple model of this historical progression, the details of which will soon become clear. Note that 2010 is marked as a possible date for the first applications of nanomedicine. These will most likely be ex vivo applications only in vivo nanomedical treatments may come much later.

Another important reason to study the history of medicine is to gain a deeper appreciation of the long, hard struggle to improve human health, a struggle that is expected finally culminate in victory in the 21st century. If the ~10 billion people that have ever lived survived an average of 40 years and spent 5 per cent of their lives in sickness or physical misery from disease, then ~200 trillion man-hours of suffering have been paid to achieve this remarkable result, a not inconsiderable price.

Prehistoric Medicine

The elementary nature of prehistorical medical practice has already been mentioned. There is direct evidence that Stone Age man used a natural fungus as a treatment for intestinal parasites. Besides basic wound-tending, the practice of circumcision is an age-old form of simple surgery with a rational, if sometimes controversial, hygienic basis.

Ancient Mesopotamian Medicine

The first written records, which came from ancient Babylon and Egypt, contain the earliest references to medical care, obviously codifying earlier practices no record of which has survived. The medicine of the Sumerians of Mesopotamia from ca. 3000 BC was primarily religious. The Mesopotamian peoples saw the hands of the gods in everything. Disease was caused by spirit invasion, sorcery, malice, or the breaking of taboos, and sickness was both judgement and punishment.

But medicine also had an empirical component, with some sicknesses being ascribed to cold, dust and dryness, putrefaction, malnutrition, venereal infection and other natural causes. With the discovery of distillation, the Mesopotamians made essence of cedar and other volatile oils. Turpentine, asafetida, henbane, myrrh, mint, poppy, fig, and mandrake are also mentioned.

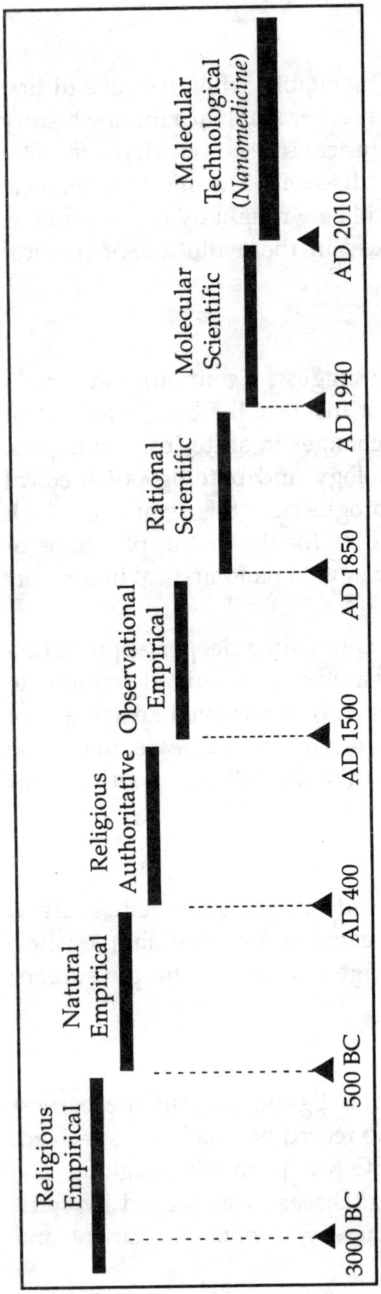

Fig. VI–A.1. Evolution of the paradigm of scientific medicine.

Surgical conditions such as wounds, fractures and abscesses were also treated by Mesopotamian surgeons. Practitioners were priests, and after 2000 BC, they were ruled by the strict laws included in the Code of Hammurabi. The Hammurabic Code mentions the Gallabu, or barber-surgeons, whose province was minor surgery, including dentistry and the branding of slaves. If Herodotus (ca. 485-425 BC) may be believed, Babylonian medicine must have declined in the 5th century BC. Herodotus states that there were no physicians, but that the people brought their sick into the marketplace in order that passers-by might make suggestions or offer cures.

Ancient Egyptian Medicine

The civilization of ancient Egypt dates from around 3000 BC. As in Mesopotamia, ancient Egyptian medicine was religious-empirical. An examination of the preserved bodies of members of the Royal family has provided much information on the diseases of ancient Egypt, including congenital deformities such as clubfoot, dental caries, gallstones, bladder and kidney stones, rheumatoidarthritis, mastoiditis, numerous eye diseases, and bone fractures, some of which show treatment by quite sophisticated splinting. Other evidence suggests that the average ancient Egyptian was extremely diseased.

The Ebers Papyrus (ca. 1550 BC), deriving from Thebes, is the principal medical document and may be the oldest surviving medical book. Over 20 meters long, it deals with scores of diseases and proposes remedies involving spells and incantations, but also includes many rational treatments.

The Edwin Smith Papyrus (ca. 1600 BC), discovered by its American namesake at Luxor in 1862, may be the world's earliest surviving surgical text, and contains material probably derived from even more ancient times.

Ancient Greek Medicine

By 1000 BC the communities later collectively known as the Greeks were emerging around the Aegean Sea. How much medical knowledge they took from Egypt remains controversial, but the contrasts between the two are striking. Little is known of Greek medicine before the appearance of written texts in the 5th century BC. Archaic, The Homeric epics (ca. 600 BC) offer glimpses of early Greek medicine. Scholars count 147 cases of battle wounds in the Iliad, including 106 spear thrusts, 17 sword slashes, 12 arrow shots, and 12 sling shots. Among the arrow wound survivors was King Menelaus of Sparta, whose physician extracted the arrow, sucked out the blood and applied a salve.

Hippocratic medicine, the foundation of Greek written medicine, explicitly grounds the art upon a quite different basis—a healing system independent of the supernatural and built upon natural philosophy. The beginnings of true medical science in the West were established when the reliance on superstition that underpinned tribal medicine was replaced by civilized and rational curiosity about the cause of illness. The growth of civilized thought allowed for argument on medical cause and cure, with great doctrinal multiplicity. The separation of medicine from religion reveals another distinctive feature of Greek healing: its openness, a quality characteristic of Greek intellectual activity in general, owing to political diversity. Those calling themselves doctors (iatroi) had to compete with bone-setters, exorcists, root-cutters, incantatory priests, gymnasts, and showmen, exposed to the quips of playwrights and the criticism of philosophers. Medicine was open to all.

Empedocles (fl. 450 BC) may have been the first to advance some of the key physiological doctrines in Greek medicine, including innate heat as the source of living processes such as digestion, the cooling function of breathing, and the notion that the liver makes the blood that nourishes the tissues. His contemporary, Alcmaeon of Croton (fl. 470 BC), believed that the brain, not the heart, was the chief organ of sensation. Alcmaeon's examination of the eyeball led him to discern the optic nerve leading into the skull, a genuine observational basis. He gave similar explanations for the sensations of hearing and smelling, because the ear and nostrils suggested passages leading to the brain.

The sixty or so works comprising the Hippocratic Corpus (ca. 440-340 BC) derive from a variety of hands, and, as with the books of the Bible, they became jumbled up, fragmented, then pasted together again in antiquity. What is now called the Corpus was gathered around 250 BC in the Library at Alexandria, with further texts added later still.

The Hippocratics promulgated the idea of "vis medicatrix naturae," or the power of nature to cure itself, and thus the belief that there was a natural tendency for things to get better on their own. This tendency could be aided by providing a beneficial environment for the patient and by improving physical function with a regimen of suitable diet, lifestyle, and exercise—the diatetica. In extreme cases, further aids to recovery could be sought. The art of diagnosis involved creating a profile of the patient's lifestyle, work and dietary habits, partly by asking questions and partly by the use of trained senses.

The technique most prized among Hippocratics was the art of prognosis—a secular version of the priestly and oracular prognostications of earlier medicine, and bearing some analogy to the 20th century weatherman, who can give a bright or gloomy forecast but is powerless to change it.

The ultimate significance of Hippocratic medicine was twofold. First, it carved out a lofty role for the selfless physician which would serve as a lasting model for professional identity and conduct. Second, it taught that an understanding of sickness required an understanding of nature.

Ancient Alexandrian Medicine

Soon after the death of Aristotle (d. 322 BC) and his most famous pupil, Alexander the Great (d. 323 BC), a great medical school was founded in Egypt at the court of King Ptolemy, at his capital, Alexandria, at the mouth of the Nile. The two earliest teachers at the Alexandrian medical school were also its greatest—Herophilus of Chalcedon and his contemporary, Erasistratus of Chios. Their writings having been lost, we know about them only through later physicians.

Herophilus was the first to dissect cadavers in public. Herophilus observed that the coats of arteries were much thicker than those of the veins, thus he speculated that the arteries were filled not with air but with blood. Herophilus wrote at least eleven treatises, discovering and naming the prostate and the duodenum (from the Greek for twelve fingers, the length of gut he found). He also wrote on the pulse as a diagnostic guide and on therapeutics, ophthalmology, dietetics, and midwifery. He recognized the brain as the central organ of the nervous system and the seat of intelligence, extending the knowledge of the parts of the brain, certain of which still bear titles translated from those given by him. He was the first to grasp the nature of the nerves, which he distinguished as motor and sensory, though he did not separate them clearly from tendons.

Erasistratus surmised that every organ is formed of a threefold system of "vessels"—veins, arteries, and nerves, dividing indefinitely. These, plaited together, were postulated to make up the tissues. In the brain, Erasistratus observed convolutions, noting that they were more elaborate in man than in animals and associating this with higher intelligence. He distinguished between cerebrum and cerebellum, and is often regarded as an early mechanist because of his model of bodily processes—for instance, digestion involved the stomach grinding food. He was opposed to intrusive remedies such as venesection, and his favorite therapeutic measures were regulated exercise, diet, and the vapor bath—very much in the Hippocratic tradition.

Ancient Roman Medicine

Roman tradition held that one was better off without doctors. According to Cato citizens had no need of professional physicians because Romans were hale and hearty, unlike the effete Greeks. Apparently Romans enjoyed bad-mouthing Greek physicians. Thus the Romans despised medicine as a profession but this did not prevent them from making use of Greek physicians or even of the services of their own slaves. In the Roman empire, the earliest scientific teacher was a Greek, Asclepiades of Bithynia.

Back in the mainstream, Galen of Pergamum provided the final medical synthesis of antiquity and the effective medical standard for the next 13 centuries. Galen developed a characteristic physiological scheme that remained in vogue until the 17th century. It supposes three types of so-called spirits associated with three types of the activity of living things. These were the natural spirits formed in the liver and distributed by the veins, the vital spirits formed in the heart and distributed by the arteries, and the animal spirits formed in the brain and distributed by the nerves. As in Greece, medicine remained personal in Rome. No medical degrees were conferred or qualifications required.

Medicine in the Middle Ages

The passage from the glorious days of Rome to the Middle Ages was often violent, especially in the West, with wave after wave of barbarian onslaughts from the East. Galen had no effective successor. Indeed, medieval medicine may be summed up as a corrupted version of Galenism. The true scientific tradition did not reappear in the West until the 16th century, after a lengthy incubation in the Islamic world.

By contrast with the naturalist focus of Hippocratic and Galenic medicine, healing became more authoritarian and intertwined with religion, for the rising Church taught that there was a supernatural plan and purpose to everything, including sickness and death. Christian and Jewish healing traditions became more prominent. Disease could be cured by prayers or by invoking the names of saints, by exorcism, by amulets or number magic, or by transferring the sickness to animals, plants, or to the soil.

Many of the distinguished surgeons of the Middle Ages were clerics, but the practice of medicine and surgery by members of the Church was not favored by the hierarchy. Many laws were passed against the practice of medicine for worldly profit. In Europe in the Middle Ages, practitioners unsuccessfully sought to cure wounds by treating only the weapons that caused them, using an ointment on the offending knife or sword known as a weapon salve.

The Middle Ages are also remarkable for their plagues, which became more numerous as people crowded closer and closer together in urban centers of increasing population density. Pre-modern medicine was powerless against bubonic plague. Before antibiotics reduced the disease to triviality in 1943, the average mortality rate was 60-70 per cent of those affected, despite all that the best hospital care could accomplish.

Renaissance and Pre-Modern Medicine

While internal medicine languished through the Middle Ages, the surgeons of the Renaissance gained wide experience during the many religious wars of the period. They had many new problems to face, including the treatment of wounds caused by firearms.

The first man to break away from this old doctrine was Ambroise Pare (1510-90). Pare came to Paris in 1532 as an apprentice to a barber-surgeon and then moved to the great Hotel Dieu as resident surgeon. In that immense medieval hospital, the only one in Paris at the time, he gained great experience, and in 1536 he began his career as a military surgeon.

In 1633 appeared the earliest book on first-aid for the injured, by one Stephen Bradwell, although the proffered advice sounds impractical and a bit odd to modern ears.

Even by the 18th century, the traditional surgeon's day-to-day business eschewed high-risk operations like amputations; rather, it was a round of minor procedures such as venesection, lancing boils, dressing skin abrasions, pulling teeth, managing whitlows, trussing ruptures, and treating skin ulcers. The fatality rates of these procedures were low, for surgeons understood their limits, and the repertoire of operations they attempted was small, because of the well-known risks of trauma, blood loss, and sepsis. Internal disorders were treated not by the knife but by medicines and management, since major internal surgery was unthinkable before anesthetics and antiseptic procedures. Improvements did occur in certain operations such as lithotomy.

The single worst malady cultivated in populous cities was tuberculosis (TB), a disease characterized by fever, night sweats, and hemoptysis (coughing up blood), called "consumption" because victims were almost literally consumed. In the continental U.S. as late as 1890, the corresponding percentage was about 13 per cent, and tuberculosis was still the leading cause of death (Table VI–A.2), though the data are partially suspect because cases of lung cancer were sometimes reported as "consumption".

One important 18th century improvement in internal medicine which decisively saved many lives was the introduction of inoculation and vaccination against smallpox. Edward Jenner, an English country doctor who performed such inoculations, noticed that cowpox, a cattle disease occasionally contracted by humans, particularly dairy maids, also conferred immunity against smallpox. Suspecting that it might be possible to produce this immunity by arm-to-arm inoculation from the cowpox pustule, and surmising it would be safer than inoculation from smallpox pustules directly, since in humans cowpox was benign, Jenner tried the experiment, and it worked.

The use of ambulances didn't catch on until the late 1800s; until then, anyone injured in the streets depended on the kindness of strangers or a nearby business shop for a place to rest until a doctor could be summoned. The first "modern" ambulance appeared in the city of Cincinnati in 1865, but the first true city ambulance system was developed in association with Bellevue Hospital in New York City in 1866, receiving 1500 requests for transport in its first three years of service.

The period also saw the development of many simple diagnostic tools that are taken for granted today. For example, a French physician, Rene Theophile Hyacinthe Laennec described pathological lesions found in the chest at autopsy and showed how they correlated with disease detected in living patients, establishing for the first time the modern concept of clinicopathological correlation, the cornerstone of modern diagnosis. Laennec also developed an instrument that he named a "stethoscope" to assist him in his examination of patients, especially female patients, against whose chests the direct placing of a male ear was socially taboo. The original device was a straight wooden tube; by mid-century rubber tubing was introduced to create a flexible monaural stethoscope, and in 1852 an American physician, George P. Cammann (1804-63), devised our familiar two-ear instrument.

Table VI–A.2. Leading Causes of Death in the United States, 1890s vs. 1990
(Crude annual death rate per 100,000 of Population)

Causes of Death	*1890 Death Rate per 100,000*	*1990 Death Rate per 100,000*
Consumption (TB)	245.4	0.7
Pneumonia	186.9	31.1
Heart disease	121.8	310.4
Diarrheal diseases	104.1	~0
Debility and atrophy	88.6	NA
Intantile cholera	79.7	~0
Bronchitis	74.4	1.7
Diphtheria	70.1	~0
Kidney diseases	59.7	8.8
Meningitis	49.1	0.4
Apoplexy (stroke)	49.0	57.9
Cancer	47.9	203.2
Typhoid Fever	46.3	~0
Old age	44.9	NA
Brain diseases	30.9	NA
Croup	27.6	~0
Malaria	19.2	0.002
Influenza	6.2	0.8
Other	~488.2	248.8
All Causes	~1840.0	863.8

The clinical thermometer is of like vintage. Galileo (1564-1642) invented the first thermometer in the late 16th century, but it was not applied to medicine. Other devices emerged later to measure pulse and blood pressure. In 1854, Karl Vierordt (1818-1884) created the sphygmograph, a pulse recorder usable for routine monitoring on humans. Blood pressure was measured using the familiar inflatable band wrapped around the upper arm, called the sphygmomanometer, whose basic design was established in 1896 by Scipione Riva-Rocci (1863-1937). The hypodermic syringe was invented in 1853.

Cells and Tissues

The doctrine of the essential cellular nature of living things was established by 1840. Modern cell theory began in botany. Jacob Henle applied cell biology to man. Physiologists stressed organ and tissue function—Claude Bernard (1813-78) emphasized the experimental method in establishing biological knowledge and urged that medical practice should be grounded in such knowledge.

Histology was raised to the status of an independent science by the Swiss microanatomist Albert von Kolliker. The medical implications of cell theory were taken up by Rudolph Virchow who dominated German biomedical research for half a century. Since Kolliker and Virchow, the study of the intimate structure and workings of the cells themselves, as distinct from the tissues, has become a separate science, cytology, further extended to the study of cells in disease, or cytopathology.

Blood Transfusions

Vague references to blood transfusion are found in medieval writings, though physicians historically were far more concerned with taking blood out of the body than with putting it back in. The first known transfusion of human blood into an already moribund person was attempted by James Blundell in 1818, and on several other occasions during the 1820s, with poor results. The cause of many of the untoward effects of blood transfusion was finally explained in 1901 when the presence of agglutinins and iso-agglutinins in the blood was demonstrated by Karl Landsteiner in Vienna, winning him the Nobel Prize in 1930; in 1907, the four main blood groups were determined by Jan Jansky of Prague. Rhesus factor was later identified by two American scientists, Philip Levine and Rufus Stetson. These advances were of fundamental importance and it became possible in the 20th century to eliminate most of the fatalities due to incompatibility, allowing blood transfusions to be practiced reliably and with uniformly good results.

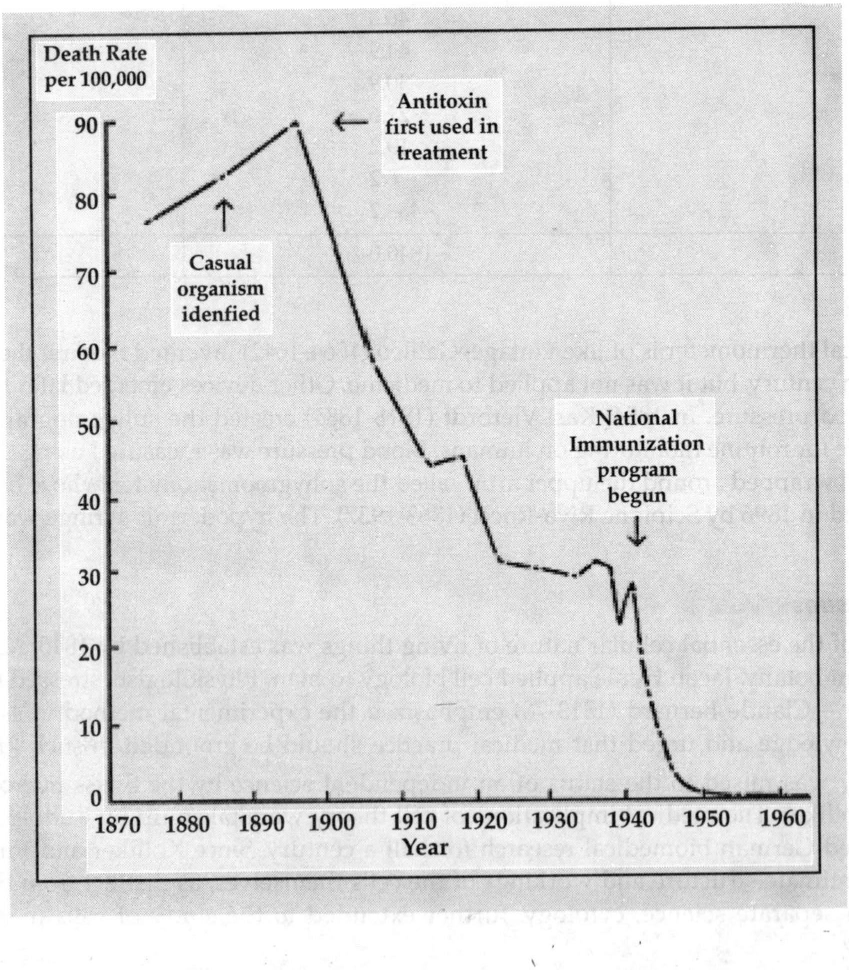

Fig. VI–A.2. Deaths due to Diphtheria in England and Wales (children under 15).

20th Century Medicine

The 20th century saw more discoveries and advances in medical science than all previous centuries combined. In this period, medicine became more powerful than ever before as scientists gained knowledge of matters and processes of illness that, at the beginning of the century, were still unknown or mysterious. Unlike the mere palliatives of earlier eras, 20th century physicians could actually cure some diseases, reverse some physical traumas, and save many lives that could not be saved before.

In the first half of the 20th century, the rational scientific paradigm that arose in the 19th century was pursued and extended. Acceptance of the germ theory of infection and the discovery of leukocytes led to the rapid emergence of immunology. This allowed medical scientists to produce protective vaccines and antisera which largely eliminated many diseases that were previously prevalent and dangerous, including whooping cough, measles, and diphtheria—the first truly effective medical treatments (Figure VI–A.2).

Acceptance of the germ theory also led to the discovery of filterable viruses (organisms that could pass through the pores of all known filters) in the 1890s, and subsequently to the identification of viruses as specific causes of yellow fever, smallpox, typhus (a Rickettsia organism), measles, poliomyelitis, rabies, and viral meningitis. Biochemists synthesized vitamins which were recognized as essential constituents of a healthy diet, thus allowing the elimination of vitamin deficiency diseases such as scurvy, rickets, osteomalacia, beriberi, pellagra, xerophthalmia, nyctalopia, and pernicious anemia, via dietary supplements. Many metabolic diseases became treatable due to biochemical investigations.

Blood group specification made transfusions convenient, facilitating dramatic advances in many branches of medicine,especially surgery. The first report of a successful autotransplant of a kidney into the neck of a dog was performed by Emerich Ullmann in 1902. The first successful kidney graft between identical twins was performed in 1954 by Joseph E. Murray, who received a Nobel Prize for his work. The first heart transplantation in man was achieved by Christiaan Barnard in 1967. Results were poor at first, but the use of new anti-rejection techniques such as the drug cyclosporin in the late 1970s, aided by advances in immunology, greatly improved the success of this and other organ transplants. By 1987, a total of ~7000 human hearts had been transplanted. At the close of the 20th century, heart, lung, heart-lung, and liver transplants were standard procedures, while grafts of small intestine and pancreas (the latter for refractory diabetes) were under active clinical investigation.

Two pivotal events transformed scientific medicine from a merely rational basis to a molecular basis, thus laying the groundwork for 21st century nanomedicine. The first pivotal event was the drug revolution, among which the most useful and spectacular were the antibiotics introduced between 1935-1945 and widely used ever since. Antibiotics are significant because they actively interfere with microbial metabolism and growth at the molecular level. The first antibiotic drugs were the sulphonamides, in 1935. Then penicillin became available in the 1940s, initially in very small quantities, then mass-produced by Pfizer during and after World War II, as a result of the research work of Alexander Fleming (1888-1955), Howard Florey (1898-1968) and Ernst Chain (1906-1979). In 1943, Selman A. Waksman (1888-1973) discovered streptomycin, the first effective anti-tuberculosis drug, for which he received the 1952 Nobel Prize.

The second pivotal event was the genetics revolution, starting with the discovery in 1953 of the information-carrying double-helix structure of DNA by Francis Crick (b. 1916) and John B. Watson (b. 1928), followed in the 1980s by the ability to chemically read the genetic code, isolate specific genes and clone them for further study. In the mid 1980s, the Human Genome Project was launched, with the objective of fully sequencing every gene in the human genome. The first phase of this project neared completion as the 20th century drew to a close.

Molecular biology became the premier scientific discipline of the latter 20th century. By 1998, the compositions of organs, tissues, cells, organelles, and membranes had been defined, and the biosynthesis and catabolism of hundreds of compounds had been elucidated. The regulation of body processes was described at progressively finer levels in biochemical language. Many pharmacologic agents were finally understood in terms of specific molecular loci and mechanisms of action. Advances were particularly rapid inimmunology, virology, cellular biology, peptide research, and structural biology. A beginning was made in explaining human behaviour in mechanistic terms, as more and more chemical mediators and pharmacologic modifiers were discovered. In biology, these disciplines developed porous boundaries with related disciplines such as physiology, pharmacology, neurosciences, biochemistry and biophysics. All entered a phase of confluence, employing the common language of chemistry.

Thus, the late 20th century is best regarded as the molecular age of basic biological science. The molecular influence pervades all the traditional disciplines underlying clinical medicine. DNA sequencing techniques and restriction endonucleases permitted precise identification of the exact structural alteration of the gene in an increasing number of hereditary diseases. Global changes in progressive aging dysfunction were shown to be strongly related to declining secretion of growth hormone by Rudman in 1990. Shortly thereafter, anti-aging medicine was recognized as a distinct discipline and was promoted by the American Academy of Anti-Aging Medicine (A^{4}M), a group that claimed >6000 physician and scientist members worldwide by 1998 and had held numerous conferences.

While biological science was vaulting forward into the molecular realm at a blistering pace, biomedical engineering lagged considerably behind, though many remarkable successes had been achieved since mid century. Radiology expanded with sophisticated radiotherapy, ultrasound, scanning and imaging techniques (e.g., CAT, PET, NMR), with submillimeter resolution in living tissues. Surgical instruments became less damaging and less invasive; minimally invasive techniques used OK (orifice and keyhole) surgery and surgery performed under the "eye" of a scan. Fetuses could be screened for abnormality and surgery then performed upon them inside the womb, or they could be removed temporarily from the womb and then returned to continue gestating, after surgery. Prospective parents with fertility difficulties could make use of a wide range of therapies including in vitro fertilization. Implantable pacemakers, defibrillators, and ventricular assist devices were commonplace, artificial wearable/implantable and full/partial replacements for lungs, heart, kidney, liver, and pancreas were either available, in clinical trials, or under development. Electron microscopes with 200,000X magnifications allowed many details of internal cellular and viral structures to be resolved as early as 1946; by 1998, atomic-force (AFM) and scanning-tunneling microscopy (STM) permitted direct tactile examination of individual biomolecules in fixed cells.

As the 20th century drew to a close, a few preliminary efforts had been made to apply the molecular approach to medical diagnostic and clinical tools. Biotechnology was one avenue being pursued, with the rational design of artificial enzymes and specified-ligand binding sites having already been achieved in certain limited cases, and the beginnings of gene therapy as noted earlier. Carbon fullerenes had been used to create a water-soluble inhibitor of HIV protease, and had other biological applications. AFM-based force-amplified biological sensors could detect defined biological species such as cells, proteins, toxins, and DNA at concentrations as low as 10^{-18} M (~1/ mm^3), and automated laboratory systems for sorting and handling individual cells and viruses were commonplace. Biotech companies such as Physiome Sciences of Princeton NJ had developed three-dimensional computer models of the heart and other organs. Physiome's heart model was based on detailed molecular, biochemical, cellular and anatomical information, including submodels of all the different cell types found in the heart embodying knowledge of the function of each cell type in healthy and diseased hearts, and information on gene function and the causes and effects of congestive heart failure, arrhythmias and heart attacks. Other cell biochemistry simulators included E-CELL and the Virtual Cell. There was also much progress and interest in nanostructure analysis and nanomaterials fabrication for medical and biological purposes, with research groups emerging at major universities. However, the greatest medical revolution of all awaits the ability to engineer and fabricate whole devices and systems at the molecular scale. Along this course lies nanotechnology and molecular manufacturing, a deep well from which nanomedicine will inevitably spring.

21st Century Medicine

It is always somewhat presumptuous to attempt to predict the future, but in this case we are on solid ground because most of the prerequisite historical processes are already in motion and all of them appear to be clearly pointing in the same direction.

Medical historian Roy Porter notes that the 19th century saw the establishment of what we think of as scientific medicine. From about the middle of that century the textbooks and the attitudes they reveal are recognizable as not being very different from modern ones. Before that, medical books were clearly written to address a different mindset.

But human health is fundamentally biological, and biology is fundamentally molecular. As a result, throughout the 20th century scientific medicine began its transformation from a merely rational basis to a fully molecular basis. First, antibiotics that interfered with pathogens at the molecular level were introduced. Next, the ongoing revolutions in genomics, proteomics and bioinformatics provided detailed and precise knowledge of the workings of the human body at the molecular level. Our understanding of life advanced from organs, to tissues, to cells, and finally to molecules, in the 20th century. In the 21st century, the entire human genome will be mapped. This map will inferentially incorporate a complete catalog of all human proteins, lipids, carbohydrates, nucleoproteins and other molecules, including full sequence, structure, and much functional information. Only some systemic functional knowledge, particularly neurological, may still be lacking by that time.

This deep molecular familiarity with the human body, along with simultaneous nanotechnological engineering advances will set the stage for a shift from today's molecular scientific medicine in which fundamental new discoveries are constantly being made, to a molecular technologic medicine in which the molecular basis of life, by then well-known, is manipulated to produce specific desired results. The comprehensive knowledge of human molecular structure so painstakingly acquired during the 20th and early 21st centuries will be used in the 21st century to design medically-active microscopic machines. These machines, rather than being tasked primarily with voyages of pure discovery, will instead most often be sent on missions of cellular inspection, repair, and reconstruction. In the coming times, the principal focus will shift from medical science to medical engineering. Nanomedicine will involve designing and building a vast proliferation of incredibly efficacious molecular devices, and then deploying these devices in patients to establish and maintain a continuous state of human healthiness.

The very earliest nanotechnology-based biomedical systems may be used to help resolve many difficult scientific questions that remain. They may also be employed to assist in the brute-force analysis of the most difficult three-dimensional structures among the 100,000-odd proteins of which the human body is comprised, or to help ascertain the precise function of each such protein. But much of this effort should be complete within the next 20-30 years because the reference human body has a finite parts list, and these parts are already being sequenced, geometered and archived at an ever-increasing pace. Once these parts are known, then the reference human being as a biological system is at least physically specified to completeness at the molecular level. Thereafter, nanotechnology based discovery will consist principally of examining a particular sick or injured patient to determine how he or she deviates from molecular reference structures, with the physician then interpreting these deviations in light of their possible contribution to, or detraction from, the general health and the explicit preferences of the patient.

In brief, nanomedicine will employ molecular machine systems to address medical problems, and will use molecular knowledge to maintain human health at the molecular scale.

Volitional Normative Model of Disease

What, exactly, is "medicine"? Dictionaries give several definitions, ranging from the very restrictive to the most general, as follows: "a drug or remedy"; "any substance used for treating disease"; "any drug or other substance used in treating disease, healing, or relieving pain"; "treatment of disease by medical, as distinguished from surgical, treatment"; "the branch of this science and art that makes use of drugs, diet, etc., as distinguished especially from surgery and obstetrics"; "the science of treating disease, the healing art"; "the art and science of preventing or curing disease"; "the act of maintenance of health, and prevention and treatment of disease and illness"; "the department of knowledge and practice dealing with disease and its treatment"; or, most generally". Here, we shall adopt the latter, maximally-inclusive, definition of "medicine"(Figure VI–A.3).

Reviewing Figure VI–A.3, the contemporary physician might at first be inclined to relegate molecular approaches to some minor subfield, perhaps "nanoanalytics," "nanogenomics," or "nanotherapeutics." This would be a serious mistake, because the application of molecular approaches to health care will significantly impact virtually every category of laboratory and clinical practice across the board. Thus we are led to the broadest possible conception of nanomedicine as "the science and technology of diagnosing, treating, and preventing disease and traumatic injury, of relieving pain, and of preserving and improving human health, using molecular tools and molecular knowledge of the human body."

This brings us to the question of "disease," a complex term whose meaning is still hotly debated among medical academics. Figure VI–A.4 shows the results of a survey of four different groups of people who were read a list of common diagnostic terms and then asked if they would rate the condition as a disease. Illnesses due to microorganisms, or conditions in which the doctor's contribution to the diagnosis was important, were most likely to be called a disease, but if the cause was a known physical or chemical agent the condition was less likely to be regarded as disease; general practitioners also had the broadest definition of disease.

No less than eight different types of disease concepts are held by at least some people currently engaging in clinical reasoning and practice, including:

1. **Disease Nominalism**—A disease is whatever physicians say is a disease. This approach avoids understanding and forestalls inquiry, rather than furthering it.

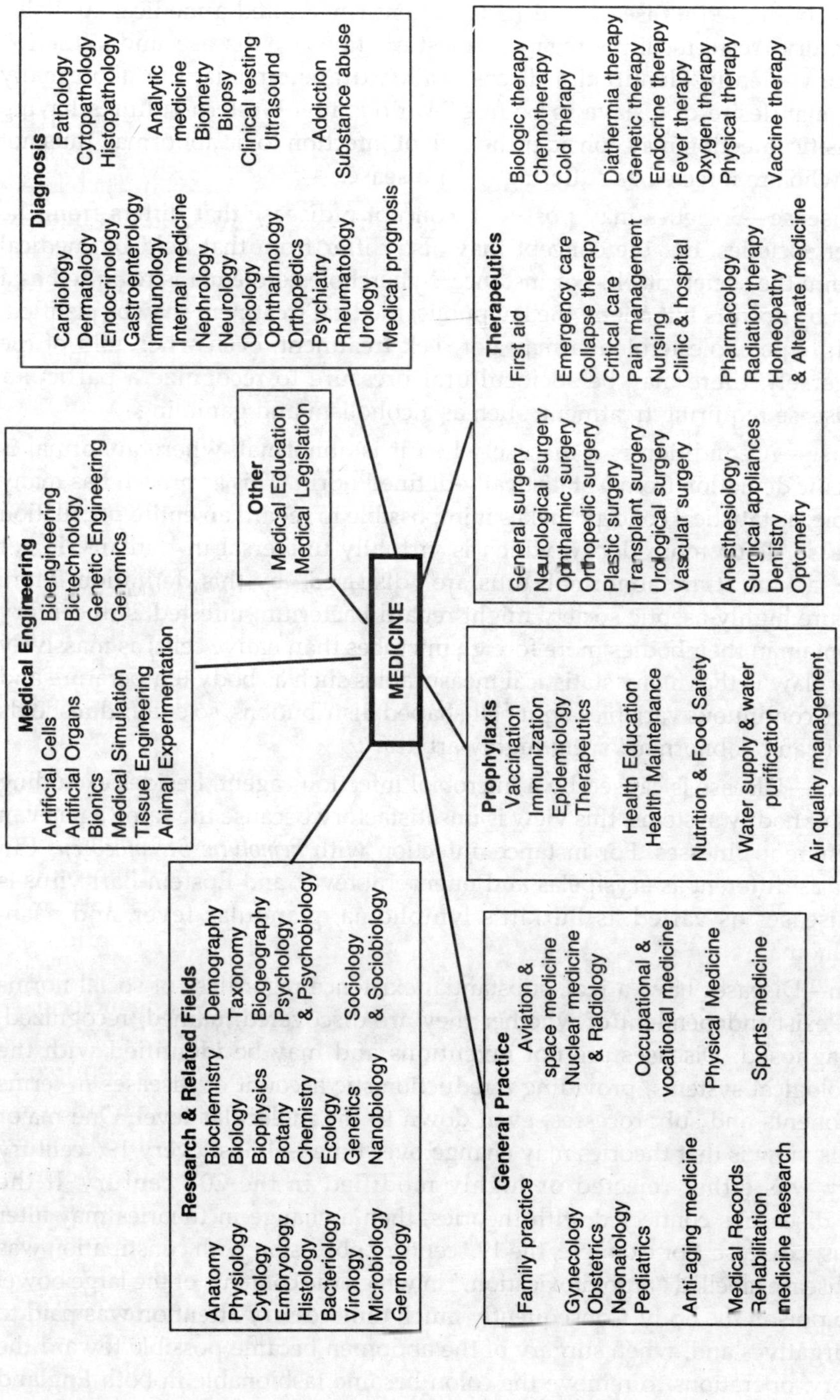

Fig. VI–A.3: Branches of medicine and associated disciplines.

2. **Disease Relativism**—A disease is identified or labeled in accordance with explicit or implicit social norms and values at a particular time. In 19th century Japan, for example, armpit odor was considered a disease and its treatment constituted a medical specialty. Similarly, 19th century Western culture regarded masturbation as a disease, and in the 18th century, some conveniently identified a disease called drapetomania, the"abnormally strong and irrational desire of a slave to be free." Various non-Western cultures having widespread parasitic infection may consider the lack of infection to be abnormal, thus not regarding those who are infected as suffering from disease.
3. **Sociocultural Disease**—Societies may possess a concept ofdisease that differs from the concepts of other societies, but the concept may also differ from that held by medical practitioners within the society itself. For instance, hypercholesterolemia is regarded as a disease condition by doctors but not by the lay public; medical treatment may be justified, but persons with hypercholesterolemia may not seek treatment, even when told of the condition. Conversely, there may be sociocultural pressure to recognize a particular condition as a disease requiring treatment, such as alcoholism and gambling.
4. **Statistical Disease**—A condition is a disease when it is abnormal, where abnormal is defined as a specific deviation from a statistically-defined norm. This approach has many flaws. For example, a statistical concept makes it impossible to regard an entire population as having a disease. Thus tooth decay, which is virtually universal in humans, is not abnormal; those lacking it are abnormal, thus are "diseased" by this definition. More reasonably, a future highly-aseptic society might regard bacterium-infested 20th century humans (who contain in their bodies more foreign microbes than native cells; as massively infected. Another flaw is that many statistical measurables such as body temperature and blood pressure are continuous variables with bell-shaped distributions, so cutoff thresholds between "normal" and "abnormal" seem highly arbitrary.
5. **Infectious Agency**—Disease is caused by a microbial infectious agent. Besides excluding systemic failures of bodily systems, this view is unsatisfactory because the same agent can produce very different illnesses. For instance, infection with *hemolytic Streptococcus* can produce diseases as different as erysipelas and puerperal fever, and Epstein-Barr virus is implicated in diseases as varied as Burkitt's lymphoma, glandular fever, and naso-pharyngeal carcinoma.
6. **Disease Realism**—Diseases have a real, substantial existence regardless of social norms and values, and exist independent of whether they are discovered, named, recognized, classified, or diagnosed. Diseases are not inventions and may be identified with the operations of biological systems, providing a reductionistic account of diseases in terms of system components and subprocesses, even down to the molecular level. One major problem with this view is that theories may change over time—almost every 19th century scientific theory was either rejected or highly modified in the 20th century. If the identification of disease is connected with theories, then a change in theories may alter what is viewed as a disease. For example, the 19th century obsession with constipation was reflected in the disease labelled "autointoxication," in which the contents of the large bowel were believed to poison the body. Consequently much unnecessary attention was paid to laxatives and purgatives and, when surgery of the abdomen became possible toward the end of the century, operations to remove the colon became fashionable in both England and America.

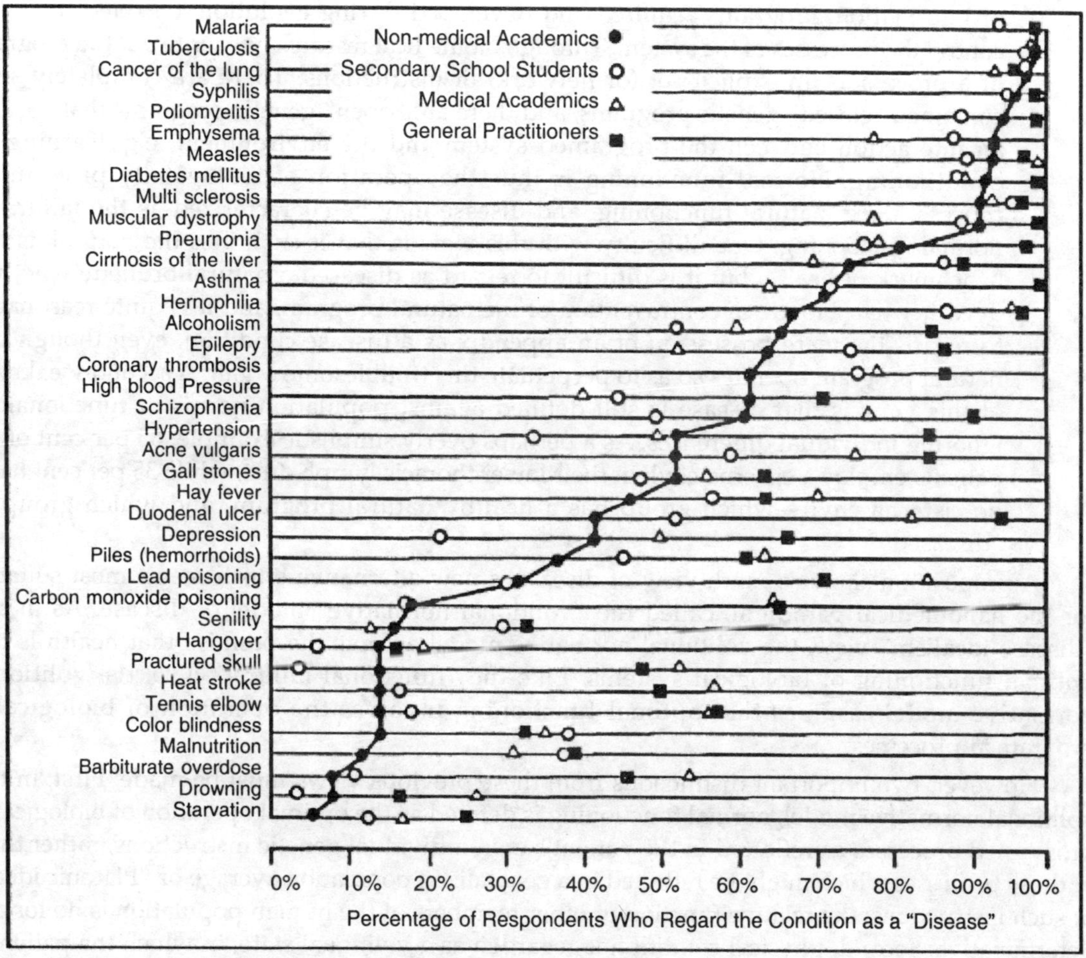

Fig. VI–A.4: Opinion survey: what is disease? (modified from Carnpbell, Scadding, and Roberts)

7. **Disease Idealism**—Disease is the lack of health, where health is characterized as the optimum functioning of biological systems. Every real system inevitably falls short of the optimum in its actual functioning. But by comparing large numbers of systems, we can formulate standards that a particular system ought to satisfy, in order to be the best of its kind. Thus "health" becomes a kind of Platonic ideal that real organisms approximate, and everyone is a less than perfect physical specimen. Since we are all flawed to some extent, disease is a matter of degree, a more or less extreme variation from the normative ideal of perfect functioning. This could be combined with the statistical approach, thus characterizing disease as a statistical variation from the ideal. But this view, like the statistical, suffers from arbitrary thresholds that must be drawn to qualify a measurable function as representing a diseased condition.
8. **Functional Failure**—Organisms and the cells that constitute them are complex organized systems that display phenomena (e.g. homeostasis) resulting from acting upon a program

of information. Programs acquired and developed during evolution, encoded in DNA, control the processes of the system. Through biomedical research, we write out the program of a process as an explicit set (or network) of instructions. There are completely self-contained "closed" genetic programs, and there are "open" genetic programs that require an interaction between the programed system and the environment, e.g. learning or conditioning. Normal functioning is thus the operation of biologically programed processes, e.g. natural functioning, and disease may be characterized as the failure of normal functioning. One difficulty with this view is that it enshrines the natural as the benchmark of health, but it is difficult to regard as diseased a natural brunette who has dyed her hair blonde in contravention of the natural program, and it is quite reasonable to regard the mere possession of an appendix as a disease condition, even though the natural program operates so as to perpetuate this troublesome organ. A second weakness of this view is that disease is still defined against population norms of functionality, ignoring individual differences. As a perhaps overly simplistic example, 65 per cent of all patients employ a cisterna chyli in their lower thoracic lymph duct, while 35 per cent have no cisterna chyli—which group has a healthy natural program, and which group is "diseased"?

A proposition shows a ninth view of disease, a new alternative which seems most suitable for the nanomedical paradigm, called the "volitional normative" model of disease. As in the "disease idealism" view, the volitional normative model accepts the premise that health is the optimal functioning of biological systems. Like the "functional failure" view, the volitional normative model assumes that optimal functioning involves the operation of biologically programed processes.

However, two important distinctions from these previous views must be made. First, in the volitional normative model, normal functioning is defined as the optimal operation of biologically programed processes as reflected in the patient's own individual genetic instructions, rather than of those processes which might be reflected in a generalized population average or "Platonic ideal" of such instructions; the relative function of other members of the human population is no longer determinative. Second, physical condition is regarded as a volitional state, in which the patient's desires are a crucial element in the definition of health. This is a continuation of the current trend in which patients frequently see themselves as active partners in their own care.

In the volitional normative model, disease is characterized not just as the failure of "optimal" functioning, but rather as the failure of either (a) "optimal" functioning or (b) "desired" functioning. Thus, disease may result from:

1. A failure to correctly specify desired bodily function (specification error by the patient).
2. A flawed biological program design that doesn't meet the specifications (programing design error).
3. Flawed execution of the biological program (execution error).
4. External interference by disease agents with the design or execution of the biological program (exogenous error).
5. Traumatic injury or accident (structural failure).

In the early years of nanomedicine, volitional physical states will customarily reflect "default" values which may differ only insignificantly from the patient's original or natural biological programing. With a more mature nanomedicine, the patient may gain the ability to substitute alternative natural programs for many of his original natural programs. For example, the genes responsible for appendix morphology or for sickle cell expression might be replaced with genes that encode other phenotypes, such as the phenotype of an appendix-free cecum or a phenotype for statistically typical human erythrocytes. Many persons will go further, electing an artificial genetic structure which, say, eliminates age-related diminution of the secretion of human growth hormone and other essential endocrines. On the other hand, a congenitally blind patient might desire, for whatever personal reasons, to retain his blindness. Hence his genetic programs that result in the blindness phenotype would not, for him, constitute "disease" as long as he fully understands the options and outcomes that are available to him. Whether the broad pool of volitional human phenotypes will tend to converge or diverge is unknown, although the most likely outcome is probably a population distribution (of human biological programs) with a tall, narrow central peak (e.g., a smaller standard deviation) but with longer tails (e.g., exhibiting a small number of more extreme outliers).

One minor flaw in the volitional normative model of disease is that it relies upon the ability of patients to make fully informed decisions concerning their own physical state. The model crucially involves desires and beliefs, which can be irrational, especially during mental illness, and people normally vary in their ability to acquire and digest information. Patients also may be unconscious or too young, whereupon default standards might be substituted in some cases.

Nevertheless, the volitional normative view of disease appears most appropriate for nanomedicine because it recognizes that the era of molecular control of biology could bring considerable molecular diversity among the human population. Conditions representing a diseased state must of necessity become more idiosyncratic, and may progressively vary as personal preferences evolve over time.

Treatment Methodology

The availability of advanced nanomedical instrumentalities should not significantly alter the classical medical treatment methodology, although the patient experiences and outcomes will be greatly improved. Treatment in the nanomedical era will become faster and more accurate, efficient, and effective. In clinical practice, patient treatment customarily includes up to six distinguishable phases: examination, diagnosis, prognosis, treatment, validation, and prophylaxis.

Evolution of Bedside Practice

The relationship between physician and patient has been evolving in response to the rapidly changing medical environment. Two of the most important trends are the decline of traditional holism (with a concomitant increase in specialization) and the rise of therapeutic customization in medical practice.

Specialization and Holistic Medicine

Holism is a philosophy which holds that individuals function as complete units that cannot be reduced merely to the sum of their parts. It is unarguable that a simplistic reductionist view of the patient which ignores the complex interactions among the many cells, tissues, organs, and systems constituting the human body is deeply flawed. For example, an understanding of the molecular basis of the contractile proteins of heart muscle will not alone tell us how heart muscle cells will look or act; knowing only about the parts of something is not sufficient to predict the behaviour of the whole. However, traditional concepts of holistic medicine go well beyond such basic systems philosophy—incorporating requirements for a consideration of all physical, emotional, social, spiritual, environmental and economic needs of the patient.

N. Jewson and others have decried the modern shift away from holistic medicine, which is asserted to have taken place in three historical phases in the West. The initial phase is identified as the practice of "bedside medicine," where wealthy fee-paying clients in the 17th and 18th centuries helped shape their own diagnosis and treatment by medical practitioners in an holistic manner. Aspects of the patient's emotional and spiritual life were seen as central by the practitioner in making a diagnosis, since most physical treatments were only palliative and so the doctor had to focus on nonphysical supportive measures. This frame of reference was progressively replaced during the 19th century with the trend toward "hospital medicine," wherein physicians concentrated on generic classifications of diseases that were manifested in the patient, moving doctors away from the earlier focus on the individual as a whole person. The 20th century saw the development of "laboratory medicine," which moved diagnosis and therapy even further away from the whole patient, "who came to be medically conceived as little more than a depersonalized object, comprised of a complex of cells."

A more charitable view is that physicians have increasingly specialized in treating those physical diseases for which effective treatments may be readily specified, leaving nonphysical and nontreatable issues for psychiatrists, social workers, fitness coaches, priests, lawyers, or other professionals to deal with. In the 21st century this operational specialization may become complete, since nanomedicine phenomenologically regards the human body as an intricately structured machine with trillions of complex interacting parts, with each part (and each subsystem of parts) subject to individual scrutiny, repair, and possibly replacement by artificial technological means. In this new medical cosmology, the concept of the whole patient almost completely dissolves into a data-intensive whirlwind of molecular detail at the cellular, tissue, organ, and systemic levels.

And yet, as in a bygone era, patients once again will help to shape their own diagnosis and treatment at the hands of medical practitioners who begin to apply the volitional normative model of disease in their practices. This may breathe new life into the age-old medical school dictum to "treat the patient as a person" and to "focus on the sick person" rather than exclusively on the body.

The availability of extremely powerful and transforming molecular technologies also argues for a return to the romantic and perhaps quaint concept of a single doctor taking care of a single patient. The potential for interactions among highly potent nanorobotic instrumentalities argues for diagnostic and therapeutic "gatekeeping" by a single trusted practitioner in whom strategic treatment responsibility is vested—in partnership, of course, with the patient.

Customized Diagnosis and Therapeutics

In science the objective is to understand the individual occurrence by means of a general law; in medical practice, knowledge of what is generally the case does not tell the physician how to treat a particular patient. Thus the problem of how to proceed from the prototypic case to the individual instance remains to be solved in a systematic manner by the practitioner.

The practitioner of "bedside medicine" in the 17th and 18th centuries had few curative and almost no customized tools at his disposal—perhaps a few dozen basic surgical techniques, performed septically, hazardously, and without anesthesia, and a drug/herbal formulary consisting of a few hundred substances. For instance, the *Edinburgh Pharmacopoeia* of 1803 listed only 222 simples while the *London Pharmacopoeia* of 1809 listed fewer than 200 items, most of which had variable, uncertain, minimal, or nonspecific potency. A few more options became available to the 19th century medical practitioner, including complex surgical techniques for specific conditions that could be performed aseptically with anesthesia and a good chance for success, a somewhat broader and more efficacious pharmacopeia, vaccines targeted to several diseases, and an improving diagnostic ability. By the 20th century, the physician could prescribe from among tens of thousands of specific drugs to target specific bacterial, viral, fungal, or parasitic infections; select from among a very precise array of anesthetic agents, chosen to avoid allergic responses in particular patients; perform a wide variety of noninvasive tests and scans for diagnostic purposes to identify very specific conditions; and perform minimally invasive surgeries directed at many arbitrary tissue masses as small as 1 mm^3 in volume. The first glimmerings of personalized genetic therapies also began to appear.

With the arrival of nanomedicine in the 21st century, the treatment paradigm will complete its transition from coarse-grained, one-size-fits-all, slow-acting methods to molecularly-precise, completely customized, speedy and highly-efficacious procedures and instrumentalities. The irregular shotgun pattern of 18th century palliatives will evolve during the early 21st century into a penetrating and perfectly tailored hail of "magic bullets" each targeting an individual cell or group of cells unique to the individual patient.

The Physician-Patient Relationship

Many other aspects of the physician-patient relationship, especially as this relationship may evolve in the coming era of nanomedicine, are important and worthy of extensive discussion. One such issue is the obligation of both parties in the partnership to tell the truth. The patient as a fellow human being has every right to know the truth about his or her biological condition, but other considerations may enter into the fulfillment of this obligation. Patients have a quite natural anxiety about their own possible death, and it has been claimed that this anxiety implies that no one can be truly objective toward his or her own body. Guttentag observes that "telling an unwelcome truth to the unprepared is as ill-conceived as trying to hide the truth from the prepared."

In the nanomedical era, the sheer number of "truths" that may become available for disclosure will increase enormously even as the terminal prognosis becomes rare. For example, each human being is believed to possess at least 4-10 potentially serious genetic defects; up to 1 per cent of human DNA is of exogenous viral origin, and as much as 10 per cent of the genome consists of transposons, discrete sequences that are positionally mobile among the chromosomes. Should something be done about this, or not? What should the average patient make of the news that his physician has discovered exactly 57 submicron-scale lamellar defects scattered throughout the compact bone of the caudal epiphysis of the patient's right humerus? In the nanomedical era, people will gain the ability to specify their own physical structure to minute detail, but many patients will not be ready, willing, or able to assume responsibility for this knowledge. Thus there is no ideal substitute for the doctor's interpretative abilities and judgements on the patient's behalf as to the personal significance of specific diagnostic information.

Perhaps the single most important aspect of the physician-patient relationship, in any century, is the humanistic quality of the good doctor. The patient seeks a physician who cares about him as a person and will diagnose and prescribe in a sensitive and compassionate manner, accepting some degree of obligation to the patient.

Changing View of the Human Body

How does a patient regard his or her own body, and how might this most intimate of all relationships change in the nanomedical era? The so-called dualist theory of the human compound, as originally developed by Descartes and widely accepted today by the ordinary person, holds that the human being consists of two separate kinds of thing: the body and the mind or soul. The body acts as a host or receptacle for the mind. The mind, often called "the ghost in the machine," is manifested by the brain, which it uses (via the bodily senses) to acquire and store information about the world and to integrate this with its genetically-driven imperative to live, thus resolving internal conflicts among action-choices and expressing these in what we (in our consciousness) experience as decisive action.

Scientific medicine has concentrated primarily on the body. The ancient Roman physician Galen first dissected and vivisected a variety of animals to increase his knowledge of anatomy and physiology, and dissection became increasingly important in the training of physicians and surgeons, and in painting and sculpture, during the Renaissance. By the late 20th century, dissection had reached the molecular level, with the insides of the human cell and nucleus being taken apart and examined by molecular biologists, literally receptor by receptor. Dissection and the mechanistic understanding it provides have led some to decry what they regard as the modern "soulless" view of the human body as a mere machine.

In the nanomedical era, even the most diehard reductionist must come to see the human body not merely as a heap of parts but rather as a finely tuned vehicle that is owned and piloted by a single human mind. As with automobiles, some body-owners will be more diligent about maintenance, regular tuneups, and paint jobs than other body-owners. Some will crave the latest upgrades, while others may prefer a more conservative model that reliably gets them around town. At either extreme, all human vices and virtues will be on full display, though one may perhaps anticipate an increasing pride of corporeal ownership if for no other reason than because maintenance and repair will become quick, convenient, and inexpensive.

From this simple analogy of body and mind to car and driver, it might at first appear that the advent of nanomedical technology will confirm and strengthen the traditional dualist conception of the body. But closer inspection reveals that the analogy is at best incomplete, and at worst deeply flawed. This is because mind, first being necessarily embedded in physical structure and relying upon that structure for its faithful execution, and second, this physical structure now being manipulable at the molecular level, enters also into the purview of our mechanic. Both car and driver may be modified in the shop. Speaking allegorically, it is as if the driver, after getting his car a tuneup, emerges from the shop no longer favoring chocolate but enjoying vanilla instead, or now preferring jazz over classical, the opposite of before. Such psychological changes may be either volitional or emergent.

Until the late 20th century, human progress was measured almost exclusively in terms of externalities. Food was gathered, then sown, then manufactured. Shelters had no running water, then gained outhouses, then indoor plumbing. Natural lighting and campfires gave way to candles, then oil lamps, then electric illumination. Finger-counting yielded first to the abacus, then the mechanical adding machine, and finally to the digital computer. But throughout all of history, the human body itself has remained largely untouched by progress. We have always regarded our bodies, evolved by natural selection, as fundamentally inviolate and immutable—subject perhaps to various natural or traumatic degradations, but rarely to any significant intrinsic improvement on the timescale of human civilization.

Now we are set to embark upon an era in which our natural physiological equipment may for the first time in history become capable of being altered, improved, augmented, or rendered more comfortable or convenient, due to advances in medical technology. The physical human body may be one of the last bastions of "naturalness". It will also be one of the last elements in our common worldview to be modernized.

Our subjective experience of reality will shift by subtle degrees. For instance, all objective information about our physical surroundings has traditionally arrived in the conscious mind via the various natural senses such as hearing, sight, and smell. In the nanomedical era, machine-mediated sensory modalities may permit direct perception of physical phenomena well removed from our bodies in both time and space, or which are qualitatively or quantitatively inaccessible to our original natural senses. Perception will gradually expand to incorporate nonphysical phenomena including abstract models of mental software, purely artificial constructs of simulated or enhanced realities, and even the mental states of others. Such new perceptions will inevitably alter the way our minds process information.

But the winds of change will sweep deeper still, into our very souls. Like ants oblivious to the collective purpose of their colony, the billions of neurons in the human brain are all busily buzzing, wholly ignorant of the emergent plan. This is the physical, mechanical world of our electrochemical hardware. People also have thoughts, feelings, emotions, and volitions, a higher level in the data processing hierarchy which in turn is equally oblivious of the brain cells. We can happily think while being totally unaware of any help from our neurons. But nanomedicine will give us unprecedented systemic multilevel access to our internal physical and mental states, including real-time operating parameters of our own organs, tissues, and cells, and, if desired, the activities of small groups of (or even individual) neurons. Diverse parts of our selves previously closed to our attention may slowly conjoin and enter our conscious awareness.

Will this access promote an integrated identity or lead to hopeless confusion, or worse?

Nanomedical alterations to the brain and other physical systems may give us vastly expanded freedom to be who we choose to be along with increased responsibility to make wise and informed choices. The ethical and legal aspects of these questions, as well as the scientific and psychological ones, are extremely important and should be thoroughly debated in the years and decades that lie ahead.

The Nanomedical Perspective

Nanomedicine and Molecular Nanotechnology

A mature nanomedicine will require the ability to build structures and devices to atomic precision, hence molecular nanotechnology and molecular manufacturing are key enabling technologies for nanomedicine. The prefix "nano-" (from the Greek root nanos, or dwarf) means one-billionth (10^{-9}) of something. The term **nanotechnology** refers generally to technology on the scale of a billionth of a meter, or a nanometer (a nanometer is ~6 carbon atoms wide). Similarly, the words "nanomachine," "nanorobot," "nanomotor" and "nanocomputer" may refer to complex engineered objects fabricated by positioning matter with molecular control.

Molecular engineering was discussed as an extension of bulk technologies in the 1960s and 1970s by von Hippel, von Foester, and Zingsheim. The phrase "Nano-Technology" was first used in print in 1974 by N. Taniguchi to refer to the increasingly precise machining and finishing of materials, progressing from larger to smaller scales and ultimately to nanoscale tolerances, following in the path of Feynman's proposed "top-down" approach, a schemata which persisted in Taniguchi's thinking throughout the 1980s and 1990s. In 1981, K.E. Drexler described a new "bottom-up" approach involving molecular manipulation and molecular engineering in the context of building molecular machines and molecular devices with atomic precision, a fundamentally different mind-set. Drexler again described molecular technology in 1982 and molecular mechanical devices in 1983, first using the word "nanotechnology" in 1985 and 1986 as synonymous with molecular technology, finally settling upon "molecular nanotechnology" in 1991 and "molecular machine systems" in 1992 to clarify that his concept involved working devices constructed with atomic precision, as distinguished from nanostructured bulk materials, micromachinery, polymeric self-assembly, pure biotechnology, nanolithography, Langmuir-Blodgett thin films, and the like. Drexler's definition—molecular nanotechnology as the three-dimensional positional control of atomic and molecular structure to create materials and devices with molecular precision—is the usage adopted in this book. The first known use of the term "nanomedicine" was in 1991 by Drexler, Peterson, and Pergamit in their popular book *Unbounding the Future.*

Is molecular nanotechnology possible? The bottom line is that molecular nanotechnology violates no physical laws and there exist many possible technical paths leading to useful results. Even by 1985, for example, G. Yamamoto had reported on molecular gears, describing "compounds that exist in conformations which are regarded as static meshed gears with two-toothed and three-toothed wheels and some of them behave as dynamic gears." H. Iwamura prepared a system that formed a chain of beveled molecular gears with ~GHz rotation rates. In 1998, it was generally accepted that molecular nanotechnology would be developed, although there was still some disagreement about how long it would take.

It is often noted that molecular biological systems are themselves nanomachines, constituting an existence proof for molecular nanotechnology. Indeed, biotechnology is one possible implementation pathway for molecular nanotechnology that is being pursued. Table VI–A.3 reveals the close functional correspondence between the macroscale components of everyday machines and the molecular components of natural biological systems. Such comparisons have a long history. For example, Marcello Malpighi (1628-94), a professor of medicine in Pisa who discovered the fine structure of the lungs and the capillaries using the microscope, once observed that "our bodies are composed of strings, thread, beams, levers, cloth, flowing fluids, cisterns, ducts, filters, sieves, and other similar mechanisms."

Contemplating applications of nanotechnology to medicine, Brian Wowk concluded that 20th century physicians were in a predicament similar to that which would be faced by 18th century engineers trying to maintain a 20th-century automobile—repairs would be crude at best, and breakdowns inevitable.

By the end of the 20th century, mainstream military (DoD), NIH, NSF, and other U.S. government and international groups had begun to seriously consider the potential future applications of molecular nanotechnology in medicine.

Here we assume that the mass production of nanomachines at modest cost is technically feasible and then explores the medical implications of this assumption. Proposed systems presented in this trilogy are intended not as final engineering blueprints but merely as points of departure for further analysis and refinement. All designs and projected capabilities are, for the most part, conservatively drawn with generous safety margins, leaving a considerable volume of design space yet to be explored by more intrepid future investigators.

Table VI–A.3. Comparison of Macroscale and Biomolecular Components and Functions

Macroscale Device	*Device Function*	*Biomecular Examples*
Struts, beams, casings	Transmit force, hold positions	Microtubules, cellulose
Cables	Transmit tension	Collagen
Fasteners, glue	Connect parts	Intermolecular forces
Solenoids actuators	Move things	Conformation-changing proteins, actin/myosin
Motors	Turn shafts	Flagellar motor
Drive shafts	Transmit torque	Bacterial flagella
Bearings	Support moving parts	Sigma bonds
Containers	Hold fluids	Vesicles
Pumps	Move fluids	Flagella, membrane proteins
Conveyor belts	Move components	RNA moved by fixed ribosome (partial analog)
Clamps	Hold workpieces	Enzymatic binding sites
Tools	Modify workpieces	Metallic complexes, functional groups
Production lines	Construct devices	Enzyme systems, ribosomes
Numerical control systems	Store and read programs	Genetic system

Nanomedicine: History of the Idea

Conclusive proof of the existence of atoms was not obtained until the close of the 19th century. This may explain why the idea of nanomedicine is an exclusively 20th century phenomenon.

Development of the concept of nanomedicine has followed two principal paths which Richard Smalley has termed "wet nanotechnology" in the biological tradition, and "dry nanotechnology" in the mechanical tradition. Both approaches were presaged in speculative fiction. The following abbreviated history focuses on nanomedicine largely to the exclusion of broader issues in molecular engineering, manufacturing and nanoscopy, and includes a number of inspirational, speculative, or fictional early references from non-refereed sources.

The Biological Tradition

The general idea of biological engineering stretches back at least to the mid-19th century, but the first science fiction story involving actual genetic engineering was "Proteus Island" (1936), written by the chemical engineer Stanley Weinbaum. Artificial engineered organisms subsequently appeared in minor roles in several stories, a notable example being the familiars employed by the fake witches in Fritz Leiber's "Gather, Darkness!" (1943), and A.E. van Vogt used "gene transformation" to create the superman in "Slan" (1940). The first artificially evolved creatures appeared in Theodore Sturgeon's "Microcosmic God" (1941), wherein a biochemist established conditions allowing accelerated artificial evolution, creating the Neoterics, a submillimeter-sized race of intelligent hypermetabolic creatures which could accomplish tasks very rapidly. The first artificially designed microcreatures appeared in James Blish's "Surface Tension" (1952). In this story, crash-landed dying human astronauts create a completely new form of humanity—tiny men and women reduced to protozoan size—who are seeded in the pools and puddles at the surface of the new planet, and who go on to master the biotechnology necessary to travel from one water puddle to another.

The scientific tradition of biological nanomachines for medical purposes began in 1964 when Robert Ettinger, an early cryonics pioneer, suggested that cellular-level or even molecular-level repair might be developed for life extension. Ettinger speculated that "... surgeon machines, working 24 hours a day for decades or even centuries, will tenderly restore the frozen brains, cell by cell, or even molecule by molecule in critical areas."

In 1965, the synthesis of artificial life was publicly proposed as a national goal by the president of the American Chemical Society, Professor Charles Price, who pointed out that many new types of life might be made, not "mere imitations" of biology as we know it.

In 1967, Isaac Asimov suggested the future possibility of "factories ... where the working machinery consists of submicroscopic nucleic acids" and that a "repertoire of hundreds or thousands of complex enzymes" could be used to "bring about chemical reactions more conveniently than any methods now used" and also for "helping to construct life."

In 1968, G.R. Taylor cited the possibilities for genetic engineering and genetic surgery: "The microsurgery of DNA may possibly be achieved by physical methods: fine beams of radiation (probably laser light or pulsed X-rays) may be used to slice through the DNA molecule at desired points." He also cited predictions that bacteria would soon be programed.

In 1969, J. White suggested that a modified virus could be used as a cell repair machine: "It has been proposed that appropriate genetic information be introduced by means of artificially constructed virus particles into a congenitally defective cell for remedy; similar means may be used for the more general case of repair. The repair program must use means such as protein synthesis and metabolic pathways to diagnose and repair any damage... [Information] can be preserved by specifying that the repair program incorporate appropriate RNA tapes into itself..."

In 1970, Jeon et al carried out "the reassembly of *Amoeba proteus* from its major components: namely nucleus, cytoplasm, and cell membrane," taken from three different cells.

In 1972, Ettinger proposed using genetic engineering to make microscopic biorobots: "Genetic engineering's most sensational impact will concern the modification of humans, but it will have other uses as well. Some of the robots that will serve us will need to be nanominiaturized." Existing organisms could be modified to make biologically-based programable biorobots for medical applications: "If we can design sufficiently complex behaviour patterns into microscopically small organisms, there are obvious and endless possibilities, some of the most important in the medical area. Perhaps we can carry guardian and scavenger organisms in the blood, superior to the leukocytes and other agents of our human heritage, that will efficiently hunt down and clean out a wide variety of hostile or damaging invaders." Computerized cellular repair machines "must use means such as protein synthesis and metabolic pathways to diagnose and repair any damage. .. [Information] can be preserved by specifying that the repair program incorporate appropriate RNA tapes into itself...." Also in 1972, Danielli escribed various possibilities for generating new life forms via "life-synthesis" and genetic engineering, noting that "macromolecular engineering" might enable the development of very powerful and compact macromolecular computer systems.

In 1974, Halacy noted "some rather inglorious ways" to use "the miracle of artificial life," including potential capabilities for growing diverse items ranging from computers to airplanes. Morowitz suggested cooling cells to cryogenic temperatures in order to analyze and determine their structure. Artificial cells could also be assembled at such temperatures and then be set in motion by thawing. Morowitz further reported that microsurgery experiments on amoebas "have been most dramatic. Cell fractions from four different animals can be injected into the eviscerated ghost of a fifth amoeba, and a living functioning organism results."

In 1975, Richard Laing described the theoretical possibility of molecular machine self-replicators using molecular (data) tapes based on the idea of universal Turing machines, examining several ways that such "artificial organisms" might replicate themselves as "a vehicle for the exploration of broad biological possibilities."

In 1976, Donaldson presented the first detailed (and quite ambitious) list of biotechnological techniques that appeared necessary to achieve cell repair and might prove feasible, writing at a time that predated many current capabilities such as automated protein/DNA sequencing and synthesis, and most knowledge of restrictions on cellular developmental pathways and genetic programs and networks. For repair at the level of the cell, Donaldson's techniques would have included:

1. The ability to design enzymes to produce specific repair functions such as renaturing denatured proteins, joining broken lipoprotein complexes, annealing broken strands of DNA or RNA, reading proteins of existing or special types onto RNA and replicating them, and giving a cell the ability to metabolize new substrates, use novel cofactors, or construct essential amino acids;
2. Specially constructed bacteria or macrophages able to replicate themselves, spread throughout a specific target tissue, and carry out specific repairs according to the programs designed into their DNA/RNA; these could be designed to operate at unnatural temperatures or to utilize metabolic pathways not presently found in nature;
3. The abilities to re-introduce lost DNA or lost organelles such as mitochondria into a cell, to introduce entirely new forms of organelles perhaps to perform specific repair functions, and to introduce new metabolic capacities into a cell;
4. The ability to modify at will the developmental program of a cell, as for instance to induce postmitotic cells such as neurons to divide according to a specific program, forming daughter cells with specific properties; and
5. Several different types of repair bacteria able to work together in an integrated fashion, and linked together by chemical means (e.g. hormones), so as to apply optimal repairs to every body cell in order to (A) diagnose the precise nature of the damage, call other repair bacteria to its location, and report to the attending doctor that new types of repair bacteria other than those already introduced are needed, and (B) identify structures which must be preserved (e.g. memory) and reconstruct them if necessary.

For repairs at the level of the whole organism, Donaldson offered the following rather aggressive biological nanotechnology techniques:

1. Understanding the physiology of aging combined with the ability to reverse it;
2. Control over growth and development, including the ability to program types of growth and development which do not naturally occur, such as growth of new eyes or other organs which have been lost or damaged, growth of an entire and well-formed body from a head alone, and regrowth of injured or lost brain tissue; and

3. Nonpermanent "substitute organs" which will take over from others which have been lost, including (A) the ability to keep a given tissue alive and healthy in vitro for an indefinite time and similar abilities for a body part, and (B) temporary replacements for any body organ which may have been lost. These would be used to support the body while new organs were growing, e.g. as "metabolic crutches." This capacity specifically includes the ability to make temporary replacements for diffuse systems such as the vascular or nervous systems. For instance, a specially created "plant" would grow an entire vascular system into the patient, down to replacement for the capillaries and venules from a single seed, always introducing its fibrils between cells and destroying none or very little of the original structure.

The last of these is a description of a (clearly speculative) "repair net," a concept which Donaldson may have been the first to propose in 1976. A whole body repair net was later termed a "chrysalis", and twelve years later, in 1988,

In 1977, Darwin further developed the theme of tissue repair and cellular repair biomachines, independently proposing a modified white blood cell to perform repair functions.

In 1981, Asimov suggested that we "consider the bacteria. These are tiny living things made up of single cells far smaller than the cells in plants and animals. . . . [We] can, by properly designing these tiniest slaves of ours, use them to reshape the world itself and build it close to our hearts' desire." More importantly, Donaldson elaborated on his earlier discussion of how cryonically suspended human beings might be repaired. He extended the earlier concepts of Ettinger and White, concluding that "with such hybrid technology as micro-miniature biological-mechanical machines the size of viruses" and related technology, "it seems unlikely that (to repair a single cell) there would be any difficulty at all in principle to carrying out any imaginable repair." He estimated that about 10 programable cell-repair biomachines could be introduced into a cell that was under repair without causing too much mechanical disruption. In 1988, Donaldson described artificial macrophages that "can carry control machinery to recognize target cells, responding only to them, or responding differently depending upon cell type or cell conditions. They can still work even if the target cell isn't functioning (unlike viruses), rebuild target cell machinery other than the genes, and transfer many more genes up to an entire copy of the patient's genome." They can also "communicate with one another [and] release diffusible chemicals to guide one another's behaviour."

By the 1990s, bioengineered viruses of various types and certain other vectors were routinely being used in experimental genetic therapies as a means to target and penetrate certain cell populations, with the objective of inserting therapeutic DNA sequences into the nucleus of human target cells in vivo. Retrovirally-altered lymphocytes (T cells) began to be injected into humans for therapeutic purposes. Another example of an engineered cell in therapeutics was the use of genetically modified cerebral endothelial cell vectors to attack glioblastoma, which was being pursued by Neurotech (in Paris) in 1998. Engineered bacteria were being pursued by Vion Pharmaceuticals in collaboration with Yale University. In their "Tumor Amplified Protein Expression Therapy" program, antibiotic-sensitive *Salmonella typhimurium* (food poisoning) bacteria were attenuated by removing the genes that produce purines vital to bacterial growth. The tamed strain (cell line VNP20009) could not long survive in healthy tissue, but quickly multiplied ~1000-fold inside tumors, which are rich in purines. The next step would be to add genes to the bacterium to produce anticancer proteins that can shrink tumors, or to modify the bacteria to deliver various enzymes, genes, or prodrugs for tumor cell growth regulation. The engineered bacteria were available in multiple serotypes to avoid potential immune response in the host. Phase I human clinical trials were expected to begin in 1999 using clinical dosages produced in 50-liter fermenters, and other possible bacterial vector species were being examined.

Micro-biorobotics was still regarded by many in the biotechnology community as a highly speculative topic in 1998. Glen A. Evans described the possible construction of synthetic genomes and artificial organisms. His proposed strategy involves determining or designing the DNA sequence for the genome, synthesizing and assembling the genome, introducing the synthetic DNA into an enucleated pluripotent host cell, then introducing the host cell into an organism. Evans foresees the high throughput "read-out" of gene products discovered through genomic sequencing, rapid construction of designer genes and genomes, automated designer vector synthesis for gene replacement/insertional mutagenesis, and finally the construction of synthetic genomes and artificial organisms. Robert Bradbury has considered requirements and costs for genome synthesis and replacement. For example, Bradbury estimates genome synthesis costs by assuming a projected ~$0.03/base for DNA synthesis (compared to ~$0.80-$1.00/base-micromole in 1999), with ~20,000 expressed genes per biorobotic cell with an average gene size of ~3000 bases, giving a ~60 megabase expressed genome per biorobotic cell (1 chromosome) and a raw synthesis cost of ~$3 million per designer cell line (excluding design costs). There has also been discussion of "chemical reaction automata" as precursors for synthetic organisms, synthetic lifeforms and "nanobiology", "biorobotics", "cell rovers", microbial engineering, lymphocyte engineering, and artificial chromosomes and cell engineering is rapidly gaining popularity.

The Mechanical Tradition

The late science fiction author Robert A. Heinlein nearly invented the concept of molecular nanotechnology in 1942 when he suggested a process for manipulating microscopic structures. Heinlein envisioned the extensive use of life-size teleoperator hands, called "waldoes," complete with sensory feedback for full, remote-controlled telepresence. His fictional hero, Waldo, used a collection of these mechanical teleoperated hands for building and operating a series of ever-smaller sets of such mechanical hands. The smallest mechanical hands, "hardly an eighth of an inch across," were equipped with micro-surgical instruments and stereo "scanners," and were used to "manipulate living nerve tissue, [to examine] its performance in situ," and to perform neurosurgery. Eric Frank Russell's 1947 story "Hobbyist" described a fabrication process with "atom fed to atom like brick after brick to build a house." In 1955, Russell's serial "Call Him Dead," also published in *Astounding Science Fiction*, had a virus-based alien intelligence that spread through contact with blood or saliva; the story features a "microforger," a man who makes "surgical and manipulatory instruments so tiny they can be used to operate on a bacillus." Also in the mechanical tradition, Isaac Asimov's "Fantastic Voyage" in 1966 took its miniaturized human crew in a miniaturized submarine through the bloodstream of a human patient on a mission of repair.

Heinlein, Russell, and Asimov overlooked the full implications of their ideas, and most scientists were unsympathetic—for example, in 1952 Erwin Schrodinger wrote that we would never experiment with just one electron, atom, or molecule. But by the early 1960s, several scientists had reinvented similar approaches to micromanipulation and miniaturization, this time extending their reach into the nanotechnology domain. The first and most famous of these scientists was the Nobel physicist Richard P. Feynman. In his remarkably prescient 1959 talk "There's Plenty of Room at the Bottom," Feynman proposed employing machine tools to make smaller machine tools, these to be used in turn to make still smaller machine tools, and so on all the way down to the atomic level. Feynman prophetically concluded that this is "a development which I think cannot be avoided." Such nanomachine tools, nanorobots and nanodevices could ultimately be used to develop a wide range of submicron instrumentation and manufacturing tools, i.e. nanotechnology. Feynman's suggested applications for these tools included producing vast quantities of ultrasmall computers and various micro- and nano-robots.

Feynman was clearly aware of the potential medical applications of the new technology he was proposing. After discussing his ideas with a colleague, Feynman offered the first known proposal for a nanomedical procedure to cure heart disease: "A friend of mine (Albert R. Hibbs) suggests a very interesting possibility for relatively small machines. He says that, although it is a very wild idea, it would be interesting in surgery if you could swallow the surgeon. You put the mechanical surgeon inside the blood vessel and it goes into the heart and looks around. (Of course the information has to be fed out.) It finds out which valve is the faulty one and takes a little knife and slices it out. Other small machines might be permanently incorporated in the body to assist some inadequately functioning organ." Later in his historic lecture, Feynman urged us to consider the possibility, in connection with biological cells, "that we can manufacture an object that maneuvers at that level!"

In 1961, K.R. Shoulders rejected the use of biological building blocks, even though biological "processes do work, and they can do so in a garbage can without supervision." He saw them as too limited environmentally and too difficult to control with available technology. Instead, he sought to directly produce much simpler, more powerful and rugged nanostructure arrays, operating at video frequency rates, which in turn could ultimately aid in their own replication. In 1965, Shoulders reported the actual operation of micromanipulators able to position tiny items with 10 nm accuracy while under direct observation by field ion microscopy.

In 1970, Volkenstein noted that "the creation of a nonmacromolecular system which would act as a model for living organisms is definitely possible" but could not arise by itself, and that the macromolecularity of present organisms is not essential, but due to their evolutionary origins. Taking some poetic license, he added: "Consequently, the cybernetic nonmacromolecular machine, which simulates life, could have been and can be created on earth only by man. Then it could perfect itself without limits." T. Nemes discussed artificial self-replicating machines and described how to construct "an automatic lathe able to reproduce itself," a concept apparently developed before von Neumann's work on machine replication.

In 1981, Drexler suggested the construction of mechanically deterministic nanodevices using biological parts; these devices could inspect cells at the molecular level and also repair cellular tissues that had been damaged during cryonic suspension. In 1982, Drexler described cell repair machines even more clearly in the mechanical tradition, in a popular publication.

By 1983, Drexler began privately circulating a draft technical paper entitled "Cell Repair Machines" which investigated for the first time, in some detail, whether an advanced mechanical-based nanotechnology would "permit construction of systems of molecular-scale sensors, computers, and manipulators able to enter and repair cells; the nature of the computational algorithms and magnitude of the computational resources needed to guide repairs; and the physical capabilities and constraints important to the repair process [in order to] sketch the conceptual design of a cell repair system based on a mature molecular technology."

Peterson notes that "medical applications were explored by Drexler during the early 1980s but the medical community was not ready for the concept." A technical paper by Drexler, invited by an editor at the *Journal of the American Medical Association*, was dismissed by a referee as "science fiction."

In 1985, G. Feinberg proposed using short-wavelength coherent laser energy to power and communicate with "nanosensors that could be implanted into the human body. They could . . . [monitor] various physiological functions from subcellular molecules up through tissues and organs . . . essential in determining some of the mechanisms involved in growth and aging."

In a 1985 book entitled *Robotics*, edited by Marvin Minsky (a well-known computer scientist and artificial intelligence pioneer), Minsky briefly described how fully-automatic cellular repair machines might work, following Drexler's vision.

R.A. Freitas Jr. also contributed a chapter to Minsky's 1985 book, offering the following suggestion for remote-controlled incisionless nanosurgery: "One possibility is the concept of remote-controlled medical mites made feasible by modern micromachinery technology. Some medical mites would be like microminiature submarines, released inside the human body for internal sensing. Other mites could float, crawl, or swim through major arteries in the human body and perform on-site repairs from within, controlled by radio link under direction of a skilled telemicrosurgeon."

In 1986, Drexler published *Engines of Creation*, a popular text with two chapters devoted to discussions of cellular repair machines. Drexler further expounded upon the topic of cellular repair machines in articles published in 1985, 1986, 1987, and 1989, and his concepts were described by Brian Wowk in 1988 and in a Time-Life book in 1989.

In 1988, A.K. Dewdney reported an early nanomedical concept of an artery-cleaning nanorobot that he attributed to Drexler, accompanied by an artist's conception with the caption "a nanomachine swimming through a capillary attacks a fat deposit."

Nanomedicine in the 1990s

Despite the tremendous importance of molecular nanotechnology, a published 1993 literature review of robotics in health care included not a single reference to nanotechnology or to nanomedicine At this writing in 1998, the list of original post-1989 technical and popular nonfiction works that deal with nanomedical topics in the mechanical tradition is short enough to permit virtually an exhaustive listing, including Beardsley, Bova, Coombs and Robinson, Crawford, Drexler, DuCharme, Emanuelson, Fahy, Fiedler and Reynolds, Freitas, Kaehler, Klatz and Kahn, Kurzweil, Lampton, Merkle, Merrill, Minsky, More, Ostman, Reifman, and Wowk and Darwin.

Biotechnology and Molecular Nanotechnology

Some pragmatic readers may be wondering what is so special about molecular nanotechnology, that existing or anticipated biotechnology could not accomplish just as well? After all, biotechnology is already an established medical capability. It has real applications and real products already on the market. Reflecting upon the future possibility of sophisticated mechanical medical nanorobots equipped with powerful nanocomputers, in 1989, one well-known cryobiologist mused that "what is not clear is just what need we would have for such devices." The simplest answer, suggested by Figure 1.5, is that each of the three contemporary branches of "nanotechnology" offers something of unique value to the practice of medicine.

Nanoscale materials technology has already found widespread use in medicine, including biocompatible materials and analytical techniques, surgical and dental practice, nerve cell research using intracellular electrodes, biostructures research and biomolecular research using near-field optical microscopy, scanning-probe microscopy and optical tweezers, and vaccine design, and also many 20th century bulk chemical and biochemical manufacturing techniques along with much of classical pharmacology.

As for "biotechnology," the original meaning of this word contemplated "the application of biological systems and organisms to technical and industrial processes". In recent times, the field has expanded to include genetic engineering and now takes as its ultimate goal no less than the engineering of all biological systems, even completely artificial organic living systems, using biological instrumentalities.

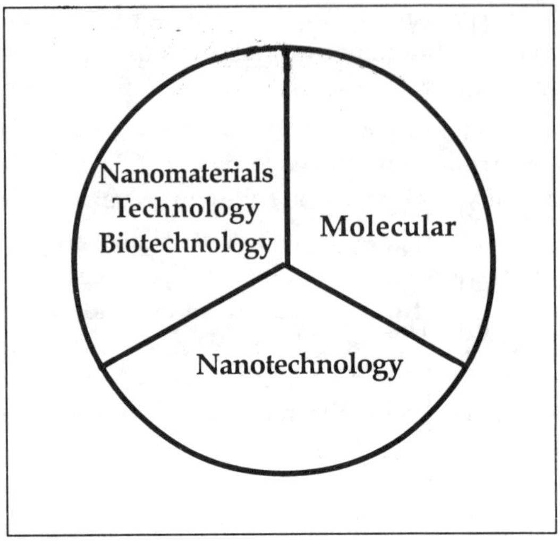

Fig. VI–A.5. Three contemporary branches of Nanotechnology.

The third branch, molecular nanotechnology, takes as its purview the engineering of all complex mechanical systems constructed from the molecular level—potentially offering new tools for medical practice, the principal subject of this book. Observes G.M. Fahy, "the difference between nanotechnologists and biotechnologists is that the former do not restrict themselves to the biological limitations of the latter, and they are much more ambitious about the kinds of accomplishments that they want to achieve."

Doctors can utilize solutions to medical problems from all three approaches. As noted earlier, 80-90 per cent of medical complaints resolve themselves via natural homeostatic processes, or without the necessary involvement of active biotechnological or molecular-nanotechnological agents. But by employing biotechnology, the range and efficacy of treatment options greatly increases. With molecular nanotechnology, the range, efficacy, comfort and speed of possible medical treatments again expands enormously. Molecular nanotechnology is essential when the damage to the human body is extremely subtle, highly selective, or time-critical (as in head traumas, burns, or fast-spreading diseases), or when the damage is very massive, overwhelming the body's natural defenses and repair mechanisms.

Table VI–A.4, inspired by an earlier discussion from G.M. Fahy, makes these options more explicit. At every difficulty level, most classes of medical problems may be resolved with varying efficacy within the homeostatic/nanomaterials, biotechnological, or molecular-nanotechnological approaches. As the chosen technology becomes more precise, active, and controllable, the range of options broadens and the quality of the options improves. Thus, the question is not whether molecular nanotechnology is required to accomplish a given medical objective. In many cases, it is not—though of course there are some things that only biotechnology and molecular nanotechnology can do, and some other things that only molecular nanotechnology can do. Rather, the important question is which approach offers a superior solution to a given medical problem, using any reasonable metric of treatment efficacy. For virtually every class of medical challenge, a mature molecular nanotechnology offers a wider and more effective range of treatment options than any other approach.

It is quite possible to imagine an advanced biotechnology that uses an engineered white cell, fibroblast, or macrophage chassis, energized by native oxygen and glucose and modified mitochondrial powerplants, driven by pseudopodia, cilia or flagella communicating and navigating via biochemical signals and even incorporating onboard digital biocomputers to make microscopic biorobots. Principal arguments favoring the biotechnology approach for medical purposes are:

(1) We are already somewhat familiar with such systems, after half a century of intensive molecular biology research.

(2) We have already "built" precursor systems, such as a whole living amoeba constructed from five distinct parts,bioengineered viruses and bacteria as DNA insertion devices, and natural replication stimulated in genetically engineered starter microbes.

(3) Biocompatibility will not be a major issue since fibroblasts (which express no HLA Class II antigens, hence stimulate no rejection response; could be used as the starter material.

(4) Both engineered viruses and bacteria are already in wide commercial and research usė.

(5) The greater complexity of self-repair in mechanical systems, should it be needed. Many believe that the development pathway to early biorobots may be considerably shorter than for the mechanical nanorobots of the molecular nanotechnology approach, for which in 1998 not a single working prototype yet existed, even in research laboratories.

It is also possible to imagine a molecular nanotechnology that uses mechanical nanorobotic systems. Such systems will have many constitutional differences from biological-based systems. For instance, mechanical systems will transport parts, materials, energy and instructions via fixed channels, whereas most (but not all) biological systems operate by diffusion. Mechanical systems will have structures constrained by specific geometries, whereas biological systems have structures defined by patterns of containment and interconnection—the shape of a membrane compartment in a cell matters less than its continuity and the contents of the volume it defines. Mechanical systems will be deterministically manufactured by operations analogous to manual construction, whereas engineered ribosomes will self-assemble via diffusion and stochastic matching of complementary parts; in other words, biology uses recipes, while mechanical systems use blueprints. Cells grow, with their parts adapting to one another; mechanical nanorobots may be constructed from parts of fixed structure. Biology uses self-repair; mechanical systems generally do not—largely because self-repair (by component exchange) will not be needed in molecular mechanical systems, whose designs may be made more simple by relying upon high component redundancy. These and other differences imply a number of important advantages that mechanical-based medical systems will enjoy over biological-based medical systems, which, taken together, strongly suggest that predominantly mechanical nanosystems may be the approach of choice for a mature medical nanotechnology. The advantages of molecular nanotechnology (e.g. the mechanical tradition) are many:

1. **Speed of Medical Treatment**—doctors may be surprised by the incredible quickness of nanorobotic action when compared to the speeds available from fibroblasts or leukocytes. Normal homeostatic processes such as dermal wound repair via natural fibroblasts may require weeks to run to completion. Typical fibroblast movements occur at 0.1-1 microns/sec, but mechanical nanomanipulators can operate at 1-10 cm/sec speeds or faster, a speed advantage of 4-5 orders of magnitude. Even the strongest biological fibers (e.g. intermediate filaments) have a failure strength 3 orders of magnitude below the strongest mechanical fibers (e.g. fullerene nanotubes). Biological cilia beat at ~30 Hz while mechanical nanocilia may cycle up to ~20 MHz, though practical power restrictions and other considerations may limit them to the ~10 KHz range for most of the time. Flexible mechanical surfaces can complete a morphing motion in ~0.1 millisec, compared to the ~100 millisec snapback time for pinched red cell membrane, again a thousandfold advantage in speed. Thus we expect that mechanical therapeutic systems can reach their

Table VI–A.4. Medical Challenges of Increasing Difficulty and Possible Approaches

Level of Difficulty	*Increasing Efficacy*			
	Specific Challenges	*Homeostatic or Nanomatorials*	*Biotechnology Approach Technology*	*Molecular nanotochnology Approach*
1.	Minor symptoms and Minor Physical Trauma	Herbal preparation, minor surgery, passago of time	Symptomatic suppression pharmacology	Pharmacyles zippocyles nospirocytes, etc.
2.	Infectious Agents	Natural immune system	Antibiotics and vaccines	Nanobiotics
3.	Mutation and Cell Disease	Natural immune system Major surgery	Molecular diagonstics and treatments; gone therapies	Nanobiotics Cell repair machines
4.	Health Maintenance and Aging	Vitamins, good diet, exercise, moderation and "clean living"	Gene therapy to enhance self-repair and immune function	Whole-body cyto-assay Cell repair machines
5.	Major Organ Replacement or Repair	Organ transplantation Major surgery	Storm cell & tissue engineering embryonic gene reactivation	Cell mills, tissue mills, organ mills, nanosurgery
6.	Morpho-Engineering	Cosmetic Surgery	Control of natural norphogenetic systems	Nanosurgery Chromatin editors
7.	Emergency Care: Restoration of Nonhomeostatic Tissue	Emergency major surgery Prosthesis or bionics	Biological repair nets Bio engineered immune cells	Tranapods Warm biostasis
8.	Augmentation of Natural Structure or Function	Prosthesis or bionics	Cross-species and artificial nonphogenetic supplementation	Cell engineering devices Autogenous control

targets up to ~10,000 times faster, all else equal, and treatments which require ~10^5 sec for a biological system may need only ~10^2 sec for a mechanical system; tachyiatria improves both patient and physician comfort. In either biological or mechanical systems, large numbers of devices of comparable physical size (e.g. ~microns) may be employed to do the work, so numbers alone cannot offset the mechanical speed advantage.

2. **Power Density and Transduction**—biological cells typically employ power densities of 10^3-10^4 W/m^3, with maximum densities of ~10^6 W/m^3 in honeybee flight muscle cells and bacterial flagellar motors. By contrast, nanomechanical power systems can produce power densities of 10^9-10^{12} W/m^3 an advantage of 10^3-10^8 for mechanical over biological systems. By 1998, conducting polymer-based actuators generated 20-100 times the force for a given cross-sectional area as mammalian skeletal muscle. Additionally, amoebic locomotion in motile cells requires diffusion-limited cytoskeletal disassembly and reassembly to achieve movement; mechanical motility systems may employ simple cable-pulling, winches, or ratchets, which are faster and more direct. Some biological energy transducers are reversible, but muscle contraction is irreversible—not only cannot muscle actively re-expand, but stretching it doesn't make it produce much useful chemical energy. In contrast, electric motors can be run backwards to generate electricity, forcing pistons makes them pump, and loudspeakers can be used as microphones.

3. **Superior Building Materials**—typical biological materials have tensile failure strengths in the 10^6-10^7 N/m^2 range, with the strongest biological materials such as wet compact bone having a failure strength of ~10^8 N/m^2, all of which compare poorly to ~10^9 N/m^2 for good steel, ~10^{10} N/m^2 for sapphire, and ~10^{11} N/m^2 for diamond and carbon fullerenes, again showing a 10^3-10^5 advantage for mechanical systems that use nonbiological materials. Nonbiological materials can be much stiffer, permitting the application of higher forces with greater precision of movement, and they also tend to remain stable over a larger range of temperature, pressure, salinity and pH. Proteins are heat sensitive in part because much of the functionality of their structure is due to noncovalent bonds involved in folding, which are broken more easily at higher temperatures; in diamond, sapphire, and many other rigid materials, structural shape is covalently fixed, hence is far more temperature-stable. Most proteins tend to become dysfunctional at cryogenic temperatures, unlike diamond-based mechanical structures. Biomaterials are not ruled out for all nanomechanical systems, but represent only a small subset of the materials that can be used in nanorobots. Mechanical systems can employ a wider variety of atoms and molecular structures in their design and construction, with novel functional forms that might be difficult to implement in a biological system such as steam engines or nuclear power. As another example, an application requiring the most effective bulk thermal conduction possible should use diamond, the best conductor available, not some biomaterial with inferior thermal performance.

4. **Nondegradation of Treatment Agents**—diagnostic and therapeutic agents constructed of biomaterials generally are biodegradable in vivo, although there is a major branch of pharmacology devoted to designing drugs that are moderately non-biodegradable—antisense DNA analogues with unusual backbone linkages and peptide nucleic acids (PNAs) are difficult to break down. However, suitably designed nanorobotic agents constructed of nonbiological materials are not biodegradable. An engineered fibroblast may not stimulate an immune response when transplanted into a foreign host, but its biomolecules are subject to chemical attack in vivo by free radicals, acids, and enzymes. Even "mirror"

biomolecules or "Doppelganger proteins" comprised exclusively of unnatural D-amino acids have a lifetime of only ~5 days inside the human body. Nonbiological materials such as diamond and sapphire are highly resistant to chemical breakdown or leukocytic degradation in vivo, and pathogenic biological entities cannot easily evolve useful attack strategies against these materials.

5. **Control of Nanomedical Treatment**—present-day biotechnological entities are not programable and cannot be switched on and off conditionally during task execution. A digital biocomputer, while possible in theory, represents a considerable conceptual departure from the usual biological paradigm. Even assuming that a digital biocomputer could be installed in a fibroblast, and that appropriate effector mechanisms could be attached, such a system would necessarily have slower clock cycles, less capacious memory per unit volume, and longer data access times, implying less diversity of action, poorer control, and less complex executable programs than would be available in nanoscale electromechanical computer systems. The mechanical approach emphasizes precise control of action, including control of physical placement, timing, strength, structure, and interactions with other (especially biological) entities. The biological approach emphasizes the use of poorly controlled natural structures, needlessly sacrificing huge blocks of the available functionality and design space.
6. **Nanodevice Versatility**—mechanical systems can readily incorporate biological elements if necessary, but artificial biological systems can incorporate nonbiological materials such as carbon nanotubes or diamond/sapphire structural elements only with difficulty, in part because biology has a more limited repertoire of "effector" mechanisms. Artificial biological systems cannot easily incorporate nonbiological materials where desired because natural biological assembly methods make no provisions for these materials either in the coded instructions in DNA or in the attachment chemistries. Rebuilding or reconstructing the human body with nonbiological components (e.g. fullerene encabled bone for bone damaged in an accident; or augmentation of human body function with unnatural abilities or features (e.g. autogenous paracrine control), will be very difficult or impossible to achieve using purely biotechnological means.
7. **Avoiding Overspecialization**—M. Krummenacker notes that one of the most glaring shortcomings of bacteria and other naturally occurring molecular machinery—when viewed as systems subject to further engineering—is the rather limited range of molecular substrates they can utilize. For example, bacterial enzymes are highly specialized devices, with very narrow substrate specificities. Thousands of different enzymes are needed in each organism, and the substance classes capable of digestion are limited to some sugars, various amino acids including proteins that can be degraded by excreted proteases, lipids, and a few other smallish oxygen-functionalized carbon molecules such as glycerol and ethanol. Some bacteria can metabolize CO_2 and a handful of aromatic compounds, but there is a vast range of organic chemicals that most bacteria cannot degrade or manufacture. A much smaller set of substantially more general molecular tools can probably be designed using the mechanical approach; mechanosynthesis can fabricate and assemble, or disassemble, a far wider range of molecular structures than are available to the cellular machinery of life.
8. **Faster and More Precise Diagnosis**—the analytic function of medical diagnosis requires rapid communication between the injected devices and the attending physician. If limited

to chemical messaging, biotechnology devices will require minutes or hours to complete each diagnostic loop. Nanomachines, with their more diverse set of input-output mechanisms, can outmessage the results of in vivo reconnaissance or testing literally in seconds. Such nanomachines can also run more tests of greater variety in less time. Mechanical nanoinstruments, including molecule-by-molecule disassemblers will make comprehensive cell mapping and cell interaction analysis possible. Bacterial resistance can be assayed at the molecular level, allowing new treatment agents to more easily be composed, manufactured and immediately deployed.

9. **More Sensitive Response Threshold for High-Speed Action**— unlike natural systems, an entire population of nanobiotic devices can be triggered globally by just a single local detection of the target antigen or pathogen. The natural immune system takes >10^5 sec to become fully engaged after exposure to a systemic pathogen or other antigen-presenting intruder. A biotechnologically enhanced immune system that can employ the fastest natural unit replication time (~10^3 sec for some bacteria) will require ~10^4 sec for full deployment post-exposure. By contrast, a nanobiotic immune system can probably be fully engaged (though not finished) in at most two blood circulation times, or ~10^2 sec.
10. **More Reliable Operation**—engineered macrophages would probably individually operate less reliably than mechanical nanorobots. Many pathogens, such as *Listeria monocytogenes* and *Trypanosoma cruzi*, are known to be able to escape from phagocytic vacuoles into the cytoplasm; while biotech drugs or cell manufactured proteins could be developed to prevent this (e.g. cold therapy drugs are entry-point blockers), nanorobotic trapping mechanisms can be more secure. Proteins assembled by natural ribosomes typically incorporate one error per ~10^4 amino acids placed; current gene and protein synthesizing machines utilizing biotechnological processes have similar error rates. A molecular nanotechnology approach will improve error rates by at least a millionfold. Mechanical systems can also incorporate sensors to determine if and when a particular task needs to be done, or when a task has been completed. Finally, it is unlikely that natural organisms will be able to infiltrate mechanical nanorobots or to co-opt their functions. By contrast, a biological-based robot could be diverted or defeated by microbes that can piggyback on its metabolism, interfere with its normal workings, or even incorporate the device wholesale into their own structures, causing the engineered biomachine to perform some new or different function than was originally intended. There are many examples of such co-option among natural biological systems, including the protozoan mixotrichs found in the termite gut that have assimilated bacteria into their bodies for use as motive engines, and the nudibranch mollusks (marine snails without shells) that steal nematocysts (stinging cells) away from coelenterates such as jellyfish (i.e. a Portuguese man-of-war) and incorporate the stingers as defensive armaments in their own skins, a process which S. Vogel has called "stealing loaded guns from the army."
11. **Verification of Progress and Treatment**—using a variety of communication modalities, nanorobots can report back to the attending physician, with digital precision, a summary of diagnostically- or therapeutically-relevant data describing exactly what was found, and what was done, and what problems were encountered, in every cell visited. A biological-based approach relying upon chemical messaging is necessarily slow with limited signaling capacity. Also unlike mechanical nanorobots, biotechnological systems generally cannot monitor their own functions while working, and, except for a few highly

specialized DNA proofreading systems, cannot directly inspect their work while it is in progress or after it is finished.

12. **Minimum Side Effects**—almost all drugs have side effects, such as conventional cancer chemotherapy which causes hair loss and vomiting, although computer-designed drugs have high specificity and relatively few side effects. Carefully tailored cancer vaccines under development in the late 1990s were expected unavoidably to affect some healthy cells. Even well-targeted drugs are distributed to unintended tissues and organs in low concentrations, although some bacteria can target certain organs fairly reliably without being able to distinguish individual cells. By contrast, mechanical nanorobots may be targeted with virtually 100 per cent accuracy to specific organs, tissues, or even individual cellular addresses within the human body. Such nanorobots should have few if any side effects, and will remain safe even in large dosages because their actions can be digitally self-regulated using rigorous control protocols that affirmatively prohibit device activation unless all necessary preconditions have been, and continuously remain, satisfied. G.M. Fahy has noted that these possibilities could transform "drugs" into "programable machines with a range of sensory, decision-making, and effector capabilities [that] might avoid side effects and allergic reactions ... attaining almost complete specificity of action. ... Designed smart pharmaceuticals might activate themselves only when, where, and if needed." Additionally, nanorobots may be programed to excuse themselves from the site of action, or even from the body, after a treatment is completed; by contrast, spent biorobotic elements containing ingested foreign materials may have more limited post-treatment mobility, thus lingering at the worksite causing inflammation when naturally degraded or removed.

13. **Reduced Replicator Danger**—Drexler points out that living systems are evolved systems, while nanomechanical replicators would be designed: "The former are shaped to serve the goal of their own survival and replication in a natural environment, whereas the latter will be shaped (whether well or poorly) to serve human goals, perhaps in an artificial environment." Genetic engineering involves not design of replicators from scratch, but tinkering with the molecular machinery of existing bioreplicators. Since bioreplicators were not designed, they are not necessarily structured in ways that lend themselves to complete understanding, and processes based on diffusion and matching allow complex nonlocal interactions that can be hard to trace. Bioreplicators can be crippled, but having evolved in nature, they resemble systems that can survive in nature. Typically, they are able to exchange genetic information with wild organisms, raising the possibility of the introduction of new, unconstrained replicators in the natural environment. Having evolved to evolve, they have a capacity for further evolution—to serve their own survival, not human goals. For example, even when stripped of key pieces of DNA to interfere with its replication powers, a live attenuated AIDS vaccine can slowly recover its virulence and can attack immune cells. R. Bradbury suggests that artificial biorobots could incorporate multiple fail-safe mechanisms including required external essential nutrients or suicide suppressors, self-destruct triggers, countdown timers like telomeres, and engineering to reject foreign DNA (the basis for restriction enzymes), but it remains logically easier for a system with inchoate capacity to evolve to resume doing so, than for a system which has never had this capability to spontaneously develop it.

In contrast, nanomechanical replicators (e.g. assemblers; will be designed from scratch and thus will differ fundamentally from biological systems. The parts and structures of designed mechanical systems will be known, and the relationships among their parts will also be designed and fixed. More important, nanoreplicators will be fundamentally alien to the biosphere, unrelated to anything that has evolved to survive in nature. Certainly the capacity to fail can appear by accident, and emergent capabilities cannot be completely ruled out, but engineering experience shows that the ability to perform complex organized activities (such as replication in a natural environment) does not normally appear spontaneously. Also, and purely as a geometrical consideration, adding a new part inside a densely organized geometric structure (as would be found in a nanomachine) typically requires changes in the relative positions of many other parts, and hence corresponding adjustments in design. Adding a part inside a densely organized topological structure (as found in living systems) typically leaves topologies unchanged—room can be made by stretching and shifting other parts, with no change in their essential design, hence permitting easier modification to biological structures by exogenous agencies.

Note that mechanical medical nanodevices need not be capable of replication. There is no requirement for replication in vivo; such replication would be needlessly dangerous, and adding this capability would reduce effectiveness in carrying out the primary medical task. Analogously, viral vectors employed in genetic therapies are modified to be "incapable" of replication.

14. **Assured Patentability**—microscopic biorobots, unavoidably derived from natural biological material, may someday be deemed unpatentable under a general prohibition on "genetic colonialism" or other emerging legal doctrines. In contrast, mechanical nanorobots, being fully-artificial and designed machines, should always be patentable provided they satisfy the customary legal criteria.

Unit VI–B

Nanomedicine Taxonomy

A Nanomedicine Taxonomy

Nanomedicine research is being funded by government sources, such as the National Institutes of Health (NIH), and by companies in various sectors, including pharmaceutical, biotechnology and medical devices. Because of the focused and shorterterm development horizon of commercial applications, nanomedicine research is more easily characterized in these areas.

The nanomedicine taxonomy summarized below and further described in this document, classifies some of the leading areas that nanotechnology tools, materials, devices, and intelligent materials and machines are currently applied in medical research. Specific company and university research activities should be interpreted as examples and are only a subset of the vast scope of nanomedicine research currently underway.

Biopharmaceutics

Drug Delivery

Nanotechnology provides a wide range of new technologies for developing customized solutions that optimize the delivery of pharmaceutical products.

To be therapeutically effective, drugs need to be protected during their transit to the target action site in the body while maintaining their biological and chemicals properties.

Some drugs are highly toxic and can cause harsh side effects and reduced therapeutic effect if they decompose during their delivery. Depending on where the drugs will be absorbed (i.e. colon, small intestine, etc), and whether certain natural defense mechanisms need to be passed through such as the blood-brain barrier, the transit time and delivery challenges can be greatly different.

Once a drug arrives at its destination, it needs to be released at an appropriate rate for it to be effective. If the drug is released too rapidly it might not be completely absorbed, or it might cause gastro-intestinal irritation and other side effects. The drug delivery system must positively impact the rate of absorption, distribution, metabolism, and excretion of the drug or other substances in the body. In addition, the drug delivery system must allow the drug to bind to its target receptor and influence that receptor's signalling and action, as well as other drugs, which might also be active in the body.

Table VI–B.1: Nanomedicine Taxonomy

Biopharmaceutics	Sensory Aids
Drug Delivery	Retina Implants
Drug Encapsulation	Cochlear Implants
Functional Drug Carriers	**Surgical Aids**
Drug Discovery	Operating Tools
Implantable Materials	Smart Instruments
Tissue Repair and Replacement	Surgical Robots
Implant Coatings	**Diagnostic Tools**
Tissue Regeneration Scaffolds	Genetic Testing
Structural Implant Materials	Ultra-sensitive Labeling and Detection Technologies
Bone Repair	High Throughput Arrays and Multiple Analyses
Bioresorbable Materials	Imaging
Smart Materials	Nanoparticle Labels
Implantable Devices	Imaging Devices
Assessment and Treatment Devices	**Understanding Basic Life Processes**
Implantable Sensors	
Implantable Medical Devices	

Drug delivery systems also have severe restrictions on the materials and production processes that can be used. The drug delivery material must be compatible and bind easily with the drug, and be bioresorbable (i.e. degrade into fragments after use which are either metabolized or eliminated via normal excretory routes). The production process must respect stringent conditions on processing and chemistry that won't degrade the drug, and still provide a cost effective product.

Nanotechnology can offer new drug delivery solutions in the following areas.

Drug Encapsulation

One major class of drug delivery systems is materials that encapsulate drugs to protect them during transit in the body. Drug encapsulation materials include liposomes and polymers (i.e. Polylactide (PLA) and Lactide-co-Glycolide (PLGA)) which are used as microscale particles. The materials form capsules around the drugs and permit timed drug release to occur as the drug diffuses through the encapsulation material. The drugs can also be released as the encapsulation material degrades or erodes in the body.

When encapsulation materials are produced from nanoparticles in the 1 to 100nm size range instead of bigger microparticles, they have a larger surface area for the same volume, smaller pore size, improved solubility, and different structural properties. This can improve both the diffusion and degradation characteristics of the encapsulation material.

In addition to liposomes and polymers, other types of nanoparticles are available for encapsulation. Materials such as silica and calcium phosphate (hydroxyapatite) have demonstrated superior properties at the nanoscale than the microscale, and can potentially be better suited for certain drug delivery challenges mentioned above.

Nanoparticle encapsulation is also being investigated for the treatment of neurological disorders to deliver therapeutic molecules directly to the central nervous system beyond the blood-brain barrier, and to the eye beyond the blood-retina barrier. Applications could include Parkinson's, Huntington's, Alzheimer's, ALS and diseases of the eye. In particular, ophthalmic disorders represent a rapidly growing area that is threatened by various diseases such as age related macular degeneration (AMD), diabetic retinopathy, glaucoma, and retinitis pigmentosa (RP).

Neurotech is developing an encapsulated cell therapy to treat eye diseases. It uses a semipermeable membrane to encapsulate cells, which also permits therapeutic agents produced by the cells to diffuse through the membrane. The membrane isolates the cells from the local environment and minimizes immune rejection. The encapsulated cells are administered by a device implanted in the eye to permit the continuous release of therapeutic molecules from living cells. This avoids direct injections into the eye, which may not be practical for regular administrations.

Advectus Life Sciences is developing a nanoparticle-based drug delivery system for the treatment of brain tumors. The anti-tumor drug doxorubicin is adhered to a nanopolymer

Poly ButylCyano Acrylate (PBCA) particle and coated with polysorbate 80. The drug is injected intravenously and circulates through the blood stream. The polysorbate 80 attracts plasma apolipoproteins and is used by the blood stream to carry lipids. This is intended to create a camouflage effect similar to LDL cholesterol, which allows the drug to pass the blood-brain barrier.

Functional Drug Carriers

Another class of drug delivery systems where nanotechnology offers interesting solutions is in the area of nanomaterials that carry drugs to their destination sites and also have functional properties. Certain nanostructures can be controlled to link with a drug, a targeting molecule, and an imaging agent, then attract specific cells and release their payload when required. Because of their nanoscale size, nanostructures have the ability to enter the cells, as cells will typically internalize materials below 100nm. Some of the leading nanostructures being used for this purpose include fullerenes, dendrimers, and nanoshells.

Fullerenes (or Buckyballs) are natural hollow spheres, one nanometer in diameter, made up of 60 carbon atoms. Fullerenes create a unique drug delivery platform that allows active pharmacopheres to be grafted to its surface in three-dimensional orientations for precise control in matching fullerene compounds to biological targets, in entrapping atoms within the fullerene cage, and for attaching fullerene derivatives to targeting agents.

C Sixty is developing fullerene-based drug delivery platforms which link fullerenes with antibodies and other targeting agents. Some of C Sixty's drug delivery systems include fullerene-decorated chemotherapeutic constructs, fullerene-radiopharmaceuticals, and fullerene-based liposome systems, called Buckysomes, for the delivery of single drug loads or multiple drug cocktails. Employing rational drug design, C Sixty has produced several drug candidates using its fullerene platform technology in the areas of HIV/AIDS, neuro-degenerative disorders and cancer.

Another nanomaterial used as a drug delivery scaffold is the dendrimer, a polymer molecule discovered by Don Tomalia of Dendritic Nanotechnologies. Researchers such as James Baker of the University of Michigan are using dendrimers to get genetic material or tumor-destroying therapies into a cell without triggering an immune response.

This is due to the dendrimer's small size and branched structure. Dendrimers can be designed to release attached compounds in response to a specific molecule or chemical reaction. A layered sphere called a nanoshell is being developed by Nanospectra for drug delivery. The nanoshell has a gold exterior layer which covers interior layers of silica and drugs.

Nanoshells can be made to absorb light energy and then convert it to heat. As a result, when nanoshells are placed next to a target area such as tumor cell, it can release tumorspecific antibodies when infrared light is administered.

Drug Discovery

Nano and micro technologies are part of the latest advanced solutions and new paradigms for decreasing the discovery and development times for new drugs, and potentially reducing the development costs.

Traditional trial-and-error methods have contributed to a discovery process lasting 10 years or more for new drugs to reach the market. In recent years, a number of new and complementary technologies have been developed which considerably impact the drug discovery process.

High-throughput arrays and ultra-sensitive labeling and detection technologies are being used to increase the speed and accuracy of identifying genes and genetic materials for drug discovery and development. These micro and nano technologies along with information technology solutions such as combinatorial chemistry, computational biology, computer-aided drug design, data mining, and data processing tools are addressing the challenges related to eliminating critical bottlenecks in drug discovery.

In addition to the capacity for processing information, the number of drug candidates that have been screened in the last ten years has increased by three orders of magnitude from 500,000 drug compounds to approximately 1.5 billion according to BCC.

Implantable Materials

Tissue Repair and Replacement

Nanotechnology provides a new generation of biocompatible nanomaterials for repairing and replacing human tissues.

Human tissue that is diseased or traumatically compromised may require synthetic materials for its repair or replacement. While most types of tissues repair the interaction of stem cells with chemical modulators, there are differences in the ways that various tissues heal.

"Hard" tissues such as bone and teeth heal by reproducing tissues indistinguishable from the original. However in cases where a dental or artificial bone implant is required, the structural material used in the implant may trigger immune rejection, corrode in the body fluids, or no longer bond to the host bone. This can require additional surgery or result in the loss of the implant's function. In many cases, the failure occurs at the tissue-implant interface, which may be due to the implant material weakening its bond with the natural material. To overcome this, implants are often coated with a biocompatible material to increase their adherence properties and produce a greater surface area to volume ratio for the highest possible contact area between the implant and natural tissue.

"Soft" tissues such as skin, muscle, nerves, blood vessels and ligaments repair damaged areas with fibrous tissue. Damaged tissue from various sources such as burns and ulcers can be self-repaired by the body, but can also result in scar formation. Graft material using artificial sheets can replace skin and other tissue with reasonable graft stability and cosmetic outcome.

In other types of tissue, notably "Ultrasoft" tissue such as cell membrane and organelles that exhibit metabolic function, tissue replacement can best occur when living cells are transplanted in a mesh-structured synthetic scaffold.

The scaffold incorporates signalling ligands or DNA fragments to elicit specific cellular responses, and molecular sensors to accept feedback from the in vivo environment. The scaffold is typically a temporary structure that is bioresorbable when the tissue is regenerated. The scaffold material needs to be fabricated into a desired shape or threedimensional structure, with specific surface properties to support the site where the organized growth of multiple cell types will take place.

Nanotechnology can new offer new solutions for tissue repair and replacement in the following areas.

Implant Coatings

Nanotechnology brings a variety of new high surface area biocompatible nanomaterials and coatings to increase the adhesion, durability and lifespan of implants. Ceramic materials such as calcium phosphate (hydroxyapatite or HAP) are made into implant coatings using nanosized particles instead of microsized particles.

In addition to the higher surface areas and improved adhesion properties of the nanoparticle coatings, improved coating techniques are also being developed. While high temperature processes such as plasma spray can melt ceramic particles and reduce their surface area and adhesion properties, new low temperature processes with electromagnetic fields can maintain the nanomaterial properties. This provides the maximum possible contact area between the implants and bone surface to improve the potential for in-growth in the host bone.

New types of nanomaterials are being evaluated as implant coatings to improve interface properties. For example, nanopolymers such as polyvinyl alcohol (PVA) can be used to coat implantable devices that are in contact with blood (e.g. artificial hearts, vascular grafts, catheters) for dispersing clots or preventing their formation.

Tissue Regeneration Scaffolds

Nanostructures are being researched for the preparation and improvement of tissue regeneration scaffolds. Research areas include the ability to develop molecularly sensitive polymers using the optical properties of nanoparticles as control systems, manipulating the stiffness and strength of scaffolds using hybrid nanostructures, and the use of nanotechnology to prepare molecular imprints to maximize long-term viability and function of cells on scaffold surfaces.

With the ultimate objective of growing large complex organs, a variety of nanomaterials and nanotechnology fabrication techniques are being investigated as tissue regeneration scaffolds that provide improved structural requirements and guide the activity of seeded cells. Some examples are as follows:

- Nanoscale polymers such as Polyvinyl alcohol (PVA) are being molded into heart valves and seeded with fibroblasts and endothelial cells.
- PVA is also being investigated for the cornea by having corneal epithelia cells seeded in a PVA hydrogel structure. This polymer material can absorb more than 20 per cent its weight in water while maintaining a distinct three-dimensional structure.
- A polyglycolic ball is being experimented with muscle cells and bladder endothelial cells.
- Polymer nanocomposites are being researched for bone scaffolds.

Commercially viable solutions are thought to be 5 to 10 years away. Scientific challenges related to a better understanding of molecular/cell biology and fabrication methods for producing large three-dimensional scaffolds are among the many obstacles yet to be overcome.

Nanostructures are also being used to study the fundamental properties of implanted tissues.

In areas of in vivo analysis, nanostructures are used as tracers for implanted cells, and to study the response of host to implanted tissues.

Structural Implant Materials

Nanotechnology provides a new generation of biocompatible materials that can be used as implants or temporary biosorbable structures.

Bone is a high strength material that is used as both weight bearing and non-weight bearing structures. Bones are more than just structural materials as they also contain interconnected pores that allow body fluids to carry nutrients and permit interfacial reactions between hard and soft tissues. In the case of bone fractures, grafts, disorders, dental applications and other types of surgery, bones may require repair or replacement. A variety of natural materials are used as bone substitutes. These include autograft from the patient's pelvis, allograft from another human, bovine material or coral blocks.

Natural materials tend to be brittle and can lose mechanical strength during sterilization. They can also cause inflammation, pain at the pelvis graft site, and potentially transmit disease.

Bone cavities can also be filled with synthetic bone cement. Current bone cements containing polymethylmethacryate (PMMA) act as a filler or grout, which is injected as a flowable paste and then hardens in vivo. While PMMA cement can offer adequate mechanical properties and bonding, it is typically recommended only for non-weight bearing bones. PMMA has also been linked to tissue damage, nerve root pain and other side effects.

Bone Repair

Nanotechnology brings a variety of new high surface area biocompatible nanomaterials that can be used for bone repair and cavity fillers.

High strength nanoceramic materials, such as calcium phosphate apatite (CPA) and hydroxyapatite (HAP), can be made into a flowable, moldable nanoparticle paste that can conform to and interdigitate with bone. As natural bone is approximately 70 per cent by weight. CPA including hydroxyapatite (HAP), biocompatibility is thought to be extremely high with minimal side effects.

As its dense surface and tight three dimensional crystalline structure will allow for a superior compressive strength to PMMA, nanoceramics may be suitable for both weight bearing and non-weight bearing bones. Nanoceramics such as CPA is nanocrystalline and can have grain size of under 50nm. Its nanocrystals can be connected to each other to form connective hard tissue required for bones.

Bone cement material being developed at the University of South Carolina and Competitive Technologies, can harden in vivo and in the presence of serum, form a solid bone-like structure capable of stabilizing a fractured bone or used as a bone substitute.

Bioresorbable Materials

Nanotechnology also brings advances in bioresorbable materials. Bioresorbable polymers are currently being used in degradable medical applications such as sutures and orthopaedic fixation devices. With new production methods, nanostructures are being fabricated which could be used as temporary implants.

Bioresorbable implants will biodegrade and do not have to be removed in a subsequent operation. In one application, nanostructured implants are being designed to degrade at a rate that will slowly transfer load to a healing bone that it is supporting.

Research is also being done on a flexible nanofiber membrane mesh that can be applied to heart tissue in open-heart surgery. The mesh can be infused with antibiotics, painkillers and medicines in small quantities and directly applied to internal tissues. The nanomaterial will degrade over time and not stick to surgeons' wet gloves, which can complicate the use of some materials currently in use.

Smart Materials

Smart materials are a class of nanomaterials that respond to changes in the environment such as a drop in temperature or pH. An environmental change could trigger a physical or chemical effect that mimics a natural mechanism in the body. For example, applications could include a smart polymer that flexes with mechanical strength as an artificial muscle, or a hydrogel that dissolves according to body chemistry to more efficiently deliver drugs.

When attached to or embedded in a structure, smart materials can potentially sense disturbances, or trigger physical reactions with force in actuators, piezo sensors, and shape memory metals. It is possible to have smart materials used in the body to exceed current human performance on multiple dimensions, and detect environmental conditions beyond current human limits.

Implantable Devices

Assessment and Treatment Devices

Nanotechnology offers sensing technologies that provide more accurate and timely medical information for diagnosing disease, and miniature devices that can administer treatment automatically if required.

Health assessment can require medical professionals, invasive procedures and extensive laboratory testing to collect data and diagnose disease. This process can take hours, days or weeks for scheduling and obtaining results.

Some medical information is extremely time sensitive such as finding out if there is sufficient blood flow to an organ or tissue after transplant or reconstructive surgery, before irreversible damage occurs.

Certain medical tests such as biopsies are subjective and can provide inconclusive or incorrect results. In a false negative result where a needle misses the tumor and then samples a normal tissue, the cancer may go untreated and can impact a patient's chances for long-term survival.

Some tests such as diabetes blood sugar levels require patients to administer the test themselves to avoid the risk of their blood glucose falling to dangerous levels. Certain users such as children and the elderly may not be able to perform the test properly, timely or without considerable pain.

People who are exposed to radiation or hazardous chemicals in their work environment are at a higher risk of illness. Occasional testing is typically done but may not detect a disease in its early stage. Early detection could initiate timely treatment with a higher chance of success, and have a worker removed from the hazardous environment to prevent further damage.

Nanotechnology can new offer new implantable and/or wearable sensing technologies that provide continuous and extremely accurate medical information. Complementary microprocessors and miniature devices can be incorporated with sensors to diagnose disease, transmit information and administer treatment automatically if required. Example applications are as follows.

Implantable Sensors

Micro and nanosized sensors can make use of a wide range of technologies that most effectively detect a targeted chemical or physical property. Researchers at Texas A&M and Penn State use polyethylene glycol beads coated with fluorescent molecules to monitor diabetes blood sugar levels. The beads are injected under the skin and stay in the interstitial fluid. When glucose in the interstitial fluid drops to dangerous levels, glucose displaces the fluorescent molecules and creates a glow. This glow is seen on a tattoo placed on the arm.

Another approach by researchers at the University of Michigan is using dendrimers attached with fluorescent tags to sense pre-malignant and cancerous changes inside living cells. The dendrimers are administered transdermally and because of their small size, pass through membranes into white blood cells to detect early signs of biochemical changes from radiation or infection. Radiation changes the flow of calcium ions within the white blood cells and eventually triggers apoptosis, or programed cell death (PCD) due to the radiation or infection. The fluorescent tags attached to the dendrimers will glow in the presence of the death cells when passed with a rentinal scanning device using a laser capable of detecting the fluorescence. A similar application is being researched with NASA for detecting radiation levels in astronauts.

Sensor microchips are also being developed to continuously monitor key body parameters including pulse, temperature and blood glucose. A chip would be implanted under the skin and transmit a signal that could be monitored continuously.

Another application uses optical microsensors implanted into subdermal or deep tissue to monitor tissue circulation after surgery. Instrumentation is linked to transmit data to a nearby sensor to provide very early indication of inadequate circulation if the surgery is unsuccessful.

Another type of implantable sensor uses MEMS devices and accelerometers for monitoring and treating paralysed limbs. Implantable MEMS sensors can measure strain, acceleration, angular rate and related parameters to determine normal and problem data.

Implantable Medical Devices

Implantable sensors can also work with a series of medical devices that administer treatment automatically if required. Tiny implantable fluid injection systems can dispense drugs electrically on demand making use of microfluidic systems, miniature pumps, and reservoirs. Initial applications may include chemotherapy that directly targets tumors in the colon and are programed to dispense precise amounts of medication at convenient times, such as after a patient has fallen asleep. Lupus, diabetes and HIV/AIDS applications are also being investigated.

Implantable sensors that monitor the heart's activity level can also work with an implantable defribulator to regulate heartbeats. These devices are used with microprocessors to deliver electricity that keeps the heart in rhythm when levels go above or below the person's normal heart range.

Implantable MEMS devices are also fitted on prostheses to mimic the stability and strain of natural limbs. An artificial leg being developed uses sensors to measure load on the foot, knee angles and motion over 50 times per second. The sensors work with an electronically controlled hydraulic knee to improve its stability.

Implantable medical devices that are greater than 1mm in diameter might unfavorably alter the functions of surrounding tissue. Smaller implantable devices with non-intrusive or minimally-intrusive systems will likely contain nanoscale materials and smaller systems approaching the nanoscale.

Functional Electrical Stimulation (FES) is a method for treating people to regain the use of their paralyzed limbs by electrically stimulating paralyzed muscles with implanted electrodes. Researchers at Aalborg University in Denmark are applying nanostructures to the electrode surfaces to improve biocompatibility and acceptance in the neural/muscle tissue. When placed within a cell membrane, the nanostructures form a bioelectric interface with the neuron or muscle cell that enables the intracellular potential of the cell to be observed and manipulated.

Researchers at Aalborg are also using nanostructures to activate denervated muscles caused by injuries to the lower motor neurons located in the spinal cord. This can result from traumatic spinal cord injury, strokes in the spinal cord, repeated vertebral subluxation, brachial plexus injuries and peripheral myopathies such as polio which destroys the nerve cells controlling muscle (i.e. denervated muscle).

The muscle fibre membrane is incorporated with potential-generating nanostructures that change the transmembrane potential of the muscle fibre and improves the extracellular electrical stimulation. Muscle fibre activation is achieved by illuminating the incorporation site on the muscle fibre to optically activate denervated muscles.

Sensory Aids

Nano and related micro technologies are being used to develop a new generation of smaller and potentially more powerful devices to restore lost vision and hearing functions. The devices collect and transform data into precise electrical signals that are delivered directly to the human nervous system.

Degenerative diseases of the retina, such as retinitis pigmentosa or age related macular degeneration, decrease night vision and can progress to diminishing peripheral vision and blindness. These retinal diseases may lead to blindness due to a progressive loss of photoreceptors (rods and cones), the light sensitive cells of the eye.

In cases where the neural wiring from the eye to the brain is still intact but the eyes' lack photoreceptor activity, photoreceptor loss could be compensated by bridging or bypassing the destroyed photoreceptors and artificially stimulating the adjacent intact cells. Artificially generated impulses could reach the brain and produce visual perception, thereby restoring some elementary vision.

In severe hearing loss, patients typically have absent or malfunctioning sensory cells in the cochlea. In a normal ear, sound energy is converted to mechanical energy by the middle ear, which is then converted to mechanical fluid motion in the cochlea. Within the cochlea, the inner and outer hair sensory cells are sensitive transducers that convert mechanical fluid motion into electrical impulses in the auditory nerve.

Cochlear implants are designed to substitute for the function of the middle ear, cochlear mechanical motion, and sensory cells. The implants transform sound energy into electrical energy that will initiate impulses in the auditory nerve. Cochlear implants include an electronic circuit that is surgically placed in the skull behind the ear on the mastoid process of the temporal bone. This circuit is attached to a bundle of tiny wires that are inserted into the cochlea. At the end of the wires are typically 8 to 24 electrodes that cause a different pitch percept when stimulated. The other part of the device is external and has a microphone, a speech processor, and connecting cables.

Current cochlear implants have a number of drawbacks. They require major surgery and can eliminate any remaining natural hearing. Because of their large size, current cochlear implants often stimulate several nerve fibers at once. This causes users to experience imprecise or distorted sound, especially with complex sounds like music.

Microtechnologies and eventually nanotechnologies are being used in a new generation of implantable devices for improving sensory loss in the following areas.

Retina Implants

Retinal implants are in development to restore vision by electrically stimulating functional neurons in the retina. One approach being developed by various groups including a project at Argonne National Laboratory is an artificial retina implanted in the back of the retina. The artificial retina uses a miniature video camera attached to a blind person's eyeglasses to capture visual signals. The signals are processed by a microcomputer worn on the belt and transmitted to an array of electrodes placed in the eye. The array stimulates optical nerves, which then carry a signal to the brain. Another approach by Optobionics makes use of a subretinal implant designed to replace photoreceptors in the retina. The visual system is activated when the membrane potential of overlying neurons is altered by current generated by the implant in response to light stimulation. The implant makes use of a microelectrode array powered by as many as 3,500 microscopic solar cells.

Cochlear Implants

A new generation of smaller and more powerful cochlear implants are intended to be more precise and offer greater sound quality. An implanted transducer is pressure-fitted onto the incus bone in the inner ear. The transducer causes the bones to vibrate and move the fluid in the inner ear, which stimulates the auditory nerve.

An array at the tip of the device makes use of up to 128 electrodes, which is five times higher than current devices. The higher number of electrodes provides more precision about where and how nerve fibers are stimulated. This can simulate a fuller range of sounds.

The implant is connected to a small microprocessor and a microphone, which are built into a wearable device that clips behind the ear. This captures and translates sounds into electric pulses which are send down a connecting wire through a tiny hole made in the middle ear.

Implant electrodes are continuously decreasing in size and in time could enter the nanoscale.

Surgical Aids

Operating Tools

Medical devices that contain nano and micro technologies will allow surgeons to perform familiar tasks with greater precision and safety, monitor physiological and biomechanical parameters more accurately, and perform new tasks that are not currently done.

Surgery can be physically demanding on both the surgeon and the patient. Open surgery such as heart operations can require a surgeon to make a large cut into the skin and underlying muscles, saw through breastbones, crack open the ribcage, then make delicate and high precision maneuvers with various surgical instruments. Not only can the patient incur severe pain, scarring, complications and extensive recovery time, the surgeon must be intensely focused for long hours hunched over an operating table. This can increase the chance of a surgical error due to fatigue in the neck and back, along with potentially significant hand tremors.

In various procedures, the surgeon's view could be limited. Judgment may be required to tell whether the tissue about to be cut has the density of the intended cartilage, or alternatively, a tendon which was grabbed by mistake.

To improve surgical results benefiting both the patient and surgeon, minimally invasive surgical procedures are increasingly being done using laparoscopic techniques. Making use of small entry ports into the area of interest, a rod shaped telescope attached to a camera and other long and narrow surgical instruments are used to perform all of the major maneuvers.

Even with its considerable benefits, laparoscopic surgery causes a major shift in surgical skills that have to be mastered as the surgeon's vision and access is severely restricted. Visual cues are obtained from a two dimensional video display instead of a three dimensional operating field. As well, the surgeon must work with long laparoscopic instruments to handle delicate tissues.

Nano and micro technologies are being applied to improve the role of the surgeon in the following areas.

Smart Instruments

Surgical tools such as scalpels, forceps, grippers, retractors and drills are being embedded with miniature sensors to provide real-time information and added functionality to aid surgeons. Surgeons can be given continuous data on the force and performance of their instruments, the tissue type about to be cut (i.e. cartilage, bone, muscle, vascular, etc) and specific tissue properties, such as density, temperature, pressure, and electrical impulses. Verimetra is developing an enhanced version of its Data Knife with logic and surgical micro-electromechanical systems that can provide added information and functionality to assist a surgeon during a procedure by stimulating electrodes, measuring and cutting with ultrasonic elements, and cauterizing.

Instruments are being developed with specific functionality like tilt and pressure to allow neurosurgeons to perform operating tasks with greater precision and safety.

Nanoparticles are also being investigated for optically guiding surgery. This can potentially allow for better removal of lesioned or diseased sites, including tumors.

Surgical Robotics

Robotic surgical systems are being developed to provide surgeons with unprecedented control over precision instruments. This is particularly useful for minimally invasive surgery.

Instead of manipulating surgical instruments, surgeons use their thumbs and fingers to move joystick handles on a control console to maneuver two robot arms containing miniature instruments that are inserted into ports in the patient. The surgeon's movements transform large motions on the remote controls into micro-movements on the robot arms to greatly improve mechanical precision and safety.

A third robot arm holds a miniature camera, which is inserted through a small opening into the patient. The camera projects highly magnified 3D images on a console to give a broad view of the interior surgical site.

The surgeon controlling the robot is seated at an ergonomically designed console with less physical stress than traditional operating room conditions.

UCI Medical Center's da Vinci Surgical System is currently approved for gall bladder, prostrate, colorectal, gynecological, esophageal and gastric bypass procedures. Clinical trials are underway for vascular repair and coronary bypass surgery. The Zeus Robotic Surgical System is presently used for cardiac surgery.

Diagnostic Tools

Genetic Testing

Nano and micro technologies provide new solutions for increasing the speed and accuracy of identifying genes and genetic materials for drug discovery and development, and for treatment-linked disease diagnostics products.

Thousands of genes and their products (i.e. RNA and proteins) function in a complicated and orchestrated manner. Genetic testing technologies used for drug discovery and in diagnostic products make it possible to measure gene sequences (i.e. genes, gene mutations) and gene expression levels (abundance), which are important indicators of growth, metabolism, development, behaviour and adaptation of living systems. With these tools, many different genes and expression patterns can be characterized as normal or diseased tissue, along with the various stage of disease.

There are many steps and processes involved in genetic testing. While the nature of the tests and nomenclature may vary, some of the key principals and related challenges are described below.

Step 1—Create or obtain DNA arrays of "known" probes

Known probes of genetic sequences are prepared or obtained commercially for a desired test. These probes are typically in the form of cDNA or an array of oligonucleotides.

While it is desirable to have as many probes as possible to maximize the search for possible results, there is currently a limited number of known specimens. This limits the range of experimentation for drug discovery and disease diagnostics. With advances in the human genome, gene cloning and molecular biology, a vast amount of new probes are increasingly available which could in time cover the entire genome, and potentially screen multiple targets in a single test.

DNA arrays are created by placing probes as individual spots in an orderly array arrangement on a glass microscope slide, a nylon membrane substrate or a silicon chip. The latter case is also referred to as a biochip.

Step 2—Collect and assay "unknown" targets

Unknown cells or molecules of interest (called "targets") need to be collected and then compared with the known probes in the DNA array. This begins with tissue, blood or other biological samples where nucleic acids (RNA or DNA) are isolated.

The nucleic acids are converted into labeled targets by tagging the DNA marker with an observable fluorescent or radioactive label. The labeled targets are incubated with the probes to permit hybridization and form stable DNA and/or RNA duplex. This is used for locating or identifying nucleotide sequences of interest. After the incubation, nonhybridized samples are washed away, and measurements are made of the signal produced when hybridization occurs at particular probe locations. The level of hybridization between a specific probe and a target indicates the level of the gene corresponding to that probe in the test solution.

Current assay technologies use fluorescent dyes to label molecules and require expensive equipment such as a laser to light up biological interactions, and an optical microscope to detect the binding sites. Fluorescent dyes are not always precise or sufficiently sensitive to detect every gene. Also, genes that are not distinguished separately can bleed together and result in a false or inconsistent result.

The hybridization may also require gene amplification or Polymerase Chain Reaction (PCR) which is used to produce millions of copies of a specific piece of DNA in a test tube from a single cell. Amplification could require a repetitive series of cycles making use of various chemical and processing stages.

Nano and micro technologies can provide new solutions in the following areas.

Ultra-sensitive Labeling and Detection Technologies

Several new technologies are being developed to improve the ability to label and detect unknown target genes. At Genicon, gold nanoparticle probes are being treated with chemicals that cling to target genetic materials and illuminate when the sample is exposed to light.

Another approach by Chad Mirkin and Nanosphere uses gold nanoparticle probes coated with a string of nucleotides that complement one end of a target sequence in the sample. Another set of nucleotides, complementing the other end, is attached to a surface between two electrodes. If the target sequence is present, it anchors the nanoprobes to the surface like little balloons, and when treated with a silver solution, they create a bridge between the electrodes and produce a current.

Quantum Dot Corp. uses quantum dots to detect biological material. Because their color can be tailored by changing the size of the dot, the potential for multiple colors increases the number of biological molecules that can be tracked simultaneously. In addition, quantum dots do not fade when exposed to ultraviolet light and the stability of their fluorescence allows longer periods of observation.

These technologies are expected to be more sensitive than fluorescent dyes and could more effectively detect low abundance and low-level expressioning genes. They may also make use of smaller and less expensive equipment to light and detect the samples.

Without the need for gene amplification, they can also provide results in less time.

High Throughput Arrays and Multiple Analyses

Arrays are usually classified by their sample spot sizes with 200 micrometer diameter as the traditional cutoff between macroarrays and microarrays. Microarrays contain thousands of spots. As the sample spot size further decreases in size, more spots can fit onto a substrate. This reduces the area and corresponding time that needs to be sampled. With continued miniaturization beyond micro, the possibility exists to greatly increase the number of spots on a single chip, with the ultimate objective of including the entire genome. At the same time, quicker analyses are being made possible by performing multiple analyses in parallel rather than in series to generate more information in less time.

By using nanomaterials as sensing particles, chips could be reduced in size. This would allow scientists to read thousands of molecules with the possibility of using cheaper equipment.

Smaller, portable, cheaper and more precise integrated diagnostic kits for diseases such as systemic lupus or other multi-marker diseases may be made possible by using microfluidic and nanofluidic technologies. DNA arrays typically perform one type of analysis thousands of times. In the event that other experiments or processes are also required, micro and nanofluidic devices such as lab-on-a-chip, can integrate mixing, moving, incubation, separation, detection and data processing in a small portable device. At the micro and nanoscale, fluids move through pipes in laminar flow, as opposed to turbulent flow at the macro level. This provides the opportunity for microfluidic devices to exploit certain physical behaviours. For example, two liquids can separately circulate through micro pipelines and valves without mixing each other.

The combination of arrays and fluidic capabilities can greatly increase the speed and accuracy of various genetic testing. Within the drug discovery and diagnostic fields, microfluidic devices in the form of nanoarrays or "lab on a chip" technologies could allow the production of more efficient and disposable DNA and protein sequencers for drug discovery and diagnostic kits.

Imaging

Nano and micro technologies offer new imaging technologies that provide high quality images not possible with current devices, along with new methods of treatment.

Malignant tumors are highly localized during the early stage of their development. If detected early, the tumors can often be surgically removed with high success.

The longer a patient has a malignant tumor the more likely the cancer will spread to neighboring lymph nodes and other anatomic structures. Aggressive surgery or chemotherapy, or very high doses of radiation may kill the cancer but at the same time severely injure normal tissue. In many patients, particularly those with breast cancer, the cancer can spread to other parts of the body even after the original cancer tumor has been removed.

Brain tumors have other challenges. Removing tumors and other cancerous cells in the brain through surgery can risk the loss of cognitive ability, hearing, speech, and other functions. Chemotherapy therapeutics can kill cancerous cells but because of their large physical size may be prevented from entering the brain by the blood-brain barrier.

Nanotechnology can offer new solutions for the early detection of cancer and other diseases. Complementary technologies can also destroy cancer cells, which are too small to locate and remove surgically.

Nanoparticle Probes

Researchers at the University of Michigan are developing nanoprobes that can be used with magnetic resonance imaging (MRI). Nanoparticles with a magnetic core are attached to a cancer antibody that attracts cancer cells. The nanoparticles are also linked with a dye which is highly visible on an MRI. When these nanoprobes latch onto cancer cells they can be detected on the MRI. The cancer cells can then be destroyed by laser or low dosage killing agents that attack only the diseased cells.

In brain tumors, the presence of cancer can weaken the blood-brain barrier. Research is being conducted to have nanoprobes cross the weakened barrier to the tumor but not cross into healthy brain tissue.

Another group at Washington University is using nanoparticles to attract to proteins emitted from newly forming capillaries that deliver blood to solid tumors. The nanoparticles circulate through the bloodstream and attach to blood vessels containing their complementary protein. Once attached, chemotherapy is released into the capillary membrane. The nanoparticles traveling in the bloodstream would be able to locate additional cancer sites which may have spread to other parts of the body.

Nanotechnology can also be used to improve the contrast agents for MRI's. Using fullerenes to encapsulate contrast agents such as gadolinium, Luna Innovations claims that traditional images can be enhanced by 50 times. If successful, this can result in the use of lower magnetic fields through smaller, cheaper and portable imaging equipment. Alternatively, smaller amounts of contrast agents could be used which would reduce the risk of allergic reaction in some patients.

Miniature Imaging Devices

Miniature wireless devices are being developed with nano and micro technologies for providing high quality images not possible with traditional devices.

Given Imaging has developed a pill containing a miniature video system. When the pill is swallowed, it moves through the digestive system and takes pictures every few seconds. The entire digestive system can be assessed for tumors, bleeding, and diseases in areas not accessible with colonoscopies and endoscopies.

Another company, Medirad is trying to develop a miniature x-ray device that can be inserted into the body. They are attempting to make carbon nanotubes into a needle shape cathode. The cathode would generate electron emissions to create extremely small x-ray doses directly at a target area without damaging surrounding normal tissue.

Nanobiotechnology

Nanotechnology, which came into its own in 2000 with the birth of the National Nanotechnology Initiative, is the next stop in the miniaturization path that gave us micro-electronics, microchips and microcircuits. The word nanotechnology derives from nanometer; which is one-thousandth of a micrometer (micron), or the approximate size of a single molecule. Nanotechnology -the study, manipulation and manufacture of ultra-small structures and machines made of as few as one molecule-was made possible by the development of microscopic tools for imaging and manipulating single molecules and measuring the electro-magnetic forces between them.

Nanobiotechnology joins the breakthroughs in nanotechnology to those in molecular biology. Molecular biologists help nanotechnologists understand and access the nanostructures and nanomachines designed by 4 billion years of engineering—cell machinery and biological molecules. Exploiting the extraordinary properties of biological molecules and cell processes, nanotechnologists can accomplish many goals that are difficult or impossible to achieve by other means.

For example, rather than build silicon scaffolding for nanostructures, DNA's ladder structure provides nanotechnologists with a natural framework for assembling nanostruct1lres; and its highly specific bonding properties bring atoms together in a predictable pattern to create a nanostructure.

Nanotechnologists also rely on the self-assembling properties of biological molecules to create nanostructures, such as lipids that spontaneously form liquid crystals.

DNA has been used not only to build nanostructures but also as an essential component of nanomachines. Most appropriately, DNA, the information storage molecule, may serve as the basis of the next generation of computers. As microprocessors and microcircuits shrink to nanoprocessors and nanocircuits, DNA molecules mounted onto silicon chips may replace microchips, with electron flow-channels etched in silicon. Such biochips are DNA-based processors that use DNAs extraordinary information storage capacity. Conceptually, they are very different from the DNA chips discussed below. Biochips exploit the properties of DNA to solve computational problems; in essence, they use DNA to do math. Scientists have shown that 1,000 DNA molecules can solve in four months computational problems that require a cent1lry for a computer to solve.

Other biological molecules are assisting in our continual quest to store and transmit more information in smaller places. For example, some researchers are using light-absorbing molecules, such as those found in our retinas, to increase the storage capacity of CDs a thousandfold.

Some more immminent applications of bio-nanotechnology include:

- Increasing the speed and power of disease diagnostics.
- Creating bio-nanostructures for getting functional molecules into cells.
- Improving the specificity and timing of drug delivery.
- Miniaturizing biosensors by integrating the biological and electronic components into a single, minute component. encouraging the development of green manufacturing practices.

Unit VI–C

Nanomedicine—The Medical Revolution

"Like primitive engineers faced with advanced technology, medicine must 'catch up' with the technology level of the human body before it can become really effective. What is the technology level? Since the human body is basically an extremely complex system of interacting molecules (i.e., a molecular machine), the technology required to truly understand and repair the body is molecular machine technology. A natural consequence of [our achieving] this level of technology will be the ability to analyze and repair the human body as completely and effectively as we can repair any conventional machine today."

—***Brian Wowk***

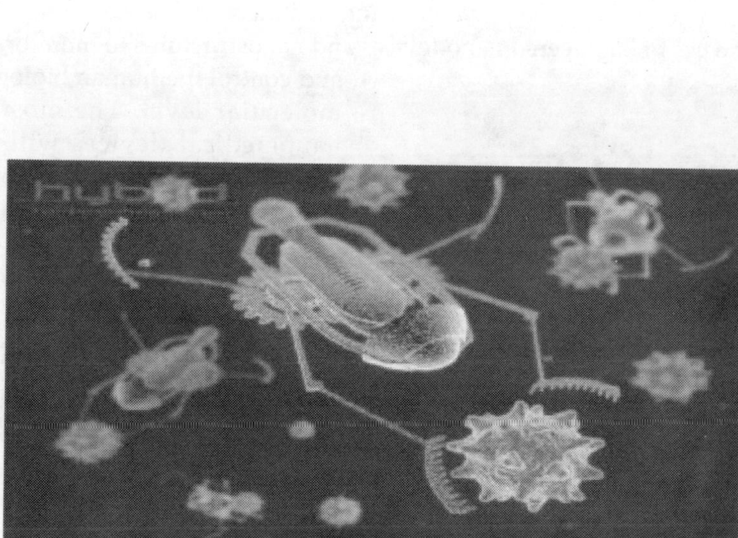

Introduction

Imagine having bones woven with a fabric such that one could fall out of a building and walk away. Imagine that in the event of a fire, microscopic vessels just ten billionths of a meter wide, pressurized with 1,000 atmospheres of pure oxygen could sense oxygen levels in the blood and provide hours of respiratory requirements for the body. Imagine medical nanites being injected into the bloodstream, consuming atherosclerotic plaques in the walls of the blood vessels; repairing cell damage caused by cancer. Or imagine nanomouthwashes that could eliminate gum disease and tooth decay—nanomachines acting as security guards and attacking any foreign entity in the body1. Sounds like something from a science fiction movie? Absolutely not. Welcome to the world of nanomedicine.

Overview of Nanomedicine

Despite advances in modern medicine over the last century, the state of current medicine has been limited both by its understanding and by its tools and in many ways is still more of an art than a science. Only within the last 50 years has medical science begun to examine disease pathology on a molecular level; thus, from a molecular viewpoint, modern medicine remains crude. For example, today's drug is essentially a single molecule with an often sophisticated but always limited repertoire. With nanomedicine, tomorrow's "smart pharmaceuticals" could essentially be programable machines with a range of "sensory," "decision-making," and "effector" capabilities. They can avoid side effects and allergic reactions by coming in generic, biocompatible housings; becoming active only upon reaching their ultimate destinations; and attaining almost complete specificity of action. They can check for overdosage before becoming active, thus preventing accidental or intentional poisoning. They may work in concert with three or four "sister" agents that together produce versatility unattainable by one agent alone. Despite the vast increase in complexity over present-day drugs, such agents via nanomedicine can be expected to be totally "pure" and predictable in their behaviour. Safety and efficacy may be inherent in the designs of such "drugs," in which case regulatory issues could be simplified rather than complicated.

So what exactly is nanomedicine? Technically it is the application of nanotechnology (engineering of tiny machines) to the prevention and treatment of disease in the human body. More specifically, it is the use of engineered nanodevices and nanostructures to monitor, repair, construct and control the human biological system on a molecular level. The most elementary of nanomedical devices will be used in the diagnosis of illnesses. A more advanced use of nanotechnology might involve implanted devices to dispense drugs or hormones as needed in people with chronic imbalance or deficiency states. Lastly, the most advanced nanomedicine involves the use of nanorobots as miniature surgeons. Such machines might repair damaged cells, or get inside cells and replace or assist damaged intracellular structures. At the extreme, nanomachines might replicate themselves, or correct genetic deficiencies by altering or replacing DNA (deoxyribonucleic acid) molecules.

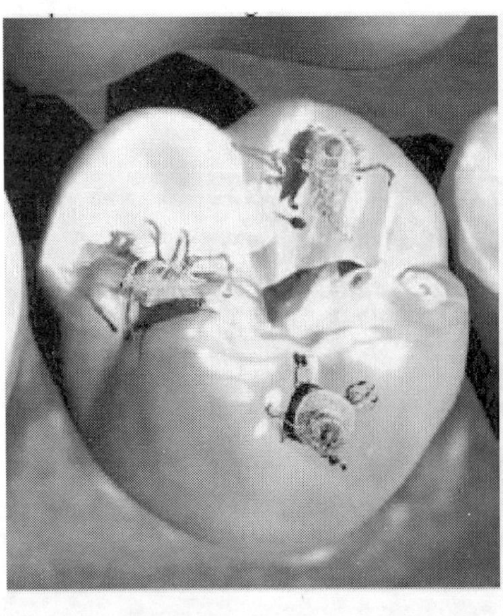

The Promise of Nanomedicine

In 1959, in a talk at CalTech titled, "There's Plenty of Room at the Bottom," acclaimed physics lecturer and Nobel laureate Richard Feynman suggested that nature could be manipulated at a nanometer scale—atom by atom.

Wiith the convergence of human gene sequencing and advances in nantotechnological engineering and a deepening understanding of the function of cellular systems, nanotechnologists believe it will be possible to design medicallyactive microscopic machines to fight disease and affect physiological repairs at the cellular level.

Invading bacteria will automatically be destroyed, since they do not "fit" within the physiological blueprint encoded in the nanorobot patrolling the body. Diagnosis of disease will no longer rely upon patient history and the results of laboratory tests, but will result from an ongoing internal examination of deviations from the encoded molecular blueprint and programed repair functions designed to correct anomalies at the molecular level6. Cell damage attributed to aging will be repaired from the inside out. The potential benefits of medical applications of nanotechnology—nanomedicine—are read more like science fiction than science fact.

In 1997, a group of health scientists with the Department of Defense concluded that if a breakthrough to a molecular assembler occurs within 10 to 15 years, an entirely new field of nanomedicine will emerge by 20207. These scientists postulated that initial applications would be focused outside the body in diagnostic and pharmaceutical manufacturing, but the most powerful uses would be within the body. Some of the applications they discussed included cell-herding machines to stimulate rapid healing and tissue reconstruction and cell repair machines to perform genetic surgery.

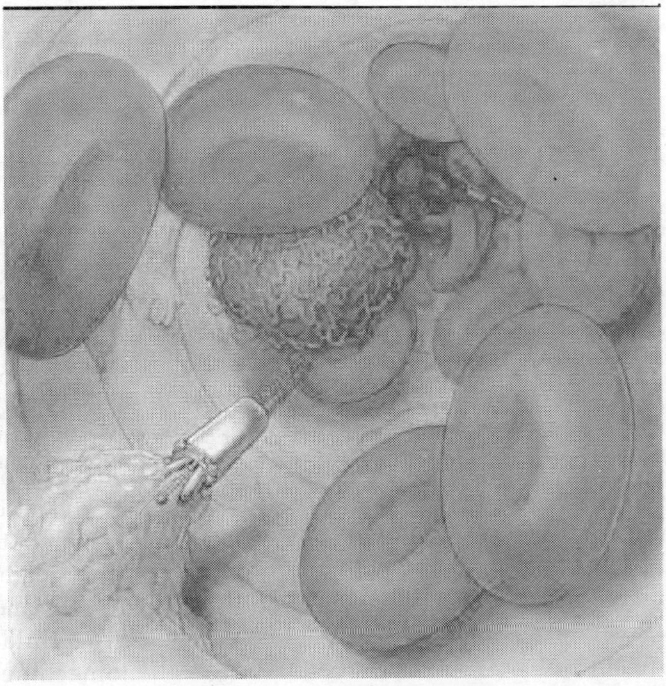

Many of these research goals may take 20 or more years to achieve, which is one reason the federal government in US is the major financial supporter of nanotechnology.

Researchers have been able to move molecules and even construct nanomotors. The first privately held nanotechnology company, Zyvex, was started in mid 1997 with the goal of building the key tool for creating molecular nanotechnology, the assembler. According the Zyvex Founder, President, and CEO James Von Ehr II, the first assembler could be available within 10 years. However, it will be some time after that before medical nanorobots will reach the market.

Technical Challenge

The problems with nanotechnology begin with assembly. Recall that nanotechnology is molecular manufacturing—building things one atom or molecule at a time with programed nanoscopic robotic arms. Atoms don't simply let themselves be pushed around. They are constantly moving, jiggling, combining, and reacting, always ready to confound the would-be molecular architect. A design that might seem perfectly reasonable to an engineer in the macroscopic world might, on a nanoscale, fold instantly into a ball of goop.

Imagine, anyway, that you have a robot hand that can assemble nanomachines, that you have managed to build a stable structure containing ten atoms, and you now want to add another piece to it. You would need to grab another atom or molecule, all the time making sure that none of the positions of the neighboring atoms are altered. Each atom's position is specified in three dimensions, so you'd have a 3D problem, (at least, if the fourth dimension—timing is not a factor). This is not easy to calculate.

The difficulties continue. Because the robot handles one atom at a time, it has to work fast, and with speed come mistakes. A free radical—a highly reactive molecule—can clip off a single atom, and thereby undo all that the robot has contrived. The robot would have to detect and correct errors as they happened, not only as it assembled a device, but also afterward, in order to maintain its integrity.

Another problem is scalability. Unlike the technology for sequencing genes, which began small and only later moved to mass production, carbon nanotubes have been produced by the thousands from the very start. Unfortunately, thousands aren't good enough if the goal is to knit the nanotubes into organs, limbs, or other cellular structure. For that, tons of nanotubes are required with no currently suitable means of mass production.

However, on the bright side, the same things were said about combinatorial chemistry, recombinant DNA, decoding the human genome, and landing to the moon.

Financial and Competency Challenges

A nanomedical company with a successful product will reap huge profits much like biotechnology and pharmaceuticals companies do. However, like biotechnology or pharmaceutical companies, they will also experience the tremendous expenditures associated with bringing a product to market. In fact, such companies will incur even higher costs than the traditional biotechnology and pharmaceutical companies because of the diversity of technology required in nanomedicine. Take a pharmaceutical company, for example. The technology that such a company has to deal with is normally confined to molecular technology. A nanomedical company, on the other hand, will require expertise in the same realm of expertise as the pharmaceutical company plus a multitude of expertise in other areas such as computer technology, nanopropullsion technology, and circuitry technology, just to name a few. The cost of developing innovative technology in just one of these areas is expensive enough, let alone multiple areas.

Furthermore, there is the basic question of can a company achieve such diversity of technologies? It seems difficult for a starting company to develop all these competencies in all these different areas at once. Companies that have a more realistic chance of pursuing nanomedicine alone are the well-established companies that are already highly diversified—companies such as GE, Cannon, Mitsubishi, IBM. These companies already have technological expertise in a variety of areas which they can leverage.

Challenge of Shorter Product Life Cycles

The difficulty with drug discovery today is that it is part science, part art, and part luck. Because of these elements, the average time to discover and win FDA approval for a prescription drug is 15 years, with 3 out of every 10 drugs failing to meet FDA approval. This long process plus the protection afforded by patents and patent extensions have allowed pharmaceutical products a relatively long product life cycle. For example, Prozac, launched in January 1988, maintained a monopoly afforded to it by patent protection until only recently when it become generic. Nanomedicine, although still protected by patents, will enjoy shorter periods of monopoly with each product. Why? Because in nanomedicine every molecule is designed—nothing is left to luck. As such, once a general level of assembly technology is understood and common in the industry, the differentiator between companies and products becomes a function of engineering skills. Take digital cameras, for example. When Sony released a 2 megapixel camera, Olympus introduced a 3.2 megapixel camera in less than a year. Sony then countered in 6 months with a 3.2 megapixel camera capable of 10 X optical zoom. Apply this same pattern to nanomedicine and you can see how the product life cycle of nanomedical devices will share a similar fate to consumer electronics. Picture Company A producing a nanopill that when swallowed will measure a variety of parameters such as blood pressure, tissue oxygenation, internal chemistry, etc. Now picture Company B introducing a different nanopill 9 months later that can measure more parameters with greater sensitivity and accuracy. Company A then launches a new version of their original nanopill within 6 months of Company B's launch. Product life cycles have been reduced and prices consequently eroded.

Challenge of Sole Proprietorship

As alluded to in the "financial and competency challenge" section, the ability for a "stand-alone" company to be a serious competitor in nanomedicine will become less likely. Because nanomedicine spans many different industries—materials, computer, circuitry, etc., the role of partnership and cross licensing of technology becomes ever more important. Companies can no longer operate in a vacuum. Imagine Pfizer branching into nanomedicine. If they were to develop a nanobot of some type, Pfizer lacks the engineering capability to develop the sensor, mechanical components or computer guidance system. The company would have to look outward for partners who have the capabilities to develop the necessary components. Development of nanomedical technology will require a cooperative effort and as result, no one company can claim one hundred percent of the profits for itself.

The challenge of sole proprietorship does not only apply to new nanomedical companies. It also applies as a challenge for existing companies rich in research and development. Few technologically wealthy companies are actively managing the licensing of their intellectual property as a profit center for the organization. The advancement of nanomedicine, and nanotechnology in general, will require such companies to rethink their business model and role of R&D within the company. As the doors of licensing open, the role of R&D centers cannot be viewed as centers of discovery and thus cost centers, but as potential profit pools and access to other company's technology through cross-licensing.

Legal Challenge

Another interesting question is how the FDA will view/classify medical applications of nanotechnology. Would nanomedicine be regulated as drugs, devices, or something else? Statutes provide that treatments that operate through mechanical means are to be treated as devices; those that rely on chemical effects are considered drugs. This distinction is not very satisfactory in the context of nanotechnology-based treatments.

Social Challenge

Like genetic engineering, the question facing nanomedicine is not "Can we?", but "Should we?" Nanomedicine can eradicate biological sources of human suffering and human limitations. For example, advocates of nanomedicine believe that the technology must be used to eliminate human vulnerability to disease. Fully developed nanorobots, in their view, will be able to correct any damage to DNA, eliminate any tumors or infections, and repair most structural damage to the body. Such technology can, in principle, make disease via inheritance or infection a thing of the past.

The same technology, however, can be used to prevent aging, remove the inherent limits on the human lifespan. Bones could be made of diamond and lungs rebuilt to breathe poisonous atmospheres. Brain enhancements by means of artificial, improved neurons will mean that limits on memory and intelligence need no longer exist. A single artificial neuron could store the entire Library of Congress, accessible to an individual on demand. Brains could have the ability to link directly via nanoengineered devices with computers, with other brains, or with the Internet. Where is the line drawn between the repair and restoration of lost function and the augmentation of "normal" capabilities? Furthermore, can we even make clear distinctions between what constitutes treatment of a disease or disability and what is "simply" self-improvement? And if some improvement is appropriate, what distinguishes legitimate improvement from manipulative alterations of people for sinister or even benign purposes?

Case Analysis

To illustrate commercialization and the challenges associated with nanomedicine, two cases will be presented. The first case examines the viability of commercializing a technology/design while still in it infancy. The second case examines a nanomedical technology that is further along in the development phase and closer to commercialization. The second case demonstrates similar challenges to that of the first case, in addition to new challenges arising from the technology being at a different stage of the commercialization and development process.

Case Study 1: Respirocytes—Mechanical Artificial Red Cell

The World Health Organization, as well as most public health organizations, unanimously agree: the world is facing a severe shortage of donor blood. The American Red Cross has 36,000 units per day to satisfy a daily demand of 80,000 units. And unfortunately, like oxygen, blood (more so the red blood cell component of blood) is essential to life and non-substitutable. Furthermore, anemia (hemoglobin—the oxygen carrying molecule of red blood cells—less than 12mg/dL) is a prevalent disorder that causes significant morbidity and mortality, and bears huge economic-costs of treatment.

Possible artificial oxygen carriers have been investigated for eight decades, starting with the first documented use of hemoglobin solutions in humans in 1916. With an estimated 4 million patients suffering from anemia as a chronic disease, the commercial potential for successful blood substitutes has been estimated at between $5-10 billion/year. Current blood substitutes include hemoglobin formulations and flurocarbon emulsions, neither of which offers performance comparable to Mother Nature's molecule.

Enter nanomedicine to offer a solution where conventional medicine has failed. Scientists at the Institute for Molecular Manufacturing have purposed a mechanical artificial red cell called a "Respirocyte". The purposed artificial Respirocyte is a spherical nanomedical device 1 micron in diameter made up of a flawless diamond or sapphire shell constructed atom by atom (Figure VI-C.1). The device utilizes an active means of conveying gas molecules into, and out of, pressurized microvessels. For this task, engineers have designed molecular sorting rotors with binding site "pockets" along the rim exposed alternately to the blood plasma and interior chamber by the rotation of the disk. Each pocket selectively binds a specific molecule when exposed to the plasma. Once the binding site rotates to expose it to the interior chamber, the bound molecules are forcibly ejected by rods and stored in tanks (Figure VI–C.2). Onboard power is provided by a mechanochemical engine that combines glucose and oxygen to generate mechanical energy to drive molecular sorting rotors and other subsystems. Sensors and an onboard computer provide the necessary controls for respiratory gas loading and unloading, rotor field and ballast tank management, glucose engine throttling, power distribution, interpretation of sensor data and commands received from the outside, self-diagnosis and activation of failsafe shutdown protocols, and ongoing revision or correction of protocols in vivo.

Are Respirocytes Commercializable?

Having developed a technology/product is one thing; developing a technology/product that has commercial demand is a completely different story. The mouse trap is the most highly designed and patented device in the U.S. patent office, and yet the market does not see a flood of mouse traps. Why? Because the market does not command the variety of mouse traps that inventors have created. Market demand is but one of many factors to consider prior to commercialization. No plan is complete without examining other factors such as financing, technology testing and validation, strategy for commercialization, transition from plan to operations, just to name a few. Considering all these elements, then, can mechanical red blood cells "cross the chasm" and become commercially viable?

Market Demand

As alluded to earlier, the treatment of anemia consists of either transfusion of natural red blood cells or red blood substitutes. Natural red blood cells are in short supply, which suggests a market demand. The next logical question is whether red blood substitutes can satisfy the remaining demand.

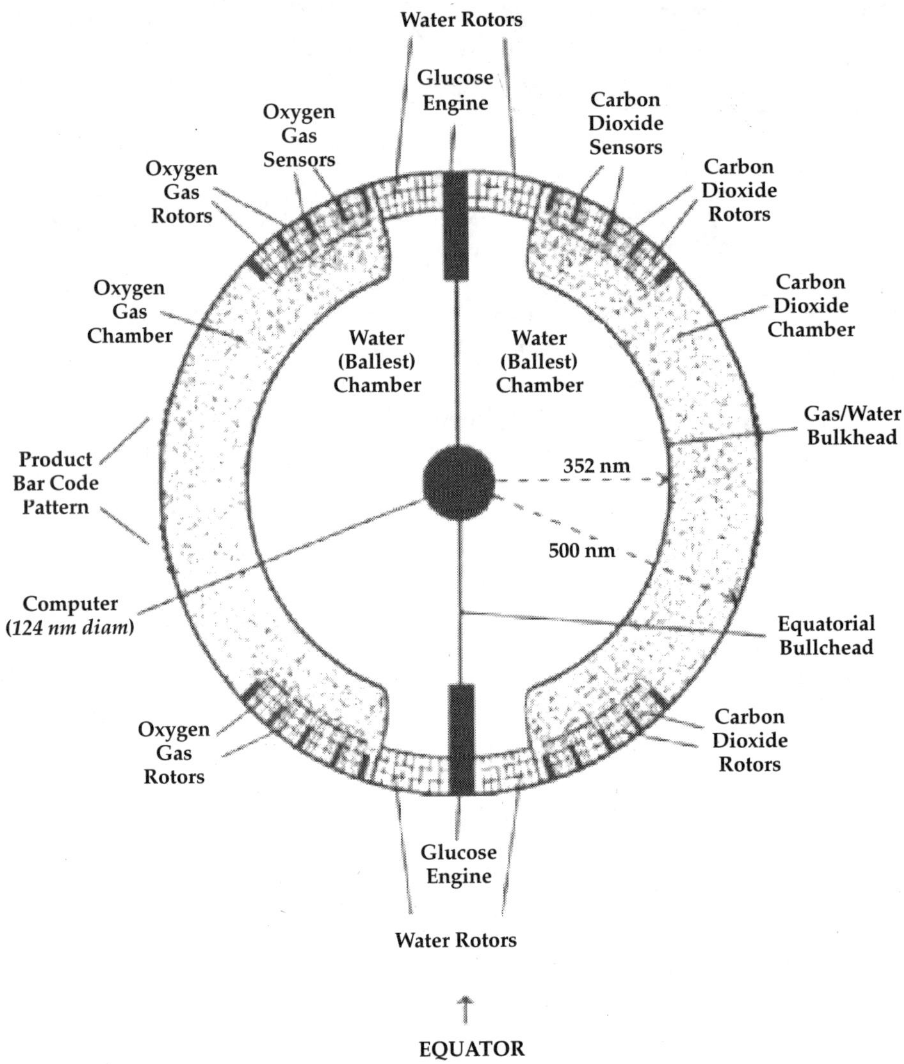

Figure VI–C.1: An example of Respirocyte

Analysis of the two forms of blood substitutes—hemoglobin solutions and fluorocarbon emulsions reveal limitations that translate into opportunities for nanomedical red blood cells. Most of the red cell substitutes under trial at present have far too short a survival time in the circulation to be useful in the treatment of chronic anemia, and are not specifically designed to regulate carbon dioxide or to participate in acid/base buffering. Several cell-free hemoglobin preparations evidently cause vasoconstriction, decreasing tissue oxygenation, and there are reports of increased susceptibility to bacterial infection. Fluosol, a widely available fluorocarbon perfusant, is approved by the FDA for use in patients at high risk of ischemic complications during percutaneous transluminal coronary angioplasty, but its experimental use in the treatment of severe anemia has been disappointing and cannot yet be recommended. The radical innovation associated with

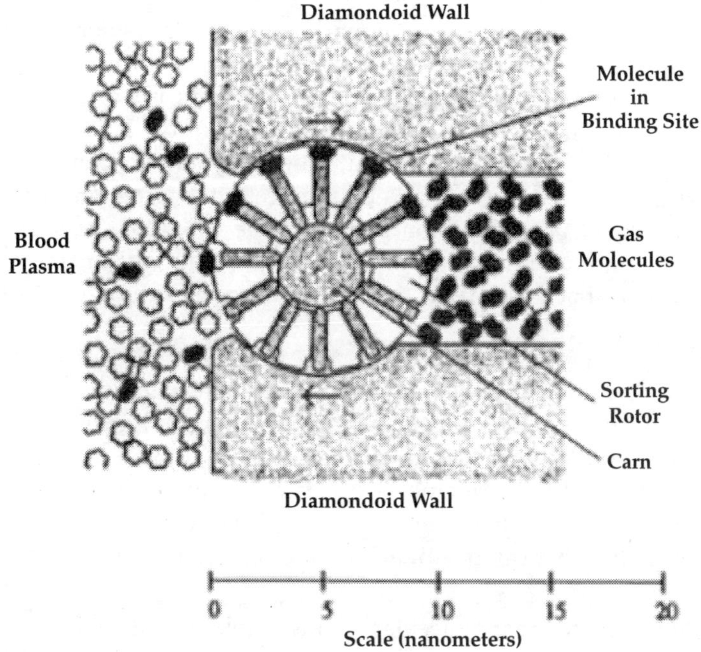

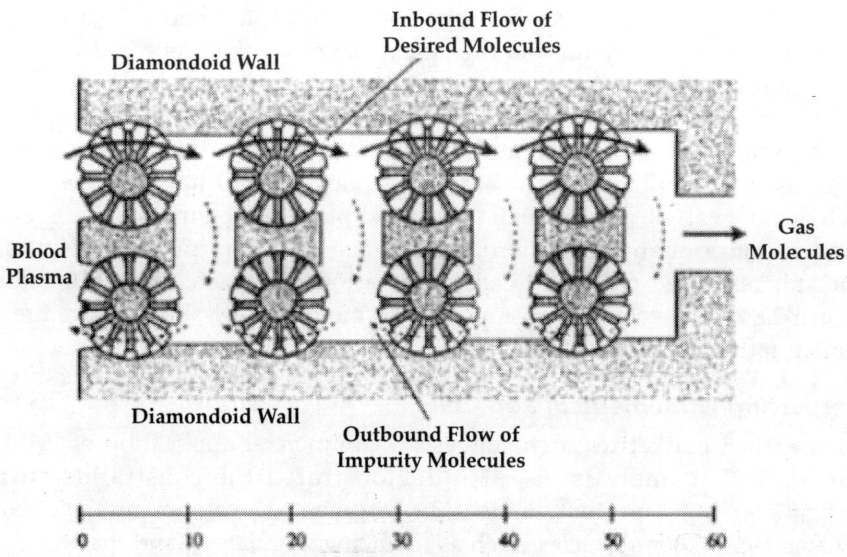

Fig. VI–C.2: Further details of 'Respirocyte'

Respirocytes addresses the "pain points" where current technology fails. Nanomedical red blood cells thus has the potential to become the standard for anemia management

How about the market size? A strong market demand does not necessarily mean a profitable business. For example, demand is present for orphan drugs such as GENOTROPIN because it treats a rare disease (Prader-Will Syndrome); however the drug is not profitable because fewer than 200,000 Americans are affected by the disease. Because of the level of technology involved with nanomedical red cells, the market size must be sufficient to warrant the capital investment necessary to bring such a product to market. In this case, the target market of Respirocytes—anemia —is certainly attractive. With 4 million people in the U.S. alone suffering from chronic anemia and an additional 1 million units (minimum) of blood required for transfusion annually, the cost of treating anemia has been estimated to be in excess of $5 billion per year.

Development

Development is one of the key challenges to the commercialization of Respirocyte. The challenge is not in manufacturing the components of the nanodevice, but in assembling the components. Molecular nanotechnology violates no physical laws. In 1959, the Nobel physicist Richard P. Feynman observed that: "The principles of physics, as far as I can see, do not speak against the possibility of maneuvering things atom by atom. It is not an attempt to violate any laws; it is something, in principle, that can be done; but, in practice, it has not been done because we are too big."

It is generally accepted that molecular nanotechnology will be developed; the proper question is no longer "if" but "when." Developmental success leading towards molecular nanotechnology has been achieved. For example, A Scanning Tunneling Microscope was used in 1989 to spell out "IBM" using 35 individual xenon atoms on a nickel surface. Atomic Force Microscopes (AFMs) have performed nanomachining operations: two rectangular slots and a 50-nm rectangular sliding member were milled from a crystal, and then the member was slid repeatedly from one slot to the other, making a reversible mechanical latch. Chemists have built a large number of potentially useful rigid nanoparts including molecular-scale rods, rings, springs, cubes, spheres, tetrahedrons, hollow tubes, propellers, and tongs, and wire-frame nanostructures of many shapes made of polymerized DNA. Micro Electro-Mechanical Systems (MEMS) has produced 100-micron-scale accelerometers, flow valves, pistons, gear trains, and piezo-driven motors that are available today off-the shelf in mass quantities; microgrippers used to manipulate individual 2.7-micron polystyrene spheres, dried red blood cells, and various protozoa; and a/1000th scale working electric car just 2.8 millimeters wide, complete with motor, wheels, body, spare tire, bumpers, and even a 10-micron thick license plate.

Challenges Facing Nanomedical Red Cells

As discussed earlier, market research suggests a commercial application of IMM's Respiorcyte design. However, the analysis has also demonstrated the constraints surrounding the commercialization of this product, namely development. Developmental challenges aside, the technology also faces other obstacles, such as financing, regulatory, and marketing. Furthermore, since nanotechnology has been achieved along so many independent pathways, the real question for scientists at the Institute of Molecular Manufacturing looking to commercialize this design is whether or not to partner with or license to other companies involved in nanotechnology.

Several compelling reasons speak to the need for IMM to consider licensing its design to other companies for development. First, the financial burden associated with commercializing the product is unlike any other consumer product on the market. Commercializing a medical product is a multi-tiered process, with each step being more expensive than the last. According to a recent study conducted by Tufts University, the average cost for a biotechnology/pharmaceutical company to develop a prescription drug is $800 million (the breakdown of these costs is approximately: basic/investigative, 10 per cent; preclinical, 15 per cent; manufacturing/process, 15 per cent; clinical, 55 per cent; post-marketing, 5 per cent). Using these numbers as a benchmark, it is reasonable to assume that the present cost of developing Respirocytes would be in excess of the average $800 million given the immaturity of technology and process. In addition, because of the novelty of this technology, regulatory bodies such as the FDA will require strong evidence of the product's efficacy and safety. In clinical trial terms this translates into bigger sample size, more metrics, and longer duration of study—all of which exponentially increases the cost of development.

Secondly, Respirocytes, unlike a drug molecule, is a combination of material science, biological science, mechanical engineering, electrical engineering, and computer science. IMM may have the competency for material design and assembly, but where will they get the expertise for computer design necessary for control of the device? Where will they get the technical know-how to build a nanomotor?

Lastly, again relating to competency, IMM's competency resides in R&D. Assuming that Respirocytes can be manufactured and does get FDA approval, the company is now faced with the unfamiliar task of marketing. Granted that the product will probably sell itself because of the promise associated with artificial mechanical red blood cells, radical technologies such as these always faced the first mover disadvantage.

Case Study 2: Using DNA as a Construction Medium

DNA is thought of as the building blocks of all living organisms—the chemical basis of heredity, the blue print of all living things. But DNA is not normally thought of as a building material. A group of chemists at NYU have developed a means of utilizing strands of DNA to construct 3-dimentional geometric structures. Utilizing the specific base paring relationship as "the glue", the group has been able to link strands of DNA together (Figure 3.1). From this, the research group has introduced bends, kinks, twists and other shapes, which serve as the building blocks for even more complex structures. See Figures VI–C.3—VI–C.9 for illustrations of this concept.

The use of DNA has several distinct advantages. The first advantage is the specificity of DNA interactions (i.e. "sticky ends") combined with branched molecules results in a system whereby the DNA intermolecular interactions are both readily programable and predictable. Therefore, it is possible to gain large amounts of control over both linking and branching topology. The second advantage of using strands of DNA as a molecular scaffold is the availability of arbitrary sequences. In other words, for a segment of DNA to "make sense" and code for a protein, a specific sequence of bases is required. For example, a sequence ATTCGT may code for a simple amino acid chain. Any deviation from this code will not be understood and produce nothing. As such, the length and therefore size of the structure can be controlled without any adverse consequences. The third advantage is that DNA can be manipulated and modified by a large battery of enzymes, including DNA ligase, restriction endonucleases, kinases and exonucleases. Lastly, the double helical DNA is a stiff polymer, stable molecule, and has an external code that can be read by proteins and nucleic acids.

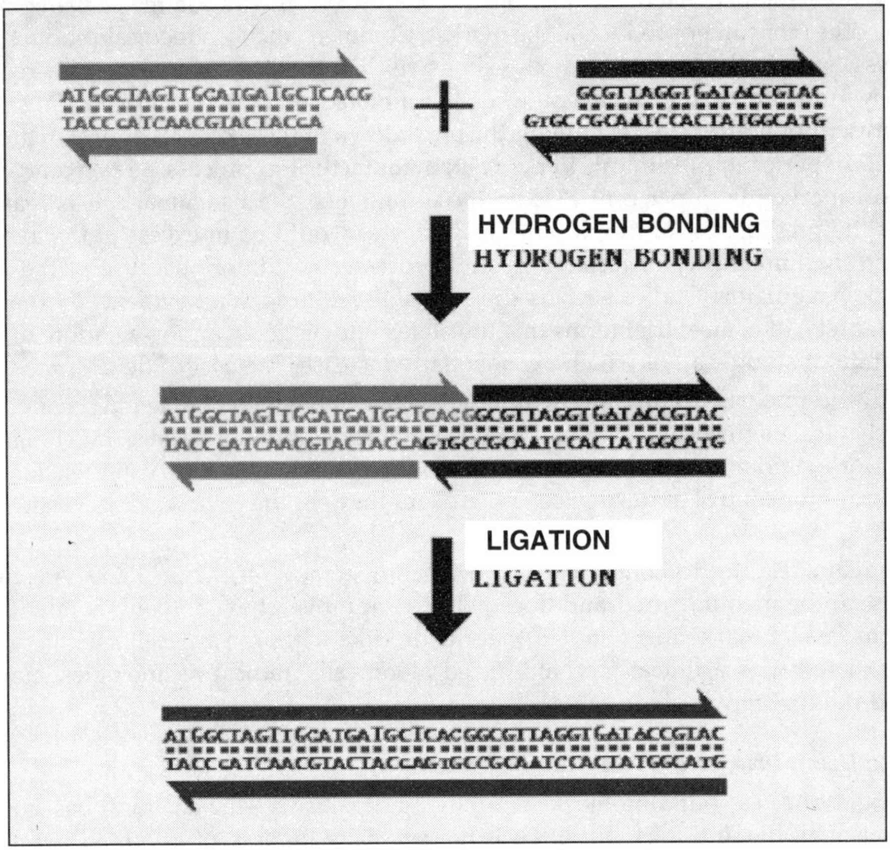

Fig. VI–C.3: ***Sticky-Ended Cohesion and Ligation.*** **Two linear double helical molecules of DNA are shown at the top of the drawing. The right end of the left molecule and the left end of the right molecule have single-stranded extensions ('sticky ends') that are complementary to each other. The middle portion shows that, under the proper conditions, these bind to each other specifically by hydrogen bonding**

Commercializing DNA Construction Medium Technology

Unlike the Case 1, the technology discussed here represents a more immediately viable commercial technology. As previously illustrated, early iterations of nanomedicine will not be in the form of nanobots with self-replicating capabilities, but in simpler forms such as diagnostic and monitoring devices or drug delivery systems. In fact, the commercial application of this technology can be immediately applied to issues such as drug delivery.

Why Drug Delivery?

Drug delivery techniques were established to deliver or control the amount, rate and, sometimes, location of a drug in the body to optimize its therapeutic effect, convenience and dose. Developing a well established formulation of a drug with a new delivery system is a relatively low risk activity

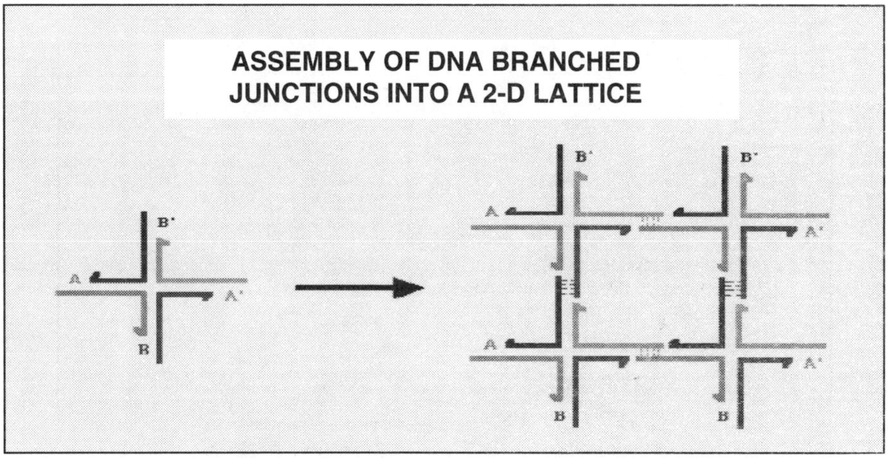

Fig. VI–C.4: Formation of a Two-Dimensional Lattice

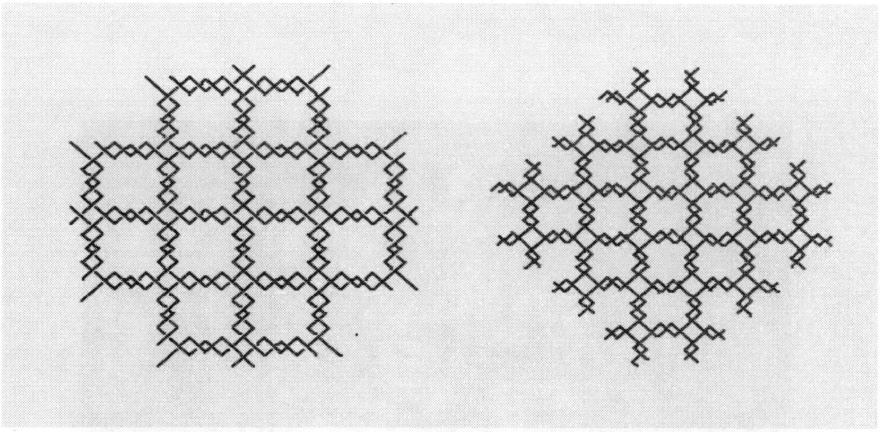

Fig. VI–C.5: More Complex Formation of a Two-Dimensional Lattice

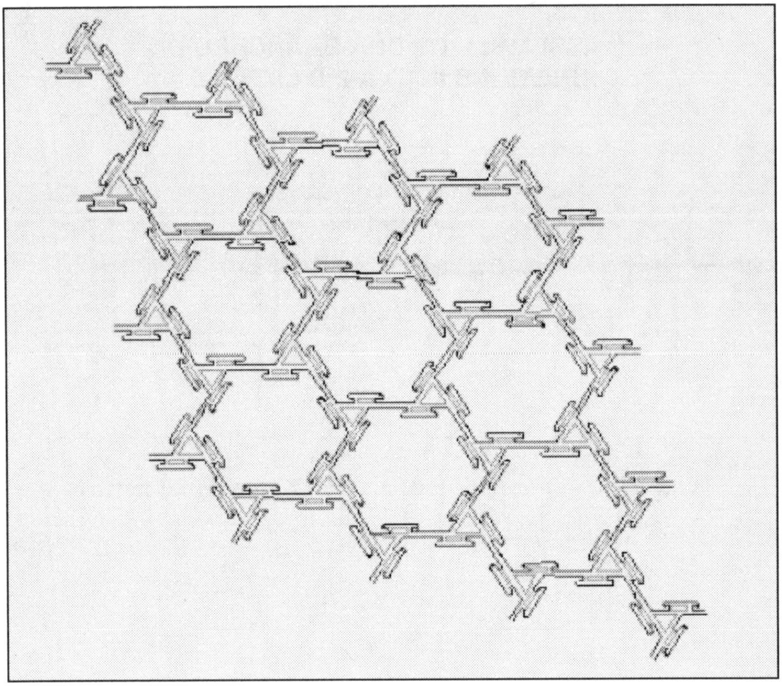

Fig. VI–C.6: DNA Triangular Lattice.

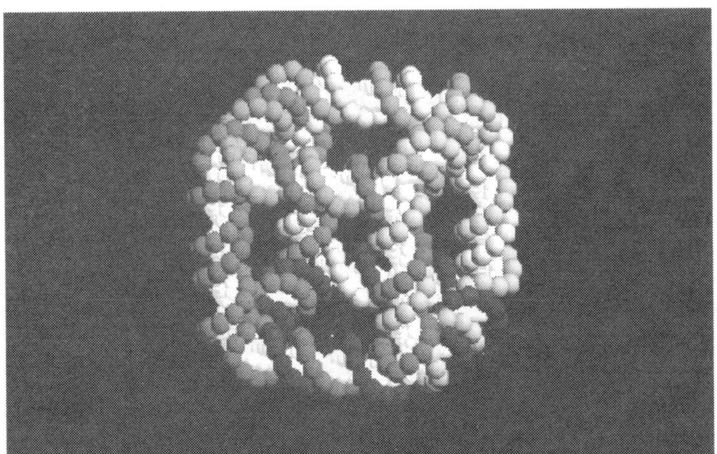

Fig. VI–C.7: A DNA Molecule with the Connectivity of a Cube.

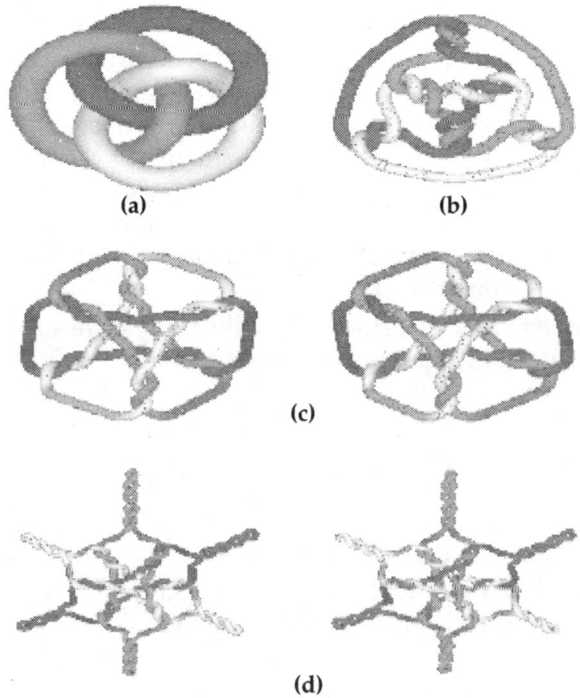

Fig. VI–C.8. The Design and Construction of Borromean Rings from DNA.

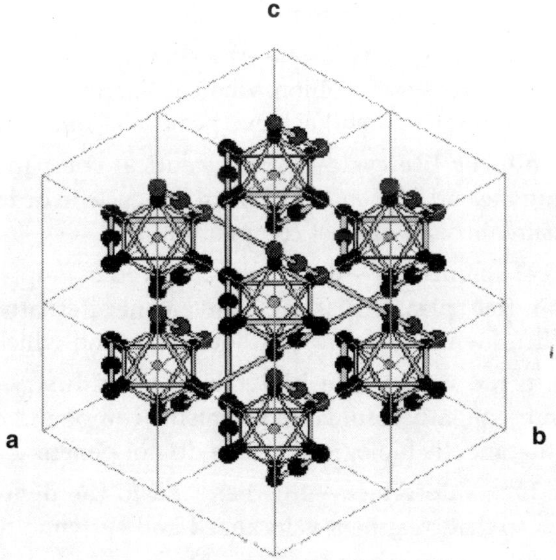

Fig. VI–C.9: Macroscopic view of a Three-Dimensional geometric structure assembled from multiple DNA scaffold.

and can be used to enhance a company's product portfolio by extending the drug's commercial life-cycle. New, therapeutic biological drugs may also depend on a novel technology to enable them to reach their target within the body. Any drug delivery system that: results in reduced side effects; is simple to use (i.e. does not require administration by a trained professional); is pain-free; is pleasant tasting; and is infrequent in its dosing schedule, will increase patients' quality of life, meaning that more patients are likely to use it and respond to the treatment. Likewise, a promising product in vitro will have limited success in vivo and on the market if it cannot: be delivered into the body with proven high bioavailability, high absorption by the body; be effective in small enough doses that cause minimal side effects; manufactured simply and costefficiently; and be tailored to the needs of the disease type being treated or prevented.

Current studies have shown that existing drug delivery systems are less than perfect. In the year 2000, 15 per cent of all hospital admissions were due to adverse drug effects with 100,000 deaths. Adverse drug effects have resulted in $136 billion in health care costs. Furthermore, from a commercial view point, drug delivery is a rapidly expanding subsection of the market for therapeutic drugs. The company MediJect estimates the current global drug delivery market to be worth $20 billion. Drug delivery is the fastest growing healthcare sector, currently growing at a rate twice that of pharmaceuticals. Drug delivery industry sales are expected to increase at an annual rate of 15 per cent, and this growth is thought likely to be sustained throughout the 21st century. There are, therefore, significant opportunities for growth and development within this market. The US drug delivery market alone is projected to grow from $14.3 billion in 1998 to $24 billion in 2003 and it is estimated that by the end of 2005, 20 per cent of pharmaceutical expenditure will be used on products using drug delivery technologies.

Current Market Drivers

Over the past 10 years there has been an increase in R&D into drug delivery technologies. This growth has been, and will continue to be, fueled by a number of factors:

- **Controlling overall healthcare costs**—the cost of developing and launching a new chemical entity can be $300–600 million, whereas the reformulation of an existing product may cost only $50 million and still achieve peak sales of $100–200 million.
- **Enhancing a product's life-cycle**—pharmaceutical companies need to enhance older products reaching the end of their commercial life-cycle in order to optimize the company portfolio and maintain sales against competition.
- **Improving patient compliance**—this can be achieved by improving the choice of delivery system with pain-free, pleasant tasting, convenient alternatives. Patients are also more likely to take a drug which has reduced side effects and which works quickly.
- **Rapid growth in the worldwide biotechnology industry**—this has increased drug discovery and subsequently resulted in promising compounds such as proteins, peptides, hormones and therapeutic biologicals which do not penetrate the body easily.
- **The need to reduce side effects**—this has lead to the development of targeted drug delivery systems so that treatment is localized and systemic side effects can be reduced.
- **The need to deliver peptides and proteins**—there is a demand for an alternative delivery system to injectable or infusion-delivered insulin for the estimated 18 million sufferers of Type 1 diabetes worldwide. A growing number of the 15 million Type 2 sufferers worldwide are also now administering insulin themselves. At present, peptide and protein

pharmaceuticals are worth approximately $10 billion worldwide and is expected to increase two to three times over the next decade.

How Will DNA Construction Medium Technology Address These Issues?

What makes this technology highly viable in terms of commercialization is the ability to satisfy such market needs. The most obvious benefit of utilizing DNA as a construction material is its ability to avoid aggravating the body's immune system and thus reducing side effects. Drug molecules, which would normally elicit immune reactions, are in effected shielded from antibodies by the inert structure of the DNA. Furthermore, aside from a protective effect, the DNA shell can also be functional. After all, the DNA shell can "code" for proteins and peptides. Again using insulin as an example, a DNA constructed structure containing molecules of insulin within it could be injected into diabetic patients28. As enzymes within the body begin reading the DNA coding on the shell, and manufacturing nascent insulin molecules, the exogenous molecules of insulin are gradually being released either through simple diffusion or a sustained release mechanism. This process has resulted in the release and production of enough insulin to last the patient 24 hours, reducing the normal dosing of insulin from three times a day to once a day, and greatly improving patient compliance. Furthermore, this process is has not mounted any immune reaction because the porousness of the engineered DNA shell is big enough for insulin molecules to escape, but small enough so that antibodies cannot enter.

Technology Commercialization

So far, the analysis has shown that the technology is viable from an engineering and scientific perspective. That a profitable market exists for this technology, and that the technology can address "pain points" and market needs. The next question to consider is how to bring the technology to market.

Unlike Case 1, the financing needs with this technology are more feasible. Average costs to develop and bring to market a drug delivery system are approximately $50 million. Given the novelty of this technology, a more liberal estimate of the cost of development should be applied, say $200 million. Is this company and technology worth $200 million? A simple calculation suggests that a company/technology such as this has a NPV well in excess of $200 million. In fact, a conservative calculation shows that this company's technology is worth $1.3 billion. Funding for this technology should not be an issue. Where the company should turn for capital is another issue that depends on multiple factors such as: how soon the company needs the funds, corporate goals, degree of control the company wishes to relinquish, among others.

Problems with Transitioning from Plan to Operation

Although the challenge of financing is less prominent for the commercialization of DNA construction medium technology, the company and technology face other challenges that must be considered as the technology is brought to market. One challenge is that of manufacturing and the scaling from laboratory to industrial. To date, the success of the experiments have been on the scale of laboratory experiments; however, it is unknown if the manufacturing process when scaled up will function as expected.

The other challenge facing the company is that of management. Who will lead the company from the commercial end? The current organization is comprised primarily of scientists and engineers who may not have the necessary business skills or experience to make the company and technology a commercial success. This issue spills over into the issue of organizational structure. As the company begins to grow, roles and responsibilities will require better definition. What degree of control will the founders be willing to concede? How will the environment and culture of the organization continue to foster innovation while maintaining competitive from a business end?

Lastly, the company must be aware of marketing risks. The marketing risk described here is not only with advertising or distribution channels, but also with the risk of myopia. It is all too easy for the company to become complacent through the capture of profits from "low hanging fruit" and miss the greater opportunity. For example, the technology has been discussed in the context of drug delivery systems, but the company should not limit its applications to drug delivery alone. This is not to say that the company must pursue other markets outside of drug delivery—its area of expertise, but that it should always consider the option of licensing or crosslicensing.

It should be cognizant of the fact that the business of licensing technology and intellectual property can be an equally profitable business proposition.

Summary of Cases

Although the two cases presented here represent nanomedical technologies that are on different planes of development, they do demonstrate fundamental difficulties common to all technology commercialization: the problem of translating ideas into practice; the problem of evolving from prototype to mass market; and the problem of environmental factor conditions that are outside the control of the entrepreneur. Both technologies were to face regulatory and legal scrutiny as well as market adoption uncertainty. The cases suggest the disadvantage of a first mover strategy and that "the first application is not the killer application".

The cases have also shown the differences in obstacles facing technologies at different stages of development. With the mechanical red blood cells, one can appreciate the magnitude of challenges facing the commercialization of the technology compared to that of the DNA construction medium. In the first case, financing and even being able to build a prototype was an issue, but in the second case, the salient issue revolved around marketing and product launch; transitioning from plan to product.

Thoughts on Nanomedicine

The field of nanomedicine will certainly take decades to develop. Perhaps before we let our imaginations run wild dreaming of the promises that nanomedicine can deliver, our first task is to decide whether the capabilities are indeed possible. Bear in mind that as we look at how to solve various problems, we notice that some that look difficult today will become easy, while others that might seem easier turn out to be more difficult. The seeming difficulty of treating disorders is always changing: Once polio was frequent and incurable, today it is easily prevented. Syphilis once caused steady physical decline leading to insanity and death; now it is cured with a shot. In the longer term, though, the story of nanotechnology in medicine will be the story of extending surgical control to the molecular level.

Unit VI–D

Nanosensors and Nanoscale Scanning

Nanosensor Technology

Medical nanorobots need to acquire information from their environment to properly execute their assigned tasks. Such acquisition is achieved using onboard nanoscale sensors, or nanosensors, of various types. Nanosensors allow for medical nanodevices to monitor environmental states at three different operational levels:

1. Internal nanorobot states.
2. Local and global somatic states (inside the human body).
3. Extrasomatic states (sensory data originating outside the human body).

The general physical limits to sensory perception are reviewed by Block and Bialek. The specific nanosensor technologies required include sensors to detect chemical substances, displacement and motion, force and mass, and acoustic, thermal, and electromagnetic stimuli.

Chemical and Molecular Nanosensors

The area of chemical microsensors is well developed, and there is increasing research interest in nanoscale chemical sensors, "chemosensors" and "biosensors". The most common nanomedical application of chemical sensors will be to measure the concentration of specific molecules and biopolymers in aqueous solvent—whether in blood serum, interstitial fluid, or cytosol. This may be accomplished using ordered arrays of receptors or by using sorting rotors to directly count molecular populations in known sample volumes. Spatial and temporal concentration gradient sensing is also essential in navigation, communications, and in mediating rapid response to environmental stimuli. Chemotactic sensors may be used to sample the chemical composition of surfaces.

Broadband Receptor Arrays

One simple broadband concentration sensor, shown schematically in Figure 4.1, consists of a graduated series of receptors having uniformly high specificity but engineered with progressively greater affinities (successively smaller equilibrium dissociation constants K_d (molecules/nm^3); for the target molecule, which is present in the test sample at concentration c_{ligand}. Exposure of the receptor array to the test sample for a time t_{EQ} necessary to reach diffusion-driven equilibrium gives a probability of receptor occupancy $P_{occ} \sim 0.91$ in a receptor with $K_d \sim 0.1c_{ligand}$, $P_{occ} \sim 0.50$ in a receptor with $K_d \sim c_{ligand}$, and $P_{occ} \sim 0.09$ in a receptor with $K_d \sim 10c_{ligand}$.

During each measurement cycle, all steric probes are simultaneously extended into their associated receptor volumes at a time t_{EQ} after presentation of sample to the array. Probes which reach full extension register an empty receptor; those which cannot fully extend register an occupied receptor. After registration, the probes are retracted and ejection rods are thrust into all receptors (typically requiring ~1 nanonewton (nN) of force for occupied receptors) to empty them while the test chamber is flushed clear in preparation for the next cycle to begin. Ideally, the probe rod pushing into the binding site should push the bound molecule further into the binding site to

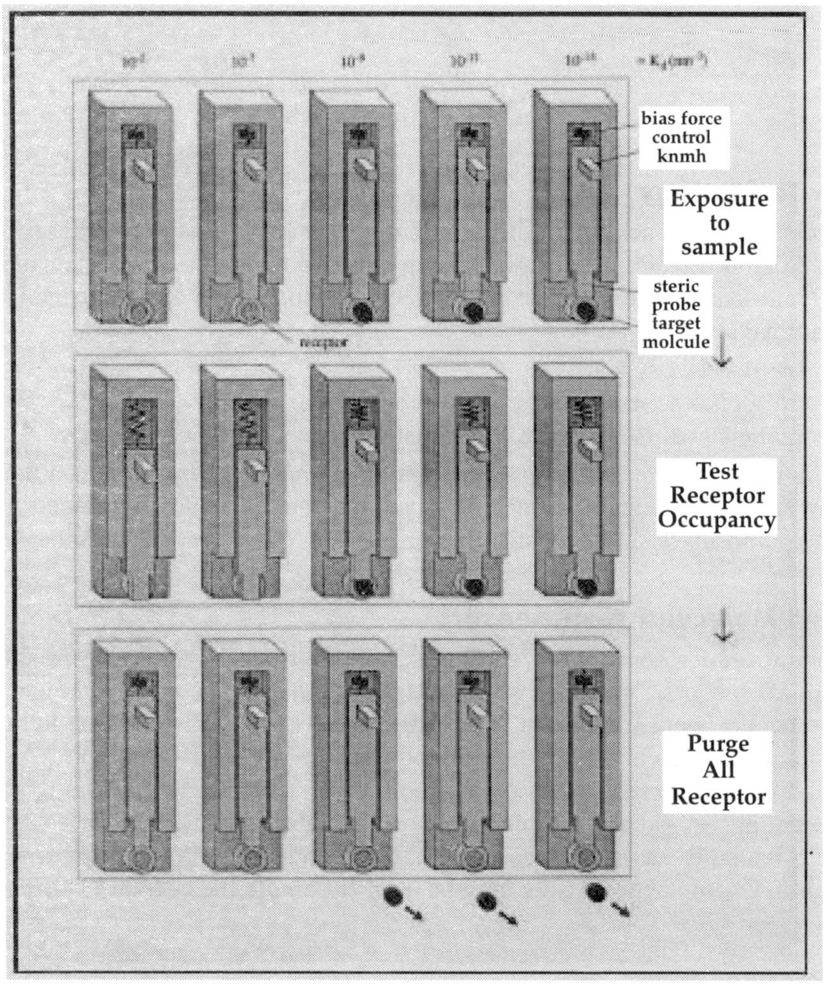

Fig. VI–D.1: Schematic of broadband chemical concentation sensor array using 5 receptor units with steric probes.

avoid complications that may arise due to competing kinetic barriers, and ejection rods should push molecules far enough to ensure that the next resample is independent.

Consider a series of N steric probe units as proposed by Drexler. Each probe measures 8 nm $\times$ $(2.5\ nm)^2 \sim 50\ nm^3$ and has mass ~ 2×10^{-22} kg. The ratio of the dissociation constants of adjacent units is $\kappa = K_{di} / K_{di+1} > 1$, with $K_{d1} \sim 10^{-4}$ molecules/nm^3 and $K_{dN} \sim 10^{-13}$ molecules/nm^3 typically in the human body, and so $N = 1 + \{\log_{10} (K_{d1}/K_{dN}) / \log_{10} (\kappa)\}$.

Minimum measurement error occurs when c_{ligand} exactly matches the K_d of a probe unit, that is, when $c_{ligand} = K_{di}$ (i.e., $P_{occupiedi} = 0.5$). Maximum measurement error occurs when c_{ligand} lies exactly midway between two probes such that $K_{di} > c_{ligand} > K_{di+1}$, or, more specifically, at the geometric midpoint $c_{ligand} = (K_{di} K_{di+1})^{1/2} = K_{di}\ \kappa^{-1/2}$. In this case,

$$P_{occupied_i} = \frac{c_{ligand}}{K_{d_i}\ \kappa^{-1/2}}\ P_{occupied_i} \tag{4.1}$$

where $P_{unoccupiedi} = 1 - P_{occupiedi}$. Additionally, sampling error is $\sim N_m^{-1/2}$ when N_m independent measurements are taken (i.e., N_m measurement cycles employing all N probe units during each cycle), for $N_m >> 1$, establishing an error bound on the probability of receptor occupancy of $P_{occupiedi} \pm N_m^{-1/2}$. If two concentrations $c_1 = P_{occupiedi}/(1 - P_{occupiedi})$ and $c_2 = (P_{occupiedi} + N_m^{-1/2})/(1 - (P_{occupiedi} + N_m^{-1/2}))$ may be distinguished, where $P_{occupiedi}$ is given by Eqn. 4.1, then the minimum detectable concentration differential $\Delta c/c$ is

$$\Delta c/c = [c_2/c_1] - 1 = \frac{(1+\kappa^{-1/2})\ N_m^{-1/2}}{1-(1+\kappa^{-1/2})^{-1} - N_m^{-1/2}} - 1 \tag{4.2}$$

For k >> 1, Eqn. 4.2 reduces to: $(\Delta c/c)_{large} \sim (1 + \kappa^{1/2})/N_m^{1/2}$. In the limit as κ approaches 1 and $N_m >> 1$, Eqn. 4.2 reduces to: $(\Delta c/c)_{small} \sim \kappa^{-1/2}\ (\Delta c/c)_{large}$.

For $\kappa = 10$ (an N = 10-probe sensor, sufficient to span the entire $K_d = 10^{-4}$ to 10^{-13} molecules nm^3 range using one probe per decade), $\Delta c/c = 0.94$ (94 per cent) at $N_m = 100$, 0.20 (20 per cent) at $N_m = 1000$, (0.06) 6 per cent at $N_m = 10{,}000$, and 0.006 (0.6 per cent) at $N_m = 10^6$ measurement cycles. Little additional discrimination is obtained by using more than N ~ 10 probes in the sensor. For $N_m = 1000$, $\Delta c/c = 0.89$ (89 per cent) at $\kappa = 178$ (N = 5 probes), 0.20 (20 per cent) at $\kappa = 10$ (N = 10 probes), 0.136 (13.6 per cent) at $\kappa = 1.23$ (N = 100 probes), and 0.135 (13.5 per cent) at $\kappa = 1.0021$ (N = 10,000 probes). Note that $\Delta c/c = 0.20$ (20 per cent) is equivalent to detecting a pH variation of ~0.08 (e.g., $10^{(0.08)} - 1 \sim 0.20$).

Sensor cycle time $\sim t_{EQ}$ may be crudely approximated by the number of random ligand-receptor encounters ($N_{encounters} \sim 100$; Drexler) necessary to ensure binding divided by the number of ligands striking the receptor surface per second, or (sec/cycle)

$$t_{EQ} \sim \frac{N_{encounters}}{A\ c_{ligand}}\ \frac{\pi\ MW_{kg}}{2\ kTN_A}\ \text{(sec/cycle)} \tag{4.3}$$

where A = active receptor cross-sectional area ~0.1 nm^2 for small molecule (MW ~100 gm/mole) receptors to ~10 nm^2 for large molecule (MW ~10^5 gm/mole) receptors; $c_{ligand} = 10^{-2}\ nm^{-3}$ for the most common molecules to $10^{-12}\ nm^{-3}$ for rare molecules in the human body; MW_{kg} = target ligand molecular weight in kg/mole, k = 0.01381 zJ/K (Boltzmann constant), T = 310 K (human body temperature), and $N_A = 6.023 \times 10^{23}$ molecules/mole (Avogadro's number).

For common small molecules, t_{EQ} ~10^{-6} sec; for common large molecules, t_{EQ} ~0.2×10^{-6} sec, but a sample chamber ~$(25\ nm)^3$ in size is diffusion-limited to a minimum t_{EQ} ~ 10^{-5} sec for c_{ligand} >~ $10^{-4}\ nm^{-3}$. For the rarest small molecules in the human body, t_{EQ} ~ 800 sec, ~200 sec for rare large molecules, and a larger chamber may also be required. Preconcentration of the test sample by dehydration using sorting rotors to rapidly remove water from the sample chamber reduces t_{EQ} to 10-40 sec for very rare molecules.

Steric probes using a driving force of ~100 pN, over 2 nm of motion per cycle require 200 zJ/measurement. If 1000 measurement cycles of a 10-probe sensor suffice to determine concentration to ~2 significant figures, this information costs ~0.002 pJ to acquire, a power consumption of from ~10 pW for large common molecules to ~10^{-9} pW (~0.6 kT/sec) for small rare molecules while the sensor is working. Power dissipation may run up to ~10 times higher, depending upon the number of occupied receptors that must be cleared via ejection rods during each cycle.

Narrowband Receptor Arrays

If the concentration of the target ligand varies only slightly within known limits, similar results may be obtained using a narrowband sensor that employs a repeating array of a single receptor type tuned to the maximum expected concentration c_{max} of the target ligand (Fig. VI–D.2). If the sensor consists of N_r receptors, choosing $K_d >\sim c_{max} N_r^{-1}/(1 - N_r^{-1})$ ensures that all receptors are occupied at $c = c_{max}$, thus avoiding array saturation at concentrations below c_{max}. At minimum sensitivity, $K_d \sim c_{min} (1 - N_r^{-1})/N_r^{-1}$, and so $(1 + (\Delta c/c)) = (c_{max}/c_{min})^{(Nr-1)}$ for the entire array. Thus the minimum detectable concentration differential $\Delta c/c$ for the array is

$$\Delta c/c = (N_r - 1)^{(Z/Nr)} - 1 \qquad (4.4)$$

If A = receptor unit active area (~50 nm^2), L = receptor unit length (~10 nm), and receptor arrays are packed in parallel sheets with separation Δx (~10 nm, for small-molecule diffusion in ~10^{-6} sec) to allow sample access, sensor volume $V_s = AL\ N_r\ (1 + (\Delta x/L))$ and sensor scale ~ $V_s^{1/3}$. Thus a sensor with N_r = 100 receptor units achieves $\Delta c/c$ = 0.10 (10 per cent) with a $(46\ nm)^3$ sensor; N_r = 1400 receptor units achieves $\Delta c/c$ = 0.01 (1 per cent) with a $(112\ nm)^3$ sensor. A dedicated micron-scale nanodevice employing a $(1.44\ micron)^3$ sensor array having 3 million receptor cells could at best distinguish 0.001 per cent concentration differentials in ~ 3 sec (from Eqn. 4.6) for common molecules like serum glucose.

The ion channel switch (ICS) biosensor (aka. the "Australian sensor") can detect small molecule concentrations as low as ~$10^{-14}\ nm^{-3}$ in a ~600 sec measurement time.

Counting Rotors

Concentration may be determined by varying exposure time to the sample. One energy-efficient design (Fig. VI–D.3) uses an input sorting rotor running at varying speeds (according to target concentration) synchronized with a counting rotor (linked by rods and ratchets to a data storage register device) to assay the number of molecules of the desired type that are present in a known volume of fluid. The counting rotor uses a steric probe, moved up and down in synchrony with the arrival of a binding site under the test probe, to count the number of molecules transferred by the sorting rotor. The fluid sample is drawn from the environment into a 4000 nm^3 reservoir ($L_{reservoir}$ ~ 16 nm) with the equivalent of 10^4 refills/sec on a flow-through basis, using paddlewheel pumps or reciprocating pistons; not shown) operated slowly enough to avoid damaging molecules in the sample volume. This sensor, which measures 45 nm × 45 nm × 10 nm comprising ~500,000 atoms

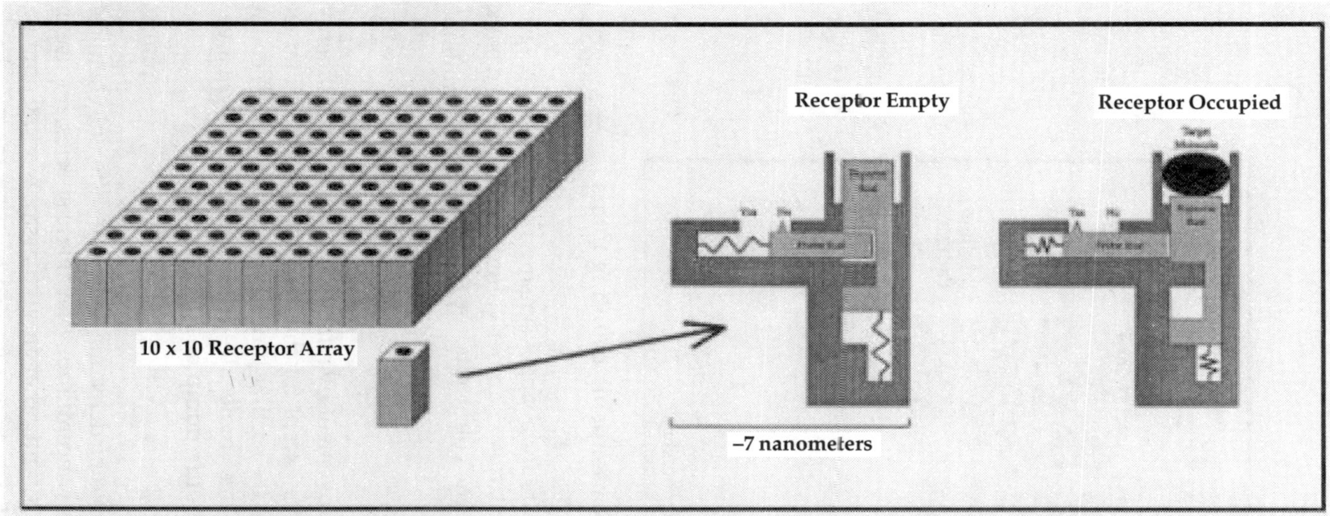

Fig. VI–D.2. Schematic of narrowband chemical concentration nanosensor array using receptors of a single type.

($\sim 10^{-20}$ kg), should count 'N ~ $10^4 - 10^5$ molecules/sec of small common molecules present at typical human serum concentrations of $10^{-3} - 10^{-2}$ nm^{-3}, such as glucose, with the rotor spinning at ~1 per cent of normal speed.

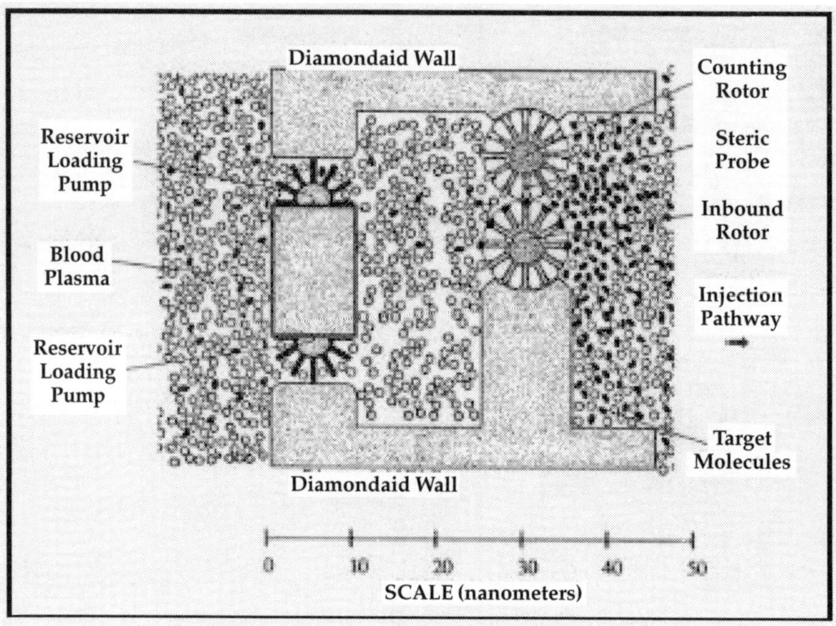

Fig. VI–D.3. Chemical concentration sensor using counting rotors.

For rare molecules with c~10^{-11} nm^{-3}, the count rate for a single sensing rotor falls to 'N~10^{-4} molecules/sec. However, a (340 nm)^3 reservoir (a ~10^4 larger chamber volume) serviced by a bank of 10^4 rotors (4×10^{-18} kg additional mass) improves the count rate for rare molecules to 'N ~ 1 molecule/sec; that is,

$$N \sim 2500\ c_{ligand}\ L^3_{reservoir}\ \text{(molecules/sec)} \tag{4.5}$$

Sample preconcentration may allow further improvement by a factor of up to 10-100. Energy requirements are comparable to those for sorting rotors except for the need to retrieve ~all target molecules from a known sample volume, which slows the sorting process, requiring ~10 zJ/count for common molecules and ~100 zJ/count for the rarest molecules in the human body. Large-molecule concentrations may be measured using a shuttle pump and a counting device.

Chemical Assay

It may also be useful to design a sensor that is capable of detecting as many different chemical species as possible, in contrast to detecting the precise concentration of a single species, for use in chemomessaging, chemonavigation, cellular diagnosis, and other in vivo assay work. Counting rotors may be used for this purpose. Each 12-receptor sorting rotor/counting rotor pair has a minimum 400 nm^3 volume. Allowing an additional ~800 nm^3 per pair to account for power, control, mechanical attachments and housings, which includes ~300 nm^3 per pair for sample chamber volume and sample access, requires ~100 nm^3 per receptor, roughly the same as for the steric probe units described earlier. A dedicated 1 $micron^3$ chemical assay nanorobot, which includes ~0.25 $micron^3$ sample volume, with ~10^7 pairs (receptors) could thus continuously scan for as many as ~10^7 different chemical species, counting 'N ~ 1 large molecule/sec at concentrations of c_{ligand} ~ 10^{-10} molecules/nm^3.

Chemical Nanosensor Theoretical Limits

Berg and Purcell estimated that an ideal concentration sensor, limited only by diffusion constraints and drawing through a spherical boundary surface of radius r_s, provides a minimum detectable concentration differential $\Delta c/c$ of

$$\Delta c/c = (1.61\ \Delta t\ D\ c\ r_{50})^{-1/2} \tag{4.6}$$

where Δt = time (sec), D is diffusion coefficient ~10^{-9} m^2/sec for small molecules (Table 3.3), and c = concentration (molecules/m^3). Using a sample chamber with r_s = 10 nm, a concentration of c = 3×10^{-3} nm^{-3} for serum glucose could be measured with 1 per cent uncertainty in Δt ~ 260 microsec, or with 0.01 per cent uncertainty in 2.6 sec. A dedicated micron-scale nanodevice using a 10-sec sampling time could at best distinguish a 0.0005 per cent concentration differential for small common molecules like serum glucose but only a 27 per cent concentration differential for large rare serum molecules like somatotropin.

Berg and Purcell also found that the capture of small molecules at nanorobot surfaces can be surprisingly efficient. Specifically, for a spherical nanodevice of radius R, across whose surface are uniformly distributed N_r receptor spots of radius r_r, the maximum diffusion current absorbed by the receptor array, J_{array}, is :

$$J_{array} = J\left(\frac{N_r\ r_r}{N_r\ r_r + \pi\ R}\right) \tag{4.7}$$

where J = maximum diffusive intake current. For r_r = 1 nm and R = 1 micron, the J_{array} is fully 50 per cent of J using only N_r = 3100 surface receptors, which occupy a mere 0.1 per cent of the total device surface area.

For a purely sensory spherical nanodevice, a uniform distribution of a relatively small number of chemoreceptors confers optimal sensitivity. However, in applications where sensors (which attract and then release) are in competition with pumps (which strongly attract and then remove) for the same target molecules at the nanodevice surface, greater sensing efficiency may be achieved by clustering the sensors, as in a "nose" organ.

Proteins may be modified as they age within biological cells. For instance, under steady state conditions, ~10 per cent of protein molecules may exhibit carbonyl (oxidation) modifications, glycosylation (non-enzymatic), and dysfunctional proteins may be ubiquitinized and phosphorylated under enzyme control. Heat shock also modifies proteins. Chemosensor receptors must be carefully designed to accommodate changes in target ligand structures or sites that are likely to be subject to such alterations.

Spatial Concentration Gradients

Medical nanodevices can detect both extremely steep and extremely shallow spatial concentration gradients. For example, a receptor array sensor consisting of 10 steric probe units has a total volume of ~1000 nm^3 including its prorated share of drive systems, housings, etc. In just a few measurement cycles, adjacent sensors can detect a maximum concentration change from 10^{-2} nm^{-3} to 10^{-11} nm^{-3} across a mean nanosensor separation of 10 nm on a nanodevice surface, a steep chemical spatial gradient of ~10^{10} per cent per nanometer. J. Soreff points out that maintaining a steep concentration gradient of a freely diffusing species implies a power dissipation of ~kT per molecule per time to diffuse a concentration-halving distance. For a common, small, tethered molecule with c_{ligand} ~ 3 × 10^{-3} molecules/nm^3 that diffuses ~1 nm in t_{diff} ~ 1 nanosec, implied power density D_{mol} ~ kT c_{ligand}/t_{diff} ~ 10^{13} watts/m^3, a power density that might possibly be encountered in the vicinity of chemoelectric transducers. (Such high power densities never occur on macroscale dimensions, and heat rapidly diffuses at nanoscale dimensions.)

Similarly, counting rotor sensors for small common serum molecules like glucose deployed at opposite ends of a 1-micron nanodevice can detect a concentration differential of ~10^{-4} over a ~1 sec measurement period, a shallow chemical spatial gradient of only ~10^{-5} per cent per nanometer. Even lower gradients may be detected by communication between physically separated cooperating nanodevices.

Temporal Concentration Gradients

Koshland and Macnab have shown that bacteria possess a mechanism for sensing small temporal concentration gradients. A bacterium detects a spatial gradient of attractant not by comparing the concentration at its head and tail, but rather by traveling through space and comparing its observations over time. *E. coli* can measure a ~0.01 per cent gradient during the time it travels one body length (~2 microns) at top speed (20-40 microns/sec), a minimum temporal gradient of 0.1 per cent per second. This appears very efficient, since the counting rotor sensor proposed above is not much better, achieving ~0.01 per cent per second. As for peak gradients, the 10-probe receptor array sensor can detect a maximum concentration change (e.g., from an already-known rare concentration) in ~10^{-4} sec, a peak temporal gradient of ~10^{15} per cent per second.

Chemotactic Sensor Pads

Besides concentration sensing, chemical sensors may be used to determine the chemical characteristics of surfaces. A binding pad coated with an array of reversible, possibly reconfigurable, artificial receptors may be pressed against test surfaces such as cell membranes, or the perimeters of very large molecules, in order to detect the presence of specific chemical ligands. If pad receptor specificity is high for the target ligand, then the degree of pad "stickiness" to the test surface is a good measure of the population of target ligands on that surface. Probing 2-D surfaces with receptors may be constrained by orientation effects, since binding may require a specific relative orientation of ligand and receptor. Compliant pad receptors can reduce this problem but may impair lateral resolution—there is a three-way tradeoff among angular tolerance, lateral resolution, and multioriented multicopy receptors. A physical contact sensor prevents confusion of surface-bound and free-solvated ligands of the same species. The selective and reversible stickiness of such pads can also be exploited for locomotion.

Molecular adhesion experiments in chemical force microscopy and interfacial force microscopy using AFM probes with functionalized tips typically detect maximum adhesive noncovalent force differentials of 40-160 piconewton (pN)/ligand for competing ligands at a peak-force distance of ~0.2 nm. Forces >10 pN are readily measured by nanosensors, so binding pad sensors should give an exact count of the number of target ligands present per unit area of test surface if (receptor areal number density) > (ligand areal number density) > (1 ligand/pad area), and if contact time is adequate. Receptors for many different ligands can be mounted on moving rotors or belts, or employed as interchangeable tool tips, thus allowing a single sensor pad to reconfigure its specificity or sensitivity under external control.

Receptor Sensors

Ligand mimics may also be used to allow a medical nanodevice to detect a particular receptor in its environment. Each such "receptor sensor" is a protruding physical structure that mimics all or part of the electrosteric structure of the ligand to which the target receptor has maximum specificity. For example, probes consisting of artificial structures that simulate antigen peptides (8-15 amino acids long) bound in an MHC-like structure could be used to sense the presence of specific T-cell receptors.

Sensing pads displaying ligand mimic structures are brought into contact with the test surface. Any target receptors present will bind to a mimic structure, thus may be counted as the sensor sweeps slowly across the surface. Immediately after recognition has occurred, the mimic structure may be reversibly fragmented, destroying receptor/mimic affinity and allowing the receptor to release the mimic. Mimics are mounted on the sensor pad with an appropriately elastic compliance to prevent structure detachment during a measurement scan.

Receptor sensing is most useful in detecting receptors for large molecules. In the case of receptors for small molecules, there may sometimes be insufficient space in the target molecular volume to match the surface electrosteric structure, to accommodate ligand fragmentation control rods or cables, or to build internal boundaries sufficient to permit ligand fragmentation without creating undesirable reactive species. Some of these problems with small-molecule receptors may be partially overcome by employing an annular repulsive ring through which a mimic ligand may be forcibly retracted without fragmentation, but receptor orientation effects will present additional constraints.

Displacement and Motion Sensors

The ability to precisely measure position and displacement, velocity and acceleration, rotation and orientation are essential requirements for nanoscale locomotion, physical manipulation, and navigation.

Displacement Sensors

Diamondoid parts may contain features, or be positioned, in picometer (1 pm = 10^{-12} m) steps, much smaller than the unavoidable RMS thermal vibrations. The nanomanipulator robot arm may be moved in picometer increments, and the RMS thermal longitudinal displacements at the tip of a 20-nm long, 10-nm wide diamondoid lever is ~1 pm. Displacement sensitivity of 10 pm is routinely achieved in STMs. It is claimed this can be improved down to the ~1 pm level; STM resolution of 2 pm has been demonstrated experimentally. A C_{60} molecule used as the active element in an electromechanical amplifier transmits ~100 times more electrical current when physically compressed by 100 pm, suggesting a minimum deformation detection limit of a few picometers. (A strain gauge and a vibration sensor using a (17,0) carbon nanotube (~2000 atoms) has been proposed.)

Thermally-induced positional uncertainty in the displacement of nanoscale components is approximated by the classical value

$$\Delta x = (kT/k_s)^{1/2} \tag{4.8}$$

where T is temperature (K) and k_s is the Hooke's law spring constant, or restoring force stiffness (N/m). At 310 K, this classical approximation is accurate to within <~10 per cent of the quantum mechanical treatment for harmonic oscillators with RMS displacements Δx >~10 pm. Spring stiffness k_s is ~0.1 N/m for nonbonded (noncovalent) interatomic interactions, ~30 N/m for covalent bond angle bending, ~400 N/m for covalent bond stretching, and ~1000 N/m for solid 1 nm^3 diamondoid blocks. At 310 K, springs of such stiffness produce minimum displacement uncertainties of 200 pm, 10 pm, 3 pm and 2 pm, respectively.

The log ratio of detection energy to noise energy, the signal/noise ratio (SNR), for a displacement sensor is derived from the harmonic potential (1/2) k_s x^2 as

$$\mathrm{SNR} = \ln\left(\frac{\text{signal energy}}{\text{noise energy}}\right) \ln\left(\frac{k_s\,\Delta x^2}{2\,kT}\right) \tag{4.9}$$

Assuming a very stiff k_s = 600 N/m, a minimal SNR – 1 gives a minimum detectable displacement of 6 pm, or 10 pm at a more reasonable SNR = 2 (20 dB). The conservative conclusion is that Δx_{min} ~ 10 pm displacements should be reliably detectable by medical nanosensors in vivo, in a measurement time t_{meas} ~ 10^{-9} sec. This compares favorably with the stereocilia of the inner ear, which can only detect 100 pm displacements in 10^{-5} sec.

Velocity and Flow Rate Sensors

Typical thermal velocities of nanoscale diamondoid components at 310 K, are ~60 m/sec for 1 nm^3 objects, ~2 m/sec for $(10\ nm)^3$ objects, 0.06 m/sec for $(100\ nm)^3$ objects, and 0.02 m/sec for 1 $micron^3$ objects. These thermal motions are in degrees of freedom whose range of motion is largely confined, by design, to sub-nanometer distances within nanoscale measuring devices. However, objects of nanomedical interest may have velocities ranging from ~10^{-9} m/sec to ~10^3 m/sec (e.g., the speed of sound in water), and may include biological elements in the environment or internal machine components. Thus it is useful to explore the fundamental limitations involved in measuring the full range of this physical variable.

There are at least two simple ways to determine a velocity. First, velocity may be measured by transferring the kinetic energy of a moving object or fluid volume to a sensor element. The movement of the sensor element is then clocked as it traverses a known distance. Second, and more efficiently, the passage of the moving object may sequentially trip two latches located a known distance apart, again revealing the moving object's velocity; alternatively, detecting the presence or absence of a protruding feature on a passing body using probe rods (e.g., Fig. VI–D.4) can determine velocity without any requirement for large kinetic energy transfers. The need for physical gating movements by sensor components and computational register-shift operations using diamondoid rods moving at ~GHz clocking speeds suggests a conservative minimum stable sensor cycle time of ~10^{-9} sec.

The two principal restrictions on the maximum detectable velocity v_{max} are the maximum linear dimension of the sensor element, or L_{sensor}, and the minimum event time that may be accurately discriminated, or Δt_{min}. For a sensor element with thermal noise energy ΔE ~ 10 kT = 43 zJ at 310 K, the quantum mechanical limit is Δt_{min} >~$\hbar/(2\,\Delta E)$ ~ 10^{-15} sec. Δt_{min} is also limited by the thermal variation in length of the individual latches. If the minimum detectable latch displacement is Δx ~ 10 pm, and the maximum speed of latch displacement is approximately the

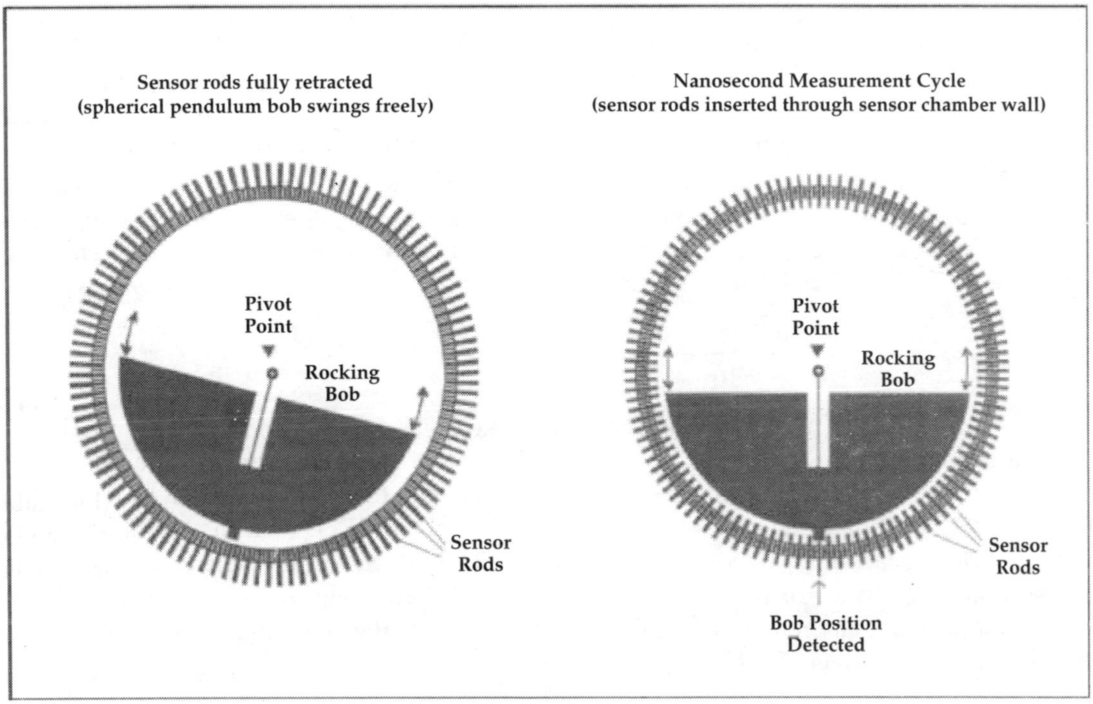

Fig. VI–D.4. Spherical nanopendulum sensor for angular displacement and rotational velocity.

speed of sound in the latch material (e.g., $v_{sound} \sim 17{,}300$ m/sec for diamond), then $\Delta t \sim \Delta x/v_{sound} \sim 10^{-15}$ sec, near the quantum mechanical limit. Δt is further restricted by the minimum detectable phase variance between latches located at either end of the sensor. An acoustically-transmitted clock pulse emitted from a centrally-located source that is Δx closer to one of the two latches can discriminate a phase variance of $>\sim \Delta x/L_{sensor} \sim 10^{-5}$ taking $L_{sensor} \sim 1$ micron, so the minimum discriminable time is again $\Delta t_{min} \sim 10^{-5}\, L_{sensor}/v_{sound} \sim 10^{-15}$ sec; hence $v_{max} \sim L_{sensor}/\Delta t_{min} \sim 10^{9}$ m/sec $\sim c$ (= 3×10^{8} m/sec, the speed of light).

However, unlike photons, physical objects moving in an aqueous medium are normally limited to the speed of sound in that medium, e.g., $v_{sound} \sim 1500$ m/sec in water at 310 K If v_{max} is taken as $\sim 10^{3}$ m/sec, then for a sensor of size $L_{sensor} \sim 1$ micron the required time discrimination is only $\Delta t_{min} \sim L_{sensor}/v_{max} \sim 10^{-9}$ sec.

The minimum detectable velocity v_{min} is limited by the minimum measurable linear displacement $\Delta x \sim 10$ pm and the limits of our patience in making the measurement. For a maximum measurement time of $t_{meas} \sim 1$ sec, $v_{min} \sim \Delta x/t_{meas} \sim 0.01$ nm/sec in vacuo. In a nanomedically-relevant aqueous medium, the systematic motion of a free-floating (untethered) object of radius R is masked by added Brownian noise, giving the limiting $v_{brownian} = \Delta X/\tau >\sim 260$ nm/sec for R = 10 micron and $v_{brownian} >\sim 2.6$ micron/sec for R = 100 nm, where $\tau - t_{meas} = 1$ sec. The thermal motions of tethered objects moving along surfaces (e.g., vesicles transported along microtubules by kinesin motors; are more restricted, hence Brownian velocity masking is reduced.

Modest accuracy may be achieved using acoustic reflection in a Doppler velocimeter.

Employing acoustic waves of frequency ν, the minimum measurable velocity change $\Delta v = v_{sound}/(\nu\ t_{meas}) = 150$ microns/sec in an aqueous medium at 310 K with $v_{sound} = 1500$ m/sec, $\nu \sim 10$ MHz, and $t_{meas} = 1$ sec, corresponding to the shifted frequency $\nu\ (1 + (\Delta v/v_{sound}))$ having enough time to make one extra cycle during t_{meas}. Even at such long wavelengths (~150 microns), significant backscattering of ultrasound from ~7.82 micron red cells is observed experimentally.

Velocity sensors may also be used to estimate volumetric flow rates in fluidic channels and blood vessels by measuring the speed at which fluid passes a nanodevice resting on the vessel wall. The parabolic velocity profile of v is a function of radial distance r from the center of a cylindrical vessel of radius R for Hagen-Poiseuille flow:

$$v = \frac{(R^2 - r^2)\ \Delta P}{4\eta\ L_v} \tag{4.10}$$

where absolute viscosity $\eta = 1.1$ centipoise (1.1×10^{-3} kg/m-sec) for plasma at 310 K and ΔP is the pressure change along the length of a vessel of length L_v. A 1-micron nanodevice obtains measurements of v at various distances from the vessel wall, then fits these data to Eqn. 4.10 to find the choice of R that minimizes the variation in estimated $\Delta P/L_v$, which should be roughly constant under nonaccelerative conditions. Once R is determined, $\Delta P/L_v$ is known as well as the cross-sectional area (πR^2) of the vessel lumen, and a simple integration of the radial velocity distribution provides the flow rate in m^3/sec. Unfortunately, the usefulness of this technique is limited by boundary layer effects.

Acceleration Sensors

Accelerations inside nanoscale devices are often measured in the trillions of g's ($g = 9.81$ m/sec^2). For example, an L = 1 nm diamondoid block ($m = 3.5 \times 10^{-24}$ kg) with thermal energy ~kT experiences random thermal accelerations of $a_{therm} \sim kT/g\ m\ L = 10^{11}$g's; a 10-nm logic rod with vibrational energy ~kT experiences ~10^{12} g oscillations; a 200-nm wide hoop spinning axially with 1000 m/sec rim speed has rotational acceleration of a ~10^{12} g's.

Box-Spring Accelerometers

As one member of a class of devices, a simple nanoscale omni-directional accelerometer is conceptualized as a reaction mass suspended by three pairs of reaction springs (one pair per orthogonal directional axis) with each spring connected to the walls of an evacuated box (Fig. 4.5). Acceleration in any direction extends some springs and compresses others, which is measured by force or displacement sensors attached to each spring. The signal/noise ratio for this sensor is

$$SNR = \ln\left(\frac{E_{spring}}{E_{noise}}\right) = \ln\left(\frac{m^2 a^2}{2\ kT\ k_s}\right) \tag{4.11}$$

For a sensor mass $m = 10^{-17}$ kg (~78 nm wide platinum or ~140 nm wide diamond cube) suspended with $k_s = 1$ N/m floppy springs at T = 310 K, the minimum detectable acceleration $a_{min} = 1.6 \times 10^6$ g's for SNR = 1 and 2.6×10^6 g's for SNR = 2 (undamped resonant acceleration ~ $L\ k_s/m \sim 10^9$ g's for L = 100 nm). Undamped resonance frequency $w_{res} = (k_s/m)^{1/2} = 3 \times 10^8$ radians/sec, which is measurable.

Sensor mass exchanges reciprocally with acceleration sensitivity in Eqn. 4.11, so the lower limit even for a dedicated micron-scale platinum-core box-spring accelerometer is a_{min} ~1000 g's at SNR ~ 2 (undamped resonant acceleration ~ 10^6 g's for L = 2 microns). The maximum detectable acceleration requires m ~ 10^{-23} kg (~1.4 nm diamond cube) and k_s ~ 1000 N/m, giving a_{max} ~5×10^{13} g's at SNR = 1.

If the accelerometer depicted in Figure VI–D.5 is installed near the center of mass of a nanorobot, the sensor may avoid confusing centrifugal and translation forces. Otherwise, information from multiple sensors must be combined to obtain an unambiguous measurement. Multiple measurements can improve the SNR for constant accelerations of long duration.

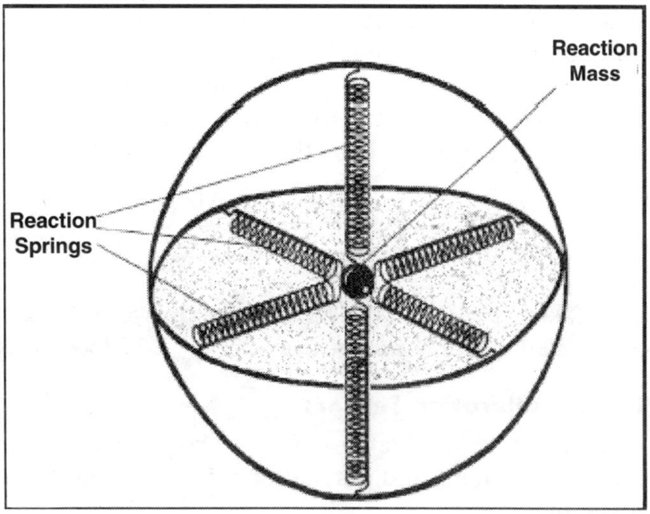

Fig. VI–D.5. Schematic of box-spring omnidirectional accelerometer.

In 1998, ~100 nm vibrating-beam silicon accelerometers were under development by Analog Devices and Daimler-Benz. Tunneling accelerometers had been fabricated with ~10^6 g resolution at a 350 Hz bandwidth or ~10^{-8} g/$Hz^{1/2}$. Silicon-based capacitive accelerometers (5-1000 g full scale) manufactured by Silicon Designs (Issaquah, WA) were widely used in automotive airbag safety systems; in 1998, Motorola shipped its ten-millionth accelerometer, and the global market for automotive silicon accelerometers was expected to reach $463 million by 2002.

Displacement Accelerometers

Formed elements in the human bloodstream (e.g., red cells, white cells, platelets) typically are buffeted by much smaller accelerations. For example at 310 K, instantaneous random thermal accelerations of $a_{thermal}$ ~ 500 g's are experienced by a 0.2-micron virus particle, ~0.05 g's by a 2-micron platelet, ~10^{-4} g's by a 10-micron neutrophil—and ~0.4 g's by a 1 $micron^3$ spherical nanorobot. Force-based sensors cannot easily detect such low accelerations, since the smaller the sensor, the larger the a_{min} for a given SNR (Eqn. 4.11). An alternative approach uses displacement sensors to determine an object's velocity twice in rapid sequence, allowing acceleration to be computed from the difference.

Consider a constantly accelerating object that triggers three clocking latches at positions x_1, x_2, and x_3, in sequence. Time and position of the object are measured as (x_1, t_1), (x_2, t_2), and (x_3, t_3). The measured constant acceleration is then given by

$$a = \frac{v_{32} - v_{21}}{t_{32} - t_{21}} \ (\mathrm{m/sec^2}) \qquad (4.12)$$

where $v_{32} = (x_3 - x_2)/(t_3 - t_2)$, $v_{21} = (x_2 - x_1)/(t_2 - t_1)$, $t_{32} = (t_3 + t_2)/2$ and $t_{21} = (t_2 + t_1)/2$.

Assume that a slowly accelerating external object passes three clocking latches that are fixed to the exterior of a nanorobot of radius R. Measurements of the object's velocity incur an unavoidable error due to nanorobot thermal motion. The object is observed to pass between latch pairs during a measurement time ~t_{meas}. During this same interval, the nanorobot translates a distance $\Delta X = (kT\ t_{meas}/3\ phR)^{1/2}$, incurring a maximum velocity measurement error of ~$\Delta X/t_{meas}$, and thus from Eqn. 4.12 a maximum acceleration measurement error of ~$2\ \Delta X/t_{meas}^2$ if $(t_3 - t_2) \sim (t_2 - t_1) \sim t_{meas}$. Thus the minimum detectable acceleration is

$$a_{min} = \left(\frac{4\ kT}{3\ \pi\ \eta\ R\ t^3_{meas}}\right)^{1/2} \ (m/sec^2) \tag{4.13}$$

Taking T = 310 K, $\eta = 1.1 \times 10^{-3}$ kg/m-sec and R = 1 micron, then $a_{min} \sim 10^{-7}$ g's for t_{meas} = 1 sec; $a_{min} \sim 0.004$ g's for a more reasonable $t_{meas} = 10^{-3}$ sec measurement time in the in vivo environment.

Fluid Acceleration Sensors

Biological fluid accelerations may also be inferred from measurements of spatial pressure gradients. Ignoring frictional and gravitational forces and assuming laminar flow, fluid acceleration a_{fluid} is given by

$$a_{fluid} = \frac{\Delta P}{\rho\ \Delta_x} \ (m/sec^2) \tag{4.14}$$

where ρ = fluid density (e.g., blood plasma is an almost incompressible Newtonian fluid, with $\rho \sim 1025$ kg/m³ at 310 K) and ΔP is the measured change in pressure over a distance Δx. The mitral and aortic valves in the heart ($\Delta P/\Delta x \sim 10^5$ N/m³, $a_{fluid} \sim 10$ g's, $\Delta x \sim$ a few cm), as well as valves in the veins and lymphatics, operate as bistable analog decelerometers employing this principle. Micron-scale medical nanorobots traveling across blood vessel surfaces could use pressure sensors capable of 10^{-6} atm resolution to infer local blood flow accelerations to an accuracy of ± 0.01 g from sequential measurements taken 1 mm apart.

Pivoted Gyroscopic Accelerometers

Consider a cylindrical gyroscope spinning at an angular velocity w (radians/sec), with radius r and thickness h, whose center of gravity lies a distance L_p from a frictionless pivot point along the rotational axis. The pivot point defines the center of a spherical sensor grid of radius R $(r^2 + 4h^2)^{1/2}$ having distinguishable grid elements located a distance $\Delta x \sim 1$ nm apart. Under uniform acceleration a, the gyroscope precesses around its spin axis at an angular velocity $\omega_p = 2\ a\ L_p/\omega\ r^2 = \Delta x/R\ t_{meas}$, where t_{meas} is measurement duration. Hence the minimum detectable acceleration is:

$$a_{min} = \frac{\Delta x\ r^2 \omega}{2\ L_p\ R\ t_{means}} \ (m/sec^2) \tag{4.15}$$

by comparing precessional motion against the grid.

In choosing an ω with which to measure a_{min}, there is an upper and a lower limit. The upper bound is the bursting strength condition :

$$\omega = \omega_{max} < \left(\frac{2\ \sigma_w}{\rho\ r^2}\right)^{1/2} \ (rad/sec^2) \tag{4.16}$$

where σ_ω is a safe working stress = 10^{10} N/m^2 (~0.2 tensile strength) for diamond and ρ is the density of the gyroscope material (~3510 kg/m^3 for diamond). The lower bound is the minimum spin angular velocity below which the pivoted gyroscope cannot spin stably about the vertical axis and begins to wobble, thus ruining the measurement. The lower bound is given by

$$\omega = \omega_{min} = \left(\frac{4\ a_{min}\ L_p\ (r^2 + {}^1/_3 h^2)}{\pi\ h\ \rho\ r^6} \right)^{1/2} \tag{4.17}$$

If $\omega_{max}/\omega_{min} > 1$, then nonwobbling spin velocities are available below the bursting speed of the gyroscope cylinder. Since this ratio scales with R, there is a minimum sensor size R_{min} below which no useful spin velocities are available. Combining Eqns. 4.15 and 4.17 and solving the quadratic in ω gives

$$R_{min} = \frac{2\ \Delta x\ (r^2 + {}^1/_3 h^2)}{\pi\ h\ t_{meas}\ r^3\ (2\sigma_w\ \rho)^{1/2}} \tag{4.18}$$

For $t_{meas} = 10^{-3}$ sec, R_{min} >~ 69 microns (optimum r = 48 microns, h = 25 microns); for t_{meas} = 1 sec, R_{min} >~ 6.9 microns (optimum r = 4.8 microns, h = 2.5 microns). Thus it appears that pivoted gyroscopic accelerometers likely will be employed only in devices >> 10 microns in size. Microscale gyroscopes implemented in silicon have been discussed by Greiff and others.

Accelerative Onset

Aside from pure acceleration, the rate of accelerative onset—sometimes called jolt, surge, or jerk—can produce additional damaging effects on macroscale mechanical and biological structures. For example, the effects of accelerative onset ('a) on the human body were studied by Col. John Paul Stapp in a series of aggressive rocket sled experiments conducted during the 1950s. For human subjects, cardiovascular shock was entirely absent for 'a >~ 500-600 g/sec but began to appear at 'a >~ 1100-1400 g/sec, at 1-40 g accelerations. Human tolerance to linear decelerative force is determined both by rate of onset and peak acceleration; the endurance limit for well-restrained healthy young males has been given as <~500 g/sec and <~50 g for >~0.2 sec durations.

By 1998, the effects of accelerative onset on nanomechanical structures had not been extensively studied, but similar limits might apply to nanomachines. For example, a diamond rod of length L_{rod} = 1 micron, cross-sectional area S_{rod} = 0.01 micron^2, mass $m_{rod} = 3.5 \times 10^{-17}$ kg, and working strength $W_{rod} \sim 5 \times 10^{10}$ N/m^2 fails when subjected to a static acceleration a_{fail} >~ $W_{rod}\ S_{rod}/m_{rod} \sim 10^{12}$ g. But accelerative shocks can produce dynamic amplification in the response spectrum of elastic systems. Dynamic stress rises to a peak of roughly twice the corresponding static value when $'a_{peak}$ >~ 4 $\nu_{res}\ a_{max}$ for a trapezoidal pulse shape, where ν_{res} is the frequency of natural vibration of the system, because the stresses from the free oscillations of the system, induced by the changes in the acceleration of the system, can add to the static stresses from the acceleration, with damage done by the total stress. Taking k_s ~ 1 N/m for the rod described above, then ν_{res} ~ 30 MHz and $'a_{peak} \sim 10^{20}$ g/sec at $a_{max} \sim 10^{12}$ g. For comparison, the teragravity nanocentrifuge has mass m ~ 10^{-17} kg and is spun up to $a_{max} = 5 \times 10^{11}$ g's in >~10^{-6} sec, giving 'a <~ 5×10^{17} g/sec << $'a_{peak}$.

Angular Displacement

There are circumstances in which it is necessary for a medical nanodevice to measure, establish or maintain a preferred orientation, as for example nanorobots with a clearly defined "top" and "bottom" or nanodevices performing lengthy intracellular operations requiring precision stationkeeping. Measurement of rotation rates may also be required.

Gimballed Nanogyroscopes

Pivoted gyroscopes of the type are impractical for orientational sensing in all but the largest medical nanodevices. Setting ω_{max} (Eqn. 4.16) equal to ω_{min} (Eqn. 4.17), taking $a_{min} = g = 9.81$ m/sec^2 (gravitational acceleration), and optimizing for minimum sensor size gives a critical nonwobbling nonbursting spin velocity $\omega_{crit} \sim 6.8 \times 10^7$ radians/sec at a minimum device radius of ~35 microns. If a pivoted gyroscope is smaller than this minimum radius, then spinning it slower than ω_{crit} causes the gyro to tumble helplessly, while spinning it faster than ω_{crit} tears the gyro apart.

However, a triaxially gimballed nanogyroscope may be regarded as having its pivot point near its center of mass, so L_p becomes very small and ω_{min} becomes very slow, making nanogyroscopes feasible in medical nanorobots. Taking h = 1 micron, r = 0.5 micron, $\rho = 3510$ kg/m^3 and $a_{min} = g$ in Eqn. 4.17, then if the gimbals are aligned such that $L_p = 1$ nm between the pivot point and the effective center of gravity along the rotation axis, then $\omega_{min} \sim 400$ rad/sec. If gimbal tolerances are improved to $L_p = 0.1$ nm, then $\omega_{min} \sim 125$ rad/sec.

How stable is a nanogyroscopic orientation standard? There are two principal considerations.

First, Brownian thermal rotation in all three angular degrees of freedom (θ, φ and ψ) gives rise to a small nutation around the nanogyroscopic precession axis of magnitude $\Delta\theta_{nutate} \sim p_\varphi/p_\psi$, where p_φ and p_ψ are angular momenta of the gyro. The minimum value for p_φ is approximated by $(1/2)\, kT \sim (1/2)\, p_\varphi^2/I_1$, where $I_1 = (1/4)\, m\, r^2 + (1/12)\, m\, h^2$, the moment of inertia around φ, and $m = \pi\, r^2\, h\, \rho$. The maximum value for p_ψ is given by the burst-strength condition $\omega < \omega_{max}$ (Eqn. 4.16) as $p_\psi \sim I_3\, \omega$, where $I_3 = (1/2)\, m\, r^2$, the moment of inertia around ψ. Hence:

$$\Delta\theta_{nutate} \sim \left(\frac{kT\, (r^2 + {}^1/_3 h^2)}{2\, \pi \sigma_w\, h\, r^4} \right)^{1/2} \quad \text{(radians)} \tag{4.19}$$

For T = 310 K, r = 0.5 micron, h = 1 micron, $\sigma_\omega \sim 10^{10}$ N/m^2, $\Delta\theta_{nutate} \sim 0.8$ microradian.

Second, the gimbal bearings have small frictional losses, exerting small torques on the gyro and causing a small precession away from the original orientation. The angular velocity of precession $\omega_{precess} \sim T_{gimbal}/(I_3\, \omega)$ (rad/sec) where T_{gimbal} is the frictional torque caused by an imperfect gimbal bearing. Assume that the gimbal is driven by external forces to oscillate at some frequency ν_{gimbal}, and that a gimbal bearing dissipates power at the rate of P_{gimbal} (watts), so that each oscillatory motion of the gimbal dissipates $E_{gimbal} = P_{gimbal}/\nu_{gimbal}$ (joules). If the angular amplitude of the gimbal motion is α_{gimbal} (radians) and the gimbal radius is r_{gimbal}, then the gimbal bearing travels $X_{gimbal} = \alpha_{gimbal}\, r_{gimbal}$ (meters) per gimbal oscillation and the force applied in dissipating an energy E_{gimbal} is $F_{gimbal} = E_{gimbal}/X_{gimbal}$. Thus, the torque applied by each gimbal oscillation event is $T_{gimbal} = F_{gimbal}\, r_{gimbal} = P_{gimbal}/(\nu_{gimbal}\, \alpha_{gimbal})$; $P_{gimbal} = k_p\, v_{gimbal}^2$ (see Drexler), where $v_{gimbal} = \alpha_{gimbal}\, \nu_{gimbal}\, r_{gimbal}$ is the sliding speed of the gimbal bearing surfaces during movement and k_p is a constant that depends solely on the geometry of the bearing.

The change in orientation angle caused by each gimbal oscillation event is $\Delta\theta_{osc} = \omega_{precess}/\nu_{gimbal}$. The number of gimbal oscillations $N = t\, \nu_{gimbal}$, where t is elapsed time since the gyro was last calibrated. If gimbal oscillations are independent and randomly distributed, then the total change in orientation angle is very roughly $\Delta\theta_{precess} \sim \Delta\theta_{osc} N^{1/2}$; assuming $r_{gimbal} \sim r$, then:

$$\Delta\theta_{precess} = \frac{2\, k_p\, \alpha_{gimbal}\, (t\, v_{gimbal})^{1/2}}{\pi\, r^2\, h\, \rho\, \omega} \text{ (radians)} \tag{4.20}$$

From Drexler, $k_p = 2.7 \times 10^{-14}$ watt-sec^2/m^2 as a conservative value for a small (~2 nm) stiff bearing. Random hydrodynamic flows induced by thermal fluctuations inside biological cells have a characteristic duration $t_{fluct} \sim 10$ millisec, which suggests $\alpha_{gimbal} \sim 0.06$ radians (~3°), for an R ~ 1 micron in cyto nanorobot. Taking r = 0.5 micron, h = 1 micron, $\rho = 3510$ kg/m^3 for diamond (giving m ~ 3×10^{-15} kg), $\omega = 1 \times 10^9$ rad/sec (just below the $\omega_{max} \sim 4.8 \times 10^9$ rad/sec bursting speed), and assuming $v_{gimbal} \sim t_{fluct}^{1} = 100$ Hz (to which the results are not terribly sensitive), then $\Delta\theta_{precess} \sim 10^{-8}\, t^{1/2}$ (radians). Thus a $\Delta\theta_{precess} \sim 1$ microradian orientation shift takes ~1 hour; a $\Delta\theta_{precess}$ ~ 4 microradian shift takes ~1 day; and a $\Delta\theta_{precess} \sim 70$ microradian (~10 arcsec) shift takes ~1 year. (Total angular error $\Delta\theta_{total} = \Delta\theta_{precess} + \Delta\theta_{nutate}$.) Thus a gimballed nanogyroscope may be sufficiently stable to serve as a readily storable, easily transportable, and highly accurate onboard orientation standard.

Nanopendulum Orientation Sensing

A nanopendulum may also be employed for orientational sensing. As one example of a large class of related devices, consider a simple rigid spherical nanopendulum of radius r and mass m, with a large hemispherical bob free to swing through the entire solid angle about a central pivot point (Fig. 4.5), with a gimballed housing to allow active avoidance of the pivot support beam. The nanopendulum bob moves along a spherical surface located far enough from a concentric spherical sensor grid of radius $R = r + z_{sep}$ to minimize van der Waals interactions, with bob and sensor surfaces electrically neutral or uniformly mutually repulsive. Much like the pin cushion receptor model, the sensor grid consists of a spherical array of diamondoid sensor rods $\Delta x \sim 1$ nm wide that are rapidly and simultaneously extended into the evacuated sensor cavity to determine bob position during a measurement cycle of duration t_{meas}. Radial sensor rod velocity should exceed typical tangential bob velocity by at least 1-2 orders of magnitude. For the example given below, a sensor stud on a 500-nm bob that is turning at $v_{turn} \sim v_{thermal} = 3.6$ mm/sec passes a ~1 nm sensor rod in ~280 nanosec; a sensor rod displacing ~1 nm in the radial direction in ~1 nanosec travels at $v_{rod} \sim 1$ m/sec. This nanopendulum may serve both as an orientation sensor and as a rotation rate sensor, using principles similar to vestibular (semicircular canal) mechanics.

If a simple nanopendulum spherical sensor housing with bob initially at rest in a uniform vertical gravity field g is rotated through the smallest detectable angle θ_{min} (radians) around the center pivot, then ignoring friction, the bob's position over the sensor grid moves a distance Δx, hence

$$\Delta\theta_{min} = \Delta x / r \tag{4.21}$$

For $\Delta x = 1$ nm and r = 500 nm, $\Delta\theta_{min} \sim 2$ milliradians (~0.1 deg). For a practical sliding interface (e.g., between bob and sensor surface) of contact area S and turning velocity v_{turn}, Drexler estimates frictional dissipation, primarily driven by shear-reflection drag and band-stiffness scattering, as $P_{drag} \sim k_1\, v_{turn}^2\, S$, where $k_1 \sim 400$ kg/m^2-sec, hence the energy dissipated between the bob and the nearest sensor rod (the one interacting with the bob) is $E_{drag} \sim k_1\, v_{turn}\, S\, \Delta x$. Bob gravitational energy $E_{bob} = m\, g\, r \sin \Delta\theta_{min} \sim m\, g\, r\, \Delta\theta_{min}$ for small displacements. The bob swings freely at the minimum displacement if $E_{bob}/E_{drag} > 1$. If $m = 10^{-15}$ kg, $r = 500$ nm, and $S = 1\ nm^2$, then $v_{turn} < 2.5$ m/sec or $\omega < 49$ megaradians/sec for the bob to swing freely (vs. $v_{thermal} = 3.6$ mm/sec and $\omega_{res} \sim 9900$ rad/sec for this pendulum).

The bob-up/bob-down energy differential is ~2 m g r = 9.8 zJ ~2.3 kT for $m = 10^{-15}$ kg and r = 500 nm. To distinguish opposite orientations ($\Delta\theta \sim \pi$) requires $N_{meas} \sim (kT/2\ m\ g\ r)^2 \sim 1$ measurement for this sensor. To distinguish $\Delta\theta_{min} \sim 2$ milliradians requires averaging $N_{meas} \sim (kT/E_{bob})^2 \sim 190{,}000$ independent measurements (of ~0.002 kT/measurement) or $t_{meas} \sim 190$ microsec using a sensor rod clock cycle of ~1 nanosec. Thus, after subtracting out the effects of translation displacements as determined by linear accelerometers, a nanodevice can measure angular displacements in its orientation to within $\pm\ \Delta\theta_{min}$ in a measurement time $t_{meas} \sim 10^{-4}$ sec. With sensor rods moving at ~1 m/sec, energy dissipation for each rod sliding in its housing is ~0.01 zJ/cycle. If only ~100 test rods need be triggered to continuously track the moving stud, then a sensor rod clock cycle of ~1 nanosec implies a continuous nanopendulum power dissipation of ~1 pW; additional drag power loss between stud and contacting rod is only $P_{drag} \sim 0.0004$ pW.

Many other configurations of this basic device are readily conceived and may be more volume-efficient, including nested spheres or ovoids, two-dimensional rocking disk sections for in-plane rotational measurements (three devices required for monitoring all angular degrees of freedom in a rigid rotator), and multiaxial circumferential tracks or tubes containing rolling balls or sliding plugs. J. Logajan suggests that a pair of linear three-dimensional translation accelerometers should be able to unambiguously extract the 3 translation and 3 rotation components of all possible acceleration vectors. Rotation about any axis will result in distinctly different mass displacements while accelerated translation along any axis will result in identical displacements.

Nanopendulum Tachometry

The smallest angular velocity detectable using a simple nanopendulum is given approximately by $\omega_{min} \sim \Delta x/r\ t_{meas} = 2$ milliradian/sec (~0.1 deg/sec) for $\Delta x = 1$ nm, r = 500 nm, and $t_{meas} = 1$ sec. The largest detectable angular velocity $\omega_{max} \sim 2\ \pi/t_{meas}$ if t_{meas} is the minimum time required to count one complete rotation of the sensor; the smallest $t_{meas} \sim 2\ \pi/\omega_{max} \sim 100$ nanosec for $\omega_{max} \sim 49$ megaradians/secor ~5 per cent of the bursting speed (Eqn. 4.16) for an r = 1 micron spherical diamondoid nanodevice.

What rotational displacements will normally be encountered by an in vivo medical nanorobot? A 1-micron device experiences instantaneous Brownian rotations of $\omega_B = (kT/4\ \pi\ \eta\ R^3\ \tau)^{1/2}$ radians/sec, where $\tau = (MW_{kg}/32\ \pi\ kT\ N_A\ c_{H2O}\ R^4)^{1/2}$, the mean time between molecular collisions with the device (mostly by water molecules), from Eqn. 4.3, where $c_{H2O} \sim 3.3 \times 10^{28}$ molecules/m^3. For R = 0.5 microns and T = 310 K, $\tau \sim 6 \times 10^{-20}$ sec/collision and $\omega_B \sim 7 \times 10^9$ rad/sec, with each displacement of $\sim \tau\ \omega_B = 4 \times 10^{-10}$ radian, none of which are detectable. However, the net of all Brownian rotational displacements is one rotation every 16 sec, ~0.4 radian/sec.

The largest measurable angular displacements in vivo are caused by tumbling due to differential shear forces in the bloodstream. Normal vessel wall shear rates in physiological bloodflow range from 100-1400 radians/sec in the larger arteries to 500-4000 radians/sec in the smaller arteries and capillaries. The measured rolling velocity of white cells during vascular "tank tread" locomotion on venule walls is ~40 micron/sec, or ~8 radians/sec, which may be encountered by nanodevices employing cytocarriage. The human vestibular mechanism has a frequency response range of 0.048-260 radians/sec.

Thus, the maximum range of physiologically relevant rotation rates is 10^{-2} to 10^{4} radians/sec. Since the natural harmonic frequency of a simple nanopendulum is $\omega_{res} = (g/r)^{1/2} = 4400$ radians/sec for r = 500 nm, the possibility of forced oscillations and resonance need be faced only when measuring the highest likely physiological rotation rates, in which case the addition of a damping force $F_{damp} \sim m\, v_{turn}\, \omega_{res}$ during resonance events should eliminate the problem. F_{damp} for critical damping peaks at ~mg ~ 0.01 pN for the m = 10^{-15} kg, r = 500 nm case (v_{turn} ~ 2 mm/sec).

Force Nanosensors

Piconewton forces may be applied and must be detected by medical nanodevices. Pendulum- and cantilever-based sensors may be used to probe nanoscale forces, to monitor the ambient gravitational field, to measure molecular-scale masses to single-proton accuracy, and to discriminate among molecules having different isotopic composition.

Minimum Detectable Force

Extremely small forces have been measured using macroscale instrumentation, particularly the atomic force microscope (AFM). In 1998, experimental forces of 1 nN in the contact regime and 1 pN in the noncontact regime were routinely measured. In 1994, D. Rugar achieved a force sensitivity of 5×10^{-16} N using an AFM with a 90-nm thick cantilever at room temperature, and in 1998 Rugar reported a force resolution of 7×10^{-18} N using a magnetic resonance force microscope (MRFM) with a 230-micron long, 60-nm thick, silicon cantilever. The theoretical limit on the force sensitivity of the AFM has been estimated as 10^{-18} – 10^{-19} N for temperatures near absolute zero using fiber optic interferometry or other techniques. However, many of these methods are impractical or impossible to employ in medical nanodevices.

What is the minimum force F_{min} likely to be detectable by nanorobot sensors? Several lines of reasoning produce consistent answers. For example, force may be measured by the displacement Δx of a beam against the known resistive pressure provided by a spring of minimum stiffness k_s; that is, $F_{min} \sim k_s\, \Delta x$. However, the energy stored in the spring must exceed thermal noise energy, with SNR ~ ln ($k_s\, \Delta x^2$ /2 kT) (Eqn. 4.9), giving $F_{min} = 2\, kT\, e^{SNR}/\Delta x$. For T = 310 K and SNR = 2, F_{min} = 100 pN at Δx = 0.6 nm, 10 pN at 6 nm, and 1 pN at 60 nm. Indeed, the natural molecular motors (kinesin, dynein, and actin-myosin) have isometric force outputs on the order of 1-14 pN per protein molecule, which are routinely observed using microscale cantilever spring sensors measuring total forces of 10-100 pN. Breaking antigen-antibody or receptor-ligand interactions generally requires rupture forces in the 50-300 pN range. A force of ~160 pN separates biotin that is seated in its avidin receptor; ~85 pN pulls iminobiotin out of the avidin receptor. A force of ~300 pN ruptures a hydrogen bond between two isolated water molecules, but >3000 pN are likely to be needed to rupture a single C-C covalent bond. The rupture strength of single covalent bonds has been directly measured experimentally by AFM as, for instance, 2000 ± 300 pN for Si-C (528 zJ/molecule dissociation enthalpy, 0.185 nm bond length) and 1400 ± 300 pN for Au-S.

Force sensing may also require using measurement devices that involve sliding interfaces of two mutually inert periodic surfaces. This motion may be characterized by the potential energy of interaction, ~0.1-1 zJ as determined by Monte Carlo simulations of typical diamondoid material at 0.2-0.5 nm separations. Assuming a minimum 0.5 zJ energy barrier, for small sliding contacts the typical force amplitude $F_{min} \sim 1.7 \times 10^{10}\, \Delta V_{barrier} \sim 8.5$ pN.

A force of 10 piconewtons across a small-molecule receptor $(0.3\ nm)^3$ of surface area 0.09 nm^2

represents an energy density of ~10^8 J/m^3 × 0.027 nm^3 = 2.7 zJ ~ 0.6 kT. And the nanorobotic manipulator arm is capable of ~picometer step sizes, representing a minimum applicable force of F_{min} ~ 10 pN assuming 0.04 m/N compliance for the entire mechanism. Thus as a conservative estimate, the smallest force that is reliably detectable (and applicable) by nanodevices is probably ~10 pN.

Finally, a nanosensor of dimension L able to distinguish two forces differing by ΔF newtons has a probability of erroneous measurement of

$$P_{err} = \exp(-L\ \Delta F/4\ kT) \qquad (4.22)$$

For ΔF = 10 pN and T = 310 K, P_{err} = 1 per cent for L = 8 nm, or 0.01 per cent for L = 16 nm, so reliable nanoscale picoforce sensors may be ~10-20 nm in size.

Nanogravimeters

Research in space medicine has discovered that micron-scale elements of the immune system have extraordinary sensitivity to gravity. One study found that bone marrow-derived (B6MP102) macrophage single cells respond after 8 seconds to exposure to a 10^{-2} g hypogravity environment as indicated by increased cell spreading, a well-known marker of cell activation. Individual immune system cells placed in microgravity exhibit enhanced growth but depressed differentiation, decreased activation by concanavalin A, a 20 per cent reduction in glucose metabolism, increased resistance to antibiotics, increased number of pseudopodia on monocytes, and changes in cytoplasmic streaming velocity and frequency.

As of 1998, the exact mechanism by which the leukocyte "gravity sensor" operates remained unknown. It has been speculated that differences in densities of several organelles, such as the nucleolus, the ribosomes, and the centrioles, could generate detectable pressures on the structure of the cytoskeleton under gravitational loading mediated through mechanosensitive stretch-activated ion channels or the extracellular matrix. While indirect gravitational effects that modify the cellular environment (such as reduced sedimentation or decreased thermal convection due to hypogravity) could also be responsible, a direct interaction of the gravitational force with cellular structures appears to be the most likely sensor mechanism. For instance, if a change in gravitational loading equivalent to the force generated by a natural molecular motor (F ~ 1 pN) is detectable by the cytoskeleton, and this force is applied to the entire mass of a (10 micron)^3 cell (m ~ 10^{-12} kg), then the minimum detectable gravitational acceleration is g ~ F/m ~ 1 m/sec^2 = 0.1 g, which should be adequate to detect the onset of hypogravity.

A nanomechanical gravity sensor of similar size to a macrophage cell may be at least 100,000 times more sensitive. For example, the energy of a tethered mass swinging freely in a uniform gravity field is approximately E_g ~ m g Δh, where Δh is the maximum vertical amplitude of the motion. To be detectable, oscillator energy must exceed thermal noise energy, so SNR ~ ln (m g Δh $N_{meas}^{1/2}$/kT), giving $g_{min} = kT\ e^{SNR}/m\ \Delta h\ N_{meas}^{1/2} \sim kT\ e^{SNR}/\rho\ L^4\ N_{meas}^{1/2}$ for a bob mass of density ρ and size L^3, N_{meas} independent measurements, and assuming Δh ~ L. If T = 310 K, SNR = 2, N_{meas} = 1, and ρ = 21,450 kg/m^3 (Pt), then g_{min} = 0.1 g for L ~ 1 micron, 0.01 g for L ~ 2 microns, 10^{-5} g for L ~ 11 microns, and 10^{-6} g for L = 20 microns, the size of the average human tissue cell.

Regarding nanogravimeter design, for a pendulum of length L in a gravity field g ~ 9.81 m/sec^2 (1 g), the resonant period of the motion $T_{res} = 2\ \pi\ (L/g)^{1/2}$ = 2 × 10^{-3} sec for L = 1 micron, 9 × 10^{-3} sec for L = 20 microns. If Δt_{min} = 1 nanosec is the minimum detectable decrease in resonant period caused by an increase in gravity from g to g + Δg, then the minimum detectable change in gravitation force is

$$\Delta g/g = 1- \left(\frac{L}{g\, k^2_g}\right) \sim \frac{10^{-9}}{L^{1/2}} \qquad (4.23)$$

for a medical nanorobot, where $k_g = (\Delta t_{min}/2\,\pi) + (L/g)^{1/2}$. For L = 1 micron, $\Delta g/g \sim 10^{-6}$; for L = 20 microns, $\Delta g/g \sim 2 \times 10^{-7}$.

Single-Proton Massometer

A nanopendulum may also be used to measure molecular-scale masses. Torsion, helical spring, and vibrating rod mechanisms all will work, but one of the most efficient designs for weighing the smallest masses may be the coiled suspension-spring cantilever massometer illustrated schematically in Figure VI–D.6, which is analogous in principle to the traditional quartz crystal microbalance. In this design, a beam of length L_{beam} and circular cross-sectional radius R_{beam} is hinged at one end and carries a sample holder of solid volume V_{sample} at the other end. The beam is supported (a variable distance L_{spring} from the hinge) by a spring consisting of a movable suspension rod of total length l_{rod}, circular cross-sectional radius r_{rod}, and spring constant $k_s \sim \pi^2 r_{rod}^4\, E\, G/(G\, l_{rod}^3 + E\, \pi\, r_{rod}^2\, l_{rod})$ using spring (rod) material of density ρ, Young's modulus E, and shear modulus G. The natural resonance period for this unloaded system is $T_0 = (2\,\pi\, L_{beam}/L_{spring})\,(m_0/k_s)^{1/2} = Z\, m_0^{1/2}$, where $Z = (2\, L_{beam}/L_{spring}\, r_{rod}^2)\,((G\, l_{rod}^3 + E\,\pi\, r_{rod}^2\, l_{rod})/E\, G)^{1/2}$ and $m_0 = \rho\,(V_{sample} + \pi\, L_{beam}\, R_{beam}^2 + \pi\, l_{rod}\, r_{rod}^2)$. Gravitational loading is negligible for suspension system mass $m << (k_s\, L_{spring}^2/g\, L_{beam}^2) \sim 10^{-15}$ kg; $m \sim 10^{-20}$ kg in the exemplar system.

After adding a test mass Δm to the sample holder, the resonance period increases to $T_m = Z\,(m_0 + \Delta m)^{1/2}$, which may be detected if $T_m - T_0 \geq \Delta t_{min} \sim 1$ nanosec, the minimum convenient time interval. Hence the minimum detectable molecular weight ΔMW, expressed in proton masses ($m_p = 1.67 \times 10^{-27}$ kg), is

$$\Delta MW = \Delta m/m_p = m_p^{-1}\,[(m^{1/2}{}_0 + \Delta t_{min}/Z)^2 - m_0]\ (\text{protons}) \qquad (4.24)$$

In Figure 4.6, beam and rod are coiled in two dimensions (although three-dimensional coiling, more difficult to depict, provides maximum compactness). L_{beam} and l_{rod} are chosen to minimize total coil area $A_{coil} = 2\, R_{beam}\, L_{beam} + 2\, r_{rod}\, l_{rod}$. Using aluminum for the rod and beam ($E = 6.9 \times 10^{10}$ N/m^2, $G = 2.6 \times 10^{10}$ N/m^2, $\rho = 2700$ kg/m^3), $L_{beam} = 1200$ nm, $l_{rod} = 1920$ nm, $r_{rod} = 0.5$ nm, $V_{sample} = 1$ nm^3, and $\Delta t_{min} = 10^{-9}$ sec, then $\Delta MW \sim 1$ proton for $L_{spring} = 1$ nm $>> \Delta x_{min} = 0.01$ nm, the minimum measurable displacement. (For $L_{spring} > 1$ nm, physiological ~1 g accelerative loads on a ~1 micron nanorobot in which the sensor is embedded produce undetectable accelerative beam displacements $< \Delta x_{min}$).

Moving the suspension rod closer to the sample holder reduces device sensitivity. For $L_{spring} = L_{beam}$, minimum $\Delta MW \sim 1200$ protons, permitting moderately large molecules to be weighed in minimum time (~10^{-6} sec). Using a very thin sensor configuration (h_{sensor}~20 nm) and coils occupying ~50 per cent of sensor cavity space, sensor dimensions are (79 nm)^2 × 20 nm giving a massometer volume ~10^5 nm^3 or ~0.01 per cent of the total volume of a 1 micron^3 nanorobot. Sensor mass is ~ 3×10^{-19} kg, including diamondoid housing and prorated share of support mechanisms. If one measurement cycle including sample loading/unloading operations can be completed every ~10 microsec, then the device processes its own mass of glucose molecules in ~10 sec. Note that biological molecules weighed in air or in vacuo may lose water of hydration, whereas weighing in water presents other complications that merit further investigation.

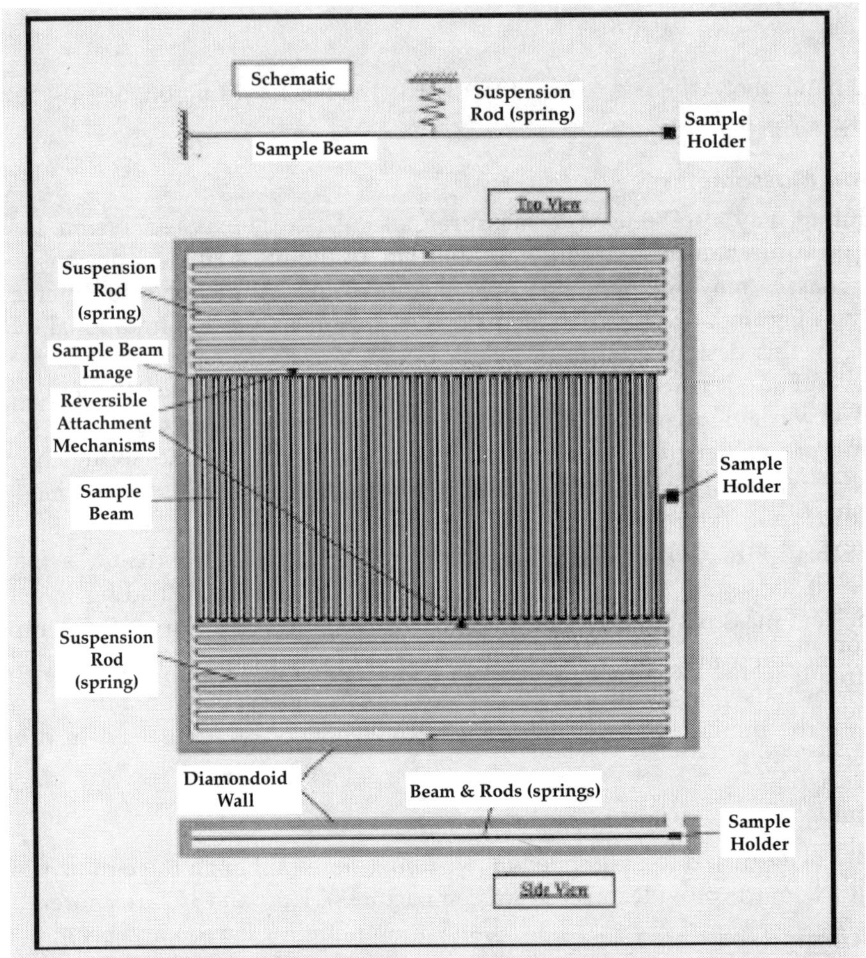

Fig. VI–D.6. Single proton massometer sensor using coiled suspension spring cantilever.

J. Soreff points out that for a molecule on a spring, the precision of the measurement is linked to the sharpness of the resonance, which is optimized to the degree the vibrational frequency of the spring/molecular mass system is kept well below the vibrational resonance frequencies of the molecule itself. The following discussion reports the results of a simple lumped-element analysis of the effects of internal vibration on the massometer which confirms the previously described performance estimates for the massometer.

Suppose that it is desired to weigh some mass M to a precision ΔM by attaching it to a weak measuring spring (spring constant k_{weak}) and measuring the shift in resonant frequency. The shift in frequency due to coupling to an internal mode of the sample must be less than $(1/2)(\Delta M/M)$ $(1/2\,\pi)\,(k_{weak}/M)^{1/2}$. If the sample has some lowest internal mode frequency ν_{int}, then approximating the sample as two M/2 masses separated by x_{strong} connected by a strong spring (k_{strong}) gives $\nu_{int} = (1/\pi)(k_{strong}/M)^{1/2}$. If this sample is attached to a weak measuring spring of length x_{weak}, which is in turn attached to a rigid support, then the equations of motion are:

$$(M/2)\ x'_{strong} = -\ k_{weak}\ x_{weak} + k_{strong}\ (x_{strong} - x_{weak}) \quad (4.25)$$

$$(M/2)\ x'_{strong} = k_{strong}\ (x_{strong} - x_{strong}) \quad (4.26)$$

Setting $x_{weak} = e^{st}$ and $x_{strong} = R\ e^{st}$ where t = time, s = time constant, and $R = -(1 + (2\ k_{weak}/M\ s^2))$, and since the angular frequencies of the two springs are $\omega_{strong} = 2\ \pi\ \nu_{int} = 2\ (k_{strong}/M)^{1/2}$ and $\omega_{weak} = (k_{weak}/M)^{1/2}$, then the lower frequency solution for $\omega_{weak} << \omega_{strong}$ may be approximated by $s = i\ \omega_{weak}\ (1 - \omega_{weak}^2/2\ \omega_{strong}^2)$ and so the (undesired) fractional change in frequency due to the flexibility of the object to be weighed is

$$\omega_{weak}/\omega_{strong} < (\Delta M/M)^{1/2} \quad (4.27)$$

For a molecule sample size of L_{sample}, $\nu_{int} \sim v_{sound}/L_{sample} \sim 500$ GHz for $v_{sound} \sim 1500$ m/sec and $L_{sample} \sim 3$ nm; $M = \rho\ L_{sample}^3 = 2.7 \times 10^{-23}$ kg for $\rho = 1000$ kg/m^3. To detect the mass difference of a single proton in a sample molecule, sensor $Q = M/m_p = 16{,}000$; from Eqn. 4.27, the sensor may operate at a factor of $Q^{1/2}$ below ν_{int}, or $\nu_{spring} \sim \nu_{int}/Q^{1/2} \sim 4$ GHz. A period of 0.25 nanosec is slightly too short to allow a phase shift of $\Delta t_{min} \sim 1$ nanosec, since the period is only one-quarter of this. However, a phase shift of π radians should accumulate in $t_{meas} = Q/2\ \nu_{spring} \sim 2$ microsec and should be detectable by looking at magnitudes of forces. Interestingly, an SPM microprobe of circular cross-section and exponential profile with a 0.1 nm^2 tip has been proposed that could also reach ~proton mass detection sensitivity at room temperature and 1 atm pressure with an eigenmode frequency of 10 GHz; an electrically-driven carbon nanotube-based resonant-beam balance demonstrated ~10^{-17} kg (~10^{10} proton or ~0.01 micron^3 mass) sensitivity in 1999.

Existing surface acoustic wave (SAW) devices are chemical sensors that can measure areal mass densities as small as 80 picogram/cm^2, or 0.01 N_2 molecule per nm^2 of surface, though current devices are limited to ~micron^2 areas. A proposed quartz microresonator microbalance could detect a 10^{-6} molecular monolayer, the equivalent of 5×10^{-9} molecule per nm^2. Detection of ~400 million protons using an MRFM vibrating rod ~60 nm thick was demonstrated by Rugar in 1998.

Isotope Discrimination

It will often be useful in nanomedicine to distinguish molecules containing different isotopes of the same chemical elements, for example to count the number of organic molecules in a sample that contain radioactive C^{14} atoms in place of stable C^{12} atoms or to make direct assays of heavy radiometal-contaminated tissues. Nanodevices may also sort isotopically heteronuclear molecules into homonuclear fractions, for example to separate deuterochemicals from hydrochemicals or to acquire isotopically pure materials with which to construct:

(a) Small, high-speed rotors with precisely balanced mass distributions.
(b) Components with extremely fine size differences.
(c) Nanostructures with maximum thermal conductivity (Table 4.1), or mixed isotopes to reduce thermal conductivity.
(d) Long-lived isotopically pure structures that generate no internal radioactivity, thus promoting longer device lifetimes.

The most reliable sensor for discriminating among isotopes within target molecules is the single-proton massometer. Isotopes of any chemical element differ in mass by at least ~1 neutron (~protonic) increments. Thus a massometer with single-proton resolution can resolve all isotopes starting from chemically pure samples. Isobars (atoms having the same mass number) and isotones (atoms having the same number of neutrons) may be finally resolved chemically; nuclear isomers differing in mass by ~0.001 amu cannot be resolved. Teragravity centrifugation and time-of-flight particle-beam deflection techniques may also prove useful.

Isotopes vary in many subtle ways due to their mass differences, although it is not yet clear how a potential isotope nanosensor might exploit some or all of these distinctions:

(1) **Line Shifts**—the electronic energy states of atoms or molecules depend on the reduced nuclear mass, causing a spectral line shift of ~1.0005 between hydrogen and deuterium, less among heavier isotopes, detectable spectroscopically.

(2) **Hyperfine Structure**—isotopes differ in electrical quadrupole moment, nuclear spin and magnetic moment, giving rise to hyperfine structure in the optical spectra, especially in the heavier elements. For example, the magnetic moments of hydrogen and deuterium (2.79268 and 0.857387 nuclear magnetons, respectively) are distinguishable via NMR. The NMR frequency in a 10 kilogauss magnetic field is 42.5759 MHz for H, 6.53566 MHz for D. Of course, not all isotopes are NMR sensitive.

(3) **Vibrations and Rotations**—molecular spectra are especially sensitive to isotopic changes, since the quanta of vibrational and rotational energies are directly dependent upon the masses of the atoms involved, and there are also some subtle effects from differences in nuclear spin. Separations are commonly achieved by laser excitation.

(4) **Cross-Sections**—isotopes have different thermal neutron capture cross sections; for instance, the cross-section of C^{12} is 3.7×10^3 barns (1 barn = 10^{-24} cm^2) compared to 9×10^{-4} barns for C^{13} and $<10^{-6}$ barns for C^{14}. (Electron scattering might be better, since the electron is less likely to cause the nucleus to fission and it is generally difficult to use neutrons to probe samples on the molecular scale.) Similarly, the neutron cross-section for H_2O is 0.6 barns compared to 9.2×10^{-4} barns for D_2O.

(5) **Other Properties**—isotopes differ slightly in various bulk properties, many of which could be measured in submicron-scale aliquots by nanosensors. The ability of both large mammals and micron-size photosynthetic plant cells to preferentially concentrate C^{12} or C^{13} in organic matter, and the predicted ability of carbon nanotubes to selectively extract tritium from low-concentration H_2-T_2 mixtures, constitute proofs of principle that at least some of the bulk properties listed in Table 4.1 may become useful in achieving nanoscale isotopic separation or purification.

Pressure Sensing

This section describes the ability of medical nanodevices to detect and measure fluid pressure variations while functioning inside the human body. Generation of acoustic signals for power transmission, communication, and navigation are considered elsewhere.

Minimum Detectable Pressure

A change in the ambient pressure in a fluid environment (ΔP) may be determined by measuring the change in volume (ΔV) of a fixed quantity of gas at constant temperature, relative to a known sensor volume at a known reference pressure. (Thermal sensors elsewhere in the nanorobot provide

Table VI–D.1. Isotopic Variation in Bulk Properties Possibly Useful in Nanomedicine

Bulk Property	*Units*	*Common Isotope*	*Rarer Isotope*	*Comments*
Absolute density	kg/m^3	997.044 (H_2O)	1194.45 (D_2O)	at 25°C
Critical density	kg/m^3	325 (H_2O)	363 (D_2O)	
Critical pressure	atm	218.5 (H_2O)	218.6 (D_2O)	
Oritical temperature	°C	374.2 (H_2O)	371.5 (D_2O)	
Temperature of minimum density	°C	3.98 (H_2O)	112 (D_2O)	
Freezing poin:	°C	0 (H_2O)	3.82 (D_2O)	
Boiling point	°C	100 (H_2O)	101.42 (D_2O)	
Triple point (solid/liquid/vapor)	K	13.96 (H_2)	18.73 (D_2)	
Triple point (Ice I/Ice III)	K	251.2 (H_2O)	254.4 (D_2O)	at 2054/2181 atm (H_2O/D_2O)
Heat of lusior.	kcal/mole	0.028 (H_2)	0.047 (D_2)	at triple point
Heat of vaporization	kcal/mole	9.70 (H_2O)	9.96 (D_2O)	
Heat of formation	kcal/mole	68.32 (H_2O)	–7041 (D_2O)	liquids at 25°C
Formation equilibrium constant	—	41.553 (H_2O)	42.654 (D_2O)	logrithm at 25°C liquid state
electrochemical reaction potential	volts	0 ($2H^+ + 2e$-H_2)	–0.044 ($2D^+ + 2e$-D_2)	at 25°C
Thermodynamic entropy	cal/K	16.72 (H_2O)	18.16 (D_2O)	liquids at 25°C
Lower limit of inflammability	% conc.	4.65% (H_2)	5.0% (D_2O)	combustion in O_2 at 1 atm
Upper limit of inflammability	% conc.	93.9% (H_2)	95.0% (D_2O)	combustion in O_2 at 1 atm
Surface tension	joule/m^2	0.07275 (H_2O)	0.0678 (D_2O)	at 20°C
Absolute viscosity	kg/m sec	0.001002 (H_2O)	0.00176 (D_2O)	at 20°C
Dielectric constant	—	80.36 (H_2O)	79.755 (D_2O)	at 20°C
Refractive index	—	1.33300 (H_2O)	1.32844 (D_2O)	
Solubilities	gm NaCl/liter	359 (H_2O)	305 (D_2O)	at 20°C
Aqueous dissociation constant	—	1.79 $\times 10^5$ (NH_4OH)	1.10$\times 10^4$ (ND_2OH)	K_0, 25°C 0.1.001N
Thermal conductivity	watts/m K	–3000 (pure C^{12})	2000 (1%C^{13})	damond at room temperature
Thermal diffusivity	cm^2/sec	18.5 (0.1% C^{13})	12.3 (1%C^{13})	damond at room temperature
Isothermal compressability	kg/m^3	1250 (H_2O)	1390 (D_2O)	at 12,000 atm 37°C
Ionization energy	—	13.9965 (H_2O)	14.869 (D_2O)	logarithm at 25°C
Ion mobilities	(mole/liter)^2	1.008 $\times 10^{34}$ (H_2O)	1.54$\times 10^{-15}$ (D_2O)	ion product at 25°C
Molar magnetic susceptability	cm^3/mole	–12.97 $\times 10^6$ (H_2O)	–12.76$\times 10^8$ (D_2O)	liquids at 20°C
Superfludity transition temperature	K	2.12 (He^4)	0.003 (He)	near absolute zero

(Contd.)

Table VI–D.1. (Contd.)

Bulk Property	*Units*	*Common Isotope*	*Rarer Isotope*	*Comments*
Superconductivity transition temperature	K	9 (PdH)	10 (PdD)	
Reversible glass/liq. transition,	K	129 (H_2O)	134 (D_2O)	
Acoietic wave speed	m/sec	1284 (H_2)	890 (D_2)	at STP
Acouptic wave speed temp. dependance	m/sec K	+2.2 (H_2)	+1.6 (D_2)	at 1 atm
Isotope-exchange chemical reaction rates	m^5/sec	6.05×10^{46} (O^{16}-$O^{32}{}_2$)	9.76×10^{-40} (O^{15}+$0^{16}{}_2$)	ozome formation
Relative crystal lattico constant	—	1 (natural Ce)	1.00002 (Ge^{15})	at 27°C
Tunneting vibrational voltage	milivolts	358 (C_2H_2)	266 (C_2D_2)	at 8K

readings by which corrections can be made for temperature variations.) If the sensor volume consists of a cylinder of fixed cross-sectional area, the change in volume is converted into a linear displacement of a piston located at the interface between gas and fluid environment. The change in Gibbs free energy is $\Delta G = \Delta P\ \Delta V$, while thermal noise ~kT. Pressure sensor signal/noise ratio SNR ~ ln $(\Delta P\ \Delta V/kT)$, so at the minimum ΔG the minimum detectable pressure ΔP_{min} is measured at the maximum possible volume change, or V_{sensor} (the sensor volume), given approximately by

$$\Delta P_{min} \sim kTe^{SNR}/V_{sensor} \tag{4.28}$$

using the classical ideal gas formulation. At SNR = 2, a $(22\ nm)^3 = 10^4\ nm^3$ sensor (~5×10^5 atoms or ~10^{-20} kg for a cube with 1 nm thick walls) can detect a minimum pressure change $\Delta P_{min} \sim 0.03$ atm; a $(680\ nm)^3 = 0.3$ micron^3 sensor (~3×10^9 atoms, ~6×10^{-17} kg) detects $\Delta P_{min} \sim 10^{-6}$ atm, probably near the practical limit for mobile in vivo medical nanodevices. Note that ΔP_{min} scales approximately as the inverse cube of mean sensor dimension, and immunity to thermal noise scales exponentially with sensor volume. Current silicon micromachined pressure sensors are ~1 mm^3 with 10^{-3} atm sensitivity, a factor of ~10^{12} poorer sensitivity than the theoretical minimum for their size. (A field-effect chemical sensor with an active sensing volume of 0.1 mm^3 that measures pressure indirectly has detected a 10^{-11} atm step in hydrogen gas with $t_{meas} \sim 3$ sec, only a factor of 1000 above the theoretical minimum for a sensor of that size.) The human ear can detect pressure waves as small as 2×10^{-10} atm.

A less sensitive approach relies on the observation that a change in pressure P alters both density and bulk modulus of a compressed fluid, variations that may be detected by measuring the change in the speed of sound v_s in the working fluid, given by

$$v_s = (B/\rho)^{1/2} \tag{4.29}$$

where B is bulk modulus of the fluid and ρ is fluid density. $(dB/dP)/B$ may be up to several times larger than $(d\rho/dP)/\rho$ for molecular fluids, but there is some $(dv_s/dP)/v_s$ for each fluid. Differentiating Eqn. 4.29 with respect to P and using $d\rho/dP = \rho/B$, then $dv_s/dP = (dB/dP - 1)/2\ \rho^{1/2}\ B^{1/2}$. For water at 1 atm, $dB/dP \sim 8.4$ atm/atm, ρ ~1000 kg/m^3 and B = 22,000 atm, giving $dv_s/dP \sim 0.25$ m/sec-atm. If Δv_s is the measured change in acoustic speed, then:

$$\Delta P_{min} = \Delta v_s/(dv_s/dP) \tag{4.30}$$

Velocity can be measured to ~0.1 mm/sec accuracy using a micron-scale velocity sensor and $t_{meas} \sim 1$ sec, so a minimum $\Delta v_s \sim 0.1$ mm/sec implies $\Delta P_{min} \sim 0.0004$ atm.

Spatial Pressure Gradients

Two 0.3-micron^3 pressure sensors located at either end of a 1 micron nanorobot could detect pressure changes as small as 10^{-6} atm per micron, a ~1 atm/m spatial gradient. The maximum detectable gradient would involve adjacent measurements of minimum (10^{-6} atm) and maximum (~10^4 atm) absolute pressure, a theoretical peak spatial gradient of ~10^{16} atm/m, though a practical device could not survive exposure to such extreme shear force.

Temporal Pressure Gradients

Driving the sensor at a frequency near ν_{res} may cause large amplitude excursions unless the motion is heavily damped. The lowest possible undamped resonance frequency ν_{res} for piston-type acoustic sensors of size ~ L with minimum spring constant k_s ~ 2 kT/L^2 for a maximally sensitive spring having an energy of ~kT when stretched by L, and piston mass m, is

$$n_{res} = \left(\frac{1}{2\pi}\right)\left(\frac{k_S}{m}\right) \tag{4.31}$$

where m = ρ L^2 h for a piston of area L^2, height h, and density ρ. Assuming h/L = 0.1 and ρ ~ 1000 kg/m^3, ν_{res} = 21 MHz for an L = 22 nm sensor (k_s = 1.8 × 10^{-5} N/m) and ν_{res} = 4 KHz for an L = 680 nm sensor (k_s = 1.9 × 10^{-8} N/m). Resonance frequency may be adjusted by selecting appropriate k_s and m. In practical devices resonance will occur at a slightly lower frequency than ν_{res} and with some broadening of the peak, because m should also include the masses of the shank of the piston plus a time-varying portion of the fluid mass occupying the piston's chamber (depending upon the time-varying position of the piston).

Consider the surface of a piston of height h, area L^2 and mass m to which a pressure spike of amplitude P (N/m^2) and frequency ν_{driven} is applied, producing a stroke length X_{stroke}. The solution for the equation of motion of an undamped forced harmonic oscillator has a maximum amplitude of $X_{stroke} = F_{spike}/\text{abs}(m(\omega_{driven}^2\ \omega_{res}^2))$ where ω = 2 π ν, F_{spike} = PL^2 and m is defined above. The driving frequency that produces stroke length X_{stroke} = L is given by:

$$\nu_{driven} = \left(\frac{1}{2\pi}\right)\left[\left(\frac{P}{\rho L h}\right) - 4\pi^2\nu^2_{ITS}\right]^{1/2} \tag{4.32}$$

For an L = 22 nm sensor driven to full throw by a minimum detectable ΔP_{min} = 0.03 atm pressure spike (again taking h/L = 0.1 and ρ ~ 1000 kg/m^3), ν_{driven} = 34 MHz; ν_{driven} = 45 GHz when the sensor is driven to full throw by the maximum possible pressure spike P_{max} = 39,000 atm consistent with water remaining in the liquid state (see below). The piston may be driven at these frequencies, or slower, using triangular-wave pulse trains of pressure spikes of these magnitudes.

For an L = 680 nm sensor driven to full throw by a minimum detectable ΔP_{min} = 10^{-6} atm pressure spike, ν_{driven} = 6 KHz; ν_{driven} = 1.5 GHz when the sensor is driven to full throw by the maximum pressure spike P_{max} = 39,000 atm.

What is the maximum tolerable pressure spike in aqueous media? The requirement of subsonic piston motion in water gives the highest maximum: $P_{max} < (1/2)\,\rho\, v_{sound}^2$ (N/m^2), if ρ = 3510 kg/m^3 for a solid diamondoid piston and the speed of sound v_s ~ 1450 m/sec in water (Eqn. 4.29) for bulk modulus B_{water} = 2.2 × 10^9 N/m^2, ρ_{water} = 1000 kg/m^3, giving P_{max} < 39,000 atm. However, rapid dissipative pressure spikes P_{max} > 26,000 atm add sufficient energy to boil 310 K water, and isothermal static compression of pure water at 310 K causes crystallization into Ice VI near ~11,500 atm. Additional restrictions on (and features of) high-frequency pressure transducers are described later.

Ullage Sensors

It will often be necessary to determine the degree to which a fluid storage tank is full or empty. In most circumstances, an ullage sensor may consist of a simple static pressure sensor embedded in the tank wall, with readout calibrated according to the van der Waals equation for gases and liquids. A fractal network of chemical sensors permeating the internal volume of the tank could provide continuous readouts of local concentration which could be integrated to estimate total storage in the tank. The force required to rotate a paddle inside the tank would increase as a tank was filled, due to the increase in net viscosity of the contents. Transmission time of thermal or acoustic pulses might also give some measure of fluid concentration and physical phase. C. Phoenix suggests storing liquids in a piston, allowing direct and precise molecular volume readout, in designs where this is convenient; in theory, a piston that slides the minimum measurable distance of ~10 pm within the 2-nm-diameter throat of a water-filled cylinder could detect the removal of a single water molecule from the column.

Thermal Nanosensors

The ability of medical nanodevices to measure both absolute temperature and changes in temperature is crucial for monitoring in vivo physiological thermoregulatory mechanisms and intracellular energy transactions. Precision thermal sensing is also important within nanoscale devices to provide corrective input for pressure, chemical, and displacement sensors, and to improve the stability of onboard clocks. General limits of thermal sensing are considered, followed by several proposals for specific implementations.

Minimum Detectable Temperature Change

A sensor system consisting of N atoms has (3N-6) internal coordinates and thus ~(3N-6) oscillators, each with an average energy of ~kT assuming no modes are high enough to be frozen out. However, the standard deviation of the energy of the oscillators around the mean energy is ~kT, and these energy fluctuations are uncorrelated (ignoring coupling between the oscillators), so (3N-6) of them will have an average fluctuation of kT $(3N-6)^{1/2}$. (By comparison, the energy in one 10-micron infrared photon is ~5 kT.) As a result, an instantaneous measurement of the total energy of all N atoms gives a minimum detectable temperature change ΔT_{min} of

$$\Delta T_{min}/T = (3N-6)^{-1/2} \sim (3N_{meas}\, n_d\, V_{sensor})^{-1/2} \qquad (4.33)$$

for a set of N_{meas} independent temperature measurements using a sensor of volume V_{sensor} constructed with material of atomic number density n_d, which might be maximized using diamond ($n_d = 1.76 \times 10^{29}$ carbon atoms/m³).

Thermal equilibration time $t_{EQ} \sim L^2 C_V/K_t$ for a sensor of size L, of heat capacity C_V (1.8×10^6 joules/m³-K for diamond) and of thermal conductivity K_t (2000 watts/m-K for diamond). Thus, $t_{EQ} \sim 10^{-13}$ sec for a sensor of size L = 10 nm, 10^{-11} sec for L = 100 nm, and 10^{-9} sec for L = 1 micron, so thermal sensors up to 1 micron in size should easily achieve thermal equilibration within a typical measurement cycle time $\Delta t_{min} \sim 10^{-9}$ sec. Sensor measurement time $t_{meas} = N_{meas}\, \Delta t_{min}$.

A (57 nm)³ sensor can detect $\Delta T_{min}/T = 10^{-4}$ (~31 millikelvins at 310 K) in a single measurement ($N_{meas} = 1$), with measurement time $t_{meas} \sim 1$ nanosec. A sensitivity of $\Delta T_{min}/T = 10^{-6}$ (~310 microkelvins at 310 K) may be achieved using either a ~1 micron³ sensor with a single measurement ($N_{meas} = 1$, $t_{meas} = 1$ nanosec) or a (124 nm)³ sensor with $N_{meas} = 1000$ independent measurement cycles and $t_{meas} = 1$ microsec. Finally, a 1 micron³ sensor can detect $\Delta T_{min}/T = 3 \times 10^{-9}$ (~1 microkelvin at 310 K) with $N_{meas} = 100{,}000$ independent sensor cycles, giving a measurement time $t_{meas} \sim 100$ microsec.

These figures are confirmed by experimental estimates of detector noise temperature, such as $\Delta T_{min}/T = (k/L^3 C_V)^{1/2} \sim 10^{-6}$ for a silicon nitride detector with $C_V = 5.2 \times 10^6$ joules/m³-K and L ~1 micron. It has been calculated that sensitivity of ~1 microkelvin is theoretically possible using quartz electronic microresonators as precision thermometers. Thermocouple probes with 100 nm tips have already shown ~100 microkelvin sensitivity, tunneling thermometers capable of measuring thermoelectric potential localized to atomic-scale dimensions have been proposed, and experiments leading toward "yoctocalorimetry" were being pursued in 1998.

For comparison, heat sensors in human skin have $\Delta T_{min}/T \sim 3 \times 10^{-4}$ (~90 millikelvins), and the infrared sensor pit of the rattlesnake is sensitive to an energy intensity of ~0.8 pJ/micron^2 in a measurement time t_{meas} ~35 millisec and achieves $\Delta T_{min}/T \sim 3 \times 10^{-6}$; mosquitos register $\Delta T_{min}/T \sim 6 \times 10^{-6}$ at a distance of ~1 cm.

Piston-Based Temperature Sensors

Consider a coiled cylinder of cross-sectional area A filled with gas at pressure P, with a piston at one end whose movement is resisted by a constant-force spring (Fig. VI–D.7). Increasing gas temperature from the coldest temperature at which the sensor will operate (piston at maximum ingress), T_0, to some warmer temperature T_1 causes the piston to move from position x_0 to position x_1 while the spring holds pressure constant at P inside the cylinder during the measurement. If Δx is the smallest measurable piston displacement and $\Delta T = T_1 - T_0$, then the poorest accuracy is

$$\Delta T/\Delta_D = \frac{x_1 - x_0}{x_0} = \frac{\Delta x}{x_0} = N_{meas}^{-1/2} (N+1)^{-1/2} \tag{4.34}$$

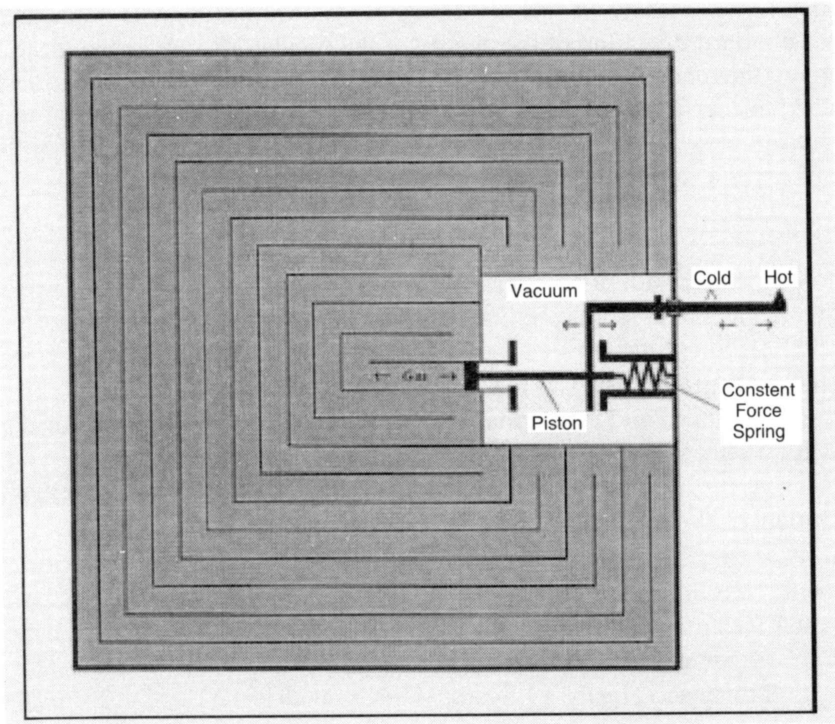

Fig. VI–D.7. Two-dimensional representation of a three-dimensional micro-kelvin piston-based coiled-cylinder temperature sensor.

because Δx has a thermal noise component as well, and the number of gas molecules $N = n_d V_{sensor}$ where $n_d \sim 10^{28}$ gas molecules/m^3 at P = 1000 atm. Sensitivity is maximized at the largest feasible

x_0; for $\Delta T/T_0 = 10^{-6}$ (~300 microkelvins at 310 K) and $\Delta x = \Delta x_{min} = 1$ nm, then $x_0 = 1000$ microns, N = 10^9 gas molecules at 1000 atm (using the van der Waals equation of state) and 310 K taking $N_{meas} = 1000$, giving $V_{sensor} = 10^{-19}$ m^3 and thus a cylinder cross-sectional area of $A = V_{sensor}/x_0 = (10$ nm$)^2$. There are ~1000 gas molecules per nanometer of cylinder length, each traveling with mean thermal velocity $v_t \sim 500$ m/sec. Tightly coiled into a cubical volume, the folded sensor size is $L_{sensor} \sim V_{sensor}^{1/3} \sim (464$ nm$)^3$ and measurement time $t_{meas} = N_{meas}\ \Delta t_{min} \sim 1$ microsec. Sensor mass is ~10^{-16} kg.

The coiled tube design, though not strictly necessary, is nonetheless convenient because it allows a continuous sensor element to be arbitrarily distributed through the nanorobot volume, to be concentrated in multiple specific interior regions, or to be placed inside nanorobot components that will be required to flex such as metamorphic protuberances. If instead the working gas is placed in a cubical box of side L_{box} with a sensor piston of area A_p, any change in temperature causes the piston to move a distance $\Delta x = (L_{box}^3/A_p)\ (\Delta T/T_0) \sim (k\ T_0/k_s)^{1/2}$, where $k_s = (A_p/L_{box}^2)$ $(N\ kT_0/L_{box}^2)$, from Drexler; hence

$$\Delta T/T_0 = (A_p/N\ L^2_{box})^{1/2} \tag{4.35}$$

which argues for a small piston, a large box, and a high operating pressure to achieve maximum sensitivity. For $A_p = 100$ nm^2, N = 10^9 molecules, and $L_{box} = 464$ nm, $\Delta T/T \sim 10^{-6}$.

Thermal-Expansion Temperature Sensors

Consider a thin rod of length x and width w, constructed as a sandwich of two dissimilar materials each of thickness h, having coefficients of linear expansion α_1 and α_2, respectively (Fig. VI–D.8). When heated, the two materials expand differentially to lengths of $x\ (1 + \alpha_i\ \Delta T)$ (neglecting the quadratic and higher-order terms in a_i), producing a cantilever deflection of $\theta_t = (x/h)\ (\alpha_1 - \alpha_2)$ (radians) from simple geometry. The minimum detectable temperature change $\Delta T_{min} = 2\ \pi\ \Delta x_{min/}$ $x\ \theta_t$ or

$$\Delta T_{min} = \frac{8\ \pi\ h^3\ w^2\ \Delta x_{min}}{(\alpha_1 - \alpha_2)\ V^2_{rod}} \tag{4.36}$$

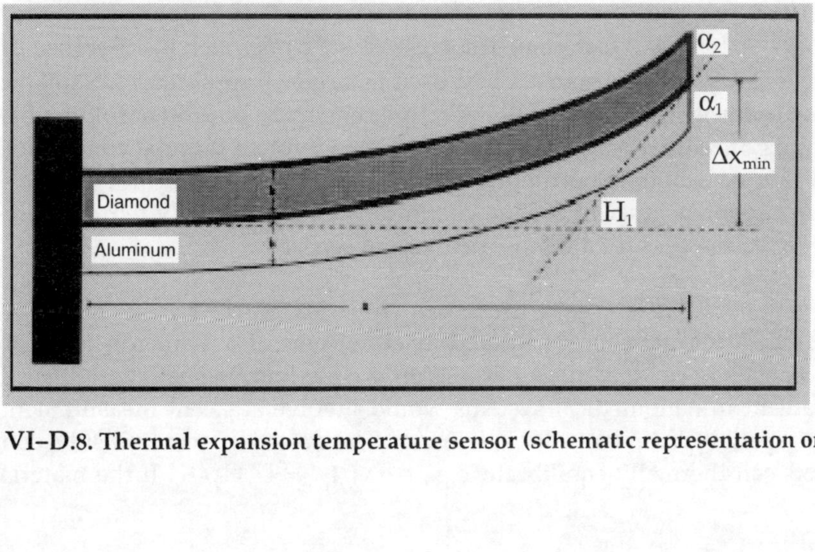

Fig. VI–D.8. Thermal expansion temperature sensor (schematic representation only).

where volume of the sandwich rod V_{rod} = 2 h w x. Taking Δx_{min} = 10 pm, $\alpha_1 = 2.38 \times 10^{-5}$ /K for aluminum, $\alpha_2 = 8 \times 10^{-7}$ /K for diamond at 310 K, h = w = 5 nm, and assuming sensor volume $V_{sensor} \sim 2 V_{rod}$ in a tightly packed three-dimensional coil configuration, ΔT_{min} = 1 microkelvin for an $L^3 \sim (227 \text{ nm})^3$ sensor ~ 1 per cent of the volume of a 1-micron nanorobot. Use of a suitable negative-coefficient material for α_2, such as the ceramics zirconium tungstate and cordierite, can further reduce L by 5-10 per cent.

Measuring the volume of the sensor is essentially like taking an inventory of total stored thermal energy, which has a kT $N^{1/2}$ fluctuation term from adding the N uncorrelated kT fluctuations; hence

$$\Delta T/T_0 \sim N_{meas}\, n_d\, V_{sensor}\,)^{-1/2} \tag{4.37}$$

Hence, to achieve ΔT ~ 1 microkelvin at T_0 = 310 K with this sensor may require N_{meas} ~ 47 million independent measurements of Δt_{min} = 1 nanosec each, or t_{meas} ~ 47 millisec.

Micromechanical silicon cantilever heat sensors (microcalorimeters) of ~20,000 micron³ volume have already achieved ΔT_{min} = 10 microkelvins at room temperature ($\Delta T_{min}/T \sim 3 \times 10^{-8}$), responding to ~1 pJ of heat in a measurement time t_{meas} ~ 1 millisec. Thermal-expansion nanosensors might also exploit the temperature sensitivity of viscoelastic materials (e.g., modulus of relaxation).

Mechanochemical Temperature Sensors

Since binding rates are temperature sensitive, a chemical concentration sensor using steric probes immersed in a sample containing a precisely known concentration of target ligands can be used as a temperature sensor. Similarly, enzyme reaction rates increase with a change in temperature according to

$$\Delta T = \frac{10 \ln (k_2/k_1)}{\ln (Q_{10})} \tag{4.38}$$

where $\Delta T = T_2 - T_1$, k_1 and k_2 are the reaction rate constants at temperatures T_1 and T_2, and Q_{10} is the well-known temperature coefficient—the ratio of the rates of reaction measured at two temperatures 10°C apart. Q_{10} ~ 2 for most enzymes, but may range up to 4 for some enzymes that catalyze reactions with unusually high activation energies. Assuming Q_{10} ~ 4, the $(112 \text{ nm})^3$ sensor with $\Delta c/c$ = 0.01(1 per cent) can distinguish temperatures 0.07 K apart; the huge $(1.44 \text{ micron})^3$ sensor with $\Delta c/c = 10^{-5}$ (0.001 per cent) resolves ΔT ~ 72 microkelvins. Because of their relative volumetric inefficiency, such sensors are best used in large ΔT applications. Similar considerations apply to thermal sensors based on the temperature sensitivity of protein folding (as in heat shock protein mediators of cellular responses to a 5 K change, protein thermal contraction, and thermal conformation rate constants for protein a helices that typically coil in ~10^{-6} sec. The quantum tunneling transmission coefficient for ion channels is also exquisitely temperature-sensitive and thus could serve as the basis of a nanoscale thermal sensor.

Spatial Thermal Gradients

Two (200-400 nm)³ thermal sensors located at either end of a 1 micron nanorobot can detect temperature changes as small as 10^6 K per micron, a ~1 kelvin/m spatial gradient. The maximum detectable gradient in nanomedical systems would involve adjacent measurements differing by ~100°C, a peak spatial gradient of ~10^8 kelvins/m. The detection event must be completed before the two sensors can thermally equilibrate, e.g., $t_{meas} < t_{EQ} = L^2 C_V/K_t$. If the material between the sensors is mostly diamondoid, then $t_{EQ} \sim 10^{-9}$ sec; if mostly water, then $t_{EQ} \sim 10^{-5}$ sec, for L = 1 micron.

Temporal Thermal Gradients

On the timescale of nanomachine operations, typically 10^{-9}-10^{-6} sec, the human body provides an essentially isothermal operating regime. However, medical nanorobots circulating in the bloodstream may encounter ΔT of ~3 K during one circulation time ~60 sec under normal circumstances, a temporal thermal gradient of ~0.1 kelvin/sec, or >10 kelvin/sec at 1 mm tissue depth for a fingertip placed on a 600 K stove. Thermal nanosensors can take independent measurements differing by as much as ~100°C in ~10^{-9} sec, giving a maximum detectable thermal gradient of 10^{11} kelvins/sec, far in excess of physiological rates. Detection of the minimum ~1 microkelvin change between two measurements taken 1 sec apart probably represents the minimum practical temperature gradient, ~1 microkelvin/sec. For comparison, the record-holder in the insect world is the eyeless cave beetle *Speophyes lucidulus,* whose antennae can detect thermal change rates as small as ~3000 microkelvins/sec.

The extraordinary thermal conductivity of diamond ensures that endogenous nanorobot waste heat is rapidly conducted to the surrounding aqueous medium. For example, a 1 micron^3 diamondoid nanorobot generating 10 picowatts of onboard power at its core experiences a temperature differential of ~10^{-8} K between core and perimeter (using thermal conductivity K_t = 2000 watts/m-K for diamond), well below minimum detectable limits.

Electric and Magnetic Sensing

Electric Fields

The electric field E (volts/m) caused by a single point charge q (e.g., an electron or singly-charged ion) is given by Coulomb's law as:

$$E = \frac{q_e}{4 \pi \varepsilon_0 \kappa_e r^2} \quad (4.39)$$

where $q_e = 1.60 \times 10^{-19}$ coul (one charge), $\varepsilon_0 = 8.85 \times 10^{-12}$ farad/m (permittivity constant), κ_e = dielectric constant of the matter traversed by the electric field (κ_e = 74.31 for pure water at 310 K, 5.7 for diamond, 1 for vacuum or air), and r = distance from the charge, in meters. The field at a distance of 1 nm from a point charge is ~10^9 volts/m in air, 2×10^7 volts/m in water. The force between two unit charges $F = q_e E$ = 230 pN in air, 3 pN in water, at 1-nm separation.

What is the magnitude of electric fields likely to be encountered in nanomedical situations? Electric fields in the human body may be generated from internal or external sources. Internally, ions and individual molecules may carry static electric fields. For example, the surface of an isolated charged amino acid has ~1 charge/nm^2. Debye-Huckel shielding due to counterion flow in salty fluids (as are found in the human body) reduces these fields very rapidly with distance. The static field from a single charged amino acid floating in human plasma falls from ~6×10^6 volts/m at a range of 1 nm to 0.7 volts/m at r = 10 nm. In plant cells, photosynthetic "receivers" detect gradients in the electrochemical potential of the order of 1 eV across distances on the order of 1 nm, an instantaneous ~10^9 volts/m.

The most important internal electrical sources are the electrochemical gradients caused by gated channel and transporter molecular pump operations at the intracellular level; muscular, membrane, digestive, and neural activity at the intercellular and organ level; and piezoelectric fields generated by movement of collagenous tissues (e.g., tendon and elastin), bones, and many other biological materials. Typically these sources generate potentials of 10-100 millivolts over distances of 0.01-10 microns, producing local fields in the range of 10^3-10^7 volts/m. For example, the 50 millivolt transmembrane potential necessary to open a sodium channel corresponds to an electric field of 5×10^6 volts/m. Interestingly, implanted electrical devices such as pacemakers may produce up to 600 volts/m one millimeter from the casing.

External sources may also contribute to detectable nanomedical electric fields within the human body, though usually to a lesser degree. Placing a hand flat between two plates charged to $E_p = 3 \times 10^6$ volts/m (near the breakdown voltage for air) induces an instantaneous field of $E \sim E_p/\kappa_{water} = 42{,}000$ volts/m transversely through the limb. (A DC field induces ion current flows which accumulate surface charge and cancel out the internal field after ~one RC time.) Electric fields directly beneath high-tension power lines are 2,000-11,000 volts/m, or 50-1000 volts/m at a lateral displacement of 25 meters along the ground. Household appliances produce 200-300 volt/m fields; 20 amperes flowing through standard Romex copper wiring in the walls of a house produces 100-600 volts/m in the middle of a room, up to 24,000 volts/m at the wall ($E = B/\mu_0 \varepsilon_0 c$; c = the speed of light, B from Eqn. 4.43); static sparks from carpet walking are ~10,000 volts/m, and ~10^5 volts/m makes hair stand on end. Quiescent atmospheric charge is ~100 volts/m, rising to 5000 volts/m during severe thunder-storms oscillating from positive to negative over a ~10,000 volt/m range. Lightning bolts (~20 coul 0.002-sec discharge) produce 60,000 volt/m spikes at a distance of 10 meters from the strike, 600 volts/m at a range of 1 kilometer. Electron-hole avalanches in semiconductors are initiated by fields >~5×10^7 volts/m. Penal electrocutions use ~1500 volts/m.

Direct force sensing of these electric fields is ineffective; from Eqn. 4.39, the minimum detectable field $E_{min} = F_{min}/q = 6 \times 10^7$ volts/m using $F_{min} = 10$ pN and a single-charge sensor. Particle deflection seems more efficient. Consider a collimated stream of singly-charged particles, each of mass m and initial velocity V_0 entering an evacuated test chamber of length L and experiencing a uniform electric field E, causing a lateral displacement $\Delta x = q\,E\,L^2/\,2\,m\,V_0^2$ measured by the sensor. For this measurement, SNR ~ln ($m\,V_0^2/2\,kT$), so the minimum detectable electric field is

$$E_{min} = \frac{4\,kT\,\Delta x_{min}\,e^{SNR}}{qL^2}\ \text{(volts/m)} \tag{4.40}$$

For $\Delta x_{min} = 10$ pm, T = 310 K and SNR = 2, $E_{min} = 10{,}000$ volts/m for a sensor of size L = 28 nm, or 100 volts/m for L = 280 nm, allowing measurement of all nanomedically relevant fields that are homogeneous across the entire device. Using the larger 280-nm sensor, time of flight is ~10^{-12} sec for electrons or ~10^{-9} sec for U^{238} ions, permitting $t_{meas} \sim 10^{-9}$ sec and dissipating at least ~30 pW waste heat (due to kinetic impact) in continuous operation. Time-varying electric fields can be measured up to $t_{meas}^{-1} \sim$ GHz frequencies, and electric current may be determined using standard techniques because current density is scale-invariant. Care must be taken to eliminate measurement error due to exogenous magnetic fields. Piezoelectric or other electromechanical transducers can convert electricity into mechanical signals appropriate for all-mechanical computation and control systems.

Another class of electric field detector of similar size and sensitivity is the electrosensitive gel sensor. For example, the negatively charged heparan sulfate proteoglycan network found in the secretory granule matrix expands 50 per cent in volume within milliseconds of exposure to a -2 volt/micron field, and the response appears linear with voltage. Thus the length of a 400-nm long gel-filled sensor of fixed cross-sectional area should expand a minimally-detectable 10 pm (= Δx_{min}) in a 100 volt/m field. Similar sensitivities to small periodic fields, analogous to shark electroreceptors (maximum sensitivity ~10^{-7} volts/m,), are found in ion-channel embedded artificial membranes that selectively amplify electric signals in noisy environments via stochastic resonance. Stochastic resonance in neurons permits detection of electric fields as weak as 10 volts/m.

The theoretical limits have been analyzed by Weaver and Astumian and by Block, who conclude that the minimum detectable static field for a hollow spherical sensor of radius r_{sen} and shell membrane thickness d_{mem} (capacitance $C_e = 4\,\pi\,\varepsilon_0\,\kappa_e\,r_{sen}^2/d_{mem}$) is

$$E_{minDC} = \left(\frac{2}{3r^2_{sen}}\right)\left(\frac{kT\,d_{mem}}{4\pi\,\varepsilon_0\,\kappa_e}\right)^{1/2} \text{(volts/m)} \tag{4.41}$$

Using κ_e – 74.31 for water at T = 310 K, r_{sen} = 0.5 micron and d_{mem} = 10 nm gives E_{minDC} = 200 volt/m. Increasing r_{sen} to 5 microns reduces E_{min} to ~2 volts/m. To detect a periodic field of frequency ν_e by sampling over ($\nu_e\,t_{meas}$) cycles for a measurement time t_{meas}, Block gives:

$$E_{minAC} = \frac{E_{minDC}}{(\nu_e\,t_{meas})^{1/2}} \text{ (volts/m)} \tag{4.42}$$

Using ν_e = 1 MHz and t_{meas} = 1 sec gives E_{minAC} = 0.2 volt/m. Cylindrical sensors are 10-100 times more sensitive, and sensitivity may possibly be further enhanced by coupling the periodic electric potential to a Michaelis-Menten type enzyme or a similar nanomechanism embedded in the sensor shell to achieve tuning. Fields as small as ~10^{-4} volts/m may be detectable by ~20-micron mammalian living cells; such detection may trigger biochemical observables which may be monitored or eavesdropped by resident nanomedical devices. Submicron-scale electrometers capable of ~0.1 electrons/$Hz^{1/2}$ (charge per unit bandwidth) have been demonstrated experimentally, and also single-electron transistors (SETs) to 10^{-4} electron/$Hz^{1/2}$, with the potential to reach 10^{-6} electron/$Hz^{1/2}$. Galvanotropic bacteria have been reported in the literature.

Magnetic Fields

The most significant source of biomagnetic fields, generated by electric charges moving as currents through the body, is neural impulses. The magnetic field generated by these currents is given by Ampere's law as

$$B = \frac{\mu_0\,\kappa_m\,i}{2\pi\,r} \tag{4.43}$$

where B is magnetic flux density (teslas), μ_0 is the permeability constant (~$4\pi \times 10^{-7}$ henry/m), relative permeability κ_m = 1 for most biological materials and in vacuo, i is electric current (~0.2 microamperes for human nerve impulses), and r = distance from the current-carrying conductor. Hence B ~ 0.04 microtesla for r = 1 micron, adjacent to an axonal conductor. For comparison, magnetic fields directly beneath high-tension power lines are 10-40 microtesla, or 2-8 microtesla at a lateral displacement of 25 meters along the ground. Household appliances produce 1-1000 microtesla at a distance of 3 cm and 0.01-2 microtesla at a 1-meter range (electric shavers produce 0.4-60 microtesla at a range of 15 cm). 20-amp Romex wiring in the walls of a house can produce 2 microtesla in mid-room, up to 80 microtesla at the wall, though in 1998 the average strength in 98 per cent of U.S. homes from artificial sources was 0.05-0.09 microtesla, typical exposures throughout the day were 0.1-1 microtesla, and the federal workplace standard was a maximum of 100 microtesla. Magnetotactic marsh bacteria, birds and insects can detect the geomagnetic field vector (identifying Earth's magnetic poles), which is 30-70 microtesla, depending on geographic position. Behavioural thresholds have been measured as 0.005-0.02 microtesla for pigeons and 0.001-0.01 microtesla for honeybees, and no more than 1000-2000 microteslas for human subjects. Some biochemical changes have been observed in human cells exposed to ~10-100 microtesla 60 Hz fields, although the data are controversial; it appears the fields pump iron atoms through cell membranes, potentially causing peroxidation or other damage to occur. In 1998, transcranial magnetic stimulation was used to expedite recovery of motor skills for stroke patients, and was being explored as a treatment for depression and a means of altering patient mood. Lightning bolts (i ~ 10,000 amp/discharge) produce 200 microtesla spikes at a distance of 10 meters from the strike, 2 microtesla at a range of 1 kilometer.

Computer simulations of a fully heterogeneous human body divided into 37,000 $(1.31 \text{ cm})^3$ voxels produced a maximum induced current of 8.84 amp/m^2 (average ~1 amp/m^2) when the body was exposed to a uniform 60 Hz 1-tesla magnetic field. Patients undergoing Echo-Planar MRI scanning report "tickling" and "pain" when exposed to field changes exceeding 60 tesla/sec; ventricular fibrillation may result from myocardial current densities of 4 amp/m^2, which is induced at 250 tesla/sec.

The simplest magnetosensor is a freely pivoting permanent magnet that experiences a restoring force when the magnet's axis is not aligned with the external field. This "compass effect" is also employed by magnetotactic bacteria, which have ~0.1 micron ferromagnets in magnetosomes in their cells that allow them to detect and track the geomagnetic field vector. Permanent magnets of cross-sectional area A_{magnet} and length l_{magnet} made of Alnico or Alcomax (ρ ~ 8000 kg/m^3) can provide B_{magnet} ~ 1.4 tesla. (In 1998, the record flux density was 13.5 tesla for dipole magnets, ~27 tesla for resistive magnets.) The force on such a magnet immersed in an external field of flux B is $F = B\, B_{magnet}\, A_{magnet}/\mu_0$, so the minimum detectable flux B_{min} is

$$B_{min} = \frac{\mu_0\, \kappa_m\, F_{min}}{B_{magnet}\, A_{magnet}} \tag{4.44}$$

If the minimum detectable force is F_{min} = 10 pN, B_{min} = 50 microtesla for A_{magnet} = 0.4 micron^2, about right for magnetotactic bacteria, and B_{min} = 2 microtesla for A_{magnet} = 2 micron^2.

The limitations of force sensing may be avoided by suspending the bar magnet from its center of mass and allowing it to oscillate in vacuo about its stable equilibrium position in the external field B, making a simple harmonic oscillator with period

$$t_{magnet} = 2\pi\, \frac{\mu_0\, \kappa_m\, I_{mag}}{B\, B_{magnet}\, A_{magnet}\, l_{magnet}} \tag{4.45}$$

where the moment of inertia $I_{mag} = m_{magnet}\, ((A_{magnet}/4) + (l_{magnet}{}^2/12))$. Oscillator energy $E_m \sim B\, B_{magnet}\, A_{magnet}\, l_{magnet}/\mu_0\, \kappa_m$, so the minimum detectable external field is

$$B_{min} = \frac{\mu_0\, \kappa_m\, k\, T\, e^{SNR}}{B_{magnet}\, A_{magnet}\, l_{magnet}} \qquad (4.46)$$

For B_{magnet} = 1.4 tesla, A_{magnet} = (660 nm)^2, l_{magnet} = 660 nm, T = 310 K and SNR = 2, then B_{min} = 0.1 microtesla and $t_{magnet} = 6 \times 10^{-4}$ sec (~t_{meas}), allowing a ~KHz sampling frequency. Magnet mass $m_{magnet} \sim 2 \times 10^{-15}$ kg. Several classes of molecular magnets have been described in the literature.

Another class of magnetosensor detects the change in dimensions when a magnetic body is magnetized, which is called magnetostriction. For instance, iron-cobalt ferrite expands by $\Delta L/L \sim 7.5 \times 10^{-4}$ at saturation magnetostriction in a ~1 tesla field, permitting detection of ~10 millitesla field changes using a 1 micron sensor and $\Delta x_{min} \sim 10$ pm. Small magnetostrictive volume changes make possible a sensor of even greater sensitivity. A field of ~0.2 tesla causes a fractional volume increase of ~10^{-6} in iron and up to 40×10^{-6} in some iron-nickel (~30 per cent Ni) alloys. A (370 nm)^3 magnetostriction sensor monitored by stretch sensors with displacement sensitivity Δx_{min} = 0.1 nm (~1 atomic radius) can in theory detect $B_{min} \sim 0.1$ microtesla. Magnetostriction is nonlinear with applied field, and there are significant hysteresis effects which may confuse data interpretation. However, multisensor arrays combining positive- and negative-magnetostriction materials should enhance signal extraction. Since the change in volume is small, temperature compensation may be required.

Hall effect probe microscopes have demonstrated 10 microtesla sensitivity with 350 nm spatial resolution, and alternative nanoscale magnetic detectors such as direct-current SQUID (superconducting quantum interference device) magnetometers, stray field devices, and magnetic spin sensors are being developed. In 1999, micro SQUID could measure magnetic moments as small as ~10^4 Bohr magnetons or ~10^{-19} J/tesla, roughly 7×10^6 protons or a 14-nm iron cube. Magnetoencephalographs using SQUID magnetometers to measure external electromagnetic fields produced by neural traffic in the brain can register 10^{-6}–10^{-3} microtesla signals. Other biomagnetic measurement techniques have helped to investigate electrophysiological disturbances in heart muscle and in the brain. By 1998, microcoils as small as 100 microns in diameter had been constructed using microcontact printing of 25 micron wide silver wires, producing magnetic flux densities >0.4 tesla with a 10 milliamp current, switchable in <10^{-3} sec.

Optical Sensing

In biosystems and nanosystems alike, the detection of an optical photon is a high-energy event. A single photon of red light at λ = 700 nm carries an energy of 280 zJ, a blue photon at 400 nm conveys 500 zJ, and an ultraviolet (UV) photon at 200 nm transmits 1000 zJ to its receptor. These energies, ranging from 65-234 kT, exceed thermal noise by a wide margin, hence are all easily detectable reliably by suitable nanosensors. Note that optical photon detection does not require a nanosensor the size of one wavelength or larger—for example, 1-nm chlorophyll molecules have no difficulty absorbing 660-nm photons.

Biological systems provide numerous examples of single-photon transducers, including photopigment molecules such as chlorophyll and rhodopsin which mediate electron transfers in <~1 picosecond by undergoing conformational changes, triggering a series of biochemical events. An estimated 100-500 cell membrane channels are affected by the detection of a single optical photon. The first documented case of a physiological role for chlorophyll in animals has been found in the fish *Malacosteus niger*, which uses a chlorophyll derivative to see far-infrared light. Mammalian cells also act as indirect UV sensors as for instance when energy absorption induces perturbations of membrane structure or a conformational change in membrane proteins, activating the JNK cascade below 300 pJ/micron^2, and centrosomes may serve as cellular directional IR sensors in single fibroblast cells. Integration times for photon sensors are briefly treated.

Artificial "bioelectronic" cyanine-quinone chromophore molecules ~1 nm in size can detect optical photons of wavelengths λ = 580-630 nm, and can trigger and reset in 1-3 picosec (governed by the tunneling times associated with reformation of the zwitterionic states of the inputs), in theory permitting ~100 GHz modulations to be received. Exciton-generating poly(p-phenylene vinylene) polymer switches have the potential to achieve ~THz switching speed; a birefringent Kerr medium also switches in a few picoseconds. Chalcogenide glass has viscosity >10^{13} kg/m-sec in darkness, but ~5×10^{11} kg/m-sec under illumination. Proposals for specific implementations of photoergic transducers and optical cables are elsewhere

Particulate and High-Energy Radiation

High energy photons including X-rays (λ ~ 2 nm, ~10^5 zJ or 0.0001 picojoule) and gamma rays (λ ~ 0.002 nm, ~0.1 picojoule), and ejecta from the decay of radioactive nuclei such as α-particles (helium nuclei, ~1 picojoule) and β- particles (free electrons, 0.01-0.1 picojoule) may be extremely destructive to nanomechanical systems and are difficult to reliably measure quantitatively. For example, a charged particle with 0.01 pJ of kinetic energy (K.E.) traversing a decelerative electrostatic field of 10^9 volts/m requires K.E./qE ~ 60 microns to slow to a halt. However, particle-induced visual sensations have been observed by dark-adapted astronauts in space, and Ruderfer has suggested that even solar neutrinos may be detectable by the human brain (though most physicists would find this implausible).

Cellular Bioscanning

The goal of cellular bioscanning is the noninvasive and non-destructive in vivo examination of interior biological structures. One of the most common nanomedical sensor tasks is the scanning of cellular and subcellular structures. Such tasks may include localization and examination of cytoplasmic and nuclear membranes, as well as the identification and diagnostic measurement of cellular contents including organelles and other natural molecular devices, cytoskeletal structures, biochemical composition and the kinetics of the cytoplasm.

The precision and speed of medical nanodevices is so great that they can provide a surfeit of detailed diagnostic information well beyond that which is normally needed in classical medicine for a complete analysis of somatic status. Except in the most subtle cases a malfunction in any one component of the cellular machinery normally provokes a cascade of pathological observables in many other subsystems. Detection of any one of these cascade observables, if sufficiently unique and well-defined, may provide adequate diagnostic information (see) to plan a proper reparative procedure.

DNA is one of the few cellular components that is regularly inspected and repaired. Most homeostatic systems adopt a more simple philosophy of periodic replacement of components regardless of functionality. Nanomedicine allows the philosophy of inspection and repair to be extended to all cellular components. The following discussion briefly describes a few of the sensory techniques that may be useful in achieving this objective.

Cellular Topographics

Tactile topographic scanning provides the most direct means for examining cellular structures in vivo. For example, ex vivo live cell scanning in air by commercially available atomic force microscopy using a 20-40 nm radius tip allows nondestructive feature resolutions of ~50 nm across the top 10 nm of the cell. Investigators have used the scanning tip to punch a hole in the cell, then pull back and scan the breach, observing the membrane heal itself via self-assembly in real time. Up to ~KHz scanning frequencies are possible, with up to 1024 data points per scan line. The nanomechanical manipulator arm achieves comparable tip velocities and positional accuracies.

Assuming a scan rate of ~10^6 pixels/sec, a micron-scale nanodevice, once securely anchored to the surface of a (~20 micron)^3 human cell, could employ a tactile scanning probe to image the 0.1 per cent of plasma membrane lying within its (1.4 micron)^2 vicinity in ~2 sec to ~1 nm^2 resolution (~1 mm/sec tip velocity), or ~50 sec to ~0.2 nm (e.g., atomic) resolution (~0.2 mm/sec tip velocity). Inside the cell, and again post-anchoring, an entire 6 micron^2 mitochondrial surface could be imaged to atomic resolution in ~100 sec; the surface of a 100 nm length of 25 nm diameter microtubule could be atomically resolved in 0.2 sec—though of course in a living cell these structures may be changing dynamically during the scanning process. The continuous power dissipation of a (0.1 micron)^2 scan head moving at 1 mm/sec through water is ~0.002 pW (~kT/pixel at ~10^6 pixels/sec), though of course the energy cost of sensing, recording, and processing each pixel must be at least ~10 kT/pixel (ignoring the possibility of pre-computational image compression), so the total scanner power draw could be as high as 0.02-0.1 pW in continuous operation. Special scanning tips and techniques should allow topographic, roughness, elastic, adhesive, chemical, electrostatic (charge density), conductance, capacitance, magnetic, or thermal surface properties to be measured.

For large (~0.1-1 micron) cellular components, identification and preliminary diagnosis of improper structure may be possible using measured surface characteristics which may be matched to entries in an extensive onboard library, perhaps combined with dynamic monitoring of anomalies in cross-membrane molecular traffic using chemical nanosensors. Most membranes are self-sealing, so it should also be possible to gently insert a telescoping member into the target organelle, which member then slowly reticulates and extrudes smaller probes with sensory tips in a fixed pattern and step size, allowing the acquisition of detailed internal structural and compositional information.

Small (~1-10 nm) protein-based components of the cell such as enzymes, MHC carriers, and ribosomes are self-assembling or require only modest assistance (e.g., molecular chaperones) to self-assemble. Reversible mechanical denaturation of these smaller proteins, using diamondoid probe structure to displace water molecules (thus reducing hydrophobic forces) and by using specialized handling tools akin to functionalized AFM tips and molecular clamps, may be followed by precise nondestructive amino acid sequencing to allow identification and diagnostic compositional analysis. Afterward, the protein molecule may be refolded back into its original minimum-energy conformation, possibly with the assistance of chaperone-like structures in some cases.

Transcellular Acoustic Microscopy

Acoustic microscopy is another noninvasive scanning technique that can provide nanomedically useful spatial resolutions. Frequencies are in the GHz range, far exceeding the relaxation time of the protoplasm. A cryogenic acoustic microscope operated at 8 GHz has demonstrated 20 nm lateral resolution in liquid helium. By 1998, the best resolution achieved with water as the coupling fluid has been 240 nm at a frequency of 4.4 GHz. Operated in water at 310 K, a nanomechanical 1.5 GHz acoustic microscope would achieve a far-field minimum lateral resolution $\lambda/2$ ~ 500 nm, ~10^5 voxels per human cell, sufficient to locate and count all major organelles and a few intermediate-scale structures. Scanning acoustic microscopes (SAM) operated at 2 GHz in reflection mode (which allows detection of interference effects) achieve 30-50 nm resolution in the direction of the acoustical axis, and the Subtraction SAM approach reveals topographical deviations of 7.5 nm at 1 GHz. It has been proposed that picosecond ultrasonics could be used to obtain an image of the cytoskeleton with detail comparable to that of conventional X-ray images of a human skeleton.

Power requirements are a significant constraint on acoustic reflection microscopy (echolocation). In the simplest case, consider an acoustic emitter of radius r_E which converts an input power of P_{in} into acoustic power of intensity $I_E \leq I_{max}$ = 1000 watts/m^2 with efficiency e per cent, conservatively taken here to be 0.50 (50 per cent). The emitter produces a train of omnidirectional pressure pulses of amplitude $A_p = (2 \rho v_{sound} I_E)^{1/2}$ (N/m^2) (Eqn. 4.52) and frequency ν which travels a distance X_{path} to a target of radius r_T. As the signal reaches the target, its amplitude has been reduced by (r_E/X_{path}) due to the $1/r^2$ dependence of intensity in spherical waves and by $e^{-\alpha tiss\ \nu\ Xpath}$ by attenuation (Eqn. 4.51) with $\alpha_{tiss} = 8.3 \times 10^{-6}$ sec/m for soft tissue.

Upon reaching the target, a fraction $f_{reflect}$ of the signal is reflected as a point source echo; if the target surface has an acoustic impedance similar to liver tissue and the medium is similar to water, then from Eqn. 4.53 and Table VI–D.2, $f_{reflect} \sim 0.05$ (5 per cent). The echo then travels a distance X_{path} back to a receiver (which may or may not be located near the emitter), the amplitude again losing (r_T/X_{path}) by geometry and $e^{-\alpha tiss\ \nu\ Xpath}$ by attenuation. The echo is finally detected by the receiver which can measure a pulse of minimum pressure amplitude $\Delta P_{min} \sim 10^{-6}$ atm.

Combining these relations gives the following result:

$$P_{in} \geq \frac{\pi \Delta P^2_{min}\ X^4_{path}}{2\rho v_{sound}\ e\ \text{per cent}\ r^{-2}_{reflect}\ r_T^2\ e^{-2\alpha_{tiss}\ \nu\ X_{path}}}\ \text{(watts)} \tag{4.47}$$

To scan the entire interior of a (20 micron)^3 cell, whether from within or without, requires X_{path} >~ 20 microns. A frequency of ν = 1.5 GHz allows $v_{sound}/\nu \sim 1$ micron spatial resolution. The fastest possible pulse repetition time is $X_{path}/v_{sound} \sim 13$ nanosec. Assuming $r_E = r_T = 0.5$ micron, P_{in} ~ 7 pW or ~5 watts/m^2, well within the safe intensity range for acoustic radiation. At maximum safe transmitter intensity I_{max}, the longest scannable path length at 1.5 GHz is $X_{path} \sim 46$ microns.

Table VI–D.2. Acoustic Impedance for Specular Reflection by Acoustic Waves Crossing a Material Interface

Body Tissue	Impedance (kg/m^2-sec)	Body Tissue	Impedance (kg/m^2-sec)
Air	400	Liver (25°C)	1.65×10^6
Lung	1.80×10^5	Blood	1.65×10^6
Fat	1.39×10^6	Nervse (optic)	1.68×10^6
Aqueous humor	1.51×10^6	Muscle	1.73×10^6
Water	1.52×10^6	Lens of eyes	1.84×10^6
Brain (25°C)	1.57×10^6	Nylon	2.9×10^6
Skin	1.6×10^6	Skull bone	7.80×10^6
Soft Tissue	1.63×10^6	Enamel	1.71×10^7
Kidney	1.63×10^6	Diamond	6.3×10^7

Power constraints are somewhat less severe in the case of acoustic transmission microscopy (acoustic tomography), a technique which requires a minimum of two physically separated components (e.g., a transmitter and a receiver). Consider an acoustic emitter radiating a series of omnidirectional pressure pulses of frequency ν which travel across a tissue cell of width X_{path}, to be detected by a receiver on the other side after a signal transit time $t_{cell} \sim X_{path}/v_{cell}$, where v_{cell} is the speed of sound in the cytosol. In the simplest case, assume that the cytosol is clear except for one target of interest along the path (e.g., an organelle) in which the speed of sound is $v_{target} \# v_{cell}$. The minimum detectable difference in travel times between a signal that passes through the target and one that does not is Δt (~10^{-9} sec for diamondoid systems), so the minimum resolvable target size is $r_{min} \sim v_{target}\, v_{cell}\, \Delta t/2\, \mathrm{abs}(v_{target} - v_{cell})$ when $r_{min} >\sim v_{cell}/\nu$. For soft tissue targets in water at 1.5 GHz, $r_{min} \sim 30$ microns; for tooth enamel in water, $r_{min} \sim 1$ micron. However, the use of sampling gates will permit the detection of phase shifts that are much less than the characteristic time constant of the system in incoming waves, hence the minimum thickness of objects in theory detectable by acoustic tomography can be far smaller than the values for r_{min} estimated above.

Computing acoustic tomographic power requirement P_{in} in a similar manner as for echolocation, except that the transmitted signal rather than the echo is detected and assuming the signals are of approximately equal strength regardless of the path taken, gives:

$$P_{in} \geq \frac{\pi \Delta P^2_{min}\, X^4_{path}}{2\rho v_{cell}\, e \text{ per cent } r_T^2 e^{-2\alpha_{tiss}\, \nu\, X_{path}}} \text{ (watts)} \tag{4.48}$$

Using the same values as the previous example, $P_{in} = 1$ pW for $X_{path} \sim 45$ microns; for $X_{path} \sim 140$ microns, $P_{in} = 1000$ pW, near I_{max}.

In either mode of operation, data gathering is accelerated by positioning more than one receiver around the target. Sensitivity is boosted by increasing emitter size, taking multiple measurements, or by illuminating the target with some large external source located elsewhere in the tissue or even outside of the organ. However, cells and body tissues are mesoscopic "junkyards"—highly heterogeneous media which may produce large numbers of nontarget scattering events, thus increasing the difficulty of extracting signal from noise. Additional complications arise due to:

1. Scattering on rough surfaces.
2. Rapid pressure variations with range in the Fresnel zone or near field of the transmitted signal.
3. Cytoplasmic viscosity inhomogeneities due to the asymmetric arrangement of cytoskeletal structure, granules, vacuoles, and the endomembrane system.
4. Variation in cytoplasmic Young's modulus due to time-varying tensions in semirandomly-distributed cytosolic fibrillar elements (e.g., a relaxed or contracted state) which alter elasticity and thus the local speed of sound.

Magnetic Resonance Cytotomography

Electrostatic scanning is largely ineffective because of Debye-Huckel shielding. Magnetic stray field probes allow resolution of 10 nm physical features, but only for materials with substantial magnetic domains—which most biological substances lack. But nuclear magnetic resonance (NMR) imaging may allow cellular tomography by creating 3-D proton (hydrogen atom) density maps. Atomic density maps of other biologically important elements with nonzero nuclear magnetic moments (including D, Li^6, B^{10}, B^{11}, C^{13}, N^{14}, N^{15}, O^{17}, F^{19}, Na^{23}, Mg^{25}, P^{31}, Cl^{35}, K^{39}, Fe^{57}, Cu^{63} and Cu^{65}) may also be compiled. For example, sodium imaging is already used clinically to assess brain damage in patients with strokes, epilepsy, and tumors.

In a hypothetical NMR cytotomographic nanoinstrument, a large permanent magnet is positioned near the surface of the cell or organelle to be examined. This creates a large static background magnetic field that polarizes the protons. The spatial gradient of this field establishes a unique resonant frequency, called the Larmor frequency, within each isomagnetic surface throughout the test volume. A second time-varying magnetic driver field (e.g., vibrating permanent magnet or weak rf field) is then scanned through the full range of resonant frequencies, exciting into resonance the protons (nuclear spins) in each isomagnetic surface, in turn. Depending on the sensor implementation chosen, each resonance detected may cause absorption of driver field energy, an increase in measured impedance, or even a return echo of magnetic energy (if the driver field is operated in pulse echo mode) as the excited protons relax to equilibrium in ~1 sec. The large polarizing magnet is then rotated to a new orientation, moving the isomagnetic surfaces to new positions within the test volume, and the scan is repeated. A 3D map of the spatial proton distribution may be computed after several scan cycles.

Only those regions within a linewidth of resonance ΔB will generate an appreciable signal, and the expected spatial resolution (say, along the z-axis) $\Delta z = \Delta B/(dB/dz)$, where dB/dz is the spatial flux gradient of B. For a polarizing field B = 1.4 tesla placed adjacent to a (20 micron)^3 cell, the cross-cell gradient dB/dz ~ 7×10^4 tesla/m; placed next to a (1 micron)^3 organelle, dB/dz ~ 1.4×10^6 tesla/m. Assuming an effective NMR linewidth ~ 0.002 tesla, minimum spatial resolutions are 29 nm and 1.4 nm, respectively.

Unfortunately, the smallest reliably detectable energy in the sensor element is ~ kT e^{SNR}. The energy required to flip a single proton is $E_{flip} = 4 \pi \nu_L L_{proton}$, where $L_{proton} = 5.28 \times 10^{35}$ joule-sec (the quantized spin angular momentum of the proton) and the Larmor resonance frequency $\nu_L = g_{proton}$ B; g_{proton} is the gyromagnetic ratio (4.26×10^7 Hz/tesla for protons) and B is the polarizing magnetic flux. However, the population difference between spin-up and spin-down nuclei in NMR is very small. For small energy differences, the Boltzmann distribution only allows a fraction $E_{flip}/2$ kT to be flipped before the upper and lower state populations are made equal and the absorption disappears. In order to flip $N_{flip} \sim kT\, e^{SNR}/4 \pi \nu_L L_{proton}$ protons, the sample volume must include at least:

$$N_{min} = 2e^{SNR} \frac{kT}{4\pi\, \nu_L\, L_{proton}} \quad \text{(protons)} \tag{4.49}$$

For T = 310 K, SNR = 2 and B = 1.4 tesla, ν_L = 60 MHz and $N_{min} \sim 1.7 \times 10^{11}$ protons.

When scanning a cell or organelle, the vast majority of the protons present are in water molecules (Table 3.2). Water at 310 K has a proton density $n_{water} = 6.7 \times 10^{28}$ protons/m^3, compared to $n_{fat} \sim 6.4 \times 10^{28}$ protons/m^3 (palmitic acid). Distinguishing fat from water thus requires a minimum sample volume of $N_{min}/(n_{water} - n_{fat})$ ~55 micron^3.

Proteins range from 4.4×10^{28} protons/m^3 (aspartic acid minus one water) to 7.6×10^{28} protons/m^3 (leucine minus one water), average $n_{protein} \sim 5.6 \times 10^{28}$ protons/m^3; $n_{carbohydrate} \sim 5.8 \times 10^{28}$ protons/m^3 (glucose minus one water), allowing NMR cellular tomography to resolve minimum feature sizes of 2-4 microns (560 micron^3). This resolution may include intracellular structures such as the endoplasmic reticulum (rough ER ~1100 micron^3, smooth ER ~400 micron^3), the Golgi complex (~500 micron^3), the nucleus (~270 micron^3), possibly the nucleolus (~50 micron^3) and even large localized ferritin granule concentrations.

In 1998, the minimum sample size for microcoil (~470 micron diameter) NMR scanning was ~10,000 micron^3 or ~10^{16} protons for a 1-minute measurement cycle. Individual T-cell vacuoles ~1 micron in diameter labeled with dextran-coated iron oxide particles have been imaged by MRI.

Near-Field Optical Nanoimaging

Electromagnetic waves of optical wavelength l that interact with an object are diffracted into two components, called "far-field" and "near-field." The propagation of electromagnetic radiation over distances $z > \lambda$ acts as a spatial filter of finite bandwidth, resulting in the familiar diffraction-limited resolution ~$\lambda/2$. Classical optics is concerned with this far-field regime with low spatial frequencies < 2/l, and conventional optical imaging will be difficult at the cellular level in vivo. (Short-wave X-rays will damage living biological cells.) Information about the high spatial frequency components of the diffracted waves is lost in the far-field regime, so information about sub-wavelength features of the object cannot be retrieved in classical microscopy.

However, for propagation over distances $z << \lambda$, far higher spatial frequencies can be detected because their amplitudes are then of the same order as the sample ($z = 0$). This second diffraction component is the "near-field" evanescent waves with high spatial frequencies $>2/\lambda$. Evanescent waves are confined to subwavelength distances from the object. Thus a localized optical probe, such as a subwavelength aperture in an opaque screen, can be scanned raster fashion in this regime to generate an image with a resolution on the order of the probe size.

The original Near-field Scanning Optical Microscope (NSOM) surpassed the classical diffraction limit by operating an optical probe at close proximity to the object. The NSOM probe uses an aluminum-coated "light funnel" scanned over the sample. Visible light emanates from the narrow end (~20 nm in diameter) of the light funnel and either reflects off the sample or travels through the sample into a detector, producing a visible light image of the surface with ~12 nm resolution at $\lambda = 514.5$ nm provided the distance between light source and sample is very short, about 5 nm, with signal intensities up to 10^{11} photons/sec (~50 nanowatts). This represents a resolution of ~$\lambda/40$. The near-field acoustic equivalent is found in the medical stethoscope, which exhibits a resolution of ~$\lambda/100$.

Applications include dynamical studies at video scanning rates, low-noise high-resolution spectroscopy, and differential absorption measurements. Optical imaging of individual dye molecules has already been demonstrated, with the ability to determine the orientation and depth of each target molecule located within ~30 nm of the scanned surface. Molecules with nanometer-scale packing densities have been resolved to ~0.4-nm diameters using STMs to create photon emission maps (an electric current generates photons in the sample), and laser interferometric NSOMs have produced clear optical images of dispersed oil drops on mica to ~1 nm resolution. Live specimen 250 nm optical sectioning for three-dimensional dynamic imaging has also been demonstrated.

Submicron laser emitters have been available since the late 1980s. It should be possible to use NSOM-like nanoprobes to optically scan the surfaces of cells or organelles to <~1 nm resolution, mapping their topography and spectroscopic characteristics to depths of tens of nanometers without penetrating the surface. However, a proper membrane-sealing invasive light funnel ~20 nm in diameter might not seriously disrupt some cellular or organelle membranes and thus could be inserted into the interiors of these bodies or through the cytoskeletal interstices to permit deeper volumetric scanning. The thermal conductivity of water at 310 K is 0.623 watts/m-K and the energy per 500 nm photon is 400 zJ, so for 1 micron^3 of watery tissue the maximum scan rate is ~35 micron^3/sec-K, if e^{SNR} photons are used to image each 1 nm^3 voxel with SNR = 2. Thus a 1-sec volumetric optical scan of an aqueous ~1-micron^3 sample volume to 1 nm^3 resolution requires a ~1 nanowatt scanner running at ~GHz bit rates, raising sample volume temperature by ~0.03 K.

NSOM permits the determination of five of the six degrees of freedom for each molecule, lacking only the optically inactive rotation around the dipole axis. In principle, it should be possible to produce <~1 nm resolution near-field optical scans of in situ protein molecules, since with atomically precise fabrication and single lines of atoms as conductors a minimum light guide (metal-dielectric-metal) is 3 atoms wide. Given a molecular laser and adequate collimation, photons passing through folded proteins will scatter according to the molecular structure. Detection of sufficient photons comprising these scattering patterns should allow the noninvasive determination of protein structure; polarized photons provide information on chirality. Absorption and fluorescent signals will be visible from phenyl rings, tryptophan, and bound cofactors such as ATP and adenine (which is fluorescent). Positions of monoclonal antibodies on virus surfaces are now identifiable experimentally using NSOM; it should be possible to map binding sites on virus and cell surfaces using fluorescently labelled antibodies. Single molecule detection has been proposed as a tool for rapid base-sequencing of DNA.

Other optically-based cellular imaging techniques must be distinguished. Optical Coherence Tomography (OCT) uses Michelson interferometry to achieve ~10 micron spatial resolutions over tissue depths of 2-3 mm in nontransparent tissue in the near infrared. However, OCT requires numerous physical components not easily implemented on a micron-size detector (e.g., beam splitter, lens-grating pair, galvanometer mirror, optical prism), femtosecond pulse shaping, and illumination power levels of ~10^7 watts/m^2 >> 100 watts/m^2 "safe" continuous limit in tissue. Bioluminescence techniques in which the light source is placed inside the tissue (e.g., a transgenic mouse with a luciferase gene in every cell of its body) has a spatial resolution limited to ~10 per cent of the depth, or ~10 microns resolution at a ~100 micron depth. Coherent anti-Stokes Raman scattering (CARS) already permits organelle imaging in living cells and can in theory create a point-by-point chemical map of a cell using two intersecting lasers.

Three-dimensional observation of microscopic biological nonliving structures by means of X-ray holography requires a high degree of spatial coherence and good contrast between target and surroundings. Good contrast may be achieved in the wavelength range between the K absorption edges of carbon (l = 4.37 nm) and oxygen (l =2.33 nm), the "water window" where carbon-containing biological objects absorb radiation efficiently but water is relatively transparent. A 50-micron diameter emitter has been tested that uses near-IR 5-femtosecond laser pulses impinging upon a helium gas target to create a well-collimated (<1 milliradian) beam of coherent soft X-rays at a 1 KHz repetition rate producing a brightness of 5×10^8 photons/mm^2-milliradian^2-sec, a peak X-ray intensity of >10^{10} watts/m^2 on the propagation axis behind the He target.

Cell Volume Sensing

Many cellular parameters need not be measured directly in order to be detected by a medical nanodevice. Cell volume sensing is a case in point. An intracellular nanodevice can indirectly monitor changes in the volume of the cell in which it resides by one of two methods. First, measurements of the mechanical deformation of the cellular membrane, stretch-activated channels, or cytoskeletal strains and structural changes are quite sensitive to alterations in total cell volume. Second, concentration or dilution of the cytoplasmic environment through cell shrinkage or swelling leads to the activation of various volume-regulation responses which may be detected by the nanorobot. Changes in the concentration of soluble cytosolic proteins may nonspecifically affect enzyme activity via "macro- molecular crowding". Minor changes in cell volume can cause severalfold changes in ion transport. Cellular signalling entities that have been linked to the transduction and amplification of the primary volume signal include Ca^{++} transients, phosphoinositide turnover, eicosanoid metabolism, kinase/phosphatase systems such as JNK and p38, cAMP, and G-proteins. These signal amplification pathways may be monitored using chemical concentration sensors aboard the medical nanodevice, allowing the nanorobot to eavesdrop on the natural sensory channel traffic of the cell.

Noninvasive Neuroelectric Monitoring

In many ways the neuron is the most nanomedically important class of cell in the human body. Nanomedical applications regarding neurons and the brain are addressed in. The following is a brief summary of noninvasive (i.e., no axonal membrane penetration) nanotechnological methods for monitoring the electrical traffic of individual neurons. Noninvasive measurement of axonal traffic within nerve bundles will require multiple sensors and greater sensitivity to compensate for shielding by the perineurium, a tight resistive sheath enclosing the bundle ~20 microns thick with resistivity ~4000 ohm-cm.

Electric Field Neurosensing

The "typical" ~20 micron human neuron discharges 5-100 times per second, moving from -60 mV potential to +30 mV potential in ~10^{-3} sec. Thus the variation in electric field at the axonal surface is ± 4500 volts/m. Since electric field sensors can detect 100 volt/m fields up to GHz frequencies, an electric sensor attached by circumaxonal cuff or pressed against the axonal surface (possibly at the node of Ranvier) should readily detect each action potential discharge. (The diameter of human nerve axons is 0.1-20 microns.) By 1998 silicon-to-neuron extracellular junctions already permitted direct stimulation of individual nerve cells in vitro without killing the cells, and extracellular electrodes were commonly used to detect neuronal electrical activity noninvasively, both in vivo and in vitro. Artificial electric fields may also be employed to trigger or moderate neural signals. Cell membrane capacitance is typically ~0.01 picofarads/micron^2, and varies with the state of the health of the cell.

Magnetic Field Neurosensing

The magnetic flux density caused by a single action potential discharge is ~0.1 microtesla at the axonal surface, which may be detected by a "compass oscillator" type magnetosensor with a ~KHz maximum sampling rate.

It might be possible for artificial magnetic fields to directly influence neural transmissions. Even a static 65 millitesla field has been shown to reduce frog skin Na^+ transport by 10-30 per cent. Each neuronal discharge develops an electrical energy of ~20 picojoule (~10^{10} kT), far smaller than the magnetic energy stored in a B = 1.4 tesla field of a permanent micromagnet traversing an $L^3 = (20 \text{ micron})^3$ volume which is $B^2 L^3/2 \mu_0 \sim 6000$ pJ. If properly manipulated, such a field may be sufficient to enhance, modulate, or extinguish a passing neural signal.

Neurothermal Sensing

While neurons come in many shapes and sizes our "exemplar" ~14,000-micron^3 neuron discharging ~90 mV into an input impedance of ~500 Kohms produces ~0.2 microampere current per pulse and generates a continuous (average) 100-300 pW of waste heat as measured experimentally. The discharge rate of 5-100 Hz can produce brief surges up to ~2000 pW during a high-frequency train, but the duty cycle of such trains is far less than 100 per cent, reducing time-averaged dissipation to the observed 100-300 pW range.

Single impulses are measured experimentally to produce 2-7 microkelvin temperature spikes in cold or room-temperature mammalian non-myelinated nerve fibers and ~23 microkelvins in non-myelinated garfish olfactory nerve fibers at an energy density ranging from 270-1670 joules/m^3-impulse from 0-20°C. In non-myelinated fibers the initial heat occurs in two temperature-dependent phases: a burst of positive heat, followed by rapid heat reabsorption (called the negative heat). The positive heat derives from the dissipation of free energy stored in the membrane capacity, and from the decrease in entropy of the membrane dielectric with depolarization. An L ~ 20-micron neuron in good thermal contact with an aqueous heat sink at 310 K has thermal conductance $L K_t \sim 10^{-5}$ watts/K, so trains of 5-100 Hz impulses lasting 1 second should raise cellular temperature by 10-30 microkelvins; up to ~200 microkelvin thermal spikes from such trains have been observed experimentally. These events are easily detectable by nanoscale thermal sensors capable of ~1 microkelvin sensitivity up to ~1 KHz. The ~microkelvin heat signature of individual impulses or very short pulse trains can probably be temporally resolved because the minimum pulse repetition time is ~10 millisec (at 100 Hz) which is much longer than the thermal time constant for an L ~ 20-micron neuron (thermal conductivity K_t = 0.528 watts/m-K and heat capacity $C_V = 3.86 \times 10^6$ joules/m^3-K for brain tissue; which is $t_{EQ} \sim L^2 C_V/K_t \sim 3$ millisec.

Direct Synaptic Monitoring

The synaptic cleft between the axonal presynaptic terminal and the dendritic postsynaptic membrane is 10-20 nm in most synapses, although in the vertebrate myoneural junction it may be as large as 100 nm. Contact area per bouton is ~1 micron^2, giving a total gap volume of ~10^7-10^8 nm^3. The density of acetylcholine receptors is highest in muscles along the crests and upper thirds of the junctional folds (~10,000/micron^2), and is lowest in the extrasynaptic regions (~5/micron^2). Each action potential discharge triggers the release of ~10^4-10^5 molecules of acetylcholine into the gap volume of an active neuromuscular junction (diffusion time ~1 microsec), raising c_{ligand} from near zero to ~3×10^{-4} molecules/nm^3 (~0.0005 M) in ~1 millisec, followed by near-complete hydrolyzation by acetylcholinesterase during the 1-2 millisec refractory period. Into the gap volume may easily be inserted a ~10^5 nm^3 neurotransmitter concentration sensor able to measure ~100 acetylcholine molecules in ~1 millisec (Eqn. 4.5), thus detecting pulses at the fastest discharge rate. A similar device could be used to precisely regulate neuro-transmitter concentration at the junction, and hence the neural signal itself, under nanodevice control. Simple electrochemical and mechanochemical artificial synapses have been demonstrated.

Other Neurosensing Techniques

Many other noninvasive neurosensing techniques are readily conceivable. For instance, counting rotors or sodium magnetic resonance imaging could detect changes in the ionic composition (e.g., Na^+, K^+) of the periaxonal fluid before, during, and after discharge.

Cellular RF and Microwave Oscillations

Starting in 1968, H. Frohlich, observing that millivolt electrical potentials maintained across cell membranes ~10 nm thick give rise to huge fields ~10^7 volts/m possibly producing an electret state, theorized that membrane molecules must be highly electrically polarized and thus could interact to produce coherent surface acoustic vibrational modes in the 10-100 GHz (microwave) frequency range; the longest wavelength is about twice the membrane thickness. Interestingly, this frequency span is very close to the maximum trigger/reset frequency for bioelectronic molecules). Note that (~100 mV) (1.6×10^{-19} coul) ~ 4 kT, so a membrane molecule with a single charge on either end should be reliably reoriented by a depolarization wave, coupling pressure waves and electrostatic field fluctuations. However, the direct detection of 10-100 GHz millimeter radiation by non-nanotechnological means is experimentally difficult and controversial because the tests must be performed in vivo in close proximity to an actively metabolizing cell in water—and water strongly absorbs microwaves over macroscale ranges (e.g., ~99 per cent absorption in 3 mm at 100 GHz).

Nevertheless, active cells have shown enhanced Raman anti-Stokes scattering, an effect ascribed to the converse of the Frohlich oscillations. In one study, the normalized growth rate of yeast cultures was enhanced or inhibited when irradiated by CW microwave fields of ~30 watts/m^2 of various frequencies; growth rate data spanning 62 separate runs revealed a repeatable frequency-dependent spectral fine structure with six distinct peaks of width ~10 MHz near 42 GHz. Investigations of related phenomena are ongoing and voluminous; the interested reader should peruse *Bioelectromagnetics*, the archival journal of this field.

100 GHz waves attenuate only ~1 per cent after passing through ~3 microns of soft tissue. A single electron injected into an integral membrane protein could act as an oscillating dipole, making a 300 volt/m signal 10 nm from the protein antenna with an energy transfer of ~0.004 kT per cycle; ~1000 oscillating electrons could produce a measurable field. A 20-micron diameter cell modeled as a nonuniform spherical dipole layer with transmembrane dipoles located 10 nm apart and embedded in a dissipative medium could produce 10^2-10^5 volt/m microwave fields 1-10 microns from the cell surface.

Thus, a variety of rf and microwave electromagnetic emanations may in theory be detectable both within and nearby living cells which could prove diagnostic of numerous internal states. Such states may include cytoskeletal dynamics, metabolic rates, plasmon-type excitations due to the collective motion of ions freed in chemical reactions, positional, rotational or conformational changes in biological macromolecules and membranes, internal movements of organelles and nerve traffic conduction, cellular pinocytosis, cellular reproduction events, cell membrane identity (e.g., distinguishing erythrocyte, Gram-positive and Gram-negative cell coat conductivities at 10 KHz), and cell-cell interactions.

Macrosensing

Macrosensing is the detection of global somatic states (inside the human body) and extrasomatic states (sensory data originating outside the human body). While the treatment here is necessarily incomplete, the discussion nevertheless gives a good feel for the kinds of environmental variables that internally-situated nanodevices could sense. Not all capabilities outlined here need be available on every nanorobot, since injection of a cocktail of numerous distinct but mutually cooperative machine species allows designers to take full advantage of the benefits of functional specialization. In many cases, a given environmental variable can be measured by several different classes of sensor device. However, since these devices are microscopic it is in theory possible to operationalize almost all of the macrosensing capabilities described below in one patient using just a billion devices (~10 mm^{-3} whole-body deployment density), a total volume of ~1 mm^3 of nanorobots or ~0.1 per cent of the typical ~1 cm^3 therapeutic dose.

Acoustic Macrosensing

Cyto-Auscultation

Can sounds generated by a single cell be detected, and thus be useful for diagnosis? Probably not, given that low frequency acoustic radiators are notoriously inefficient. For example, mitochondrion organelles of the giant amoeba *Reticulomyxa* are shuttled back and forth by 1-4 cytoplasmic dynein motors while riding on the outside of a bundle of 1-6 microtubules. Each dynein motor generates 2.6 pN of force and drives the mitochondria at up to ~10 micron/sec, developing 0.3-1.0 × 10^{-16} watts of mechanical power within each 320-nm diameter organelle. Taking each organelle as a cylindrical acoustic radiator with a mechanical input power of P_{in} ~ 10^{-16} watts at ν ~ 1000 Hz, the output acoustic pressure at the organelle surface is only ~10^{-9} atm an acoustic power intensity I ~ 10^{-15} watts/m^2 (Eqn. 4.52) which is not detectable by micron-sized nanorobots. Nevertheless, cells and intracellular elements are capable of vibrating in a dynamic manner with complex harmonics that can be altered by growth factors and by the process of carcinogenesis, so the possibility cannot be completely ruled out.

Blood Pressure and Pulse Detection

Blood pressure ranges from 0.1-0.2 atm in the arteries to as low as 0.005 atm in the veins. The systolic/diastolic differential ranges from 0.05-0.07 atm in the aorta and 0.01-0.02 atm in the pulmonary artery, falling to 0.001-0.003 atm in the microvessels, or 0.003-0.005 atm if the precapillary sphincter is dilated. In venous vessels, pulse fluctuations are 0.002-0.010 atm in the superior vena cava, 0.004-0.006 atm in the subclavian vein, ~0.004 atm in venules generally, and ~0.0005 atm in the brachial vein. There is also a ~0.05 Hz random fluctuation in the microvessels with amplitude on the order of 0.004-0.007 atm. Both blood pressure and pulse rate can be reliably monitored by a medical nanodevice virtually anywhere in the vascular system using a (68 nm)^3 pressure sensor with ~0.001 atm sensitivity.

Pulse propagation through body tissue is somewhat muted due to absorption in compressible fatty membranes, but most cells lie within 1-3 cell-widths of a capillary so the cardiac acoustic signal should still be measurable using more sensitive detectors. The time-averaged interstitial pressure in subcutaneous tissue is 0.001-0.004 atm.

Arterial pulse waves (vascular oscillations) carry subtle messages about the health of internal organs and the arterial tree. The idea of using pulse waves for diagnosis has a long history dating back 2000 years in China. For example, in the *Book on Pulse Waves* by Wang Shu-He (201-285 AD), waves detected by manual probing are classified using such subjective and qualitative descriptors as floating, deep, hidden, rapid, slow, moderate, feeble, replete, full, thready, faint, weak, soft, slippery, hesitant, hollow, firm, long, short, swift, running, intermittent, uneven, taut, string-tight, gigantic, or tremulous. Abnormal waves were empirically related to disease states. Wave data gathered by nanodevices could make possible a theoretically sound, quantitative system of noninvasive observation, classification, and diagnosis as a supplement to other nanomedical tools.

Respiratory Audition

The variation of mechanical pressure over a complete respiratory cycle is ~0.003 atm in the pleura, ~0.002 atm at the alveoli, detectable by nanomedical pressure sensors positioned in the vicinity of the respiratory organs. Holding a deep breath further stretches the pulmonary elastic tissue, up to 0.02 atm.

However, turbulent flows at Reynolds number N_R > 2300 in the trachea, main bronchus and lobar bronchus produce a whooshing noise that may be the loudest noncardiac sound in the human torso during conventional auscultation. The energy dissipation for turbulent flow in a tube is:

$$P_{turb} = P_{lam} Z = 8 \pi \eta_{air} v^2 LZ \text{ (watts)} \quad (4.50)$$

where P_{lam} is the dissipation for laminar (Poiseuillean) flow in a long circular cylindrical tube of length L, v is the mean flow velocity, and turbulence factor $Z = 0.005 (N_R^{3/4} - (2300)^{3/4})$, a well-known empirical formula. For $\eta_{air} = 1.83 \times 10^{-5}$ kg/m-sec for room-temperature air (20°C), P_{turb} = 0.87 milliwatts for the trachea (L = 0.12 m, v = 3.93 m/sec and N_R = 4350 at 1 liter/sec volume flow); P_{turb} = 0.66 milliwatts for the main bronchus (L = 0.167 m, v = 4.27 m/sec and N_R = 3210); and P_{turb} = 0.09 milliwatts for the lobar bronchus (L = 0.186 m, v = 4.62 m/sec and N_R = 2390), totalling ~1.6 milliwatts acoustic emission from a ~120 cm^3 upper tracheobroncheal volume. This is a power density of 13 watts/m^3 corresponding to a pressure of 4×10^{-5} atm assuming a 300 millisecond measurement window at the maximum respiration rate.

The amplitude of an acoustic plane wave propagating through tissue attenuates exponentially with distance due to absorption, scattering and reflection. The amplitude is given approximately by

$$A_x = A_0 e^{-\alpha F x} \quad (4.51)$$

where A_0 is the initial wave amplitude in atm, A_x is the amplitude a distance x from the source, and α is the amplitude absorption coefficient. The function F expresses the frequency dependence of the attenuation. For pure liquids, $F = F_{liq} = \nu^2$ (Hz^2); for example, $\alpha_{liq} = 2.5 \times 10^{-14}$ sec^2/m for water at room temperature. However, for soft tissues, $F = F_{tiss} \sim \nu$ (Hz). Values for α_{tiss} (in sec/m) are in Table VI–D.3; the value for diamond was estimated from the acoustic line discussion in. Assuming <10 KHz bronchial turbulence noise of initial amplitude $A_0 = 4 \times 10^{-5}$ atm, $A_x \sim 3.7 \times 10^{5}$ atm through 1 meter of typical soft tissue; $A_x \sim 3.4 \times 10^{-5}$ atm even if 0.1 meter of bone is interposed in the acoustic path. Either A_x is reliably detected from anywhere in the body using a >(210 nm)^3 pressure sensor (Eqn 4.28).

Table VI–D.3. Amplitude Absorption Coefficients for Acoustic Waves in Human Body Tissue

Body Tissue	Coefficient α (sec/m)	Body Tissue	Coefficient α (sec/m)
For α_{kg} (sec^2/m):		Fat	7.0×10^{-6}
Water	2.5×10^{-14}	Sett tissue (avg.)	8.3×10^{-6}
Castor oil	1.2×10^{-11}	Liver	1.0×10^{-5}
Air (SIP)	1.4×10^{-10}	Nerves	1.0×10^{-5}
For α_{tiss} (sec/m):		Brain (adult)	1.1×10^{-5}
Diamond (est.)	-2×10^{-15}	Kidney	1.2×10^{-5}
Aqueous humor	1.1×10^{-6}	Muscle	2.3×10^{-5}
Vitreous humor	1.2×10^{-6}	Eye lens	2.6×10^{-5}
Blood	2.1×10^{-6}	Polythene (plastic)	5.8×10^{-5}
Brain (intant)	3.4×10^{-6}	Bone	1.6×10^{-4}
Abdomen	5.9×10^{-6}	Lung	4.7×10^{-4}

Mechanical Body Noises

Many other mechanical body noises should be globally audible to properly instrumented medical nanodevices. If normal chewing motions (of hard foods) release 1-10 milliwatts in a ~100 cm^3 oral volume with a ~1 sec jawstroke, power density is ~10 watts/m^3 or ~10^{-4} atm of tooth-crunching noise. A stomach growl registering 45 dB (vs. 30 dB whisper, 60 dB normal conversation) at 2 meters has a source power of 160 milliwatts; released from a 10 cm^3 gastric sphincter volume gives a ~2×10^{-6} atm acoustic wave, detectable throughout the body. Walking and running releases 20-100 joules/footfall for a 70 kg man; assuming the energy is absorbed within a ~1 cm thickness or within ~1 second by the sole of the foot, Eqn. 4.52 implies an upward-moving planar compression wave of 0.4-2.0 atm, easily detectable by acoustically instrumented nanodevices body-wide. (Shoe insoles dissipate energy and alter the shock wave pulse shape.) Hand-clapping generates 0.02-0.2 atm pulses, also easily detectable.

Lesser noises including ~30 millisec hiccups at (4-60)/min, intestinal and ureteral peristalsis, sloshing of liquid stomach contents, heart murmurs, a tap on the shoulder by a friend, nasal sniffling and swallowing, clicks from picking or drumming fingernails, crepitations, manustuprations and ejaculations, the rustling noise of clothing against the skin, flapping eyelids, anal towelling, bruits (including murmurs and thrills) due to vascular lesions, dermal impact of water while showering, copulatory noises, urethral flow turbulence during urination, transmitted vibrations from musical instruments, creaking joints, and squeaking muscles can be detected locally if not globally. Implantation of significant interconnected in vivo diamondoid structures may produce increased sensitivity to internal noises, due to the extremely low acoustic absorption coefficient of diamond (Table VI–D.2).

Vocalizations

Average source power for conversational speech in air is ~10 microwatts at the vocal cords (60 dB), up to ~1000 microwatts for shouting (90 dB) and as little as 0.1 microwatts (30 dB) for whispering. Vocal cord surface area ~1 cm^2, giving an acoustic intensity I ~ 0.001-10 watts/m^2. (Using the decibel notation, dB = 10 $\log_{10}$ (I/I_0), where $I_0 \sim 5 \times 10^{-13}$ watts/m^2 in air, $I_0 \sim 1 \times 10^{-16}$ watts/m^2 in water.) In a planar traveling wave, pressure amplitude A_p (N/m^2) is related to power intensity I by

$$A_p = (2\pi \, v_{sound} \, I)^{1/2} \ (N/m^2) \tag{4.52}$$

For water at 310 K, ρ = 993.4 kg/m^3 and v_{sound} = 1500 m/sec, therefore A_p = 0.0005-0.05 atm for speech, detectable by nanodevices throughout the body due to minimal attenuation at audible frequencies. Other easily detectable vocalizations include whistling, humming, coughing, sneezing, rales, wheezing, expectorating, eructations, flatus, vomiting, hawking and noseblowing.

In the case of spherical waves diverging from an omnidirectional transmitter of radius r at a distance X from the transmitter, average intensity declines inversely as the square of the distance and so I in Eqn. 4.52 must be replaced by $(I\, r^2/X^2)$.

Environmental Sources

Can in vivo nanodevices directly detect sounds emanating from the environment outside of the body, such as other people talking in the same room or a door slamming? The waves from an external acoustic source of power P_R watts travel through the air and, upon arriving at the air/skin interface a distance x_R from the source with amplitude $A_{incident}$, are transmitted through the interface with amplitude $A_{transmit}$. For a specular reflector—interface dimensions (human body ~ 2

m) > acoustic wavelength (~0.03-3.0 m for typical audible sounds in air)—with acoustic impedance Z_1 and Z_2 on either side of the interface and perpendicular incidence,

$$A_{transmit} = A_{incident}\left[1 - abs\left(\frac{Z_1 - Z_2}{Z_1 + Z_2}\right)\right] \tag{4.53}$$

Acoustic impedance, like the speed of sound, is essentially frequency-independent over the nanomedically-relevant range of ultrasonic frequencies. For Z_{air} = 400 kg/m²-sec and assuming Z_{skin} ~ 1.6 × 10⁶ kg/m²-sec from Table 4.3, then $A_{transmit} = (5 \times 10^{-4})\ A_{incident}$. In other words, there is ~99.95 per cent reflection from the air-skin interface, which is why coupling mediums like gels and oils are commonly employed in ultrasound imaging. If an immediately subdermal nanodevice can detect a minimum $A_{transmit}$ ~ 10^{-6} atm, then from Eqn. 4.52 and simple geometry:

$$P_R = \frac{2\pi x_R^2 A^2_{incident}}{\rho\ v_{sound}} \tag{4.54}$$

For STP (1 atm, 0°C) air, ρ = 1.29 kg/m³ and v_{sound} = 331 m/sec. If the minimum detectable pressure ~10^{-6} atm ~$A_{transmit}$, then at a distance of x_R = 2 meters the acoustic source must have a power of ~2000 watts, far exceeding the ~1 milliwatt output of a person loudly shouting. To hear normal conversation at x_R = 2 m, minimum nanodevice detector sensitivity falls to 7 × 10^{-11} atm requiring a subdermal pressure nanosensor ~(17 micron)³ in size (Eqn. 4.28), roughly the dimensions of a single human cell; other methods may prove more efficient.

Of course, an ex vivo acoustic nanosensor may receive sound that has passed through no interface, hence may detect pressure waves ~3 orders of magnitude lower in amplitude. Assuming $A_{transmit} = A_{incident}$, P_R ~ 600 microwatts, so ex vivo nanorobots with a 0.3 micron³ sensor could hear people shouting at x_R = 2 meters. To hear talking (~10 microwatt source) requires a 2.4 micron³ ex vivo pressure sensor (limit ~10^{-7} atm), from Eqn. 4.28.

Optimally positioned and calibrated nanomedical pressure sensors could directly measure changes in the ambient barometric pressure to within ± 10^{-6} atm. Normal atmospheric variation due to weather ranges from 0.94-1.05 atm; such slow moving changes are readily monitored. Very near the Earth's surface, the air pressure P at altitude h above sea level is approximated by $P = e^{-k_p h}$ (atm), where k_p = 1.16 × 10^{-4} m⁻¹ at 20°C; at sea level, a 10^{-6} atm change in pressure reflects a change in altitude of only ~1 cm. However, the opening or closing of a door inside a (~5 m)³ room that displaces >125 cm³ of air also causes a minimally detectable >10^{-6} atm pressure pulse. Other sources of environmental pressure variation such as infrasonic (~0.2 Hz) microbaroms from offshore ocean storms, wind entering through open windows, forced-air currents from central heating or A/C systems, or even the movements of nearby people and pets may be detectable and thus, may further confuse the measurement, reducing absolute accuracy unless suitable corrections are made.

Proprioceptive Macrosensing

Kinesthetic Macrosensing

Using a navigational transponder network, a population of ~10^{11} nanodevices each spaced an average ~100 microns apart throughout the body tissues can determine relative location to a positional accuracy of ~3 microns and an angular accuracy of ~2 milliradian, with the data updatable once every millisecond. Thus, dispersed, the network can continuously monitor and record the relative positions of all limbs with worst-case (cumulative error) ~0.8 mm accuracy over a 2-meter span. These devices can prepare high-resolution dynamic maps of body position, velocity, acceleration and rotation, allowing precise real-time digital kinesthesia during sports or artistic activities such as gymnastics, pole vaulting, or ballet; during transportative activities such as driving cars around hairpin turns, roller-coaster rides, military aircraft maneuvers and space launches; during precision tool-using such as needle-threading, antique watch repair, or while using the fingers as measurement calipers; during self-defense activities requiring complex motions such as karate or judo; and during emergency situations such as automobile crashes and tumbling motions during falls from great heights. All such sensory data is readily outmessaged to the patient or user in real time.

Nanorobots could assist in path integration or the reconstruction of experienced limb trajectories. Given the ability to determine rotation rates and to measure applied forces and torques, the network should also be able to forecast the anticipated future positions of limbs. By comparing these projections to actual results, the network can then infer the viscosity of the medium in which the activity is taking place (e.g., air or water), whether the environment is stable or is translating or rotating in some direction, and whether the human user is physically supported or in free-fall. Internal nanodevices can directly measure if a body is sitting, standing, laying prone, inverted, falling, floating or diving, and then communicate that information directly to the human user. Detection of patient activity states—e.g., the patient is sitting, standing, or walking—can be used to control nanorobot behaviours, activate or deactivate outmessaging displays, and so forth. A simple macroscale wearable tactile "compass belt" that would convey desired geographical directions has already been proposed.

Orientational Macrosensing

Since nanopendular sensors can reliably determine the direction of the local gravity field vector to within ~2 milliradians in $\sim 10^{-4}$ sec, the navigational network can poll its members and arrive at an accurate consensus on which direction is up. Nanodevices affixed to relatively stable hard body parts will exhibit more consistent orientational readings. Measurement of the gravity vector allows the vertical orientation of the rest of the body to be precisely fixed in space, especially useful for gymnasts, trapeze artists, underwater divers in murky lakes, and vestibular-impaired individuals, to whom this information may be outmessaged directly.

The gravitational vector may also be indirectly measured, albeit more slowly and less accurately, by monitoring the body's natural reactions to changes in the axis of gravitational loading. For instance, the variation in head-to-toe hydrostatic pressure for a 1.7 m tall, 70 kg human standing in a 1-g gravity field is ~0.17 atm; the ability to measure a systematic cross-body differential of 10^{-6} atm allows the detection of a $\sim 10^{-5}$ g change along the lengthwise aspect, or $\sim 10^{-4}$ g change along the transverse aspect. Thus a continuously updated whole-body barostatic map allows the human body to serve as a three-dimensional gravity/orientation sensor.

Body Weight Measurement

A network of nanodevices can inventory the volume and density of each of the body's $\sim 10^8$ (1 mm)^3 voxels using a combination of acoustic ranging, somatic mapping, and flowmetry. Each voxel is identified as fat, muscle tissue, bone mass, interstitial fluid, and so forth. These measurements allow the body's weight to be precisely computed as volume times density, rather than the usual method of determining weight using gravitational force data. The existence of a natural physiological "ponderostat," a crudely analogous humoral body-mass detector, has been proposed; indeed, the insulin-leptin system has been found to serve a related purpose, and is easily eavesdropped by medical nanorobots.

Gravitational Geographic Macrosensing

Medical nanodevices can measure variations in the gravity field to $\sim 10^{-6}$ g's for L = 20 micron gravimeters in a measurement time t_{meas} = 2-9 millisec. This implies that in vivo nanodevices can take precise measurements of their latitude and altitude relative to sea level ~100 times every second. Gravity increases toward the poles and at lower altitudes. Specifically, using the formula of Cassinis (accounting for rotational and polar flattening effects on the Earth) with the Bouguer correction to the free air variation by altitude (assuming flat topography), measured gravity g_{meas} is given approximately by

$$g_{meas} = g_0 \left[1+k_1 \sin^2(\theta_L) - k_2 \sin^2(2\theta_L)\right] - k_3 h + k_4 h \rho_{earth} \tag{4.55}$$

where θ_L = terrestrial latitude (equator = 0°), h = height above sea level in meters, g_0 = 9.78039 m/sec^2 (equatorial sea-level value ofg), $k_1 = 5.2884 \times 10^{-3}$, $k_2 = 5.9 \times 10^{-6}$, $k_3 = 3.086 \times 10^{-6}$ sec^{-2}, $k_4 = 4.185 \times 10^{-7}$, and ρ_{earth} = 5522 kg/m^3.

Since sea-level g varies from 9.78039 m/sec^2 at the equator to 9.83217 m/sec^2 at the north pole, a 20-micron gravimeter ($\Delta g = 10^{-6}$ g) detects a change in position of 1 arcmin of latitude or ~1900 meters north/south along the Earth's surface. Similarly, since at 45° latitude g varies from 9.806 m/sec^2 at sea level to 9.803 m/sec^2 at 1000 meters altitude, a 20-micron gravimeter detects a change in altitude of ~3.3 meters (e.g., upstairs vs. downstairs in a house). For comparison, in 1998 high-quality commercial gravity gradiometers measured gradients of $\sim 10^{-9}$ g/meter and allowed the compilation of micro-g (~1 milligal) resolution aerial gravity maps; atom interferometers measured the gravitational acceleration of atoms to a precision of 10^{-10}.

To achieve such phenomenal positional accuracies, the nanodevice must be able to computationally resolve several complicating factors. First, localized mass concentrations representing nonuniformities in crustal density produce residuals of up to ±0.0006 m/sec^2, which may be removed from the data using a standard map of known terrestrial isostatic variations and anomalies. Indeed, matching observations to such a map could provide useful longitudinal information as well. Another complication is the variation in gravity due to tidal forces amounting to $\sim 3 \times 10^{-7}$ g's, twice daily, which lies at the limits of detectability for a 20-micron gravimeter. Other minor geodesic and terrain-related corrections, too complicated for discussion here, may also need to be applied in certain circumstances. Note that the presence of nearby heavy objects does not influence measurement accuracy: a 100-ton building 10 meters away adds a lateral acceleration of only 7×10^{-9} g's to a human body.

One final complication is that patient movements create kinematic accelerations that must be distinguished from the gravitational accelerations. Gravity readings can be corrected by taking derivatives of the signals from kinesthetic monitoring to give gravity in the reference frame of the patient's room, and many individual measurements may be averaged to improve accuracy since the gravity vector normally changes only very slowly.

Electric/Magnetic Macrosensing

Vascular-Interstitial Closed Electric Circuits

In vivo studies of the electric properties of blood vessels shows that the walls of veins and arteries present a specific electric resistance ~200 times greater than the conductive medium of blood (plasma)—roughly 200 ohm-m vs. 0.7 ohm-m. Thus blood vessels are properly regarded as relatively insulated conducting cables which can electrically connect an injured tissue with surrounding noninjured tissue. Capillaries form an electric junction with the interstitial fluid, so the electric gradient can be canceled by ionic transport. Cell membranes are insulating dielectrics with resistance and capacitance, penetrated by ionic channels or gates for ionic transport. Additionally, in 1941 Szent-Gyorgyi suggested the possibility of semi-conduction in proteins, a theory which has since been elaborated by many investigators.

B. Nordenstrom has proposed that the above schema represents a system for selective electrogenous mass transport of material between blood and tissue which he calls Vascular-Interstitial Closed (electric) Circuits (VICCs)—in effect, an additional electro-circulatory system operating in parallel with the well-known diffusive, osmotic, and hydrodynamic mechanisms of regular bloodstream materials transport. As long as ions can leak through open pores and migrate through ion channels, long distance transport cannot take place and the VICC system remains primarily a local (e.g., within tissues), not global, electrical circuit. Connections also exist with conductive media including cerebrospinal fluid, bile, and urine.

Nanorobots capable of monitoring the status of local VICCs may rapidly and efficiently acquire a wealth of systemic information without the need for direct inspection of the affected tissues. For example, when a working muscle produces catabolic products such as lactic acid, an electrochemical potential gradient develops between the muscle and surrounding tissue in a known manner. Injured tissues become polarized in relation to surrounding tissue, as initial catabolic degradation products acidify the tissue. Malignant neoplasms, benign neoplasms, and internally necrotic granulomas all may polarize electrically in relation to surrounding tissue; a vascular thrombus also contains ionized material. Large deflections in the diffusion potential of blood occur if blood is deoxygenated or is infected with Gram-negative bacteria. Blood changes its electric potential from +500 mV to +1000 mV during spontaneous coagulation, and the spontaneous electric potential at the site of a crush injury of the iliac crest in rats oscillated several times between +190 mV and -60 mV over a four day experimental trial.

Nordenstrom proposes that in vivo electrophoresis via the VICC system may also play a role in mediating leukotaxis. When tissue is artificially polarized by electrodes, accumulation of granulocytes with margination and development of pseudopods is the result. In one experiment, exposure of a mesenterial membrane to a 1 microampere 1 volt field for 30 minutes stimulated diapedetic bleeding near the anode. At higher power, the arterio-capillaries contracted, narrowed, and emptied of blood cells, while the veno-capillaries and venules widened and filled with granulocytes. Monitoring such actions of local VICCs experiencing natural voltage fluctuations could alert medical nanorobots to the presence of local injuries, tissue changes, or pathologies that otherwise might go unnoticed for a considerable time.

Electric/Magnetic Geographic Macrosensing

Detection of electric fields to ~100 volts/m, the normal vertical atmospheric gradient, in theory might allow determination of aboveground altitude to ~1 m accuracy if the time between recalibrations is very brief. However, a human body (which is a good conductor), standing on level ground, acquires a slight negative surface charge and becomes "part of the ground" electrically. This distorts upward the equipotentials that usually run parallel to the ground, thoroughly corrupting an altitude measurement. Nearby lightning hits and storms are detectable, which is useful.

However, medical nanodevices with appropriate magnetometers can measure variations in magnetic field to ~0.1 microtesla using a $(660 \text{ nm})^3$ permanent magnet sensor, in a measurement time t_{meas} ~ 0.6 millisec. Geomagnetic field maps of the Earth's surface are highly nonisotropic in both latitudinal and longitudinal directions; knowledge of the field plus the absolute direction of true north (e.g., using a nanogyroscope) gives longitude information from the separation of the planetary magnetic and spin poles. This implies that properly equipped in vivo nanodevices can establish their latitude and longitude on Earth's surface ~1000 times every second by this means. In particular, the horizontal component of the geomagnetic field vector ranges from 0-41 microtesla from magnetic pole to magnetic equator; the independent vertical field component ranges from 0-70 microtesla. Thus a 0.1-microtesla sensor should resolve each component to an accuracy of ~11 arcmin or ~20 km in both latitude or longitude. Terrestrial surface magnetic anomalies (e.g., iron deposits) produce local variations of 0.03-30 microtesla and are thus detectable. Additional complications to these measurements include artificial magnetic field sources, the 0.01-0.1 microtesla solar daily variation, occasional magnetic storms producing erratic geomagnetic fluctuations of 0.01-5 microtesla, sudden-commencement ionospheric electrojets up to 0.3 microtesla, and various long-term secular variations in the geomagnetic field.

Radio, television, and direct broadcast satellite signals probably are not detectable by individual nanodevices, but such detection may be indirectly achieved in vivo using nanorobot-accessible dedicated macroscale antennas implanted in the body.

Piezoelectric Stress Macrosensing

The piezoelectric effect is "the production of electrical polarization in a material by the application of mechanical stress". Many materials in the human body are piezoelectric, including tendon and elastin, dentin and bone. This polarization, or surface charge, varies as the physical stress imposed on the material changes over time. Thus, measurement of the piezoelectric effect in bone or tendon can provide information on the physical loads that are being carried by these materials. For example, a shear stress applied along the long axis of a bone alters the polarization voltage that appears on a surface at right angles to the axis. In another experiment, piezoelectric surface charges on a loaded human femur were measured to range from -131 to +207 picocoulombs/cm^2 (-8 to +13 charges/micron^2) from one end of the bone to the other end, depending upon position along the shaft. Real-time monitoring of this data would permit specific inferences as to the amount of load the bone was carrying and from what direction, including bending, shearing, and twisting forces, from which whole-body activity states could subsequently be inferred. Knowledge of these surface charge variations might also be exploited in osteographic or functional navigation.

Optical Macrosensing

In vivo medical nanodevices can gather little useful optical information from the external environment. Here's why.

In biological soft tissues, scattering dominates absorption except in the pigmented layers of the epidermis and stratum corneum. Thus the propagation of light in tissues may be regarded as occurring in two steps.

In the first step, optical photons of intensity I_0 falling perpendicularly on skin are transmitted according to Beer's law through tissue to a depth z with a transmitted intensity of

$$I_x = I_0 (1 - r_{sp}) e^{-\sigma_t z} \text{ (watts/m}^2\text{)} \qquad (4.56)$$

where r_{sp} = specular reflection coefficient for visible light (Fresnel reflection at the air-tissue surface) ~ 4-7 per cent, or 0 per cent if the original light source lies within the body; and the transmission coefficient $\sigma_t = \sigma_a$ (absorption coefficient ~ 300 m^{-1}) + σ_s (scattering coefficient ~30,000 m^{-1}) for various human soft tissues at optical wavelengths. (Coefficient values for the most heavily pigmented skin layers may be 5-7 times higher.) Typical values for $\sigma_t \sim$ 10,000-100,000 m^{-1}, average ~30,000 m^{-1} for soft tissue, although exceptionally clear tissues with σ_t = 1000 m^{-1} have been reported. Thus the mean free path of an optical photon in human soft tissue is 10-100 microns, average ~ 30 microns (~1.5 tissue cell-widths), up to an extreme maximum of ~1 mm for the most transparent tissues known. For $\sigma_t \sim$ 30,000 m^{-1}, at z = 150 microns, $I_z/I_0 \sim$ 0.01 and ~99 per cent of all photons have been scattered at least once from their initial path.

In the second step, in an unbounded medium the patch of fully scattered photons continues to propagate through the tissue via diffusion until all photons are absorbed. The complicated governing diffusion equation has not yet been been completely solved analytically, but the asymptotic diffusive fluence is given roughly by:

$$I_d = I_i\, e^{-\sigma_d x}\ (\text{watts/m}^2) \tag{4.57}$$

where σ_d is the diffusion exponent or effective attenuation coefficient (averaging ~900 m^{-1} for typical soft tissues but with an extremely wide range reported over the ultraviolet, visible, and near-infrared wavelengths, from 10-1,000,000 m^{-1}.) As a crude approximation, the initial intensity of the fully scattered photon patch $I_i \sim I_0 (1 - r_{sp})\, a_{scat}$, where $a_{scat} = \sigma_s/(\sigma_s + \sigma_a) \sim 0.987$ is the albedo for single particle scattering.

Over the optical band from 400-700 nm, incident intensity I_0 = 100-400 watts/m^2 when standing in direct sunlight; artificial lighting in homes and offices is typically 0.1-10 watts/m^2; moonlight provides only $I_0 \sim 10^{-4}$ watts/m^2; and the absolute threshold for human vision is ~10^{-8} watts/m^2. For $\sigma_d \sim$ 900 m^{-1}, the intensity ratio of transmitted/incident visible light falls to I_d/I_0 = 0.1 at z = 2.5 mm depth (~eyelids), ~10^{-4} at z = 1 cm depth, and ~10^{-11} at z = 2.8 cm—a depth at which the tissue would be completely dark to the human eye even under direct sunlight illumination at the outermost skin surface.

An optical nanosensor with N_{sensor} receiver elements, with each receiver element having an area A_e = 1 nm^2 and capable of single-photon detection, requires an integration time for reliable detection of e^{SNR} photons given by

$$t_{meas} = \frac{h\,\nu\, e^{SNR}}{I_d\, N_{sensor}\, A_e}\ (\text{sec}) \tag{4.58}$$

where h = 6.63 × 10^{-34} joule-sec and ν = 4.3-7.5 × 10^{14} Hz for optical photons. Let us assume that a patient is standing in $I_0 \sim$ 400 watt/m^2 direct unfiltered sunlight, and that t_{meas} = 1 sec, SNR = 2, N_{sensor} = 25,000 elements giving a nanorobot eyespot of $A_e N_{sensor}$ = 0.025 micron^2, and that σ_d = 900 m^{-1}. Then from Eqns. 4.57 and 4.58 the maximum tissue depth for optical photon detection is z_{max} ~ 17 mm ($I_d \sim 10^{-4}$ watt/m^2) which includes a volume of tissue comprising up to ~30 per cent of total body volume. Changes in illumination at the minimum indoor artificial level of ~0.1 watts/m^2 are visible to z_{max} ~ 7 mm depth. This eyespot, if exposed to air on the outermost surface of the skin, is just sensitive enough to detect full moonlight.

Our general conclusion is that variations in normal indoor lighting may be directly measurable by nanorobots stationed within the outermost ~1 cm of body tissues. Imaging, as opposed to the simple illumination detection system described here, is a much more difficult design challenge.

Neural Macrosensing

The ability to detect individual neural cell electrical discharges noninvasively in many different ways, coupled with the abilities (A) to recognize and identify specific desired target nerve cells and (B) to pool data gathered independently by spatially separated nanodevices in real time, offers the possibility of indirect neural macrosensing of complex environmental stimuli by eavesdropping on the body's own regular sensory signal traffic.

For example, facultatively mobile nanodevices may swim into the spiral artery of the ear and down through its bifurcations to reach the cochlear canal, then position themselves as neural monitors in the vicinity of the spiral nerve fibers and the nerves entering the epithelium of the organ of Corti (cochlear or auditory nerves) within the spiral ganglion. These monitors can detect, record, or rebroadcast to other nanodevices in the communications network all auditory neural traffic perceived by the human ear. Advanced speech recognition systems may permit recovery of spoken words and identification of individual speakers (including vocalizations by the user), recognition of background noises, reception and validation of cues or commands spoken directly to in vivo nanosystems by authorized medical personnel, and so forth. Since properly configured monitors can also modulate or stimulate nerve impulses, these devices may add audible signals to the audio traffic, thus may be employed as hearing aids (using feedback loops), real-time language translation mechanisms, continuous vocalization/audition recorders, voice-stress analyzers, or nanodevice-user communications links.

Nanomonitors positioned at the afferent nerve endings emanating from hair cells located in the otolithic membrane of the utricle and saccule and in the cristae ampullaris of the semicircular canals allow medical nanodevices to directly record, amplify, attenuate, or modulate the body's own sensations of gravity, rotation, and acceleration, although kinesthetic sensory management might also be required for complete control. Motor neurons likewise can be monitored to keep track of limb motions and positions, or specific muscle activities, and even to exert control. Neuron-resident nanorobots may detect auditory effects induced in the brain by pulsed microwaves from external sources. Feline cochlear neurons sensitive to audible frequencies of >300 Hz respond to single microwave pulses at a threshold-specific absorption rate of 6-11 watts/kg-pulse, while in humans the threshold for effect, which depends on energy per pulse, may be as low as ~0.02 joule/m^2-pulse for people with low hearing threshold.

Olfactory and gustatory sensory neural traffic similarly may be eavesdropped by nanosensory instruments. Nerve taps in the medulla oblongata or at the phrenic nerve that drives the diaphragmatic muscles allow direct monitoring of respiratory activity. Pain signals may be recorded or modified as required, as can mechanical and temperature nerve impulses from other receptors located in the skin. Even psychological variables such as emotionality, vigilance, and mental workload may be directly monitored by measuring ANS activity in sympathetic efferent fibers outside of the brain.

The most complex and difficult challenge in neural macrosensing will be optic nerve taps. The retina is thoroughly vascularized, permitting ready access to both photoreceptor (rod, cone, bipolar and ganglion) and integrator (horizontal, amacrine, and centrifugal bipolar) neurons. However, the optic nerve bundle itself has ~10^6 tightly bunched individual nerve fibers, a 10-100 MHz signal bandwidth, and significant natural data compression techniques which all must be untangled in real time. Developing algorithms capable of interpreting raw optical nerve traffic, say, to recognize a specific human face or a specific scene in the vision field, would prove a significant research challenge. (Direct monitoring of photo-receptors or retinal membrane potentials and other techniques may simplify untangling of the signal compression.) Rapid visual field identification using artificial neural nets is also a subject of much current research interest. Eyeball rotations (e.g., via resident intradevice nanogyroscopes, eyelid position, pupil aperture, and lens accommodation under ciliary muscle control must be monitored to supplement the vision field analysis.

Other Macrosensing

We have only scratched the surface of the potential for nanorobot macrosensing. Medical nanodevices can quantitatively monitor variables not normally accessible to human consciousness, such as:

1. Hormone and neurotransmitter levels.
2. Gastroelectric oscillations and skin conductivity.
3. Pupil dilations.
4. Drug and alcohol (and breakdown product) concentrations in the blood.
5. Internal organ damage or malfunction.
6. Digital real-time performance data on vital organs or limbs (e.g., continuous kidney volume throughput, pancreatic insulin output, cholesterol metabolism in the liver, or lactic acid production in specific muscles during exercise).
7. Continuous mapping of thermoregulatory isotherms from which the temperature and heat capacity of the external medium (e.g., swimming pool water, cool night air) and the presence of sunlight or shade on the skin (via dermal thermal differentials) may be inferred.

Various chemical substances cyclically increase and decrease in serum concentration with time of day, permitting crude temporal macrosensing. Other substances (e.g., aldosterone; Appendix B) significantly vary in serum concentration depending upon whether the patient is supine or standing, thus permitting limited biochemical postural macrosensing.

If nanorobots can leave and re-enter the body, then macrosensing may include direct sampling of the external environment upon demand and the possibilities for remote data acquisition become virtually limitless.

Unit VI–E

Six Opportunities in Nano-Enabled Drug Delivery Systems

Nanomedicine is widely expected to become a multi-billion dollar business over the next decade. Although some of the applications for nanomedicine are quite futuristic, others are on the verge of generating significant revenues. One of these is nano-enabled drug delivery systems which promise to increase the bioavailability of drugs and reduce the likelihood of toxicity. Researches pinpoint several areas where nano-engineering of drug delivery systems is leading to greater control over active compounds, resulting in the physician's ability to send the smallest dose possible in order to achieve the desired result in the patient.

This chapter reviews the main opportunities that are emerging for nano-enabled drug delivery systems. In most cases, these are based on delivering drugs in nanoparticulate form. Most drugs perform better as nanoparticles, because they can be targeted better and there are fewer side effects. In addition, by using smaller amounts of drugs, treatment costs may be reduced. For example, according to researches, patients show a strong preference for nanoparticle inhalers as an alternative to the widely used injectable methods, which may permit far lower doses of expensive protein-based drugs like insulin.

Opportunities

It is expected that the dosing benefits of nano-enabled drug delivery systems will be extended to compounds used in treating both infectious disease and cancer, and has identified six types of drug delivery systems in which nanotechnology is likely to have an impact.

Injectable Delivery Systems

Although injectable delivery systems have some inherent discomfort for the patient and other types of delivery may be preferred by both the patient and the physician, nanotechnology may be applied to produce improvements and additional revenue streams for the drug company. A case in point is Johnson & Johnson, which, at the beginning of 2005, announced that its Élan NanoCrystal technology would be used in a Phase III clinical trial for an injectable formulation of paliperidone palmitate, an anti-schizophrenia drug. This is a new "nano formulation" of an older drug which overcomes the original's insolubility, by reducing the particle size to under 200 nm. This new formulation enables the product life cycle to be extended by a significant amount.

Implantable Delivery Dystems

Implantable drug delivery systems are often preferable to the use of injectable drugs, because the latter frequently display first-order kinetics (e.g. initial concentration is elevated, which then drops exponentially over time). This may cause difficulties with toxicity, if and when the peak concentration remains above the therapeutic range. Conversely, drug efficacy often diminishes as the drug concentration falls below the targeted range. By contrast, implantable "time release" systems may help minimize peak plasma levels and reduce the risk of adverse reactions; allow for more predictable and extended duration of action; reduce the frequency of re-dosing; and improve patient acceptance and compliance.

To these advantages nanotechnology can add the benefits listed at the beginning of this White Paper. Implantable drug delivery systems are expected to be used for the delivery of proteins, hormones, pain medications, and other pharmaceutical compounds. An example of how nanotechnology is already making its impact felt in this part of the drug delivery business is provided by pSivida's BioSilicon product. BioSilicon is a novel nanostructured material that effectively stores an active compound in nanosized pockets that controllably release minute amounts of drug as the silicon dissolves. pSivida is currently exploring biodegradable implantable methods for tissue engineering and ophthalmic delivery.

Nano-implants will also be used in the not-too-distant future for treating cancer. Among the first nanoscale devices to show promise in anti-cancer therapeutics and drug delivery are structures called "nanoshells," which it is believed may afford a degree of control never before seen in implantable drug delivery products. Nanoshells typically have a silicon core that is sealed in an outer metallic core. By manipulating the ratio of wall to core, the shells can be precisely tuned to scatter or absorb very specific wavelengths of light. For example, Gold encased nanoshells have been used to convert light into heat, enabling the destruction of tumors by selective binding to malignant cells. A physician can use infrared rays to pass harmlessly through soft tissue, while initiating a lethal application of heat when the nanoshells are excited. Some researchers are experimenting with temperature-sensitive drug delivery control methods, using nanoshells that release their payload only when illuminated with the proper infrared wavelength.

Oral Delivery Systems

The vast majority of consumers prefer an oral drug delivery system to implantables or injectables. With this in mind, various development groups are working to enhance traditional oral delivery systems with nanoengineered improvements. There are some areas where nano-enhanced drugs could make a big difference in increasing oral bioavailability and reducing undesirable side effects.

By increasing bioavailability, nanoparticles can increase the yield in drug development and more importantly may help treat previously untreatable conditions. Because of the blood brain barrier (BBB) many new chemical entities aimed at treating brain disorders have proved not to be clinically useful. By contrast, nanoparticles have been demonstrated to cross the BBB with little difficulty. Companies such as Germany's NanoPharm have developed systems capable of reaching the brain for anesthesia (Dalargin, an analgesic), cancer drugs, and various therapeutics. The company claims several advantages over existing systems, including:

(1) no requirement to open the BBB,

(2) the ability to use potentially any drug, whether hydrophilic or hydrophobic, and
(3) the drug does not need to be modified.

Meanwhile, researchers at the University of Texas at Austin have described a means of using nanospheres for oral drug delivery. These nanosphere carriers are derived from hydrogels, which are highly stable organic compounds that swell when their environment becomes more acidic. They have been successfully formulated into controlled-release tablets and capsules, which release active compounds when the hydrogel body swells.

Topical Delivery Systems

NanoMarkets' research indicates that nanomaterials provide a unique opportunity for rapid topical delivery of active compounds. Given their very small size, nanoparticles are able to enter human tissues and cells quickly. Companies such as Novavax have already developed regulated topical systems that take advantage of the unique properties of micellar nanoparticles. It has two hormone replacement therapies dubbed Estrasorb (which received FDA approval in October 2003) and Androsorb (which successfully completed Phase I human trials in 2003).

Transdermal Systems

The number of FDA-approved polymers available for use on-skin is increasing rapidly. With this in mind, the industry will be presented with opportunities to create new transdermal platform designs that improve on-skin properties and diffusion of active molecules from patches.

This interesting trend is expected to result in smaller and less invasive patches that increase the universe of available drug candidates. In some cases, electronics are even being integrated into patch-like platforms involving wound care, monitoring, and diagnostic methods. As such, patch-like platforms will play an important role in healthcare to treat, measure, diagnose, and generally improve quality of life.

Toxin Removal

Colloidal dispersions have been demonstrated to remove potentially lethal compounds from the bloodstream, including high concentrations of lipophilic therapeutics, illegal drugs, and chemical and biological agents. A team of scientists at the University of Florida and Clarkson University in Potsdam, N.Y., has demonstrated favorable results to this end, using biocompatible microemulsions. These oil-in-water systems have a rapid and efficient absorption capacity for many target molecules that are frequently overdosed, whether this be intentional or accidental. The microemulsions use a polymeric surfactant, in combination with an ionic co-surfactant.

Other oil compositions have been made with varying absorption characteristics for different toxins, in all cases avoiding hemolysis and unwanted blood coagulation. The mechanism by which nanoparticles trap toxins involves aromatic or cyclodextrin receptors with electron deficiencies, which form complexes with selected toxins, sometimes referred to as "benzenoid moieties.

Some Conclusions

The analysis above establishes that the impact of nanotechnology on the drug delivery business is likely to be very broad in three senses:

- Nanotechnological improvements promise to impact almost every common mode of drug delivery oral, implantable, injectable, etc.

- Such improvements will themselves cover a lot of ground, including lower drug toxicity, lowering cost of treatments, improved bioavailability and extension of the economic life of proprietary drugs.
- Nano enabled drug delivery will provide improvements in the treatment of some of the most widespread diseases, such as cancer and diabetes.

This is an impressive list. Also impressive is the fact that many of the categories of nano-enabled drug delivery systems are already close to or at the point of marketing. It is believed that not only will the nano-enabled drug delivery market be one of the first true nanomedicine markets to evolve, but as it does so, the revenues from nano-enabled drug delivery systems will be quite large.

Unit VI–F

Nano-Enabled Drug Discovery: Seven Technologies to Watch

The pharmaceutical industry faces increasingly challenging market conditions that are leading to an intensified search for better drug discovery technologies. Nanotechnology can help with that search. This chapter reviews seven nanotechnologies that are believed will make serious contributions to this search.

Introduction

The pharmaceutical industry has to discover and develop innovative medicines for a wide range of diseases in a marketplace that is (1) likely to experience growing regulatory and pricing pressures and (2) that is increasingly targeted towards diseases that have been traditionally resistant to pharmacotherapy (e.g., solid tumors). This chapter profiles a number of tools that are emerging from nanotechnology that are believed will be important aids in the process of drug discovery in this environment.

Today, nearly all pharmaceutical companies follow common technology processes for discovering drugs. These include cloning and expressing human receptors and enzymes in formats that allow high-throughput, automated screening and the application of combinatorial chemistry. Thus, random screening can now be achieved with libraries sufficiently large and diverse to have a relatively high probability of finding a novel molecule. These libraries are possible because they can be generated by the techniques of combinatorial chemistry (combichem). Backing this are important developments in the application of genetics and genomics to understand associations between diseases and gene products. Importantly, bioinformatics are beginning to identify putative targets for a number of diseases.

An important consequence of today's approach is that more lead molecules are being discovered for diverse targets, giving the medicinal chemist more scope to find a candidate molecule. Yet the probability of success of launching a candidate molecule into the market remains basically unchanged at around 8 percent, and the perception is that the time to market is not reducing as might be expected with greater automation. Numerous problem areas remain:

- Current methodologies for compound synthesis.
- Storage and retrieval are inefficient and inadequate.
- Better information is needed to improve target selection and validation.
- More predictive screens are essential.
- More emphasis must be placed on acquiring good ADMET data earlier in the discovery process.
- Novel screening methodologies are needed that dramatically increase throughputs and reduce reagent consumption.

It is becoming clear that many of the important solutions will be nanotechnological in nature. This is because the size range that holds the most interest in drug discovery is more or less the same as that which defines nanotechnology from 100 nm down to the atomic level (approximately 0.2 nm). Nanotechnology cannot provide the ultimate solutions to all these problems. However, it is becoming clear that many of the tools developed to pursue nanotechnology and nanoscience will have important roles in drug discovery.

Atomic Force Microscopy

While AFM microscopes are invaluable for imaging objects at the nanoscale, until recently they haven't been able to see how components of a cell react in biological processes, such as their response to a specific chemical or compound. New imaging techniques involve attaching antibodies specific to individual proteins to the tip of an atomic force microscopes' probe. When an antibody reacts with the protein to which it is targeted, it creates a variance in the microscope's reading compared to a reading with a bare tip, showing the presence of a protein or other specific material in the region being scanned.

This technique leads to a greater understanding of the chemical dynamics involved in how cells react to stimuli, and could prove particularly significant for drug discovery. However, although NanoMarkets believes that AFM is a useful tool to improve the understanding of molecular interactions, our research also indicates that the widespread use of this technology in bioanalytical applications has been hindered by the limited throughput of these techniques and the high experimental burden imposed by complicated and expensive instrumentation. Highly parallel micrometer and submicrometer cantilever arrays, which are currently being developed, might increase the throughput of AFM-based force spectroscopy.

Near-field Scanning Optical Microscope

NSOMs allow the study of optical properties of the sample surface with a resolution better than the wavelength of the light. By scanning the optical probe at very small distances from the sample (a few nanometers) "evanescent waves" from the surface are detected by the probe. The use of evanescent waves allows bypassing of the wavelength limitation of traditional optical techniques (the traditional limit of resolution being half the wavelength in use).

NSOM can be used to image biological samples and could ultimately be a significant contributor to drug discovery. However, there is an obstacle to such deployment at the present time manufacturing the probes reliably, the requirement for a very small hole through which the light passes being problematic.

Surface Plasmon Resonance

SPR is a phenomenon that occurs when light is reflected off thin metal films and a small amount interacts with electrons in the film, reducing the light intensity. The refractive index of the materials sandwiching the film dictates the angle at which the light reduction (essentially a shadow) occurs. Using an SPR-based approach, the interaction of biomolecules can be detected in real time, offering applications largely for observing biological systems in action on a very small scale, but also with potential for biosensors.

Surface plasmons were the subject of some interesting general research in late 2002 and early 2003. There is still a lot to be learned in this field, but NanoMarkets believes that there will be important improvements for imaging systems (and thus for drug discovery) as this learning process proceeds.

Nano Mass Spectrometry

Current technologies used for proteomic studies are based on a variety of separation techniques followed by identification of the separated proteins and proteolytic peptides using mass spectrometry (MS). One popular technique is high-resolution two-dimensional (2D) gel electrophoresis, which is capable of resolving 2500 or more distinct protein spots from complex samples. The in-gel protein spots are then identified using sensitive MS and sequence database searching. An alternative approach to 2D gel electrophoresis is chromatographic separation of peptides with electrospray ionization (ESI)-MS or tandem MS (MS/MS) detection.

The need for relatively large sample volumes is a major challenge for conventional ESI-MS analysis. There are now systems that address this challenge by applying nanotechnology to the electrospray nozzles. Thus, Advion's ESI Chip contains an array of nanoelectrospray nozzles, each one-fifth the diameter of a human hair, etched in a silicon wafer.

Dip-pen Nanolithography

Nanolithography is the art and science of etching, writing, or printing at the nanoscopic level. DPN is a scanning probe nanopatterning technique in which an AFM tip is used to deliver molecules to a surface via a solvent meniscus, which naturally forms in the ambient atmosphere. This direct-write technique offers high-resolution patterning capabilities for a number of molecular and biomolecular "inks" on a variety of substrates, such as metals, semiconductors, and monolayer functionalized surfaces.

DPN is becoming a significant tool for the scientist interested in fabricating and studying soft- and hard-matter on the nanoscale. DPN enables precise multiple patterns with near-perfect registration. It's both a fabrication and imaging tool, as the patterned areas can be imaged with clean or ink-coated tips. The ability to achieve precise alignment of multiple patterns is an additional advantage earned by using an AFM tip to write, as well as read, nanoscopic features on a surface. NanoMarkets believes that together, all these attributes make DPN a valuable tool for studying fundamental issues on colloid chemistry, surface science, and nanobiotechnology.

Nano-Arrays

The central paradigm in proteomics studies has been to identify differential protein levels in healthy and diseased cells, characterize these proteins and determine the protein's role in biochemical pathways. These proteins can then serve as diagnostic markers and potential drug targets.

Traditionally, 2D gel electrophoresis has been the workhorse in proteomics on the front end of mass spectroscopy. This technology, however, has been described as a bottleneck in high-throughput proteomic studies. As a result there has been a significant push to develop alternative technologies that serve to alleviate this rate failing step in protein separation prior to further analysis. At the forefront of these emerging and growing technologies are protein and DNA microarrays that allow the highly specific capture and analysis of a large number of proteins expressed in various cell types exposed to given perturbations in a high throughput manner.

Microarrays are generally defined as a substrate and surface chemistry onto which a biomolecule or capture agent has been immobilized for the purpose of expression analysis and functional characteristic. The potential to decrease the time required for the critical steps or protein separation and characterization are serving as a considerable driver in the protein microarray market. Microarray technologies can be considered platforms for nanoscale bioanalysis, and these products have already proven their value in the marketplace. However, currently available microarray technologies suffer from certain limitations that prohibit the exploitation of the full range of drug discovery applications. These limitations are already being addressed at the nano level with "nanoarrays," which are ultra-miniaturized versions of the traditional microarray that can measure interactions between individual molecules down to resolutions of as little as one nanometer. Nanoarrays are being touted as the next evolutionary step in the miniaturization of bioaffinity tests for proteins, nucleic acids, and receptor-ligand pairs.

It is believed that nanometer-scale resolution capabilities of nanoarray technology offer many advantages in the emerging field of functional proteomics. Unlike nucleic acids, which can easily be multiplied by amplification methods, individual proteins cannot be easily increased in quantity. Using nanoarray technology, very small quantities of individual proteins can be effectively screened against a large set of drug targets. In addition, nanoarrays can be incorporated as sensors in ways that would be impossible with larger microarrays. In addition, unlike traditional microarrays, a nanoarray can be used "in solution." Microarrays, once exposed to the target protein, must be removed from solution and rinsed before they are scanned to see whether or not there have been any "hits." Not only does this take time, but the rinsing process can also easily result in the degradation of the array itself.

Quantum Dots

QDots are nanometer-sized semiconductor crystals, or electrostatically confined electrons. While other bio-imaging tools such as fluorescent organic dyes have been available for many years. QDots hold the promise of a true revolution in bio-imaging quality and sensitivity.

Since 1914, researchers studying everything from blood tests to the effects of cancer drugs on the workings of a cell have used fluorescent dyes to tag cells. The dyes, however, can be problematic. Each dye molecule requires a laser of the same wavelength to cause it to illuminate. The dyes are also imprecise and have a tendency to bleed together. They can only be lit up for very short periods of time, typically just a few seconds after a light source is applied. QDots take advantage of the quantum confinement effect predicted by quantum mechanical theory to fluoresce extremely brightly when excited by a light source such as a laser. And they don't have any of the disadvantages of conventional technologies. By varying the size of the crystals one can cause a rainbow of colors to fluoresce. In addition, QDots stay lit for much longer periods of time than conventional dyes often hours or days.

QDots enable the tagging of a variety of different biological components like proteins or different strands of DNA with specific colors. QDots are designed to bond with and illuminate individual biological targets of choice whether genes, nucleic acids, proteins, small molecules, cancer cells, or even entire blood vessels. It may be noted that some of the world's largest pharmaceutical and biotech companies, such as GlaxoSmithKline, Pfizer, and AstraZeneca, and leading biotech firm Genentech are applying QDots in high-content drug screening and have completed initial drug screens using QDots as the biological read-out. Genentech is applying QDots to the detection of breast cancer.

Conclusions

This unit demonstrates how nanotechnology offers several improvements in drug discovery technology. These include:

- Improved imaging (including real-time imaging) that enables researchers to better understand how biological processes are affected by a new compound.
- The use of dip pen technology to improve fundamental research techniques for drug discovery applications.
- Major improvements in today's microarray and fluorescent dye technologies.

As the drug industry strives to meet the increasingly difficult task of producing new drugs and especially new blockbuster drugs, it is increasingly seeking new drug discovery technologies that can improve R&D success rates and time to market. It is therefore believed that these improvements are likely to lead to substantial new business revenues over the next few years.

Appendices

APPENDIX A

Research Programs and Centres Worldwide: A Select Listing

PROGRAMS

Scientific breakthroughs combined with recent research programs in the United States, Japan, and Europe, and various initiatives in Australia, Canada, China, Korea, Singapore, and Taiwan highlight the international interest in nanoscale science and technology. Definitions of nanotechnology vary somewhat from country to country. Nanotechnology as defined for the projects reviewed in this chapter arises from the exploitation of the novel and improved physical, chemical, mechanical, and biological properties, phenomena, and processes of systems that are intermediate in size between isolated atoms/molecules and bulk materials, where phenomena length and time scales become comparable to those of the structure. It implies the ability to generate and utilize structures, components, and devices with a size range from about 0.1 nm (atomic and molecular scale) to about 100 nm (or larger in some situations) by control at atomic, molecular, and macromolecular levels. Novel properties occur compared to bulk behavior because of the small structure size and short time scale of various processes. Nanotechnology's size range and particularly its new phenomena set it apart from the technologies referred to as microelectromechanical systems (MEMS) in the United States or microsystems technologies (MST) in Europe.

It is estimated that nanotechnology is presently at a level of development similar to that of computer/information technology in the 1950s. As indicated in the preceding chapters and as evidenced by the WTEC panel's research and observations during the course of this study, the development of nanoscale science and technology is expected by most scientists working in the field to have a broad and fundamental effect on many other technologies. This helps to explain the phenomenal levels of R&D activity worldwide. This chapter presents an overview of most of the significant nanotechnology research programs in the world. Where possible, a general picture is given of the funding levels of the programs, based on site interviews and publications.

The United States

Various U.S. public and private funding agencies; large companies in chemical, computer, pharmaceutical, and other areas; as well as small and medium-size enterprises provide support for precompetitive research programs on nanotechnology. Most of the supported programs are evolving out of disciplinary research programs, and only some are identified as primarily dealing with nanotechnology. U.S. government agencies sponsored basic research in this area at a level estimated at about $116 million in 1997 (Siegel et al., 1998), as shown in Table 8.1. the National

Science Foundation (NSF) has the largest share of the U.S. government investment, with an expenditure of about $65 million per year, or about 2.4% of its overall research investment in 1997. In 1998 it expanded its research support to functional nanostructures with an initiative in excess of $13 million.

NSF activities in nanotechnology include research supported by the Advanced Materials and Processing Program; the Ultrafine Particle Engineering initiative dedicated to new concepts and fundamental research to generate nanoparticles at high rates; the National Nanofabrication User Network (NNUN); and Instrument Development for Nano-Science and Engineering (NANO-95) to advance atomic-scale measurements of molecules, clusters, nanoparticles, and nanostructured materials. A current activity is the initiative, Synthesis, Processing, and Utilization of Functional Nanostructures (NSF 98-20 1997).

In the United States, a number of large multinational corporations, small enterprises, and consortia are pursuing nanotechnology-related research and development activities. Dow, DuPont, Eastman Kodak, Hewlett-Packard (HP), Hughes Electronics, Lucent, Motorola, Texas Instruments, Xerox, and other multinationals have established specialized groups in their long-term research laboratories, where the total research expenditure for nanotechnology research is estimated to be comparable to the U.S. government funding. Computer and electronics companies allocate up to half of their long-term research resources to nanotechnology programs. HP spends 50% of long-term (over 5 years) research on nanotechnology (Williams, 1998). Small business enterprises, such as Aerochem Research Laboratory, Nanodyne, Michigan Molecular Institute, and Particle Technology, Inc., have generated an innovative competitive environment in various technological areas, including dispersions, coatings, structural materials, filtration, nanoparticle manufacturing processes, and functional nanostructures (sensors, electronic devices, etc.). Small niches in the market as well as support from several U.S. government agencies through the Small Business for Innovative Research (SBIR) program have provided the nuclei for high-tech enterprises. the university-small business technology transfer (STTR) program at NSF is dedicated to nanotechnology in fiscal year 1999. Two semiconductor processing consortia, The Semiconductor Manufacturing and Technology Institute (Sematech) and the Semiconductor Research Corporation (SRC), are developing significant research activities on functional nanostructures on inorganic surfaces. A series of interdisciplinary centers with nanotechnology activities has been established in the last few years at many U.S. universities, creating a growing public research and education infrastructure for this field. Examples of such centers are:

- Brown University, Material Research Science and Engineering Center
- Rice University, Richard Smalley's Center for Nanoscale Science and Technology (CNST)
- University of California-Berkeley, nanoelectronics facilities
- University of Illinois at Urbana, The Engineering Research Center on Microelectronics in collaboration with the Beckman Institute, a private foundation
- University of North Carolina
- University of Texas-Austin
- Rensselaer Polytechnic Institute
- University of Washington (focus on nanobiotechnology)
- University of Wisconsin at Madison (focus on nanostructured materials)

NNUN, mentioned above, is an interuniversity effort supported by NSF at five universities: Cornell, Stanford, University of California-Santa Barbara (UCSB), Penn State, and Howard. It has focused on nanoelectronics, optoelectronics, electromechanical systems, and biotechnology. the Center for Quantized Electronic Structures (QUEST) at UCSB is a national facility developing expertise on underlying physics and chemistry aspects. Hundreds of graduate students have completed their education in connection with these centers in the last few years.

Current interest in nanotechnology in the United States is broad-based and generally spread into small groups. the research themes receiving the most attention include:

1. metallic and ceramic nanostructured materials with engineered properties
2. molecular manipulation of polymeric macromolecules
3. chemistry self-assembling techniques of "soft" nanostructures
4. thermal spray processing and chemistry-based techniques for nanostructured coatings
5. nanofabrication of electronic products and sensors
6. nanostructured materials for energy-related processes such as catalysts and soft magnets
7. nanomachining
8. miniaturization of spacecraft systems

In addition, neural communication and chip technologies are being investigated for biochemical applications; metrology has been developed for thermal and mechanical properties, magnetism, micromagnetic modeling, and thermodynamics of nanostructures; modeling at the atomistic level has been established as a computational tool; and nanoprobes have been constructed to study material structures and devices with nanometer length scale accuracy and picosecond time resolution. While generation of nanostructures under controlled conditions by building up from atoms and molecules is the most promising approach, materials restructuring and scaling-down approaches will continue. Exploratory research includes tools of quantum control and atom manipulation, computer design of hierarchically structured materials (e.g., Olson 1997), artificially structured molecules, combination of organic and inorganic nanostructures, biomimetics, nanoscale robotics, encoding and utilization of information by biological structures, DNA computing, interacting textiles, and chemical and bioagent detectors.

Commercially viable technologies are already in place in the United States for some ceramic, metallic, and polymeric nanoparticles, nanostructured alloys, colorants and cosmetics, electronic components such as those for media recording, and hard-disk reading, to name a few. the time interval from discovery to technological application varies greatly. For instance, it took several years from the basic research discovery of the giant magnetoresistance (GMR) phenomenon in nanocrystalline materials (Berkowitz et al., 1992) to industry domination by the corresponding technology by 1997. GMR technology has now completely replaced the old technologies for computer disk heads, the critical components in hard disk drives, for which there is a $20+ billion market (Williams, 1998). All disk heads currently manufactured by IBM and HP are based on this discovery. In another example, nanolayers with selective optical barriers are used at Kodak in more than 90% of graphics black and white film (Mendel, 1997) and for various optical and infrared filters, which constitute a multibilliondollar business. Other current applications of nanotechnology are hard coatings, chemical and biodetectors, drug delivery systems via nanoparticles, chemical-mechanical polishing with nanoparticle slurries in the electronics industry, and advanced laser technology. Several nanoparticle synthesis processes developed their scientific bases decades ago, but most processes are still developing their scientific bases (Roco 1998). Most of the technology base development for nanoparticle work is in an embryonic phase, and industry alone cannot sustain the research effort required for establishing the scientific and technological infrastructure. This is the role of government (e.g., NSF and NIH) and private agency (e.g., Beckman Institute) support for fundamental research.

Nanotechnology research in the United States has been developed in open competition with other research topics within various disciplines. This is one of the reasons that the U.S. research efforts in nanotechnology are relatively fragmented and partially overlapping among disciplines, areas of relevance, and sources of funding. This situation has advantages in establishing competitive paths in the emerging nanotechnology field and in promoting innovative ideas; it also has some disadvantages for developing system applications. An interagency coordinating "Group on Nanotechnology" targets some improvement of the current situation. the group was established in 1997 with participants from twelve government funding/research agencies to enhance communication and develop partnerships among practicing nanoscience professionals.

Canada

Canada's National Research Council supports nanotechnology through the Institute for Microstructural Science, which has the mission to interact with industry and universities to develop the infrastructure for information technology. the main project, the Semiconductor Nanostructure Project, was initiated in 1990. It provides support for fundamental research at a series of universities, including Queen's, Carleton, and Ottawa Universities.

Japan

The term "nanotechnology" is frequently used in Japan specifically to describe the construction of nanostructures on semiconductors/inorganic substrates for future electronic and computer technologies, and to describe the development of equipment for measurement at nanometer level (Sienko, 1998). There are, however, Japanese programs in a number of other areas related to nanotechnology in the broader definition used in this report.

Government agencies and large corporations are the main sources of funding for nanotechnology in Japan; small and medium-size companies play only a minor role. Research activities are generally grouped in relatively large industrial, government, and academic laboratories. the three main government organizations sponsoring nanotechnology in Japan are the Ministry of International Trade and Industry (MITI), the Science and Technology Agency (STA), and Monbusho (the Ministry of Education, Science, Sports, and Culture). Funding for nanotechnology research should be viewed in the context of an overall increased level of support for basic research in Japan since 1995 as a result of Japan's Science and Technology Basic Law No. 130 (effective November 15, 1995), even if the law has not been fully implemented. the data presented below are based on information received from Japanese colleagues during the WTEC visit in July 1997. All budgets are for the fiscal year 1996 (1 April 1996 to 31 March 1997) and assume an exchange rate of ¥115 = $1, unless otherwise stated. the first five-year program on ultrafine particles started in 1981 under the Exploratory Research for Advanced Technologies (ERATO) program; an overview of the results of that program was published in 1991 (Uyeda, 1991).

It is estimated that the Agency of Industrial Science and Technology (AIST) within MITI had a budget of approximately $60 million per year for nanotechnology in 1996/97 (roughly 2.2% of the AIST budget). the National Institute for Advancement of Interdisciplinary Research (NAIR) hosts three AIST projects:

1. Joint Research Center for Atom Technology (JRCAT), with a ten-year budget of about $220 million for 1992-2001 ($25 million per year in 1996)
2. Research on Cluster Science program, with about $10 million for the interval 1992-1997
3. Research on Bionic Design program, with $10 million for 1992-1997, about half on nanotechnology

Other efforts supported to various degrees by MITI include the following:

- the Electrotechnical Laboratory in Tsukuba, which allocates about 17% (or $17 million per year) of its efforts on advanced nanotechnology projects
- the Quantum Functional Devices Program, funded at about $64 million for 1991-2001 (about $6.4 million in 1996)
- the Osaka National Research Institute and the National Industrial Research Institute of Nagoya, which each spend in the range of $2.5-3 million per year for nanotechnology
- the Association of Super-Advanced Electronics Technologies (ASET), a relatively new MITI-sponsored consortium with partial interest in nanotechnology; it has similarities with the U.S. Ultra Electronics program of DARPA

It is estimated that STA investment in nanotechnology research was about $35 million in 1996, mainly within five organizations:

1. Institute of Physical and Chemical Research (RIKEN), where nanotechnology is included in the Frontier Materials Research initiative
2. National Research Institute for Metals (NRIM)
3. National Institute for Research in Inorganic Materials (NIRIM)
4. Japan Science and Technology Corporation (JST—formerly called JRDC), which manages the ERATO program, including four nanotechnology-related projects, each with total budgets of $13-18 million for five years:
   - Quantum Wave Project (1988-93)
   - Atomcraft Project (1989-94)
   - Electron Wavefront Project (1989-94)
   - Quantum Fluctuation Project (1993-98)

Monbusho supports nanotechnology programs at universities and national research institutes, as well as via the Japan Society for Promotion of Science (JSPS). the most active programs are those at Tokyo University, Kyoto University, Tokyo Institute of Technology, Tohoku University, and Osaka University. the Institute of Molecular Science and the Exploratory Research on Novel Artificial Materials and Substances program promote new research ideas for next-generation industries (5-year university-industry research projects). the "Research for the Future" initiative sponsored by JSPS has a program on "nanostructurally controlled spin-dependent quantum effects" (1996-2001) at Tohoku University. Monbusho's funding contribution to nanotechnology programs is estimated at ~ $25 million.

In total, MITI, STA, and Monbusho allocated ~$120 million for nanotechnology in 1996.

Large companies also drive nanotechnology research in Japan. Important research efforts are at six institutions: Hitachi (Central R&D Laboratories, where nanotechnology is ~25% of long-term research) ~Hitachi has ~$70 billion per year in sales; NEC (Fundamental Research Laboratories, where nanotechnology is estimated to be ~50% of the precompetitive research)—NEC has ~$40 billion per year in sales; NTT (Atsugi Lab); Fujitsu (Quantum Electron Devices Lab); Sony; and Fuji Photo Film Co. An allocation of 10% of sales for research and development is customary in these companies, with ~10% of this for long-term research. Some Japanese nanostructured products already have considerable market impact. Nihon Shinku Gijutsu (ULVAC) produces over $4 million per year in sales of particles for electronics, optics, and arts. Also, there are in Japan, at the United States, private consortia making an increased contribution to nanotechnology R&D:

- The Semiconductor Industry Research Institute of Japan (SIRI), established in 1994, focuses on long-term research with partial government funding.
- Semiconductor Leading Edge Technologies, Inc. (SELETE), established by ten large

Japanese semiconductor companies in 1996, focuses on applied research and development with an estimated budget of $60 million in 1997.

- Semiconductor Technology Academic Research Center (STARC) promotes industry-university interactions.

Strengthening of the nanotechnology research infrastructure in the last years has been fueled by both the overall increase of government funding for basic research and by larger numbers of academic and industry researchers choosing nanotechnology as their primary field of research. Potential industrial applications provide a strong stimulus. A systems approach has been adopted in most laboratory projects, including multiple characterization methods and processing techniques. A special Japanese research strength is instrumentation development. the university-industry interaction is stimulated by the new MITI projects awarded to universities in the last few years that encourage temporary hiring of research personnel from industry. Other issues currently being addressed are more extensive use of peer review, promoting personnel mobility and intellectual independence, rewarding researchers for patents, promoting interdisciplinary and international interactions, and better use of the physical infrastructure.

China

Nanoscience and nanotechnology have received increased attention in China since the mid-1980s. Approximately 3,000 researchers there now contribute to this field (Bai, 1996). the ten-year "Climbing Project on Nanometer Science" (1990-99) and a series of advanced materials research projects are the core activities. the Chinese Academy of Sciences sponsors relatively large groups, while the China National Science Foundation (CNSF) provides support mainly for individual research projects. Areas of strength are development of nanoprobes and manufacturing processes using nanotubes. the Chinese Physics Society and the Chinese Society of Particuology are societies involved in the dissemination of nanotechnology research.

India

India's main research activities are on nanostructured materials and electronic devices (Sikka, 1995). These involve a combination of research institutions (the Central Electronics Engineering Research Institute in Pilani, the Space Application Center in Ahmedabad, and others), funding organizations (the Centre for Development of Materials in Pune, the Indian Institute of Science in Bangalore, and others), and industry.

Taiwan

Taiwan's major nanotechnology research effort is conducted in the area of miniaturization of electronic circuits. the research is conducted in academic institutions and at the Industrial Technology Research Institute. Government funds for fundamental research are channeled via the National Science Council.

South Korea

A national research focus on nanotechnology was established in Korea in 1995. the Electronics and Telecommunications Research Institute (ETRI) in Taejon, Korea's science city, targets advanced technologies for information and computer infrastructures, with a focus on nanotechnology (ATIP, 1998). the emphasis is on nanoscale semiconductor devices and particularly on semiconductor quantum nanostructures and device applications (lasers, modulators, switches and logical devices, resonant tunneling devices, self-assembled nanosize dots, single-electron transistors, and quantum wires).

Singapore

Nanotechnology research received a considerable boost in Singapore by the initiation of a national program in this area in 1995.

Australia

The National Research Council (NRC) of Australia has sponsored R&D in nanotechnology since 1993 (ASTC, 1993). Research groups work on synthesis of nanoparticles for membranes and catalysts (University of New South Wales), nanofiltration (UNESCO Center for Membrane Science and Technology), and use of nanoparticles in processing minerals for special products (the Advanced Mineral Products Research Center at the University of Melbourne). AWA Electronics in Homebush has the largest industrial research facility for nanoelectronics in Australia.

European Community (EC)

The term "nanotechnology" is frequently defined in Europe as "the direct control of atoms and molecules" for materials and devices. A more specific definition from H. Rohrer (1997) is a "one-to-one relationship between a nano-object or nano-part of an object and another nano-, micro- or bulk object." the nanotechnology field as defined in this WTEC report includes these aspects, with the clarification that only the specific, distinctive properties and phenomena manifesting at length scales between individual atoms/molecules and bulk behaviour are considered.

There are a combination of national programs, collaborative European (mostly EC) networks, and large corporations that fund nanotechnology research in Europe.

Multinational European programs include the following:

1. The ESPRIT Advanced Research Initiative in Microelectronics and the BRITE/EURAM projects on materials science in the EC are partially dedicated to nanotechnology.
2. The PHANTOMS (Physics and Technology of Mesoscale Systems) program is a network created in 1992 with about 40 members to stimulate nanoelectronics, nanofabrication, optoelectronics, and electronic switching. Its coordinating center is at the IMEC Center for Microelectronics in Leuven, Belgium.
3. The European Science Foundation has sponsored a network since 1995 for Vapor-phase Synthesis and Processing of Nanoparticle Materials (NANO) in order to promote bridges between the aerosol and materials science communities working on nanoparticles. the NANO network includes 18 research centers and is codirected by Duisburg University and Delft University of Technology.
4. The European Consortium on NanoMaterials (ECNM) was formed in 1996, with its coordinating center in Lausanne, Switzerland. This group aims at fundamental research to solve technological problems for nanomaterials and at improved communication between researchers and industry.
5. NEOME (Network for Excellence on Organic Materials for Electronics) has had some programs related to nanotechnology since 1992.
6. The European Society for Precision Engineering and Nanotechnology (EUSPEN) was designed in 1997 with participation from industry and universities from six EC countries.
7. The Joint Research Center Nanostructured Materials Network, established in 1996, has its center in Ispra, Italy.

It is expected that the European Framework V will introduce additional programs on nanotechnology, particularly by adding a new dimension in nanobiology in the next four-year plan.

Germany

The Federal Ministry of Education, Science, Research, and Technology (BMBF) in Germany provides substantial national support for nanotechnology. the Fraunhofer Institutes, Max Planck Institutes, and several universities have formed centers of excellence in the field. It is estimated that in 1997 BMBF supported programs on nanotechnology with a budget of approximately $50 million per year. Two of the largest upcoming projects are "CESAR," a $50 million science center in Bonn equally sponsored by the state and federal governments with about one-third of its research dedicated to nanoscience, and a new institute for carbon-reinforced materials near Karlsruhe ($4 million over 3 years, 1998-2001). BMBF is establishing five "centers of competence in nanotechnology" in Germany starting in 1998, with topics ranging from molecular architecture to ultraprecision manufacturing.

U.K.

A network program (LINK Nanotechnology Programme) was launched in the United Kingdom in 1988 with an annual budget of about $2 million per year. the Engineering and Physical Sciences Research Council (EPSRC) is funding materials science projects related to nanotechnology with a total value of about $7 million for a five-year interval (1994-1999). About $1 million is specifically earmarked for nanoparticle research. the National Physical Laboratory established a forum called the National Initiative on Nanotechnology (NION) for promoting nanotechnology in universities, industry, and government laboratories.

France

The Centre National de la Récherche Scientifique (CNRS) has developed research programs on nanoparticles and nanostructured materials at about 40 physics laboratories and 20 chemistry laboratories in France. Synthesis methods include molecular beam and cluster deposition, lithography, electrochemistry, soft chemistry, and biosynthesis. Nanotechnology activity has grown within a wide variety of research groups, including ones focused on molecular electronics, large gap semiconductors and nanomagnetism, catalysts, nanofilters, therapy problems, agrochemistry, and even cements for ductile nanoconcretes. It is estimated that CNRS spends about 2% of its budget and dedicates 500 researchers in 60 laboratories (or about $40 million per year) on projects related to nanoscience and nanotechnology. Companies collaborating to research and produce nanomaterials include Thompson, St. Gobain, Rhone Poulenc, Air Liquide, and IEMN. Also, there is the "French Club Nanotechnologie," aimed at promoting interactions in this field in France.

Sweden

The estimated total expenditure for research on nanotechnology in Sweden is ~ $10 million per year. There are four materials research consortia involved in this field: 1. Ångstrom Consortium in Uppsala, with a budget of ~ $0.8 million per year in 1998 for surface nanocoatings, 2. Nanometer Structures Consortium in Lund with a budget of ~ $3.5 million per year partially supported by ESPRIT (~ $1 million per year), 3. Cluster-based and Ultrafine Particle Materials in the University of Uppsala and Royal Institute of Technology with a budget of ~$0.4 million per year in 1998, 4. Brinell Center at the Royal Institute of Technology.

Switzerland

There is a Swiss national program on nanotechnology with a special strength in instrumentation. the most advanced research centers are focused on nanoprobes and molecule manipulation on surfaces (IBM Research Laboratory in Zurich), devices and sensors (Paul Scherrer Institute), nanoelectronics (ETH Zürich), and self-assembling on surfaces in patterns determined by the substrate or template (L'École Polytechnique Fédérale de Lausanne). See the site reports on all of these institutions.

The Netherlands

The most active research centers in the Netherlands are the DIMES institute at Delft University of Technology, which receives one-third support from industry, and the Philips Research Institute in Eindhoven, which researches self-assembling monolayers and patterning on metallic and silicon surfaces. the SST Netherlands Study Center for Technology Trends is completing a study on nanotechnology and aims at promoting increased funding and research networking in the Netherlands (ten Wolte, 1997).

Finland

The Academy of Finland and the Finnish Technology Development Center began a three-year nanotechnology program in 1997. the program involves sixteen projects with funding of $9 million for a three-year period (1997-99) for nanobiology, functional nanostructures, nanoelectronics, and other areas. Research on actuators and sensors is the Finnish area of strength.

Belgium and Spain

Since about 1993 both Belgium and Spain have established nanotechnology programs, centers of excellence, and university-industry interactions.

Multinational Efforts

Large multinational companies with significant nanotechnology research activities in Europe include IBM (Zurich), Philips, Siemens, Bayer, and Hitachi. Degussa Co., with headquarters in Germany, is a commercial supplier since 1940 of microparticles and, now, nanoparticles.

Western Europe has a variety of approaches to funding research on nanotechnology. These are discussed in detail in other studies. IPTS (the Science and Technology Forecast Institute) has conducted a study on nanotechnology research in the EC (Malsh, 1997). Other European studies published recently include those by VDI (1996), UNIDO (1997), the U.K. Parliamentary Office of Science and Technology (1997), and NANO network (Fissan and Schoonman, 1997). the overall expenditure for nanotechnology research within the EC was estimated in 1997 to be over $128 million per year.

Russia and Other FSU Countries

Support for generation of nanoparticles and nanostructured materials has a tradition in Russia and other countries of the former Soviet Union (FSU) dating back to the mid-1970s; before 1990 an important part of this support was connected to defense research. the first public paper concerning the special properties of nanostructures was published in Russia in 1976. In 1979 the Council of the Academy of Sciences created a section on "Ultra-Dispersed Systems." Research strengths are in the areas of preparation processes of nanostructured materials and in several basic scientific aspects. Metallurgical research for special metals, including those with nanocrystalline structures, has received particular attention; research for nanodevices has been relatively less developed. Due to funding limitations, characterization and utilization of nanoparticles and nanostructured materials requiring costly equipment are less advanced than processing.

Russian government funds are allocated mainly for research personnel and less for infrastructure. Funding for nanotechnology is channeled via the Ministry of Science and Technology, the Russian Foundation for Fundamental Research, the Academy of Sciences, the Ministry of Higher Education, and other ministries with specific targets. the Ministry of Higher Education has relatively little research funding. Overall, 2.8% of the civilian budget in Russia in 1997 was planned for allocation to science. There is no centralized program on nanotechnology; however, there are components in specific institutional programs. Currently, about 20% of science research in Russia is funded via international organizations. the significant level of interest in the FSU can be identified by the relatively large participation at a series of Russian conferences on nanotechnology, the first in 1984 (First USSR Conference on Physics and Chemistry of Ultradispersed Systems), a second in 1989, and a third in 1993.

The Ministry of Science and Technology contributes to nanotechnology through several of its specific programs related to solid-state physics, surface science, fullerenes and nanostructures, and particularly "electronic and optical properties of nanostructures." This last program involves a network of scientific centers: the Ioffe Institute in St. Petersburg, Lebedev Institute in Moscow, Moscow State University, Novgograd Institute of Microstructures, Novosibirsk Institute of Semiconductor Physics, and others. This research network has an annual meeting on nanostructures, physics, and technology, and has developed interactions with the PHANTOMS network in the EC. the U.S. Civilian Research and Development Foundation has provided research funds in the FSU for several projects related to nanotechnology, including "Highly Non-Equilibrium States and Processes in Nanomaterials" at the Ioffe Institute (1996-98).

Russian government and international organizations are the primary research sponsors for nanotechnology in Russia. However, laboratories and companies privatized in the last few years, such as the Delta Research Institute in Moscow, are under-development. With a relatively lower base in characterization and advanced computing, the research focus is on advanced processing and continuum modeling. Research strengths are in the fields of physico-chemistry, nanostructured materials, nanoparticle generation and processing methods, and applications for hard materials, purification, and the oil industry, and biologically active systems (Siegel et al. n.d.). There are related programs in Ukraine, Belarus, and Georgia, mostly under the direction of the respective academies of sciences in these countries, that are dedicated to crystalline nanostructures and advanced structural and nanoelectronic materials. Several innovative processes, such as diamond powder production by detonation synthesis at SINTA in Belarus, are not well known abroad.

Nanotechnology in the United States, Japan, and Western Europe is making progress in developing a suitable research infrastructure. the promise of nanotechnology is being realized through the confluence of advances in two fields: (1) scientific discovery that has enabled the atomic, molecular, and supramolecular control of material building blocks, and (2) manufacturing that provides the means to assemble and utilize these tailored building blocks for new processes and devices in a wide variety of applications. Technology programs cannot be developed without strong supporting science programs because of the scale and complexity of the nanosystems. the overlapping of discipline-oriented research with nanotechnology-targeted programs seems appropriate at this point in time. Highly interdisciplinary and multiapplication nanotechnology provides generic approaches that enable advances in other technologies, from dispersions, catalysts, and electronics to biomedicine. Essential trends include the following:

- learning from nature (including templating, self-assembly, multifunctionality)
- building up functional nanostructures from molecules
- convergence of miniaturization and assembly techniques

- novel materials by design
- use of hierarchical/adaptive simulations

A characteristic of discovery in nanotechnology is the potential for revolutionary steps. the question "what if?" is progressively replaced by "at what cost?" the road from basic research to applications may vary from a few months to decades. Research and development is expensive, and the field needs support from related areas. the R&D environment should favor multiapplication and international partnerships. Based on the data for 1996 and 1997 collected during this WTEC study, 1997 government expenditures for nanotechnology research were at similar absolute levels in the United States, Japan, and Western Europe. (Estimated OECD data for 1997 indicated GDPs of $4.49 billion for Japan, $7.76 billion for the United States, and $7.00 billion for Western Europe.) the largest funding opportunities for nanotechnology are provided by NSF in the United States (approximately $65 million per year for fundamental research), by MITI in Japan (approximately $50 million per year for fundamental research and development), and by BMBF in Germany (approximately $50 million per year for fundamental and applied research). Large companies in areas such as dispersions, electronics, multimedia, and bioengineering contribute to research to a larger extent in Japan and the United States than in Europe.

While multinational companies are pursuing nanotechnology research activities in almost all developed countries, the presence of an active group of small and medium-size companies introducing new processes to the market is limited to the United States. In the United States, individual and small-group researchers as well as industrial and national laboratories for specialized topics have established a strong position in synthesis and assembly of nanoscale building blocks and catalysts, and in polymeric and biological approaches to nanostructured materials. the Japanese large-group research institutes, and more recently academic laboratories, have made particular advances in nanodevices and nano-instrumentation. the European "mosaic" provides a diverse combination of university research, networks, and national laboratories with special performance in dispersion and coatings, nanobiotechnology, and nanoprobes.

With a relatively lower base in characterization and computing infrastructure, the research focus in Russia is on physico-chemistry phenomena, advanced processing, and continuum modeling. Interest and economic support, particularly for device-related research, is growing in China, Australia, India, Taiwan, Korea, and Singapore.

The pace of revolutionary discoveries that we are witnessing now in nanotechnology is expected to accelerate in the next decade worldwide. This will have a profound impact on existing and emerging technologies in almost all industry sectors, in conservation of materials and energy, in biomedicine, and in environmental sustainability.

Research Centers

- Institute for Molecular Manufacturing—nonprofit foundation carrying out research aimed at developing molecular manufacturing (molecular nanotechnology).
- Rice University—Center for Nanoscale Science and Technology—devoted to nurturing science and technology at the nanometer scale.
- Cornell NanoScale Science & Technology Facility (CNF) —subjects of research encompass physical sciences, life sciences, and engineering, particularly with inter-disciplinary emphasis.
- Cornell University—Nanobiotechnology Center (NBTC) —devoted to studying the convergence of nano-microfabrication and biosystems.
- University of Washington—Center for Nanotechnology University of Birmingham (United

Kingdom) —Nanoscale Physics Research Laboratory—advancing the frontiers of the physics, chemistry, and technology of nanometre-scale structures, devices, and processes.
- University of Newcastle upon Tyne—Centre for Nanoscale Science and Technology—multidisciplinary research centre developing micro- and nano-devices for biomedical applications.
- University of California, Santa Barbara Nanotech—offers expertise in compound semiconductor-based device fabrication providing a full range of processes to the scientific and research communities.
- Center for Nanoscience and Nanotechnology—Georgia Institute of Technology—working to improve the human condition by providing fundamental and applicable knowledge to nanotechnology and its related fields of science.
- University of Michigan—Center for Biologic Nanotechnology—dedicated to the application of nano-molecular architecture to medicine.
- University of Notre Dame Nanofabrication Facility (NDNF) —provides information on the nanoelectronics fabrication facility.
- Arizona State University—Graduate Research Training Program in Biomolecular Nanotechnology—offers training in biomolecular nanotechnology, molecular electronics, and molecular devices.
- University of Nebraska-Lincoln—Center for Electro-Optics—specializes in optical/electromagnetic measurement and remote sensing technologies research.
- Schubert-Group: Macromolecular Chemistry and Nanoscience—Eindhoven University of Technology—devoted to the combinatorial materials research efforts.
- California Nanosystems Institute (CNSI)—R&D institute working on nanotechnologies and its commercial applications. Jointly established by University of California, Los Angeles, and University of California, Santa Barbara.
- Center for Nanotechnology—NASA Ames Research Center—focuses on experimental research and development in nano and biotechnologies as well as computational nanotechnology, nanoelectronics, and optoelectronics.
- Institute for Soldier Nanotechnologies—intent to create a University Affiliated Research Center (UARC), with industry partners, to develop nanometer-scale science and technology solutions for the soldier.
- National Institute of Standards and Technology—Nanometer-Scale Metrology—research projects and measurement services in nanometer scale metrology.

World-wide Nanoelectronics Research

U.S. Government Labs

- Department of Defense (DOD) Laboratories
  DOD Labs conduct some of the most advanced and imaginative research in nanofabrication, nanoelectronics, and nanocomputing. Not all of this research is well represented on the World-Wide Web, however. the Nanoelectronics & Nanocomputing Home Page has attempted to list some of this research, at least by title, on the Who's Who in Nanoelectronics page, as well as other pages. For further information on DOD Labs and an index of DOD technology departments and projects, see the TechTRANSIT Page. TechTRANSIT connects technology transfer resources and activities to meet the requirements of the Department of Defense Office of Technology Transition.

- Naval Research Laboratory (NRL)
  The Naval Research Laboratory (NRL) is the Navy's corporate research and development laboratory. NRL is the leader among Government Laboratories in nanotechnology and nanocomputing research. This includes very strong efforts in the application of proximal probes, modeling of nanosystems, novel approaches to nanofabrication, as well as other areas. However, only limited information about NRL's nanoelectronics activities is available on their WWW pages. For some information on nanoelectronics work at the NRL, see the NRL Electronics Science & Technology Division.
- Los Alamos National Laboratory (LAND).
  Investigators at LANL are deeply involved in the study of nanoelectronics and nanofabrication as part of the Theory, Modeling, and High Performance Computing core competency at the laboratory. However, only limited information about LANL's nanoelectronics activities is available on their WWW pages. Some of the LANL investigators are represented, however, on the Who's Who in Nanoelectronics page.

U.S. Universities

- UCLA Nanoelectronics Lab
- The University of Cincinnati Nanoelectronics Lab
  The Nanoelectronics Laboratory is engaged in research at (or around) the nanometer scale in semiconductors and other electronic materials. Specific research areas include focused ion beam processing, integrated optoelectronic devices and waveguides, plasma assisted etching, and deposition.
- Cornell University Nanofabrication Facility (CNF)
  The Cornell Nanofabrication Facility (CNF) has served the US research community for more than 15 years by providing nanofabrication services to over 500 research projects. Subject areas of projects completed are as diverse as Astronomy to Plant Pathology, as well as the more conventional Micro-electronics, Physics and Materials Research, and emerging fields of Integrated Optics and Micro-Electro-Mechanical Systems (MEMS).
- Massachusetts Institute of Technology (MIT) NanoStructures Laboratory
  The NanoStructures Laboratory (NSL) at MIT (formerly the Submicron Structures Laboratory) develops techniques for fabricating surface structures with feature sizes in the range from nanometers to micrometers, and uses these structures in a variety of research projects.
- University of Minnesota Nanostructure Lab
  The Nanostructure Lab's goal is to develop new nanotechnologies for fabricating structures and devices substantially smaller than current technology permits, and to explore innovative nanoscale electronic, optoelectronic, and magnetic storage devices. Research at the lab is headed by Stephen Chou.
- Notre Dame Microelectronics Lab
  The Microelectronics Laboratory is equipped to fabricate integrated circuits and devices in both silicon and compound semiconductors with geometries as small as 0.01 microns (10 nm).
- Penn State Electronic Materials and Processing Research Laboratory (EMPRL)
- Purdue University Nanoscale Physics Laboratory
  This Laboratory is a part of an interdisciplinary Center of Excellence in nanotechnology that has been developed at Purdue 'University as part of a University Research Initiative

funded via the Army Research Office. Among the accomplishments of the interdisciplinary, interdepartmental Purdue team is the fabrication of arrays of functioning nanometer-scale circuit elements and wires using techniques of chemical self-assembly.

- Stanford University Nanofabrication Facility
  The Stanford Nanofabrication Research Facility is located in the Center for Integrated Systems. This state-of-the-art research laboratory consists of over $10M worth of modern processing equipment in a 10,000 sq.ft. clean room, extensive computing and networking capability, established processes for silicon, gallium arsenide and other material, a highly skilled technical staff and some of the most advanced research projects in the world.

U.S. Government Sponsored Research and Development Programs

- Ultra Dense, Ultra Fast Computing Components/Nanoelectronics (ULTRA) Program
  Sponsored by the Defense Advanced Research Projects Agency, the goals of the ULTRA Nanoelectronics program are to explore and develop material, processing technologies, quantum and conventional devices and device architectures for a next generation of information processing systems. the ULTRA program seeks improved speed, density, and functionality beyond that which can be achieved by scaling transistors.
- Nanoelectronics Modeling (NEMO) Program
  The U.S. Government's Project NEMO is developing an advanced software tool for modeling nanoelectronic devices.

U.S. Government Sponsoring Agencies

- Defense Advanced Research Projects Agency (DARPA)
  The Defense Advanced Research Projects Agency (DARPA) is the advanced research and development organization for the U.S. Department of Defense. More information on DARPA's nanoelectronics work is available at the DARPA Electronics Technology Office (ETO) Home Page. the DARPA ULTRA Program (for the development of Ultra-dense and Ultra-fast computers) is the premier nanoelectronics research program in the world.
- Air Force Office of Scientific Research (AFOSR)
  The Air Force Office of Scientific Research (AFOSR) is charged with directing the Air Force's basic research program, which includes extensive efforts in next-generation electronics and related technologies.
- Army Research Office (ARO)
  The Army Research Office sponsors a diverse set of research and development projects in nanoelectronics. ARO is led by Dr. Gerry Iafrate, who was a very early supporter and investigator in the fields of quantum electronics and nanoelectronics.
- Office of Naval Research (ONR)
  The Office of Naval Research (ONR) coordinates, executes, and promotes the science and technology programs of the United States Navy and Marine Corps through universities, government laboratories, and nonprofit organizations. the ONR Electronics Division has a nanoscale electronics program. the objective of this program is to study the electronic properties of solid state materials and develop ultra small devices for microelectronic applications.
- The National Science Foundation (NSF)
  NSF's mission is to promote the progress of science; to advance the national health, prosperity, and welfare; and to secure the national defense. NSF funds research at U.S.

universities in nanoelectronics, and it also has considerable involvement in the funding of DNA/biomolecular nanocomputation.

Federally Funded Research and Development Centers (FFRDCs)

- The Aerospace Corporation
  Aerospace operates a Federally Funded Research and Development Center serving as an architect-engineer for advanced space systems. Aerospace has been concerned with developing and applying nanoelectronics and nanotechnology to ease the manufacture and, especially, to reduce the size and weight of spacecraft. This could have dramatic impact in reducing launch costs, as well as in improving reliability. Aerospace has worked with NASA to sponsor several meetings on these topics, and they have published technical reports and proceedings in connection with these activities. to learn more about nanoelectronics at the Aerospace Corporation, see the Web pages for their Electronic Systems Division.
- MIT Lincoln Laboratory
  Dr. Gerhard Sollner of the MIT Lincoln Lab is a well-known investigator at the forefront of nanoelectronics research.
- The MITRE Corporation
  Nanoelectronics research at the MITRE Corporation is performed by the MITRE Nanosystems Group. the MITRE Corporation is a Federally Funded Research and Development Center (FFRDC) for systems engineering and integration serving in the public interest. MITRE applies leading-edge technology to some of the most important systems in the world, including U.S. Military command and control systems and air traffic control systems, as well as similar systems around the world.
- The RAND Corporation
  RAND is a nonprofit institution that helps improve public policy through research and analysis. In 1995, RAND published a report on "The Potential of Nanotechnology for Molecular Manufacturing" by Max Nelson and Calvin Shipbaugh.

Private Foundations

- The Mayo Foundation/Mayo Clinic, Rochester, Minnesota
  Nanoelectronics research at the Mayo Foundation is focused in the Mayo Special Purpose Processor Development Group. This group is led by Dr. Barry Gilbert, who is one of the world's preeminent experts in the field of quantum-effect electronic devices nanoelectronics.

US Industry

- Hughes Electronics
  Hughes is engaged in nanoelectronics research. However, only limited information about Hughes' nanoelectronics activities is available on their corporate WWW pages.
- International Business Machines (IBM)
  IBM is one of the leaders world-wide in nanotechnology and nanoelectronics research. Some of their capabilities are on view in the • IBM Almaden Scanning Tunneling Microscope Image Gallery. This page contains spectacular images of atoms being manipulated with an STM.
  Additional IBM nanotechnology research is performed at the IBM-Zurich Lab. One of the most important tools of nanotechnology, the scanning tunneling electron microscope

(STM), the first proximal probe device, was invented at IBM-Zurich. the Lab's Web site has a description of an IBM-Zurich molecular positioning experiment that uses an STM.

- Motorola, Inc.
  Motorola is among the leaders in nanoelectronics research. However, only limited information about Motorola's nanoelectronics activities is available on their corporate WWW pages.
- Texas Instruments, Inc. (TI)
  The Texas Instruments Nanoelectronics Group is one of the world's leaders in nanoelectronics research. However, only limited information about TI's nanoelectronics research activities is available on their corporate WWW pages. Several of the TI nanoelectronics investigators are listed, however, on the Who's Who in Nanoelectronics page.

Japanese Laboratories

- ERATO Single Quantum Dot Project
  Five year, $20 million interdisciplinary project to investigate the basic properties and fabrication of quantum dots. Begun in October 1995. Sponsored by Japan's ERATO (Exploratory Research for Advanced Technology) Initiative, under the auspices of Japan's Science and Technology Agency.
- Fasol Lab at the University of Tokyo
  The Fasol Lab home page has an in-depth nanotechnology index.
- Hitachi Corp. Research and Development, Tokyo, Japan
  Hitachi is engaged in an extensive nanoelectronics research effort, involving a number of preeminent scientists and engineers. However, only limited information about Hitachi's nanoelectronics activities is available on their corporate WWW pages.

European Laboratories

- Cavendish Laboratory, Cambridge University, Great Britain
  As noted by Dr. William Tolles in his 1994 report "Nanoscience and Nanotechnology in Europe," important research on the subject of nanoscience and nanoengineering is ongoing at the Cavendish Laboratory. However, little information about Cavendish's nanoelectronics activities is available on their laboratory WWW pages.
- Project on Nanometre Structures for Future Optoelectronic Applications sponsored by the European Union's Espirit Program for Information Technologies Research
- Project on Organic Materials for Electronics (and Nanostructures) sponsored by the European Union's Espirit Program for Information Technologies Research
- Project on Resonant Tunneling for Novel Electronic Applications sponsored by the European Union's Espirit Program for Information Technologies Research
- University of Glasgow Nanoelectronics Research Center, Glasgow, Scotland
- IBM-Zurich Laboratory, Zurich, Switzerland
  Much of the IBM's nanotechnology research is performed at the IBM-Zurich Lab. the Lab's Web site has a description of an IBM-Zurich molecular positioning experiment.
- National Physical Laboratory, Teddington, Great Britain
  As noted by Dr. William Tolles in his 1994 report "Nanoscience and Nanotechnology in Europe," important research on the subject of nanometrology is ongoing at the National Physical Laboratory. However, little information about the National Physical Laboratory's

nanoelectronics activities is available on their laboratory WWW pages.

- University College London, UK
  Professors Terry Fountain and Michael Duff at the University College London are leading experts in the field of architectures for nanoelectronic computers. However, little information about the University College London's nanoelectronics activities is available on their university WWW pages.

Nanotechnology Research—Lab Groups

- Massachusetts Institute of Technology (MIT)—Microsystems Technology Laboratories—explores microsystems, microfabrication, and manufacturing technologies.
- Princeton University—NanoStructures Laboratory—explores and develops new nanotechnologies to fabricate structures; innovates nanoscale electronic, optoelectronic, and magnetic devices.
- New York University (NYU)—Seeman Lab-investigating unusual DNA molecules in model systems that use synthetic molecules, with a major effort in DNA nanotechnology.
- University of Cincinnati—Nanoelectronics Laboratory—research at or around the nanometer scale in semiconductors and other electronic materials.
- Academia Sinica—Surface Nanostructure Laboratory—creates, characterizes, and exploits nanostructures (1-50 nm) on solid surfaces.
- Waseda University—Ohdomari Lab—research on the process of nanostructure fabrication, and the physics of silicon surfaces and interfaces.
- University of Pune—Scanning Tunneling Microscopy Group—performs research on nanoparticles, nanolithography, and noise in STM.

NIH Nanotechnology and Nanoscience Information

This page contains information on (1) currently active NIH and BECON research and training opportunities, and (2) listings of funded grants for NIH and BECON program announcements related to nanotechnology and nanoscience. Additional information on the NIH bioengineering funding opportunities can be found at http://www.becon.nih.gov/becon_funding.htm.

While much of biology is grounded in nanoscale phenomena, NIH has not re-classified most of its basic research portfolio as nanotechnology. Only those studies that use nanotechnology tobls and concepts to study biology; that propose to engineer biological molecules toward functions very different from those they have in nature; or that manipulate biological systems by methods more precise than can be done by using molecular biological, synthetic chemical, or biochemical approaches that have been used for years in the biology research community, are classified as nanotechnology projects. NIH participates in the National Nanotechnology Initiative (www.nano.gov), the U.S. federal multi-agency R&D program for nanoscale science, engineering, and technology.

A list of technical and financial contacts for Nanoscience and Nanotechnology related projects is given at: http://www.becon.nih.gov/nano_contacts.htm.

A list of technical and financial contacts for PA-02-125, Bioengineering Nanotechnology Initiative (SBIR) is given at: http://www.becon.nih.gov/nano_contacts_sbir.htm.

Nanomedicine NIH Roadmap Initiatives

- Nanomedicine Roadmap Homepage

NIH Reports Related to Nanotechnology

- NIBIB/DOE Workshop on Biomedical Applications of Nanotechnology, March 17-18, 2005
- Nanoscience and Nanotechnology Symposium—June 25-26, 2000
- NHLBI Nanotechnology Working Group Report
- NCI Cancer Nanotechnology Brochure
- Recent Slideshow on NIH Nanotechnology Funding Opportunities/Examples

Other Trans-NIH Initiatives

- Nanoscience and Nanotechnology in Biology and Medicine
- Bioengineering Nanotechnology Initiative
- Bioengineering Research Partnerships
- Exploratory/Developmental (R21) Bioengineering Research Grants

National Heart Lung and Blood Institute

- Innovative Technologies for Engineering Small Blood Vessels
- NHLBI Programs of Excellence in Nanotechnology

National Cancer Institute

- NCI Alliance for Nanotechnology in Cancer (including current funding opportunities)
- Innovative Technologies for Molecular Analysis of Cancer
- Application of Emerging Technologies for Cancer Research

National Institute of General Medical Sciences

- Single Molecule Biophysics and Nanoscience—Research on, and development of, new and improved instruments, methods, and technologies for nanoscience, and for the analysis of single protein and nucleic acid molecules and their complexes in vivo and in vitro. Current approaches include optical and fluorescent spectroscopies, scanning probe microscopy, and biomechanical techniques to analyze the behavior and heterogeneity of single molecules and sub-cellular structures at the nanometer scale. Examples of targets for study include protein or RNA folding, enzyme catalysis, signaling, molecular machines, and the assembly and dynamics of complex cellular structures. A major goal is to develop and enhance existing methods and reagents for the 3-D visualization of cellular processes in living cells in real time at high resolution.

National Toxicology Program (NTP)

Three agencies form the core of the NTP:

- National Institute of Environmental Health Sciences of the National Institutes of Health (NIEHS/NIH)
- National Institute for Occupational Safety and Health of the Centers for Disease Control and Prevention (NIOSH/CDC)
- National Center for Toxicological Research of the Food and Drug Administration (NCTR/FDA)
- The NTP is conducting studies on the toxicology of nanoscale materials
- Report of the first nanotoxicology workshop http://www.nanotoxicology.ufl.edu/aimed at developing appropriate experimental strategies for evaluating the safety/toxicity of nanomaterials

Additional NIH Nanoscience and Nanotechnology Research Funding Opportunities

- DEVELOPMENT OF CELL-SELECTIVE TOOLS FOR STUDIES OF THE BLADDER, PROSTATE, AND GENITOURINARY TRACT—Application Receipt Dates: February 1, 2005
- NOVEL APPROACHES TO CORNEAL TISSUE ENGINEERING—Application Receipt Dates: February 1, June 1, October 1, ending June 1, 2005
- NINDS EXPLORATORY/DEVELOPMENTAL PROJECTS IN TRANSLATIONAL RESEARCH—Application Receipt Dates: February 1, June 1, October 1, ending July 31, 2005
- SMALL BUSINESS BIODEFENSE PROGRAM—Application Receipt Dates: April 1, August 1, December 1, ending Aug. 2, 2005
- CUTTING-EDGE BASIC RESEARCH AWARDS (CEBRA)—Application Receipt Dates: February 1, June 1, October 1, ending Oct. 5, 2005
- NEUROTECHNOLOGY RESEARCH, DEVELOPMENT, AND ENHANCEMENT—Application Receipt Dates: February 1, June 1, October 1, ending October 2006

Information on Funded Grants and Projects for NIHIBECON Program Announcements

- Summary of Funded NIH Bioengineering Nanotechnology Initiative (SBIR)
- Summary of Funded Nanoscience and Nanotechnology in Biology and Medicine (PAR-03-045)

NSF-NSES

The potential impact of research on the nanoscale is well recognized, and yet to remain nationally and internationally competitive formidable challenges must be met. Through a unique partnership of academic institutions, industrial partners, and national labs, the NSF Nanoscale Science and Engineering Center (NSEC) for Integrated Nanopatterning and Detection Technologies, headquartered at Northwestern University under the leadership of Chad A. Mirkin, is helping to meet these challenges by providing a cross-disciplinary research program well integrated with innovative educational outreach programs.

The NU-NSEC brings together recognized leaders from Northwestern University, University of Chicago, University of Illinois/Urbana-Champaign, and Argonne National Laboratory, who, building on the nanopatterning tools and synthetic methods unique to this Center, are making advances in the development of advanced nanopatterning techniques and nanoscale sensors. Potential applications in disease and biological/chemical detection systems arena are expected to be profound. However, this research will also directly impact new technological directions outside of the existing purview of the Center including molecular electronics, catalysis, information storage, and therapeutics.

In a field as dynamic as nanotechnology, collaborations are a vital means to transfer knowledge and an important component of the Center mission. NSEC researchers have numerous meaningful collaborations with industry, national laboratories, and academia. Additionally, the Nanotechnology Corporate Partners (NCP) program provides a forum for companies and business groups with allied interests to develop partnerships with the NSEC and help transition its technology into the private sector. Working with the Kellogg School of Management, the NSEC also offers the Small Business Evaluation and Entrepreneurs (SBEE) program, which gives scientists and engineers with new technologies assistance with development of comprehensive business plans for presentation to potential investors. A $10 million seed fund was established by Lurie Investment

Fund to initiate business development based on technology developed at the Center, and three companies have already been successfully launched.

Excellent in-house user facilities are available to NSEC researchers. Additionally, the State of Illinois committed $5,000,000 to support the development of next-generation instrumentation and to establish a multi-user, multi-purpose, open-access Nanoscale Imaging, Fabrication, Testing and Instrumentation (NIFTI) facility to provide soft lithographic writing, imaging, and analysis capabilities.

The NU-NSEC is also committed to educating the scientists, technicians, and teachers of tomorrow. This commitment is realized not only in the research of graduate students, undergraduates, and post-doctoral associates, but also through the Center's innovative *educational programs*. Some of these programs build on current educational strengths and infrastructure, while others launch new initiatives both internally and through exciting new partnerships with the Museum of Science and Industry, Chicago, and Harold Washington College (a primarily minority institution).

Undergraduates and minority undergraduates from around the country have opportunities to engage in research at the Center. High school science teachers, many from schools with large minority populations, participate in an innovative program which combines hands-on research with curriculum development. NSEC researchers are collaborating with pre-college educators on the development of a "Nanoscience Module" to introduce nanoscience concepts at the high school level through guided hands-on activities and design challenges. the NSEC is also collaborating with the Museum of Science & Industry, Chicago, to produce a design concept for a 13,000 square foot exhibit in nanotechnology, which is expected to open in 2006 and educate millions of museum visitors each year. An interactive web site for children entitled "Mission Spec-tacular" will be launched in spring 2003, and a series of dramatic educational demonstrations entitled "Frontiers of Nanotechnology" has already been premiered with pre-college and lay audiences. NSEC researchers and their students are also actively involved and highly visible in the broader community, through lectures, radio, and television.

Through synergistic research of the highest quality, strong and active collaborations, and innovative outreach programs, the NU-NSEC is advancing scientific knowledge and understanding in the area of integrated nanopatterning and nanoscale detection technologies, while advancing formal education as well as public understanding of nanoscale science and engineering.

Nanoscience and Technology at US—DOE Labs

The Role of DOE in the National Nanotechnology Initiative:

> "The DOE has a stunning portfolio of research and scientific user facilities devoted to visualizing, characterizing, and controlling the nanoworld—from atoms and molecules to bulk materials—which makes the Department's research capabilities unique in the world. the DOE is currently making a broad range of contributions in these areas. For example, the enhanced properties of nanocrystals for novel catalysts, tailored light emission and propagation, nanocomposites and supercapacitors are all being explored. Nanocrystals and layered structures offer unique opportunities for tailoring the optical, magnetic, electronic, mechanical and chemical properties of materials, and DOE researchers are have synthesized layered structures for electronics, novel magnets, and surfaces with tailored hardness.

Specific examples of past accomplishments at DOE include:

- Addition of aluminum oxide nanoparticles that converts aluminum metal into a material with wear resistance equal to that of the best bearing steel
- Novel optical properties of semiconducting nanocrystals that are used to label and track molecular processes in living cells
- Nanoscale layered materials that can yield a four-fold increase in the performance of permanent magnets
- Layered quantum well structures to produce highly efficient, low-power light sources and photovoltaic cells
- Novel chemical properties of nanocrystals that show promise as photocatalysts to speed the breakdown of toxic wastes
- Meso-porous inorganic hosts with self-assembled organic monolayers that are used to trap and remove heavy metals from the environment

The DOE also maintains a large array of major national user facilities that are ideally suited to nanoscience discovery and to developing a fundamental understanding of nanoscale processes. Large computational facilities at DOE will also be key contributors in nanoscience discovery, modeling and understanding.

National Nanotechnology Initiative Research at the DOE

Major new efforts in nanoscale science, engineering, and technology at the Department of Energy will take advantage of opportunities afforded by recent advances. These efforts will be part of the Basic Energy Sciences (BES) program and have the following broad goals:

1. to attain a fundamental scientific understanding of nanoscale phenomena, particularly collective phenomena;
2. to achieve the ability to design and synthesize materials at the atomic level to produce materials with desired properties and functions;
3. to attain a fundamental understanding of the processes by which living organisms create materials and functional complexes to serve as a guide and a benchmark by which to measure our progress in synthetic design and synthesis; and
4. to develop experimental characterization tools and theory/modeling/simulation tools necessary to drive the nanoscale revolution.

The principal missions of DOE in science, energy, defense, and environment will benefit greatly from developments in these areas. For example, nanoscale synthesis and assembly methods will result in significant improvements in solar energy conversion; more energy-efficient lighting; stronger, lighter materials that will improve efficiency in transportation; greatly improved chemical and biological sensing; use of low-energy chemical pathways to break down toxic substances for environmental remediation and restoration; and better sensors and controls to increase efficiency in manufacturing."

Sandia National Labs

It's no wonder that venturing into the tiny domain of atoms and molecules is catching on. Amazing discoveries are being made, inspiring scientists the world over.

Throughout the scientific community, including Sandia National Laboratories, researchers say building things atom-by-atom or molecule-by-molecule will revolutionize the production of virtually every human-made object. *Exciting* prospects—but they also point out the promise of nanotechnology can *only* be realized if we learn to understand the special rules that control behavior at this small scale and develop the skill needed to integrate these concepts into practical devices.

The excitement stems from the understanding that the behavior of materials at the nanoscale is *nothing* like that at the large scale. Only recently have the necessary tools, such as powerful new microscopes, been developed to let researchers see these surprising behaviors. Sandia National Labs' scientists and engineers are among the leading architects and builders of these tools. to appreciate their unique design features, it helps to get an idea of the size of things at the atomic level. In nanoscience objects are measured in nanometers, 1 billionth of a meter. For comparison, the smallest features on current computer chips, measure about 200 nanometers. And a human hair is 100,000 nanometers thick.

Moving beyond observation, scientists are now poised to make exciting advances in nanotechnology, the creation of materials, devices and systems through the *control* of matter at the atomic level. By understanding and controlling the way molecules organize into nanoscale patterns, scientists are discovering new phenomena and learning to design materials with vastly different sets of properties. As one Sandia scientist put it "Design possibilities are limited only by one's imagination."

Sandia Labs continues to enhance its proficiency in many fields, all in keeping with its Department of Energy mission to unite science and engineering to serve national needs. And along with world class capabilities in materials science, micro fabrication—including 40,000 square feet of clean room space—high performance computing and systems engineering—Sandia is uniquely positioned to be the integrating center for new discoveries in nanoscience.

Sandia is uniquely positioned to be the intergrating center for new discoveries in nanoscience. Nano materials are already helping to attain scientist's vision for new technology such as, the *micro-chemlab*, by increasing a thousandfold the ability to collect the chemicals of interest. Currently in the development stage, this handheld device will carry out all of the functions of a full chemistry laboratory allowing first responders, military, and emergency personnel to rapidly access chemical and biological hazards.

ESSENTIAL NANOSCIENCE INSTRUMENTS

X-rays and Neutrons: Essential Tools for Nanoscience Research

Virtually all of the grand challenges in nanotechnology have as overriding themes the need to determine the structure over a wide range of length scales and to understand how the combination of the structure and dynamics leads to their unique properties. Both x-rays and neutrons cover all the length scales of interest in nanoscience, from the atomic structure of individual building blocks to the configuration of assembled, functional structures, making them essential tools for the elucidation of these challenges.

Tools for Nano-technology

Electron Microscopy

In 1924 L. de BROGLIE discovered the wave-character of electron rays thus giving the prerequisite for the construction of the electron microscope. the prototype was built by M. KNOLL and E. RUSKA (Technische Universitat Berlin, 1932). One of the first biological objects depicted was the tobacco mosaic virus (TMV). the first picture of a cell was published in 1945 by K.R. PORTER, A. CLAUDE and E.F. FULLAM (Rockefeller Institute, New York).

The Transmission Electron Microscope (TEM)

The conventional electron microscopy is nowadays called TEM (transmission electron microscopy). We will therefore start with its construction. the ray of electrons is produced by a pin-shaped cathode heated up by current. the electrons are vacuumed up by a high voltage at the anode. the acceleration voltage is between 50 and 150 kV. the higher it is, the shorter are the electron waves and the higher is the power of resolution. But this factor is hardly ever limiting. the power of resolution of electron microscopy is usually restrained by the quality of the lens-systems and especially by the technique with which the preparation has been achieved. Modern gadgets have powers of resolution that range from 0,2-0,3 nm. the useful resolution is therefore around 300,000 x.

The accelerated ray of electrons passes a drill-hole at the bottom of the anode. Its following way is analogous to that of a ray of light in a light microscope. the lens-systems consist of electronic coils generating an electromagnetic field. the ray is first focused by a condenser. It then passes through the object, where it is partially deflected. the degree of deflection depends on the electron density of the object. the greater the mass of the atoms, the greater is the degree of deflection. Biological objects have only weak contrasts since they consist mainly of atoms with low atomic numbers (C, H, N, O). Consequently it is necessary to treat the preparations with special contrast enhancing chemicals (heavy metals) to get at least some contrast. Additionally they are not to be thicker than 100 nm, because the temperature is raising due to electron absorption. This again can lead to destruction of the preparation. It is generally impossible to examine living objects.

After passing the object the scattered electrons are collected by an objective. Thereby an image is formed, that is subsequently enlarged by an additional lens-system (called projective with electron microscopes). the thus formed image is made visible on a fluorescent screen or it is documented on photographic material. Photos taken with electron microscopes are always black and white. the degree of darkness corresponds to the electron density (= differences in atom masses) of the candled preparation.

The Scanning Electron Microscope (SEM)

The path of the electron beam within the scanning electron microscope differs from that of the TEM. the technology used is based on television techniques. the method is suitable for the depiction of preparations with conductive surfaces. Biological objects have thus to be made conductive by coating with a thin layer of heavy metal (usually gold is taken). the power of resolution is normally smaller than in transmission electron microscopes, but the depth of focus is several orders of magnitude greater. Scanning electron microscopy is therefore also well-suited for very low magnifications. Numerous examples will be given in the following.

The surface of the object is scanned with the electron beam point by point whereby secondary electrons are set free. the intensity of this secondary radiation is dependent on the angle of inclination of the object's surface. the secondary electrons are collected by a detector that sits at an angle at the side above the object. the signal is then enhanced electronically. the magnification can be chosen smoothly (depending on the model) and the image appears a little later on a viewing screen.

The properties of the light microscope as opposed to that of transmission and scanning electron microscopes are collected in a table.

Finally some outlines of new and further developments are given. the high voltage electron microscope: it operates with an accelerating voltage of 700—3000 kV. Its power of resolution is greater, the preparation can be thicker, the strain on the preparation is smaller. But the enormous technical expenditure is disadvantageous. Only few gadgets exist. New results concerning botany have not been gained. the scanning transmission electron microscope (STEM): In this development of the SEM do the electrons pass through the preparation and the secondary radiation thus generated is used for image formation. Here, too, the expenditure is large, but it is still worthwhile, since large molecules like nucleic acids or proteins or molecular complexes like viruses can be depicted much better and gentler than with the TEM. No news for botany, though. the interpretation of images gained with electron microscopy is increasingly done with computerized interpretation programs. But they are usually only suitable for the reconstruction of regularly recurring patterns and these, again, are found more often on a molecular level than on a cellular one.

Atomic Force Microscopy—AFM Tools

The Company: Since the late 1990's Novascan has been a nanotechnology company that specializes in atomic force microscope products (AFM) and AFM/SPM services. Our tools are used worldwide to visualize, characterize and manipulate microscopic and nanoscopic environments.

Our Mission: We are committed to developing and producing innovative and state of the art devices that facilitate our clients projects in nanotechnology. These improvements range from specialized AFM instruments to the probes and sensors used to characterize samples.

Technology: At Novascan we strive to develop forward looking devices with the hope that we can help you make new scientific discoveries in your work and projects. Want to learn more?

Scanning Tunneling Microscopy

Quantum Corrals

Scientists discovered a new method for confining electrons to artificial structures at the nanometer lengthscale. Surface state electrons on Cu(111) were confined to closed structures (corrals) defined by barriers built from Fe adatoms. the barriers were assembled by individually positioning Fe adatoms using the tip of a low temperature scanning tunneling microscope (STM). A circular corral of radius 71.3 Angstrom was constructed in this way out of 48 Fe adatoms. STM image shows the direct observation of standing-wave patterns in the local density of states of the Cu(111) surface. These spatial oscillations are quantum-mechanical interference patterns caused by scattering of the two-dimensional electron gas off the Fe adatoms and point defects.

NanoBiology/Life Science

Virtually all the breathtaking advances now taking place throughout the life sciences are directly linked to the ability to resolve and work with nanoscale biologic structures—and no one has provided more leadership to exploring the nanoscale than FEI. the company's pioneering work and continual progress in imaging and understanding the basic building blocks of life is enabling some of the biggest ambitions of NanoBiology, from the rapid characterizations of viruses, cells, DNA and proteins and the development of smart drugs and delivery systems, to dreams of developing in-vivo devices for fighting life-threatening diseases. FEI's opening up the nanoscale is not only inspiring the century's biggest ideas, it is also encouraging the convergence of engineering and scientific disciplines in order to achieve them. the reach of life science and structural biology has vastly expanded thanks to FEI's tools for nanotech and their grasp of molecular and atomic details.

Systems

DualBeam™ /Lab-Based

- Quanta™ 200 3D
- Nova™ 200 NanoLab

Sample Preparation & Transfer Solutions

- Vitrobot™

Scanning Electron Microscopes (SEMs)

- Nova NanoSEM
- Quanta 200
- Quanta 200 FEG
- Quanta 400
- Quanta 400 FEG
- Quanta 400 HV
- Quanta 600
- Quanta 600 FEG
- Quanta 200 HV
- Sirion 200
- Sirion 400

Transmission Electron Microscopes (TEMs)

- Morgagni™ 268(D)
- Tecnai™ G^2 Spirit TWIN
- Tecnai™ G^2 Spirit BioTWIN
- Tecnai™ G^2 20 TWIN
- Tecnai™ G^2 20 X-TWIN
- Tecnai™ G^2 F20 TWIN
- Tecnai™ G^2 F20 S-TWIN
- Tecnai™ G^2 F20 S-TWIN TMP
- Tecnai™ G^2 F20 X-TWIN
- Tecnai™ G^2 F20 X-TWIN TMP
- Tecnai™ G^2 30 TWIN
- Tecnai™ G^2 30 S-TWIN
- Tecnai™ G^2 F30 TWIN
- Tecnai™ G^2 F30 S-TWIN
- Tecnai™ G^2 Sphera
- Tecnai™ G^2 Polara

FEI's Tools for Nanotech bring the nanoscale within the reach of researchers and developers who are working towards the biggest nanotechnology innovations of the 21st century. Our scanning and transmission electron microscopes and DualBeam (FIB/SEM) systems are enabling tools developed to meet the rigorous technology roadmaps of our customers in the diverse markets we serve. Below you can view our entire listing of electron microscopy and focused ion beam systems for NanoResearch, NanoElectronics and NanoBiology.

Software

FEI's software is specifically written to operate with FEI's hardware, guaranteeing a true balance and perfect conjunction between the two. It uses independent and internally consistent calibrations, resulting in data that are reproducible and more accurate. FEI software offers a *total solution,* guiding the user through the entire process by means of a clear workflow with fail-safe routines, that are optimized for ease-of-use and faster time-to-result. FEI software builds on expertise from user networks and leading experts and embodies the user's needs. FEI Company provides full support on all software, and provides dedicated customer training and workshops.

Software for Tecnai Instruments

To guide you through the software and related products

- Low-Dose Exposure Technique
- Tecnai™ Software
- Xplore3D™
- TrueImage™

FEI Systems—Tools for Nanotech

DualBeam™/Fab-Based

System	*Description*
Certus-3D™	The Certus-3D is a full wafer autoloading and fully digital DualBeam system for critical 3D analysis of data storage devices in a clean room environment.
Accura™ XT+	The Accura XT+ is an advanced DualBeam™ mask repair system designed with both focused ion beam (FIB) and environmental SEM (ESEM) columns for repairing defects on today's most complex masks and also for tomorrow's next generation mask technologies. the Accura XT+ can recover any type of defect, enabling advanced masks to reach production more rapidly at reduced cost.
CLM-3D™	The CLM-3D 300 mm DualBeam (FIB/SEM) system provides automated 3D metrology in a clean room environment, delivering real-time process characterization or validation.
Defect Analyzer™ 300 HP	The Defect Analyzer 300 HP is a fully automated in-fab 300 mm DualBeam (FIB/SEM) critical for yield-limiting defect identification, in-line module qualification, and defect and process marginality correlation.

DualBeam™/Lab-Based

Nova™ 600 NanoLab	The Nova 600 NanoLab is a versatile, high-performance DualBeam (FIB/SEM) designed to support the diverse imaging, analytical and sample preparation/modification demands of the high-end lab market. It is equipped with a high-precision, 6" (150 mm) piezo stage that allows investigation of 4", 5" and 6" wafers.
Quanta™ 200 3D	The Quanta 200 3D is a DualBeam (FIB/SEM) system with a tungsten electron column which utilizes ESEM technology to section, image and analyze a wide range of conducting and non-conducting samples.
Altura™ 855	The Altura 855 is a 200 mm, manual loading DualBeam (FIB/

System	*Description*
	SEM) for defect characterization, failure analysis, and automated TEM sample preparation on piece parts and patterned and unpatterned wafers.
Expida 1285	The Expida 1285 is a 300 mm DualBeam (FIB/SEM) for defect characterization, failure analysis, and automated TEM sample preparation on piece parts and patterned and unpatterned wafers.
Nova™ 200 NanoLab	The Nova 200 NanoLab is a versatile, high-performance DualBeam (FIB/SEM) designed to support the diverse imaging, analytical and sample preparation/modification demands of the high-end lab market.
Strata™ 400	The Strata 400 high-performance DualBeam (FIB/SEM) supports analytical laboratories' increasing need for high-resolution analytical capabilities as device geometries shrink and new material systems are introduced.
Strata™ 400 STEM	The Strata 400 STEM high-performance DualBeam (FIB/SEM) supports analytical laboratories' increasing need for highest-resolution analytical capabilities as device geometries shrink and new material systems are introduced. It includes a FlipStage for in-situ foil preparation and 30 kV STEM imaging.
Focused Ion Beam (FIBs)	
VectraVision™	The VectraVision is an advanced FIB (focused ion beam) system for circuit edit applications at the 90 nm process node and beyond.
Laser MicroChemical Etch and Deposition	
Vectra LMC 9900	The Vectra LMC 9900 provides rapid access to backside circuitry for faster, more efficient editing. It's also a micromachining tool for nanotechnology devices such as MEMs and microoptics.
Sample Preparation & Transfer Solutions	
NanoLift™	The NanoLift is an in-situ sample lift-out, load-lock, and transport mechanism used for transferring cartridges containing multiple TEM samples.
Ultra View™	From Wafers to Atoms in Minutes-UltraView is a complete sample transfer and analysis suite (FIB/SEM/TEM) that rapidly extracts samples from in-process wafers and transfers the samples to near-fab labs for ultra-high resolution STEM and atomic-level TEM imaging and analysis.
Vitrobot™	The Vitrobot™ enables automated cryo fixation of TEM samples at constant and user-definable physical and mechanical conditions (e.g. temperature, relative humidity, blotting conditions and freezing velocity). This ensures reproducible, high-quality cryo fixation results and a high sample preparation throughput prior to cryo observation in a TEM.

System	*Description*
Scanning Electron Microscopes (SEMs)	
Nova NanoSEM	World's only true high-resolution, low-vacuum FEG-SEM for characterization of charging and/or contaminating nano-materials or -devices.
Quanta 200	The Quanta 200 SEM is a versatile high performance scanning electron microscope, with three modes (high-vacuum, low-vacuum and ESEM) to accommodate the widest range of samples of any SEM system.
Quanta 200 FEG	The Quanta 200 FEG SEM is a versatile high resolution scanning electron microscope with three modes (high vacuum, low vacuum and ESEM) to accommodate the widest range of samples of any SEM system.
Quanta 400	The Quanta 400 SEM is a versatile high resolution scanning electron microscope with three modes (high-vacuum, low-vacuum and ESEM) to accommodate the widest range of samples of any SEM system.
Quanta 400 FEG	The Quanta 400 FEG SEM is a versatile high resolution scanning electron microscope with three modes (high vacuum, low vacuum and ESEM) to accommodate the widest range of samples of any SEM system.
Quanta 400 HV	The Quanta 400 HV SEM is a high performance, entry-level scanning electron microscope with a tungsten electron source that fulfils all needs for conventional imaging and microanalysis.
Quanta 600	The Quanta 600 is a versatile high resolution SEM (scanning electron microscope) with three modes (high vacuum, low vacuum and ESEM) to accommodate the widest range of samples of any SEM system.
Quanta 600 FEG	The Quanta 600 FEG SEM is a versatile high resolution scanning electron microscope with three modes (high vacuum, low vacuum and ESEM) to accommodate the widest range of samples of any SEM system.
Quanta 200 HV	The Quanta 200 HV SEM is a high performance, entry-level scanning electron microscope with a tungsten electron source that fulfils all needs for conventional imaging and microanalysis.
Sirion 200	The Sirion 200 SEM is an ultra-high resolution scanning electron microscope with immersion lens electron optics and in-lens detector technology for nanocharacterization and analysis.
Sirion 400	The Sirion 400 SEM is an ultra-high resolution scanning electron microscope with immersion lens electron optics and in-lens detector technology for nanocharacterization and analysis.
Stylus NanoProfilometer (SNP)	
SNP XT	Stylus Nanoprofilometry (SNP) for Automated, 3D Mask Metrology and Defect Inspection for 90 nm Technology and Beyond

System	*Description*
Transmission Electron Microscopes (TEMs)	
Morgagni™ 268(D)	The Morgagni 268(D) TEM is an adaptable, easy-to-use digital transmission electron microscope that delivers a high sample throughput and excellent image quality for cell biology applications, diagnostic screening, and chemical and pharmaceutical development.
Titan™ 80-300 S/TEM	FEI's Titan™ 80-300 kV transmission electron microscope is a high-end imaging and analytical S/TEM instrument dedicated to corrector and monochromator technologies. Designed for optimum stability which enables sub-Angstrom imaging of nanostructures in both TEM and STEM modes.
Tecnai™ G^2 Spirit TWIN	Tecnai G^2 Spirit TEM is designed to meet the stringent imaging and analysis demands of Life Science and Soft Matter research. Spirit produces excellent 2D and 3D images. the TWIN lens configuration is designed for general high-resolution purposes.
Tecnai™ G^2 Spirit BioTWIN	Tecnai G^2 Spirit TEM is designed to meet stringent imaging and analysis demands of Life Science and Soft Matter research. Spirit produces excellent 2D and 3D images. the BioTWIN lens configuration is designed specifically for high-contrast imaging.
Tecnai™ G^2 20 TWIN	The Tecnai G^2 20 TEM is a highly automated 200 kV transmission electron microscope equipped with a TWIN lens. It can be equipped with dedicated software and hardware for superb TEM imaging, both in 2D and in 3D.
Tecnai™ G^2 20 S-TWIN	The Tecnai G^2 20 S-TWIN is a 200 kV TEM (transmission electron microscope) for high spatial resolution imaging and microanalysis with a tungsten or LaB6 filament source.
Tecnai™ G^2 20 X-TWIN	The Tecnai G^2 20 X-TWIN is a 200 kV transmission electron microscope (TEM) for high spatial resolution imaging and microanalysis with a tungsten or LaB6 filament source.
Tecnai™ G^2 20 U-TWIN	High-resolution TEM (transmission electron microscope) for imaging at atomic resolution.
Tecnai™ G^2 F20 TWIN	The Tecnai G^2 F20 TWIN TEM combines excellent imaging performance with excellent analytical capability using the Schottky FEG electron source.
Tecnai™ G^2 F20 S-TWIN	The Tecnai G^2 F20 S-TWIN TEM (transmission electron microscope) provides excellent imaging performance and versatile capabilities making it the perfect general analytical laboratory tool.
Tecnai™ G^2 F20 S-TWIN TMP	The Tecnai G^2 F20 S-TWIN TMP TEM (transmission electron microscope) is an analytical system with excellent high-resolution imaging and atomic resolution microanalysis capabilities. It has a totally oil-free vacuum system, utilizing turbo and ion-getter pumps throughout, without any need for cold traps or liquid nitrogen.

System	Description
Tecnai™ G^2 F20 X-TWIN	The Tecnai G^2 F20 X-TWIN TEM (transmission electron microscope) is an analytical system with excellent high-resolution imaging and atomic resolution microanalysis capabilities. the system is optimized for high spatial resolution analysis and has an EDS solid angle of 0.3 steradian.
Tecnai™ G^2 F20 X-TWIN TMP	The Tecnai G2 F20 X-TWIN TMP TEM (transmission electron microscope) is an analytical system with excellent high-resolution imaging and atomic resolution microanalysis capabilities. It has a totally oil-free vacuum system, utilizing turbo and ion-getter pumps throughout, without any need for cold traps or liquid nitrogen.
Tecnai™ G^2 F20 U-TWIN	The Tecnai G^2 F20 U-TWIN TEM is a very high-resolution transmission electron microscope popular in the semiconductor and data storage industries.
Tecnai™ G^2 30 TWIN	The Tecnai G^2 30 TWIN TEM is a highly automated 300 kV transmission electron microscope equipped with a TWIN lens. It can also be equipped with dedicated software and hardware for superb TEM imaging, both in 2D and in 3D.
Tecnai™ G^2 30 S-TWIN	This high-resolution transmission electron microscope (TEM) has good analytical capabilities and features excellent performance in imaging of a complete range of materials samples.
Tecnai™ G^2 30 U-TWIN	The Tecnai G^2 30 U-TWIN TEM is the thermal emission transmission electron microscope with the highest resolution. It has excellent sample penetration characteristics and can be very successfully equipped with a GIF for elemental mapping
Tecnai™ G^2 F30 TWIN	The Tecnai 300 kV field emission TEM is the workhorse transmission electron microscope for all laboratory types. Especially valuable in multi-user laboratories, these systems utilize advanced windows protocols and registry to allow every user to complete their research or analysis in a timely fashion.
Tecnai™ G^2 F30 S-TWIN	The Tecnai G^2 F30 S-TWIN TEM (transmission electron microscope) is a perfect high-end analytical laboratory tool with excellent and versatile capabilities for high resolution imaging as well as extremely good analytical performance. Ultra-stable high-tension technology ensures reliable and excellent long-term energy stability for electron energy loss experiments.
Tecnai™ G^2 F30 U-TWIN	Tecnai G^2 F30 U-TWIN TEM (transmission lectron microscope) is the ultimate high-resolution imaging tool for the top-laboratory needs.
Tecnai™ G^2 Sphera	The Tecnai G^2 Sphera TEM is a 200 kV transmission electron microscope equipped with a TWIN lens and a high level of automation. It has dedicated software and hardware for superb TEM imaging, both in 2D and in 3D.

System	*Description*
Tecnai™ G^2 Polara	The Tecnai G^2 Polara TEM is a 300 kV electron transmission electron microscope dedicated for cryo applications at liquid nitrogen and liquid helium temperatures.

NANOTECHNOLOGY INFORMATION RESOURCES, LINKS AND WEBSITES

AZoNano Information

Why AZoNano

A totally new concept in the field of Nanotechnology publishing and information provision—AZoNano.com (the A to Z of Nanotechnology) was formed with the primary aim of increasing the use of Nanotechnology by the science, engineering and design community worldwide.

Over 21 million engineers and scientists now use the internet as part of their daily lives, yet the vast majority of them are still unaware what can be achieved by the use of Nanotechnology. AZoNano.com is here to change that.

AZoNano's Mission

The aim of AZoNano.com is to become the primary Nanotechnology information source for the science, engineering and design community worldwide. It also aims to be the primary publicist of news, views and developments within the Nanotechnology community.

However, unlike many other Nanotech related organisations AZoNano.com is totally focussed on the needs of the end users of Nanotechnology. to achieve this aim, all of the educational, informative and news content on AZoNano.com is easy to access and search and is provided on a free of charge, no subscription, no charge per article, totally free basis.

The Idea

The idea to launch AZoNano.com originated from discussions held in 2002 between AZoM.com and the Institute of Nanotechnology.

Both parties to the AZoNano.com collaboration were aware that Nanotechnology had a big future, but both organisations were frustrated by the lack of awareness of the end-user industrial community.

Consequently, a decision was taken to build a Nanotechnology knowledgebase that would really help to translate Nanotechnology into tangible industrial benefits. Making it very easy for the layman and non-Nanotech specialist to discover how Nanotechnology could benefit them and bringing together the providers of Nanotechnology with organisations that can benefit from implementing Nanotechnologies.

The construction of AZoNano.com began in 2003 using the already proven and successful AZoM.com internet portal and technology platform. the AZoM.com site having successfully demonstrated how materials technologies and information can be distributed to millions of end-users via the web.

Currently, AZoM.com attracts over 440,000 visitor sessions per month (over 6 million hits per month) and is the premier on-line materials destinations.

AZoNano's Team

AZoNano.com is fortunate to have a first rate team of scientists, software engineers, editors and

Nanotech specialists, these include:

- Ottilia Saxl—CEO, the Institute of Nanotechnology
- Dr. Ian Birkby—CEO, AZoM.com
- Dr. Cameron Chai—Editor in Chief
- David Freund—Features Editor
- Joel Scanlon—CTO
- Rolf Easto—Sales Manager
- Grant Millar—Editor/Researcher

Site Listing of Nanotechnology

- *Nanotechnology Now*—offers a resource portal for nano science and nano tech news with event information, interviews, directories, glossary, and a tutorial on the basics.
- *NanotechNews.com*—news for nanotechnologists.
- *NanoComputer Dream Team*—desires to directly influence the design and development of a nanometer scale computer.
- *Nanomedicine*—analyzes developments in molecular machine systems to address medical problems.
- *Scientific American: Nanotechnology*—features current coverage, highlights, and article archives.
- *A to Z of Nanotechnology*—offering nanotechnology information, news, new products, suppliers, and expert directory.
- *Nanotech 2005:2005 NSTI Nanotechnology Conference and Trace Show*—May 8-12, 2005; Anaheim, California, U.S.A.
- *Nanotechnology Gallery*—images, presentations, reports, and webcasts from NASA's Center for Nanotechnology.
- *NanoTechWire.com*—provides news and information on nanotechnology resources, research, patents, materials, investments, industry events, and optics.
- *Nanotech 2003: NanoTechnology Conference & Trade Show*—February 23-27, 2003; San Francisco, CA.
- *ANTS: Automomous NanoTechnology Swarm*—NASA mission architecture and AI paradigm for space exploration based on large numbers of autonomous specialized spacecraft working together as inspired by social insect colonies.
- *Nanoword.net*—online distributor of publications, with a focus on nano-science, technology, and education. Includes the Encyclopedia Nanotech.
- *Nanotechnology in Australia*—information for nanotechnology capability companies in Australia.
- *Nanotechnology Engineering*—collection of articles presented by the Institute of Atomic-Scale Engineering.
- *Societal Implications of Nanoscience and Nanotechnology*—final report from the September 2000 workshop held at the National Science Foundation.
- *Foresight Institute vs. Scientific American: Debate on Nanotechnology*
- *Exploring the Nanoworld*—images, K-12 teaching material, and general information about nanotechnology.
- *Nanodot*—news and discussion of emerging technologies.
- *Nanotechnology & the Motley Fool*—Download this free report from the Motley Fool and discover a handful of rocket stocks with spectacular profit potential. Don't miss out on

the next round of market-busting growth stocks, www.fool.com

- *Nanotechnology Consulting*—Nanotech Plus is a nanotechnology consulting firm providing information on nanotechnology products, markets and applications with a realistic viewpoint, www.nanotechplus.net
- *Subscribe—Forbes/Wolfe Nanotech Report*—Intelligent nanotechnology research based on a sound understanding of science. Get 12 issues of the Forbes/Wolfe Nanotech Report for $195—a 67% savings—plus a free report. Click here, www.forbesnanotech.com

Open Directory—Science Technology and Nanotechnology

- *The Foresight Institute*—Non-profit institute focused on nanotechnology, the coming ability to build materials and products with atomic precision, and systems to aid knowledge exchange and critical discussion, thus improving public and private policy decisions.
- *Nanotechnology Now*—Comprehensive and informative nanotechnology portal. Provides introduction to nanotechnology, general information, images, interviews, news, events, research, books, glossary and links.
- *Applied Computational Research Society*—Nonprofit organization fostering interdisciplinary approaches to modeling and simulation in the advancement of computational science and high technology. Focused on nanoscience and MEMS applications.
- *The ASME Nanotechnology Institute*—American Society of Mechanical Engineers: nanotechnology resources, events, programs, activities, online courses, webcasts, news and links.
- *CMP Cientifica*—News, networks, conferences and nanotechnology resources in Europe.
- *Euspen*—European Society for Precision Engineering and Nanotechnology—contains society news, conference information, discussion forums and expertise search. Euspen is a not-for-profit company based at Cranfield University in the UK, with members around the world.
- *Exploring the Nanoworld*—Movies of nano-structured materials: ferrofluids, memory metals, amorphous metals, LEDs, re-entrant foams, self-assembly, DNA, Magnetic Resonance Imaging, atomix imaging, Lego models, Giant Magneto Resistance, silicon, carbon nanotubes, polymers, and semiconductor devices. National Science Foundation supported Materials Research Science and Engineering Center on Nanostructured Materials and Interfaces at the University of Wisconsin-Madison.
- *The gRobots Project*—An open source project to build and run a globally distributed supercomputer for the simulation of evolvable nanotechnology.
- *How Stuff Works: How Nanotechnology Will Work*—Animated narrative shows how Nanotechnology has the potential to totally change manufacturing, health care and many other areas.
- *IBM Almaden STM Molecular Art*—Some of the famous images of atoms and molecules made with IBM's scanning tunneling microscope.
- *Institute for Atomic-Scale Engineering*—Personal page of Forrest Bishop: published and unpublished concept articles.
- *Institute of Nanotechnology (UK)*—The Institute of Nanotechnology has been created to foster, develop and promote all aspects of science and technology in those domains where dimensions and tolerances in the range of 0.1 nm to 100 nm play a critical role.
- *Iran Nanotechnology Policy Studies Committee*—A committee for studying nanotechnology trends in Iran and the world.

- *Minatec*—Centre for innovation in micro and nanotechnology. Teaching, research, and industrial development.
- *Minatec*—Centre for Innovation in Micro and Nanotechnology in Grenoble. Teaching, research, and industrial development.
- *MITRE Corp. Nanosystems Group*—Reviews current nanotechnology research and people behind it, downloadable copies of several widely-used articles. Nanoelectronics focus.
- *Molecular Manufacturing Shortcut Group*—Promotes developing nanotechnology as a way to facilitate space exploration and settlement.
- *Moletronics and Nanotechnology references*—Links to sites with information mainly on molecular electronics.
- *Nano Science & Technology Institute*—Events, publications and news in nanotechnology, microtechnology and biotechnology.
- *Nano Space*—Nano technology as it applies now and in the future to Space Exploration and Studies.
- *Nano Apex*—Nanotechnology portal.
- *Nanobase*—Loyola College, Maryland. National Science Foundation sponsored site presenting up-to-date sources of information on nanotechnology in the areas of major research centers, funding agencies, major reports and books.
- *NanoComputer Dream Team*—Multidisciplinary project to make Earth's first nanometer supercomputer via Internet.
- *Nanodot*—News and Discussion of Coming Technologies: Weblog devoted to nanotechnology-related news stories, with attached discussion forum, (based on Slashdot software)
- *Nanoforum.org: European Nanotechnology Gateway*—Offers background information, news and developments in industry and research and a selection of publications.
- *Nanogloss*—Online dictionary of nanotechnology.
- *Nanoguys*—Nanotechnology Career Consultants. Recruiting and placement of staff in nanotechnolgy-related positions such as Organic, Inorganic, Biological, Project Management, Research Scientist, Engineer, and Lab Technician.
- *Nanomagazine*—Interviews with people in the Nanotechnology field.
- *NanoManipulator: 3rdTech*—Commercial version of the NanoManipulator system originally from University of North Carolina, Chapel Hill.
- *NanoManipulator: University of North Carolina*—The nanoManipulator provides an improved, natural interface to SPMs (STMs and ATMs) by coupling the microscope to a virtual-reality interface giving a virtual telepresence on the surface, scaled by a factor of about a million, to one.
- *Nanomedicine*—Bevoted to the medical applications and implications of molecular nanotechnology. Nanomedicine may be defined as the monitoring, repair, construction and control of human biological systems at the molecular level, using engineered nanodevices and nanostructures. Includes the Nanomedicine Art Gallery. (From the Foresight Institute.)
- *Nanoscale Science and Engineering*—National Science Foundation. Activities, Information & Education, Research & Development, Connections, and NSF & NNI (National Nanotechnology Initiative) Reports.
- *NanoScout*—Portal with forums, and links sorted by topic and popularity.
- *NanoSpace*—The Center for NanoSpace Technologies is a Texas-based nonprofit scientific

research and education foundation chartered to conceive, establish, and conduct cutting-edge technology, research, and development in the areas of aerospace, education, energy, life sciences, and shipping and transportation.

- *NanotechNews*—News for nanotechnologists and investors: regular updates, many links to other nanotechnology websites, archives, search function, newswire.
- *Nanotechnology*—Institute of Physics quarterly journal.
- *Nanotechnology*—Institute of Physics monthly journal for aspects of nanoscale science and technology
- *The NanoTechnology Group*—A consortium of nano companies, universities and organizations developing a nanoscience curriculum for K-12.
- *Nanotechnology in Manufacturing*—What Next? the Coming Revolution In Manufacturing: Autodesk Technology Forum Presentation by John Walker, May 10th, 1990. Somewhat dated but well worth reading.
- *Nanotechnology Journals Database*—References, and links to, papers about nanotechnology, from selected scientific journals (e.g. PRB, PRL, JACS, JAP, JPCB).
- *Nanotechweb.org*—A portal for nanoscience and nanotechnology online resources for academics, industrialists and investors, as well as for laymen interested in nanotechnology.
- *NanoTech Wire.com*—Provides information for nanotechnology and research.
- *Nanotek*—Nanotechnology: articles online, news, pictures, cvs.
- *nanoTEN*—Nanotechnology consulting firm. Services include nanotechnology matchmaking, assessment of trends, feasibility studies, materials and products analysis, and presentation assistance.
- *nanoTITAN*—Nanotechnology Software, Data & Services: nanoTITAN offers Java libraries and applications, a nanodevice database, the nanoML specification, custom simulation and visualization to the nanotechnology and general scientific communities.
- *Nanotube Patent Search*—Results from a search of nanotube or nanotube-related patents at the United States Patent and Trademark Office
- *The Nanotube Site*—Michigan State University's Library of Links for the Nanotube Research Community
- *NanoWave*—An MIT startup, Nanowave offers SPPE (Scanning Probe Position Encoder) technology for picometer-order (subnanometer) resolution over ranges greater than several inches.
- *National Science Foundation*—Partnership in Nanotechnology (US). Recent technical and policy reports, funding opportunities, profiles of projects currently supported, and links to other nanoscience/technology sites.
- *Phantoms*—European Union sponsored nanoelectronics network. Contains comprehensive information about European nanotechnology, and links to groups and events worldwide.
- *Polymer Simulation*—Computer simulation of a polymer sticking to a surface.
- *Scanning Tunneling Microscope (STM)*—Describes the invention of the topografiner, a precursor instrument, between 1965 and 1971, and also tells of the STM's development.
- *Scientific American*—Nanotechnology articles, some free and some archived.
- *Self-Replicating Systems*—NASA and Self-Replicating Systems: Implications for Nanotechnology.
- *Small Times*—Daily articles covering MEMS, nanotechnology, and microsystems, with a business angle.
- *Stony Brook Buckyballs*—A virtual tour of fullerenes in Laszlo Mihaly's laboratory at the

Physics Department at SUNY, Stony Brook.

- *STT: Netherlands Study Center for Technology Trends*—Somewhere in this confusing site is good material in English, though it is often labeled in Dutch, and Dutch material is often labeled in English. STT hopes to bring their project to the attention of a wide audience, thereby contributing to the development of nanotechnology.
- *Technology Review: Nanotechnology and Materials*—Articles, discussion and services for nanotech built on MIT's award-winning monthly magazine.
- *Thermodynamics Research Laboratory*—Classical, Quantum, Nanoscale & Statistical Thermodynamics & Mechanics data and property calculation sites.
- *Virtual Journal of Nanoscale Science & Technology*—A weekly multi-journal compilation of the latest research on nanoscale systems. Published by the American Institute of Physics and the American Physical Society.
- *VISION Online*—Virtual Institute supported by the EC, committed to promotion and training in the precision engineering and nanotechnologies. Contains discussion forums, training programmes and patent library.
- *Xerox/Zyvex Nanotechnology*—Former Xerox Palo Alto Research Center Nanotechnology page: brief introduction to core concepts of molecular nanotechnology (MNT), and links for further reading.
- *Yahoo Groups: Nanotechnology*—Active nanotechnology discussion group.
- *Yashnanotech*—Provides marketing and consultancy services for products developed by nanotechnology. Promotes nanotech awareness through training, educational publications, seminars, workshops and conferences.

Nanoword Feature Articles

Interview with Rocky Rawstern

We've heard plenty about the potential of nanotechnology from scientists, engineers, capitalists and entrepreneuers, but what does the unbiased public want and expect from nanotechnology? Rocky Rawstern addresses public expectations, media spin and 'the singularity'.

Trends in NanoTechnology (TNT) 2001

The TNT 2001 conference took place September 3-7, 2001 in Segovia, Spain. Some of the more trendy aspects of nanotechnology were: Electronics, Photonics, Nanotubes, Mechanical Devices, Theoretical Models, Self-Assembly, Networks and Funding.

Molecular Electronics

With our increasing understanding of electron transfers within and between individual molecules, it may only be a matter of time before a lucky researcher discovers the right parameters to cheaply integrate moles of addressable molecular circuits.

Interview with Tim Harper

This interview with Tim Harper, CEO of CMP Cientifica discusses what makes a product 'nano', revolutionary products and the potential of nanotechnology.

Alchemy, the Next Generation

Nanotechnology can be qualitatively understood as an approach to alchemy that is based on the current model of matter. Only time will tell whether its dreams will be realized or face the same fate as the philosopher's stone and the elixir of life.

Electrons: from Microscopy to Lithography

The electron is a most convenient particle (or wave) for microscopy and lithography. Whether they are being focused via e-beams or tunneled through STM tips, without the electron there would be no nanotech.

Mechanical VLSI

The miniaturization of mechanical devices (MicroElectroMechanical Systems—MEMS) is gaining increasing attention due to numerous commercial applications. Integration of basic MEMS components is bridging the gap between micro- and nanotechnology.

Frontiers in Optical Lithography

Several methods are available to the chip designer who wishes to use light to carve nanoscale features. the diffraction barrier can be broken in various ways and novel photosensitive materials promise a bright future for photolithography.

Nanotech Education

Which subjects should a student learn in preparation for nanotechnology? Biology, chemistry, physics, engineering and computer science each provide a unique perspective of the nanoscale. Various combinations of these fields are where the most progress is being made, thus resulting in interdisciplinary nanoscience.

Surface Wetting

Understanding surface wetting has had important implications for preventing wear, for medicine and for self-cleaning products. A model for surface tension based on a two-dimensional frozen film is proposed.

Nanoscale Chemistry

While chemistry deals with molecules in a statistical sense, nanotechnology deals with them as discrete entities, each requiring special attention. For instance, in order to understand and fabricate molecular assembly lines or carry out single molecule reactions in vessicles it is useful to tweak some fundamental units such as temperature.

Quantum Biology

At a time when numerous physicists are racing to construct quantum computers, molecular biologists may unknowingly be racing to dismantle them.

Biomolecular Self-Assembly

Living systems have a vast array of molecular components that are specially designed for self-assembly. Why build your own components when you can borrow DNA, protein and lipids from nature?

Seeing Things Smaller Than Light

Scanning Probe Microscopy (SPM) is a family of techniques that allows one to observe and create features far smaller than the wavelength of visible light. STM, SFM (AFM), SNOM (NSOM), MFM and SCM are a few of the variations.

Quantum Dots

Quantum dots have zero dimensional electronic properties and could be used as the qubits in quantum computers. Quantum dots are nothing new to biological systems, which use them for everything from photosynthesis to the electron transport chain.

DNA Mediated Energy Transfer

The ability of DNA to transfer electrons through the one-dimensional double helix has been experimentally observed. As biologists figure out what this means for medicine, nanoscientists are using DNA to assemble nanoelectronic devices.

Nanolinks

Nanoelectronics Planet

Nanoelectronics Planet provides news and events for the nanotechnology business communities of interest: business development, investment.technology transfer, and government.

NanoTechnology Magazine

Excellent all-around source for news and information about nanotechnology. Good for beginners as well as seasoned veterans.

Ralph C. Merkle's Nanotechnology Web Site

Good introduction to nanotechnology with lots of links to additional information. the site is maintained by Ralph C. Merkle, a scientist at Xerox PARC.

CSC Scientists Discover Ways to Build Chips From the Atom Up

This article from the Computer Sciences Corporation details how they reduced the size of existing chips but by building them up from individual molecules.

Feynman's 1959 talk

Nobel Laureate Richard P. Feynman's 1959 talk to the American Physical Society, in which he discusses minituarization at the molecular level. It is widely viewed as the seminal speech that launched nanotechnology as an independent field of research.

The Nanoelectronics and Nanocomputers Home Page

Focuses on the electronics and computer applications of nanotechnology. An excellent source for top-down information.

Books

These and other books can be ordered from *the Foresight Institute's book store.* the best technical introduction to this area is:

- *Nanosystems: molecular machinery, manufacturing, and computation* by K. Eric Drexler (Wiley, 1992).

A technical introduction to medical applications of nanotechnology:

- *Nanomedicine*, currently being written by Robert A. Freitas, is a technical introduction to medical applications of nanotechnology. Volume I is available. A web page on *Nanomedicine* has an overview. the full text is also available.

Further reading:

- *Engines of Creation* (full text in html) by K. Eric Drexler (Anchor 1986) discusses both the technology and its possible applications and consequences.
- *Unbounding the Future*, by K. Eric Drexler, Chris Peterson and Gayle Pergamit (Quill, 1991) provides a non-technical discussion of what nanotechnology should let us do, using technically feasible scenarios to clearly illustrate the possibilities.

- *Nano!* by Ed Regis (Little, Brown, 1995) is an engaging and entertaining book that describes the researchers involved in this area, particularly Drexler, and the reactions of different members of the scientific community to the concept.

Journals, Publications and Newsgroups

- The Foresight Update is a newsletter published by the Foresight Institute and is an excellent way to keep abreast of developments and events in this rapidly moving area.
- *Foresight lists a variety of resources about nanotechnology.*
- *sci.nanotech* is a net news discussion group that covers nanotechnology and related areas.
- The *journal Nanotechnology* covers nanotechnology both in the specific sense used here and in the broader sense. It has had a special issue for each of the Foresight Conferences on Molecular Nanotechnology and is well worth reading. Ask your library to subscribe! *Nanotechnology* is published by the *Institute of Physics Publishing.*

General Information and Policy

- Future Trends in Microelectronics Device Simulation (JPL) (May 1999)
- *International Comparative Study on Applying Molecular and Mat erials Modelling* [World Technology Division (WTEC), *International Technology Research Institute (ITRI)*] (March 2000)
- *CAD Framework initiative* standards for CAD tools and systems.
- *DOE Initiative in Computational Materials Science* (May 1998)
- *DOE Strategic Simulation Plan: Basic Science Component* (October 1998)
- *IDA MEMS* ("identifying transition opportunities for MEMS technology into defence systems") (2000)
- *Future & Emerging Technologies (FET) Proactive Initiative: Nanotechnology Information Devices (NID)* (European Commission, Information Society Technologies (IST))
- *PHANTOMS*—Nanotechnology for Information Processing and Storage (European Commission, Information Society Technologies (1ST)) *http:// www.phantomsnet.com*
- *Technology Roadmap for Nanoelectronics* (MELARI NANO, ed. 1999)
- *U.S. National Nanotechnology Initiative* (July 2000)
- *Nanotechnology Research Directions: Vision for Nanotechnology R&D in the Next Decade* (NSTC IWGN Workshop Report, Sept. 1999)
- *NSF National Nanotechnology Initiative* (| Crosscutting | NANO) (2000)
- *Notes on Funding of Nanotechnology in Japan* (1999)
- *WTEC Panel Report on Nanostructure Science & Technology* (August 1999)
- *Introduction to Nanotechnology* (from Zyvex)
- *Molecular Simulation and Nanotechnology*
- *Nanoelectronic Modelling* (JPL) (November 1998)

Associations

Directories of Associations

- *Professional Societies and Associations in Metals and Materials Engineering* listed at the University of British Columbia Library:

International

- *IEEE*—Institute of Electrical and Electronic Engineers
- *Semiconductor Equipment and Materials International*

- *International Society for Molecular Electronics & BioComputing (ISMEBC)*
- *Materials Research Society (MRS)*
- *Microelectricalmechanical Systems (MEMS) Clearinghouse*
- *Applied Computational Research Society (ACRS)*
- *Foresight Institute (FI)* (non-profit organization devoted to nanotechnology)
- *Institute for Molecular Manufacturing (IMM)*
- *Institute of Nanotechnology*

Regional /National

- *CORDIS* Community Research and Development Information Service the European Commission's Research home page
- *IST* the European Commission's fifth framework Information Society Technologies Program
- *FET* Future and Emerging Technologies; a key action within the IST program
- *FET Initiative Quantum Information Processing and Communication*
- *Psi-k* (or *try here*). Psi-k is an EU network dealing with quantum chemical calculations of solids and materials for the semiconductor industry
- *ATOMCAD* An EU network, the objective of which is to integrate new atomistic TCAD tools with existing conventional packages with the aim of introducing atomic scale modelling into the microelectronics industry
- *EMRS* the European Materials Research Society
- *NEXUS User-Supplier Club (USC) on CAD for Microsystems* (European Microsystems Technology On-Line (EMSTO))
- *National Microelectronics Application Centre* (product development, design, information technology)
- *IEE* Institute of Electrical Engineers (UK)
- *Institut fuer Nanotechnologie.* Forschungszentrum Karlsruhe
- *DARPA* Defence Advanced Research Projects Agency (USA)
- *DARPA Microsystems Technology Office,* quantum computing. Involves theory and technology demonstration
- Use of the spin, degree of freedom for new devices, *SPINS,* including quantum information processing
- *NSF* National Science Foundation (USA)
- *Electronics, Photonics and Device Technologies* Program run by the NSF
- *DesCartES,* Distributed Center for Electronic Simulation, is a program run by the NSF in the field of atomic/quantum scale simulation in electronics devices down to the nanoscale
- *NIST Office of Microelectronics Programs* (USA)
- *Computational Material Science Network (CMSN)* (USA)
- *DoD Computational Electronics & Nanoelectronics (CEN)* (USA)
- *MITI* the Ministry of Trade and Industry (Japan)
- *Taiwan Semiconductor Industry Association*
- *Laboratory of Semiconductor Materials science.* Institute of semiconductors, Chinese academy of sciences
- *Institute of Software,* Chinese Academy of Sciences
- *Center for Nanoscale Science and Technology* (Rice University, USA)
- *Center for Simulating Dynamic Response of Materials* (USA, Caltech)
- *Institute for Nanotechnology* (Northwestern University, USA)
- *Interagency Working Group on Nano Science, Engineering and Technology* (USA)

- *Metallurgy Division* of the U.S. National Institute of Standards (NIST) ("developing measurement and standards infrastructure for nanotechnology") (USA)
- *National Nanofabrication Users' Network (NNUN)* ("provides users with access to some of the most sophisticated nanofabrication technologies in the world with facilities open to all users from academia, government, and industry") (USA)
- *Japan Society for Simulation Technology on NACSIS* (Japan)
- *Atom Technology Group, National Institute for Advanced Interdisciplinary Research (NAIR)* (Japan)

Companies

- *Alcatel Microelectronics* (Belgium)
- *Infineon Technologies AG* (Germany)
- *Philips Semiconductors* (Holland)
- *STMicroelectronics* (France)
- *TDK Semiconductor Europe*
- *Advanced Micro Devices Inc.* (USA)
- *Analog Devices Inc.* (USA)
- *CMP Cientmfica* (Spain)
- *Dallas Semlconuductor* (USA)
- *International Business Machines (IBM) Corporation* (USA)
- *Intel Corporation* (USA)
- *Lucent Technologies Microelectronics Group—Agere Systems* (USA)
- *MEMX, Inc.* Offering turn-key MEMS and Nanotechnology Design and Product realization capabilities
- *Motorola Semiconductor Products Sector* (USA)
- *Semiconductor Research Corp.* (USA)
- *SRI International* (USA)
- *Texas Instruments* (USA)
- *Fujitsu* (Japan)
- *FECIT*—Fujitsu European Centre for Information Technology
- *Hitachi* (Japan)
- *NEC Global* (Japan)
- *Sony Semiconductor* (Japan)
- *Toshiba Corporation* (Japan)
- *Hyundai Heavy Industries* (Korea)
- *Samsung Semiconductor* (Korea)
- *California Molecular Electronics*
- *HP Labs*
- *Innovations for High Performance (IHP) Microelectronics*
- *List of Company Links* (from Nanotechnology Industries)
- *List of Company Links* (from IDA MEMS, DARPA)
- *List of Organizations* (from Mitre)
- *Molecular Manufacturing Enterprises, Inc. (MMEI)* (a seed capital firm)
- *Nanochip* ("Develops mass storage devices, based on molecular scale technology")
- *NanoMagnetics, Ltd.*
- *Nanomat* ("manufacturing nanomaterials and nanotechnology for structural, non-structural, microelectronic, and biomedical applications")

- *NanoWorld Projects*
- *SciCo* ("innovation for materials design—data, simulation, analysis")
- *Zyvex* ("The First Molecular Nanotechnology Company")

Academic Institutes and Groups

- *NMRC* at *University College Cork* Ireland
- *Computational Modelling* at the NMRC.
- *IMEC* Belgium
- *DIMES* Delft Institute for Microelectronics and Sub-Micron Technology at the technical university of Delft, Holland
- *Institute for Microelectronics* at the Technical University of Vienna
- *Nanoscale Science Laboratory* at the Cambridge University Engineering Department
- *Silicon Technology Research at the University of Edinburgh*
- *Scottish Microelectronics Centre* at the *department of electronic and electrical engineering,* university of Edinburgh, Scotland
- *TCAD* Research at the University of Stanford
- *TCAD Training and Research* at Louisiana Tech University
- *Microsystem Technology laboratory at MIT.* Includes Microsystems TCAD
- *Software and Analysis of Advanced Materials Processing Center* in the college of engineering at the university of Florida.
- *NASA Semiconductor Device Modelling*
- *NASA Nanoelectronic modelling*
- *Electronic Engineering at University of California, Berkeley*
- *University of Texas at Austin Microelectronics Research Center*
- *S. Dunham's research group at Washington* Atomistic simulation of deep sub-micron fabrication

Constants and Conversion Factors

- *Constants/Conversion Factors*

Databases

- *Database of materials studied using CRYSTAL.* From the theoretical chemistry group at Torino.
- *NSF Nanotechnology Database*
- *U. Idaho's VLSI Database*

Software Links

- *Silvaco*
- *Avant! Software* Formerly Technology Modelling Associates
- *Integrated Systems Engineering AG*
- *Microelectronics Software* (from Institute for Microelectronics)
- *Siborg System Inc.* MicroTec: 2D process and devices simulator for the PC
- *Los Alamos Grid Toolbox* Grid generation and optimisation software for device modelling
- *Purdue University Network Computing Hubs (PUNCH) Online Computing for Engineering & Science* (Purdue Simulation Hub: Computational Electronics, Computer Architecture, Parallel Programming, VLSI Design)
- *BIPSIM Inc.* BIPOLE3: Quasi 3D device simulation

- *PDF Solutions Inc.*
- *Coventor* formerly MEMCAD
- *Domain Software Solutions Inc.* Statistical Software for Process Simulation
- *Celestry* formerly BTA Technology Inc.
- *The Computer Physics Communications* software library at Queen's University Belfast
- *GAMESS* a free quantum chemical code
- *GAMESS-UK* a quantum chemical code
- *Gaussian Inc.*
- *Accelrys.* Formerly Molecular Simulations Inc.
- *Q-Chem Inc.*
- *Q-Chem Inc.—ab initio and DFT molecular modeling*
- *Schroedinger Inc.*
- *TURBOMOLE*
- *SPARTAN* available from Wavefunction Inc.
- *FHI96MD* from the Fritz Haber Institut, Berlin
- *Nanotechnology Software links* (from Nanotechnology Industries)
- *Nanoelectronics Modelling (NEMO)*

Comprehensive Sites

- *Harris Lexicon of semiconductors*
- *Semiconductor Subway* at MIT is a comprehensive resource providing links to information related to microelectronics and microsystems technology.
- The *Semiconductor Manufacturing Technology* consortium of semiconductor manufacturing companies in the US and, with international SEMATECH, the world.
- The *International Technology Roadmap for Semiconductors* home page
- *MEMS Virtual Learning Cybercenter* (by Abdelkader Tayebi; "a distance-learning educational web page")
- *NANOMAT*
- *NANONET*
- *About.com's Nanotechnology category* (articles, list of centers, etc.)
- *NanoPhiles* (extremely wide-ranging nanotechnology site—even includes a section on careers)
- *Nanotechnology Industries*
- *The Nanoelectronics & Nanocomputing Home Page* (by Mitre's Nanosystems Group; Internet gateway to nanoelectronics research & development information and resources from around the world)
- *Computational Chemistry List (CCL)*

Journals

Technical Journals

IEEE Publications. Some of the more pertinent journals include:

- TCAD Journal of Technology Computer Aided Design, an electronic journal covering Technology Computer Aided Design
- IEEE Transactions on Electron Devices
- IEEE Transactions on Computer Aided Design of Circuits and Systems
- IEEE Electron Device Letters

- IEEE Journal of Quantum Electronics
- IEEE Transactions on Semiconductor Manufacturing
- Elsevier Science or ScienceDirect a gateway direct to the Journals published by Elsevier Science
- Microelectronic Engineering
- Microelectronics Journal
- Materials Science in Semiconductor Processing
- Solid State Electronics
- Computer Physics Communications
- Online Journal Publication Service from the American Institute of Physics
- Applied Physics Letters
- Journal of Applied Physics
- Journal of Chemical Physics

Electronic Journals, Articles, etc.

- EE Times.
- Nanotech Bulletin.
- Semiconductor Magazine
- Computational Methods for Atomistic Simulations of Materials
- Small Times Daily news, mostly in the area of MEMS
- Yahoo! Semiconductor news
- Wired Daily News
- Integrated System Design Magazine
- VLSI Design: An International Journal of Custom-Chip Design, Simulation, and Testing
- Archive: Nanotechnology Papers, sci.nanotech
- Year 2067.com (Emerging Technologies E-Magazine)
- List of Relevant Electronic Journals (from Sean Morgan's Nanotechnology Pages)
- Computational Science Education Project (for >= advanced undergrads)
- Daily Diffs Nanotechnology Folder
- Yahoo! Nanotechnology News (Full Coverage)
- Journal of Nanoparticle Research
- NanoTechnology Magazine

Electronic Lists

- CCP5 Collaborative Computational Project; Computer simulation of condensed matter.

Upon subscribing, one can receive mailings regarding job and studentship postings and upcoming conference announcements

- List of Nanotechnology Mailing Lists (from Sean Morgan's Nanotechnology Pages)
- Nanotechnology Conference/Discussion (Council of Chemical Research's Net Web Board)

Lists of Links

- NanoLink ("Key Nanotechnology Sites on the Web")
- Webring for Nanotechnology (you can add your site to the webring)

Related Keywords

Microelectronics

- nanoelectronics, molecular electronics, molectronics

- molecular electronic devices, molecular nanoelectronic devices, molecular scale devices, nanoscale electronic devices, single-electron devices
- molecular engineering, nanotechnology, microsystems
- nanoscience, nanophysics
- VLSI circuits, nanoscale integrated circuits
- semiconductor physics, solid-state physics, device physics

Design

- computational materials science, computational electronics, computational nanoelectronics, computational nanoscience, computational nanotechnology
- modelling, modelling, simulation

Conferences and events

- See the Foresight web page listing of events.

The Feynman Prizes

- *The Foresight Institute has a web page which gives information on the prizes they sponsor,* most notably the Feynman prizes in nanotechnology.

Some articles on the web

- *There's plenty of room at the bottom,* by Richard P. Feynman, is a classic 1959 article which discusses the limits of miniaturization and forecast the ability to "...arrange the atoms the way we want; the very atoms, all the way down!"
- *Molecular engineering: an approach to the development of general capabilities for molecular manipulation,* by K. Eric Drexler. the first journal article on molecular nanotechnology.
- A summary of *Advanced automation for space missions,* a 1980 NASA study which provides a good introduction to self-replicating systems
- *Atomistic design and simulations of nanoscale machines and assembly*
- *That's impossible: how good scientists reach bad conclusions*
- *Nanotechnology: what will it mean?,* IEEE Spectrum, January 2001
- *Nanotechnology is coming,* Frankfurter Allgemeine Zeitung, September 11, 2000
- *Nanotechnology: Designs for the Future*
- Several papers on nano design and related issues (including several mpeg movies of simulated nanostructures) and a paper discussing NASA applications of molecular nanotechnology from the computational nanotechnology project at NASA Ames.
- *A proposed "metabolism" for a hydrocarbon assembler.*
- *Overview of self-replication.*
- *Self-replicating systems and molecular manufacturing.*
- *Self-replicating systems and low cost manufacturing.*
- *Molecular manufacturing: adding positional control to chemical synthesis.*
- *It's a small, small, small, small world,* published in *MIT's Technology Review,* provides a general introduction to nanotechnology.
- *Nanotechnology and the next 50 years,* by Richard Smalley.
- *A new family of six degree of freedom positional devices* discusses the Stewart platform, a simple robotic arm, and a new proposal: the double tripod. It then analyzes and compares their positional accuracy in the face of thermal noise at room temperature.
- *Steps towards molecular manufacturing* discusses the design of molecular building blocks that could be used in conjunction with positional assembly in solution (no vacuum) to

build a useful range of non-diamondoid molecular structures, including early assemblers.

- *Computational nanotechnology* discusses the idea of using computer simulation to speed the development of this new technology.
- *Theoretical studies of a hydrogen abstraction tool for nanotechnology* is an ab initio study of a proposed molecular tool.
- *A proof about molecular bearings.*
- *Design considerations for an assembler* discusses the design of a "simple" diamondoid assembler.
- *Convergent assembly* can make meter scale or larger products starting with nanometer scale parts.
- *Nanotechnology and medicine* discusses some of the possible medical applications of nanotechnology.

Drexler and Smalley debate feasibility of molecular nanotechnology in Chemical & Engineering News cover story

- Foresight issues *press release.* "[Smalley] offers vehement opinions and colorful metaphors but no relevant, defensible scientific arguments..."
- Kurzweil analyzes the issues. "Smalley's position, which denies both the promise and the peril of molecular assembly, will ultimately backfire."
- Howard Lovey's nano blog covers *Clash of the nanotech titans.* "...I've covered local and national government enough to confidently question the motives of those who side with the Smalley camp."
- The Center for Responsible Nanotechnology (CRN) issued a press release. "If Smalley's goal is to demonstrate that machine-phase chemistry is fundamentally flawed, he has not been effective..."
- The *New York Times*: "The debate has caught widespread attention among nanotechnology researchers..."
- A bibliography on mechanosynthesis and proposal for further research. Computational chemistry can validate the feasibility of mechanosynthesis, what's needed is funding.
- Lawrence Lessig in Wired says: "Should science tell the truth? You'd think that question would need no answer. But in the vortex known as Washington, DC, the obvious too often gets bent."

Sites with many links

- The Foresight Institute
- KurzweilAI.net
- The National Nanotechnology Initiative
- Nanotechnology Industries
- UCLA's chemistry page has many links to chemistry-related topics.
- The NanoLink list of nanotechnology web sites.

Some groups focused on nanotechnology

- The Foresight Institute. Motto: *Preparing for future technologies. A* nonprofit organization, the Foresight Institute has played a pivotal role in educating both the general public and the research community about the potential impact of nanotechnology. Address: Foresight Institute, Box 61058, Palo Alto CA 94306 USA; phone: 415-324-2490; fax: 415-324-2497; e-mail: inform@foresight.org; WWW: http://www.foresight.org.

- The Institute for Molecular Manufacturing, a nonprofit foundation formed to carry out research aimed at developing molecular manufacturing.
- The National Nanotechnology Initiative.
- The Center for Responsible Nanotechnology, focuses on nano policy research.
- Molecular Manufacturing Enterprises, Incorporated (MMEI) was founded to help accelerate advancements in the field of nanotechnology. They provide seed capital and other support to those developing key advances.
- The Numerical Aerodynamic Simulation Systems Division (NAS) at NASA Ames Research Center, in partnership with the Ames computational chemistry branch and others, has pursued substantial research in computational molecular nanotechnology, and funded a grant program in the field.
- The Center for Nanoscale Science and Technology at Rice University is devoted to nurturing science and technology at the nanometer scale.
- The Molecular Manufacturing Shortcut Group (MMSG) is a chapter of the National Space Society (NSS). MMSG's motto: *promote the development of nanotechnology as a means to facilitate the settlement of space.* NSS has a position paper on molecular nanotechnology.

Other pages

- The House Science Committee, Subcommittee on Basic Research, held hearings on nanotechnology on June 22nd, 1999. Testimony was heard from Eugene Wong (NSF), Paul McWhorter (Sandia), Richard Smalley (Rice), and Ralph Merkle.
- NSF is funding research in nanotechnology. Neal Lane, the Director of NSF, said: "If I were asked for an area of science and engineering that will most likely produce the breakthroughs of tomorrow, I would point to nanoscale science and engineering, often called simply "nanotechnology"."
- James Gimzewski (formerly at IBM Zurich) made the world's smallest abacus as well as positioned individual molecules at room temperature.
- The Rice University Nanotechnology Initiative.
- The Laboratory for Molecular Robotics at USC is run by Aristides Requicha and is investigating the precise manipulation of atoms and molecules.
- The NASA Institute for Advanced Concepts is interested in revolutionary new ideas that "leap-frog" the evolution of current aerospace systems in a 10-40 year time horizon.
- Wilson Ho and his group show their atomically resolved and precise work in pictures.
- Charles Lieber's group at Harvard.
- Scanning tunneling microscopy at IBM Almaden includes images of several structures built by positioning individual atoms.
- The Materials and Process Simulation Center at Caltech, run by Bill Goddard, has computationally modeled a broad range of structures, including those relevant to the development of nanotechnology. For example, Charles Musgrave and Jason Perry, then with Goddard's group, used *ab initio* quantum chemistry to analyze a molecular tool which should be useful in the synthesis of diamondoid structures (*Theoretical studies of a hydrogen abstraction tool for nanotechnology,* Musgrave et. al., *Nanotechnology* 2 (1991) pages 187-195).
- NRL (Naval Research Laboratory) has several groups pursuing various aspects of nanotechnology. the Chemistry Division (among others) pursues research in nanostructures and nanofabrication.

- Ned Seeman's lab is working on nanotechnological applications of DNA, including (for example) a truncated octahedron. the *Stewart platform*, a well known positional device, is basically an octahedron six of whose struts can be adjusted in length. While DNA is not as stiff as might be desired for molecular robotics applications, the ability to synthesize an octahedral structure suggests that the self-assembly of a simple positional device is possible.
- Links to information about diamond CVD (Chemical Vapor Deposition).
- An Alpha release of Carol Shaw's Molecular Assembly Sequence Software (MASS) for the Macintosh.
- Ralph C. Merkle's talk on the need for continued nanotechnology research at the Stanford
- Spiritual Robots Symposium.
- Ralph C. Merkle's brief video introduction to nanotechnology (from his appearance on
  - Big Thinkers).
- Some constants, conversion factors, etc. that are useful in nanotechnology.
- Geoff Leach's nano directory with information on Crystal Clear, a crystal editor with a graphical user interface.
- University of North Carolina at Chapel Hill's Virtual Reality Nanomanipulator Project.
- A group at Oak Ridge National Lab (ORNL) is modeling molecular components.
- Some reactions to nanotechnology from the technical community.
- RAND has issued a report on *the Potential of Nanotechnology for Molecular Manufacturing*.
- Nanotechnology in manufacturing by John Walker, part of a talk he gave in 1990 at the Autodesk technology forum.
- NIST has an interest in nanomanufacturing of atom-based standards.
- A macroscopic modular reconfigurable robot has been designed modeled and a prototype built at Stanford.
- HotWired has a page of scenarios about the future, including some speculative scenarios about nanotechnology.
- A new version of the planetary gear illustrated in *Nanosystems* on pages 311 and 312.
- MITRE has a web page on nanoelectronics and nanocomputing.
- Reversible computing is also an important issue if we are to continue improving computer performance. Molecular manufacturing will let us put a very large number of logic elements into a very small volume, so if we are to avoid creating a *great* deal of heat we'll need to keep the energy dissipation per logic operation *very* low indeed!
- Visual images of some proposed molecular machines.
- The slides for some talks on nanotechnology are available.

NANOSCIENCE AND NANOTECHNOLOGY REPORTS: WORLDWIDE OR INTERNATIONAL STUDIES

1. *Small Wonders, Endless Frontiers: A Review of the National Nanotechnology Initiative*

The chapter on history has a table indicating planned goals until 2007. Stupp, S.I., *Small Wonders, Endless Frontiers: A Review of the National Nanotechnology Initiative*. 2002, National Academy Press: Washington, D.C., 2002.

2. *Nanotechnology-Size Matters*

Researched by 3i, Economist Intelligence Unit and the Institute of Nanotechnology, this report examines input from the leading experts in the field of nano-science and technology around the world and summarizes the key pointers that help understand the dynamics of the field as a whole. 3i. "Nanotechnology-Size Matters", *Building a Successful Nanotechnology Company,* http://www.3i.com/, July 2002

3. Nanostructure Science and Technology: A Worldwide Study

This report reviews the status of research and development in nanoparticles, nanostructured materials, and nanodevices worldwide, with particular focus on comparisons between the United States and other leading industrialized countries. Topics covered include particle synthesis and assembly, dispersions and coatings of nanoparticles, high surface area materials, functional nanoscale devices, bulk behavior of nanostructured materials, and biological methods and applications. the final chapter is a review of related government funding programs around the world. Siegel, R.W., E. Hu, and M.C. Roco, eds., *Nanostructure Science and Technology: A Worldwide Study,* NSTC/CT/IWGN, 1999.

4. Nanotech: the Tiny Revolution

This white paper is a preview of the Nanotechnology Opportunity Report™ which provides a thorough analysis of the global nanotechnology market and of the key technologies coming to market now and in the near future. It also includes an extensive guide to the 455 public and private companies, 95 investors, and 271 academic institutions, and government entities that are involved in the near-term applications of the technology worldwide. Analysis is provided on both markets & technologies, at a price of $1995. Hollister, P., *Nanotechnology—the Tiny Revolution,* July 2002, CMP cientifica: www.cientifica.com.

5. Economic Perspectives of Nanotechnology: Enormous Markets for Tiny Particles

Growth forecast of the world market of nanomaterials from the stand point of BASF. Being one of the world leaders in chemical manufacturing, BASF's point of view may be valuable in assessing the market for nanomaterials. Ebenau, A., *Economic Perspectives of Nanotechnology: Enormous Markets for Tiny Particles,* BASF Future Business GmbH, Ludwigshafen: Mannheim, October 2002.

6. CORDIS

This web site provides almost all required information on the EU's efforts and achievements in nanotechnology and science and technology in general, http:// www. cordis.lu/nanotechnology/

7. Sci-Tech Focus Report

Concise discussion of the growth of nanotechnology in Korea. "Survey of Korea's Nanotechnology", *Sci-Tech Focus;* 2002, http://nr.stic.gov.tw/ejournal/SciTechFocus/ SciTechFocus_e.htm. p. 5.

8. Asia Pacific Nanotech Weekly

Publishes weekly reports, commentary and analyses on nanotechnology activities in Asia Pacific. the reports are in PDF format and freely accessible. Liu, L., Asia Pacific Nanotech Weekly, http://www.nanoworld.jp/apnw/.

9. Nanoscale Chemicals and Materials—An Overview on Technology, Products and Applications

Compiled by SRI Consulting, a subsidiary of SRI international, this by far is the most informative report particularly on Asia; it provides data on current and future budget plans of Asian and European countries, along with the core emphasis technologies. Fink, U., et al., *Nanoscale Chemicals and Materials—An Overview on Technology, Products and Applications.* December 2002, http://

scup.sric.sri.com/ Enframe/Report.html?report=NANOT000&show=Navigaton.html

10. Nanotechnology Market and Company Report, 2003

The definition of the term "nanotechnology" was discussed with an international advisory board. the nanotechnology world market is predicted for specific product categories in each of the four nanotechnology segments. the company profiles are identified via various databases, conferences and internet researches. These profiles include information such as founding year, revenue, number of staff and short descriptions of key business activities with respect to nanotechnology. For an analysis of the present status and future visions of nanotechnology, interviews with 48 leading experts are carried out. the experts' responses on the future visions and future products serve as an indicator of future developments from the present point of view. Stark, D., *Finding Hidden Pearls: Nanotechnology Market and Company Report, 2003*. 2003, http://www.nanoreport.de/home.html.

11. Current Medical Technologies, Applications & Market Opportunities in Nanotechnology & MEMS, 2002-12

While much of the enormous potential in nanotechnology and micro-electromechanical systems (MEMS) is at least a few years from beginning to materialize, the existence of substantial current revenue, current profits and a wide array of current activity demands scrutiny now by any entities seeking to truly capitalize on this market. This report on nanotechnology and MEMS is a detailed market and technology assessment and forecast of the products and technologies involved in the medical application of MEMS and nanotechnology. the Report details the current and emerging products, technologies and markets in the medical use of MEMS and nanotechnology. Specifically, the report comprehensively details the clinical applications, the status of products & technologies under development, the current/forecast market for each major application, the current or likely competitors and their positions in the market, and the opportunities for current or hopeful competitors in this field. Driscoll, P., *Current Medical Technologies, Applications & Market Opportunities in Nanotechnology & MEMS, 2002-12*. 2003: http://www.mediligence.com;rpt601.htm/

12. Implications of Emerging Micro- and Nanotechnologies

An evaluation of the current trends in micro and nanotechnology for the Air Force. Increased information capabilities, miniaturization, new materials and increased functionality and autonomy emerged as the four main themes for emerging technologies. Brueck, S.R.J., *Implications of Emerging Micro- and Nanotechnologies*. 2002: Washington D.C.

13. Societal Implications of Nanoscience and Nanotechnology

This report was created right before the commencement of the NNI in 2001. Current status is documented in section I.2.2.5 of this report. Roco, M.C. and W.S. Bainbridge, eds. *Societal Implications of Nanoscience and Nanotechnology*. NSET Workshop Report, NSET: Arlington, Virginia, March 2001.

14. the National Nanotechnology Initiative: the Initiative and Its Implementation Plan

The purpose of the initiative was to support long-term nanoscale research and development leading to potential breakthroughs in areas such as materials and manufacturing, nanoelectronics, medicine and healthcare, environment, energy, chemicals, biotechnology, agriculture, information technology, and national security. the effect of nanotechnology on the health, wealth, and lives of people could be at least as significant as the combined influences of microelectronics, medical imaging, computer-aided engineering, and man-made polymers developed in this century. Downey, M.L., ed. *National Nanotechnology Initiative: the Initiative and Its Implementation Plan.* NSTS/CT/NSET: Washington D.C. 144. 2000.

15. the National Nanotechnology Initiative FY2004 Budget Request

NNI FY2004 Budget, in http://www.nano.gov/fy2004_budget_ostp03_0204.pdf.

16. National Nanotechnology Initiative: From Vision to Commercialization

This meeting was designed to update attendees on the most promising nanotech advances. the meeting would feature status reports by the participating U.S. Government Departments and Agencies and the Nanoscale Science and Engineering Centers addressing:

- The main focus areas of nanotech research for FY2002 and FY2003
- The primary research objectives and associated timelines
- Highlights of research to-date and examples of the most promising developments
- Important challenges that need to be overcome
- The potential for near-term commercial applications and their estimated time-frames
- Other commercial applications that could be derived from the research and their estimated time-frames
- Efforts to transition such developments into the private sector
- Solicitations, proposals, and grants

National Nanotechnology Initiative. From Vision to Commercialization, April 2003, http://www.ccst.us/www.infocastinc.com/NNI/home.asp

17. Nanotechnology's Power Brokers

The team at Forbes/Wolfe Nanotech Report surveyed leading investors, scientists, corporate executives and high-ranking government officials providing a jump on who the top movers and shakers are in nanotech. Survey responses were combined with quantitative criteria to arrive at the final rankings. Wolfe, J., "Nanotechnology's Power Brokers," *Nanotech Report*. March 2003, pp. 1-8, http://www.forbesnanotech.com/

18. U.S. Nanotechnology 2002: Technologies, Application Development, M&A, Patent Disputes, Markets and Business Opportunity

Gives an insightful overview of nanotechnology. the report analyses how this technology works and how it will apply to the consumer market in the near future. the report also details the activities of private and government innovators involved in the highly budgeted nanotechnology R&D arena. This technology, which uses the most miniscule of building-blocks as a base for technological innovations will soon revolutionize the commercial world. MRG, U.S. Nanotechnology 2002: *Technologies, Application Development, M&A, Patent Disputes, Markets and Business Opportunity,* 2002, http://www.mrgco.com/TOC_Nanotechnology_2002.html.

19. 21st Century Nanotechnology Research and Development Act

This legislation, signed on 12/3/2003, puts into law programs and activities supported by the NNI. the authroization bill calls for $3.68 Billion for nanotechnology R&D for FY 2005-2008. http://www.nano.gov/html/news/PresSignsNanoBill.htm.

20. Preparing for the Next Silicon Valley: Opportunities and Choices

The Next Silicon Valley Initiative is working to shape both a framework for understanding and communicating what is happening in the Valley economy, and a process for engaging leaders in a regional discussion of the opportunities and choices for the next wave of innovation. This paper identifies the economic opportunities and risks associated with the evolving convergence of biotechnology, nanotechnology and information technology. the purpose of this paper is to stimulate discussion and action in preparing for the next wave of social and technological innovation. Henton, D., et al., *Preparing for the Next Silicon Valley: Opportunities and Choices*, Joint Venture: June 2002, http://www.jointventure.org/

21. State Technology and Science Index: Comparing and Contrasting California

Clusters of existing technologies expand and emerging science-based technologies formed will be critical factors in determining economic winners and losers in the first half of the 21st century. As economic activity is based more on intangible assets, those states with vibrant technology clusters will experience superior economic growth. It is imperative for state and local development officials and business leaders to promote high-tech expansion and cluster formation. Although high tech is not the only development strategy to pursue, it will be the key distinguishing feature of regional vitality in the 21st century. States that recognize these changes and alter course quickly will be ahead in the economic development game. DeVol, R.C., R. Koepp, and F. Fogelbach, State Technology and Science Index: Comparing and Contrasting California. Milken Institute: September 2002, http://www.milkeninstitute.org/publicactions/publicaltions.taf?function =detail&ID =163&cat=ResRep

22. Nanotechnology Yellow Pages

Generated by Larta, this is a somewhat out of date description of some of the research being carried out in academic and commercial setups in the Southern California region. This report portrays the most prominent themes of California's endeavors in the field of nanotechnology. It does not highlight the diverse scientific developments, restricted primarily to academic research laboratories, that gives the southern California region its paramount position in the global nanotechnology map at these early stages. Foster, L., *Nanotechnology Yellow Pages*. Larta: 2001, http://www.larta.org/.

Worldwide Studies

23. Promoting Japanese-Style Nanotechnology Enterprise

An analysis of Japan's strategy in developing nanotechnology. the presentation emphasizes that strategic alliances across national boarders may help Japan acquire the leading role in the global market. It gives a useful comparison of key technologies owned by the US and Japan. KAMEI, S.-i., Promoting Japanese-style Nanotechnology Enterprises (The Need for Strategic International Alliances). 2002: http://www.intv.net/ftmrw/DrKamei022602/pdf.

24. Proceedings of EPA Nanotechnology and the Environment: Applications and Implications (STAR Progress Review Workshop)

Collection of papers presented in the EPA conference addressing the impact of the growth and development of nanomaterials and nanotechnology from EPA's standpoint. Karn, B. and N. Savage, *EPA Nanotechnology and the Environment: Applications and Implications STAR Progress Review Workshop*, the Office of Research and Development's National Center for Environmental Research, 2002.

25. Nanotechnology Innovation for Chemical, Biological, Radiological, and Explosive (CBRE): Detection and Protection

A workshop on May 2-3, 2002, was organized by the AVS and coordinated with a meeting of the AVS Science and Technology Society held in Monterey, California on May 1 and 2, 2002. the workshop was supported in part by the NNCO. This report summarizes the workshop's conclusions concerning the opportunities and challenges for nanoscience and technology research as applied to the NNI Grand Challenge on Chemical, Biological, Radiological, and Explosive (CBRE) Detection and Protection. Baker, J., et al., *Nanotechnology Innovation for Chemical, Biological, Radiological, and Explosive (CBRE): Detection and Protection*, November 2002, p. 32.

26. Nanoscale Science, Engineering and Technology in DOE's Office of Basic Energy Sciences. Research Directions and Nanoscale Science Research Centers

This brochure briefly discusses the challenges of science at the nanoscale and DOE's role in addressing them. It includes a brief description of the Basic Energy Sciences (BES) research program, and the new Nanoscale Science Research Centers are introduced. These user facilities will provide the Nation's research community with world-class resources for the synthesis, processing, fabrication, and analysis of materials at the nanoscale. BES, D., *Nanoscale Science, Engineering and Technology in DOE's Office of Basic Energy Sciences: Research directions and nanoscale science research centers*, 2003, p. 20.

27. Nanobusiness 2003

The NanoBusiness Conference brings together leaders from business, government, academia and the financial community, providing attendees with all the tools and information required to push nanotechnology research and application development toward commercialization, 2003, N., the official conference of the Nanobusiness Alliance: http://www.nanobusiness2003.com/index.shtml

28. Molecular Manufacturing: Societal Implications of Advanced Nanotechnology

In this U.S. House of representatives committee on science hearing, more research into the societal impacts and ethical responsibilities of molecular manufacturing was urged. Witnesses praised H.R. 766, the Nanotechnology Research and Development Act of 2003, sponsored by Committee Chairman Sherwood Boehlert (R-NY) and Rep. Mike Honda (D-CA), as an important step toward achieving that goal. the full Committee hearing is available at [http://www.house.gov/science/hearings/full03/index.html] Peterson, April 9, 2003, 2003; Tringe, 2003 #24; April 9, 2003, 2003.

29. Nano Republic

As in 2002, the conference will be held on July 10, 2003 in Pasadena, California, and will be chaired by Dr. Jim Heath of Caltech. the conference is produced by LARTA, a prominent think tank for technology businesses, in collaboration with Caltech, the California NanoSystems Institute, Rockwell Scientific Company, UCSB, UCLA, USC, and UCI. the Nano Republic Conference will discuss recent developments in creating actual products and services based on nanotechnologies. the program features presentations and discussion by world leaders from industry and academia on topics such as materials, design and instrumentation, components (simple devices), and systems (complex devices). LARTA, Nano Republic. 2003: http://www.larta.org/nanorepublic/index.htm.

30. California NanoSystems Institute (CNSI)

The vision of the CNSI is to establish a coherent and distinctive organization that serves California and national purposes and that is embedded on the UCLA and UCSB campuses. the CNSI will be a world-class intellectual and physical environment that supports collaboration among California's university, industry and national laboratory scientists. UCLA and UCSB, *California NanoSystems Institute*, http://www.cnsi.ucla.edu/mainpage.html.

31. Center for Nanoscale Science and Engineering (CNSE)

UC Riverside in its current phase of strong growth is building a pre-eminent program in Nanotechnology through the establishment of the CNSE. Nanomedicine, where nanomaterials and nanodevices are brought to bear on biological processes and medical ailments is CNSE's thrust area. the effort that UCR is mounting in nanotechnology and its outreach in the Inland Empire will start to catalyze the location of high-tech industry in the Inland Empire.

32. Center for Nanoscience Innovation for Defense (CNID)

In addition to the campus center, UCR joined UCLA and UCSB in July 2002 to form the Center for Nanoscience Innovation for Defense (CNID), an alliance created to facilitate a rapid transition of research innovation in the nanosciences into applications for the defense sector. the three UC institutions will equally share U.S. government allocations of about $20 million. the Defense Advanced Research Project Agency (DARPA) and Defense MicroElectronics Activity (DMEA) sponsor the three-campus CNID. the center aims at the control and understanding of nanoscale materials, with applications in information and communications technology.

33. Northern California Nanotechnology Initiative

The NCNano Northern California Nanotechnology Initiative is a regional economic development program committed to building the world's leading Nanotechnology Cluster in Northern California. Their major organizational goals are to bring $6 billion in Nanotechnology investment and grant money into the Northern California Region and to create 150,000 new local jobs. Through creation of the Northern California Nanotechnology Institute NCNano will provide the unifying fabric integrating Universities, Research Labs, Businesses, Capital, Local and Regional Governments and Entrepreneurs. the challenge is fostering the creation of a dynamic environment where new ventures, spinouts and established companies can mutually prosper in the new, nanotechnology-driven, global technology economy.

Initiative, T.N.C.N., NC nano. 2003: http://www.ncnano.org/index.php.

Jr. R.P., *House Approves Boehlert-Honda $2.36 Billion Nanotechnology Legislation Bill to Boost Silicon Valley Economy*. 2003, May 7: http://www.ncnano.org/ index.php?module =ContentExpress&func=display&ceid=28

34. Bay Area Nanotechnology Initiative (BANI)

BANI was initiated in April by the San Francisco Center for Economic Development in conjunction with leaders from industry and academia in the Bay Area and Silicon Valley. BANI is currently undertaking a regional assets assessment to assist the region to retain its pre-eminent position as the leader in research and development of nanoscience and nanotechnology. BANI has been building partnerships with industry, academia, research laboratories, venture capitalists and existing.

APPENDIX B

Brief Listing of Nanotechnology Applications

Scientists Learn to Prevent Nano 'merging' 2005-10-07

Controling the size of the particles is important because size is strongly linked to properties like electronic structure and melting temperature.

Nano World: Nanowires Help Spot Cancer 2005-09-30

Arrays of silicon nanowires with biomolecular coatings can spot molecular traces of cancer.

Coventor & Mems Exchange: Mems Tools Initiative Partnership 2005-09-30

Creating access to the most comprehensive array of foundry processes and software tools

Nano World: Nano-tool Markets Rising 2005-09-21

The market for tools used to fabricate emerging nanotechnology could grow from less than $20 million in 2004 to nearly $235 million by 2010.

Tiny Robot Imitates Caterpillar 2005-09-16

Researchers at Dartmouth College have invented a robot so small that an entire army of 200 of them could march across an M&M.

Rensselaer Researchers to Study Nano Springs, Rods, Beams 2005-09-15

Researchers exploring the potential of making and testing springs, rods, and beams at the nanoscale.

Living With Cancer 2005-09-15

Collaboration is critical for getting nano solutions into clinics.

Joint Project Produces Unique Bacterial Diagnosis Application 2005-09-15

Lab-on-a-chip application allows a more accurate detection of infectious diseases.

2005-10-07

- Scientists Learn to Prevent Nano 'merging': Controling the size of the particles is important because size is strongly linked to properties like electronic structure and melting temperature

2005-09-30

- Nano World: Nanowires Help Spot Cancer: Arrays of silicon nanowires with biomolecular coatings can spot molecular traces of cancer

- Coventor & Mems Exchange: Mems Tools Initiative Partnership: Creating access to the most comprehensive array of foundry processes and software tools

2005-09-21

- Nano World: Nano-tool Markets Rising: the market for tools used to fabricate emerging nanotechnology could grow from less than $20 million in 2004 to nearly $235 million by 2010

2005-09-16

- Tiny Robot Imitates Caterpillar: Researchers at Dartmouth College have invented a robot so small that an entire army of 200 of them could march across an M&M

2005-09-15

- Rensselaer Researchers to Study Nano Springs, Rods, Beams: Researchers exploring the potential of making and testing springs, rods, and beams at the nanoscale
- Living With Cancer: Collaboration is critical for getting nano solutions into clinics
- Joint Project Produces Unique Bacterial Diagnosis Application: Lab-on-a-chip application allows a more accurate detection of infectious diseases

2005-09-14

- Like Fireflies And Pendulum Clocks, Nano-oscillators Synchronize Their Behavior: Nanoscale oscillators could replace much bulkier and expensive components in microwave circuits
- Nano World: Public Attitudes Toward Nano: Public has low trust in government and industry regarding the health risks of nanotechnology

2005-09-09

- Nano World: Nano For Artificial Kidneys: Nano-filter mimics the function of the human kidney

2005-09-06

- Nano Shuttles Suggest Lifting Things May Become Thing of the Past: Microlitre drops moved up a one millimetre 12 degree slope against the force of gravity

2005-09-01

- Nanospheres Block Pain of Sensitive Teeth: Nanospheres could help dentists fill the tiny holes in our teeth that make them incredibly sensitive and that cause severe pain
- Mems Accelerometers Pick Up Speed: New sensors can discern more-complex and subtle three-dimensional movements
- Ce Growth Supercharging Mems Market: Pushing this growth are recent refinements in MEMS processing

2005-08-31

- Argonne Researchers Create New Diamond-nanotube Composite Material: New process "grows" diamond and carbon nanotubes together

2005-08-30

- Gold Bowties May Shed Light On Molecules And Other Nano-sized Objects: "Bowtie nanoantenna" greatly improves the optical mismatch between nanoscale objects and light
- Nano-tech Boffins Haven't the Foggiest: Polymer coating made of silica nanoparticles that can create clear glass surfaces that never become foggy
- New Microfluidics Device Can Deliver Targeted Anti-cancer Drugs to Brain Tumors: MEMS device is a candidate to deliver multiple drugs into the tumor microenvironment

2005-08-29

- Virginia Tech Researcher Reports Nano-particle Dispersion Technique Improves Polymers: Supercritical fluid carbon dioxide used; melt properties provide monitor

- Physicists Pose 'chip Dip' Nano Process: Nanotubes are deposited by dipping a chip covered with a gluelike substance into the nanotube solution

2005-08-24

- Nano-pen Writes In Tiny Letters: sharpened tip the size of a single atom applies an electric charge through a thin film of oil molecules onto a target surface
- Nano Diamonds Serve As Circuitry-writing Pens: Low wear and writing capability of ultra nanocrystalline tips with chemical inks is very promising

2005-08-22

- Nano Steel Will Light Up Batteries And Fuel Cells: Researchers have made an advance in nanotechnology by producing transparent carbon nanotube sheets that are stronger than the same-weight steel sheets

2005-08-19

- Nano Coalition Unveils Environmental, Health And Safety Database: Database marks the first effort to integrate the vast and diverse scientific literature on the impacts of nanoparticles

2005-08-17

- Nano-metrology to Probe Chip Structures At Atomic Level: Engineers are investigating a nanoscale approach to metrology that will allow them to examine new semiconductor structures at the atomic level

2005-08-16

- Nano-switches Could Yield Even Smaller Gadgets: Tiny Y-shaped tubes of carbon that act like electrical switches

2005-08-11

- Diamonds Are A Scientist's Best Friend: Tribological properties of diamond far exceed those of silicon
- Nano-sized Bomb Targets Tumors: Drug-delivering nanocell used to pinpoint and obliterate disease without harming healthy cells

2005-08-10

- Photons Are Exciting For Mems Disk: A micro-mechanical structure that vibrates when pumped with light has been developed at the California Institute of Technology

2005-08-09

- Nano World: A Semiconductor Nanotools Boom: Semiconductor industry tools and instruments that work on the nanoscale could form a $5.5 billion market by 2012
- Mems Work Uses Polysilicon-germanium Over Cmos: Gyroscope demonstrates combining a MEMS sensor and a signal conditioning circuit

2005-08-07

- Missed Opportunities In Nano: Nanotechnology companies are missing opportunities to help corporate buyers integrate nanoscale components into advanced products

2005-08-04

- Study May Expand Applied Benefits of Super-hard Ceramics: Discovery could speed the design of materials that approach the hardness of diamond yet remain supple enough to be worked like metal

2005-08-03

- Magnetic Nano-oscillator Addresses Limitations of Integrated Oscillators In Wireless Devices: Project aims at demonstrating the concept of spin torque in a nano-scale microwave integrated oscillator

2005-08-02

- Stress Testing At the Nano Scale: Material stress testing tool incorporates a flexure-based piezoelectric positioner, a capacitive probe displacement sensor, and a high-resolution load cell for testing of thin films under stress

2005-08-01

- Nano Silver Fights Infections: Silver nanoparticles could help fight hospital-related infections that afflict 2 million patients and lead to 90,000 deaths in the United States each year
- Micro, Nano Wrestle With Similar Challenges: Micro and Nanotechnology are both facets of science that address matter, devices, and products on a scale where the phenomena that govern their performance are different from traditional science.

2005-07-28

- DNA-based Molecular Nano-wires: Wires use derivatives of DNA with a greater electronic potential than DNA itself

2005-07-26

- Nano Detector Fingers Pathogens: A portable nano detection tool could be used by processors to ensure food safety

2005-07-21

- Ornl Mirrors Powerful Tools For Studying Micro/Nano-materials: Precision mirrors to focus X-rays and neutron beams could speed the path to new materials and perhaps help explain why computers, cell phones and satellites go on the blink.

2005-07-20

- Nano-surgeons Break the Atomic Bond: the ultimate in surgery has been carried out in a vibration-free bunker
- New Mems Design Smokes Out False Alarms: New generation of fire detectors could reduce the rate of false alarms in the cargo and baggage compartments of commercial airliners.
- Two Steps Forward, One Back?: the nanotechnology concept is as old as the constituents of the pottery glazes used by the ancient Romans and as new as some of the coatings and composites used by automakers.
- Startups Rebuild Amid Telecom's Rubble: MEMS and nanotech startups re-emerge with overhauled products and business plans

2005-07-19

- Nanomemory: Commercial Opportunities for Nano-based Memory and Storage Technologies

2005-07-15

- UCLA Chemists Create Nano Valve: UCLA chemists have created the first nano valve that can be opened and closed at will to trap and release molecules
- Nano-graphite May Store H_2 Gas: Graphite films only nanometers or billionths of a meter thick could help store hydrogen in an inexpensive, easily manufactured, lightweight and nontoxic manner

2005-07-06

- U.S. Risks Losing Nano Lead: the U. S. has fallen behind Asian competitors when spending levels are corrected to reflect exchange rates
- Nano-imprint Makes Its Mark: Imprint lithography used to operating circuits with a width of 30 nanometers

2005-07-01

- Nano World: Wiring Up Single Molecules: Carving of infinitesimal gaps into nanowires soon could help scientists connect electronics to single molecules

2005-06-23

- Buckyball Aggregates Are Soluble, Antibacterial: Research offers clues about C60 behavior in natural environments

2005-06-20

- Teleporting Over the Internet: Eventually, the teleported objects will be built with "nano-dust"—tiny objects that can be programmed to bind to each other and move

2005-06-17

- Nano Cluster Devices Unveils Hydrogen Sensor Prototype: Sensors could be used in a wide variety of applications

2005-06-13

- Nano World: Nano For Stem-cell Research: Nanotechnology can help regenerate organs with stem-cell biology and help people walk again and recover after strokes and heart attacks
- Brush Up On Your Nanotechnology: the world's smallest brushes have bristles more than a thousand times finer than a human hair

2005-06-12

- Nanotechnology Could Speed the Adoption of RFID Tags: Nanotechnology could really help accomplish the goal of five-cent RFIDs for ubiquitous use

2005-06-09

- Hp Reveals Design For Future Nano-electronic Circuits Packaging: Future nano-electronic circuits to use coding theory

2005-06-08

- Mit's Nanoprinter Could Mass-produce Nano-devices: Nature's most efficient printing technique: the DNA/RNA information transfer

2005-06-03

- New Self-assembling Technique Provides Path to Manufacturing Complex Nano-electronic Devices: Molecules arrange themselves to replicate the underlying pattern without imperfections

2005-06-02

- Interstate Effort to Boost Nano In Chesapeake Bay Region: Initiative will help make the Mid-Atlantic region a leader in the research, development and commercialization of nanotechnology

2005-06-01

- Magnetic Resonance Goes Nano: MRI device is small enough to fit on a computer chip
- Going Nano Boosts Thermoelectrics: Researchers use nanomaterials that allow only cold electrons to flow in order to minimize the amount of heat transfer
- Nano Leds Made Easier: Each lamp in the moth-eye array is 220 nanometers in diameter

2005-05-31

- Army's New Sensors Rock: the U.S. military is developing miniature electronic sensors disguised as rocks

2005-05-27

- The Future of Nano-biology: Regenerating Tissue And Artificial Proteins: Every tissue from head to toe is being regenerated somewhere across the planet

2005-05-25

- Nasa Goes Nano For Air Purification: NASA is working to shrink the size of a device used to detect and measure pollutants in the air
- Nano-grating Dvds Could Store 100 Times More: reflective nano-structures to be used to encode data in a highly multi-level format

2005-05-23

- Nano Technology Tracks Cancer Spread: Microscopic new technology checks for the spread of cancer using magnetic resonance imaging machines found at most hospitals

2005-05-20

- Silicon Valve Holds Promise of Fluid Controls Revolution: MEMS chip capable of controlling the refrigerant system in a car's air-conditioner

2005-05-19

- Nems Device Detects the Mass of A Single Dna Molecule: Technology could be combined with microfluidics to perform genetic analysis of very small samples of DNA
- Report: U.S. On Right Path With Nano: the U.S. program to promote nanotechnology got a thumbs-up from its federal oversight body, which nevertheless said the NNI needs to take further steps to communicate and establish links to U.S. industry
- Quantum Dots: Nano-probes of the Future: Scientists have created a procedure to render quantum dots water-soluble and non-toxic

2005-05-09

- Mems Startups Still Attractive, Says Vc Executive: MEMS design and manufacturing processes have yet to stabilize sufficiently to allow the MEMS industry to divide into the fabless designers and the fabbed makers of components
- New Research Raises Questions About Buckyballs And the Environment: Buckyballs dissolve in water and could have a negative impact on soil bacteria
- Motorola Unveils Nano Emissive Flat Screen Technology: Optimized for large-screen high-definition televisions less than 1-inch thick, the prototype employs CNTs to change the design and fabrication of flat panel displays

2005-05-05

- Washington Kicks Off A Nano Initiative of Another Stripe: Nanotechnology finally has something to offer the state's industries

2005-05-04

- Motor Transport In Bio-nano Systems: Micrometer bead is pulled by molecular motors, which are too small to be visible, along parallel filaments, which are immobilized on a substrate surface
- Nano Pyramids Boost Fuel Cells: One way to improve fuel cells that generate hydrogen on-the-fly is to increase the amount of surface area in a cell that can host the necessary chemical reactions

2005-05-02

- Nano World: Nano Could Hasten Hydrogen: Nanotechnology-assisted solar energy could help bring the dream of hydrogen-fueled vehicles to reality more quickly and cheaply

2005-04-28

- Researchers Devise Nano-scale Method For Investigating Living Systems: By observing how tiny specks of crystal move through the layers of a biological membrane, a team of electrical and computer engineers and biologists has devised a new method for investigating living systems on the molecular level

2005-04-27

- Nano Investors Facing 'implosion': Early investors in nano-technology may get their fingers burned

2005-04-20

- Nano Won't Ko Silicon, Says Moore: Nanotechnology will not replace silicon in the near future, according to Gordon Moore

2005-04-19

- Nano Startups Skip Across Chasm: If early 2005 is any guide, nano funding is headed up

2005-04-18

- Adaptive Optics Industry Finds New Markets Through Mems: Each mirror in micro-mirror array can independently tip, tilt and change position to smooth out aberrations in an optical beam
- Water, Water Everywhere: Nano Filtration: One of the single biggest applications of nanotechnology could be solving the global shortage of pure water

2005-04-15

- Wall Street: When Nano Firms Are Ready They'll Move Quickly: If Nanotech ever takes on Wall Street in a big way, it will appear to happen suddenly.
- Invasion of the Poverty-fighting Nano-bots: Nano technologies that purify drinking water, produce energy and grow food can benefit poor countries

2005-04-13

- Mems Advances From A Bit-role Player to A Star Performer At Consumer Electronics Show: What a difference a year makes
- Tiny Engines Could One Day Be Used to Repair Damaged Cells: Relaxation oscillator is now the world's smallest motor

2005-04-12

- Nanotechnology's Miniature Answers to Developing World's Biggest Problems: Lab-on-a-chip could transform the lives of billions of the world's most vulnerable inhabitants

2005-04-08

- Tire Pressure: Ruling Seems Clear, But Read the Fine Print: A new tire pressure ruling appears to favor MEMS but over the long-term could accommodate other technologies
- Nano World: Nano For Quantum Computers: Possibilities for a nanotech quantum computer include: quivering nanotubes, superconducting nanocircuits and quantum dots

2005-04-07

- Ask the Analyst: Four micro/nano experts show what life's like inside market research
- Nano-agents That Strip For Action: New nanotechnology improves fuel efficiency

2005-04-04

- A Nano Drug's Giant Promise: After a decade-long slog to FDA approval, APP's Abraxane, a novel, less-toxic cancer treatment, has doctors and investors hopeful

2005-03-31

- Nano-probes Stay Inside A Cell's Nucleus For Days: Research could help biologists better understand DNA replication, genomic alterations, and cell cycle control

2005-03-29

- Toshiba's 'nanobattery' Recharges In Only One Minute: Battery fuses Toshiba's latest advances in nano-material technology for the electric devices sector with cumulative know-how in manufacturing lithium-ion battery cells

2005-03-28

- Mems Makers' Efforts Bear Fruit As Apple Embraces Sensor: Laptop computers will come equipped with a sudden motion sensor designed to prevent damage to sensitive components
- Nano-based Chips Won't Easily Replace Silicon: ... according to former Intel co-founder Moore

2005-03-18

- Nano Hazards: Exposure to Minute Particles Harms Lungs, Circulatory System: Several dozen reports unveiled detail how nanopollutants interact with the body
- Nano-probes Allow Inside Look At Cell Nuclei: Technique could be used to determine whether a drug has arrived where it is supposed to, and if it is having the desired impact
- Nano World: Water, Water Everywhere Nano: One of the single biggest applications of nanotechnology could be solving the global shortage of pure water
- Bacteria Bridges New Nano-circuit: Scientists in Wisconsin have used bacteria to make tiny bio-electronic circuits, a step towards new sensors and perhaps new nano-manufacturing techniques

2005-03-16

- Researchers Study How to Make Nanomaterial Industry Environmentally Sustainable: Study is the first to show what factors affect the size of aggregate particles
- Data Storage May Enter New Nanotech Phase: Itsy bitsy rods and disks may be able to store vast numbers of data bits
- Hp Plots Its Nano Course: HP is looking at nanotechnology to solve its future computing needs

2005-03-14

- In Solution, Tiny Magnetic Wires Scatter Light: Researchers believe "Nanowires" will one day become critical components in ever-shrinking electronic circuits

2005-03-12

- 'millipede' Small Scale Mems Prototype Shown At Cebit: Technology capable of achieving data storage densities of more than one terabit (1000 gigabit) per square inch

2005-03-10

- Mems Motor Means No More Dead Batteries: Tiny engines that generate power may one-day replace batteries
- Nano Particles Could Solve Ground Water Problem: Using nanotechnology, researchers were able to maximize the number of palladium atoms that come in contact with TCE molecules and improve efficiency by several orders of magnitude over bulk palladium catalysts

2005-03-08

- Sampling 'small Atmospheres' In the Tiny New Worlds of Mems: Sandia gas sampling device rapidly determines whether MEMS seals are effective
- Nanobiotech: Nano Poised For Liftoff: First 'nano' technologies yield fruit in the lab and clinic with the promise of more to come

2005-03-07

- Investing In Nano: Venture Capitalists Tell Their Side of the Story: Mark Brandt, Jeff Fagnan, Matthew McCall, Waqar Qureshi answer questions posed by Jeff Karoub

2005-03-06

- Early Cancer Detection Is Treatment of the Future, Nano-medicine: the metallic nanoparticles are made especially for medical applications in a patented process

2005-03-04

- Nano World: Nano to Speed Drug Discovery: Nanotechnology could help more candidate drugs make it to market through clinical trials

2005-03-02

- Tracking In Locations Where GPS Won't Go: to improve accuracy, MEMS gyroscope uses three sensors: one on each leg and a third worn at the waist

2005-02-28

- Nano 'ruler' Could Provide Microchip Benchmark: A nanoscopic measuring device that uses atomic lattices to gauge tiny distances could enable microelectronics engineers to build better components

2005-02-25

- Nano World: Edible Nanotech On the Horizon: Edible forms of nanotechnology could help make smart programmable drinks and more effective drugs
- Nano-enabled Drug Discovery Set to Dominate: Nanotechnology has been singled out as a tool, which could make the difference in an industry

2005-02-23

- The Worries Over Nano No-nos: Could the same properties that make the tiny particles so effective also turn them into efficient troublemakers inside the human body?

2005-02-22

- Tiny Data Switch May Store More: Mechanical device is a completely new approach to improving data storage

2005-02-21

- Successful Test of Single Molecule Switch Opens the Door to Biomolecular Electronics: Developments due to approach based on measurements made with the molecules suspended in solution
- Startup Uses Tiny Probes to Store Data: Technology could take the $7 billion flash memory market in an entirely new direction

2005-02-18

- A Mason Finds His Muse In Nano's Building Blocks: NanoTechno series consists of images of molecular structures and materials—quantum dots, nanowires, carbon nanotubes and thin films

2005-02-17

- Nano Startups Consolidate IP Positions: Small tech patenting activities—of which nanotechnology is a part—have skyrocketed in recent years

2005-02-15

- Nano Mechanism to Control Protein May Lead to New Protein Engineering: UCLA scientists have created a mechanism at the nanoscale to externally control the function and action of a protein
- Outfit Greases Wheels of Nano Commercialization: Particles have a unique structure of nested spheres that lubricate by acting as miniscule roller balls

2005-02-14

- Nanobusiness Alliance: Budget Likely to Give Nano A Trim: Group says more focus needs to be put on funding the gap between the research stage and the point where most venture capital firms are willing to provide support for bringing a product to the marketplace

2005-02-11

- Nano World: Nanotech May Cut Pharma Waste: Much of the waste the pharmaceutical industry generates could be cut dramatically by using nanotechnology reactors
- Engineers Develop Biowarfare Sensing Elements: Devices Permit Mass Production of Highly Sensitive and Stable Nerve-Gas Detectors
- Devising Nano Vision For An Optical Microscope: Phase-sensitive, scatter-field optical imaging could extend life of conventional light-based imaging methods

2005-02-09

- Mems Rf Front-end Filters Needed: MEMS-based solutions have quality factors that exceed 10,000, well beyond conventional ceramic filters
- Side-impact Regulations Could Drive More Demand For Mems Sensors: Airbags like the side and curtain ones are expected to provide enhanced protection in passenger vehicles
- Mems Industry Group Releases '05 Report: Group noted a critical lack of MEMS characterization tools

2005-02-08

- Nano-sized Probes Allow Researchers to See Tumors Through Flesh And Skin: Technology will be less costly and more accessible than MRI-based methods and free of the harmful side effects associated with radioactivity

2005-02-07

- Nano And Chips: Uneasy Ties: Semiconductor execs bridle at the suggestion that this new technology could unseat them. It is, however, a distinct possibility

2005-02-04

- The Business of Nanotech: There's still plenty of hype, but nanotechnology is finally moving from the lab to the marketplace
- New York's Big Hopes For Nano: With fabs and labs sprouting around the capital in Albany, the state is shaping up as a power in the emerging technology

2005-02-03

- Nano Surfaces Could Slash Cost of Solar Energy: Light-trapping technologies could reduce the thickness of semiconductor materials needed in solar panels
- Battery Uses 'nano Needles': Nanotechnology provides a unique mechanism to combine the chemicals and control the characteristics of the chemical reaction

2005-02-02

- '04 Nano Funding Report: Less Money: Dollars were down, but nano deals increased from 34 in 2003 to 45 in 2004

2005-02-01

- Channeling Nano's Future: Startups will plot multiple courses to public markets this year
- Nanometer Schmanometer: Keep your eye on MEMS

2005-01-31

- Phased-array Antennas Get Mems Relays: MEMS components could go into lower-cost phased-array antennas for collision avoidance and adaptive cruise control systems in automobiles
- Florida Professor Leads Drive For Nano-enhanced Armor: Ballistic shields will be ready for field testing in three to six months

2005-01-28

- Nano's Road to the Future: 5-year-old National Nanotechnology Initiative keeps U.S. efforts on course...for now

- Danish Researchers Design Virtual Nano-catalyst: Research offers new opportunities in the fields of renewable energy, pollution control and in the chemical industry
- Nano Paint Could Boost Antiterrorism, Rescue Efforts: New technology may be used to detect cancer in the first cells to become malignant

2005-01-25

- Lab-on-a-chip Finally Catching On In Mems Field: Key applications include life science research areas of genomics, pharmacogenomics, and proteomics

2005-01-21

- 2005 May Be A Momentous Year For Mems: Or maybe just a momentum-building year
- Lotus Plant Inspiring Mems Technology: Researchers at Ohio State University have been taking a look at the lotus plant in a quest to reduce friction

2005-01-18

- A Giant Leap For Nano: American Pharmaceutical Partners, Bristol Myers Squibb battle for market share forcasted to be $3 billion
- Mems Researchers Make Supersensitive Motion Detector: Sandia researchers created a MEMS sensor that can detect motions smaller than one nanometer
- Musclebots Crawling In Petri Plate: Device is said to be better than micromotors as it does not requires any external power

2005-01-14

- Study Finds Advantages to Iron Nanoparticles For Environmental Clean Up: Iron nanoparticles may be effective in cleaning up carbon tetrachloride in contaminated groundwater

2005-01-12

- Nano Gas Turbine Designed: Nanoscale gas turbine made from nanotubes

2005-01-11

- New Nano-plastic Can 'see' In the Dark: Semiconductor crystals were dispersed in everyday solvents similar to how particles in paint behave
- Pins & Vias: Mems Enter the Mainstream: Use of devices based on MEMS in electronic products is expected to grow

2005-01-10

- Regional Recap: States Fund, Flaunt Nano Competitiveness: Nanotech appears to be on lawmakers' minds as state governors and legislators set their agendas for 2005

2005-01-07

- 'nano-scissors' Laser Shows Precise Surgical Capability: Tool opens new frontier for biologists studying nerve regeneration
- Nano World: Molecular Electronics Rising: Molecular electronics will be capable of doing things that silicon cannot
- Nano-drug Bombs Target Tumors: Nanoscale capsules may one day deliver cancer drugs directly to tumors
- Nano-propellers Sent For A Spin: Metallic rods about 500 times smaller than the width of a human hair have been turned into tiny "propellers"
- Fund Manager Advises Balancing Risk: EVEN IN A MICRO-NANO PORTFOLIO

2005-01-02

- Breakthrough In 'nano' Tracking of Cancer Hailed: Cancer successfully predicted in 90% of early experiments

- UCSB Scientists Build Nanoscale 'jigsaw' Puzzles Made of RNA: Nine different RNA fabrics were generated by tectosquares self-assembly

2004-12-29

- Chip-scale Magnetic Sensor Draws on Mini Clock Design: Magnetometer is about as tall as a grain of rice

2004-12-28

- Nano Firms Tie the Knot: the future of carbon nanotubes just got a little more exciting

2004-12-26

- Nano-lubricant Could Mean No More Oil Changes: Lubricant to be based on spherical inorganic nanoparticles

2004-12-22

- Simmons Remakes Bed With Nano-enhanced Fabric: Bed features a removable mattress top that can be washed to rid the mattress of germs

2004-12-17

- Nano World: Nanomaterials Buyers Beware: More than 200 companies worldwide sell nanomaterials. As a group, they have a poor track record
- Polymer Science Enables New Breakthroughs In It And Nano Medicine

2004-12-14

- Tight Twist Toughens Nano Fiber: Multiwall nanotubes contain several layers of successively larger tubes
- Tiny Tubes for the Big Picture: Nano-materials could be used in medical imaging

2004-12-13

- Physicists Discover Potential In Ferroelectric Nanodisks And Nanorods: New phenomenon could open door to nano-memory

2004-12-10

- Survey Shows Public can Discern Nano's Benefits: Nano advances have reached the point at which potential is becoming reality

2004-12-06

- Spilling the beans on new technology: You can now laugh at any drink thrown in your face because of nanotechnology
- Software, tools key markets in China: Chinese government selects Coventor design tools software for its MEMS program

2004-12-02

- Uc-berkeley to build nano research center: Center to be built from $11.9 million grant from the National Science Foundation

2004-12-01

- Multiphysics analysis speed mems development: In the absence of a superphysicist, all members of the team can help solve these multiphysics problems

2004-11-30

- Flash market offers mems ray of hope: Companies have developed storage technologies that couple a grid of MEMS cantilevers with atomic force microscope tips

2004-11-29

- Polymer made in a nano test tube: Fullerene epoxide polymerizes to form a linear chain inside a carbon nanotube

2004-11-28

- Model prospects: accelrys announces nano consortium: Goal is to accelerate the development of software for designing nanomaterials and nanodevices

2004-11-25

- Infineon shrinks transistors with nanotubes: Researchers grew nanotubes measuring no more than 1.1 nanometers in diameter in a controlled process

2004-11-24

- Mini generator has enough power to run electronics: New microengines would be smaller, last 10 times longer than batteries

2004-11-23

- Gold nano anchors put nanowires in their place: Rows of horizontal zinc-oxide nanowires grown on a sapphire surface are held in place with gold nanoparticles

2004-11-22

- Siemens begins work on silicon-based mems devices for monitoring building climate: Prototypes will measure the CO_2 and ammonia content in mouse cages

2004-11-19

- Nano fabric may make computers thinner: Researchers in Russia and England who claim they have discovered the world's first single-atom-thick fabric
- Mems-makers tune in to digital TV market: Among the contenders is a MEMS-based technology pioneered by Texas Instruments known as DLP

2004-11-18

- Nanotech could put a new spin on sports: "It's all about controlling the physics of how the ball spins"

2004-11-15

- Tunable takes a simpler tack: Tunable lasers should replace all fixed-wavelength lasers in communications applications
- New technology breathes life into your food: Samsung and LG devising micro and nano technology to prolong the shelf life of frozen foods

2004-11-12

- EPA Backs Nanomaterial Safety Research: Activists Say $4 Million Is Too Little for Studies

2004-11-10

- Gold nano anchors put nanowires in their place: Researchers at NIST have demonstrated a technique for growing single-crystal nanowires in place

2004-11-08

- Twosomes Become Troublesome If One Side Goes Under: Partnership inherently is a risky proposition, given the high percentage of MEMS firms that fail.
- CMOS-compatible SiGe micromachine: Belgian researchers have developed a deposition sequence of poly-SiGe at CMOS-compatible temperatures

2004-11-05

- Nanotechnology may change energy industry: Technology can make coal a cleaner source of energy

2004-11-03

- Molecules form nano containers: Research shows the self-assembly of minuscule multicompartment structures.

2004-11-02

- Tiny tools carve glass: Researchers develop microscopic carving instruments

2004-10-28

- Nano-product sales to be $2.6 trillion by 2014: Sizing Nanotechnology's Value Chain

- Nanotech Group's Invitations Turned Down by Environmentalists: None of the three invited representatives of environmental groups has agreed to join the newly created International Council on Nanotechnology

2004-10-27

- Nano Central: A VC looks at how nanotechnology is putting new life into old economy industries

2004-10-25

- Nano-Fabric Reveals Unique Properties: Electrons can travel without any scattering over submicron distances, an important property for making very-fast-switching transistors
- Kerry May Pursue More Nano Funding; Bush Expected to Continue Support: John Kerry favors increased funding for science and technology initiatives
- Hitachi devises MEMS chip to quickly mix liquids at set ratios: Chip is capable of processing tens of milliliters of liquid per minute

2004-10-21

- Serial Entrepreneurs: True tales from industry veterans
- Nano-bullet for non-invasive treatment of cancers: the advantage of using smaller particles is that they can be inserted into any part of the human body and treat cancer cells in their infancy

2004-10-20

- Researchers refine nano self-assembly: Device organizes itself into predictable patterns

2004-10-19

- Stabilizing Mems System Lifts Hovering Vehicle's Chances: MEMS devices help the organic air vehicle remain stable as it hovers above surveillance targets

2004-10-18

- Dreaming of a Single-Chip Mobile Phone: Integration ultimately leads to lower manufacturing costs

2004-10-12

- Trying Nano On for Size: Don't be alarmed, but nanotechnology is in your pants
- Smart shirt that can call for help: Jacket is fitted with a sensor and a Bluetooth transmitter that communicates with a receiver

2004-10-11

- HP bets inkjets will open new markets: Inkjet printing is based on a manufacturing technology that is a close cousin to MEMS

2004-10-06

- Mems, Aka 'big Nano,' Seeks to Elevate Its Political Profile: Director of Berkeley Sensor and Actuator Center sees technical and political convergence for micro and nanotechnology

2004-10-05

- High-tech tweezers enable nano-assembly lines: Technique used to put things together at the microscopic or atomic level
- Virus forms nano template: Self-assembly mechanisms bring together charged membranes and oppositely charged polymers

2004-10-04

- SEMI Publishes Six New Technical Standards: New standards were developed by equipment suppliers, device manufacturers and other companies
- Firms Get Over $20 Million For Micro And Nano Products: Funding makes it easier for companies to take on otherwise fiscally risky projects

- Device has keen sense for mini-motions: Sensor claimed to be 1,000 times better than existing technology

2004-10-01

- For Sensors, MEMS the Word: Get started experimenting with a MEMS sensor!

2004-09-30

- Unifying Approaches Sought At Mems Conference: Standards and reliability discussed at MEMS Industry Group meeting

2004-09-28

- Nano proponents square off against specter of 'gray goo': Experts say Nano field will contribute $1 trillion to the U.S. economy by 2015

2004-09-27

- Buckyballs: Study Finds Toxicity, And A Cure: Researchers at Rice have demonstrated a way to reduce the toxicity of water-soluble buckyballs by a factor of ten million
- Samsung Sets Nano Benchmark: Congomerate earned an estimated $779 million in revenue since its 2003 nano launch

2004-09-20

- Radant Takes Lead In Race to Bring Mems Rf to Market: Firm has a $5-million government grant to verify its switches
- Light excites nano-antenna: Antenna captures visible light like radio antennas capture radiowaves
- Emerging Nanophotonics Markets: New Revenues from Processing Light with Nano Devices

2004-09-17

- Mems May Make Smart Homes More Affordable: Thanks to MEMS, the basic building blocks actually are there for bringing intelligence into housing for the masses

2004-09-13

- Cancer Institute Starts Nanotechnology Drive: Nanotechnology offers new ways to detect, diagnose and to treat cancer at its earliest stages and with minimal side effects

2004-09-08

- DARPA Contracts Lucent for Maskless Litho: Lucent to design, develop and demonstrate MEMS-based Spatial Light Modulators to enable maskless optical lithography
- Nano memory scheme handles defects: Researchers from Hongik University in Korea have devised a memory architecture designed for nanoscale crossbar electronics

2004-09-07

- Nanotech in Fashion: the Trend in New Fabrics: A new wave of nanotechnology allows clothes to resist spills and wrinkles, and wear longer between washings
- Molecular Imaging Focuses on Nano-level Life Sciences: Atomic force microscopes (AFMs) allow scientists to peer at the nano world

2004-09-03

- Bookham adds overload protection to 10 Gbit/s receiver: Integrated MEMS Variable Optical Attenuator controls the power reaching the optical receiver

2004-09-02

- Corns 2004 Kicks Off With Upbeat Look At State of Industry: Killer MEMS applications remain an elusive target, a thriving nano/MEMS community continues to grow

2004-08-31

- The Daintiest Dynamos: By harvesting energy from radioactive specks, nuclear microbatteries could power tomorrow's cellphones

- Mems Researchers Perfect Fabrication of Atomic Clock: Researchers at NIST have developed the core technology for a MEMS-based atomic clock
- Cell phones could keep atomic time: Atomic clocks may be headed into cell phones, thanks to a breakthrough by federal researchers.

2004-08-30

- The Players and Pretenders of Nanotech: Investment opportunities in the nanotech world
- Environmental Fate of Nanoparticles: A pressing question remains as to the environmental impact of manufactured nano-sized materials

2004-08-26

- Four startups gunning for TI's microdisplay business: Engineers foresee significant growth for microdisplay-based TVs

2004-08-25

- Nanotech will tap nature's potential: Humans can learn a lot from animals, particularly the microscopic life forms living in thermal vents

2004-08-19

- Nano-coating cuts phthalate migration: Researchers say that the material cuts phthalate migration in medical applications.

2004-08-18

- Nano Discovery to Make Internet 1000 Times Faster: Canadian researchers have shown that nanotechnology can be used to pave the way to a supercharged Internet-based entirely on light.
- Wireless Sensors: How Not to Replace 1000 Batteries: Wireless microsensors will be used to monitor everything from nesting conditions for endangered birds to stress-induced cracks in large structures. the big question? How to make batteries for them that last for years.

2004-08-05

- Nanosys Says No to IPO: On second thought, never mind.
- MEMS makes a nonvolatile-memory match: the MEMS memory works by flexing a microbeam.
- Embedding A Mobile Phone In Your Ear: If you thought people walking down the street talking into their bluetooth headsets looked odd enough, just wait until mobile phones are completely embedded in your ear.

2004-08-02

- Princeton Researchers Use Nanoimprinting for Mass Producing Nano Scale Devices: Low-cost production method may lead to greater memory capacity and lower costs for computers, digital cameras and other devices
- Science of the small could create 'nano-divide': New technologies could increase the divide between rich and poor nations

2004-07-30

- Federal funds go to national underdogs in nano research: Scientists and entrepreneurs working together to promote Oregon take note of $5 million in likely federal funding and a $1.3 million grant
- Nanotechnology Precaution Is Urged: Minuscule Particles in Cosmetics May Pose Health Risk, British Scientists Say

2004-07-29

- Myths and realities of nano futures: Ever since John Dalton proved the existence of atoms in 1803, scientists have wanted to do things with them

2004-07-27

- The promise and perils of the nanotech revolution: Possibilities range from disaster to advances in medicine, space
- Advancements in micro technology catapult MEMS-based applications: the hunt for the 'next big thing' is on

2004-07-21

- VIA to launch Nano-ITX embedded motherboards in 2H: Nano-ITX motherboards will have dimensions of 12 × 12 cm
- Small Fuel Cells 2004, an event review: Small Fuel Cells 2004, the sixth in a series of annual symposia concentrating on portable applications, took place in Arlington VA, USA, in May 2004.

2004-07-19

- Green Goo: the New Nano-Threat: Nanobiotech carries the power to create completely new organisms that have never existed
- Three new standards for MEMS devices: Standards help researchers measure more accurately material characteristics used in construction

2004-07-16

- IBM claims nano-scale imaging breakthrough: 'Micro-cantilever' is 1,000 times thinner than a human hair
- USC Scientist Invents Technique to Grow Superconducting and Magnetic 'Nanocables': Nanocables manufactured from a potent new class of substances with extraordinary properties called Transition Metal Oxides
- Mems Update: 2004 Likely to Be Banner Year: the first half of 2004 is looking a lot like the last half of 2003 for the MEMS industry, only better

2004-07-12

- Konarka: Giant Leap With Nano-based Solar Power: Semiconducting particles of titanium dioxide coated with light-absorbing dyes used to convert light into electricity
- Active Spectrum Switches Gears With Wireless: Using MEMS technology, Active Spectrum wants to make a multi-frequency transceiver

2004-07-09

- A Nickel Investment For Future's Grid Will Pay Off: Energy is the single most important challenge facing humanity today
- Nano-team spins tomorrow's yarn: A method to continuously spin nano-material has been developed by Cambridge-MIT Institute scientists

2004-07-08

- University Develops 12 Tbyte Nano Memory: Crystals are 10 nm across and can be deposited in a single crystal layer on a substrate
- Is Telecom Ready to Embrace Mems-based Solutions?: Cautious optimism—the market is now finally beyond bottom

2004-07-06

- European Union Earmarks Billions to Prepare For A Nano Future: Technology is crucial to the development of mobile communications device-makers

2004-07-01

- Oregon Blazes Nano Trail With More Money, Grants And Events: ONAMI in the running for $10 million in federal defense grants for 2005

- Tech Company Gets Hypersensitive: Sensor technology claimed 10,000 times more sensitive than what's available today

2004-06-29

- MEMS Industry Finding Success One Step at a Time Reports In-Stat/MDR: Venture capital funding in the first half of 2004 alone has already exceeded that provided in all of 2003
- Hubble's Early Demise Sends Nanotech Back to Waiting List: Cancelled mission to the Hubble Space Telescope would have been first applied nanotech in space

2004-06-17

- MEMS Enjoying Revenue Growth, VC Attention: 2003 revenues in the MEMS industry reached $5.3 billion, up 35.7 percent year over year.
- MEMS-Driven Wearable Display For Auto Techs In the Works: Head-up display lets an automobile service technician consult reference manuals without stepping away from the vehicle
- Mems Sensors May Speed Work of Jet Engine Designers: Sensors measure the fluctuating pressure as air passes through the inlets and mixes in the engine chambers.

2004-06-10

- Kodak's New Image: Century-old Firm Develops Nano Strategy: Company has developed pigment nanoparticles as small as 10 nanometers

2004-06-07

- IEEE and SEMI Sign Agreement to Foster Nanotechnology and MEMS Standards: the agreement marks the first standards collaboration between the two organizations.
- MEMS Come to Oz Wine Industry: Sensors are distributed across a vineyard and send their information by wireless link

2004-06-04

- Amateur Scientists Study Microscopy In Basement Nanolabs: Amateur scientists satisfy their curiosity through kitchen chemistry, backyard stargazing and garage physics

2004-06-03

- Too Little Too Late?: Report Gives Brits Failing Nano Grade

2004-05-28

- GI Joe Goes Nano: Soldier's items could be lightened, combined or integrated
- Scientists use nano-sized particles: Particles one-billionth of a meter in size provide a "stable imaging marker" during surgery to remove brain tumors

2004-05-25

- Scaling Friction Down to the Nano/Micro Realm: Friction measurements made at the micro- and nano-scale can differ substantially due to changes in applied load

2004-05-21

- Keystone cops: MEMS accelerometers outlaw projector flaw

2004-05-18

- The Way We Live: Nanotechnology and Medicine

2004-05-17

- Gyricon Is Xerox's Poster Child For Copy Company's New Nano Image: Etch-a-Sketch examplifies commercial small tech from the "The Document Company"
- Language of Science Lags Behind Nanotech: Burgeoning Field in Need of Universal Way to Describe Creations

2004-05-14

- Big Blue Says Breakthrough Means Millipede May Crawl Out of Lab: the Millipede chip holds 4,096 miniature read-write heads

2004-05-11

- Intel Debuts 90 Nano Mobile Chip: the 90 nanometer process produces smaller, faster transistors

2004-05-10

- Chiral Photonics Brings A New Twist to Optical Devices, Lasers: 15,000 helical twists per inch in the core of the fiber are each about 2 microns in width

2004-05-09

- Nano's Great Leap: A safer approach is to find the companies that sell "enabler" technologies

2004-05-07

- Nano-scale trees created at Lund Institute of Technology: Research demonstrates the possibility of producing more complex nano-scale structures

2004-05-03

- Postal Service Delays Rollout of Cepheid's Anthrax Detector: Genexpert testing system detects trace levels of DNA.

2004-04-30

- Nano Wires Make Tiny Compasses: Compass needles are just under half the size of the previously smallest known compasses
- Nano Weapons Join the Fight Against Cancer: "Nanoshells" and other ultrasmall tools could improve diagnosis and treatment of tumors

2004-04-29

- High-speed Nanotube Transistors Could Lead to Better Cell Phones, Faster Computers: Findings add to mounting enthusiasm about nanotechnology's revolutionary potential
- A Conveyor Belt for the Nano-Age: Someday, nanoscale conveyor belts could expedite the atom-by-atom construction of the world's smallest devices

2004-04-26

- Vermont's Seldon Labs Wants to Keep Soldiers' Water Pure: Firm to develop a water filter that uses carbon nanotubes

2004-04-19

- Living On A Bootstrap Budget, Luna Is Proud Parent to Five Spinoffs: Prototype products made into commercial development with outside partners
- China develops first nano-satellite: Super-small satellite weighs 25 kg

2004-04-16

- Immunicon's Nano IPO Gets 'sane' Response From Investors: Cytometer that enumerates and differentiates between the immuno-magnetically selected cells based on the fluorescence signature of the cells
- Technion Develops Nano-Sized Vehicles to Detect Gas Contamination: Device was modeled on dandelion seeds enabling them to fly in the wind

2004-04-13

- Nanomaterials Production Weaves Job Growth In Former Textile Town: 400,000 square-foot tobacco warehouse converted for production of nanomaterials

2004-04-12

- Nano Ribbons Coil into Rings: Tiny rings are potentially useful as sensors, resonators, transducers and actuators

- Nanoparticles Player Stirs the Pot With Room Temperature Process: ANW's process can yield 2-nanometer zinc oxide particles.

2004-04-09

- Environmental Journal Publishes Controversial Buckyball Study: Rates of brain damage to be 17 times higher in nine large-mouth bass

2004-04-07

- 'Stentennas' Could Signal Change In How Arteries Are Monitored: A "stentenna" keeps arteries free of plaque and monitors blood flow and pressure.

2004-04-06

- Mems Technology Magnifies Opportunities for Low-cost Sem: NOVESEM uses MEMS to make less expensive components for a tabletop SEM.

2004-04-05

- Ancient Greeks help scientists build environmentally friendly nano devices: Technique inspired by the lost-wax casting process used by ancient Greeks for sculpture
- Mems Gyroscopes Making Strides In Replacing Entrenched Technologies: MEMS gyroscopes quickly matching the performance parameters of older gyrotechnologies
- New Kyocera Company Shows Industry's Oled Intent: the LG4050 has an OLED subdisplay outside and a conventional screen inside.

2004-04-02

- Latest Gadgets and Future must Haves' Seen at Consumer Show: Texas Instruments remains leader with DLP technology for digital TV

2004-04-01

- 'Nano-Lightning' Could Be Harnessed to Cool Future Computers: "micro-scale ion-driven airflow" device uses a sort of nano-lightning to create tiny wind currents

2004-03-31

- Tiny machines need even tinier lubricants: Gases to provide microscale thin films of slippery coating.
- Mems Doesn't Spell Auto Opportunity, Despite Market Report: French report shows MEMS gyros and accelerometers market is growing

2004-03-29

- Millennial Net Hopes to Hook Up 21st Century Sensor Networks: Radio-connected sensors self-organize into wireless networks
- Nanoparticles Toxic in Aquatic Habitat, Study Finds: "nanoparticles" can trigger organ damage and other toxic effects

2004-03-26

- A Few Billion Years in Development, Biomimetics Emerge at Last: Velcro is the quintessential example of biomimetics

2004-03-25

- Businesses Begin to Pay Attention to Nanoparticle Health Debates: Nanotox raises questions about the toxicity of nanoparticles

2004-03-24

- 'Nano-lightning' could cool future computers: New technology creates tiny wind currents
- DNA has Nano Building in Hand: Researchers used DNA branch migration to construct the DNA hand.

2004-03-23

- Where's the Road to the 'last Mile'? Perhaps In the Packaging: Industry watchers predict 2004 will see progress in getting "fiber to the home"

2004-03-22

- Technology Conceived in the Cold War Changes With Times at Sionex: Sensor product can be calibrated to identify explosives, germs or other materials

2004-03-19

- Hey, is that a Nanocapsule on your Face?: Line of cosmetics contains nanocapsules, which help active ingredients get to the skin's deeper layers

2004-03-18

- Atomic microscope spots viruses: Nanotechnology used to develop a very small silicon chip to catch and identify viruses
- Hernia Patient Can Continue Active Life With Nano-enhanced Device: TiMESH, contains a 30-nanometer-thick coating on a standard plastic mesh implanted to repair damaged tissue

2004-03-15

- Small Times Names Top 10 Small Tech Hot Spots: Which states lead the race to become hot spots in nanotechnology, MEMS and microsystems?
- The Small and the Beautiful: Max Planck researchers use nanotechnology to visualize cellular processes crucial for the development of new cancer drugs

2004-03-12

- Effects of nanotechnology on the environment: There is still uncertainty about the impact of releasing nanoparticles into the environment

2004-03-11

- Little things could mean a Lot: Nestle is screaming for smoother ice cream...

2004-03-10

- Intel's research chief tells nano crowd to think even smaller: Think very small: 10 nanometers or less, says Intel's David Tennenhouse.

2004-03-09

- Nanolab Knows that Profits are not Made on Nanotubes Alone: and, yes, the company does make a profit.

2004-03-03

- Precisely Manipulating Millions of Atoms: Method may soon help scientists build nanoscale devices much more rapidly
- Sliding Without Friction: Friction can be controlled at the nanometer scale

2004-03-01

- One Startup, Two Employees and one Really Smart Membrane: Filaments can be customized to bind to specific targets

2004-02-25

- Profiting From Nanotechnology: Overview of what the science is and where opportunity may lie for investors

2004-02-24

- New Index to Track Nano Stocks, But Large-caps Stay off for now: Investors interested in nanotechnology finally get what they want: a stock index.

2004-02-23

- Nanomagnetics has New Materials in Store for Memory Market: DataInk captures metal granules in a shell of ferritin protein with an 8-nanometer inner cavity.

2004-02-20

- Gyros to Go: SENSORS WILL KEEP MOVING FOR MILITARY, AUTO

- Quantum Cryptography Companies Tap into Nanoscale's Quirky Core: Curious physics of individual photons, puts unique power of nanoscale to work.

2004-02-17

- 3D fabrication uses light-activated molecules to create complex microstructures: Technique could compete with existing processes for fabricating microfluidic devices

2004-02-16

- Water-based Nano-reagent for Molecular Biology Research Tool Kits: Neowater (TM) enhances performance of molecular biology tool kits
- STMicro Announces 3-Axis MEMS-based Accelerometer: Device provides both three-axis sensing in a single package and a digital output
- Personal medicine in waiting room: U.S. Genomics finds smaller market

2004-02-12

- Nano-origami: Scientists at Scripps research create single, clonable strand of DNA that folds into an octahedron
- Mems Usa Hires Executive, Going Public: Reverse merger is seen as a quick and efficient way of going public
- For Nanoart to Imitate Real Life, Exhibition Goes Back to Basics: Nano exhibit inspired by the "cell", capable of self-sustaining existence and replication.

2004-02-11

- Nanomaterials Could Soup up Supercapacitors if Price is Right: It's a battery! It's a power pump! It's Super Capacitor!

2004-02-10

- Stents rivalry to intensify as Boston Scientific enters market: Rival developers of drug-releasing stents have been battling for supremacy on the patent front. Now, there are issues of safety on the patient front as the two-way tussle moves to the marketplace.

2004-02-09

- Optical switches primed for test debut: DirectLight platform to target telecommunications, manufacturing, as well as fiber distribution, production and storage
- Nano Patterning: IBM brings closer to reality chips that put themselves together
- New Microscope Shows Nano-Fibre Formation: Microscopy probes nanofiber growth process with angstrom resolution

2004-02-06

- Industries await dawning of nanotechnology age: Stain-resistant fabrics and fresh food packaging only the beginning
- Incorporate Disassembly into every self-assembled Nanotech Product: Molecular nanotechnology may clean up the toxic mess left by older technologies

2004-02-04

- Treatment Could Remove Toxins in Blood before Damage Occurs: Nanoscale magnetic particles being developed for medical applications.

2004-02-03

- President thinks a billion of Nano, but a bit less than Allowed by Law: Proposed 2005 budget would give nano programs a bit less than the amount authorized under the bill.

2004-02-02

- Evolved Nanomaterial Sciences Separates the Bad From the Good: While most nanotech companies attempt to build things up from the bottom, Evolved Nanomaterial Sciences Inc. is approaching the nanoscale from the opposite end.

- Flexible Display Screens Readied for Production: Roll 'Em, Fold 'Em, Stick 'Em in Your Pocket: Long-Envisioned Plastic Sheets will make E-Newspapers, E-Books a Reality

2004-02-01

- For Science, Nanotech Poses Big Unknowns: Studies show nanoparticles can act as poisons in the environment and accumulate in animal organs.

2004-01-30

- Nanotennis Anyone? Tiny, Sporty Materials have their day on Court: VS Nanotube racket is responsive when you are out of position and can barely get your frame on a ball.

2004-01-29

- Coatings company wants consumers to look for the U-right Label: NANOECO sportswear is treated with Texcote technology, which places a layer of protective coating on the surface.

2004-01-28

- Self-assembly Technique Shines even if flash Device was for Show: Technique for making nanocrystals could become a cornerstone in IBM's future miniaturized components.

2004-01-27

- Quantum Control holds the key to a Shining LED Lighting Market: Market for high-brightness LEDs projected to grow from $1.53 to $2.85 billion

2004-01-22

- Asia-pacific Governments Invest in Nano Labs and Research Centers: Singapore's Biopolis research park will host a mix of public research institutes and private biomedical companies.

2004-01-20

- Molecular memory not a dream as ZettaCore attracts real money: Tip of the molecular electronics wedge may ultimately evolve into mainstream nanoelectronics

2004-01-19

- Pursuit of a Small Tech Building Block Goes Down to the Nanowire: When is a wire more than a wire? When you add the "nano" prefix, of course.

2004-01-15

- British Scientist says Nanoparticles might move from Mom to Fetus: Vyvyan Howard study could flag up a new, particular, hazard of nanoparticles

2004-01-13

- Small Tech Tackles Concussion Syndrome with Mems Helmets: MEMS accelerometers to create an "impact history" for each wearer.

2004-01-12

- Nanomed Pharmaceuticals' Co-founders Invade Inner Space: Nanotechnology is used to package drugs in tiny spheres and send them to specific cells

2004-01-08

- Europe's MEMS Research Program Surprisingly Spawns Startups: Licensed technologies create new businesses.

2004-01-07

- Nanosphere's newest detector zeros in on specific diseases: Latest gadget could someday detect AIDS and Alzheimer's disease.

2004-01-06

- Philips finds an e-paper technology that's quicker on the draw: Thin electronic-paper display uses a process called "electrowetting"

2003-12-30

- Hitachi set to plant its own 'nanostamp' on the medical market: Process uses a silicon "stamp" to produce long, thin "nanopillars."

2003-12-22

- Universal display wants to shine its OLED light on the world: Organic light-emitting device technology to see the light of day in 2004

2003-12-16

- Geeky garments keep you comfy, and just might save your life: Who says geeks can't look good?
- IntelliSense Software Launched: IntelliSense Software Announces Acquisition of Software Business Unit Assets of Corning IntelliSense

2003-12-15

- Here's a new spin (cycle) on small tech: Smart appliances: Over the next several years, "intelligent" appliances will hit home

2003-12-10

- Ibm's Directed Self-assembly One Facet In Broad Nanotech Program: Chuck Black and Kathryn Guarini demonstrated their nanotech breakthrough with strands of purple and red pipe-cleaner wire
- Nano Europe Goes Down to the Crossroads Between East And West: TRIESTE, Italy has become the temporary home for more than 1,000 nanophiles this week.

2003-12-08

- Microreactors could redefine chemistry, nanodrip by drop: Move over lab-on-a-chip, make room for chemical-plant-in-a-box.

2003-12-04

- EV Group and AMO GmbH to Cooperate on Nanoimprint Lithography: Agreement on low cost, high resolution, large area patterning process

2003-12-03

- IPD library from PHS MEMS supports Agilent's EDA software: New Integrated Passive Devices library designed for Agilent's electronic design automation software

2003-12-02

- Boob tubes in your Buick?: RF MEMS-based antenna arrays may make it possible
- New IBM ThinkPad retains data even if owner is thoughtless: MEMS accelerometers may soon protect data

2003-11-26

- MEM Research Releases EM3DS 6.2 Beta: Now including 3D and 2.5D engines

2003-11-20

- Small tech VC in Q3: the story remains staid and sustainable

2003-11-19

- Investors take a Serious look at Nanotech as Senate Passes Nanotechnology Bill: Investorideas.com features News, Research, Links, Company Profiles For Nanotechnology Stocks
- India selects Coventor MEMS software for national program: India's MEMS Initiative Program chooses CoventorWare? for new design centers

2003-11-17

- MEMS/MST magazine free of charge: by Yole Developpement
- DALSA selects EV Group as Strategic Supplier for MEMS Equipment: Wafer Bonding & Thick Polymer Lithography Production Equipment to be provided

2003-11-14

- DELFMEMS: the MEMS solution!: DELFMEMS provides services in Micro/ Nanotechnologies.

2003-11-13

- PHS MEMS Integrated Passive Devices (IPD) confirm expectations: OEMs have recognized improved performance and testability by using this state of the art technology for several applications

2003-11-12

- Microfabrica to Enhance Microdevice Developments in Japanese Market: Secures strategic distribution alliances with Itochu Corporation and Sumitomo Corporation

2003-11-11

- Southeast France plants seeds for a fertile small tech future: France is beginning to look more like the French version of Silicon Valley

2003-11-10

- Small times magazine recognizes small tech in today's world: Small Times announces award winners
- McGill University selects EV Group equipment to build nanotools: McGill University's Nanotools Facility will be formally inaugurated on November 10th, 2003

2003-11 07

- Nanotechnology educational materials: available for free online

2003-11-04

- Gold nano bullets shoot down cancer tumours: Tests on human breast cancers, both in the test tube and in tumours in mice, were highly successful

2003-11-03

- Play it Smart: Sensor makers create 'cool' and cut costs to pry open dog-eat-dog toy market
- MEMS industry recognizes packaging can make or break a product: the technical difficulties involved in packaging can be considerable
- Partnering makes difference for packaging, testing, assembly
- Company's IPO shows good form, but it will be a tough act to follow
- Bantamweights vs. giants: Is lab-on-a-chip brawl a mismatch?
- Molybdenum products and others: Molybdenum products and others
- Silicon Carbide heating element: SiC heating elements
- Molybdenum disilicide heating elements: $MoSi_2$ heating elements

2003-10-31

- EV Group and SiGen forge wafer bonding alliance on Next-Generation Plasma Bonding

2003-10-30

- FLX Micro Secures Additional Venture Financing: November 3, 2003
- Tronic's Microsystems And Sercel Present the Results of Their Collaboration In Mems-based Seismic Sensor Production: Tronic's industrial and technological expertise enables a new generation of geophones for land-based oil and gas exploration
- Tronic's Microsystems Opens New Mems Manufacturing Facility
- Polyimide film from China: insulating plastic films

2003-10-29

- EV Group's aligner and bonder enhance biochip research at ASU

2003-10-26

- Integrated Micro Devices Corporation provides rapid polymer MEMS prototyping

2003-10-22

- Indian fab to pursue smart-card, MEMS manufacturing

2003-10-20

- New Suss nano Prep TEchnology first to Revolutionize Direct Wafer Bonding
- IMEC and EV Group will jointly develop new wafer-level packaging and MEMS wafer bonding techniques

2003-10-14

- IMT Welcomes Dr. Ian Johnston From STS

2003-10-13

- Coventor to Assist Intellisense Customers with Software Upgrade: CoventorWare? conversion package provides MEMS focused training and support
- China Selects Coventor As Preferred Vendor For Government Mems Development Program: CoventorWare software to be supplied to 30 universities through China's MEMS/ 863 program

2003-10-10

- MEMtronics Receives $3.69 Million DARPA Contract for RF MEMS Switch Development
- Numbers may not Gauge it, but MEMS Industry is Moving At Midyear

2003-10-09

- Block MEMS Receives $2 Million Government Contract to Develop Miniaturized Chemical Agent Detector (ChemPen) for Battlefield and Homeland Defense

2003-10-07

- RM in Life Science Applications: Breakthrough in micro fluidic devcies

2003-09-30

- MICRO SYSTEM Technologies 2003 Best Paper and Best Poster Awards:
- MICRO SYSTEM Technologies 2003 Conference and Exhibition, October 7-8, Munich, Germany

2003-09-28

- $MoSi_2$ heating element and SiC heating elements: $MoSi_2$ heating element and SiC heating elements

2003-09-26

- DISCERA Selects DALSA for Manufacture of Quartz Replacement Technology: Relationship Leverages Low-Cost, Low-Power Benefits of RF MEMS
- Silicon Microstructures (SMI) announces miniature pressure sensor chip
- Silicon Microstructures (SMI) announces new pressure sensor chip with improved media compatibility
- Silicon Microstructures (SMI) releases first in its family of co-integrated sensors
- Fraunhofer Institute Selects Substrate Bonder from SUSS to Support MEMS Commercialization Program

2003-09-17

- SUSS MicroTec's New Substrate Bonders Offer Superior Process Control: Bonders enable high yields and ensure functionality of MEMS, compound semiconductors

2003-09-16

- Flow Science Announces Release of Flow-3d® Version 8.2

- Silicon Microstructures (SMI) completes expansion and renovation of its MEMS wafer fab: Previously announced multi-million dollar plan completed 2 months ahead of schedule
- MEMS-based Explosive Detector Could Catch Future 'shoe Bombers'
- Micronit Microfluidics Signed Agreement on High Volume Lab-on-chip Foundry Manufacturing

2003-09-06

- FLX Micro Enters Exclusive Licensing Agreement with case Western Reserve University

2003-09-03

- Memgen changes name to Microfabrica

2003-08-26

- MEMS Standards, while small, may mean much for the Industry
- Molybdenum Products: Molybdenum Powder, Molybdenum electrodes...
- $MoSi_2$ heating elements and others

2003-08-25

- Axsun Isn't Sitting by the Phone, Waiting for Telecom to Call Again

2003-08-22

- PhoeniX invests in MST software: PhoeniX develops software for Micro System Technology to provide the bridge between the design team and the real cleanroom environment

2003-08-20

- Tire Sensor Makers Gauging Pressures of Court Decision

2003 08 18

- FLX Micro Names Bob Lynch as Chief Executive Officer: August 20, 2003

2003-08-12

- Thick Negative Photoresist which can be removed: SU8 Alternative

2003-08-11

- Flow Science Announces New Associate
- Palomar Hopes to Dominate Assembly With Its One-stop Shop

2003-08-08

- EV Group presents new company website www.EVGroup.com

2003-08-04

- STS Sharpens its MEMS Tools to Carve a Niche in Industry

2003-08-01

- Not Exactly Quicker than a Ray of Light, but Optical MEMS will Fly

2003-07-29

- Integrated Micromachines Equipment Auction

2003-07-28

- Atomic IC and nanocomposites: atomic electromechanical (AEM) systems, nanocomposites
- Jmar's in Motion to Grab a Corner of the Microchips Bag

2003-07-21

- Anti-stiction layer deposition in a vapor phase: samples solicitation

2003-07-16

- Flow Science Announces New Associate
- Dutch MEMS: Strong Research, but no Mass Production . . . Yet

2003-07-15

- Advanced Microsystem for RF and Millimeter wave Communications: AMICOM "Network of Excellence" Within 6th Framework

2003-07-14

- With New Cash, Facility and Focus, Tronic's Tries for New Markets

2003-07-09

- EV Group Sells Production Lithography Package to Silex

2003-07-08

- A Delicious Idea—But it may be too Pricey to be Palatable

2003-07-07

- Amkor Provides Glimpse of Future IC-packages
- Hybrid MEMS 3D Dynamics Analyser
- Cea-leti Selects EV Group's Hot Embossing System to Conduct Nanotechnology Research

2003-07-01

- New Purpose Built 12" MEMS Foundry In Nickel: Photo Electroformed Nickel MST

2003-06-30

- Olympus' Partnership Across the Pacific Expands Its Mems Borders

2003-06-23

- Agency Launches Full-service MEMS Practice

2003-06-18

- Micralyne, Canada Collaborate to Mass-produce Mems on the Cheap

2003-06-17

- Silex Microsystems has attracted USD 9 million

2003-06-16

- Nexense:: Direct Digital Measurement

2003-06-13

- Coal-Mine Canaries on a Chip
- Ultrasound Works Like Bat's ears to Make Real-time 3-D Images

2003-06-12

- MEMGen Unveils '3-D MEMS Design Challenge' Winners: Top three prizes include free prototypes and up to $10,000 cash

2003-06-11

- Researchers create potential toxic sensor chip by combining electronics with living cell

2003-06-10

- EV Group introduces a new wafer edge inspection system at SEMICON West

2003-06-09

- STS positions platform for MEMS volume etch
- SiSonic surface mount microphone to appear in mobile devices soon
- RF MEMS tools and IP to grow from European project
- Immense potential lies in amazingly small things

2003-06-06

- XACTIX to Unveil Xetch eXplorerTM: Will make breakthrough MEMS release process obtainable for a wider range of programs

2003-06-05

- Hitachi Enters a House Haunted with Ghosts of Optical Past

2003-06-04

- Discera Begins Sampling Wireless Industry's First Micro-oscillator: First Single-Die Multi-Frequency Device Will Replace Multiple Quartz Crystals

2003-06-03
- Discera Joins Center for Wireless Integrated Microsystems (WIMS): Microcomponent Developer Strengthens Ties to Innovative University Group
- Discera Joins Berkeley Sensor & Actuator Center: BSAC Affiliation Provides Access to Latest Research Results and Technology Commercialization Opportunities

2003-05-29
- Software Firms Keep Hard Times at Bay with new Business Models

2003-05-23
- DARPA funds "smart-dust" research
- Onstream Bankrupt again, Leaving Mems Projects Homeless

2003-05-22
- Neonode Multimedia Phone to Use Emkay Sisonic MEMS Mike

2003-05-20
- Innovative Micro Technology Welcomes Dr. Amir R. Mirza from GE NovaSensor
- EV Group's temporary bonding and debonding technology enables thinned wafer processing in high volume manufacturing

2003-05-19
- Read/Write Heads Teach Advanced Microsensors Small Tech 'rithmatic

2003-05-17
- MEMS/NEMS suppliers database available on the website of Intelligent MEMS Design, Inc.

2003-05-16
- LioniX moves to new premises
- LioniX Forms Alliance with Indes: Companies to develop and integrate microsystems

2003-05-14
- PHS MEMS Launches Custom Integrated Passive Devices For Wireless Applications: Design Services and Volume Manufacturing Process Enable Miniaturization and Cost Reduction within RF Modules
- The Latest Developments In Microsystems, MEMS, MOEMS And Microfluidics At Micro System Technologies 2003 in Munich: Conference program now decided
- Xerox Runs this up the (Telephone) Pole: MEMS For Magical 'Last Mile'

2003-05-07
- TRONIC's Microsystems unveils new MEMS production facility
- Small tech battle-tested in Iraq in prelude to warfare's new wave

2003-05-05
- Abtech Scientific uses small tech to sense the future of medicine

2003-05-01
- What can Europe teach America? a little self-help, for starters

2003-04-30
- FLX Micro issues "call for designs" for its third MUSiC run

2003-04-29
- Small tech companies outduel all comers at Purdue competition

2003-04-28
- Its cells may be thin, but Cymbet says they pack a big punch
- MEM Research announces FBAR modeling capabilities: Ver. 6.1 of EM3DS includes full-wave modeling of Bulk Acoustic Wave Resonators

- ALCATEL signs exclusive agreement with CANON SALES on Japanese MEMS market

2003-04-25

- As privacy vs. security debate heats up, NSF primes sensor pump

2003-04-23

- Discera Speeds to Commercialization of Wireless Microcomponents with New Hires and California Office: Product and Manufacturing Managers to Help Launch Miniaturized Low-Power Products

2003-04-22

- Here's a new idea: the Internet; customers demand e-commerce

2003-04-21

- FlyTimer jettisons half of 'a wing and a prayer' for small planes

2003-04-17

- IMMI introduces a new low cost fluxless wafer-level MEMS packaging

2003-04-15

- MEMGen Extends Deadline for MEMS Design Challenge to April 30, 2003

2003-04-11

- Officially, war kills nano summit; but some suspect it went AWOL

2003-04-10

- OAI Installs New Back-side Mask Aligner at TFI: Automated, versatile aligner facilitates manufacturing

2003-04-09

- Motors of light may resolve dilemma of how to power MEMS
- Developing motors of light to power small electronic devices
- Memsic finalizes plan for China facility

2003-04-07

- IMMI is switching from telecom to more promising products

2003-04-02

- Small Times Media names Crosby as new president and publisher
- Dutch company positions itself for the growing MEMS market

2003-04-01

- STMicroelectronics puts some marketing muscle into BioMEMS

2003-03-27

- Emkay's SiSonic Microphone Designed in Neonode's State-of-the-Art Mobile Phone

2003-03-25

- Deadline Nears for MEMGen's MEMS Design Challenge: Prizes include free prototypes and up to $10,000 cash
- MEM Research releases EM3DS 6.0: the Free limited and the full trial versions are available for download

2003-03-24

- EnOcean Proffers Piezo Power so Low-energy Products see the Light
- EV Group supports Waseda University on Nanotechnology Research

2003-03-19

- Coventor DESIGNER Trial License Program Announced: Layout and modeling software gives design teams MEMS capability
- CoventorWare 2003 MEMS Design Training Announced: Course fee Waived through July 2003

- New Motown MEMS Lab Will Steer Delphi Beyond the Automobile

2003-03-18

- WSU unveils MEMS lab

2003-03-17

- Small Times Magazine Names Top 10 Small Tech Hot Spots
- Nanoledge builds on nanotube knowledge to make macro material
- LNL mans the photonic torpedoes to prepare for telecom's return

2003-03-13

- U.S. nanotech funding expected to hit $1 billion
- Good things, small packages: Binghamton University electronics engineering center charts new directions at 'micro' and 'nano' scale

2003-03-12

- California, Massachusetts Locked in Bicoastal Clash for Supremacy

2003-03-07

- MEMS Worn to be Wild, But Air Bag Vests a Hard Sell to Easy Riders

2003-03-06

- OMM Dies in Telecom's Waiting Room; Customers Mourn While IP's on Ice

2003-03-05

- Hannover Fair 2003: Micro-Scale High-Tech for Competitive Products: 37 International Exhibitors on the IVAM-Pavilion Present Product and System Solutions

2003-03-04

- MEMSCAP and DOLPHIN partner to bring high level language solutions for MEMS simulation to market: MEMSMaster deploys MEMSCAP modeling expertise and powerful design environment enabled by Dolphin's SMASH single engine, mixed-signal, multi-level and multi-language simulator.

2003-03-03

- Flow Science Announces New Associate
- Bioprocessors Breaks a Bottleneck In the Drug-discovery Process
- EV Group Celebrates Opening of New Joint Venture Company in Taiwan

2003-02-27

- New Opportunities for Optical MEMS Are Emerging Outside of Telecom, Says New CIR Report

2003-02-26

- Not Content to Gather Dust in the Lab, Pioneer Brings Motes to Market

2003-02-25

- Nanotech to pave way for micro-machines

2003-02-20

- Nanoimprint Equipment Companies Set to Make Market Impression

2003-02-19

- Intel donates 150 mm wafer line to New Mexico MEMS project

2003-02-17

- EVG highlights 300 mm Automated IQ Aligner at SEMICON Europa 2003

2003-02-14

- New MEMS Startup Offers Design, Modeling and Fabrication Services as Well as MEMS Equipment and Patent Services: MEMS and ASIC service, MEMS equipment, IP service
- After Columbia: Small tech can help make Space Travel Safer

2003-02-12

- New MEMS Startup Offers Design, Modeling and Fabrication Services
- MEMS to remain a niche technology?
- Kionix Partners with Kefico to Supply MEMS Accelerometers
- SoCs are "dead," Intel manager declares
- Survey Finds "disconnects" Between MEMS Developers and Fabs
- EV Group ships volume MEMS production equipment to Asia Pacific Microsystems Inc.

2003-02-11

- Calient Networks to provide Optical Switches for "OptIPuter"
- Coventor Teams up with Cadence
- Agilent Invests in MEMX

2003-02-10

- Inventor throws curve at screens
- Corning IntelliSense, Northrop Grumman team on RF MEMS

2003-02-04

- TRONIC's Microsystems Appoints Peter Pfluger as CEO: Former CEO Stephane Renard Takes on New Role as Executive Chairman
- Small Tech's Voice Heard Amid Clamor of Competing Agendas

2003-01-31

- Brewer Science increases focus on MEMS market
- Emkay Releases SiSonic—the World's First Surface Mount Microphone: Based on MEMS Semiconductor Technology
- EV Group strengthens its activities in the advanced packaging market

2003-01-30

- Patent Promises to Place Smart Scalpels in Surgeons' Hands

2003-01-29

- In Telecom's Long, Dark Winter, MEMS Survivors Keep it Simple

2003-01-23

- MEMS/Nano LIGA Division Launched by Digital Matrix

2003-01-22

- Small Tech Auto Applications begin their Trip in High-end Vehicles

2003-01-20

- Silicon Microstructures (SMI) embarks on major MEMS facility expansion: Relocation of Los Angeles etch operations to Milpitas facility; Multi-million dollar expansion of Milpitas MEMS facility to increase capacity and add capability
- Miniature Tool & Die Is Helping to Mold the Micro Market

2003-01-17

- EV Group provides industry first 300 mm volume production wafer bonding system

2003-01-16

- CoventorWare 2003 MEMS Design Software Available

2003-01-15

- MEMGen Ignites Micro-Device Innovation: Kicks off MEMS Design Challenge; Prizes include free prototypes and up to $10,000 cash
- Massive Data Storage in Tiny Devices
- MEMS meet semiconductors in beginning of a beautiful friendship

2003-01-14

- New Penn State Summer Institute for Biomaterials and Bionanotechnology: A summer research opportunity for undergraduates and graduate students

2003-01-10

- MEMS industry hopes to hire, while veterans see growing pains

2003-01-09

- Companies put their MEMS together to drum up more customers

2003-01-08

- How to start a nanotechnology (or MEMS) business

2003-01-06

- Iolon's MEMS-Driven Lasers Reduce Need for Spare Telecom 'Tires'

2003-01-03

- Tiny radio signal switches set to turn on RF MEMS market

2002-12-30

- PamGene says it's found the fast track to gene research products

2002-12-27

- European study says only the big photonics players will survive

2002-12-23

- Microlab seeks funding as it prepares launch of RF MEMS switch

2002-12-20

- Companies Race to Develop Ultimate Detection Technology in Terror War

2002-12-19

- Swedish MEMS Fab SILEX MICROSYSTEMS installs ALCATEL AMS 200 ICP Silicon Etcher: Annecy, France (19th December 2002).
- Belgian center spins off company to sell drug-screening technology

2002-12-18

- MUSiC Call for Designs

2002-12-13

- Tanner Research Announces Availability of MEMS R&D and Prototyping Services

2002-12-12

- MUMPs Update from MEMSCAP: New worldwide academic pricing, design rules, run schedules available for MUMPs

2002-12-11

- Innovative Micro Technology wins $1.8 m DARPA Award for Developing MEMS Cell Purifier
- SUSS MicroTec introduces Embossing Equipment for Nanoimprinting Technology

2002-12-10

- UK Govt. Requests Proposals for Spectrum Efficiency Scheme
- Small tech gets a larger piece of the U.S. defense research pie

2002-12-05

- Inventor's long, strange trip to market fueled by MEMS
- Japanese Wafer Manufacturers Produce with SOI Wafer Production Bonding Systems from EVG

2002-12-03

- Alcatel introduces AMS 200 DE: New Deep Oxide Etching ICP Production Tool for MEMS & MOEMS

2002-11-26

- Startups selected for DARPA RF MEMS program

2002-11-22

- Swiss MEMS fab navigates new funding while it guides missiles

2002-11-21

- For MEMS, the key to lowering cost is all in the packaging

2002-11-20

- Trikon Intros Deep Silicon Etch Module for MEMS Devices
- FLX Micro Advances the World's First Micromachining Process in Silicon Carbide
- OnStream MST and Coventor Announce MEMS Partnership: RF MEMS Volume Manufacturing Ramp to Begin in 2003

2002-11-19

- New MEMS Market Research Report: MEMS Technology: Where To?
- MEMS really is the word at Munich electronics trade show

2002-11-18

- Cepheid poised for postal contract, plows on with life science plans

2002-11-15

- Sandia National Laboratories MEMS/LIGA Research Selects Digital Matrix

2002-11-13

- OMM to Supply Optical Switching Modules to ZTE of China
- Small Times magazine announces Best of Small Tech Award winners
- Few firms have figured out how to make MEMS quickly and cheaply

2002-11-12

- IVAM: Positive Signs for Hannover Fair: Industrial Solutions "en miniature" are much sought-after

2002-11-11

- STMicro shows dual-function DNA analysis chip
- Coventor co-invents products while it reinvents itself

2002-11-10

- MEMS Etcher: An economical XeF_2 etching system

2002-11-07

- EV Group Annouces New EVG®40 Series Top-to-Bottom Side Alignment Accuracy Measurement Systems

2002-11-05

- MEMSCAP Finalizes JDSU/Cronos Acquisition and Fab Financing
- Radioactive isotopes fuel microscopic battery
- Thomson Derwent Launches MEMS Patents Profile: New Resource Targeted At Engineers, Scientists and Researchers Working in Micro Devices

2002-11-04

- MEMS Supplier Silicon Microstructures (SMI) hires Mike Dunbar for Strategic Marketing
- Telecom is next on the continuum for 'smart ceramic' MEMS company

2002-11-01

- STMicro prototypes DNA analysis chip
- Display start-up projects new image
- Next Generation Data Storage on the Nanometer-Scale

2002-10-31
- Photonicsourcing launches a powerful search engine for photonic-related products

2002-10-29
- MEMS give farmers the straight story on productive planting

2002-10-28
- Corning IntelliSense and Technovena Pte. Ltd. Partner to Develop Chinese CAD for MEMS Market: Market distribution partnership to strengthen support for Chinese customers of IntelliSuite®

2002-10-25
- 'MEMS' the word for new Star Center project: Research aims to build miniature instruments

2002-10-24
- OAI Installs Back-side Aligner: Automated, versatile aligner facilitates MEMS manufacturing

2002-10-22
- Thousand-chamber biochip debuts
- Breast cancer diagnostic tool attracts interest, but needs cash

2002-10-21
- Support for EUV lithography grows
- Flow Science Announces Release of FLOW-3D® Version 8.1

2002-10-18
- Akustica Raises $2.25 Million in Series A Funding: MEMS Technology Company Spun-off from Carnegie Mellon University
- Tiny atomic battery developed at Cornell: Could run for decades unattended, powering sensors or machines

2002-10-16
- Corning IntelliSense Corporation Awarded $2 Million from NIST Advanced Technology Program.: Research to advance development of digitally controlled MEMS.

2002-10-15
- Generics forms alliances with Coventor, Epigem and LioniX: Launching a new microsystems development service, MST-charged(tm).
- With prototype, funding in hand, Arradial ready to discover drugs

2002-10-14
- Chips & Drivers: MEMS are making memorable strides toward ubiquity
- Teravicta teams with Dow-Key on RF MEMS switches
- EVG highlights Production Resist Processing System for Spin Coating and Developing at SEMICON Japan 2002

2002-10-10
- University of Florida researchers make progress on tiny battery
- Expansion of Silicon Microstructures, Inc. MEMS Fabrication facility and equipment.
- U.S. announces technology funding recipients

2002-10-09
- MEMS event calendar now available in iCalendar format: Usable from Apple's iCal, Mozilla calendar.

2002-10-08
- "Terahertz" camera prototype developed
- Veeco Came, saw, Acquired Majority of the AFM Market

2002-10-03
- SEMI's Oasis provides respite from GDSII

2002-10-02
- TRONIC's Microsystems Announces Agreements
- IMT Partners with L-3 Communications for MEMS-based Inertial Measurement Unit
- For Triad, it's not the MEMS, it's the Method of Making Them

2002-10-01
- Cadence buys IBM's design-for-test tools business
- Analog Devices spins a monolithic MEMS gyro

2002-09-30
- MEMS chipmaker makes $1.9M in series C funding
- MEMS poised to move into the mainstream
- Expanding MEMS with low-cost SOI wafers
- RF MEMS Market to Exceed US$ 1 Billion by 2007: Market study by WTC

2002-09-27
- CFDStudio/ANSWER Version 5.5 is released

2002-09-26
- Smart Artillery Shells Promise a Major MEMS Device Market

2002-09-25
- MEMS Foundry Service Prospers In Southern California

2002-09-24
- XinTec Inc. Expands Shellcase Chip Scale Packaging Technology with EV Group's Production Mask Aligner and Resist Coater

2002-09-20
- Philips puts small tech to the test: It needs to beat other technologies

2002-09-17
- Newsweek on New Technology Jobs: "Fields of Dreams: New Job Opportunities"

2002-09-11
- Chicago area small tech firms are getting their act together
- Military calling all MEMS as U.S. beats Iraq war drums
- Small tech provides some answers to a nation questioning its safety

2002-09-10
- Michigan will help lead the drive toward small tech, governor says

2002-09-09
- Silicon Light Machines Introduces Reconfigurable Blocking Filter: Wavelength-Agile Optical Module Boosts Flexibility and Reduces Costs for Service Providers
- Michigan is host to COMS 2002, gathering of microsystems leaders
- Warning: Loose lips sink IP; shut up and patent, say experts

2002-09-06
- New CVD technique produces thick, high-quality diamond films

2002-09-05
- MEMSIC Debuts ADI pin-for-pin Compatible Accelerometer: New temperature compensated accelerometers offer 50% cost savings and industry-best 50,000 g shock survivability rating
- New technology to seal MEMS could speed integration with chips
- Where have all the brainiacs gone? Ph.D. dearth may stymie small tech

2002-09-04

- Coventor Launches Teragenics: Spin-off Will Combine Distinctive Microfluidics Technology With Biotech Intellectual Property

2002-08-30

- Nanotechnology creates ideal optical medium
- Wireless MEMS are loud and clear while telecom suffers static

2002-08-29

- OLEDs get ready to light up the market for flexible screens

2002-08-27

- Kodak tries to sneak up on TI with its digital cinema strategy

2002-08-26

- CALMEC hopes to build a company a single-molecule switch at a time

2002-08-23

- Team has many obstacles to leap before the Doctor-on-a-Chip is in

2002-08-15

- Agere to dump optical business, leaving MEMS in state of limbo

2002-08-14

- Rise and Decline of Standard MEMS: Was It Just 'Too Much, Too Early'?

2002-08-13

- MEMS Micromirrors Reflecting the Big Picture in Home Theater

2002-08-12

- IMT Receives Investment from L-3 Communications for MEMS: A Strategic Partnership Formed
- Flow Science Announces New Associate
- Harris & Harris Nixes 'Tiny' Transformation
- Proteomics and Promises, Promises: Drug Makers Not Easily Convinced
- What Can Small Tech Do for Its Country? Just Ask, Says Top Soldier

2002-08-01

- MEMS Exchange: Number of Enlisted Fab Sites Keeps Climbing
- MEMSCAP to Acquire Cronos Mems unit from JDS Uniphase: Company to Become Exclusive Supplier of MEMS-Based Products to JDS Uniphase

2002-07-31

- N.J. Governor Funds Small Tech Consortium Based at Bell Labs
- High-speed Data Transfer with Microsystems Technology: IVAM Workshop Micro-optics/Opto-electronics at Dortmund

2002-07-30

- Coventor Announces Enhanced Electrostatic Solver: MemCap Turbo Module Now Shipping
- Swiss School Keeps Spinning out Small Tech Firms at Dizzying Pace

2002-07-29

- Small Tech Funding Pitches Sought for COMS 2002 Conference

2002-07-26

- Amersham to Expand Microarrays after Buying System from Motorola
- EV Group installs wafer bonding system at Analog Devices: EV Group's SOI wafer bonding system provides volume, quality and yield enhancements at Analog Devices' SOI wafer production fab

2002-07-25

- India Ready to Launch its First Privately Funded MEMS Lab

2002-07-24

- Firm Finds New Polymer Process

2002-07-22

- PHS MEMS Counts on Growth in Market for Mobile Handsets

2002-07-16

- Cronos Introduces Two New Multi-User MEMS Processes: MUMPs Program Expands to Include SOI and MetalMUMPs

2002-07-15

- Partech International Investment to Boost MEMGen Growth: Leads Series Al financing

2002-07-12

- Gaming may be the 'Trojan Horse' that sneaks MEMS into your Palm
- Microstrain's orientation sensor gets the attention of the U.S. Navy

2002-07-10

- Teravicta Technologies to Offer RF MEMS-Based Products: Company's RF MEMS Switch Currently in Evaluation at 12 Leading Wireless Companies

2002-07-01

- Coventor Demonstrates MEMS Switches
- EV Group showcases Automated Spray and Spin Coating System at Semicon West

2002-06-10

- PowerZyme says it has the juice to play in portable power market
- MicroMD opens as part of a technology initiative

2002-06-05

- MEMSIC Awarded Best of Show at Sensors Expo: Ultra Low Noise sensor recognized for innovation by leading sensor publication

2002-06-04

- Small Times Launches Small Tech Census: the Most Comprehensive Measurement of the Small Tech Industry
- TRONIC'S Microsystems SA and the CEA have signed a MEMS R&D agreement
- U.S. military's research agency gets a new microsystems chief: Clark Nguyen joins DARPA
- BigBangwidth sets out to create a new universe of smart switches

2002-05-30

- Micralyne Reports Large MEMS Order from Automotive Sector
- APLAC Introduces MEMS Library

2002-05-29

- Micralyne may be a model for making money in MEMS

2002-05-28

- Brits give firm license to share

2002-05-24

- MEMSCAP Strengthens its Position in Asia: Long Term Deal with BL Marketing Services

2002-05-21

- Laser Micro-machining for MEMS and MEOMS

2002-05-20

- MEMS focused issue of Materials Today magazine: Materials Today for tomorrow's technology

2002-05-16

- MEMSIC Accelerometers Selected by Master Brake: Leading brake systems supplier cuts costs and improves performance of lightweight trailers using MEMSIC dual axis accelerometers
- MEMSIC Delivers Accelerometer with Lowest Noise Floor: MEMSIC only vendor to integrate circuit and sensor on one chip based on standard CMOS process to lower cost and maximize reliability

2002-05-07

- Critical Point Dryer for MEMS: Gentle and economic method for drying MEMS
- New Tight Genes Look Good as Better Disease Treatments
- Genefluidics Counts on Glass to Break into Nanobio Market
- Biodegradable Nanofiber could Prevent Scar Tissue
- Time For Physical Sciences to Get Its Act Together, says D.C. Counsel
- French Venture Capitalists Keen on Nano, But Pinch Their Euros
- Electronic Nose Sniffs out new markets for California Firm
- Plasma Process for Particles Could Fuel Explosive Growth

2002-04-26

- Mr. Micromachine: Before they Called it MEMS, there was Kurtz

2002-04-22

- Key to Life Science Success is Market, not Technology

2002-04-19

- Better Living Through Small Tech: MEMS and Nanotechnology Meet Real Life
- New Advanced Deep Plasma Etching Tool for High Volume Production of MEMS & MOEMS: Alcatel introduces AMS 200 "I-Speeder" at Semicon Europa

2002-04-17

- Solutions Come in Small Packages for German Firm with New Process

2002-04-16

- Hannover Fair 2002: Microtechnology Road Maps Point in Different Directions

2002-04-15

- Report from Hannover Fair 2002: Microtech Section is Small, But it's 'Where the Action Is'
- Lion Photonix and 3T MST group join their efforts: Lion Photonix Technologies BV and the business-unit Micro System Technologies of 3T BV, both based in Enschede, the Netherlands have joined forces for the development of Lab-on-a-chip systems.

2002-04-09

- Sound Holding changes its name to Sonion

2002-04-08

- New Microsystems Applications Drive Fast Growth, Study Says

2002-04-05

- Flow Science Announces New Associate
- MEM Research Releases Ver 5.2 of EM3DS: EM3DS 5.2 available for free download. Also available a free MEMS Gallery

2002-04-04

- Firms Lay Financial Foundation Resting on Nanoclay Composites

2002-04-02

- MEMS Fabrication Services: RF and Optical MEMS Manufacturing

2002-04-01
- Three-part series on higher ED programs in small tech

2002-03-26
- Relax, your Car's Sensors Soon will be Better than your Senses
- German Nano, Bio Join Together to Try Building a Better Biochip

2002-03-21
- You've Got MEMS under your Skin? Sensor to Track Your Every Move
- New Lasers Tuned with MEMS may make Nets more Efficient
- Program Tries to Kick-Start European Microtech Industry

2002-03-19
- 1×10 MEMS Mirror Array Available from Coventor and Colibrys

2002-03-14
- Small Times Analysis: Silicon Valley Could Rise Again as the Small Tech Hub

2002-03-13
- FiberLead Launches High Precision Optical Fiber Arrays

2002-03-12
- What Telecom Slowdown? MEMS Maker Debuts Product
- MEMS Robots to Give Surgeons Sense of Touch, Long-Distance

2002-03-11
- Flow Science, Inc. Announces New Associate
- Nanotech Reality Check: New Report Tries to Cut Hype, Keep Numbers Real: Nanotechnology Opportunity Report released today

2002-03-09
- microTEC presents New Solutions for Packaging and Interconnection Technologies

2002-03-08
- U.S. Regulators Want to Know Whether Nanotech Can Pollute

2002-03-07
- Micralyne Enhances Microfluidics Consumables Offering with Launch of ProtoChip

2002-03-06
- Noted Photonics Conference to Honor Nobel Laureates

2002-02-28
- For MEMS Only: An Evening at the Micromachines Club

2002-02-27
- Investors Put Faith, Money Into Microlab's MEMS Switch

2002-02-25
- LLNL Develops Powerful New Rechargable Battery
- Handheld Disease Detector Could Be Developed in 3 Years

2002-02-22
- Dispatch from the Telecom Shakeout: Auction Sells Remnants of MEMS Firm

2002-02-21
- French System Ensures Good Research Translates Into Marketable Products

2002-02-20
- Microvision Piles up Patents in Retinal Display Technology

2002-02-19
- Swiss MEMS Company Sees Profitability in Flexibility: Colibrys profile; Corning Intellisense also highlighted

2002-02-18

- Altair Hopes for Lift from Pigment Patent

2002-02-15

- U.S. Law on Tire Sensors Skids Between Agencies

2002-02-14

- SphereTeK Solder Microball Technology: New low cost solder microballs for $3.00 per wafer

2002-02-13

- IMT Expands Fab and Capabilities
- Innovative Micro Technology Appoints New VP of Marketing and Sales
- Making MEMS: the Working Life of Kionix's New Microdevice Fab

2002-02-04

- Applied MEMS Teams with Kistler Instruments: Low Noise Accelerometer to be Offered

2002-02-01

- Umachines introduces Digital 1×2 MEMS Switch: Umachines is currently sampling this 1×2 characterized by low insertion loss, low power consumption, low PDL and lifetimes greater than 100 million cycles.

2002-01-29

- Flow Science, Inc. Announces Release of Version 8.0 With Multi-Block Functionality
- Analog Devices Selects Coventor to Integrate Design Tools into Optical iMEMS® Development Environment: New MEMS development tools will reduce ADI's optical iMEMS® product design cycles

2002-01-23

- Micro Industry Involved in Power Struggle; Small, Cheap Energy will Emerge Victorious

2002-01-22

- Record Numbers at MEMS Conference 'Means the Industry is Really Forming'

2002-01-21

- Coventor Introduces Low-Cost Variable Optical Attenuator Integration Alternative
- MEMS Market Continues to Grow, Says Industry Group's New Report

2002-01-17

- MEMS Optical Names Angelo V. Ugge as CEO

2002-01-15

- Lion Photonix and OnStream MST combine their efforts

2002-01-14

- New structuring (USHD) and Polishing method (MDF) for glass and fused silica wafers

2002-01-10

- MEMSCAP Still Thinks Large Despite Telecom's Troubles: MEMSCAP building world's biggest MEMS factory

2002-01-09

- Coventor and SensFab Team for MEMS Product Design and Manufacturing: Relationship Gives MEMS Integrators Access to Proven MEMS Production Capability

2001-12-27

- MEMSCAP Acquires CAPTO. New MEMSCAP Unit Brings Medical and Aerospace Business

2001-12-26

- First Book on Microfluidics: New Textbook/monograph

2001-12-17
- BTG and Coventor to Commercialize MEMS IP

2001-12-10
- MEM Research Releases EM3DS 5.0x: Trial version available for download

2001-12-07
- Alcatel Micro Machining Systems expands in Japan: New application laboratory in Tokyo

2001-12-05
- PHS MEMS Chooses Coventor Architect Tool for MEMS Development: Selection Key to Reducing Time to Market for Optical and Wireless Products

2001-12-04
- Teen Inventors Earn Scholarships by Making MEMS Devices Run Better: Siemens Westinghouse Science and Technology Competition Winners Announced

2001-12-03
- Primaxx Single Wafer MEMS Etch Release System: Vapor phase "no stiction" MEMS release

2001-11-29
- Introduction to MEMS Short Course: Offered by Carnegie Mellon University; Pittsburgh, PA

2001-11-28
- Beauty is Skin-deep with High-Tech Sensors: MEMSCAP and La Licorne Enter Multi-Million Dollar Partnership

2001-11-26
- MEMS Optical, Inc. Announces the Completion of a Major Expansion
- MEMS Collaboration Accelerates Time to Market: Agreement between Colibrys and Coventor will yield evaluation samples.

2001-11-12
- Kionix, Inc. Announces Completion of MEMS Fabrication Facility: Press Release 12 Nov, 2001

2001-11-10
- Microstructures by "click": Online Quotations

2001-10-17
- LIGA Now Available Through MEMS Exchange

2001-10-15
- Coventor Named One of Fastest Growing Companies by Inc Magazine: Coventor is First MEMS Company to Receive the Award
- NJIT Aiding Development of DNA Microarray Technology

2001-10-13
- 1 Micron Precision Automation System

2001-10-12
- Micro-Map 5000: New interferometer available

2001-10-10
- MEMS: Mechanics and Measurements Call for Papers: Event Date: June 10-12, 2002 Abstracts Due: October 2001

2001-10-05
- Alcatel opens new North American MEMS Application Lab in Boston USA: For immediate release

2001-09-27

- Tronic's Microsystems has Leveraged 10.5 Million Euros to Increase its Manufacturing Capacities

2001-09-20

- Training Course: Start-up creation in the field of Microsystems: October 12 at Munich (DE)
- Training Course: CAD Tools for MEMS: October 8-9, 2001 at Munich (DE)

2001-09-19

- Advion BioSciences Selects IntelliSense to Manufacture ESI Chip?: Leading Mass Spectrometry Laboratory Transfers Microfluidic Chip Production to IntelliSense

2001-09-06

- Microsystems Technology Short Course: 28-30 November 2001 @ Imperial College, London, UK

2001-09-05

- Randall Wolf of Ohio State University joins Verimetra's Strategic Advisory Board

2001-08-29

- New Gold Deposition Service

2001-08-28

- David Hill Joins IntelliSense as President: Seasoned Semiconductor Executive takes Helm at Industry-Leading MEMS Company
- MEM Research released releases version 4.2 of EM3DS: A completely free version of the program is also available

2001-08-27

- IntelliSense and MM Solutions, Ltd. Partner to Develop European CAD for MEMS? Market: New Distribution partnership to strengthen support for European customers of IntelliSuite®

2001-08-08

- Expertech Announces New Reduced Footprint Line of Horizontal Furnaces Targeted for Emerging Device Market: Scotts Valley, CA, August 8, 2001

2001-08-06

- 2001 International Semiconductor Device Research Symposium (ISDRS): Sessions on MEMS Materials and Processes, Optical MEMS, Microfluidics and Power MEMS

2001-08-01

- Microtechnology at Hannover Fair 2002: USA to have greater presence

2001-07-31

- MEMS Technology Workshop: October 16-19, 2001, Boston, MA, USA

2001-07-27

- Verimetra appoints two leading surgeons to its Strategic Advisory Board

2001-07-26

- Ardesta Launches Discera: to Supply Miniaturized Communications Subsystems to Wireless Industry

2001-06-22

- IntelliSense Corporation co-sponsors Singapore MEMS course

2001-06-11

- microTEC rapid manufacturing without tooling
- ISSYS Inc. Receives $2.75 Million in Funding to Commercialize Advanced Micromachined Biofluidic Chips

- MEMSCAP Establishes Finnish Subsidiary for Active Optical Components
- MEMSCAP Offers Active Optical Component Products for POF
- PHS MEMS raises 31 million Euros in a major funding round
- MEM Research has released EM3DS 4.0, a new software well suited for the electromagnetic analysis of MEM switches and capacitors

2001-06-01

- Verimetra, Inc. closes first round

2001-05-19

- IntelliSense Corporation co-sponsors Singapore MEMS course
- Fully automated wafer mounting and demounting

2001-04-29

- Walsin Lihwa Chooses MEMSCAP in Multimillion Dollar Partnership for MEMS Design and Production

2001-04-25

- Ardesta Announces Translume
- New Source for Daily Microsystems News goes Live: www.smalltimes.com
- Speakers list for MEMS Conference (Nick Ortyl ...)

2001-04-19

- New Conference and Trade Show Created for Emerging Small Tech Industry
- Kokusai Completes Customer Development and Demonstration of MEMS Thermal Processing Suite Using the Kokusai BDF-41 Horizontal Furnace Reactor

2001-04-14

- Tystar Secures Large Equipment Order For MEMS Device Production

2001-04-08

- Microtronic A/S (DK) and microFAB Bremen GmbH (D) join forces to set up production of silicon microphones

2001-03-16

- MEMSCAP Offers Integrated MEMS Design Platform for Cadence Users
- Berkeley MEMS Conference picking up steam

2001-03-03

- TRONICS Microsystems and DEBIOTECH close Collaboration Agreement

2001-02-22

- Mezzo Systems Announces Availability of LIGA Foundry Services
- IntelliSense Certified to ISO 9001 Quality Standard
- Conference on Health Engineering
- Nano Alumina Fiber—Product La\unch
- Memscap's Mems Comm RF-MEMS Library Available for Popular Agilent Technologies Ads Tool Suite
- Umech Technologies Introduces First Commercial Full-Field 3D MEMS Motion Measurement Tool
- MEMSCAP Expands Optical Mems Product Portfolio
- XACTIX spins off bio-MEMS business
- DISCO is at the Cutting Edge of Tiny Technology
- Microcosm Technologies and JENOPTIK Mikrotechnik team to support optical and microfluidics MEMS
- IEEE Starts Sensors Journal

- Argonide Announces New Copper Coating Process
- GT Equipment Technologies, Inc. (GTi) announces new website
- National Nanotechnology Initiative
- New paint capable of metallizing substrates with thick or thin films.
- ADC Teams with MEMSCAP for MEMS-Based Optical Communications Components
- NASA Hosts First IntelliSuite, CAD for MEMS, User Group Meeting
- Free Trial of CFD-Micromesh, a Fast 3D Model Builder from Layouts
- ELSEVIER launches Smart Materials Bulletin

APPENDIX C

Nanotechnology Development: Focus on Social, Ethical, Legal and Geopolitical Implications

(i) Societal Implications of Nanotechnology

Scientific discoveries by themselves rarely create change. It is the confluence of old and new technologies with old and emerging social needs that creates change. There are dynamic interactions and "tipping points" that are difficult to foresee (e.g., the capabilities that emerge when computers get one more order of magnitude faster, coupled with a clever new idea for software, leveraged by increasingly computer-literate 20-year olds).

Thus, the problem for prediction is not only to track the advances of nanotechnology, but also to track other advances and changes in society at the same time. Visionaries and science fiction writers have bold visions, but researchers must also take a role in producing a useful understanding of possible societal directions.

Perhaps the best we can do is to open our thinking and be more aware of the kinds of changes that may occur, how to spot them as early as possible, and how to prepare to influence the course of changes when deemed necessary or desirable. Of course, it is important to consider who would be doing the "deeming." History suggests that those in power tend to suppress or co-opt new technologies (e.g., RCA's success in subverting Farnsworth's patents on TV and convincing the public that they were the inventor, Fisher and Fisher 1997), and the newly empowered try to do the same (cf., Bill Gates).

So, I accept my more modest role to suggest social science research methods for studying societal implications. I begin with some assumptions about research and nanotechnology, and then present goals for social science approaches to societal implications of nanotechnology. I then suggest sources of measures and kinds of indicators that can be helpful, including "leading indicators" that might provide early hints of change. Finally, I offer possibilities for types of research designs and some conclusions.

Some Assumptions

Nanotechnology, as a family of tools and techniques, is a source of products and other "stuff." It is these products and stuff that will be fought over, demanded, and used in ways that will bring layers of change. Each of these bits has the potential to be mundane variations with a bit more zip or a bit less cost, or to be dramatic advances. We are not likely to know which is which, and the nanotech equivalent of the hoola hoop may turn out to be critically important when combined with something else.

Initially, the impacts of nanotechnology will be via specific products and innovations.

Such primary effects would be to make things work better, cheaper, with more features, etc. This might, for example, increase food yields, generate new textiles for clothing, improve power production, cure certain diseases, or whatever. Secondary effects might be shifts in demand for products and services, so that people come to expect different kinds of food, medical care, entertainment, etc. the required infrastructure for nanotechnology may create interdisciplinary research centers, new educational programs to supply nanoscientists and nanotechnologists, etc. Later, tertiary effects would move upstream in our social structures and cultural patterns, such as shifts in education and career patterns, family life, government structure, and so forth. Will nanotechnology extend our lifetimes more rapidly than it extends our health, or *vice versa*? Will nanotechnology enable so much connectivity to information and to each other that it revolutionizes society (cf., Star Trek's the Borg) or shift us more toward solitary lifestyles of "distance experiences" and "virtual experiences" rather than personal contacts (cf., Asimov's *the Naked Sun)*? Research, including social science research, is an activity intended to generate and validate knowledge, based on systematic rules agreed to by a community of scientists.

Since there are many subcommunities of scientists, the rules vary somewhat from discipline to discipline, place to place, and time to time. It is important to realize that the activities of social science research also influence policy-makers and the general public, and thereby change the way society thinks and acts (Gergen, 1973). to measure something is also to shape the future, by changing what we pay attention to, expect, reward, and punish. Therefore, it is important to measure both what is desired, and also what is feared.

Research Goals

In light of the above assumptions, it seems reasonable to pursue the following goals for social science research on the societal impacts of nanotechnology. First, we need to define and measure "societal impacts." Second, we need to find leading indicators or first signs of impacts. Third, we need to develop theories that explain impacts, identify causal mechanisms and contingent conditions (e.g., under what circumstances would particular products have particular impacts), relate various advances and impacts together in more comprehensive systems models, and permit (tentative) extrapolation to possible futures. Finally, we would like to assist policy development on the basis of what is known from our research and what is known about desires and values, i.e., what are "society's" goals and how will these goals change over time as technology advances?

Measures and Indicators of Societal Impacts

There are a huge number of potential measures for societal impacts. Our standard measures of the social world are all useful and all insufficient. From a methodological viewpoint, there are various techniques for measuring societal phenomena, including self-report surveys and interviews; diary studies (e.g., new software that allows Palm Pilots to interview their owners) and think-aloud protocols; social and economic statistics; direct observation (of, for example, meetings, point of sale, point of use); bibliometric and content analysis (of patents, citations, email, Web sites and hits, chat room content, 10Ks); network analysis of relationships across people and organizations; accretion measures of what piles up (e.g., trash heaps); and erosion measures of what gets worn out (e.g., repair calls). For some general principles and examples, see Babbie, 1989; Judd et al., 1991; Webb et al., 1981).

It may be helpful to sort through potential measures by using some conceptual criteria or categories. For example, we could distinguish *process* from *content*. the temporal *process* might include scientific or social visionaries who first imagine possibilities, nanoscience and nanotechnology discoveries that connect with these possibilities, actual products that are developed and marketed, public acceptance by scientists, consumers, and policy-makers, substitutions of usage patterns as new products replace old products, interactions of new product capabilities with existing technical and social arrangements that reshape demand and usage, and finally transformations of social institutions and associated infrastructures. the *content* domains might include health, wealth, food, fuel, productivity, education, employment, national security, happiness, social capital (networks of relationships), political participation, ethical thought, etc.

From the process categories, we can generate more specific indicators of innovation and change. Visions are likely to appear in the media, at meetings, and on the internet. Discoveries can be tracked through patent applications and new products and services. Investments are visible in budgets, grants, strategic plans, projects, job titles, hiring results, and educational programs. Public acceptance is indicated by attitude measures, purchases, usage, and even charitable contributions. Interactions show up in studies of professional and other social networks, fields of study, and communities. Substitutions are evident in the decline of old industries. Transformations can be measured in indicators of lifestyle and social institutions such as deurbanization and demassification.

Other conceptual criteria underlie the objectives and recommendations in the IWGN Workshop report. These constitute a nascent "theory" of the nanotechnology process and impacts. the report divides its objectives and recommendations into sectors or classes of participants such as academe, private sector, government labs, funding agencies, and professions. For each sector, the objectives suggest process and/or content measures of change. In academe, the workshop endorsed interdisciplinary work, new courses, fellowships, information flow, and regional coalitions. For the private sector, the focus was on investments, startups, and coalitions. For government labs, the report looked at budgets, equipment, standards, and coalitions. For funding agencies, the priorities were on new initiatives, databases, and centers. the professions were expected to create new forums, symposia, and job fairs where interdisciplinary topics and careers could flourish.

Of particular importance would be indicators that provide early signs of change. In our process indicators, we would probably look upstream to the beginning of the chain of events and examine allocations of effort in academic and corporate R&D labs and analyses of patent applications to see the emerging technological trends. More generally, we might consider who, what, and where we could find "bellwethers" or first movers.

California seems to be the first place that social change occurs in the United States. Finland and Singapore are examples of countries that have embraced new technologies and undergone rapid change. Within both rapidly changing and slowly changing societies, there may be classes of "early changers" such as 15 year olds who are at the forefront. In some cases, such as in the health care domain, it may be the 90-year-olds who show the first signs of change. Science enthusiasts, nerds, and hackers may be first movers in a nanotechnology-rich world. Start-up companies, university labs, and Internet chat rooms may be places to look for changes and impacts. Particular social strata, such as groups "on the margin" of society, may also exhibit the early signs of changes (things that seem weird or bad) that will later spread to mainstream society.

Research Design

Research is more effective when it is designed into the process being studied, rather than having to explain what happened after the fact. It is not surprising that many of our best studies of the impact of technologies are retrospective analyses from 10 or 100 years ago (e.g., Bijker, 1995; Fischer, 1992) but we can't afford to wait that long for knowledge of current developments and impacts. If social scientists can be introduced earlier into partnerships and collaborations with nanoscientists and nanotechnologists, there is a better chance to learn more and learn quickly enough to guide policy. Research designs are typically stronger on measures and mechanisms if introduced early, so that measures can be made over time and informative controls can be established to strengthen causal understanding.

I have listed some typical design types in the order in which I would guess they will be used. In other words, designs that are easier to execute are likely to be used first, and those that require great skill and/or great control over circumstances may come later.

Surveys, expert panels, and statistical summaries of socioeconomic data are relatively easy to do and provide useful snapshots. Simulations (computer-based or game-like) allow us to project behavior in hypothetical scenarios (e.g., Sterman, 1989). Case studies (Yin, 1989) permit a rich set of information gathered around particular cases of inventions, products, companies, communities, supply chains, etc. Quasi-experiments (Cook and Campbell, 1979) allow focused comparisons that strengthen causal inference, such as time series and econometric designs and control groups based upon diffusions of new technologies (some groups naturally get it earlier, some later). Ethnographies (Van Maanen, 1988) and in-depth immersion in real-world sites allow a rich understanding of complex interdependencies and subtle phenomena. True experiments allow us to answer precise questions, but are difficult to arrange in real-world settings.

In conclusion, we must remember that we cannot easily predict the products and impacts of nanotechnology on society. We are not studying a monolithic "nanotechnology" but rather a host of varied technologies, products, services, and other interventions. Some implications and impacts are relatively easy to predict and to study, but others will be emergent and surprising. However, everything in society will be changing (not just what nanotechnology has touched directly), and everything will be connected together (loosely or tightly), so predictions will be uncertain and causal explanations will be difficult to validate.

Research is needed to help us understand changes and to plan action. A wide range of indicators will be needed because we do not know what will emerge or what will turn out to be important. Helpful theories are especially important to focus attention on key issues and processes, to guide research, and to represent the results of research. Society needs theories and system models to understand how changes in one part of the system, whether a particular type of technology or a particular element of society, spread out to create intended and unintended effects throughout the system. This includes understanding the impact of society on technology. This will enhance our ability to educate everyone earlier and to improve the quality of public debate. It will also enable scenario analyses, strategic planning, and simulated public debates to be more informative and rich.

At its best, the research attitude is one of openness, curiosity, sharing, and constant improvement. This is a model for increasing the capability of the public and the scientific communities to plan intelligently, communicate effectively, respond to emergent circumstances, and understand themselves and the broader society and world in which they live and work. From an ecological perspective, there is no guarantee that "progress" (however defined) has any particular consequences for the human race; nor is there a guarantee that more science and technology will always find an answer to human problems. We must improve ourselves as a thoughtful and ethical human society at the same time that we improve our mastery over the physical world.

(ii) Ethical Issues in Nanotechnology

Because we are concerned with issues of practical and professional ethics, not theoretical ethics, we must examine nano initiatives along with associated social practices, institutions, organizations, and the choices and actions of individuals within them. In nano science and engineering, ethical and responsibility issues connected with various initiatives will reflect many complexities—those from the multidisciplinary and multiinstitutional character of research and development, as well as from more technical aspects.

To survey ethical implications and identify risks it is necessary to step back from captivating visions of profound transformations of the material world and of society. the focus must be on specific proposed initiatives in their institutional environments.

Normative questions, that is, questions about what it is right and appropriate to do, gain a foothold at the macro-level in relation to societal and organizational policies and activities and, at the micro-level, in relation to ethical standards, responsibilities, decisions, and actions of individuals.

An important aim of ethical investigation is to anticipate ethical problems preventable harms, conflicts about justice and fairness, and issues concerning respect for persons likely to arise from specific nano initiatives. A second important aim is to foster sensitivity to ethical issues and responsibilities at every level of decision-making by both technical and policy people.

How to Proceed

At the outset, it is essential to clarify the term "nanotechnology." There is no disagreement about the oversimplifying and misleading character of that term. It is a catch all that has caught on because of its convenience and market appeal, its usefulness as a rallying point. Though aware of its deficiencies, interested parties are not inclined to reject use of the term. It is therefore necessary to make clear what are the initiatives, disciplines, and institutions that "nanotechnology" embraces.

(1) Initiatives with implications for advances in medicine, computing, space exploration, energy conversion and storage, optics, and materials, including catalysts, are among those often cited.

(2) A large number of disciplines—and specialities within them—in science, engineering, mathematics, and computing are encompassed.

(3) Institutions of government, academe, and industry, and organizations and practices within these institutions are also included.

Government agencies act as promoters of initiatives, and they have begun to fund university research and graduate training. Universities are gearing up for and launching new programs, instituting graduate studies, renovating and putting up buildings, and forming alliances with other universities, national laboratories, and private firms. New collaborative arrangements between universities and private companies to carry forward research and development are being, and will continue to be, forged. Of use in forming these new collaborations is the experience recently gained from crafting such arrangements to advance work in information technology and biotechnology.

That experience suggests a second task: to examine recent history. Study of our experience with biotechnology and information technology may help to locate nodes of ethical concern. Caution in drawing parallels is necessary, however, in light of claims made for the uniqueness of nanoscience and nanotechnology. It may, nevertheless, be useful to direct attention, from an ethical perspective, to questions that these earlier technologies suggest. Included are questions about how government plays a role in promoting, launching, and supporting technological developments and about how projects to produce products are selected and by whom. Government agencies' promotional discourse to boost the NNI is already available for scrutiny, and there may be opportunities to study how nano projects are currently selected. Our history with earlier technologies suggests the need to devise processes and settings for information exchange with and wider participation by members of the public in order to promote transparency. (See Wynne, 1991 for an especially insightful discussion.) So compelling are the ethical and practical benefits of building in openness, disclosure, and public participation from the outset that efforts towards those ends should begin without delay.

Recent history with other technologies indicates that the obstacles to achieving these aims are formidable. Observers have already expressed concern that multidisciplinary meetings present special risks to open exchange of information (NSTC, 2000, p. 32). Similar concern is appropriate regarding the flow of information across institutional boundaries, between academe, government, and the private sector (Blumenthal, 1992). Some problematic patterns relating to the flow of information between those engaged in technological development and the public are familiar from technological developments of the recent past. For example, by becoming locked into a definition of the public as the "other"—the enemy, uncomprehending, standing in the way of advance—those with authority over information may withhold it and thereby cut themselves off from public reactions. Failing to nourish genuine information exchange, they may invite the very opposition they wish to ward off. Such patterns from the past point to the need for care and caution in framing "the communication problem", to avoid seeing it as a problem of one-way communication downward (Wynne, 1991).

A third task at this stage is to try to identify ethical issues that have already arisen or are likely to arise. As elsewhere in practical and professional ethics, it is necessary to disaggregate, to look case by case at specific nano science and technology options and their consequences—current, foreseen, foreseeable, or speculative. Initiatives and options should not be divorced from their institutional contexts, however complex the latter may be. At times technical people fail to foresee what is foreseeable and within their sphere of responsibility. Sometimes they claim to foresee what is not foreseeable, given available knowledge. Nanotechnology options afford opportunities for both kinds of foreseeability problems to arise regarding the consequences of specific developments.

The recent history of rapidly developing technologies suggests that we should be alert to unintended consequences. Consider, for instance, the complex issues that have arisen about privacy in connection with information technologies (Johnson and Nissenbaum, 1995). the privacy debates also illustrate how ethical issues concerning respect for persons are generated with the propagation of new technologies. Biotechnology offers another cautionary example relating to unintended consequences. When investigators in the new biotechnology achieved confidence in control of their products in the laboratory, vigilance regarding unintended consequences did not extend to the new products in the field. Information about interactions of new bioengineered organisms with other members of an actual ecosystem has been scarce and slow to appear. Accordingly, concern about unintended consequences from agricultural products of biotechnology remains high in many places (Weil, 1996).

While trying to stay alert to unintended consequences, we should also try to avoid taking it for granted that there is wide agreement on the desirable consequences of various nanotechnology options. It is essential to obtain a diversity of perspectives on the desirability of particular options. For eliciting a range of perceptions, conversation between people in nano research and development and members of the public is necessary. It may be essential to create mechanisms, such as citizen/scientist panels, to bring different perspectives into the conversation. In the exchanges, people will have to learn to identify interests that are in play, their own and those of other parties to the conversation. Experience with biotechnology shows the costliness of proceeding with mistaken assumptions about what are desirable outcomes and products (Crow, 2001).

Identifying Ethical Issues

To go further in identifying ethical issues, we need concrete points of departure, actual examples or cases that pose ethical questions, quandaries, or conflicts. This is how we proceed in practical and professional ethics when looking at issues in other social practices. Lacking specific nano examples at this time, we may note features of nanotechnology that, because of their alleged novelty, may be sources of ethical concern.

For example, forecasts of development of new catalysts stress the creation of new production processes. the latter are likely to raise ethical questions, for instance, about the need for safeguards for workers that specific new processes might generate.

Questions of these kinds can provide points of entry to the institutional, organizational settings in which potential problems are embedded and in which they must be examined.

The new processes are confidently predicted to produce workforce changes (NSTC, 2000, 20). Anticipated impacts on the "human resource infrastructure" will surely bring benefits to some and harms to others. When there are winners and losers, issues about equity cannot be avoided. Again, on a case by case basis, it will be important to identify winners and losers so as not to inflict preventable harms or lose opportunities for mitigating harms or compensating losers.

Ethical issues associated with intellectual property protection are virtually certain to arise (NSTC, 2000, 31). They will be novel insofar as novel features of nanotechnologies and their social environments introduce new complexities relating to intellectual property protection. In environments where patents and trade secrets are generated, there will be implications for open exchange among technical people and communication with the public. the rationales justifying ownership are likely to be as vigorously debated and contested as those associated with information technology and biotechnology. Equity issues raised by intellectual property protection should generate debate as well. Vigorous and extensive public discussion could even lead to reexamination and revisions of intellectual property policies.

Ethical questions about university/industry relationships are hardly novel, but they are virtually certain to arise. These questions have engendered discussion, literature, and large-scale empirical studies since at least the mid-1980s (Blumenthal et al. 1986). By now, some specialists contend that institutional accommodations to new relationships with private companies have transformed universities, bringing significant changes in university values and practices (Webster and Etzkowitz, 1991). For instance, many in universities now accept a need to allow faculty members to accumulate great wealth through their research (Forest, 2000). Yet there is a clear understanding on both sides that universities and private sector enterprises are valuable to each other as partners precisely because of their differences (Weil, 1988; Weil, 2000).

The close association of university research with the private sector has brought problems of conflict of interest to the forefront. For example, questions arise about whether a university researcher's ties to a for-profit firm threaten reliable judgment in university research. Observers have suggested that universities as institutions can have conflicts of interests (Frankel, 1996; Pritchard, 1996). A strong program of research and development in nanoscience and nanotechnology will subject university values and practices to new pressures, and universities will have to continue to make accommodations preserving core university values. They will have great need for ethical guidelines as they make those accommodations and review their policies in the light of experience.

The focus on institutions, organizations, and practices should not obscure the need to focus on the individual responsibility of engineers, scientists, and others involved in the processes of producing new technologies "with unprecedented control over the material world". to focus on individual responsibility, university programs of graduate study and research in nano areas should include attention to ethical issues specific to their own nano areas. They should give attention to these issues in their training of students in scientific research ethics and in their technical training of students. Professional societies have a role to play in affording opportunities for debate and discussion and helping to devise, for individuals and organizations, guidelines that incorporate ethical principles responsive to emerging issues.

Addressing the Issues

This survey of ethical dimensions suggests three main fronts of activity for responding to issues. One is activity within research and development initiatives funded by government agencies and carried out in universities. Government agencies should require an ethics component in grants for graduate training and in grants for research. In this way, they can provide for attention to ethical issues by individual researchers and principal investigators within research groups, by research groups as collectives, and by university departments and centers in which the nano research and training goes on. It is necessary to connect specialists in ethics and behavioral sciences with such projects from their outset and maintain the specialists' association with ongoing nano projects of research and development. National Science Foundation proposal guidelines already incorporate provisions requesting an ethics component. Such provisions should be standard components of requirements for submitting proposals in the nano domain to government agencies. Over time, the provisions can be further detailed or improved in other ways, as experience indicates.

The second front of activity has already been suggested—a fresh and energetic effort to devise opportunities for genuine conversation with members of the public about current and proposed initiatives. In view of the ethical and practical reasons for commitment to disclosure and incorporation of democratic processes in advancing nano initiatives, government agencies that are promoting nanoscience and nanotechnology should accept an obligation to create appropriate channels and fora. Concerted activity on these two fronts could be innovative and capable of transforming research and development in step with the innovative and transforming features of nano processes, structures, and products.

A third front of activity is education. While some ethics specialists are already available to help in addressing the ethics component in research, development, and education related to nano initiatives, there are not enough. Mainstream graduate training in philosophy and behavioral science is not yet oriented to respond to this need. the intellectual interest and acknowledged importance of issues raised by the NNI justifies government initiatives to address the lack of qualified ethics and behavioral science specialists.

It should be possible to devise training programs led by existing specialists in collaboration with faculty who direct graduate and postgraduate study in appropriate disciplines. Postdoctoral studies may be especially suitable for providing the training and forming the collaborations across disciplines that are needed. A program to train people for the ethics work within nanotechnology and nanoscience initiatives is essential, and feasible, with government support. This education should be part of a concerted endeavor to improve education at all levels, in scientific and engineering disciplines, and areas that cross disciplines, as well as in philosophy, ethics, and behavioral sciences. For the NNI to begin to fulfil its promise of carrying forward research and development initiatives that bring benefits to society fairly distributed, concerted efforts on all three fronts are needed.

(iii) Social Acceptance of Nanotechnology

The notion of "social acceptance" of technology is prevalent throughout both scholarly and social science studies of technology and in popular literature, yet it is, in an obvious way, a very obscure idea. On the one hand, it connotes empirical content, perhaps even measurable criteria, so that whether or not a technology has been socially accepted appears to be decidable question, a matter of fact about social relations or how things stand in the world. On the other hand, the phrase "social acceptability" suggests a normative judgment in a way that makes social acceptance come to involve inherently contentious characterizations of "society's values." Here I want to make a few general remarks about the social acceptance of new technology, to note some lessons learned and yet to be learned from the ongoing saga of biotechnology, and to make some suggestions about how these lessons might apply to nanotechnology.

Here are some of the indices that might be taken to measure social acceptance of technology.

- *Geographical Parameters:* Where is the technology used—and "where" can be characterized in terms of location, demography, culture, class, etc.?
- *Economic Parameters:* What is the market penetration for the technology (what percentage of the potential purchasers of the technology actually purchase it), what is the price sensitivity, etc.?
- *Psycho-Social Parameters:* What do surveys indicate when people are asked how what they think about a given technology?
- *Affective Parameters:* What's the "comfort level" of users? Do they feel a sense of regret? Do they feel a sense of moral disapproval? (One could argue, for example, that despite widespread use, chemical pesticides are not a socially accepted technology).
- *Cognitive Parameters:* What's the level of awareness that a technology is being used? Are people presented with clear opportunities to accept or reject a technology? (Arguably, many of the technologies in daily use have not been "socially accepted" simply because they are, for the most part, wholly unknown.)
- *Technical Administrative Parameters:* Where is a given technology in the process of regulatory review and approval? Are there regulatory or court decisions that actively sanction the use of the technology?
- *Political Parameters:* What is the level of debate, contention and organized opposition to a technology? What is the potential to mobilize opposition at any given time, and at what cost?

There are also normative and quasi-normative parameters for the social acceptability of technology. Here are a few:

- *Religious Acceptability:* What do religious doctrines or religious authorities have to say about the technology? There are cases (such as whether cheese made with recombinant chymosin meets Kosher standards) where these are rather straightforward questions.

- *Cultural Acceptability:* Given the fact that many cultural norms are implicit and veiled, the question of whether a given technology is culturally acceptable is often vexed and contentious.
- *Ethical Acceptability:* I take this to be the broadest and most irreducibly normative category. *Should* society accept a given technology, and if so, under what constraints or qualifying conditions? (So, one might argue that a technology is acceptable only if its advantages and benefits will be distributed fairly across all economic classes).

Many of the social sciences employ forms of analysis that generate evaluations that are at least hypothetically ethical, in the above sense. Standard economic cost-benefit analysis is an obvious example. Though saying whether the costs of a technology outweigh its benefits is not a normative judgement in the sense of actually recommending for or against, such analyses nevertheless suggest qualified normative criteria that might be (and indeed commonly are) used for such a judgement. I hope it goes without saying that what "cost," "benefit" or "fairly" means is itself an ethical issue, that what it means for society to accept a technology (in a normative sense) is an ethical issue, and that whether there even *are* such normative criteria is an ethical issue. One can debate terminology, but the category of ethical acceptability is logically inescapable, since to deny the relevance or possibility of ethical criteria is to make a contestable normative claim.

Criteria for descriptive parameters for social acceptance tend to become entangled with normative criteria for social acceptability. the reasons have little to do with the study of technology *per se.* For example, behavioral social science may require framing assumptions about rationality in order to structure data bases or to characterize a behavioral phenomenon as an instance of "choice". But calling a form of judgment or behavior *irrational* is generally taken to imply a normative judgment about it. This opens out into both philosophical and highly tendentious debates, especially about the kinds of social science analysis noted above.

While it is tempting to dismiss talk about the social acceptability of technology and society's values as lacking proper rigor (Is society really the sort of thing that can have values?), it is useful to recognize some important things that are going on in such talk.

Clearly, players—and we are all players here—try to "spin" the social acceptance of a technology such as nuclear power, food irradiation, e-commerce, or biotechnology in an effort to influence opinion. That is, social acceptance is a thoroughly reflexive phenomenon. Given the multiple ways (above) that a technology might or might not be characterized as socially accepted (or not), it is generally possible for those who wish to quiet or aggravate acceptance as measured by any given parameter to say (with some degree of truth) that the technology is/is not/may/may not be/will be/won't be socially accepted (with respect to some other parameter).

Given the economic, political and cultural and professional interests that are (or might potentially be) at stake, anyone may be an interested party. This is *not* to say that there are no facts with respect to the indices that might be taken to measure social acceptance of nanotechnology, nor to say that everyone wants to influence the social acceptance of a technology that they study. But it does suggest that any reported observation regarding these indices may with fairness be subjected to a normative critique. For my money, the reports of sources that disclose their interests will be more credible.

Most scientists (including social scientists) are not trained at clearly understanding their own interests, much less at disclosing them openly to others. Unfortunately, the arguments that are proffered in enduring philosophical debates, edifying in their own right and crucial to rigorous social science, are quite likely to be deployed in the strategic gamesmanship of players seeking to advance or retard the social acceptance of a given technology.

Has biotechnology been socially accepted? I will confine myself to food and agricultural biotechnology, which is my primary area of expertise. Certainly there are published reports that bear on each of the indices noted above.

- *Geographical Parameters:* One can measure the number of hectares sown in biotech crops, the countries in which they are grown, and the demographic characteristics of farmers who use biotech seeds.
- *Economic Parameters:* There are many economic studies of the adoption of biotech, as well as studies of the competitiveness of the industry.
- *Psycho-Social Parameters:* There have been repeated surveys of public opinion on biotech. Levels of acceptability vary in degree and in change of direction across national groups.
- *Affective Parameters:* Surveys and focus group research provide some indication of affective parameters. This is an ongoing area of data collection and qualitative analysis.
- *Cognitive Parameters:* Surveys indicate a general measure of awareness, generally quite low, it is worth noting, for a technology that has become a model case study for a contested technology.
- *Technical Administrative Parameters:* Since the early nineties there have been regular reports of regulatory approvals. the weakness in this area would appear to be the lack of a general theory of what might have been thought to be an obvious question: what constitutes legal or administrative acceptance?
- *Political Parameters:* A substantial amount of qualitative research exists. There do not appear to be standard methodologies for developing indices of political contentiousness and acceptability.

Some Lessons: Though a fair amount can be said about each of these parameters as they relate to food and agricultural biotechnology, what do we know about the social acceptance of food and agricultural biotechnology? I would be skeptical of anyone who professed to know whether this technology *has been* accepted or *will be* accepted, both on a global or a regional basis. While it seems unlikely that agricultural biotechnology will simply disappear, it is anyone's guess as to whether the levels measured in these indices are stable, or will increase or decrease over time.

To me, this suggests that we don't know very much about how these parameters affect one another, about the dynamics of social acceptance. the notion of social acceptance seems intuitively clear, and even demonstrable for some historical cases. For example, can we really question whether electrification has been socially accepted in the industrialized world? Yet the question whether we really have a very good sense of what real-time social acceptance amounts to. the strategic, normative and reflexive dimensions of social acceptance may account for the open-ended nature of the real-time index for social acceptance. But it would be useful to have a theoretically clear and rich statement of why this is the case.

Has the battle of strategic positioning for the social acceptance of nanotechnology already begun? It is inevitable that, whatever the motives of their authors, the documents which already exist (the documents we are generating at this workshop) will be spun by players down the road. Because I have ideas about how which criteria and procedures should lead to the acceptance, rejection or qualification of any new technology, I would argue that several things should be done at the earliest opportunity. Most importantly, science funders should invest in the creation of fora that are shielded from strategic actors to the extent possible. There are, of course, limits to the extent that this is possible, but several of the following things would help. One is clear, open and ongoing multi-disciplinary clarification of the interests that are advanced and retarded by the development of nanotechnology. A second goal would be to increase scientists' capacity to reflectively understand the sense in which they are interested parties, and encourage disclosure of those interests. Here, career, disciplinary and ego-based interests can be as decisive as pecuniary ones. One should not shy away from development of explicitly normative studies and position papers, but such efforts should aspire to high standards of transparency, clarity of analysis and to the creation of a public record.

It would also be useful to build on the literature of social acceptance as it has been developed with respect to technologies such as biotechnology, to design studies that would compare and integrate the indices described above, and to examine analogies that might suggest a basis for normative evaluations of the social acceptability of nanotechnology. the "technology out of control theme" or the precautionary principle as applied to biotechnology might, for example, present a suggestive starting point for normative studies on the acceptability of nanotechnologies.

The scientists and engineers behind nanotechnology must be involved in all the above, and to me this suggests that there will be a need for an ongoing series of conferences, workshops, seminars and publications with a fairly high level of visibility. Such people will exclude themselves from the standard run of social science disciplinary research outlets, so something else is needed. These activities need to be visible enough so that someone who gets interested in the ethics or social acceptance of nanotechnology fifteen years from now won't need extraordinary good luck to find the public record. That's sorely missing with respect to agricultural biotechnology, where the social acceptance wheel is constantly being reinvented by players and naive researchers alike. Perhaps that points to a national center, or perhaps to a program like the Ethical, Legal, and Social Implications of Human Genetics Research (ELSI) at the National Institutes of Health.

(iv) Social, Ethical, and Legal Implications of Nanotechnology

The term "nanotechnology" may include such products as mono-layered materials, nanocomponents in smarter MEMS, and Fullerene-based computers. Some decades from now, it may also include communicating and/or programmable molecular machines. the implications discussed herein assume that some decades from now, the latter capabilities are achieved.

In the short-term, incremental improvements in processes and materials can come about by improved knowledge and skills in the nano-realm. the additive effects of these changes can have secondary and tertiary impacts that are transformational. For example, the social, ethical, and legal impacts of the World Wide Web were not originally thought to be considerable. the Web's enormous impacts are still not fully understood. In a similar way, nanotechnology may appear gradually and yet have a revolutionary effect. In the longer-term, the risks and rewards of nanosystems will certainly be exaggerated as our technological capabilities improve.

Virtually every "millennium survey" of the future poses some social, ethical, or legal questions about nanotechnology. But there is little so far in the way of serious study. Bill Joy's recent article, "Why the Future Doesn't Need Us" in *Wired* is a widely read example. Mr. Joy suggests that since nanotechnology is potentially dangerous, we should relinquish our study of it. His article, though intriguing, misses a most critical point. the United States doesn't have a monopoly on nanotechnology research. the rest of the world is spending over $1 billion per year in the field. Relinquishment by friendly governments, even in the unlikely event that it could be enforced, does not ensure that all researchers will make the same decision.

Others have also encouraged the public's fears about nanotechnology and biotechnology.

Just as fear of cloning could slow efforts in biomedicine and fear of genetically modified foods could contribute to hunger, fear of futuristic nanobots running amok could delay the benefits offered by nanotechnology. When discourse is founded on emotional prejudices, its unreasonableness discredits the legitimate need to identify and assess risks.

Nanotechnology could ultimately turn out to be risky, but the prudent way to assess the risks is not the abandonment of the field. Asilomar demonstrated that the scientific community could design systems to contain high-risk technologies; the Shelter Island meetings showed us we could put great minds towards thinking about risk; the space program showed us we could have a dangerous program with almost no fatal missions. Notwithstanding the negativists, we have proven that we can manage the risks of powerful technologies. This does not suggest that we are safe—merely that we are not inevitably doomed to the worst of possible technological outcomes. We should hear from the pessimists, but we should not hear *only* from them.

Here are some examples of possible social, ethical, and legal implications that bear some thought and consideration. You have a handout that contains a more robust list.

Short-term (3-5 years)

- The NNI will fund careers for scientists and graduate students; commercial firms will invest in nanotechnology R&D. These efforts will be widely dispersed politically, geographically, and scientifically.
- Nanotechnology research will both require and produce enabling technologies that will have beneficial spin-offs.
- There will be unintended consequences, both good and bad.
- We will have to balance the opportunity costs of studying assembler-based nanotechnology (with its huge potential payoffs) against research in fields where advances might have nearer-term but smaller payoffs.
- Designs for future nanosystems will be created, challenged, modeled, improved.

Medium-term (5-15 years)

- Nanotechnology research may allow an otherwise moribund Moore's Law to continue operating.
- Fullerene-based computer chips may require enormously expensive fabrication facilities that turn out chips by the millions—needed for smart packaging, foods, Bluetooth devices, etc. This could result in a proliferation of inexpensive unit costs but prohibitively (and anti-competitively) expensive initial capital costs.
- We may discover and perfect nano-sized sensors and tools that can diagnose disease much sooner than ever before—perhaps long before we have cures.
- We should expect revolutionary advances in materials, MEMS, etc., resulting in abundant markets but also in an upheaval in global financial and manufacturing systems.
- Public education may be needed to balance the views expressed by anti-technology writers and press.

Long-term (over 20 years)

- Nanosystems may help solve problems of disease and aging, pollution and scarcity, overpopulation and starvation, and could create revolutionary changes unlike any ever seen.
- Nanosystems could help produce alternatives to fossil fuels and their high environmental price and reliance on foreign sources.

- The transition from a pre-nano to a post-nano world could be very traumatic and could exacerbate the problem of haves *vs.* have-nots. Have-nots do not easily obtain access to new technologies; the difference between the lives of the nano-rich and the nano-poor will likely be striking.
- Potential harmful uses—intentional and unintentional—need to be studied well in advance: nano weapons; intelligence-gathering devices; nanotechnology combined with Artificial Intelligence to form super-intelligent but virtually invisible devices; artificial viruses to which humans have no immunity, etc.

Recommended Social and Ethical Research Areas

- We should follow a broad path in the ethical studies of nanotechnology including utilitarian ethics (in its many forms), virtue ethics, communitarian ethics, deontological and religious studies, and views of administrative and distributive justice.
- We need to look at the broadest reasonable set of worldviews if we are to do justice to the problem.
- Scientists and engineers like those at Rice University, New York University, MITRE Corporation, and DARPA can inform us on what is feasible.
- Technology pessimists can probe areas of risk that might escape a less vigorous review.
- Optimists from organizations like the Foresight Institute can provide insights into the kinds of systems that seem to be within the realm of the possible if assemblers can be made to work.
- Organization and process professionals like those from ITRI and the Institute for Alternative Futures can coordinate experts who would typically not relate to each other.
- Technology assessment professionals like those with Coates & Jarratt, Inc. can apply time-tested tools to estimate a range of possible circumstances and assess the secondary and tertiary effects of incremental and radical changes.
- We need formal risk analysis because risk is one of the most important ethical and social issues we could imagine:
  - Risk of physical harm—Risk to economic and social systems
  - Risk to political and financial power bases
  - Risk of the haves fighting to keep the upper hand over the have-nots
- We need experts from several legal disciplines to understand the legal implications:
  - Litigators
  - Patent attorneys
  - Privacy law specialists
  - Constitutional lawyers
  - Academic legal scholars
  - Attorneys who understand the process of using the law as a weapon to retard progress

Most importantly, a multidisciplinary approach is critical to a satisfactory understanding of the social, ethical, and legal implications of nanotechnology. For this reason, there needs to be a gathering of all the PIs and other listed key personnel from each grant at least once every six to nine months with mandatory attendance and at least one paper presentation by each sponsored grantee institution (firm, individual, or university). This will not be inexpensive or easy to administer, but it will increase the odds of researchers understanding the issues.

At some point over the course of the next several years, we will have to think about how —scientifically—we are going to achieve a nanotechnology that includes communicating or computing capability. In order to assess the unintended consequences—secondary, tertiary, quaternary uses and effects of new technologies—we must consider what approaches are used. Otherwise, we can't possibly think forward to what unintended consequences might be or how they might be manifested. Nanotechnology arrived at through chemistry may have quite different characteristics than nanotechnology arrived at through biology.

(v) Envisioning Life on the Nano-frontier

In the spirit of Bonnie Nardi's paper in this volume (Nardi, 2001), I would like to "envision" some of the legal and organizational challenges that might arise from the widespread introduction of nanotechnology into the U.S. economy. Obviously, true prediction currently lies beyond the capabilities of even the most informed and foresightful experts in the field, let alone an outside observer like myself. Nonetheless, social scientific research on previous technological revolutions suggests at least a few tentative hypotheses. In particular, as we speculate on policy implications, we might be well advised to distinguish between two types of nanotechnology—which I will call nano-materials and nano-mechanisms, respectively—and to explore the implications of each separately. I suspect that the social disruptions and governance challenges stemming from the former are likely to be much more localized and manageable than those stemming from the latter.

Two Flavors of Nanotechnology

Accounts of the coming nanotechnology revolution seem to involve two somewhat separable agendas—each with its own distinctive social implications. the first (and, in the short run, probably more significant) agenda links nanotechnology to chemical engineering and material science. This variety of nanotechnology focuses on controlling the nano-scale organization of macro-scale substances. Examples of such "nanomaterials"—or "nanates"—might include wear-resistant nanostructured polymers for tires and drive belts, super-hard nanostructured ceramics for drill bits and cutting surfaces, or ultra-fine nanostructured membranes for filters and seals.

The second (and, in the long-run, probably more disruptive) agenda links nanotechnology to mechanical engineering and robotics. This variety focuses on constructing nano-scale devices for operation in macro-scale environments. Examples of such "nanomechanisms"—or "nanites"—might include ultra-small *in-vivo* medical devices, miniaturized surveillance systems, or lilliputian mining and manufacturing equipment.

Needless to say, these two varieties of nanotechnology blur together at their boundaries: One could, for example, imagine a nano-material filter that employed interconnected nano-mechanism "turnstiles"; or, conversely, one could imagine a nano-mechanism robot that assembled, inspected or repaired nano-material compounds. However, the broad distinction between nano-materials and nano-mechanisms considered as archetypes serves to highlight important "scope conditions" for predictions about how nanotechnology will (or will not) transform the social order.

Nano-Materials as Discrete Technological Discontinuities

For many industries and many aspects of social life, nanostructured materials are likely to represent profoundly important technological developments. Indeed, if nanates live up to their early billing, the resulting societal transformations could equal the transformations that attended the development of bronze implements at the end of the stone age or the introduction of nuclear weapons in the 20th century. Given this, it may seem odd to predict that the impacts of nano-materials will be "localized and manageable," as asserted above.

My premise, however, is that nano-materials with enhanced performance characteristics do not, in and of themselves, pose unprecedented challenges to social organization.

Admittedly, particular new materials may engender the development of particular new products that profoundly transform particular areas of social life—and the ripples from such transformations may spread through society in complex and unpredictable ways. (One need but imagine the implications of a bullet coating that allowed small arms fire to penetrate tank armor, or of a photovoltaic cell that eliminated the need for fossil fuels.) But such transformative potential is hardly unique to nanoscience. Nanotechnology may be more likely to yield these radically new materials, but humanity has developed quite a few other transformative compounds—from glass to gasoline to plastic—without the aid of nano-scale understanding. Thus, although the introduction of new nanates may be revolutionary, it promises to be revolutionary in relatively familiar ways.

Recent years have generated a substantial literature on the industrial impact of technological discontinuities (for a survey, see Tushman and Anderson 1996). In brief, the central finding of this literature is that the introduction of any radically new technology initiates a period of ferment within the affected industrial sectors. During this ferment, businesses, public agencies and other "technology champions" jockey for position, attempting to mobilize various economic, cultural and political forces to frame and tame the discontinuity, and to thereby establish a new "dominant design." Once such a dominant design emerges, however, the affected sectors restabilize—although often in significantly different configurations than before. Presumably, a similar dynamic would attend the introduction of new nano-materials, albeit perhaps at a higher intensity and a broader scale.

Of course, this hardly obviates the need for foresightful policy attention. The changes initiated by even a relatively modest technological discontinuity can be quite farreaching, and the framing and taming process can be intensely political, both within the affected industries themselves and at the level of the larger polity. However, the differences between previous technological discontinuities and the discontinuities resulting from nano-materials seem likely to be matters of degree, rather than of kind.

Industries will certainly be transformed, and some of these transformations will be wrenching and risky, but the involvement of nanotechnology will not, in itself, make the transformations any more wrenching and risky than the transformations that we have seen in the past. It is in this sense that the impact of nano-materials will be "localized and manageable": the policy issues, even if large, will arise from the particular performance characteristics of particular products, not from the inherent nature of nanotechnology *per se*. If so, case-by-case planning would appear to represent an appropriate and, arguably, sufficient response.

(As an aside, it may be worth-noting that nano-materials could produce unprecedented disruptions if they instigated technological discontinuities in an unusually large number of industries simultaneously. Without entirely ruling out such simultaneity, however, the practical limits on human resources and attention strongly mitigate against this prospect. In all likelihood, applications of nanotechnology will be neither more nor less staggered in their arrival than applications of previous multi-purpose technologies, like semiconductors, synthetic polymers, or wireless telecommunications.)

Nano-mechanisms as Unique Governance Challenges

In contrast to the relatively familiar challenges of nano-materials, nano-mechanisms promise (or threaten) to confront society with policy issues that are as unprecedented as they are profound. By allowing humans to manipulate the world at a previously unattainable scale, nano-mechanisms open a genuinely new frontier, beyond the contemplation of traditional legal and governmental regimes. While, as Richard Feynman put it, "there is plenty of room at the bottom," (Feynman, 1961) at the moment there are very few sheriffs at the bottom, to keep that room safe and productive.

As currently envisioned, nano-mechanisms are likely to possess at least three properties that will generate novel safety and governance challenges: invisibility, micro-locomotion, and self-replication.

1. Invisibility is, of course, an inherent property of nano-scale objects, whether natural or artificial. Artificial nanites, however, would be among the first complex constructions intentionally engineered to accomplish human purposes at a microscopic (or sub-microscopic) level, and their introduction into the technological armamentarium would dramatically increase the potential for orchestrated covert activities.
2. Micro-locomotion is less inherent in nanotechnology than is invisibility, since some nano-mechanisms will undoubtedly work best when anchored to substrates. Nonetheless, free ranging nanites have clear advantages for certain tasks, and as such devices are introduced, they will radically challenge traditional understandings of macro-boundaries and barriers. Fences, walls and even human skin are largely open space, at the nano-scale.
3. Self-replication is clearly not an inherent property of nano-mechanisms, and indeed, creating self-replicating nanites may prove to be one of the most difficult technical hurdles of the nanotechnology revolution. Self-replication, however, is essential for the economical production of complex nano-mechanisms in useful quantities, and thus it seems likely that by the time such nano-mechanisms become socially significant, self-replication will in fact be a fairly common attribute. Unfortunately, self-replication also poses profound challenges to human foresight and control, since a population of carelessly designed self-replicating nanites could grow exponentially, without a ready "off switch."

Needless to say, the hazards of invisibility, micro-locomotion and self-replication would be magnified if nanites also possessed a capacity for autonomous operation and self-modification; however, such higher-order artificial intelligence is not a prerequisite for the envisioned crisis of nano-governance. Even "dumb" nano-mechanisms would require a dramatic rethinking of society's current legal and normative structures.

Three governance issues promise to be particularly acute: monitoring, ownership, and control. Moreover, these issues may play out not merely with respect to nanomechanisms themselves, but also with respect to the "nanospace" that such mechanisms render accessible:

1. *Monitoring:* Invisible, micro-locomoting, self-replicating nanites will severely test society's established assumptions about what can and should be monitored, and by whom. On the one hand, the lay public may find itself inundated (or even interpenetrated) by complex devices whose presence, provenance and purpose remain undetectable without sophisticated technical assistance. On the other hand, those actors that can produce or purchase nanites will find themselves able to monitor their worlds (including their social worlds) in more profound and surreptitious ways than ever before. Contrary to traditional assumptions, ordinary people will no longer be able to observe all the socially relevant activities in their own surroundings; yet public authorities will be hard pressed to provide policing assistance without further endangering individual privacy. In the absence of trustworthy institutions to regulate this newly-opened invisible frontier, the potential for rampant abuse will be matched only by the potential for rampant paranoia.

2. *Ownership:* Nano-mechanisms will also severely test society's assumptions about what can and should be owned, and by whom. Self-replication raises difficult questions of whether property

rights persist from one generation of mechanisms to the next—not to mention difficult practical problems of "branding" proprietary nanites and policing "nanite rustling." Equally importantly, micro-locomotion raises difficult questions of who owns the nanospace through which nanites pass. Controversies akin to historical debates over airspace and rights-of-way may emerge over the nanospaces within macroscopically "solid" objects. If your nanites get sucked into my air conditioning system, are you trespassing or am I kidnapping? Can public health authorities claim an easement for the passage of biomedical nano-sensors through my gut? If nano-robots in my tap water "harvest" a microscopic quantity of copper from my plumbing, is their owner guilty of breaking and entering?

3. *Finally*, and perhaps most significantly, nano-mechanisms will test established assumptions about responsibility and control. Traditionally, legal liability for mechanical devices can outlast ownership, but can be attenuated by unforeseeable circumstances or by the intervention of a third party's will. the viability of these principles, however, becomes unclear in a world of self-replicating nano-mechanisms that operate invisibly in an only partially understood micro-environment. One might imagine governing the control of nano-mechanisms by analogy to defective products, or to toxic emissions, or to speeding vehicles, or to straying livestock—but the implications of these analogies are not entirely congruent with one another. Moreover, a parallel set of complexities and contradictions surround the control of nanospace: Who is responsible for policing the boundaries between "open range" and "enclosed territory" at the nanoscale? Can negligently maintained nanospace pose an "attractive nuisance" to passing nanites? Does a state maintain territorial rights to the nanospace within the bodies of citizens traveling abroad?

Although the distinction between nano-materials and nano-mechanisms is to some extent an artificial one, it highlights an important dimension of differentiation in the impact of nanotechnology: Nanostructured materials may pose serious practical and ethical challenges for particular policy domains, but these challenges will arise at a familiar macro scale, for which we have numerous rules, institutions and historical precedents. In contrast, nano-engineered mechanisms will force us to reformulate our rules and institutions to govern an unfamiliar setting with which we have no prior experience. By allowing human will to operate in the previously unreachable interstices of macro-reality, the nanotechnology revolution promises to open vast new frontiers for exploration, exploitation and settlement. Taming those frontiers, however, will require us not only to control physical matter, but also to control ourselves.

(vi) Societal Implications of Nanotechnology

In the past my thoughts and ideas about nanotechnology have flowed easily, as they were prompted by the excitement and prospects of a coming new wave of scientific challenges and opportunities. Putting together a statement on the societal implications of nanotechnology has proven more difficult. During the past six months I have asked a number of people what they thought nanotechnology meant, and what they thought the societal implications might be. While most responses were positive, they varied depending on the degree of technical background. Most scientists and engineers saw nanotechnology as an encompassing technology that would broadly impact on microelectronics, data storage and the computer area, energy conversion and chemical processing, defense and national security, biotechnology (including agriculture), and the biomedical areas of therapeutics and diagnostics. While some scientists and engineers felt that nanotechnology was not quite as "new" as perceived and presented by the popular media, they did agree that a new level of much higher activity and broader applications was definitely happening. For the most part, this group felt that the overall implications of nanotechnology would be very good. In fact, some felt it might even be vital that we progress as rapidly as possible into this new area. the feeling was that nanotechnology (as well as other related technologies) may provide solutions to some of the more intractable problems related to eliminating disease and famine in developing countries, and helping to improve the economy and general living standard for the rest of the world.

In general, scientists and engineers that I talked with tended to feel nanotechnology would carry risks similar to those seen in the development of other previous technology revolutions, e.g., microelectronics. the feeling was that, while nanotechnology could certainly be used for some insidious purposes, the overall benefits would far outweigh the risks. If there was concern about nanotechnology, it centered more on the need to be careful not to oversell or make inflated promises about nanotechnology, as this could lead to loss of credibility and confidence by the general public. Responses from non-technical professional people were different from those of the scientists and engineers. In general, this group was not as well aware of the potentially more far-reaching applications of nanotechnology (and this would not be unexpected).

This group tended to equate nanotechnology primarily with the further miniaturization of microelectronic devices and very advanced computers. Some from the non-technical group related to nano-type medical devices, which might for example be put into the blood stream of humans to treat cancer or other diseases. While this group may not really understand nanotechnology, they generally thought that it was both very interesting and important, and that it might provide many future benefits. Some from both non-technical and technical groups do have an underlying uneasiness with how quickly technology in general is advancing. This uneasiness could most certainly reflect on nanotechnology.

Some concerns about nanotechnology (and technology in general) are not just related to deliberate or insidious misuse of these technologies, but also include problems which could result from carelessness or unexpected adverse effects which appear later in implementing the technology. By way of previous examples, the dangers from nuclear power plant breakdowns, the pollution from chemical processing and toxic reactions to certain therapeutic agents. While these types of concerns are justified, most people still believe that technology will of course continue to advance in the future and will most certainly provide benefits for all of mankind.

(vii) Socio-economic Research on Nanoscale Science and Technology: A European Overview and Illustration

This is true even more so in a field where two different systems of science come together. Natural scientists and technologists have to understand elementary notions of social science, and social scientists in turn have to learn about the scientific and technological object of their study to do a meaningful analysis. the basic vocabularies have to be learned yet.

The objective of this paper is to provide a European point of view on the social implications of NST. Past and ongoing efforts in the social sciences that are related to NST are introduced. Part I of this paper will give an overview of mainly European studies in this area. Over the past few years, a modest amount of activity of social scientific research has been carried out in Europe that specifically addressed economic and social issues of nanotechnology. A review of these activities might provide a useful platform for further discussions.

Part II will give an illustration of a quantitative social scientific approach to NST. This is done to clarify where social sciences can contribute and where inputs from actors in the field are needed. Finally, a number of conclusions are drawn and some suggestions are made with respect to where future social scientific research on NST might be of interest.

Part I: An Overview of Ongoing Socio-Economic Research

Economic and Social Aspects of Nanotechnology: One should make a distinction between activities in the social science arena that are directed at describing the emergence of nanoscience and technology as it happens and those activities that deal with issues that might become relevant once nanotechnology has entered a more mature stage in its development.

Studies on Current NST Activities or Near-Term Applications: Often and especially in a European context, nanotechnology or nanoscience is discussed in terms of potential paths towards commercialization, niche markets for already or soon to be available products. This is also the major focus of ESANT, the European working group for Economic and Social Aspects of Nanotechnology, a small community of researchers and practitioners, mostly with a technological background. the group has published a report (http://www.nano.org.uk/ESANT99.htm). In their report, potential avenues of near-term commercial development of nanotechnology are outlined.

Other social science research focuses on how governments support nano-scientific research and the implications this has for the development of nanotechnology. One example is the work commissioned by the Finnish National Technology Agency (Meyer, 2000b). Nanotechnology is addressed in many different ways. the different perspectives observed at the country-level seem to coincide with the strengths of the respective national innovation system. There is no common definition as to what nanotechnology comprises. It appears that specific scientific and technological communities have reached a common basis as to what "nano" means with respect to their particular area. the current state of development allows government actors to play a substantial role in developing a novel technological area. In some European countries, governments rely to a considerable extent on technology foresight studies in their long-term science and technology policy planning. These exercises provide governments with estimates about future technological developments and the implications they might have for society. Some of these studies also addressed NST. We shall have a closer look at them in the following section.

Mid- and Long-Term Issues: Foresight studies attempt to depict an image of a possible future using a variety of techniques. An important foresight method is the Delphi survey. A Delphi survey basically is a tool to create consensus and detect areas of conflicting expert opinions. In a first round, experts are confronted with a number of topics they have to evaluate with respect to time of realization, implication on wealth creation, quality of life, and similar issues. In another round the results of the previous round are introduced to the experts who then have a chance to re-evaluate the topic. An overview of the nanotechnology-related topics covered in the 1995 German Mini-Delphi survey. Kuusi and Meyer try to distill five major areas of developments *("leitbilds")* drawing on this Delphi exercise and information from other technology studies (Kuusi, 1999). *Leitbild* is a loan word from German, meaning a guiding image. Within the context of the social studies of science and technology, it can denote developments in unfolding areas that might lead to the emergence of a novel paradigm or technological trajectory. Its major function is to provide a platform integrating actors from different areas by providing a common goal, which can be an envisaged technique or product.

1. Nano-resolution methods of analysis. One *leitbild* is that of nano-resolution analytical methods (topics 20, 22). the aim here is to further improve existing tools by adding new functions to analysis tools. Here, realized techniques are further generalized into promising tools. the development of analytical tools and techniques is closely related to progress in other areas. Often nano-resolution tools are identified with nano-resolution optical microscopy since these methods combine the possibility of measurement with that of manipulation, which makes them more versatile as tools for nanotechnology. A number of scanning probe microscopes have been developed and some of these methods do not require complicated sample preparation; e.g., the ability to work on *in vivo* substrates and determine structure function relationships is the main reason why the AFM, the atomic force microscope, is so popular amongst biologists. It is said that even (nearly) two decades after its invention, the implication of the scanning-tunneling microscope (STM) and its follow-ups is still growing. In conjunction with continuous technical further development of this method, researchers are discovering new phenomena in the fields of physics, chemistry, and biology. At the same time the STM methods are used as a nano-tool rather than a nanoprobe. the idea is to modify surfaces and tailor their structures on the nanoscale, down to the manipulation of individual atoms (Frenken, 1998). This case has shown that realized techniques—such as AFM and STM—can provide a basis for developing more sophisticated "promising" techniques that will go beyond mere measurement. Ultimately, they might facilitate large-scale manipulation at the nanometer level.

2. Nano-materials. Another *leitbild* ranges around the notion of nanostructured materials. Nano-materials can be seen as structures that have particular properties owing to their size. the *leitbild* in this context is to manufacture a nano-material in a way that allows predetermination of its properties. This generalization pattern describes the transition from realized techniques to promising targets. Paired with a better scientific understanding of the subject matter, a variety of already realized techniques allow us to develop rather concrete ideas of improved materials. the idea here is to take advantage of nanoscale characteristics of structures and substances to create new materials with enhanced properties, such as those common to polymers, composites, or other materials (topics 16, 17, 19). the aim is not necessarily direct control of individual atoms; bulk operations suffice to exploit the nanoscale properties. One example to illustrate the idea of bulk-processing nano-materials is that of colloidal dispersions (Philipse, 1998). Colloidal science deals with the physics and chemistry of finely dispersed matter. Colloids are generally understood as particles or other objects with at least one dimension roughly in the sub-micron range. Nano-particles then are viewed as colloids smaller than 100 nanometer. It is pointed out that colloid science has a long tradition involving nano-particles and that not all that is nano is necessarily new (Philipse, 1998). In this sense, colloids encompass gold colloids, colloidal silica, and aluminum oxide powders. Due to their small dimensions, colloids exhibit Brownian Motion. Owing to their large surface area, surface forces and repulsions determine the interaction between colloidal particles in the liquid phase. the balance between these forces depends critically on the details of the particle surface and the liquid composition. Colloids aggregate easily to form large floes, networks or gels. the control and understanding of these aggregation processes is the major challenge of colloidal nanoscience. Suspensions of colloids, or dispersions, are, for example, milk, blood, cosmetics, such as toothpaste, or ink. Examples for inorganic colloids are clay particles, iron oxides on computer disks and in magnetic fluids, pigments in paints, or powders for technical ceramics. Colloidal systems are studied from a (bulk) chemical synthesis perspective. Such research integrates the study of physical properties of dispersions and gelation phenomena, obtaining important input from computer simulations and statistical mechanics. It is said that this wide scope of colloidal science is unavoidable, since finely dispersed matter can be encountered in many disciplines and applications.

The nano-materials *leitbild* provides a focus point for integrating this and other bulk chemistry topics with related activities.

3. Ultra-thin layers. A number of topics also revolve around thin layers (topics 14, 18). Here, efforts appear to be directed at characterizing these structures. Thin-film technologies constitute a considerably well developed field. An exact ultra-fine production of thin films is a necessary condition for the subsequent characterization. Designing ultra-thin layers can be associated with a small number of technical aims, such as atomically exact delineations of layers, quantumized potential distribution, defined pore distribution in layers, ultra-thin separation and protection layers, improved layer function by way of multilayer structuring. These technical targets are related to a greater number of applications. Examples are information storage layers, films with quantum effects, optical layers, multilayer piles for quantum/semiconductor laser and x-ray optical compounds, displays, sensor layers, tribologic films, biocompatible films, photovoltaic films, membrane films, and chemically active surfaces (Bachmann, 1998). One might find that activities within the context of this *leitbild* follow a generalization pattern where promising targets are used to develop promising techniques. Realized techniques permit already sufficiently exact operations at the nano level to convey an idea of future products. the idea of future products, in turn, requires even more exact and precise tools.

4. Bottom-up techniques. While the preceding *leitbild* is still to be positioned in the field of top-down nanotechnology, the subsequent *leitbild* addresses the nanoscale from a different, "bottom-up" perspective. This perspective attempts to simulate nature to develop materials with novel properties by way of self-organization. These approaches are often dubbed "biomimetics" (topics 19, B). In this case, an integration of various techniques is necessary. It has been shown that one can create structures by way of self-organization in a biomimetic process. However, our technological means are inadequate to fully utilize the potential this *leitbild* offers us. Being aware of the general feasibility—thanks to already realized artifacts—we can make reasonable assumptions about the requirements of the techniques necessary to pursue this path of development further.

5. Direct atomic control. Another *leitbild* focuses on direct control of atoms, to rearrange them into new atomic structures (topic 15). This could lead to novel materials. the difference between the materials approach mentioned above and this one is related to how one controls the process (bulk reactions *vs.* atomic control). Atomic control is also strongly related to the idea of atoms being effectively used as carriers of certain functions, such as data storage, etc.

Societal Implications of Nanotechnology: the workshop's efforts are chiefly directed at identifying areas where mature nanotechnology can influence society. In Europe, the ESANT working group has undertaken the effort to outline possible avenues of societal implications of nanotechnology. the idea was to "foresee some beneficial aspects of nanotechnology for society in general" that reach beyond mere economic benefits, a topic a number of studies focus on currently.

Even though the ESANT study stresses the beneficial role nanotechnology is expected to play in economic and societal development, the group felt it is necessary to point to a possible detrimental effects and risks that may be associated with nanotechnology.

APPENDIX D

FAQs: Frequently Asked Questions

What is Absolute Zero?

Lowest possible temperature equal to zero Kelvins (i.e. –273.15 Celsius degrees, centigrades). Completely empty (void of matter and radiation) space can be considered to have such a temperature. Even at absolute zero, the atoms and molecules do not 'stand still' but still move around due to the Heisenberg uncertainty principle. This is called 'zero-point motion'. In laboratories, extremely low temperatures are achieved by using liquid nitrogen or liquid helium as cooling media. These are in fact very common in laboratories and various industries, especially liquid nitrogen. Temperature is a physical quantity which gives us an idea of how hot or cold an object is. The temperature of an object depends on how fast the atoms and molecules which make up the object can shake, or oscillate. As an object is cooled, the oscillations of its atoms and molecules slow down. For example, as water cools, the slowing oscillations of the molecules allow the water to freeze into ice. In all materials, a point is eventually reached at which all oscillations are the slowest they can possibly be. The temperature which corresponds to this point is called absolute zero. Note that the oscillations never come to a complete stop, even at absolute zero. There are three temperature scales. Most people are familiar with either the Fahrenheit or the Celsius scales, with temperatures measured in degrees Fahrenheit (°F) or degrees Celsius (°C) respectively. On the Fahrenheit scale, water freezes at a temperature of 32° Fahrenheit and boils at 212°F. Absolute zero on this scale is not at 0° Fahrenheit, but rather at –459° Fahrenheit. The Celsius scale sets the freezing point of water at 0° Celsius and the boiling point at 100° Celsius. On the Celsius scale, absolute zero corresponds to a temperature of –273° Celsius. Scientists—especially those who study what happens to things when they become very, very cold—commonly use the Kelvin scale, with temperatures measured in Kelvin (K). This scale uses the same temperature steps as the Celsius scale, but is shifted downward. On this scale, water freezes at 273 K and boils at 373 K. Only on the Kelvin temperature scale does absolute zero actually fall at 0 K.

What is Backscattered Electron Imaging?

The production of backscattered electrons varies directly with the specimen's atomic number. This differing production rates causes higher atomic number elements to appear brighter than lower atomic number elements. This interaction is utilized to differentiate parts of the specimen that have different average atomic number. An example is shown in the SEM output section, specifically the mechanically alloyed specimen micrograph. Detection of Backscattered Electrons is achieved by using a "donut" shaped solid state semiconductor device mounted on the bottom of the objective lens. When Backscattered Electrons strike the detector electron-hole pairs are created which are then counted. This quantity is translated into a pixel intensity and displayed on the CRT, forming the image.

What is Fibrinogen?

Plasma glycoprotein clotted by thrombin, composed of a dimer of three non-identical pairs of polypeptide chains (alpha, beta, gamma) held together by disulfide bonds. Fibrinogen clotting is a sol- gel change involving complex molecular arrangements: whereas fibrinogen is cleaved by thrombin to form polypeptides A and B, the proteolytic action of other enzymes yields different fibrinogen degradation products. A plasma protein that, when activated by thrombin, becomes fibrin, the principle component of a blood clot. Plasma glycoprotein clotted by *thrombin*, composed of a dimer of three non-identical pairs of polypeptide chains (alpha, beta, gamma) held together by disulfide bonds. Fibrinogen clotting is a sol-gel change involving complex molecular arrangements: whereas fibrinogen is cleaved by thrombin to form *polypeptides* A and B, the proteolytic action of other enzymes yields different fibrinogen degradation products.

What is Bacteria?

Single-celled microorganisms, about one micrometer (one thousand nanometers) across. Microscopic cells present everywhere. Many types of bacteria cause diseases in humans, but many others are very useful to us. Bacteria are very small living organisms made of only one cell. They are present just about everywhere: the air, the soil, and the skin, for example. Many of them are microbes that cause diseases (rhinitis, listeriosis, and others), but others are very helpful to humans. For example, bacteria in the intestine help digestion and we often use bacteria to make food products (yoghurt, sauerkraut, and others). Bacteria responsible for diseases can be easily eliminated thanks to antibiotics. And it is also possible to prevent certain bacterial infections (tuberculosis, diphtheria, and others) by using vaccines that prepare the immune system for the bacterial intrusion. Bacteria have, in their cytoplasm, a circular molecule of DNA. They also contain other, smaller DNA molecules, called plasmids, which can carry, for example, the genes for resistance to antibiotics. Researchers use these plasmids to study genes and proteins. By integrating a gene into a plasmid, they can obtain a large quantity of the gene because it is very easy to multiply bacteria. What's more, it is simple to isolate these plasmids and recuperate the gene under study. One-celled living organisms, typically about one micron in diameter. Bacteria are among the oldest, simplest, and smallest types of cells. Single-celled microorganisms, about one micrometer (one thousand nanometers) across.

What is Genomic Technologies?

One of the primary reasons for the success of the Human Genome Project has been the development and use of high- throughput strategies for data generation, and the placement of the data immediately in the public domain. Most of the sequence data, the underlying maps and the sequence assemblies were generated through the use of large-scale automated processes. Now, methods such as sequence analysis of whole genomes, DNA microarray technology and mass spectrometry have been or are being developed as high-throughput approaches for additional types of genomic analyses, such as determining the parameters of gene expression or the location of gene products by the thousands at a time instead of individually. High-throughput methods to determine the location of cis- regulatory elements and, to a lesser extent, other sequence elements, are also beginning to be developed. However, at present, there is no single approach or compilation of approaches that can accurately and efficiently identify every sequence feature in genomic DNA. Determination of all Functional Elements in Human DNA.

What is Fermi Paradox?

"If there are other intelligent beings in the Universe, why aren't they here?" Since it appears to be quite possible for a technological species to spread across the galaxy in less than 10 million years (using von Neumann machines) or otherwise change things on such a large scale that it would be very visible (see Kardaschev types), the lack of such evidence is puzzling or implies that other technological civilizations doesn't exist. There have been many attempts to explain this, for example the "Wildlife Preserve" idea (the aliens doesn't want to interfere with younger civilizations), that they transcend and become incomprehensible, that they hide or that they are actually here, hidden on the nanoscale, but the problem with these attempts is that most of them just explain why some aliens would not be apparent.

What is Universal Assembler?

This is a term that was introduced by Eric Drexler. In his terminology, an assembler is enzyme or catalyst that brings two molecules together and connects them. DNA can be used to program the assembly of a wide variety of proteins. Drexler imagines a universal assembler to be something that could be programmed to fabricate anything. The following quote was taken from Eric Drexler's book Engines of creation in the section where he discusses universal assemblers. These second-generation nanomachines—built of more than just proteins—will do all that proteins can do, and more. In particular, some will serve as improved devices for assembling molecular structures. Able to tolerate acid or vacuum, freezing or baking, depending on design, enzyme-like second-generation machines will be able to use as "tools" almost any of the reactive molecules used by chemists—but they will wield them with the precision of programmed machines. They will be able to bond atoms together in virtually any stable pattern, adding a few at a time to the surface of a workpiece until a complex structure is complete. Think of such nanomachines as assemblers. Because assemblers will let us place atoms in almost any reasonable arrangement (as discussed in the Notes), they will let us build almost anything that the laws of nature allow to exist. In particular, they will let us build almost anything we can design—including more assemblers. The consequences of this will be profound, because our crude tools have let us explore only a small part of the range of possibilities that natural law permits. Assemblers will open a world of new technologies. Advances in the technologies of medicine, space, computation, and production—and warfare—all depend on our ability to arrange atoms. With assemblers, we will be able to remake our world or destroy it. So at this point it seems wise to step back and look at the prospect as clearly as we can, so we can be sure that assemblers and nanotechnology are not a mere futurological mirage. Uses raw atoms and molecules to construct consumer goods, and is pollution free. Can be programmed to build anything that is composed of atoms and consistent with the rules of chemical stability. Eric Drexler talks about these assemblers as nanorobots with telescoping manipulator arms that are capable of picking up individual atoms, and combining them however they are programmed.

What is Triac?

A triac is an electronic component equivalent to two silicon controlled rectifiers joined end to end or back to back with their gates connected together. This results in a two-directional electronic switch which can conduct current in both directions when it is triggered (turned on). It can be triggered by either a positive or a negative voltage. The device turns off again when the current through it drops below a certain threshold value, such as the end of a half-cycle of alternating current. This makes the triac a convenient switch for AC circuits. Applying a trigger pulse at a controllable point in an AC cycle allows one to control the percentage of current that flows through the triac. Lower power triacs are used in lots of applications, especially domestic electrical appliances, such as light dimmer.

What is UV Embossing?

Cold embossing uses UV light to polymerize and solidify materials after forming by the stamp. The starting materials are thin films of UV-curable materials. Typical materials may be monomers, or inorganic-organic hybrid polymers, which are polymerized by UV to polymers with desirable optical properties. The films may be deposited onto silicon, gallium arsenide, or similar wafers. The UV-transmissive patterned master template is pressed against the film, imprinting the pattern, which is UV cured before pressure is released. UV embossing allows accurately aligning and overlaying multiple levels of features. UV embossing has a clear advantage over hot embossing in allowing multi-level applications, since the first level is polymerized after exposure, making it irrevocably formed and dimensionally stable, regardless of subsequent layer applications and exposures. UV embossing is a low pressure, room temperature process, with little shrinkage of the pattern. It also allows partial UV casting, by limiting the illumination to maintain clear areas on the wafer. Cold embossing may be done on one or both sides of the wafer. Double-sided embossing allows micromolding.

What is Quantum Well?

A P-N-P junction in which the "N" layer is ~10 nm (where traditional physics leaves off and quantum effects take over) and an "electron trap" is created. "If one makes a heterostructure with sufficiently thin layers, quantum interference effects begin to appear prominently in the motion of the electrons. The simplest structure in which these may be observed is a *quantum well*, which simply consists of a thin layer of a narrower-gap semiconductor between thicker layers of a wider-gap material." See Center for Quantum Electronics U of Dallas.

What is Quantum Teleportation?

'Teleportation' (or copying) of the quantum state from one particle to another. Not really spectacular as in Star Trek serial where macroscopic objects (humans, weapons and similar) are 'teleported'. Present technology can 'teleport' quantum objects such as photons, electrons as similar. It is important to note that the object itself is not teleported during this process. What actually happens is that the quantum state (e.g. spin number) is exactly copied from one object to another, both spatially and temporally present *prior* to 'teleportation'. The technique used is based on the Einstein-Podolsky-Rosen effect (EPR) or paradox, which relies on the so-called nonlocality of the wavefunction. An EPR source produces pair of entangled particles, one of which serves as a 'carbon copy'. The third particle (original, source particle to be copied, introduced besides the EPR pair) is 'entangled' with one of the particles from the EPR pair which enables the copying of its state to another EPR particle that has never interacted with the original, yet it becomes its copy. The experimental setup requires one measuring apparatus that entangles the original with one of the EPR particles and another apparatus that acts on another EPR particle which becomes the copy. The experimentalists operating the two apparatuses must be in contact i.e. the one that acts on the copy particle must know the results of the entanglement measurement on the source (original) particle and another EPR particle which is 'waisted' in the process. It should also be noted here that none of the experimentalists does not know the quantum state of the original particle, even when the whole process finishes, neither is it measured during the process.

What is Waste Recycling?

An advanced nanotechnology might be able to build filters that dynamically modify themselves to attract the contaminant molecules detected by the air and water quality sensors. Once attached to the filter, the filter could in principle move the offending molecules to a molecular laboratory for modifications to useful or at least inert products. A swarm might implement such an active filter if it was able to dynamically manufacture proteins that could bind contaminant molecules. The protein and bound contaminant might then be manipulated by the swarm for transportation. With a sufficiently advanced nanotechnology it might even be possible to directly generate food by non-biological means. Then agriculture waste in a self-sufficient space colony could be converted directly to useful nutrition. Making this food attractive will be a major challenge.

What is Space Elevator?

Issacs and Pearson proposed a space elevator—a cable extending from the Earth's surface into space with a center of mass at geosynchronous altitude. If such a system could be built, it should be mechanically stable and vehicles could ascend and descend along the cable at almost any reasonable speed using electric power (actually generating power on the way down). The first incredibly difficult problem with building a space elevator is strength of materials. Maximum stress is at geosynchronous altitude so the cable must be thickest there and taper exponentially as it approaches Earth. Any potential material may be characterized by the taper factor—the ratio between the cable's radius at geosynchronous altitude and at the Earth's surface. For steel the taper factor is tens of thousands—clearly impossible. For diamond, the taper factor is 21.9 including a safety factor. Diamond is, however, brittle. Carbon nanotubes have a strength in tension similar to diamond, but bundles of these nanometer-scale radius tubes shouldn't propagate cracks nearly as well as the diamond tetrahedral lattice. Thus, if the considerable problems of developing a molecular nanotechnology capable of making nearly perfect carbon nanotube systems approximately 70,000 kilometers long can be overcome, the first serious problem of a transportation system capable of truly large scale transfers of mass to orbit can be solved. The next immense problem with space elevators is safety—how to avoid dropping thousands of kilometers of cable on Earth if the cable breaks. Active materials may help by monitoring and repairing small flaws in the cable and/or detecting a major failure and disassembling the cable into small elements.

What is Wavelength Routing?

A concept developed by Cisco Systems where at each intermediate node between the end nodes, light coming in on one port at a given wavelength gets routed out of one and only one port. It allows carriers to leverage their core network technologies to support the massive growth of data services. It avoids wasting transmitted power, by channeling the energy transmitted by each node along a restricted path instead of letting it spread over the entire network.

What is Thermodynamical Stability?

Stability of a certain system with respect to changes in thermodynamical parameters. For nanoscopic systems, the most important parameter is temperature, i.e. stability of a system with respect to increase of temperature. Some of nanoscopic structures are stable only at sufficiently low temperatures, while at higher temperatures their structure tends to degrade. At higher temperatures, the atoms of which a nanostructure is built vibrate with larger amplitudes which may result in degradation of the structure. Additionally, nanostructured material, or a nanoparticle may be unstable when exposed to chemical influences from the surrounding, i.e. atoms and molecules. Some nanostructures are stable only in ultra-high vacuum conditions, which very much limits their use. Thermodynamical and chemical stability is a major concern in many applications involving nanoparticles and nanostructured materials.

What is X-ray Crystallography?

X-ray crystallography involves firing X-rays through the crystal of a molecule to produce a diffraction pattern. This pattern provides information on the structure of that crystal. For example, X-ray crystallography helped scientists discover that the DNA molecule exists as a double helix. For more information see Crystallography 101 (Lawrence Livermore National Laboratory, USA) and X-ray crystallography (The British Biophysical Society, UK).

What is Zephyr?

A common term used to describe lightweight fabrics. The Zephyr philosophy is *building compilers from parts*. Parts might include front ends, back ends, optimizers, and the glue that holds all these pieces together. You might even *generate parts automatically* from compact specifications. If you *describe your intermediate forms* using Zephyr's Abstract Syntax Description Language (ASDL), we can generate data-structure definitions in C, C++, *Java, Standard ML, and Haskell*. Your IR can be serialized on disk and *freely exchanged* among compiler passes written in these languages; asdlGen creates the glue that holds your compiler together. You can also *visualize your IR* using our graphical browser. To *compile code for multiple targets*, the state of the art is to use a code-generator generator, which lets you specify a translation from an intermediate code to a target instruction set. We're trying to separate the intermediate code from the instruction set, so we can develop *machine descriptions that can be reused*, not just with compilers, but also with other tools. The idea is to *generate the machine-dependent parts* from descriptions of instructions' semantics, of binary representations, or of other properties. Zephyr's Computer Systems Description Languages (CSDL) let you *describe as much or as little as you need* for your application. Getting high-quality code requires optimization. Zephyr's very portable optimizer (vpo) provides *instruction selection*, *instruction scheduling*, and classical *global optimization*. Vpo operates on programs represented as register-transfer lists (RTLs) in SSA form. All vpo optimizations are "plug and play," and it is easy to add new ones. Vpo provides a level playing field for *evaluating different optimizations*.

What is meant by Rosette Nanotubes?

Rosette nanotubes are tubes of nanometric sizes obtained through the self-assembly of organic molecular components. The basic unit for the self-assembly of rosette nanotubes is a base borrowing the structural properties of guanine and cytosine. Thus, the basic unit is hydrophobic and has complementary donor-donor-acceptor and acceptor-acceptor-donor H-bond arrays on its sides. The self-assembly process takes place in two stages. At first, six basic units self-assemble into a supermacrocycle ring. In the second stage, supermacrocycle rings stack forming a nanotube with a central hollow pore. Although very recently discovered (2001) rosette nanotubes allow us to envision interesting applications due to their tubular structure combined with the flexibility given by organic chemistry, which opens up a large range of possibilities for their functionalization. It is thought that rosette nanotubes could one day be used for orthopaedic implants since one can attach growth factors and specific bone recognition peptide sequences that will preferentially attract bone cell adhesion. Similarly to the peptide nanotubes, rosette nanotubes could act as antibiotics by inserting in to the membrane of bacteria and sucking up nutrients.

What are Fluorescent Semiconductor Nanocrystals?

Fluorescent semiconductor nanocrystals are nanometer sized pieces of crystal comprising from a few tens to several thousands or more of atoms. Due to a quantal effect, quantum confinement, their electronic and optical properties are size dependent. In particular, when shined with UV light they emit narrow spectrum fluorescence light whose "colour" shifts towards shorter wavelengths as the size of the crystallites decreases. Nanocrystals can be synthesized dispersed in a solid matrix (SiO_2, ZrO_2,...), by sol-gel method, or they can be prepared as colloids by wet chemistry techniques. This latter method yields the best samples in terms of size homogeneity and purity of the fluorescence color. Applications making use of the nanocrystals take advantage of their robust and tunable electronic/optical properties. One rapidly growing application of nanocrystals is in the fields of biology and medicine. Several recent studies have shown the potential of nanocrystals for multicolor fluorescent labelling of live cells, in vivo imaging of growing organisms, single particle tracking in cell traficking, intracellular staining, etc. Other fields of application is as light absorbing or emitting centers in photovoltaic devices or electroluminescent diodes. Of interest also is using the nanocrystals as tags in microbeads to prepare molecular barcodes.

Fluorescence: fluorescence corresponds to the fast emission of a photon (quantum of light) by a molecule/particle following the absorption of an excitation photon, of higher energy. Fluorescence stops when the excitation is turned off, in contrast to phosphorescence. Molecules/particles which possess this property are fluorophores. The probability of this process is termed the quantum yield (maximum value 1).

What is Thermodynamical Stability?

Stability of a certain system with respect to changes in thermodynamical parameters. For nanoscopic systems, the most important parameter is temperature, i.e. stability of a system with respect to increase of temperature. Some of nanoscopic structures are stable only at sufficiently low temperatures, while at higher temperatures their structure tends to degrade. At higher temperatures, the atoms of which a nanostructure is built vibrate with larger amplitudes which may result in degradation of the structure. Additionally, nanostructured material, or a nanoparticle may be unstable when exposed to chemical influences from the surrounding, i.e. atoms and molecules. Some nanostructures are stable only in ultra-high vacuum cónditions, which very much limits their use. Thermodynamical and chemical stability is a major concern in many applications involving nanoparticles and nanostructured materials.

What is Tensiometer?

Force measuring instrument for determining surface/interfacial tensions of liquids, and dynamic contact angles of (regular shaped) solids, fibers, powders and porous materials. The measurement of surface/interfacial tension of liquids and contact angles of liquids against solids as performed by a tensiometer is based on force measurements of the interaction of a solid probe with the interface of a liquid or between 2 immiscible liquids. In these experiments a probe is hung on a balance and brought into contact with the liquid interface. The forces experienced by the balance as the probe interacts with the surface of the liquid can be used to calculate surface/interfacial tension or contact angles. The forces present in this situation depend on the following factors; size and shape of the probe, contact angle of the liquid/solid interaction and surface tension of the liquid. The size and shape of the probe are easily controlled. For surface / interfacial tension measurements the contact angle of the liquid towards the probe is controlled to be zero (complete wetting). This is achieved by using probes with high energy surfaces. For example, probes made of a platinum/iridium alloy insure complete wetting and they are very easily cleaned in a reliable way. The mathematical interpretation of the force measurements depends on the shape of the probe used. Two types of probes are commonly used for surface/interfacial tension measurements, the Wilhelmy plate and DuNouy Ring. The Wilhelmy plate method utilizes the interaction of a platinum plate with the liquid interface being tested. The calculations for this technique are based on the geometry of a fully wetted plate in contact with, but not submerged in, the liquid. In this method the position of the probe relative to the surface is significant. Therefore the measurement is made at the so called zero depth of immersion of the probe, and the force acting on the probe at this position is registered and can directly be used to calculate the surface tension of the liquid when the perimeter of the plate is accurately known. If only limited quantity of the liquid to be tested is available one may consider using a thin round platinum rod as the probe. In such a case the measurement is exactly the same as with the Wilhelmy plate, but the probe dimensions of the probe will be smaller which affects the accuracy of the measurement and hence also might affect the reproducibility of the end results. The Du Nouy ring method utilizes the interaction of a platinum ring with the liquid interface being tested. The ring is submerged below the interface and subsequently raised upwards. As the ring moves upwards it raises a meniscus of the liquid. Eventually this meniscus tears from the ring and returns to it's original position. Prior to this event, the volume, and thus the force exerted, of the meniscus passes through a maximum value and begins to diminish prior to the actually tearing event. The calculation of surface or interfacial tension by this technique is based on the measurement of this maximum force. As an additional volume of liquid which is raised due to the proximity of one side of the ring to the other mathematical corrections are needed in order to obtain the correct surface/interfacial tension values. He tensiometric method for measuring dynamic contact angles (Advancing and Receding contact angles) measures the forces that are present when a sample of solid is brought into contact with a test liquid. If the forces of interaction, geometry of the solid and surface tension of the liquid are known the contact angle may be calculated. The solid sample has to have a symmetrical and regular shape (rod, cube, round rod, rectangle, wire etc.) in order to use this method. The advantages of this method are that the actual measurement is very simple to perform, one measurement gives an average for the whole sample and that wetting characteristics of single fibers can easily be obtained. He Tensiometer instrument also enables to determine the wettability characteristics i.e. contact angles of powders and porous solids by using the Washburn method (Powder Wettability Measurement).

What is Swarms?

Active materials can theoretically be made entirely of machines. These are sometimes called swarms since they consist of large numbers of identical simple machines that grasp and release each other and exchange power and information to achieve complex goals. Swarms change shape and exert force on their environment under software control. Although some physical prototypes have been built, at least one patent issued, and many simulations run, swarm potential capabilities are not well analyzed or understood. We briefly discuss some concepts here. Michael proposes brick-shaped machines of various sizes that slide past each other to assume a variety of shapes. He has generated a large number of videos showing computer simulations of simple motions. Although his web site contains rather extravagant claims, this work has received a U.K. patent. Yim built a small swarm with macroscopic (size in inches) components called polypod, built a simulator of polypod, and programmed it to move in various ways to study locomotion. There are two brick shaped components in polypod, one of which has two prismatic joints linked by a revolute joint. The second component is a cubic connector with no mechanical motion. Polypod is programmed by tables for each member of the swarm. Each member is programmed to move at various speeds in each degree of freedom for certain amounts of time. The swarm components are implicitly synchronized so there is no clock signal. Hall proposes a swarm with 10 micron dodecahedral components each with 12 arms that can move in and out, rotate a little, and grab and release each other. This concept is called the "utility fog." Hall estimates that the utility fog would have a density of 0.2, tensile strength of 1000 psi in action and 100,000 psi in a passive mode, and have a maximum shear rate of 100 km/second/meter. Bishop proposes a swarm consisting of 100 nanometer brick-shaped components that slide past each other to change shape. Globus proposes a swarm with two kinds of components—edges and nodes. The terms "node" and "edge" are chosen to correspond to those in graph theory. The roughly spherical nodes are capable of attaching to five edges (for a tetrahedral geometry with one free edge per node) and rotating each edge in pitch and yaw. The rod-like edges are capable of changing length, rotating around their long axis, and attaching/detaching to/from nodes.

What is Rebuilding Tissues?

Again, skin provides easy examples and may be a natural place to start in practice. People often want hair where they have bare skin, and bare skin where they have hair. Cell herding machines could move or destroy hair follicle cells to eliminate an unwanted hair, or grow more of the needed cells and arrange them into a working follicle where a hair is desired. By adjusting the size of the follicle and the properties of some of the cells, hairs could be made coarser, or finer, or straighter, or curlier. All these changes would involve no pain, toxic chemicals, or stench. Cell-herding devices could move down into the living layers of skin, removing unwanted cells, stimulating the growth of new cells, narrowing unnaturally prominent blood vessels, insuring good circulation by guiding the growth of any needed normal blood vessels, and moving cells and fibers around so as to eliminate even deep wrinkles. At the opposite end of the spectrum, cell herding will revolutionize treatment of life-threatening conditions. For example, the most common cause of heart disease is reduced or interrupted supply of blood to the heart muscle. In pumping oxygenated blood to the rest of the body, the heart diverts a portion for its own use though the coronary arteries. When these blood vessels become constricted, we speak of coronary-artery disease. When they are blocked, causing heart muscle tissue to die, we speak of someone "having a coronary," another term for heart attack. Devices working in the bloodstream could nibble away at atherosclerotic deposits, widening the affected blood vessels. Cell herding devices could restore artery walls and artery linings to health, by ensuring that the right cells and supporting structures are in the right places. This would prevent most heart attacks. But what if a heart attack has already destroyed muscle tissue, leaving the patient with a scarred, damaged, and poorly functioning heart? Once again, cell-herding devices could accomplish repairs, working their way into the scar tissue and removing it bit by bit, replacing it with fresh muscle fiber. If need be, this new fiber can be grown by applying a series of internal molecular stimuli to selected heart muscle cells to "remind" them of the instructions for growth that they used decades earlier during embryonic development. Cell-herding capabilities should also be able to deal with the various forms of arthritis. Where this is due to attacks from the body's own immune system, the cells producing the damaging antibodies can be identified and eliminated. Then a cell-herding system would work inside the joint where it would remove diseased tissues, calcified spurs, and so forth, then rework patterns of cells and intercellular material to form a healthy, smoothly working, and pain-free joint. Clearly, learning to repair hearts and learning to repair joints will have some basic technologies in common, but much of the research and development will have to be devoted to specific tissues and specific circumstances. A similar process—but again, specially adapted to the circumstances at hand—could be used to strengthen and reshape bone, correcting osteoporosis. In dentistry, this sort of process could be used to fill cavities, not with amalgam, but with natural dentin and enamel. Reversing the ravages of periodontal disease will someday be straightforward, with nanomedical devices to clean pockets, join tissues, and guide regrowth. Even missing teeth could be regrown, with enough control over cell behavior.

What is Quantum Mechanics?

A largely computational physical theory explaining the behavior of quantum phenomena, which incorporates the theory of special relativity. Despite dilignet attempts, general relativity has not been successfully incorporated into quantum mechanics. A physical model of chemical and optical phenomena, as well as the behaviour of matter in general on a small scale. Quantum mechanics describes a system of particles in terms of a wave function defined over the configuration space of the system. Although the concept of particles having distinct locations is implicit in the potential energy function that determines the wave function (e.g., of a ground-state system), the observable dynamics of the system cannot be described in terms of the motion of such particles from point to point. In describing the energies, distributions, and behaviors of electrons in nanometer-scale structures, quantum mechanical methods are necessary. Electron wave functions help determine the potential energy surface of a molecular system, which in turn is the basis for classical descriptions of molecular motion. Nanomechanical systems can almost always be described in terms of classical mechanics, with occasional quantum mechanical corrections applied within the framework of a classical model. A physical model of chemical and optical phenomena, as well as the behaviour of matter in general on a small scale. Quantum mechanics describes a system of particles in terms of a *wave function* defined over the *configuration space* of the system. Although the concept of particles having distinct locations is implicit in the *potential energy* function that determines the wave function (e.g., of *aground-state* system), the observable dynamics of the system cannot be described in terms of the motion of such particles from point to point. In describing the energies, distributions, and behaviors of electrons in nanometer-scale structures, quantum mechanical methods are necessary. Electron *wave functions* help determine the *potential energy surface* of a molecular system, which in turn is the basis for *classical* descriptions of molecular motion. *Nanomechanical* systems can almost always be described in terms of classical mechanics, with occasional quantum mechanical corrections applied within the framework of a classical model.

What is Resonant Tunneling Devices (RTDs)?

Here we talk about the future nanoelectronic devices using quantum effects employed in semiconductor devices for fabrication of computer devices and high frequency oscillators. The underlying principles used for defining the perpendicular transport of electrons in low dimensional structures up to room temperature and its current and voltage characteristics are explained using energy band diagrams. MOSFETS with its special characteristic features serves two purposes. First case its acts as an excellent conductor device even with its scaling properties and even with small changes in its bias voltage results in large changes in its conductivity, hence its acts as an amplifier device. Though it functions as a two state device there some factors need to be considered for its implementation. The obstacles associated with MOSFET are high electric field, heat power dissipation, vanishing bulk properties, shrinkage of depletion layers and unevenness of the depletion layer. Corrections have to be applied for these devices. Once the devices approach the nanoelectric properties, the bulk properties of semiconductor devices can be replaced by quantum physics which depend on the electron properties, energy quantization effects and tunneling come into effect. When quantum effects are applied to semiconductor devices the consideration of doped materials no longer considered as major issue. These devices are of great importance for high speed electronics. Here we discuss the quantum properties, general used terms such a quantum dot, tunneling effect, island, subband and electron behavior in quantized energy levels to deal with the resonant tunneling devices and the application of the quantum properties in various devices such as Resonant Tunneling Diodes (RTDs), Resonant Tunneling Transistors (RTTs). Here we explain the theory related to tunneling and the mathematical derivation for transmission probability of the particle in the double barrier junction.

What is Superconductor?

An element, inter-metallic alloy, or compound that will conduct electricity without resistance below a certain temperature. However, this applies only to direct current (DC) electricity and to finite amounts of current. All known superconductors are solids. None are gases or liquids. And all require extreme cold to enter a superconductive state. Once set in motion, current will flow forever in a closed loop of superconducting material—making it the closest thing to perpetual motion in nature. Scientists refer to superconductivity as a "macroscopic quantum phenomenon". In addition to being classified Type 1 and Type 2, superconductors can be categorized further by their dimensionality. Most are 3-D. But some compounds, like surface-doped $NaWO_3$ and some organic superconductors are 2-D. Li_2CuO_2 and single-walled carbon nano-tubes have shown rare 1-D superconductivity. In addition to repelling magnetic fields, enhanced thermal conductivity, higher optical reflectivity and reduced surface friction are also properties of superconductors. The term "superconductor" is also used in some instances to refer to materials that have near infinite thermal conductivity-such as carbon nanotubes. However, on this website it is used in the context of electrical conductivity only.

An object or substance that conducts electricity with zero resistance. Superconductors, materials that have no resistance to the flow of electricity, are one of the last great frontiers of scientific discovery. Not only have the limits of superconductivity not yet been reached, but the theories that explain superconductor behavior seem to be constantly under review. In 1911 superconductivity was first observed in mercury by Dutch physicist Heike Kamerlingh Onnes of Leiden University (shown above). When he cooled it to the temperature of liquid helium, 4 degrees Kelvin (–452F, –269C), its resistance suddenly disappeared. The Kelvin scale represents an "absolute" scale of temperature. Thus, it was necessary for Onnes to come within 4 degrees of the coldest temperature that is theoretically attainable to witness the phenomenon of superconductivity. Later, in 1913, he won a Nobel Prize in physics for his research in this area. The next great milestone in understanding how matter behaves at extreme cold temperatures occurred in 1933. German researchers Walter Meissner (above) and Robert Ochsenfeld discovered that a superconducting material will repel a magnetic field (below graphic). A magnet moving by a conductor induces currents in the conductor. This is the principle upon which the electric generator operates. But, in a superconductor the induced currents exactly mirror the field that would have otherwise penetrated the superconducting material—causing the magnet to be repulsed. This phenomenon is known as strong diamagnetism and is today often referred to as the "Meissner effect" (an eponym). The Meissner effect is so strong that a magnet can actually be levitated over a superconductive material.

In subsequent decades other superconducting metals, alloys and compounds were discovered. In 1941 niobium-nitride was found to superconduct at 16 K. In 1953 vanadium-silicon displayed superconductive properties at 17.5 K. And, in 1962 scientists at Westinghouse developed the first commercial superconducting wire, an alloy of niobium and titanium (NbTi). High-energy, particle-accelerator electromagnets made of copper-clad niobium-titanium were then developed in the 1960s at the Rutherford-Appleton Laboratory in the UK, and were first employed in a superconducting accelerator at the Fermilab Tevatron in the US in 1987. The first widely-accepted theoretical understanding of superconductivity was advanced in 1957 by American physicists John Bardeen, Leon Cooper, and John Schrieffer (above). Their *Theories of Superconductivity* became know as the BCS theory-derived from the first letter of each man's last name—and won them a Nobel prize in 1972. The mathematically-complex BCS theory explained superconductivity at temperatures close to absolute zero for elements and simple alloys. However, at higher temperatures and with different superconductor systems, the BCS theory has subsequently become inadequate to fully explain how superconductivity is occurring.

Another significant theoretical advancement came in 1962 when Brian D. Josephson (above), a graduate student at Cambridge University, predicted that electrical current would flow between 2 superconducting materials—even when they are separated by a non-superconductor or insulator. His prediction was later confirmed and won him a share of the 1973 Nobel Prize in Physics. This tunneling phenomenon is today known as the "Josephson effect" and has been applied to electronic devices such as the SQUID, an instrument capable of detecting even the weakest magnetic fields. The 1980's were a decade of unrivaled discovery in the field of superconductivity. In 1964 Bill Little of Stanford University had suggested the possibility of organic (carbon-based) superconductors. The first of these theoretical superconductors was successfully synthesized in 1980 by Danish researcher Klaus Bechgaard of the University of Copenhagen and 3 French team members. $(TMTSF)_2PF_6$ had to be cooled to an incredibly cold 1.2K transition temperature (known as Tc) and subjected to high pressure to superconduct. But, its mere existence proved the possibility of "designer" molecules-molecules fashioned to perform in a predictable way.

Then, in 1986, a truly breakthrough discovery was made in the field of superconductivity. Alex Müller and Georg Bednorz (above), researchers at the IBM Research Laboratory in Rüschlikon, Switzerland, created a brittle ceramic compound that superconducted at the highest temperature then known: 30 K. What made this discovery so remarkable was that ceramics are normally insulators. They don't conduct electricity well at all. So, researchers had not considered them as possible high-temperature superconductor candidates. The Lanthanum, Barium, Copper and Oxygen compound that Miiller and Bednorz synthesized, behaved in a not-as-yet-understood way. The discovery of this first of the superconducting copper-oxides (cuprates) won the 2 men a Nobel Prize the following year. It was later found that tiny amounts of this material were actually superconducting at 58 K, due to a small amount of lead having been added as a calibration standard—making the discovery even more noteworthy.

Müller and Bednorz' discovery triggered a flurry of activity in the field of superconductivity. Researchers around the world began "cooking" up ceramics of every imaginable combination in a quest for higher and higher Tc's. In January of 1987 a research team at the University of Alabama-Huntsville substituted Yttrium for Lanthanum in the Miiller and Bednorz molecule and achieved an incredible 92 K Tc. For the first time a material (today referred to as YBCO) had been found that would superconduct at temperatures warmer than liquid nitrogen—a commonly available coolant. Additional milestones have since been achieved using exotic—and often toxic—elements in the base perovskite ceramic. The current class (or "system") of ceramic superconductors with the highest transition temperatures are the mercuric-cuprates. The first synthesis of one of these compounds was achieved in 1993 by Prof. Dr. Ulker Onbasli at the University of Colorado and by the team of A. Schilling, M. Cantoni, J.D. Guo, and H.R. Ott of Zurich, Switzerland. The world record Tc of 138 K is now held by a thallium-doped, mercuric-cuprate comprised of the elements Mercury, Thallium, Barium, Calcium, Copper and Oxygen. The Tc of this ceramic superconductor was confirmed by Dr. Ron Goldfarb at the National Institute of Standards and Technology-Colorado in February of 1994. Under extreme pressure its Tc can be coaxed up even higher-approximately 25 to 30 degrees more at 300,000 atmospheres. The first company to capitalize on high-temperature superconductors was Illinois Superconductor (today known as ISCO International), formed in 1989. This amalgam of government, private-industry and academic interests introduced a depth sensor for medical equipment that was able to operate at liquid nitrogen temperatures (~77K). While no significant advancements in superconductor Tc's have been achieved in recent years, other discoveries of equal importance have been made. In 1997 researchers discovered that at a temperature very near absolute zero an alloy of gold and indium was both a superconductor and a natural magnet. Conventional wisdom held that a material with such properties could not exist! Since then, over a half-dozen such compounds have been found. Recent years have also seen the discovery of the first high-temperature superconductor that does NOT contain any copper (2000), and the first all-metal perovskite superconductor (2001). Also in 2001 a material that had been sitting on laboratory shelves for decades was found to be an extraordinary new superconductor. Japanese researchers measured the transition temperature of magnesium diboride at 39 Kelvin—far above the highest Tc of any of the elemental or binary alloy superconductors. While 39 K is still well below the Tc's of the "warm" ceramic superconductors, subsequent refinements in the way MgB_2 is fabricated have opened the door to its use in industrial applications. Laboratory testing has found MgB_2 will outperform NbTi and Nb_3Sn wires in high magnetic field applications like MRI. Startling—and serendipitous—discoveries like these are forcing scientists to continually re-examine longstanding theories on superconductivity and to consider heretofore-unimagined combinations of elements. Magnetic-levitation is an application where superconductors perform extremely well. Transport vehicles such as trains can be made to "float" on strong superconducting magnets, virtually eliminating friction between the train and its tracks. Not only would conventional electromagnets waste much of the electrical energy as heat, they would have to be physically much larger than superconducting magnets. A landmark for the commercial use of MAGLEV technology occurred in 1990 when it gained the status of a nationally-funded project in Japan.

What is Electron-beam Lithography?

Electron beam lithography is a process that can be used to write fine patterns with an electron beam. An electron beam can be generated by extracting electrons from a sharp needle with an electric field. If the electrons are then accelerated towards a metal plate with a small hole in it, a narrow beam of electrons emerges from this hole. The electrons are typically accelerated through a potential of several kilovolts and are traveling at a good fraction of the speed of light. An electron beam can be deflected by electric and magnetic field which makes it possible to write with the beam. Such electron beams are used in televisions and computer monitors. It is possible to focus an electron beam to have a very small diameter of about 1 nanometer. Such narrowly focused beams are used in scanning electron microscopes, transmission electron microscopes, and electron-beam pattern generators. Typically in an electron-beam pattern generator, the electron beam writes a pattern in a thin organic film that is coated over a wafer. Usually the wafer is a single crystal of silicon that is about 0.5 mm thick and 10-30 cm in diameter. The thin (~100 nm) organic film contains long molecules that wrap around each other like cooked spagetti. The energetic electrons in the electron beam cut these molecules up into small pieces. The film is then dipped in developer which dissolves the short pieces but leaves the long unexposed sections unaffected. A pattern is thereby defined in the resist. The pattern can be transfered to the substrate by putting the wafer in a gas or liquid that dissolves the substrate material. Patterns with features of a few tens of nanometers can be made this way.

What is Secondary Electron Imaging?

Production of secondary electrons is very topography related. Due to their low energy, 5eV, only secondaries that are very near the surface (<10nm, see Interaction Volume) can exit the sample and be examined. Any changes in topography in the sample that are larger than this sampling depth will change the yield of secondaries due to collection efficiencies. Collection of these electrons is aided by using a "collector" in conjunction with the secondary electron detector. The collector is a grid or mesh with a +100V potential applied to it which is placed in front of the detector, attracting the negatively charged secondary electrons to it which then pass through the grid-holes and into the detector to be counted. When a Secondary Electrons collide with the solid-state semiconductor detector an electron-hole pairs are created which are then counted. This quantity is translated into a pixel intensity and displayed on the CRT, forming the image. See the SEM page for details on image formation.

What is Energy Storage?

Buckytubes have the intrinsic characteristics desired in material used as electrodes in batteries and capacitors, two technologies of rapidly increasing importance. Buckytubes have a tremendously high surface area (~1000 m^2/g), good electrical conductivity, and very importantly, their linear geometry makes their surface highly accessible to the electrolyte. Research has shown that buckytubes have the highest reversible capacity of any carbon material for use in lithium-ion batteries. In addition, buckytubes are outstanding materials for supercapacitor electrodes and are now being marketed. Buckytubes also have applications in a variety of fuel cell components. They have a number of properties including high surface area and thermal conductivity that make them useful as electrode catalyst supports in PEM fuel cells. They may also be used in gas diffusion layers as well as current collectors because of their high electrical conductivity. Buckytubes' high strength and toughness to weight characteristics may also prove valuable as part of composite components in fuel cells that are deployed in transport applications where durability is extremely important.

What is Thrombogenesis?

Blood coagulation involves a complex series of reactions in which various proteins are enzymatically activated in a sequential manner, transforming liquid blood into a gel-like clot which is then stabilized to form a thrombus (clot) consisting of platelets, fibrin, and red cells. The series of reactions is classically divided into two pathways-extrinsic and intrinsic -involving more than a dozen factors that converge on a single common final pathway, resulting in clot formation. Since, these factors are carried by the blood, the vasculoid can regulate their local concentration and thus, prevent thrombogenic pathways from proceeding to completion. Additionally, platelets must undergo adhesion and activation for coagulation to occur. The adhesion of platelets to exposed collagen in injured blood vessels is mediated by a bridging molecule called von Willebrand's factor that is secreted by endothelial cells into plasma, which prevents platelets from detaching under the high shearing stresses developed near vessel walls.

The activation of normally quiescent platelets is a complex phenomenon that includes changes in cell shape, increased movement, release of the contents of their granules (containing nucleotidyl phosphates, serotonin, various factors, enzymes and plasma proteins), and aggregation. The most potent activator of platelets in vivo is thrombin, which interacts with a receptor on the platelet plasma membrane, followed by transmembrane signaling and subsequent activation of the cell. Collagen is the other most important platelet activator; ADP can stimulate aggregation but not granule release. In principle, the blood-contacting surfaces of a nanoorgan, or of nanorobots in sufficient bloodstream numbers and concentrations, could activate platelets (and thus either of the two coagulation pathways), but careful choices of materials and of allowable mechanical motions should reduce or eliminate inherent nanodevice thrombogenicity. For example, DLC diamond-coated stents, heart valves and other blood-contacting LVAD surfaces or substrates generally show reduced thrombogenicity and no platelet activation. Sapphire (alumina ceramic) has low thrombogenicity and both platelet adhesion and activation are low. Hemolysis is near-zero for diamond and alumina powders. Additionally, in the vasculoid plateletogenesis and release of platelets could be actively regulated and reduced to minimal levels. Future experiments must determine if ordinary diamondoid surfaces will have to be supplemented with additional antithrombogenic coatings in order to achieve vasculoid performance objectives. If such coatings are required, one simple possibility is surface-immobilized heparin, a ~15 kD straight-chain anionic (acidic) mucopolysaccharide (glycosaminoglycan) that forms polymers of various lengths. Heparin, first discovered in 1916, is produced naturally by human liver mast cells and basophil leukocytes, and inhibits coagulation primarily by accelerating the interaction between antithrombin and thrombin. Nanorobot exteriors can be "heparinized", and thereby rendered thromboresistant by immobilized heparin on all blood-contacting surfaces at ~monolayer surface concentration (e.g., 7-10 pmol/cm^2). Cellulose membranes coated with 3.6 pmol/cm^2 of endothelial-cell-surface heparan sulphate show complete inhibition of platelet adhesion. If satisfactory passive nonthrombogenic surfaces cannot be found, nanorobots might employ any of several active strategies to prevent iatrogenic coagulation.

What is Thyristor?

The thyristor (also called a silicon-controlled rectifier or SCR) is a solid-state semiconductor device similar to a diode, with an extra terminal which is used to turn it on. Once turned on, the thyristor will remain on (conducting) as long as there is a significant current flowing through it. If the current falls to zero, the device switches off. The thyristor is a four-layer semiconducting device, with each layer consisting of an alternately N or P-type material, for example N-P-N-P. The main terminals, labelled anode and cathode, are across the full four layers, and the control terminal, called the gate, is attached to one of the middle layers. The operation of a thyristor can be understood in terms of a pair of tightly coupled transistors, arranged to cause the self-latching action. Thyristors are mainly used where high currents and voltages are involved, and are often used to control alternating currents, where the change of sign of the current causes the device to automatically switch off. This is known as *synchronous operation* or Zero Cross operation. This principle is used to control the desired loading by adjusting the frequency of the sinusoidal input. The range of frequencies is great because there is no limit to the number of cycles a thyristor can perform, and exhibits no "wear out" modes. This is a frequency domain method of control. With phase angle control a thyristor is turned on at a specific and adjustable portion of the cycle of the controlling sinusoidal input. Moving the point at which the thyristor is turned on regulates power output. An example of this method of control is a dimmer switch for lights. The turn on point of a thyristor is controlled to occur at a particular point on the sine curve of the AC supply. The thyristor stays on for the remainder of that cycle and the longer the thyristor stays on, the brighter the light. Fine resolution of output is possible with this method and is suitable for fast responding loads such as tungsten filament lamps or temperature variable resistance loads. Phase-angle control is also essential for inductive loads. The drawback of a thyristor is that, like a diode, it only conducts in one direction. A similar self-latching 5-layer device, called a triac, is able to work in both directions. An earlier gas filled tube device called a Thyratron provided a similar electronic switching capability, where a small control voltage could switch a large current. Modern thyristors can switch large amounts of power (up to megawatts). In the realm of very high power applications, they are still the primary choice. However, in low and medium power (from few tens of watts to few tens of killowatts) they have almost been replaced by other devices with superior switching characteristics like MOSFETs or IGBTs. One major problem of thyristor is that is it not a fully controllable switch in the sense that triggering current direction need to be reversed to switch it off. GTO (Gate Turn-off Thyristor) is another related device which addresses this problem. In high-frequency applications, thyristors are poor candidates due to large switching times arising out of bipolar conduction. MOSFETs, on the other hand, has much faster switching capability because of its unipolar conduction (only majority carriers carry the current).

What is Thruster?

A way of controlling a satellite's attitude. Thrusters usually contain compressed gas that when sent out of the end of the thruster will move the satellite in space. The force of the compressed gas (the action) causes the satellite to move in the opposite direction (the reaction). This is Newton's third law of motion—for every action there is an equal and opposite reaction.

What is Electron Microscopy?

Visual and photographic microscopy in which electron beams with wavelengths thousands of times shorter than visible light are used in place of light, thereby allowing much greater magnification. In high-resolution electron microscopy one can begin to do "crystallography without crystals", averaging thousands of images of single molecules or other assemblies to reveal near atomic level structure. These methods demand intense computing hardware, software and algorithm development.

What is Electrolyte/Insulator/Silicon?

Structures at the heart of a broad family of potentiometric silicon sensors. The best-known member of the family is the ion-sensitive field effect transistor, known as the ISFET or CHEMFET and the light-addressable potentiometric sensor LAPS. The principle of operation of devices using such structures is as follows. A potential with respect to a reference electrode is generated at the interface between the liquid solution and the insulator. The surface potential ($ø_0$) is determined by that ionic species which has the fastest exchange rate (i_o) with the membrane covering the insulator. If no intentional membrane is deposited on an oxide covered insulator that species will be H^+. Surface potential changes in turn change the Si flat-band voltage V_{FB}. The flat-band voltage is the potential one needs to apply to the Si in order to have the bands flat throughout the semiconductor. The flat band voltage of an EIS structure has been shown to be given by: $V_{FB} = E_{REF} - Ö^{Si}/q - ø_0 - Q_{ins}/C_{inss}$, where V_{FB} stands for the flat-band voltage of the structure, E_{REF} for the reference electrode potential, $Ö^{Si}$ for the work function of silicon, $ø_0$ for the surface potential at the insulator/electrolyte interface, Q_{ins} for the charge at the insulator/silicon interface and C_{ins} for the insulator capacitance. At least two terms in the above equation are not known with a precision greater than a few hundred millivolts. This is true for E_{REF} as well as for Q_{ins}/C_{ins} which can vary from device to device by several hundred millivolts. For a given EIS sensor, these inaccurately known quantities are constant, and variations in flat-band voltage can be equated to variations of the surface potential.

What is Scanning Transmission Electron Microscopy?

A special TEM-technique in which an electron transparent sample is bombarded with a finely focused electron beam (typically of a diameter of less than 10 nm) which can be scanned across the specimen or rocked across the optical axis and transmitted secondary, backs scattered and diffracted electrons as well as the characteristic X-ray spectrum can be observed. STEM essentially provides high resolution imaging of the inner microstructure and the surface of a thin sample (or small particles), as well as the possibility of chemical and structural characterization of micrometer and sub-micrometer domains through evaluation of the X-ray spectra and the electron diffraction pattern.

What is Hot Embossing?

Hot embossing of wafers uses the cloned master die as the wafer stamp. Wafers are first coated with a thermoplastic polymer, that is, a polymer that softens when heated. The wafer is positioned in a controlled-atmosphere stamping chamber. After heating the wafer, the stamp is brought down with pre-determined pressure and temperature to soften the polymer and simultaneously form the image on the whole wafer. After allowing time for polymer flow during heating, the stamp and wafer are cooled to harden the image. To maintain the pattern, pressure must continue to be applied until both stamp and substrate are cooled below the thermoplastic softening point. The hot-embossed wafer may be used for its embossed features, or it may be post-processed as a resist for etching, material deposition, and lift-off patterning. For resist use, curing may be followed by a light etch to remove the remaining traces of materials from the clear areas.

What is High Surface Area Materials?

Materials which have large surface to volume ratio. Consider for example a chunk of material in the shape of a sphere of radius R. Its surface to volume ratio is 3/R. Thus, for smaller pieces of materials, the surface to volume ratio is larger. For pieces of matter on a nanometer scale (nanoparticles), the contribution of the surface to the overall properties of such matter becomes very important. For example, in a nanoscopic particle with a diameter of 5 nm, about half of its atoms are on its surface (low-coordination-number atoms). This is why the techniques and methods of surface science are needed to describe properties of nanoscopic matter. Even the macroscopic pieces of nanostructured materials can have high surface area, since their nanoscopic constituents can arrange in such a way to still have a large exposed surface. For example, single-walled carbon nanotube materials consist of carbon nanotube bundles. The bundles are made of several tens of carbon nanotubes and thus have a large exposed surface. Bundles are in a macroscopic material organized in such a way that their large surface still remains exposed (a particular bundle is not shielded by other bundles in the material). High surface area materials are of interest for catalysis, due to the fact that catalysis proceeds on the surface of materials. Porous materials such as zeolites also have a large area of exposed surface. Such materials can also be used for separation of different gases (purification of gases, removal of ultra-fine contaminants from air and water) since, due to geometrical constraints, molecules of a particular gas can enter the nanoscopic pores, while those of another gas cannot due to the fact that they are to big to fit in. High surface are materials can also be used for storage of gases (adsorption/desorption materials) and as a battery/ capacitor elements.

What is Quantum Wire?

Another form of quantum dot, but unlike the single-dimension "dot," a quantum wire is confined only in two dimensions—that is it has "length," and allows the electrons to propagate in a "particle-like" fashion. Constructed typically on a semiconductor base, and (among other things) used to produce very intense laser beams, switchable up to multi-gigahertz per second.

What is Crude Technology?

As the Silicon Valley Faire Scenario shows, molecular manufacturing will work much like ordinary manufacturing, but with devices built so small that a single loose molecule of pollutant would be like a brick heaved into a machine tool. John Walker of Autodesk, a leading company in computer-aided design, observes that nanotechnology and today's crude methods are very different: "Technology has never had this kind of precise control; all of our technologies today are bulk technologies. We take a big chunk of stuff and hack away at it until we're left with the object we want, or we assemble parts from components without regard to structure at the molecular level." Molecular manufacturing will orchestrate atoms into products of symphonic complexity, but modern manufacturing mostly makes loud noises. These figurative noises are sometimes all too literal: A crack in a metal forging grows under stress, a wing fails, and a passenger jet crashes from the sky. A chemical reaction goes out of control, heat and pressure build, and a poisonous blast shakes the countryside. A lifesaving product cannot be made, a heart fails, and a hospital's heart-monitoring machine signals the end with a high-pitched wail. Today, we make many things from metal, by machining. From the perspective of our standard, simulated molecular world, a typical metal part is a piece of terrain many days' journey across. The metal itself is weak compared to the bonds of the protein chain or other tough nanomechanisms: solid steel is no stronger than your simulated fingers, and the atoms on its surface can be pushed around with your bare hands. Standing on a piece of metal being machined in a lathe, you would see a cutting blade crawl past a few times per year, like a majestic plough the size of a mountain range. Each pass would rip up a strip of the metal landscape, leaving a rugged valley broad enough to hold a town. This is machining from a nanotechnological perspective: a process that hacks crude shapes from intrinsically weak materials. Today, electronics are made from silicon chips. We have already seen the landscape of a finished chip. During manufacturing, metal features would be built up by a centuries-long drizzle of metal-atom rain, and hollows would be formed by a centuries-long submergence in an acid sea. From the perspective of our simulation, the whole process would resemble geology as much as manufacturing, with the slow layering of sedimentary deposits alternating with ages of erosion. The term *nanotechnology* is sometimes used as a name for small-scale microtechnology, but the difference between molecular manufacturing and this sort of microlandscaping is like the difference between watchmaking and bulldozing. Today, chemists make molecules by solution chemistry. We have seen what a liquid looks like in our first simulation, with molecules bumping and tumbling and wandering around. Just as assemblers can make chemical reactions occur by bringing molecules together mechanically, so reactions can occur when molecules bump at random through thermal vibration and motion in a liquid. Indeed, much of what we know today about chemical reactions comes from observing this process. Chemists make large molecules by mixing small molecules in a liquid. By choosing the right molecules and conditions, they can get a surprising measure of control over the results: only some pairs of molecules will react, and then only in certain ways. Doing chemistry this way, though, is like trying to assemble a model car by putting the pieces in a box and shaking. This will only work with cleverly shaped pieces, and it is hard to make anything very complex. Chemists today consider it challenging to make a precise, three-dimensional structure having a hundred atoms, and making one with a thousand atoms is a great accomplishment. Molecular manufacturing, in contrast, will routinely assemble millions or billions. The basic chemical principles will be the same, but control and reliability will be vastly greater. It is the difference between throwing things together blindly and putting them together with a watchmaker's care. Technology today doesn't permit thorough control of the structure of matter. Molecular manufacturing will. Today's technologies have given us computers, spacecraft, indoor plumbing, and the other wonders of the modern age. Tomorrow's will do much more, bringing change and choices.

What is Dendrimer?

A dendrimer is a spherical, highly branched polymer molecule. Dendrimers are made from two different monomers—a reactive amine and an acrylic acid - and are assembled in discrete steps, which allow them to be constructed with an exact size that depends on the number of enlargement steps. This picture shows a dendrimer that has undergone three "generations" of enlargements. Dendrimers are of particular interest for cancer applications because of their defined and reproducible size, but more importantly, because it is easy to add a variety of other molecules to the surface of a dendrimer. Such molecules could include tumor-targeting agents (including but not restricted to monoclonal antibodies), imaging contrast agents to pinpoint tumors, drug molecules for delivery to a tumor, and reporter molecules that might detect if an anticancer drug is working. A dendrimer is a tree-like highly branched polymer molecule (Greek *dendra = tree).* Dendrimers are synthesized from monomers with new branches added in discrete steps ("generation") to form a tree-like architecture. A high level of synthetic control is achieved through step-wise reactions and purifications at each step to control the size, architecture, functionality and monodispersity. Several different kinds of dendrimers have been synthesized utilizing different monomers and some are commercially available. This picture shows a "3rd generation" polyamidoamine (PAMAM) dendrimer. Dendrimers are of particular interest for cancer applications because of their defined and reproducible size, but more importantly, because it is easy to attach a variety of other molecules to the surface of a dendrimer. Such molecules could include tumor-targeting agents (including but not restricted to monoclonal antibodies), imaging contrast agents to pinpoint tumors, drug molecules for delivery to a tumor, and reporter molecules that might detect if an anticancer drug is working. (3) A dendrimer is a tree-like highly branched polymer molecule (Greek *dendra = tree).* Dendrimers are synthesized from monomers with new branches added in discrete steps ("generation") to form a tree-like architecture. A high level of synthetic control is achieved through step-wise reactions and purifications at each step to control the size, architecture, functionality and monodispersity. Several different kinds of dendrimers have been synthesized utilizing different monomers and some are commercially available. This picture shows a "3rd generation" polyamidoamine (PAMAM) dendrimer. Dendrimers are of particular interest for cancer applications because of their defined and reproducible size, but more importantly, because it is easy to attach a variety of other molecules to the surface of a dendrimer. Such molecules could include tumor-targeting agents (including but not restricted to monoclonal antibodies), imaging contrast agents to pinpoint tumors, drug molecules for delivery to a tumor, and reporter molecules that might detect if an anticancer drug is working. (4) A polymer having a regular branched structure; If suitably functionalized may be used as a soluble support, in which case the desired, dendrimer-supported, material may be isolated by size- exclusion chromatography. Dendrimers may also be attached to a polymer and used as a *solid support,* with significantly increased loading over the initial resin. [IUPAC COMBINATORIAL CHEMISTRY] Dendrimers consist of interconnected monomeric subunits that hybridize to form a tree- like structure. Each monomer is a double- stranded DNA molecule where the two strands share a region of sequence complementarity in the middle of molecule. [CHI Microarrays report] A dendrimer (from Greek *dendra* for tree) is an artificially manufactured or synthesized molecule built up from branched units called monomers. Such processes involve working on the scale of nanometers (a nanometer is 10^{-9} meter or a millionth of a millimeter). Technically, a dendrimer is *a polymer,* which is a large molecule comprised of many smaller ones linked together. Dendrimers have some proven applications, and numerous potential applications. [whatis.com] Also known as *"cascade molecules"*. Polymeric molecule about the size of an average protein that has a branching shape. There is a dendrimeric polymer whose monomers are acrylic acid and a diamine. Depending on the dendrimer core (central part), the dendrimer can start with 3 to 8 (or more) branches, with 3 and 4 being the most common numbers observed. There is a large number of molecules which can be used as cores. It has been suggested that dendrimers should exhibit strong catalytic properties. Another suggestion for their use is for drug delivery and release. The 'lego-like' nature of such molecules could perhaps be used to engineer and optimize their structure to suit specific needs. The name 'dendrimer' is derived from the ancient Greek word 'dendron' (tree), and the suffix 'mer' (segment).

What is Early Attrition?

Of poor drug candidates is central to the new drug discovery paradigm. The major trend in lead optimization is the movement toward in silico and high- throughput in vitro approaches. Computational methods can be applied to chemical structures to predict ADMET properties even before the compound is synthesized so that only favorable compounds need be synthesized for screening. Whole cells are also increasingly being used in high-content screening mode to provide selectivity information along with other valuable data concerning the effects of compounds on cell function. Another area of growing impact involves the use of cells equipped with reporter gene constructs for high- throughput analysis of the expression of specific, predetermined genes in response to compound administration. New approaches for predicting toxicity are also examined in this report, particularly the efforts of some groups to develop better animal model systems and to look at ways to introduce these assays earlier in the process. CHI Advances in Lead Optimization: Accelerating Drug Discovery and Development report, 2003.

What is Energy Dispersive Spectroscopy?

A technique that is based on the characteristic X-ray peaks that are generated when an energetic electron beam interacts with the specimen. Each element produces characteristic x-rays that may be used to identify the presence of that element in the region being examined. Comparison of the relative intensities of x-ray peaks may be used to determine the relative concentrations of each element in the specimen. Elements with an atomic number less than that of carbon carbon (Z=5) are not generally detectable.

What is Diamondoid?

Hydrocarbon molecule with a three-dimensional structure whose carbon-carbon framework constitutes the fundamental repeating unit in the diamond lattice. Structures that resemble diamond in a broad sense, strong stiff structures containing dense, three dimensional networks of covalent bonds; diamondoid materials could be as much as 100 to 250 times as strong as titanium, and far lighter. Structures that resemble diamond in a broad sense, strong stiff structures containing dense, three dimensional networks of covalent bonds, formed chiefly from first and second row atoms with a valence of three or more. Many of the most useful diamondoid structures will in fact be rich in tetrahedrally coordinated carbon. [NTN] Materials with superior strength to weight ratio, as much as 100 to 250 times as strong as Titanium, and much lighter. Possibly used to build stronger lighter rockets and space components, or a variety of other earth-bound articles for which weight and strength are a consideration.

What is Effective Fuel Catalyst?

In order for a catalyst to be effective when additised into fuel three principal properties are required. Firstly, complete hydrocarbon oxidation should be promoted, secondly formation of NOx should not be favoured, and finally the catalyst should remain thermally stable. As has been discussed, the ability of cerium oxide to donate oxygen allows complete hydrocarbon and soot burning in principle. In practice the activation energy of the cerium oxide, i.e. the lowest temperature at which oxygen donation occurs, is a crucial factor.

Although the gas temperature of a diesel engine is high (approximately 1700°C), a low catalyst 'switch-on' temperature will clearly promote a more complete fuel burn in the milliseconds during which combustion occurs. A lower overall burn temperature, notwithstanding transient higher maximum temperatures, also mitigates against NOx production due to the high activation energy of nitrogen oxidation.

What is Density Functional Theory?

A property of a material defined as the ratio of mass to volume of a sample. A branch of quantum theory of many electron systems. Density functional theory (DFT) rests on the two basic theorems by Kohn and Hohenberg which state that the ground state energy of the N-electron systems is a unique functional of the electron density distribution. The theorems are quite important since they allow one to concentrate on the search for the electron density of the system, rather than the many-body quantum wave function, which is a much more complicated object. DFT had great success in explaining and calculating the electronic structure of molecules and solids, and is used extensively in chemistry and physics. DFT provides an information on the electron density distribution in a system of interest and its ground state energy. There are many variants of DFT, possibly the most promising ones being those dealing with time dependent implementation of DFT. Walter Kohn and John A. Pople received a Nobel Prize in chemistry (1998) for the development of density functional theory and computational methods in quantum chemistry.

What is Fullerene?

A large molecule composed exclusively of carbon atoms and having shape of an empty cage (full name buckminsterfullerene). There are fullerenes with different number of carbon atoms and different sizes. The carbon atoms are arranged in pentagons and hexagons within a fullerene molecule. Some consider carbon nanotube as a special form of fullerene. Fullerenes were discovered by R.F. Curl Jr., H.W. Kroto and R.E. Smalley. For this discovery they were awarded Nobel Prize (chemistry, 1996). A class of cage-like carbon compounds composed of fused, pentagonal and/or hexagonal sp^2 carbon rings. A new allotrope of carbon characterized by a closed cage structure consisting of an even number of three coordinate carbon atoms devoid of hydrogen atoms. This class was originally limited to closed-cage structures with twelve isolated five-membered rings, the rest being six- membered rings. [IUPAC Provisional Recommendations for the Nomenclature for the C_{60}-I_h and C_{70}-$D_{5h(6)}$ Fullerenes, 2001] "Though there tends to be some variation in definitions of fullerenes and related structures, generally buckminsterfullerene, or the buckyball, is taken to refer only to C_{60}, an approximately spherical, hollow, carbon molecule containing 60 carbon atoms arranged in interlocking hexagons and pentagons, akin to the geodesic dome created by architect Buckminster Fuller, hence the name. The structure also looks very much like a soccer ball. The other fullerenes are similar cage-like molecules with more or less carbon atoms, but still composed of pentagons and hexagons. The remaining fullerenes contain more or less carbon atoms, again arranged in pentagons and hexagons." "A closed, convex cage molecule containing only hexagonal and pentagonal faces. (This definition intentionally leaves out possible heptagons, which are responsible for the concave parts and are treated as defects.)". Fullerene: A class of cage-like carbon compounds composed of fused, pentagonal and/or hexagonal sp^2 carbon rings. Fullerenes were first discovered in 1985 when the soccer ball shaped C_{60} (Buckminsterfullerene) was synthesized. The carbon atoms in C_{60} are arranged in a geometric shape consisting of 12 pentagons and 20 hexagons. Other spherical fullerenes (collectively known as buckyballs) were subsequently synthesized with a different number of hexagonal faces. The smallest possible fullerene is the dodecahedral C_{20}, a shape consisting of 12 pentagonal faces and no hexagonal faces. Larger, fullerenes have been found to exist in nature. Nanotubes, nanohorns and buckybowls are other examples of fullerenes. More Definitions of Fullerene: "Though there tends to be some variation in definitions of fullerenes and related structures, generally buckminsterfullerene, or the buckyball, is taken to refer only to C_{60}, an approximately spherical, hollow, carbon molecule containing 60 carbon atoms arranged in interlocking hexagons and pentagons, akin to the geodesic dome created by architect Buckminster Fuller, hence the name. The structure also looks very much like a soccer ball. The other fullerenes are similar cage-like molecules with more or less carbon atoms, but still composed of pentagons and hexagons. The remaining fullerenes contain more or less carbon atoms, again arranged in pentagons and hexagons." "A closed, convex cage molecule containing only hexagonal and pentagonal faces. (This definition intentionally leaves out possible heptagons, which are responsible for the concave parts and are treated as defects.)".

What is Fractal Robots?

AKA: Fractal Shape Shifting Robots and Programmable "Digital Matter", are programmable machines that can do unlimited tasks in the physical world, the world of matter. Load the right software and the same "machines" can vacuum the carpet, paint your car, or construct an office building and later, wash that building's windows. This is the beginning of "Digital Matter". Fractal Shape Shifting Robots look like "Rubic's Cubes" that can "slide" over each other on command, changing and moving in any overall shape desired for a particular task. These cubes communicate with each other and share power through simple internal induction coils (or surface contacts in some models), have batteries, a small computer and various kinds of internal magnetic and electric inductive motors (depending on size) used to move over other cubes. When sufficiently miniaturized (below 0.1mm) and fabricated using photolithography and E-Beam methods, the machines may exceed human manual dexterity and could then be programmed to assemble complex fractal aggregates or even to maintain the photolithographic and E-Beam equipment itself! The ultimate goal is self sustaining systems and "self-assembly" features that can drop cost dramatically and enable successive generations of robots exhibiting greater utility and value, to be built along the way.

What is Herding Cells?

Destroying invaders will be helpful, but injuries and structural problems pose other problems. Truly advanced medicine will be able to build up and restructure tissues. Here, medical nanodevices can stimulate and guide the body's own construction and repair mechanisms to restore healthy tissue. What is healthy tissue? It consists of normal cells in normal patterns in a normal matrix all organized in a normal relationship to the surrounding tissues. Surgeons today (with their huge, crude tools) can fix some problems at the tissue level. A wound disrupts the healthy relationship between two different pieces of tissue, and surgical glues and sutures can partly remedy this problem by holding the tissues in a position that promotes healing. Likewise, coronary artery bypass surgery brings about a more healthy overall configuration of tissues—one that provides working plumbing to supply blood to the heart muscle. Surgeons cut and stitch, but then they must rely on the tissue to heal its wounds as best it can. Healing establishes healthy relationships on a finer scale. Cells must divide, grow, migrate, and fill gaps. They must reorganize to form properly connected networks of fine blood vessels. And cells must lay down materials to form the structural, intercellular matrix-collagen to provide the proper shape and toughness, or mineral grains to provide rigidity, as in bone. Often, they lay down unwanted scar tissue instead, blocking proper healing. With enough knowledge of how these processes work (and nanoinstruments can help gather that knowledge) and with good enough software to guide the process—a more difficult challenge-medical nanomachines will be able to guide this healing process. The problem here is to guide the motion and behavior of a mob of active, living cells—a process that can be termed *cell herding*. Cells respond to a host of signals from their environment: to chemicals in the surrounding fluids, to signal molecules on neighboring cells, and to mechanical forces applied to them.

Cell-herding devices would use these signals to spur cell division where it is needed and to discourage it where it is not. They would nudge cells to encourage them to migrate in appropriate directions, or would simply pick them up, move them along, and deliver them where needed, encouraging them to nestle into a proper relationship with their neighbors. Finally, they would stimulate cells to surround themselves with the proper intercellular-matrix materials. Or—like the owner of a small dog who, on a cold day, wraps the beast in a wool jacket—they would directly build the proper surrounding structures for the cell in its new location. In this way, cooperating teams of cell-herding devices could guide the healing or restructuring of tissues, ensuring that their cells form healthy patterns and a healthy matrix and that those tissues have a healthy relationship to their surroundings.

What is Dry Nanotechnology?

Derives from surface science and physical chemistry, focuses on fabrication of structures in carbon (e.g. fullerenes and nanotubes), silicon, and other inorganic materials. Unlike the "wet" technology, "dry" techniques admit use of metals and semiconductors. The active conduction electrons of these materials make them too reactive to operate in a "wet" environment, but these same electrons provide the physical properties that make "dry" nanostructures promising as electronic, magnetic, and optical devices. Another objective is to develop "dry" structures that possess some of the same attributes of the self-assembly that the wet ones exhibit. [Rice University]

What is DNA Tweezers?

Dr. Bernard Yurke and colleagues made an artificial DNA based molecular machine that also accepted DNA as a fuel. The machine, called DNA tweezers, consisted of three strands of DNA labeled A, B and C. Strands B and C are partially hybridized on to the central strand A with overhangs on both ends. This conformation of the machine is the open conformation. When an auxiliary fuel strand F is introduced, that is designed to hybridize with both overhang regions, the machine attains a closed conformation. The fuel strand is then removed from the system by the introduction of its exact complement, leaving the system to go back to its original open conformation. This way, a reversible motion is produced, which can be observed by attaching fluorescent tags to the two ends of the strand A. In this case the 5′ end was labeled with the dye TET (tetrachloro-fluorescein phosphoramidite) and the 3′ end was labeled with TAMRA (carboxy-tertamethylrhodamine). Aside from the creation of a completely new molecular machine, this showed a way of selective fueling of such machines. The fuel strands are sequence-specific, so they will work on only those machines towards which they are directed, and not trigger other machines surrounding them. This machine was later improved to form a Three-State Device which had two robust states and one flexible intermediate state. A variation of the tweezers came about as the DNA-scissors.

What is Aluminum Nanopositioning Stages?

Nanopositioning stages are constructed from a variety of materials to take advantage of their thermal characteristics. While aluminum alloys are inexpensive and easily machined, they exhibit a fairly high coefficient of thermal expansion (2.3×10^{-5} mm/mm per °C). If a positioning application requires nanometer level resolution, temperature changes and time become important factors. In general, if the entire measurement or procedure can be completed in a "short" timeframe (i.e. minutes vs. hours) then average laboratory heating/ cooling systems provide adequate temperature control and aluminum nanopositioners can be used.

What is Clay-Nylon Nanocomposites?

The earliest example of the in situ polymerization method was work by Toyota on synthesis of clay-nylon nanocomposites and this remains probably the most studied system, including work by Bayer and Ube. In a typical synthesis ADA-modified clay is dispersed in the monomer caprolactam, which is polymerized to form the nylon-6-clay hybrid as an exfoliated composite. Complete exfoliation may be preceded by intercalation of the monomer in the clay. Generally, low concentration of clay (a few per cent) are incorporated in these nanocomposites, partly because this is often sufficient to modify the desired properties significantly, but also because higher levels of clay can adversely increase the system viscosity leading to poor processability, although the viscosity increase is shear rate dependent. Other nylons and copolyamides (e.g. nylon-6/6,6) have also been incorporated in clay nanocomposites. Functionality such as hydroxyl groups can be introduced into the onium salt modifiers to improve compatibility with the nylon via hydrogen bonding - this can lead to an enhancement of desirable nanocomposite properties. Similar modification of the chemistry may be required for successful exfoliation with other types of polymer system. In the case of ethylene-vinyl alcohol (EVOH) copolymers, use of hydroxylated quaternary ammonium ions improves compatibility between the clay and the EVOH by introducing favourable hydroxyl group interactions. In polypropylene (PP) nanocomposites, maleic anhydride grafted PP is used as a compatibiliser. Polymerization initiators can be anchored to the clay platelet surface and this approach has been extended tp living free radical polymerization of styrene, where an initiating species bonded to the TEMPO free radical is attached to the surface of the clay platelets via ion exchange.

What is DNA Probes?

A DNA or nucleic acid probe is a short strand of DNA that locates and binds to its complementary sequence in samples containing single strands of DNA or RNA enabling identification of specific sequences. Nucleic acid probe assays exploit the fundamental hybridization reaction that occurs spontaneously between two complementary DNA:DNA or DNA:RNA strands. As in immunoassays, detection of the hybrid requires that the probe be labeled. Various direct and indirect methods have been devised for the detection of the hybrid. Direct labeling involves attaching the label directly to the probe sequence; indirect labeling binds an antibody to the DNA:DNA or DNA:RNA hybrid. As in immunoassays, non-isotopically-labeled probes are preferred over radio-labeled probes primarily because of radiation hazards, disposal problems, and short reagent shelf life. In addition, the factors determining the detection limits of hybridization assays based on labeled probes

are similar to those in immunoassays. Therefore, the development of a simple, inexpensive and sensitive direct detection system which eliminates the use of labels is highly desirable.

What is Integrated Drug Discovery?

Drug discovery today faces more difficult hurdles than ever before. More money, and resources are delivering fewer clinical candidates and drugs on the market than a decade ago, and high- throughput screening and combinatorial chemistry are perceived as not having lived up to their promises. Today however, discovery departments are being tightly integrated through cheminformatics and discovery informatics. These integrated discovery efforts, together with targeted and directed high- throughput synthesis and screening efforts, are changing the impact of these high- throughput disciplines and changing the nature of the drug discovery process. Cheminformatics, machine learning and data mining methods are being directed toward the simultaneous and multivariate optimization of potency, efficacy, selectivity and more importantly, ADMET and drug- like properties, with significant impact. Cheminformatics, Data Mining and Virtual Screening in Integrated Drug Discovery, Charles Manly, Ph.D., V.P., Discovery Technologies, Neurogen.

What is Catenanes?

The catenanes are also special type of interlocked structures that represent a growing family of molecular machines. They are synthesized by supramolecular assistance to molecular synthesis. The general structure of a catenane is that of two interlocked ring-like components that are non-covalently linked via a mechanical bond, that is, they are held together without any valence forces. Both the macrocyclic components have recognition sites that are atoms of groups that are redox-active or photochemically reactive. It is possible to have both rings with similar recognition sites. In such a scenario, one of the rings may rotate inside the other with the conformations stabilized by noncovalent interactions, but the two states of the inner ring differing by 180° will be undistinguishable (degenerate). For better control and distinguishable molecular conformations, it is desirable to have different recognition sites within the macrocycles. Then they can be controlled independently through their own specific stimuli. The stereo-electronic property of one recognition site within a macrocycle can be varied such that at one point it has more affinity to the sites on the other ring. At this instant, the force balance will guide the rotating macrocycle for a stable conformation that requires that particular site to be inside the other macrocycle. Similarly, with other stimulus, this affinity can be turned off, or even reversed along with the affinity of the second recognition site on the rotating macrocycle increased towards those on the static one. There is a need for computational modeling, simulation and analysis of such molecular machine motion. Like rotaxanes, catenanes also can be designed for chemical, photochemical or electrochemical control.

What is Ciliary Distribution Subsystem?

The interior volume of the vasculoid is lined with a sapphire surface (comprised of watertight adjacent installed plates) from which protrude trillions of mechanical cilia (arranged in patterns designed to maximize transport speed and reliability. A secondary role is to assist the vasculocytes (small mobile nanorobots) in cleaning up after accidental internal spills, component malfunctions, or other pathological events. Even by 1997, prototype MEMS ciliary arrays of up to 1024 cilia had already demonstrated transport speeds up to 200 microns/sec with ~3 micron positional accuracy.

What is Hydrogen Storage?

Studies of H_2 absorption and packing in carbon nanotubes and nanoropes are in progress at NASA Ames and elsewhere. Nanotubes provide large pore sizes and nanoropes have different pore sizes depending on interstitial and other locations. Dillon estimated that the single walled nanotubes in their sample contained 5 to 10 per cent by weight of H_2. The nanotubes were about 0.1 to 0.2 per cent by weight of the total sample. Computational studies at Ames suggest that to store 7-10 per cent H_2 in single walled nanotubes at room temperature the H_2s must be stored inside the tubes, not merely absorbed on the walls. This work suggests that carbon nanotubes might be developed into an excellent H_2 storage medium within 3-5 years

What is Biomedical Nanotechnology?

The comprehensive monitoring, control, construction, repair, defense, and improvement of all human biological systems, working from the molecular level, using engineered nanodevices and nanostructures; the science and technology of diagnosing, treating, and preventing disease and traumatic injury, of relieving pain, and of preserving and improving human health, using molecular tools and molecular knowledge of the human body; the employment of molecular machine systems to address medical problems, using molecular knowledge to maintain and improve human health at the molecular scale. A rapidly expanding field that includes many potential technologies and approaches. The key to this definition is that phenomena and materials at the nanometer scale are known to have properties that are uniquely attributable to that scale length. Nanomedicine could similarly be defined [as nanotechnology] as the design, synthesis, or application of materials, devices, or technologies in the nanometer-scale for the basic understanding, diagnosis, and/or treatment of disease. Canadian Institute of Health Research, Regenerative Medicine and Nanomedicine, RFA, 2003. The monitoring, repair, construction and control of human biological systems at the molecular level, using engineered nanodevices and nanostructures.

What is Stem Cell Research?

Stem cell research is a relatively new technology that uses human embryo cells to treat disease. The embryo cells are called "stem cells" because they can be stimulated to develop into any type of body cell or system. Human embryo cells are especially valuable because they can be used to treat and possibly cure many diseases such as heart disease, diabetes, and Alzheimers. Stem cells have been identified in placental tissue, blood, and fat but these sources are limited in their efficacy. Embryonic stem cells have much more potential and are harvested when the embryos are no more than a few days old. The process destroys the embryos. The biggest source of human embryos is those left over from in-vitro fertilization procedures. Fertility treatments usually result in multiple embryos. Since only the most viable are selected for implantation, the remaining embryos are frozen. Costs are incurred with the cryogenic (freezing) process and with ongoing maintenance. There is a great deal of controversy over the ethics of stem cell research. Some say that life begins at conception and that the use of humans (even immature ones) for research purposes is immoral. Others claim that the embryos are only tiny amounts of undifferentiated tissue and since they are already scheduled for destruction, they should be used to help others. Adoption programs have resulted in the use of some of these excess embryos and live births have occurred. Since these leftover embryos are typically less viable, they may not result in healthy children. Stem cell research using these leftover embryos or embryos created solely for scientific purposes is currently banned in the United States. As an alternative, the government has set up a Federal Registry of 60 stem cell "lines"—cells whose initial embryos have already been destroyed. Not all of the lines have been approved by the National Institutes of Health (NIH) and scientists have complained that their quality could be inferior. Federal funding is available only for research using these lines. Other countries have embraced the idea and embryonic stem cell research is currently being conducted all over the world. Vast financial rewards await the developers of viable disease treatments. Recent polls have shown increased public support for the idea.

What is Laser Scanning Microscopy?

There are two major forms of laser scanning microscopy, namely confocal laser scanning microscopy (CLSM) and multiphoton laser scanning microscopy (MPLSM). The two forms are very similar at the illumination side (as opposed to the detection side). ... MPLSM is more sensitive that CLSM because all the light generated to make an image is sent directly to the photon multiplier tube. This contrasts with CLSM where a pin hole is required to select the light from the focal plane. In CLSM there is considerable loss of signal in the optics required to direct the light to the pin hole. MPLSM gives a sharper image than CLSM because of the lack of extraneous light and improved geometry of detection. In MPLSM the photon multiplier tube can be placed very close to the specimen whereas CLSM has all the intervening optics and the pin hole.

What is Kinesin Linear Motor?

Kinesin and Dynein family of proteins are involved in cellular cargo transport along microtubules as opposed to actin in the case of myosin. Microtubules are 25 nm diameter tubes made of protein tubulin and are present in the cells in an organized manner. Microtubules have polarity; one end being the plus (fast growing) end while the other end is the minus (slow growing) end. Kinesins move from minus end to plus end, while dyneins move from plus end to the minus end of the microtubules. Microtubule arrangement varies in different cell systems. In nerve axons, they are arranged longitudinally in such a manner that their plus ends point away from the cell body and into the axon. In epithelial cells, their plus end points towards the basement membrane. They deviate radially out of the cell center in fibroblasts and macrophages with the plus end protruding outwards. Like myosin, kinesin is also an ATP-driven motor. One unique characteristic of kinesin family of proteins is their processivity—they bind to microtubules and literally 'walk' on it for many enzymatic cycles before detaching. Also, each of the globular heads/motor domains of kinesin is made of one single polypeptide unlike myosin (heavy and light chains and dynein heavy, intermediate and light chains).

What is Disruptive Technologies?

Some technologies are improved in a linear fashion or incrementally. Others truly change the paradigm. Clayton Christensen writes about these in *The Innovator's Dilemma*. What is particularly interesting about Christensen's analysis (based on data from the disk drive industry) is that he found disruptive technologies tended to be much cheaper than existing technologies. Existing companies were quite capable of developing the technologies (and had). What they couldn't do was figure out how to market them and whether it made sense to devote sufficient resources to them (which in many cases would not have been the responsible thing to do.)

What is Finite Fermion Systems?

Low-dimensional nanometer-sized systems have defined a new research area in condensed-matter physics within the last 20 years. Modern semiconductor processing techniques allowed the artificial creation of quantum confinement of only a few electrons. Such finite fermion systems have much in common with atoms, yet they are man-made structures, designed and fabricated in the laboratory. Usually they are called "quantum dots", referring to their quantum confinement in all three spatial dimensions. A common way to fabricate quantum dots is to restrict the two-dimensional electron gas in a semiconductor heterostructure laterally by electrostatic gates, or vertically by etching techniques. This creates a bowl-like potential in which the conduction electrons are trapped. In addition to the many possible technological applications, what makes the study of these "artificial atoms" or "designer atoms" interesting are the far-reaching analogies to systems that exist in nature and have defined paradigms of many-body physics: atoms, nuclei, and, more recently, metallic clusters or trapped atomic gases. Quantum dots added another such paradigm. Their properties can be changed in a controlled way by electrostatic gates, changes in the dot geometry, or applied magnetic fields. Their technological realization gave access to quantum effects in finite low- dimensional systems that were largely unexplored. After the initial success in the fabrication and control of mesoscopic semiconductor structures, which are typically about one hundred nanometers in size and confine several hundred electrons, many groups focused on the further miniaturization of such devices. A breakthrough to the "atomic" regime was achieved with the experimental discovery of shell structure in fluctuations of the charging energy spectra of small, vertical quantum dots: the borderline between the physics of bulk condensed matter and few-body quantum systems was crossed. Much of the many-body physics that was developed for the understanding of atoms or nuclei could be applied. In turn, measurements on artificial atoms yielded a wealth of data from which a fundamental insight into the many-body physics of low-dimensional finite fermion systems was obtained. With further progress in experimental techniques, artificial atoms will continue to be a rich source of information on many-body physics and undoubtedly will hold a few surprises."

What is Dip Pen Nanolithography?

A method for nanoscale patterning of surfaces by the transfer of a material from the tip of a scanning probe microscope onto the surface. Dip Pen Nanolithography (DPN) is a method developed at Northwestern University for depositing thin organic films in patterns with feature sizes of around 10 nm (about 20 times better than the best optical lithography). By coating an Atomic Force Microscope (AFM) tip with an ink, the scientists are able to deposit well defined lines of the ink in a manner similar to a traditional ink pen. While such molecular scale precision is quite a feat, similar precision has been achieved with other methods. However, this particular method for patterned film deposition can be operated in parallel because the deposition is independent of the applied force. New AFM-based soft-lithography technique which was invented in our labs. Dip-Pen Nanolithography (DPN) is a scanning probe nanopatterning technique in which an AFM tip is used to deliver molecules to a surface via a solvent meniscus, which naturally forms in the ambient atmosphere. This direct- write technique offers high- resolution patterning capabilities for a number of molecular and biomolecular 'inks' on a variety of substrates Chad Mirkin group, Dip Pen Nanolithography, Dept of Chemistry, Northwestern Univ., US.

What is Fluorophore?

The categories of greatest need [in single molecule studies] are improving the photophysical properties of fluorophores used for single molecule spectroscopy There is a need for synthesis of probes with desirable spectral and luminescent characteristics, such as small size, high quantum yield, high extinction, reduced photobleaching, blinking, and photoisomerization. The best probes will be compatible with conditions inside the cell and will move freely in the cell. Emerging technologies have made use of silicon and lanthanide nanocrystals (Quantum dots), which emit enough photons to be detected at very low concentrations, plasmon and Raman probes, and G/C/Y/R-fluorescent proteins, but there is still much to be done to optimize these probes. High throughput screening and combinatorial approaches need to be applied to this problem.

What is Cerium Oxide?

There are numerous commercial applications for cerium including metallurgy, glass and glass polishing, ceramics, phosphors and catalysts. In catalysis, cerium is used in the form of cerium (IV) oxide, CeO_2. Cerium oxide is a highly stable, nontoxic, refractory ceramic material with a melting point of 2600°C and a density of 7.13 g.cm^3. The crystal structure is fluorite face centred cubic with a lattice constant of 5.11Å. The rare earth oxides are generally some of the most thermally stable of all known materials and as such may be used in extremely high temperature applications without decomposition of the oxide.

What is Chemical Synthesis?

Both metals and ceramics can be produced using a variety of chemical approaches such as sol-gel or thermal decomposition. These methods provide large quantities of nano-sized agglomerates at a low cost. Chemical processes also allow effective control of the stoichiometry of the final product. However, the precursor chemicals required may leave residual contamination on the particle surfaces, which leads to difficulties in compaction and sintering. Additionally, powders produced through wet chemical techniques often have difficulties with agglomeration. Other methods which implement chemical reactions include Examples: A common method of producing nanostructured powders is through mechanical deformation, i.e. milling or shock deformation. These processes produce nanostructured materials through severe mechanical deformation of a coarse grained precursor material. The nanometer sized grains nucleate within the dislocation cell structures located in the shear bands. The final grain size is a function of the amount of energy input during the milling as well as time, temperature during milling and milling atmosphere. The precursor materials can be crystalline or amorphous materials in the form of powders or tapes. Through mechanical milling, it is possible to produce nanostructured powders of systems which are otherwise immicsible. A major disadvantage of this method is the possibility of contamination from the milling media due to the large forces and energies involved in the milling process. Other synthesis methods based on *mechanical deformation:* friciton/sliding wear. Examples: Three dimensional nanostructured materials are also synthesized through thermal crystallization of an amorphous material. By controlling the nucleation and growth during annealing of an amorphous material, one can produce a bulk material with average grain sizes of less than 20 nm without the need for consolidation and sintering steps. However, this process is limited to material composition which are readily available in the form of a metallic glass, which has an amorphous microstructure.

What is Fluidic Self Assembly?

A novel technique for accurately assembling large numbers of very small devices. The small size, planarity, and accuracy of the assembly also result in very low parasitic interconnects, comparable to on die traces. This massively parallel assembly process combines the capability and flexibility of assembly with the cost effectiveness of integration. Invented by Mr. Mark Hadley and was part of his Ph.D. dissertation while he was studying at University of California, Berkley. The FSA process became the foundation for the origins of a new company named Alien Technology Corporation. In the FSA process, specifically shaped semiconductor devices ranging in size from 10 microns to several hundred microns are suspended in liquid and flowed over a surface which has correspondingly shaped "holes" or receptors on it and into which the devices settle. The shape of the devices and of the holes is designed so that the devices fall easily into place and are selfaligning. Alien has successfully demonstrated the assembly of tens of thousands of devices in a single process step.

What is Flagella Motors?

Unicellular organisms, such as, E. coli have an interesting mode of motility (for reviews see. They have a number of molecular motors, about 45 nm in diameter, that drive their 'feet' or the flagella that help the cell to swim. Motility is critical for cells, as they often have to travel from a less favorable to a more favorable environment. The flagella are helical filaments that extend out of the cell into the medium and perform a function analogous to what the oars perform to a boat. The flagella and the motor assembly are called a flagellum. The flagella motors impart a rotary motion into the flagella. In addition to a rotary mechanism, the flagella machines consist of components such as rate meters, particle counters, and gearboxes. These are necessary to help the cell decide which way to go, depending on the change of concentration of nutrients in the surroundings. The rotary motion imparted to the flagella needs to be modulated to ensure the cell is moving in the proper direction as well as all flagella of the given cell are providing a concerted effort towards it. When the motors rotate the flagella in a counterclockwise direction as viewed along the flagella filament from outside, the helical flagella create a wave away from the cell body. Adjacent flagella subsequently intertwine in a propulsive corkscrew manner and propel the bacteria. When the motors rotate clockwise, the flagella fly apart, causing the bacteria to tumble, or change its direction. These reversals occur randomly, giving the bacterium a 'random walk', unless of course, there is a preferential direction of motility due to reasons mentioned earlier. The flagella motors allow the bacteria to move at speeds of as much as 25 mm/s with directional reversals occurring approximately 1 per second. A number of bacterial species in addition to E. coli., depend on flagella motors for motility. Some of these are Salmonella enterica serovar Typhimurium (Salmonella), Streptococcus, Vibrio spp., Caulobacter, Leptospira, Aquaspirrilum serpens and Bacillus. The rotation of flagella motors is stimulated by a flow of ions through them which is a result of a build-up of a transmembrane ion gradient. There is no direct ATP-involvement; however the proton gradient needed for the functioning of flagella motors can be produced by ATPase.

What is Anechoic Chamber?

Also called an "echo free" chamber. A room at the David Florida Laboratory where satellites are tested to make sure they are sending and receiving clear signals. Anechoic chambers duplicate the silence of space, where there is no medium (like air or water) through which the sound waves can travel. The walls of the anechoic chambers are made of carbon compound pyramids which absorb any microwave and radio signals that hit them.

What is Electronic Tongues?

For liquid analysis, based on the organizational principles of biological sensory systems, developed rapidly during the last decade. ... The exciting possibility of establishing a correlation between the output from an electronic tongue and human sensory assessment of food flavour, thereby enabling quantification of taste and flavour, is described. Application areas of electronic tongue systems including foodstuffs, clinical, industrial, and environmental analysis are discussed in depth.

What is Zener Diodes?

Diodes that can be made to conduct backwards. This effect, called Zener Breakdown, occurs at a precisely defined voltage, allowing the diode to be used as a precision voltage reference. Some devices labelled as high-voltage Zener diodes are actually avalanche diodes. Two (equivalent) Zeners in series and in reverse order, in the same package, constitute a transient absorber (or *Transorb*, a registered trademark). They are named for Dr. Clarence Melvin Zener of Southern Illinois University, inventor of the device. Semiconductor diode that has a well-defined turn-on voltage for conduction in the reverse direction.

Appendix E

Superconductor Terminology and Naming Schemes

Anneal: To heat and then slowly cool a material to reduce brittleness.
Annealing of ceramic superconductors usually follows sintering and is done in an oxygen-rich atmosphere to restore oxygen lost during calcination. The oxygen content of a ceramic superconductor is critical. For example, YBCO with 6.4 atoms of oxygen will *not* superconduct. But YBCO with 6.5 atoms *will*.

Anti-ferromagnetism: A state of matter where adjacent ions in a material are aligned in opposite or "anti-parallel" arrays. Such materials display almost no response to an external magnetic field at low temperatures and only a weak attaction at higher temperatures. There is evidence that anti-ferromagnetism in the cuprate oxides plays a role in the formation of Cooper pairs and, thus, in facilitating a superconductive state in some compounds.

BCS Theory: The first widely-accepted theory to explain superconductivity put forth in 1957 by John Bardeen, Leon Cooper, and John Schreiffer. The theory asserts that, as electrons pass through a crystal lattice, the lattice deforms inward towards the electrons generating sound packets known as "phonons". These phonons produce a trough of positive charge in the area of deformation that assists subsequent electrons in passing through the same region in a process known as phonon-mediated coupling. This is analogous to rolling a bowling ball up the middle of a bed. 2 people, one lying on each side of the bed, will tend to roll toward the center of the bed, once the ball has created a depression in the mattress. And, a 2nd bowling ball, placed at the foot of the bed, will now, quite easily, roll toward the middle.

Borocarbides: Superconducting borocarbides are compounds containing both boron and carbon in combination with rare-earth and transition elements; some of which exhibit the unusual ability to return to a normal, non-superconductive state at temperatures below Tc.

BSCCO: An acronym for a ceramic superconductor system containing the elements Bismuth, Strontium, Calcium, Copper and Oxygen. Typically, a small amount of lead is also included in these compounds to promote the highest possible Tc. BSCCO has probably found the widest acceptance among high-Tc superconductor applications due to its unique properties. BSCCO compounds exhibit both an intrinsic Josephson Effect and anisotropic (directional) behavior.

Ceramics: Ceramic superconductors are inorganic compounds formed by reacting a metal with oxygen, nitrogen, carbon or silicon. The best-known of these are the copper-perovskites. Ceramics are typically hard, brittle, heat-resistant materials formed by a process known as solid-state reaction.

Charge Reservoirs: In superconductors, charge reservoirs are the layers that may control the oxidation state of adjacent superconducting planes (even though they themselves are not superconducting). In the layered cuprates, these consist of copper-oxide chains.

Chevrel (phases): A class of molybdenum chalcogenides (compounds containing Group VI elements S, Se or Te along with molybdenum and a positively charged metal ion) - named for Roger Chevrel of the University of Rennes, whose research brought them to the attention of the scientific community in the early 1970's. Recently, the Superconductivity Group at the University of Durham (UK) reported a novel fabrication technique that increases Bc2 (upper critical field) in the chevrel $PbMo_6S_8$ from 50 T in bulk materials up to > 100 T.

Coherence Length: The size of a cooper pair - representing the shortest distance over which superconductivity can be established in a material. This is typically on the order of 1000Å; although it can be as small as 30Å in the copper oxides.

Cooper Pair: Two electrons that appear to "team up" in accordance with theory—BCS or other—despite the fact that they both have a negative charge and normally repel each other. (Named for Leon Cooper.) Below the superconducting transition temperature, paired electrons form a condensate - a macroscopically occupied single quantum state - which flows without resistance. However, since only a small fraction of the electrons are paired, the bulk does not qualify as being a "bose-einstein condensate".

DAC: An acronym for "diamond anvil cell". Often the Tc of a superconductor can be coaxed upward with the application of high pressure. The DAC is used to accomplish this in the laboratory. A DAC is composed of 2 specially-cut diamonds and a stainless steel gasket. The gasket goes between the diamonds and seals a small chamber in which a fluid is placed. Since neither the diamonds nor the liquid will compress, hydrostatic forces in excess of a million atmospheres can be brought to bear on a sample suspended within the fluid.

Diamagnetism: The ability of a material to repel a magnetic field. Many naturally-occurring substances (like water, wood and paraffin, and many of the elements) exhibit weak diamagnetism. Superconductors exhibit strong diamagnetism below Tc. In a few rare compounds, a material may become superconductive at a higher temperature than the point at which diamagnetism appears. But, as a rule, the onset of strong diamagnetism is one of the most reliable ways to ascertain when a material has become superconductive.

D-Wave: A form of electron pairing in which the electrons travel together in orbits resembling a four-leaf clover. Wave functions help theoreticians describe (and predict) electron behavior. The d-wave models have gained substantial support recently over s-wave pairing as the mechanism by which high-temperature superconductivity might be explained.

Energy Gap: This is the energy required to break up a pair of electrons. According to BCS theory, the formula for determining the energy gap (in meV) is $E_g=7/2\ KT_c$. Where K = Boltzmann's constant (8.62e-5 eV/K). And where T_c is the critical transition temperature in Kelvin. Since electron-pairing is universally agreed to be the method by which superconductivity occurs, this is the amount of energy required to disrupt the superconducting state.

ESR: An acronym for "Electron Spin Resonance" (also EPR: Electron Paramagnetic Resonance). This is another mechanism by which superconductivity might be explained in some materials. Simply put, ESR is the response of electrons to electromagnetic radiation or magnetic fields at discrete frequencies. Electrons, as they move, create tiny magnetic moments. Nearby electrons are influenced either beneficially or adversely. When the moments are complementary, the electrons become paired and can help each other move through a crystal lattice.

Ferrite: Ferrites are ceramics with magnetic properties. They are included on this page because

many of the same elements used in ferrites (e.g. Ba, Sr, Tm, O) are also key constituents in ceramic superconductors. This may be an important clue in understanding high-temperature superconductivity.

Ferromagnetism: A state wherein a material exhibits magnetization through the alignment of internal ions (neighboring magnetic moments). This contrasts with paramagnetism, which is temporary, much weaker and results from unpaired electrons.

Flux-Lattice: A configuration created when flux lines from a strong magnetic field try to penetrate the surface of a Type 2 superconductor. The tiny magnetic moments within each resulting vortex repel each other and a periodic lattice results as they array themselves in an orderly fashion.

Fluxon: The smallest magnetic flux (flux quantum) that exists in nature. Just as electrons are quantized charge, fluxons are a quantized flux. The term is used in association with vortices, which result from magnetic fields penetrating Type 2 superconductors in single fluxon quanta.

Flux-Pinning: The phenomenon where a magnet's lines of force (called flux) become trapped or "pinned" inside a superconducting material. This pinning binds the superconductor to the magnet at a fixed distance. Flux-pinning is only possible when there are defects in the crystalline structure of the superconductor (usually resulting from grain boundaries or impurities). Flux-pinning is desirable in high-temperature ceramic superconductors in order to prevent "flux-creep", which can create a pseudo-resistance and depress Jc and Hc.

Four-point Probe: The most common method of determining the Tc of a superconductor. Wires are attached to a material at four points with a conductive adhesive. Through two of these points a voltage is applied and, if the material is conductive, a current will flow. Then, if any resistance exists in the material, a voltage will appear across the other two points in accordance with Ohm's law (voltage equals current times resistance). When the material enters a superconductive state, its resistance drops to zero and no voltage appears across the second set of points. By using the four-point method, instead of just two points, resistance in the adhesive and wires can be ignored; as the second set of points do not themselves conduct any current and can, therefore, only reflect what voltage exists across the body of the material.

Hall Effect: When a magnetic field is applied perpendicularly to a thin metal film or semiconductor film that is conducting an electric current, a small voltage will appear perpendicular to the axis of both the film and the magnetic field. This voltage is proportional to the strength of the applied field. However, the output is typically not linear. The Hall resistance (the ratio of the Hall voltage to the current) changes in steps, pursuant to the laws of quantum mechanics. This is known as the Integral Quantum Hall Effect or just Quantum Hall Effect. Discovered in 1879, the Hall effect was named for its discoverer Edwin H. Hall, a graduate student at Johns Hopkins University.

Hc: The scientific notation representing the "critical field" or maximum magnetic field that a superconductor can endure before it is "quenched" and returns to a non-superconducting state. Usually a higher Tc also brings a higher Hc.

Heavy Fermions: Compounds containing the elements cerium, ytterbium or uranium; whose (inner shell) conduction electrons often have effective masses (called quasiparticle masses) several hundred times as great as that of a "free" (normal) electron mass. This gives them what's known as a low "Fermi energy" and makes them unlikely—and unusual—superconductors. Research suggests cooper-pairing in heavy fermion systems arises from the magnetic interactions of the electron spins.

Hole: A positively-charged vacancy within a crystal lattice resulting from the shortage of an electron

in that region. Holes are typically induced by doping a material with an impurity. However, they can also be synthesized electronically with devices like the field-effect transistor (FET). Modern electronic devices rely heavily on holes (as p-type semiconductors) to function. There is evidence that the holes of hypocharged oxygen in charge-reservoirs are, in fact, what makes possible high-temperature superconductivity in the layered cuprates.

HTS: An acronym for "High-Temperature Superconductor" (or Superconductivity). There is no widely-accepted temperature that separates HTS from LTS (Low-Temperature Superconductors). However, all the superconductors known before the 1986 discovery of the superconducting oxocuprates would be classified LTS. The barium-lanthanum-cuprate fabricated by Müller and Bednorz, with a Tc of 30K, is generally considered to be the first HTS material. Certainly any compound that will superconduct above the boiling point of liquid nitrogen (77K) would be HTS.

Hysteresis (loop): Hysteresis, as it applies to a superconductor, relates to the dynamic response of a superconductor to a strong magnetic field impinged upon it. As the strength of a nearby magnetic field (H) increases, the critical transition temperature (Tc) of a superconductor will decrease. And, at some point superconductivity will completely disappear, as it becomes "quenched". However, as the magnetic field is gradually withdrawn, the superconductor may *NOT* immediately return to a superconductive state. Herein lies the hysteresis. The graph of H-vs-Tc is different retreating than it is advancing (creating a "loop" shape). This fact must be weighed carefully in high-current applications where the superconductor Hc may, even briefly, be exceeded; as significant power losses can result.

Infinite layer: Infinite layer compounds have no clear separation between molecules. Rather than electrostatic bonding between discrete molecules to form a bulk crystalline aggregate, all the atoms are bound together by covalent or co-ionic bonding to form the equivalent of one huge molecule. $(Ba,Sr)CuO_2$ and $Na_2Ba_6Si_{46}$ are examples of "infinite layer" or "infinite network" superconductor compounds.

Isotope Effect: The influence atomic mass contributes to the critical transition temperature of a superconductor. For example, $^{203.4}Hg$ has a Tc of 4.126K. While ^{198}Hg has a Tc of 4.177K. Since both forms of mercury have the same lattice structure, this difference in Tc can be attributed solely to the difference in mass.

Jc: The scientific notation representing the "critical current density" or maximum current that a superconductor can carry. Also note that, as the current flowing through a superconductor increases, the Tc will usually decrease.

Josephson Effect (also DC Josephson Effect): A phenomenon named for Cambridge graduate student Brian Josephson, who predicted that electrons would "tunnel" through a narrow (<10 angstroms) non-superconducting region, even in the absence of an external voltage. In a normal conductor, electrical current only flows when there's a voltage differential and contiguous electrical connection. It has been theorized that the Josephson Effect arises from the incoherent phase relationships between superconducting electrons in the two (separated) superconductors. The AC Josephson Effect is where the current flow oscillates as an external magnetic field impinged upon it increases beyond a critical value. [at a frequency of 2eV/h, where e is the electron charge, V is the voltage that appears, and h is Planck's constant] Sidebar: This oscillation frequency has, in fact, resulted in an upward revision of Planck's constant from 6.62559e-34 to 6.626196e-34.

Josephson Junction: A thin layer of insulating material sandwiched between 2 superconducting layers. Electrons "tunnel" through this non-superconducting region in what is known as the

"Josephson effect" (see above). Sidebar: The standard volt is now defined as the voltage required to produce a frequency of 483,597.9 GHz in a Josephson Junction.

Kelvin: A scale of temperature measurement that starts at "absolute zero", the coldest theoretical temperature attainable. (Named for Lord William Thomson Kelvin.)

Meissner Effect: Exhibiting diamagnetic properties to the total exclusion of all magnetic fields. (Named for Walter Meissner.) This is a classic hallmark of superconductivity and can actually be used to levitate a strong rare-earth magnet.

Mott Transition: The high- temperature copper oxides are composed of CuO_2 planes that are separated from each other by ionic "blocking layers". Although it has one conduction electron (or hole) per Cu site, each CuO_2 plane is originally insulating because of the large electron correlation. That behavior is typical of the Mott insulator state, in which all the conduction electrons are tied to the atomic sites. The superconducting state emerges when holes from the blocking layers dope the CuO_2 layers in a way that alters the number of conduction electrons and triggers the Mott transition. Researchers believe that the strong antiferromagnetic correlation, which originates in the Mott-insulating CuO_2 sheets and persists into the metallic state, [could be] a possible mechanism of high-temperature superconductivity.

Organics: Organic superconductors are a sub-class of organic conductors that include molecular salts, polymers and pure carbon systems (including carbon nanotubes and C_{60} compounds). They may also be referred to as "molecular" superconductors. They are typically large, carbon-based molecules of 20 or more atoms, consisting of a planar organic molecule and a non-organic anion.

Penetration Depth (also London Penetration Depth): This term relates to how deeply a magnetic field will penetrate the surface of a superconductor. An external magnetic field impinged upon a Type 2 superconductor will decay exponentially into the surface based on the paired electron density within the superconductor (only a small fraction of the electrons are in a superconductive state). The "London" name comes from brothers F. London and H. London, who in 1935 created a theoretical model of superconductivity.

Perovskites: A large family of crystalline ceramics that derive their name from a mineral known as a perovskite. They are the most abundant minerals on earth and have a metal-to-oxygen ratio of approximately 2-to-3. Copper-oxide superconductors are layered perovskites. The perovskite name comes from Russian mineralogist Count Lev Aleksevich von Perovski.

Phase-Slip (also Quantum Phase-Slip): A point where a material in a superconductive state spontaneously changes from one state to another, generating a topological "defect". This defect causes paired electrons to become "out of step" with each other, producing a voltage and, ergo, non-zero electrical resistance. This phenomenon has been observed in ultra-thin wires less than a few tens-of-nanometers in diameter. Though bulk superconductivity may persist ($T<T_c$), one consequence of phase slip is a lower current-carrying state. A similar phenomenon occurs in Josephson Junctions.

Proximity Effect: The phenomenon where a thin film of non-superconductive material in close proximity with a superconductor takes on superconductive properties. The Josephson junction is a device that takes advantage of this phenomenon. The Inverse Proximity Effect is where just the opposite occurs. A non-superconductive metal can enhance the Tc of an adjacent superconductor. This inverse effect has been observed with silver and lead.

P-Wave: A rare form of electron pairing in which two electrons travel together in spherical orbits; with both having the same direction of rotation. (See "D-Wave" explanation above.)

Quasiparticle: A bare particle that is "dressed" or "clothed" by a cloud of other surrounding particles.

Quasiparticles behave similarly to bare (normal) particles, but usually have a larger effective mass due to this cloud moderating interactions with other particles.

Quench: The phenomenon where superconductivity in a material is suppressed; usually by exceeding the maximum current the material can conduct (Jc) or the maximum magnetic field it can withstand (Hc).

Re-entrant (behavior): A condition where a material retreats from its superconductive state and then re-enters it. This can be caused by a strong external magnetic field that dynamically exceeds the Hc of the material and/or is mis-aligned (in the case of some organic superconductors), a discordant temperature below Tc (in the case of some borocarbides), or by Jc hysteresis (momentarily exceeding the critical current density, causing the Tc to shift downward).

Resistance: The opposition of a material to the flow of electrical current through it. Energy lost due to resistance is a result of vibrations at the molecular level and manifests itself as heat in proportion to the square of the current flow. In a superconductor all resistance disappears below a certain temperature. However, this applies only to direct current (DC) electricity. Other types of losses result when transporting alternating current (AC). Examples of this include hysteresis, reactive-coupling and radiational losses. In the new high temperature ceramic superconductors, the power loss in applications like transmission lines is inversely proportional to the critical current density for low magnetic field applications. This limitation can be compensated for to some degree by increasing the ratio of voltage to current. In Type 2 superconductors carrying high-frequency alternating current, "skin effect" losses also result as the energy tends to migrate to the surface where the conductive medium is incontiguous, producing a pseudo-resistance. In some materials the amount of resistance may also depend on the direction of current flow (anisotropic resistivity) and/or presence of an external magnetic field (hall effect).

Room-temperature Superconductor: There are NO confirmed room-temperature superconductors (as was once reported for lithium-beryllium-hydride and for lead-silver-carbonate). However, it has been theorized that a metallic form of hydrogen might be a room-temperature superconductor. In 1996 physicists at Lawrence Livermore Laboratory were able to briefly create metallic hydrogen. But, its existence was fleeting and no measurement of the Meissner effect was possible. Zero resistance *has* been observed at room temperatures in ballistic quantum wire. However, having one-dimensional geometry, this wire does not exhibit the Meissner effect, except when configured as a closed loop.

Sinter: The process of heating a material to just below its melting point. An extended period of sintering is the method by which the constituent components of a ceramic superconductor are combined in a solid-state reaction. Since ceramic superconductors are inherently brittle, sintering helps promote intergranular bonding and hardness.

SQUID: A superconducting loop interrupted in 2 places by Josephson junctions. When sufficient electrical current is conducted across the squid body, a voltage is generated proportional to the strength of any nearby magnetic field. The SQUID, an acronym for Superconducting QUantum Interference Device, is the most sensitive detector known to science.

Stripes: Stripes are microscopic rivers of charge that flow across the surface of a Type 2 superconductor. It is theorized that stripes encourage "holes" to pair up and, as such, may play a role in facilitating charge transfer. Recently, at the Stripes 2000 conference in Rome, Italy, it was shown that there exists a critical value of micro-strain that must be exerted upon the CuO_2 planes for stripes to form.

Superconductor: An element, inter-metallic alloy, or compound that will conduct electricity without resistance below a certain temperature. However, this applies only to direct current (DC) electricity and to finite amounts of current. All known superconductors are solids. None are gases or liquids. And all require extreme cold to enter a superconductive state. Once set in motion, current will flow forever in a closed loop of superconducting material - making it the closest thing to perpetual motion in nature. Scientists refer to superconductivity as a "macroscopic quantum phenomenon". In addition to being classified Type 1 and Type 2, superconductors can be categorized further by their dimensionality. Most are 3-D. But some compounds, like surface-doped $NaWO_3$ and some organic superconductors are 2-D. Li_2CuO_2 and single-walled carbon nano-tubes have shown rare 1-D superconductivity. In addition to repelling magnetic fields, enhanced thermal conductivity, higher optical reflectivity and reduced surface friction are also properties of superconductors. The term "superconductor" is also used in some instances to refer to materials that have near infinite thermal conductivity - such as carbon nanotubes. However, on this website it is used in the context of electrical conductivity only.

Susceptibility: A measure of the relative amount of induced magnetism in a material. Magnetic susceptibility is often used in lieu of resistance measurements to determine the transition temperature of a superconductor. Although, on occasion, the two techniques produce very different Tc's. In a typical superconductor, the (arbitrary) value of susceptibility will change from zero to a negative number as the temperature drops through Tc. However, in some materials it changes from positive to negative, as paramagnetism yields to diamagnetism.

S-Wave: A form of electron pairing in which the electrons travel together in spherical orbits, but in opposite directions.

Tc: The scientific notation representing the critical transition temperature below which a material begins to superconduct. The sudden loss of resistance in a superconductive medium may occur across a range as small as 20 millionths of a degree or, in the case of some stoichiometrically imperfect compounds, tens of degrees.

Thin Film (Deposition): A method of fabricating ceramic superconductors to more precisely control the growth of the crystalline structure to eliminate grain boundaries and achieve a desired Tc. This can involve Pulsed-Laser Deposition (PLD) or Pulsed-Electron Deposition (PED) of the material. A variation of this technique can be used to increase the Tc of a superconductor by growing it on a supporting material with a smaller interatomic spacing. The supporting material acts as a molecular "girdle" to compress the atomic lattice of the superconductor, thereby raising its transition temperature. Superconductive tape is made using thin film deposition technology.

Translational Symmetry: As it applies to superconductivity, translational symmetry is where the process of charge transfer is repeated exactly as the charge carriers (paired electrons) traverse the solid. In a normal conductor, latent heat continuously vibrates the atomic lattice, deflecting mobile free electrons and preventing "perfect" translational symmetry. In a superconductor this scattering tendency is overcome.

Tungsten-bronze: A nebulous term used to describe alkali metal tungstenates, vanadates, molybdates, titanates and niobates. The term was originally coined to describe Na_xWO_3 compounds; the crystals of which look much like the copper-tin alloy known as bronze. There have been reports of superconductivity as high as 91K for a surface-doped sodium tungsten-bronze. This material was the first high-temperature superconductor discovered that does not contain any copper.

Ultraconductor: Materials known as ultraconductors™ display room-temperature resistance many orders of magnitude lower than the best metallic conductors. Examples of these materials include oxidized atactic polypropylene (OAPP) and other polymers. Since ultraconductor™ is a colloquial term, these materials might better be described as "hyperconductors". The Meissner effect cannot be confirmed in them, but strong (giant) diamagnetism *is* in evidence. Some of them may actually find acceptance in high-current applications ahead of superconductors as a result of their low losses at ambient temperatures and pressures.

Undressing: The process by which a quasiparticle becomes more like a bare (normal) particle. It is theorized this may be a driving force behind superconductivity, as undressed electrons are significantly lighter and can, thus, conduct current more readily.

Unit Cell: A unit cell is the smallest assemblage of atoms, ions, or molecules in a solid, beyond which the structure repeats to form the 3-dimensional crystal lattice.

Vortices (plural of vortex): Swirling tubes of electrical current induced by an external magnetic field into the surface of a superconducting material that represent a topological singularity in the wavefunction. These are particularly evident in Type 2 superconductors during "mixed-state" behavior when the surface is just partially superconducting. Superconductivity is completely suppressed within these volcano-shaped structures. Recent research suggests that flux vortices may NOT possess quantum values (equal to multiples of Planck's constant divided by 2 times electron charge). But may instead have but a tiny fraction of the basic unit of magnetism. The movement of vortices can produce a pseudo-resistance and, as such, is undesirable. While superconductivity is a "macroscopic" phenomenon, vortices are a "mesoscopic" phenomenon.

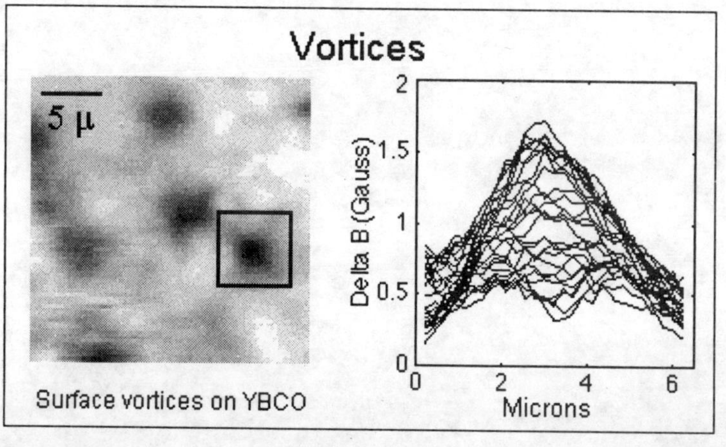

YBCO: An acronym for a well-known ceramic superconductor composed of Yttrium, Barium, Copper and Oxygen. This was the first truly "high temperature" ceramic superconductor discovered; having a transition temperature well above the boiling point of liquid nitrogen—a commonly available coolant. Its actual molecular formula is $YBa_2Cu_3O_7$, making it a "1-2-3" superconductor. YBCO compounds exhibit d-wave electron pairing. The patent for YBCO is held by Lucent Technologies.

With the discovery of the ceramic superconductors, came a need for a classification system to describe structure types. The cuprate superconductors all have blocks of conducting CuO_2 planes, alternating with insulating, spacing and separating layers. This makes possible a systematic Naming Scheme that allows for identification and comparison. The scheme chosen uses four numbers. The first denotes the number of insulating layers between adjacent conducting blocks. The second represents the number of spacing layers between identical CuO_2 blocks. The third gives the number of layers that separate adjacent CuO_2 planes within the conducting block. And, the fourth is the number of CuO_2planes within a conducting block.

Using the $TlBa_2Ca_2Cu_3O_9$ molecule depicted at left as an example, there is 1 insulating TlO layer, 2 spacing BaO layers, 2 separating Ca layers, and 3 conducting CuO_2 planes-making it a "1223" type.

There are also occasionally letters of the alphabet in the 5th and 6th positions. Suffixes like "T*", "C", and "F" describe to researchers structure detail in the oxygen positions.

To further complicate matters, several of the more popular structures have names that do NOT follow the 4-number scheme. For example, the compound $Y_1Ba_2Cu_3O_7$ is often referred to by the 3-number name "123". This delineates the number of metal atoms (the stoichiometry)—without regard to atom location—within the structure. Using the 4-number scheme, it would be classified a "1212".

Bibliography

Adalsteinsson, D. and J. Sethian. 1997. An overview of level set methods for etching, deposition, and lithography development. *IEEE Transactions on Semiconductor Devices*. 10(1), 167-184.

Adeyeye, A., G. Lauhoff, J. Bland, C. Daboo, D. Hasko, and H. Ahmed. 1997. *Appl. Phys. Lett.* 70:1046.

Adleman, L., "Molecular Computation of Solutions to Combinatorial Problems," Science, Vol. 266, 11 November 1994, pp. 1021-1023. *This is THE landmark paper that started the new field of DNA computing.*

Adleman, L.M. 1994. Molecular computation to solutions of combinatorial problems. *Science* 266:1021-1024.

Aizawa, M. 1994. Molecular interfacing for protein molecular devices and neurodevices. *IEEE Engineering in Medicine and Biology* (Feb./March):94-102.

Ajayan, P.M., and S. Iijima. 1993. *Nature*. 361:333. Arakawa, Y., H. Sakaki. 1982. *Appl. Phys. Lett.* 40: 939.

Akis, R., D. Ferry, C. Musgrave, "Kinetic Lattice Monte Carlo Simulations of Processes on the Silicon (100) Surface," *Physica E-Low Dimensional Systems and Nanostructures* 19, 183-187 (2003).

Aksay, I.A., et al., eds. 1992. Hierarchical structured materials. *MRSProc.* 255. Angstrom Technology Partnership. 1995. *Challenging the angstrom.* Tokyo.

Aksay, I.A., et al., eds. 1992. Hierarchically structured materials. *MRS Proceeding* 255.

Aksay, I.A., M. Trau, S. Marine, I. Honma, N. Yao, L. Zhou, P. Fenter, P.M. Eisenberger, and S.M. Gruner. 1996. Biomimetic pathways for assembling inorganic thin films. *Science* 273:892-898.

Aksay, Ilhan. 1998. Nanostructured ceramics through self-assembly. In *R&D status and trends*, ed. Siegel et al.

Al Globus, Charles W. Bauschlicher Jr., Jie Han, Richard L. Jaffe, Creon Levit, Deepack Srivastava, "Machine phase fullerene nanotechnology," *Nanotechnology* 9(1998):192-199.

Albert Einstein, "Investigations on the Theory of Brownian Movement," with notes by R. Furth, translated into English from German by A.D. Cowper, Methuen, London (1926), Dover Publications (1956); original paper in *Ann. Phys.* 17(1905):549.

Albrecht, T.R., Akamine, S., Carver, T.E., and Quate, C.F. (1990) Microfabrication of cantilever styli for the atomic force microscope. *J. Vac. Sci. Tcchnol.* A 8(4), 3386-3396.

Albrecht, T.R., Grütter, P., Home, D., and Rugar, D. (1991) Frequency modulation detection using high-Q cantilevers for enhanced force microscope sensitivity. *J. Appl. Phys.* 69(2), 668-673.

Albus, J.S., 2003, 'Engineering of Mind for Enhancing Human Productivity', in: M.C. Roco, W.S.

Bainbridge (eds.), *Converging Technologies for Improving Human Performance*, Kluwer, Dordrecht, pp. 281-293.

Albus, J.S.; Bainbridge, W.S.; Banfield, M.; Dastoor, C.A.; Murray, K.; Carley, M.; Hirshbein, T.; Masciangioli, T.; Miller, R.; Norwood, R.; Price, P.; Rubin, J.; Sargent, G.; Wallace, W.A.: 2003, 'Enhancing Group and Societal Outcomes: Theme D Summary', in M.C. Roco, W.S. Bainbridge (eds.), *Converging Technologies for Improving Human Performance*, Kluwer, Dordrecht, pp. 275-277.

Alexander, S., Hellemans, L., Marti, O., Schneir, J., Elings, V., Hansma, P.K., Longmiro, M., and Gurley, J. (1989) An atomic-resolution atomic-force microscope implemented using an optical lever. *J. Appl. Phys.* 65(1), 164-167.

Alexandra Goho, "Protein Power: Solar cell produces electricity from spinach and bacterial proteins," *Science News Online*, week of June 5, 2004: Vol. 165, No. 2, p. 355.

Alibek, K.A., Testimony of Dr. Kenneth Alibek, Chief Scientist at Hadron, Inc., before the House Armed Services Committee, October 20,1999. Ref: http://www.house.gov/ hasc/testimony/ 106thcongress/99-10-20alibek.htm.

Alivisatos, A.P. 1996. Perspectives on the physical chemistry of semiconductor nanocrystals. *Journal of Physical Chemistry* 100:13226-13239.

Allara, D.L. 1995. "Critical Issues in Applications of Self-Assembled Monolayers", *Biosensors & Bioelectronics*, 10, 771-783.

Allara, D.L. 1996. Nanoscale structures engineered by molecular self-assembly of functionalized monolayers. In *Nanofabrication and biosystems*, ed. Hoch et al. Birge, R.R. 1995. Protein-based computers. *Sci. Am.* (Mar.):90-95.

Altmann, J. & Gubrud, M.A.: 2004, 'Military, Arms Control, and Security Aspects of Nanotechnology', in: D. Baird, A. Nordmann & J. Schummer (eds.), *Discovering the Nanoscale*, Amsterdam: IOS Press, pp. 269-277.

Altmann, J. Military Uses of Microsystem Technologies. Dangers and Preventive Arms Control. Munster: Agenda Verlag, 2001.

Altmann, J.: 2004, 'Military Uses of Nanotechnology: Perspectives and Concerns', *Security Dialogue*, 35 (1): 61-79.

Alves, H., M. Ferreira, U. Koster, and B. Muller. 1996. *Materials Science Forum*, 225-227: 769.

Amann, K: 1994, 'Menschen, Mause und Fliegen', *Zeitschrift fur Soziologie*, 23, 22-40.

Andrea G. Bodnar, Michel Ouellette, *et al.*, "Extension of life-span by introduction of telomerase into normal human cells," *Science* 279 (16 January 1998):349-352.

Andres, R.P., S. Datta, D.B. Janes, C.P. Kubiak, and R. Reifenberger. 1998. the design, fabrication and electronic properties of self-assembled molecular nanostructures. In *the Handbook of Nanostructured Materials and Nanotechnology*, ed. H.S. Nalwa, San Diego: Academic Press.

Andrew Pollack, *Atom by Atom, Scientists Build 'Invisible' Machines of the Future*, the New York Times Science Section, Tuesday, November 26, 1991, page B7.

Anne Eisberg, "Benign Viruses Shine on the Silicon Assembly Line," *New York Times*, February 12, 2004.

Anonymous: 2002, *Small Wonders, Endless Frontiers: A Review of the National Nanotechnology Initiative*, National Academies Press, Washington, DC.

Antonio Regalado, "Nanotechnology Patents Surge as Companies Vie to Stake Claim," *Wall Street Journal*, June 18, 2004, p. 1.

Arap W., R. Pasqualini, E. Ruoslahti. 1998. "Cancer Treatment by Targeted Drug Delivery to Tumor Vasculature in a Mouse Model", *Science*, Jan. 16; 279 (5349): pp. 377-380.

Arendt, H. the Human Condition, the University of Chicago Press, 1958.

Arias, T.A. and Sohrab Ismail-Beigi. 2000. Novel algebraic formulation of density-functional calculation. *Computer Physics Communications.* In press.

Arriagada, F.J., and K. Osseo-Asave. 1995. *J. Colloid Interface Sci.* 170:8.

Arribart, H.: 2004, 'Les nanomatériaux autres que ceux des technologies de l'information et des communications (TICS)', in: Académie des sciences, Academie des technologies (eds.), *Nanosciences, nanotechnologies,* editions Tec & Doc, Paris, pp. 361-382.

Arthur C., Clarke, *Profiles of the Future,* Harper & Row, Publishers, New York, 1962.

Asada, M., Y. Mayamoto, and Y. Suematsu. 1986. *IEEE Journ. Quantum Electronics* QE-22(9): 1915-1921.

Asimov, I. *the Naked Sun.* Garden City, NY: Doubleday, 1957.

Asimov, I., A *Choice of Catastrophes,* Simon and Schuster, NY (1979).

Atkins, P.: 2003 'The Future of Matter', in: J. Brockman (ed.), *the Next Fifty Years: Science in the First Half of the Twenty-First Century,* Vintage Books, New York, pp. 194-205.

Attard, G.S., et al. 1997. *Science* 278:838. Averback, R.S., J. Bernholc, and D.L. Nelson. 1991. *MRS Symposium Proceedings,* Vol. 206.

Attebery, B.: 2004, 'Dust, Lust, and Other Messages from the Quantum Wonderland', in: N.K. Hayles (ed.), *Nanoculture: Implications of the New Technoscience,* Bristol, UK: Intellect Books, pp. 161-172.

Averback, R.S., J. Bernholc, and D.L. Nelson. 1991. *MRS Symposium Proceedings,* Vol. 206.

Axelbaum, R.L. 1997. Developments in sodium/halide flame technology for the synthesis of unagglomerated non-oxide nanoparticles. In *Proc. of the Joint NSF-NIST Conference on Nanoparticles: Synthesis, Processing into Functional Nanostructures and Characterization* (May 12-13, Arlington, VA).

B. Ding, R. Sha & N.C. Seeman, Pseudohexagonal 2D DNA Crystals from Double Crossover Cohesion, *J. Am. Chem. Soc.* 126, 10230-10231 (2004).

B. Liu, N.B. Leontis and N.C. Seeman, Bulged 3-arm DNA Branched Junctions as Components for Nanoconstruction, Nanobiology 3, 177-188 (1994). N.C. Seeman, Molecular Craftwork with DNA, the Chemical Intelligencer 1(3), 38-47 (1995).

B.W. Matthews, H. Nicholson, and W. J. Becktel, "Enhanced protein thermostability from site-directed mutations that decrease the entropy of unfolding." *Proc. Nat, Acad. Sci., 84:* 6663-6667 (1987), and included references.

B. Yurke, A.J. Turberfield, A.P. Mills, Jr., F.C. Simmel, J.L. Neumann, "A DNA-fuelled molecular machine made of DNA," *Nature,* 406(10 August 2000):605-608.

B.H. Robinson and N.C. Seeman, the Design of a Biochip: A Self-Assembling Molecular-Scale Memory Device. Protein Engineering 1, 295-300 (1987).

Babbie, E. *the Practice of Social Research,* 5th ed. Belmont, CA: Wadsworth, 1989. Bachand, G. and CD. Montemagno. 2000. Constructing organic/inorganic NEMS devices powered by biomolecular motors. *Journal of Biomedical Microdevices.* 2, No.3, 179-184.

Bachelard, G.: 1984, *the New Scientific Spirit,* Beacon Press, Boston.

Bai, Q., K.D. Wise, and D.J. Anderson. 2000. A high-yield microassembly structure for three-dimensional microelectrode arrays. *IEEE Trans. Biomed Eng.* 47(3): p. 281-9.

Baibich, M..N., J.M. Broto, A. Fert, F. Nguyen Van Dau, F. Petroff, P. Etienne, G. Creuzet, A Friederich, and J. Chazelas. 1998. *Phys. Rev. Lett.* 61:2472.

Bainbridge, W.S.: 2002, 'Public Attitudes Toward Nanotechnology', *Journal of Nanoparticle Research,* 4, 561-570.

Baird, D. & Shew, A.: 2004, 'Probing the History of Scanning Tunneling Microscopy', in: D. Baird, A. Nordmann & J. Schummer (eds.), *Discovering the Nanoscale,* Amsterdam: IOS Press, pp. 145-156.

Baird, D.: 1993, 'Analytical Chemistry and the Big Scientific Instrumentation Revolution', *Annals of Science,* 50, 267-290.

Baird, D.: 2004, *Thing Knowledge: A Philosophy of Scientific Instruments,* University of California Press, Berkeley.

Baird, D.; Nordmann, A. & Schummer, J. (eds.): 2004, *Discovering the Nanoscale,* Amsterdam: IOS Press.

Baker, R.T.K. Synthesis, properties and applications of graphite nanofibers. In *R&D status and trends,* ed. Siegel et al.

Baker, R.T.K. Synthesis, properties and applications of graphite nanofibers. In *R&D status and trends,* ed. Siegel et al.

Ball, P.: 2002, 'Natural Strategies for the molecular engineer', *Nanotechnology,* 13, 15-28.

Ball, P.: 2002, 'Tutorial: Natural Strategies for the Molecular Engineer', *Nanotechnology,* 13, 15-28.

Ball, P.: 2003, 'Nanotechnology in the firing line', *Nanotechweb.org,* 23 December [http://www.nanotechweb.org/articles/society/2/12/1/1].

Barber, Benjamin. 1996. *Jihad* vs. *McWorld: How Globalism and Tribalism are Reshaping the World* (New York: Ballantine Books).

Baschnagel, J., K. Binder, K. Kremer et al. 2000. Bridging the gap between atomistic and coarse-grained models of polymers: status and perspectives. *Advances in Polymer Science.* Volume 152. Springer-Verlag.

Bass, T.A., "Gene Genie," Wired Magazine, August 1995, p. 114. *Well-informed and well-written survey of the technical developments and people in the new field ofDNA computing.*

Bate, R., Frazier, G., Frensley, W., Reed, M., "An Overview of Nanoelectronics," *Texas Instruments Technical Journal,* July-August 1989, pp. 13-20.

Baum, R.; Drexler, K.E.; Smalley, R.: 2003, 'Nanotechnology: Drexler and Smalley Make the Case For and Against 'Molecular Assemblers', *Chemical and Engineering News,* 81, 37-42.

Beck, J.S., J.C. Vartuli, W.J. Roth, M.E. Leonowicz, C.T. Kresge, K.D. Schmitt, C.T.-W. Chu, D.H. Olsen, E.W. Shepard, S.B. McCullen, J.B. Higgins, and J.L. Schlenker. 1992. *J. Am. Chem. Soc.* 114:10834.

Becker, M.F., J.R. Brock, H. Cai, N. Chaudhary, D. Henneke, L. Hilsz, J.W. Keto, J. Lee, W.T. Nichols, and H.D. Glicksman. 1997. Nanoparticles generated by laser ablation. In *Proc. of the Joint NSF-NIST Conf. on Nanoparticles.*

Ben-David, Joseph and Randall Collins, 1966. "Social Factors in the Origins of a New Science: the Case of Psychology." *American Sociological Review.* Vol. 31 pp 451-465.

Bennett, C.H., "Quantum Information and Computation," *Physics Today,* October.

Bensaude-Vincent, B.: 2001, 'The Construction of a Discipline: Materials Science in the United States', *Historical Studies in the Physical and Biological Sciences,* 31, 223-248.

Bensaude-Vincent, B.: 2004, 'Two Cultures of Nanotechnology?', *Hyle: International Journal for Philosophy of Chemistry,* 10(2), 65-82.

Bensaude-Vincent, B.; Arribart, B.Y.; Sanchez, C: 2002, 'Chemists at the School of Nature', *New Journal of Chemistry,* 26, 1-5.

Berger, R., C. Gerber, H.P. Lang, and J.K. Gimzewski. 1996. Micromechanics: A Toolbox for Femtoscale Science: "Towards a laboratory on a tip." *Microelectronic Engineering,* 35:373-9. (International Conference on Micro- and Nanofabrication, Glasgow, U.K., 22-25 Sept. 1996).

Berkowitz, A.E., J.R. Mitchell, M.J. Carey, A.P. Young, S. Zhang, F.E. Spada, F.T. Parker, A. Hutten, and G. Thomas. 1992. *Phys. Rev. Lett.* 68:3745.

Berndt, C.C., J. Karthikeyan, T. Chraska, and A.H. King. 1997. Plasma Spray Synthesis of Nanozirconia Powder. In *Proc. of the Joint NSF-NIST Conf. on Nanoparticles.*

Berube, D.M.: 2004, 'The Rhetoric of Nanotechnology', in: D. Baird, A. Nordmann & J. Schummer (eds.), *Discovering the Nanoscale,* Amsterdam: IOS Press, pp. 173-192.

Bijker, W.E.: 1995a, *Of Bicycles, Bakelite, and Bulbs: Toward a Theory of Sociotechnical Change,* MIT Press, Cambridge, MA.

Bijker, W.E.: 1995b, 'Sociohistorical Technology Studies', in: S. Jasanoff, G.E. Markle, J. Petersen, T. Pinch (eds.), *Handbook of Science and Technology Studies,* Sage, London, pp. 229-256.

Bijker, W.E.; Law, J.: 1992, 'Do Technologies Have Trajectories?', in: W.E. Bijker, J. Law (eds.), *Shaping Technology/Building Society: Studies in Sociotechnical Change,* MIT Press, Cambridge, MA, pp. 1-16.

Bijker, W.E.; Pinch, T.: 1987, 'The Social Construction of Facts and Artifacts: Or How the Sociology of Science and the Sociology of Technology Might Benefit Each Other', in: W.E. Bijker, T.P. Hughes, T. Pinch (eds.), *the Social Construction of Technological Systems: New Directions in the Sociology and History of Technology,* MIT Press, Cambridge, MA, pp. 17-50.

Billings P.R., M.A. Kohn, M. de Cuevas, J. Beckwith, J.S. Alper, M.R. Natowicz. 1992. "Discrimination as a Consequence of Genetic Testing," *American Journal of Human Genetics,* Mar; 50 (3): pp. 476-82. Drews, J. 1996. "Genomic Sciences and the Medicine of Tomorrow," *Nature Biotechnology,* Nov., 14 (11), pp. 1516-1518.

Bimber, B., and Guston, D., eds., *Technology Assessment: the End of OTA,* special issue of *Technology Forecasting and Social Change,* Vol. 54, No. 2 and 3, 1997.

Bimber, B.: 1994, 'Three Faces of Technological Determinism', in: M.R. Smith, L. Marx (eds.), *Does Technology Drive History?: the Dilemma of Technological Determinism,* MIT Press, Cambridge, MA, pp. 79-100.

Bimberg, D., N. Kirstaedter, N.N. Ledentsov, Zh.I. Alferov, P.S. Kopev, and V.M. Ustinov. 1998. In GaAs-GaAs quantum-dot lasers. *IEEE J. of Selected Topics in Quant. Electronics* 3:196-205.

Binnig, G., Quate, C.F., and Gerber, Ch. (1986) Atomic Force Microscope. *Phys. Rev. Lett.* 56(9), 930-933

Baker, R.T.K. Synthesis, Properties and Applications of Graphite Nanofibers. In *R&D status and trends,* ed. Siegel et al.

Ball, P.: 2002, 'Natural Strategies for the Molecular Engineer', *Nanotechnology,* 13, 15-28.

Ball, P.: 2002, 'Tutorial: Natural Strategies for the Molecular Engineer', *Nanotechnology,* 13, 15-28.

Ball, P.: 2003, 'Nanotechnology in the Firing Line', *Nanotechweb.org,* 23 December [http://www.nanotechweb.org/articles/society/2/12/1/1].

Barber, Benjamin. 1996. *Jihad vs. McWorld: How Globalism and Tribalism are Reshaping the World* (New York: Ballantine Books).

Baschnagel, J., K. Binder, K. Kremer et al. 2000. Bridging the gap between atomistic and coarse-grained models of polymers: status and perspectives. *Advances in Polymer Science.* Volume 152. Springer-Verlag.

Bass, T.A., "Gene Genie," Wired Magazine, August 1995, p. 114. *Well-informed and well-written survey of the technical developments and people in the new field of DNA computing.*

Bate, R., Frazier, G., Frensley, W., Reed., M., "An Overview of Nanoelectronics," *Texas Instruments Technical Journal,* July-August 1989, pp. 13-20.

Baum, R.; Drexler, K.E.; Smalley, R.: 2003, 'Nanotechnology: Drexler and Smalley Make the Case

For and Against 'Molecular Assemblers', *Chemical and Engineering News*, 81, 37-42.

Beck, J.S., J.C. Vartuli, W.J. Roth, M.E. Leonowicz, C.T. Kresge, K.D. Schmitt, C.T.-W. Chu, D.H. Olsen, E.W. Shepard, S.B. McCullen, J.B. Higgins, and J.L. Schlenker. 1992. *J. Am. Chem. Soc.* 114:10834.

Becker, M.F., J.R. Brock, H. Cai, N. Chaudhary, D. Henneke, L. Hilsz, J.W. Keto, J. Lee, W.T. Nichols, and H.D. Glicksman. 1997. Nanoparticles generated by laser ablation. In *Proc. of the Joint NSF-NIST Conf. on Nanoparticles.*

Ben-David, Joseph and Randall Collins, 1966. "Social Factors in the Origins of a New Science: the Case of Psychology." *American Sociological Review.* Vol. 31 pp 451-465.

Bennett, C.H., "Quantum Information and Computation," *Physics Today*, October

Bensaude-Vincent, B.: 2001, 'The Construction of a Discipline: Materials Science in the United States', *Historical Studies in the Physical and Biological Sciences*, 31, 223-248.

Bensaude-Vincent, B.: 2004, 'Two Cultures of Nanotechnology?', *Hyle: International Journal for Philosophy of Chemistry*, 10(2), 65-82.

Bensaude-Vincent, B.; Arribart, B.Y.; Sanchez, C., 2002, 'Chemists at the School of Nature', *New Journal of Chemistry*, 26, 1-5.

Berger, R., C. Gerber, H.P. Lang, and J.K. Gimzewski. 1996. Micromechanics: a toolbox for femtoscale science: "Towards a laboratory on a tip." *Microelectronic Engineering*, 35:373-9. (International Conference on Micro- and Nanofabrication, Glasgow, U.K., 22-25 Sept. 1996).

Berkowitz, A.E., J.R. Mitchell, M.J. Carey, A.P. Young, S. Zhang, F.E. Spada, F.T. Parker, A. Hutten, and G. Thomas. 1992. *Phys. Rev. Lett.* 68:3745.

Berndt, C.C., J. Karthikeyan, T. Chraska, and A.H. King. 1997. Plasma spray synthesis of nanozirconia powder. In *Proc. of the Joint NSF-NIST Conf. on Nanoparticles.*

Berube, D.M.: 2004, 'The Rhetoric of Nanotechnology', in: D. Baird, A. Nordmann & J. Schummer (eds.), *Discovering the Nanoscale,* Amsterdam: IOS Press, pp. 173-192.

Bijker, W.E.: 1995a, *Of Bicycles, Bakelite, and Bulbs: Toward a Theory of Sociotechnical Change,* MIT Press, Cambridge, MA.

Bijker, W.E.: 1995b, 'Sociohistorical Technology Studies', in: S. Jasanoff, G.E. Markle, J. Petersen, T. Pinch (eds.), *Handbook of Science and Technology Studies,* Sage, London, pp. 229-256.

Bijker, W.E.; Law, J.: 1992, 'Do Technologies Have Trajectories?', in: W.E. Bijker, J. Law (eds.), *Shaping Technology I Building Society: Studies in Sociotechnical Change,* MIT Press, Cambridge, MA, pp. 1-16.

Bijker, W.E.; Pinch, T.: 1987, 'The Social Construction of Facts and Artifacts: Or How the Sociology of Science and the Sociology of Technology Might Benefit Each Other', in: W.E. Bijker, T.P. Hughes, T. Pinch (eds.), *the Social Construction of Technological Systems: New Directions in the Sociology and History of Technology,* MIT Press, Cambridge, MA, pp. 17-50.

Billings P.R., M.A. Kohn, M. de Cuevas, J. Beckwith, J.S. Alper, M.R. Natowicz. 1992. "Discrimination as a Consequence of Genetic Testing," *American Journal of Human Genetics,* Mar; 50 (3): pp. 476-82. Drews, J. 1996. "Genomic Sciences and the Medicine of Tomorrow," *Nature Biotechnology,* Nov., 14 (11), pp. 1516-1518.

Bimber, B., and Guston, D., eds., *Technology Assessment: the End of OTA,* special issue *of Technology Forecasting and Social Change,* Vol. 54, No. 2 and 3, 1997.

Bimber, B.: 1994, 'Three Faces of Technological Determinism', in: M.R. Smith, L. Marx (eds.), *Does Technology Drive History?: the Dilemma of Technological Determinism,* MIT Press, Cambridge, MA, pp. 79-100.

Bimberg, D., N. Kirstaedter, N.N. Ledentsov, Zh.I. Alferov, P.S. Kopev, and V.M. Ustinov. 1998. In

GaAs-GaAs quantum-dot lasers. *IEEE J. of Selected Topics in Quant. Electronics* 3:196-205.

Binnig, G., Quate, C.F., and Gerber, Ch. (1986) Atomic force microscope. *Phys. Rev. Lett.* 56(9), 930-933.

Bishop, A.R., and R.G. Nuzzo. 1996. Self-assembled monolayers: Recent developments and applications. *Current Opinion in Colloid & Interface Sci.* 1:127-136.

Bishop, J. et al. 1990. *Surface modified drug nanoparticles.* U.S. Patent application. Docket 61894 Filed 9/17/90.

Black, C.T.; Murray, C.B.; Sandstrom, R.L.; Sun, S.H. 2000. "Spin-dependent tunneling in self-assembled cobalt-nanocrystal superlattices", *Science,* 290, 1131-1134.

Blackman, M. 1994. *An evaluation of the Link Nanotechnology Program and the National Initiative on Nanotechnology.* Cambridge.

Blumenthal, D. 1992. "Academic-Industry Relationships: Extent, Consequences, and Management," *the Journal of the American Medical Association,* 268:3, December, pp. 3344-3349.

Blumenthal, D.M., M. Gluck, K.S. Louis, M. Stoto, and D. Wise. 1986. "University-Industry Research Relationships in Biotechnology: Implications for the University," *Science* 232: pp. 1361-1366 (June 13).

BMBF. 1996. "Delphi-Bericht 1995 zur Entwicklung von Wissenschaft und Technik -Mini-Delphi." Bonn, author's translation.

Bockrath, M., D.H. Cobden, P.L. McEuen, N.G. Chopra, A. Zettl, A. Thess, and R.E. Smalley. 1997. "Single-Electron Transport in Ropes of Carbon Nanotubes." *Science,* 275.

Bohn, R., T. Haubold, R. Birringer, and H. Gleiter. 1991. *Scripta Metall. Mater.* 25:811.

Bonnemain, B. 1998. "Superparamagnetic agents in magnetic resonance imaging: Physicochemical characteristics and clinical applications—A review", *Journal of Drug Targeting,* 6, 167-174.

Bowes, C.L., A. Malek, and G.A. Ozin. 1996. *Chem. Vap. Deposition* 2: 97.

Brandt, A. 2000. *Multiscale Scientific Computation: 2000.* Weizmann Institute preprint. October.

Braun, P.V., P. Osenar, and S.I. Stupp. 1996. *Nature.* 368: 2. Brinker, C.J. 1996. *Curr. Opin. Solid State Mater. Sci.* 1:798. Chen, C-Y., S.L. Burkett, H.-X. Li, and M.E. Davis. 1993. *Microporous Mater.* 2:27.

Braun, T.; Schubert, A. & Zsindely, S.: 1997, 'Nanoscience and nanotechnology on the balance', *Scientometrics,* 38, 321-325.

Brave, Ralph. 2000. "Genome Hell is Revealed in Many Forms." *Baltimore Sun,* July 2, p. 1C.

Brinker, C.J. 1996. *Curr. Opin. Solid State Mater. Sci.* 1:798.

Broderick, D. the Spike, Forge, New York, 2001. Brooks, L.J.: 2003, *the Institute for the Future: A Site for Mapping, Maintaining, and Transforming Images of the Future (Ph.D.* dissertation) University of California, San Diego, La Jolla.

Brotzman, R. 1998. Nanoparticle Dispersions. In *R&D status and trends,* ed. Siegel et al. Bu, X., P. Feng, and G.D. Stucky. 1998. Large-cage zeolite structures with multidimensional 12-ring channels. *Science* 278:2080-2085.

Brotzman, R. 1998. Nanoparticle dispersions. In *R&D status and trends,* ed. Siegel et al. Brus, L.E. 1996. Theoretical metastability of semiconductor crystallites in high pressure phases with applications to beta tin structures of silicon. *J. Am. Chem. Soc.* 118:4834-38.

Brown, J.S. and Duguid, P. Don't Count Society Out: A Response to Bill Joy. Paper prepared for the Societal Implications of Nanotechnology Conference, 2000.

Brown, J.S. and P. Daguid. 2001. "Don't Count Society Out: A Response to Bill Joy" in this volume: *Societal Implications of Nanoscience and Nanotechnology.* National Science Foundation, Arlington, VA.

Brown, Lawrence A. 1981. *Innovation Diffusion: A New Perspective.* New York: Methuen.

Brown, N.: 2003, 'Hope Against Hype—Accountability in Biopasts, Presents and Futures', *Science Studies*, 16(2), 3-21.

Brown, N.: 2004, 'Needle on the Real: Technoscience and Poetry at the Limits of Fabrication', in: N.K. Hayles (ed.), *Nanoculture: Implications of the New Technoscience*, Bristol, UK: Intellect Books, pp. 173-190.

Bruchez, M.; Moronne, M.; Gin, P.; Weiss, S.; Alivisatos, A.P. 1998. "Semiconductor nanocrystals as fluorescent biological labels", *Science*, 281, 2013-2016.

Brus, L. 1996. Semiconductor colloids: Individual nanocrystals, opals and porous silicon. *Current Opinion in Colloid & Interface Science* 1:197-201. *Chemical Engineering News.* 1997. Particulate matter health studies to be reanalyzed (August 18):33.

Bueno, O.: 2004, 'The Drexler-Smalley Debated on Nanotechnology: Incommensurability at Work?', *Hyle: International Journal for Philosophy of Chemistry*, 10(2), 83-98.

Bueno, O.: 2004, 'Von Neumann, Self-Reproduction and the Constitution of Nanophenomena', in: D. Baird, A. Nordmann & J. Schummer (eds.), *Discovering the Nanoscale*, Amsterdam: IOS Press, pp. 101-115.

Bush, V.: 1945, *Science, the Endless Frontier*, US Government Printing Office, Washington, DC.

Bush, Vannevar. 1945. *Science, the Endless Frontier.* Washington D.C. U.S. Government Printing Office.

Byl, J., Physica D 34, 295-299 (1989). *Self-Reproduction in Small Cellular Automata.*

C. Mao, T. LaBean, J.H. Reif and N.C. Seeman, Logical Computation Using Algorithmic Self-Assembly of DNA Triple Crossover Molecules, *Nature* 407, 493-496 (2000).

C. Mao, W. Sun and N.C. Seeman, Designed Two-Dimensional DNA Holliday Junction Arrays Visualized by Atomic Force Microscopy, *Journal of the American Chemical Society* 121, 5437-5443 (1999).

C. Mao, W. Sun, Z. Shen and N.C. Seeman, A DNA Nanomechanical Device Based on the B-Z Transition, *Nature* 397, 144-146 (1999).

C. Mao, W. Sun, Z. Shen, N.C. Seeman, "A nanomechanical device based on the B-Z transition of DNA," *Nature* 397(14 January 1999):144-146.

C.D. Montemagno, G.D. Bachand, "Constructing nanomechanical devices powered by biomolecular motors," *Nanotechnology* 10(1999):225-231; G.D. Bachand, C.D. Montemagno, "Constructing organic/inorganic NEMS devices powered by biomolecular motors," *Biomedical Microdevices* 2(2000): 179-184.

C.T. Kuo, P.J. Bostick, R.F., Irie, D.L. Morton, A.J. Conrad, D.S. Hoon, "Assessment of messenger RNA of beta l→4-N-acetylgalactosaminyl-transferase as a molecular marker for metastatic melanoma," *Clin. Cancer Res.* 4(February 1998):411-418.

Caflisch, R.E., W.E., M.F. Gyure, B. Merriman, and C. Ratsch. 1999. Kinetic model for a step edge in epitaxial growth. *Phys. Rev. E.* Vol 59. 6879.

Calcote, H.F., and D.G. Keil. 1997. Combustion synthesis of silicon carbide powder. In *Proc. of the Joint NSF-NIST Conf. on Nanoparticles.*

Callon, M.: 1986, 'Some Elements of a Sociology of Translation', in J. Law (ed.), *Power, Action and Belief: A New Sociology of Knowledge?*, Routledge, London, pp. 196-223.

Canguilhem, G.: 1952, *Machine et organisme' in La connaissance de la vie*, Hachette, Paris [quoted from the fourth edition Vrin, Paris, 1971].

Carroll, J.S.: 2001, 'Social Science Research Methods for Assessing Societal Implications of Nanotechnology', in: M.C. Roco and W.S. Bainbridge (eds.), *Societal Implications of Nanoscience and Nanotechnology*, Dordrecht: Kluwer, pp. 188-192.

Carroll, J.S.: 2001, 'Social Science Research Methods for Assessing Societal Implications of Nanotechnology', in: M.C. Roco, W.S. Bainbridge (eds.), *Societal Implications of Nanoscience and Nanotechnology,* National Science Foundation, pp. 188-192.

Carsley, J.E., R. Shaik, W.W. Milligan, and E.C Aifantis. 1997. In *Chemistry and physics of nanostructures and related non-equilibrium materials,* ed. E. Ma, B. Fultz, R. Shull, J. Morral, and P. Nash. Warrendale, PA: TMS.

Casey, Jr., H.C., and M.B. Panish. 1978. *Heterostructure Lasers, Part A: Fundamental Principles.* (Academic Press, New York), pp. 1-9.

Cerf, Christopher, and Victor Navasky. 1984. *the Experts Speak.* New York: Pantheon Books.

Cetron, M., and Davies, O. 1997. *Probable Tomorrows: How Science and Technology Will Transform Our Lives in the Next Twenty Years* (New York: St. Martin's Press).

Chan, W.C.W. and S. Nie. 1998. Quantum dot bioconjugates for ultrasensitive nonisotopic detection. *Science.* 281, 2016 (25 September).

Chance, B., Mueller, P., DeVault, D. & Powers, L. (1980) *Phys. Today* 33 (10), 32-38.

Chang, K. "Smaller Computer Chips Built Using DNA as Template." *New York Times,* November 21, 2003.

Chantrell. 1994. *J. Appl. Phys.* 76:6811. Erb, U., G. Palumbo, R. Zugic, and K.T. Aust. 1996. In *Processing and properties of nanocrystalline materials,* ed. C. Suryanarayana, J. Singh, and F.H. Froes. Warrendale, PA: T.M.S. Gertsman, V.Y., M. Hoffman, H. Gleiter and R. Birringer, 1994. *Ada Metall. Mater.* 42:3539-3544.

Charles B. Musgrave, Jason K. Perry, Ralph C. Merkle, William A. Goddard III, "Theoretical studies of a hydrogen abstraction tool for nanotechnology," *Nanotechnology* 2(1991): 187-195; http://www.zyvex.com/nanotech/Habs/Habs.html

Charles B. Musgrave, Jason K. Perry, Ralph C. Merkle, William A. Goddard III, "Theoretical studies of a hydrogen abstraction tool for nanotechnology," *Nanotechnology* 2(1991):187-195; http://www.zyvex.com/nanotech/Habs/Habs.html

Chaudhry, L., *Valley to Bill Joy: 'Zzzzzzz',* WiredNews, Apr. 5,2000. Ref: http://piglet.ex.ac.uk/mail/cybersociety.2000/0305.html.

Chen, C.-Y., K. Ishihara, N. Nakabayashi, E. Tamiya, and I. Karube. 1999. Multifunctional biocompatible membrane and its application to fabricate a miniaturized glucose sensor with potential for use *in vivo. Journal of Biomedical Microdevices.* 1, No. 2, 155-166.

Chen, J. and N.C. Seeman. 1991. the synthesis from DNA of a molecule with the connectivity of a cube. *Nature* 350:631-633.

Chianelli, R.R. 1998. Synthesis, fundamental properties and applications of nanocrystals, sheets, and fullerenes based on layered transition metal chalcogenides. In *R&D status and trends,* ed. Siegel et al.

Chianelli, R.R., M. Daage, and M.J. Ledoux. 1994. *Advances in Catalysis* 40:177. Chidsey, E.E.D., and R.W. Murray. 1986. *Science,* 231: 25.

Chondroudis, K. and D.B. Mitzi. 1999. Electroluminescence from an organic-inorganic perovskite incorporating a quaterthiophene dye within lead halide perovskite layers. *Chemistry of Materials.* Vol 11. p. 3028.

Chopra, N.G., R.J. Luyken, K. Cherrey, H. Crespi, M.L. Cohen, S.G. Louie, and A. Zettl. 1995. *Science* 269:966.

Christensen, Clayton M. *the Innovator's Dilemma: When New Technologies Cause Great Firms to Fail.* New York: Harperbusiness, 2000.

Cipra, Barry A. 2000. the best of the 20th century: editors name top 10 algorithms. *SIAM News*. vol. 33(4).

Claeson, T., and Likharev, K., "Single Electronics," *Scientific American*, June 1992, pp. 80-85.

Clasen, R. 1990. *Int. Journal of Glass and Science Technology* 63: 291. Crandall, B.C. and J. Lewis, (eds.). 1992. *Nanotechnology: Research and perspectives*. Cambridge: MIT Press.

Clausen-Schaumann, H.; Seitz, M.; Krautbauer, R.; Gaub, H.E.: 2000. 'Force Spectroscopy with Single Bio-Molecules', *Current Opinion in Chemical Biology*, 4, 524-530.

Clifford Lau of the US Defense Department to Barnaby Feder, "Frontier of Military Technology is the Size of a Molecule," *New York Times*, April 8, 2003, p. C2.

Coenen, C., 2004. 'Nanofuturismus: Anmerkungen zu seiner Relevanz, Analyse und Bewertung', *Technikfolgenabschatzung' Theorie und Praxis*, 13 (2), 78-85.

Cohen, J.B. 1992. *Metall. Trans. A*. 23A:2685.

Colbert, D.T. and R.E. Smalley. 1999. Fullerene nanotubes for molecular electronics. *Trends in Biotechnology*. Vol. 17. 46-50.

Collier, C.P., E.W. Wong, M. Belohradsky, F.M. Raymo, J.F. Stoddart, P.J. Kuekes, R.S. Williams, and J.R. Heath. 1999. "Electronically Configurable Molecular-Based Logic Gates." *Science*, 285.

Computing in Science and Engineering, A.I.P and IEEE Press, also covered by http://www.siam.org/siamnews/05-00/topten.pdf, Jan/Feb.

Constant, E.W., II: 1980, *the Origins of the Turbojet Revolution*, Johns Hopkins University Press, Baltimore.

Cook, T.K. and Campbell, D.T. *Quasi-Experimentation*. Rand-McNally, 1979 Fischer, C.S. *America Calling: A Social History of the Telephone to 1940*. Berkeley: U. of California, 1992.

Crandall, B.: 1999, 'Molecular Engineering', in: B. Crandall (ed.), *Nanotechnology: Molecular Speculations on Global Abundance*, MIT Press, Cambridge, MA, pp. 1-46.

Crandall, B.C. and Lewis, J., *Nanotechnology: Research and Perspectives*, MIT Press, Cambridge, 1992.

Crommie, M.F.; Lutz, C.P.; Eigler, D.M.; Heller, E.J. 1996. "Quantum interference in 2D atomic-scale structures," *Surface Science*, 362, 864-869.

Crow, M. & Sarewitz, D.: 2001, 'Nanotechnology and Societal Transformation', in: M.C. Roco and W.S. Bainbridge (eds.), *Societal Implications of Nanoscience and Nanotechnology*, Dordrecht: Kluwer, pp. 45-54.

Crow, M. & Sarewitz, D.: 2001, 'Nanotechnology and Societal Transformation', in: M.C. Roco, W.S. Bainbridge (eds.), *Societal Implications of Nanoscience and Nanotechnology*, National Science Foundation, pp. 45-54.

Crow, M.M. 2001, "Harnessing Science to Benefit Society," *the Chronicle of Higher Education, Chronicle.com*. 9 March, 2001: 17 pars. March 8, 2001 (http:// chronicle.com/weekly/v471/i26/26b02001.htm).

Crowe, W. J., Statement of Admiral William J. Crowe, Jr., before the House Committee on Government Reform, October 12, 1999. Ref: http://www.house.gov/reform/ hearings/healthcare/99.10.12/crowe.htm.

Csicsery-Ronay, I. Jr.: 1991, 'The SF of Theory: Baudrillard and Haraway', *Science Fiction Studies*, 18, 387-404.

Csicsery-Ronay, I. Jr.: 1996, 'The Seven Beauties of Science Fiction', *Science Fiction Studies*, 23, 385-392.

Czekai, D., et al. 1994. *Use of smaller milling media to prepare nanoparticulate dispersions*. U.S. Patent application. Docket 69802(02) Filed 2/25/94.

D.B. Dusenbery, "Minimum size limit for useful locomotion by free-swimming microbes," *Proc. Natl. Acad. Sci. USA* 94(30 September 1997): 10949-10954.

D.T. Dennis et al, "Tularemia as a biological weapon: medical and public health management," *J. Amer. Med. Assoc.* 285(6 June 2001):2763-2773.

D.W. Brenner, S.B. Sinnott, J.A. Harrison, O.A. Shenderova, "Simulated engineering of nanostructures," *Nanotechnology* 7(1996):161-167; http://www.zyvex.com/nanotech/nano4/brennerPaper.pdf

Dai, H.J. 2000. "Controlling nanotube growth," *Physics World*, 13, 43-47.

Dai, H., E.W. Wong, Y.Z. Lu, S. Fan, and C.M. Leiber. 1995. *Nature* 375:769. deHeer, W.A., W.S. Bacsa, A. Chatelain, T. Gerfin, R. Humphrey-Baker, L. Forro, and D. Ugarte. 1995. *Science* 268:845.

Dangl, J.L., 1998, Plants just say NO to pathogens, *Nature.* Vol. 394, 525-526.

Darden, T., L. Perera, L. Li, L. Pedersen, 1999, New tricks for modelers from the crystallization toolkit: the particle mesh Ewald algorithm and its use in nucleic acid simulations, *Structure.* Vol. 7, R55-R60.

Darken, L.S. 1961. *Trans. Am. Soc. Met.* 54:599.

David, Paul. 1991. "Computer and Dynamo: the Modern Productivity Paradox in a Not-too-Distant Mirror," *Technology and Productivity: the Challenge for Economic Policy* (Paris: OECD).

de Gennes, P.G. 1974. *the Physics of Liquid Crystals.* Oxford University Press.

de la Mora, J.F., I.G. Loscertales, J. Rosell-Llompart, K. Serageldin, and S. Brown. 1994.

de Souza e Silva, A.: 2004, 'The Invisible Imaginary: Museum Spaces, Hybrid Reality and Nanotechnology', in: N.K. Hayles (ed.), *Nanoculture: Implications of the New Technoscience*, Bristol, UK: Intellect Books, pp. 27-82.

Decker, M.; Fiedeler, U.; Fleischer, Th.: 2004, 'Ich sehe was, was Du nicht siehst... zur Definition von Nanotechnologie', *Technikfolgenabschatzung' Theorie und Praxis*, 13 (2), 10-16.

Delaney, S.R.: 1984, *Starboard Wine: More Notes on the Language of Science Fiction*, Dragon Press, New York.

Delaney, S.R.: 1994, *Silent Interviews*, Univ. Press of New England, Hanover, NH.

DeSimone, J.M., E.T. Samulski, et al. 2000. Critical micelle density for the self-assembly of block copolymer surfactants in supercritical carbon dioxide. *Langmuir.* Vol.16. 416-421.

Dewey, J.: 1958, *Experience and Nature*, Dover, New York.

Di Paolo, E.A., *Artificial Life Bibliography of On-line Publications*, June 28, 2000.

Dimitrakopoulos, C.D., S. Purushothaman, J. Kymissis, A. Callegari, and J.M. Shaw. 1999. Low-voltage organic transistors on plastic comprising high dielectric constant gate insulators. *Science.* Vol. 283, 5 February, pp. 822-824.

DiVincenzo, D.P., "Quantum Computation," Science, Vol. 270, 13 October 1995, pp. 255-261

Doering, R.R. and P.K. Chatterjee. 1998. "The Future of Microelectronics." *Proceedings of the IEEE*, 86, 1.

Doering, Robert and Yoshio Nishi. 2000. "Limits of Integrated-Circuit Manufacturing." *Proceedings of the IEEE.* September.

Doi, M. and S.F. Edwards. 1986. *the Theory of Polymer Dynamics.* Oxford University Press (Clarendon), London: New York.

Donald, A.M. and A.H. Windle. 1992. Liquid crystalline polymers. *Cambridge Solid State Science Series.* Cambridge University Press.

Drake, J. (1969) *Nature (London)* 221, 1132.

Drell, S.D., A.D. Sofaaer, and G.D. Wilson, Hoover Digest 2000, No. 1, *the Present Threat.*

Drell, S.D., A.D. Sofaaer, and G.D. Wilson, *the New Terror, Facing the Threat of Biological and Chemical Weapons*, Hoover Institution Press, Stanford University (1999).

Dresselhaus, M., and G. Dresselhaus. 1995. *Ann. Rev. Mat. Sci.* 25:487.

Dresselhaus, M.S. 1998. Carbon-based nanostructures. In *R&D status and trends*, ed. Siegel et al.

Dresselhaus, M.S., G. Dresselhaus, and P. Eklund. 1996. *Science offullerenes and carbon nanotubes.* San Diego: Academic Press.

Dresselhaus, M.S., G. Dresselhaus, and P. Eklund. 1996. *Science offullerenes and carbon nanotubes.* San Diego: Academic Press. Gleiter, H. 1989. *Prog. Mater. Sci.* 33:223.

Drexler K.E.: 1995, 'Introduction to nanotechnology', in: Krummenacker, M. & Lewis, J. (eds.), *Prospects in Nanotechology. Proceedings of the 1st general conference on nanotechnology: developments, applications, and opportunities, November 11-14, 1992, Palo-Alto*, John Wiley & Sons, New York, pp. 1-20.

Drexler, E., Phoenix, C: 2004, 'Self-Replication in Nanotechnology: Feasible, Potentially Safe, and Unnecessary' (unpublished paper, in preparation).

Drexler, E.: 1986, *Engines of Creation: the Coming Era of Nanotechnology*, Anchor Books, New York (expanded edition with a new afterword, 1990).

Drexler, E.: 1990 [1986], *Engines of Creation*, Anchor Books, New York.

Drexler, E.: 1992, *Nanosystems: Molecular Machinery, Manufacturing, and Computation*, John Wiley & Sons, New York.

Drexler, E.: 2003a, 'Open Letter to Richard Smalley', *Chemical & Engineering News*, 81, 38-39.

Drexler, E.: 2003b, 'Drexler Counters', *Chemical & Engineering News*, 81, 40-41.

Drexler, E.; Smalley, R.E.: 2003, 'Point-Counterpoint: Nanotechnology', *Chemical and Engineering News*, 81(48), 37-42.

Drexler, K. Eric. 1986. *Engines of Creation*. Garden City, N.Y.: Anchor Press/Doubleday.

Drexler, K.E.: 1981, 'Molecular Engineering—an Approach to the Development of General Capabilities for Molecular Manipulation', *Proceedings of the National Academy of Sciences of the United States of America—Physical Sciences*, 78, 5275-5278.

Drexler, K.E.: 1981, 'Molecular engineering: An approach to the development of general capabilities for molecular manipulation', *Proceedings of the National Academy of Sciences*, 78, No. 9, Chemistry section, 5275-78.

Drexler, K.E.: 1986, *Engines of Creation*, Anchor Books, New York.

Drexler, K.E.: 1990 [1986], *Engines of Creation: the Coming Era of Nanotechnology*, Anchor Books, New York.

Drexler, K.E.: 1992, *Nanosystems: Molecular Machinery, Manufacturing, and Computation*, Wiley, New York.

Drexler, K.E.: 2001, 'Machine-Phase Nanotechnology', *Scientific American*, (Sept.), 66-67.

Drexler, K.E.: 2003a, 'Open Letter', *Chemical & Engineering News*, 81, no. 48, 37-42.

Drexler, K.E.: 2003b, 'An open letter to Richard Smalley', April 16, [published on KurzweilAI.net].

Drexler, K.E.: 2004, 'Nanotechnology: From Feynman to Funding', *Bulletin of Science, Technology & Society*, 24, 21-27.

Dudenbostel, D., K.-L. Krieger, C. Candler and R. Laur. 1997. A new passive CMOS telemetry chip to receive power and transmit data for a wide range of sensor applications. *Proceedings of Transducers '97*, an International Conf. On Solid-State Sensors and Actuators, Chicago, June (published by the IEEE), p. 995.

Dupuy, J.P. "Common Knowledge, Common Sense." *Theory and Decision* 27 (1989): 37-62.

Dupuy, J.P. "Two Temporalities, Two Rationalities: A New Look at Newcomb's Paradox." in Bourgine, P. & Walliser, B. (eds.), Economics and Cognitive Science. New York: Pergamon. 1992, 191-220.

Dupuy, J.P.: 2000, *the Mechanization of the Mind,* Princeton University Press, Princeton N.J.

Dupuy, J.-P.: 2004, 'Complexity and Uncertainty: A Prudential Approach to Nanotechnology', in: European Commission (Community Health and Consumer Protection): *Nanotechnologies: A Preliminary Risk Analysis on the Basis of a Workshop, Brussels, 1-2 March 2004,* pp. 71-93 (www.europa.eu.int/comm/health/ph_risk/ documents/ev_20040301_en.pdf).

Dupuy, J.P.: 2004, 'Pour une evaluation normative du programme nanotechnologique', *Annales des Mines,* (February 2004), 27-32.

Dym, H. and H.P. McKean,. 1972. *Fourier Series and Integrals.* Academic Press.

E. Francis J. Ring, Barbara Phillips, eds., *Recent Advances in Medical Thermology,* Plenum Press, New York, 1984.

E. Winfree, F. Liu, L.A. Wenzler, and N.C. Seeman, Design and Self-Assembly of Two-Dimensional DNA Crystals, *Nature* 394, 539-544 (1998).

E. Winfree, X. Yang and N.C. Seeman, Universal Computation via Self-assembly of DNA: Some Theory and Experiments, In: DNA Based Computers II, ed. by L.F. Landweber and E.B. Baum, *Am. Math. Soc. Providence,* pp. 191-213 (1998).

E.M. Purcell, "Life at low Reynolds number," *Am. J. Physics* 45 (January 1977):3-11.

E.T. Foley, A.F. Kam, J.W. Lyding, P.H. Avouris, P.H. (1998), "Cryogenic UHV-STM study of hydrogen and deuterium desorption from Si(100)," *Phys. Rev. Lett.* 80(1998):1336-1339.

Ehrlich, Paul R., Anne H. Ehrlich, and John P. Holdren. 1977. *Ecoscience: Population, Resources, Environment*. San Francisco: W.H. Freeman.

Eigler, D.M.; Schweizer, E.K.: 1990, 'Positioning Single Atoms with a Scanning Tunneling Microscope', *Nature,* 344, 524-526.

Einsiedel, E.F.; Goldenberg, L.: 2004, 'Dwarfing the Social? Nanotechnology Lessons from the Biotechnology Front', *Bulletin of Science, Technology & Society,* 24, 28-33.

Elkins, C: 1979, 'Science Fiction *versus* Futurology: Dramatic *Versus* Rational Models',
Science Fiction Studies, 6, 20-31. Ellsberg, D. "Risk, Ambiguity and the Savage Axioms." *Quarterly Journal of Economics* 75 (1961): 643-669. Ellul, J. 1964. *the Technological Society.* New York: Alfred Knopf. (First published 1954).

Dupuy, J.P. "Two Temporalities, Two Rationalities: A New Look at Newcomb's Paradox." in Bourgine, P. & Walliser, B. (eds.), Economics and Cognitive Science. New York: Pergamon. 1992, 191-220.

Dupuy, J.P.: 2000, *the Mechanization of the Mind,* Princeton University Press, Princeton N.J.

Dupuy, J.-P.: 2004, 'Complexity and Uncertainty: A Prudential Approach to Nanotechnology', in: European Commission (Community Health and Consumer Protection): *Nanotechnologies: A Preliminary Risk Analysis on the Basis of a Workshop, Brussels, 1-2 March 2004,* pp. 71-93 (www.europa.eu.int/comm/health/ph_risk/ documents/ev_20040301_en.pdf).

Dupuy, J.P.: 2004, 'Pour une evaluation normative du programme nanotechnologique', *Annales des Mines,* (February 2004), 27-32.

Dym, H. and H.P. McKean,. 1972. *Fourier Series and Integrals.* Academic Press.

E. Francis J. Ring, Barbara Phillips, eds., *Recent Advances in Medical Thermology,* Plenum Press, New York, 1984.

E. Winfree, F. Liu, L.A. Wenzler, and N.C. Seeman, Design and Self-Assembly of Two-Dimensional DNA Crystals, *Nature* 394, 539-544 (1998).

E. Winfree, X. Yang and N.C. Seeman, Universal Computation via Self-assembly of DNA: Some Theory and Experiments, In: DNA Based Computers II, ed. by L.F. Landweber and E.B. Baum, *Am. Math. Soc, Providence*, pp. 191-213 (1998).

E.M. Purcell, "Life at low Reynolds number," *Am. J. Physics* 45(January 1977):3-11.

E.T. Foley, A.F. Kam, J.W. Lyding, P.H. Avouris, P.H. (1998), "Cryogenic UHV-STM study of hydrogen and deuterium desorption from Si(100)," *Phys. Rev. Lett.* 80(1998):1336-1339.

Ehrlich, Paul R., Anne H. Ehrlich, and John P. Holdren. 1977. *Ecoscience: Population, Resources, Environment*. San Francisco: W.H. Freeman.

Eigler, D.M.; Schweizer, E.K.: 1990, 'Positioning Single Atoms with a Scanning Tunneling Microscope', *Nature*, 344, 524-526.

Einsiedel, E.F.; Goldenberg, L.: 2004, 'Dwarfing the Social? Nanotechnology Lessons from the Biotechnology Front', *Bulletin of Science, Technology & Society*, 24, 28-33.

Elkins, C: 1979, 'Science Fiction *versus* Futurology: Dramatic *versus* Rational Models', *Science Fiction Studies*, 6, 20-31.

Ellsberg, D. "Risk, Ambiguity and the Savage Axioms." *Quarterly Journal of Economics* 75 (1961): 643-669.

Ellul, J. 1964. *the Technological Society*. New York: Alfred Knopf. (First published 1954).

Empedocles, S., Bawendi, M. 1999. "Spectroscopy of Single CdSe Nanocrystallites," *Accounts of Chemical Research*, 32, 389-396.

Entingh, D., Dunn, A., Glassman, E., Wilson, J.E., Hogan, E. and Damstra, T. (1975) in *Handbook of Psychobiology*, eds. Gazzinga, M.S. & Blakemore, C. (Academic, New York), pp. 201-238.

Esant. 1999. "Economic and Social Aspects of Nanotechnology." First draft report of working group 6 of the Euroconferences on Nanoscience for Nanotechnology. Also at http://www.nano.org.uk/ESANT99.htm.

Estermann, M., L.B. McCusker, C. Baerlocher, A. Merrouche, and H. Kessler. 1991. *Nature* 331:698.

Esumi, A., A. Suzuki, N. Aihara, K. Uswi, and K. Torigoe. 1998. *Langmuir*, 14:3157.

ETC: 2003a, *the Big Down: Atom Tech' Technologies Converging at the Atomic Scale*. Winnipeg, Canada: Action Group on Erosion, Technology and Concentration [www.etcgroup. org/documents/TheBigDown.pdf].

ETC: 2003b, 'No Small Matter II: The Case for a Global Moratorium 'Size Matters!' *Occasional Paper Series* 7(1) [www.etcgroup.org/documents/ Occ.Paper_Nanosafety.pdf].

Etzkowitz, H., 2001, 'Nano-Science and Society: Finding a Social Basis for Science Policy', in: M.C. Roco and W.S. Bainbridge (eds.), *Societal Implications of Nanoscience and Nanotechnology*, Dordrecht: Kluwer, pp. 121-128.

Etzkowitz, H., M. Gulbrandsen and J. Levitt. 2000. *Public Venture Capital: Government Funding Sources for Technology Entrepreneurs*. New York: Harcourt.

Etzkowitz, Henry, 2001. "The Endless Transition: Finding a Social Basis for Science Policy." *Scipolicy: the Journal of Science and Health Policy*. Vol. 1. No. 2, Spring 2001.

Etzkowitz, Henry, Carol Kemelgor and Brian Uzzi. 2000. *Athena Unbound: the Advancement of Women in Science and Technology*. Cambridge: Cambridge University Press.

European Comission: *European Workshop on Social and Economic Research on Nanotechnologies and Nanosciences, Brussels, 14-15 April 2004* [www.stage-research.net/STAGE/PAGES/Nano.html].

European Commission (Community Health and Consumer Protection): 2004, *Nanotechnologies: A Preliminary Risk Analysis on the Basis of a Workshop, Brussels, 1-2 March 2004* [www.europa.eu.int/comm/health/ ph_risk/documents/ev_2004030l_en.pdf].

European Commission: 2004, *Converging Technologies: Shaping the Future of European Societies (Report of the High Level Expert Group "Foresighting the New Technology Wave"),* Brussels: European Commission Research [http://europa.eu.int/comm/ research/conferences/2004/ntw/index_en.html].

Evans, D.F.; Wennerstrom, H.: 1999, *the Colloidal Domain: Where Physics, Chemistry, Biology and Technology Meet,* John Wiley & Sons, New York.

Evans, J.H.: 2002, *Playing God? Human Genetic Engineering and the Rationalization of Public Bioethical Debate,* Chicago University Press, Chicago.

F. Buot, "Mesoscopic Physics and Nanoelectronics: Nanoscience and Nanotechnology," *Physics Reports,* pp. 73-174, 1993.

F. Capasso and S. Datta, "Quantum Electron Devices," *Physics Today,* pp. 74-82, February 1990.

F. Liu, H. Wang and N.C. Seeman, Short Extensions to Sticky Ends for DNA Nanotechnology and DNA-Based Computation, *Nanobiology* 4, 257-262 (1999).

F. Liu, R. Sha and N.C. Seeman, Modifying the Surface Features of Two-Dimensional DNA Crystals, *Journal of the American Chemical Society.* 121, 917-922 (1999).

F. Mathieu, S. Liao, C. Mao, J. Kopatsch, T. Wang, and N.C. Seeman, Six-Helix Bundles Designed from DNA, *NanoLetters* 5, 661-665 (2005).

Fantel, Hans. 1987. "Portable CD Players Advance." *New York Times,* May 17, Section 2, p. 33, col. 1.

Faraday, M.: 1857, 'Experimental relations of gold (and other metals) to light', *Philosophical Transactions of the Royal Society,* London, 147, 145-181.

Fedor N. Dzegilenko, Deepak Srivastava, Subhash Saini, "Simulations of carbon nanotube tip assisted mechano-chemical reactions on a diamond surface," *Nanotechnology* 9 (December 1998):325-330.

Fennema, O.R. (1973) in *Low-Temperature Preservation of Foods and Living Matter,* eds. Fennema, O.R., Powrie, W.D. & Marth, E.H. (Dekker, New York), pp. 150-239, 436-503.

Fermi, E., J. Pasta, and S. Ulam. 1974. Studies of Nonlinear Problems, I. *Los Alamos Report.* LA1940. 1955.

Feyerabend, P. 1981, *Realism, Rationalism and Scientific Method* (Philosophical Papers, volume 1), Cambridge University Press, Cambridge.

Feynman, R. "There's Plenty of Room At the Bottom." Talk given on at the annual meeting of the American Physical Society at the California Institute of Technology, 1959.

Feynman, R. (1961) in *Miniaturization,* ed. Gilbert, H.D. (Reinhold, New York), pp. 282-296.

Feynman, R., "There's Plenty of Room at the Bottom: An invitation to Enter a New Field of Physics," Talk at the Annual Meeting of the American Physical Society, 29 December 1959. Reprinted in Appendix B of Crandall and Lewis. See citation above.

Feynman, R.: 1960, 'There's Plenty of Room at the Bottom', *Engineering and Science,* 23, 22-36.

Feynman, R.: 1999 [1959], 'There's Plenty of Room at the Bottom: An Invitation to Enter a New Field of Physics', in: *the Pleasure of Finding Things Out,* Perseus, Cambridge, MA, pp. 117-139.

Feynman, R.P. 1961. "There is plenty of room at the bottom." In *Miniaturization.* New York: Reinhold.

Fielder, F.A.; Reynolds, G.H.: 1994, 'Legal Problems of Nanotechnology: An Overview', *Southern California Interdisciplinary Law Journal,* 3, 593-629.

Filler, M., C. Mui, G. Wang, C. Musgrave, and S. Bent, "Competition and Selectivity in the Reaction of Nitriles on Ge (100)-2xl," *Journal of the American Chemical Society* 125, 4928-4936 (2003).

Fisher, D.E. and Fisher, M.J. *Tube: the Invention of Television.* Harcourt Brace, 1997.

FitzGerald, Frances. 2000. *Way Out There in the Blue: Reagan, Star Wars, and the End of the Cold War.* New York: Simon & Schuster.

Fleischer, Th.; Decker, M. & Fiedeler, U. (eds.): 2004, *Grofie Aufmerksamkeit fur kleine Welten' Nanotechnologie und ihre Folgen,* special issue of *Technikfolgenabschdtzung' Theorie und Praxis,* 13 (2), 5-85 [www.itas.fzk.de/ tatup/042/inhalt.htm].

Fleming, D. "The Economics of Taking Care: An Evaluation of the Precautionary Principle." in Freestone, D. & Hey, E. (eds.), the Precautionary Principle and International Law. La Haye: Kluwer Law International, 1996.

Foerster, H.V.: 1962, 'Bio-Logic', in: E.E. Bernard & M.R. Kare (eds.), *Biological prototypes and synthetic systems,* Plenum Press, New York, Vol. 1., pp. 1-12

Fogelberg, H. & Glimell, H. 2003, 'Molecular Matters: In Search of the Real Stuff, in: H. Fogelberg & H. Glimell, *Bringing Visibility to the Invisible: Towards A Social Understanding of Nanotechnology,* Gdteborg: Gb'teborg University, pp. 5-32 [www.sts.gu.se/publications/STS_report_6.pdf].

Fogelberg, H. & Glimell, H.: 2003, *Bringing Visibility to the Invisible: Towards A Social Understanding of Nanotechnology,* Goteborg: Gb'teborg University [www.sts.gu.se/ publications/ STS_report_6.pdf].

Fogelberg, H.: 2003, '"Science" and "Technology" as Knowledge Cultures', in: H. Fogelberg, H. Glimell, *Bringing Visibility to the Invisible: Towards a Social Understanding of Nanotechnology,* Goteborg University, Goteborg, pp. 77-85.

Fogelberg, H.: 2003, 'The Grand Politics of Technoscience: Contextualizing Nanotechnology', in: H. Fogelberg & H. Glimell, *Bringing Visibility to the Invisible: Towards A Social Understanding of Nanotechnology,* Goteborg: Goteborg University, pp. 33-54 [www.sts.gu.se/publications/ STS_report_6.pdf].

Fogelberg, H.: 2003, 'The Material Culture of Nanotechnology', in: H. Fogelberg & H. Glimell, *Bringing Visibility to the Invisible: Towards A Social Understanding of Nanotechnology,* Goteborg: Goteborg University, pp. 99-114.

Fogelberg, H.; Glimell, H.: 2003, *Bringing Visibility to the Invisible: Towards a Social Understanding of Nanotechnology,* Goteborg University, Goteborg.

Forest, G. 2000. Comment at Conference on Societal Implications of Nanoscience and Nanotechnology, September 29, 2000, NSF, Arlington, VA.

Forest, M.G., Q. Wang, and H. Zhou. 2000a. Exact banded patterns from a Doi-Marrucci-Greco model of nematic liquid crystal polymers. *Physical Review E.* Vol. 61(6). 6665-6672.

Forest, M.G., Q. Wang, and H. Zhou. 2000b. Homogeneous pattern selection and director instabilities of nematic liquid crystal polymers induced by elongational flows. *Physics of Fluids.* Vol. 12(3). 490-498.

Forum Proceedings of Sigma Xi, the Scientific Research Society, *Ethics, Values, and the Promise of Science,* Feb. 25-26, 1993.

Foucault, M. 1977. *Power/Knowledge: Selected Interviews and Other Writings 1972-1977.* New York, Pantheon Books.

Foucault, M.: 1992, *the Archaeology of Knowledge,* Routledge, London.

Fourier, J. 1878. *the Analytical Theory of Heat,* translated by A. Freeman. Cambridge University Press. London and New York.

Francoeur, E.: 1997, 'The Forgotten Tool: the Design and Use of Molecular Models', *Social Studies of Science,* 27, 7-40.

Frankel, M. 1996. "Perception, Reality, and the Political Context of Conflict of Interest in University-Industry Relationships," *Academic Medicine,* 71:12, December, pp. 1297-1304.

Franklin, N.R.; Dai, H.J. 2000. "An enhanced CVD approach to extensive nanotube networks with directionality", *Advanced Materials*, 12, 890-894.

Frazier, G., "An Ideology For Nanoelectronics," in *Concurrent Computations: Algorithms, Architecture, and Technology*, Plenum Press, New York, 1988.

Freitas, L. *Nanomedicine*. Bioscience, 1999.

Frenken, J.W.M. 1998. "Scanning Tunneling Microscopy", section 5.2, A. ten Wolde (ed.), *Nanotechnology: Towards a molecular construction kit*. STT Report #60, the Hague, 289-299.

Friedlander, S.K. 1993. Controlled synthesis of nanosized particles by aerosol processes. *Aerosol Sci. Technol.* 19:527.

Friedlander, S.K. 1998. Synthesis of nanoparticles and their agglomerates: aerosol reactors. In *R&D status and trends*, ed. Siegel et al.

Friedlander, S.K. 1998. Synthesis of nanoparticles and their agglomerates: Aerosol reactors. In *R&D status and trends*, ed. Siegel et al.

Friedlander, S.K., H.D. Jang, and K.H. Ryu. 1998. *Appl. Phy. Lett.* 72(2):173.

G. Serrano & N.C. Seeman, Nanotecnologia basada en ADN, Revista de Quimica 19, 11-20 (2004), in Spanish.

Gacoin, T., L. Malier, and J.P. Boilot. 1997. *Chem. Mater.* 9:1502.

Galison, P.: 1996, 'Computer Simulations and the Trading Zone', in: P. Galison, D.J. Stump (eds.), *the Disunity of Science: Boundaries, Contexts, and Power*, Stanford University Press, Stanford, pp. 118-157.

Galison, P.: 1997, *Image and Logic: A Material Culture of Microphysics*, University of Chicago Press, Chicago.

Galison, P.; Stump, D.J. (eds.): 1996, *the Disunity of Science: Boundaries, Contexts, and Power*, Stanford University Press, Stanford.

Gallego-Juarez, J.A. (1989) Piezoelectric ceramics and ultrasonic transducers. *J. Phys. E: Sci. Instrum.* 22, 804-816

Gates, B.; Park, S.H.; Xia, Y.N. 2000. "Tuning the photonic bandgap properties of crystalline arrays of polystyrene beads by annealing at elevated temperatures", *Advanced Materials*, 12, 653-656.

Geneva Convention, *Protocol for the Prohibition of the Use in War of Asphyxiating, Poisonous or Other Gases, and of Bacteriological Methods of Warfare*, signed at Geneva on June 17,1925.

George D. Skidmore, Eric Parker, Matthew Ellis, Neil Sarkar, Ralph Merkle, "Exponential assembly," paper presented at the 8th Foresight Conference, Washington DC, November 2000; http://www.foresight.org/Conferences/MNT8/Papers/Skidmore/ index.html

George M. Whitesides, "The Once and Future Nanomachine," *Scientific American* 285 (September 2001):78-83.

Gergen, K.J. Social psychology as history. *Journal of Personality and Social Psychology*, 1973, 26, 309-20.

Gerlach, N.; Hamilton, S.: 2000, 'Telling the Future, Managing the Present: Business Restructuring Literature as SF', *Science Fiction Studies*, 27, 461-477.

Gieryn, T.: 1999, *Cultural Boundaries of Science*, Univ. of Chicago Press, Chicago.

Gieryn, T.F.: 1999, *Cultural Boundaries of Science: Credibility on the Line*, University of Chicago Press, Chicago.

Gieryn, T.F.; Figert, A.E.: 1986, 'Scientists Protect Their Cognitive Authority: the Status Degradation Ceremony of Sir Cyril Burt', in: G. Bb'hme, N. Stehr (eds.), *the Knowledge Society*, Reidel, Dordrecht, pp. 67-89.

Gleiter, H. 1989. *Prog. Mater. Sci.* 33:223. Goddard, W.A. 1998. Nanoscale theory and simulation. In *R&D status and trends*, ed. Siegel et al.

Gleiter, H. 1990. *Progress in Materials Science* 33:4. Guinier, A. 1938. *Nature* 142:569; Preston, G.D. 1938. *Nature* 142:570.

Glimell, H.: 2001, 'Challenging Limits' Excerpts from an Emerging Ethnography of Nano Physicists', in: H. Glimell & O. Johlin (eds.), *the Social Production of Technology: On the Everyday Life with Things*, Goteburg, SE: BAS Publisher, chapter 7, pp. 111-131 (reprinted in: H. Fogelberg & H. Glimell, *Bringing Visibility to the Invisible: Towards A Social Understanding of Nanotechnology*, Göteborg: Göteborg University, pp. 115-139.

Glimell, H.: 2001, 'Dynamics of the Emerging Field of Nanoscience', in: M.C. Roco and W.S. Bainbridge (eds.), *Societal Implications of Nanoscience and Nanotechnology*, Dordrecht: Kluwer, pp. 156-160.

Glimell, H.: 2003, 'A Nano Narrative: the Mircopolitics of a New Generic Enabling Technology', in: H. Fogelberg & H. Glimell, *Bringing Visibility to the Invisible: Towards A Social Understanding of Nanotechnology*, Göteborg: Göteborg University, pp. 55-77 [www.sts.gu.se/publications/STS_report_6.pdf].

Glimell, H.: 2003, 'Dynamics of the Emerging Field of Nanoscience', in: H. Fogelberg & H. Glimell, *Bringing Visibility to the Invisible: Towards A Social Understanding of Nanotechnology*, Göteborg: Göteborg University, pp. 79-85 [www.sts.gu.se/publications/STS_report_6.pdf].

Glimell, H.: 2004, 'Grand Visions and Lilliput Politics: Staging the Exploration of the 'Endless Frontier'', in: D. Baird, A. Nordmann & J. Schummer (eds.), *Discovering the Nanoscale*, Amsterdam: IOS Press, pp. 231-246.

Goddard, W.A. 1998. Nanoscale theory and simulation. In *R&D status and trends*, ed. R. Siegel et al.

Godsell, D.: 2003, *Living Machinery. Bionanotechnology: Lessons from Nature*, Wiley-Liss, New-York.

Goldbeck-Wood, G. and A.H. Windle. 1999. Lattice modeling of nematodynamics. *Rheol. Ada.* Vol. 38. 548-561.

Goldhaber-Gordon, D., Montemerlo, M.S., Love, J.C., Opiteck, G.J., and Ellenbogen, J.C., "Overview of Nanoelectronic Devices," submitted to the *Proceedings of the IEEE*, February 1997. For more information, please send e-mail to nanotech@mitre.org.

Gorman, M.E.; Groves, J.F. & Shrager, J.: 2004, 'Societal Dimensions of Nanotechnology as a Trading Zone: Results from a Pilot Project', in: D. Baird, A. Nordmann & J. Schummer (eds.), *Discovering the Nanoscale*, Amsterdam: IOS Press, pp. 63-73.

Graham, P.: 2002, 'Hypercapitalism: Language, New Media and Social Perceptions of Value', *Discourse and Society*, 13(2), 227-249.

Gref, Ruxandra, Yoshiharu Minamitake, Maria Teresa Peracchia, Vladimir Trubetskoy, Vladimir Torchilin, Robert Langer. 1994. "Biodegradable Long Circulating Polymeric Nanospheres", *Science*, Vol. 263, 18 Mar, pp 1600-1603.

Griliches, Zvi. 1957. "Hybrid Corn: An Exploration in the Economics of Technological Change," *Econometrica*, 25 (4), pp 501-522.

Griliches, Zvi. 1960. "Hybrid Corn and the Economics of Innovation," *Science*, 132, pp 275-280. Mass, William. 1989. "Mechanical and Organizational Innovation: the Case of the Draper Loom," *Business History Review*, 63 (Winter).

Grinbaum, A. & Dupuy, J.-P.: 2004, 'Living with Uncertainty: Toward the Ongoing Normative Assessment of Nanotechnology', *Techne: Research in Philosophy and Technology*, 8(3) (forthcoming).

Grinbaum, A.: 2004, 'La condition de l'homme moderne et les nanotechnologies', in: G. Nivat (ed.), *Les limites de l'humain. 39émes Rencontres Internationales de Genéve,* L'Age d'Homme, Geneve, p. 141.

Grondahl B., W. Puppe, A. Hoppe, I. Kühne, J.A. Weigl, H.J. Schmitt. 1999. "Rapid Identification of Nine Microorganisms Causing Acute Respiratory Tract Infections by Single-Tube Multiplex Reverse Transcription-PCR: Feasibility Study." *Journal of Clinical Microbiology,* Jan; 37 (1): pp. 1-7.

Grosberg, A.Y. and A.R. Khokhlov. 1994. *Statistical Physics of Macromolecules,* AIP Series in Polymers and Complex Materials, AIP Press, New York.

Gross, M.: 1999, *Travels to the Nanoworld: Miniature Machinery in Nature and Technology,* Plenum, New York.

Grunwald, A.: 2004, 'Ethische Aspekte der Nanotechnologie. Eine Felderkundung', *Technikfolgenabschdtzung' Theorie und Praxis,* 13 (2), 71-78.

Guarnieri, F., M. Fliss, and C. Bancroft. 1996. Making DNA Add. *Science* 273:220-223. Hanes, J., J.L. Cleland, and R. Langer. 1997.

Gubrud, M. A., *Nanotechnology and International Security,* Fifth Foresight Conference on Molecular Nanotechnology, November 5-8, 1997. Ref: http://www.foresight.org/ Conferences/MNT05/ Papers/Gubrud/.

Guerlac, H.: 1961, *Lavoisier—The Crucial Year,* Cornell University Press, Ithaca.

Gund, P., Andose, J.D., Rhodes, J.B. & Smith, G.M. (1980) *Science* 208, 1425-1431.

Giinther, B., A Baalmann, and H. Weiss. 1990. *Mater. Res. Soc. Symp. Proc.* 195:611-615.

Gupta, V.K. & Pangannaya, N.B.: 2000, 'Carbon Nanotubes: Bibliometric Analysis of Patents', *World Patent Information,* 22, 185-189.

Gurley, Bill. 2000. Digital music: the real law is Moore's law. *Fortune.* 2 October, page 268. Joy, Bill. 2000. Why the future doesn't need us. *Wired.* April, pp. 238-262.

Guthold, M.; Falvo, M.R.; Matthews, W.G.; Paulson, S.; Washburn, S.; Erie, D.A.; Superfine, R.; Brooks, F.P.; Taylor, R.M.: 2000, 'Controlled Manipulation of Molecular Samples with the NanoManipulator', *IEEE-ASME Transactions on Mechatronics,* 5, 189-198.

Gutte, B., Dannigen, M. & Wittschieber, E. (1979) *Nature (London)* 281, 650-655.

H. Qiu, J.C. Dewan and N.C. Seeman, A DNA Decamer with a Sticky End: the Crystal Structure of d-CGACGATCGT. *J. Mol. Biol.* 267, 881-898 (1997).

H. Yan and N.C. Seeman, Edge-Sharing Motifs in DNA Nanotechnology. *Journal of Supramolecular Chemistry* 1, 229-237 (2003).

H. Yan, X. Zhang, Z. Shen and N.C. Seeman, A Robust DNA Mechanical Device Controlled by Hybridization Topology, *Nature* 415, 62-65 (2002).

Hacking, I.: 1983, *Representing and Intervening: Introductory Topics in the Philosophy of Natural Science,* Cambridge University Press, Cambridge, UK.

Hacking, I.: 1992, 'The Self-Vindication of the Laboratory Sciences', in: A. Pickering (ed.), *Science as Practice and Culture,* University of Chicago Press, Chicago, pp. 29-64.

Hadjiconstantinou, N.G. 1999. Combining atomistic and continuum simulations of contact-line motion. *Physical Review E.* Vol. 59(2).2, 2475-2478.

Hadjipanayis, C.G. 1998. Nanostructured magnetic materials. In *R&D Status and Trends,* ed. Siegel et al.

Hamilton, S.: 2003, 'Biotechnology, SF and the Media', *Science Fiction Studies,* 30, 267-282.

Han, W., S. Fam, O. Li, and Y. Hu. 1998. Synthesis of gallium nitride nanorods through a carbon nanotube-confined reaction. *Science* 277:1287-1289.

Hansson, S.O. "The False Promises of Risk Analysis." *Ratio* 6 (1993): 16-26.

Hansson, S.O.: 2004, 'Great Uncertainty about Small Things', *Techne: Research in Philosophy and Technology*, 8(3) (forthcoming).

Haraway, D.J.: 1997, *Modest Witness@Second Millenium.FemaleMand Meets OncoMoused*, Routledge, London and New York.

Harold J. Morowitz, "A Model of Reproduction," *American Scientist* 47(1959):261-263. Hayes, B., "The Square Root of NOT," *American Scientist*, July-August 1995, pp. 304-308.

Hayles, N.K. (ed.): 2004, *Nanoculture: Implications of the New Technoscience*, Bristol, UK: Intellect Books.

Hayles, N.K.: 2004, 'Connecting the Quantum Dots: Nanotechscience and Culture', in: N.K. Hayles (ed.), *Nanoculture: Implications of the New Technoscience*, Bristol, UK: Intellect Books, pp. 11-26.

Hayward, R.C.; Saville, D.A.; Aksay, L.A. 2000. "Electrophoretic assembly of colloidal crystals with optically tunable micropatterns", *Nature*, 404, 56-59.

Hecht, G., Allen, M.T.: 2001, 'Introduction: Authority, Political Machines, and Technology's History', in: M.T. Allen, G. Hecht (eds.), *Technologies of Power: Essays in Honor of Thomas Parke Hughes and Agatha Chipley Hughes*, MIT Press, Cambridge, MA, pp. 1-23.

Hecht, G.: 1998, *the Radiance of France: Technology, Politics, and National Identity after World War II*, MIT Press, Cambridge, MA.

Heidegger, M.: 1962, *Being and Time*, trans. J. Macquarrie, E. Robinson, Harper & Row, New York.

Heidegger, M.: 1977, *the Question Concerning Technology and Other Essays*, Harper Colophon Books, New York.

Held, R., T. Heinzel, A.P. Studerus, K. Ensslin, and M. Holland. 1997. Semiconductor quantum point contact fabricated by lithography with an atomic force microscope. *Appl. Phys. Lett.* 71:2689-91.

Hellemans, A. 1998. X-rays find new ways to shine. *Science* 277:1214-15.

Helmreich, S.: 1998, *Silicon Second Nature: Culturing Artificial Life in a Digital World*, University of California Press, Berkeley, CA.

Hemeren, G.: 2004, 'Nano Ethics Primer', in: European Commission (Community Health and Consumer Protection): *Nanotechnologies: A Preliminary Risk Analysis on the Basis of a Workshop, Brussels, 1-2 March 2004* (www.europa.eu.int/comm/ health/ ph_risk/documents/ ev_20040301_en.pdf), pp. 95-102.

Hendrickson, C.T., A. Horvath, S. Joshi, and L.B. Lave. 1998. Economic input-output models for environmental life-cycle assessment. *Environmental Science & Technology*, p. 32.

Hennig, J.: 2004, 'Changes in the Design of Scanning Tunneling Microscopic Images from 1980 to 1990', *Techne: Research in Philosophy and Technology*, 8(3) (forthcoming).

Hessenbruch, A.: 2004, 'Nanotechnology and the Negotiation of Novelty', in: D. Baird, A. Nordmann & J. Schummer (eds.), *Discovering the Nanoscale*, Amsterdam: IOS Press, pp. 135-144.

Hietpas, P.B., S.D. Gilman, R.A. Lee, M.R. Wood, N. Winograd, and A.G. Ewing. 1996. Development of votammetric methods, capillary electrophoresis and tof sims imaging for constituent analysis of single cells. *In Nano fabrication and biosystems*, ed.

Higgins, R.J. 1997. An economical process for manufacturing of nano-sized powders based on microemulsion-mediated synthesis. In *Proc. of the Joint NSF-NIST Conf. on Nanoparticles.*

Hiruma, K., M. Yazawa, T. Katsoyama, K. Ogawa, K. Haraguchi, M. Koguchi, and H. Kakibayashi. 1995. *J. Appl. Phys.* 77(2):476.

Hoch et al. Ho, S.V,, P.W. Sheridan, and E. Krupetsky. 1996. Supported polymeric liquid membranes for removing organics from aqueous solutions.

Hoch, H.C., L.W. Jelinski, and H.G. Craighead, eds. 1996. *Nanofabrication and biosystems.* New York: Cambridge University Press.

Hoh, J.H. and Hansma, P.K. (1992) Atomic force microscopy for high-resolution imaging in cell biology. *Trends Cell Bio.* 2, 208-213

Homer Jacobson, "On Models of Reproduction," *American Scientist* 46(1958):255-284. Hopwood, J., and S. Mann. 1997. *Chem. Mater.* 9:1819.

Howland, H.: 1962, Structural, hydraulic and economic aspects of leaf venation and shape', in: E.E. Bernard & M.R. Kare (eds.), *Biological prototypes and synthetic systems*, Plenum Press, New York, Vol. 1, pp. 183-192.

Hoyningen-Huene, P.: 1993, *Reconstructing Scientific Revolutions: Thomas S. Kuhn's Philosophy of Science* (trans, by A. Levin), University of Chicago Press, Chicago.

Huang, J.Y., Y.K. Wu, and H.Q. Ye. 1996. *Ada Mater.* 44:1211. Inoue, A. 1997. Private communication. Inturi, R.B., and Z. Szklavska-Smialowska. 1992. *Corrosion* 48:398. Karch, J., R. Birringer, and H. Gleiter. 1987. *Nature* 330:556.

Hubbell, J.A., and R. Langer. 1995. Tissue engineering. *Chem. Eng. News* (March 13): 42-54.

Hughes, P.M., *Global Threats and Challenges: the Decades Ahead*, Lt. Gen. Patrick M. Hughes, Director of the Defense Intelligence Agency, in Statement for the Senate Committee on Intelligence, January 28, 1998. Ref:http://www.fas.org/ irp/congress/1998_hr/s980128h.htm.

Hughes, T.P.: 1983, *Networks of Power: Electrification in Western Society, 1880-1930*, Johns Hopkins University Press, Baltimore.

Hughes, T.P.: 1994, 'Technological Momentum', in: M.R. Smith, L. Marx (eds.), *Does Technology Drive History? the Dilemma of Technological Determinism*, MIT Press, Cambridge, MA, pp. 101-114.

Huijer, M.: 2003, 'Reconsidering Democracy: History of the Human Genome Project', *Science Communication*, 24(4), 479-502.

Hullmann, A. & Meyer, M.: 2003, 'Publications and Patents in Nanotechnology. An overview of previous studies and the state of the art', *Scientometrics*, 58 (3) 507-527.

Huntington, Samuel P. 1996. *the Clash of Civilizations and the Remaking of the World Order* (New York: Simon and Schuster).

Huo, Q., D.I. Margolese, U. Ciesla, P. Feng, T:E. Gier, P. Sieger, R. Leon, P.M. Petroff, F. Schuth, and G.D. Stucky. 1994. *Nature* 368:317.

Huo, Q., D.I. Margolese, U. Ciesla, P. Feng, T.E. Gier, P. Sieger, R. Leon, P.M. Petroff, F. Schuth, and G.D. Stucky. 1994. *Nature* 368:317.

Huxley, A. *Brave New World*. New York: Harper and Row, 1946.

Ihde, D.: 1991, *Instrumental Realism: the Interface between Philosophy of Science and Philosophy of Technology*, Indiana University Press, Bloomington.

Imae, Y., and T. Atsumi. 1989. T. Na+-driven bacterial flagellar motors: A mini-review. *J. Bioenergetics and Biomembranes* 21:705-716.

Itakura, K. & Riggs, A.D. (1980) *Science* 209, 1401-1405.

J. Chen and N.C. Seeman, the Synthesis from DNA of a Molecule with the Connectivity of a Cube, *Nature* 350, 631-633 (1991).

J.D. Le, Y. Pinto, N.C. Seeman, K. Musier-Forsyth, T.A. Taton and R.A. Kiehl, Self-Assembly of Nanoelectronic Component Arrays by *In Situ* Hybridization to 2D DNA Scaffolding, *NanoLetters* 4, 2343-2347 (2004).

J. Qi, X. Li, X.P. Yang and N.C. Seeman, the Ligation of Triangles Built from Bulged Three-Arm DNA Branched Junctions, *Journal of the American Chemical Society*, 118, 6121-6130 (1996).

J. Storrs Hall, "Architectural considerations for self-replicating manufacturing systems," *Nanotechnology* 10 (1999):323-330; http://www.foresight.org/Conferences/MNT6/ Papers/Hall/index.html

J. Storrs Hall, "Architectural considerations for self-replicating manufacturing systems," *Nanotechnology* 10 (1999):323-330; http://www.foresight.org/Conferences/MNT6/ Papers/Hall/index.html

J. Travis, "Tick, tock, enzyme rewinds cellular clock," *Science News* 153 (17 January 1998):37.

J.-H. Chen, N.R. Kallenbach and N.C. Seeman, A Specific Quadrilateral Synthesized from DNA Branched Junctions. *Journal of the American Chemical Society* 111, 6402-6407 (1989).

J.H. Reif, T.H. LaBean & N.C. Seeman, Challenges and Applications for Self-Assembled DNA Nanostructures, Sixth International Workshop on DNA-Based Computers, DNA 2000, Leiden, the Netherlands, (June, 2000) ed. A. Condon, G. Rozenberg. Springer-Verlag, Berlin Heidelberg, *Lecture Notes in Computer Science* 2054, 173-198, (2001).

J.K. Gimzewski, C. Joachim, R.R. Schlittler, V. Langlais, H. Tang, I. Johannsen, "Rotation of a Single Molecule Within a Supramolecular Bearing," *Science* 281(24 July 1998):531-533.

J.W. Lyding, K. Hess, G.C. Abeln, D.S. Thompson, J.S. Moore, M.C. Hersam, E.T. Foley, J. Lee, Z. Chen, S.T. Hwang, H. Choi, P.H. Avouris, I.C. Kizilyalli, "UHV-STM nanofabrication and hydrogen/deuterium desorption from silicon surfaces: implications for CMOS technology," *Appl. Surf. Sci.* 130 (1998):221-230.

J.W. Lyding, K. Hess, G.C. Abeln, D.S. Thompson, J.S. Moore, M.C. Hersam, E.T. Foley, J. Lee, Z. Chen, S.T. Hwang, H. Choi, P.H. Avouris, I.C. Kizilyalli, "UHV-STM nanofabrication and hydrogen/deuterium desorption from silicon surfaces: implications for CMOS technology," *Appl. Surf. Sci.* 130(1998):221-230.

Jacobs, K.M., et al. 2000. Postlesional Epilepsy: the Ultimate Brain Plasticity. *Epilepsia.* 41 (Suppl 6): p. S153-61.

James, E.: 1994, *Science Fiction in the 20th Century*, Oxford University Press, Oxford.

Jameson, F.: 1982, 'Progress Versus Utopia; or, Can We Imagine the Future', *Science Fiction Studies*, 9, 147-158.

Janet Raloff, "Nano Hazards: Exposure to minute particles harms lungs, circulatory system," Science News Online, Week of March 19, 2005; Vol. 167, No. 12.

Jaworek, T., D. Deher, G. Wegner, R.H. Wieringa, and A.J. Schouten. 1998. Electromechanical properties of an ultrathin layer of directionally aligned helical polypeptides. *Science* 279:57-60.

Jena, P., S.N. Khanna, and B.K. Rao. 1996. In *Science and technology of atomically engineered materials*, ed. P. Jena. River Edge, NJ: World Scientific.

Jena, P., S.N. Khanna, and B.K. Rao. 1996. In *Science and technology of atomically engineered materials*, ed. P. Jena. River Edge, NJ: World Scientific.

Johansson, M.: 2003, 'Plenty of room at the bottom: Towards an anthropology of nanoscience', *Anthropology Today*, 19 (No. 6), 3-6.

John Cumings, A. Zettl, "Low-friction nanoscale linear bearing realized from multiwall carbon nanotubes," *Science* 289(28 July 2000):602-604.

John von Neumann, *Theory of Self-Reproducing Automata*, A.W. Burks, ed., University of Illinois Press, Urbana IL, 1966.

Johnson, A.: 2004, 'The End of Pure Science: Science Policy from Bayh-Dole to the NNI', in: D. Baird, A. Nordmann & J. Schummer (eds.), *Discovering the Nanoscale*, Amsterdam: IOS Press, pp. 217-230.

Johnson, D.G. and H. Nissenbaum (eds). 1995. *Computers, Ethics and Social Values.* Prentice Hall:Englewood Cliffs, NJ. pp. 262-393.

Johnson, P.C.: 2003, 'Implications of the Continuum of Bioinformatics', in: M.C. Roco; W.S. Bainbridge (eds.), *Converging Technologies for Improving Human Performance,* Kluwer, Dordrecht, pp. 207-212.

Jonas, H. the Imperative of Responsibility. In Search of an Ethics for the Technological Age. Chicago: University of Chicago Press, 1985.

Jones, C.W., and W.J. Koros. 1994. *Carbon* 32:1419.

Joshua Lederberg, "Infectious History," *Science* 288(14 April 2000):287-293.

Jounet, C, W.K. Maser, P. Bernier, A. Loiseau, M. Lamy de la Chapelle, S. Lefrant, P. Deniard, R. Lee, and J.E. Fischer. 1997. *Nature* 388:756.

Jounet, C, W.K. Maser, P. Bernier, A. Loiseau, M. Lamy de la Chapelle, S. Lefrant, P. Deniard, R. Lee, and J.E. Fischer. 1997. *Nature* 388:756.

Joy, B.: 2000, 'Why the Future Doesn't Need us', *Wired,* 8, 1-11 [www.wired.com/wired/archive/8.04/joy.html].

Joy, Bill, *Why the future doesn't need us,* http://www.wired.eom/wired/archive/8.04/joy_pr.html.

Joy, Bill. "Why the Future Doesn't Need Us." *Wired,* April 8, 2000.

Joy, W. 2000. Why the Future Doesn't Need Us. *Wired,* April.

Judd, C, Smith, E. and Kidder, L. *Research Methods in Social Relations,* (6th Edition). Holt, Rinehart, Winston, 1991.

Junno, T., S.-B. Carlsson, H. Xu, L. Montelius, and L. Samuelson. 1998. Fabrication of quantum devices by angstrom-level manipulation of nanoparticles with an atomic force microscope. *Appl. Phys. Lett.* 72:548-550.

Justus, B.L., R.J. Tonucci, and A.D. Berry. 1992. *Appl. Phys. Lett.* 61:3151.

K.E. Drexler, "Molecular engineering: An approach to the development of general capabilities for molecular manipulation." *Proc. Nat. Acad. Sci.,* 78: 5275-5258 (1981).

K.E. Drexler, *Engines of Creation,* Anchor/Doubleday (New York, 1986).

K. Eric Drexler, "Molecular Engineering: An Approach to the Development of General Capabilities for Molecular Manipulation," *Proc. Natl. Acad. Sci. (USA)* 78(1981):5275-5278.

K. Eric Drexler, "Molecular Nanomachines: Physical Principles and Implementation Strategies," *Annu. Rev. Biophys. Biomol. Struct.* 23(1994):377-405.

K. Eric Drexler, "Nanomachinery: Atomically precise gears and bearings," *IEEE Micro Robots and Teleoperators Workshop,* Hyannis, Cape Cod, November 1987.

K. Eric Drexler, David Forrest, Robert A. Freitas Jr., J. Storrs Hall, Neil Jacobstein, Ralph Merkle, Christine Peterson, "On Physics, Fundamentals, and Nanorobots: A Rebuttal to Smalley's Assertion that Self-Replicating Mechanical Nanorobots Are Simply Not Possible"; http://www.imm.org/SciAmDebate2/smalley.html

K. Eric Drexler, *Engines of Creation: the Coming Era of Nanotechnology,* Anchor Press/Doubleday, New York, 1986; http://www.foresight.org/EOC/

K. Eric Drexler, *Nanosystems: Molecular Machinery, Manufacturing, and Computation,* John Wiley & Sons, New York, 1992; http://www.zyvex.com/nanotech/nanosystems.html; http://www.foresight.org/NanoRev/Bookstore.html#anchor 1025139

K. Eric Drexler, *Nanosystems: Molecular Machinery, Manufacturing, and Computation,* John Wiley & Sons, New York, 1992, Chapter 8.

K.E. Drexler, *Nanosystems: Molecular Machinery, Manufacturing, and Computation,* Wiley, New York, 1992.

Kalman B., F.D. Lublin. 1999. "The Genetics of Multiple Sclerosis. A Review." *Biomedicine and Pharmacotherapy*, Sep. 53 (8): pp. 358-370.

Kang, J. and C. Musgrave, "A Quantum Chemical Study of the Self-Directed Growth Mechanism of Styrene and Propylene Molecular Nanowires on the Silicon (100) 2x1 Surface," *Journal of Chemical Physics* 116, 9907-9913 (2002).

Kang, J. and C. Musgrave, "The Mechanism of Atomic Layer Deposition of SiO_2 on the Silicon (100)-2xl Surface Using $SiCl_4$ and H_2O as Precursors," *Journal of Applied Physics* 91, 3408-3414 (2002).

Karak, N., and S. Maiti. 1997. Dendritic polymers: A class of novel material. *J. Polym. Mater.* 14:105.

Karger, J., and D.M. Ruthven. 1992. *Diffusion in zeolites.* New York: J. Wiley.

Karger, J., and D.M. Ruthven. 1992. *Diffusion in zeolites.* New York: J. Wiley. Kratschmer, W., L.D. Lamb, K. Fostiropoulps, and D.R. Huffman. 1990. *Nature* 347:354.

Karplus, M. & Weaver, D.L. (1976) *Nature* (London) 260, 404-406.

Kasuga, T., M. Hiramatsu, A. Hoson, T. Sekino, and K. Niihara. 1998. *Langmuir* 14:3160.

Katari, J.E.B., V.L. Colvin, and A.P. Alivisatos. 1994. *J. Phys. Chem.* 98:4109.

Kay, L.: 2000, *Who Wrote the Book of Life? A History of the Genetic Code*, Stanford University Press, Stanford.

Kay, L.: 2000, *Who Wrote the Book of Life? A History of the Genetic Code*, Stanford University Press, Stanford.

Ke, M., S.A. Hackney, W.W. Milligan, and E.C. Aifantis. 1995. *Nanostructured Mater.* 5:689.

Kear, B., and G. Skandan. 1998. Nanostructural bulk materials: Synthesis, processing, properties and performance. *In R&D status and trends*, ed.

Kear, B.H., R.K. Sadangi, and S.C. Liao. 1997. Synthesis of WC/Co/diamond nanocomposites. In *Proc. of the Joint NSF-NIST Conf. on Nanoparticles.*

Keiper, A.: 2003, 'The Nanotechnology Revolution', *the New Atlantis*, 2, 17-34.

Keller, D.J. and Chih-Chung, C. (1992) Imaging steep, high structures by scanning force microscopy with electron beam deposited tips. *Surf. Sci.* 268, 333-339.

Kennedy, P., *the Rise and Fall of the Great Powers, Economic Change and Military Conflict from 1500 to* 2000, Random House (1987).

Kennedy, Paul. *the Rise and Fall of the Great Powers, Economic Change and Military Conflict from 1500 to 2000.* New York: Random House, 1987.

Keyssar, Alexander. 1986. *Out of Work: the First Century of Unemployment in Massachusetts* (Cambridge: Cambridge University Press).

Khushf, G.: 2004, 'A Hierarchical Architecture for Nano-scale Science and Technology: Taking Stock of the Claims About Science Made By Advocates of NBIC Convergence', in: D. Baird, A. Nordmann & J. Schummer (eds.), *Discovering the Nanoscale*, Amsterdam: IOS Press, pp. 21-33.

Khushf, G.: 2004, 'Systems Theory and the Ethics of Human Enhancement: A Framework for NBIC Convergence', *Annals of the New York Academy of Sciences*, 1013, 124-149.

Kirschvink, J.L., A. Koyayashi-Kirschvink, and B.J. Woodford. 1992. Magnetite biomineralization in the human brain. *Proc. Nat'l. Acad. Sci.* USA 89:7683-7687.

Kishida, M., T. Fujita, K. Umakoshi, J. Ishiyama, H. Nagata, and K. Wakabayashi. 1995. *Chem. Commun.* 763.

Klein, J.D., et al. 1993. *Chem. Mater.* 5:902. Koch, C.C. 1989. Materials synthesis by mechanical alloying. *Annual Review of Mater. Sci.* 19:121-143.

Kline, R.: 1992, *Steinmetz: Engineer and Socialist*, Johns Hopkins University Press, Baltimore.

Kline, R.: 1995, 'Construing 'Technology' as 'Applied Science": Public Rhetoric of Scientists and Engineers in the United States, *Isis*, 86, 194-221.

Kline, R.: 2000, 'The Paradox of'Engineering Science': A Cold War Debate about Education in the United States', *IEEE Technology and Society Magazine*, 19, 19-25.

Kline, R.; Pinch, T.: 1996, 'Users as Agents of Technological Change: the Social Construction of the Automobile in the Rural United States', *Technology and Culture*, 37, 763-95.

Knight, F.H. Risk, Uncertainty and Profit. London School of Economics and Political Science, London, 1933.

Knorr-Cetina, K: 1992, 'The Couch, the Cathedral, and the Laboratory: On the Relationship between Experiment and Laboratory in Science', in: A. Pickering (ed.), *Science as Practice and Culture*, University of Chicago Press, Chicago, pp. 113-138.

Knorr-Cetina, K.: 1999, *Epistemic Cultures: How the Sciences Make Knowledge*, Harvard University Press, Cambridge, MA.

Koch, C.C. 1998. Bulk behavior. In *R&D status and trends*, ed. Siegel et al. 32 *Evelyn L. Hu and David T. Shaw.*

Kolata, G., "A Vat of DNA May Become a Vast Computer of the Future," *N.Y. Times*, 11 April 1995, p. Cl. *Large, much cited spread at the top of the front page of the Tuesday "Science Times" Section that first brought the Adelman discovery to wide public attention. Beautiful illustrations that explain how it works.*

Kortan, A.R., R. Hull, R.L. Opila, M.G. Bawendi, M.L. Steigerwald, P.J. Carroll, and L.E. Brus. 1990. *J. Am. Chem. Soc.* 112:1327.

Kourilsky, Ph. & Viney, G. Le Principe de precaution. Report to the Prime Minister, Paris, Éditions Odile Jacob, 2000.

Kranakis, E.: 1997, *Constructing a Bridge: An Exploration of Engineering Culture, Design, and Research in Nineteenth-Century France and America*, MIT Press, Cambridge, MA.

Krätschmer, W., L.D. Lamb, K. Fostiropoulos, and D.R. Huffman. 1990. *Nature* 347:354.

Krejchi, M.T., E.D.T. Atkins, A.J. Waddon, M.J. Fournier, T.L. Mason, and D.A. Tirrell. 1994. Chemical sequence control of beta-sheet assembly in macromolecular crystals of periodic proteins. *Science* 265:1427-1432.

Krejchi, M.T., S.J. Cooper, Y. Deguchi, E.D.T. Atkins, M.J. Fournier, T.L. Mason, and D.A. Tirrell. 1997. Crystal structures of chain-folded antiparallel beta-sheet assemblies from sequence-designed periodic polypeptides. *Macromolecules* 30:5012-5024.

Kresge, C.T., M.E. Leonowicz, W.J. Roth, J.C. Vartuli, and J.S. Beck. 1992. *Nature* 359:710.

Kresge, C.T., M.E. Leonowicz, W.J. Roth, J.C. Vartuli, and J.S. Beck. 1992. *Nature* 359:710. Kupperman, A., S. Nadimi, S. Oliver, G. Ozin, J. Garces, and M. Olken. 1993. *Nature* 365:239.

Kroes, P. & Meijers, A. "The Dual Nature of Technical Artifacts—Presentation of a New Research Programme." *Techne* 6, 2 (2002): 4-8.

Krstic, V., U. Erb, and G. Palumbo. 1993. *Scripta Metall. et Mater.* 29:1501.

Krumhansl, J. A. & Pao, Y. H. (1979) *Phys. Today* 32 (11), 25-32.

Kruskal, M., and N. Zabusky. 1965. Interaction of solitons in a collisionless plasma and the recurrence of initial states, *Phys. Rev. Lett.* Vol. 15, 240-243.

Kuhn, T.: 1970, *the Structure of Scientific Revolutions* (second edition), University of Chicago Press, Chicago.

Kuhn, T.S.: 1996, *the Structure of Scientific Revolutions*, University of Chicago Press, Chicago.

Kumar, A., and G.M. Whitesides. 1993. Features of gold having micrometer to centimeter dimensions can be formed through a combination of staming with an elastomeric stamp and an alkanethiol ink followed by chemical etching. *App. Phys. Lett.* 63:2002-2004.

Kupferman, R., M. Kawaguchi, and M.M. Denn. 2000. Emergence of structure in models of liquid crystalline polymers with elastic coupling. *J. Non-Newtonian Fluid Mech.* Vol. 91. 255-271.

Kupperman, A., S. Nadimi, S. Oliver, G. Ozin, J. Garces, and M. Olken. 1993. *Nature* 365:239.

Kurzweil, R. 1999. *the Age of Spiritual Machines.* Penguin Books.

Kurzweil, R.: 1998, *the Age of Spiritual Machines, How We Will Live, Work and Think in the New Age of Intelligent Machines,* Phoenix, New York.

Kuusi, O. 1999b. *Expertise in the future use of generic technologies.* VATT Research Reports #59, Helsinki.

Kuusi, O. and M. Meyer. 2001. "Technology Generalizations and Leitbilds—The Anticipation of Technological Opportunities." Manuscript submitted to *Technological Forecasting and Social Change.*

Kuusi, O., 1999. "Epistemic value analysis—a tool to evaluate technological options." Personal communication.

Kuusi, O.; Meyer, M.: 2002, 'Technological generalizations and leitbilder' the anticipation of technological opportunities', *Technological Forecasting & Social Change,* 69, 625-639.

Kyprianidou-Leodidou, T., W. Caseri, and V. Suter. 1994. *J. Phys. Chem.* 98:8992. Kung, H.H., and E.I. Ko, 1996. *Chem. Eng. J.* 64:203.

L. Chong, B. van Steensel, D. Broccoli, H. Erdjument-Bromage, J. Hanish, P. Tempst, T. de Lange, "A Human Telomeric Protein," *Science* 270 (8 December 1995): 1663-1667; D. Broccoli, A. Smogorzewska, L. Chong, T. de Lange, "Human telomeres contain two distinct Myb-related proteins, TRF1 and TRF2," *Nature Genet.* 17 (October 1997):231-235.

L.J. Perry and R. Wetzel, "Disulfide bond engineered into T4 lysozyme: stabilization of the protein toward thermal inactivation." *Science,* 226: 555-557 (1984).

L. Regan and W. F. DeGrado, "Characterization of a helical protein designed from first principles." *Science,* 241: 976-978 (1988).

L. Zhu, P.S. Lukeman, J. Canary & N.C. Seeman, Nylon/DNA: Single-Stranded DNA with Covalently Stitched Nylon Lining, *J. Am. Chem. Soc.* 125,10178-10179 (2003).

L.J. Lauhon, W. Ho, "Inducing and observing the abstraction of a single hydrogen atom in bimolecular reaction with a scanning tunneling microscope," *J. Phys. Chem.* 105(2000):3987-3992.

L.J. Lauhon, W. Ho, "Inducing and observing the abstraction of a single hydrogen atom in bimolecular reaction with a scanning tunneling microscope," *J. Phys. Chem.* 105(2000):3987-3992.

L.S. Penrose, "Self-Reproducing Machines," *Scientific American* 200 (June 1959):105-114.

Laibinis, P.E.; Hickman, J.J.; Wrighton, M.S.; Whitesides, G.M. 1989. "Orthogonal Self-Assembled Monolayers—Alkanethiols On Gold and Alkane Carboxylic-Acids On Alumina", *Science,* 245, 845-847.

Lambright, W. Henry. 1995. *Powering Apollo: James E. Webb of NASA.* Baltimore: Johns Hopkins University Press.

Landon, B.: 2004, 'Less is More: Much Less is Much More: the Insistent Allure of Nanotechnology Narratives in Science Fiction', in: N.K. Hayles (ed.), *Nanoculture: Implications of the New Technoscience,* Bristol, UK: Intellect Books, pp. 131-146.

Landon, B.: 2004, 'Less is More: Much Less is Much More: the Insistent Allure of Nanotechnology Narratives in Science Fiction', in: K. Hayles (ed.), *NanoCulture,* Intellect Books, Bristol, pp. 59-78.

Lapham, E.V., C. Kozma, J.O. Weiss. 1996. "Genetic Discrimination: Perspectives of Consumers." *Science,* Oct 25; 274 (5287): pp. 621-624.

Larrére, R.: 2002, 'Agriculture: artificialisation ou manipulation de la nature?', *Cosmpolitiques,* 1 (June), 158-173.

Larson, R.G. 1999. *the Structure and Rheology of Complex Fluids.* Oxford University Press.

Laszlo, P.: 2004, 'Is There Life After Partington?', *Hyle: International Journal for Philosophy of Chemistry,* 10(2), 169-178.

Latour, B. & Woolgar, S.: 1986, *Laboratory Life: the Social Construction of Scientific Facts,* 2nd edn., Princeton University Press, Princeton.

Latour, B.: 1983, 'Give Me a Laboratory and I Will Raise the World', in: K. Knorr-Cetina, M. Mulkay (eds.), *Science Observed,* Sage, London, pp. 141-170.

Latour, B.: 1986, 'Visualisation and Cognition: Thinking With Eyes and Hands', *Knowledge and Society,* 6, 1-40.

Latour, B.: 1996, *Aramis or the Love of Technology,* Harvard University Press, Cambridge, MA.

Latour, B.: 2002, 'Morality and Technology: the End of the Means', *Theory, Culture and Society,* 19(5/6), 247-260.

Laudan, L.: 1984, *Science and Values: the Aims of Science and their Role in Scientific Debate,* University of California Press, Berkeley.

Lave, L.B. & H.L. MacLean. 2000. *An Environmental-Economic Evaluation of Hybrid Electric Vehicles: Toyota's Prius vs. Its Conventional Internal Combustion Engine Corolla,* working paper. August.

Lave, L.B., A. Russell, F.C. McMichael, and C.T. Hendrickson. 1996. Environmental implications of battery powered vehicles as a transitional technology: ozone *vs.* lead. *Environmental Science & Technology,* p. 30.

Lave, L.B., C.T. Hendrickson, and F.C. McMichael. 1995b. Environmental implications of electric cars. *Science,* Vol. 268: 992-5. May 19.

Lave, L.B., E. Cobas-Flores, C.T. Hendrickson, and F.C. McMichael. 1995a. Using input-output analysis to estimate economy wide discharges. *Environmental Science & Technology,* 29(9).

Lax, Peter D. 1989. the flowering of applied mathematics in America, in *A Century of Mathematics in America—Part II,* edited by Peter Duren et al. American Mathematical Society. 455-466.

Layton, E.T., Jr.: 1971, 'Mirror-Image Twins: the Communities of Science and Technology in 19th-century America', *Technology and Culture,* 12, 562-580.

Leah Eisenberg, "Medicating Death Row Inmates So They Qualify for Execution," AMA Case in Health Law, Vol. 6, No. 9, September 2004.

Ledley, Fred D. 1995. "Nonviral Gene Therapy: the promise of Genes as Pharmaceutical Products," *Human Gene Therapy,* Vol. 6, Sep, pp. 1129-1144.

Lee, M.S. and M. Head-Gordon. 2000. Absolute and relative energies from polarized atomic orbital selfconsistent field calculations and a second order convergence with size and composition of the secondary basis. *Computers and Chemistry.* Vol. 24(3,4). 295-301.

Lee, S.C. 1998. the nanobiological strategy for construction of nanodevices. *Biological molecules in nanotechnology: the convergence of biotechnology, polymer chemistry and materials science,* edited by S.C. Lee and L. Savage (IBC Press, Southborough, MA).

Lehn, J.M., 2004: 'Une Chimie Supramoleculaire Foisonnante', *La lettre de l'Academie des Sciences,* 10, 12-13.

Lehn, J.M.: 1995, *Supramolecular Chemistry,* VCH, Weinheim.

Lemann, Nicholas. 1991. *the Promised Land: the Great Black Migration and How it Changed America* (New York: Random House).

Lenhard, J.: 2004, 'Nanoscience and the Janus-Faced Character of Simulations', in: D. Baird, A. Nordmann & J. Schummer (eds.), *Discovering the Nanoscale*, Amsterdam: IOS Press, pp. 93-100.

Lenhard, J.: 2004, 'Nanoscience and the Janus-Faced Character of Simulations', in: D. Baird, A. Nordmann, J. Schummer (eds.), *Discovering the Nanoscale*, IOS Press, Amsterdam, pp. 93-100.

Lent, C.S., Tougaw, P.D., Porod, W., Bernstein, G.H., "Quantum Cellular Automata," *Nanotechnology*, Vol. 4, p. 49, 1993.

Leonard, D., M. Krishnamurthy, C.M. Reaves, S.P. Denbaars, and P.M. Petroff. 1993. Direct formation of quantum-sized dots from uniform coherent islands of In GaAs on GaAs surfaces. *Appl. Phys. Lett*. 63:3203-5.

Leslie-Pelecky, D.L., and R.D. Reike. 1996. *Chem. Mater.* 8:1770.

Lewak, S.E.: 2004, "What's the Buzz? Tell Me What's A-Happening: Wonder, Nanotechnology, and Alice's Adventures in Wonderland', in: N.K. Hayles (ed.), *Nanoculture: Implications of the New Technoscience*, Bristol, UK: Intellect Books, pp. 201-.

Liang, G., Z. Li, and E. Wang. 1996. *J. Mater. Sci.* Makino, A., A. Inoue, T. Hatanai, and T. Bitoh. 1997. *Materials Science Forum* 235-238: 723.

Lide, D.R., ed. 1993-1994. *CRC Handbook of.Chemistry and Physics*, 74th ed. Lide, D.R., ed. 1993-1994. *CRC Handbook of Chemistry and Physics*, 74th ed. Lin, H.-P., and C.-Y. Mou. 1996. *Science* 273:765.

Lin-Easton, P.C.: 2001, 'It's Time for Environmentalists to Think Small Real Small: A Call for the Involvement of Environmental Lawyers in Developing Precautionary Policies Molecular Nanotechnology', *Georgetown International Law Review*, 14, 106-134.

Lipowsky, R. 1991. the conformation of membranes. *Nature*. Vol. 349. 475.

Lipson, H. and J.B. Pollack, *Nature*, 406, 974-978, (2000). *Automatic design and manufacture of robotic life forms*. See, also, several additional articles on this subject in contemporary issues of Nature.

Lloyd, S., "A Potentially Realizable Quantum Computer," *Science*, 17 September 1993, pp. 1569-1571.

Lloyd, S., "Quantum-Mechanical Computers," *Scientific American*, October 1995,

López, J.: 2003, *Society and Its Metaphors: Language Social Theory and Social Structure*, Continuum, New York and London.

López, J.: 2004, 'Bridging the Gaps: Science Fiction in Nanotechnology', *Hyle: International Journal for Philosophy of Chemistry*, 10(2), 129-152.

López, J.: 2004, 'How Sociology Can Save Bioethics...Maybe', *Sociology of Health and Illness*, 26(7), 1-22.

Lósch, A.: 2004, 'Nanomedicine and Space: Discursive Orders of Mediating Innovations', in: D. Baird, A. Nordmann & J. Schummer (eds.), *Discovering the Nanoscale*, Amsterdam: IOS Press, pp. 193-202.

Lu, H.P.; Xun, L.Y.; Xie, X.S. 1998. "Single-molecule enzymatic dynamics", *Science*, 282, 1877-1882.

Luce, R.D. & Raiffa, H. Games and Decisions, Wiley, New York, 1957. Lukens, R.: 2004, 'Historic Nanotechnology', *Chemical Heritage*, 22, 29.

Lynn, David M., Robert Langer. 2000. "Degradable Poly(b-amino esters): Synthesis, Characterization, and Self-Assembly with Plasmid DNA," *Journal of the American Chemical Society*, Vol. 122.

M.C. Hersam, G.C. Abeln, J.W. Lyding, "An approach for efficiently locating and electrically contacting nanostructures fabricated via UHV-STM lithography on Si(100)," *Microelectronic Engineering* 47(1999):235-.

M.C. Hersam, G.C. Abeln, J.W. Lyding, "An approach for efficiently locating and electrically contacting nanostructures fabricated via UHV-STM lithography on Si(100)," *Microelectronic Engineering* 47(1999):235-.

M.C. Roco, "National Nanotechnology Initiative: Overview," September 20, 2004.

M.L. Petrillo, C.J. Newton, R.P. Cunningham, R.-I. Ma, N.R. Kallenbach and N-C. Seeman. the Ligation and Flexibility of 4-Arm DNA Junctions. *Biopolymers*, 27, 1337-1352 (1988).

Mackenzie, D.: 1996a, 'Economic and Sociological Explanations of Technological Change', in: *Knowing Machines: Essays on Technical Change*, MIT Press, Cambridge, MA, pp. 49-65.

Mackenzie, D.: 1996b, 'Marx and the Machine', in: *Knowing Machines: Essays on Technical Change*, MIT Press, Cambridge, MA, pp. 23-47.

Mahmood, U., C.H. Tung, A. Bogdanov Jr., R. Weissleder. 1999. "Near-Infrared Optical Imaging of Protease Activity for Tumor Detection." *Radiology*, Dec; 213 (3): pp. 866-870.

Majetich, S.A. and A.C. Canter. 1993. *J. Phys. Chem.* 97:8727.

Malinow, R., Z.F. Mainen, and Y. Hayashi. 2000. LTP mechanisms: from silence to four-lane traffic. *Curr. Opin. Neurobiol.* 10(3): p. 352-7.

Mao, C., W. Sun, and N.C. Seeman. 1997. Construction of Borromean rings from DNA. *Nature* 386:137-138.

Mark A. Reed, James M. Tour, "Computing with molecules," *Scientific American*, 282 (June 2000): 86-93.

Mark, H., Statement of the Honorable Hans Mark, Director, Defense Research and Engineering, before the House Armed Service Committee, October 20, 1999. Ref: http://www.house.gov/hasc/testimony/106thcongress/99-10-20mark.htm.

Marshall, K.: 2004, 'Atomizing Risk Technology', in: K. Hayles (ed.), *NanoCulture*, Intellect Books, Bristol, pp. 79-102.

Marshall, K.: 2004, 'Atomizing the Risk Technology', in: N.K. Hayles (ed.), *Nanoculture: Implications of the New Technoscience*, Bristol, UK: Intellect Books, pp. 147-160.

Martin, T.P., N. Malinowski, U. Zimmerman, U. Naher, and H. Schaber. 1993. *J. Chem. Phys.* 99:4210.

Martin, T.P., U. Naher, H. Schaber, U. Zimmerman. 1993. *Phys. Rev. Lett.* 70:3079. Prigogine, I., and S. Rice. 1988. *Advances in Chemical Physics*, Vol. 70, Parts 1 & 2. New York: J. Wiley.

Marty, A.: 2003, *Testimony before the Committee on Science on HR 766, Nanotechnology Research and Development Act*, US Government Printing Office, Washington, DC.

Marx, K.: 1963 [1847], *the Poverty of Philosophy*, International Publishers, New York.

Matsuki, H., K. Ofuji, N. Chubachi, and S. Nitta. 1996. Signal transmission for implantable medical devices using figure-of-eight coils. *IEEE Trans, on Magnetics.*, Vol. 32, No. 5, 5121. September.

Matsuki, H., Y. Yamakata, N. Chubachi, S. Nitta, and H. Hashimoto. 1996. Transcutaneous DC-DC converter for totally implantable artificial heart using synchronous rectifier. *IEEE Trans. On Magnetics*. 32, No.5, 5118 (Sept.).

Mayer, S.: 2002, 'From genetic modification to nanotechnology: the dangers of 'sound science", in: T. Gilland (ed.), *Science: Can We Trust the Experts?*, London: Hodder and Stoughton, pp. 1-15.

McCain, L.: 2002, 'Informing Technology Policy Decisions: the US Human Genome Project's Ethical, Legal and Social Implications Programs as a Critical Case', *Technology and Society*, 24, 111-132.

McCammon, J.A., Gelin, B.R. & Karplus, M. (1977) *Nature (London)* 267, 585-590.

McConnell, H.M. 1996. Light-addressable potentiometric sensor: Applications to drug discovery. In *Nanofabrication and biosystems*, ed. Hoch et al.

McConnell, S.A., ed. 1998. *Pharmaceuticals.* Dun and Bradstreet Industry Reference Handbooks, Gale, Detroit, pp. 30-34.

McCulloch, W.S.: 1962, 'The imitation of one form of life by another—Biomimesis', in: E.E. Bernard & M.R. Kare (eds.), *Biological prototypes and synthetic systems,* Plenum Press, New York, Vol. 1, pp. 393-97.

McHale, B.: 1987, *Postmodern Fiction,* Methuen, New York & London. Meadows, D.L., et al. 1972. *the Limits to Growth,* New York: University.

Mehl, R.F., and R.W. Cahn. 1983. Historical development. In *Physical metallurgy.* North Holland. Milligan, W.W., S.A. Hackney, M. Ke, and E.C. Aifantis. 1993. *Nanostructured Materials* 2:267.

Mehta, M.D.: 2002, 'Nanoscience and Nanotechnology: Assessing the Nature of Innovation in These Fields', *Bulletin of Science, Technology & Society,* 22(4), 269-273.

Mehta, M.D.: 2004, 'From Biotechnology to Nanotechnology: What Can We Learn from Earlier Technologies?, *Bulletin of Science, Technology & Society,* 24, 34-39.

Merz, J.L. 1986. "The Optoelectronics Joint Research Laboratory: Light Shed on Cooperative Research in Japan." *Scientific Bulletin* (ONR Far East Office). 11 (4), 1-30 (Oct.-Dec).

Methods of Molecular Quantum Mechanics, by R. McWeeny, Second Edition, Academic Press 1989.

Meyer, G. and Amer, N.M. (1988) Novel optical approach to atomic force microscopy. *Appl. Phys. Lett.* 53(12), 1045-1047.

Meyer, G. and Amer, N.M. 1990. Simultaneous measurement of lateral and normal forces with an optical-beam-deflection atomic force microscope. *Appl. Phys. Lett.* 57(20), 2089-2091.

Meyer, M. & Kuusi, O. 2004, 'Nanotechnology: Generalizations in an Interdisciplinary Field of Science and Technology', *Hyle: International Journal for Philosophy of Chemistry,* 10(2), 153-168.

Meyer, M. 2000. "Patent citations in a novel field of technology: What can they tell about interactions of emerging communities of science and technology?" *Scientometrics,* XXXXIIX (2), 151-178.

Meyer, M. 2000b. "Hurdles on the way to growth. Commercializing novel technologies." Institute of Strategy and International Business, Working Papers Series 2000/1, Helsinki University of Technology, Espoo.

Meyer, M. and O. Persson. 1998. "Nanotechnology—interdisciplinarity, patterns of collaboration and differences in application," *Scientometrics,* XXXXII (2), 195-205.

Meyer, M.: 2000a, 'Does science push technology? Patents citing scientific literature', *Research Policy,* 29, 409-434.

Meyer, M.: 2000b, 'Patent citations in a novel field of technology: What can they tell about interactions of emerging communities of science and technology?', *Scientometrics,* 48, 151-178.

Meyer, M.: 2001a, 'Patent citations in a novel field of technology: An exploration of nano-science and nano-technology', *Scientometrics,* 51, 163-183.

Meyer, M.: 2001b, 'Socio-economic Research on Nanoscale Science and Technology: A European Overview and Illustration', in: M.C. Roco and W.S. Bainbridge (eds.), *Societal Implications of Nanoscience and Nanotechnology,* Dordrecht: Kluwer, pp. 217-241.

Meyer, M.; Persson, O.: 1998, 'Nanotechnology interdisciplinarity, patterns of collaboration and differences in application', *Scientometrics,* 47, 195-205.

Michael, E.J. Holwill, Peter Satir, "II.4 Generation of Propulsive Forces by Cilia and Flagella," in J. Bereiter-Hahn, O.R. Anderson, W.-E. Reif, eds., *Cytomechanics: the Mechanical Basis of Cell Form and Structure,* Springer-Verlag, Berlin, 1987, pp. 120-130.

Michael Page, Donald W. Brenner, "Hydrogen abstraction from a diamond surface: *Ab initio* quantum chemical study using constrained isobutane as a model," *J. Am. Chem. Soc.* 113(1991):3270-3274.

Michael Page, Donald W. Brenner, "Hydrogen abstraction from a diamond surface: *Ab initio* quantum chemical study using constrained isobutane as a model," *J. Am. Chem. Soc.* 113(1991):3270-3274.

Michalet, X. and D. Bensimon. 1988. Vesicles of toroidal topology: observed morphology and shape transformations. *J. Phys. II.* France. Vol. 5. 598.

Mihail Roco and William Sims Bainbridge, eds., Converging Technologies for Improving Human Performance, NSF/DOC Report, June 2002.

Miksanek, T.: 2001, 'Microscopic Doctors and Molecular Black Bag: Science Fiction's Prescription for Nanotechnology and Medicine', *Literature and Medicine,* 21(1), 55-70.

Milburn, C: 2002, 'Nanotechnology in the Age of Post-human Engineering: Science Fiction as Science', *Configurations,* 10, 261-295 [reprinted in: N.K. Hayles (ed.), *Nanoculture: Implications of the New Technoscience,* Bristol, UK: Intellect Books, 2004, pp. 109-130].

Milburn, C: 2002, 'Nanotechnology in the Age of Posthuman Engineering: Science Fiction as Science', *Configurations,* 10, 261-295.

Milburn, C: 2004, 'Nano/Splatter: Disintegrating the Postbiological Body', *New Literary History,* 35 (in print).

Minsky, M.: 1995, 'Virtual Molecular Reality', in: Krummenacker, M. & Lewis, J. (eds.), *Prospects in Nanotechology. Proceedings of the 1st general conference on nanotechnology: developments, applications, and opportunities, November 11-14, 1992, Palo-Alto,* John Wiley & Sons, New York, pp. 187-205.

Mirsky, Steve. 2000. "Tantalizing Tubes." *Scientific American.* June.

Misa, T.: 1988, 'How Machines Make History and How Historians (and Others) Help Them to Do So', *Science, Technology, and Human Values,* 13, 308-331.

Mishra, R.S., and A.K. Mukherjee. 1997. Oral presentation at TMS meeting, Indianapolis, Indiana, 16-18 September 1997, to be published in proceedings of Symp. "Mechanical Behavior of Bulk Nano-Materials."

Mishra, R.S., R.Z. Valiev, and A.K. Mukherjee. 1997. *Nanostructured Materials* 9:473. Morris, D.G., and M.A. Morris. 1991. *Ada Metall. Mater.* 39:1763-1779.

Mizrach, S., *Should there be a limit placed on the integration of humans and computers and electronic technology?* Ref: http://www.limmat.ch/koni/texte/cyborg-ethics.html.

Mnyusiwalla, A.; Abdallah, S.D.; Singer, P.A.: 2003, 'Mind the gap: science and ethics in nanotechnology', *Nanotechnology,* 14, R9-R13.

Mnyusiwalla, A.; Daar, A.S.; Singer, P.A.: 2003, '"Mind the Gap": Science and Ethics in Nanotechnology', *Nanotechnology,* 14, R9-R13.

Mody, C.C.M.: 2000, "A New Way of Flying': Differance, Rhetoric, and the Autogiro in Interwar Aviation', *Social Studies of Science,* 30, 513-543.

Mody, C.C.M.: 2001, 'A Little Dirt Never Hurt Anyone: Knowledge-Making and Contamination in Materials Science', *Social Studies of Science,* 31, 7-36.

Mody, C.C.M.: 2004, 'How Probe Microscopists Became Nanotechnologists', in: D. Baird, A. Nordmann & J. Schummer (eds.), *Discovering the Nanoscale,* Amsterdam: IOS Press, pp. 119-133.

Mody, C.C.M.: 2004, 'Instruments in Training: the Growth of American Probe Microscopy in the 1980s', in: D. Kaiser (ed.), *Pedagogy and the Practice of Science: Producing Physical Scientists, 1800-2000,* Cambridge, MA: MIT Press (forthcoming).

Mody, C.C.M.: 2004, 'Small, but Determined: Technological Determinism in Nanoscience', *Hyle: International Journal for Philosophy of Chemistry*, 10(2), 99-128.

Mody, C.C.M.: 2004, *Crafting the Tools of Knowledge: the Invention, Spread, and Commercialization of Probe Microscopy, 1960-2000*, Ph.D. dissertation, Cornell University.

Molecular engineering: An approach to the development of general capabilities for molecular manipulation, by K. Eric Drexler, Proceedings of the National Academy of Sciences, U.S.A. 78(9): 5275-5278, 1981.

Molecular Mechanics, by Ulrich Burkert and Norman L. Allinger, ACS Monograph 177, American Chemical Society, 1982.

Molecular Mechanics: the Art and Science of Parameterization, by J. Phillip Bowen and Norman L. Allinger, page 81-97 of reference [15].

Montemerlo, M.S., Love, J.C., Opiteck, G.J., Goldhaber, D. J., and Ellenbogen, J.C., "Technologies and Designs for Electronic Nanocomputers," MITRE Technical Report 96W0000044, the MITRE Corporation, McLean, VA, July 1996. For more information, send e-mail to nanotech@mitre.org.

Moor, J.H. & Weckert, J.: 2004, 'Nanoethics: Assessing the Nanoscale From an Ethical Point of View', in: D. Baird, A. Nordmann & J. Schummer (eds.), *Discovering the Nanoscale*, Amsterdam: IOS Press, pp. 301-310.

Moore, J.C., H.M. Jin, O. Kuchner, and F.H. Arnold. 1997. Strategies for the *in vitro* evolution of protein function: Enzyme evolution by random recombination of improved sequences. *J. Mol. Biol.* 272:336-347.

Moronne, M.M.; Ben Dahan, M.; Bruchez, M.; Hamamoto, D.J.; Weiss, S.; Alivisatos, A.P. 1999. "Nanocrystal phosphors as biological probes for fluorescence microscopy", *Biophysical Journal*, 76, A450-A450.

Morris, D.G., and M.A. Morris. 1997. *Materials Science Forum* 235-238:861. Nagpal, P., and I. Baker. 1990. *Scripta Metall. Mater.* 24:2381.

Moshe Sipper, "Fifty Years of Research on Self-Replication: An Overview," *Artificial Life* (Summer 1998): 237-257, in "Special Issue on Self-Replication." See also: Moshe Sipper, "The Artificial Self-Replication Page," http://lslwww.epfl.ch/-moshes/selfrep/

Mui, C. and C. Musgrave, "The Hydroxylation and Oxidation of the Ge(100)-2xl Surface by Hydrogen Peroxide and Water," *Langmuir*, 20, 7604-7609 (2004).

Mui, C., S. Bent and C. Musgrave, "A Quantum Chemistry Based Statistical Mechanical Model of Hydrogen Deposition from Si(100)-2xl, Ge(100)-2xl, and SiGe Alloy Surfaces," *Journal of Physical Chemistry B.*, 108, 18243-18253, (2004).

Mui, C., Y. Widjaja, J. Kang, and C. Musgrave, "Surface Reaction Mechanisms for Atomic Layer Deposition of Silicon Nitride," *Surface Science*, 557, 159-170, (2004).

Mumford, Lewis. 1934. *Technics and Civilization* (New York: Harcourt Brace Jovanovich, Inc.). National Science and Technology Council (NSTC). 1999. *Nanotechnology: Shaping the World Atom by Atom* (Washington, DC: September; available at http://www.nano.gov).

Munn Sanchez, E.: 2004, 'The Expert's Role in Nanoscience and Technology', in: D. Baird, A. Nordmann & J. Schummer (eds.), *Discovering the Nanoscale*, Amsterdam: IOS Press, pp. 257-266.

Murray, C.B.; Kagan, C.R.; Bawendi, M.G. 2000. "Synthesis and characterization of monodisperse nanocrystals and close-packed nanocrystal assemblies", *Annual Review of Materials Science*, 30, 545-610.

N. Hanai, K. Nakamura, K. Shitara, "Recombinant antibodies against gangliosides expressed on tumor cells," *Cancer Chemother. Pharmacol.* 46(2000):S13-S17.

N. Jonoska, P. Sa-Ardyen and N.C. Seeman, Computation by Self-Assembly of DNA Graphs, *Journal of Genetic Programming and Evolvable Machines* 4, 123-137 (2003).

N. Jonoska, S. Liao, & N.C. Seeman, Transducers with programmable input by DNA Self-assembly, Aspects of Molecular Computing, Lecture Notes in Computer Science 2340, Springer Verlag, Berlin, 219-240 (2004).

N. Koumura, R.W. Zijlstra, R.A. van Delden, N. Harada, B.L. Feringa, "Light-driven monodirectional molecular rotor," *Nature* 401(9 September 1999):152-155.

N.C. Seeman and A.M. Belcher, Emulating Biology: Nanotechnology from the Bottom Up, Proceedings of the National Academy of Sciences (USA) 99 (supp. 2), 6451-6455 (2002).

N.C. Seeman and N.R. Kallenbach, Design of Immobile Nucleic Acid Junctions. *Biophysical Journal* 44, 201-209 (1983).

N.C. Seeman and N.R. Kallenbach, Nucleic Acid Junctions, the Tensors of Life? In: Nucleic Acids: the Vectors of Life, ed. by B. Pullman and J. Jortner, D. Reidel, Dordrecht, 183-200 (1983).

N.C. Seeman and N.R. Kallenbach, Nucleic Acid Junctions: A Successful Experiment in Macromolecular Design. In: Molecular Structure: Chemical Reactivity and Biological Activity, ed. by J.J. Stezowski J.-L. Huang and M.-C. Shao, Oxford University Press, Oxford, (1988) pp. 189-194.

N.C. Seeman and P.S. Lukeman, Nucleic Acid Nanostructures, *Reports on Progress in Physics* 68, 237-270 (2005).

N.C. Seeman, At the Crossroads of Chemistry, Biology and Materials: Structural DNA Nanotechnology, *Chemistry & Biology* 10, 1151-1159 (2003).

N.C. Seeman, B. Ding, S. Liao, T. Wang, W.B. Sherman, P.E. Constantinou, J. Kopatsch, C. Mao, R. Sha, F. Liu, H. Yan and P.S. Lukeman, Experiments in Structural DNA Nanotechnology: Arrays and Devices, Proc. SPIE; Nanofabrication: Technologies, Devices and Applications 5592, 71-81 (2005).

N.C. Seeman, Biochemistry and Structural DNA Nanotechnology: An Evolving Symbiotic Relationship, *Biochemistry* 42, 7259-7269 (2003).

N.C. Seeman, Branched DNA: A 3-D Structural Design System, *Clinical Chemistry* 39, 722-724(1993).

N.C. Seeman, C. Mao, F. Liu, R. Sha, X. Yang, L. Wenzler, X. Li, Z. Shen, H. Yan, P. Sa-Ardyen, X. Zhang, W. Shen, J. Birac, P. Lukeman, Y. Pinto, J. Qi, B. Liu, H. Qiu, S.M. Du, H. Wang, W. Sun, Y. Wang, T.-J. Fu, Y. Zhang, J.E. Mueller and J. Chen, Nicks, Nodes, and New Motifs for DNA Nanotechnology, Frontiers of Nano-Optoelectronic Systems, ed. by L. Pavesi & E. Buzanova, Kluwer, Dordrecht, 177-198 (2000).

N.C. Seeman, De Novo Design of Sequences for Nucleic Acid Structure Engineering, *Journal of Biomolecular Structure and Dynamics* 8, 573-581 (1990).

N.C. Seeman, Design and Engineering of Nucleic Acid Nanoscale Assemblies, *Current Opinion in Structural Biology*, 6, 519-526 (1996).

N.C. Seeman, Directing the Structure of Matter through Nanotechnology, Proceedings of IEEE International Joint Symposia on Intelligence and Systems IEEE Computer Society, Los Alamitos, CA, 146-150 (1998).

N.C. Seeman, DNA Components for Molecular Architecture. *Accounts of Chemical Research* 30, 357-363 (1997).

N.C. Seeman, DNA Engineering and its Application to Nanotechnology, *Trends in Biotechnology* 17, 437-443 (1999).

N.C. Seeman, DNA in a Material World, *Nature* 421, 33-37 (2003).

N.C. Seeman, DNA Nanostructures for Mechanics and Computing: Nonlinear Thinking with

Life's Central Molecule. NanoBiotechnology. Editors, Chad Mirkin and Christof Niemeyer, Wiley-VCH Verlag GmbH & Co., KgaA, Weinheim, Chapter 20, pp 308-318 (2004).

N.C. Seeman, DNA Nanostructures for Mechanics and Computing: Nonlinear Thinking with Life's Central Molecule. NanoBiotechnology. Editors, Chad Mirkin and Christof Niemeyer, Wiley-VCH Verlag GmbH & Co., Weinheim, Chapter 20, pp 308-318 (2004).

N.C. Seeman, DNA Nanotechnology, Encyclopedia of Supramolecular Chemistry, Marcel Dekker, New York, 475-483 (2004).

N.C. Seeman, DNA Nanotechnology, In: WTEC Workshop Report on R & D Status and Trends in Nanoparticles, Nanostructured Materials, and Nanodevices in the United States, ed. by R.W. Siegel, E. Hu and M.C. Roco, International Technology Research Institute, Baltimore (1998).

N.C. Seeman, DNA Nanotechnology: From Topological Control to Structural Control, in Pattern Formation in Biology, Vision and Dynamics, ed. by A. Carbone, M. Gromov, P. Pruzinkiewicz, World Scientific Publishing Company, Singapore, 271-309 (2000).

N.C. Seeman, DNA Nanotechnology: Life's Central Performer in a New Role, *Biological Physics Newsletter* 2(1) 2-6 (2002).

N.C. Seeman, DNA Nanotechnology: Novel DNA Constructions. *Annual Review of Biophysics and Biomolecular Structure* 27, 225-248 (1998).

N.C. Seeman, DNA Nicks and Nodes and Nanotechnology, *NanoLetters* 1, 22-26 (2001).

N.C. Seeman, DNA Structural Engineering Utilizing Immobile Junctions, *Current Opinion in Structural Biology* 1, 653-661 (1991).

N.C. Seeman, DNA: Beyond the Double Helix, *Macromolecular Symposia* 201, 237-244 (2003).

N.C. Seeman, F. Liu, C. Mao, X. Yang, L. A. Wenzler, R. Sha, W. Sun, Z. Shen, X. Li, J. Qi, Y. Zhang, T.-J. Fu, J. Chen and E. Winfree, Two Dimensions and Two States in DNA Nanotechnology, Proceedings of the 11th Conversation in Biomolecular Stereodynamics, ed. by R.H. Sarma and M.H. Sarma, Adenine Press, New York, 253-262 (2000).

N.C. Seeman, From Genes to Machines: DNA Nanomechanical Devices, *Trends in Biochemical Sciences* 30, 119-235 (2005).

N.C. Seeman, H. Wang, B. Liu, J. Qi, X. Li, X. Yang, F. Liu, W. Sun, Z. Shen, Y. Wang, R. Sha, C. Mao, S. Zhang, T.-J. Fu, S. M. Du, J.E. Mueller, Y. Zhang and J. Chen, the Perils of Polynucleotides: the Experimental Gap Between the Design and Assembly of Unusual DNA Structures, In: DNA Based Computers II, ed.by L.F. Landweber and E.B. Baum, *Am. Math. Soc, Providence*, pp. 215-233 (1998).

N.C. Seeman, H. Wang, J. Qi, X.J. Li, X.P. Yang, Y. Wang, H. Qiu, B. Liu, Z. Shen, W. Sun, F. Liu, J.J. Molenda, S.M. Du, J. Chen, J.E. Mueller., Y. Zhang, T.-J. Fu, and S. Zhang, DNA Nanotechnology and Topology. In: Biological Structure and Dyinamics, ed. by R.H. Sarma and M.H. Sarma, Adenine Press, New York, Vol. 2, pp. 319-341(1996).

N.C. Seeman, H. Wang, X. Yang, F. Liu, C. Mao, W. Sun, L. Wenzler, Z. Shen, R. Sha, H. Yan, M.H. Wong, P. Sa-Ardyen, B. Liu, H. Qiu, X. Li, J. Qi, S.M. Du, Y. Zhang, J.E. Mueller, T.-J. Fu, Y. Wang, and J. Chen, New Motifs in DNA Nanotechnology, Nanotechnology 9, 257-273 (1998).

N.C. Seeman, In the Nick of Space: Generalized Nucleic Acid Complementarity and the Development of DNA Nanotechnology, Synlett 2000, 1536-1548 (2000).

N.C. Seeman, It Started with Watson and Crick, But it Sure Didn't End There: Pitfalls and Possibilities beyond the Classic Double Helix, *Natural Computing* 1, 53-84 (2002).

N.C. Seeman, J. Chen, S.M. Du, J.E. Mueller, Y. Zhang, T.-J. Fu, H. Wang, Y. Wang, and S. Zhang, Synthetic DNA Knots and Catenanes, *New Journal of Chemistry* 17, 739-755 (1993).

N.C. Seeman, J. Chen, Y. Zhang, B. Liu, H. Qiu, T.-J. Fu, Y. Wang, X. Li, X. Yang, J. Qi, F. Liu, L.A. Wenzler, S, Du, J.E. Mueller, H. Wang, C. Mao, W. Sun, Z. Shen, M.H. Wong, H, Yan and R. Sha, A Bottom-Up Approach to Nanotechnology Using DNA, In: Biological Molecules in Nanotechnology, ed. by S.C. Lee, IBC Libraries, Southborough, MA, Chapter 8 (1998).

N.C. Seeman, J. Qi, X. Li, X. Yang, N.B. Leontis, B. Liu, Y. Zhang, S.M. Du and J. Chen, the Control of DNA Structure: From Topological Modules to Geometrical Modules. In: Modular Chemistry, ed. by J. Michl, Kluwer, 95-104 (1997).

N.C. Seeman, Key Experimental Approaches in DNA Nanotechnology, Current Protocols in Nucleic Acid Chemistry, Unit 12.1, John Wiley & Sons, New York (2002).

N.C. Seeman, Macromolecular Design, Nucleic Acid Junctions and Crystal Formation. *Journal of Biomolecular Structure and Dynamics* 3, 11-34 (1985).

N.C. Seeman, Nanoengineering with DNA, Biomolecular Materials: Materials Research Society Symposium Proceedings 292, 123-134 (1993).

N.C. Seeman, Nanoscale Assembly and Manipulation of Branched DNA: A Biological Starting Point for Nanotechnology, NANOCON Proceedings, ed. by J. Lewis and J.L. Quel, NANOCON, P.O. Box 40176, Bellevue, WA 98004, pp. 101-123; transcript of oral presentation, pp. 30-36 (1989).

N.C. Seeman, Nucleic Acid Junctions and Lattices. *Journal of Theoretical Biology* 99, 237-247 (1982).

N.C. Seeman, Nucleic Acid Junctions: Building Blocks for Genetic Engineering in Three Dimensions. In: Biomolecular Stereodynamics, ed. by R.H. Sarma, Adenine Press, New York (1981), pp. 269-277.

N.C. Seeman, Nucleic Acid Nanostructures and Topology. *Angewandte Chemie.* 110, 3408-3428 (1998); Angewandte Chemie International Edition 37, 3220-3238 (1998).

N.C. Seeman, Physical Models for Exploring DNA Topology. *Journal of Biomolecular Structure and Dynamics* 5, 997-1004 (1988).

N.C. Seeman, Structural Control and Engineering of Nucleic Acids, In: Concepts in Protein Engineering and Design, P. Wrede and G. Schneider, eds., Walter-de-Gruyter, Berlin, 319-343 (1994).

N.C. Seeman, Structural DNA Nanotechnology: A New Organizing Principle for Advanced Nanomaterials, *Materials Today* 6 (7), 24-29 (2003).

N.C. Seeman, Synthetic DNA Topology, Molecular Catenanes, Rotaxanes and Knots, ed. by J.-P. Sauvage and C. Dietrich-Buchecker, Wiley-VCH, Weinheim, pp. 323-356 (1999).

N.C. Seeman, the Construction of 3-D Stick Figures from Branched DNA, *DNA and Cell Biology* 10, 475-486 (1991).

N.C. Seeman, the Interactive Manipulation and Design of Macromolecular Architecture Utilizing Nucleic Acid Junctions. *Journal of Molecular Graphics* 3, 34-39 (1985).

N.C. Seeman, the Use of Branched DNA for Nanoscale Fabrication, *Nanotechnology* 2, 149-159 (1991).

N.C. Seeman, Y. Zhang, and J. Chen, DNA Nanoconstructions, *Journal of Vacuum Science and Technology* A12, 1895-1903 (1994).

N.C. Seeman, Y. Zhang, S.M. Du and J. Chen, Construction of DNA Polyhedra and Knots Through Symmetry Minimization, In: Supramolecular Stereochemistry, ed. by J. Siegel, NATO Advanced Research Workshops Series, Kluwer, Dordrecht, 27-32 (1995).

N.C. Seeman, Y. Zhang, S.M. Du, H. Wang, J.E. Mueller, Y. Wang, B. Liu, J. Qi and J. Chen, the Control of DNA Structure and Topology: An Overview., Molecularly Designed Ultrafine/ Nanostructured Materials: *Materials Research Society Symposium Proceedings* 351, 57-66 (1994).

N.C. Seeman, Y. Zhang, T.-J. Fu, S. Zhang, Y. Wang and J. Chen, Chemical Synthesis of Nanostructures, Biomolecular Materials by Design: *Materials Research Society Symposium Proceedings* 330, 45-56 (1994).

N.C. Seeman. Nanotechnology and the Double Helix, Scientific American, 290, (6) 64-75 (2004).

N.C. Seeman. Structural DNA Nanotechnology: An Overview. Methods in Molecular Biology 303: Bionanotechnology Protocols, Editors, Sandra J. Rosenthal and David W. Wright, Humana Press, Totowa, NJ, pp. 143-166 (2005).

N.C. Seeman, J. Chen, Y. Zhang, T.-J. Fu, X. Li, X. Yang, Y. Wang, J. Qi, B. Liu and F. Liu, Control of Structure and Topology in DNA Nanotechnology. In: Molecular Nanotechnology: Biological Approaches and Novel Applications, ed. by S.A. Minden, IBC Libraries, Southborough, MA, Chapter 2.2 (31 pages) (1997).

N.R. Kallenbach, R.-I. Ma and N.C. Seeman, An Immobile Nucleic Acid Junction Constructed from Oligonucleotides. *Nature* 305, 829-831 (1983).

Nardi, B., O'Day, V. 1999. *Information Ecologies: Using Technology with Heart.* MIT Press.

National Coordination Office for Information Technology, *Ethical, Legal, and Societal Implications of Information Technology Research and Development.* Available at: http://www.itrd.gov, October 1999.

National Human Genome Research Institute, *Ethical, Legal, and Social Implications of Human Genetics Research of the National Institutes of Health.*

National Science and Technology Council (NSTC). 2000. *National Nanotechnology Initiative: Leading to the Next Industrial Revolution* (Supplement to the President's FY 2001 Budget) (Washington, DC: February; available at http://www.nano.gov).

National Science and Technology Council (NSTC). 2000. *National Nanotechnology Initiative: Leading to the Next Industrial Revolution.* February, (http://www.nano.gov/ nni.pdf).

Naughton, G. 1999. the Advanced Tissue Sciences story. *Sci. Am.* 280(4): p. 84-5.

Nauta, K. and R.E. Miller. 1999. Non-equilibrium self-assembly of long chains of polar molecules in superfluid helium. *Science.* Vol. 283. 1895-1897.

New Approaches to Empirical Force Fields, by Uri Dinur and Arnold T. Hagler, pages 99-164.

Nicholas Wade, "A DNA Success Raises Bioterror Concern," *New York Times,* January 12, 2005.

Nieman, G.W., J.R. Weertman, and R.W. Siegel. 1991a. *Mater. Res. Soc. Symp. Proc.* 206:581-586.

Nomura, M. & Held, W. (1974) in *Ribosomes,* eds. Nomura, M., Tissiers, A. & Lengyel, P. (Cold Spring Harbor Laboratory, Cold Spring Harbor, NY), pp. 193-203.

Nordmann, A. (rapp.) Converging Technologies—Shaping the Future of European Societies, European Commission report, 2004.

Nordmann, A.: 2003, 'Shaping the World Atom by Atom: Eine nanowissenschaftliche WeltBildanalyse', in: A. Grunwald (ed.), *Technikgestaltung zwischen Wunsch und Wirklichkeit,* Berlin: Springer, pp. 191-199.

Nordmann, A.: 2004, 'Molecular Disjunctions: Staking Claims at the Nanoscale', in: D. Baird, A. Nordmann & J. Schummer (eds.), *Discovering the Nanoscale,* Amsterdam: IOS Press, pp. 51-62.

Nordmann, A.: 2004, 'Molecular Disjunctions', in: D. Baird, A. Nordmann, J. Schummer (eds.), *Discovering the Nanoscale,* IOS Press, Amsterdam, pp. 51-62.

Nordmann, A.: 2004, 'Nanotechnology: Convergence and Integration', Presentation at *EuroNanoForum, Trieste, December 10, 2003* (Proceedings in preparation).

Nordmann, A.: 2004, 'Nanotechnology's WorldView: New Space for Old Cosmologies', *IEEE Technology and Society Magazine,* 23 (forthcoming).

Nordmann, A.: 2004, 'Social Imagination for Nanotechnology', in: European Commission (Community Health and Consumer Protection): *Nanotechnologies: A Preliminary Risk Analysis on the Basis of a Workshop, Brussels, 1-2 March 2004*, pp. 111-113 [www.europa.eu.int/comm/health/ph_risk/documents/ev_20040301_en.pdf].

Nordmann, A.: 2004, 'Was ist TechnoWissenschaft?' Zum Wandel der Wissenschaftskultur am Beispiel von Nanoforschung und Bionik', in: T. Rossmann & C. Tropea (eds.), *Bionik' Neue Forschungsergebnisse aus Natur-, Ingenieur- und Geisteswissenschaften*, Berlin: Springer, 2004 (forthcoming).

Nowotny, H.; Scott, P.; Gibbons, M.: 2001, *Re-thinking Science*, Polity, Cambridge.

NSET. 2000. *National Nanotechnology Initiative: Leading to the Next Industrial Revolution*. 2000. A report by the Interagency Working Group on Nanoscience, Engineering and Technology (NSET), of the National Science and Technology Council's Committee on Technology. Washington, D.C., February.

NSTC (National Science and Technology Council): 1999, *Nanotechnology: Shaping the World Atom by Atom* [www.wtec.org/loyola/nano/IWGN.Public.Brochure/ IWGN.Nanotechnology.Brochure.pdf, accessed: 15 Nov. 2003].

Nye, D.E.: 1994, *American Technological Sublime*, MIT Press, Cambridge, MA.

Nye, D.E.: 2003, *America as Second Creation: Technology and Narratives of New Beginnings*, MIT Press, Cambridge, MA.

P. Livingston, "Ganglioside vaccines with emphasis on GM2," *Semin. Oncol.* 25(December 1998):636-645.

P. Sa-Ardyen, A.V. Vologodskii and N.C. Seeman, the Flexibility of DNA Double Crossover Molecules. *Biophysical Journal* 84, 3829-3837 (2003).

P. Sa-Ardyen, N. Jonoska and N.C. Seeman, Self-Assembling DNA Graphs, DNA-Based Computers VIII, LNCS 2568, Springer-Verlag, Berlin, 1-9 (2003).

P. Sa-Ardyen, N. Jonoska and N.C. Seeman, Self-Assembling DNA Graphs, Natural Computing, 2, 427-438 (2003).

P. Sa-Ardyen, N. Jonoska and N.C. Seeman, the Construction of Graphs Whose Edges are DNA Helix Axes, *J. Am. Chem. Soc.*, 126, 6648-6657 (2004).

P.S. Lukeman, A. Mittal, & N.C. Seeman, Two Dimensional PNA/DNA Arrays: Estimating the Helicity of Unusual Nucleic Acid Polymers, Chemical Communications 2004, 1694-1695 (2004).

Parikh, A.N.; Schivley, M.A.; Koo, E.; Seshadri, K.; Aurentz, D.; Mueller, K.; Allara, D.L. 1997. "nalkylsiloxanes: From single monolayers to layered crystals. the formation of crystalline polymers from the hydrolysis of n-octadecyltrichlorosilane", *Journal of the American Chemical Society*, 119, 3135-3143.

Parker, J.C., et al. 1995. U.S. Patent 5,460,701. POST (Parliamentary Office of Science and Technology). 1995.

Parr, R.G. and W. Yang. 1989. *Density-Functional Theory of Atoms and Molecules*. Oxford University Press. Oxford.

Paschen, H.; Coenen, C; Fleischer, T.; Grünwald, R.; Oertel, D. & Revermann, C: 2004, *Nanotechnologie: Forschung, Entwicklung, Anwendung*, Berlin: Springer (*Nanotechnologie, TAB-Arbeitsbericht 92*, Berlin: Büro für Technikfolgen-Abschatzung beim Deutschen Bundestag).

Paul A. Schulte, Frederica P. Perera, eds., *Molecular Epidemiology: Principles and Practices*, Academic Press, San Diego, CA, 1993.

Peng, X.G.; Manna, L.; Yang, W.D.; Wickham, J.; Scher, E.; Kadavanich, A.; Alivisatos, A.P. 2000. "Shape control of CdSe nanocrystals", *Nature*, 404, 59-61.

Perkal, M., C. Marks, M.I. Lorber, W.H. Marks. 1992. "A three-year experience with serum anodal trypsinogen as a biochemical marker for rejection in pancreatic allografts. False positives, tissue biopsy, comparison with other markers, and diagnostic strategies." *Transplantation,* Feb; 53 (2): pp. 415-419.

Perspectives on the Computer Revolution, Zenon W. Pylyshyn, Prentice Hall, 1970.

Philip Ball, "Synthetic Biology: Starting from Scratch," *Nature,* 431, pp. 624-626, 7 October 2004.

.Philip G. Collins, Michael S. Arnold, Phaedon Avouris, "Engineering carbon nanotubes and nanotube circuits using electrical breakdown," *Science* 292(27 April 2001):706-709.

Philip G. Collins, Phaedon Avouris, "Nanotubes for electronics," *Scientific American* 283 (December 2000):62-69.

Philipse, A.P. 1998. "Colloidal Dispersions," section 3.4, A. ten Wolde (ed.), *Nanotechnology: Towards a molecular construction kit.* STT Report #60, the Hague, 171-178.

Phoenix, C.; Drexler, E.: 2004, 'Safe Exponential Manufacturing', *Nanotechnology,* 15, 869-72.

Pierce, N.A. and M.B. Giles. 2000. Adjoint reconstruction of superconvergent functionals from PDE approximations. *SIAM Rev.* Vol. 42(2). 247-264.

Pinch, T.: 1986, *Confronting Nature,* Reidel, Dordrecht.

Pinch, T.: 1993, "Testing—One, Two, Three ... Testing!': Toward a Sociology of Testing', *Science, Technology, and Human Values,* 18, 25-41.

Pitt, J.C.: 2004, 'The Epistemology of the Very Small', in: D. Baird, A. Nordmann & J. Schummer (eds.), *Discovering the Nanoscale,* Amsterdam: IOS Press, pp. 157-163.

Piveteau, Laurent-Dominique, Anthe S. Zandvliet and Robert Langer. 2001. "Nitroxide and Galactose Grafted Dendrimer: A Specifically Targeted Contrast Agent for MRI and EPR Imaging of the Liver," *in preparation.*

Polanyi, Karl. 1944. *the Great Transformation: the Political and Economic Origins of Our Time* (Boston: Beacon Press, 1957; orig. ed. 1944).

Polanyi, M.: 1962, *Personal Knowledge: Towards a Post-Critical Philosophy,* Harper Torchbooks, New York.

Poldrack, R.A. 2000. Imaging brain plasticity: conceptual and methodological issues—a theoretical review. *Neuroimage.* 12(1): p. 1-13.

Pollock, C.R. 1995. *Fundamentals of Optoelectronics.* Irwin, Inc. pp. 3-5.

Porod, W., C.S. Lent, G.H. Bernstein, A.O. Orlov, I. Amlani, G.L. Snider, and J.L. Merz. 1999. Quantumdot cellular automata: computing with coupled quantum dots. *Int. Journal of Electronics* 86, No. 5, 549-590, pp.140-145.

Pressman, J.: 2004, 'Nano Narrative: A Parable from Electronic Literature', in: N.K. Hayles (ed.), *Nanoculture: Implications of the New Technoscience,* Bristol, UK: Intellect Books, pp. 191-200.

Prigogine, I., and S. Rice. 1988. *Advances in chemical physics,* Vol. 70, Parts 1 & 2. New York: J. Wiley.

Pritchard, M.S. 1996. "Conflicts of Interest: Conceptual and Normative Issues," *Academic Medicine,* 71:12, December, pp. 1305-1313.

Putman, C.A.J., De Grooth, B.G., Van Hulst, N.F., and Greve, J. (1992) A detailed analysis of the optical beam deflection technique for use in atomic force microscopy. *J. App. Phys.* 72(1), 6-212

Putnam, D., Gentry, C.A., Pack, D.W., Langer, R. 2001. "Polymer-based Gene Delivery with Low Cytotoxicity by a Unique Balance of Side-chain Termini." *Proceedings of the National Academy of Sciences,* U.S.A., Volume 98, Number 3, pages 1200-1205.

R. Sha, F. Liu and N.C. Seeman, Atomic Force Measurement of the Interdomain Angle in Symmetric Holliday Junctions, *Biochemistry* 41, 5950-5955 (2002).

R. Sha, F. Liu, D.P. Millar, N.C. Seeman, Atomic Force Microscopy of Parallel DNA Branched Junction Arrays, *Chemistry & Biology* 7, 743-751 (2000).

R. Sha, F. Liu, M. Bruist and N.C. Seeman, Parallel Helical Domains in DNA Branched Junctions Containing 5', 5' and 3', 3' Linkages, *Biochemistry* 38, 2832-2841 (1999).

R.T. Bate, "Nanoelectronics," *Nanotechnology*, Vol. 1, pp. 1-7, 1990.

R.G. Postier, "Antibiotic-resistant organism infection," *Am. Surg.* 66 (February 2000):112-116.

R.-I. Ma, N.R. Kallenbach, R.D. Sheardy, M.L. Petrillo and N.C. Seeman, 3-Arm Nucleic Acid Junctions Are Flexible. *Nucleic Acids Research* 14, 9745-9753 (1986).

R.J. Jackson, A.J. Ramsay, C.D. Christensen, S. Beaton, D.F. Hall, I.A. Ramshaw, "Expression of mouse interleukin-4 by a recombinant ectromelia virus suppresses cytolytic lymphocyte responses and overcomes genetic resistance to mousepox," *J. Virol.* 75 (February 2001):1205-1210.

R.N. Jones, "Resistance patterns among nosocomial pathogens: trends over the past few years," *Chest* 119 (February 2001): 397S-404S.

Rademann, K., B. Kaiser, U. Even, F. Hensel. 1987. *Phys. Rev. Lett.* 59: 2319.

Rao, C.N.R., B.C. Satishkumar, and A. Govindaraj. 1997. *Chem. Commun.* 1581.

Ralph C. Merkle, "Anew family of six degree of freedom positional devices," *Nanotechnology* 8(1997):47-52; http://www.zyvex.com/nanotech/6dof.html

Ralph C. Merkle, "A Proof About Molecular Bearings," *Nanotechnology* 4(1993):86-90; http :// www. zy vex. com/nanotech/bearingProof.html

Ralph C. Merkle, "A proposed 'metabolism' for a hydrocarbon assembler," *Nanotechnology* 8(1997):149-162; http://www.zyvex.com/nanotech/hydroCarbonMetabolism.html

Ralph C. Merkle, "Binding sites for use in a simple assembler," *Nanotechnology* 8(1997):23-28; http://www.zyvex.com/nanotech/bindingSites.html

Ralph C. Merkle, "Casing an assembler," *Nanotechnology* 10(1999):315-322; http:// www.zyvex.com/nanotech/casing

Ralph C. Merkle, "Casing an assembler," *Nanotechnology* 10(1999):315-322; http:// www.zyvex.com/nanotech/casing

Ralph C. Merkle, "Convergent assembly," *Nanotechnology* 8(1997):18-22; http://www. zyvex. com/ nanotech/convergent.html

Ralph C. Merkle, "Design considerations for an assembler," *Nanotechnology* 7(1996):210-215; http:/ /www.zyvex.com/nanotech/nano4/merklePaper.html

Ralph C. Merkle, "Design-ahead for nanotechnology," in Markus Krummenacker, James Lewis, eds., *Prospects in Nanotechnology: Toward Molecular Manufacturing*, John Wiley & Sons, New York, 1995, pp. 23-52.

Ralph C. Merkle, "Molecular building blocks and development strategies for molecular nanotechnology," *Nanotechnology* II (2000):89-99; http://www.zyvex.com/ nanotech/mbb/ mbb.html

Ralph C. Merkle, "Self-replicating systems and molecular manufacturing," *J. Brit. Interplanet. Soc.* 45(1992):407-413; http://www.zyvex.com/nanotech/selfRepJBIS.html

Ralph C. Merkle, "Self-replicating systems and low cost manufacturing," in M.E. Welland, J.K. Gimzewski, eds., the Ultimate Limits of Fabrication and Measurement, Kluwer, Dordrecht, 1994, pp. 25-32; http://www.zyvex.com/nanotech/selfRepNATO.html

Ramanan, V.R. 1998. Nanocrystalline soft magnetic alloys for applications in electrical and electronic devices. In *R&D Status and Trends*, ed. Siegel et al.

Ramos, L.; Lubensky, T.C.; Dan, N.; Nelson, P.; Weitz, D.A. 1999. "Surfactant-mediated two-dimensional crystallization of colloidal crystals", *Science*, 286, 2325-2328.

RAND Corporation. *the Potential of Nanotechnology for Molecular Manufacturing*. 1995.

Rao, C.N.R., B.C. Satishkumar, and A. Govindaraj. 1997. *Chem. Commun.* 1581.

Rao, M.B., and S. Sircar. 1993. *Gas Separation and Purification* 7: 279.

Reetz, M.T., et al. 1995. *Science* 267:367.

Ratner, M.; Ratner, D.: 2003, *Nanotechnology*, Prentice Hall, New Jersey.

Ratner, M.; Ratner, D.: 2003, *Nanotechnology: A Gentle Introduction to the Next Big Idea*, Prentice Hall, Upper Saddle River, NJ.

Ravetz, J.: 1999, 'What is Post-normal Science?', *Futures*, 31, 647-653.

Rechard, R.P. "Historical Relationship Between Performance Assessment for Radioactive Waste Disposal and Other Types of Risk Assessment." *Risk Analysis* 19, 5 (1999): 763-807.

Reed, M.A., "Quantum Dots," *Scientific American*, January 1993, pp. 118-123.

Reed, Mark A. and James M. Tour. 2000. "Computing with Molecules." *Scientific American*. June. SIA (Semiconductor Industry Association). 1999. "1999 International Technology Roadmap for Semiconductors."

Regis, E.: 1995, *Nano: the Emerging Science of Nanotechnology*, Little, Brown, Boston.

Reinhardt, C: 2004, 'Chemistry in a Physical Mode: Molecular Spectroscopy and the Emergence of NMR', *Annals of Science*, 61, 1-32.

Reviews in Computational Chemistry, Vol. 2, Kenny B. Lipkowitz and Donald B. Boyd, VCH Publishers, 1991.

Rheinberger, H.-J.: 1997, *Toward a History of Epistemic Things: Synthesizing Proteins in the Test Tube*, Stanford University Press, Stanford.

Richard A. Laing, "Automation introspection," *J. Comp. System Sci.* 13(1976):172-183.

Richard A. Laing, "Automaton models of reproduction by self-inspection," *J. Theoret. Biol.* 66(1977):437-456.

Richard A. Laing, "Some alternative reproductive strategies in artificial molecular machines," *J. Theoret. Biol.* 54(1975):63-84.

Richard E. Smalley, "Of chemistry, love, and nanobots," *Scientific American* 285 (Sept. 2001):76-77.

Richard P. Feynman, "There's Plenty of Room at the Bottom," *Engineering and Science* (California Institute of Technology), February 1960, pp. 22-36. Reprinted in B.C. Crandall, James Lewis, eds., *Nanotechnology: Research and Perspectives*, MIT Press, 1992. pp. 347-363, and in D.H. Gilbert, ed., *Miniaturization*, Reinhold, New York, 1961, pp. 282-296; http://www.zyvex.com/nanotech/ feynman.html

Rietman, E.A.: 2001, 'Drexler hypothesis of a universal assembler is supported not by theoretical arguments alone but by existence proof in the form of biological life', in: *Molecular Engineering of Nanosystems*, Springer, New York & Berlin.

Rinzler, A.G.; Liu, J.; Dai, H.; Nikolaev, P.; Huffman, C.B.; Rodriguez-Macias, F.J.; Boul, P.J.; Lu, A.H.; Heymann, D.; Colbert, D.T.; Lee, R.S.; Fischer, J.E.; Rao, A.M.; Eklund, P.C.; Smalley, R.E. 1998.

Rip, A., Misa, Th. J., & Schot, J.W. (eds.). Managing Technology in Society. the Approach of Constructive Technology Assessment. London: Pinter Publishers, 1995.

Robert A. Freitas Jr., "Clottocytes: Artificial Mechanical Platelets," Foresight Update No. 41, 30 June 2000, pp. 9-11; http://www.imm.org/Reports/Rep018.html

Robert A. Freitas Jr., "Exploratory design in medical nanotechnology: A mechanical artificial red cell," *Artificial Cells, Blood Substitutes, and Immobil. Biotech.* 26(1998):411-430; http://www.foresight.org/Nanomedicine/Respirocytes.html

Robert A. Freitas Jr., "Microbivores: Artificial Mechanical Phagocytes using Digest and Discharge Protocol," Zyvex preprint, March 2001; http://www.zyvex.com/ Publications/articles/

Microbivores.html. See also: Robert A. Freitas Jr., "Microbivores: Artificial Mechanical Phagocytes," Foresight Update No. 44, 31 March 2001, pp. 11-13; http://www.imm.org/Reports/Rep025.html.

Robert A. Freitas Jr., "Nanodentistry," *J. Amer. Dent. Assoc.* 131 (November 2000):1559-1566.

Robert A. Freitas Jr., "Some Limits to Global Ecophagy by Biovorous Nanoreplicators, with Public Policy Recommendations," Zyvex preprint, April 2000; http:// www.foresight.org/NanoRev/Ecophagy.html

Robert A. Freitas Jr., *Nanomedicine, Volume I: Basic Capabilities,* Landes Bioscience, Georgetown, TX, 1999; http://www.nanomedicine.com/

Robert A. Freitas Jr., *Nanomedicine, Volume I: Basic Capabilities,* Landes Bioscience, Georgetown, TX, 1999, Section 2.4.2.

Robert A. Freitas Jr., *Nanomedicine, Volume I: Basic Capabilities,* Landes Bioscience, Georgetown, TX, 1999, Sections 6.5.3 and 9.3.5.3.4.

Robert A. Freitas Jr., William P. Gilbreath, eds., *Advanced Automation for Space Missions, NASA Conference Publication CP-2255,*1982; http://www.islandone.org/MMSG/aasm/

Robert F. Service, "Is nanotechnology dangerous?" *Science* 290 (24 November 2000): 1526-1527.

Roberts, J.A.: 2004, 'Deciding the Future of Nanotechnologies: Legal Perspectives on Issues of Democracy and Technology', in: D. Baird, A. Nordmann & J. Schummer (eds.), *Discovering the Nanoscale,* Amsterdam: IOS Press, pp. 247-255.

Robinson, C.: 2004, 'Images in NanoScience/Technology', in: D. Baird, A. Nordmann & J. Schummer (eds.), *Discovering the Nanoscale,* Amsterdam: IOS Press, pp. 165-169.

Robison, W.L.: 2004, 'Nano-Ethics', in: D. Baird, A. Nordmann & J. Schummer (eds.), *Discovering the Nanoscale,* Amsterdam: IOS Press, pp. 285-300.

Robot Factory, http://www.fanuc.co.jp/eannai/servo.htm; citation courtesy of Tihamer Toth-Fejel, 2000.

Roco, M.C., R.S. Williams, and P. Alivisatos, *Nanotechnology Research Directions: Vision for Nanotechnology in the Next Decade,* Kluwer Academic Publishers (2000).

Roco, M. and Bainbridge, W. (eds.). Converging Technologies for Improving Human Performances, National Science Foundation report, 2002.

Roco, M.C. & Bainbridge, W.S. (eds.): 2001, *Societal implications of nanoscience and nanotechnology,* (Proceedings of a workshop organized by the National Science Foundation, September 28-29, 2000), Kluwer: Dordrecht [available online at http://itri.loyola.edu/nano/societalimpact/nanosi.pdf]

Roco, M.C. & Bainbridge, W.S., (eds.): 2001, *Societal Implications of Nanoscience and Nanotechnology,* National Science Foundation [online: www.wtec.org/loyola/ nano/societalimpact/nanosi.pdf, accessed: 15 November 2003].

Roco, M.C. & Bainbridge, W.S., (eds.): 2003, *Converging Technologies for Improving Human Performance: Nanotechnology, Biotechnology, Information Technology and Cognitive Science,* Kluwer, Dordrecht.

Roco, M.C. & Tomellini, R. (eds.): 2002, *Nanotechnology: Revolutionary Opportunities and Societal Implications, Workshop, Lecce (Italy), 31 January—1 February 2002,* Luxemburg: Office for Official Publications of the European Communities, [200 pp.]

Roco, M.C, R.S. Williams, and P. Alivisatos (eds). 1999. *Nanotechnology Research Directions: IWGN Workshop Report, Vision for Nanotechnology R&D in the Next Decade.* (Kluwer Academic Publishers.

Roco, M.C: 2003, 'Broader Societal Issues of Nanotechnology', *Journal of Nanoparticle Research,* 5, 181-189.

Roco, M.C: 2003b, 'Coherence and Divergence in Megatrends in Science and Engineering', in: M.C. Roco, W.S. Bainbridge (eds.), *Converging Technologies for Improving Human Performance*, Kluwer, Dordrecht, pp. 79-96.

Roco, M.C: 2003c, 'Fundamentally New Manufacturing Processes and Products', in: Roco, M.C. & Bainbridge, W.S., (eds.), *Converging Technologies for Improving Human Performance*, Kluwer, Dordrecht, pp. 300-302.

Roco, M.C: 2004, 'The Emergence and Policy Implications of Converging Technologies' [www.nsf.gov/home/crssprgm/nano/nbic_roco_04_0422_@aaas_57sl.pdf, accessed May 2004].

Roco, M.C; Bainbridge, W.S. (eds.): 2002, *Converging Technologies for Improving Human Performance: Nanotechnology, Biotechnology, Information Technology and the Cognitive Science*, Arlington, VA: National Science Foundation.

Rod Logic and Thermal Noise in the Mechanical Nanocomputer, by K. Eric Drexler, in *Proceedings of the 3rd International Symposium on Molecular Electronic Devices*, F.L. Carter, R.E. Siatkowski, H. Wohltjen, eds., North-Holland 1988.

Rodney Brooks, "The Merger of Flesh and Machines," the Next Fifty Years: Science in the First Half of the Twenty-First Century, ed. by John Brockman, 2002, p. 191.

Rofagha, R., R. Langer, A.M. El-Sherik, U. Erb, G. Palumbo, and K.T. Aust. 1991. *Scr. Metall. Mater.* 25:2867.

Roher, H. 1993. *Jpn. J. Appl. Phys.* 32:1335. Rohlfing, E.A., D.M. Cox, and A. Kaldor. 1984. *J. Chem. Phys.* 81:3846.

Rohlfing, E.A., D.M. Cox, and A. Kaldor. 1984. *J. Chem. Phys.* 81:3846.

Rohrer, H.: 1995, 'The Nanometer Age—Challenge and Chance', *Microelectronic Engineering*, 27, 3-15.

Roland Pease, "'Living' robots powered by muscle," *BBC News*, January 17, 2005.

Romanov, A.E., V.I. Vladimirov. 1992. In *Dislocations in solids*, ed. F.R.N. Nabarro, Vol. 9. Amsterdam: North-Holland. Salishekev, G.A., O.R. Valiakhmetov, V.A. Valitov, and S.K. Mukhtarov. 1994. *Materials Science Forum.* 170-172:121.

Rosen, A. 1998. A periodic table in three dimensions: A sightseeing tour in the nanometer world. In *Advances in quantum chemistry.* In press.

Rosen, P.: 1993, 'The Social Construction of Mountain Bikes: Technology and Postmodernity in the Cycle Industry', *Social Studies of Science*, 23, 479-513.

Rosenberg, N. "Why Technology Forecasts Often Fail." *Futurist*, July/August 1995, 16-21.

Rosenberg, Nathan. 1994. *Exploring the Black Box: Technology, Economics, and History.* New York: Cambridge University Press.

Rothman, B.K.: 1998, *Genetic Maps and Human Imagination*, W.W. Norton, New York & London.

Ruthven, D.M., S. Farooq, K.S. Knaebel. 1994. *Pressure swing adsorption.* New York: VCH Publishers.

Ruthven, D.M., S. Farooq, K.S. Knaebel. 1994. *Pressure swing adsorption.* New York: VCH Publishers. Sayari, A. 1996. *Chem. Mater.* 8:1840.

S. Liao and N.C. Seeman, Translation of DNA Signals into Polymer Assembly Instructions, *Science* 306, 2072-2074 (2004).

S.P. Ho and W.F. DeGrado, "Design of a 4-Helix bundle protein: Synthesis of peptides which self-associate into a helical protein." *J. Am. Chem. Soc*, 109: 6751-6758 (1987).

S. Sasaki, I. Karube, "The development of microfabricated biocatalytic fuel cells," *Trends Biotechnol.* 17 (February 1999):50-52.

S. Xiao, F. Liu, A. Rosen, J.F. Hainfeld, N.C. Seeman, K.M. Musier-Forsyth & R.A. Kiehl, Self-Assembly of Nanoparticle Arrays by DNA Scaffolding, *J. Nanoparticle J. Research* 4, 313-317 (2002).

S. Zhang, C. Cordon-Cardo, H.S. Zhang, V.E. Reuter, S. Adluri, W.B. Hamilton, K.O. Lloyd, P.O. Livingston, "Selection of tumor antigens as targets for immune attack using immunohistochemistry: I. Focus on gangliosides," *Int. J. Cancer* 73 (26 September 1997):42-49.

S.B. Levy, "The future of antibiotics: facing antibiotic resistance," *Clin. Microbiol. Infect.* 6(2000):101-106 (Suppl. 3).

S.P. Walch, W.A. Goddard III, R.C. Merkle, "Theoretical studies of reactions on diamond surfaces," Fifth Foresight Conference on Molecular Nanotechnology, 1997; http://www.foresight.org/Conferences/MNT05/Abstracts/Walcabst.html

San Francisco Chronicle, *Scientist Is Fearful of Computer Mutiny; Sun Micro co-founder says replicating robots could replace humans*, March 13, 2000.

Sanders, P.G., J.A. Eastman, and J.R. Weertman. 1996. In *Processing and properties of nanocrystalline materials*, ed. Suryanarayana et al.

Sanders, P.G., M. Rittner, E. Kiedaisch, J.R. Weertman, H. Kung, and Y.C. Lu. 1997. *Nanostructured Mater.* 9:433.

Sankey, H.: 1994, *the Incommensurability Thesis*, Avebury, Aldershot.

Santini, John T., Jr., Michael J. Cima and Robert Langer. 1999. "A Controlled-release Microchip," *Nature*, Vol. 397, 28 Jan. pp. 335-338.

Sarewitz, D.; Woodhouse, E.: 2003, 'Small is Powerful', in: A. Lightman, D. Sarewitz & Chr. Desser, (eds.), *Living with the Genie: Essays on Technology and the Quest for Human Mastery*, Washington, DC: Island Press, pp. 63-83.

Sarikaya, M.; Aksay, I. (eds.), 1995, *Biomimetics, Design and Processing of Materials*, AIP Press, Woodbury, New York.

Scanlan, R.M., W.A. Fietz, and E.F. Koch. 1975. *J. Appl. Phys.* 46:2244.

Scheraga, H.A. (1978) in *Versatilty of Proteins*, ed. Li, C.H. (Academic, New York), pp. 119-132.

Schiefsky, M.J.: forthcoming, 'Art and Nature in Ancient Mechanics', in: W.R. Newman & B. Bensaude-Vincent (eds.), *The Artificial and the Natural: An Ancient Debate and its Modern Descendants*, MIT Press, Cambridge, MA.

Schiemann, G.: 2004, 'Dissolution of the Nature-Technology Dichotomy? Perspectives on Nanotechnology from an Everyday Understanding of Nature', in: D. Baird, A. Nordmann & J. Schummer (eds.), *Discovering the Nanoscale*, Amsterdam: IOS Press, pp. 209-213.

Schmidt, J.C.: 2004, 'Unbounded Technologies: Working Through Technological Reductionism of Nanotechnology', in: D. Baird, A. Nordmann & J. Schummer (eds.), *Discovering the Nanoscale*, Amsterdam: IOS Press, pp. 35-50.

Schuler, E.: 2004, 'Perception of Risks and Nanotechnology', in: D. Baird, A. Nordmann and J. Schummer (eds.), *Discovering the Nanoscale*, Amsterdam: IOS Press, pp. 279-284.

Schultz, L., J. Wecker, and E. Hellstern. 1987. *J. Appl. Phys.* 61:3583.

Schultz, L., K. Schnitzke, and J. Wecker. 1989. *J. Magn. Mater.* 80:115.

Schummer, J.: 2001, 'Ethics of Chemical Synthesis', *Hyle: International Journal for Philosophy of Chemistry*, 7, 103-124 [online: www.hyle.org/journal/issues/7/ schummer.htm].

Schummer, J.: 2003, 'The Notion of Nature in Chemistry', *Studies in History and Philosophy of Science*, 34 (2003), 705-736.

Schummer, J.: 2004, 'Interdisciplinary Issues in Nanoscale Research', in: D. Baird, A. Nordmann & J. Schummer (eds.), *Discovering the Nanoscale*, Amsterdam: IOS Press, pp. 9-20.

Schummer, J.: 2004, 'Multidisciplinarity, Interdisciplinarity, and Patterns of Research Collaboration in Nanoscience and Nanotechnology', *Scientometrics*, 59, 425-465.

Schummer, J.: 2004, 'Naturverhaltnisse in der modernen Wirkstoff-Forschung', in: K. Kornwachs (ed.), *Technik ' System ' Verantwortung*, Minister: LIT, pp. 629-638.

Schummer, J.: 2004, 'Why do Chemists Perform Experiments?', in: D. Sobczynska, P. Zeidler, E. Zielonacka-Lis (eds.), *Chemistry in the Philosophical Melting Pot*, Peter Lang (Frankfurt/M.), 2004, pp. 395-410.

Schummer, J.: 2005, 'Reading Nano: the Public Interest in Nanotechnology as Reflected in Book Purchase Patterns,' *Public Understanding of Science*, 14 (2) (forthcoming).

Schummer, J.; Baird, D. (eds.): 2004-5, *Nanotech Challenges*, joint special issue of *Hyle: International Journal for Philosophy of Chemistry & Techne: Research in Philosophy and Technology* (forthcoming).

Schumpeter, Joseph Alois. 1942. *Capitalism, Socialism, and Democracy*. New York: Harper.

Schwarz, A.E.: 2004, 'Shrinking the 'Ecological Footprint' with NanoTechnoScience?', in: D. Baird, A. Nordmann & J. Schummer (eds.), *Discovering the Nanoscale*, Amsterdam: IOS Press, pp. 203-208.

Schwarz, R.B. 1998. Storage of hydrogen in powders with nanosized crystalline domains. In *R&D status and trends*, ed. Siegel et al.

Senosiain, J., C. Musgrave and D. Golden, "Temperature and Pressure Dependence of the Reaction of OH and CO: Master Equation Modeling on a High Level Potential Energy Surface," *International Journal of Chemical Kinetics* 35, 464-474 (2003).

Senturia, S., N. Aluru, and J. White. 1997. Simulating the behavior of MEMS devices: computational methods and needs. *IEEE Computational Science and Engineering*. Vol. 4(1). 30-43.

Sha, R., Zhang, X., Liao, S., Constantinou, P.E., Ding, B., Wang, T., Garibotti, A.V., Zhong, H., Israel, L.B., Wang, X., Wu, G., Chakraborty, B., Chen, J., Zhang, Y., Mao, C., Yan, H., Kopatsch, J., Zheng, J., Lukeman, P.S., Sherman, W.B. and Seeman, N.C., Motifs and Methods in Structural DNA Nanotechnology, *Proc. Intl. Conf. Nanomaterials*, NANO 2005, July 13-15, 2005, Mepco Schlenk Engineering College, Srivakasi, India, V. Rajendran, ed., pp. 3-10 (2005).

Shen, T.D., C.C. Koch, T.Y. Tsui, and G.M. Pharr. 1995. *J. Mater. Res.* 10:2892.

Shor, P., *Proc. 35th Ann. Symp. Foundations of Computer Science*, 1994.

Shull, R.D. 1998. NIST activities in nanotechnology. In *R&D Status and Trends*, ed.

Shull, R.D., R.D. McMichael, and J.J. Ritter. 1993. *Nanostructured Mater.* 2:205.

Shultz, W.: 2000, 'Nanotechnology: the Next Big Thing', *Chemical and Engineering News*, 78(18), 41-47.

Siegel et al. Ogunnaike, B and W. Ray. 1994. *Process dynamics, modeling and control.* Oxford University Press, pp 5-21; 1033-48.

Siegel et al. Siegel, R.W. 1997. *Materials Science Forum* 235-238:851.

Siegel R.W. and G.E. Fougere. 1994. In *Nanophase materials*, ed. G.C. Hadjipanayis and R.W. Siegel. Netherlands: Kluwer Acad. Publ.

Siegel, H.: 1980, 'Objectivity, Rationality, Incommensurability, and More', *British Journal for the Philosophy of Science*, 31, 359-384.

Siegel, R.W. 1991. *Ann. Rev. Mater. Sci.* 21:559.

Siegel, R.W., E. Hu, and M.C. Roco, eds. 1998. *R&D status and trends in nanoparticles, nanostructured materials, and nanodevices in the United States.* Baltimore: Loyola College, International Technology Research Institute. NTIS #PB98-117914.

Simon, Julian L. (ed). 1995. *the State of Humanity*, Oxford: Blackwell.

Simos, P.G., et al. 2000. Insights into brain function and neural plasticity using magnetic source imaging.

Smalley, R.: 2001, 'Of Chemistry, Love and Nanobots', *Scientific American*, 285, 76-77.

Smalley, R.: 2003a, 'Smalley Responds', *Chemical & Engineering News,* 81, 39-40.

Smalley, R.: 2003b, 'Smalley Concludes', *Chemical & Engineering News,* 81, 41-42.

Smalley, R.E. 2000. Hearing of the Senate Science and Technology Caucus on Nanotechnology. April 5, (unpublished).

Smalley, R.E.: 1999, 'Prepared written statement and supplemental material', Rice University, 22 June [http//www.house.gov/science/smalley_062299.htm].

Smalley, R.E.: 2001, 'Of Chemistry, Love and Nanobots', *Scientific American,* (Sept.), 76-77.

Smith, A., *the Wealth of Nations: An Inquiry into the Nature and Causes,* Random House, New York, 1994.

Smith, C.S. 1981. A *search for structure.* Cambridge, Mass.: MIT Press.

Smith, D., H. Babcock, and S. Chu. 1999. Single-polymer dynamics in steady shear flow. *Science.* Vol. 283. 1724-1727. March.

Smith, M.R.; Marx, L. (eds.): 1994, *Does Technology Drive History? the Dilemma of Technological Determinism,* MIT Press, Cambridge, MA.

Snider, G.L.; Orlov, A.O.; Amlani, I.; Zuo, X.; Bernstein, G.H.; Lent, C.S.; Merz, J.L.; Porod, W. 1999. "Quantum-dot cellular automata: Review and recent experiments (invited)", *Journal of Applied Physics,* 85, 4283-4285.

Snow, E.S., P.M. Campbell, and F.K. Perkins. 1997. Nanofabrication with proximal probes. *Proceedings of the IEEE* 85:601-11.

Soffer, A., J.E. Koresh, S. Saggy. 1987. United States Patent 4,685,940.

Soong, R.; Bachand, G.; Neves, H.; Olkhovets, A.; Craighead, H.; Montemagno, C: 2000, 'Powering an Inorganic Nanodevice with a Biomolecular Motor', *Science,* 290, 1555-1558.

Staden, H.V.: forthcoming, '*Physis* and *Techne* in Greek Medicine', in: W.R. Newman & B. Bensaude-Vincent (eds.), *the Artificial and the Natural: An Ancient Debate and its Modern Descendants,* MIT Press, Cambridge, MA.

Star, S.L.; Griesemer, J.R.: 1989, 'Institutional Ecology, 'Translations' and Boundary Objects: Amateurs and Professionals in Berkeley's Museum of Vertebrate Zoology, 1907-1939', *Social Studies of Science,* 19, 387-420.

Steelman, John R. 1947. *Science and Public Policy* (Washington, DC: US Government Printing Office). Stephenson, Neal. 1995. *the Diamond Age, or, Young Lady's Illustrated Primer* (New York: Bantam Books).

Stephen P. Walch, Ralph C. Merkle, "Theoretical studies of diamond mechanosynthesis reactions," *Nanotechnology* 9(1998):285-296.

Sterling, B., *When Robots Act Like Rabbits, The Wall Street Journal,* September 6, 2000.

Sterman, J. D. (1989b). Modeling Managerial Behavior: Misperceptions of Feedback in a Dynamic Decision Making Experiment. *Management Science,* 35(3), 321-339.

Steve, Jurvetson, "Transcending Moore's Law with Molecular Electronics," *Nanotechnology Law & Business Journal,* Vol. 1, No. 1, article 9, p. 9.

Stipe, B.C.; Rezaei, M.A.; Ho, W. 1998. "Single-molecule vibrational spectroscopy and microscopy", *Science,* 280, 1732-1735.

Stix Gary, "Toward 'Point One'," *Scientific American,* February 1995, pp. 90-95.

Stix, G. 1996. Trends in nanotechnology: waiting for breakthroughs. *Sci. Am.* April.

Stix, G.: 1996, 'Trends in Nanotechnology: Waiting for Breakthroughs', *Scientific American,* 274(4), 94-99.

Stix, G.: 2001, 'Little Big Science: Nanotechnology Is All the Rage. But Will It Meet Its Ambitious Goals? And What the Heck Is It?', *Scientific American* 285, 32-37

Stokes, Donald E. 1997. *Pasteur's Quadrant: Basic Science and Technological Innovation*. Washington, D.C.: Brookings Institution Press.

Storrs, J., Hall, *Flying Saucers?* http://www.nanocentral.com/NanoWorld/Perspectives/ JoshHall/ askjosh_aug96.html.

Stranick, S.J.; Atre, S.V.; Parikh, A.N.; Wood, M.C.; Allara, D.L.; Winograd, N.; Weiss, P.S. 1996. "Nanometer-scale phase separation in mixed composition self-assembled monolayers", *Nanotechnology*, 7, 438-442.

Stuckless, J.T, D.E. Starr, D.J. Bald, C.T. Campbell. 1997. *J. Chem. Phys.* 107:5547.

Stupp, S., and P.V. Braun. 1997. Molecular manipulation of microstructures: biomaterials, ceramics, and semiconductors. *Science* 277:1242-1248.

Suchman, M.: 2001, 'Envisioning Life on the Nano-Frontier', in: M.C. Roco and W.S. Bainbridge (eds.), *Societal Implications of Nanoscience and Nanotechnology*, Dordrecht: Kluwer, pp. 211-216.

Suchman, M.: 2002, 'Social Science and Nanotechnologies', M. Roco & R. Tomellini (eds.), *Nanotechnology ' Revolutionary Opportunities and Societal Implications*, Luxembourg: European Communities, pp. 95-99.

Sullivan, D.: 1994, 'Exclusionary Epideictic: *NOVA's* Narrative Excommunication of Fleischmann and Pons', *Science, Technology, and Human Values*, 19, 283-306.

Sun, S., C.B. Murray, D. Weller, L. Folks, and A. Moser. 2000. Monodisperse FePt nanoparticles and ferromagnetic FePt nanocrystal superlattices. *Science.* Vol. 287, 17 March, pp. 1989-1992.

Sun, S.H.; Murray, C.B. 1999. "Synthesis of monodisperse cobalt nanocrystals and their assembly into magnetic superlattices", *Journal of Applied Physics*, 85, 4325-4330.

Sun, S.H.; Murray, C.B.; Weller, D.; Folks, L.; Moser, A. 2000. "Monodisperse FePt nanoparticles and ferromagnetic FePt nanocrystal superlattices", *Science*, 287, 1989-1992.

Sunstrom, J.E., IV, W.R. Moser, and B. Marshik-Guerts. 1996. *Chem. Mater.* 8:2061.

Surowiecki, J.: 2004, 'Bring on the Nanobubble', *the New Yorker*, March 15, p. 68.

Susan B. Sinnott, Richard J. Colton, Carter T. White, Donald W. Brenner, "Surface patterning by atomically-controlled chemical forces: molecular dynamics simulations," *Surf. Sci.* 316(1994):L1055-L1060.

Susan Smith, Izabela Giriat, Anja Schmitt, Titia de Lange, "Tankyrase, a poly (ADP-ribose) polymerase at human telomeres," *Science* 282(20 November 1998): 1484-1487.

Suslick, K.S., T. Hyeon, and F. Fang. 1996. *Chem. Mater.* 8:2172.

Suvin, D.: 1979, *Metamorphoses of Science Fiction*, Yale University Press, New Haven.

Swami, N.: 2002, *Testimony before Senate Committee on Commerce, Science, and Transportation*, US Government Printing Office, Washington, DC.

Sweeney, A.E.; Seal, S. & Vaidyanathan, P.: 2003, 'The promises and perils of nanoscience and nanotechnology: Exploring emerging social and ethical issues', *Bulletin of Science, Technology & Society*, 23 (4), 236-245.

Swiss Re: 2004, *Nanotechnology: Small Matter, Many Unknowns*, Zurich: Swiss Re [www. swissr e. com].

T.E. Creighton, "The protein-folding problem." *Science*, 240: 267, 344 (1988).

T. LaBean, H. Yan, J. Kopatsch, F. Liu, E. Winfree, J.H. Reif and N.C. Seeman, the Construction of DNA Triple Crossover Molecules, *Journal of the American Chemical Society* 122, 1848-1860 (2000).

T.R. Kelly, H. De Silva, R.A. Silva, "Unidirectional rotary motion in a molecular system," *Nature* 401 (9 September 1999):150-152.

Tadmor, E., R. Phillips, and M. Ortiz. 2000. Hierarchical modeling in the mechanics of materials. *Int. J. Solids Struct.* Vol. 37. 379.

Tanev, P.T., M. Chibwe, and T.J. Pinnavaia. 1994. *Nature* 368:321.

Taylor, K.J., C.L. Pettiette-Hall, O. Cheshnovsky, and R.J. Smalley. 1992. *J. Chem. Phys.* 96:3319.

Ten Wolde, A. (ed.): 1998, *Nanotechnology: Towards a Molecular Construction Kit,* the Hague: the Netherlands Study Centre for Technology Trends (STT).

Tenner, E.: 2001, 'Nanotechnology and Unintended Consequences', in: M.C. Roco and W.S. Bainbridge (eds.), *Societal Implications of Nanoscience and Nanotechnology,* Dordrecht: Kluwer, pp. 241-246.

Thacker, E.: 2000, 'SF, Technoscience, Net.art: the Politics of Extrapolation (science fiction)', *Art Journal,* 59(3), 64-73.

Thacker, E.: 2004, *Biomedia,* University of Minnesota Press, Minneapolis.

The Nuclear Motion Problem in Molecular Physics, by B.T. Sutcliffe, in Advances in Quantum Chemistry, Vol. 28,1997, pages 65-80. (This reference added 97.08.31).

The Royal Society and The Royal Academy of Engineering: 2004, *Nanoscience and nanotechnologies: opportunities and uncertainties,* London [www.nanotec.org.uk/finalReport.htm].

Theis, T.: 2001, 'Information Technology Based on a Mature Nanotechnology', in: M.C. Roco, W.S. Bainbridge (eds.), *Societal Implications of Nanoscience and Nanotechnology,* National Science Foundation, pp. 60-67.

Theoretical studies of a hydrogen abstraction tool for nanotechnology, by Charles B. Musgrave, Jason K. Perry, Ralph C. Merkle, and William A. Goddard III, *Nanotechnology.*

Theory of Self-Reproducing Automata, by John von Neumann, edited and completed by Arthur W. Burks, University of Illinois Press, 1966.

There's Plenty of Room at the Bottom, a talk by Richard Feynman at an annual meeting of the American Physical Society given on December 29, 1959.

Thomas, E.L., D.M. Anderson, C.S. Henkee, and D. Hoffman. 1988. Periodic area-minimizing surfaces in block copolymers. *Nature.* Vol. 334. 598.

Thompson, D.: 1992 [1942], *On Growth and Forms,* Cambridge University Press, Cambridge.

Thompson, P.B.: 2001, 'Social Acceptance of Nanotechnology', in: M.C. Roco, W.S. Bainbridge (eds.), *Societal Implications of Nanoscience and Nanotechnology,* online: National Science Foundation, pp. 198-202.

Tighe, T.S., J.M. Worlock, and M.L. Roukes. 1997. Direct thermal conductance measurements on suspended monocrystalline nanostructures. *Appl. Phys. Lett.* 70:2687-9.

Tihamer Toth-Fejel, "Modeling kinematic cellular automata: An approach to self-replication," Environmental Research Institute of Michigan, Ann Arbor, MI, 4 October 2000.

Tilton, John E. (ed). 1991. *World Metal Demand,* Washington, DC: Resources for the Future.

Tobio, M., R. Gref, A. Sanchez, R. Langer and M.J. Alonso. 1998. "Stealth PLA-PEG Nanoparticles as Protein Carriers for Nasal Administration", *Pharmaceutical Research,* Vol. 15, No. 2, pp. 270-275.

Tohru Watanabe, WatanabefV@aol.com, Department of Robotics, Ritsumeikan University; personal communication to Tihamer Toth-Fejel, 21 March 2000.

Tourney, C.: 2004, 'Anticipating Public Reactions to Nanotechnology', *Techne: Research in Philosophy and Technology,* 8(3) (forthcoming).

Tour, J.M.: 2003, *Molecular Electronics: Commercial Insights, Chemistry, Devices, Architecture and Programming,* World Scientific, Singapore.

Tramontin, A.D. and E.A. Brenowitz. 2000. Seasonal plasticity in the adult brain. *Trends Neurosci.* 23(6): pp. 251-8.

Trivelli, A., and W.F. Smith. 1939. *Photog. J.* 79:330,463,609. As referred to in T.H. James. 1977. *the theory of the Photographic process.* New York: MacMillan (p. 100).

Trudeau, M.L., and J.Y. Ying. 1996. *Nanostructured Materials* 7:245.

Tschope, A.S., W. Liu, M. Flyzani-Stephanopoulos, and J.Y. Ying 1995. *J. Catal.* 157:42.

Tschope, A.S., and J.Y. Ying. 1994. *Nanostructured Materials* 4:617. Yeo, Y.Y., C.E. Wartnaby, and D.A. King. 1995. *Science* 268:1731.

Tsuji, T. and A. Rey. 1998. Orientation mode selection mechanisms for sheared nematic liquid crystalline materials. *Phys. Rev. E.* Vol. 57(5). 5609-5625.

Turnbull, D.: 2000, *Masons, Tricksters and Cartographers*, Harwood, Amsterdam.

Turner, F.: forthcoming, 'When the Counterculture Met Computer Networks and the New Economy: Revisiting the WELL and the Origins of Virtual Community', *Technology and Culture.*

Turton, R., *the Quantum Dot*, Oxford University Press, 1995.

Tushman, Michael and Philip Anderson (1996), Managing Strategic Innovation and Change: A Collection of Readings. London: Oxford University Press.

Two Types of Mechanical Reversible Logic, by Ralph C. Merkle.

Uyeda, R. 1991. *Prog. in Mater. Sci.* 35:1. Van de Zande, B.M.I., M.R. Bohmer, L.G.J. Fokkink, and C. Shonenberger. 1997. *J. Phys. Chem.* 101:852.

Vaia, R., D. Tomlin, M. Schulte, and T. Bunning. 2000. Two-phase nanoscale morphology of polymer/LC composites. *Polymer.* In press.

Vale, R.D.; Milligan, R.A. 2000. "The way things move: Looking under the hood of molecular motor proteins", *Science*, 288, 88-95.

Valenti. Plaza y Valdes, S.A. de C.V. Instituto de Investigaciones Sociales. Universidad Nacional Autonoma de Mexico. Universidad Autonoma Metropolitana, pp. 81-93.

Van Maanen, J. *Tales of the Field.* Chicago: U. Chicago, 1988.

Verbeek, A. & Callaert, J.: 2002, *Detailed analysis of the science-technology interaction in the field of Nanotechnology*, Luxemburg: Office for Official Publications of the European Communities (Linking Science to Technology. Bibliographic References in Patents, Vol. 9) [79 pp.].

Vermaas, P.E.: 2004, 'Nanoscale Technology: A Two-Sided Challenge for Interpretations of Quantum Mechanics', in: D. Baird, A. Nordmann & J. Schummer (eds.), *Discovering the Nanoscale*, Amsterdam: IOS Press, pp. 77-91.

Vettiger, P.; Brugger, J.; Despont, M.; Drechsler, U.; Durig, U.; Haberle, W.; Lutwyche, M.; Rothuizen, H.; Stutz, R.; Widmer, R.; Binnig, G. 1999. "Ultrahigh density, high-data-rate NEMS-based AFM data storage system", *Microelectronic Engineering*, 46, 11-17.

Vettiger, P.; Despont, M.; Drechsler, U.; Durig, U.; Haberle, W.; Lutwyche, M.I.; Rothuizen, H.E.; Stutz, R.; Widmer, R.; Binnig, G.K. 2000. "The "Millipede"—More than one thousand tips for future AFM data storage", *IBM Journal of Research and Development*, 44, 323-340.

Viani, M.B.; Schaffer, T.E.; Chand, A.; Rief, M.; Gaub, H.E.; Hansma, P.K.: 1999, 'Small Cantilevers for Force Spectroscopy of Single Molecules', *Journal of Applied Physics*, 86, 2258-2262.

Vicki Glaser, "Carbohydrate-Based Drugs Move Closer to Market," *Genetic Engineering News*, 15 April 1998, pp. 1, 12, 32, 34.

Vidyasankar, S., M. Ru, and F.H. Arnold. 1997. Molecularly imprinted ligand-exchange adsorbents for the chiral separation of underivatized amino acids. *J. Chromatography A* 775:51-63.

Vincenti, W.G.: 1990, *What Engineers Know and How They Know It: Analytical Studies from Aeronautical History*, Johns Hopkins University Press, Baltimore.

Vogel, S.: 1998, *Cats' paws and catapults*, Norton & Cy, New York, London.

von Neumann, J. "The General and Logical Theory of Automata." Talk given at the Hixon Symposium at the California Institute of Technology, 1948.

Voter, A.F. 1997. Hyperdynamics: accelerated molecular dynamics of infrequent events. *Phys. Rev. Lett.* Vol. 78. 3908.

W. Shen, M. Bruist, S. Goodman & N.C. Seeman, A Nanomechanical Device for Measuring the Excess Binding Energy of Proteins that Distort DNA, *Angew. Chem. Int. Ed.* 43, 4750-4752 (2004); Angew.Chem. 116, 4854-4856 (2004).

W.B. Sherman and N.C. Seeman, A Precisely Controlled DNA Bipedal Walking Device, *NanoLetters* 4, 1203-1207 (2004); Erratum, 4, 1801-1801.

W.C. Albrich, M. Angstwurm, L. Bader, R. Gartner, "Drug resistance in intensive care units," *Infection* 27(1999):S19-S23.

Wald, C.A.: 2004, 'Working Boundaries on the NANO Exhibition', in: N.K. Hayles (ed.), *Nanoculture: Implications of the New Technoscience,* Bristol, UK: Intellect Books, pp. 83-108.

Walsh, C. (1979) *Enzymatic Reaction Mechanisms* (Freeman, San Francisco), pp. 33, 38. Wang, C.C., Z. Zhang, and J.Y. Ying. 1997. *Nanostructured Mater.* 9:583.

Warner, M. and E.M. Terentjev. 1996. Nematic elastomers: a new state of matter? *Progress in Polymer Science.* Vol. 21(5). 853.

Warner, M., E.M. Terentjev, R.B. Meyer, and Y. Mao. 2000. Untwisting of a cholesteric elastomer by a mechanical field. *Phys. Rev. Lett.* Vol. 85(11). 2320.

Washington, London, and Moscow Convention, Convention on the prohibition of the development, production and stockpiling of bacteriological (biological) and toxin weapons and on their destruction, signed Washington, London, and Moscow, April 10, 1972, entered into force March 26, 1975.

Webb, E.J., Campbell, D.T., Schwartz, R.D., Sechrest, L., and Grove, J.B. *Unobtrusive Measures: Nonreactive Research in the Social Sciences* (2nd ed.). Boston: Houghton Mifflin, 1981.

Webster, A. and H. Etzkowitz. 1991. "Academic-industry Relations: the Second Academic Revolution?" Science Policy Support Group, London, 1991.

Weckert, J.: 2001, 'The Control of Scientific Research: the Case of Nanotechnology', *Australian Journal of Professional and Applied Ethics,* 3, 29-44.

Weil, V. 1988. "Afterword," *Biotechnology: Professional Issues and Social Concerns.* (AAAS Publication) co-edited with Paul DeForest, Mark Frankel, and Jeanne Poindexter.

Weil, V. 1996. "Biotechnology: Societal Impact and Quandaries," in *Biotechnology and Ethics: A Blueprint for the Future, Report of NSF Workshop,* Northwestern University, Center for Biotechnology.

Weil, V. 2000. "Como Pueden Convivir Los Valores Universitarios Y Las Normes Empresariales?" in *Dos Ejes en La Vinculacion de Las Universidades a la Produccion.* ed. by Rosalba Casas and Giovanna

Weil, V.: 2001, 'Ethical Issues in Nanotechnology', in: M.C. Roco and W.S. Bainbridge (eds.), *Societal Implications of Nanoscience and Nanotechnology,* Dordrecht: Kluwer, pp. 193-198.

Weisenhorn, A.L., Hansma, P.K., Albrecht, T.R., and Quate, C.F. (1989) Forces in atomic force microscopy in air and water. *Appl. Phys. Lett.* 54(26), 2651-2653

Weiss, S. 1999. "Fluorescence spectroscopy of single biomolecules", *Science,* 283, 1676-1683.

Wenger, E.: 1998, *Communities of Practice: Learning, Meaning, and Identity,* Cambridge University Press, Cambridge, UK.

White, J.G. 1986. *Philos. Trans. R. Soc. Lond. B Biol. Sci.* 314: pp. 1-340.

White, L.: 1962, *Medieval Technology and Social Change,* Clarendon, Oxford.

White, Lynn Jr. 1974. Technology assessment from the stance of a medieval historian. *the American Historical Review.* 79(1): 1-13.

White, Lynn T. 1962. *Medieval Technology and Social Change* (London: Oxford University Press).

Whitesides, G. 2001. 'The Once and Future nanomachines', *Scientific American,* (Sept.), 78-83.

Whitesides, G. 1995. 'Self-Assembling Materials,' *Scientific American*, (Sept.), 146-9.

Whitesides, G. 1998. 'Nanotechnology: Art of the Possible', *Technology, MIT Magazine of Innovation*, (Nov.-Dec), pp. 8-13.

Whitesides, G.; Love, J.C. 2001. 'The Art of Building Small', *Scientific American*, (Sept.), 38-47.

Whitesides, G.M. "Self-Assembling Materials.

Whitesides, G.M. 2001. 'The Once and Future Nanomachine', *Scientific American*, 285, 78-83.

Whitesides, G.M.; Ismagilov, R.F. 1999. 'Complexity in Chemistry', *Science*, 284, 89-92.

Whitesides, G.M.; Love, J.C. 2001. 'The Art of Building Small: Researchers Are Discovering Cheap, Efficient Ways to Make Structures Only a Few Billionths of a Meter Across', *Scientific American*, 285, 39-47.

Whitesides, G.M.; Mathias, J.P.; Seto, C.T. 1991. "Molecular Self-Assembly and Nanochemistry—a Chemical Strategy For the Synthesis of Nanostructures", *Science*, 254, 1312-1319.

Widjaja, Y. and C. Musgrave, "Atomic Layer Deposition of Hafnium Oxide: A Detailed Reaction Mechanism from First Principles," *Journal of Chemical Physics* 117, 1931-1934 (2002).

Widjaja, Y. and C. Musgrave, "Indirect Adsorbate-Adsorbate Interactions Mediated Through the Surface Electronic Structure of the Si(100)-(2xl) Surface," *Journal of Chemical Physics* 120, 1555-1559 (2004).

Widjaja, Y. and C. Musgrave, "Quantum Chemical Study of the Elementary Reactions in Zirconium Oxide Atomic Layer Deposition," *Applied Physics Letters* 81, 304-306 (2002).

Wilbur, J.L.; Whitesides, G.M. 1999. *Self-Assembly and Self-Assembled Monolayers in Micro- and Nanofabrication*; Timp, G., Ed.; Springer-Verlag, New York, Inc.: New York, pp. 331-370.

Wilder, K., H.T. Soh, T. Soh, A. Atalar, and C.F. Quate. 1997. Hybrid Atomic Force/Scanning Tunneling Lithography. *Journal of Vacuum Science & Technol. B.* 15:1811-17.

Wildoer, J.W.G., L.C. Venema, A.G. Rinzler, R.E. Smalley, and C. Dekker. 1998. Electronic Structure of Atomically Resolved Carbon Nanotubes. *Nature* 391:59-62.

William-Jones, B.; Corrigan, O.: 2003, 'Rhetoric and Hype: Where's the Ethics in Pharmacogenomics?', *American Journal of Pharmacogenomics*, 3(6), 375-383.

Williams, R.S. 1998. Functional Nanostructures. In *R&D Status and Trends*, ed. Siegel et al. Wiltzius, P. 1998.

Wilson Ho, Hyojune Lee, "Single Bond Formation and Characterization with a Scanning Tunneling Microscope," *Science* 286 (26 November 1999):1719-1722; http:// www.physics.uci.edu/~wilsonho/stm-iets.html

Winner, L. 1993, 'Upon Opening the Black Box and Finding It Empty: Social Constructivism and the Philosophy of Technology', *Science, Technology, and Human Values*, 18, 362-378.

Winner, Langdon, ed., *Democracy in a Technological Society, in Philosophy and Technology*, Vol. 9, Kluwer Academic Publishers, Dordrecht, the Netherlands, 1992.

Winningham, M.J. and D.Y. Sogah. 1997. A modular approach to polymer architecture control via catenation of prefabricated biomolecular segments: Polymers containing parallel beta-sheets templated by a phenoxathiin-based reverse turn mimic. *Macromolecules* 30: 862-876.

Wolfson, J.R.: 2003, 'Social and Ethical Issues in Nanotechnology: Lessons from Biotechnology and Other High Technologies', *Biotechnology Law Report*, 22 (4), 376-96.

Wong, H.-S.P., D.J. Frank, and P.M. Solomon. 1998. "Device Design Considerations for Double-Gate, Ground-Plane, and Single-Gated Ultra-Thin SOI MOSFETs at the 25-nm Channel-Length Generation." *Technical Digest of the 1998 International Electron Devices Meeting.*

Wood, S., Jones, R.; Geldart, A.: 2003, *the Social and Economic Challenges of Nanotechnology*, UK Economic and Social Research Council [www.esrc.ac.uk/esrccontent/DownloadDocs/Nanotechnology.pdf, accessed: 3 Feb. 2004].

Woods, S.; Jones, R. & Geldart, A.: 2003, *the Social and Economic Challenges of Nanotechnology*, Report to the Economic and Social Research Council (ESRC), Swindon, UK [www.esrc.ac.uk/esrccontent/DownloadDocs/Nanotechnology.pdf].

Wu, M.K., R.S. Windeler, C.K. Steiner, T. Bors, and S.K. Friedlander. 1993. Controlled Synthesis of Nanosized Particles by Aerosol Processes. *Aerosol Sci. Technol.* 19: 527.

Wynne, B. 1991. "Knowledges in Context." *Science, Technology, and Human Values*, 16:1, Winter 1991, pp. 111-121.

X. Yang, B. Liu, A. Vologodskii, B. Kemper and N.C. Seeman, Torsional Control of Double Stranded DNA Branch Migration. *Biopolymers* 45, 69-83 (1998).

X. Yang, L.A. Wenzler, J. Qi, X. Li and N.C. Seeman, Ligation of DNA Triangles Containing Double Crossover Molecules, *Journal of the American Chemical Society* 120, 9779-9786 (1998).

X. Zhang, H. Yan, Z. Shen and N.C. Seeman, Paranemic Cohesion of Topologically-Closed DNA Molecules, *J Am. Chem. Soc.* 124, 12940-12941 (2002).

X.J. Li, X.P. Yang, J. Qi, and N.C. Seeman, Antiparallel DNA Double Crossover Molecules as Components for Nanoconstruction, *Journal of the American Chemical Society*, 118, 6131-6140 (1996).

Xie, R.H., and J. Jiang. 1998. *J. Appl. Phys.* 83(6):3001.

Xu, Y. and C. Musgrave, "A DFT Study of the Atomic Layer Deposition of Al_2O_3 on SAMs: the Effect of SAM Termination," *Chemistry of Materials*, 16, 646 (2004).

Y. Wang, J.E. Mueller, B. Kemper, and N.C. Seeman, the Assembly and Characterization of 5-Arm and 6-Arm DNA Junctions, *Biochemistry* 30, 5667-5674 (1991).

Y. Zhang and N.C. Seeman, A Solid-Support Methodology for the Construction of Geometrical Objects from DNA, *Journal of American Chemical Society* 114, 2656-2663 (1992).

Y. Zhang and N.C. Seeman, the Construction of a DNA Truncated Octahedron, *Journal of American Chemical Society* 116, 1661-1669 (1994).

Yeo, Y.Y., C.E. Wartnaby, and D.A. King. 1995. *Science* 268:1731.

Yin, H., M.D. Wang, K. Svoboda, R. Landick, J. Gelles, and S.M. Block. 1995. Transcription against an Applied Force. *Science* 270: 1653-1657.

Yin, R.K. *Case Study Research* (2nd ed.). Newbury Park, CA: Sage, 1989.

Ying, J.Y. 1998. Nanostructure Processing of Advanced Catalytic Materials. In *R&D Status and Trends*, ed. Siegel et al.

Ying, J.Y., and T. Sun. 1997. Research Needs Assessment on Nanostructured Catalysts. *Journal of Electroceramics* 1:219-238.

Yuan, Y., J. Fendler, and I. Cabasso. 1992. *Chem. Mater.* 4:312.

Z. Shen, H. Yan, T. Wang and N.C. Seeman, Paranemic Crossover DNA: A Generalized Holliday Structure with Applications in Nanotechnology, *J. Am. Chem. Soc.* 126, 1666-1674 (2004).

Zachariah, M.R. 1994. Flame Processing, *in-situ* Characterization, and Atomistic Modeling of Nanoparticles in the Reacting Flow Group at NIST. In *Proc. of the Joint NSF-NIST Conf. on Ultrafine Particle Engineering* (May 25-27, Arlington, VA).

Zachary, G. Pascal. 1997. *Endless Frontier: Vannevar Bush, Engineer of the American Century.* New York: Free Press.

Zener, C. 1948. *Elasticity and Anelasticity of Metals.* Chicago: University of Chicago Press. Zhang, Z., M.L. Roukes, and P.C. Hammel. 1996.

Nanomedicine Glossary

Aerobots (aerobotics)—aerial (flying) robots.

Autogenous control—in medical nanorobotics, the conscious control of in vivo nanorobotic systems by the human user or patient; in biochemistry, the action of a gene product that either inhibits (negative autogenous control) or activates (positive autogenous control) expression of the gene coding for it.

Baronatation—in medical nanorobotics, locomotion through a frozen fluid by applying mechanical pressure along the path travelled to induce melting ahead, followed by regelation (refreezing) behind.

Cell engineering—deliberate artificial modifications to biological cellular systems on a cell-by-cell basis.

Cell surgery—in medical nanorobotics, modifying cellular structures using medical nanomachines.

Chemotactic nanosensor—in medical nanorobotics, a nanosensor used to determine the chemical characteristics of surfaces, possibly configured as a pad coated with an array of reversible, perhaps reconfigurable, artificial molecular receptors.

Chronocyte—in medical nanorobotics, a theorized mobile, mass-storage (nanorobotic) device, similar to a communicyte, that may be used as a mobile source of precisely synchronized universal time inside the human body.

Clottocytes Artificial Mechanical Platelets" (artificial platelets).

Communicyte—in medical nanorobotics, a theorized mobile, mass-storage (nanorobotic) device that can be used for information transport throughout the human body.

Conjugation—in medical nanorobotics, the docking of two or more nanorobots for the purpose of exchanging information, energy or materials, or to establish a larger multirobotic structure; in biology, the union of two unicellular organisms accompanied by an interchange of nuclear material, as in Paramecium.

Crystallescence—in medical nanorobotics, the crystallization of solid solute that is offloaded by nanorobot sorting rotors at a concentration that exceeds the solvation capacity of the surrounding solvent.

Cytocarriage—in medical nanorobotics, the commandeering of a natural motile cell, by a medical nanorobot, for the purposes of in vivo transport (of the nanorobot), or to perform a herding function (of the affected cell), or for other purposes.

Cytocide—the killing of living cells.

Cytography—a physical description (and mapping) of the living cell.

Cytoidentification—identification of cell type.

Cytometrics—the quantitative measurement of cell sizes, shapes, structures, and numbers.

Cytonatation—in medical nanorobotics, swimming around inside a living cell.

Cytonavigation—in medical nanorobotics, navigation inside the cell; cellular navigation.

Cytopenetration—in medical nanorobotics, entry into cells by penetrating the plasma membrane.

Cytoskeletolysis—in medical nanorobotics, purposeful destruction of the cellular cytoskeleton by a nanorobot, for cytocidal purposes.

Cytotomography—tomographic imaging of an individual cell.

Cytovehicle—in medical nanorobotics, a living cell that has been commandeered by a medical nanorobot for use during cytocarriage.

Demarcation—in medical nanorobotics, a crude form of functional navigation in which artificial conditions detectable by in vivo medical nanorobots are created at or near the target treatment site, such as warm or cold spots, pressure spots, or injected chemical plumes.

Dermal Zippers—see **zippocytes**.

Diamondoid—structures that resemble diamond in a broad sense; strong, stiff structures containing dense, three-dimensional networks of covalent bonds, formed chiefly from first and second row atoms with a valence of three or more. Many of the most useful diamondoid structures will be rich in tetrahedrally coordinated carbon.

Diamondophagy—eating diamond.

Disassembler—in molecular nanotechnology, a nanomachine or system of nanomachines able to take an object apart while at each step recording the structure and composition of that object at the molecular level.

Disequilibration—in medical nanorobotics, maintenance or inducement of a state of perpetual ionic, chemical, or energetic disequilibrium in a living cell by a medical nanorobot, usually for the purpose of inducing cytocide.

Effervescence—in medical nanorobotics, bubble formation by a gaseous solute that is offloaded by nanorobot sorting rotors at a concentration that exceeds the solvation capacity of the surrounding solvent.

Engulf formation—in medical nanorobotics, a configuration that may be adopted by a metamorphic nanorobot, in which the nanorobot reshapes itself to create an interior cavity capable of trapping a living cell, virion, or other biological particle.

Exploratory engineering—design and analysis of systems that are theoretically possible but cannot be built yet, owing to the limitations of currently available tools.

Functional navigation—in medical nanorobotics, a form of nanorobotic navigation in which nanodevices seek to detect subtle variations in their environment, comparing sensor readings with target tissue/cell profiles and then congregating wherever a precisely defined set of preconditions exists.

Histonatation—in medical nanorobotics, locomotion (swimming) through tissues by a nanorobot.

Histonavigation—in medical nanorobotics, navigation through tissues by a nanorobot.

Hypsithermal limit—the maximum amount of energy that may be released at Earth's surface, as a result of human technological activities, without significantly altering the natural global energy balance; estimated as 10^{13}-10^{15} watts.

In cyto—within a biological cell.

Inmessaging—in medical nanorobotics, conveyance of information from a source external to the human body, or external to working nanodevices, to a receiver located inside the human body.

In nucleo—within the nucleus of a cell.

In sanguo—within the bloodstream.

Macrosensing—in medical nanorobotics, the detection of global somatic states (inside the human body) and extrasomatic states (sensory data originating outside of the human body) by in vivo nanorobots.

Massometer—in medical nanorobotics, a nanosensor device for measuring the mass of individual molecules or small physical objects to single-proton resolution.

Messenger molecule—a chemically recognizable molecule which can convey information after it is received and decoded by an appropriate chemical sensor.

Metamorphic—in medical nanorobotics, capable of adopting multiple physical configurations via smooth changes from one configuration to another.

Microbiotagraphics—mapping the microbiotic populations present in the human body.

Microbivores Artificial Mechanical Phagocytes" (artificial white cells).

Molecular assembler—a general-purpose device for molecular manufacturing, able to guide chemical reactions by positioning individual molecules to atomic accuracy (e.g. mechanosynthesis) and to construct a wide range of useful and stable molecular structures according to precise specifications.

Molecular manufacturing—manufacturing using molecular machinery, giving molecule-by molecule control of products via positional chemical synthesis, to produce complex molecular structures manufactured to precise specifications.

Molecular medicine—a variety of pharmaceutical techniques and gene therapies that address specific molecular diseases or molecular defects in biological systems.

Molecular nanotechnology—thorough, inexpensive control of the structure of matter based on molecule-by-molecule control of products and byproducts; the products and processes of molecular manufacturing, including molecular machinery; a technology based on the ability to build structures to complex, atomic specifications by mechanosynthesis or other means; most broadly, the engineering of all complex mechanical systems constructed from the molecular level.

Molecular sorting rotor—a class of nanomechanical device capable of selectively binding (or releasing) molecules from (or to) solution, and of transporting these bound molecules against significant concentration gradients.

Molecular surgery (molecular repair)—in medical nanorobotics, the analysis and physical correction of molecular structures in the body using medical nanomachines.

Monkeywrenching—in medical nanorobotics, the mechanical or chemical jamming of cellular equilibrium processes, with a cytocidal objective. See also disequilibration.

Nanapheresis—in medical nanorobotics, the removal of bloodborne medical nanorobots from the body using aphersis-like processes.

Nanocentrifuge—in medical nanorobotics, a proposed nanodevice that can spin materials at very high speed, imparting rotational accelerations of up to one trillion gravities (g's), thus permitting rapid sortation.

Nanochronometer—in medical nanorobotics, a proposed clock or timing mechanism constructed of nanoscale components.

Nanocrit (Nct)—in medical nanorobotics, volume-fraction or bloodstream concentration of medical nanorobots, expressed as a percentage.

Nanomedicine—(1) the comprehensive monitoring, control, construction, repair, defense, and improvement of all human biological systems, working from the molecular level, using engineered nanodevices and nanostructures; (2) the science and technology of diagnosing, treating, and preventing disease and traumatic injury, of relieving pain, and of preserving and improving human health, using molecular tools and molecular knowledge of the human body; (3) the employment of molecular machine systems to address medical problems, using molecular knowledge to maintain and improve human health at the molecular scale.

Nanorobot—a computer-controlled robotic device constructed of nanometer-scale components to molecular precision, usually microscopic in size (often abbreviated as "nanobot").

Nanosensor—a chemical or physical sensor constructed using nanoscale components, usually microscopic or submicroscopic in size.

Nanosieving—in medical nanorobotics, a nanodevice that can sort molecules or other nanoscale objects by physical sieving.

Naturophilia—an exclusive love of Nature, disdaining everything that is artificial or technological.

Navicyte—in medical nanorobotics, a mobile, mass-storage (nanorobotic) device, similar to a communicyte, that may be used to establish a navigational network inside the human body.

Outmessaging—in medical nanorobotics, conveyance of information from a transmitter located inside the human body, especially from working nanodevices, to the patient or to a recipient external to the human body.

Pharmacyte—in medical nanorobotics, a theorized (nanorobotic) device capable of delivering precise doses of biologically active chemicals to individually-addressed human body tissue cells (e.g. cell-by-cell drug delivery).

Positional navigation—in medical nanorobotics, a form of nanorobotic navigation in which nanodevices know their exact location inside the human body to ~micron accuracy continuously at all times.

Presentation semaphore—in medical nanorobotics, a mechanical device used to display specific antigens, chemical ligands, or other molecular objects to the external environment, with the purpose of selectively modifying the chemical or other surface characteristics of a nanorobot exterior.

Replicator—any system that can build copies of itself when provided with the appropriate raw materials and energy.

Respirocrit—in medical nanorobotics, the volume-fraction or bloodstream concentration of respirocyte nanorobots, expressed as a percentage.

Respirocytes Artificial Mechanical Red Cell.

Respirocyte—in medical nanorobotics, a theorized bloodborne spherical 1-micron (nanorobotic) device having a 1000-atm pressure vessel with active pumping powered by endogenous serum glucose, that serves as a mechanical artificial red blood cell.

Sanguinatation—in medical nanorobotics, locomotion (especially swimming by a nanorobot) through the bloodstream.

Sapphirophagy—eating sapphire (corundum).

Thermogenic limit—in medical nanorobotics, the maximum amount of waste heat that may safely be released by a population of in vivo medical nanorobots that are operating within a given tissue volume.

Transtegumental—crossing or passing through the skin or covering of a body.

Vasculocyte—in medical nanorobotics, a theorized (nanorobotic) device capable of performing repairs of an injured vascular luminal surface.

Vasculoid—is a single, complex, multisegmented nanotechnological medical robotic system capable of duplicating all essential thermal and biochemical transport functions of the blood, including circulation of respiratory gases, glucose, hormones, cytokines, waste products, and cellular components.

Vasculography—a physical description (and mapping) of the human vascular system.

Vitamins (engineering)—in machine replication theory, vitamin parts are components of a self-replicating machine which the machine is incapable of producing itself, therefore these vital parts must be supplied from an external source.

Volitional normative model of disease—in medical nanorobotics, disease is said to be present in a human being upon either (1) the failure of optimal physical (e.g. biological) functioning, or (2) the failure of desired (by the patient) functioning.

Zippocytes—in medical nanorobotics, a theorized medical nanorobot that can rapidly perform incision-wound repairs to the dermis and epidermis; dermal zippers.

Index

N

O

T

U

V